T0318578

EPIGENETIC BIOMARKERS AND DIAGNOSTICS

EPIGENETIC BIOMARKERS AND DIAGNOSTICS

Edited by

JOSÉ LUIS GARCÍA-GIMÉNEZ
Center for Biomedical Network Research on Rare Diseases (CIBERER), Madrid, Spain; Medicine and Dentistry School; Biomedical Research Institute INCLIVA, University of Valencia, Spain

AMSTERDAM • BOSTON • HEIDELBERG • LONDON
NEW YORK • OXFORD • PARIS • SAN DIEGO
SAN FRANCISCO • SINGAPORE • SYDNEY • TOKYO
Academic Press is an imprint of Elsevier

Academic Press is an imprint of Elsevier
125 London Wall, London EC2Y 5AS, UK
525 B Street, Suite 1800, San Diego, CA 92101-4495, USA
225 Wyman Street, Waltham, MA 02451, USA
The Boulevard, Langford Lane, Kidlington, Oxford OX5 1GB, UK

Notices
Knowledge and best practice in this field are constantly changing. As new research and experience broaden our understanding, changes in research methods, professional practices, or medical treatment may become necessary.

Practitioners and researchers must always rely on their own experience and knowledge in evaluating and using any information, methods, compounds, or experiments described herein. In using such information or methods they should be mindful of their own safety and the safety of others, including parties for whom they have a professional responsibility.

To the fullest extent of the law, neither the Publisher nor the authors, contributors, or editors, assume any liability for any injury and/or damage to persons or property as a matter of products liability, negligence or otherwise, or from any use or operation of any methods, products, instructions, or ideas contained in the material herein.

ISBN: 978-0-12-801899-6

British Library Cataloguing-in-Publication Data
A catalogue record for this book is available from the British Library

Library of Congress Cataloging-in-Publication Data
A catalog record for this book is available from the Library of Congress

For information on all Academic Press publications
visit our website at http://store.elsevier.com/

Publisher: Mica Haley
Acquisition Editor: Catherine Van Der Laan
Editorial Project Manager: Lisa Eppich
Production Project Manager: Chris Wortley
Designer: Mark Rogers

Typeset by TNQ Books and Journals
www.tnq.co.in

Printed and bound in the United States of America

Contents

6. Genome-Wide Techniques for the Study of Clinical Epigenetic Biomarkers

ENEDA TOSKA AND F. JAVIER CARMONA SANZ

7. Sequenom MassARRAY Technology for the Analysis of DNA Methylation: Clinical Applications

ENRIQUE J. BUSÓ AND MARISA IBORRA

8. The Role of Methylation-Specific PCR and Associated Techniques in Clinical Diagnostics

FANG ZHAO AND BHARATI BAPAT

9. Pyrosequencing and Its Application in Epigenetic Clinical Diagnostics

ANA-MARIA FLOREA

10. Mass Spectrometry for the Identification of Posttranslational Modifications in Histones and Its Application in Clinical Epigenetics

ROBERTA NOBERINI, ALESSANDRO CUOMO AND TIZIANA BONALDI

11. High-Throughput Analysis of Noncoding RNAs: Implications in Clinical Epigenetics

VALERIO COSTA, MARIA R. MATARAZZO, MIRIAM GAGLIARDI, ROBERTA ESPOSITO AND ALFREDO CICCODICOLA

12. Circulating Noncoding RNAs as Clinical Biomarkers

FRANCESCO RUSSO, FLAVIA SCOYNI, ALESSANDRO FATICA, MARCO PELLEGRINI, ALFREDO FERRO, ALFREDO PULVIRENTI AND ROSALBA GIUGNO

13. DNA Methylation Biomarkers in Lung Cancer

JUAN SANDOVAL AND PAULA LOPEZ SERRA

14. DNA Methylation Alterations as Biomarkers for Prostate Cancer

JOÃO RAMALHO-CARVALHO, RUI HENRIQUE AND CARMEN JERÓNIMO

15. DNA Methylation in Breast Cancer

RAIMUNDO CERVERA, ALBERTO RAMOS, ANA LLUCH AND JOAN CLIMENT

16. DNA Methylation in Obesity and Associated Diseases

ANA B. CRUJEIRAS AND ANGEL DIAZ-LAGARES

17. DNA Methylation Biomarkers in Asthma and Allergy

AVERY DeVRIES AND DONATA VERCELLI

18. DNA Methylation for Prediction of Adverse Pregnancy Outcomes

DEEPALI SUNDRANI, VINITA KHOT AND SADHANA JOSHI

19. Epigenetics in Infectious Diseases

GENEVIEVE SYN, JENEFER M. BLACKWELL AND SARRA E. JAMIESON

20. DNA Methylation in Neurodegenerative Diseases

SAHAR AL-MAHDAWI, SARA ANJOMANI VIRMOUNI AND MARK A. POOK

21. The Histone Code and Disease: Posttranslational Modifications as Potential Prognostic Factors for Clinical Diagnosis

NICOLAS G. SIMONET, GEORGE RASTI AND ALEJANDRO VAQUERO

22. Histone Posttranslational Modifications and Chromatin Remodelers in Prostate Cancer

FILIPA QUINTELA VIEIRA, DIOGO ALMEIDA-RIOS, INÊS GRAÇA, RUI HENRIQUE AND CARMEN JERÓNIMO

23. Histone Posttranslational Modifications in Breast Cancer and Their Use in Clinical Diagnosis and Prognosis

LUCA MAGNANI, ANNITA LOULOUPI AND WILBERT ZWART

24. Histone Variants and Posttranslational Modifications in Spermatogenesis and Infertility

JUAN AUSIO, YINAN ZHANG AND TOYOTAKA ISHIBASHI

25. Circulating Histones and Nucleosomes as Biomarkers in Sepsis and Septic Shock

JOSÉ LUIS GARCÍA GIMÉNEZ, CARLOS ROMÁ MATEO, MARTA SECO CERVERA, JOSÉ SANTIAGO IBAÑEZ CABELLOS AND FEDERICO V. PALLARDÓ

26. Chromatin Landscape and Epigenetic Signatures in Neurological Disorders: Emerging Perspectives for Biological and Clinical Research

PAMELA MILANI AND ERNEST FRAENKEL

27. MicroRNA Deregulation in Lung Cancer and Their Use as Clinical Tools

PAULA LOPEZ-SERRA AND JUAN SANDOVAL

28. Circulating Nucleic Acids as Prostate Cancer Biomarkers

CLAIRE E. FLETCHER, AILSA SITA-LUMSDEN, AKIFUMI SHIBAKAWA AND CHARLOTTE L. BEVAN

29. MicroRNAs in Breast Cancer and Their Value as Biomarkers

OLAFUR ANDRI STEFANSSON

30. MicroRNAs in Bone and Soft Tissue Sarcomas and Their Value as Biomarkers

TOMOHIRO FUJIWARA, YU FUJITA, YUTAKA NEZU, AKIRA KAWAI, TOSHIFUMI OZAKI AND TAKAHIRO OCHIYA

31. miRNAs in the Pathophysiology of Diabetes and Their Value as Biomarkers

EOIN BRENNAN, AARON McCLELLAND, SHINJI HAGIWARA, CATHERINE GODSON AND PHILLIP KANTHARIDIS

List of Contributors

Carolina Abril-Tormo IBSP-CV Biobank, FISABIO, Valencia, Spain

Sahar Al-Mahdawi Department of Life Sciences, College of Health & Life Sciences, Brunel University London, Uxbridge, UK; Synthetic Biology Theme, Institute of Environment, Health & Societies, Brunel University London, Uxbridge, UK

Diogo Almeida-Rios Cancer Biology and Epigenetics Group – Research Center, Portuguese Oncology Institute, Porto, Portugal; Department of Pathology, Portuguese Oncology Institute, Porto, Portugal

Sara Anjomani Virmouni Department of Life Sciences, College of Health & Life Sciences, Brunel University London, Uxbridge, UK; Synthetic Biology Theme, Institute of Environment, Health & Societies, Brunel University London, Uxbridge, UK

Juan Ausio Department of Biochemistry and Microbiology, University of Victoria, Victoria, BC, Canada

Bharati Bapat Lunenfeld Tanenbaum Research Institute, Mount Sinai Hospital, Toronto, ON, Canada; Department of Laboratory Medicine and Pathobiology, University of Toronto, Toronto, ON, Canada; Department of Pathology, University Health Network, Toronto, ON, Canada

Charlotte L. Bevan Department of Surgery & Cancer, Imperial Centre for Translational & Experimental Medicine, Imperial College London, Hammersmith Hospital Campus, London, UK

Jenefer M. Blackwell Telethon Kids Institute, The University of Western Australia, Subiaco, WA, Australia

Tiziana Bonaldi Department of Experimental Oncology, European Institute of Oncology, Milano, Italy

Abdelhalim Boukaba Drug Discovery Pipeline, Guangzhou Institutes of Biomedicine and Health, Chinese Academy of Science, Guangzhou, People's Republic of China

Eoin Brennan Diabetic Complications Division, Baker IDI Heart and Diabetes Institute, Melbourne, VIC, Australia

Enrique J. Busó Unidad Central de Investigación, University of Valencia, Valencia, Spain

F. Javier Carmona Sanz Human Oncology & Pathogenesis Program (HOPP), Memorial Sloan-Kettering Cancer Center (MSKCC), New York, NY, USA; Cancer Epigenetics and Biology Program (PEBC), Bellvitge Institute for Biomedical Research (IDIBELL), Barcelona, Spain

Raimundo Cervera Hematology and Oncology Unit, Biomedical Research Institute INCLIVA, Valencia, Spain

Alfredo Ciccodicola Institute of Genetics and Biophysics "Adriano Buzzati-Traverso", CNR, Naples, Italy; Department of Science and Technology, University Parthenope of Naples, Italy

Joan Climent Hematology and Oncology Unit, Biomedical Research Institute INCLIVA, Valencia, Spain

Valerio Costa Institute of Genetics and Biophysics "Adriano Buzzati-Traverso", CNR, Naples, Italy

Ana B. Crujeiras Laboratory of Molecular and Cellular Endocrinology, Instituto de Investigación Sanitaria (IDIS), Complejo Hospitalario Universitario de Santiago (CHUS) and Santiago de Compostela University (USC), Santiago de Compostela, Spain; CIBER Fisiopatología de la Obesidad y la Nutrición (CIBERobn), Madrid, Spain

Alessandro Cuomo Department of Experimental Oncology, European Institute of Oncology, Milano, Italy

Avery DeVries Department of Cellular and Molecular Medicine, Arizona Respiratory Center and Arizona Center for the Biology of Complex Diseases, University of Arizona, Tucson, AZ, USA

Angel Diaz-Lagares Cancer Epigenetics and Biology Program (PEBC), Bellvitge Biomedical Research Institute (IDIBELL), Barcelona, Spain

Roberta Esposito Institute of Genetics and Biophysics "Adriano Buzzati-Traverso", CNR, Naples, Italy

Alessandro Fatica Department of Biology and Biotechnology Charles Darwin, Sapienza University of Rome, Rome, Italy

Alfredo Ferro Department of Clinical and Experimental Medicine, University of Catania, Catania, Italy

Claire E. Fletcher Department of Surgery & Cancer, Imperial Centre for Translational & Experimental Medicine, Imperial College London, Hammersmith Hospital Campus, London, UK

Ana-Maria Florea Department of Neuropathology, Heinrich-Heine-University Düsseldorf, Dusseldorf, Germany

Ernest Fraenkel Department of Biological Engineering, Massachusetts Institute of Technology, Cambridge, MA, USA

Yu Fujita Division of Molecular and Cellular Medicine, National Cancer Center Research Institute, Tokyo, Japan

Tomohiro Fujiwara Department of Orthopaedic Surgery, Okayama University Graduate School of Medicine, Dentistry, and Pharmaceutical Sciences, Okayama, Japan; Center for Innovative Clinical Medicine, Okayama University Hospital, Okayama, Japan; Division of Molecular and Cellular Medicine, National Cancer Center Research Institute, Tokyo, Japan

Miriam Gagliardi Institute of Genetics and Biophysics "Adriano Buzzati-Traverso", CNR, Naples, Italy

José Luis García-Giménez Center for Biomedical Network Research on Rare Diseases (CIBERER), National Institute of Health Carlos IIII, Spain; Department of Physiology, Medicine and Dentistry School, University of Valencia, Valencia, Spain; Biomedical Research Institute INCLIVA, University of Valencia, Valencia, Spain

Rosalba Giugno Department of Clinical and Experimental Medicine, University of Catania, Catania, Italy

Catherine Godson Conway Institute, Diabetes Complications Research Centre, University College Dublin, Dublin, Ireland

Inês Graça Cancer Biology and Epigenetics Group – Research Center, Portuguese Oncology Institute, Porto, Portugal; School of Allied Health Sciences (ESTSP), Polytechnic of Porto, Porto, Portugal

Kirsten Grønbæk Department of Hematology, Rigshospitalet, Copenhagen, Denmark

Shinji Hagiwara Diabetic Complications Division, Baker IDI Heart and Diabetes Institute, Melbourne, VIC, Australia

Rui Henrique Cancer Biology & Epigenetics Group, IPO-Porto Research Center (CI-IPOP), Portuguese Oncology Institute, Porto, Portugal; Department of Pathology, Portuguese Oncology Institute, Porto, Portugal; Department of Pathology and Molecular Immunology, Institute of Biomedical Sciences Abel Salazar – University of Porto (ICBAS-UP), Porto, Portugal

José Santiago Ibañez Cabellos Center for Biomedical Network Research on Rare Diseases, Medicine and Dentistry School, University of Valencia, Valencia, Spain; Biomedical Research Institute INCLIVA, Valencia, Spain

Marisa Iborra Gastroenterology Department, Hospital Universitari i Politècnic La Fe, Valencia, Spain

Toyotaka Ishibashi Division of Life Science, Hong Kong University of Science and Technology, Kowloon, Hong Kong, HKSAR; Department of Biomedical Engineer, Hong Kong University of Science and Technology, Kowloon, Hong Kong, HKSAR

Sarra E. Jamieson Telethon Kids Institute, The University of Western Australia, Subiaco, WA, Australia

Carmen Jerónimo Cancer Biology & Epigenetics Group, IPO-Porto Research Center (CI-IPOP), Portuguese Oncology Institute, Porto, Portugal; Department of Pathology and Molecular Immunology, Institute of Biomedical Sciences Abel Salazar – University of Porto (ICBAS-UP), Porto, Portugal

Sadhana Joshi Department of Nutritional Medicine, Interactive Research School for Health Affairs, Bharati Vidyapeeth University, Pune, Maharashtra, India

Phillip Kantharidis Diabetic Complications Division, Baker IDI Heart and Diabetes Institute, Melbourne, VIC, Australia

Akira Kawai Department of Musculoskeletal Oncology, National Cancer Center Hospital, Tokyo, Japan

Vinita Khot Department of Nutritional Medicine, Interactive Research School for Health Affairs, Bharati Vidyapeeth University, Pune, Maharashtra, India

Lasse Sommer Kristensen Department of Hematology, Rigshospitalet, Copenhagen, Denmark

Ana Lluch Hematology and Oncology Unit, Biomedical Research Institute INCLIVA, Valencia, Spain

José Antonio López-Guerrero Laboratory of Molecular Biology and Biobank, Fundacion Instituto Valenciano de Oncologia, Valencia, Spain

Paula Lopez-Serra Epigenetic and Cancer Biology Program (PEBC), Bellvitge Biomedical Research Institute (IDIBELL), Barcelona, Spain

Annita Louloupi Division of Molecular Pathology, The Netherlands Cancer Institute, Amsterdam, The Netherlands

Luca Magnani Department of Surgery and Cancer Imperial Centre for Translational and Experimental Medicine, Imperial College Hammersmith, London, UK

Jacobo Martínez-Santamaría IBSP-CV Biobank, FISABIO, Valencia, Spain; Valencian Biobank Network, FISABIO, Valencia, Spain

Maria R. Matarazzo Institute of Genetics and Biophysics "Adriano Buzzati-Traverso", CNR, Naples, Italy

Aaron McClelland Diabetic Complications Division, Baker IDI Heart and Diabetes Institute, Melbourne, VIC, Australia

Pamela Milani Department of Biological Engineering, Massachusetts Institute of Technology, Cambridge, MA, USA

Yutaka Nezu Division of Molecular and Cellular Medicine, National Cancer Center Research Institute, Tokyo, Japan

Roberta Noberini Center of Genomic Science, Istituto Italiano di Tecnologia, Milano, Italy

Takahiro Ochiya Division of Molecular and Cellular Medicine, National Cancer Center Research Institute, Tokyo, Japan

Toshifumi Ozaki Department of Orthopaedic Surgery, Okayama University Graduate School of Medicine, Dentistry, and Pharmaceutical Sciences, Okayama, Japan

Federico V. Pallardó Center for Biomedical Network Research on Rare Diseases, Medicine and Dentistry School, University of Valencia, Valencia, Spain; Biomedical Research Institute INCLIVA, Valencia, Spain

Lorena Peiró-Chova INCLIVA Biobank, INCLIVA Biomedical Research Institute, Valencia, Spain

Marco Pellegrini Laboratory of Integrative Systems Medicine (LISM), Institute of Informatics and Telematics (IIT) and Institute of Clinical Physiology (IFC), National Research Council (CNR), Pisa, Italy

Tandy L.D. Petrov Department of Biology, The University of Alabama at Birmingham, Birmingham, AL, USA

Olga Bahamonde Ponce INCLIVA Biobank, INCLIVA Biomedical Research Institute, Valencia, Spain

Mark A. Pook Department of Life Sciences, College of Health & Life Sciences, Brunel University London, Uxbridge, UK; Synthetic Biology Theme, Institute of Environment, Health & Societies, Brunel University London, Uxbridge, UK

Alfredo Pulvirenti Department of Clinical and Experimental Medicine, University of Catania, Catania, Italy

João Ramalho-Carvalho Cancer Biology & Epigenetics Group, IPO-Porto Research Center (CI-IPOP), Portuguese Oncology Institute, Porto, Portugal

Alberto Ramos Hematology and Oncology Unit, Biomedical Research Institute INCLIVA, Valencia, Spain

George Rasti Chromatin Biology Laboratory, Cancer Epigenetics and Biology Program (PEBC), Bellvitge Biomedical Research Institute (IDIBELL), Barcelona, Spain

Nicole C. Riddle Department of Biology, The University of Alabama at Birmingham, Birmingham, AL, USA

Peter H.J. Riegman Department of Pathology, Erasmus Medical Center, Rotterdam, The Netherlands

Carlos Romá Mateo Center for Biomedical Network Research on Rare Diseases, Medicine and Dentistry School, University of Valencia, Valencia, Spain; Biomedical Research Institute INCLIVA, Valencia, Spain

Francesco Russo Laboratory of Integrative Systems Medicine (LISM), Institute of Informatics and Telematics (IIT) and Institute of Clinical Physiology (IFC), National Research Council (CNR), Pisa, Italy; Department of Computer Science, University of Pisa, Pisa, Italy

Fabian Sanchis-Gomar Department of Physiology, Medicine and Dentistry School, University of Valencia, Valencia, Spain; Biomedical Research Institute INCLIVA, University of Valencia, Valencia, Spain

Juan Sandoval Epigenetic and Cancer Biology Program (PEBC), Bellvitge Biomedical Research Institute (IDIBELL), Barcelona, Spain

Flavia Scoyni Department of Biology and Biotechnology Charles Darwin, Sapienza University of Rome, Rome, Italy

Marta Seco Cervera Center for Biomedical Network Research on Rare Diseases, Medicine and Dentistry School, University of Valencia, Valencia, Spain; Biomedical Research Institute INCLIVA, Valencia, Spain

Akifumi Shibakawa Department of Surgery & Cancer, Imperial Centre for Translational & Experimental Medicine, Imperial College London, Hammersmith Hospital Campus, London, UK

Nicolas G. Simonet Chromatin Biology Laboratory, Cancer Epigenetics and Biology Program (PEBC), Bellvitge Biomedical Research Institute (IDIBELL), Barcelona, Spain

Ailsa Sita-Lumsden Department of Surgery & Cancer, Imperial Centre for Translational & Experimental Medicine, Imperial College London, Hammersmith Hospital Campus, London, UK

Olafur Andri Stefansson Cancer Research Laboratory, Faculty of Medicine, University of Iceland, Reykjavik, Iceland

Deepali Sundrani Department of Nutritional Medicine, Interactive Research School for Health Affairs, Bharati Vidyapeeth University, Pune, Maharashtra, India

Genevieve Syn Telethon Kids Institute, The University of Western Australia, Subiaco, WA, Australia

Trygve O. Tollefsbol Comprehensive Cancer Center, Center for Aging, Comprehensive Diabetes Center, Nutrition Obesity Research Center, Cell Senescence Culture Facility, University of Alabama at Birmingham, Birmingham, AL, USA

Eneda Toska Human Oncology & Pathogenesis Program (HOPP), Memorial Sloan-Kettering Cancer Center (MSKCC), New York, NY, USA

Marianne B. Treppendahl Department of Hematology, Rigshospitalet, Copenhagen, Denmark

Toshikazu Ushijima Chief of Division of Epigenomics, National Cancer Center Research Institute, Tokyo, Japan

Alejandro Vaquero Chromatin Biology Laboratory, Cancer Epigenetics and Biology Program (PEBC), Bellvitge Biomedical Research Institute (IDIBELL), Barcelona, Spain

Donata Vercelli Department of Cellular and Molecular Medicine, Arizona Respiratory Center and Arizona Center for the Biology of Complex Diseases, University of Arizona, Tucson, AZ, USA

Filipa Quintela Vieira Cancer Biology and Epigenetics Group – Research Center, Portuguese Oncology Institute, Porto, Portugal; School of Allied Health Sciences (ESTSP), Polytechnic of Porto, Porto, Portugal

Yinan Zhang Division of Life Science, Hong Kong University of Science and Technology, Kowloon, Hong Kong, HKSAR

Fang Zhao Lunenfeld Tanenbaum Research Institute, Mount Sinai Hospital, Toronto, ON, Canada; Department of Laboratory Medicine and Pathobiology, University of Toronto, Toronto, ON, Canada

Wilbert Zwart Division of Molecular Pathology, The Netherlands Cancer Institute, Amsterdam, The Netherlands

Preface

Epigenetics is an emerging frontier of biology, and its definition is continuously being adapted based on new scientific findings. In fact, the NIH Roadmap Epigenomics Project has recently defined epigenetics as "the heritable changes in gene activity and expression (in the progeny of cells or of individuals) and also stable, long-term alterations in the transcriptional potential of a cell that are not necessarily heritable." In this regard, epigenetics includes DNA methylation, noncoding RNAs, and histone posttranslational modifications. This integrative definition of epigenetics reflects the potential of the discipline to expand beyond the control of a particular gene expression program for each cell type, defining the cellular and developmental identity and function of cells, and, finally, translating this potential to health and disease conditions in human beings.

Due to the rapid progress in the field of epigenetics, new and promising methodologies to advance biomedical research are being developed. Epigenetic research and epigenetic pharmaceutical drug development are now considered areas of great interest and promise in the biomedical scene. The advantage of human epigenetics compared with human genetics and genomics is that it provides vital information about gene function in individual cell types, while incorporating information from the environment and lifestyle, and unlike most genetic defects causative of human disease, epigenetic alterations are modulable and reversible. The volume *Epigenetic Biomarkers and Diagnostics* is intended to describe both epigenetic biomarkers that can be adopted into clinical routine as well as advanced technologies and tools for their analysis. In this regard, epigenetic biomarkers provide clinicians valuable information about the presence or absence of a disease (diagnostic value), the patient prognosis (prognostic value), the response to a specific treatment (predictive value), the effects of ongoing treatment (therapy-monitoring biomarkers), and the future risk of disease development (risk prediction). Furthermore, several advantages may arise from the use of epigenetic biomarkers versus gene expression in clinical practice, such as higher stability, for example, in biofluids. They can also fill clinical gaps by bridging genetic information, mRNA transcription, and protein translation. In consequence, the associations between epigenome alterations and diseases become clearer, providing a way to act directly on gene expression by developing specific drugs or even by adopting healthy lifestyles.

Many methodologies, including classical methods and next-generation-sequencing-based technologies, are available to clinicians and researchers to identify new epigenetic biomarkers and analyze them from several biological sources. In this context, genome-wide methylation analysis, chromatin immunoprecipitation coupled with high-throughput platforms, and noncoding RNA sequencing are described in this volume. Furthermore, some techniques for DNA methylation analysis are more likely to be rapidly adopted in clinical laboratories, such as EpiTYPER MassARRAY, methyl specific PCR (MSP), and pyrosequencing. On the other hand, immunoassays are well established in clinical laboratories for the analysis of histones (i.e., inflammatory and autoimmune diseases). However, it is expected that the incorporation of mass spectrometry technologies into laboratories for clinical diagnostics (replacing routine immunoassays) will be the tendency in the coming years, as will be the analysis of histone posttranslational modifications associated with pathological states.

Although it is not possible to cover all epigenetic markers, this volume includes chapters describing the most promising biomarkers for cancer (i.e., breast, lung, colon, etc.), metabolic disorders (i.e., diabetes and obesity), autoimmune diseases, infertility, allergy, infectious diseases, and neurological disorders; and, where possible, we will focus our attention on those which are feasible to be adopted for clinical use.

This book was written in a comprehensive manner by outstanding experts in their corresponding fields for a broad target audience such as advanced students, basic scientists, biomedical and biotechnological companies, as well as clinical researchers, clinicians (i.e., pathologists, immunologists, oncologists, endocrinologists, etc.) and analysts from clinical laboratories who can adopt these potential biomarkers into clinical practice.

In the coming years, epigenetics will continue to provide an exciting future in biomedicine and clinical practice. The chapters covered in *Epigenetic Biomarkers and Diagnostics* highlight the unprecedented impact of epigenetics in clinical diagnostics and will contribute to the discovery and development of new epigenetic biomarkers in the future.

José Luis García-Giménez
Valencia, Spain

CHAPTER

1

Epigenetic Biomarkers: New Findings, Perspectives, and Future Directions in Diagnostics

José Luis García-Giménez[1], *Toshikazu Ushijima*[2], *Trygve O. Tollefsbol*[3]

[1]Department Physiology, Center for Biomedical Network Research on Rare Diseases, National Institute of Health Carlos IIII, Institute of Health Research INCLIVA, Medicine and Dentistry School, University of Valencia, Valencia, Spain; [2]Chief of Division of Epigenomics, National Cancer Center Research Institute, Tokyo, Japan; [3]Comprehensive Cancer Center, Center for Aging, Comprehensive Diabetes Center, Nutrition Obesity Research Center, Cell Senescence Culture Facility, University of Alabama at Birmingham, Birmingham, AL, USA

OUTLINE

Epigenetic Biomarkers and Diagnostics
http://dx.doi.org/10.1016/B978-0-12-801899-6.00001-2

1. INTRODUCTION

The literal meaning of the term epigenetic is "above or on top of genetics," although it has had many different definitions over the years. Conrad Hal Waddington was the first to define epigenetics, in 1942, as "the branch of biology which studies the causal interaction between genes and their products, which bring the phenotype into being" [1]. In 1990, Holiday defined epigenetics as "the study of the mechanisms of temporal and spatial control of gene function during the development of organisms" [2]. A few years later, epigenetics was defined in a narrower manner, as the different epigenetic modifications or mechanisms that produce heritable changes affecting gene expression without affecting the DNA sequence. In 2001, Jenuwein and Allis proposed the histone code as an epigenetic mechanism which is a critical feature of a genome-wide mechanism of information storage and retrieval and that considerably expands the information potential of the genetic code. Therefore, they pointed out that epigenetics imparts a fundamental regulatory system beyond the sequence information of our genetic code [3]. By 2007, the definition of epigenetics had changed yet again, when Bird defined epigenetics as "the structural adaptation of chromosomal regions so as to register, signal or perpetuate altered activity states" [4]. The same year, Goldberg, Allis, and Bernstein defined epigenetics as "the study of any potentially stable and, ideally, heritable change in gene expression or cellular phenotype that occurs without changes in Watson-Crick base pairing of DNA" [5]. It is evident that both definitions proposed by Bird and Goldberg et al. were inclusive of transient chemical modifications of DNA and histones; modifications that have not always been universally accepted and that are still a subject of debate. Therefore, in 2008, a consensus definition was made at a Cold Spring Harbor meeting, giving a more integrative definition for epigenetics as a "stably heritable phenotype resulting from changes in a chromosome without alterations in the DNA sequence" [6]. It is evident that epigenetics is an emerging frontier of science, so its definition is continuously being adapted based on new scientific findings. In fact, the NIH Roadmap Epigenomics Project defines epigenetics as "the heritable changes in gene activity and expression (in the progeny of cells or of individuals) and also stable, long-term alterations in the transcriptional potential of a cell that are not necessarily heritable" [7] (www.roadmapepigenomics.org). In this regard, epigenetics includes DNA methylation, noncoding RNAs (ncRNAs), and histone posttranslational modifications (PTMs). This last integrative definition of epigenetics reflects the potential of epigenetics to go beyond the control of a particular gene expression to produce a unique gene expression program of each cell type, defining the cellular and developmental identity [8], as well as potential health and disease outcomes [9].

Our epigenome is characterized by its ability to dynamically respond to intra- and extracellular stimuli, so that epigenetic changes are reversible and in consequence are potential contributors to health and disease. Recent progress in the field of epigenetics opens promising ways to advance biomedical research. Epigenetic research and epigenetic pharmaceutical drug development are now considered a bright spot in the biomedical research field. This research will contribute to biomarker discovery, new therapy development, computational and bioinformatics training, and the development of new next-generation sequencing (NGS) technologies and applications. The advantage of epigenetics compared to genetics is that it provides vital information about gene function in individual cell types and incorporates information from the environment. Furthermore, several advantages may arise from the use of epigenetic biomarkers versus gene expression (by measuring mRNAs). Epigenetic biomarkers have shown higher stability in fluids and formalin-fixed paraffin-embedded (FFPE) biospecimens compared to mRNAs. In addition, epigenetic biomarkers can benefit biomedical research and fill clinical research gaps by bridging genetic information and mRNA expression. In consequence, this makes the association between epigenome alterations and diseases clearer, and also provides a way to act directly on them by developing specific drugs or adopting healthy lifestyles. One of the most important issues is that the associations that have been found between epigenetics and certain diseases will have a synergistic effect on the development of personalized medicine. In recent years, epigenetics has aroused the interest of different biomedical and scientific fields, and it also has impacted society. In 2010, *TIME Magazine* published "Why Your DNA Isn't Your Destiny" opening the public's eyes to the latest research and changing the general view about DNA and its direct role in our lives.

2. EPIGENETIC MECHANISMS

Three major events are mainly involved in epigenetic regulation and chromatin structure control: DNA methylation, histone PTMs, and ncRNAs (i.e., microRNAs (miRNAs), long noncoding RNAs (lncRNAs)). Disruption of one or more of these epigenetic mechanisms can lead to inappropriate expression of genes, resulting in an alterated state of cell homeostasis or disease. In this regard, epigenetic-based biomarkers are an important new research area. With the potent technologies now available, diagnostic tools can be created to analyze these biomarkers and therefore contribute to the study of human diseases. Here, we summarize the three most relevant mechanisms and discuss the technologies available to analyze them.

2.1 DNA Methylation

DNA methylation is the most-studied DNA modification since its discovery in the late 1940s. It consists of the addition of a methyl group at the 5′ position of the cytosine base (5-methylcytosine (5mC)), which protrudes into the major groove of DNA, representing a potential recognition site for protein binding without changing the Watson–Crick base pairing. The methyl group at the 5′ position of cytosine is donated by S-adenosylmethionine (SAM) via Dnmt1, Dnmt3A, and Dnmt3B (DNA(cytosine-5-)methyltransferases (DNMTs)). Importantly, CpG sites are underrepresented and unevenly distributed across the human genome, giving rise to vast low-density CpG regions interspersed with CpG clusters located mainly in CpG islands [10]. Generally speaking, 5mCs play essential roles in maintaining cellular function and genome stability.

In cancers, global DNA hypomethylation and regional hypermethylation are almost always observed and have been associated with genome instability, altered chromatin conformation, and chromosome fragility [11–13]. At the same time, hypermethylation of CpG islands [14] and their flanking regions, called CpG shores [15], is also observed in specific CpG islands, including those in promoters of tumor suppressor genes (driver methylation) and other genes methylated in association with cancer development (passenger methylation). Importantly, DNA methylation can be detected by a wide range of sensitive and cost-effective techniques [16,17], as described in Chapters 6–9 of Section II in this volume. Moreover, DNA methylation is a stable chemical mark that is not easily altered, which makes it a feasible biomarker for diagnostics, prognostics, and treatment monitoring. The field of DNA methylation has expanded with the identification of multiple cytosine variants thanks to the design of new methodologies. Ten-eleven translocation (Tet) family of cytosine oxygenase enzymes (TETs) are responsible for oxidizing 5mCs into 5-hydroxymethylcytosine (5hmC), 5-formylcytosine (5fC), and 5-carboxylcytosine (5caC) by means of an α-ketoglutarate- and O_2-Fe(II)-dependent reactions (Figure 1). Recent evidence reveals that these DNA demethylation

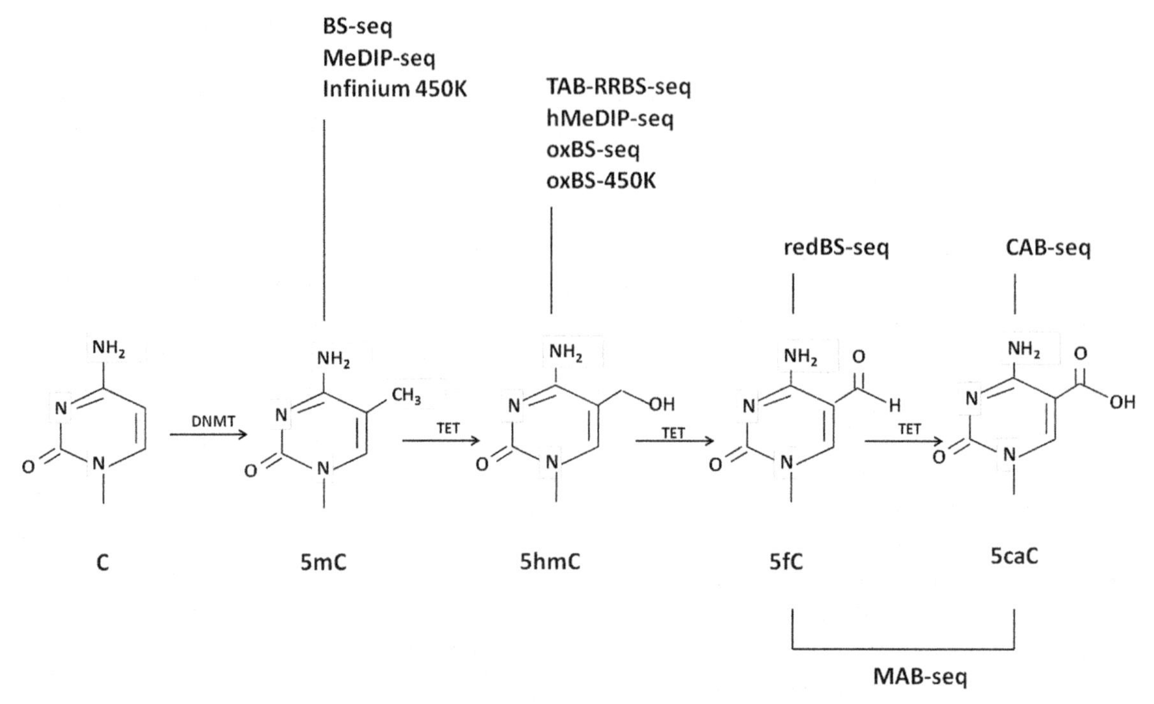

FIGURE 1 Dynamic DNA methylation and demethylation catalyzed by DNA(cytosine-5-)methyltransferases (DNMTs) and ten-eleven translocase (TET) proteins. 5-methylcytosine (5mC) is produced by the activity of DNMTs using S-adenosylmethionine as methyl donor. 5mC can be converted to 5-hydroxymethylcytosine (5hmC), 5-formylcytosine (5fC), and 5-carboxylcytosine (5caC) by iron-dependent dioxygenases TET proteins. Some methods exist to identify the cytosine modifications at single-base resolution in whole-genome studies. Bisulfite genomic sequencing (BS-seq), Infinium 450K DNA methylation (Illumina), and direct immunoprecipitation of methylated DNA (MeDIP-seq) are used for 5mC studies. DNA pull-down using 5hmC antibodies in combination with genome-wide sequencing (hMeDIP-seq), Tet-assisted (bisulfite) reduced representation bisulfite sequencing (TAB-RRBS), and oxidative bisulfite coupled to 450K BeadChip (oxBS-450K) to study 5hmC. The method methylation-assisted bisulfite sequencing (MAB-seq) serves for mapping genome-wide 5fC and 5caC. However, using the reduced bisulfite sequencing (redBS-seq) or chemical modification-assisted bisulfite sequencing it is possible to identify 5fC or 5caC, respectively.

intermediates are dynamic and participate in key regulatory functions in distal gene regulatory elements in mammalian genomes and also in biological processes, such as demethylation in zygote formation and in germ cell lineage (for a review see Ref. [18]).

Novel experimental designs consisting of chemical oxidation or chemical reduction of the modified cytosines allow the identification of 5mC, 5hmC, 5fC, and 5caC at single-base resolution [19–21] and, when implemented in DNA sequencing protocols, will allow the analysis of whole genomes and the exploitation of these procedures in biomedicine and clinical research. In this regard, high-throughput methods for analyzing cytosine variants have appeared in recent years. Bisulfite sequencing (BS-seq), Infinium HumanMethylation 450K BeadChip (Illumina), and immunoprecipitation of methylated DNA followed by sequencing (MeDIP-seq) have been developed to analyze 5mC. On the other hand, Tet-assisted bisulfite sequencing (TAB-seq), 5hmC immunoprecipitation coupled to sequencing, oxidative bisulfite sequencing (oxBS-seq), and oxidative bisulfite hybridization in the Infinium 450K BeadChip (oxBS-450K) have been designed to map 5mC and 5hmC residues at single-base resolution on a genome-wide scale [22]. Using named methylation-assisted bisulfite sequencing (MAB-seq), the quantitation of the 5fC and 5caC residues at single-base resolution is possible [23], and using chemical modification-assisted bisulfite sequencing (CAB-seq) the specific whole-genome analysis of 5caC is possible [24]. By using reduced bisulfite sequencing (redBS-seq), it is possible to identify 5fC [25].

As shown in Figure 1, these different directed assays have been developed to study the cytosine variants which participate in the regulation of DNA methylation. Therefore, these methodologies will contribute to explaining the intriguing epigenetic regulation underlying human pathologies [26]. It is obvious that the direct application of these high-throughput procedures in clinical routine seems to be so far from reality. However, these techniques will contribute to the identification of passenger methylation underlying several human diseases and also to develop clinical epigenetics.

2.2 Histone PTMs and Histone Variants

Chromatin is composed of repeating arrays of nucleosomes which comprise the essential units of organization in eukaryotic chromosomes. Each nucleosome is formed by 145 bp of DNA wrapped around a histone octamer. Each histone octamer consists of two copies each of canonical histone H4, H3, H2A, and H2B or their variants. Histone can be chemically modified and when it is, chromatin state and gene expression are considered to be modified as well. The histone PTMs produced (i.e., acetylation, methylation, phosphorylation, butyrylation, hydroxybutyrylation, crotonylation, citrullination, formylation, glycosylation, O-GlcNAcylation [27], carbonylation, parsylation [28], and glutathionylation [29]) mainly occur on amino acids in the N-terminal tail domains (Figure 2). These chemical modifications change the nucleosome structure and spread to different regions of the genome. The hypothesis of the histone code proposes that the control of the chromatin state is exerted by the PTMs [30]. Evidence accumulated in recent years supports the idea that the PTM signature is altered in a wide range of diseases, such as cancer [31,32], neurological syndromes [33,34], and rare diseases such as Rubinstein–Taybi syndrome [35] and Cofin–Lowry syndrome [36]. In addition, clinicians have become increasingly interested in histone PTMs in recent years because it is possible to analyze them in specific regulatory domains of genes to obtain valuable information for the diagnostics of disease [17].

Another additional level in the supramolecular structural organization of chromatin involves the participation of histone variants (Figure 2). Histone variants differ in their primary sequence of amino acids [37]. Variants for all of the histone protein families have been described and have emerged as important elements involved in chromatin dynamics and organization, and have also served as scaffolds for other proteins that create silenced regions in chromosomes [38,39]. Therefore, the incorporation of histone variants in specific domains is very important, and their regulation plays a crucial role in cellular processes such as differentiation, proliferation, and nuclear reprogramming. For this reason, mutations in specific histone variants are involved in human disease [40] like cancer, as recently reviewed by Vardabasso et al. [41]. Due to the importance of histone variants in cell differentiation and reprogramming it is not surprising that they are closely related to male infertility, as described later in this volume by Ausió, Zhang, and Ishibashi in Chapter 24.

Interestingly, histones can also be found in biological fluids such as blood, serum, and plasma. Therefore, they have been proposed as clinical biomarkers since their presence in body fluids suggests tissue damage, inflammation, and cellular apoptosis [42]. Recent evidence supports the idea that extracellular histones contribute to human disease. Circulating nucleosomes were recently associated with disease

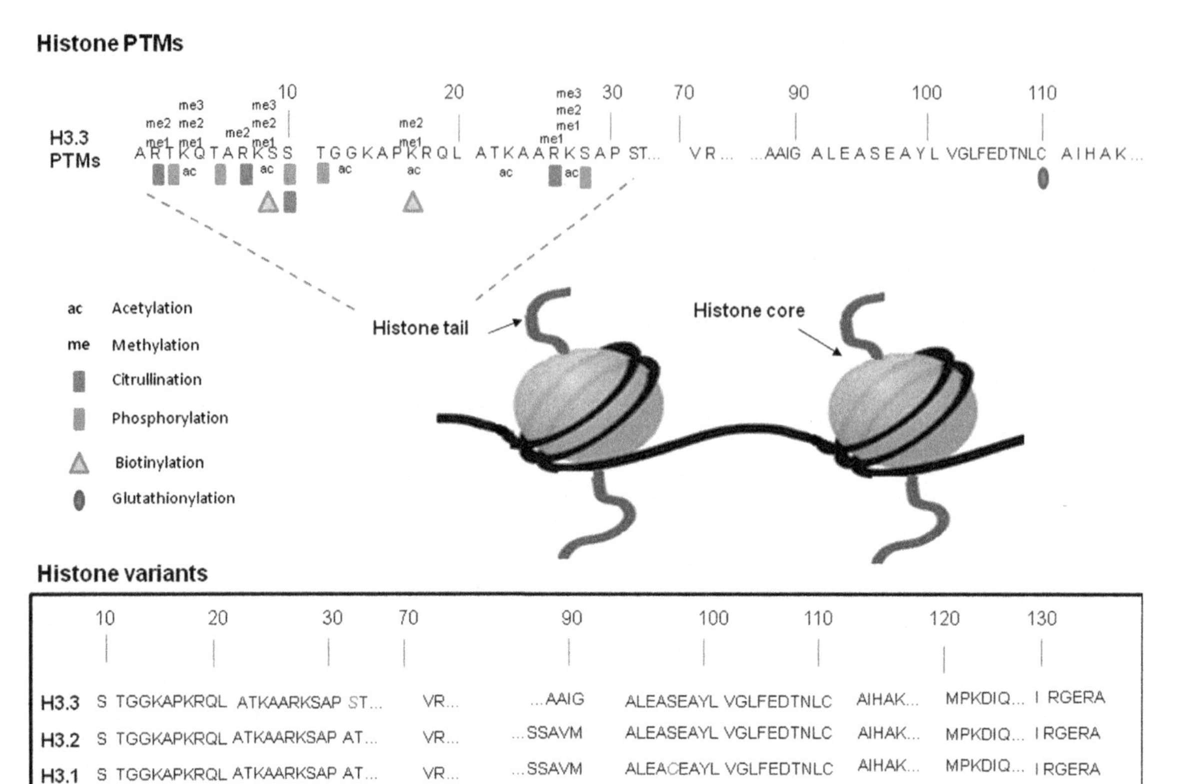

FIGURE 2 **Graphical representation of nucleosomes showing histone octamer and histone tails for histone H3.3.** On the top of the figure is shown partially the primary sequence of histone H3.3 in which main posttranslational modifications (PTMs) in amino acids (acetylation, ac; monomethylation, me1; dimethylation, me2; trimethylation, me3; biotinylation, yellow triangle; citrullination, red square; phosphorylation, green square; and glutathionylation, blue oval) are represented. H3 variants are showed into the box at the bottom of the figure, in which is compared the alignment of human noncentromeric histone H3 variants. Differences in amino acid sequence among H3.3, H3.2, H3.1, H3.1t (testicular variant), and H3.Y are shown in italics (in online version they are shown colored in red).

progression in patients with thrombotic microangiopathies [43]. Ekaney et al. detected higher levels of histone H4 in septic patients compared to patients with multiple organ failure without infection or to patients with minor trauma [44]. Zeerleder et al. found increased nucleosome levels in systemic inflammation and septic shock [45,46] and studied how these levels correlated with the severity of the inflammatory response and mortality in children affected by meningococcal sepsis [46]. Unfortunately, there exists no consensus on the levels of circulating histones and nucleosomes that discriminate patients from healthy subjects, due in part to the diversity of procedures used in their determination.

Currently, different commercial kits are available for the identification of circulating nucleosomes and histones in blood samples of autoimmune disease patients, mainly in drug-induced systemic lupus erythomatosus [47]. Many of them are based on enzyme immunoassays [48]. Furthermore, immunohistochemistry and immunofluorescence-based procedures can be used to analyze histone PTMs. However, a trend toward the use of mass spectrometry (MS)-based methodologies is gaining interest in clinical diagnostic laboratories. MS methodologies can be employed in the analysis of histone PTMs and histone variants. In this volume, Noberini et al. discuss the potential of MS-based technologies in biomarker discovery and their applications in clinical epigenetics, focusing on novel techniques to dissect the histone code in clinical samples (Chapter 10). Importantly, MS holds great promise for clinical epigenetics because its potential allows histone PTM combinations to be dissected.

2.3 Noncoding RNAs

Only a small percentage of the transcribed genes encode proteins, so these genome regions were described as "dark matter RNAs" [49]. ncRNAs are considered active participants in controlling a wide range of biological processes, such as the regulation transcription of single genes, as well as entire transcriptional programs [50]. Many thousands of regulatory nonprotein-coding RNAs were demonstrated to be transcribed in genome-wide studies, including miRNAs, small RNAs, PIWI-interacting RNAs, and various classes of lncRNAs [51]. Most clinical applications are being found for miRNAs and lncRNAs, so we will briefly describe these ncRNAs. Later chapters of this volume are focused on the ncRNA species and their potential as clinical biomarkers.

miRNAs consist of a large family of short ncRNAs (17–25 nucleotides). These ncRNAs are involved in many biological processes [52] (i.e., cellular development, differentiation, apoptosis, proliferation, tumor growth, metastatic dissemination, and resistance to therapy, among others) [53,54] as a consequence of their ability to control the expression (downregulation or upregulation) of numerous genes. High stability and low susceptibility to degradation are important characteristics of miRNAs [17,55,56]. This could be due to their short length, particular biogenesis, and strong association with proteins, or membrane-bound vesicles, such as exosomes, microvesicles, or apoptotic bodies, etc. Importantly, miRNA expression is frequently altered in cancer and has shown promise as a tissue-based, circulating biomarker for cancer classification and prognostics [56]. An interesting pioneer study on this was performed by Volinia et al. They identified many overexpressed miRNAs in several solid tumors by performing a large-scale miRNome analysis and found specific miRNA signatures for each tumor type [57]. Of great clinical importance is that unique miRNA expression patterns can distinguish tumors from different anatomical locations, and also differentiate various tumor subtypes at a single anatomic locus in kidney cancer [58], breast cancer [59], and papillary renal carcinoma [60], among others.

lncRNAs are generally defined as transcripts longer than 200 nucleotides that can be processed like mRNA, i.e., spliced and polyadenylated [61]. These kinds of ncRNAs have been associated with neurological disorders [62], cancer [63], and complex metabolic disorders [64].

There are currently a number of studies showing the interconnection between distinct epigenetic events. For example, a subgroup of ncRNAs are additionally classified as epi-miRNAs because they can directly or indirectly regulate the expression of several components of the epigenetic machinery, such as DNA methyltransferases or histone deacetylases, creating a very well-orchestrated mechanism [65,66] in such a way that the joint activities of different epigenetic modifications result in a common outcome. An altered balance of these processes leads to pathological conditions, so it is important to evaluate the levels of ncRNAs (miRNAs and lncRNAs) in human biospecimens. High-throughput sequencing technologies, computational pipelines, and bioinformatics algorithms allow us to identify the profile of dysregulated ncRNAs in human diseases. In Chapter 11, Costa et al. describe some of these approaches and tools for ncRNA analysis. In addition, Russo et al. (Chapter 12) discuss recent discoveries and controversial topics on circulating ncRNAs and describe several public resources recently developed for ncRNA analysis. In Table 1 the most relevant properties and information of miRNAs and lncRNAs are summarized.

Epigenetic regulation does not always produce gene regulation in a binary system that is either ON or OFF. Epigenetic mechanisms can be propagated over multiple cell divisions in somatic cells. In addition, epigenetic information is modified during cellular differentiation and partially erased in the germ line and the early embryo [67]. We also know that lifestyle, environment, especially chronic inflammation, and nutrition induce epigenetic changes during our life [68–73] and that aberrant placement of epigenetic marks and epigenetic dysregulation (produced by mutations in genes that codify for epigenetic machinery) are involved in disease. Thus, DNA methylation, histone PTMs, and ncRNAs play a critical role in diseases such as cancer, infectious diseases, infertility, and metabolic and neurodegenerative disorders. Some examples are shown in section III of Epigenetic Biomarkers and Diagnostics. Therefore, the identification of these epigenetic changes serves to define healthy or disease states in humans and to set the basis for the identification of epigenetic biomarkers.

3. EPIGENETIC BIOMARKERS AND IN VITRO DIAGNOSTICS

A biomarker is a characteristic that is objectively measured and evaluated as an indicator of normal biologic processes, pathogenic processes, or pharmacologic responses to a therapeutic intervention. The United Nations World Health Organization (WHO) defines a biomarker as any substance, structure, or process that can be measured in the body or its products and which influences or predicts the incidence of outcome of diseases (*Biomarkers in Risk Assessment: Validity and Validation, Environmental Health Criteria Series, No. 222, WHO*). In this regard, the goal of clinical biomarkers is to provide clinicians

TABLE 1 General Information for Noncoding RNAs

Properties	miRNAs	lncRNAs
Length (nt)	17–25	>200
Poly-A tail	No	Yes
No. of sequences described in human	1881[a]	111685[b]
Found in	Cell, tissues, body fluids/exosomes	Cell, tissues, body fluids/exosomes
Stability in body fluids and FFPE biospecimens	High	Low
Isolation	RNA conserving miRNAs or miRNA purification protocols	RNA purification protocols
Procedures for the analysis	RNA-seq, arrays, qRT-PCR	RNA-seq, arrays, qRT-PCR

[a] *miRbase v21 (Kozomara A, Griffiths-Jones S. miRBase: annotating high confidence microRNAs using deep sequencing data. Nucl Acids Res 2014 January; 42(D1): D68–D73).*

[b] *LNCipedia.org v3.1 (Volders P-J, Verheggen K, Menschaert G, Vandepoele K, Martens L, Vandesompele J and Mestdag P. An update on LNCipedia: a database for annotated human lncRNA sequences. Nucl Acids Res 2014 2015 January; 43(D1): D174-D180).*

miRNAs, microRNAs; lncRNAs, long noncoding RNAs; FFPE, formalin-fixed paraffin-embedded; qRT-PCR, quantitative real-time polymerase chain reaction.

with valuable information about the presence or absence of a disease (diagnostic biomarker), the patient prognosis (prognostic biomarker), the response to a specific treatment (predictive biomarker), the effects of ongoing treatment (therapy monitoring biomarkers), or a future risk of disease development (risk markers) [74]. Based on these descriptions, recommended properties for clinical biomarkers are (1) they should be specific, sensitive, and stable; (2) they should be validated by different institutions in a large number of samples followed by approval from the (US Food and Drug Administration) FDA and/or (European Medicines Agency) EMA; and (3) although there exist excellent single markers that serve for diagnostics, the use of biomarker signatures instead of only one biomarker is preferable because the combination of biomarkers increases the sensitivity and specificity [75].

One of the most important properties of epigenetic marks is that they are highly stable (methylated DNA and miRNA) in multiple biospecimens (i.e., urine, blood, plasma, FFPE tissues, etc.). This is the case of miRNAs, which are very stable molecules in blood [56], urine [76], and also in FFPE [55], different from mRNA and proteins, as discussed by Peiró-Chova et al. in Chapter 2.

Based on these precedents, an epigenetic biomarker can be defined as "any epigenetic mark or altered epigenetic mechanism (1) that can be measured in the body fluids or tissues and (2) that defines a disease (detection), predicts the outcome of diseases (prognostic) or response to a therapy (predictive), monitors treatment responses (therapy monitoring), or predicts risk of future disease development (risk)."

In general, epigenetic biomarkers may represent the effect of the environment and natural history on the particular evolution of disease in each patient. Therefore, one of the most promising properties of epigenetic biomarkers is that they will contribute to the improvement of precision medicine. The advantage offered by epigenetic biomarkers versus genetic biomarkers is that the former is a dynamic one which changes based on disease evolution and intra- or extraenvironmental cellular conditions while the latter, in contrast, is a static biomarker since our gene sequence generally does not change.

The potential of epigenetic biomarkers in clinical practice is currently being demonstrated. The in vitro diagnostics (IVD) market is seeking new diagnostic and prognostic biomarkers which contribute to personalized medicine, and epigenetics is currently contributing to this [77,78]. Recent advances in NGS have allowed epigenetics to be used more easily both by researchers and clinicians [79–81].

The exponential growth of the IVD market is evident at the moment, since the identification of biomarkers and the design of IVD tests facilitate diagnostic, prognostic, and treatment monitoring. Another point to consider is that the USFDA encourages the integration of biomarkers into drug development and their appropriate use in clinical practice. In that way, the effective integration of biomarkers into clinical development programs may facilitate new medical product development and promote personalized medicine [82,83]. In recent years, many epigenetic drugs have been discovered, so the development of new, predictive, and sensitive biomarkers for clinical practice is expected to reduce the time and cost of drug development and also contribute to overcoming problems during the evaluation of new therapies during clinical trials [84]. Therefore, the use of epigenetic biomarkers has tremendous potential to affect the success rate of clinical trials and drug discovery. Proof of this is the growth observed in the number of clinical trials using epigenetic drugs during last decade (Figure 3). In fact, epigenetic therapy will be improved considerably after identification of good pretreatment biomarkers predicting response. Many large clinical trials in combination with novel high-throughput screening methods have contributed to these improvements, as described by Treppendahl et al. in Chapter 5.

Specifically, epigenetic biomarkers are coevolving and have reached a critical point

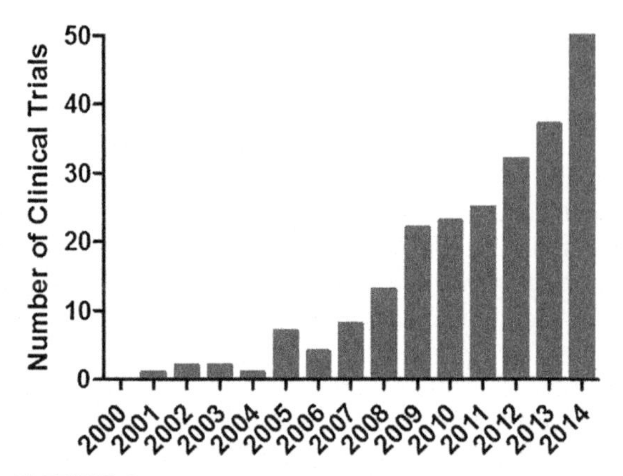

FIGURE 3 **Number of clinical trials using epigenetic drugs and biomarkers reported by ClinicalTrials.gov from 2000 to 2014.** The figure presents the growth rate for clinical trials in different statuses (i.e., not yet recruiting, recruiting, enrolling by invitation, active but not recruiting, completed). The shape of the curve indicates a research area that is reaching amount of interest in clinical practice.

wherein enough signatures have been identified across various disease classes, most notably in cancer, that some of these signatures and expression patterns can be validated and used in the clinic [85,86]. The potential of epigenetic biomarkers in clinical practice has directed them to their use in IVD. In fact, the global epigenetics market was valued at an estimated $413.24 million in 2014 and it is expected to reach $783.17 million in 2019. The IVD market is expected to experience an enormous expansion in the coming years as well, and epigenetics will clearly contribute to that.

4. EPIGENETIC BIOMARKERS AND THE CLINICAL LABORATORY

Many methods, including classical methods and NGS-based methods, are available to clinicians and researchers to identify new epigenetic biomarkers, as reviewed by Petrove and Riddle in Chapter 4. In recent years, technological improvements have allowed researchers to examine the epigenome on a genome-wide fashion.

In Section II of this volume we also provide a description of the most relevant technologies that have contributed to epigenetic biomarker identification and analysis and outline which are feasible and useful technologies to be adopted in clinical laboratories.

These new technological improvements have set the basis for new biomarker identification and analysis. In this context, genome-wide methylation analysis and chromatin immunoprecipitation with high-throughput platforms, including DNA microarrays and genome sequencing, have permitted the performance of comprehensive studies regarding the epigenetic bases of several diseases. These technologies are well described by Toska and Carmona-Sanz in Chapter 6. NGS has also revolutionized the analysis of miRNAs and lncRNAs by allowing the application of NGS-based methods for clinical epigenetics, as described in Chapter 11 by Costa et al. A recent example of this application was published by Keller et al. In this study, the authors analyzed 863 miRNA signatures from 454 human blood biospecimens. The samples were obtained from patients suffering from 14 different diseases, including lung cancer, pancreatic ductal adenocarcinoma, prostate cancer, ovarian cancer, Wilms tumor, multiple sclerosis, periodontitis, sarcoidosis, myocardial infarction, and chronic obstructive pulmonary disease [87]. By utilizing the data of dysregulated miRNAs in the blood and developing mathematical algorithms, Keller et al. were able to accurately predict disease in more than two-thirds of individuals [87]. It is obvious that more ncRNAs are being identified as research advances, so data integration in public databases such as the human miRNA-associated disease database (HMDD) [88] (www.cuilab.cn/hmdd) or miRandola for extracellular circulating miRNAs [89] serve as resources for researchers screening miRNA and lncRNA profiles for a wide range of diseases. These databases are therefore fundamental tools for biomedical research as described by Russo et al. in Chapter 12.

Since DNA methylation was the first epigenetic mark to be widely studied, further progress has been made in the field of clinical diagnostics. Some techniques for DNA methylation analysis have the potential to be rapidly well established in clinical laboratories. In this regard, Buso and Iborra describe the use of Sequenom MassARRAY technology in the analysis of DNA methylation in Chapter 7. This MS-based method provides differences in DNA methylation at the individual CpG level. The EpiTYPER MassARRAY system of Sequenom is a candidate to be adopted in clinical laboratories, although there exist some inconveniences to this technology, such as the fact that it requires expensive equipment and trained technicians. Furthermore, other procedures that can be more rapidly adopted in clinical laboratories are described by Zhao and Bapat in Chapter 8. They describe MethyLight and multiplex MethyLight assays using TaqMan® fluorescent probes. These easy-to-use procedures allow the analysis of the specific methylated gene/s in a single or multiplex fashion. These procedures are based on quantitative real-time (qRT) MSP (methylation-specific polymerase chain reaction (PCR)) which is able to differentiate methylated from unmethylated cytosines in DNA by a bisulfite treatment and allele-specific primers. Because qRT-PCR technologies are well adopted in clinical diagnostics laboratories, MSP-based procedures and their required equipment are very reliable technologies to be integrated into clinical laboratories. In addition, pyrosequencing is another available methodology to be adopted in clinical laboratories. As described in Chapter 9, pyrosequencing is a quantitative sequence-based detection technology which is applicable, for example, in toxicity tests to investigate the modification of DNA methylation levels upon exposure to chemicals. In this regard, a number of assays for CpG island analysis are available that have proven to be highly reproducible and highly sensitive in clinical practice.

As we described in the previous section, PTMs have also led to a revolution in clinical diagnostics. As an alternative to traditional antibody-based methods in order to increase sensitivity and reproducibility, MS methods have emerged as a powerful analytical tool to identify PTMs. In Section II, Bonaldi and her team describe different approaches to analyze PTMs by MS, offering a perspective on how this technology and its methods can be integrated into clinical practice (Chapter 10).

5. EPIGENETIC BIOMARKERS: AN OVERVIEW OF RECENT ADVANCES

Recent technical advances have contributed by making epigenome mapping increasingly cost-efficient and less unmanageable. Therefore, research on disease-specific biomarkers will continue over the following years. Furthermore, the development of bioinformatic tools, which will increase the efficiency of the analysis of data generated in epigenetic research and epigenome-wide association studies (EWAS), will help to accomplish epigenetic biomarker development projects [90,91]. Exciting advances are rapidly occurring in this field, providing new biomarkers for diagnostics, prognostics, and clinical monitoring for several diseases.

In the field of DNA methylation, diverse methylation-derived chemical modifications have recently been discovered, such as 5hmC, 5fC in mouse stem cells (ES) and brain cortex, and 5caC in mice ES, [19]. The discovery of these kinds of new chemical modifications reveals the necessity for further investigations to elucidate their role in cell physiology and gene regulation, so as to provide new epigenetic biomarkers. Emerging innovative methylation and hydroxymethylation detection strategies are focused on addressing the validation of DNA methylation-based biomarkers in order to provide potential applications for these kinds of biomarkers in diverse clinical settings [92].

Histone modifications as well as the enzymes that catalyze them have been explored for their potential to be used as biomarkers and also as therapeutic targets and biomarkers in cancer [93] and other diseases. However, some questions regarding histone PTMs should be raised by researchers, as pointed out by Boukaba A. et al. in Chapter 3. The first question is why so many PTMs? Are they truly PTMs or just histone chemical moieties of adducts that are catalyzed by distinct enzymatic reactions? Although not real epigenetic PTMs introduced by histone modifiers, some of them could set the basis for the discovery of new PTMs involved in pathological processes, thus serving as biomarkers. In the advent of new MS procedures and recent achievements of MS-based proteomics for qualitative and quantitative characterization of histone PTMs and histone variants, MS has been converted into a well-established method in epigenetics research [94].

On the other hand, it has been shown that miRNAs of cancer cells modulate the microenvironment via noncell-autonomous mechanisms such as angiogenesis, tumor immune invasion, and tumor–stromal interaction, favoring the acquisition of hallmark cancer traits in neighboring cells to initiate cancer progression [95]. The understanding of these molecular processes may yield novel solutions to treat cancer tissues and design better therapeutic responses through the identification of new miRNA-based targets or their use in treatment monitoring. New technologies have already contributed to our understanding of miRNA-based mechanisms not only in cancer but also in the central nervous system and brain processing, including learning, memory, and cognition, setting the basis for the study of incurable neurological disorders [96]. Moreover, miRNAs are also known as major constituents of exosomes, extracellular vesicles that are proposed to transmit signals from cell to cell. In this regard, knowing the function of specific miRNAs and in which particles can they be found in biological fluids can contribute to improving the knowledge of ncRNAs for better biomarker design.

Section III of this volume presents a number of epigenetic biomarkers of cancer (i.e., lung cancer, prostate cancer, and breast cancer), metabolic syndrome, infertility, pregnancy complications, allergy and respiratory diseases, and also neurodegenerative disorders. This section is focused on the description of DNA methylation, histone PTMs, and ncRNA-based biomarkers, thus offering the most reliable biomarkers to be used in clinical practice and serve as a useful manual for clinicians and biomedical scientists. In some chapters, a number of epigenetic biomarkers with clinical value and that are used in clinical diagnostics are mentioned. In other cases, although some of the epigenetic biomarkers discussed are not used in clinical practice yet, they show promising diagnostic value and are candidates to be adopted promptly into the clinical setting.

6. EPIGENETICS: PERSPECTIVE OF IMPLANTATION IN CLINICAL LABORATORIES

The use of epigenetic biomarkers in molecular diagnostics needs to reach important milestones. The knowledge of new methodologies and bioinformatic tools and the consideration of financial, regulatory, and bioethical aspects are important issues that must be taken into account before the incorporation of epigenetic biomarkers into clinical practice [17]. For biomarker validation, it is necessary to perform clinical validation by enrolling eligible and clinically relevant participants in a research study. For example, a "detection" epigenetic biomarker should have the power to differentiate between liver cancer patients and liver cirrhosis patients, not between liver cancer patients and age-matched healthy volunteers.

While there are a number of published reports identifying strategies to recruit participants for genetic research, there is scarce information regarding the recruitment strategy for epigenetic research [97]. In this regard, the ethical conduct of epigenetic research in clinical trials is extremely important. As we know, some epigenetic changes have an intergenerational influence [98]. Indeed, it is of notable interest to reflect on ethics not only in epigenetic diagnostics but also in epigenetic clinical trials. In light of this, Jallo et al. discusses the strategy to be developed and implemented to enroll subjects for epigenetic studies [97]. Following the performance of epigenetic studies and once the epigenetic biomarker has been validated, biomedical research is able to deliver a reliable and useful clinical biomarker. However, it is important to take into account that clinical laboratories need easy-to-use and low-cost technologies to analyze these biomarkers and prepare clinical reports [17], so it is evident that providing reliable epigenetic biomarkers is difficult and several obstacles may have to be solved along the way.

To overcome the most important barriers for the implementation of epigenetic biomarkers in diagnostic laboratories, some issues regarding the clinical utility and clinical validity of epigenetic biomarkers and regulatory issues should be taken into account. In fact, to become a clinically epigenetic biomarker approved by regulatory agencies, the biomarker should be confirmed and validated, and it should be reproducible, specific, and sensitive. Therefore, it is of high importance to completely characterize the epigenetic biomarkers, first by analytical validation which ensures the consistency of the method, test, or technology used to measure the epigenetic biomarker, and second to establish the clinical validity which relates to the consistency and accuracy of the biomarker for diagnostics, predicting the clinical outcome or monitoring the effect of a treatment.

In addition, other issues require solving before using epigenetic biomarkers in clinical routine, including changing the paradigm of clinical diagnostics based exclusively on genetics. In this regard it is necessary to (1) *change conventional diagnostic approaches*. The adoption of new biomarkers by the medical community will improve success in diagnostic and also therapeutic interventions by minimizing the secondary events produced by nonpersonalized therapy. To introduce this concept to health professionals we need specialists to educate the health community about novel diagnostic approaches based on epigenetics. (2) *Increase knowledge about new epigenetic-based biomarkers*. Direct communication with health professionals and medical societies will help change the paradigm and adopt these epigenetic-based biomarkers into clinical routine. Epigenetics is a relative young biomedical discipline. So many health professionals do not know the potential of epigenetics and its clinical utility. So, meetings, scientific journals, and clinical reviews and new edited volumes such as *Epigenetic Biomarkers and Diagnostics* may contribute to this task.

Regarding the use of technologies required for epigenetic studies it is important to (3) *introduce new technologies into clinical laboratories*. It would be very useful to introduce NGS and MS spectrometry technologies into clinical laboratories. NGS has transformed genomic medicine, because it has reduced the cost of large-scale sequencing. By using NGS, it is possible to analyze an individual's near-complete exome, genome, methylome, miRNome and also to analyze complete histone posttranslational maps to assist in the diagnosis of a wide array of clinical scenarios. For NGS, one of the most promising applications is based on the advances in Nano-ChIP-Seq, which will allow the analysis of specific chromatin regions from far fewer cells for embryology and development studies. (4) *Replace classical methods for other cost-effective and time-effective technologies*. For example, the potential of MS-based assays instead of those based on immunoassays for the analysis of histone PTMs shows several advantages, including quantification, high sensitivity, specificity and low cost, among others [99]. However, there is an associated problem consisting of the high cost of this kind of technology and the requirement of specialists in MS [99]. Importantly, it has been estimated that the incorporation of MS in laboratories for clinical diagnostics (replacing routine immunoassays) would save about $250,000 per year in each tertiary hospital laboratory [100]. So, if the future tendency of clinical laboratories is to replace immunoassays with MS-based procedures, it would be better that efforts were focused on the improvement of MS-based procedures for the identification of histone PTMs. (5) *Implement new applications for conventional methodologies*. It is also possible to adapt conventional technologies used in laboratories for clinical diagnostics to epigenetic applications. In this regard, well-established qPCR can be used to analyze DNA methylation by MSP assays in clinical diagnostic.

Finally, other strategic considerations need to be taken into account for the adoption of epigenetic biomarkers in clinical diagnostics, including (6) *government regulations*. It would be very relevant to work with different stakeholders to overcome regulatory hurdles in different countries around the world to achieve the implementation of epigenetic diagnostics in clinical routine. It is possible to use "in-house developed" tests in a Clinical Laboratory Improvement Amendments (CLIA)-certified reference laboratory to perform different epigenetic tests. It is also possible to apply for a less rigorous 510(k) process or for a more extensive premarket approval application until full FDA or EMA review is performed. The "in-house developed" tests in CLIA-certified laboratories may contribute to adopt epigenetic assays in tertiary hospitals and increase the knowledge of epigenetics in clinical practice. (7) *Third party*. The adoption of new technologies or epigenetic biomarkers by physicians and clinicians (prescribers) is subjected to decisions made by national/regional public health administrators and insurance companies, which are ultimately responsible for the adoption of new technologies. Therefore, it is very important to perform clinical trials to validate the potential of epigenetic biomarkers and also to align patient and clinician demands with the decision-making of administrators. It is obvious that epigenetics has revolutionized biomedicine, and its potential applications in diagnostics and disease treatment make it easy to be recognized by public health administrators as a useful tool in clinical practice. (8) *Ethical issues*. It is important to address the possible ethical issues regarding the adoption of epigenetics into clinical routine. Rothstein et al. have identified a number of issues in which epigenetics differs from genetics. Therefore, epigenetic research which leads to new ethical issues addressing a number of potential legal and ethical implications, including legal and moral responsibility, intergenerational considerations, and issues regarding access to health care [101,102].

7. PERSPECTIVES OF EPIGENETICS IN DIAGNOSTICS

Since the first human methylome was published in 2009 [103], technologies and biomedical research have offered a wide range of clinical applications for epigenetics. No doubt in the coming years epigenetics will overcome the limitations of traditional genetics and genomics by providing new tools based on epigenetic biomarkers and new epigenetic drugs able to control the function of our genome. Epigenetics will help address some unresolved questions in our understanding of personalized medicine. Epigenetic biomarkers (DNA methylation, PTM in histones, and ncRNAs) serve in the dynamic study of physiopathological conditions, and therefore, epigenetic biomarkers will serve to predict the evolution of disease and to monitor the effect of treatments on diseases. Furthermore, the development of single-cell epigenetic assays or technologies such as Nano-ChIP-Seq, which analyzes just a small number of cells, will have a profound impact on clinical epigenetics, because they solve the problem of limited tissue availability and they also discern between different cellular populations, or even cell heterogeneity, among other advantages. However, the advances in epigenetics and the identification of new epigenetic biomarkers have also produced new ethical issues, regarding causal and moral responsibility, due to how epigenetics and the exposome (which covers factors, including environment, contamination, nutrition, drugs, etc.) produce gene expression changes that in turn affect the health of people [104]. With the rapid advances of NGS technologies and their potential in clinical diagnostics it is important to validate new technologies and epigenetic biomarkers for diagnostics and prognostics and to provide professional standards and guidelines in a similar way to those provided for the American College of Medical Genetics and Genomics for genomic medicine [105].

If we are to address the current increasing burden to national health-care systems from diseases such as cancer, metabolic diseases, and neurological disorders, among others, we need to develop a wide array of biomarkers (diagnostic, prognostic, predictive, and for treatment monitoring) which may contribute to improving precision medicine and helping rationalize health-care funding and resources. The outbreak of epigenetic biomarkers into clinical diagnostics will contribute not only to improving the health of people but also to increasing the sustainability of health-care systems.

LIST OF ABBREVIATIONS

ChIP Chromatin immunoprecipitation
CLIA Clinical Laboratory Improvement Amendments
DNMTs DNA(cytosine-5-)methyltransferases
EMA European Medicines Agency
EWAS Epigenome-wide association studies
FDA Food and Drug Administration
FFPE Formalin-fixed paraffin-embedded tissues
IVD In vitro diagnostics
MS Mass spectrometry
NGS Next-generation sequencing
PTMs Posttranslational modifications
TETs Ten-eleven translocases

References

[1] Waddington CH. The epigenotype. Endeavour 1942;1:18–20.

[2] Holliday R. Mechanisms for the control of gene activity during development. Biol Rev Cambr Phil Soc 1990;65:431–71.

[3] Jenuwein T, Allis CD. Translating the histone code. Science August 10, 2001;293(5532):1074–80.

[4] Bird A. Perceptions of epigenetics. Nature 2007; 447(7143):396–8.

[5] Goldberg ADA, Allis CD, Bernstein E. Epigenetics: a landscape takes shape. Cell 2007;128:635–8.

[6] Berger SL, Kouzarides T, Shiekhattar R, Shilatifard A. An operational definition of epigenetics. Genes Dev April 1, 2009;23(7):781–3.

[7] Bernstein BE, Stamatoyannopoulos JA, Costello JF, Ren B, Milosavljevic A, Meissner A, et al. The NIH Roadmap Epigenomics Mapping Consortium. Nat Biotechnol October 2010;28(10):1045–8.

[8] Kanherkar RR, Bhatia-Dey N, Csoka AB. Epigenetics across the human lifespan. Front Cell Dev Biol 2014;2:49.

[9] Rivera CM, Ren B. Mapping human epigenomes. Cell September 26, 2013;155(1):39–55.

[10] Sandoval J, Esteller M. Cancer epigenomics: beyond genomics. Curr Opin Genet Dev February 2012; 22(1):50–5.

[11] Ehrlich M. DNA methylation in cancer: too much, but also too little. Oncogene August 12, 2002;21(35): 5400–13.

[12] Deng G, Nguyen A, Tanaka H, Matsuzaki K, Bell I, Mehta KR, et al. Regional hypermethylation and global hypomethylation are associated with altered chromatin conformation and histone acetylation in colorectal cancer. Int J Cancer June 15, 2006;118(12):2999–3005.

[13] Kondo T, Bobek MP, Kuick R, Lamb B, Zhu X, Narayan A, et al. Whole-genome methylation scan in ICF syndrome: hypomethylation of non-satellite DNA repeats D4Z4 and NBL2. Hum Mol Genet March 1, 2000;9(4): 597–604.

[14] Jones PA, Baylin SB. The epigenomics of cancer. Cell February 23, 2007;128(4):683–92.

[15] Irizarry RA, Ladd-Acosta C, Wen B, Wu Z, Montano C, Onyango P, et al. The human colon cancer methylome shows similar hypo- and hypermethylation at conserved tissue-specific CpG island shores. Nat Genet February 2009;41(2):178–86.

[16] Esteller M, Corn PG, Baylin SB, Herman JG. A gene hypermethylation profile of human cancer. Cancer Res April 15, 2001;61(8):3225–9.

[17] Sandoval J, Peiro-Chova L, Pallardo FV, Garcia-Gimenez JL. Epigenetic biomarkers in laboratory diagnostics: emerging approaches and opportunities. Expert Rev Mol Diagn June 2013;13(5):457–71.

[18] Song CX, He C. Potential functional roles of DNA demethylation intermediates. Trends Biochem Sci October 2013;38(10):480–4.

[19] Booth MJ, Raiber EA, Balasubramanian S. Chemical methods for decoding cytosine modifications in DNA. Chem Rev August 5, 2014;115.

[20] Song CX, Szulwach KE, Dai Q, Fu Y, Mao SQ, Lin L, et al. Genome-wide profiling of 5-formylcytosine reveals its roles in epigenetic priming. Cell April 25, 2013;153(3):678–91.

[21] Booth MJ, Branco MR, Ficz G, Oxley D, Krueger F, Reik W, et al. Quantitative sequencing of 5-methylcytosine and 5-hydroxymethylcytosine at single-base resolution. Science May 18, 2012;336(6083):934–7.

[22] Booth MJ, Ost TW, Beraldi D, Bell NM, Branco MR, Reik W, et al. Oxidative bisulfite sequencing of 5-methylcytosine and 5-hydroxymethylcytosine. Nat Protoc October 2013;8(10):1841–51.

[23] Neri F, Incarnato D, Krepelova A, Rapelli S, Anselmi F, Parlato C, et al. Single-base resolution analysis of 5-formyl and 5-carboxyl cytosine reveals promoter DNA methylation dynamics. Cell Rep February 4, 2015;5(10):674–83.

[24] Lu X, Song CX, Szulwach K, Wang Z, Weidenbacher P, Jin P, et al. Chemical modification-assisted bisulfite sequencing (CAB-Seq) for 5-carboxylcytosine detection in DNA. J Am Chem Soc June 26, 2013;135(25): 9315–7.

[25] Booth MJ, Marsico G, Bachman M, Beraldi D, Balasubramanian S. Quantitative sequencing of 5-formylcytosine in DNA at single-base resolution. Nat Chem May 2014;6(5):435–40.

[26] Ulahannan N, Greally JM. Genome-wide assays that identify and quantify modified cytosines in human disease studies. Epigenetics Chromatin 2015;8:5.

[27] Huang H, Sabari BR, Garcia BA, Allis CD, Zhao Y. SnapShot: histone modifications. Cell October 9, 2014;159(2):458-e1.

[28] Garcia-Gimenez JL, Ledesma AM, Esmoris I, Roma-Mateo C, Sanz P, Vina J, et al. Histone carbonylation occurs in proliferating cells. Free Radic Biol Med April 15, 2012;52(8):1453–64.

[29] Garcia-Gimenez JL, Olaso G, Hake SB, Bonisch C, Wiedemann SM, Markovic J, et al. Histone h3 glutathionylation in proliferating mammalian cells destabilizes nucleosomal structure. Antioxid Redox Signal October 20, 2013;19(12):1305–20.

[30] Strahl BD, Allis CD. The language of covalent histone modifications. Nature January 6, 2000;403(6765):41–5.

[31] Hake SB, Xiao A, Allis CD. Linking the epigenetic 'language' of covalent histone modifications to cancer. Br J Cancer February 23, 2004;90(4):761–9.

[32] Chi P, Allis CD, Wang GG. Covalent histone modifications–miswritten, misinterpreted and mis-erased in human cancers. Nat Rev Cancer July 2010;10(7):457–69.

[33] Gray SG. Epigenetic treatment of neurological disease. Epigenomics August 2011;3(4):431–50.

[34] Ren R-JD, Eric B, Wang G, Seyfried NT, Levey AI. Proteomics of protein post-translational modifications implicated in neurodegeneration. Transl Neurodegener 2014;3:23.

[35] Park E, Kim Y, Ryu H, Kowall NW, Lee J. Epigenetic mechanisms of Rubinstein-Taybi syndrome. Neuromol. Med March 2014;16(1):16–24.

[36] Pereira PM, Schneider A, Pannetier S, Heron D, Hanauer A. Coffin-Lowry syndrome. Eur J Hum Genet June 2010;18(6):627–33.

[37] Talbert PB, Henikoff S. Histone variants–ancient wrap artists of the epigenome. Nat Rev Mol Cell Biol April 2010;11(4):264–75.

[38] Henikoff S, Furuyama T, Ahmad K. Histone variants, nucleosome assembly and epigenetic inheritance. Trends Genet July 2004;20(7):320–6.

[39] Ausio J. Histone variants–the structure behind the function. Brief Funct Genomic Proteomic September 2006;5(3):228–43.

[40] Maze I, Noh KM, Soshnev AA, Allis CD. Every amino acid matters: essential contributions of histone variants to mammalian development and disease. Nat Rev Genet April 2014;15(4):259–71.

[41] Vardabasso C, Hasson D, Ratnakumar K, Chung CY, Duarte LF, Bernstein E. Histone variants: emerging players in cancer biology. Cell Mol Life Sci February 2014;71(3):379–404.

[42] Xu J, Zhang X, Pelayo R, Monestier M, Ammollo CT, Semeraro F, et al. Extracellular histones are major mediators of death in sepsis. Nat Med November 2009;15(11):1318–21.

[43] Fuchs TA, Kremer Hovinga JA, Schatzberg D, Wagner DD, Lammle B. Circulating DNA and myeloperoxidase indicate disease activity in patients with thrombotic microangiopathies. Blood August 9, 2012;120(6):1157–64.

[44] Ekaney ML, Otto GP, Sossdorf M, Sponholz C, Boehringer M, Loesche W, et al. Impact of plasma histones in human sepsis and their contribution to cellular injury and inflammation. Crit Care 2014;18(5):543.

[45] Zeerleder S, Zwart B, Wuillemin WA, Aarden LA, Groeneveld AB, Caliezi C, et al. Elevated nucleosome levels in systemic inflammation and sepsis. Crit Care Med July 2003;31(7):1947–51.

[46] Zeerleder S, Stephan F, Emonts M, de Kleijn ED, Esmon CT, Varadi K, et al. Circulating nucleosomes and severity of illness in children suffering from meningococcal sepsis treated with protein C. Crit Care Med December 2012;40(12):3224–9.

[47] Burlingame RW. The clinical utility of antihistone antibodies. Autoantibodies reactive with chromatin in systemic lupus erythematosus and drug-induced lupus. Clin Lab Med September 1997;17(3):367–78.

[48] Kavanaugh A, Tomar R, Reveille J, Solomon DH, Homburger HA. Guidelines for clinical use of the antinuclear antibody test and tests for specific autoantibodies to nuclear antigens. American College of Pathologists. Arch Pathol Lab Med January 2000;124(1):71–81.

[49] Johnson JM, Edwards S, Shoemaker D, Schadt EE. Dark matter in the genome: evidence of widespread transcription detected by microarray tiling experiments. Trends Genet February 2005;21(2):93–102.

[50] Kugel JF, Goodrich JA. Non-coding RNAs: key regulators of mammalian transcription. Trends Biochem Sci April 2012;37(4):144–51.

[51] Taft RJ, Pang KC, Mercer TR, Dinger M, Mattick JS. Non-coding RNAs: regulators of disease. J Pathol January 2010;220(2):126–39.

[52] He L, Hannon GJ. MicroRNAs: small RNAs with a big role in gene regulation. Nat Rev Genet July 2004;5(7):522–31.

[53] Hwang HW, Mendell JT. MicroRNAs in cell proliferation, cell death, and tumorigenesis. Br J Cancer March 27, 2006;94(6):776–80.

[54] Garzon R, Marcucci G, Croce CM. Targeting microRNAs in cancer: rationale, strategies and challenges. Nat Rev Drug Discov October 2010;9(10):775–89.

[55] Peiro-Chova L, Pena-Chilet M, Lopez-Guerrero JA, Garcia-Gimenez JL, Alonso-Yuste E, Burgues O, et al. High stability of microRNAs in tissue samples of compromised quality. Virchows Arch December 2013;463(6):765–74.

[56] Mitchell PS, Parkin RK, Kroh EM, Fritz BR, Wyman SK, Pogosova-Agadjanyan EL, et al. Circulating microRNAs as stable blood-based markers for cancer detection. Proc Natl Acad Sci USA July 29, 2008;105(30):10513–8.

[57] Volinia S, Calin GA, Liu CG, Ambs S, Cimmino A, Petrocca F, et al. A microRNA expression signature of human solid tumors defines cancer gene targets. Proc Natl Acad Sci USA February 14, 2006;103(7):2257–61.

[58] Petillo D, Kort EJ, Anema J, Furge KA, Yang XJ, Teh BT. MicroRNA profiling of human kidney cancer subtypes. Int J Oncol July 2009;35(1):109–14.

[59] Blenkiron C, Goldstein LD, Thorne NP, Spiteri I, Chin SF, Dunning MJ, et al. MicroRNA expression profiling of human breast cancer identifies new markers of tumor subtype. Genome Biol 2007;8(10):R214.

[60] Wach S, Nolte E, Theil A, Stohr C, TR T, Hartmann A, et al. MicroRNA profiles classify papillary renal cell carcinoma subtypes. Br J Cancer August 6, 2013;109(3):714–22.

[61] Guil S, Esteller M. Cis-acting noncoding RNAs: friends and foes. Nat Struct Mol Biol November 2012;19(11):1068–75.

[62] Salta E, De Strooper B. Non-coding RNAs with essential roles in neurodegenerative disorders. Lancet Neurol February 2012;11(2):189–200.

[63] Kunej T, Obsteter J, Pogacar Z, Horvat S, Calin GA. The decalog of long non-coding RNA involvement in cancer diagnosis and monitoring. Crit Rev Clin Lab Sci December 2014;51(6):344–57.

[64] Beltrami C, Angelini TG, Emanueli C. Noncoding RNAs in diabetes vascular complications. J Mol Cell Cardiol December 20, 2014.

[65] Iorio MV, Piovan C, Croce CM. Interplay between microRNAs and the epigenetic machinery: an intricate network. Biochim Biophys Acta Oct-Dec 2010;1799(10–12):694–701.

[66] Murr R. Interplay between different epigenetic modifications and mechanisms. Adv Genet 2010;70:101–41.

[67] Reik W. Stability and flexibility of epigenetic gene regulation in mammalian development. Nature May 24, 2007;447(7143):425–32.

[68] Alegria-Torres JA, Baccarelli A, Bollati V. Epigenetics and lifestyle. Epigenomics June 2011;3(3):267–77.

[69] Sanchis-Gomar F, Garcia-Gimenez JL, Perez-Quilis C, Gomez-Cabrera MC, Pallardo FV, Lippi G. Physical exercise as an epigenetic modulator: Eustress, the "positive stress" as an effector of gene expression. J Strength Cond Res December 2012;26(12):3469–72.

[70] Pareja-Galeano H, Sanchis-Gomar F, Garcia-Gimenez JL. Physical exercise and epigenetic modulation: elucidating intricate mechanisms. Sports Med April 2014;44(4):429–36.

[71] Woldemichael BT, Bohacek J, Gapp K, Mansuy IM. Epigenetics of memory and plasticity. Prog Mol Biol Transl Sci 2014;122:305–40.

[72] Zheng J, Xiao X, Zhang Q, Yu M. DNA methylation: the pivotal interaction between early-life nutrition and glucose metabolism in later life. Br J Nutr December 2014;112(11):1850–7.

[73] Ushijima T, Hattori N. Molecular pathways: involvement of *Helicobacter pylori*-triggered inflammation in the formation of an epigenetic field defect, and its usefulness as cancer risk and exposure markers. Clin Cancer Res Official J Am Assoc Cancer Res February 15, 2012;18(4):923–9.

[74] Bock C. Epigenetic biomarker development. Epigenomics October 2009;1(1):99–110.

[75] Mishra A, Verma M. Cancer biomarkers: are we ready for the prime time? Cancers (Basel) 2010;2(1):190–208.

[76] Mall C, Rocke DM, Durbin-Johnson B, Weiss RH. Stability of miRNA in human urine supports its biomarker potential. Biomark Med August 2013;7(4):623–31.

[77] Weber WW. The promise of epigenetics in personalized medicine. Mol Interv 2011;10(6):363–70.

[78] Budiman MAS, Smith SW, Ordway JM. DNA methylation in personalized medicine. Pers Med 2011;8(1):35–43.

[79] Schones DE, Zhao K. Genome-wide approaches to studying chromatin modifications. Nat Rev Genet March 2008;9(3):179–91.

[80] Li JZQ, Bolund L. Computational methods for epigenetic analysis: the protocol of computational analysis for modified methylation-specific digital karyotyping based on massively parallel sequencing. NY, USA: Humana Press; 2011.
[81] Statham AL, Robinson MD, Song JZ, Coolen MW, Stirzaker C, Clark SJ. Bisulfite sequencing of chromatin immunoprecipitated DNA (BisChIP-seq) directly informs methylation status of histone-modified DNA. Genome Res June 2012;22(6):1120–7.
[82] Marrer E, Dieterle F. Promises of biomarkers in drug development – a reality check. Chem Biol Drug Des June 2007;69(6):381–94.
[83] Amur S, Frueh FW, Lesko LJ, Huang S-M. Integration and use of biomarkers in drug development, regulation and clinical practice: a US regulatory perspective. Biomarkers Med 2008;2(3):305–11.
[84] Sistare FD, DeGeorge JJ. Preclinical predictors of clinical safety: opportunities for improvement. Clin Pharmacol Ther August 2007;82(2):210–4.
[85] Hayes J, Peruzzi PP, Lawler S. MicroRNAs in cancer: biomarkers, functions and therapy. Trends Mol Med August 2014;20(8):460–9.
[86] Cited; Available from: https://www.academia.edu/7174962/Epigenetics_and_MicroRNA_Biomarkers_Market_Status_and_Trends.
[87] Keller A, Leidinger P, Bauer A, Elsharawy A, Haas J, Backes C, et al. Toward the blood-borne miRNome of human diseases. Nat Methods 2011;8(10):841–3.
[88] Lu M, Zhang Q, Deng M, Miao J, Guo Y, Gao W, et al. An analysis of human microRNA and disease associations. PLoS One 2008;3(10):e3420.
[89] Russo F, Di Bella S, Nigita G, Macca V, Lagana A, Giugno R, et al. miRandola: extracellular circulating microRNAs database. PLoS One 2012;7(10):e47786.
[90] Rakyan VK, Down TA, Balding DJ, Beck S. Epigenome-wide association studies for common human diseases. Nat Rev Genet August 2011;12(8):529–41.
[91] Paul DS, Beck S. Advances in epigenome-wide association studies for common diseases. Trends Mol Med October 2014;20(10):541–3.
[92] Olkhov-Mitsel E, Bapat B. Strategies for discovery and validation of methylated and hydroxymethylated DNA biomarkers. Cancer Med October 2012;1(2):237–60.
[93] Gezer U, Holdenrieder S. Post-translational histone modifications in circulating nucleosomes as new biomarkers in colorectal cancer. Vivo May-Jun 2014;28(3):287–92.
[94] Soldi M, Cuomo A, Bremang M, Bonaldi T. Mass spectrometry-based proteomics for the analysis of chromatin structure and dynamics. Int J Mol Sci 2013;14(3):5402–31.
[95] Suzuki HI, Katsura A, Matsuyama H, Miyazono K. MicroRNA regulons in tumor microenvironment. Oncogene August 18, 2014;34.
[96] Wang W, Kwon EJ, Tsai LH. MicroRNAs in learning, memory, and neurological diseases. Learn Mem September 2012;19(9):359–68.
[97] Jallo N, Lyon DE, Kinser PA, Kelly DL, Menzies V, Jackson-Cook C. Recruiting for epigenetic research: facilitating the informed consent process. Nurs Res Pract 2013;2013:935740.
[98] Whitelaw NC, Whitelaw E. Transgenerational epigenetic inheritance in health and disease. Curr Opin Genet Dev June 2008;18(3):273–9.
[99] Wu AH, French D. Implementation of liquid chromatography/mass spectrometry into the clinical laboratory. Clin Chim Acta May 2013;420:4–10.
[100] Hetu PO, Robitaille R, Vinet B. Successful and cost-efficient replacement of immunoassays by tandem mass spectrometry for the quantification of immunosuppressants in the clinical laboratory. J Chromatogr B Anal Technol Biomed Life Sci February 1, 2012;883–884:95–101.
[101] Rothstein MA, Cai Y, Marchant GE. The ghost in our genes: legal and ethical implications of epigenetics. Winter Health Matrix Clevel 2009;19(1):1–62.
[102] Rothstein MA, Cai Y, Marchant GE. Ethical implications of epigenetics research. Nat Rev Genet April 2009;10(4):224.
[103] Lister R, Pelizzola M, Dowen RH, Hawkins RD, Hon G, Tonti-Filippini J, et al. Human DNA methylomes at base resolution show widespread epigenomic differences. Nature November 19, 2009;462(7271):315–22.
[104] Chadwick R, O'Connor A. Epigenetics and personalized medicine: prospects and ethical issues. Pers Med 2013;10(5):463–71.
[105] Rehm HL, Bale SJ, Bayrak-Toydemir P, Berg JS, Brown KK, Deignan JL, et al. ACMG clinical laboratory standards for next-generation sequencing. Genet Med September 2013;15(9):733–47.

CHAPTER

2

The Importance of Biobanks in Epigenetic Studies

Lorena Peiró-Chova[1], Olga Bahamonde Ponce[1], Carolina Abril-Tormo[2], Jacobo Martínez-Santamaría[2,3], José Antonio López-Guerrero[4], Peter H.J. Riegman[5]

[1]INCLIVA Biobank, INCLIVA Biomedical Research Institute, Valencia, Spain; [2]IBSP-CV Biobank, FISABIO, Valencia, Spain; [3]Valencian Biobank Network, FISABIO, Valencia, Spain; [4]Laboratory of Molecular Biology and Biobank, Fundacion Instituto Valenciano de Oncologia, Valencia, Spain; [5]Department of Pathology, Erasmus Medical Center, Rotterdam, The Netherlands

OUTLINE

Epigenetic Biomarkers and Diagnostics
http://dx.doi.org/10.1016/B978-0-12-801899-6.00002-4

1. BIOBANKS

1.1 Definition and Types of Biobanks

Although there is no generally accepted definition for "biobank," and current definitions often have slightly different meanings, the following definition encompasses its essence: "A biobank is a type of biorepository that stores biological samples and data for use in research" [1].

In 2001, the Organization for Economic Cooperation and Development (OECD) defined biobanks as Biological Resources Centres (BRCs) "that consist of service providers and repositories of the living cells, genomes of organisms, and information relating to heredity and functions of biological systems...that must meet the high standards of quality and expertise demanded by the international community of scientist and industry for the delivery of biological information and materials,...that must provide access to biological resources on which R&D in the life sciences and the advancement of biotechnology depends" [2].

According to the International Society for Biological and Environmental Repositories (ISBER), a biobank is "an entity that receives, stores, processes and/or disseminates specimens, as needed. It encompasses the physical location as well as the full range of activities associated with its operation" [3].

In these two last definitions, the term biobank refers to any type of collection of biological samples (plant, animal, microbe, human, etc.), but in medical sciences the term is reserved for human specimens. In this sense, increasingly accepted are definitions like the one by the Public Population Project in Genomics and Society (P3G): "an organized collection of human biological material and associated information stored for one or more research purposes" [4]. This review will focus on this kind of biobank.

Since the late 1990s, biobanks have become an important resource for biomedical research, supporting many types of contemporary research like genomics and personalized medicine, and may be classified according to different criteria. When sorted by purpose, two large groups of biobanks can be discerned: *population-based biobanks* (PBs) and *hospital-integrated biobanks* (HIBs). PBs developed first. They collect samples from a large number of individual volunteers representing different population cohorts. This type of biobank stores both biomaterials and the associated information from these individuals, mainly lifestyle habits, disease history, and environmental data obtained for the most part from participant questionnaires. Research performed on these biomaterials is usually devoted to the epidemiological, exposition, and environmental risk factors for certain types of diseases in relation to the genotype of the participant [5,6]. HIBs are usually located at hospitals in which samples representing a variety of diseases are collected. Research performed with these samples focuses mainly on biomarkers associated with the progression of the disease or the treatment response [7]. In Spain, the development of a new multidisciplinary model of biobanking has been promoted in which all human samples collected for biomedical research purposes from different hospital departments are managed by a single networked infrastructure created as a service for supporting and enhancing cooperative biomedical research. Nowadays many intermediate forms of biobanks can be found, making the distinction between PBs and HIBs less pronounced.

In addition, biobanks can host two main types of collections: *project-driven* and *general or archival* collections [8]. *A project-driven* collection means that specimens are collected and distributed to answer specific research questions. The advantage this kind of collection has is that the investigator receives exactly what is requested. Nevertheless, its disadvantage is that it may take a long period of time (years/decades) to collect enough samples, especially if follow-up data are required [8]. To the contrary, *archival* or *general* collections are devoted to establishing reference collections. They are not meant to meet particular research goals, but to be available in order to respond to multiple requests for an assortment of research uses. Hence, a large number of samples can be provided to researchers immediately, although the specifically required data may not always be available. In *archival* collections, usually the data collected are less detailed than in *project-driven* collections in which objectives of specific research projects are previously defined [8]. In addition, in *archival* collections from hospitals, the sample quality needs only to be fit for diagnostic testing purposes. This certainly does not imply that these samples would be useful in any technique available for biomedical research.

1.2 General Processes in Biobanks

Biobanks, understood as technical units, define their workflows into general operating procedures for obtaining, handling, and distributing human specimens, along with their relevant associated information. An operating procedure is defined as each activity or set of activities using resources. It is managed, so the elements of the input process can be transformed into results at the output level (ISO 9000:2005—Quality Management Systems—Fundamentals and Vocabulary). In turn, procedures are interrelated in a more or less sequential and multidimensional way. Thus, the application of a system based on processes within the organization of a biobank, and the identification and definition of the interactions between them, provides a holistic view of the operation of a biobank called the "process approach" (ISO 9001:2008—Quality Management System—Requirements) (Figure 1).

These processes are mainly defined in the quality management system adopted by a biobank [9]. They are organized into three main levels: *strategic processes*, related to management efficiency; *support processes*, necessary to ensure control over the available resources; and *key processes*. The last are defined in terms of the

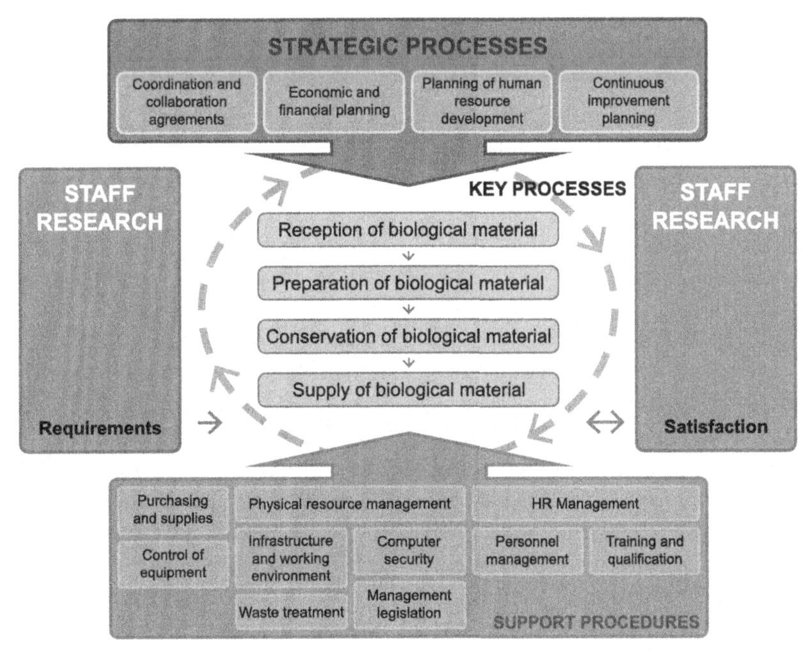

FIGURE 1 A typical process model that could be implemented in biobanks according to a *process approach* in response to the requirements of ISO 9001:2008.

nature of the biobank and the types of collections it hosts. Key processes essentially refer to four sequential common or fundamental processes for any type of biobank. These common key processes are, first, the *reception of the biological material*, followed by its *processing* and *preservation*, and finally, the *distribution* of that biological material and associated data to researchers.

Despite the fact that there are so many different types of biobanks, the preparation of biological material remains a key process. Such key processes should be described in the standard operating procedures (SOPs). These SOPs describe how to manipulate, for example, peripheral blood to obtain blood components (serum, plasma, PBMCs, DNA, RNA, micro RNAs (miRNAs), histones, etc.) or other types of samples obtained without invasive procedures, such as saliva, urine, feces, and tears. To improve the exchangeability of the samples needed to be shared in multicenter research projects, these SOPs need to be standardized. Within an environment where the biobanks are efficiently designed to answer different research questions, the handling of samples often differs. The variety may be of importance when exchanging samples, and this needs to be considered before the materials are used. Moreover, there are biobanks with very specific materials arising from surgical or clinical procedures, such as tumor and peritumoral tissues, fresh, fixed, or postmortem tissues, cerebrospinal fluid, and bone marrow.

In order to harmonize the variations that may arise from the different SOPs used by different biobanks, several initiatives have emerged across best practice guidelines [10,2,11]. One of these is the SPREC (Standard PREanalytical Code) [12,13], which is understood as a tool for describing, using a seven-position code, the possible preanalytical variables associated with the preparation and processing of the biological samples (centrifugation, precipitation, phase separation, etc.) without the intervention of chemical agents. This tool is complemented by other initiatives that have a global character, but with a more general and less explicit purpose than SPREC: BRISQ (Biospecimen Reporting for Improved Study Quality) [14] and MIABIS (Minimum Information About BIobank data Sharing) [15]. These initiatives have a greater capacity to cover the different variables that may be related to biobanking material and its associated clinical and epidemiological information. In the case of BRISQ, these variables are organized into different levels, and they aim to characterize part of the SOPs of the four key processes of a biobank. The purpose of BRISQ is to publish the conditions in which the used samples were preserved in order to enhance the reproducibility of the results obtained. This can also have a normalizing effect and decrease the variability associated with the biomaterial and increase the feasibility of the results obtained, for example, in large-scale epigenomic studies.

Overall, the main objective of biobanks is to be guarantors of the traceability and reproducibility of everything concerning the essence of biobanks. It is to provide the scientific community with biological material of high quality and appropriately characterized in order to enhance excellence in biomedical research.

1.3 Ethical Issues in Biobanking

Biobanks have provoked questions on research and medical ethics, and have opened widespread discussion. Since Chapter 3 describes the ethical concerns regarding epigenetic studies, attention here is focused on the ethical issues related to biobanking. The collection, storage, distribution, and use of biological materials and associated data should be conducted in a way that respects individuals, ensuring their privacy and confidentiality. Therefore, the following are the fundamental ethical issues for biobanks: the *informed consent* (IC) of the donors with the right to revoke their permission (including the informed opt-out systems when using residual materials); *data protection;* and *anonymization or pseudoanonymization procedures* of the biological material for research.

Protecting autonomy through the process of obtaining consent is important as it shows respect for the individual. The principle of IC is largely recognized and considered a pillar in the practice of bioethics. Although it does not in itself protect a person, IC allows individuals to exercise their fundamental right to decide whether and how their biomaterials and associated data will be used in research [1].

In recent years, biobank ethical debate has focused especially on the validity of general or broad consent. Some would argue that *the more general the consent is, the less informed it becomes*. Others would say that if *the information provided covers all aspects relevant to the person's choice, then that person's consent is appropriately informed*. For those biobanks following the *general* collection model, a trend toward broad consent for future research is now regarded as appropriate [16,17].

Another discussion reflects the risks of the experiment and the conditions that need to be adhered to by stakeholders using the biomaterials for medical research. A trial in which a patient is subjected to a potentially life-altering intervention is completely different from the situation where residual material and anonymized clinical data are used for medical research purposes. In the latter case, an (informed) opt-out system is very capable of both safeguarding the unwanted use of biomaterials, and providing for the later revocation of the permission when the only risk concerns privacy. This method is especially valuable when using archival materials for medical research and saves a lot of scarce medical research resources, especially when the patient contact with the pathologist involved is not included in the normal routine [18].

The right to revoke is mentioned in all consent forms for biomedical research. That notwithstanding, how it can be implemented in practice is a challenge because biomaterial is exchanged, data are distributed in complex databases, and sometimes samples undergo transformation into cell lines that can be exchanged and duplicated. Therefore, although biobanks should ensure the availability of donor ICs, current consent forms used in biobanking must be reviewed to ensure all these issues are carefully addressed and explained by establishing practical and feasible procedures. Otherwise, not only the act of consenting but also the right to revoke may be only concepts with no real significance [1].

Still, protecting the donor against research risks is a key responsibility for biobanks. In the case of biobanking and the associated noninterventional research, the main risk for patients/donors is that identifiable personal data fall into the wrong hands. Hence, safeguarding confidentiality is a paramount aspect of the protection of individuals and groups participating in biobanks [17].

The existence of many different levels of identifiability, coupled with their many different interpretations, has posed a major problem for discussions on confidentiality issues. Now, a nomenclature proposed by the European Medicines Agency (EMA) has been adopted by the International Conference on Harmonization of Technical Requirements (ICH), which brings together the regulatory authorities in Europe, Japan, and the United States. Briefly, the nomenclature is as follows. *Identified* data and samples are labeled with personal identifiers such as name or identification numbers (e.g., social security or national insurance number). *Coded* data and samples are labeled with at least one specific code and do not carry any personal identifiers. *Anonymized* data and samples are initially single or double coded, but there is a link with the subjects through the coding key(s). *Anonymous* data and samples are never labeled with personal identifiers when originally collected, neither is a coding key generated. Therefore, there is no potential to trace clinical data and samples to individual subjects [17].

It is also recognized that the degree of data protection is closely linked to questions of withdrawal from research, dissemination of results to participants (generally or individually), follow-up of participants, and third-party access to research data. There are a variety of situations ranging from identifying personal data (for which there is significant protection, very limited access, and the possibility of withdrawal) to anonymized data (for which there is free access for research purposes, with no possibility of recontacting participants and no possibility of returning the results or of withdrawing from the project) [19]. With the realization that absolute data safety is an illusion, the idea of *open consent* has been proposed. This type of consent refrains from any promises of anonymity, privacy, or confidentiality [20].

In genetic studies, including epigenetics, working with nonanonymized or even anonymous samples and data is of special importance since the analysis of methylomes and miR-Nomes, for example, could lead to incidental findings affecting future generations, making it necessary to recontact participants.

2. BIOBANKS IN BIOMEDICAL RESEARCH

The value of biobanks to translational medical science is widely recognized [21,22]. Promoting translational research and the application of innovative technology requires easy access support infrastructure to facilitate agile experimental demonstrations of a hypothesis or simulated models. In this scenario, biobanks constitute one of the most attractive alternatives contributing to building bridges between all the biomedical research specialties.

Among the situations that have occurred in the last decade of biomedical research and that have resulted in the establishment of biobanks, the following deserve highlighting:

- the vertiginous development of the "-omics" disciplines (metabolomics, proteomics, transcriptomics, genomics, epigenomics, etc.) in obtaining large amounts of data from high-quality samples [23],
- the need of large quantities of representative biological samples with their associated clinical information from different scenarios where the "-omics" technologies can be applied, and
- the "globalization of biomedical research" through an increasingly abundant and close collaboration among different research groups around the world that implies the interchange of biological samples and data.

Hence, biobanks constitute a research infrastructure committed to providing facilities, resources, and services to research groups in an open, transparent, and generous manner that benefits not only science but also participants (patients, healthy donors, etc.).

2.1 Biobanks as Research Infrastructures

Biobanks constitute an exclusive research infrastructure that requires different governance mechanisms than do project-based mechanisms. This infrastructure necessitates a specialized level of organization focused not only on ensuring the quality of the biological samples and their associated information but also on the access rules and confidence for both end users: researchers, as well as participants. Because of this operating philosophy, biobanks need to be dynamic and efficient in meeting the requirements of researchers, while preserving the rights and the confidentiality of the donors. Overcoming this challenge requires the involvement of many actors (clinicians, data managers, technicians, pathologists, etc.) and institutional support. Trust between the biobank and the investigators using the biobank is of major importance. The biomaterials and data must be available for their projects. In addition, the trust between the biobank and the participants is crucial. Within this force field, the biobank manager needs to find the solution that best fits each situation. In order to guarantee a minimum of functionality within the framework described above, at least the following parameters should be considered.

2.1.1 IC and Governance

Rather than providing IC for a specific project, participants in most biobanks offer an *open consent* for multiple future biomedical projects, the details of which cannot be provided at the time of enrollment. For this reason, the governance mechanism must balance the needs of the scientific community and the participants with an emphasis on the recognition of participants, trustworthiness, and adaptive management [24,25]. Hence, formal governance structures are a common and necessary component of biobanks. Although the institutional review board (IRB) is an essential component for oversight and safety, most biobanks utilize a formal access or oversight committee to approve the use of samples and data [26,27,28]. These committees may serve to review the scientific content of the research proposals as well as provide management for finite biological samples. Members of these committees should be skilled in scientific, ethical, and clinical domains to provide an additional level of safety and rigor to projects using biobanks.

In many countries, as well as in some collaborative research consortiums, the role of participants is increasingly taken into account. In some contexts, informal governance structures, such as community advisory boards (CABs), can be an important component of biobank governance [25]. CABs may provide advice about the efficacy of the IC process and the implementation of research protocols [29], and they are representative of the community participating in the research being reviewed (for instance, cancer patient coalitions, rare disease advocacy groups, special and minority groups). Input from the community can add insight into the perspectives of participants when questions arise. Issues with return of results, academic/industry partnership, and privacy will evolve over the life of the biobank. Hence, participants are key stakeholders in these issues and CABs can serve as a representative voice of the community [30].

2.1.2 Associated Data

Most prevalent diseases are considered to be caused by a large number of genetic and environmental effects or modest gene–environment interactions [31]. These interactions may also affect the response to treatments and patient follow-up after clinical interventions. In these scenarios, the scientific value of biobanks is greatly enhanced when they also have lifestyle, risk factor data, and other clinical information available [5]. Clinically useful data commonly include demographic information, general health and functioning, personal and family medical history, health behaviors (e.g., diet, physical activity, and smoking), medication use (both prescription and over-the-counter medications), diagnosis (type of disease, biomarkers, etc.), type of treatment, and follow-up. The systematic collection of family history data can also be highly useful, as it is generally stored as unstructured text in the electronic health record (EHR), making it difficult to cost-effectively retrieve it for research studies [32].

2.1.3 Sustainability

Biobanks are often launched without a long-term plan of sustainability [5,17]. While a large component of the cost is the upfront collection and processing of samples, there are significant costs to maintaining samples, data, and access to a biobank. Cost recovery models vary from institutional support to complete support through user fees, although the latter are hard to set, given large initial costs and the life cycle of a biobank over decades [5]. More recently, it has been suggested that biobanking in a clinical context might be incorporated into the cost of business and embedded into the fee and insurance reimbursement structure [33].

2.2 Biobanks as Centers for Enhancing Security and Quality of Samples and Data

To be successful, biobanks must pay close attention to collecting high-quality specimens and relevant associated information. An important goal is to ensure a stable and comparable preanalytical phase, therefore SOPs and adherence to the SOPs should be implemented with sufficient sample handling information so that the history of each sample is completely retraceable [5]. This is especially the case when the results show unexpected deviations. Quality metrics are key indicators of the usefulness of biobank specimens. For instance, DNA is fundamental to epigenetic studies; numerous methods to estimate quality have been developed, including total DNA yield and DNA amplification by polymerase chain reaction (PCR) [34].

The value of biobank samples is enhanced by the presence of a high-quality information management system (IMS) to track overtime data concerning enrollment and consent; sample acquisition, processing, storage, and distribution; quality assurance/quality control; collection and/or linkage to subject data (such as clinical data); data security and access; and reporting functions. The IMS plays a critical role in providing sample and data accountability and in tracking a sample from collection to processing, storage, use, and final disposal. Use of barcodes to enhance the tracking is strongly encouraged [35]. A robust IMS will be able to integrate large volumes of data from multiple sources, including both clinical and research data. Use of recognized standards (like SPREC) [13,36] enhances the ability to harmonize with other biobanks for the pooling of projects.

Biobanks with an IMS that can be linked to EHRs have an especially rich resource from which to draw a wealth of data. Because of this, the development of methods to rapidly extract phenotype data from the EHR is an active area of investigation [37]. Still, there are several challenges in using EHRs for research, including data that are often incomplete, inaccurate, conflicting, highly complex, and potentially biased [25].

Another data source for biobanks are the data observations from completed studies conducted using the biobank's samples. The return of the research data to the biobank at the completion of the study should become a normal procedure. This enables the secondary use of existing research data which can provide many opportunities for new discoveries beyond the scope of the original study, leading to the optimization of the resources. For example, infrastructures like the Mayo Clinic Biobank require all the data generated using their materials to be deposited into a secure central database for future use [27]. In this context, the biobank builds again on trust and acts as an *honest broker* safeguarding privacy and confidentiality; hence, any study using returned data would need approval from the Access Committee and a separate IRB approval [38]. Popular data types for reuse are genome-wide association and whole-exome sequencing data, which can be used to reduce genotyping costs for subsequent studies and improve the characterization of genetic variants that are clinically relevant and actionable. Another model would be that the data become part of the hospital data, which also harbor the biobank and clinical data. This way, the research is brought close to both the patient and to the biobank for reuse. Clinicians are able to look into the research data and can contact the researchers about clinical implications.

Thus, biobanks have the capacity to generate a large amount of genetic and epigenetic data, some of which may have health implications for the participants and their relatives. This raises the need to address the return of both primary research results and incidental findings. In countries like Spain, this type of information is considered a right of the participants of the study and is guaranteed by law.

It is in the IC form where participants express their desire to be informed of the research results. Others recommend the results be returned to the biobank participants if certain criteria are met [39,40], although the problem of what to do with genes with pleiotrophic effects is yet to be resolved [41]. The process of returning results to participants incurs significant costs to biobanks and must be taken into account when planning for result disclosure [42]. Input from a CAB can also provide important insights when developing return-of-results policies and procedures. Furthermore, it is necessary that biobank participants receive appropriate clinical follow-up with a geneticist and/or genetic counselor after receiving research results, which may incur costs to participants [39].

3. EPIGENETICS AND BIOBANKING

Epigenetics has recently emerged as a new and promising field within the genetics discipline. Lifestyle, nutrition, stress, diseases, and pharmacological interventions have a great impact on the epigenetic code of the cells by altering DNA methylation, miRNA expression, and histone posttranslational modifications (PTMs). These changes are inherently more plastic and dynamic than genetic mutations and they contribute to the pathology of several diseases. Moreover, there is an increasing need to find new biomarkers for diagnosis, prognosis, and therapy monitoring, hence epigenomic maps of cell types and tissues in specific diseases or physiological situations will provide a valuable resource for the identification of promising biomarkers and will also facilitate the characterization of targets for epigenetic drug development [43].

In this scientific context, the recruitment of biological samples must be entrusted to dedicated biobanks with dedicated personnel. This will help ensure the adequate collection, processing, and preservation of comparable, high-quality samples.

3.1 Epigenetic Biomarkers: Stability in Different Sample Sources

More and more studies are focused on the identification of biomarkers that may be of clinical utility. To deliver reliable clinical biological markers, researchers will have to look for those characterized by high stability, reproducibility, sensitivity, and specificity. Epigenetic biomarkers, which are based on methylation of specific genes, miRNA signatures, and specific PTMs in histones, reflect all of these properties. Consequently, they are promising candidates for delivering potential clinical biomarkers.

When searching for clinical markers, researchers need to focus on the study of stable molecules. Therefore, it is necessary to assess the stability of epigenetic markers under various, clinically relevant conditions. The characterization of the degree of stability of epigenetic biomarkers and the identification of determinants that may affect their integrity can provide critical information for the correct design and conduction of clinical trials and validation of biomarkers for clinical practice. Therefore, well-designed studies evaluating the stability of epigenetic biomarkers under the different sample handling processes (collection, preparation of samples, extraction procedures, preservation, etc.) are required. In this regard, biobanks can implement subprocesses with SOPs that ensure the stability of the preanalytical phase of the biological samples they host.

Interestingly, intrinsic properties characterize the biomolecules that can be used for epigenetic studies, making them suitable substrates for these purposes. Methylated DNA has been shown to be more stable in vivo and in vitro than unmethylated DNA [44]. Furthermore, it has been demonstrated that methylated DNA circulates in blood longer than unmethylated DNA does. Thus, cell-free circulating methylated DNA can be a good candidate for diagnostic biomarkers [44]. In fact, methylated DNA has been determined in serum and plasma [45–49], urine and semen [50–52], bronchoalveolar lavage fluid [53], saliva [54], sputum [55,56], ductal lavage fluid [57], and fine-needle aspirates [58]; a fact which gives an idea of the high stability of methylated DNA.

miRNAs are a large family of short noncoding RNAs (17–25 nucleotides) [59], which are involved in many biological processes and cancer [60,61]. The miRNAs have been described as highly stable molecules, and some authors have attributed their stability to nuclease resistance due to smaller nucleic acid size [62] and/or microvesicular containment [63]. Therefore, the study of these stable small RNA molecules involved in a broad spectrum of biological processes may lead to the discovery of promising new biomarkers with a diagnostic, prognostic, or therapeutic value [43,64].

Mitchell et al. [65] and Chen et al. [66] were the first to report the stability of miRNAs in serum and their potential use as biomarkers for disease. Since then, many other investigations have analyzed the stability of miRNAs over time, taking into account preanalytical conditions in body fluids, such as serum and plasma [67] and urine [68].

Additionally, specific miRNA signatures with clinical value have been identified in the body fluids of cancer patients, making cell-free miRNAs good candidates for minimally invasive biomarkers [69]. The miRNAs and miRNA signatures from serum, urine, saliva, cerebrospinal fluid [70,71], and exosomes [72] have been described as promising diagnostic biomarkers. As has been pointed out before, some authors have attributed the unusual stability of miRNA in several biofluids, as well as in tissues, to their location within exosomes [73], since these vesicles act as a barrier to nucleases.

Exosomes containing miRNAs have been described as playing a role in cell-to-cell communication [74,75]. Recent studies have described the isolation of the exosomes from various body fluids, including plasma, malignant ascites, urine, amniotic fluid, breast milk, and saliva [76,77]. The stability of exosomes and exosomal miRNA in body fluids and their accurate detection by quantitative PCR further support their potential as noninvasive, or at least minimally invasive, biomarkers for disease [78]. In this regard, it is important to point out that procedures for isolating exosomes and miRNAs, along with methodologies for quantifying miRNA species from biological material, are relatively recent developments. Thus, SOPs for the collection, processing, and preservation of biological specimens from which exosomes and miRNAs can be isolated must be carefully evaluated before these molecules are used for further research as candidate biomarkers. In this regard, biobanks can play a crucial role, fine-tuning these procedures and ensuring the quality of samples.

The miRNAs can be readily isolated not only from body fluids but also from fresh or fixed tissues. The stability of these small molecules in tissue samples of compromised quality, such as formalin-fixed paraffin-embedded (FFPE) samples, has also been evaluated [79]. Formalin fixation and the paraffin-embedding process cause enzymatic degradation and even a chemical modification of RNA, giving rise to crosslinks with proteins and making RNA extraction difficult [80,81,82]. However, small RNAs are easily extracted after proteinase K digestion. Therefore, miRNAs from FFPE samples are similarly recovered than miRNAs from optimally preserved frozen samples, thus supporting the high degree of stability of miRNAs [79]. Thus, tissue samples, mainly found in HIBs, could be the starting material for early biomarker studies. Retrospective studies using tissue samples for the identification of miRNA signatures in diseases may be the first step in the discovery of candidate biomarkers. Consequently, once potential biomarkers are identified, methods can be optimized in specific SOPs so as to detect and validate these specific miRNAs in body fluids in order to obtain useful minimally invasive epigenetic biomarkers to be used in everyday clinical practice.

Histones are other biomolecules that can be used as epigenetic biomarkers, and both the presence of circulating histones and altered PTMs in histones in specific tissue or cell lines are indicative of pathological conditions.

Extracellular histones have been demonstrated to function as endogenous danger signals. Levels of circulating histones as well as nucleosomes are increased in patients with cancer, inflammation, and infection [83,84]. Hence, detecting the levels of histones and nucleosomes in biological fluids could set the basis for the design of new biomarkers. EDTA (10 mmol/L) avoids further digestion of nucleosomes through the Ca^{+2}- and Mg^{+2}-dependent endonucleases present in blood. These processed samples have shown good stability for at least 6–12 months at −80 and −20 °C [85]. In addition, a 5-year storage at −80 °C revealed a decrease of about 32% of the nucleosome values, suggesting that long-term stability of these kinds of proteins in blood derivatives can be compromised during the storage time [86]. Clarifying what are the optimal processes for preserving the integrity of these proteins in body fluids, thus avoiding the degradation by endogenous proteases, is a challenge in which biobanks can contribute actively by providing SOPs that ensure the integrity of these nuclear proteins.

Histone PTMs play an important role in the epigenetic regulation, and irregular patterns of histone global acetylation and methylation have frequently been observed in various diseases [87,88]. As a consequence, quantitative analysis of the PTMs can be very useful in identifying new biomarkers, looking for therapeutic targets, or monitoring treatment. Mass spectrometric (MS) approaches, such as selected reaction monitoring, can quantify histone PTMs directly from core histone samples [89]. Therefore, it is very important to characterize the source from which histones have been purified. In this context, nonfixed samples are further recommended for MS studies because protein extraction from FFPE tissues faces challenges due to deparaffinization and cross-link reversion. Despite the advances in proteomics, factors affecting the stability and utility of FFPE biomaterials will require detailed evaluation and validation, and PTM enrichment procedures would be required. In conclusion, the PTM-directed analysis in FFPE would be complicated by dynamic changes during the fixation process and due to complications of affinity enrichment procedures [90]. Since the optimal starting material for MS studies, such as fresh-frozen tissues, are not always readily available, the optimization of methods for the extraction of proteins from different sources of biological material to carry out these studies is also needed [91].

Other important experimental procedures used in epigenomic studies are chromatin immunoprecipitation (ChIP) experiments followed by real-time PCR or sequencing analysis [92]. ChIP experiments consist of the immunoprecipitation of the chromatin using specific antibodies that recognize proteins bound to chromatin and specific PTMs in histones. Although FFPE tissue samples comprise a potentially valuable resource for retrospective biomarker discovery studies, these biological samples have drawbacks, as described earlier. These samples often display degradation and loss of antigenicity due to harsh fixation conditions and prolonged storage. These factors can affect the ease of chromatin preparation for experiments of ChIP-seq and require optimized protocols for good results. Some commercial kits partially solve this problem. For example, Active Motif, Inc. has developed a ChIP-IT® FFPE Chromatin Preparation Kit that contains specifically formulated reagents and an improved protocol to extract high-quality chromatin from histological slides or tissue sections. Moreover, there also exists literature where optimized ChIP-seq procedures to obtain good-quality chromatin from FFPE samples are described [93].

Overall, the stability of candidate biomarkers is a prerequisite for the adequate identification and validation of epigenetic biomarkers which may be used in clinical practice to improve the accuracy of diagnosis, predict prognosis, and monitor disease progression and response to therapy. The nature and essence of biobanks make them a feasible source for providing researchers this appreciated biomaterial and associated data of high quality.

3.2 How Can Biobanks Contribute to Epigenetic and Epigenomic Studies?

Large-scale epigenomic studies have revolutionized three major areas in science: basic gene regulatory processes, cellular differentiation and reprogramming, and the role of epigenetic regulation in disease [94]. In a similar way, epigenetics has accelerated biomedical research in three major areas: the identification of pathogenic pathways and targets participating in disease, the identification of new epigenetic biomarkers, and the development of epigenetic drugs for treating diseases. The optimal conditions for the collection, processing, and long-term preservation of biological samples need to be defined before performing biomarker discovery studies. Furthermore, taking into account the preanalytical variables associated with the above-mentioned processes is very important for the success of epigenetic and epigenomic studies. Additionally, there is a need for adequate isolation and detection procedures for the specific biomolecules that are to be analyzed as candidate biomarkers, as described in the previous sections. Moreover, an adequate selection of cell types, tissues, and body fluids of high quality to carry out biomedical research should be ensured. Consequently, biobanks play a crucial role in ensuring the availability to researchers of high-quality biological samples that have been collected, processed, and preserved the best way, following standardized operating procedures.

Furthermore, biobanks can contribute to the systematization, protocolization, standardization, and innovation of technical procedures that could facilitate researchers with optimal starting material for their investigations. The implementation and standardization of technical procedures in biobanks may contribute to decrease the amount of resources needed in order to acquire the biomaterial required for epigenetic studies [43]. Moreover, as previously mentioned, the return of research data to biobanks at the completion of a study could lead to an optimization of the resources, avoiding the need for researchers to reanalyze the same parameters while reducing the cost and time required for subsequent studies.

Large-scale epigenomic projects and studies focusing on the identification of epigenetic biomarkers require a large number of biological samples with their associated clinical data. Nevertheless, a single biobank may not always be able to provide the number of samples needed to provide statistical power in this kind of studies. Thus, there is a need for biobanks to network. The standardization of procedures between biobanks plays a key role in these networks [95]. The participation of several biobanks in multicenter projects requires all of them to adopt common SOPs [15]. Hence, networking between biobanks also requires a multidirectional flow of information. Consequently, the proper management and traceability of preanalytical variables associated with samples are crucial to ensuring effective interconnection and interoperability among biobanks [96]. Thus, the coordination and integration of biobanks in a way that allows them to share SOPs and information about the samples they hold may facilitate large-scale collaborative projects. For the shared information, it is crucial to look carefully at the annotations used and translate them to a common format wherever necessary.

Finally, genes, epigenetic regulation, and environment are important contributors in the development of human diseases [97]. Individuals have different genomes and environmental exposure histories, the latter also defined as *exposomes* [98], affecting their epigenomic signature. Epigenetics and exposomes are tightly associated, and it is surprising how environment can affect the human epigenome. Regarding the impact of environment on epigenetics, biobanks interested in human biomonitoring that host samples from humans exposed to different environmental conditions (exposure to chemicals, for example) have appeared recently in great quantity. The number of this type of biobank is exploding. *Environmental specimen biobanks* can be integrated into PBs. These biobanks may make possible the development of interesting epigenomic projects that shed light on how environment affects our health.

Overall, the identification and analysis of candidate biomarkers to be translated to clinics require a proper research ecosystem that can be mediated by biobanks. Before performing biomarker discovery studies, optimal material with the associated data needs to be used, and adequate isolation and detection procedures need to be defined. To a large degree, biobanks can cover the area of regulatory and ethical issues, as well as that of social engagement and focus on the implications of epigenetic studies. Biobanks would act as providers of high-quality biological samples with well-annotated data to the scientific community, facilitating the deep research collaborations needed for large-scale epigenomic projects (Figure 2). Therefore, dedicated biobanks are a key step in the development of personalized/precision medicine [5,99,100].

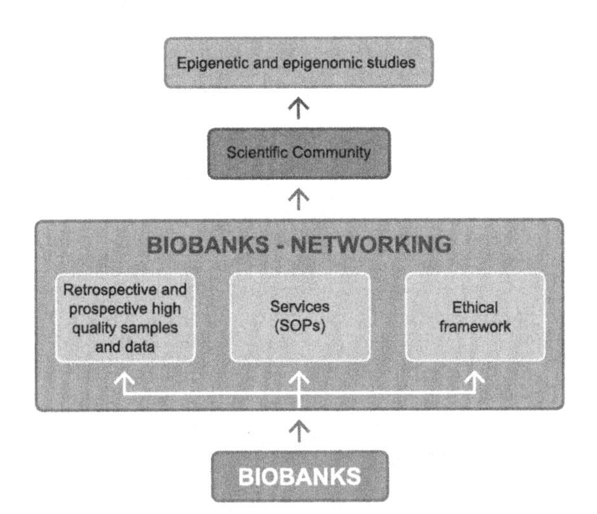

FIGURE 2 A schematic representation of how biobanks contribute to epigenetic and epigenomic studies, providing high-quality samples, data, and services to the scientific community in an ethical framework, facilitating deep research collaborations through biobanking networks.

4. CONCLUSIONS

To conclude, the following are some of the most important aspects that have been discussed in this chapter.

- During the last few years, biobanks have become an important resource for biomedical research, constituting an exclusive infrastructure that requires specific governance mechanisms.
- Biobanks must pay close attention to the collection of high-quality specimens and their relevant associated information. They must conduct their activities in a way that respects individuals by ensuring their privacy and confidentiality. Biobanks also need to be dynamic and efficient in order to meet the requirements of researchers.
- Epigenetic biomarkers, which are based on the methylation of specific genes, miRNA signatures, and specific PTMs in histones, are promising clinical biomarkers because they are highly stable under different conditions.
- Biobanks provide critical research and infrastructure support for biomedical research in the field of personalized medicine, facilitating the discovery and validation of epi- and genetic associations with exposome interactions. These interactions will shed light onto disease pathogenesis which can ultimately be translated into risk assessment/stratification schemes, new diagnostic and pharmacogenomic tools, and new drugs.

- All biobanks involved in health-care research should encourage a learning environment and a dynamic adaptation to ensure the rapid and valid translation of epi- and genetics results to patients and public health systems.

LIST OF ABBREVIATIONS

BRCs Biological resources centers
BRISQ Biospecimen Reporting for Improved Study Quality
CABs Community advisory boards
EHR Electronic health record
EMA European Medicines Agency
FFPE Formalin-fixed paraffin-embedded
HIB Hospital-integrated biobanks
IMS Information management system
ICH International Conference on Harmonization of Technical Requirements
IC Informed consent
ISBER International Society for Biological and Environmental Repositories
IRB Institutional review board
MS Mass spectrometry
MIABIS Minimum Information About BIobank data Sharing
OECD Organization for Economic Co-operation and Development
P3G Public Population Project in Genomics and Society
PB Population-based biobanks
R&D Research and development
SPREC Standard PREanalytical Code
SOPs Standard operating procedures

Acknowledgments

The authors would like to offer thanks for the financial support of the Ministerio de Economía y Competitividad, ISCIII, FEDER, under grants PT13/0010/0004, PT13/0010/0037, and PT13/0010/0064, the European Union Seventh Framework Programme FP7 under grant agreement n° 260791: EurocanPlatform "A European Platform for Translational Cancer Research" and the persons working in WP10, and EC H2020-ICT-2014-1-644242 SAPHELY.

References

[1] Cambon-Thomsen A, Rial-Sebbag E, Knoppers BM. Trends in ethical and legal frameworks for the use of human biobanks. Eur Respir J August 2007;30(2):373–82.
[2] OECD. Best practice guidelines on biosecurity for biological resource centers. 2007.
[3] ISBER. Best practices for repositories: collection, storage, retrieval and distribution of biological materials for research. Cell Preserv Technol 2008;6(1):3–58.
[4] Kauffmann F, Cambon-Thomsen A. Tracing biological collections: between books and clinical trials. JAMA May 21, 2008;299(19):2316–8.
[5] Riegman PH, Morente MM, Betsou F, de Blasio P, Geary P. Biobanking for better healthcare. Mol Oncol October 2008;2(3):213–22.
[6] Labant MA. Biobank diversity facilitates drug & diagnostic development. Genet Eng Biotechnol News 2012;32(2):42–4.
[7] Bevilacqua G, Bosman F, Dassesse T, Hofler H, Janin A, Langer R, et al. The role of the pathologist in tissue banking: European Consensus Expert Group Report. Virchows Arch April 2010;456(4):449–54.
[8] Riegman PH, Dinjens WN, Oosterhuis JW. Biobanking for interdisciplinary clinical research. Pathobiology 2007;74(4):239–44.
[9] Carter A, Betsou F. Quality assurance in cancer biobanking. Biopreserv Biobank June 2011;9(2):157–63.
[10] 2012 best practices for repositories collection, storage, retrieval, and distribution of biological materials for research international society for biological and environmental repositories. Biopreserv Biobank April 2012;10(2):79–161.
[11] NCI. National Cancer Institute best practices for biospecimen resources. Bethesda: NCI; 2007.
[12] Betsou F, Lehmann S, Ashton G, Barnes M, Benson EE, Coppola D, et al. Standard preanalytical coding for biospecimens: defining the sample PREanalytical code. Cancer Epidemiol Biomarkers Prev April 2010;19(4):1004–11.
[13] Lehmann S, Guadagni F, Moore H, Ashton G, Barnes M, Benson E, et al. Standard preanalytical coding for biospecimens: review and implementation of the Sample PREanalytical Code (SPREC). Biopreserv Biobank August 2012;10(4):366–74.
[14] Moore HM, Kelly AB, Jewell SD, McShane LM, Clark DP, Greenspan R, et al. Biospecimen reporting for improved study quality (BRISQ). Cancer Cytopathol April 25, 2011;119(2):92–101.
[15] Norlin L, Fransson MN, Eriksson M, Merino-Martinez R, Anderberg M, Kurtovic S, et al. A minimum data set for sharing biobank samples, information, and data: MIABIS. Biopreserv Biobank August 2012;10(4):343–8.
[16] Hansson MG. Ethics and biobanks. Br J Cancer January 13, 2009;100(1):8–12.
[17] Hewitt RE. Biobanking: the foundation of personalized medicine. Curr Opin Oncol January 2011;23(1):112–9.
[18] Riegman PH, van Veen EB. Biobanking residual tissues. Hum Genet September 2011;130(3):357–68.
[19] Porteri C, Togni E, Pasqualetti P. The policies of ethics committees in the management of biobanks used for research: an Italian survey. Eur J Hum Genet February 2014;22(2):260–5.

[20] Gottweis HaL G. Biobank governance in the postgenomic age. Personal Med 2010;7:187–95.
[21] Watson PH, Wilson-McManus JE, Barnes RO, Giesz SC, Png A, Hegele RG, et al. Evolutionary concepts in biobanking - the BC BioLibrary. J Transl Med 2009;7:95.
[22] Ginsburg GS, Burke TW, Febbo P. Centralized biorepositories for genetic and genomic research. JAMA March 19, 2008;299(11):1359–61.
[23] Hawkins AK. Biobanks: importance, implications and opportunities for genetic counselors. J Genet Couns October 2010;19(5):423–9.
[24] O'Doherty KC, Burgess MM, Edwards K, Gallagher RP, Hawkins AK, Kaye J, et al. From consent to institutions: designing adaptive governance for genomic biobanks. Soc Sci Med August 2011;73(3):367–74.
[25] Olson JE, Bielinski SJ, Ryu E, Winkler EM, Takahashi PY, Pathak J, et al. Biobanks and personalized medicine. Clin Genet July 2014;86(1):50–5.
[26] Yuille M, Dixon K, Platt A, Pullum S, Lewis D, Hall A, et al. The UK DNA banking network: a "fair access" biobank. Cell Tissue Bank August 2010;11(3):241–51.
[27] Olson JE, Ryu E, Johnson KJ, Koenig BA, Maschke KJ, Morrisette JA, et al. The Mayo Clinic Biobank: a building block for individualized medicine. Mayo Clin Proc September 2013;88(9):952–62.
[28] Lopez-Guerrero JA, Riegman PH, Oosterhuis JW, Lam KH, Oomen MH, Spatz A, et al. TuBaFrost 4: access rules and incentives for a European tumour bank. Eur J Cancer November 2006;42(17):2924–9.
[29] Strauss RP, Sengupta S, Quinn SC, Goeppinger J, Spaulding C, Kegeles SM, et al. The role of community advisory boards: involving communities in the informed consent process. Am J Public Health December 2001;91(12):1938–43.
[30] Haga SB, Zhao JQ. Stakeholder views on returning research results. Adv Genet 2013;84:41–81.
[31] Ogino S, Lochhead P, Chan AT, Nishihara R, Cho E, Wolpin BM, et al. Molecular pathological epidemiology of epigenetics: emerging integrative science to analyze environment, host, and disease. Mod Pathol April 2013;26(4):465–84.
[32] Kho AN, Pacheco JA, Peissig PL, Rasmussen L, Newton KM, Weston N, et al. Electronic medical records for genetic research: results of the eMERGE consortium. Sci Transl Med April 20, 2011;3(79):79re1.
[33] McDonald SA, Watson MA, Rossi J, Becker CM, Jaques DP, Pfeifer JD. A new paradigm for biospecimen banking in the personalized medicine era. Am J Clin Pathol November 2011;136(5):679–84.
[34] Betsou F, Gunter E, Clements J, DeSouza Y, Goddard KA, Guadagni F, et al. Identification of evidence-based biospecimen quality-control tools: a report of the International Society for Biological and Environmental Repositories (ISBER) Biospecimen Science Working Group. J Mol Diagn January 2013;15(1):3–16.
[35] Troyer D. Biorepository standards and protocols for collecting, processing, and storing human tissues. Methods Mol Biol 2008;441:193–220.
[36] Nussbeck SY, Benson EE, Betsou F, Guadagni F, Lehmann S, Umbach N. Is there a protocol for using the SPREC? Biopreserv Biobank October 2013;11(5):260–6.
[37] Hripcsak G, Albers DJ. Next-generation phenotyping of electronic health records. J Am Med Inf Assoc January 1, 2013;20(1):117–21.
[38] Mayol-Heath DN, Woo P, Galbraith K. Biorepository regulatory considerations: a detailed topic follow-up to the blueprint for the development of a community-based hospital biorepository. Biopreserv Biobank December 2011;9(4):321–6.
[39] Wolf SM, Lawrenz FP, Nelson CA, Kahn JP, Cho MK, Clayton EW, et al. Managing incidental findings in human subjects research: analysis and recommendations. J Law Med Ethics Summer 2008;36(2): 219–48, 1.
[40] McGuire AL, Caulfield T, Cho MK. Research ethics and the challenge of whole-genome sequencing. Nat Rev Genet February 2008;9(2):152–6.
[41] Kocarnik JM, Fullerton SM. Returning pleiotropic results from genetic testing to patients and research participants. JAMA February 26, 2014;311(8):795–6.
[42] Black L, Avard D, Zawati MH, Knoppers BM, Hebert J, Sauvageau G. Funding considerations for the disclosure of genetic incidental findings in biobank research. Clin Genet November 2013;84(5):397–406.
[43] Sandoval J, Peiro-Chova L, Pallardo FV, Garcia-Gimenez JL. Epigenetic biomarkers in laboratory diagnostics: emerging approaches and opportunities. Expert Rev Mol Diagn June 2013;13(5):457–71.
[44] Skvorstova TE. In: Grahan PB, editor. Circulating nucleic acids in plasma and serum. Springer; 2011.
[45] Esteller M, Hamilton SR, Burger PC, Baylin SB, Herman JG. Inactivation of the DNA repair gene O6-methylguanine-DNA methyltransferase by promoter hypermethylation is a common event in primary human neoplasia. Cancer Res February 15, 1999;59(4):793–7.
[46] Esteller M, Sanchez-Cespedes M, Rosell R, Sidransky D, Baylin SB, Herman JG. Detection of aberrant promoter hypermethylation of tumor suppressor genes in serum DNA from non-small cell lung cancer patients. Cancer Res January 1, 1999;59(1):67–70.
[47] Sanchez-Cespedes M, Esteller M, Wu L, Nawroz-Danish H, Yoo GH, Koch WM, et al. Gene promoter hypermethylation in tumors and serum of head and neck cancer patients. Cancer Res February 15, 2000;60(4):892–5.
[48] Kawakami K, Brabender J, Lord RV, Groshen S, Greenwald BD, Krasna MJ, et al. Hypermethylated APC DNA in plasma and prognosis of patients with esophageal adenocarcinoma. J Natl Cancer Inst November 15, 2000;92(22):1805–11.

[49] Usadel H, Brabender J, Danenberg KD, Jeronimo C, Harden S, Engles J, et al. Quantitative adenomatous polyposis coli promoter methylation analysis in tumor tissue, serum, and plasma DNA of patients with lung cancer. Cancer Res January 15, 2002;62(2):371–5.

[50] Goessl C, Krause H, Muller M, Heicappell R, Schrader M, Sachsinger J, et al. Fluorescent methylation-specific polymerase chain reaction for DNA-based detection of prostate cancer in bodily fluids. Cancer Res November 1, 2000;60(21):5941–5.

[51] Cairns P, Esteller M, Herman JG, Schoenberg M, Jeronimo C, Sanchez-Cespedes M, et al. Molecular detection of prostate cancer in urine by GSTP1 hypermethylation. Clin Cancer Res September 2001; 7(9):2727–30.

[52] Jeronimo C, Usadel H, Henrique R, Silva C, Oliveira J, Lopes C, et al. Quantitative GSTP1 hypermethylation in bodily fluids of patients with prostate cancer. Urology December 2002;60(6):1131–5.

[53] Ahrendt SA, Chow JT, Xu LH, Yang SC, Eisenberger CF, Esteller M, et al. Molecular detection of tumor cells in bronchoalveolar lavage fluid from patients with early stage lung cancer. J Natl Cancer Inst February 17, 1999;91(4):332–9.

[54] Rosas SL, Koch W, da Costa Carvalho MG, Wu L, Califano J, Westra W, et al. Promoter hypermethylation patterns of p16, O6-methylguanine-DNA-methyltransferase, and death-associated protein kinase in tumors and saliva of head and neck cancer patients. Cancer Res February 1, 2001;61(3):939–42.

[55] Belinsky SA, Nikula KJ, Palmisano WA, Michels R, Saccomanno G, Gabrielson E, et al. Aberrant methylation of p16(INK4a) is an early event in lung cancer and a potential biomarker for early diagnosis. Proc Natl Acad Sci USA September 29, 1998;95(20):11891–6.

[56] Palmisano WA, Divine KK, Saccomanno G, Gilliland FD, Baylin SB, Herman JG, et al. Predicting lung cancer by detecting aberrant promoter methylation in sputum. Cancer Res November 1, 2000;60(21):5954–8.

[57] Evron E, Dooley WC, Umbricht CB, Rosenthal D, Sacchi N, Gabrielson E, et al. Detection of breast cancer cells in ductal lavage fluid by methylation-specific PCR. Lancet April 28, 2001;357(9265):1335–6.

[58] Jeronimo C, Costa I, Martins MC, Monteiro P, Lisboa S, Palmeira C, et al. Detection of gene promoter hypermethylation in fine needle washings from breast lesions. Clin Cancer Res August 15, 2003;9(9):3413–7.

[59] He L, Hannon GJ. MicroRNAs: small RNAs with a big role in gene regulation. Nat Rev Genet July 2004;5(7): 522–31.

[60] Hwang HW, Mendell JT. MicroRNAs in cell proliferation, cell death, and tumorigenesis. Br J Cancer March 27, 2006;94(6):776–80.

[61] Garzon R, Marcucci G, Croce CM. Targeting microRNAs in cancer: rationale, strategies and challenges. Nat Rev Drug Discov October 2010;9(10):775–89.

[62] Hanke M, Hoefig K, Merz H, Feller AC, Kausch I, Jocham D, et al. A robust methodology to study urine microRNA as tumor marker: microRNA-126 and microRNA-182 are related to urinary bladder cancer. Urol Oncol November-December 2010;28(6):655–61.

[63] Schwarzenbach H, Hoon DS, Pantel K. Cell-free nucleic acids as biomarkers in cancer patients. Nat Rev Cancer June 2011;11(6):426–37.

[64] Lawrie CH. MicroRNA expression in lymphoid malignancies: new hope for diagnosis and therapy? J Cell Mol Med September-October 2008;12(5A):1432–44.

[65] Mitchell PS, Parkin RK, Kroh EM, Fritz BR, Wyman SK, Pogosova-Agadjanyan EL, et al. Circulating microRNAs as stable blood-based markers for cancer detection. Proc Natl Acad Sci USA Jul 29, 2008;105(30):10513–8.

[66] Chen X, Ba Y, Ma L, Cai X, Yin Y, Wang K, et al. Characterization of microRNAs in serum: a novel class of biomarkers for diagnosis of cancer and other diseases. Cell Res October 2008;18(10):997–1006.

[67] Kroh EM, Parkin RK, Mitchell PS, Tewari M. Analysis of circulating microRNA biomarkers in plasma and serum using quantitative reverse transcription-PCR (qRT-PCR). Methods April 2010;50(4):298–301.

[68] Mall C, Rocke DM, Durbin-Johnson B, Weiss RH. Stability of miRNA in human urine supports its biomarker potential. Biomark Med August 2013;7(4):623–31.

[69] Zandberga E, Kozirovskis V, Abols A, Andrejeva D, Purkalne G, Line A. Cell-free microRNAs as diagnostic, prognostic, and predictive biomarkers for lung cancer. Genes Chromosom Cancer April 2013;52(4):356–69.

[70] Tzimagiorgis G, Michailidou EZ, Kritis A, Markopoulos AK, Kouidou S. Recovering circulating extracellular or cell-free RNA from bodily fluids. Cancer Epidemiol December 2011;35(6):580–9.

[71] Weber JA, Baxter DH, Zhang S, Huang DY, Huang KH, Lee MJ, et al. The microRNA spectrum in 12 body fluids. Clin Chem November 2010;56(11):1733–41.

[72] Michael A, Bajracharya SD, Yuen PS, Zhou H, Star RA, Illei GG, et al. Exosomes from human saliva as a source of microRNA biomarkers. Oral Dis January 2010;16(1):34–8.

[73] Valadi H, Ekstrom K, Bossios A, Sjostrand M, Lee JJ, Lotvall JO. Exosome-mediated transfer of mRNAs and microRNAs is a novel mechanism of genetic exchange between cells. Nat Cell Biol June 2007;9(6):654–9.

[74] Montecalvo A, Larregina AT, Shufesky WJ, Stolz DB, Sullivan ML, Karlsson JM, et al. Mechanism of transfer of functional microRNAs between mouse dendritic cells via exosomes. Blood January 19, 2012;119(3):756–66.

[75] Wahlgren J, De LKT, Brisslert M, Vaziri Sani F, Telemo E, Sunnerhagen P, et al. Plasma exosomes can deliver exogenous short interfering RNA to monocytes and lymphocytes. Nucleic Acids Res September 1, 2012;40(17):e130.

[76] Keller S, Ridinger J, Rupp AK, Janssen JW, Altevogt P. Body fluid derived exosomes as a novel template for clinical diagnostics. J Transl Med 2011;9:86.

[77] Gu Y, Li M, Wang T, Liang Y, Zhong Z, Wang X, et al. Lactation-related microRNA expression profiles of porcine breast milk exosomes. PLoS One 2012;7(8):e43691.

[78] Lv LL, Cao Y, Liu D, Xu M, Liu H, Tang RN, et al. Isolation and quantification of microRNAs from urinary exosomes/microvesicles for biomarker discovery. Int J Biol Sci 2013;9(10):1021–31.

[79] Peiro-Chova L, Pena-Chilet M, Lopez-Guerrero JA, Garcia-Gimenez JL, Alonso-Yuste E, Burgues O, et al. High stability of microRNAs in tissue samples of compromised quality. Virchows Arch December 2013;463(6):765–74.

[80] Srinivasan M, Sedmak D, Jewell S. Effect of fixatives and tissue processing on the content and integrity of nucleic acids. Am J Pathol December 2002;161(6):1961–71.

[81] Masuda N, Ohnishi T, Kawamoto S, Monden M, Okubo K. Analysis of chemical modification of RNA from formalin-fixed samples and optimization of molecular biology applications for such samples. Nucleic Acids Res November 15, 1999;27(22):4436–43.

[82] Korga A, Wilkolaska K, Korobowicz E. Difficulties in using archival paraffin-embedded tissues for RNA expression analysis. Postepy Hig Med Dosw (Online) 2007;61:151–5.

[83] Allam R, Kumar SV, Darisipudi MN, Anders HJ. Extracellular histones in tissue injury and inflammation. J Mol Med Berl May 2014;92(5):465–72.

[84] Chen R, Kang R, Fan XG, Tang D. Release and activity of histone in diseases. Cell Death Dis 2014;5:e1370.

[85] Holdenrieder S, Stieber P, Bodenmuller H, Fertig G, Furst H, Schmeller N, et al. Nucleosomes in serum as a marker for cell death. Clin Chem Lab Med July 2001;39(7):596–605.

[86] Holdenrieder S, Kolligs FT, Stieber P. New horizons for diagnostic applications of circulating nucleosomes in blood? Clin Chem July 2008;54(7):1104–6.

[87] Greer EL, Shi Y. Histone methylation: a dynamic mark in health, disease and inheritance. Nat Rev Genet May 2012;13(5):343–57.

[88] Leroy G, Dimaggio PA, Chan EY, Zee BM, Blanco MA, Bryant B, et al. A quantitative atlas of histone modification signatures from human cancer cells. Epigenetics Chromatin 2013;6(1):20.

[89] Tang H, Fang H, Yin E, Brasier AR, Sowers LC, Zhang K. Multiplexed parallel reaction monitoring targeting histone modifications on the QExactive mass spectrometer. Anal Chem June 3, 2014;86(11):5526–34.

[90] Sprung Jr RW, Brock JW, Tanksley JP, Li M, Washington MK, Slebos RJ, et al. Equivalence of protein inventories obtained from formalin-fixed paraffin-embedded and frozen tissue in multidimensional liquid chromatography-tandem mass spectrometry shotgun proteomic analysis. Mol Cell Proteomics August 2009;8(8):1988–98.

[91] Lai X, Schneider BP. Integrated and convenient procedure for protein extraction from formalin-fixed, paraffin-embedded tissues for LC-MS/MS analysis. Proteomics August 4, 2014. http://dx.doi.org/10.1002/pmic.201400110.

[92] Park PJ. ChIP-seq: advantages and challenges of a maturing technology. Nat Rev Genet October 2009;10(10):669–80.

[93] Fanelli M, Amatori S, Barozzi I, Minucci S. Chromatin immunoprecipitation and high-throughput sequencing from paraffin-embedded pathology tissue. Nat Protoc December 2011;6(12):1905–19.

[94] Satterlee JS, Schubeler D, Ng HH. Tackling the epigenome: challenges and opportunities for collaboration. Nat Biotechnol October 2010;28(10):1039–44.

[95] Simeon-Dubach D, Burt AD, Hall PA. Quality really matters: the need to improve specimen quality in biomedical research. J Pathol October 1, 2012. http://dx.doi.org/10.1002/path.4117.

[96] Asslaber M, Zatloukal K. Biobanks: transnational, European and global networks. Brief Funct Genomic Proteomic September 2007;6(3):193–201.

[97] Hunter DJ. Gene-environment interactions in human diseases. Nat Rev Genet April 2005;6(4):287–98.

[98] Bjornsson HT, Sigurdsson MI, Fallin MD, Irizarry RA, Aspelund T, Cui H, et al. Intra-individual change over time in DNA methylation with familial clustering. JAMA June 25, 2008;299(24):2877–83.

[99] Compton C. Getting to personalized cancer medicine: taking out the garbage. Cancer October 15, 2007;110(8):1641–3.

[100] Vaught J, Rogers J, Myers K, Lim MD, Lockhart N, Moore H, et al. An NCI perspective on creating sustainable biospecimen resources. J Natl Cancer Inst Monogr 2011;2011(42):1–7.

CHAPTER

3

Epigenetic Mechanisms as Key Regulators in Disease: Clinical Implications

Abdelhalim Boukaba[1], *Fabian Sanchis-Gomar*[2,3], *José Luis García-Giménez*[2,3,4]

[1]Drug Discovery Pipeline, Guangzhou Institutes of Biomedicine and Health, Chinese Academy of Science, Guangzhou, People's Republic of China; [2]Department of Physiology, Medicine and Dentistry School, University of Valencia, Valencia, Spain; [3]Biomedical Research Institute INCLIVA, University of Valencia, Valencia, Spain; [4]Center for Biomedical Network Research on Rare Diseases (CIBERER), Spain

OUTLINE

Epigenetic Biomarkers and Diagnostics
http://dx.doi.org/10.1016/B978-0-12-801899-6.00003-6

1. INTRODUCTION

1.1 The Cross talk between Metabolism and Epigenetics

Modulation of gene expression is of paramount importance for the survival of organisms and cells in an internal milieu that undergoes continuous changes. Cell growth, cell differentiation, migration, transport regulation of amino acids, regulation of enzymatic activity (i.e., telomerase, proteasome, etc.), DNA synthesis, and regulation of the different cell death mechanisms are processes that control not only the fate of cells but also the organismal proneness to pathological situations like cancer, neurodegeneration, and impairment of immunity. Numerous metabolites fuel the enzymatic activities of different epigenetic machineries. Hence, epigenetic gene regulation and metabolism participate in a perfect intricate coordination in which the deregulation of any component could have detrimental consequences for the cell and ultimately for the organism.

Originally it was proposed that epigenetics is the study of heritable changes in gene expression that are not due to modifications in the DNA sequence [1]. Nowadays, epigenetics is the study of all regulatory aspects of chromatin-associated processes no matter if they are heritable or not [2], as described in detail in Chapter 1.

1.2 Epigenetic Mechanisms in Gene Regulation

1.2.1 General Overview

Epigenetic gene regulation refers to how a specific structural and chemical configuration of chromatin translates into a defined outcome on its transcriptional status. That is, it examines the question, how do the combinatorial epigenetic signatures that portray any gene unit drive its expression pattern? The set of instructions that chromatin receives locally could be regarded as a set of signal inputs which are converted into effector signal outputs through complex molecular interactions engaging multicomponent protein complexes. The nucleosomal nature of the chromatin imposes a barrier to transcription but the nucleosome is emerging as a signaling unit on which complex biological transactions take place [3]. These transactions are difficult to dissect individually, but they integrate into the very nature of any chromatin-templated process.

Chromatin is composed of repeating arrays of nucleosomes, which are the building block of eukaryotic chromosomes [4]. Each nucleosome is characterized by 147 bp of DNA wrapped around a histone octamer. A histone octamer consists of two copies each of canonical histone H4, H3, H2A, and H2B or their variants that are organized in a special configuration to accommodate ~1.6 superhelical turns of DNA [5]. Both the tails and the globular domains of the histones and their variants are subjected to over 100 posttranslational modifications (PTMs). Conceptually, a modification can either disrupt the association of histones with the DNA, hence dictate the topology of the chromatin, define chromosome territories, or act as scaffolds to influence the recruitment of transcription factors and other chromatin-associated proteins.

Epigenetic alterations can produce the same phenotypic defect as a null mutant when they induce the improper silencing of a gene. Or, the contrary could result in overexpression of one particular gene and its protein product. However, the existence of multilevel gene control can act at the mRNA or protein level just to add to the complexity of this scenario. In consequence, one expects that the interpretation of these multilevel gene controls mediated by epigenetic regulation can be used for the comprehension of pathological phenotypes which in turn, could translate into clinical applications. Hence, it is important for clinicians to use specific epigenetic signature alterations for diagnosis, prognosis, personalized therapy, and treatment monitoring. It is crucial for translational epigenetics to understand how alterations in the epigenetic signatures of a gene can and do occur, and result in transcriptional defect(s). Intermediary metabolites, by-products of the metabolism, diet, environmental insults, lifestyle, etc., have the potential to alter the epigenetic context with immediate outcome on our health status.

The genetic information is not always expressed in a binary system being either OFF or ON (0 or 1). Rather, its expression displays varying degrees of intensities even within the same homogeneous population of cells as demonstrated by single cell transcriptome analysis. Transcriptional activation within a population of identical cells obeys a stochastic model due to intrinsic and extrinsic parameters inherent to a multistep process which requires the hierarchical coordination of multicomponent protein machineries [6]. In addition, the processing of nucleosomal DNA, itself, may cause stalling of the polymerase at nicks, for example, caused by the elongation process, or by the oxidative activities of some chromatin modifiers which uses H_2O_2 as a cofactor [7].

1.2.2 DNA Methylation

DNA methylation is the best studied epigenetic modification since its discovery [8]. It affects the 5′ carbon position of cytosine mostly in the context of CpG dinucleotide. DNA methylation plays crucial roles in all chromatin-templated processes and diseases as illustrated in some chapters dedicated to it in the present book. Hence, not surprisingly it is implicated in embryonic development, cell differentiation, parental gene imprinting, gene transcriptional regulation, and genome stability, as well as chromatin structure, only to mention classical aspects of its functionalities. Some aspects are only starting to emerge such as the control of the rate of elongation or the modulation of splicing [9,10], which dictates the frequency and the prevalence of protein variants. Another aspect is the deregulation of splicing, which is associated with a huge number of human diseases [11].

Initially, DNA methylation was regarded as a static modification whose pattern is stably inherited, qualifying it for the status of a true epigenetic mark. This dogma has been reversed by the discovery that ten-eleven translocases (TETs) proteins are able to oxidize the 5-methylcytosine (5mC) to 5-hydroxymethylcytosine (5hmC), and to 5-formylcytosine (5fC), and 5-carboxylcytosine (5caC), as reviewed by Wu and Zhang [12]. DNA methylation is catalyzed by the maintenance methylase, DNMT1 (DNA(cytosine-5-)-methyltransferase 1), and by the de novo methylases, DNMT3A and 3B (DNA (cytosine-5-)-methyltransferase 3 alpha/beta). These enzymes are essential for embryonic development (reviewed in Ref. [13]) and also for genomic stability as we describe for ICF syndrome (immunodeficiency, centromeric region instability, and facial anomalies syndrome) in the following sections.

Two ways by which DNA methylation affects gene regulation have been postulated. It can alter the binding of transcription factor(s) to their cognate sequence(s), especially for those bearing the CpG dinucleotide. Alternatively, CpG methylation is recognized by specific protein factors (readers), which recruit chromatin-modifying complexes to target sequences. CpG methylation is recognized by MeCP2 (methyl CpG-binding protein 2) and MBD2 (methyl CpG-binding domain protein 2) (reviewed in Ref. [13]). Interestingly, mutations in MeCP2 produce one of the most frequent syndromes related to intellectual disability in women, called Rett syndrome (RTT), as we describe next. Unmethylated CpGs are also recognized by specific modules containing proteins such as CXXC zinc finger. Readers of 5hmC have also been identified in recent screenings, highlighting that this modification is an epigenetic mark and not an intermediary product during the demethylation of cytosines [14].

Mapping DNA methylation signatures is relatively easy compared with histone PTMs profiling. Hence clinically, they harbor great potential value as diagnostic and prognostic biomarkers. The possibility to screen blood-circulating DNA for alterations in DNA methylation also adds to its value. Alterations in DNA methylation patterns have been linked to a growing number of human pathologies. These issues are discussed in detail in this chapter.

1.2.3 *Histone Modifications*

The repertoire of histone modifications as well as the enzymes implicated in the modification process exceeds largely the scope of any review article, not only because of their number but also mainly because of the diversity and the complexity of the processes in which they are involved. Not surprisingly, a few chapters have been devoted to this topic in the present book. Readers are referred to the histone modification database [15], or to snapshot of histone modifications [16]. For the scope of this chapter, histone modifications as well as the enzymes that catalyze them have been explored for their potential as therapeutic targets and biomarkers in cancer [17]. The first question is why so many PTMs in histones? Next, are they truly PTMs or just transient histone-chemical moiety adducts that are catalyzed by illicit enzymatic reactions? Indeed, some HATs (histone acetyltransferases) may use acetyl-CoA or propionyl-CoA to either acetylate or propionylate histones. The enzymes that remove acetyl groups are also able to remove propionyl moieties [18]. Other additions include butyrylation, hydroxybutyrylation, crotonylation, citrulination, formylation, glycosylation, O-GlcNAcylation [19], carbonylation, PARsylation, and glutathionylation [20,21]. What are the functional roles of such modifications? It is clear that epigenetic regulation is inherently linked to metabolism and it is important not only for basic science to uncover the functionalities of such modifications but also for translational epigenetics, as it provides new biomarkers to diagnose or identify druggable targets to treat diseases.

The chromatin is inherently defined by its epigenetic state. Epigenetic state refers to the pattern of histone modifications, DNA methylation, and nucleosome positioning including histone variants. Locally chromatin domains are defined by epigenetic signatures that distinguish them structurally and functionally from other domains. Namely, the pattern of histone modifications has served to redefine, biochemically, the subdivision of chromatin domains and extended its classical cytological partition in heterochromatin and euchromatin. Thus, at any given moment, any human cell has a defined epigenome that is portrayed by the epigenetic modifications of each one of the ~30 million nucleosomes that make up human chromosomes, and by the epigenetic modifications of cytosine of the underlying DNA, mainly in the context of CpG dinucleotide. The epigenome is highly dynamic and it is subjected to continuous fluctuations as a result of the action of chromatin-modifying and demodifying enzymes, which make a steady state virtually nonexistent. The acetyl group turnover on histones, for example, has been shown to last for less than 30 min in yeast. Knowing that most histone modifications could be removed by countermodifying enzymes or erasers, it is logical to postulate that they are subjected to either fast or slow turnover. Thus, even single-cell epigenomics is only a snapshot of the epigenomic landscape of that cell at that moment. Even though the number of histone PTMs exceeds astronomically the number of total histones that can exist in a unique configuration, the exploitation of the unimaginable number of possible epigenomes a single cell can adopt may still need quantic computing power that is yet to come.

In 2000, Brian Turner [22] early on posited that histone acetylation on different residues may configure an epigenetic code that is recognized by specific protein readers. In the same year, this hypothesis was further extended by Strahl and Allis under the name of the histone code hypothesis [23], stating that a histone modification can either enhance or impede another giving rise to specific combinatorial configurations to histone tails which are recognized by specific module-containing proteins. However, the finding that different reader modules can recognize the same histone PTM or the same reader module can recognize a PTM regardless of its sequence context casted a serious shadow on its validity [24,25]. Moreover, readers have very low affinity to their target(s) when singularly considered in vitro, which contrasts with their anticipated specificity in a chromatin context [26]. This hypothesis has been conceptually reformulated to accommodate further players and to gain wider perspective [27]. The rationale is that different modules combine to create diverse interfaces of recognition, which confer increasing specificity. The thermodynamics of this process is favored because it engages more interactions. Thus, the readout of histone PTMs is combinatorial and multivalent, and explains per se the logics underpinning the spatial and temporal coordination observed in vivo on all chromatin-templated processes.

The findings of protein domains, which interact specifically with definite histone modifications and with methylated cytosine (readers), support the existence of an epigenetic code. Moreover, an early study provided a mechanistic workflow of how the epigenetic code dictates the directionality to the chain of events during transcriptional activation [28]. Chromatin proteins bearing distinct combinations of epigenetic module readers have been described. Moreover, chromatin proteins are normally part of macromolecular complexes. To fulfill the combinatorial readout requirement of the epigenetic code, only a few subunits have to bear at least one module reader. Chromatin-targeting complexes with such characteristics are known to be widespread.

In summary, the dynamic nature of all chromatin-associated processes requires the coordinated interplay between writers, erasers, and readers to drive forward the process. While drugs targeting some epigenetic erasers have been approved by the US Food and Drug Administration (FDA) to be used in antitumor treatment, and others are in clinical trial, a milestone has been achieved by targeting a chromatin reader to treat cancer [29,30], namely, the BET protein family (bromodomain and extraterminal: BRD2, BRD3, BRD4, and BRDT (bromodomain 2/3/4/testis-specific, respectively)), which shares a common structural configuration featuring tandem bromodomains at their N-terminus. The BET proteins play a fundamental role in transcriptional elongation and cell-cycle progression by regulating the transcription of key oncogenes. Recurrent translocations involving *BRD3/4* are associated with the aggressive form of nuclear protein in testis (NUT)-midline carcinoma (NMC) [31]. NMC is primarily a chromosomal rearrangement involving the fusion of most of the coding region of NUT protein located on chromosome 15q14 with BRD4 (75% of cases) located on chromosome 19p13.1 or BRD3 [32]. This rearrangement results in aggressive forms of carcinomas. Targeting the BET bromodomains has proven to be promising against NMC, as well as in other hematological malignancies (reviewed in Ref. [29]).

1.2.4 MicroRNAs and Long Noncoding RNAs

Epigenetic regulation is also mediated by noncoding RNAs (ncRNAs) which are functional, but not protein-coding RNAs. In general, ncRNAs can regulate gene expression at the transcriptional and posttranslational level. Those ncRNAs include microRNAs (miRNAs) and long noncoding RNAs (lncRNAs), among others. miRNAs are a large family of short RNA molecules, with an average size of 17–25 nucleotides [33]. lncRNA transcripts are longer than 200 nucleotides and poorly conserved. lncRNAs are transcribed from all over the genome, including intergenic regions, domains overlapping one or more exons of another transcript on the same strand (sense) or on the opposite strand (antisense), or intronic regions of protein-coding genes [34].

ncRNAs contribute to the dynamics of the epigenome, so they are key players in the regulation of gene expression and cellular activity, and both miRNAs and lncRNAs play a role in heterochromatin formation, histone PTMs, DNA methylation, and gene silencing. miRNAs are involved in several biological processes, such as cellular development, differentiation, apoptosis, cell proliferation, and cancer [35,36]. These small RNA molecules regulate gene expression mainly by base-pairing to the 3′-untranslated region (3′-UTR) of target mRNAs, although evidence suggests that miRNAs can also bind to coding regions or to the 5′-untranslated region (5′-UTR) of some mRNAs [37,38]. miRNAs can downregulate gene expression by inhibiting mRNA translation or by degradation of mRNA molecules, as well as increasing mRNA translation of some targets [39].

The role of lncRNAs in epigenetic regulation is dual because they can serve as scaffolds for chromatin-modifying complexes and they can guide other epigenetic machinery to the target site of the chromatin [40,41]. It has also been reported that some lncRNAs show enhancer-like activity, and they regulate alternative splicing and other posttranscriptional RNA modifications [42] that determine the activity of our genome. Many lncRNAs act by forming complexes with chromatin-modifying proteins and recruiting them to specific sites in the genome, thereby modifying chromatin states and influencing gene expression [43].

2. EPIGENETICS AND METABOLISM INTERSECTION

2.1 Relevant Genes in the Methionine Cycle Contribute to Epigenetic Regulation

The methionine cycle and transsulfuration pathway have special relevance in the control of S-adenosylmethionine (SAM), which is the most important methyl donor for DNA methyltransferases (DNMTs) and histone methyltransferases (HMTs) (Figure 1). In the methionine cycle, the methylenetetrahydrofolate reductase gene (*MTHFR*) is the enzyme that reduces 5,10-methylenetetrahydrofolate to 5-methyltetrahydrofolate, necessary for the transference of a methyl group to the methionine. Mutations in the *MTHFR* gene produce low levels of SAM [44]. Genetic polymorphisms in the *MTHFR* gene results in high homocysteine (HCys) levels and low methionine production, which in turn may lead to accumulation of S-adenosylhomocysteine (SAH) and low levels of SAM in human population [45]. The low levels of SAM and accumulation of SAH alter the SAM/SAH methylation ratio inducing global hypomethylation of DNA and histones [45], which contribute to genetic instability and chromosome aberrations [46]. Vitamin B12 and SAM/SAH are essential in regulating the activity of the methionine synthase (MS) and it has been found that a deficit in MS produces alterations in genome-wide methylation [47], so affecting the epigenetic program.

The transsulfuration pathway is a very important pathway for the production of cysteine (Cys)-containing amino acids and the synthesis of glutathione, one of the most important mammalian nonenzymatic antioxidants (Figure 1). The transsulfuration pathway involves the

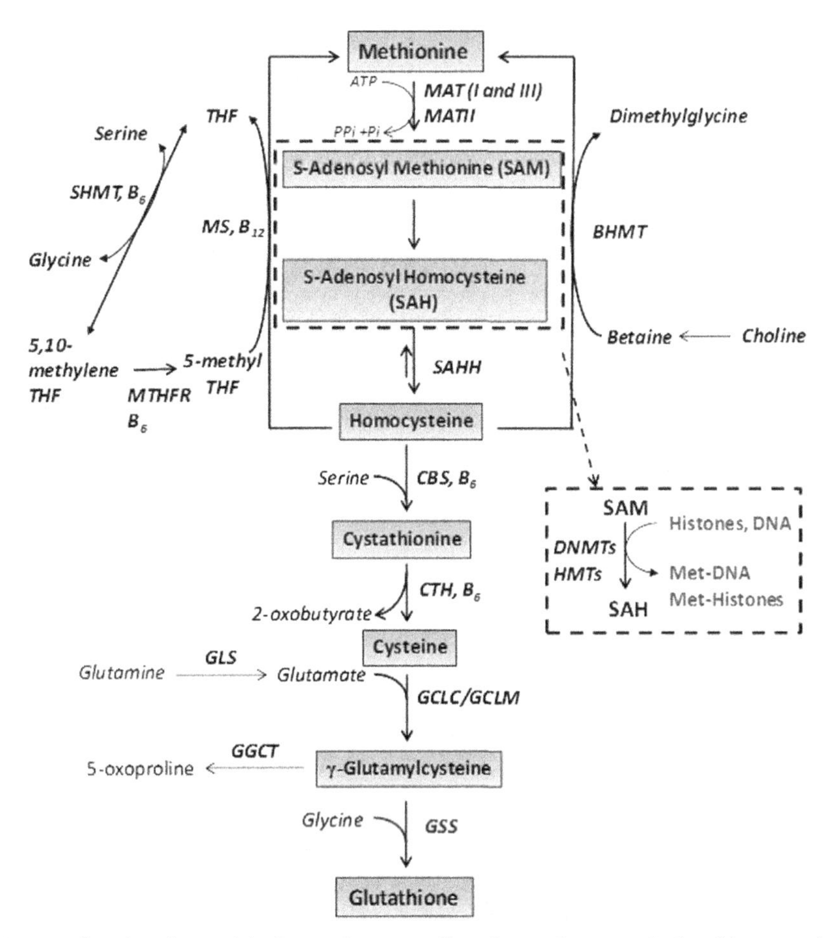

FIGURE 1 Diagram showing the methionine cycle, transsulfuration pathway, and glutathione synthesis (GSH) pathway and its connection to epigenetic metabolites. Abbreviations in the diagram: B6 and B12, vitamins B6 and B12, respectively; BHMT, betaine homocysteine methyltransferase; CBS, cystathionine β-synthase; CTH, cystathionase; DNMTs, DNA methyltransferases; GCLC and GCLM, catalytic and modulatory subunits of γ-glutamylcysteine synthetase; GGT, γ-glutamyltranspeptidase; GLS, glutaminase; GNMT, glycine N-methyltransferase; GSS, glutathione synthetase; HMT, histone methyltransferases; MAT, methionine adenosyltransferase; MS, methionine synthase; SAHH, S-adenosylhomocysteine hydrolase; and THF, tetrahydrofolate.

interconversion of Cys and HCys, through the intermediate cystathionine. The production of homocysteine through transsulfuration allows the conversion of HCys to methionine, which as described above is catalyzed by MS [48]. Following the steps of the transsulfuration pathway, HCys is transformed into cystathionine by CBS (cystathionine β-synthase) (Figure 1). Mutations in the *CBS* gene are found in patients with a rare disease called cystathionine β-synthase deficiency (ICD-10 E72.1; ORPHA394) that produces an accumulation of HCys. Hyperhomocysteinemia induces oxidative stress, which has been postulated as a mechanism that produces hypomethylation of several substrates [49,50] including DNA [51]. In patients with cystathionine β-synthase deficiency, the levels of HCys, SAM, and SAH were significantly elevated compared with controls. But, differences in global DNA methylation were not observed [52], probably because the SAM/SAH ratio was not altered and DNMTs were not down- nor upregulated.

2.2 The Tricarboxylic Acid Cycle and Epigenetics

The tricarboxylic acid cycle (TCA), or Kreb's cycle, is the main pathway for the metabolism of sugars, lipids, and amino acids (Figure 2). The TCA cycle provisions cell with reductive power in the form of NADH and $FADH_2$ (reduced nicotinamide adenine dinucleotide, and reduced flavin adenine dinucleotide, respectively), which endow their electrons to the mitochondrial respiratory chain for ATP (adenosine triphosphate) synthesis, which is essential for cellular metabolism and survival. The TCA cycle begins with the addition of acetyl-CoA to oxaloacetate to form citrate. This reaction is catalyzed by citrate synthase. Subsequently, citrate can be exported to the cytoplasm (where it is used as a precursor for lipid biosynthesis) or remains in the mitochondria (where it is converted to isocitrate by aconitase, an iron–sulfur protein that functions as an electron carrier which employs a dehydration–hydration mechanism for citrate to isocitrate conversion). The oxidative decarboxylation of isocitrate catalyzed by isocitrate dehydrogenase (IDH) produces

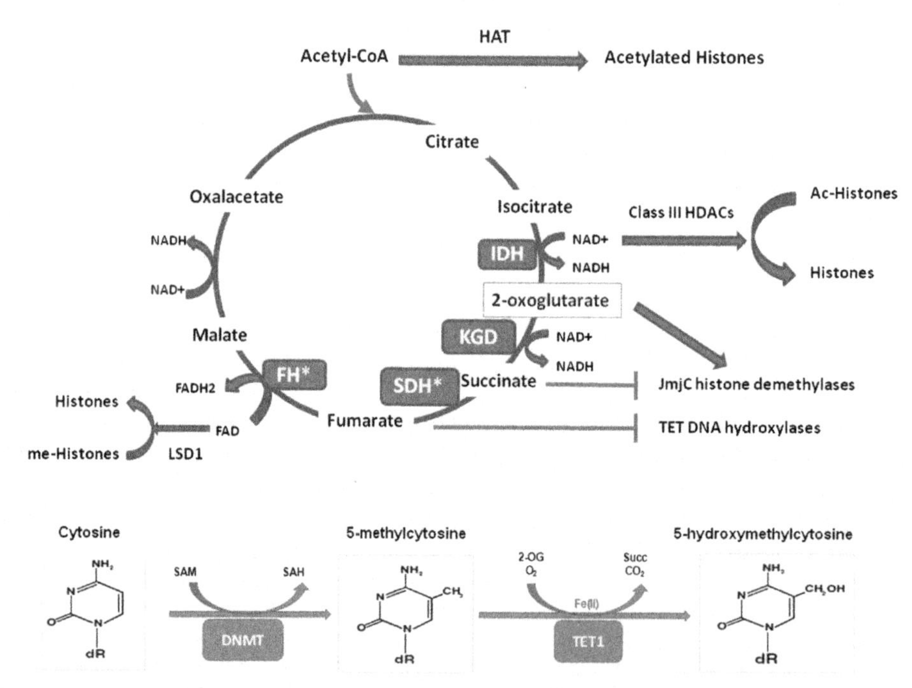

FIGURE 2 Scheme presenting the tricarboxylic acid (TCA) cycle and the production of cofactors and substrates for different epigenetic enzymes. Histones and DNA-modifying enzymes sense the metabolic changes because they depend on cofactors and substrates produced in the TCA cycle. With the acetyl-CoA cofactor, these enzymes link glycolysis, fatty acid, and amino acid metabolism with epigenetic gene regulation because acetyl-CoA is used by acetyltransferases (histone acetyltransferases (HAT)) for the acetylation of lysine residues. HDACs (histone deacetylases) are the counterpart of HATs and require, in the particular case of Class III HDACs, NAD+ (nicotinamide adenine dinucleotide, oxidized form) as a cofactor to break the bond between lysine and the acetyl group. Furthermore, as described in the text, 2-oxoglutarate is one of the cofactors along with O_2 required by JmjC histone demethylases and TET DNA hydroxylases, which are inhibited by succinate and fumarate, which are intermediates of the TCA.

α-ketoglutarate (α-KG) whose decarboxylation by the α-KG dehydrogenase (α-KGDH) complex yields succinyl-CoA (Figure 2). Succinyl-CoA is then converted to succinate by the succinyl-CoA synthetase whose oxidation by succinyl dehydrogenase complex (SDH) produces fumarate which is then hydrated by fumarate hydrolase (FH) to yield malate. The oxidation of malate by malate dehydrogenase regenerates oxaloacetate, henceforth ensuring the completion of the TCA cycle (Figure 2) (reviewed in Ref. [53]).

Emerging studies have demonstrated that TCA intermediates participate in the control of the epigenetic machinery, and some reviews on this subject have been published recently [53–55]. The key roles of TCA intermediates in epigenetic regulation have been observed at different levels. For example, acetyl-CoA is used by HATs [56,57], the cofactor NAD^+ is generally used by the class III histone deacetylases (HDACs, also called Sirtuins) [57,58], and FAD^+ is a cofactor for LSD1, the lysine-specific demethylase 1A [59], which can remove mono- and dimethylated groups from H3K4 and H3K9, through an FAD-dependent oxidative mechanism (Figure 2). Citrate can enhance histone acetylation. Furthermore, α-KG (alpha-ketoglutaric acid), succinate, and fumarate can regulate the level of DNA and histone methylation (Figure 2). In this regard, the demethylases of DNA, the Fe(II)-O_2-dependent TETs (TET1, TET3) and Jumonji C histone lysine demethylases (KDM2-7), are dependent on Fe(II), oxygen (O_2), and the TCA intermediate α-KG [60,61] (Figure 2). Interestingly, recent studies have revealed the dependence of TET1/3 and KDM2-7 of these cofactors, being both enzymes inhibited by succinate and fumarate [54,55,62].

3. LIFESTYLE, NUTRITION, AND PHYSICAL ACTIVITY IN EPIGENETIC REGULATION

Epigenetic regulation not only depends on the flux of metabolites provided by the methionine cycle and the TCA, but also on the interaction of our organism with the environment, so we will review the implications of lifestyle, nutrition, and physical activity (PA) in epigenetic regulation, as well as its potential clinical relevance.

There is increasing evidence for the involvement of epigenetics in human diseases such as cancer, metabolic diseases (including obesity, metabolic syndrome (MetS), and diabetes mellitus), inflammatory disease, and cardiovascular disease [63], and neurological disorders such as Alzheimer's disease (AD) [64]. Lifestyle factors that include PA, nutrition, behavior, alcohol consumption, and smoking are emerging as critical factors of epigenetic regulatory mechanisms [65–68]. Among the aforementioned factors and conditions, PA, nutrition, and obesity are those with abundant evidence on their implication in the epigenetic mechanisms regulation [65,69–71] (*Note:* Obesity is a condition not a lifestyle). Long-term lifestyle has influence in regulating epigenetic mechanisms described above [66] (see Figure 3). PA positively affects health, reduces/slows the aging process, and decreases the incidence of cancer through induced "eustress" or "positive stress" and epigenetic mechanisms [71]. Thus, stress may stimulate genetic adaptations through epigenetics, modulating the connection between the environment, human lifestyle factors, and genes [71]. In addition, PA induces several metabolic adaptations, while epigenetic modifications and many epigenetic enzymes are potentially dependent on changes in the concentrations of metabolites, such as oxygen, TCA intermediates, 2-oxoglutarate, 2-hydroxyglutarate, and β-hydroxybutyrate, as we described in the previous section. Exercise induces fluctuations in these enzymes in a tissue-dependent manner. Therefore, these substrates and signaling molecules, regulated by exercise, affect some of the most important epigenetic mechanisms which ultimately control the gene expression involved in metabolism [72]. In addition, epigenetic alterations induced by physical exercise can alter the

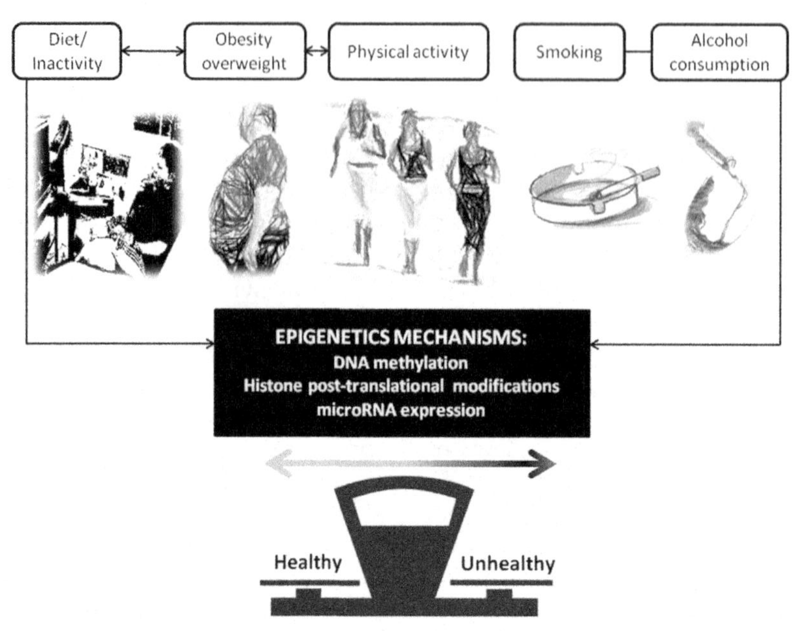

FIGURE 3 How lifestyle affects the epigenetic mechanisms. Diet and inactivity are linked to obesity and overweight, which are prevented by physical activity. Furthermore, alcohol consumption and smoking are unhealthy habits that deregulate epigenetic mechanisms, so producing epigenetic alterations associated to diseases such as cardiovascular disease and obesity, metabolic syndrome and diabetes, neurodegenerative disorders, and cancer.

level of transcription of various muscle genes in order to adapt muscles to mechanical loads [73]. Physical exercise mainly causes hypomethylation of the whole genome in skeletal muscle cells mainly related to promoters for metabolic genes, such as peroxisome proliferator-activated receptor gamma coactivator 1-alpha (*PGC-1α*), mitochondrial transcription factor A (*TFAM*), peroxisome proliferator-activated receptor delta (*PPAR-δ*), pyruvate dehydrogenase lipoamide kinase isozyme 4 (*PDK4*), and citrate synthase [73], as well as regulates the transcription of the myosin heavy chain genes (*MHCs*) [74], and reduces the expression of various types of miRNAs in skeletal muscle that regulates genes such as runt-related transcription factor (RUNX1), paired box 3 (PAX3), and sex-determining region Y-box 9 (SOX9) [75]. Physical exercises also induced changes in miRNAs expression profiles, thus modulating the expression of important metabolic genes, such as *PGC-1α* and *PDK4* [76,77], and affecting the activation of angiogenesis that occurs in skeletal muscle [78].

Expression of cancer-associated genes in the majority of sporadic cancers is also controlled by epigenetics [79]. Physical exercise and the lifestyle are usually associated with higher levels of global genomic DNA methylation restoration in cancer [80,81]. It has been demonstrated that maintaining a normal weight, participating in regular physical exercise, and eating a healthy diet improve survival outcomes in cancer [82]. The possible mechanisms of colorectal cancer that can be positively influenced by changing lifestyle are directly connected with epigenetics [82]. Important modifications in DNA methylation (hypomethylation and hypermethylation) patterns have been also reported during aging [83]. Likewise, epigenetic alterations are also associated with gene–diet and gene–environment interactions affecting lipid metabolism, inflammation, and other metabolic imbalances leading to cardiovascular disease and obesity [84]. Polyunsaturated fatty acids-rich diets could affect metabolism thus inducing epigenetic alterations through mechanisms linked to free radicals and oxidative stress [85,86]. Polyphenols have been shown to modify the activity of DNMTs, HATs, and HDACs [87,88]. Genes involved in adipogenesis (e.g., suppressor of cytokine signaling 1/3, *SOCS1/SOCS3*), inflammation genes, intermediary metabolism, and insulin signaling pathway genes as well as the methylation pattern of obesity-related genes (e.g., fibroblast growth factor 2 (FGF2), phosphatase and tensin homolog (PTEN), cyclin-dependent kinase inhibitor 1 (CDKN1A), and estrogen receptor 1 (ESR1)) could be implemented in clinical settings as epigenetic biomarkers of obesity, thus helping to predict susceptibility to and prevention of obesity [89]. Accordingly, nutritional effects on epigenetic programming have a major relevance to cardiovascular disease risk.

On the other hand, cigarette smoke condensate also decreases levels of H4K16 acetylation and H4K20 trimethylation [90], which are similar to changes found in lung cancer tissues. For example, demethylation of *H19* and *IGF2* (insulin-like growth factor 2 receptor) happened before the DNA hypermethylation-mediated silencing of p16 (also known as CDKN2A), MGMT (O-6-methylguanine-DNA methyltransferase), DAPK (death-associated protein kinase 1), E-cadherin (CDH1), and cadherin-13 (CDH13) tumor suppressor genes in the first stages of the lung carcinogenesis induced by tobacco smoke [91]. Hypomethylation of p53 in exon 5–8 (hypermutable regions) has been also reported in peripheral blood lymphocytes of smoker-type lung cancer patients [92].

To conclude, epigenetic alterations may also imply clinical relevance, playing a key role in the prevention, diagnosis, prognosis, and even treatment of MetS, diabetes, and neurological disorders like AD, aging, cancers, and other diseases.

4. EPIGENETIC MECHANISMS UNDERLYING COMPLEX DISEASES

A huge effort has been made in recent years to understand the causes and mechanisms of complex non-Mendelian diseases. Nonetheless, our comprehension of some of these diseases remains, in some aspects, elusive. Although numerous molecular genetic linkage and genome-wide association studies (GWAS) have been conducted in order to explain the predisposition to complex diseases, the data obtained are quite often controversial. A new approach to advance the comprehension of these diseases comes from the introduction of epigenetics and how the genes interact with the environment, or how the lifestyle alters our epigenome. The advantage of epigenetics in complex diseases relies in the ability to integrate and match a variety of apparently unrelated mechanisms and molecular pathways. Here, we describe two well-known complex diseases (MetS and AD) as paradigms of epigenetic contributions in disease.

4.1 MetS and Diabetes

The MetS (MIM #605552, where the MIM# is the Online Mendelian Inheritance in Man database accession number) is represented by a cluster of disorders that include obesity, hyperglycemia, hyperinsulinemia, hyperlipidemia, hypertension, and insulin resistance [93]. The MetS is not a disease per se but its core components contribute individually or collectively to the risk of developing cardiovascular diseases or type 2 diabetes mellitus [94]. In addition, the MetS has been associated with increased risks of developing other diseases such as chronic kidney disease (or CKD) [95], cancer [96], and AD [97]. The concept of MetS has generated numerous debates and has undergone constant redefinitions over decades. The current pathophysiological criteria established by different health organizations for its diagnosis have been comprehensively reviewed by Kaur et al. [98].

Several hypotheses have been postulated to explain the etiology of the MetS. The *thrifty gene hypothesis* proclaims that humans have selected genes to promote insulin resistance [99]. This allows individuals to store fat in a period of food abundance in order to secure supply during periods of food shortage. The argument raised against this hypothesis is the acquisition of obesity by humans in a relatively short period of time, which led to the proposition of the *thrifty phenotype hypothesis* by Hales and Barker [100]. This hypothesis proclaims that insulin resistance/deficiency is an adaptive response to the intrauterine environment nutrient limitations. Building upon this model, the predictive adaptive response hypothesis by Gluckman and Hanson states that the fetus makes changes in its developmental program according to predictive postnatal environment [101]. In this scenario, MetS results from a mismatch between the predicted and actual environment. The importance of the latter two hypotheses is that they highlight that the interaction between the fetus and the environment results in developmental plasticity [101].

The alteration of developmental programming in the fetus by the environment results in the alteration of the transcriptional status of a subset of metabolic genes, among others, through epigenetic mechanisms [93,102,103]. In addition, as described in the previous section, our lifestyle also affects the transcriptional status of several metabolic genes. There is evidence from animal models that the artificial manipulation of diet induces DNA methylation changes on the promoter of a subset of genes implicated in the metabolism, among others, of the offspring which persist in adults (reviewed in Ref. [104]). Mice deficient in JHDM2a (lysine (K)-specific demethylase 3A, a H3K9 histone demethylase) develop MetS [105]. Thus, the evidence supports the role of epigenetic factors in the etiology of MetS and the notion that mutations in epigenetic factors recapitulate complex human diseases [106]. In humans, it has been shown that epigenetic differences in adult offspring, prenatally exposed to the Dutch Hunger Winter in 1944–1945, present with less DNA methylation of the imprinted *IGF2* gene as compared with their unexposed, same-sex siblings [107]. This evidence supports the idea that the uterine environment can induce epigenetic changes in the fetus. However, a second study did not find a prevalence of MetS in these adults other than HDL levels [108], yet these adults did manifest other metabolic pathologies [109]. These studies suggest that other variables aside from nutrient limitations might contribute to the development of MetS. A posteriori studies have documented persistent changes in DNA methylation on several examined loci implicated in growth and metabolic diseases as a consequence of prenatal famine exposure and that these changes were depending on the sex of the exposed individual and the gestational timing of the exposure [110].

The involvement of the circadian clock in the etiology of a growing number of metabolic diseases [111], including MetS [112], further complicates both basic and clinical investigations into MetS etiology. Interestingly, sleep fragmentation, a common disorder in pregnant women, has been shown in a mouse model to induce epigenetic changes in visceral white adipose tissue of offspring by affecting the expression status of key epigenetic enzymes. Namely, changes include the upregulation of the expression of DNMT3A/B with concomitant increases in global 5mC, reduction of HAT activity, and downregulation of the expression of TET1/2/3. These alterations were found to affect the epigenetic signatures of the adiponectin gene (*ADIPOQ*) promoter and its enhancer whose expression negatively correlates with obesity and insulin resistance, and are characteristics of MetS [113].

In summary, integrative approaches are needed to gain insights into the multidimensionality of fetal reprogramming that predispose the fetus to postnatal metabolic disorders [114]. MetS is cataloged as a type of noncommunicable disease that can be induced in utero or inherited transgenerationally through epigenetic mechanisms. Notably, and given the socioeconomic impact of MetS worldwide, it provides molecular clues as to how to deal with the disease at early stages of life. In this regard, translational epigenetics could prevent the manifestation of MetS by correcting both maternal and paternal nutritional imbalances. Studies have also shown that paternal nutritional habits [115] and lifestyle, such as PA [71,72], could induce or reverse epigenetic changes in epigenetic control, affecting the offspring [116].

4.2 Alzheimer's Disease

AD (MIM #104300) is a complex neurodegenerative disease that leads to dementia. This pathobiological condition is characterized by the deposition and accumulation of amyloid plaques, tangles of intracellular hyperphosphorylated tau, gliosis, and synaptic dysfunction followed by cell death. It can affect different parts of the brain. The disease affects over 20 million individuals worldwide, according to conservative estimates [117]. Presently, AD has no cure and the socioeconomic burden is huge, as AD requires continuous care of elderly individuals who experience progressive degeneration/impairment of cognitive faculties. AD has a complex etiology as it has been linked to innumerable genetic factors with concomitant alterations in the expression of a huge number of genes operating through different pathways. However, none of the factors identified in genetic screens has been delineated as a direct cause of the disease (reviewed in Refs [117,118]). Population genetics of affected individuals identified two classes of genes. The first class is thought to play a direct role in the pathogenesis of the disease. The second class has been linked to increased risks in developing the disease. GWAS are difficult to interpret as differences such as SNPs (single nucleotide polymorphisms), though statistically significant, may not be biologically meaningful given our limited understanding of the process. The number of individuals analyzed may also limit the power of such approaches.

Early studies with HDAC inhibitors in animal models of AD established a link between epigenetic determinants and the disease (reviewed in Ref. [118]). Recently, two independent studies have assessed the alterations of the epigenetic landscapes associated with AD in a large number of individuals [119,120]. The full potential harbored by such studies will only unfold in the future. However, differential DNA methylation on a number of CpGs has confirmed the identity of a number of genes that were identified in previous genetic screens as susceptibility genes. Thus, epigenetic approaches to uncover risk factors and susceptibility genes are strong experimental tools that may hold the key to advance our understanding of complex diseases. Of note, the reported differential DNA methylation (5mC) was highly correlated with the expression data, arguing that these changes were not related to an age-dependent increase in methylation. Both studies have sufficient overlap to mutually support their findings. These differentially methylated regions were unlikely to be driven by SNPs, arguing that they may represent genuine alterations in epigenetic signatures and hence they could be used as biomarkers for AD (especially if they occur during the onset of the disease). The problem that still has to be addressed is how both of these studies distinguish between DNA methylation and hydroxymethylation (5hmC) given the limitation of the technologies used. It is very difficult to discriminate between 5mC and 5hmC by means of mapping techniques based on bisulfite conversion. However, there exist alternative methods for 5hmC sequencing. Recent modifications in the protocol for the Infinium BeadChip (450K, Illumina) method allow the detection of 5hmC. This method has been called oxBS-450K [121]. Other methods available are the oxidative bisulfite sequencing (oxBS-seq) [122] and the Tet-assisted bisulfite sequencing (TAB-seq) [123]. Notably, the brain is enriched in hydroxymethylation marks, which is being regarded not only as an oxidative intermediate but also, contrarily to methylation, has an activating role [124]. A priori in this setting, it is not easy to assimilate how gain in methylation could increase the expression status of some AD-associated genes [119].

How alterations in DNA methylation mechanistically happen is still not well understood, but alteration in metabolic pathways, diet, lifestyles, and exposure to environmental toxic compounds may provide a hint. However, the brain is characterized by mosaic copy number variation which complicates further the approach to understand neurodegenerative diseases [125]. It is possible that, at different stages during differentiation, neurons undergo stochastic genomic alterations that will condition their future epigenetic landscape and susceptibility to diseases. Hence, only some areas of the brain are affected. Recently, a new focus intending to target rare mosaic somatic mutations that only affect a small subset of cells has been implemented to gain insights into neurodegenerative diseases [126,127]. The results of these new studies show that there is a need to improve deep sequencing techniques in order to not miss vital data that may, in time, prove to be a determinant in solving the puzzle behind the pathogenesis of neurodegenerative diseases. This focus should also be applied to epigenetic signature profiling as some alterations may only affect small populations of cells, especially if mutations drive epigenetic alterations. Again, single-cell analysis could shed light on the pathogenesis of neurodegenerative diseases such as AD.

In summary, the potential that epigenetic signature profiling may harbor to diagnose disease at an early stage could provide complementary alternatives to genetic tests. This has become an area of particular importance, since it is now widely accepted that complex diseases could be driven by epigenetic changes.

5. EPIGENETICS IN RARE DISEASES—CLINICAL IMPLICATIONS

The relevance of epigenetics in maintaining normal development, biology, cell signaling, and function is reflected by the observation that many diseases (monogenic and complex diseases) appear or develop when an aberrant pattern of epigenetic marks is introduced at the wrong time or at the wrong place [128,129]. Over the last years, it has become evident that mutation of genes which codify for epigenetic machinery, such as DNMTs, methyl-binding domain proteins, HDACs, and HMTs are not only associated to human rare diseases, such as ICF syndrome 1 (ICF1); RTT; Rubinstein–Taybi syndrome (RSTS1); Weaver syndrome (WVS); etc., but also other diseases such as neurological disorders and cancer.

In this part of the chapter, we describe different rare syndromes as paradigms of other complex diseases associated to epigenetic dysregulation. Therefore, these rare syndromes can help us to the comprehension of fundamental epigenetic mechanisms. In that way, the acquired knowledge from the study of these diseases has already contributed to develop epigenetic biomarkers, mainly in cancer. In addition, we also describe the rare neuropathy Friedreich's ataxia (FRDA), in which are found some epigenetic signature alterations. Therefore, research focused on this disease will contribute to design therapies targeted at epigenetic modulation, which hold promise for many of trinucleotide repeat expansions-related disorders (i.e., Huntington's disease, spinocerebellar ataxia, myotonic dystrophy, etc.) (Table 1).

5.1 ICF Syndrome

The ICF1 (ORPHA2268, MIM #242860), where ORPHA# (Orphanet classification of diseases accession number) is a very rare syndrome which follows an autosomal recessive pattern of inheritance. It has been described in about 50 patients worldwide [130]. This very rare syndrome is characterized by immunodeficiency and centromere instability (mainly at chromosomes 1 and 16, and sometimes 9) along with frequent somatic recombination events [131]. Other distinguishing feature of the

TABLE 1 Epigenetics Associated with Complex and Rare Diseases

Disease/disorder	Disease ID#	Brief description	Primary gene(s)	Protein product	Epigenetic markers	References
Immunodeficiency, centromeric region instability, and facial anomalies syndrome 1 (ICF1)	ORPHA2268, MIM242860	Immunodeficiency with instability in centromeres of Chr 1, 9, and 16, and to a lesser extent 21, and having facial anomalies	DNMT3B (20q11.2)	DNA methyltransferase 3B	DNA hypomethylation in satellites 2 and 3, subtelomeric regions, and transposable Alu sequences	[132,133,136,138]
Rett syndrome (RTT)	ORPHA778, MIM312750	Severe neurological and neurodevelopmental disorder	MeCP2 (Xq28), CDKL5 (Xp22), Netrin G1 (1p13.3)	Methyl CpG-binding protein	Mutations in MeCP2	[145,147,151,152]
Weaver syndrome (WVS2)	ORPHA3447, MIM615521	Accelerated overgrowth and osseus maturation with craniofacial and limb anomalies	EZH2 (7q36.1)	Enhancer of zeste homolog 2	Mutations in EZH2, low levels of H3K27me3	[175,178]
Rubinstein–Taybi syndrome (RSTS1)	ORPHA783, MIM180849	Growth deficiency syndrome with facial malformations and intellectual disability	Microdeletion in 16p13.3, CREBBP (16p13.3), EP300 (22q13.2)	CREB-binding protein mutations	Mutations in p300/CBP	[161–163,167,168]
				E1A-binding protein mutations	Low levels of acetylated H4K5, H3K14, H3K18, H3K27, H3K56	

Continued

TABLE 1 Epigenetics Associated with Complex and Rare Diseases—cont'd

Disease/disorder	Disease ID#	Brief description	Primary gene(s)	Protein product	Epigenetic markers	References
Friedreich's ataxia (FRDA)	ORPHA95, MIM229300	Neurodegenerative disorder with progressive loss of myelin in peripheral sensory neurons. Cardiomiopathy is the main cause of death in these patients	Frataxin (FXN) (expansion of trinucleotide GAA repeats) (9q21.11)	Frataxin protein	Hypermethylation of DNA and H3K9 in FXN promoter. High levels of H3K9me2, H3K9me3, H3K27me3, and HP1 at GAA repeat miR-155(rs5186) associated to cardiac phenotype Dysregulation of 27 miRNAs miR-886-3p found in blood from FRDA patients	[183,188,189,191,192, 194–197]

disorder is the mild facial dysmorphism. Other relevant features found in the ICF patients are growth retardation and abnormal psychomotor development, which compromise seriously the patients to look after oneself [130]. Sixty percent of ICF1 patients carry mutations in the de novo methyltransferase DNMT3B [132,133]. These mutations in DNMT3B produce DNA hypomethylation in ICF1, mainly in specific repetitive sequences, such as satellites 2 (localized primarily on chromosomes 1 and 16, and to a lesser extent on 2 and 10) and 3 (localized to chromosomes 1 and 9), subtelomeric regions [134], very repetitive and transposable Alu sequences [135], and in imprinted genes located at heterochromatin regions [136]. That hypomethylation produces chromatin decondensation and chromosome instability which facilitate deletions, chromosome breaks, fusions, and multiradial chromosome junctions in the above-mentioned chromosomes, and usually occurs in lymphocytes [131,137], which may explain the immunodeficiency found in these patients.

Recently, whole-genome bisulfite sequencing of one ICF1 patient and one matched healthy control have been performed to analyze DNA methylation at base pair of resolution [138]. Although the study was performed in only one patient, the authors concluded that ICF1 patient has about 42% less global DNA methylation, mainly in inactive heterochromatin regions. It is likely that the global DNA hypomethylation may induce genetic instability and gene expression deregulation. In fact, more than 700 genes with deregulated gene expression have been found, especially genes related to immune system, development, and neurogenesis [139].

In contrast to the dysregulation produced by hypomethylation of protein-coding genes, no changes in histone PTMs associated with heterochromatic genes were found, as described by Brun et al. [136]. These authors proposed that gene activation (1) due to euchromatinization does not occur in the whole chromatin, but rather, (2) is restricted to cell lines which are affected in the ICF1 syndrome (i.e., lymphocytes). In addition, defects in DNA replication have also been described, including shortening of the S-phase and early replication of heterochromatin genes [133], suggesting that DNMT3B function is essential during DNA replication.

In summary, when considering the combination of the striking loss of DNA methylation not globalized to all chromatin regions, chromosome instability and fusions, and defects in DNA replication fork, together with their variable aberrant expression, the ICF1 syndrome stands as an ideal model to investigate the pathological consequences of defects in the coherent pattern of DNA methylation. So, after the characterization of DNMT3B function it is possible to look for mutations in other pathologies which can be used as biomarkers for clinical diagnosis or prognosis. In this regard, heterozygous carriers of an SNP at −149 bp in DNMT3 promoter have decreased survival expectancy in patients with small cell carcinoma of the head and neck [140]. In addition, the identification of DNMT3 mutations can serve to predict the disease onset of hereditary nonpolyposis colorectal cancer [141]. Finally, DNMT3B high levels have been associated as a poor prognostic marker in acute myeloid leukemia [142].

5.2 Rett Syndrome

RTT (ORPHA778, MIM #312750) is a postnatal neurological and neurodevelopmental, but not a neurodegenerative, disorder. RTT affects the central nervous system development. RTT is characterized by normal neurological features during the first months of life but surprisingly neurodevelopmental arrest during the first year of life suddenly occurs [143]. When RTT syndrome initiates, then regression of acquired skills, loss of speech, unusual stereotyped movements, and intellectual disability are detected [144].

RTT occurs almost exclusively in women during the early years of life being the most frequent intellectual disability in girls. Prevalence is about 1 affected in 10,000 in girls [143]. Interestingly, male patients with RTT have been also described showing variability in their phenotypes [145]. Men affected by RTT are characterized by severe to moderate congenital encephalopathy, infantile death, or heterogeneous psychiatric manifestations.

A wide range of gene mutations, including missense, frameshift, and nonsense mutations, and intragenic deletions are found in RTT patients [143]. In 96% of RTT patients mutations in the X-linked gene MeCP2 is detected [145]. Additional mutations in other genes have been suggested to produce RTT clinical phenotype. In this regard, the other genes proposed are cyclin-dependent kinase like (CDKL5) and Netrin G1 [143]. Interestingly, the increase in MeCP2 dosage due to duplications of the locus and surrounding areas produces Lubs X-linked mental retardation syndrome (MRXSL, ORPHA85281, MIM #300260), which is also a neurological disorder characterized by hypotonia and recurrent infections [146].

MeCP2 is a gene that encodes for the protein named methyl CpG-binding protein 2 which is essential for the normal function of nerve cells. Importantly, MeCP2 protein is involved in the binding to methylated cytosines, which in turn contribute to repress or silence several genes by recruiting other silencing protein complexes (e.g., mSin3a (mammalian SIN3 transcription regulator family A), N-CoR (nuclear receptor corepressor), etc.) which bear chromatin-modifying activities such as HDACs, histones demethylases, etc., which shut down transcription [147]. There exists further evidence of the involvement of MeCP2 in regression from a normal mature brain to a Rett-like brain. MeCP2 has been found in high concentrations in neurons [144,148] and in glia cells [149], so it is proposed that its defect is the origin of RTT. Recently, it has been observed that MeCP2 participates not only in transcriptional regulation but also in RNA splicing [150], chromatin condensation [151], and the silencing of repetitive elements [152]. Therefore, the role of MeCP2 goes beyond a proposed function in neuron cells.

Taking together the implications of MeCP2 in epigenetic regulation and also in neurological disorders, it must be stressed that both down- and overexpression of MeCP2 result in altered neuron function, an aspect that must be especially considered for therapeutic purposes since it opens the door for new opportunities for phenotypic reversion by means of MeCP2 restoration.

A milestone has been accomplished by Adrian Bird's lab showing that the artificial introduction of MeCP2 gene into a mice laboratory model of RTT could revert all symptoms associated with the disease [153]. However, even if this breakthrough holds huge promise for translational medicine and several labs are engaged to exploit these results, several aspects on MeCP2 functionalities are still not well understood together with the presence of two isoforms that are not totally redundant as thought initially (reviewed in Ref. [154]) impose several hurdles in the quest for a cure. Not to add, as said above, that the levels of MeCP2 have to be balanced for the normal functioning of the brain.

As described, MeCP2 has been mainly associated to RTT. However, MeCP2 family proteins are also involved in cancer (for a review, see Ref. [155]). The dysregulation of the protein product as consequence of genetic polymorphism in MeCP2 genes has been also observed in systemic lupus erythematosus (SLE) patients [156–158]. So, the identification of SNPs in MeCP2 genes may contribute to identify such patients.

5.3 RSTS1 Syndrome

The RSTS1 syndrome (RSTS1, ORPHA783, MIM #180849) is a rare syndrome characterized by growth deficiency during life and congenital anomalies (microcephalia, specific facial malformations which become more prominent with age, broad thumbs, big toes, skin anomalies, joint hypermobility, and halluces) [159,160]. RSTS1 patients are also characterized by intellectual disability. Patients suffered from a variety of heart malformations (e.g., ventricular and atrial septal defect, patent ductus arteriosus, etc.) [159,160]. RSTS1 patients can suffer from tumors, such as leukemia during early ages and meningioma when patients become adults, which are the main causes of death in these patients [160].

Although the exact molecular etiology of RSTS1 is not clearly understood, it is widely accepted that microdeletion at chromosome 16p13.3, in which is located CREB-binding protein mutations (CREBBP or CBP), occurs in RSTS1 patients. Additionally, mutations in E1A-binding protein mutations (EP300) located at 22q13.2 have been found in some RSTS1 patients. Recent investigations employing larger series of RSTS1 patients have detected CBP mutations in about 50% of patients, and about 3% of them have mutations in p300 [161,162]. However, since the cytogenetic and molecular tests only found about 65% of patients with associated mutations to p300/CBP, the other one-third of patients without a "classic" genetic abnormality remains without identification of genetic causes. CBP and p300 constitute the KAT3 family of histone lysyl acetyltransferases, which interact physically and functionally with over 400 different proteins [163]. They can acetylate H4K5, H3K14, H3K18, H3K27, and H3K56, and probably other lysines in histone tails [164–166]. RSTS-associated mutations have been found to affect CBP enzymatic activity, suggesting that reduced HAT function underlies the syndrome [167,168].

CBP and p300 show a high degree of homology (63%) and both play important roles as global transcriptional coactivators. CBP was first described as a nuclear transcription coactivator that binds specifically to CREB when it is phosphorylated [169], while p300 was originally described by protein-interaction assays with the adenoviral E1A oncoprotein [170].

The study of such syndrome has increased the knowledge of the function of CBP and p300, so contributing to the characterization of these proteins as valuable biomarkers in cancer. In this regard, overexpression of p300 predicts increased risk of recurrence and poor survival in non-small-cell lung cancer (NSCLC) [171], and overexpression of both CBP and p300 is associated with increased risk of cancer recurrence of small-cell lung carcinoma [172].

5.4 Weaver Syndrome 2

WVS2 (ORPHA3447, MIM #615521) is a very rare autosomal dominant syndrome. WVS2, also known as Weaver–Smith syndrome, is characterized by overgrowth, accelerated osseous maturation associated with craniofacial and limb anomalies. Craniofacial manifestations such as broad forehead and face, ocular hypertelorism, large ears and large bifrontal diameter, and long and prominent wide philtrum are also found in WVS2 patients [173]. Anthropometric limb anomalies found consist of prominent finger pads, deep horizontal chin growth, thin deep-set nails, camptodactyly, wide distal long bones, foot deformities [173], and scoliosis. WVS2 patients suffer from learning and intellectual disabilities. One of the most important clinical manifestations of WVS2 patients is that they are predisposed to hematological malignancies [174].

Heterozygous mutations in the histone methyltransferase EZH2 gene (Chr 7q36.1) were identified in WVS2 patients. EZH2 protein is a member of the PRC2 (polycomb repressive complex 2) which catalyzes the formation of H3K27me3, a histone mark that is found in heterochromatin regions [175]. Mammalian EZH2 has a relevant role in X-chromosome inactivation, genomic imprinting during germ line development, and cell lineage determination, including osteogenesis, myogenesis, and hematogenesis [176,177], and cancer metastasis [178].

Prognosis in WVS2 patients is variable; one of the reasons is that the epigenetic effect of EZH2 mutations may involve different epigenetic alterations throughout life of the patients. There does not exist therapeutic interventions or drugs approved for this rare disease. However, epigenetic therapy development can offer a therapeutic alternative for these patients, because it is possible to regulate the activity of EZH2.

Importantly, it has been recently found that EZH2 overexpression is associated with tumor grade, growth, and poor prognostic in glioblastoma [179]. Furthermore, it has been found that patients with NSCLC in which EZH2 is overexpressed in FFPE tumor samples exhibited chemoresistance to cysplatin chemotherapy [180].

5.5 Friedreich's Ataxia

FRDA (ORPHA95, MIM #229300), an autosomal recessive neurodegenerative disease, is the most prevalent of the inherited ataxias in Caucasians with a prevalence of 2 patients for each 100,000 people. This neuropathy is characterized by a progressive loss of myelinated axons, particularly in the dorsal root ganglia, the degeneration of posterior columns of the spinal cord, and the loss of peripheral sensory neurons. FRDA starts at early childhood and culminates in gait and limb ataxia, absent tendon reflexes, and dysarthria. Life-threatening hypertrophic cardiomyopathy is found in two-thirds of the patients at the time of diagnosis [181,182]. Scoliosis and diabetes are additional features of FRDA patients [181].

Most FRDA patients are homozygous for large expansions of GAA triplet repeats in the first intron of the gene encoding for the nuclear-encoded mitochondrial protein frataxin (FXN). While normal alleles have less than 36 repeats, FRDA-affected people have 2 FXN alleles each with >70 repeats and reaching up to 1700 repeats [183].

The absence or reduction of the protein product FXN produces abnormalities in respiratory chain Fe–S proteins (e.g., aconitase and the mitochondrial respiratory chain complexes I–III), which result in the accumulation of iron into the mitochondria [184]. High iron levels in the heart have been described in FRDA patients [185]. The higher iron levels and the mitochondrial dysfunction and alterations in mitochondrial biogenesis produce oxidative damage in the mitochondrial DNA, proteins, and tissues of FRDA patients [186,187] compared to healthy individuals.

It is not known what causes the lengthening of the GAA repeat but it is believed that errors in DNA replication, DNA recombination, and/or DNA repair are the primary determinants. The expansion of the GAA triplet repeats alters the epigenetic program of the FXN gene, affecting negatively its expression status. The finding that GAA triplet expansion increases both 5mC and 5hmC on the FXN gene is somewhat intriguing given that they are not synonymous in terms of their impact on transcription. Possibly, the increase in 5mC would induce increase in 5hmC as a result of the action of the DNA demethylases but whether this interplay is coordinated or spurious remains to be investigated. The potential contribution of epigenetics in FRDA disease was first suggested by Saveliev et al. [188]. Afterward, Greene and colleagues observed high hypermethylation of specific CpG sites upstream of the GAA repeat [189]; Castaldo et al. observed how the degree of CpG methylation at the FXN gene correlates with the length of the GAA repeats and inversely correlates with the age of disease onset [190]; and more recently, Al-Mahdawi et al. have showed how 5hmC is enriched upstream to GAA repeat region [191]. Regarding histone PTMs at the FXN gene, it was described that high levels of the repressive marks H3K9me2, H3K9me3, and H3K27me3 together with high levels of the repressive heterochromatin protein 1 (HP1) are present upstream and downstream to GAA repeat region in lymphoblastoid cell line [189,192] and fibroblasts from FRDA patients [193]. In addition, reduction of histone modifications associated with a more open chromatin such as H3K4me3, H3K36me3, and H3K79me3 flanking the GAA repeat regions of the FXN gene were found in FRDA cells [194,195].

The mechanisms described above refer to the epigenetic events that reduce the expression of FXN. However, other epigenetic mechanisms are involved in the physiopathology of FRDA. Two studies have suggested the role of miRNAs in the FRDA phenotype. The first study demonstrated how the binding site in angiotensin-II type-1 receptor gene (AGTR1) for the regulatory miR-155 is altered by the introduction of one polymorphism (rs5186), contributing to the cardiac phenotype in FRDA [196]. The second study compared the miRNA expression profiles in fibroblasts and lymphoblasts from FRDA patients and healthy subjects [197] and found the dysregulation of 27 miRNAs. Particularly, miR-886-3p was found in blood from FRDA patients and was corroborated that the reduction of this miRNA increases the levels of FXN [197]. These results open new opportunities for the management of these patients by implementing clinical prognostic in FRDA patients.

In summary, there is good evidence that epigenetic silencing at FXN genes produces the reduction of FXN protein in FRDA. Unfortunately, FRDA has no specific treatment and most studies have focused on controlling and ameliorating the symptoms associated with this disease. For example, iron chelation and antioxidant therapy have centered much attention during years. In addition, current treatment strategies aim to increase FXN levels. The identification of epigenetic events in the pathophysiology of FRDA opens a new door for epigenetic therapy. In this regard, epigenetic-based transcriptional activation of the FXN gene is an attractive therapeutic strategy which will contribute to the generation of new epigenetic drugs [198,199].

6. EPIGENETIC BIOMARKERS AND CLINICAL LABORATORY

The recent characterization of human DNA methylome map at the single-nucleotide level of resolution, identification of new histone variants and PTMs, the unveiling of genome-wide nucleosome positioning maps, the identification of miRNA signatures, and the characterization of the function of epigenetic machinery have all contributed to our understanding of how aberrant placement of the epigenetic marks or defects in the epigenetic machinery is involved in disease [129].

To facilitate the incorporation of biomarkers in clinical practice, this biomarker should define altered physiological situations, thus it should be able to discern between altered and normal status, predict clinical outcome, disease progression, and individual response to defined therapy. Furthermore, the biomarker would show feasibility; thereby it should not show variability as a consequence of sample processing, or methods used for its identification. The biomarker would have good performance (i.e., high sensibility and good specificity). In addition, it is preferable using noninvasive procedures to obtain the source (biomaterial) in which biomarkers should be analyzed. So, the measurability in body fluids or excreted substance is preferable. Finally, the biomarkers should be time and cost-effective. These aspects and how to solve these issues have been deeply discussed in Chapter 1.

The ultimate tendency is to move toward personalized medicine "the right drug to the right patient at the right time" [200]. In February 2015, the National Institutes of Health (NIH) has started a visionary research initiative to accelerate progress in precision medicine, which aims to achieve more individualized molecular approach for prevention, diagnostics, prognostic, and selection of more effective therapies [201]. Importantly the identification of proper epigenetic biomarkers and the implementation of new technologies in clinical laboratories are crucial for this purpose. Epigenetic biomarkers are likely to be introduced in routine diagnostics in clinical laboratories in the next future, thanks to the development of innovative technologies [202,203].

7. CONCLUSIONS

A comprehensive understanding of epigenetic mechanisms (DNA methylation, PTMs of histones, and ncRNAs) in terms of their interactions and alterations that define health and disease has become a priority in medicine. The study of some rare diseases, such as ICF syndrome, RTT, Coffin–Lowry syndrome, and others, has allowed for the characterization of the role of epigenetic machinery, thereby offering new answers for human diseases.

Great progress has been made in the description of aberrant methylation modifications in diseased tissues; new PTMs have been linked to disease; and, promising ncRNA signatures have been associated to disease diagnostic and prognostic. These areas offer a new perspective in biomedical research for the study of complex diseases such as MetS, AD, and cancer, among others.

Many key questions still remain unresolved, and the exact role of new PTMs and ncRNAs are yet to be described in most diseases. However, advances in technological developments are contributing to epigenomic analyses on a large scale enabling us to identify methylomes and histone maps across the human genome. In addition, bioinformatic tools and basic and clinical investigations around the world are contributing to increasing the knowledge in this field. So, the release of whole epigenomic data (methylomes, histone modification maps, and miRNA signatures) into public databases will contribute to the improvement of our knowledge of the epigenetic basis of diseases, the finding of new biomarkers, and the design of new epigenetic drugs. In this regard, in a few years, it should be possible to obtain a patient's methylome, histone map, or miRNA signature profile, and compare them with that of healthy or disease-related epigenomes to efficiently diagnose diseases, or, for example, identify a primary tumor, or establish a highly specific treatment regime and prognosis. Therefore, we foresee the potential of epigenetics to implement the new Precision Medicine Initiative.

LIST OF ABBREVIATIONS

5caC 5-Carboxylcytosine
5fC 5-Formylcytosine
5hmC 5-Hydroxymethylcytosine
5mC 5-Methylcytosine
α-KG α-Ketoglutarate
α-KGDH α-KG dehydrogenase
AD Alzheimer's disease
AGTR1 Angiotensin-II type-1 receptor gene
ATP Adenosine triphosphate
BET Bromodomain and extraterminal: BRD2, BRD3, BRD4
BRDT Bromodomain 2/3/4/testis-specific
CBS Cystathionine β-synthase
CDH1 E-cadherin
CDH13 Cadherin-13
CDKL5 Cyclin-dependent kinase like
CDKN1A Cyclin-dependent kinase inhibitor 1
CKD Chronic kidney disease
CNV Copy number variation
CREBBP CREB-binding protein mutations
DAPK Death-associated protein kinase 1
DNMT1 DNA (cytosine-5-)-methyltransferase 1
DNMT3A and 3B DNA (cytosine-5-)-methyltransferase 3 alpha/beta
ESR1 Estrogen receptor 1
$FADH_2$ Reduced flavin adenine dinucleotide
FGF2 Fibroblast growth factor 2
FH Fumarate hydrolase
FRDA Friedreich's ataxia
FXN Frataxin gene
GWAS Genome-wide association studies
HATs Histone acetyltransferases
HCys Homocysteine
HDACs Histone deacetylases
HMTs Histone methyltransferases
ICF syndrome Immunodeficiency, centromeric region instability, and facial anomalies syndrome
IDH Isocitrate dehydrogenase
IGF2 Insulin-like growth factor 2 receptor
lncRNAs Long noncoding RNAs
MBD2 Methyl CpG-binding domain protein 2
MeCP2 Methyl CpG-binding protein 2
MetS Metabolic syndrome
MGMT O-6-methylguanine-DNA methyltransferase
miRNAs MicroRNAs
MS Methionine synthase
MTHFR Methylenetetrahydrofolate reductase gene
NADH Reduced nicotinamide adenine dinucleotide
N-CoR Nuclear receptor corepressor
ncRNAs Noncoding RNAs
NSCLC Non-small-cell lung cancer
PA Physical activity

PAX3 Paired box 3
PDK4 Pyruvate dehydrogenase lipoamide kinase isozyme 4
PGC-1α Peroxisome proliferator-activated receptor gamma coactivator 1-alpha
PPAR-δ Peroxisome proliferator-activated receptor delta
PRC2 Polycomb repressive complex 2
PTEN Phosphatase and tensin homolog
PTMs Posttranslational modifications
RSTB1 Rubinstein–Taybi syndrome
RTT Rett syndrome
RUNX1 Runt-related transcription factor
SAH S-adenosylhomocysteine
SAM S-adenosylmethionine
SDH Succinyl dehydrogenase complex
SLE Systemic lupus erythematosus
SNPs Single-nucleotide polymorphisms
SOCS1/SOCS3 Suppressor of cytokine signaling 1/3
SOX9 Sex-determining region Y-box 9
TCA Tricarboxylic acid cycle
TETs Ten-eleven translocases
TFAM Mitochondrial transcription factor A
WVS2 Weaver syndrome 2

Acknowledgments

Abdelhalim Boukaba work was supported by the grant from Guangdong Province to the Molecular Epigenetics Laboratory.

José Luis García-Giménez is supported by the Center for Biomedical Network Research on Rare Diseases (CIBERER) from the Instituto de Salud Carlos III-ISCIII (Spain) ACCI2014, the regional grant GV/2014/132 and PI12/02263 from the Ministerio de Economía y Competitividad and ISCIII.

References

[1] Bird A. Perceptions of epigenetics. Nature May 24, 2007;447(7143):396–8.
[2] Dawson MA, Kouzarides T. Cancer epigenetics: from mechanism to therapy. Cell July 6, 2012;150(1):12–27.
[3] Turner BM. Nucleosome signalling; an evolving concept. Biochim Biophys Acta August 2014;1839(8):623–6.
[4] Kornberg RD, Lorch Y. Twenty-five years of the nucleosome, fundamental particle of the eukaryote chromosome. Cell August 6, 1999;98(3):285–94.
[5] Luger K, Mader AW, Richmond RK, Sargent DF, Richmond TJ. Crystal structure of the nucleosome core particle at 2.8 A resolution. Nature September 18, 1997;389(6648):251–60.
[6] Elowitz MB, Levine AJ, Siggia ED, Swain PS. Stochastic gene expression in a single cell. Science August 16, 2002;297(5584):1183–6.
[7] Perillo B, Ombra MN, Bertoni A, Cuozzo C, Sacchetti S, Sasso A, et al. DNA oxidation as triggered by H3K9me2 demethylation drives estrogen-induced gene expression. Science January 11, 2008;319(5860):202–6.
[8] Holliday R, Pugh JE. DNA modification mechanisms and gene activity during development. Science January 24, 1975;187(4173):226–32.
[9] Shukla S, Kavak E, Gregory M, Imashimizu M, Shutinoski B, Kashlev M, et al. CTCF-promoted RNA polymerase II pausing links DNA methylation to splicing. Nature November 3, 2011;479(7371):74–9.
[10] Gelfman S, Cohen N, Yearim A, Ast G. DNA-methylation effect on cotranscriptional splicing is dependent on GC architecture of the exon-intron structure. Genome Res May 2013;23(5):789–99.
[11] Wang GS, Cooper TA. Splicing in disease: disruption of the splicing code and the decoding machinery. Nat Rev Genet October 2007;8(10):749–61.
[12] Wu SC, Zhang Y. Active DNA demethylation: many roads lead to Rome. Nat Rev Mol Cell Biol September 2010;11(9):607–20.
[13] Goll MG, Bestor TH. Eukaryotic cytosine methyltransferases. Ann Rev Biochem 2005;74:481–514.
[14] Spruijt CG, Gnerlich F, Smits AH, Pfaffeneder T, Jansen PW, Bauer C, et al. Dynamic readers for 5-(hydroxy) methylcytosine and its oxidized derivatives. Cell February 28, 2013;152(5):1146–59.
[15] Zhang Y, Lv J, Liu H, Zhu J, Su J, Wu Q, et al. HHMD: the human histone modification database. Nucleic Acids Res January 2010;38(Database issue):D149–54.
[16] Waldmann T, Schneider R. Targeting histone modifications–epigenetics in cancer. Curr Opin Cell Biol April 2013;25(2):184–9.
[17] Gezer U, Holdenrieder S. Post-translational histone modifications in circulating nucleosomes as new biomarkers in colorectal cancer. In Vivo May–June 2014;28(3):287–92.
[18] Liu B, Lin Y, Darwanto A, Song X, Xu G, Zhang K. Identification and characterization of propionylation at histone H3 lysine 23 in mammalian cells. J Biol Chem November 20, 2009;284(47):32288–95.
[19] Huang H, Sabari BR, Garcia BA, Allis CD, Zhao Y. SnapShot: histone modifications. Cell October 9, 2014;159(2):458-e1.
[20] Garcia-Gimenez JL, Ledesma AM, Esmoris I, Roma-Mateo C, Sanz P, Vina J, et al. Histone carbonylation occurs in proliferating cells. Free Radic Biol Med April 15, 2012;52(8):1453–64.
[21] Garcia-Gimenez JL, Olaso G, Hake SB, Bonisch C, Wiedemann SM, Markovic J, et al. Histone h3 glutathionylation in proliferating mammalian cells destabilizes nucleosomal structure. Antioxid Redox Signal October 20, 2013;19(12):1305–20.
[22] Turner BM. Histone acetylation and an epigenetic code. Bioessays September 2000;22(9):836–45.

[23] Strahl BD, Allis CD. The language of covalent histone modifications. Nature January 6, 2000;403(6765):41–5.
[24] Becker PB. Gene regulation: a finger on the mark. Nature July 6, 2006;442(7098):31–2.
[25] Henikoff S. Histone modifications: combinatorial complexity or cumulative simplicity? Proc Natl Acad Sci USA April 12, 2005;102(15):5308–9.
[26] Ptashne M. On the use of the word 'epigenetic'. Curr Biol April 3, 2007;17(7):R233–6.
[27] Ruthenburg AJ, Li H, Patel DJ, Allis CD. Multivalent engagement of chromatin modifications by linked binding modules. Nat Rev Mol Cell Biol December 2007;8(12):983–94.
[28] Agalioti T, Lomvardas S, Parekh B, Yie J, Maniatis T, Thanos D. Ordered recruitment of chromatin modifying and general transcription factors to the IFN-beta promoter. Cell November 10, 2000;103(4):667–78.
[29] Dawson MA, Kouzarides T, Huntly BJ. Targeting epigenetic readers in cancer. N Engl J Med August 16, 2012;367(7):647–57.
[30] Filippakopoulos P, Knapp S. Targeting bromodomains: epigenetic readers of lysine acetylation. Nat Rev Drug Discov May 2014;13(5):337–56.
[31] Filippakopoulos P, Qi J, Picaud S, Shen Y, Smith WB, Fedorov O, et al. Selective inhibition of BET bromodomains. Nature December 23, 2010;468(7327):1067–73.
[32] French CA. NUT midline carcinoma. Cancer Genet Cytogenet November 2010;203(1):16–20.
[33] He L, Hannon GJ. MicroRNAs: small RNAs with a big role in gene regulation. Nat Rev Genet July 2004;5(7):522–31.
[34] Ponting CP, Oliver PL, Reik W. Evolution and functions of long noncoding RNAs. Cell February 20, 2009;136(4):629–41.
[35] Hwang HW, Mendell JT. MicroRNAs in cell proliferation, cell death, and tumorigenesis. Br J Cancer March 27, 2006;94(6):776–80.
[36] Garzon R, Marcucci G, Croce CM. Targeting microRNAs in cancer: rationale, strategies and challenges. Nat Rev Drug Discov October 2010;9(10):775–89.
[37] Tay Y, Zhang J, Thomson AM, Lim B, Rigoutsos I. MicroRNAs to Nanog, Oct4 and Sox2 coding regions modulate embryonic stem cell differentiation. Nature October 23, 2008;455(7216):1124–8.
[38] Lytle JR, Yario TA, Steitz JA. Target mRNAs are repressed as efficiently by microRNA-binding sites in the 5′ UTR as in the 3′ UTR. Proc Natl Acad Sci USA June 5, 2007;104(23):9667–72.
[39] Vasudevan S, Tong Y, Steitz JA. Switching from repression to activation: microRNAs can up-regulate translation. Science December 21, 2007;318(5858):1931–4.
[40] Pandey RR, Mondal T, Mohammad F, Enroth S, Redrup L, Komorowski J, et al. Kcnq1ot1 antisense noncoding RNA mediates lineage-specific transcriptional silencing through chromatin-level regulation. Mol Cell October 24, 2008;32(2):232–46.
[41] Wang KC, Yang YW, Liu B, Sanyal A, Corces-Zimmerman R, Chen Y, et al. A long noncoding RNA maintains active chromatin to coordinate homeotic gene expression. Nature April 7, 2011;472(7341):120–4.
[42] Wang KC, Chang HY. Molecular mechanisms of long noncoding RNAs. Mol Cell September 16, 2011;43(6):904–14.
[43] Mercer TR, Mattick JS. Structure and function of long noncoding RNAs in epigenetic regulation. Nat Struct Mol Biol March 2013;20(3):300–7.
[44] Castro R, Rivera I, Ravasco P, Camilo ME, Jakobs C, Blom HJ, et al. 5,10-methylenetetrahydrofolate reductase (MTHFR) 677C-->T and 1298A-->C mutations are associated with DNA hypomethylation. J Med Genet June 2004;41(6):454–8.
[45] Thomas P, Fenech M. Methylenetetrahydrofolate reductase, common polymorphisms, and relation to disease. Vitam Horm 2008;79:375–92.
[46] Tuck-Muller CM, Narayan A, Tsien F, Smeets DF, Sawyer J, Fiala ES, et al. DNA hypomethylation and unusual chromosome instability in cell lines from ICF syndrome patients. Cytogenet Cell Genet 2000;89(1–2):121–8.
[47] Ulrey CL, Liu L, Andrews LG, Tollefsbol TO. The impact of metabolism on DNA methylation. Hum Mol Genet April 15, 2005;14(1):R139–47.
[48] Aitken SM, Lodha PH, Morneau DJ. The enzymes of the transsulfuration pathways: active-site characterizations. Biochim Biophys Acta November 2011;1814(11):1511–7.
[49] Perna AF, Ingrosso D, De Santo NG. Homocysteine and oxidative stress. Amino Acids December 2003;25(3–4):409–17.
[50] Vanzin CS, Biancini GB, Sitta A, Wayhs CA, Pereira IN, Rockenbach F, et al. Experimental evidence of oxidative stress in plasma of homocystinuric patients: a possible role for homocysteine. Mol Genet Metab September–October 2011;104(1–2):112–7.
[51] Valinluck V, Tsai HH, Rogstad DK, Burdzy A, Bird A, Sowers LC. Oxidative damage to methyl-CpG sequences inhibits the binding of the methyl-CpG binding domain (MBD) of methyl-CpG binding protein 2 (MeCP2). Nucleic Acids Res 2004;32(14):4100–8.
[52] Heil SG, Riksen NP, Boers GH, Smulders Y, Blom HJ. DNA methylation status is not impaired in treated cystathionine beta-synthase (CBS) deficient patients. Mol Genet Metab May 2007;91(1):55–60.
[53] Cardaci S, Ciriolo MR. TCA cycle defects and cancer: when metabolism tunes redox state. Int J Cell Biol 2012;2012:161837.
[54] Salminen A, Kauppinen A, Hiltunen M, Kaarniranta K. Krebs cycle intermediates regulate DNA and histone methylation: epigenetic impact on the aging process. Ageing Res Rev July 2014;16:45–65.

[55] Salminen A, Kaarniranta K, Hiltunen M, Kauppinen A. Krebs cycle dysfunction shapes epigenetic landscape of chromatin: novel insights into mitochondrial regulation of aging process. Cell Signal July 2014;26(7):1598–603.
[56] Wellen KE, Hatzivassiliou G, Sachdeva UM, Bui TV, Cross JR, Thompson CB. ATP-citrate lyase links cellular metabolism to histone acetylation. Science May 22, 2009;324(5930):1076–80.
[57] Iacobazzi V, Infantino V. Citrate–new functions for an old metabolite. Biol Chem April 2014;395(4):387–99.
[58] Sack MN, Finkel T. Mitochondrial metabolism, sirtuins, and aging. Cold Spring Harb Perspect Biol December 2012;4(12).
[59] Shi Y, Lan F, Matson C, Mulligan P, Whetstine JR, Cole PA, et al. Histone demethylation mediated by the nuclear amine oxidase homolog LSD1. Cell December 29, 2004;119(7):941–53.
[60] Loenarz C, Coleman ML, Boleininger A, Schierwater B, Holland PW, Ratcliffe PJ, et al. The hypoxia-inducible transcription factor pathway regulates oxygen sensing in the simplest animal, *Trichoplax adhaerens*. EMBO Rep January 2011;12(1):63–70.
[61] Hausinger RP. FeII/alpha-ketoglutarate-dependent hydroxylases and related enzymes. Crit Rev Biochem Mol Biol January–February 2004;39(1):21–68.
[62] Chen H, Dzitoyeva S, Manev H. Effect of aging on 5-hydroxymethylcytosine in the mouse hippocampus. Restor Neurol Neurosci 2012;30(3):237–45.
[63] Whayne TF. Epigenetics in the development, modification, and prevention of cardiovascular disease. Mol Biol Rep September 10, 2014.
[64] Nicolia V, Lucarelli M, Fuso A. Environment, epigenetics and neurodegeneration: focus on nutrition in Alzheimer's disease. Exp Gerontol October 14, 2014.
[65] van Dijk SJ, Molloy PL, Varinli H, Morrison JL, Muhlhausler BS. Epigenetics and human obesity. Int J Obes (London) January 2015;39(1):85–97.
[66] Alegria-Torres JA, Baccarelli A, Bollati V. Epigenetics and lifestyle. Epigenomics June 2011;3(3):267–77.
[67] Denham J, Marques FZ, O'Brien BJ, Charchar FJ. Exercise: putting action into our epigenome. Sports Med February 2014;44(2):189–209.
[68] Ntanasis-Stathopoulos J, Tzanninis JG, Philippou A, Koutsilieris M. Epigenetic regulation on gene expression induced by physical exercise. J Musculoskelet Neuronal Interact June 2013;13(2):133–46.
[69] Vickers MH. Early life nutrition, epigenetics and programming of later life disease. Nutrients June 2014;6(6):2165–78.
[70] Susiarjo M, Bartolomei MS. Epigenetics. You are what you eat, but what about your DNA? Science August 15, 2014;345(6198):733–4.
[71] Sanchis-Gomar F, Garcia-Gimenez JL, Perez-Quilis C, Gomez-Cabrera MC, Pallardo FV, Lippi G. Physical exercise as an epigenetic modulator: eustress, the "positive stress" as an effector of gene expression. J Strength Cond Res December 2012;26(12):3469–72.
[72] Pareja-Galeano H, Sanchis-Gomar F, Garcia-Gimenez JL. Physical exercise and epigenetic modulation: elucidating intricate mechanisms. Sports Med April 2014;44(4):429–36.
[73] Barres R, Yan J, Egan B, Treebak JT, Rasmussen M, Fritz T, et al. Acute exercise remodels promoter methylation in human skeletal muscle. Cell Metab March 7, 2012;15(3):405–11.
[74] Baar K. Epigenetic control of skeletal muscle fibre type. Acta Physiol (Oxford, England) August 2010;199(4):477–87.
[75] Keller P, Vollaard NB, Gustafsson T, Gallagher IJ, Sundberg CJ, Rankinen T, et al. A transcriptional map of the impact of endurance exercise training on skeletal muscle phenotype. J Appl Physiol (1985) January 2011;110(1):46–59.
[76] Aoi W, Naito Y, Mizushima K, Takanami Y, Kawai Y, Ichikawa H, et al. The microRNA miR-696 regulates PGC-1{alpha} in mouse skeletal muscle in response to physical activity. Am J Physiol Endocrinol Metab April 2010;298(4):E799–806.
[77] Safdar A, Abadi A, Akhtar M, Hettinga BP, Tarnopolsky MA. miRNA in the regulation of skeletal muscle adaptation to acute endurance exercise in C57Bl/6J male mice. PLoS One 2009;4(5):e5610.
[78] Fernandes T, Magalhaes FC, Roque FR, Phillips MI, Oliveira EM. Exercise training prevents the microvascular rarefaction in hypertension balancing angiogenic and apoptotic factors: role of microRNAs-16, -21, and -126. Hypertension February 2012;59(2):513–20.
[79] Toden S, Goel A. The importance of diets and epigenetics in cancer prevention: a hope and promise for the future? Altern Ther Health Med October 2014;20(Suppl. 2):6–11.
[80] Zhang FF, Cardarelli R, Carroll J, Zhang S, Fulda KG, Gonzalez K, et al. Physical activity and global genomic DNA methylation in a cancer-free population. Epigenetics March 2011;6(3):293–9.
[81] Voisin S, Eynon N, Yan X, Bishop DJ. Exercise training and DNA methylation in humans. Acta Physiol (Oxford, England) January 2015;213(1):39–59.
[82] Lee J, Jeon JY, Meyerhardt JA. Diet and lifestyle in survivors of colorectal cancer. Hematol Oncol Clin North Am February 2015;29(1):1–27.
[83] Lillycrop KA, Hoile SP, Grenfell L, Burdge GC. DNA methylation, ageing and the influence of early life nutrition. Proc Nutr Soc August 2014;73(3):413–21.
[84] Ordovas JM, Robertson R, Cleirigh EN. Gene-gene and gene-environment interactions defining lipid-related traits. Curr Opin Lipidol April 2011;22(2):129–36.

[85] Bartsch H, Nair J. Oxidative stress and lipid peroxidation-derived DNA-lesions in inflammation driven carcinogenesis. Cancer Detect Prev 2004;28(6):385–91.
[86] Arsova-Sarafinovska Z, Eken A, Matevska N, Erdem O, Sayal A, Savaser A, et al. Increased oxidative/nitrosative stress and decreased antioxidant enzyme activities in prostate cancer. Clin Biochem August 2009;42(12):1228–35.
[87] Fini L, Selgrad M, Fogliano V, Graziani G, Romano M, Hotchkiss E, et al. Annurca apple polyphenols have potent demethylating activity and can reactivate silenced tumor suppressor genes in colorectal cancer cells. J Nutr December 2007;137(12):2622–8.
[88] Link A, Balaguer F, Goel A. Cancer chemoprevention by dietary polyphenols: promising role for epigenetics. Biochem Pharmacol December 15, 2010;80(12):1771–92.
[89] Campion J, Milagro FI, Martinez JA. Individuality and epigenetics in obesity. Obes Rev July 2009;10(4):383–92.
[90] Marwick JA, Kirkham PA, Stevenson CS, Danahay H, Giddings J, Butler K, et al. Cigarette smoke alters chromatin remodeling and induces proinflammatory genes in rat lungs. Am J Respir Cell Mol Biol December 2004;31(6):633–42.
[91] Liu F, Killian JK, Yang M, Walker RL, Hong JA, Zhang M, et al. Epigenomic alterations and gene expression profiles in respiratory epithelia exposed to cigarette smoke condensate. Oncogene June 24, 2010;29(25):3650–64.
[92] Woodson K, Mason J, Choi SW, Hartman T, Tangrea J, Virtamo J, et al. Hypomethylation of p53 in peripheral blood DNA is associated with the development of lung cancer. Cancer Epidemiol Biomarkers Prev January 2001;10(1):69–74.
[93] Wang J, Wu Z, Li D, Li N, Dindot SV, Satterfield MC, et al. Nutrition, epigenetics, and metabolic syndrome. Antioxid Redox Signal July 15, 2012;17(2):282–301.
[94] Samson SL, Garber AJ. Metabolic syndrome. Endocrinol Metab Clin North Am March 2014;43(1):1–23.
[95] Prasad GV. Metabolic syndrome and chronic kidney disease: current status and future directions. World J Nephrol November 6, 2014;3(4):210–9.
[96] Mendonca FM, de Sousa FR, Barbosa AL, Martins SC, Araujo RL, Soares R, et al. Metabolic syndrome and risk of cancer: which link? Metabolism February 2015;64(2):182–9.
[97] Rios JA, Cisternas P, Arrese M, Barja S, Inestrosa NC. Is Alzheimer's disease related to metabolic syndrome? A Wnt signaling conundrum. Prog Neurobiol October 2014;121:125–46.
[98] Kaur J. A comprehensive review on metabolic syndrome. Cardiol Res Pract 2014;2014:943162.
[99] Neel JV. The "thrifty genotype" in 1998. Nutr Rev May 1999;57(5 Pt 2):S2–9.
[100] Hales CN, Barker DJ. The thrifty phenotype hypothesis. Br Med Bull 2001;60:5–20.
[101] Gluckman PD, Hanson MA. The developmental origins of the metabolic syndrome. Trends Endocrinol Metab May–June 2004;15(4):183–7.
[102] Burdge GC, Lillycrop KA. Nutrition, epigenetics, and developmental plasticity: implications for understanding human disease. Annu Rev Nutr August 21, 2010;30:315–39.
[103] Symonds ME, Sebert SP, Hyatt MA, Budge H. Nutritional programming of the metabolic syndrome. Nat Rev Endocrinol November 2009;5(11):604–10.
[104] Seki Y, Williams L, Vuguin PM, Charron MJ. Minireview: epigenetic programming of diabetes and obesity: animal models. Endocrinology March 2012;153(3):1031–8.
[105] Inagaki T, Tachibana M, Magoori K, Kudo H, Tanaka T, Okamura M, et al. Obesity and metabolic syndrome in histone demethylase JHDM2a-deficient mice. Genes Cells August 2009;14(8):991–1001.
[106] Berdasco M, Esteller M. Genetic syndromes caused by mutations in epigenetic genes. Hum Genet April 2013;132(4):359–83.
[107] Heijmans BT, Tobi EW, Stein AD, Putter H, Blauw GJ, Susser ES, et al. Persistent epigenetic differences associated with prenatal exposure to famine in humans. Proc Natl Acad Sci USA November 4, 2008;105(44):17046–9.
[108] de Rooij SR, Painter RC, Holleman F, Bossuyt PM, Roseboom TJ. The metabolic syndrome in adults prenatally exposed to the Dutch famine. Am J Clin Nutr October 2007;86(4):1219–24.
[109] Schurno A, Eggers H, Dahl D. Hearing disorders in morbus haemolyticus neanatorum after treatment with blood transfusions. Kinderarztl Prax October 1965;33(10):457–64.
[110] Tobi EW, Lumey LH, Talens RP, Kremer D, Putter H, Stein AD, et al. DNA methylation differences after exposure to prenatal famine are common and timing- and sex-specific. Hum Mol Genet November 1, 2009;18(21):4046–53.
[111] Gamble KL, Berry R, Frank SJ, Young ME. Circadian clock control of endocrine factors. Nat Rev Endocrinol August 2014;10(8):466–75.
[112] Turek FW, Joshu C, Kohsaka A, Lin E, Ivanova G, McDearmon E, et al. Obesity and metabolic syndrome in circadian clock mutant mice. Science May 13, 2005;308(5724):1043–5.
[113] Khalyfa A, Mutskov V, Carreras A, Khalyfa AA, Hakim F, Gozal D. Sleep fragmentation during late gestation induces metabolic perturbations and epigenetic changes in adiponectin gene expression in male adult offspring mice. Diabetes October 2014;63(10):3230–41.

[114] Sookoian S, Gianotti TF, Burgueno AL, Pirola CJ. Fetal metabolic programming and epigenetic modifications: a systems biology approach. Pediatr Res April 2013;73(4 Pt 2):531–42.

[115] DelCurto H, Wu G, Satterfield MC. Nutrition and reproduction: links to epigenetics and metabolic syndrome in offspring. Curr Opin Clin Nutr Metab Care July 2013;16(4):385–91.

[116] Mutskov V, Khalyfa A, Wang Y, Carreras A, Nobrega M, Gozal D. Early life physical activity reverses metabolic and Foxo1 epigenetic misregulation induced by gestational sleep disturbance. Am J Physiol Regul Integr Comp Physiol January 7, 2015. ajpregu 00426 2014.

[117] Ballard C, Gauthier S, Corbett A, Brayne C, Aarsland D, Jones E. Alzheimer's disease. Lancet March 19, 2011;377(9770):1019–31.

[118] Mastroeni D, Grover A, Delvaux E, Whiteside C, Coleman PD, Rogers J. Epigenetic mechanisms in Alzheimer's disease. Neurobiol Aging July 2011; 32(7):1161–80.

[119] De Jager PL, Srivastava G, Lunnon K, Burgess J, Schalkwyk LC, Yu L, et al. Alzheimer's disease: early alterations in brain DNA methylation at ANK1, BIN1, RHBDF2 and other loci. Nat Neurosci September 2014;17(9):1156–63.

[120] Lunnon K, Smith R, Hannon E, De Jager PL, Srivastava G, Volta M, et al. Methylomic profiling implicates cortical deregulation of ANK1 in Alzheimer's disease. Nat Neurosci September 2014;17(9):1164–70.

[121] Stewart SK, Morris TJ, Guilhamon P, Bulstrode H, Bachman M, Balasubramanian S, et al. oxBS-450K: a method for analysing hydroxymethylation using 450K BeadChips. Methods August 28, 2014.

[122] Booth MJ, Branco MR, Ficz G, Oxley D, Krueger F, Reik W, et al. Quantitative sequencing of 5-methylcytosine and 5-hydroxymethylcytosine at single-base resolution. Science May 18, 2012;336(6083):934–7.

[123] Yu M, Hon GC, Szulwach KE, Song CX, Zhang L, Kim A, et al. Base-resolution analysis of 5-hydroxymethylcytosine in the mammalian genome. Cell June 8, 2012;149(6):1368–80.

[124] Song CX, Szulwach KE, Fu Y, Dai Q, Yi C, Li X, et al. Selective chemical labeling reveals the genome-wide distribution of 5-hydroxymethylcytosine. Nat Biotechnol January 2011;29(1):68–72.

[125] McConnell MJ, Lindberg MR, Brennand KJ, Piper JC, Voet T, Cowing-Zitron C, et al. Mosaic copy number variation in human neurons. Science November 1, 2013;342(6158):632–7.

[126] Jamuar SS, Lam AT, Kircher M, D'Gama AM, Wang J, Barry BJ, et al. Somatic mutations in cerebral cortical malformations. N Engl J Med August 21, 2014;371(8):733–43.

[127] Cai X, Evrony GD, Lehmann HS, Elhosary PC, Mehta BK, Poduri A, et al. Single-cell, genome-wide sequencing identifies clonal somatic copy-number variation in the human brain. Cell Rep September 11, 2014;8(5):1280–9.

[128] Straussman R, Nejman D, Roberts D, Steinfeld I, Blum B, Benvenisty N, et al. Developmental programming of CpG island methylation profiles in the human genome. Nat Struct Mol Biol May 2009;16(5):564–71.

[129] Portela A, Esteller M. Epigenetic modifications and human disease. Nat Biotechnol October 2010;28(10): 1057–68.

[130] Ehrlich M, Jackson K, Weemaes C. Immunodeficiency, centromeric region instability, facial anomalies syndrome (ICF). Orphanet J Rare Dis 2006;1:2.

[131] Ehrlich M. The ICF syndrome, a DNA methyltransferase 3B deficiency and immunodeficiency disease. Clin Immunol October 2003;109(1):17–28.

[132] Xu GL, Bestor TH, Bourc'his D, Hsieh CL, Tommerup N, Bugge M, et al. Chromosome instability and immunodeficiency syndrome caused by mutations in a DNA methyltransferase gene. Nature November 11, 1999;402(6758):187–91.

[133] Lana E, Megarbane A, Tourriere H, Sarda P, Lefranc G, Claustres M, et al. DNA replication is altered in Immunodeficiency Centromeric instability Facial anomalies (ICF) cells carrying DNMT3B mutations. Eur J Hum Genet October 2012;20(10):1044–50.

[134] Yehezkel S, Segev Y, Viegas-Pequignot E, Skorecki K, Selig S. Hypomethylation of subtelomeric regions in ICF syndrome is associated with abnormally short telomeres and enhanced transcription from telomeric regions. Hum Mol Genet September 15, 2008;17(18):2776–89.

[135] Miniou P, Bourc'his D, Molina Gomes D, Jeanpierre M, Viegas-Pequignot E. Undermethylation of Alu sequences in ICF syndrome: molecular and in situ analysis. Cytogenet Cell Genet 1997;77(3–4): 308–13.

[136] Brun ME, Lana E, Rivals I, Lefranc G, Sarda P, Claustres M, et al. Heterochromatic genes undergo epigenetic changes and escape silencing in immunodeficiency, centromeric instability, facial anomalies (ICF) syndrome. PLoS One 2011;6(4):e19464.

[137] Ehrlich M, Buchanan KL, Tsien F, Jiang G, Sun B, Uicker W, et al. DNA methyltransferase 3B mutations linked to the ICF syndrome cause dysregulation of lymphogenesis genes. Hum Mol Genet December 1, 2001;10(25):2917–31.

[138] Heyn H, Vidal E, Sayols S, Sanchez-Mut JV, Moran S, Medina I, et al. Whole-genome bisulfite DNA sequencing of a DNMT3B mutant patient. Epigenetics June 1, 2012;7(6):542–50.

[139] Jin B, Tao Q, Peng J, Soo HM, Wu W, Ying J, et al. DNA methyltransferase 3B (DNMT3B) mutations in ICF syndrome lead to altered epigenetic modifications and aberrant expression of genes regulating development, neurogenesis and immune function. Hum Mol Genet March 1, 2008;17(5):690–709.

[140] Wang L, Rodriguez M, Kim ES, Xu Y, Bekele N, El-Naggar AK, et al. A novel C/T polymorphism in the core promoter of human de novo cytosine DNA methyltransferase 3B6 is associated with prognosis in head and neck cancer. Int J Oncol October 2004;25(4):993–9.

[141] Jones JS, Amos CI, Pande M, Gu X, Chen J, Campos IM, et al. DNMT3b polymorphism and hereditary nonpolyposis colorectal cancer age of onset. Cancer Epidemiol Biomarkers Prev May 2006;15(5):886–91.

[142] Hayette S, Thomas X, Jallades L, Chabane K, Charlot C, Tigaud I, et al. High DNA methyltransferase DNMT3B levels: a poor prognostic marker in acute myeloid leukemia. PLoS One 2012;7(12):e51527.

[143] Williamson SL, Christodoulou J. Rett syndrome: new clinical and molecular insights. Eur J Hum Genet August 2006;14(8):896–903.

[144] Zachariah RM, Rastegar M. Linking epigenetics to human disease and Rett syndrome: the emerging novel and challenging concepts in MeCP2 research. Neural Plast 2012;2012:415825.

[145] Moretti P, Zoghbi HY. MeCP2 dysfunction in Rett syndrome and related disorders. Curr Opin Genet Dev June 2006;16(3):276–81.

[146] Van Esch H, Bauters M, Ignatius J, Jansen M, Raynaud M, Hollanders K, et al. Duplication of the MECP2 region is a frequent cause of severe mental retardation and progressive neurological symptoms in males. Am J Hum Genet September 2005;77(3):442–53.

[147] Chahrour M, Jung SY, Shaw C, Zhou X, Wong ST, Qin J, et al. MeCP2, a key contributor to neurological disease, activates and represses transcription. Science May 30, 2008;320(5880):1224–9.

[148] Luikenhuis S, Giacometti E, Beard CF, Jaenisch R. Expression of MeCP2 in postmitotic neurons rescues Rett syndrome in mice. Proc Natl Acad Sci USA April 20, 2004;101(16):6033–8.

[149] Ballas N, Lioy DT, Grunseich C, Mandel G. Non-cell autonomous influence of MeCP2-deficient glia on neuronal dendritic morphology. Nat Neurosci March 2009;12(3):311–7.

[150] Young JI, Hong EP, Castle JC, Crespo-Barreto J, Bowman AB, Rose MF, et al. Regulation of RNA splicing by the methylation-dependent transcriptional repressor methyl-CpG binding protein 2. Proc Natl Acad Sci USA December 6, 2005;102(49):17551–8.

[151] Ishibashi T, Thambirajah AA, Ausio J. MeCP2 preferentially binds to methylated linker DNA in the absence of the terminal tail of histone H3 and independently of histone acetylation. FEBS Lett April 2, 2008;582(7):1157–62.

[152] Muotri AR, Marchetto MC, Coufal NG, Oefner R, Yeo G, Nakashima K, et al. L1 retrotransposition in neurons is modulated by MeCP2. Nature November 18, 2010;468(7322):443–6.

[153] Guy J, Gan J, Selfridge J, Cobb S, Bird A. Reversal of neurological defects in a mouse model of Rett syndrome. Science February 23, 2007;315(5815):1143–7.

[154] Liyanage VR, Rastegar M. Rett syndrome and MeCP2. Neuromolecular Med June 2014;16(2):231–64.

[155] Parry L, Clarke AR. The roles of the methyl-CpG binding proteins in Cancer. Genes Cancer June 2011;2(6):618–30.

[156] Sawalha AH, Webb R, Han S, Kelly JA, Kaufman KM, Kimberly RP, et al. Common variants within MECP2 confer risk of systemic lupus erythematosus. PLoS One 2008;3(3):e1727.

[157] Liu K, Zhang L, Chen J, Hu Z, Cai G, Hong Q. Association of MeCP2 (rs2075596, rs2239464) genetic polymorphisms with systemic lupus erythematosus: a meta-analysis. Lupus August 2013;22(9):908–18.

[158] Webb R, Wren JD, Jeffries M, Kelly JA, Kaufman KM, Tang Y, et al. Variants within MECP2, a key transcription regulator, are associated with increased susceptibility to lupus and differential gene expression in patients with systemic lupus erythematosus. Arthritis Rheum April 2009;60(4):1076–84.

[159] Pagon RA, Adam MP, Ardinger HH. Rubinstein–Taybi syndrome. 2002.

[160] Hennekam RC. Rubinstein–Taybi syndrome. Eur J Hum Genet September 2006;14(9):981–5.

[161] Roelfsema JH, White SJ, Ariyurek Y, Bartholdi D, Niedrist D, Papadia F, et al. Genetic heterogeneity in Rubinstein–Taybi syndrome: mutations in both the CBP and EP300 genes cause disease. Am J Hum Genet April 2005;76(4):572–80.

[162] Tsai AC, Dossett CJ, Walton CS, Cramer AE, Eng PA, Nowakowska BA, et al. Exon deletions of the EP300 and CREBBP genes in two children with Rubinstein–Taybi syndrome detected by aCGH. Eur J Hum Genet January 2011;19(1):43–9.

[163] Bedford DC, Kasper LH, Fukuyama T, Brindle PK. Target gene context influences the transcriptional requirement for the KAT3 family of CBP and p300 histone acetyltransferases. Epigenetics January 1, 2010;5(1):9–15.

[164] Das C, Lucia MS, Hansen KC, Tyler JK. CBP/p300-mediated acetylation of histone H3 on lysine 56. Nature May 7, 2009;459(7243):113–7.

[165] Jin Q, Yu LR, Wang L, Zhang Z, Kasper LH, Lee JE, et al. Distinct roles of GCN5/PCAF-mediated H3K9ac and CBP/p300-mediated H3K18/27ac in nuclear receptor transactivation. EMBO J January 19, 2011;30(2):249–62.

[166] Bedford DC, Brindle PK. Is histone acetylation the most important physiological function for CBP and p300? Aging (Albany NY) April 2012;4(4):247–55.

[167] Kalkhoven E, Roelfsema JH, Teunissen H, den Boer A, Ariyurek Y, Zantema A, et al. Loss of CBP acetyltransferase activity by PHD finger mutations in Rubinstein–Taybi syndrome. Hum Mol Genet February 15, 2003;12(4):441–50.

[168] Murata T, Kurokawa R, Krones A, Tatsumi K, Ishii M, Taki T, et al. Defect of histone acetyltransferase activity of the nuclear transcriptional coactivator CBP in Rubinstein–Taybi syndrome. Hum Mol Genet May 1, 2001;10(10):1071–6.

[169] Chrivia JC, Kwok RP, Lamb N, Hagiwara M, Montminy MR, Goodman RH. Phosphorylated CREB binds specifically to the nuclear protein CBP. Nature October 28, 1993;365(6449):855–9.

[170] Eckner R, Ewen ME, Newsome D, Gerdes M, DeCaprio JA, Lawrence JB, et al. Molecular cloning and functional analysis of the adenovirus E1A-associated 300-kD protein (p300) reveals a protein with properties of a transcriptional adaptor. Genes Dev April 15, 1994;8(8):869–84.

[171] Hou X, Li Y, Luo RZ, Fu JH, He JH, Zhang LJ, et al. High expression of the transcriptional co-activator p300 predicts poor survival in resectable non-small cell lung cancers. Eur J Surg Oncol June 2012;38(6):523–30.

[172] Gao Y, Geng J, Hong X, Qi J, Teng Y, Yang Y, et al. Expression of p300 and CBP is associated with poor prognosis in small cell lung cancer. Int J Clin Exp Pathol 2014;7(2):760–7.

[173] Weaver DD, Graham CB, Thomas IT, Smith DW. A new overgrowth syndrome with accelerated skeletal maturation, unusual facies, and camptodactyly. J Pediatr April 1974;84(4):547–52.

[174] Basel-Vanagaite L. Acute lymphoblastic leukemia in Weaver syndrome. Am J Med Genet A February 2010;152A(2):383–6.

[175] Kirmizis A, Bartley SM, Kuzmichev A, Margueron R, Reinberg D, Green R, et al. Silencing of human polycomb target genes is associated with methylation of histone H3 Lys 27. Genes Dev July 1, 2004; 18(13):1592–605.

[176] Chou RH, Yu YL, Hung MC. The roles of EZH2 in cell lineage commitment. Am J Transl Res May 15, 2011;3(3):243–50.

[177] Wyngaarden LA, Delgado-Olguin P, Su IH, Bruneau BG, Hopyan S. Ezh2 regulates anteroposterior axis specification and proximodistal axis elongation in the developing limb. Development September 2011;138(17):3759–67.

[178] Cao R, Zhang Y. The functions of E(Z)/EZH2-mediated methylation of lysine 27 in histone H3. Curr Opin Genet Dev April 2004;14(2):155–64.

[179] Zhang J, Chen L, Han L, Shi Z, Pu P, Kang C. EZH2 is a negative prognostic factor and exhibits pro-oncogenic activity in glioblastoma. Cancer Lett January 28, 2015; 356(2 Pt B):929–36.

[180] Xu C, Hao K, Hu H, Sheng Z, Yan J, Wang Q, et al. Expression of the enhancer of zeste homolog 2 in biopsy specimen predicts chemoresistance and survival in advanced non-small cell lung cancer receiving first-line platinum-based chemotherapy. Lung Cancer November 2014;86(2):268–73.

[181] Durr A, Cossee M, Agid Y, Campuzano V, Mignard C, Penet C, et al. Clinical and genetic abnormalities in patients with Friedreich's ataxia. N Engl J Med October 17, 1996;335(16):1169–75.

[182] Delatycki MB, Paris DB, Gardner RJ, Nicholson GA, Nassif N, Storey E, et al. Clinical and genetic study of Friedreich ataxia in an Australian population. Am J Med Genet November 19, 1999;87(2):168–74.

[183] Silva AM, Brown JM, Buckle VJ, Wade-Martins R, Lufino MM. Expanded GAA repeats impair FXN gene expression and reposition the FXN locus to the nuclear lamina in single cells. Hum Mol Genet March 26, 2015.

[184] Huynen MA, Snel B, Bork P, Gibson TJ. The phylogenetic distribution of frataxin indicates a role in iron–sulfur cluster protein assembly. Hum Mol Genet October 1, 2001;10(21):2463–8.

[185] Delatycki MB, Camakaris J, Brooks H, Evans-Whipp T, Thorburn DR, Williamson R, et al. Direct evidence that mitochondrial iron accumulation occurs in Friedreich ataxia. Ann Neurol May 1999;45(5):673–5.

[186] Emond M, Lepage G, Vanasse M, Pandolfo M. Increased levels of plasma malondialdehyde in Friedreich ataxia. Neurology December 12, 2000; 55(11):1752–3.

[187] Garcia-Gimenez JL, Gimeno A, Gonzalez-Cabo P, Dasi F, Bolinches-Amoros A, Molla B, et al. Differential expression of PGC-1alpha and metabolic sensors suggest age-dependent induction of mitochondrial biogenesis in Friedreich ataxia fibroblasts. PLoS One 2011;6(6):e20666.

[188] Saveliev A, Everett C, Sharpe T, Webster Z, Festenstein R. DNA triplet repeats mediate heterochromatin-protein-1-sensitive variegated gene silencing. Nature April 24, 2003;422(6934):909–13.

[189] Greene E, Mahishi L, Entezam A, Kumari D, Usdin K. Repeat-induced epigenetic changes in intron 1 of the frataxin gene and its consequences in Friedreich ataxia. Nucleic Acids Res 2007;35(10):3383–90.

[190] Castaldo I, Pinelli M, Monticelli A, Acquaviva F, Giacchetti M, Filla A, et al. DNA methylation in intron 1 of the frataxin gene is related to GAA repeat length and age of onset in Friedreich ataxia patients. J Med Genet December 2008;45(12):808–12.

[191] Al-Mahdawi S, Sandi C, Mouro Pinto R, Pook MA. Friedreich ataxia patient tissues exhibit increased 5-hydroxymethylcytosine modification and decreased CTCF binding at the FXN locus. PLoS One 2013;8(9): e74956.
[192] Herman D, Jenssen K, Burnett R, Soragni E, Perlman SL, Gottesfeld JM. Histone deacetylase inhibitors reverse gene silencing in Friedreich's ataxia. Nat Chem Biol October 2006;2(10):551–8.
[193] De Biase I, Chutake YK, Rindler PM, Bidichandani SI. Epigenetic silencing in Friedreich ataxia is associated with depletion of CTCF (CCCTC-binding factor) and antisense transcription. PLoS One 2009; 4(11):e7914.
[194] Punga T, Buhler M. Long intronic GAA repeats causing Friedreich ataxia impede transcription elongation. EMBO Mol Med April 2010;2(4):120–9.
[195] Kumari D, Biacsi RE, Usdin K. Repeat expansion affects both transcription initiation and elongation in Friedreich ataxia cells. J Biol Chem February 11, 2011;286(6):4209–15.
[196] Kelly M, Bagnall RD, Peverill RE, Donelan L, Corben L, Delatycki MB, et al. A polymorphic miR-155 binding site in AGTR1 is associated with cardiac hypertrophy in Friedreich ataxia. J Mol Cell Cardiol November 2011;51(5):848–54.
[197] Mahishi LH, Hart RP, Lynch DR, Ratan RR. miR-886-3p levels are elevated in Friedreich ataxia. J Neurosci July 4, 2012;32(27):9369–73.
[198] Sandi C, Sandi M, Anjomani Virmouni S, Al-Mahdawi S, Pook MA. Epigenetic-based therapies for Friedreich ataxia. Front Genet 2014;5:165.
[199] Soragni E, Miao W, Iudicello M, Jacoby D, De Mercanti S, Clerico M, et al. Epigenetic therapy for Friedreich ataxia. Ann Neurol October 2014;76(4):489–508.
[200] Ma Q, Lu AY. Pharmacogenetics, pharmacogenomics, and individualized medicine. Pharmacol Rev June 2011;63(2):437–59.
[201] Collins FS, Varmus H. A new initiative on precision medicine. N Engl J Med February 26, 2015;372(9):793–5.
[202] Garcia-Gimenez JL, Sanchis-Gomar F, Lippi G, Mena S, Ivars D, Gomez-Cabrera MC, et al. Epigenetic biomarkers: a new perspective in laboratory diagnostics. Clin Chim Acta October 9, 2012;413(19–20):1576–82.
[203] Sandoval J, Peiro-Chova L, Pallardo FV, Garcia-Gimenez JL. Epigenetic biomarkers in laboratory diagnostics: emerging approaches and opportunities. Expert Rev Mol Diagn June 2013;13(5):457–71.

CHAPTER

4

The Evolution of New Technologies and Methods in Clinical Epigenetics Research

Tandy L.D. Petrov, Nicole C. Riddle

Department of Biology, The University of Alabama at Birmingham, Birmingham, AL, USA

OUTLINE

Epigenetic Biomarkers and Diagnostics
http://dx.doi.org/10.1016/B978-0-12-801899-6.00004-8

1. INTRODUCTION

Traditionally, clinical researchers and medical practitioners consider genetic predispositions, environmental risk factors, and the interactions between these two forces as causes of human disease. Most conditions encountered by physicians and clinical researchers fall somewhere between these two extremes. A small number of human disease conditions are due to genetic mutations, i.e., hemophilia A, and thus fall on one end of the spectrum. Others are mostly due to environmental conditions, i.e., type 2 diabetes, and thus are on the other end of the spectrum. However, for many conditions, genetics predisposes a person to a disease condition if the right combination of environmental factors arise, and thus, they are somewhere in the middle between these extreme cases. While this model of disease, as a combination of genetic and environmental factors, has a long history, recently, it has become clear that this model is no longer sufficient and that epigenetic effects have to be incorporated into the model as well.

Epigenetics is the field of research that studies heritable changes in phenotype that are not caused by a change in deoxyribonucleic acid (DNA) sequence (genotype) [1]. Thus, epigenetics provides a second information system in addition to DNA that can store information in what is called the epigenotype. Phenotypes linked to epigenetics range from molecular phenotypes such as gene expression to organism-level characteristics such as coat color, eye color, or flower shape [2–5]. Well-known examples of epigenetic phenomena include variegated gene expression of transgenes due to their position in the genome (position effect variegation (PEV) in *Drosophila* [6], yeast [7], and mouse [8]), parent-of-origin-specific expression of imprinted genes in mammals and plants [9,10], and X-inactivation in female mammals [11]. Just as the phenotypes affected by epigenetic processes are very diverse, so are the molecular mechanisms that serve as underpinnings for these phenomena. These molecular mechanisms include DNA methylation, chromatin structure—including histone modifications and chromosomal proteins—and noncoding ribonucleic acids (ncRNAs) to name a few [1]. Together, these epigenetic mechanisms contribute to an organism's phenotypes in ways that we are only now beginning to appreciate.

Generally speaking, epigenetic processes tend to impact gene expression, which in turn leads to changes in protein levels and eventually the organism-level phenotypes. Because it is malleable—and not as stable as the DNA sequence—epigenetic information provides a link between the genotype and the environment. When gene expression changes in response to environmental cues, the epigenetic information or epigenotype changes, reflecting the new gene expression status [12–15]. However, it is not clear if the epigenotype changes precede the gene expression changes, or if the gene expression changes occur prior to the changes in the epigenotype. Regardless of the order of events, the epigenotype has the potential to contain useful information about an organism's current gene expression state and past expression history, possibly reflecting environmental exposures, age, and disease state.

Researchers have begun to investigate the correlation between various external life events and specific epigenetic information because of the links between epigenotype, gene expression, and environment. Clinical researchers are interested specifically in epigenetic changes that distinguish normal and diseased tissues, in how the epigenome responds to environmental stimuli, and in how epigenetic information might be used for the diagnosis, therapy, or prevention of human diseases. These questions are the focus of clinical epigenetics studies, which have increased immensely since 2000—searching the PubMed database reveals only two citations in a search for "clinical epigenetics" prior to 2000, but more than 420 in the first 10 months of 2014 (Figure 1). This increase in publications partially reflects the intensifying interest in epigenetics that occurred over the last 15 years. However, it also reflects the astonishing development in methods and technologies for the study of epigenetic systems over the same time period; methods that for the first time make studies with many human samples possible.

In this chapter, we will review the major epigenetic information systems and discuss the various methods used to study them. While epigenetics has been the focus of study for decades, the number and types of methods available have changed dramatically over the last 15 years, possibly due to the increased interest in the field (see Figure 2, time line). We will describe the new technologies used for the study of epigenetic information systems and compare them to classical approaches. The utility of each method will be illustrated with relevant examples, and their use in clinical research and whenever applicable in the clinical setting will be discussed. We will end by giving an outlook on how we expect these technologies will move to the clinic, focusing on remaining obstacles and highlighting relevant opportunities.

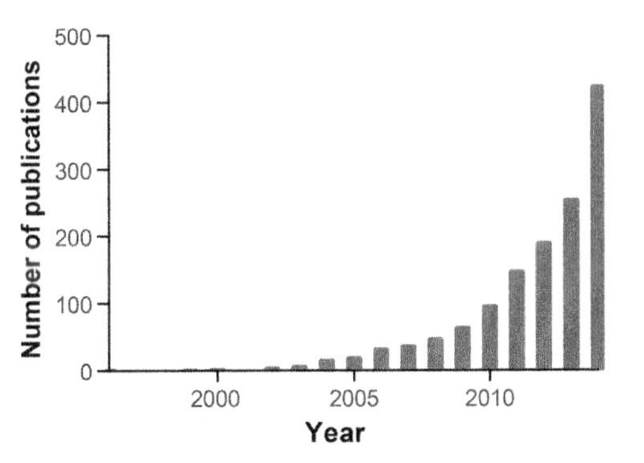

FIGURE 1 Publications matching the search term "Clinical Epigenetics Research" in PubMed have increased dramatically over the last 15 years.

2. DNA METHYLATION

One of the most highly studied epigenetic modifications is cytosine methylation, referred to often simply as DNA methylation [16]. There are other bases of the DNA that can be methylated, such as adenine in bacteria, but in mammals, cytosine is the most often modified DNA base [17,18]. Cytosine methylation is produced by the addition of a methyl group to the 5-position of cytosine residues through the action of DNA methyltransferase enzymes. Removal of methyl group can be achieved by several different pathways [19]. These pathways include DNA glycosylases, which remove the methylated base by base excision repair [20–23], and TET proteins, which produce various derivatives of 5-methylcytosine (5mC) such as 5-hydroxymethylcytosine (5hmC), 5-formylcytosine (5fC), and 5-carboxylcytosine (5caC) [24,25]. These 5mC derivatives have been discovered only recently, and thus, older studies mostly provide a composite picture of all 5mC derivatives.

DNA methylation in general is well characterized, and it plays essential roles in gene expression regulation, development, and cellular identity [19]. In mouse, for example, DNA

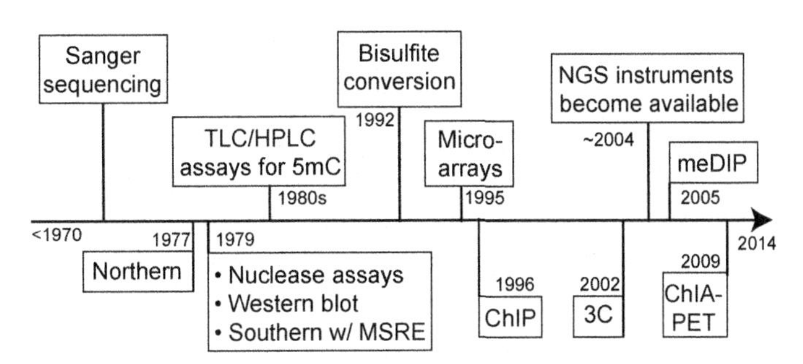

FIGURE 2 Time line for the methods development of commonly used techniques in epigenetics research from the 1970s to the present. For abbreviations, please see the list of abbreviations.

methylation is essential for survival, as mutants die during early embryonic development [26]. Because of its role in gene regulation, DNA methylation can serve as indicator of genome activity [27]. 5mC patterns inherent in the DNA contain information about the expression status of individual genes, which can be assayed instead of RNA. For clinical researchers, identifying sites of DNA methylation that are predictive for disease is of great interest as often the biological samples available to them are not suitable for RNA studies.

2.1 Composite Measures of DNA Methylation

Because DNA methylation has been studied since the 1970s [28], there are many methods to assay it. These methods differ significantly in the types of data they generate, in their cost, and in their sample requirements. The most basic type of data that can be obtained is a composite measure of the amount of 5mC that is found in the genome. Classically, this information was obtained by thin-layer chromatography (TLC) or high-performance liquid chromatography (HPLC) analyses of DNA digested into single nucleotides [29,30]. Alternatively, this information can also be obtained by mass spectrometry methods (for example, see Ref. [31]). More recently, the percent of 5mC present in a genome is assayed using antibodies to 5mC in enzyme-linked immunosorbent assays (ELISAs) (for an early example, see Ref. [32]). Due to their popularity, kits are available from several vendors. This popularity is due to the simple, short (~3h) protocol, which uses detection with a spectrophotometer, the possibility of high throughput in a 96 well format, and the relatively low cost. In addition, these new ELISAs only require about 50–100 ng of DNA, which is a significant improvement compared to the traditional TLC and HPLC assays. Thus, with chromatography, mass spectrometry, and ELISA, there are three reliable methods available to determine bulk DNA methylation levels from samples of interest.

Due to the improvements of these methods, bulk DNA methylation assays have become more attractive for clinical research. In particular, the lower amount of starting material and the potential for high-throughput analysis are important for clinical research applications. While information about composite 5mC levels is somewhat limited, it can be useful. Various studies use these assays to determine the impact of environmental exposures or drug treatment in general on the epigenotype or investigate the correlation between disease phenotypes/prognosis with global methylation levels. Thus, often these methods are used as screening tools and for follow-up experiments to investigate the source of any changes observed globally.

2.2 Restriction Enzyme-Based DNA Methylation Assays

The second class of DNA methylation assays evaluates DNA methylation at specific genomic loci. Several of these assays exploit the properties of methylation-sensitive restriction enzymes. Most of these enzymes will be inhibited by the presence of DNA methylation at their recognition site; however, some require the presence of DNA methylation for successful cutting of the DNA. Southern blots are the traditional way to assay single loci for DNA methylation [33]. These assays use methylation-sensitive (inhibited by 5mC) and methylation-insensitive isoschizomer enzymes to determine if DNA methylation is present at a specific locus. DNA is digested with both enzymes, separated according to size by gel electrophoresis, and transferred to a membrane. The membrane is then probed for a locus of interest, and if no DNA methylation is present, the restriction pattern for the two enzymes is identical. However, if 5mC is present, the pattern will differ, with the sensitive enzyme showing less digestion and larger DNA fragments than the insensitive enzyme [34]. This method is very reliable, but labor-intensive, requires large amounts of DNA (usually ~5 μg or more), and is low throughput as a single locus is assayed at a time. Southern blot analysis with methylation-sensitive restriction enzymes was used, for example, to screen for mutations affecting 5mC in *Arabidopsis thaliana* [35], and the technique was instrumental for the early work on DNA methylation.

After the advent of polymerase chain reaction (PCR), a new method combining methylation-sensitive enzymes with PCR was developed to assay 5mC [36]. Here, PCR primers flank the restriction site one wants to assay for the presence of DNA methylation. DNA digested with either a methylation-sensitive or methylation-insensitive restriction enzyme isoschizomer serves as template. The amount of amplification observed for the two reactions is compared to judge the amount of methylation present. Differences in amplification are observed between the two samples only if a site is methylated. The great advantage to this assay is that it is easily carried out with large numbers of samples, and that only 100–300 cells are required [36]. This assay was used in combination with Southern blot analysis to demonstrate the presence of a methylation imprint at the *Igf2r* locus in mouse [37]. Despite its advantages, however, the assay was still restricted to the analysis of a single locus.

A more comprehensive analysis of 5mC is possible with MSAP (methylation-sensitive amplification polymorphism) assays [38]. MSAP does not require a sequenced genome and is derived from amplified fragment length polymorphism (AFLP) protocols. It utilizes digestion by methylation-sensitive restriction enzymes, followed by PCR fragment amplification after adapter ligation. With this method, a large number of fragments can be assayed simultaneously. Due to the fact that MSAP does not require a sequenced genome, it is used often in nonmodel species. MSAP has thus been applied to species ranging from octopus [39] to maize [40] and will likely remain popular as a cost-efficient tool in nonmodel organisms.

2.3 Traditional Bisulfite Conversion-Based Assays

An immense step forward came with the development of the bisulfite conversion method by Frommer and colleagues in 1992 [41]. This method exploits the fact that treatment of DNA with bisulfite leads to the deamination of cytosines and their conversion to uracil, and that the methyl group of 5mC inhibits this reaction (the reaction is about two order of magnitudes slower [42]). Thus, treated DNA can be amplified by PCR, the fragments cloned and subsequently sequenced, which allows one to compare the sequence from treated DNA to that of untreated DNA and identify individual cytosines that are methylated (C in both sequences) or unmethylated (C in the untreated sequence, T/U from the bisulfite-treated sample) [41]. By sampling different clones, this method reports percent methylation for each individual cytosine. One drawback is that the original protocols required large amounts of DNA; however, improved protocols and kits today allow for the use of as little as 100 pg. This method made it possible for the first time to assay any chosen cytosine residue at single-base resolution.

From this basic method, numerous other techniques have been derived. Methylation-specific PCR (MSP) employs primers specific to the DNA sequences expected after bisulfite treatment for methylated versus unmethylated cytosines to determine the presence of methylated versus unmethylated sequences within a sample [43]. Another method based on the bisulfite conversion is COBRA (combined bisulfite restriction analysis) [44], which uses unbiased primers to PCR amplify DNA fragments from bisulfite-treated samples independent of their methylation status. The amplified fragments are then assayed by restriction enzyme digestion assaying for either retention of restriction sites due to the presence of methylation or the generation of a new restriction site due to the cytosine to uracil conversion [44]. A third method, dependent first on a bisulfite treatment, is methylation-sensitive single-nucleotide primer extension (Ms-SNuPE) [45]. As in COBRA, a region of interest is amplified by PCR from bisulfite-treated PCR using unbiased primers. Then, the PCR fragments are denatured and annealed to a primer adjacent to the cytosine to be assayed. Next, the primer is extended by a single, radiolabeled nucleotide, either cytosine or thymine. If the cytosine was methylated in the original sample, extension will occur with the cytosine; if no DNA methylation was present, amplification will occur with the thymine. Intermediate levels of DNA methylation at a site will result in partial signal in both reactions [45]. Pyrosequencing has been used also to examine bisulfite-treated DNA to determine levels of methylation at individual cytosines [46], as has MS-HRM (methylation-sensitive high-resolution melting) [47,48]. In MS-HRM, the melting behavior of DNA fragments amplified from bisulfite-treated samples is observed using fluorescent detection similar to the detection used for quantitative real-time PCR [47]. Compared to the original bisulfite conversion sequencing method, developed by Frommer and colleagues, the derivative methods are less labor-intensive, more suitable for high throughput, and tend to require less starting materials, characteristics that are particularly important for clinical research.

2.4 Genome-Wide DNA Methylation Assays

A new level in DNA methylation analysis was reached with the development of genome-wide, single cytosine resolution methods that utilize either array technology or next-generation sequencing (NGS). Two major types of genome-wide DNA methylation assays are currently in use, those methods based on bisulfite conversion, and those based on immunoprecipitation reactions. The simplest genome-wide bisulfite-based method is BS-seq (bisulfite sequencing), bisulfite treatment of DNA followed by NGS [41]. This method allows the researcher to obtain methylation information for every cytosine in the genome, but is very expensive and some feel wasteful, as in human samples, for example, the majority of cytosines are unmethylated, and thus a large amount of money is spent getting no new information. This concern has lead to the development of RRBS (reduced representation bisulfite sequencing), where the sample is enriched for methylated regions of the genome prior to NGS analysis [49]. By digestion of genomic DNA and size selection, a smaller fraction (1–5%) of the genome can be assayed by NGS while still obtaining results representative of the genome as a whole. Another improvement is that this method can be adjusted to work with as little as 30 ng of genomic DNA to obtain genome-wide methylation profiles from clinical samples [50].

In addition, several additional methods based on the BS-seq concept have been developed that begin to address the problem that 5mC, 5hmC, 5fC, and 5caC cannot be distinguished in the traditional bisulfite protocols. In 2012, Booth and colleagues describe a technique called oxBS-seq (oxidative bisulfite sequencing), which can distinguish 5mC and 5hmC [51]. While 5mC and 5hmC are resistant to bisulfite, 5fC is converted to uracil. By comparing the results from regular BS-seq and a DNA sample that has undergone a specific oxidation reaction to convert 5hmC to 5fC, one can precisely map 5hmC with single-base pair resolution [51,52]. An alternative technique, TAB-seq (Tet-assisted bisulfite sequencing) was developed at the same time by Yu and colleagues [53]. In 2014, Booth and colleagues also described a method for the precise mapping of 5fC, redBS-seq (reduced BS-seq), which first reduces 5fC in a DNA sample to 5hmC prior to bisulfite treatment and sequencing (redBS-seq [52]). By combining BS-seq, oxBS-seq, and redBS-seq, it is possible to map 5mC,

5hmC, and 5fC in any genome [52]. Thus, these technologies combining bisulfite treatment with NGS method provide unprecedented opportunities to investigate DNA methylation.

While the NGS methods provide the most comprehensive view of DNA methylation, they are costly, especially for large mammalian genomes. As in mammals, much of the DNA methylation is nonrandomly distributed in the genome, several array-based methods remain popular to assay DNA methylation. These include the Illumina Infinium methylation assay [54] and the Illumina GoldenGate methylation assay (recently reviewed in Ref. [55]). The Illumina GoldenGate methylation assay is an adaption of a high-throughput single nucleotide polymorphism (SNP) genotyping system that allows for the assessment of 1536 CpG sites to be assayed simultaneously and can be carried out with as little as 200ng of bisulfite-converted DNA [56,57]. The CpG sites assayed are customizable, and this assay is a good option if an intermediate number of loci needs to be assayed. In contrast, the Illumina Infinium methylation arrays, which are also an adaptation of a high-throughput genotyping system, interrogate a much larger number of loci [57–59]. The current array, the HM-450k BeadChip, assays more than 485,000 methylation sites and requires approximately 500ng of bisulfite-converted DNA [60]. Both of these array-based methods provide a cost-efficient alternative for large genomes, while allowing researchers to still monitor a significant portion of the genome.

The second type of genome-wide DNA methylation assays is based on immunoprecipitation. Antibodies against DNA modifications can be used to isolate DNA fragments containing the modification. This idea is the basis for meDIP (methylated DNA immunoprecipitation), where DNA fragments containing 5mC are isolated by immunoprecipitation with an anti-5mC antibody [57,59,61]. Specific regions of interest are assayed by quantitative PCR (qPCR), arrays, or NGS [62]. This method has been successfully used for 5mC, 5hmC, and also has used antibodies to methyl-binding domain proteins to isolate regions of high 5mC content [63–65]. Recently, this method has been adapted for small cell numbers [66]. One limitation of this technique is its resolution—immunoprecipitated DNA fragments are typically several hundred base pairs, and thus contain numerous cytosines. Thus, the immunoprecipitation-based methods for the analysis of DNA methylation cannot provide single-nucleotide resolution.

2.5 Applications in Clinical Research and Beyond

In clinical research, many of the methods for DNA methylation analyses are used, depending on the research question and goals. The main limitations are often the amount of sample available for analysis and cost. Many cancer studies conduct genome-wide surveys of DNA methylation to identify aberrant methylation that contributes to tumorigenesis. This approach is illustrated by a recent publication from the Cancer Genome Atlas Research Network, which profiled lung adenocarcinomas integrating messenger RNA, microRNA, DNA sequencing, copy number, methylation, and proteomic analyses to gain a comprehensive picture of the molecular changes occurring in these cancer tissues [67]. In contrast to this large-scale study, many human studies focus on a select number of target loci and thus utilize MSP, the Illumina platforms, or HS-MRS to study DNA methylation.

Additional limitations exist that constrain the use of the above technologies in the clinic. Cost and the amount of sample required are relevant here as well, but in addition, any test used in the clinical setting has to be easily interpretable. At this point, DNA methylation assays are not done routinely in the clinical setting. The recently FDA-approved Cologuard noninvasive colon cancer screening test assays two methylated DNA targets, *NDRG4* and *BMP3* [68]. As far as we know, this test represents the first widely applied DNA methylation assay that has progressed into the clinical setting. Given the prevalence of DNA methylation biomarkers in a variety of human disease conditions and particularly cancer [69], we anticipate that DNA methylation assays will become much more prevalent in the near future.

3. HISTONE MODIFICATIONS

Posttranslational modifications (PTMs) of histones, the protein components of the nucleosome that eukaryotic DNA is wrapped around, are another set of molecular marks encoding epigenetic information. These PTMs were first discovered in the 1960s by the lab of Vincent Allfrey [70]. Since then, an impressive catalog of histone modifications has been compiled [71]. Histones can be modified by methylation, acetylation, phosphorylation, and many other modifications [72,73]. PTMs of histones primarily affect the nucleosome and chromatin structure, but the impacts are far-reaching, influencing cellular processes such as DNA repair, replication, recombination, and gene regulation [74]. As suggested by their many functions, improper localizations of histone modifications are observed in a variety of human diseases [75]. Understanding their impact on gene expression and other cellular processes is an important topic of study in epigenetics research and of critical interest to clinical researchers focused on understanding the contribution of epigenetics to various disease processes.

3.1 Global Levels of Histone Modifications

The first available techniques to study histone modifications were designed to report bulk histone modification levels from the genome. One of these techniques is Western blot analysis, developed during 1979 by Harry Towbin [76]. Western blot analysis detects specific proteins using antibodies to the protein of interest after the proteins have been separated by size using SDS-PAGE and transferred to a membrane [77]. The Western blot method can be applied to any protein or protein modification for which a reliable antibody is available. In epigenetics research, antibodies to a variety of chromosomal proteins, including histone-modifying enzymes, are used in addition to antibodies against histone modifications. In particular for histone modifications, antibody availability and quality has been an issue in the past [78]. Today, reliable, well-characterized monoclonal antibodies are available for the study of many histone modifications by Western blot analysis. They are used extensively in the research setting investigating a wide variety of research questions including studies regarding the response of histone PTMs to genetic mutations (for example, see Ref. [79]) or environmental factors (for example, see Ref. [80]). If information about bulk levels of histone PTMs is needed, Western blot analysis remains a popular technique as it does not require specialized equipment and has proven to be reliable and efficient.

A second technique for the detection of histone modifications is immunohistochemistry, where antibodies to the PTM of interest are used to stain cells or tissue sections. Levels of PTMs can be estimated using this technique by determining the percentage of cells for which a positive signal is observed [81]. Immunohistochemistry is of particular interest to clinical epigenetics researchers, as it can be applied to paraffin-embedded tissue samples [82]. While this technique can be applied to samples often readily available, it has lower sensitivity than other methods and accurate quantification of PTMs is difficult. In addition, immunohistochemical methods tend to be time-consuming. Overall, immunohistochemistry is a good initial screening tool; however, additional studies with other methods are often required.

Mass spectrometry is another technique that has been used to evaluate bulk histone PTMs. However, while mass spectrometry can provide quantitative information, in the study of histone PTMs, it has often been used for discovery purposes. Mass spectrometry is an analytical technique that allows for identification of the types of molecules in a sample by measuring mass-to-charge ratios and abundance of gas-phase ions [83,84]. Mass spectrometry analysis first revealed that individual histones can have multiple modifications on their N-terminal tails [85–87]. Mass spectrometry also first detected the subtle differences between mono-, di-, or trimethylation at individual histone residues [88–90]. Improvements in mass spectrometry techniques have led to the continued discovery of additional histone modifications with histone lysine crotonylation being one of the most recent [91]. Thus, mass spectrometry has been instrumental in producing a comprehensive catalog of histone PTMs.

Despite the power of mass spectrometry analysis, some drawbacks do exist, the foremost being the equipment required. Mass spectrometry requires large, very expensive equipment, a large amount of computing power, and highly trained technical experts. However, in contrast to Western blot analysis, it does not require specialized reagents such as antibodies, it is unbiased in the analysis of PTMs, and it typically requires relatively small samples (pg of protein). With mass spectrometry, it is possible to assay many PTMs simultaneously. This is particularly important for clinical research, as it allows for one to gain a large amount of information from relatively small samples.

In recent years, ELISA kits have become available to obtain global levels of histone PTMs. These kits determine the amount of modified histones present in the sample via the interaction with an antibody and quantification by fluorometry or colorimetry. These assays can be carried out in a 96 well format, are quick (less than 5h), require relatively small samples (1–2μg of total protein), and fairly cost-efficient if large numbers of samples need to be evaluated. These assays are essentially the same used for the detection of global levels of 5mC described in Section 2.1 and are often used due to their convenience. Depending on their specific needs, researchers have four reliable methods to assay global levels of histone PTMs: Western blot analysis, immunohistochemistry, mass spectrometry, and ELISA-based assays.

3.2 Assays for Histone Modifications that Provide Location Information

In addition to the techniques reporting global levels of histone PTMs, several methods exist that will provide information for specific loci or provide location information in addition to quantity. The most basic of these methods is immunohistochemistry [92]. In addition to serving as screening tool to evaluate overall levels of PTMs as described in the previous section, immunohistochemistry can be used to determine the nuclear localization of the PTM by examining the distribution of fluorescent signal in the nucleus (for example, see Ref. [93]). This method has been exploited effectively in *Drosophila*, where large polytene chromosomes from the salivary gland can be examined in great detail, revealing localization of histone PTMs to specific chromosome arms or regions of the chromosomes [94]. While this method works well for *Drosophila*, the resolution is fairly low in other types of cells, and so this method is most often used as an initial screening method to be followed up with a higher resolution assay.

Chromatin immunoprecipitation (ChIP) is such a high-resolution assay. ChIP in general is used to assay the specific DNA sequences associated with a protein of interest and can be applied to histone modifications and chromosomal proteins [95,96]. ChIP protocols involve the cross-linking of DNA to its associated proteins with formaldehyde followed by sonication to produce small chromatin fragments (usually 200–500bp). The cross-linked DNA–protein complexes that contain the protein of interest are then isolated using an antibody specific to the protein of interest. Following this immunoprecipitation, the DNA associated with the protein of interest is isolated from the immunoprecipitated DNA–protein complexes after reversing the cross-linking and removal of the protein [95,96]. The resulting DNA sample contains all regions of the genome that were associated with the protein of interest. Thus, ChIP can provide specific location information for a protein of interest such as a posttranslationally modified histone.

The DNA recovered from ChIP can be analyzed by several methods, depending on the research question. If only few loci need to be assayed, usually, ChIP is followed by qPCR, resulting in a quantitative measure of the assayed protein [97]. If genome-wide data are desired, ChIP can be followed by microarray analysis (ChIP-chip [98]) or NGS (ChIP-seq [99–101]). Both of these assays can be used to produce a genome-wide landscape of individual histone modifications, and they have been used extensively to address a variety of research questions. ChIP has also been used by several large consortium projects, among them ENCODE and modENCODE, to characterize the histone modification landscape in detail in multiple tissues and species [102–105]. A recent publication based on the data from these consortia, for example, compares the distribution of histone modifications in *Caenorhabditis elegans*, *Drosophila melanogaster*, and humans [106]. They find that while many histone modifications are used in a similar manner by the species examined, some species-specific usages occur [106]. These studies reveal the power of the ChIP approach and the detailed knowledge that can be gained from these experiments about histone modifications.

Just as other methods, ChIP does have some limitations; ChIP depends on antibodies, and thus antibody quality [78]. Results from early uses of ChIP are especially difficult to interpret as our knowledge of histone modifications was incomplete, and thus antibody testing typically was insufficient to reveal all cross-reactivity. Another limitation is the large number of cells required for the most robust protocols (often ~1×10^7 cells). However, newer protocols exist that work for smaller number of cells. Microchip protocols are reported to perform well with ~10,000 cells [107–109], and a recent 2014 publication reports ChIP-seq data from samples with only 500 cells [110]. With the newer protocols optimized for small numbers of cells, ChIP is an excellent method for determining the location of specific histone modifications even if only small samples are available.

3.3 Applications in Clinical Research and Beyond

Today, Western blot analysis, immunohistochemistry, mass spectrometry, ELISAs for histone modifications, and ChIP analysis are routinely used to address a wide range of research questions, including the field of clinical epigenetics research. While mass spectrometry analyses typically are performed at local core facilities that serve entire academic institutions, the other assays for histone modifications are usually performed in individual laboratories. The choice of method typically depends on the research question, i.e., if it is important to capture a large number of histone PTMs or if one specific modification is of interest, if locus-specific information is sufficient, or if genome-wide data are required.

To address questions in clinical epigenetics research, ChIP profiles have been generated for a variety of histone modifications from numerous disease states [111,112]. Especially from cancer tissues, a large number of genome-wide ChIP profiles for histone modifications have been generated [112]. But the role of histone modifications has also been investigated for a number of other disease conditions. For example, in a recent study of autoimmune disease, genetic and epigenetic data were combined to get a more comprehensive understanding. Specifically, ChIP data for H3K27ac, a mark of active enhancers and promoters, were used to reveal the role of SNPs detected in the genetic analysis. The ChIP data allowed the authors to classify their candidate SNPs and revealed that many of the noncoding, disease-associated SNPs are associated with active enhancers [113]. This example illustrates the utility of histone modification information for clinical research, and the powerful analyses that can be performed if the epigenetics data are integrated with other data types.

While histone modifications are studied extensively in clinical research, genome-wide ChIP analysis from clinical samples is often impossible due to the limited amount of tissue/cells available. The newer micro-ChIP protocols mentioned above will alleviate this problem, but currently, they are not employed widely. In addition, genome-wide analyses are costly, and processing a large number of human samples is often not possible. Until the micro-ChIP protocols become more robust and costs drop further, histone modification analyses are unlikely to be employed in the clinical setting. Currently, if histone modification analyses are required in the clinical setting, they often rely on semiquantitative analyses by immunohistochemistry [81,82].

4. CHROMATIN STRUCTURE

Chromatin structure has been the subject of study for over 100 years, since the discovery that a fraction of most genomes decondenses and stains lightly during interphase, while the remainder of the genome remains highly condensed and stains darkly [114]. This initial characterization of two chromatin structures, heterochromatin and euchromatin, based on their response to DNA-specific dyes, has been augmented by additional studies using a variety of techniques. The techniques to study chromatin structure are varied and include genetic assays (PEV), biochemical assays, immunohistochemistry, biophysical studies, and microscopy [115]. Here, we will introduce the most commonly used methods that work well independent of the organism under study.

4.1 Accessibility Assays

Chromatin structure regulates the accessibility of DNA sequences by the transcriptional machinery. Similar to this process, chromatin structure also modulates accessibility of DNA sequences to other compounds such as nucleases and various chemicals. For example, the DNA that is wrapped around the nucleosome is protected from nucleases while the DNA that is present in the linker regions between nucleosomes can be degraded [116]. Thus, accessibility can serve as a tool to study chromatin structure.

In a typical accessibility assay, isolated chromatin—not just DNA—is exposed to a nuclease. This nuclease can be a specific restriction endonuclease, or a more general nuclease such as micrococcal nuclease or deoxyribonuclease (DNase) I [115]. The nuclease is then allowed to partially digest the chromatin, with the timing optimized to reach intermediate digestion levels that can reveal differences in accessibility across the genome. The classical method assays DNA from the digested sample and various control samples (e.g., undigested chromatin, isolated DNA digested with the nuclease) with Southern blots, probing for a single specific locus of interest. Comparing a locus in euchromatin to a locus in heterochromatin, more digestion (and smaller fragments) is seen for the euchromatic than the heterochromatic locus (for example, see Ref. [117]). Thus, this type of assay reveals relative measures of accessibility.

While this classical, Southern-based assay has been used very successfully, one is usually limited to assaying a small number of loci. Extension of the accessibility assay to the entire genome by combining the assay with NGS (DNase-seq [118]) or array (DNase-chip [119]) analysis has made these assays much more powerful in addition to decreasing the sample amount required. DNase-seq uses DNase I as the nuclease, which preferentially digests so-called hypersensitive sites that lack nucleosomes. DNA isolated after digesting with DNase I is subjected to NGS analysis, revealing lower read counts in hypersensitive regions, as the DNA there has been digested away [120,121]. Applied to the whole genome, this assay is very powerful and can be used to reveal regulatory regions in any genome from relatively small samples.

A slightly different strategy is taken by the formaldehyde-assisted isolation of regulatory elements (FAIRE)-seq [122,123]. To produce a FAIRE sample for NGS analysis, chromatin is treated with formaldehyde to cross-link DNA and proteins. After sonication to fragment the chromatin, DNA is extracted from the samples by phenol–chloroform extraction. Because most DNA is associated with histones and other proteins, the majority of sequences will be retained in the phenol fraction. Only sequences such as the DNase I hypersensitive sites that lack histones will be enriched in the aqueous phase and extracted for sequencing, thus revealing regulatory regions of the genome [122]. Thus, FAIRE-seq experiments produce high read counts in areas of open chromatin, but are depleted for sequence reads from regions of closed chromatin such as the centromeres.

Overall, the FAIRE-seq approach is complementary to the DNase-seq method, and they have been used successfully together to produce a comprehensive picture of the accessible regions in the human genome [124]. A third method of note that is similar to FAIRE-seq and DNase-seq is sonication of cross-linked chromatin sequencing (Sono-seq), where DNA isolated from formaldehyde cross-linked, sonicated chromatin samples is analyzed by NGS, which also reveals regions of open chromatin [125]. All three methods, DNase-, FAIRE-, and Sono-seq have the advantage that they do not depend on antibodies or other organism-specific resources, but are applicable to any species under study.

4.2 3D Chromatin Packaging

Accessibility measures only provide one measure of chromatin structure. In a nucleus, chromatin fibers are interacting with each other in a 3D structure, which is difficult to assay. The most commonly used methods to detect contacts between chromatin fibers are chromatin confirmation capture (3C) methods [126] and its derivatives 4C, 5C, Hi-C [127,128]. 3C uses formaldehyde cross-linked chromatin as the starting material. After cross-linking, the chromatin is digested with an appropriate restriction enzyme and then undergoes intramolecular ligation. This ligation reaction will result in the preferential ligation of DNA fragments that are part of the same cross-linked DNA–protein complex, irrespective of their location on the linear DNA strands. If two far-away genomic loci interact, this can now be detected by qPCR on the new, ligated DNA fragments [126]. Thus, by examining the frequency of new ligation products, one can estimate the frequency of any two loci of interest.

The original 3C protocol was developed to interrogate interactions between specific loci of interest. Since then, several derivative methods have been developed that allow for the assessment of DNA interactions on larger scales—up to the entire genome. Two types of 4C protocols exist, 3C-on-chip [129] and circular chromosome conformation capture [130]. Both protocols extend the 3C technology to larger portions of the genome, screening the genome for sequences that interact with one specific locus ("one versus all" strategy [127]). 5C (3C-carbon-copy) combines 3C with highly multiplexed ligation-mediated amplification, which then allows for the assessment of the interaction libraries with array or NGS [131,132]. Specifically, 5C allows the assessment of interactions between many different loci ("many versus many" strategy [127]).

The HiC method is a modification of the 3C method that utilizes NGS to implement a genome-wide interaction screen [133] ("all versus all" strategy [127]). Thus, with these various methods based on 3C, researcher has options that allow them to study chromatin structure at various levels, from a single locus to genome-wide, at much higher resolution than possible with microscopy-based techniques.

ChIA-PET, chromatin interaction analysis using paired end tag sequencing, is one other commonly used method to assay chromatin structure at high resolution [134]. This method is similar to the 3C-derived methods, but prior to the ligation step, a ChIP step is introduced in the protocol. After this immunoprecipitation, only chromatin interactions mediated by one specific protein are assayed, leading to the recovery of a specific subset of interactions compared to the 3C-type methods. Thus, ChIA-PET experiments provide structural data as well as some mechanistic insights into what might be causing the chromatin interactions observed. Together, ChIA-PET and 3C with its derivatives give epigenetics researchers unprecedented new opportunities to gain insights into higher level chromatin packaging.

4.3 Applications in Clinical Research and Beyond

The use of chromatin structure assays in clinical research is not as widespread as the study of DNA and histone modifications, possibly due to the fact that many of the protocols are somewhat more complex and convenient kits are not available, making it more challenging for new researcher to enter this area of research. In addition, chromatin structure assays tend to require more tissue than DNA or histone modification assays, and thus can be difficult for researchers to justify with limited sample availability. Despite these challenges, chromatin structure studies can generate important insights for clinical researchers. A recent ChIA-PET study, profiling the physical interactions mediated by six factors in human cells, revealed that these interactions included over 90% of transcription start sites, suggesting that specific interactions are widespread and involve significant portions of the genome [135]. These results—and those from other pioneering studies—suggest that the integration of chromatin structure studies into clinical epigenetics research will lead to novel insights that are relevant to human disease.

5. NONCODING RNAs

The last two decades have seen the discovery of an ever-increasing array of ncRNAs, which have been generally included in the field of epigenetics [136–138]. These ncRNAs include a variety of small RNAs, but also RNAs that are longer than 200 nucleotides (long noncoding RNAs (lncRNAs)). These ncRNAs carry out diverse functions during development. For example, they participate in the regulation of gene expression, they can serve as signal molecules (both intra- and extracellular), and they are important for a number of epigenetic phenomena such as X-inactivation, and the control of imprinted gene expression [139]. Due to these roles in diverse cellular pathways, ncRNAs are of significant interest for clinical research.

5.1 Small ncRNAs

Small RNAs range in size from ~20 to 30 nucleotides, and they include small interfering RNAs (short interfering RNAs, siRNAs), microRNAs (miRNAs), and PIWI-interacting RNAs (piRNAs) [140]. siRNAs, miRNAs, and piRNAs are produced by distinct molecular pathways, and they also carry out different functions. The main function of miRNAs is in developmental gene control, which commonly occurs by translational inhibition [141]. piRNAs function in the germ line and are especially important for the silencing of transposable elements [142,143]. siRNAs are involved in posttranscriptional gene silencing; they target sequence-matched mRNAs for degradation [144]. While other species of small RNAs exist, siRNAs, miRNAs, and piRNAs are the three species most often studied.

The methods used to examine these small RNAs are similar to the methods used for the study of larger RNAs. The classical method is Northern blot analysis [145], the separation of the RNAs by size using agarose gel electrophoresis, the transfer of these RNAs to a membrane, and the probing of this membrane for a single sequence. Northern blot analysis is very low throughput, labor-intensive, and requires a lot of RNA starting material. For example, in an early study from *A. thaliana*, Hamilton and Baulcombe report the discovery of ~25 nucleotide RNAs required for posttranscriptional gene silencing and document the siRNAs by Northern blot [146]. Given the labor-intensive nature of the experiments required to characterize small RNAs, prior to the advent of the NGS technologies, few small RNA species were described in detail.

The NGS methods that have become prevalent over the last 10 years made large-scale analysis of small RNAs possible (for example, see Ref. [147]). These experiments revealed that the populations of small RNAs were vast, much larger than previously anticipated. When Brennecke and colleagues investigate different classes of small RNAs in *Drosophila* with 454 sequencing technology (Roche) in an early study, they sequenced approximately 60,000 small RNAs and approximately 80% of these sequences were recovered only once [148]. While the number of sequences examined are small by today's standard (usually millions of reads are generated today per sample), they illustrate the complexity of the small RNA pools and demonstrate that the capacity of the new NGS technologies is necessary for us to gain an understanding of these highly variable RNA populations.

Specialized protocols—and often-commercial kits—are now available for the preparation of libraries from various small RNA sequences [149]. Small RNA sequencing is now routinely carried out by academic genomics core facilities and reliable analysis pipelines exist as well (for example, see Ref. [150,151]). With the increase in the number of small RNA samples analyzed, our basic comprehension of the small RNA pathways has improved greatly; and as these protocols require much less starting materials than Northern blots, they are feasible for clinical samples.

5.2 Long Noncoding RNAs

lncRNAs have been known to exist for over two decades, as several of them—*Xist*, *H19*, and *AIR*—were identified in genetic studies [152–154]. *Xist*, the first lncRNA discovered, is an approximately 17-kb transcript required for X-inactivation in female mammals [152]. Genome-wide transcript profiling technologies have brought the importance of lncRNAs to attention of biologists [137]. Early array studies as well as later RNA sequencing (RNA-seq) data generated by the various NGS technologies revealed a variety of novel transcripts, including an unexpectedly high number of lncRNAs (for example, see Refs [155,156]). As RNA-seq datasets from different developmental time points and tissues accumulate in public databases and are scrutinized in depth, more lncRNAs will likely be discovered; as they tend to be expressed at lower levels, they are restricted in their expression pattern, and are problematic to predict computationally. Due to these difficulties, it can be challenging to determine if a novel candidate lncRNA is indeed a functional, bona fide lncRNA [152]. ChIP data are often used to confirm the identification of a new lncRNA, as RNA polymerase and certain histone modifications should be present only at functional loci [139]. Using these methods, thousands of lncRNAs have been identified in mammalian genomes, and most remain to be characterized.

Among the few lncRNAs that are well studied are illustrious examples such as *Xist*, the regulatory RNA that coats the inactive X chromosome in female mammals [157]. *Rox1* and *rox2*, the two ncRNAs, involved in dosage compensation in Drosophila also fall into this well-studied category, as does the RNA integral to the eukaryotic telomerase complex, TR [158,159]. More recently characterized lncRNAs include *Braveheart (Bvht)*, an lncRNA important for cardiovascular development [160]. *Bvht* was identified from a set of candidate lncRNAs that showed on average lower expression levels in differentiated, committed cells than in embryonic stem cells. Testing expression levels in various heart tissues, *Bvht* was found to have the highest expression levels in differentiated heart tissues, making it a candidate for a lineage commitment lncRNA. Extensive additional characterization of *Bvht* demonstrated that it was indeed essential for the development of cells in the cardiovascular lineage [160]. The example of *Bvht* illustrates the extensive characterization required to go from a candidate lncRNA identified based on array or RNA-seq data to a well-characterized lncRNA. To date, such careful descriptions of lncRNAs and their function are available for only a small fraction of total lncRNAs, and most lncRNAs are in need of further study.

5.3 Applications in Clinical Research and Beyond

Overall, our knowledge about ncRNAs has progressed to where they are now of significant interest to clinical epigenetics researchers. One of the reasons for the interest in miRNAs is due to their roles in gene regulation. As miRNAs' expression is developmentally regulated, they might serve as biomarkers for disease states. miRNAs are being investigated as biomarkers for several cancers, and they show significant promise. For example, a recent study of miRNAs in fluid from pancreatic cyst, a benign lesion often preceding pancreatic cancer, discovered 13 up- and 2 downregulated miRNAs that deserve further study as a potential prognostic biomarker [161,162]. Of specific interest are circulating miRNAs that are thought to be present in the blood due to certain disease conditions such as cancer [163]. While currently no miRNA biomarkers are in use, they show great promise, especially if the disease biomarkers can be simply detected from a blood sample as circulating miRNAs.

6. CONCLUSIONS AND OUTLOOK

As our review shows, many methods are available to the clinical researchers interested in incorporating epigenetics into their research (Table 1). These methods allow for the assessment of all major molecular mechanisms contributing to the epigenotype, including DNA methylation, histone modifications, chromatin structure and composition, and ncRNAs. For most of the techniques, extensively tested protocols have been published in the various "methods" collections, and the use of some techniques such as ChIP is so widespread that protocols are now included in standard laboratory protocol collections such as "Molecular Cloning" and "Advances in Epigenetic Technology" [164,165]. This wide availability of detailed protocols illustrates the high level of interest in these methods and eases entry into epigenetics research by scientists new to the field.

While the availability of well-tested protocols—and also the availability of convenient kits—has facilitated the incorporation of epigenetics research into the field of clinical research, some obstacles do persist. Foremost among them is the limitation by sample amount requirements for some of the epigenetics assays. Typically, tissue samples available for clinical research are small, much smaller than what is often required as starting materials for the most well-established protocols. This limitation then forces clinical researchers to work with more difficult protocols such as the micro-ChIP protocols [166],

TABLE 1 Commonly Used Methods in Epigenetics Research

Method	Data type	Cost	Sample required	Ease of use
HPLC	Bulk DNA modification	Low	++	Requires specialized equipment
Southern blot	Local DNA methylation	Low	++	Easy
BS-PCR assays	DNA methylation levels	Low	–	Easy
BS genome-wide	Methylome	High	–	Requires specialized equipment
ELISA	Bulk DNA modification or histone PTM	Low	––	Easy
Mass spectrometry	Bulk DNA modification or histone PTM	Moderate	––	Requires specialized equipment
Western blot	Bulk histone PTM	Moderate	++	Easy
Immunohistochemistry	Overall histone PTM levels	Low	–––	Easy
ChIP	Genome-wide histone PTM profile	High	++	Requires specialized equipment
Accessibility assays	Relative accessibility	Moderate	++	Easy
3C and derivatives	Chromatin interaction profiles	High	++	Requires specialized equipment
Northern blot	RNA levels	Moderate	+	Easy
RNA-seq	Transcriptome	High	+	Requires specialized equipment

+ and – are used to indicate the amount of sample required for a given assay (+: large amount of sample required; –: small amount of sample required). HPLC, high-performance liquid chromatography; BS-PCR, bisulfite polymerase chain reaction; DNA, deoxyribonucleic acid; PTM, posttranslational modification; RNA, ribonucleic acid; RNA-seq, RNA sequencing; ChIP, chromatin immunoprecipitation; 3C, chromatin confirmation capture.

or might require the development of new methods. Storage of clinical samples can also be an issue, especially when fragile molecules such as RNAs need to be assayed. However, given the high interest from researchers and companies, many of the technical limitations will likely be overcome in the near future.

Especially the integration of classical epigenetics methods with NGS technologies has opened up research avenues that were in the realm of dreams just a short while ago. The ability to investigate epigenetic modifications on a genome-wide scale is unprecedented and has given epigenetics researchers the tools to answer research questions about the molecular mechanisms controlling epigenetic information, the distribution of epigenetic information in the genome, the changes in epigenetic information with aging or environmental exposures, and many more. New technology developments that are on the horizon include single-cell assays for histone modification, chromosomal proteins, DNA methylation, ncRNAs, and chromatin structure [166–169], and advances in sequencing technology that would allow for the detection of modified DNA bases without bisulfite conversion [170–172]. Thus, with the advent of single-cell assays for epigenetic modifications the evolution of new methods for epigenetics research continues. We anticipate that the single-cell technologies currently in development will have an even more profound impact on clinical epigenetics research, as they will circumvent the limitations of tissue availability.

LIST OF ABBREVIATIONS

3C methods Chromatin confirmation capture
5caC 5-Carboxylcytosine
5fC 5-Formylcytosine
5hmC 5-Hydroxymethylcytosine
5mC 5-Methylcytosine
AFLP Amplified fragment length polymorphism
BS-seq Bisulfite sequencing
CATCH-IT Covalent attached tags to capture histones and identity turnover
ChIA-PET Chromatin interaction analysis using paired end tag sequencing
ChIP Chromatin immunoprecipitation
COBRA Combined bisulfite restriction analysis
DNA Deoxyribonucleic acid
DNAme DNA methylation
DNase Deoxyribonuclease
ELISA Enzyme-linked immunosorbent assay
FAIRE-seq Formaldehyde-assisted isolation of regulatory elements sequencing
FDA Food and Drug Administration
FISH Fluorescent in situ hybridization
HAT Histone acetyltransferase
HDAC Histone deacetyltransferase
HiC Hydrophobic interaction chromatography
HPLC High-performance liquid chromatography
lncRNA Long noncoding RNA
MBD proteins Methyl-binding domain proteins
meDIP Methylated DNA immunoprecipitation
miRNA MicroRNA
MSAP Methylation-sensitive amplified polymorphism
MS-HRM Methylation-sensitive high-resolution mapping
MSP Methylation-specific PCR
Ms-SNuPE Methylation-sensitive single-nucleotide primer extension
ncRNA Noncoding RNA
oxBS-seq Oxidative bisulfite sequencing
PCR Polymerase chain reaction
PEV Position effect variegation
piRNA PIWI-interacting RNA
PTM Posttranslational modification
qPCR Quantitative PCR
redBS-seq Reduced BS-seq
RNA Ribonucleic acid
RNA-seq RNA sequencing
RRBS Reduced representation bisulfite sequencing
SDS–acrylamide Sodium dodecyl sulfate–acrylamide
siRNA Short interfering RNA
Sono-seq Sonication of cross-linked chromatin sequencing
TAB-seq Tet-assisted bisulfite sequencing
TET proteins Ten-eleven translocation proteins
TLC Thin-layer chromatography

References

[1] Felsenfeld G. A brief history of epigenetics. Cold Spring Harb Perspect Biol January 3, 2014;6(1).
[2] Argeson AC, Nelson KK, Siracusa LD. Molecular basis of the pleiotropic phenotype of mice carrying the hypervariable yellow (Ahvy) mutation at the agouti locus. Genetics February 4, 1996;142(2):557–67.
[3] Eissenberg JC, James TC, Foster-Hartnett DM, Hartnett T, Ngan V, Elgin SC. Mutation in a heterochromatin-specific chromosomal protein is associated with suppression of position-effect variegation in *Drosophila melanogaster*. Proc Natl Acad Sci USA December 6, 1990;87(24):9923–7.
[4] Cubas P, Vincent C, Coen E. An epigenetic mutation responsible for natural variation in floral symmetry. Nature September 4, 1999;401(6749):157–61.
[5] Morgan HD, Sutherland HG, Martin DI, Whitelaw E. Epigenetic inheritance at the agouti locus in the mouse. Nat Genet November 1, 1999;23(3):314–8.
[6] Elgin SC, Reuter G. Position-effect variegation, heterochromatin formation, and gene silencing in *Drosophila*. Cold Spring Harb Perspect Biol August 4, 2013;5(8):a017780.
[7] Allis CD, Jenuwein T, Reinberg D, Caparros ML. Epigenetics 2007. Available from: http://library.wur.nl/WebQuery/clc/1831507.
[8] Blewitt M, Whitelaw E. The use of mouse models to study epigenetics. Cold Spring Harb Perspect Biol November 5, 2013;5(11):a017939.
[9] Cleaton MA, Edwards CA, Ferguson-Smith AC. Phenotypic outcomes of imprinted gene models in mice: elucidation of pre- and postnatal functions of imprinted genes. Annu Rev Genomics Hum Genet January 3, 2014;15:93–126.
[10] McKeown PC, Laouielle-Duprat S, Prins P, Wolff P, Schmid MW, Donoghue MT, et al. Identification of imprinted genes subject to parent-of-origin specific expression in *Arabidopsis thaliana* seeds. BMC Plant Biol January 6, 2011;11:113.
[11] Gendrel A-VV, Heard E. Noncoding RNAs and epigenetic mechanisms during x-chromosome inactivation. Annu Rev Cell Dev Biol October 6, 2014;30:561–80.

[12] Li E. Chromatin modification and epigenetic reprogramming in mammalian development. Nat Rev Genet September 2002;3(9):662–73.

[13] Wolff GL, Kodell RL, Moore SR, Cooney CA. Maternal epigenetics and methyl supplements affect agouti gene expression in Avy/a mice. FASEB J August 6, 1998;12(11):949–57.

[14] Angel A, Song J, Dean C, Howard MA. Polycomb-based switch underlying quantitative epigenetic memory. Nature August 4, 2011;476(7358):105–8.

[15] Jaenisch R, Bird A. Epigenetic regulation of gene expression: how the genome integrates intrinsic and environmental signals. Nat Genet March 6, 2003; 33(Suppl. 245–54).

[16] Bird A. DNA methylation patterns and epigenetic memory. Genes Dev January 2, 2002;16(1):6–21.

[17] Zemach A, Zilberman D. Evolution of eukaryotic DNA methylation and the pursuit of safer sex. Curr Biol September 2, 2010;20(17):R780–5.

[18] Guibert S, Weber M. Functions of DNA methylation and hydroxymethylation in mammalian development. Curr Top Dev Biol January 2, 2013;104:47–83.

[19] Li E, Zhang Y. DNA methylation in mammals. Cold Spring Harb Perspect Biol May 4, 2014;6(5):a019133.

[20] Jost JP. Nuclear extracts of chicken embryos promote an active demethylation of DNA by excision repair of 5-methyldeoxycytidine. Proc Natl Acad Sci USA May 6, 1993;90(10):4684–8.

[21] Gehring M, Huh JH, Hsieh T-FF, Penterman J, Choi Y, Harada JJ, et al. DEMETER DNA glycosylase establishes MEDEA polycomb gene self-imprinting by allele-specific demethylation. Cell February 5, 2006; 124(3):495–506.

[22] Gong Z, Morales-Ruiz T, Ariza RR, Roldán-Arjona T, David L, Zhu JK. ROS1, a repressor of transcriptional gene silencing in *Arabidopsis*, encodes a DNA glycosylase/lyase. Cell December 5, 2002;111(6):803–14.

[23] Kohli RM, Zhang Y. TET enzymes, TDG and the dynamics of DNA demethylation. Nature October 4, 2013;502(7472):472–9.

[24] Tan L, Shi YG. Tet family proteins and 5-hydroxymethylcytosine in development and disease. Development June 5, 2012;139(11):1895–902.

[25] Wu H, Zhang Y. Mechanisms and functions of Tet protein-mediated 5-methylcytosine oxidation. Genes Dev December 4, 2011;25(23):2436–52.

[26] Li E, Bestor TH, Jaenisch R. Targeted mutation of the DNA methyltransferase gene results in embryonic lethality. Cell June 5, 1992;69(6):915–26.

[27] Zemach A, McDaniel IE, Silva P, Zilberman D. Genome-wide evolutionary analysis of eukaryotic DNA methylation. Science May 5, 2010;328(5980):916–9.

[28] Razin A, Riggs AD. DNA methylation and gene function. Science November 5, 1980;210(4470):604–10.

[29] Kuo KC, McCune RA, Gehrke CW, Midgett R, Ehrlich M. Quantitative reversed-phase high performance liquid chromatographic determination of major and modified deoxyribonucleosides in DNA. Nucleic Acids Res October 5, 1980;8(20):4763–76.

[30] Kuchino Y, Hanyu N, Nishimura S. Analysis of modified nucleosides and nucleotide sequence of tRNA. Meth Enzymol January 4, 1987;155:379–96.

[31] Song L, James SR, Kazim L, Karpf AR. Specific method for the determination of genomic DNA methylation by liquid chromatography-electrospray ionization tandem mass spectrometry. Anal Chem January 6, 2005;77(2):504–10.

[32] Achwal CW, Ganguly P, Chandra HS. Estimation of the amount of 5-methylcytosine in *Drosophila melanogaster* DNA by amplified ELISA and photoacoustic spectroscopy. EMBO J February 3, 1984;3(2):263–6.

[33] Southern EM. Detection of specific sequences among DNA fragments separated by gel electrophoresis. J Mol Biol November 3, 1975;98(3):503–17.

[34] Cedar H, Solage A, Glaser G, Razin A. Direct detection of methylated cytosine in DNA by use of the restriction enzyme MspI. Nucleic Acids Res January 1, 1979;6(6):2125–32.

[35] Vongs A, Kakutani T, Martienssen RA, Richards EJ. *Arabidopsis thaliana* DNA methylation mutants. Science June 5, 1993;260(5116):1926–8.

[36] Singer-Sam J, LeBon JM, Tanguay RL, Riggs AD. A quantitative HpaII-PCR assay to measure methylation of DNA from a small number of cells. Nucleic Acids Res February 1990;18(3):687.

[37] Stöger R, Kubicka P, Liu CG, Kafri T, Razin A, Cedar H, et al. Maternal-specific methylation of the imprinted mouse Igf2r locus identifies the expressed locus as carrying the imprinting signal. Cell April 5, 1993;73(1):61–71.

[38] Reyna-López GE, Simpson J, Ruiz-Herrera J. Differences in DNA methylation patterns are detectable during the dimorphic transition of fungi by amplification of restriction polymorphisms. Mol Gen Genet February 4, 1997;253(6):703–10.

[39] Díaz-Freije E, Gestal C, Castellanos-Martínez S, Morán P. The role of DNA methylation on *Octopus vulgaris* development and their perspectives. Front Physiol January 3, 2014;5:62.

[40] Lauria M, Piccinini S, Pirona R, Lund G, Viotti A, Motto M. Epigenetic variation, inheritance, and parent-of-origin effects of cytosine methylation in maize (*Zea mays*). Genetics March 6, 2014;196(3):653–66.

[41] Frommer M, McDonald LE, Millar DS, Collis CM, Watt F, Grigg GW, et al. A genomic sequencing protocol that yields a positive display of 5-methylcytosine residues in individual DNA strands. Proc Natl Acad Sci USA March 1992;89(5):1827–31.

[42] Hayatsu H, Shiragami M. Reaction of bisulfite with the 5-hydroxymethyl group in pyrimidines and in phage DNAs. Biochemistry February 2, 1979;18(4):632–7.
[43] Herman JG, Graff JR, Myöhänen S, Nelkin BD, Baylin SB. Methylation-specific PCR: a novel PCR assay for methylation status of CpG islands. Proc Natl Acad Sci USA September 2, 1996;93(18):9821–6.
[44] Xiong Z, Laird PW. COBRA: a sensitive and quantitative DNA methylation assay. Nucleic Acids Res June 1997;25(12):2532–4.
[45] Gonzalgo ML, Jones PA. Rapid quantitation of methylation differences at specific sites using methylation-sensitive single nucleotide primer extension (Ms-SNuPE). Nucleic Acids Res June 1997;25(12):2529–31.
[46] Colella S, Shen L, Baggerly KA, Issa JP, Krahe R. Sensitive and quantitative universal pyrosequencing methylation analysis of CpG sites. BioTechniques July 2, 2003;35(1):146–50.
[47] Wojdacz TK, Dobrovic A. Methylation-sensitive high resolution melting (MS-HRM): a new approach for sensitive and high-throughput assessment of methylation. Nucleic Acids Res January 1, 2007;35(6):e41.
[48] Wojdacz TK, Dobrovic A, Algar EM. Rapid detection of methylation change at H19 in human imprinting disorders using methylation-sensitive high-resolution melting. Hum Mutat October 3, 2008;29(10):1255–60.
[49] Meissner A, Gnirke A, Bell GW, Ramsahoye B, Lander ES, Jaenisch R. Reduced representation bisulfite sequencing for comparative high-resolution DNA methylation analysis. Nucleic Acids Res January 6, 2005;33(18):5868–77.
[50] Gu H, Bock C, Mikkelsen TS, Jäger N, Smith ZD, Tomazou E, et al. Genome-scale DNA methylation mapping of clinical samples at single-nucleotide resolution. Nat Methods February 1, 2010;7(2):133–6.
[51] Booth MJ, Branco MR, Ficz G, Oxley D, Krueger F, Reik W, et al. Quantitative sequencing of 5-methylcytosine and 5-hydroxymethylcytosine at single-base resolution. Science May 5, 2012;336(6083):934–7.
[52] Booth MJ, Marsico G, Bachman M, Beraldi D, Balasubramanian S. Quantitative sequencing of 5-formylcytosine in DNA at single-base resolution. Nat Chem May 4, 2014;6(5):435–40.
[53] Yu M, Hon GC, Szulwach KE, Song C-XX, Zhang L, Kim A, et al. Base-resolution analysis of 5-hydroxymethylcytosine in the mammalian genome. Cell June 5, 2012;149(6):1368–80.
[54] Bibikova M, Barnes B, Tsan C, Ho V, Klotzle B, Le JM, et al. High density DNA methylation array with single CpG site resolution. Genomics October 6, 2011;98(4):288–95.
[55] Bibikova M, Le J, Barnes B, Saedinia-Melnyk S, Zhou L, Shen R, et al. Genome-wide DNA methylation profiling using Infinium® assay. Epigenomics October 4, 2009;1(1):177–200.
[56] Bibikova M, Lin Z, Zhou L, Chudin E, Garcia EW, Wu B, et al. High-throughput DNA methylation profiling using universal bead arrays. Genome Res March 3, 2006;16(3):383–93.
[57] Fan J-BB, Gunderson KL, Bibikova M, Yeakley JM, Chen J, Wickham Garcia E, et al. Illumina universal bead arrays. Meth Enzymol January 2006;410:57–73.
[58] Steemers FJ, Chang W, Lee G, Barker DL, Shen R, Gunderson KL. Whole-genome genotyping with the single-base extension assay. Nat Methods January 2006;3(1):31–3.
[59] Steemers FJ, Gunderson KL. Whole genome genotyping technologies on the BeadArray platform. Biotechnol J January 1, 2007;2(1):41–9.
[60] Weisenberger DJ. Comprehensive DNA methylation analysis on the illumina infinium assay platform Technical report. 2008.
[61] Weber M, Davies JJ, Wittig D, Oakeley EJ, Haase M, Lam WL, et al. Chromosome-wide and promoter-specific analyses identify sites of differential DNA methylation in normal and transformed human cells. Nat Genet August 1, 2005;37(8):853–62.
[62] Mohn F, Weber M, Schübeler D, Roloff T-CC. Methylated DNA immunoprecipitation (MeDIP). Methods Mol Biol January 4, 2009;507:55–64.
[63] Serre D, Lee BH, Ting AH. MBD-isolated genome sequencing provides a high-throughput and comprehensive survey of DNA methylation in the human genome. Nucleic Acids Res January 5, 2010;38(2):391–9.
[64] Nestor CE, Meehan RR. Hydroxymethylated DNA immunoprecipitation (hmeDIP). Methods Mol Biol January 3, 2014;1094:259–67.
[65] Stroud H, Feng S, Morey Kinney S, Pradhan S, Jacobsen SE. 5-Hydroxymethylcytosine is associated with enhancers and gene bodies in human embryonic stem cells. Genome Biol January 6, 2011;12(6):R54.
[66] Borgel J, Guibert S, Weber M. Methylated DNA immunoprecipitation (MeDIP) from low amounts of cells. Methods Mol Biol January 2012;925:149–58.
[67] Comprehensive molecular profiling of lung adenocarcinoma. Nature July 4, 2014;511(7511):543–50.
[68] Imperiale T, Ransohoff D, Itzkowitz S, Levin T, Lavin P, Lidgard G, et al. Multitarget stool DNA testing for colorectal-cancer screening. N Engl J Med 2014;370.
[69] Mikeska T, Craig JM. DNA methylation biomarkers: cancer and beyond. Genes (Basel) January 3, 2014;5(3):821–64.
[70] Allfrey VG, Faulkner R, Mirsky AE. Acetylation and methylation of histones and their possible role in the regulation of RNA synthesis. Proc Natl Acad Sci USA May 5, 1964;51:786–94.
[71] Butler JS, Koutelou E, Schibler AC, Dent SY. Histone-modifying enzymes: regulators of developmental decisions and drivers of human disease. Epigenomics April 2012;4(2):163–77.

[72] Kouzarides T. SnapShot: histone-modifying enzymes. Cell November 5, 2007;131(4):822.

[73] Martin C, Zhang Y. Mechanisms of epigenetic inheritance. (Internet). Elsevier Curr Opin Cell Biol 2007;19(3):266–72. Available from: http://www.sciencedirect.com/science/article/pii/S0955067407000543.

[74] Deribe YL, Pawson T, Dikic I. Post-translational modifications in signal integration. Nat Struct Mol Biol June 2, 2010;17(6):666–72.

[75] Arnaudo AM, Garcia BA. Proteomic characterization of novel histone post-translational modifications. Epigenet Chromatin January 2, 2013;6(1):24.

[76] Towbin H, Staehelin T, Gordon J. Electrophoretic transfer of proteins from polyacrylamide gels to nitrocellulose sheets: procedure and some applications. Proc Natl Acad Sci USA September 6, 1979;76(9):4350–4.

[77] Seligman AM, Karnovsky MJ, Wasserkrug HL, Hanker JS. Nondroplet ultrastructural demonstration of cytochrome oxidase activity with a polymerizing osmiophilic reagent, diaminobenzidine (DAB). J Cell Biol July 1, 1968;38(1):1–14.

[78] Egelhofer TA, Minoda A, Klugman S, Lee K, Kolasinska-Zwierz P, Alekseyenko AA, et al. An assessment of histone-modification antibody quality. Nat Struct Mol Biol January 6, 2011;18(1):91–3.

[79] Brower-Toland B, Riddle NC, Jiang H, Huisinga KL, Elgin SC. Multiple SET methyltransferases are required to maintain normal heterochromatin domains in the genome of *Drosophila melanogaster*. Genetics April 3, 2009;181(4):1303–19.

[80] Yang H, Howard M, Dean C. Antagonistic roles for H3K36me3 and H3K27me3 in the cold-induced epigenetic switch at *Arabidopsis* FLC. Curr Biol August 1, 2014;24(15):1793–7.

[81] Seligson D, Horvath S, Shi T, Yu H, Tze S, Grunstein M, et al. Global histone modification patterns predict risk of prostate cancer recurrence. Nature 2005;435(7046):1262–6.

[82] Tzao C, Tung H-J, Jin J-S, Sun G-H, Hsu H-S, Chen B-H, et al. Prognostic significance of global histone modifications in resected squamous cell carcinoma of the esophagus. Mod Pathol 2008;22(2):252–60.

[83] Karch KR, Denizio JE, Black BE, Garcia BA. Identification and interrogation of combinatorial histone modifications. Front Genet January 2, 2013;4:264.

[84] Moradian A, Kalli A, Sweredoski MJ, Hess S. The top-down, middle-down, and bottom-up mass spectrometry approaches for characterization of histone variants and their post-translational modifications. Proteomics March 6, 2014;14(4–5):489–97.

[85] Young NL, DiMaggio PA, Plazas-Mayorca MD, Baliban RC, Floudas CA, Garcia BA. High throughput characterization of combinatorial histone codes. Mol Cell Proteomics October 4, 2009;8(10):2266–84.

[86] DiMaggio PA, Young NL, Baliban RC, Garcia BA, Floudas CA. A mixed integer linear optimization framework for the identification and quantification of targeted post-translational modifications of highly modified proteins using multiplexed electron transfer dissociation tandem mass spectrometry. Mol Cell Proteomics November 2009;8(11):2527–43.

[87] Johnson L, Mollah S, Garcia BA, Muratore TL, Shabanowitz J, Hunt DF, et al. Mass spectrometry analysis of *Arabidopsis* histone H3 reveals distinct combinations of post-translational modifications. Nucleic Acids Res January 4, 2004;32(22):6511–8.

[88] Jung HR, Pasini D, Helin K, Jensen ON. Quantitative mass spectrometry of histones H3.2 and H3.3 in Suz12-deficient mouse embryonic stem cells reveals distinct, dynamic post-translational modifications at Lys-27 and Lys-36. Mol Cell Proteomics May 6, 2010;9(5):838–50.

[89] Zhang K, Sridhar VV, Zhu J, Kapoor A, Zhu J-KK. Distinctive core histone post-translational modification patterns in *Arabidopsis thaliana*. PLoS One January 1, 2007;2(11):e1210.

[90] Beck HC, Nielsen EC, Matthiesen R, Jensen LH, Sehested M, Finn P, et al. Quantitative proteomic analysis of post-translational modifications of human histones. Mol Cell Proteomics July 6, 2006;5(7):1314–25.

[91] Tan M, Luo H, Lee S, Jin F, Yang J, Montellier E, et al. Identification of 67 histone Marks and histone lysine crotonylation as a new type of histone modification. Cell 2011;146(6).

[92] Zhang K, Li L, Zhu M, Wang G, Xie J, Zhao Y, et al. Comparative analysis of histone H3 and H4 post-translational modifications of esophageal squamous cell carcinoma with different invasive capabilities. J Proteomics September 2, 2014;112C:180–9.

[93] Haery L, Lugo-Picó JG, Henry RA, Andrews AJ, Gilmore TD. Histone acetyltransferase-deficient p300 mutants in diffuse large B cell lymphoma have altered transcriptional regulatory activities and are required for optimal cell growth. Mol Cancer January 3, 2014;13:29.

[94] Langer-Safer PR, Levine M, Ward DC. Immunological method for mapping genes on *Drosophila* polytene chromosomes. Proc Natl Acad Sci USA July 4, 1982;79(14):4381–5.

[95] Orlando V, Strutt H, Paro R. Analysis of chromatin structure by in vivo formaldehyde cross-linking. Methods February 6, 1997;11(2):205–14.

[96] Orlando V. Mapping chromosomal proteins in vivo by formaldehyde-crosslinked-chromatin immunoprecipitation. Trends Biochem Sci March 3, 2000;25(3):99–104.

[97] Taneyhill LA, Adams MS. Investigating regulatory factors and their DNA binding affinities through real time quantitative PCR (RT-QPCR) and chromatin immunoprecipitation (ChIP) assays. Methods Cell Biol January 2, 2008;87:367–89.

[98] Lieb JD. Genome-wide mapping of protein-DNA interactions by chromatin immunoprecipitation and DNA microarray hybridization. Methods Mol Biol January 3, 2003;224:99–109.
[99] Johnson DS, Mortazavi A, Myers RM, Wold B. Genome-wide mapping of in vivo protein-DNA interactions. Science June 5, 2007;316(5830):1497–502.
[100] Robertson G, Hirst M, Bainbridge M, Bilenky M, Zhao Y, Zeng T, et al. Genome-wide profiles of STAT1 DNA association using chromatin immunoprecipitation and massively parallel sequencing. Nat Methods August 3, 2007;4(8):651–7.
[101] Barski A, Cuddapah S, Cui K, Roh T-YY, Schones DE, Wang Z, et al. High-resolution profiling of histone methylations in the human genome. Cell May 5, 2007;129(4):823–37.
[102] Landt SG, Marinov GK, Kundaje A, Kheradpour P, Pauli F, Batzoglou S, et al. ChIP-seq guidelines and practices of the ENCODE and modENCODE consortia. Genome Res September 6, 2012;22(9):1813–31.
[103] Gerstein MB, Lu ZJ, Van Nostrand EL, Cheng C, Arshinoff BI, Liu T, et al. Integrative analysis of the *Caenorhabditis elegans* genome by the modENCODE project. Science December 5, 2010;330(6012):1775–87.
[104] Gerstein MB, Kundaje A, Hariharan M, Landt SG, Yan K-KK, Cheng C, et al. Architecture of the human regulatory network derived from ENCODE data. Nature September 4, 2012;489(7414):91–100.
[105] Roy S, Ernst J, Kharchenko PV, Kheradpour P, Negre N, Eaton ML, et al. Identification of functional elements and regulatory circuits by *Drosophila* modENCODE. Science December 5, 2010;330(6012):1787–97.
[106] Ho JW, Jung YL, Liu T, Alver BH, Lee S, Ikegami K, et al. Comparative analysis of metazoan chromatin organization. Nature August 4, 2014;512(7515):449–52.
[107] Dahl JA, Collas P. MicroChIP–a rapid micro chromatin immunoprecipitation assay for small cell samples and biopsies. Nucleic Acids Res February 5, 2008;36(3):e15.
[108] Dahl JA, Collas P. A rapid micro chromatin immunoprecipitation assay (microChIP). Nat Protoc January 2, 2008;3(6):1032–45.
[109] Adli M, Bernstein BE. Whole-genome chromatin profiling from limited numbers of cells using nano-ChIP-seq. Nat Protoc October 6, 2011;6(10):1656–68.
[110] Lara-Astiaso D, Weiner A, Lorenzo-Vivas E, Zaretsky I, Jaitin DA, David E, et al. Immunogenetics. Chromatin state dynamics during blood formation. Science August 5, 2014;345(6199):943–9.
[111] O'Geen H, Echipare L, Farnham PJ. Using ChIP-seq technology to generate high-resolution profiles of histone modifications. Methods Mol Biol January 6, 2011;791:265–86.
[112] Lund AH, van Lohuizen M. Epigenetics and cancer. Genes Dev October 5, 2004;18(19):2315–35.
[113] Farh KK, Marson A, Zhu J, Kleinewietfeld M, Housley WJ, Beik S, et al. Genetic and epigenetic fine mapping of causal autoimmune disease variants. Nature October 3, 2014;518.
[114] Heitz E. Das Heterochromatin der Moose. Jahrbuecher fuer Wiss Bot 1928;69:762–818.
[115] Sajan SA, Hawkins RD. Methods for identifying higher-order chromatin structure. Annu Rev Genomics Hum Genet January 2012;13:59–82.
[116] Wu C, Bingham PM, Livak KJ, Holmgren R, Elgin SC. The chromatin structure of specific genes: I. Evidence for higher order domains of defined DNA sequence. Cell April 1979;16(4):797–806.
[117] Wallrath LL, Elgin SC. Position effect variegation in *Drosophila* is associated with an altered chromatin structure. Genes Dev May 1, 1995;9(10):1263–77.
[118] Crawford GE, Holt IE, Whittle J, Webb BD, Tai D, Davis S, et al. Genome-wide mapping of DNase hypersensitive sites using massively parallel signature sequencing (MPSS). Genome Res January 2006;16(1):123–31.
[119] Crawford GE, Davis S, Scacheri PC, Renaud G, Halawi MJ, Erdos MR, et al. DNase-chip: a high-resolution method to identify DNase I hypersensitive sites using tiled microarrays. Nat Methods July 6, 2006;3(7):503–9.
[120] Song L, Crawford GE. DNase-seq: a high-resolution technique for mapping active gene regulatory elements across the genome from mammalian cells. Cold Spring Harb Protoc February 1, 2010;2010(2). pdb.prot5384.
[121] He HH, Meyer CA, Hu SS, Chen M-WW, Zang C, Liu Y, et al. Refined DNase-seq protocol and data analysis reveals intrinsic bias in transcription factor footprint identification. Nat Methods January 3, 2014;11(1):73–8.
[122] Giresi PG, Kim J, McDaniell RM, Iyer VR, Lieb JD. FAIRE (Formaldehyde-Assisted Isolation of Regulatory Elements) isolates active regulatory elements from human chromatin. Genome Res June 5, 2007;17(6):877–85.
[123] Giresi PG, Lieb JD. Isolation of active regulatory elements from eukaryotic chromatin using FAIRE (Formaldehyde Assisted Isolation of Regulatory Elements). Methods July 3, 2009;48(3):233–9.
[124] Song L, Zhang Z, Grasfeder LL, Boyle AP, Giresi PG, Lee B-KK, et al. Open chromatin defined by DNaseI and FAIRE identifies regulatory elements that shape cell-type identity. Genome Res October 6, 2011;21(10):1757–67.
[125] Auerbach RK, Euskirchen G, Rozowsky J, Lamarre-Vincent N, Moqtaderi Z, Lefrançois P, et al. Mapping accessible chromatin regions using Sono-Seq. Proc Natl Acad Sci USA September 2, 2009;106(35):14926–31.
[126] Dekker J, Rippe K, Dekker M, Kleckner N. Capturing chromosome conformation. Science February 5, 2002;295(5558):1306–11.

[127] De Wit E, de Laat W. A decade of 3C technologies: insights into nuclear organization. Genes Dev January 2012;26(1):11–24.
[128] Van de Werken HJ, de Vree PJ, Splinter E, Holwerda SJ, Klous P, de Wit E, et al. 4C technology: protocols and data analysis. Meth Enzymol January 2012;513: 89–112.
[129] Simonis M, Klous P, Splinter E, Moshkin Y, Willemsen R, de Wit E, et al. Nuclear organization of active and inactive chromatin domains uncovered by chromosome conformation capture-on-chip (4C). Nat Genet November 3, 2006;38(11):1348–54.
[130] Zhao Z, Tavoosidana G, Sjölinder M, Göndör A, Mariano P, Wang S, et al. Circular chromosome conformation capture (4C) uncovers extensive networks of epigenetically regulated intra- and interchromosomal interactions. Nat Genet November 3, 2006;38(11):1341–7.
[131] Dostie J, Richmond TA, Arnaout RA, Selzer RR, Lee WL, Honan TA, et al. Chromosome Conformation Capture Carbon Copy (5C): a massively parallel solution for mapping interactions between genomic elements. Genome Res October 2006;16(10): 1299–309.
[132] Van Berkum NL, Dekker J. Determining spatial chromatin organization of large genomic regions using 5C technology. Methods Mol Biol January 4, 2009;567:189–213.
[133] Lieberman-Aiden E, van Berkum NL, Williams L, Imakaev M, Ragoczy T, Telling A, et al. Comprehensive mapping of long-range interactions reveals folding principles of the human genome. Science October 5, 2009;326(5950):289–93.
[134] Fullwood MJ, Liu MH, Pan YF, Liu J, Xu H, Mohamed YB, et al. An oestrogen-receptor-alpha-bound human chromatin interactome. Nature November 4, 2009; 462(7269):58–64.
[135] Heidari N, Phanstiel DH, He C, Grubert F, Jahanbani F, Kasowski M, et al. Genome-wide map of regulatory interactions in the human genome. Genome Res September 2, 2014;24(12).
[136] Stuwe E, Tóth KF, Aravin AA. Small but sturdy: small RNAs in cellular memory and epigenetics. Genes Dev March 6, 2014;28(5):423–31.
[137] Yang L, Froberg JE, Lee JT. Long noncoding RNAs: fresh perspectives into the RNA world. Trends Biochem Sci January 3, 2014;39(1):35–43.
[138] Guil S, Esteller M. Cis-acting noncoding RNAs: friends and foes. Nat Struct Mol Biol November 4, 2012;19(11):1068–75.
[139] Rinn JL, Chang HY. Genome regulation by long noncoding RNAs. Annu Rev Biochem January 2012;81: 145–66.
[140] Aalto AP, Pasquinelli AE. Small non-coding RNAs mount a silent revolution in gene expression. (Internet). Elsevier Curr Opin Cell Biol 2012;24(3):333–40. Available from: http://www.sciencedirect.com/science/article/pii/S0955067412000373.
[141] Bartel DP. MicroRNAs: target recognition and regulatory functions. Cell January 5, 2009;136(2):215–33.
[142] Watanabe T, Lin H. Posttranscriptional regulation of gene expression by piwi proteins and piRNAs. Mol Cell October 4, 2014;56(1):18–27.
[143] Ishizu H, Siomi H, Siomi MC. Biology of PIWI-interacting RNAs: new insights into biogenesis and function inside and outside of germlines. Genes Dev November 4, 2012;26(21):2361–73.
[144] Svoboda P. Renaissance of mammalian endogenous RNAi. FEBS Lett August 5, 2014;588(15):2550–6.
[145] Alwine JC, Kemp DJ, Stark GR. Method for detection of specific RNAs in agarose gels by transfer to diazobenzyloxymethyl-paper and hybridization with DNA probes. Proc Natl Acad Sci USA December 4, 1977;74(12):5350–4.
[146] Hamilton AJ, Baulcombe DC. A species of small antisense RNA in posttranscriptional gene silencing in plants. Science October 5, 1999;286(5441):950–2.
[147] McGinn J, Czech B. Small RNA library construction for high-throughput sequencing. Methods Mol Biol January 3, 2014;1093:195–208.
[148] Brennecke J, Aravin AA, Stark A, Dus M, Kellis M, Sachidanandam R, et al. Discrete small RNA-generating loci as master regulators of transposon activity in *Drosophila*. Cell March 5, 2007;128(6):1089–103.
[149] Malone C, Brennecke J, Czech B, Aravin A, Hannon GJ. Preparation of small RNA libraries for high-throughput sequencing. Cold Spring Harb Protoc October 1, 2012; 2012(10):1067–77.
[150] Giurato G, De Filippo MR, Rinaldi A, Hashim A, Nassa G, Ravo M, et al. iMir: an integrated pipeline for high-throughput analysis of small non-coding RNA data obtained by smallRNA-Seq. BMC Bioinforma January 2, 2013;14:362.
[151] Han BW, Wang W, Zamore PD, Weng Z. piPipes: a set of pipelines for piRNA and transposon analysis via small RNA-seq, RNA-seq, degradome- and CAGE-seq, ChIP-seq Genomic DNA Sequencing. Bioinforma October 5, 2014;31(4).
[152] Brown CJ, Ballabio A, Rupert JL, Lafreniere RG, Grompe M, Tonlorenzi R, et al. A gene from the region of the human X inactivation centre is expressed exclusively from the inactive X chromosome. Nature January 4, 1991;349(6304):38–44.
[153] Bartolomei MS, Zemel S, Tilghman SM. Parental imprinting of the mouse H19 gene. Nature May 4, 1991;351(6322):153–5.

[154] Lyle R, Watanabe D, te Vruchte D, Lerchner W, Smrzka OW, Wutz A, et al. The imprinted antisense RNA at the Igf2r locus overlaps but does not imprint Mas1. Nat Genet May 1, 2000;25(1):19–21.

[155] Kapranov P, Cawley SE, Drenkow J, Bekiranov S, Strausberg RL, Fodor SP, et al. Large-scale transcriptional activity in chromosomes 21 and 22. Science May 5, 2002;296(5569):916–9.

[156] Rinn JL, Euskirchen G, Bertone P, Martone R, Luscombe NM, Hartman S, et al. The transcriptional activity of human chromosome 22. Genes Dev February 6, 2003;17(4):529–40.

[157] Brown SD. XIST and the mapping of the X chromosome inactivation centre. Bioessays November 5, 1991;13(11):607–12.

[158] Shippen-Lentz D, Blackburn EH. Functional evidence for an RNA template in telomerase. Science February 5, 1990;247(4942):546–52.

[159] Theimer CA, Feigon J. Structure and function of telomerase RNA. Curr Opin Struct Biol June 4, 2006;16(3):307–18.

[160] Klattenhoff CA, Scheuermann JC, Surface LE, Bradley RK, Fields PA, Steinhauser ML, et al. Braveheart, a long noncoding RNA required for cardiovascular lineage commitment. Cell January 4, 2013;152(3):570–83.

[161] Wang J, Paris PL, Chen J, Ngo V, Yao H, Frazier ML, et al. Next generation sequencing of pancreatic cyst fluid microRNAs from low grade- benign and high grade-invasive lesions. Cancer Lett October 2, 2014;356.

[162] Wang C, Caron M, Burdick D, Kang Z, Auld D, Hill WA, et al. A sensitive, homogeneous, and high-throughput assay for lysine-specific histone demethylases at the H3K4 site. Assay Drug Dev Technol April 2012;10(2):179–86.

[163] Witwer KW. Circulating MicroRNA biomarker studies: pitfalls and potential solutions. Clin Chem November 3, 2014;61(1).

[164] Green MR, Sambrook J. Molecular cloning: a laboratory manual. 1012. p. 2,028.

[165] Tollefsbol TO. Epigenetics protocols. Springer Science & Business Media; 2004; 287. Available from: http://books.google.com/books?hl=en&lr=&id=C8S4LfEmZU8C&oi=fnd&pg=PR5&ots=APFwQ-YVUt&sig=DwAZmfge_5mbO_BFMD-OfqbcD1o.

[166] Adli M, Zhu J, Bernstein BE. Genome-wide chromatin maps derived from limited numbers of hematopoietic progenitors. Nat Methods August 2010;7(8):615–8.

[167] Saliba A-EE, Westermann AJ, Gorski SA, Vogel J. Single-cell RNA-seq: advances and future challenges. Nucleic Acids Res January 3, 2014;42(14):8845–60.

[168] Ting DT, Wittner BS, Ligorio M, Vincent Jordan N, Shah AM, Miyamoto DT, et al. Single-cell RNA sequencing identifies extracellular matrix gene expression by pancreatic circulating tumor cells. Cell Rep September 4, 2014;8(6):1905–18.

[169] Smallwood SA, Lee HJ, Angermueller C, Krueger F, Saadeh H, Peat J, et al. Single-cell genome-wide bisulfite sequencing for assessing epigenetic heterogeneity. Nat Methods August 5, 2014;11(8):817–20.

[170] Laszlo AH, Derrington IM, Brinkerhoff H, Langford KW, Nova IC, Samson JM, et al. Detection and mapping of 5-methylcytosine and 5-hydroxymethylcytosine with nanopore MspA. Proc Natl Acad Sci USA November 2, 2013;110(47):18904–9.

[171] Schreiber J, Wescoe ZL, Abu-Shumays R, Vivian JT, Baatar B, Karplus K, et al. Error rates for nanopore discrimination among cytosine, methylcytosine, and hydroxymethylcytosine along individual DNA strands. Proc Natl Acad Sci USA November 2, 2013;110(47):18910–5.

[172] Wescoe ZL, Schreiber J, Akeson M. Nanopores discriminate among five C5-cytosine variants in DNA. J Am Chem Soc November 5, 2014;136(47).

CHAPTER

5

Biomarkers and Methodologies for Monitoring Epigenetic Drug Effects in Cancer

Marianne B. Treppendahl, Lasse Sommer Kristensen, Kirsten Grønbæk

Department of Hematology, Rigshospitalet, Copenhagen, Denmark

OUTLINE

Epigenetic Biomarkers and Diagnostics
http://dx.doi.org/10.1016/B978-0-12-801899-6.00005-X

1. INTRODUCTION

Cancer has long been regarded as a genetic disease since mutations that drive malignant transformation have been identified in almost all human cancers. However, it is now clear that most cancers are much more complex at the molecular level and cannot be explained in purely genetic terms. Molecular modifications of DNA and histone proteins, which are stably inherited through somatic cell divisions, have been shown to influence the transcription of genes without changing the DNA sequence. These alterations are known as epigenetic modifications and may be much more abundant in cancer cells than genetic changes. In particular, tumor suppressor genes are often silenced due to hypermethylated promoter regions and/or loss of histone acetylation, and oncogenes may be overexpressed due to specific histone modifications at promoters and enhancers such as hyperacetylation and bromodomain protein occupancy. Consequently, drugs aiming at resetting cancer-associated epigenetic alterations have been developed. Currently, two classes of epigenetic drugs are approved for clinical use: inhibitors of the DNA methyltransferases (DNMTs) and inhibitors of the histone deacetylases (HDACs). DNMT inhibitors (DNMTis) are approved by the US Food and Drug Administration (FDA) and the European Medicines Agency (EMA), while HDAC inhibitors (HDACis) are only approved by the FDA. In preclinical settings and clinical trials, many second-generation epigenetic drugs are currently tested, as reviewed elsewhere [1]; however, solid clinical data are still lacking.

Since the first-generation epigenetic drugs were approved, much focus has been on predicting response to epigenetic therapy. Only a few single-center studies, with a limited amount of patients, have identified biomarkers for predicting response to therapy with HDACis, probably due to the limited effect of HDACis as single-agent therapy. We have recently reviewed this subject [2], and no new potential biomarkers have been published since. Accordingly, in this contribution we have chosen to focus only on biomarkers predicting response to DNMTis.

1.1 DNMTis in the Treatment of Hematological Malignancies

Recent multicenter studies demonstrated that DNMTis have significant efficacy in the treatment of hematological malignancies [3–7]. Based on these studies, two different DNMTis, 5-azacytidine (azacytidine) and 5-aza-2′-deoxycytidine (decitabine), were approved by the FDA and EMA, however, not with the same indications. The FDA has approved both azacytidine and decitabine for all types of myelodysplastic syndrome (MDS, #614286), chronic myelomonocytic leukemia (CMML, MIM#607785) with 10–29% bone marrow blasts, and acute myeloid leukemia (AML, MIM#601626) with 20–30% bone marrow blasts. The EMA has approved azacytidine for MDS with IPSS (International Prognostic Scoring System) int-2 and high groups, CMML with 10–29% bone marrow blasts, and AML with 20–30% bone marrow blasts, while decitabine is approved for AML patients above 65 years who are not candidates for standard induction chemotherapy. Though azacytidine and decitabine significantly improve the survival of patients with MDS and AML, only about 50% of the patients achieve a clinical response [8], and prolonged therapy of 4–6 months is often required before treatment failure becomes evident. Accordingly, predictive biomarkers would be of great clinical value, as nonresponders could be spared 4–6 months of ineffective treatment with potential side effect, needless costs, and waste of precious time before considering other treatment options, e.g., alternative regimens currently available in clinical trials. The median duration of response is 12–18 months and relapse unfortunately seems to be almost unavoidable [4]. The outcome after treatment failure is exceedingly poor with a median survival of only 5.6 months [9]. Therefore, a better understanding of how these drugs work and why they fail in some patients is urgently needed in order to develop new therapies or combining DNMTis with other drugs.

1.2 How Do DNMTis Work?

Early in vitro studies indicate that DNMTis can reprogram somatic cells by DNA demethylation of aberrantly silenced genes. Under normal circumstances, the DNMTs copy the methylation pattern of the parental DNA strand after replication, ensuring that methylation patterns are maintained during cell division. During treatment, the DNMTis covalently bind and thus inactivate DNMTs. After successive cell divisions, the original DNA methylation pattern is lost (Figure 1) [10]. While decitabine is directly incorporated into DNA, azacytidine needs to be reduced at the diphosphate level by ribonucleotide reductase in order to be incorporated into

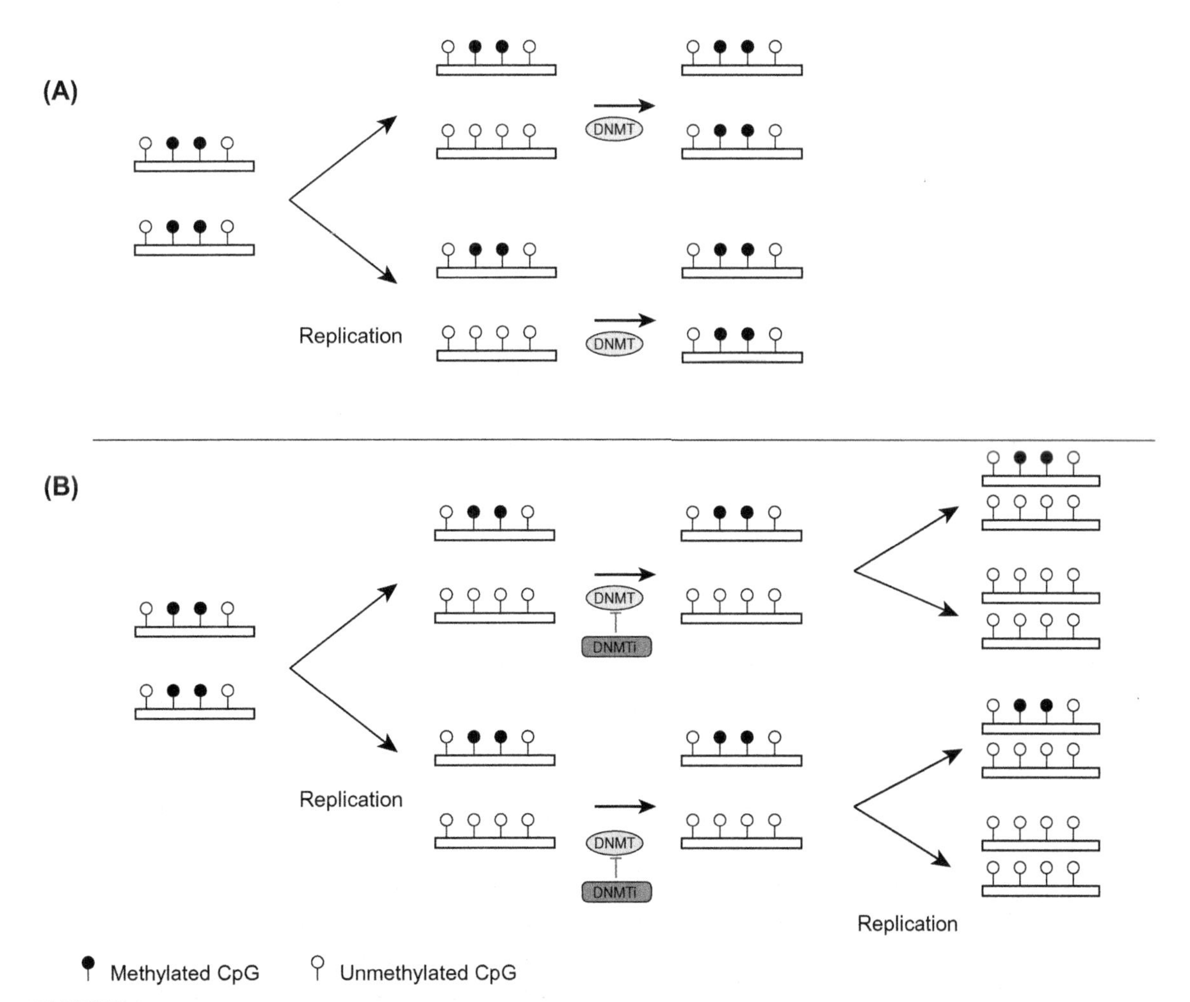

FIGURE 1 **Mechanism of action of DNMTis.** A) Under normal circumstances, the DNA methyltransferases (DNMTs) copy the methylation pattern of the parental DNA strand after replication, ensuring that methylation patterns are maintained during cell division. B) During treatment, DNMT inhibitors (DNMTis) are incorporated into DNA and RNA, where they covalently bind and thus inactivate DNMTs. After successive cell divisions, the original DNA methylation pattern is lost.

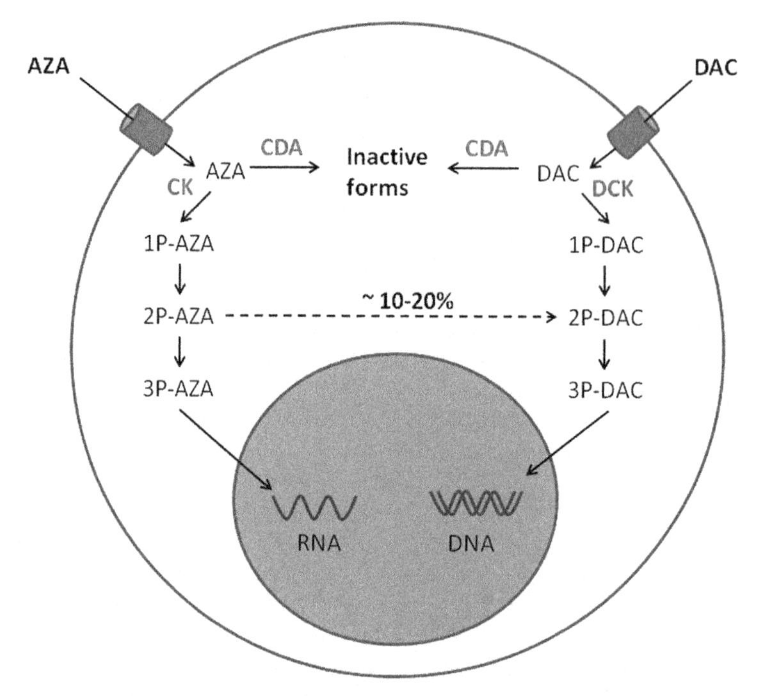

FIGURE 2 Cellular uptake and intracellular processing of azacytidine and decitabine. Azacytidine (AZA) and decitabine (DAC) enter the cell by the human nucleoside transporters. Once inside the cell, the drugs undergo an initial rate-limiting phosphorylation step by cytidine kinase (CK) or deoxycytidine kinase (DCK), which leads to the production of a monophosphate metabolite. A second and third phosphorylation step is then performed by nucleoside monophosphate kinase and nucleoside diphosphate kinase, respectively. At the diphosphate metabolite stage, approximately 10–20% of 2P-AZA is reduced to 2P-DAC by a ribonucleotide reductase. Triphosphates can be incorporated in DNA or RNA, in competition with their normal counterparts. Excess azanucleotides are rapidly deaminated to uracil by cytidine deaminase (CDA) and excreted by the kidneys [11].

DNA. It is estimated that about 10–20% of azacytidine undergoes this reduction, while the main part of azacytidine is incorporated into RNA (Figure 2) [11]. Nevertheless, the exact mechanisms of action of DNMTis in patients are currently unknown. The most obvious would be the reactivation of epigenetically silenced tumor suppressor genes and genes involved in normal differentiation as suggested by Gore et al. [12], but data are contradictory. We, and others, have shown that DNMTis can render the malignant cells immunogenic by induction of cancer/testis antigens (CTAs) [13–15]. This suggests that immune stimulation may be an important contributor to the clinical effect of these drugs. Inhibition of ribonucleotide reductase, leading to a reduced deoxyribonucleotide pool and impaired DNA synthesis and repair, has also been suggested as a possible mechanism of action for azacytidine [16], since 80–90% of the drug is incorporated into RNA. Several cellular pathways are affected by DNMTis leading to miscellaneous cellular responses (Figure 3).

Second-generation DNMTis are designed to improve the pharmacological profile. One of these, SGI-110, is currently in phase III clinical trial for AML (NCT02348489), in phase II for MDS (NCT01261312), ovarian cancer (MIM#167000) (NCT01696032), and advanced hepatocellular carcinoma (MIM#114550) (NCT01752933), while others are mainly being investigated in preclinical settings [17].

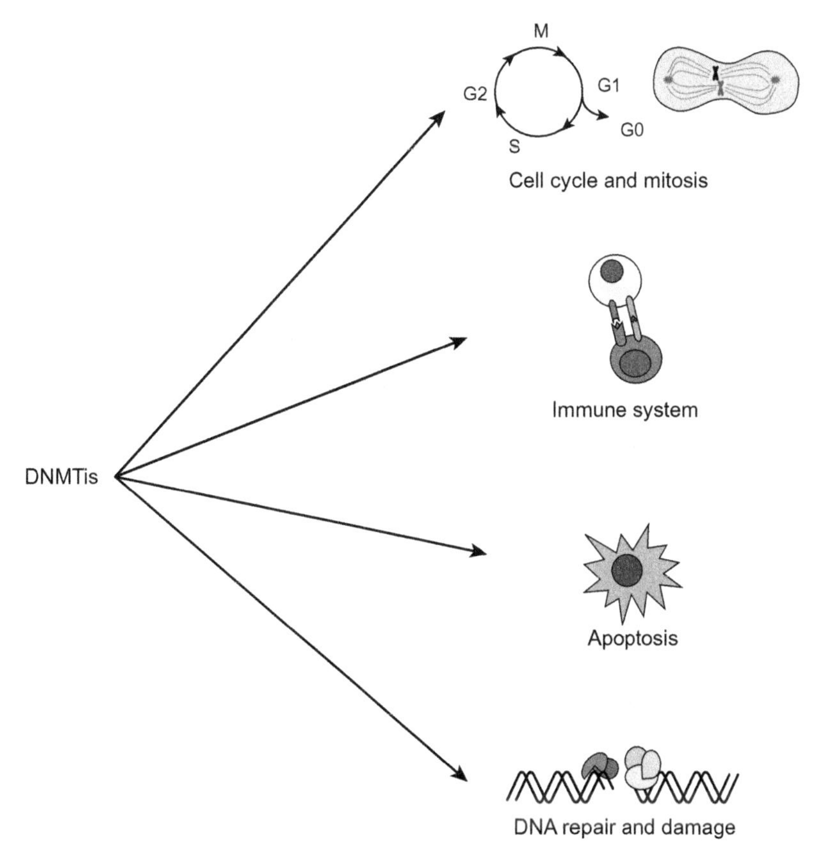

FIGURE 3 **Cellular pathways affected by DNMTis.** DNA methylation plays important roles in multiple cellular pathways, which all may be affected by DNMTis. Accordingly, the inhibition of DNMTs can lead to miscellaneous responses, each of which may require a different biomarker.

2. PREDICTING RESPONSE TO DNMTi TREATMENT

2.1 Pharmacological Factors with Potential Impact on DNMTi Resistance

Appropriate cellular uptake and intracellular processing are essential for the efficacy of azanucleosides (Figure 3). Both azacytidine and decitabine enter the cells by human nucleoside transporters (hNTs). Azacytidine and decitabine are transported by different hNTs and their cytotoxicity is dependent on the type of hNT present on the cells [18,19]. These observations suggest that hNTs may be useful biomarkers for the efficacy of DNMTis, however, clinical data are still not available.

Once inside the cell, the next crucial step in DNMTi processing is the initial monophosphorylation of azacytidine and decitabine by cytidine kinase (CK) and deoxycytidine kinase (DCK), respectively. It has been demonstrated in the HL60 cell line that disruption of DCK function by a point mutation results in resistance to decitabine [20]. While DCK mutations are rarely detected in patients [21], a borderline significant lower expression of DCK has been observed in nonresponders to decitabine [22].

TABLE 1 Molecular Markers for Response to DMNTi—Pharmacological Factors

Predictor	Patients types included	Treatment	Number of patients	Predict overall survival	Predict therapy response	References
CDA	MDS (not otherwise specified)	Azacytidine or decitabine	90	X	N.e	[23]
CDA/DCK ratio	MDS (all IPSS groups)	Decitabine	32	N.e	X	[22]

X, correlation; N.e, correlation not examined; CDA, cytidine deaminase; DCK, deoxycytidine kinase; MDS, myelodysplastic syndrome; IPSS, International Prognostic Scoring System.

Both azacytidine and decitabine are inactivated by cytidine deaminase (CDA) through an irreversible hydrolytic deamination of cytidine/deoxycytidine to uridine/deoxyuridine counterparts, and accordingly, a high CDA expression/activity decreases the half-life of the drugs. The expression level and enzymatic activity of CDA have been reported to influence on overall survival (OS) in patients treated with DNMTis. Males have been shown to have higher CDA expression/activity compared to women, and among 90 MDS patients treated with DNMTis, females had significantly better OS [23]. Although the observations in this study are indirect, this may suggest that CDA is involved in a gender-specific response. Nevertheless, conflicting results exist, with one other study showing gender-specific differences in OS [24], while in others, including the AZA-001 trial, a negative impact of male gender was not observed [4,6,25,26]. The novel drug SGI-110 is designed to overcome the effects of CDA [27], and it will be interesting to see if male patients do relatively better in the SGI-110 trial. Only a few studies have investigated the effect of combined disruption of DNMTi metabolic enzymes, however, one study of 32 MDS patients shows that the CDA/DCK ratio is negatively correlated with clinical response to decitabine [22] (Table 1).

2.2 Clinical Predictors

2.2.1 The French Prognostic Scoring System for MDS Patients

The French Prognostic Scoring System (FPSS) has been shown to separate azacytidine-treated, higher risk patients into three prognostic groups with significantly different OS. In total, 282 higher risk MDS patients (IPSS int-2 and high risk groups) treated with azacytidine were evaluated, and it was shown that bone marrow blasts >15%, abnormal karyotype, and previous treatment with low-dose cytarabine independently predicted poor response to azacytidine including Eastern Cooperative Oncology Group (ECOG) performance status ≥2, presence of circulating blasts, red blood cell transfusion dependency ≥4 units/8 weeks, and intermediate- or high-risk cytogenetics (according to IPSS), independently predicted poorer OS. Based on these factors, the FPSS was developed [28,29]. This prognostic score was later validated in the 161 higher risk MDS patients treated in the AZA-001 trial [4], which represent an independent, but highly selected, patient cohort. The prognostic score has recently been further validated in two independent patient cohorts of 60 and 90 patients, respectively [30,31]. In addition, the FPSS could identify patients with potential of obtaining complete response (CR), as all CRs were observed in the low- or int-1 and int-2 risk groups [30].

The FPSS includes the original IPSS cytogenetic risk groups, however, after the revised IPSS (IPSS-R) [32] was introduced in 2012, the initial patient cohort was reclassified according to the IPSS-R. The study shows that both the IPSS-R cytogenetic risk groups and the overall IPSS-R group have a strong prognostic value in MDS patients treated with azacytidine [33]. ECOG and Northern American Leukemia Intergroup have

TABLE 2 Clinical Markers for Response to DMNTi

Predictor	Patients types included	Treatment	Number of patients	Predict overall survival	Predict therapy response	References
French Prognostic Scoring System	MDS (int-2 and high risk)	Azacytidine	282 161	X	X	[28,29]
	MDS (int-2 and high risk) and CMML	Azacytidine	60	X	X	[30]
	MDS (int-1, int-2, and high risk), CMML and AML	Azacytidine	90	X	N.e	[31]
Splenomegaly bone marrow blast count	CMML	Azacytidine	76	X	%	[6]
Platelet doubling time	MDS (int-1, int-2, and high risk), CMML and AML	Azacytidine	90	X	N.e	[31]
	MDS, CMML, and AML	Azacytidine or azacytidine and entinostat	102	X	X	[35]
IPSS-R	MDS (int-2 and high risk)	Azacytidine	282	X	%	[33]
	MDS, CMML, and AML	Azacytidine or azacytidine and entinostat	150	X	%	[34]

X, correlation; %, no correlation; N.e, correlation not examined; MDS, myelodysplastic syndrome; IPSS-R, Revised International Prognostic Scoring System; CMML, chronic myelomonocytic leukemia; AML, acute myeloid leukemia.

subsequently compared the prognostic utility of the IPSS-R and the FPSS, and it was shown that neither of the scores predict response, but both discriminate patients with dissimilar OS. The FPSS was superior in OS prediction compared to the IPSS-R [34].

2.2.2 *Platelet Doubling Time*

In a cohort of 90 MDS, CMML, and AML patients treated with azacytidine, it was observed that an, at least, two-fold increase in platelet counts at the initiation of the second treatment cycle, as compared to the pretreatment values, was associated with significantly better OS [31]. This result has later been validated by the ECOG and Northern American Leukemia Intergroup in 102 MDS and AML patients treated with azacytidine. This study, however, shows that additional platelet doubling after first cycle of azacytidine does not improve the survival prediction by either the IPSS-R or the FPSS [35].

2.2.3 *Clinical Predictors in CMML Patients*

The impact of different clinical factors has been evaluated in a study of 76 CMML patients treated with azacytidine [6]. No predictive factors for clinical response were identified, however, increased bone marrow blast counts, splenomegaly, and high white blood cell counts were associated with significantly shorter OS. On the other hand, only bone marrow blast counts and splenomegaly retained impact on OS in multivariate analysis (Table 2).

2.3 Molecular Predictors

2.3.1 *Chromosomal Abnormalities*

Poor-risk cytogenetics has been associated with shorter response duration and shorter OS (see Section 2.2) [28,30,31]. However, among the patients with poor-risk cytogenetics, several groups have observed that patients with deletion or loss of chromosome 7 have better clinical response rates and a relatively favorable outcome after DNMTi treatment [4,5,26,28,36–38]. Karyotyping of blasts in metaphase is still the method of choice for cytogenetic risk stratification of MDS and AML patients, but recently whole-genome scanning technologies are being implemented as a supplement to classical cytogenetic analyses. Cluzeau et al. have investigated molecular genetics by single nucleotide polymorphism microarray (SNP-A) in 51 MDS and AML patients treated with azacytidine. All 51 patients had genomic modifications detected by SNP-A, 100% had genomic gains, 84% had genomic losses, and uniparental disomy was detected in 82% of the patients. Patients with more than 100 Mb genomic gains or losses had a significantly worse OS compared to patients with less genomic aberrations. A significant difference in overall response rate could not be shown [39] (Table 3).

TABLE 3 Molecular Markers for Response to DMNTi—Chromosomal Abnormalities

Predictor	Patients types included	Treatment	Number of patients	Predict overall survival	Predict therapy response	References
Poor-risk cytogenetics	MDS (int-2 and high risk) and CMML	Azacytidine	60	X	X	[30]
	MDS (int-1, int-2 and high risk), CMML and AML	Azacytidine	90	X	N.e	[31]
	MDS (int-2 and high risk)	Azacytidine	282 161	X	X	[28,29]
Isolated chromosome 7 abnormalities	MDS (int-2 and high risk), CMML and AML<30% blasts	Azacytidine	358	X	N.e	[4]
	MDS (all IPSS groups), AML<30% blasts	Azacytidine	34	N.e	X	[26]
	MDS (int-2 and high risk), CMML and AML<30% blasts	Decitabine	124	N.e	X	[36]
	MDS (int-1, int-2 and high risk), CMML and AML<30% blasts	Decitabine	170	N.e	X	[5]
	AML	Decitabine	23	N.e	X	[37]
SNP	MDS and AML	Azacytidine	51	X	%	[39]

X, correlation; %, no correlation; N.e, correlation not examined; MDS, myelodysplastic syndrome; IPSS, International Prognostic Scoring System; CMML, chronic myelomonocytic leukemia; AML, acute myeloid leukemia; SNP, single-nucleotide polymorphism.

2.3.2 *Point Mutations*

2.3.2.1 MUTATIONS IN EPIGENETIC REGULATORS

During the past decade mutations in epigenetic regulators have been identified in most cancers. In hematological cancers, mutations in enzymes that are involved in the regulation of DNA methylation are particularly frequent [40]. Even though it seems logical that mutations in these enzymes would influence the response to DNMTis, and therefore be of prognostic importance, the results have been contradictive. A correlation between clinical response and mutations in ten-eleven translocation 2 (*TET2*) (which is involved in DNA demethylation) has been reported in two studies. Itzykson et al. examined *TET2* mutations by Sanger sequencing in 86 MDS and AML patients treated with azacytidine [41]. Significantly more patients with *TET2* mutations responded to azacytidine. Bejar et al. examined 213 MDS patients treated with azacytidine or decitabine using targeted sequencing capable of detecting even small mutated subclones [42]. They found, however, that only the presence of *TET2* mutations at a >10% allelic burden predicted response to DNMTis. Neither of the two studies showed a difference in OS between cases with and without *TET2* mutations. Meanwhile, no correlation between *TET2* mutational status and clinical response or OS was observed in studies of 38 higher risk MDS patients treated with a combination of azacytidine and valproic acid (an HDACi) [43] and also in 39 CMML patients treated with decitabine [44].

A positive correlation between mutations in the *DNMT3A* and clinical response has been observed in 46 AML patients treated with decitabine. However, this response did not translate into an OS benefit [45]. On the other hand, a correlation between *DNMT3A* mutations and clinical response could not be shown in a cohort of 68 AML patients treated with azacytidine or decitabine with 62% of the patients receiving concomitant therapy with either vorinostat or valproic acid [46]. In the same study, a lack of correlation between *IDH1/2* mutations and outcome was also observed.

The impact of several point mutations on the response has been examined in two studies. Traina et al. examined 10 genes in 92 MDS, MDS/MPN (myeloproliferative neoplasm), and MDS/AML patients treated with either azacytidine, azacytidine and lenalidomide, decitabine, or decitabine and azacytidine [47]. *TET2* and/or *DNMT3A* mutations were associated with a better overall response rate and progression-free survival, but not with OS. Mutations of the putative polycomb-associated protein *ASXL1* were correlated to poor OS, while mutations of the splice factor 3B, *SF3B1*, were associated with better OS. However, these data need confirmation, since this patient cohort was heterogeneous with regard to both diagnosis and choice of treatment modalities, and only around 50% of the examined samples were collected before the initiation of DNMTi treatment. In the more recent study (Bejar et al.), targeted sequencing of 40 genes was conducted. As already mentioned, >10% mutated *TET2* alleles correlated to a better clinical response, particularly in patients not carrying *ASXL1* mutations. Most importantly, no particular mutation pattern correlated to DNMTi resistance, and responses were observed even in patients with mutations that are known to be associated with very poor prognosis, e.g., *TP53* (see Section 2.3.2.1) [42].

2.3.2.2 *TP53* MUTATIONS

Mutation of the tumor suppressor gene *TP53* is clearly associated with poor OS and drug resistance [48–50]. The impact of mutated *TP53* on response to azacytidine treatment has recently been examined in four different cohorts of MDS, AML, CMML, and MDS/MPN patients, consisting of 44, 62, 100, and 213 patients, respectively [42,48–50]. All four studies show that the presence of *TP53* mutations does not have a

TABLE 4 Molecular Markers for Response to DMNTi—Point Mutations

Predictor	Patients types included	Treatment	Number of patients	Predict overall survival	Predict therapy response	References
TET2 mutation	MDS (int-1, int-2 and high risk) and AML	Azacytidine	86	%	X	[41]
	MDS (int-2 and high risk) and CMML	Azacytidine	38 27	%	%	[43]
	CMML	Decitabine	39	%	%	[44]
	MDS (all IPSS groups)	Azacytidine or decitabine	213	%	X	[42]
DNMT3A mutation	AML	Decitabine	46	%	X	[45]
	AML	Azacytidine + vorinostat or valproric acid or decitabine ± vorinostat or valproric acid	68	%	%	[46]
TP53 mutation	MDS, AML, CMML, and MDS/MPN	Azacytidine	44	N.e	%	[49]
	MDS, AML, and CMML	Azacytidine	62	X	%	[48]
	MDS (int-2 and high risk) AML and MDS/MPN	Azacytidine	100	%	%	[50]
	MDS (all IPSS groups)	Azacytidine or decitabine	213	X	%	[42]

X, correlation; %, no correlation; N.e, correlation not examined; MDS, myelodysplastic syndrome; IPSS, International Prognostic Scoring System; CMML, chronic myelomonocytic leukemia; AML, acute myeloid leukemia; MPN, myeloproliferative neoplasm; TET2, ten-eleven translocation 2; DNMT3A, DNA methyltransferase 3A; TP53, tumor protein p53.

negative impact on treatment response. Müller-Thomas et al. actually found that MDS patients with *TP53* mutations have a significantly higher overall response rate to azacytidine, but this was not reflected in OS, where a non-significant trend towards poorer OS for patients with *TP53*-mutated MDS was observed [50]. Bally et al. and Bejar et al. show poorer OS for patients with *TP53* mutations, which is probably associated with the shorter response duration in the *TP53*-mutated cases [42,48,50] (Table 4).

2.3.3 *Gene Expression*

2.3.3.1 *CJUN* AND *CMYB* EXPRESSION

One study has identified expression of two proto-oncogenes, *CJUN* and *CMYB*, as potential predictive biomarkers in a cohort of 36 CMML patients treated with decitabine [44]. *CJUN* has previously been shown to promote aberrant monocyte transformation [51]. *CJUN* was significantly downregulated in monocytes from responding patients, and high *CJUN* expression was correlated with shorter survival [44].

Deregulation of *CMYB* has been implicated in leukemia [52], and higher *CMYB* expression was also associated with short survival [44].

2.3.3.2 *MLL5* EXPRESSION

The human trithorax-group gene *MLL5* was originally identified from molecular mapping of the frequently deleted region of chromosome 7q22 in patients with myeloid malignancies [53]. Other trithorax-group proteins have well-documented histone lysine methyltransferase (HKMT) activity; however, the role of *MLL5* as a new HKMT has long been debated [54]. High *MLL5* expression levels have been associated with favorable outcome in core-binding factor AML and cytogenetic normal AML [54]. In a study of 57 AML patients treated with decitabine, it was shown that patients with high *MLL5* expression levels in the leukemic blasts had better OS than patients with low *MLL5* expression levels. Furthermore, it was shown that low *MLL5* expression was associated with significantly lower genome-wide promoter DNA methylation levels, suggesting a correlation between global DNA methylation levels and *MLL5* expression [54].

2.3.3.3 PHOSPHOINOSITIDE-PHOSPHOLIPASE C β1

Phosphoinositide-phospholipaseCβ1(PLCβ1) is a key enzyme in lipid signaling pathways and is involved in cell proliferation and differentiation. PLCβ1 is also highly expressed in the early stages of normal hematopoietic differentiation [55]. In higher risk MDS patients, *PLCβ1* is hypermethylated and could be a specific target for azacytidine. An increase in *PLCβ1* expression and a decrease in *PLCβ1* methylation were observed in 9 out of 10 patients with hematological response, while a decrease in *PLCβ1* expression and an increase in *PLCβ1* promoter methylation were observed in all eight nonresponding patients [56]. In later studies, the same group observed a similar association in two cohorts of 32 and 26 azacytidine-treated, low-risk MDS patients, respectively [57,58]. In the latter cohort, the PLCβ1 target cyclin D3 was induced in responding patients, supporting the notion that the PLCβ1 pathway is activated during azacytidine treatment [58]. Due to the involvement of PLCβ1 in early hematopoietic differentiation, it is hypothesized that PLCβ1 upregulation by promoter demethylation leads to differentiation of the abnormal hematopoietic progenitor cells in MDS.

2.3.3.4 MICRORNA-29b

MicroRNA-29b (miR-29b) is involved in the regulation of DNA methylation by targeting the DNMTs, DMNT3A/3B and DNMT1 [59,60]. A positive correlation between the clinical response and high pretreatment levels of miR-29b was observed in a phase II clinical trial in older AML patients treated with decitabine [37]. Recently, in vitro studies from the same group have shown that priming of AML cell lines and primary AML blasts with a new HDACi (AR-42) leads to upregulation of miR-29b expression and enhanced antileukemic effect of subsequently administered decitabine [61]. However, a lack of association between pretreatment miR-29b expression levels and clinical responses to azacytidine in AML patients was subsequently reported [62]. The conflicting results obtained by these studies may be explained by the different sources used for miR analysis (peripheral blood vs bone marrow) and the use of different DNMTis in the two studies.

2.3.3.5 BCL-2 FAMILY PROTEINS

The anti-apoptotic BCL-2 family proteins have recently, in an RNA interference screen, been identified as important modulators of the antileukemic effect of azacytidine. In vitro studies of myeloid cell lines and primary AML, MDS, and MPN samples showed a synergistic effect of different BCL-2 family inhibitors and azacytidine. In 22 AML, MDS, and MPN patients treated with azacytidine, BH3 profiling was done to functionally interrogate the overall balance of pro- and anti-apoptotic BCL-2 family proteins [63]. The underlying principle of BH3 profiling is that mitochondrial depolarization following BH3 peptide exposure serves as a functional biomarker for cellular response to pro-apoptotic cues, described as "primedness" [64]. Patients who obtained a clinical response to azacytidine treatment had a significantly higher level of BH3 profiling "primedness" compared to nonresponding patients [63].

2.3.3.6 FAS EXPRESSION

Expression of the pro-apoptotic protein Fas in $CD45^{lo}/CD34^{+}$ bone marrow cells from low- and high-risk MDS and secondary AML patients has been positively correlated to response to azacytidine. A correlation between promoter methylation and Fas expression was also observed. Among 63 patients, low Fas expression at diagnosis (presumably due to promoter hypermethylation) was correlated to clinical response, while no association between Fas expression and OS was observed [65]. For a subgroup, consisting of 38 patients, Fas expression was examined before and after ≥3 cycles of azacytidine, and significant increase in Fas expression was observed in responding patients (23/38).

2.3.3.7 CD25 ANTIGEN EXPRESSION

Expression of the α-chain of the interleukin-1 receptor, CD25 antigen, has been associated with poor outcome in adults with AML [66] or acute lymphoblastic leukemia [67], and it has been suggested that $CD25^{+}$ blasts are enriched in chemoresistant leukemia stem cells. Patients with AML developed from MDS, more often express CD25 compared to de novo AML patients. In a cohort of 61 higher risk MDS and CMML patients treated with azacytidine, CD25 expression was associated with significantly shorter median OS and median event-free survival [68].

2.3.3.8 ABERRANT MYELOID PROGENITORS

The predictive value of immune phenotyping, by flow cytometry, for response to azacytidine was recently investigated in 42 MDS, CMML, or AML patients. It has previously been shown that a flow cytometric scoring system (FCSS) was able to estimate survival and relapse after allogeneic hematopoietic stem cell transplantation in MDS patients [69]. This score is generated by weighed scoring of the number of aberrations in the maturing myelomonocytic compartment and the percentage of progenitor cells, and in each category a maximum of 4 points can be obtained. A significantly worse OS was observed in patients with high pretreatment FCSS (6–8) and/or aberrant immunophenotype of the myeloid progenitors. Furthermore, a decline in FCSS during treatment with azacytidine was correlated to clinical response [70].

2.3.3.9 REGULATORY T CELLS

Regulatory T cells (Tregs) are central mediators of immunosuppression and crucial for the maintenance of self-tolerance, but are, however, also believed to play an important role in maintenance of immune tolerance in cancer. Tregs are characterized by a high expression of the transcription factor FOXP3, which has been shown to be essential for their function [71]. The FOXP3 expression in T cell is regulated by DNA methylation, with promoter hypomethylation in Tregs and promoter hypermethylation in naïve and effector T cells [72]. It has previously been shown that the number of Tregs is significantly increased in higher risk MDS patients [73]. In a recent study of 68 MDS patients treated with azacytidine, it was shown that the percentage and the absolute numbers of Tregs at diagnosis were higher in nonresponders compared to the responders. The total number of Tregs decreased during continued treatment with azacytidine; but despite this reduction in numbers, the remaining Tregs expressed increased levels of FOXP3 compared to untreated cells [74] (Table 5).

TABLE 5 Molecular Markers for Response to DMNTi—Gene Expression

Predictor	Patients types included	Treatment	Number of patients	Predict overall survival	Predict therapy response	References
CJUN CMYB	CMML	Decitabine	36	X	X	[44]
MLL5	AML	Decitabine	57	X	N.e	[54]
BLC2 familiy	MDS, AML, MPN	Azacytidine	22	N.e	X	[63]
CD25	MDS, CMML	Azacytidine	61	X	N.e	[68]
Aberrant progenitors	MDS, AML and CMML	Azacytidine	42	X	X	[70]
Treg	MDS	Azacytidine	68	N.e	X	[74]
Fas	MDS (all IPSS), AML	Azacytidine	38	%	X	[65]
PI-PLCβ1	MDS (int-2 and high risk) and AML	Azacytidine	18	N.e	X	[56]
	MDS (low and int1 risk)	Azacytidine	26	N.e	X	[58]
	MDS (low and int-1 risk)	Azacytidine	32	N.e	X	[57]
Mir-29b	AML	Decitabine	23	N.e	X	[37]
	AML	Azacytidine, valporic acid, and ATRA	45	%	%	[62]

X, correlation; %, no correlation; N.e, correlation not examined; MDS, myelodysplastic syndrome; IPSS, International Prognostic Scoring System; CMML, chronic myelomonocytic leukemia; AML, acute myeloid leukemia; MPN, myeloproliferative neoplasm; TET2, ten-eleven translocation 2; DNMT3A, DNA methyltransferase 3A; TP53, tumor protein p53.

2.3.4 DNA Methylation

A substantial amount of data highlights the importance of aberrant DNA methylation in carcinogenesis [75]. As already mentioned, the most obvious mechanism of action of DNMTis is demethylation of aberrantly methylated promoters, for example, tumor suppressor genes or genes involved in normal differentiation [12]. This demethylation will lead to a reexpression of the silenced genes. Accordingly, numerous studies have examined whether responses to DNMTis can be predicted by pretreatment DNA methylation levels at individual gene promoters, by combinations of gene promoters, or by global DNA methylation screening.

2.3.4.1 *CDKN2B* METHYLATION

One of the best-studied epigenetic events in MDS is hypermethylation of *CDKN2B*. *CDKN2B* encodes the cell cycle inhibitor p15, which controls the progression from G1 to S phase; p15 is expressed selectively during myeloid and megakaryocytic differentiation [26]. Hypermethylation of *CDKN2B* is frequently observed in MDS and is an independent prognostic factor [26]. Accordingly, the relation between the clinical response to DNMTi and *CDKN2B* methylation status has been extensively studied in MDS [26,76–80]. However, the conclusions of these studies diverge. Some reported a positive correlation between low-level pretreatment *CDKN2B* methylation and clinical response, while others observed a correlation between *CDKN2B* demethylation/expression during decitabine treatment and clinical response [12,76,79]. Yet, other groups did not detect any correlation at all [25,80,81]. These variations can probably be explained by variation in patient groups, combinations of epigenetic therapies, and methodologies for monitoring DNA methylation.

2.3.4.2 *BCL2L10* METHYLATION

Methylation of another gene, *BCL2L10*, which is an anti-apoptotic BCL-2 family member, has also been studied with varying results. Both a negative correlation between *BCL2L10* promoter methylation and response to azacytidine, and association with significantly poorer OS in patients with more than 50% *BCL2L10* methylation, have been reported in several studies. These results were based on an initial analysis of 38, and a validation in 27, azacytidine-treated, higher risk MDS patients [43]. By contrast, others showed that azacytidine-resistant MDS/AML patients have an increased fraction of BCL2L10 positive cells in the bone marrow, and that patients with low BCL2L10 expression had significantly better OS [82].

2.3.4.3 METHYLATION SIGNATURES

In a large study of 317 MDS patients, a methylation signature consisting of 10 hypermethylated genes (*CDH1*, *CDH13*, *ERα*, *NOR1*, *NPM2*, *OLIG2*, *CDNK2B*, *PGRA*, *PGRB*, and *RIL*) has been identified among 24 genes previously shown to be hypermethylated in MDS/AML. Reduction of methylation after more than 4 months of treatment was positively correlated to clinical response in a subgroup of 34 patients treated with decitabine [25]. However, pretreatment methylation levels of these genes were not correlated to clinical response.

In non-small-cell lung cancer (NSCLC) promoter methylation of four genes (*APC*, *RASSF1A*, *CDH13*, and *CDKN2A*) has been shown to correlate negatively with survival. Methylation analysis of these four genes in plasma samples from 26 NSCLC patients before treatment with azacytidine and the HDACi entinostat was associated with higher clinical response rates in patients with methylation of two or more genes [83].

Abáigar et al. examined the methylation level of 24 known tumor suppressor genes in 63 MDS/AML patients treated with azacytidine and found that patients with methylation of ≥2 of the genes had a shorter OS [38]. However, the number of methylated genes did not correlate with the treatment response to azacytidine.

2.3.4.4 PROGRAMMED DEATH-1 METHYLATION

Programmed death-1 (PD-1) is an immunoinhibitory receptor mainly expressed on activated T cells, which is regulated by DNA methylation [84]. The major role of PD-1 is to limit T cell effector responses in peripheral tissues during infection and inflammation and to limit autoimmunity. In addition, PD-1 plays an important role in tumor immunology [85]. Hypomethylation of the PD-1 promoter was observed in $CD8^+$ T cells with inhibited function, the so-called exhausted T cells. In a cohort of 18 MDS, CMML, and AML patients treated with 5-aza in combination with the HDACi vorinostat, it was shown that the treatment led to a demethylation of the *PD-1* promoter in peripheral blood mononuclear cells (PBMNCs) and the pretreatment methylation levels of *PD-1* promoter in PBMNCs were significantly higher in nonresponders [86]. Recently, our group has shown that *PD-1* demethylation occurs in both $CD8^+$ and $CD4^+$ T cells and is followed by PD-1 mRNA and protein upregulation during treatment with azacytidine, in a study of 27 MDS, AML, and CMML patients. Patients without *PD-1* promoter demethylation during treatment showed a significantly higher overall response rate and a trend toward better OS [87]. Thus, it is likely that PD-1 gene reactivation by demethylation during treatment is more important for response than the actual baseline methylation levels.

2.3.5 Global Methylation

Examination of the methylation status of repetitive elements in PBMNCs during treatment has been another approach to monitor treatment efficacy by DNMTi. This surrogate marker is attractive since the sampling of peripheral blood during treatment is easy, the analysis is independent of the presence of tumor cells in the sample, and the assay is feasible for any routine laboratory. Significant demethylation of LINE-1 and *Alu* elements during treatment has been shown by several studies for both azacytidine and decitabine [78–80]. However, a prognostic effect of neither pretreatment methylation levels nor methylation changes during treatment has been substantially documented.

A recent study analyzed the global DNA methylation levels by deep sequencing analysis of methylated DNA captured by methyl-binding protein (MBD2) (MethylCap-seq) in 16 AML patients treated with decitabine. A trend towards a higher baseline methylation level and more pronounced methylation decrease during treatment was observed among responding patients [88].

For the time being no predictive markers based on methylation arrays have been published (Table 6).

TABLE 6 Molecular Markers for Response to DMNTi—Methylation

Predictor	Patients types included	Treatment	Number of patients	Predict overall survival	Predict therapy response	References
CDKN2B methylation	MDS (all IPSS), AML<30% blasts	Azacytidine	34	N.e	X	[26]
	MDS (int-2 and high risk), CMML, and AML	Azacytidine and entinostat	30	N.e	-	[80]
	MDS (int-1, int-2 and high risk), CMML and AML	Decitabine	23	N.e	X	[76]
	AML, CML	Decitabine	41	N.e	X	[78]
	MDS, AML	Decitabine and valproic acid	54	N.e	X	[77]
	MDS, CMML	Decitabine	95	N.e	%	[79]
	AML, MDS, CML, ALL	Decitabine	50	N.e	%	[81]
PD-1 methylation/ expresssion	MDS, AML, and CMML	Azacytidine and vorinostat	18	N.e	X	[86]
	MDS (int-2 and high risk), AML, and CMML	Azacytidine	27	%	X	[87]
BCL2L10 methylation/ expresssion	MDS (int-2 and high risk) and CMML	Azacytidine	38 27	X	X	[43]
	MDS (int-2 and high risk), and AML<30% blasts	Azacytidine	77	X	X	[82]

Continued

TABLE 6 Molecular Markers for Response to DMNTi—Methylation—cont'd

Predictor	Patients types included	Treatment	Number of patients	Predict overall survival	Predict therapy response	References
10-Gene methylation signature	MDS (all IPSS), CMML	Decitabine	34	X	%	[25]
4-Gene methylation signature	NSCLC	Azacytidine and entinostat	26	N.e	X	[83]
≥2 Hypermethylated TSG	MDS (all IPSS), AML	Azacytidine	63	X	%	[38]
Global methylation	AML	Decitabine	16	N.e	X	[88]

X, correlation; %, no correlation; N.e, correlation not examined; MDS, myelodysplastic syndrome; IPSS, International Prognostic Scoring System; CMML, chronic myelomonocytic leukemia; AML, acute myeloid leukemia; NSCLC, non-small-cell lung cancer.

3. MONITORING SINGLE-LOCUS DNA METHYLATION BIOMARKERS IN CLINICAL SAMPLES

Monitoring DNA methylation changes in clinical samples is not straightforward compared to analyses of genetic changes. This is mainly because the potentially methylated regions, such as CpG islands, often contain a large number of CpG sites that may not be uniformly methylated. Along the same DNA molecule, some CpG sites may be methylated while others may remain unmethylated and, in addition, CpG sites may be differentially methylated between the two copies of the genome within a cell. Finally, the methylation status of CpG sites may also differ between different cell types, and even cells of the same type may be differentially methylated (Figure 4). Thus, different results may be obtained depending on the specific CpG sites analyzed, in particular, when analyzing clinical samples composed of more than one clone. The choice of method for monitoring single-locus DNA methylation biomarkers in clinical samples is not always straightforward. For some biomarkers, such as

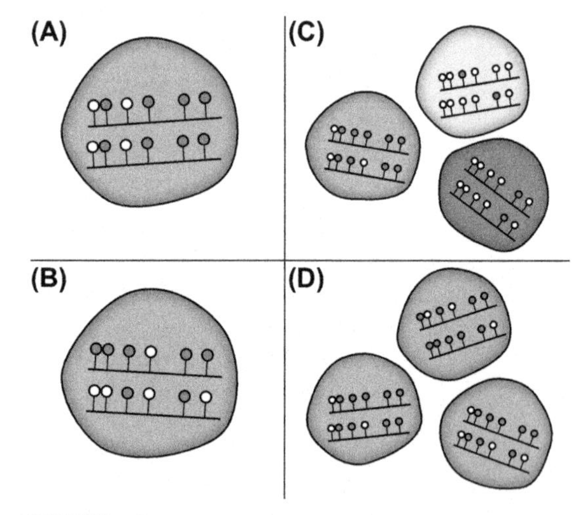

FIGURE 4 **Heterogeneous methylation.** Potentially methylated regions, such as cytosine–phosphate–guanine (CpG) islands, often contain a large number of CpG sites (denoted as lollipops), which may not be uniformly methylated. (A) Along the same DNA molecule, some CpG sites may be methylated (denoted as dark circles) while others may remain unmethylated (denoted as open circles). (B) CpG sites may also be differentially methylated between the two copies of DNA within a cell. (C) The methylation status of CpG sites often differs between different cell types. (D) Even cells of the same type may be differentially methylated.

LINE-1 methylation (PyroMark Q24 CpG LINE-1, from Qiagen), commercial kits are available, which have been thoroughly evaluated and validated. However, for the vast majority of potential DNA methylation biomarkers, commercial kits are not available, and a proper method should be selected for the analysis based on the information needed (e.g., qualitative/quantitative data), sample requirements, equipment needed, costs, hands-on-time, sensitivity, specificity, reproducibility, etc.

3.1 Polymerase Chain Reaction-Based Methods Rely on Sodium Bisulfite Treatment of the DNA

Single-locus DNA methylation biomarkers may be analyzed using methylation-sensitive restriction enzymes, antibodies specific for methylcytosine (5mC), or using polymerase chain reaction (PCR). There are several advantages of PCR-based methods including superior analytical sensitivity and specificity, and, currently, the most popular methods for the analysis of DNA methylation at single loci are based on PCR [89]. However, since DNA polymerases do not distinguish between 5mC and unmethylated C it is necessary to pretreat the DNA in order to conserve the DNA methylation information during PCR. Sodium bisulfite treatment is the method of choice for this purpose as it converts unmethylated Cs to uracil (U) while 5mCs remain as C. During the subsequent PCR, U will be amplified as thymine (T). Therefore, it can be inferred that CpG sites, which have been changed to TpG sites after bisulfite conversion, were unmethylated [90]. This is based on the assumption that the bisulfite treatment has converted all unmethylated Cs to Us. Many commercial suppliers produce kits for this purpose, which generally show very good conversion rates. Nevertheless, the sodium bisulfite treatment may still be expected to introduce some variability in DNA methylation analysis. Finally, it should be mentioned that 5mC may be further oxidized to hydroxymethylcytosine (5hmC) by TET proteins [91]. Although this epigenetic mark is found at very low levels in most cell types, with the exception of embryonic stem cells and Purkinje neurons [92], particular genomic regions such as enhancers are enriched for 5hmC. This is important to keep in mind when using bisulfite-based methods, since they cannot distinguish 5hmC from 5mC and its oxidized derivatives [93].

3.2 Many Different PCR-Based Methods Have Been Developed

During the last few decades, a plethora of different PCR-based methods have been developed for the analysis of DNA methylation [94]. Depending on the method of choice, different results may be obtained. Therefore, it is important to select the most optimal method for the detection of a particular DNA methylation biomarker, as the use of a suboptimal method may compromise the diagnostic sensitivity and specificity of the biomarker. Most importantly, the method of choice should be cost-efficient, user friendly, based on a closed-tube system to avoid PCR contamination, and highly reproducible while not producing false-positive or false-negative results. The ability to quantitatively assess the methylation levels within clinical samples may also be helpful in order to establish a cutoff, which provides the highest possible diagnostic sensitivity and specificity of the biomarker. Quantitative data are also instrumental for the analysis of loci undergoing DNA hypomethylation during cancer progression [95]. Finally, it is also important that the method of choice meets the particular demand for throughput and is compatible with the samples to be analyzed. More detailed information on the role of methylation-specific PCR can be found in Chapter 8 "The role of Methylation-specific PCR and associated techniques in clinical diagnostics" by Zhao and Bapat.

3.3 Challenges of Analyzing Clinical Samples

When studying potential DNA methylation biomarkers in cancer, the clinical samples may often contain significant infiltrates of normal cells if no micro- or macrodissections or cell sorting are performed before DNA extraction. In hematological malignancies, peripheral blood and bone marrow samples are composed of a wide range of different cells that may be differentially methylated with respect to the DNA methylation biomarker analyzed. Thus, in case of monitoring the biomarker in consecutive samples, for instance, during treatment, changes in methylation may reflect a mere change in the composition of the cell types within the samples. Other clinical samples derived from body fluids, such as plasma, serum, sputum, or urine, may also contain large proportions of normal cells making it difficult to detect a signal from the cancer cells within the sample. In these situations, it may be advisable to choose a very sensitive method.

Another potential problem is that clinical samples often have been conserved as formalin-fixed paraffin-embedded (FFPE) tissues. DNA extracted from FFPE tissues is often highly degraded, and it is important to be aware that the quality of DNA may vary substantially depending on time of fixation and storage time. Assessing the quality and quantity of the extracted DNA before bisulfite conversion is important to exclude samples of very poor quality before further analyses. In our experience, the DNA quantity measured with a spectrophotometer may not be comparable with measurements provided by chip-based capillary electrophoresis or intercalating dyes, as these methods have different sensitivities toward degraded DNA. The quality may be measured using chip-based capillary electrophoresis or using a multiplex PCR assay [96]. Nevertheless, DNA methylation can be studied successfully in FFPE samples stored for decades by PCR-based methods when small amplicons are designed [97]. In addition, it may often be useful to adjust the number of PCR cycles and/or the DNA input in the reactions when using FFPE tissues, and it may be expected that the results differ somewhat when comparing couples of matched cryopreserved and FFPE samples [98]. In addition, caution should be taken when interpreting data derived from very low amounts of input DNA, since formalin fixation may lead to hydrolytic deamination of mC to T and C to U, which may give rise to erroneous results after PCR amplification [99].

3.4 PCR Primer Design Strategies

Two different strategies have been employed for the design of PCR primers for amplification of bisulfite-treated DNA. Methylation-independent PCR (MIP) primers are designed to amplify methylated DNA and unmethylated DNA proportionally, whereas methylation-specific PCR (MSP) primers are designed to amplify only methylated DNA (Figure 5). Since the sense and antisense strands are no longer complementary after bisulfite conversion, the PCR primers may be designed for either of the two strands. Thus, if primer design is difficult for the selected strand, for instance, due to long stretches of T, it may be possible to select a more suitable region for the primers if designed for the other strand.

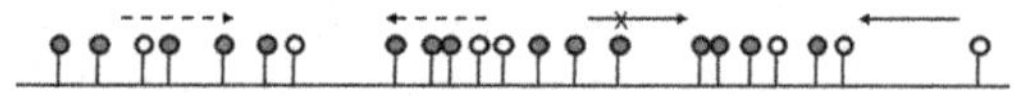

FIGURE 5 **Primer design for bisulfite-converted DNA.** Two different strategies have been employed for the design of polymerase chain reaction (PCR) primers for bisulfite-converted DNA. Methylation-independent PCR (MIP) primers (denoted as solid arrows) are designed to amplify methylated DNA and unmethylated DNA proportionally. If CpG sites cannot be avoided in MIP primers, the position in the primer overlaying the C (denoted as X) should be a base that pairs with none of the possible methylation states. Methylation-specific PCR primers (denoted as dashed arrows) are designed to amplify only methylated DNA by overlaying several CpG sites. CpG sites are denoted as lollipops, methylated Cs as dark circles, and unmethylated Cs as open circles.

Common for both MIP and MSP primers is that non-CpG Cs should be included in the primer sequences to select against the amplification of incompletely converted molecules. A list of the main criteria for the design of primers for DNA methylation studies can be found in Table 7.

3.5 MSP-Based Methods

Generally, methods based on MSP primers provide the most sensitive assays for DNA methylation detection, but cannot provide methylation information on individual CpG sites. The specificity for methylated template is achieved by including many CpG sites in the primer sequences, preferably at or close to the 3′ end. Conventional MSP uses gel electrophoresis to confirm the presence of DNA methylation in a sample. If no band is present on the gel, this may imply that the sample was unmethylated or that the DNA was lost during bisulfite treatment. Therefore, it is important to design a second set of primers specific for unmethylated DNA to confirm that the DNA was not lost during bisulfite treatment [100]. The use of real-time PCR has improved MSP-based assays significantly, as quantitative methylation information may be obtained using a closed-tube system, and false-positive results can be reduced to a minimum [101,102]. Also, the use of locked nucleic acids (LNAs) in primers and probes increases the sensitivity and specificity of MSP-based assays as they demonstrate increased affinity and specificity for their cognate DNA sequences [103]. Nevertheless, it is always important to include appropriate positive and negative controls when using MSP-based methods [94]. In particular, bisulfite-treated DNA, which is completely unmethylated, is a crucial negative control when using MSP. Finally, MSP has also been coupled with pyrosequencing to limit false-positive results [104] and to gain information on the methylation status of individual alleles if a heterozygous SNP is present in between

TABLE 7 List of the Main Criteria for the Design of Primers for DNA Methylation Studies

	MSP primers	MIP primers	Useful links
In silico bisulfite converts the template	✓	✓	http://biq-analyzer.bioinf.mpi-inf.mpg.de/tools/BiConverter/index.php
Match the melting temperatures of the forward and reverse primers	✓	✓	http://www.basic.northwestern.edu/biotools/oligocalc.html http://eu.idtdna.com/calc/analyzer
Single-base mismatch DNA thermodynamics		✓[a]	http://eu.idtdna.com/calc/analyzer
Avoid primer dimers	✓	✓	http://eu.idtdna.com/calc/analyzer
Avoid SNPs in primer sequences	✓	✓	http://www.ncbi.nlm.nih.gov/snp
Include non-CpG cytosines in primer sequences	✓	✓	
Include CpG sites in primer sequences	✓		
Avoid long stretches of A and T bases in primer sequences	✓	✓	
Reverse complement the reverse primer	✓	✓	http://www.bioinformatics.org/sms/rev_comp.html

[a] *Only applicable for primers where CpG sites cannot be avoided.*

the primers [105]. Since low-level methylation and/or allele-specific methylation may not always lead to gene silencing [106], the use of a purely qualitative method such as conventional MSP may compromise the sensitivity and specificity of many DNA methylation biomarkers.

3.6 MIP-Based Methods

On the other hand, MIP-based methods may provide a more detailed assessment of the methylation status of the locus under investigation, depending on the method used for the methylation analysis. Melting curve analysis is useful for screening samples, and provides a semiquantitative estimate of the methylation levels, but does not give information on individual CpG sites [107]. However, PCR products can be subjected to Sanger sequencing or pyrosequencing, if one of the primers has been labeled with biotin, after the melting analysis. Direct Sanger sequencing of MIP products can be used to gain information on individual CpG sites. However, due to a limited sensitivity and quantitative accuracy, Sanger sequencing is often used to sequence cloned PCR products [90]. While this approach has been considered the gold standard for DNA methylation analysis, it is labor-intensive, cost-inefficient, not performed in a closed-tube system, and may therefore not be suitable for use in clinical settings. Pyrosequencing may provide more sensitive and quantitatively accurate DNA methylation data [108], but common to all MIP-based methods is that it is difficult to achieve amplification completely independent of DNA methylation status [109]. Therefore, we recommend that all MIP-based PCR assays are thoroughly evaluated using a standard dilution series of methylated DNA into unmethylated DNA to characterize the PCR bias and determine the limit of detection. It may under some circumstances be possible to overcome PCR bias by raising the annealing temperature of the PCR [110], and if a PCR bias is observed towards unmethylated DNA, it may be overcome by introducing a limited number of CpG sites in the primer sequence toward the 5′ end [111]. However, if standards of known methylation level are analyzed in parallel with the samples of interest, some PCR bias may be acceptable. Finally, next-generation sequencing (NGS) may be used if the sensitivity and quantitative accuracy need to be further increased or if many loci need to be analyzed in parallel [112], and in the future it may be possible to detect 5hmC and 5mC without the need for bisulfite treatment and amplification of the target [113]. Detailed information regarding applications and methodologies for epigenetic studies using NGS can be found in Chapter 6 "Genome-wide techniques for the study of clinical epigenetic biomarkers" by Toska and Carmona-Sanz. An overview of the methods discussed here can be found in Table 8.

4. MONITORING GLOBAL DNA METHYLATION IN CLINICAL SAMPLES

Several different techniques have been developed for measuring global DNA methylation levels, including high-performance liquid chromatography (HPLC), two-dimensional thin layer chromatography, and high-performance capillary electrophoresis. Common to these methods is that they require a large amount of starting material, and thus may often not be compatible with clinical samples where the amount of DNA may be limited and/or of poor quality. However, because the human genome constitutes more than 50% repetitive sequences [114], these can be used as surrogates for measuring global DNA methylation levels. In particular, PCR assays targeting Alu and LINE-1 repeats are more suitable to be adopted by clinical laboratories since they require less input DNA, are compatible with FFPE samples, and have less demanding protocols. Different methods for downstream detection have been applied, such as fluorescent probes [115], pyrosequencing [116], and

TABLE 8 Overview of the Methods Discussed Within This Chapter

Method	Primer design	Closed-tube	Analytical sensitivity	Analytical specificity	Quantitative accuracy	Cost-effectiveness	References
Conventional MSP	MSP	No	High	Low	Low	High	Herman [100]
qMSP (fluorescent probes)	MSP	Yes	High	High	High	Medium	Eads [101]
qMSP (HRM dyes)	MSP	Yes	High	High	High	Medium	Kristensen [102]
LNA-MSP	MSP	Yes[a]	High	High	Depending on method used for detection	Medium	Gustafson [103]
MSP-pyrosequencing	MSP	No	High	High	High[a]	Medium	Shaw [104] Kristensen [105]
Direct Sanger sequencing	MIP	No	Low	High	Medium	Medium	
Sanger sequencing of single clones	MIP	No	Low	High	Depending on number of clones	Low	Clark [90]
Melting curve analysis	MIP	Yes	Medium	High	Medium	Medium	Guldberg [107]
Pyrosequencing	MIP	No	Medium	High	High	Medium	Tost [108]

[a] *When combined with fluorescent probes or HRM.*
HRM, high-resolution melting; MIP, methylation-independent PCR; MSP, methylation-specific PCR; LNA, locked nucleic acids.

high-resolution melting (HRM) [117]. Alternatively, assays based on methylation-sensitive restriction enzymes, such as the luminometric methylation assay (LUMA), may be used as surrogates for measuring global DNA methylation levels [118]. However, it is important to keep in mind that these surrogate assays may not always correlate well with global methylation measurements by HPLC [119].

5. CONCLUSIONS

Despite the development of a plethora of advanced technologies for measuring epigenetic changes and increasing insight into the mechanisms of action of epigenetic therapy, biomarkers for the efficacy of epigenetic therapy are still lacking.

There may be numerous reasons why the identification of biomarkers for epigenetic therapy has been less successful. Most studies have been performed in relatively small and miscellaneous patient cohorts, and the currently available data need confirmation in larger, independent studies. Furthermore, individual biomarkers have been analyzed by various methodologies in different types of patient material, e.g., PBMNCs, bone marrow MNCs, or sorted cells. Obviously, strictly stratified conditions for sampling, technology, and analyses are needed in order to identify reliable biomarkers. The epigenetic drugs are often used in combinations either with other epigenetics drugs or conventional chemotherapy complicating the search for relevant biomarkers. Each class of epigenetic drugs comprises several agents with different pharmacological function, metabolism, and elimination pathways. Accordingly, it is likely that each individual drug will require a specific biomarker. In addition, the revolution in molecular genetics has unraveled a plethora of diverse subsets of MDS/AML with distinct molecular profiles. Thus it is quite likely that no single biomarker can be used in these miscellaneous diseases, and that we, in addition to individualized therapy, also will need individualized biomarkers.

In conclusion, there is still lack of a straightforward relation between the molecular mechanisms of action and the applied biomarkers. Thus, for the time being, there is still much to uncover before the responses to epigenetic therapy can be consistently predicted, but hopefully, many large multicenter clinical trials in combination with novel high-throughput screening methods and upcoming individualized therapeutic approaches may enable us to identify relevant biomarkers of value for an individual patient. Whether these biomarkers will be genetic or epigenetic in nature, we will have to wait and see.

LIST OF ABBREVIATIONS

5hmC 5-Hydroxymethylcytosine
5mC 5-Methylcytosine
AML Acute myeloid leukemia
APC Adenomatous polyposis coli
ASXL1 Additional sex combs like transcriptional regulator 1
BCL-2 B cell CLL/lymphoma 2
C Cytosine
CDA Cytidine deaminase
CDH1 Cadherin 1
CDH13 Cadherin 13
CDKN2B Cyclin-dependent kinase 4 inhibitor B
CJUN Jun proto-oncogene
CK Cytidine kinase
CMML Chronic myelomonocytic leukemia
CMYB V-myb avian myeloblastosis viral oncogene homolog
CpG Cytosine—phosphate—guanine
CTA Cancer/testis antigens
DCK Deoxycytidine kinase
DNA Deoxyribonucleic acid
DNMT DNA methyltransferase
DNMTi DNA methyltransferase inhibitor
ECOG Eastern Cooperative Oncology Group
EMA European Medicines Agency
ERα Estrogen receptor α
Fas Fas cell surface death receptor
FCSS Flow cytometric scoring system
FDA U.S. Food and Drug Administration
FFPE Formalin-fixed paraffin embedded
FOXP3 Forkhead box P3
FPSS French Prognostic Scoring System
HDAC Histone deacetylases
HDACi Histone deacetylases inhibitors
HKMT Histone lysine methyltransferase
hNTs Human nucleoside transporters
HPLC High-performance liquid chromatography
HRM High-resolution melting
IDH1/2 Isocitrate dehydrogenase 1/2
IPSS International Prognostic Scoring System
IPSS-R Revised International Prognostic Scoring System
LINE-1 Long interspersed elements
LNAs Locked nucleic acids
LUMA Luminometric methylation assay
MBD2 Methyl-binding protein 2
MDS Myelodysplastic syndrome
MIP Methylation-independent PCR
MLL5 Mixed lineage leukemia 5
MNC Mononuclear cells
MPN Myeloproliferative neoplasm
MSP Methylation-specific PCR
NGS Next-generation sequencing
NOR1 Oxidored-nitro domain-containing protein isoform 1
NPM2 Nucleoplasmin 2
NSCLC Non-small-cell lung cancer
OLIG2 Oligodendrocyte transcription factor 2
OS Overall survival
PBMNC Peripheral blood mononuclear cells
PCR Polymerase chain reaction
PD-1 Programmed death-1
PGRA Progesterone receptor A
PGRB Progesterone receptor B
PLCβ1 Phosphoinositide-phospholipase C β1
RASSF1A Ras association (RalGDS/AF-6) domain family member 1
RIL PDZ and LIM domain 4
RNA Ribonucleic acid
RNAi RNA interference
SF3B1 Splice factor 3B1
SNP Single nucleotide polymorphism
SNP-A Single nucleotide polymorphism microarray
T Thymine
TET2 Ten-eleven translocation 2
TGFβ Transforming growth factor
TP53 Tumor protein p53
Tregs Regulatory T cells
U Uracil

References

[1] Di Costanzo A, Del Gaudio N, Migliaccio A, Altucci L. Epigenetic drugs against cancer: an evolving landscape. Arch Toxicol August 2, 2014:1651–68.

[2] Treppendahl MB, Kristensen LS, Grønbæk K. Predicting response to epigenetic therapy. J Clin Invest January 2, 2014;124(1):47–55.

[3] Fenaux P, Mufti GJ, Hellström-Lindberg E, Santini V, Gattermann N, Germing U, et al. Azacitidine prolongs overall survival compared with conventional care regimens in elderly patients with low bone marrow blast count acute myeloid leukemia. J Clin Oncol February 1, 2010;28(4):562–9.

[4] Fenaux P, Mufti GJ, Hellstrom-Lindberg E, Santini V, Finelli C, Giagounidis A, et al. Efficacy of azacitidine compared with that of conventional care regimens in the treatment of higher-risk myelodysplastic syndromes: a randomised, open-label, phase III study. Lancet Oncol March 2009;10(3):223–32.

[5] Kantarjian H, Issa J-PJ, Rosenfeld CS, Bennett JM, Albitar M, DiPersio J, et al. Decitabine improves patient outcomes in myelodysplastic syndromes: results of a phase III randomized study. Cancer April 15, 2006;106(8):1794–803.

[6] Adès L, Sekeres Ma, Wolfromm A, Teichman ML, Tiu RV, Itzykson R, et al. Predictive factors of response and survival among chronic myelomonocytic leukemia patients treated with azacitidine. Leuk Res June 12, 2013;37(6):609–13.

[7] Kantarjian HM, Thomas XG, Dmoszynska A, Wierzbowska A, Mazur G, Mayer J, et al. Multicenter, randomized, open-label, phase III trial of decitabine versus patient choice, with physician advice, of either supportive care or low-dose cytarabine for the treatment of older patients with newly diagnosed acute myeloid leukemia. J Clin Oncol July 20, 2012;30(21):2670–7.

[8] Lee Y-G, Kim I, Yoon S-S, Park S, Cheong JW, Min YH, et al. Comparative analysis between azacitidine and decitabine for the treatment of myelodysplastic syndromes. Br J Haematol May 2013;161(3):339–47.

[9] Prébet T, Gore SD, Esterni B, Gardin C, Itzykson R, Thepot S, et al. Outcome of high-risk myelodysplastic syndrome after azacitidine treatment failure. J Clin Oncol August 20, 2011;29(24):3322–7.

[10] Jones PA, Taylor SM. Cellular differentiation, cytidine analogs and DNA methylation. Cell May 1980;20(1):85–93.

[11] Li LH, Olin EJ, Buskirk HH, Reineke LM. Cytotoxicity and mode of action of 5-azacytidine on L1210 leukemia. Cancer Res November 1970;30(11):2760–9.

[12] Gore SD, Baylin S, Sugar E, Carraway H, Miller CB, Carducci M, et al. Combined DNA methyltransferase and histone deacetylase inhibition in the treatment of myeloid neoplasms. Cancer Res June 15, 2006;66(12):6361–9.

[13] Qiu X, Hother C, Ralfkiær UM, Søgaard A, Lu Q, Workman CT, et al. Equitoxic doses of 5-azacytidine and 5-aza-2′deoxycytidine induce diverse immediate and overlapping heritable changes in the transcriptome. PLoS One January 2010;5(9):e12994.

[14] Almstedt M, Blagitko-Dorfs N, Duque-Afonso J, Karbach J, Pfeifer D, Jäger E, et al. The DNA demethylating agent 5-aza-2′-deoxycytidine induces expression of NY-ESO-1 and other cancer/testis antigens in myeloid leukemia cells. Leuk Res July 2010;34(7):899–905.

[15] Goodyear O, Agathanggelou A, Novitzky-Basso I, Siddique S, McSkeane T, Ryan G, et al. Induction of a $CD8^+$ T-cell response to the MAGE cancer testis antigen by combined treatment with azacitidine and sodium valproate in patients with acute myeloid leukemia and myelodysplasia. Blood September 16, 2010;116(11):1908–18.

[16] Aimiuwu J, Wang H, Chen P, Xie Z, Wang J, Liu S, et al. RNA-dependent inhibition of ribonucleotide reductase is a major pathway for 5-azacytidine activity in acute myeloid leukemia. Blood May 31, 2012;119(22):5229–38.

[17] Fahy J, Jeltsch A, Arimondo PB. DNA methyltransferase inhibitors in cancer: a chemical and therapeutic patent overview and selected clinical studies. Expert Opin Ther Pat December 2012;22(12):1427–42.

[18] Rius M, Stresemann C, Keller D, Brom M, Schirrmacher E, Keppler D, et al. Human concentrative nucleoside transporter 1-mediated uptake of 5-azacytidine enhances DNA demethylation. Mol Cancer Ther January 2009;8(1):225–31.

[19] Damaraju VL, Mowles D, Yao S, Ng A, Young JD, Cass CE, et al. Role of human nucleoside transporters in the uptake and cytotoxicity of azacitidine and decitabine. Nucleosides Nucleotides Nucleic Acids January 2012;31(3):236–55.

[20] Qin T, Jelinek J, Si J, Shu J, Issa J-PJ. Mechanisms of resistance to 5-aza-2′-deoxycytidine in human cancer cell lines. Blood January 15, 2009;113(3):659–67.

[21] Van den Heuvel-Eibrink MM, Wiemer Ea, Kuijpers M, Pieters R, Sonneveld P. Absence of mutations in the deoxycytidine kinase (dCK) gene in patients with relapsed and/or refractory acute myeloid leukemia (AML). Leukemia May 2001;15(5):855–6.

[22] Qin T, Castoro R, El Ahdab S, Jelinek J, Wang X, Si J, et al. Mechanisms of resistance to decitabine in the myelodysplastic syndrome. PLoS One January 2011;6(8):e23372.

[23] Mahfouz RZ, Jankowska A, Ebrahem Q, Gu X, Visconte V, Tabarroki A, et al. Increased CDA expression/activity in males contributes to decreased cytidine analog half-life and likely contributes to worse outcomes with 5-azacytidine or decitabine therapy. Clin Cancer Res February 15, 2013;19(4):938–48.

[24] Bally C, Thépot S, Quesnel B, Vey N, Dreyfus F, Fadlallah J, et al. Azacitidine in the treatment of therapy related myelodysplastic syndrome and acute myeloid leukemia (tMDS/AML): a report on 54 patients by the Groupe Francophone Des Myelodysplasies (GFM). Leuk Res June 14, 2013;37(6):637–40.

[25] Shen L, Kantarjian H, Guo Y, Lin E, Shan J, Huang X, et al. DNA methylation predicts survival and response to therapy in patients with myelodysplastic syndromes. J Clin Oncol February 2010;28(4):605–13.

[26] Raj K, John A, Ho A, Chronis C, Khan S, Samuel J, et al. CDKN2B methylation status and isolated chromosome 7 abnormalities predict responses to treatment with 5-azacytidine. Leukemia September 2007;21(9):1937–44.

[27] Yoo CB, Jeong S, Egger G, Liang G, Phiasivongsa P, Tang C, et al. Delivery of 5-aza-2′-deoxycytidine to cells using oligodeoxynucleotides. Cancer Res July 1, 2007;67(13):6400–8.

[28] Itzykson R, Thépot S, Quesnel B, Dreyfus F, Beyne-Rauzy O, Turlure P, et al. Prognostic factors for response and overall survival in 282 patients with higher-risk myelodysplastic syndromes treated with azacitidine. Blood January 13, 2011;117(2):403–11.

[29] Itzykson R, Thépot S, Quesnel B, Dreyfus F, Recher C, Wattel E, et al. Long-term outcome of higher-risk MDS patients treated with azacitidine: an update of the GFM compassionate program cohort. Blood June 21, 2012;119(25):6172–3.

[30] Breccia M, Loglisci G, Cannella L, Finsinger P, Mancini M, Serrao A, et al. Application of French prognostic score to patients with International Prognostic Scoring System intermediate-2 or high risk myelodysplastic syndromes treated with 5-azacitidine is able to predict overall survival and rate of response. Leuk Lymphoma May 2012;53(5):985–6.

[31] Van der Helm LH, Alhan C, Wijermans PW, van Marwijk Kooy M, Schaafsma R, Biemond BJ, et al. Platelet doubling after the first azacitidine cycle is a promising predictor for response in myelodysplastic syndromes (MDS), chronic myelomonocytic leukaemia (CMML) and acute myeloid leukaemia (AML) patients in the Dutch azacitidine compassionate named p. Br J Haematol December 2011;155(5):599–606.

[32] Greenberg PL, Tuechler H, Schanz J, Sanz G, Garcia-Manero G, Solé F, et al. Revised international prognostic scoring system for myelodysplastic syndromes. Blood September 20, 2012;120(12):2454–65.

[33] Lamarque M, Raynaud S, Itzykson R, Thepot S, Quesnel B, Dreyfus F, et al. The revised IPSS is a powerful tool to evaluate the outcome of MDS patients treated with azacitidine: the GFM experience. Blood December 13, 2012;120(25):5084–5.

[34] Zeidan AM, Lee J-W, Prebet T, Greenberg P, Sun Z, Juckett M, et al. Comparison of the prognostic utility of the revised International Prognostic Scoring System and the French Prognostic Scoring System in azacitidine-treated patients with myelodysplastic syndromes. Br J Haematol August 9, 2014;166(3):352–9.

[35] Zeidan AM, Lee J-W, Prebet T, Greenberg P, Sun Z, Juckett M, et al. Platelet count doubling after the first cycle of azacitidine therapy predicts eventual response and survival in patients with myelodysplastic syndromes and oligoblastic acute myeloid leukaemia but does not add to prognostic utility of the revised IPSS. Br J Haematol October 4, 2014;167(1):62–8.

[36] Lübbert M, Wijermans P, Kunzmann R, Verhoef G, Bosly A, Ravoet C, et al. Cytogenetic responses in high-risk myelodysplastic syndrome following low-dose treatment with the DNA methylation inhibitor 5-aza-2′-deoxycytidine. Br J Haematol August 2001;114(2):349–57.

[37] Blum W, Garzon R, Klisovic RB, Schwind S, Walker A, Geyer S, et al. Clinical response and miR-29b predictive significance in older AML patients treated with a 10-day schedule of decitabine. Proc Natl Acad Sci USA April 20, 2010;107(16):7473–8.

[38] Abáigar M, Ramos F, Benito R, Díez-Campelo M, Sánchez-del-Real J, Hermosín L, et al. Prognostic impact of the number of methylated genes in myelodysplastic syndromes and acute myeloid leukemias treated with azacytidine. Ann Hematol November 6, 2013;92(11):1543–52.

[39] Cluzeau T, Moreilhon C, Mounier N, Karsenti J-M, Gastaud L, Garnier G, et al. Total genomic alteration as measured by SNP-array-based molecular karyotyping is predictive of overall survival in a cohort of MDS or AML patients treated with azacitidine. Blood Cancer J January 2013; 3(September):e155.

[40] Issa J-PJ. The myelodysplastic syndrome as a prototypical epigenetic disease. Blood May 9, 2013;121(19):3811–7.

[41] Itzykson R, Kosmider O, Cluzeau T, Mansat-De Mas V, Dreyfus F, Beyne-Rauzy O, et al. Impact of TET2 mutations on response rate to azacitidine in myelodysplastic syndromes and low blast count acute myeloid leukemias. Leukemia July 2011;25(7): 1147–52.

[42] Bejar R, Lord A, Stevenson K, Bar-Natan M, Pérez-Ladaga A, Zaneveld J, et al. TET2 mutations predict response to hypomethylating agents in myelodysplastic syndrome patients. Blood October 23, 2014;124(17):2705–12.

[43] Voso MT, Fabiani E, Piciocchi a MC, Brandimarte L, Finelli C, et al. Role of BCL2L10 methylation and TET2 mutations in higher risk myelodysplastic syndromes treated with 5-azacytidine. Leukemia December 2011;25(12):1910–3.

[44] Braun T, Itzykson R, Renneville A, de Renzis B, Dreyfus F, Laribi K, et al. Molecular predictors of response to decitabine in advanced chronic myelomonocytic leukemia: a phase 2 trial. Blood October 6, 2011;118(14):3824–31.

[45] Metzeler KH, Walker A, Geyer S, Garzon R, Klisovic RB, Bloomfield CD, et al. DNMT3A mutations and response to the hypomethylating agent decitabine in acute myeloid leukemia. Leukemia May 2012;26(5):1106–7.

[46] DiNardo CD, Patel KP, Garcia-Manero G, Luthra R, Pierce S, Borthakur G, et al. Lack of association of IDH1, IDH2 and DNMT3A mutations with outcome in older patients with acute myeloid leukemia treated with hypomethylating agents. Leuk Lymphoma August 2014;55(8):1925–9.

[47] Traina F, Visconte V, Elson P, Tabarroki A, Jankowska AM, Hasrouni E, et al. Impact of molecular mutations on treatment response to DNMT inhibitors in myelodysplasia and related neoplasms. Leukemia September 18, 2013 (August):1–10.

[48] Bally C, Adès L, Renneville A, Sebert M, Eclache V, Preudhomme C, et al. Prognostic value of TP53 gene mutations in myelodysplastic syndromes and acute myeloid leukemia treated with azacitidine. Leuk Res March 23, 2014;38(7):751–5.

[49] Kulasekararaj AG, Smith AE, Mian SA, Mohamedali AM, Krishnamurthy P, Lea NC, et al. TP53 mutations in myelodysplastic syndrome are strongly correlated with aberrations of chromosome 5, and correlate with adverse prognosis. Br J Haematol March 2013;160(5):660–72.

[50] Müller-Thomas C, Rudelius M, Rondak I-C, Haferlach T, Schanz J, Huberle C, et al. Response to azacitidine is independent of p53 expression in higher-risk myelodysplastic syndromes and secondary acute myeloid leukemia. Haematologica October 27, 2014;99(10). e179–81.

[51] Yang Z, Kondo T, Voorhorst CS, Nabinger SC, Ndong L, Yin F, et al. Increased c-Jun expression and reduced GATA2 expression promote aberrant monocytic differentiation induced by activating PTPN11 mutants. Mol Cell Biol August 2009;29(16):4376–93.

[52] Ramsay RG, Gonda TJ. MYB function in normal and cancer cells. Nat Rev Cancer July 2008;8(7):523–34.

[53] Emerling BM, Bonifas J, Kratz CP, Donovan S, Taylor BR, Green ED, et al. MLL5, a homolog of *Drosophila trithorax* located within a segment of chromosome band 7q22 implicated in myeloid leukemia. Oncogene July 18, 2002;21(31):4849–54.

[54] Yun H, Damm F, Yap D, Schwarzer A, Chaturvedi A, Jyotsana N, et al. Impact of MLL5 expression on decitabine efficacy and DNA methylation in acute myeloid leukemia. Haematologica September 3, 2014;99(9):1456–64.

[55] Follo MY, Finelli C, Mongiorgi S, Clissa C, Chiarini F, Ramazzotti G, et al. Synergistic induction of PI-PLCβ1 signaling by azacitidine and valproic acid in high-risk myelodysplastic syndromes. Leukemia February 2011;25(2):271–80.

[56] Follo MY, Finelli C, Mongiorgi S, Clissa C, Bosi C, Testoni N, et al. Reduction of phosphoinositide-phospholipase C beta1 methylation predicts the responsiveness to azacitidine in high-risk MDS. Proc Natl Acad Sci USA September 29, 2009;106(39):16811–6.

[57] Filì C, Malagola M, Follo MY, Finelli C, Iacobucci I, Martinelli G, et al. Prospective phase II study on 5-days azacitidine for treatment of symptomatic and/or erythropoietin unresponsive patients with Low/INT-1-risk myelodysplastic syndromes. Clin Cancer Res June 15, 2013;19(12):3297–308.

[58] Follo MY, Russo D, Finelli C, Mongiorgi S, Clissa C, Filì C, et al. Epigenetic regulation of nuclear PI-PLCbeta1 signaling pathway in low-risk MDS patients during azacitidine treatment. Leukemia May 28, 2012;26(5):943–50.

[59] Fabbri M, Garzon R, Cimmino A, Liu Z, Zanesi N, Callegari E, et al. MicroRNA-29 family reverts aberrant methylation in lung cancer by targeting DNA methyltransferases 3A and 3B. Proc Natl Acad Sci USA October 2, 2007;104(40):15805–10.

[60] Garzon R, Liu S, Fabbri M, Liu Z, Heaphy CE, Callegari E, et al. MicroRNA-29b induces global DNA hypomethylation and tumor suppressor gene reexpression in acute myeloid leukemia by targeting directly DNMT3A and 3B and indirectly DNMT1. Blood June 18, 2009;113(25):6411–8.

[61] Mims A, Walker AR, Huang X, Sun J, Wang H, Santhanam R, et al. Increased anti-leukemic activity of decitabine via AR-42-induced upregulation of miR-29b: a novel epigenetic-targeting approach in acute myeloid leukemia. Leukemia April 26, 2013;27(4):871–8.

[62] Yang H, Fang Z, Wei Y, Hu Y, Calin GA, Kantarjian HM, et al. Levels of miR-29b do not predict for response in patients with acute myelogenous leukemia treated with the combination of 5-azacytidine, valproic acid, and ATRA. Am J Hematol February 2011;86(2):237–8.

[63] Bogenberger JM, Kornblau SM, Pierceall WE, Lena R, Chow D, Shi C-X, et al. BCL-2 family proteins as 5-azacytidine-sensitizing targets and determinants of response in myeloid malignancies. Leukemia August 23, 2014;28(8):1657–65.

[64] Del Gaizo Moore V, Letai A. BH3 profiling–measuring integrated function of the mitochondrial apoptotic pathway to predict cell fate decisions. Cancer Lett May 28, 2013;332(2):202–5.

[65] Ettou S, Audureau E, Humbrecht C, Benet B, Jammes H, Clozel T, et al. Fas expression at diagnosis as a biomarker of azacitidine activity in high-risk MDS and secondary AML. Leukemia October 2012;26(10):2297–9.

[66] Terwijn M, Feller N, van Rhenen A, Kelder A, Westra G, Zweegman S, et al. Interleukin-2 receptor alpha-chain (CD25) expression on leukaemic blasts is predictive for outcome and level of residual disease in AML. Eur J Cancer June 2009;45(9):1692–9.

[67] Nakase K, Kita K, Miwa H, Nishii K, Shikami M, Tanaka I, et al. Clinical and prognostic significance of cytokine receptor expression in adult acute lymphoblastic leukemia: interleukin-2 receptor alpha-chain predicts a poor prognosis. Leukemia February 2007;21(2):326–32.

[68] Miltiades P, Lamprianidou E, Vassilakopoulos TP, Papageorgiou SG, Galanopoulos AG, Vakalopoulou S, et al. Expression of CD25 antigen on CD34$^+$ cells is an independent predictor of outcome in late-stage MDS patients treated with azacitidine. Blood Cancer J January 2014;4:e187.

[69] Wells DA, Benesch M, Loken MR, Vallejo C, Myerson D, Leisenring WM, et al. Myeloid and monocytic dyspoiesis as determined by flow cytometric scoring in myelodysplastic syndrome correlates with the IPSS and with outcome after hematopoietic stem cell transplantation. Blood July 1, 2003;102(1):394–403.

[70] Alhan C, Westers TM, van der Helm LH, Eeltink C, Huls G, Witte BI, et al. Absence of aberrant myeloid progenitors by flow cytometry is associated with favorable response to azacitidine in higher risk myelodysplastic syndromes. Cytom B Clin Cytom May 2014;86(3):207–15.

[71] Pardoll DM. The blockade of immune checkpoints in cancer immunotherapy. Nat Rev Cancer April 2012;12(4):252–64.

[72] Lal G, Zhang N, van der Touw W, Ding Y, Ju W, Bottinger EP, et al. Epigenetic regulation of Foxp3 expression in regulatory T cells by DNA methylation. J Immunol January 2009;182(1):259–73.

[73] Kordasti SY, Ingram W, Hayden J, Darling D, Barber L, Afzali B, et al. CD4$^+$CD25high Foxp3$^+$ regulatory T cells in myelodysplastic syndrome (MDS). Blood August 1, 2007;110(3):847–50.

[74] Costantini B, Kordasti SY, Kulasekararaj AG, Jiang J, Seidl T, Abellan PP, et al. The effects of 5-azacytidine on the function and number of regulatory T cells and T-effectors in myelodysplastic syndrome. Haematologica August 2013;98(8):1196–205.

[75] Baylin SB, Jones PA. A decade of exploring the cancer epigenome – biological and translational implications. Nat Rev Cancer October 2011;11(10):726–34.

[76] Daskalakis M, Nguyen TT, Nguyen C, Guldberg P, Köhler G, Wijermans P, et al. Demethylation of a hypermethylated P15/INK4B gene in patients with myelodysplastic syndrome by 5-Aza-2′-deoxycytidine (decitabine) treatment. Blood October 15, 2002;100(8):2957–64.

[77] Garcia-Manero G, Kantarjian HM, Sanchez-Gonzalez B, Yang H, Rosner G, Verstovsek S, et al. Phase 1/2 study of the combination of 5-aza-2′-deoxycytidine with valproic acid in patients with leukemia. Blood November 15, 2006;108(10):3271–9.

[78] Yang AS, Doshi KD, Choi S-W, Mason JB, Mannari RK, Gharybian V, et al. DNA methylation changes after 5-aza-2′-deoxycytidine therapy in patients with leukemia. Cancer Res May 15, 2006;66(10):5495–503.

[79] Kantarjian H, Oki Y, Garcia-Manero G, Huang X, O'Brien S, Cortes J, et al. Results of a randomized study of 3 schedules of low-dose decitabine in higher-risk myelodysplastic syndrome and chronic myelomonocytic leukemia. Blood January 1, 2007;109(1):52–7.

[80] Fandy TE, Herman JG, Kerns P, Jiemjit A, Sugar Ea, Choi S-H, et al. Early epigenetic changes and DNA damage do not predict clinical response in an overlapping schedule of 5-azacytidine and entinostat in patients with myeloid malignancies. Blood September 24, 2009;114(13):2764–73.

[81] Issa J-PJ, Garcia-Manero G, Giles FJ, Mannari R, Thomas D, Faderl S, et al. Phase 1 study of low-dose prolonged exposure schedules of the hypomethylating agent 5-aza-2′-deoxycytidine (decitabine) in hematopoietic malignancies. Blood March 1, 2004;103(5):1635–40.

[82] Cluzeau T, Robert G, Mounier N, Karsenti JM, Dufies M, Puissant A, et al. BCL2L10 is a predictive factor for resistance to azacitidine in MDS and AML patients. Oncotarget April 2012;3(4):490–501.

[83] Juergens RA, Wrangle J, Vendetti FP, Murphy SC, Zhao M, Coleman B, et al. Combination epigenetic therapy has efficacy in patients with refractory advanced non-small cell lung cancer. Cancer Discov December 2011;1(7):598–607.

[84] Youngblood B, Oestreich KJ, Ha S-J, Duraiswamy J, Akondy RS, West EE, et al. Chronic virus infection enforces demethylation of the locus that encodes PD-1 in antigen-specific CD8(+) T cells. Immunity September 23, 2011;35(3):400–12.

[85] Okazaki T, Honjo T. PD-1 and PD-1 ligands: from discovery to clinical application. Int Immunol July 2007;19(7):813–24.

[86] Yang H, Bueso-Ramos C, DiNardo C, Estecio MR, Davanlou M, Geng Q-R, et al. Expression of PD-L1, PD-L2, PD-1 and CTLA4 in myelodysplastic syndromes is enhanced by treatment with hypomethylating agents. Leukemia June 2014;28(6):1280–8.

[87] Ørskov AD, Treppendahl MB, Skovbo A, Holm MS, Friis LS, Hokland M, et al. Hypomethylation and up-regulation of PD-1 in T cells by azacytidine in MDS/AML patients: a rationale for combined targeting of PD-1 and DNA methylation. Oncotarget April 20, 2015;6(11):9612–26.

[88] Yan P, Frankhouser D, Murphy M, Tam H-H, Rodriguez B, Curfman J, et al. Genome-wide methylation profiling in decitabine-treated patients with acute myeloid leukemia. Blood September 20, 2012; 120(12):2466–74.

[89] Kristensen LS, Hansen LL. PCR-based methods for detecting single-locus DNA methylation biomarkers in cancer diagnostics, prognostics, and response to treatment. Clin Chem August 2009;55(8):1471–83.

[90] Clark SJ, Harrison J, Paul CL, Frommer M. High sensitivity mapping of methylated cytosines. Nucleic Acids Res August 11, 1994;22(15):2990–7.

[91] Tahiliani M, Koh KP, Shen Y, Pastor WA, Bandukwala H, Brudno Y, et al. Conversion of 5-methylcytosine to 5-hydroxymethylcytosine in mammalian DNA by MLL partner TET1. Science May 15, 2009;324(5929):930–5.

[92] Kriaucionis S, Heintz N. The nuclear DNA base 5-hydroxymethylcytosine is present in Purkinje neurons and the brain. Science May 15, 2009;324(5929): 929–30.

[93] Huang Y, Pastor WA, Shen Y, Tahiliani M, Liu DR, Rao A. The behaviour of 5-hydroxymethylcytosine in bisulfite sequencing. PLoS One January 2010;5(1):e8888.

[94] Kristensen LS, Treppendahl MB, Grønbæk K. Analysis of epigenetic modifications of DNA in human cells. Curr Protoc Hum Genet April 2013; Chapter 20:Unit20.2.

[95] Soes S, Daugaard IL, Sorensen BS, Carus A, Mattheisen M, Alsner J, et al. Hypomethylation and increased expression of the putative oncogene ELMO3 are associated with lung cancer development and metastases formation. Oncoscience 2014;1(5):367–74.

[96] Søes S, Sørensen BS, Alsner J, Overgaard J, Hager H, Hansen LL, et al. Identification of accurate reference genes for RT-qPCR analysis of formalin-fixed paraffin-embedded tissue from primary non-small cell lung cancers and brain and lymph node metastases. Lung Cancer August 2013;81(2):180–6.

[97] Kristensen LS, Wojdacz TK, Thestrup BB, Wiuf C, Hager H, Hansen LL. Quality assessment of DNA derived from up to 30 years old formalin fixed paraffin embedded (FFPE) tissue for PCR-based methylation analysis using SMART-MSP and MS-HRM. BMC Cancer January 2009;9:453.

[98] Tournier B, Chapusot C, Courcet E, Martin L, Lepage C, Faivre J, et al. Why do results conflict regarding the prognostic value of the methylation status in colon cancers? the role of the preservation method. BMC Cancer January 2012;12:12.

[99] Wong SQ, Li J, Tan AY-C, Vedururu R, Pang J-MB, Do H, et al. Sequence artefacts in a prospective series of formalin-fixed tumours tested for mutations in hotspot regions by massively parallel sequencing. BMC Med. Genomics January 2014;7(1):23.

[100] Herman JG, Graff JR, Myöhänen S, Nelkin BD, Baylin SB. Methylation-specific PCR: a novel PCR assay for methylation status of CpG islands. Proc Natl Acad Sci USA September 3, 1996;93(18):9821–6.

[101] Eads Ca, Danenberg KD, Kawakami K, Saltz LB, Blake C, Shibata D, et al. MethyLight: a high-throughput assay to measure DNA methylation. Nucleic Acids Res April 2000;28(8):E32.

[102] Kristensen LS, Mikeska T, Krypuy M, Dobrovic A. Sensitive Melting Analysis after Real Time- Methylation Specific PCR (SMART-MSP): high-throughput and probe-free quantitative DNA methylation detection. Nucleic Acids Res April 2008;36(7):e42.

[103] Gustafson KS. Locked nucleic acids can enhance the analytical performance of quantitative methylation-specific polymerase chain reaction. J Mol Diagn January 2008;10(1):33–42.

[104] Shaw RJ, Akufo-Tetteh EK, Risk JM, Field JK, Liloglou T. Methylation enrichment pyrosequencing: combining the specificity of MSP with validation by pyrosequencing. Nucleic Acids Res January 2006;34(11):e78.

[105] Kristensen LS, Treppendahl MB, Asmar F, Girkov MS, Nielsen HM, Kjeldsen TE, et al. Investigation of MGMT and DAPK1 methylation patterns in diffuse large B-cell lymphoma using allelic MSP-pyrosequencing. Sci Rep January 2013;3:2789.

[106] Kristensen LS, Nielsen HM, Hager H, Hansen LL. Methylation of MGMT in malignant pleural mesothelioma occurs in a subset of patients and is associated with the T allele of the rs16906252 MGMT promoter SNP. Lung Cancer February 2011;71(2):130–6.

[107] Guldberg P, Worm J, Grønbaek K. Profiling DNA methylation by melting analysis. Methods June 2002;27(2):121–7.

[108] Tost J, Gut IG. DNA methylation analysis by pyrosequencing. Nat Protoc January 2007;2(9):2265–75.

[109] Warnecke PM, Stirzaker C, Melki JR, Millar DS, Paul CL, Clark SJ. Detection and measurement of PCR bias in quantitative methylation analysis of bisulphite-treated DNA. Nucleic Acids Res November 1, 1997;25(21):4422–6.

[110] Shen L, Guo Y, Chen X, Ahmed S, Issa J-PJ. Optimizing annealing temperature overcomes bias in bisulfite PCR methylation analysis. Biotechniques January 2007;42(1). 48, 50, 52 passim.

[111] Wojdacz TK, Hansen LL. Reversal of PCR bias for improved sensitivity of the DNA methylation melting curve assay. Biotechniques September 2006;41(3). 274, 276, 278.

[112] Taylor KH, Kramer RS, Davis JW, Guo J, Duff DJ, Xu D, et al. Ultradeep bisulfite sequencing analysis of DNA methylation patterns in multiple gene promoters by 454 sequencing. Cancer Res September 15, 2007;67(18):8511–8.

[113] Korlach J, Turner SW. Going beyond five bases in DNA sequencing. Curr Opin Struct Biol June 2012;22(3):251–61.

[114] De Koning APJ, Gu W, Castoe TA, Batzer MA, Pollock DD. Repetitive elements may comprise over two-thirds of the human genome. PLoS Genet December 2011;7(12):e1002384.

[115] Weisenberger DJ, Trinh BN, Campan M, Sharma S, Long TI, Ananthnarayan S, et al. DNA methylation analysis by digital bisulfite genomic sequencing and digital MethyLight. Nucleic Acids Res August 2008;36(14):4689–98.

[116] Yang AS, Estécio MRH, Doshi K, Kondo Y, Tajara EH, Issa J-PJ. A simple method for estimating global DNA methylation using bisulfite PCR of repetitive DNA elements. Nucleic Acids Res January 2004;32(3):e38.

[117] Newman M, Blyth BJ, Hussey DJ, Jardine D, Sykes PJ, Ormsby RJ. Sensitive quantitative analysis of murine LINE1 DNA methylation using high resolution melt analysis. Epigenetics January 1, 2012;7(1):92–105.

[118] Karimi M, Johansson S, Stach D, Corcoran M, Grandér D, Schalling M, et al. LUMA (LUminometric Methylation Assay)–a high throughput method to the analysis of genomic DNA methylation. Exp Cell Res July 1, 2006;312(11):1989–95.

[119] Lisanti S, Omar WAW, Tomaszewski B, De Prins S, Jacobs G, Koppen G, et al. Comparison of methods for quantification of global DNA methylation in human cells and tissues. PLoS One January 2013;8(11):e79044.

CHAPTER

6

Genome-Wide Techniques for the Study of Clinical Epigenetic Biomarkers

Eneda Toska[1], *F. Javier Carmona Sanz*[1,2]

[1]Human Oncology & Pathogenesis Program (HOPP), Memorial Sloan-Kettering Cancer Center (MSKCC), New York, NY, USA; [2]Cancer Epigenetics and Biology Program (PEBC), Bellvitge Institute for Biomedical Research (IDIBELL), Barcelona, Spain

OUTLINE

1. INTRODUCTION

Despite intense investigation carried out in the epigenetics field during the last decades, it has not been until recently that technological improvements have permitted researchers to interrogate the epigenome in a genome-wide fashion and uncover aberrant methylation changes associated with specific pathological contexts. Genomic platforms have expanded

http://dx.doi.org/10.1016/B978-0-12-801899-6.00006-1

initial discoveries based on candidate gene approaches that interrogated the methylation status of known gene promoters on different tumor types or stages of progression and provided a much broader picture of DNA methylation in human cancer and other diseases [1]. Also, despite 5-methylcytosine (5mC) is overrepresented at cytosine–phosphate–guanine (CpG) islands, mainly within gene promoters, researchers started looking at different genomic compartments where DNA methylation also took place. Specifically, new terms as CpG shores—sequences up to 2kb distant from the transcription start site—or CpG shelves—sequences flanking outward from a CpG shore—were found to be equally correlated with gene expression patterns [2]. Additionally, gene-body DNA methylation—occurring at introns and exons—has also gained importance due to the impact on regulating transcription elongation and alternative splicing [3,4].

Until very recently, DNA methylation research focused on determining the distribution of the most abundant mark of DNA methylation, 5mC, across the genome. However, recently discovered oxidative derivatives of 5mC including 5-hydroxymethylcytosine (5hmC), 5-formylcytosine (5fC), and 5-carboxylcytosine (5caC) have been proposed to be demethylation intermediaries linked to the maintenance of DNA methylation-free regions, or to the active DNA demethylation of specific loci in genomic reprogramming processes occurring both in normal development and in disease [5]. Active DNA demethylation has long remained a matter of debate. However, the discovery of a sequential 5mC oxidation cascade mediated by the ten-eleven translocation (TET) family enzymes provided convincing evidence for an active DNA demethylation process in vertebrates [6] that will be further discussed in this text (Figure 1). Currently, a plethora of methodologies to profile the DNA methylome of the cell is available (reviewed in Ref. [7]), each one with advances and limitations depending on the context (Figure 2; Table 1).

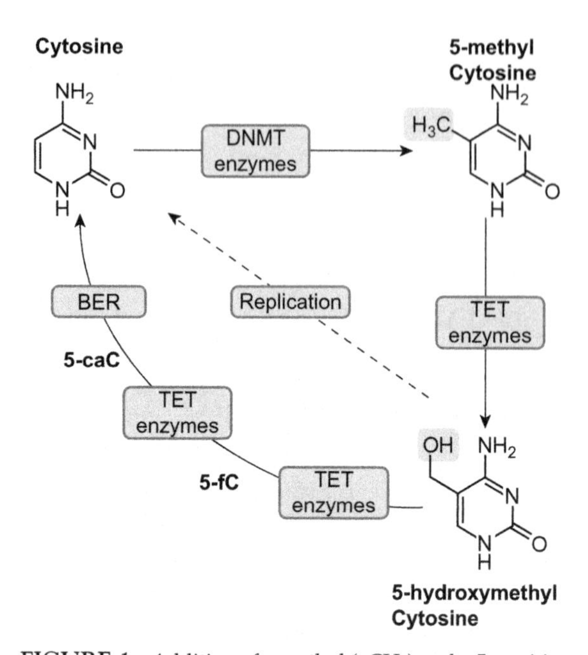

FIGURE 1 Addition of a methyl (–CH_3) at the 5′ position of cytosine by the DNA methyltransferases (DNMT) in the presence of a methyl-donor cofactor generates 5-methylcytosine (5mC). 5mC can then be converted by the ten-eleven translocation (TET) family of dioxygenases to generate 5-hydroxymethylcytosine (5hmC). This molecule can then follow the active DNA demethylation pathway through subsequent oxidization reactions mediated by the TET enzymes that convert 5hmC to 5-formylcytosine (5fC) and then 5-carboxylcytosine (5caC) that is converted back into cytosine through the base excision repair (BER). Alternatively, passive DNA demethylation is postulated to occur through DNA replication and lack of maintenance of DNA methylation.

Genome activity is deeply impacted by both changes in DNA methylation and posttranslational histone modifications, which mainly occur at the protruding histone tails [1,8]. Both processes are interconnected and determine the chromatin accessibility of DNA-binding proteins. Regarding the study of histone modifications, major contributions have been made after the development of genome-wide approaches that combine chromatin immunoprecipitation (ChIP) with high-throughput platforms including DNA microarrays and genomic sequencing (ChIP-seq). Furthermore, comprehensive studies integrating the data on DNA methylation

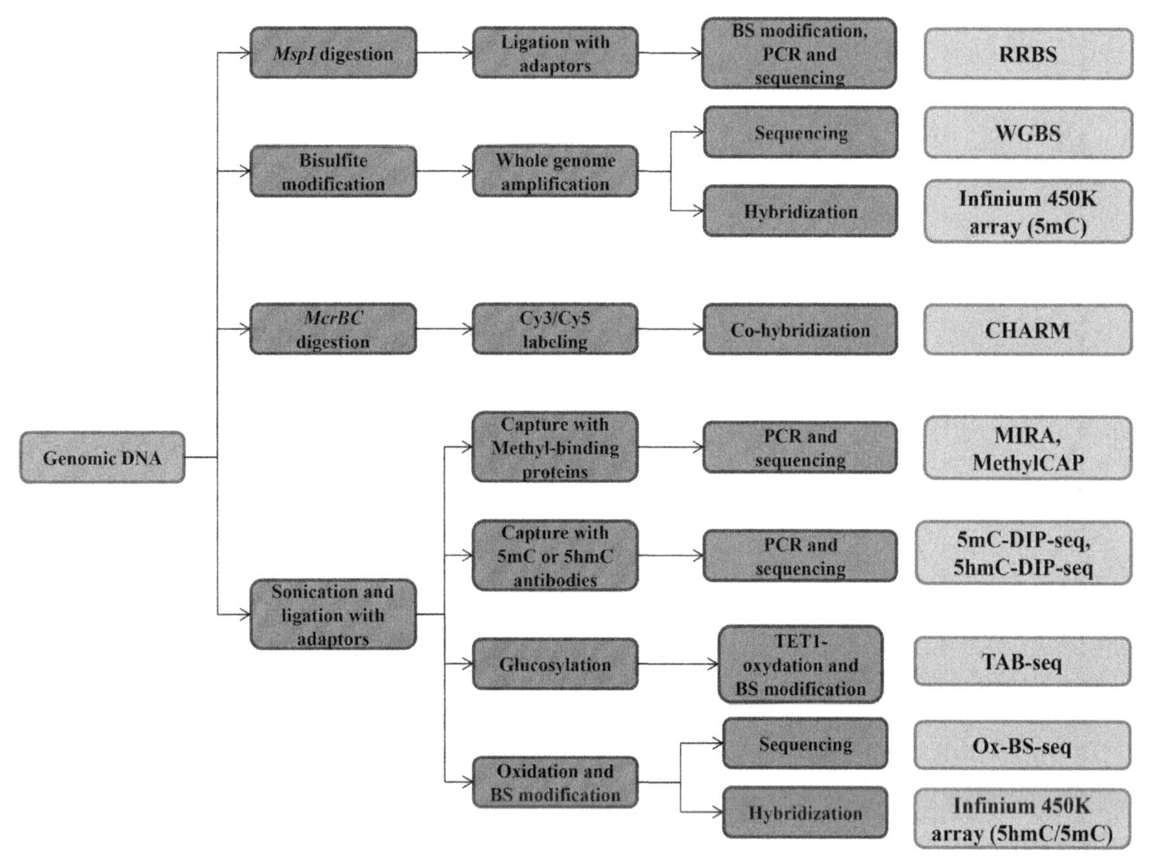

FIGURE 2 Methods for quantifying 5mC and its oxidative derivatives *(Adapted from Ref. [7].)*. Many methodologies are currently available to quantify DNA methylation genome wide involving selective modification of DNA by bisulfite (BS) modification or digestion using specific restriction enzymes.

profiles and histone modifications in combination with transcriptome studies have uncovered the intricacies of epigenetic modifications in delineating the ultimate identity and phenotype of the cell and how their deregulation contributes to malignant transformation and progression.

2. FIRST APPROACHES TO ANALYZE 5-METHYLCYTOSINE LEVELS AT A GENOMIC SCALE

Early reports that attempted to study 5mC changes in a genome-wide manner relied on restriction enzyme-based assays that examined a large series of samples to depict aberrant DNA methylation events in a qualitative manner.

These methods included restriction landmark genomic scanning (RLGS), which was the first large-scale method for investigating DNA methylation [9,10], combining DNA methylation-sensitive restriction digestion and two-dimensional electrophoresis; differential methylation hybridization (DMH), which was the first array method to be optimized for identifying novel methylated loci in cancer [11]; and amplification of intermethylated sites (AIMS) [12]. The principle of these techniques relied on the differential sensitivity of specific restriction enzymes for DNA depending on its methylation

TABLE 1 Comparison of Targeted or Whole-Genome Assays Is Currently Used to Measure DNA Methylation Genome Wide and Applicability for Clinical Diagnostics

	Assay	Input	Throughput	Coverage	Advantages	Disadvantages	Clinical Applicability
Targeted approach	Infinium 450K	200–500 ng	12 samples per chip	2% of genomic CGs	DNA repair and ligation protocols for degraded samples Automated processing Easiness of analysis pipeline High coverage	Restricted to promoter regions	High
	CHARM	2 μg	one sample per chip	19% of genomic CGs	Highly quantitative Measures CGs at promoter-distal regions	Large amount of input DNA required	Medium
Whole-genome approach	WGBS	0.5–1 μg	96-reaction format	>90%	Whole-genome coverage	Requires high-quality DNA Data analysis High cost Biased by antibody affinity	Low
	MeDIP-seq	100 ng –5 μg	96-reaction format	50–90%	Whole-genome coverage	High sequence coverage required Data analysis	Low
	RRBS	100 ng	96-reaction format	20–60%	High sensitivity at reduced cost	Does not captures all promoter CGs	Medium
	ChIP-seq	10–50 ng	N/A	>90%	Whole-genome coverage High sensitivity	Requires large amount of input Lack of precision in the binding events	Medium

CGs, cytosine–guanine sites; CHARM, comprehensive high-throughput array for relative methylation; WGBS, whole-genome bisulfite sequencing; MeDIP-seq, 5mC immunoprecipitation with high-throughput sequencing; RRBS, reduced representation bisulfite sequencing; ChIP-seq, chromatin immunoprecipitation with high-throughput sequencing.

status, enabling to compare the methylation patterns of different cell populations. These provided the first lists of DNA methylation alterations in colon cancer and also allowed to infer the degree of distortion of DNA methylation and subsequent gene activity using genetically manipulated colon cancer cell lines [13]. Specifically, the disruption of the two major DNA methyltransferases (DNMT1 and DNMT3b) in the colorectal cancer cell line HCT-116 [14] provided a useful tool for identifying differentially methylated loci on a global genomic scale by comparing the methylation profile of the DNMT1/3b double knockout and the unmodified HCT-116 cell lines [15]. These original approaches were not exempt of technical limitations, including laborious and time-consuming protocols; their dependence on inaccurate linker ligation and linker polymerase chain reaction (PCR) amplification, as is the case with DMH, for example, the limited number of sequences that can be interrogated due to the deficient or biased activity of restriction enzymes; the high rate of false-positive and false-negative results; and the low reproducibility due to the great complexity of the methods. Despite all, these pioneering studies opened a field of research in cancer by identifying and exposing the great amount of aberrantly methylated loci in cancer and provided the first clues about tumor-specific DNA methylation patterns that have been subsequently exploited in the clinic.

The discovery of sodium bisulfite modification of DNA (BS-DNA) as a method to differentially alter the sequence of genomic cytosines depending on their methylation status was another milestone in epigenetic research [16]. BS treatment converts unmethylated cytosine residues to uracil, but leaves 5mC residues unaffected. Genomic sequencing of BS-DNA or amplification of specific sequences using methylation-specific primers allowed researchers to discriminate the methylation status of single DNA molecules in a fast and sensitive way. Over 20 years after the first demonstrations of bisulfite conversion to detect methylated cytosines were provided by Marianne Frommer and Susan Clark, this protocol has been adapted to several array platforms and next-generation sequencing (NGS).

3. DNA METHYLATION ARRAYS

Genome-wide platforms quantifying DNA methylation based on the discrimination of bisulfite-induced C to T conversion have proven superior to other methods relying on methylation-sensitive restriction enzymes or isolating the methylated fraction of the genome by affinity enrichment (Table 1).

The first systematic high-throughput platform available to study 5mC profiles was the Golden Gate Cancer Panel I released by Illumina [16]. It examined the DNA methylation status of 1536 targeted CpG sites on the promoter of 807 candidate genes previously associated with aberrant DNA methylation in cancer, including tumor suppressor genes, oncogenes, and genes involved in cancer development. The assay includes one to three probes on most of the genes interrogated, and two probe pairs are designed for each CpG site: an allele-specific oligo (ASO), and a locus-specific oligo (LSO) pair for the methylated and unmethylated state of the CpG [16,17]. Subsequently, the ligated products are amplified using fluorescently labeled primers that enable the identification of methylated and unmethylated loci by calculating the ratio of fluorescent signals from the two alleles. This platform achieves considerable reproducibility between technical replicates and is sensitive enough to reliably detect differences in methylation of >20%. Moreover, experimental validation through independent techniques has proven the robustness of the assay. Golden Gate technology has been extensively used in cancer research to profile DNA methylation alterations between normal and cancer tissue [18], primary and metastatic tumors [19], and to examine DNA methylation patterns characterizing diverse normal tissues [20]. For example, this platform was applied to identify DNA methylation biomarkers for colorectal cancer diagnosis by examining DNA extracted from feces [21]. In this study, researchers identified the most frequently hypermethylated genes between normal mucosa and paired primary tumors by Golden Gate profiling and designed pyrosequencing validation assays to confirm their findings. As a result, a panel of biomarkers was found that allowed for sensitive and early diagnosis of colorectal cancer. However, despite the advantages of this platform over previous methods, limitations included the reduced number of genes assayed; the limited number of probes per gene; and the limited applicability to study human cancer samples, given the criteria to select the genes assayed.

Aiming to extend genome-wide DNA methylation studies to a different context, a custom DNA methylation assay was designed to include target genes involved in neurological degeneration in mice. This platform was used to examine the epigenetic landscapes of different brain regions and identified aberrant DNA methylation patterns occurring in mouse models of Alzheimer's disease [22]. However, despite its usefulness, few additional epigenomic platforms are available to interrogate nonhuman samples.

An extension to the Golden Gate DNA methylation platform that incorporated the same chemistry was designed to increase the coverage and extend the analysis to additional genes. The HumanMethylation27 panel from Illumina targets more than 27,000 cytosine–guanine (CG) sites covering over 14,000 genes and allowed to expand DNA methylation research to other areas of investigation [23]. As DNA methylation research increased, new concepts emerged and regulatory functions were attributed to the noncanonical CG-rich regions. Following this line of evidence, Andrew Feinberg's group designed tiling arrays to study DNA methylation in genomic DNA regions [2,24]. Employing a comprehensive high-throughput array for relative methylation (CHARM), authors identified additional promoter-distal sequences regulating gene activity by DNA methylation termed CpG shores and CpG shelves [24]. These are regions located between 200 and 2000 kilobases away from the canonical islands, and genes associated with these CpG shores showed a clear inverse correlation between transcription and methylation levels of these newly identified regions, rather than with that of the gene promoter CpG island [24]. These new regions were subsequently incorporated in the design of the HumanMethylation450 BeadChip by Illumina (Figure 2). This comprehensive platform integrates a novel chemistry using degenerate oligonucleotide probes to examine more than 450,000 CG sites across the human genome, including gene promoters, shores, and shelves, as well as CG sites present at gene bodies and intergenic sequences [25,26]. Moreover, this method was designed as a clinical diagnostic platform that allowed the screening of paraffin-embedded tissue [27] showing great reproducibility as compared to paired fresh-frozen tissue samples. This methodology has been successfully applied in phase I non-small-cell lung cancer to interrogate aberrant DNA methylation changes informative of disease outcome [28] and resulted in the identification of a panel of biomarkers (HIST1H4F, PCDHGB6, NPBWR1, ALX1, and HOXA9) whose methylation status can be studied in the clinic with PCR-based techniques. Genome-wide DNA methylation analysis was also applied to identify DNA methylation biomarkers capable of predicting the primary tumor of origin of carcinomas of unknown primary (CUP). Using the DNA methylation array technology, authors extracted a panel of DNA methylation biomarkers exclusive of particular tumor types and built algorithms that were able to predict the origin of tumor samples of unknown origin based on their DNA methylation profile [20]. This holds promising applications in the clinic, since the survival of around 3% of patients that are annually diagnosed with a CUP do not exceed 3–4 months in most cases.

Moreover, The Cancer Genome Atlas (http://cancergenome.nih.gov) has profiled thousands of samples representative of the major cancer types using this methodology, and an international initiative, the BLUEPRINT consortium (http://www.blueprint-epigenome.eu/), is moving ahead to interrogate the epigenome on a variety of reference samples from healthy and diseased individuals aiming to identify epigenetic markers for diagnostic use.

However, despite the high reproducibility of the genome-wide platforms aforementioned, single-locus validation using independent techniques is required to confirm the results derived from such studies (discussed in Chapter 8 "Epigenetic Biomarkers and Diagnostics"). The standard is bisulfite sequencing-based methods, such as pyrosequencing [29] and EpiTYPER [30], that give precise quantification of DNA methylation values and can guide clinical decision making.

4. QUANTIFICATION OF 5mC OXIDATIVE DERIVATIVES

With the discovery of 5hmC abundance in embryonic stem cells and terminally differentiated neurons, researchers wondered about the functional role of this chemical modification on the genome and on the role of TET dioxygenase enzymes in the active removal of DNA methylation (Figure 1).

Several techniques were adapted or developed for quantification of 5hmC in the genome. Some pioneering methods for global quantification include liquid chromatography and mass spectrometry, as well as DNA pull-down using 5hmC antibodies in combination with genome-wide sequencing (hMeDIP-seq). However, these protocols are technically laborious, cost-inefficient, require large amounts of input DNA, or lack of single-base resolution (Table 1). Traditional bisulfite modification protocols failed to distinguish between 5mC and 5hmC [31], since 5hmC was also found to be resistant to deamination [32]. However, the development of oxidative bisulfite sequencing (oxBS) provided an opportunity to implement bisulfite-based 5mC profiling platforms to detect 5hmC. This is especially relevant in mammalian brain tissue where 5hmC levels are high [33]. The recent coupling of oxidative bisulfite-based chemistry to the Infinium 450K DNA methylation platform allowed genome-wide profiling of 5hmC and provides a more accurate quantification of 5mC by removing the confounding factor of 5hmC [34]. Moreover, this protocol is also extensible to single-loci validation techniques as pyrosequencing.

An additional method for genome-wide analysis of 5hmC is TET-assisted bisulfite sequencing [34], which has been used to study the role of 5hmC in cancer [35], normal neural development in human and nonhuman organisms [36], as well as neurodevelopmental disorders [37]. This method is based on the glucosylation-mediated protection of genomic 5hmC and subsequent treatment with recombinant TET1 that mediates the oxidation of 5mC to 5caC while leaving the glucosylated 5hmC unaltered. Following bisulfite treatment of DNA, 5caC is converted to thymine (T), whereas cytosine's reads in the resulting sequence are interpreted as 5hmC (reviewed in Ref. [6]) (Figure 2).

5. AFFINITY ENRICHMENT OF METHYLATED DNA

Enrichment-based technologies are based on capturing methylated DNA after shearing, either by using specific antibodies or recombinant proteins that do not require bisulfite conversion of DNA.

Direct immunoprecipitation of methylated DNA (MeDIP) using a monoclonal antibody against 5-methylcytidine resulted a suitable technique for the parallel comparison of two populations in the search for differentially methylated loci. This antibody does not recognize 5hmC, hence overcomes the inability of bisulfite conversion methods to do so. In addition, coupling this application with standard bisulfite genomic sequencing has enabled the identification of a large number of genes with hypermethylated CpG islands in many tumor types [38–40]. One approach to extend the analysis is to combine MeDIP with DNA microarrays to assess relative methylation levels at the loci included on the array. Some years ago, a cross-platform algorithm for the quantitative analysis of the MeDIP data generated using arrays (MeDIP-chip) or genomic sequencing platforms (MeDIP-seq) was reported [41], but its practical implementation on a routine basis remains to be established. Alternatively, combination with NGS (MeDIP-seq) can be used to quantify a much broader proportion of mappable CpGs to the genome, including repetitive sequences. However, MeDIP-seq yields a complexity that requires an excessive amount of sequence reads to give in sufficient coverage at unique genomic regions.

Quantification methods based on isolation of methylated DNA were also optimized using recombinant proteins binding to methylated cytosines. Assays such as methylated CpG island recovery assay (MIRA) and MethylCap rely on methyl-CpG-binding domains (MBD) or proteins like MBD1, MBD2, or MBD3L1 to capture methylated DNA after DNA fractionation either by restriction digestion or sonication. These methods reduce the complexity as compared with the full genome and can also be combined with microarray or NGS technologies (MethylCap-seq) to identify biomarkers for cancer diagnosis and DNA methylation maps of cancer genomes [42]. These approaches have been applied to a variety of sample types from diverse organisms and human tumors [43,44] and have shown a remarkable degree of reproducibility as compared with other methodologies to quantify DNA methylation genome wide [45]. However, these methods present also some limitations, such as the low resolution based on the size of immunoprecipitated DNA fragments and the bias toward the level of enrichment of methylated DNA depending on the abundance of CpGs.

6. READING THE DNA METHYLOME: GENOME SEQUENCING APPLICATIONS

Commercial second-generation platforms emerged in 2005 aiming to increase the throughput and lower the cost of Sanger sequencing methods. Specifically, in the field of epigenetics the emergence of powerful and cost-effective sequencing technologies have allowed a much more comprehensive interpretation of the DNA methylation traits across the genome. Reduced representation bisulfite sequencing (RRBS) was introduced in 2005 by Meissner et al. [46] as an efficient method for absolute quantification of the methylation status of more than one million CpG sites at single base-pair resolution, covering regions of moderate to high CpG density [47], and reducing the cost of shotgun sequencing. This method uses a methylation-insensitive restriction enzyme (MspI) that reduces the coverage to 1% of the genome and allows researchers to perform epigenomic studies requiring the analysis of large numbers of samples starting from as little material as 10–300 ng of genomic DNA in a cost-efficient manner [48]. An improved pipeline optimizing RRBS for performing whole-genome DNA methylation profiling on clinical samples has been recently published, that uses a double enzymatic digestion (*MspI* and *TaqαI*) to reduce the bias associated to CG density [47].

In the last years, the cost of whole-genome sequencing has been dramatically reduced, and bioinformatic protocols for data analysis have been optimized to systematically interrogate the

DNA methylome. Whereas previously described techniques only cover around 5% of the CpGs in the DNA, whole-genome bisulfite sequencing (WGBS) allows for an unbiased assessment of DNA methylation at single-base resolution with full coverage of more than 28 million CpG sites in the human genome. After initial DNA shearing and library preparation, more than 500 million reads are required to provide high coverage of the genome. Using bisulfite conversion of DNA, NGS provided the first high-resolution DNA methylome of a living organism, *Arabidopsis thaliana* [49,50]. Subsequently, WGBS has been used to analyze DNA methylation variation in a variety of contexts, including colorectal and breast cancer, and certain types of leukemia [4,51,52], and permitted the identification of clinically relevant DNA methylation biomarkers. Moreover, such comprehensive analyses of the DNA methylation changes have uncovered unprecedented roles for gene-body DNA methylation in transcriptional activity [53].

6.1 Third-Generation Sequencing

First- and second-generation sequencing technologies have contributed to a comprehensive characterization of the DNA sequence composition and variation and have revolutionized our knowledge about genomic regulation. However, newly developed technologies are emerging that allow the sequencing of single DNA molecules with massive throughput at a modest cost. Third-generation sequencing (TGS) technologies can be classified by (1) nanopore sequencing, initially developed by Bayley and coworkers for sequencing large amounts of DNA using a nanopore-based device coupled to an exonuclease enzyme [54]; (2) direct imaging of individual DNA molecules using transmission electron microscopy [55]; and (3) DNA sequencing-by-synthesis technologies where DNA polymerases catalyze the biochemical reaction for deriving template sequence information [56]. These single-molecule sequencing platforms reduce bias by reducing the involvement of enzymes and PCR-based protocols for DNA preparation and library construction and allow the identification of nucleotide modifications independently of upstream protocols used currently to differentiate the diverse base variants on the genome. There are several TGS platforms progressively available and incorporated at research facilities for genomic analyses. The sequencing cost is among the most outstanding attribute of third-generation sequencers relative to their next-generation counterparts, followed by ease of workflow and read length, however, their implementation for the routine analysis of genomic DNA with diagnostic purposes is still to come.

6.2 ChIP-Seq Experiments: Advantages and Challenges

Genome-wide mapping of protein–DNA interactions and epigenetic marks is crucial to understand gene regulation. Precise mapping of the genomic location of transcription factor binding, core transcriptional machinery, and histone modifications is essential to fully understand the gene regulatory networks that underlie numerous biological processes such as development and differentiation processes and disease states [57]. There is growing recognition that the interplay between transcription, which is at the heart of gene regulation, and chromatin is dynamic and more complex than previously thought. Hence, understanding how genomic information is translated into gene regulation has been the subject of rigorous research over the past decades. Methods for mapping transcription factor binding across the genome by ChIP were developed more than a decade ago [58–61]. In ChIP, antibodies are used to enrich the association of the protein of interest with DNA. Initially, the DNA sites were identified by DNA hybridization to a microarray (ChIP–chip) [58–61]. Owing to the rapid technological developments in NGS, the arsenal of the genomic assays available has been revolutionized. One of the earliest applications of NGS was ChIP followed by sequencing (ChIP-seq), which was first published in 2007 and remains the main tool to investigate the protein–DNA binding in vivo [62–65]. The goal of ChIP-seq is to map the binding sites of a target protein with maximal completeness genome wide (Table 1).

ChIP-seq has been widely used to map the binding for many transcription factors, chromatin-modifying enzymes, histone modifications, and chromatin-associated proteins in a wide variety of organisms and biological states [66]. In a ChIP experiment, cells or tissues are treated with a cross-linking agent, usually formaldehyde, which cross-links proteins covalently to DNA in vivo. This is followed by cell lysis and sonication of chromatin into small fragments of a target size of 100–300 bp for ChIP-seq and 200–1000 bp range for ChIP (Figure 3). An antibody specific to the protein of interest such as a transcription factor, modified histone, RNA polymerase, etc., is used to immunoprecipitate the DNA–protein complex. After immunoenrichment, the cross-links are reversed and the enriched DNA is purified and prepared for analysis. Detailed protocols can be obtained from the Encyclopedia of DNA Elements (ENCODE) at the UCSC genome browser: http://encodeproject.org/ENCODE/. In ChIP–chip, the DNA is fluorescently labeled and hybridized to a DNA microarray [58,59], while in ChIP-seq the DNA is analyzed by high-throughput DNA sequencing.

ChIP-seq offers many advantages compared to ChIP–chip. Its base resolution and maximal signal-to-noise ratio may be the greatest improvement over ChIP–chip. The maximum resolution of ChIP–chip is array dependent and it is generally 30–100 bp while ChIP-seq maximal resolution is a single nucleotide. Microarrays can be tiled to a high density but this needs a large number of probes, which can be

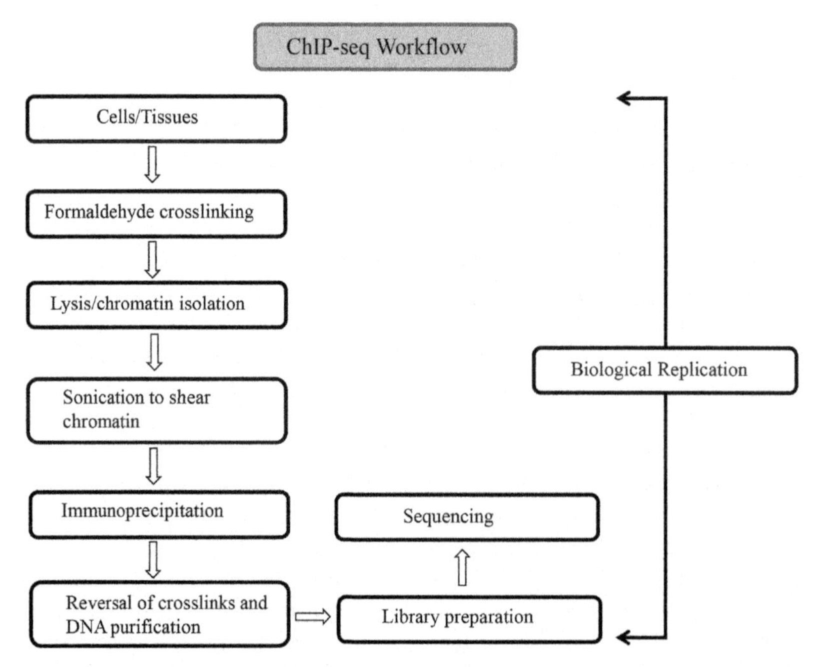

FIGURE 3 An introduction to the ChIP-sequencing workflow. Chromatin is sheared to about 100–500 bp fragments. DNA and proteins are cross-linked and bound DNA is analyzed by massively parallel short-read sequencing.

expensive for mammalian genomes [67]. Moreover, the noise generated by the hybridization process is eliminated in ChIP-seq experiments. Hybridization depends on many variables such as secondary structure of the target and probe sequences and their length, concentration, and GC content. Hence, cross-hybridization between incorrect sequences contributes to noise [57]. Finally, ChIP-seq genome coverage is not limited by the repertoire of the probes of the array. This proves to be important for the analysis of repetitive genomic regions such as heterochromatin or microsatellites, which are usually masked by ChiP–chip assays [57]. Nevertheless, there are few disadvantages with ChIP-seq as well as it is with all profiling technologies, which can produce artifacts. However, sequencing accuracy has been improved recently also with the help of alignment algorithms and computational analysis. When the technology was introduced, one of the main disadvantages was cost and availability. But as more researchers are utilizing ChIP-seq and the sequencing cost continues to decline, ChIP-seq has become the most widely used procedure to map genome-wide DNA–protein interactions.

Even though ChIP-seq is now widely used, there has been much diversity in the way ChIP-seq experiments have been conducted and reported. The ENCODE Consortium [66] has performed more than a thousand individual ChIP-seq experiments for more than 140 different factors and histone modifications in hundreds of cell lines in four different organisms (*Drosophila melanogaster*, *Caenorhabditis elegans*, mouse, and human). This experience has been used to develop a set of working standards and guidelines to follow. It is worth to note that given the diversity of cell types, factors, conditions, and biological questions to answer, it is hard to define common practices appropriate for all situations. From a technical standpoint, the challenge and the success of ChIP-seq are governed by the quality and the specificity of the antibody to the DNA-binding protein or modification. Prior validations to check whether the antibody has poor reactivity against the target or cross-reactivity with other DNA-associated proteins are necessary and described on the ENCODE guidelines [66]. Moreover, typically a large number of cells, around 10 million, are needed for ChIP-seq and this can be a challenge for small model organisms. New techniques such as nano-ChIP-seq have been carried out on as few as 10,000 cells for histone modifications. In this assay, the small amount of DNA enriched following ChIP is PCR amplified using custom primers [68]. This technique can be particularly useful for abundant histone modifications or transcription factors but it will probably need further optimization for the ones that are less abundant. Additionally, ChIP-seq experiments use sonication to shear the chromatin and build libraries containing DNA molecules that are ~200 bases long, while the binding site for most proteins is only a few base pairs. This contamination of DNA has been resolved with ChIP-exo, which uses lambda phage nuclease to digest the 5′ end of protein-bound and formaldehyde-cross-linked DNA fragments to a fixed distance from the bound protein [69]. Thus, there are technical challenges with ChIP-seq assays addressing specific biological questions and possible solutions have emerged.

6.3 The Analysis of ChIP-Seq Data

There has also been a large effort to improve the analytical tools required to interpret the sequence data output from ChIP-seq experiments. Different protein classes have diverse modes of interaction with the genome and demand different analytical approaches [66,70]. They are classified into three groups: point-source factors, broad-source factors, and mixed-sources factors [66,70]. Point-source factors such as transcription factors and their cofactors are localized at specific positions in the genome and produce highly localized signals. Broad-source factors are associated with large genomic domains and examples include some histone modifications such as H3K36me3. Mixed-source factors can bind in point-source way in some location and in broad-source way in others, such as RNA polymerase II [66]. Each of these requires distinct detection strategies as some protocols are specialized on one type of peak rather than others.

Effective analysis of ChIP-seq data also requires sufficient sequencing depth, which depends for the most part on the size of the genome and the number and size of the binding sites of the protein. For a typical point-source DNA-binding factor, the number of peaks typically increases with the number of sequence reads [51]. However, according to ENCODE, a 20-million-mapped reads is currently used as a minimum for all ChIP experiments targeting point-source transcription factors. Proteins that have more binding sites and mixed peaks (e.g., RNA Pol II or certain histone marks) will require more reads, up to 60 million for mammalian ChIP-seq. Sometimes reads from technical replicate experiments can be combined [71].

After mapping reads to the genome, there are few pivotal analyses for ChIP-seq to successfully identify regions of the genome where the protein is enriched with significant number of mapped reads (peaks). Based on sensitivity and specificity criteria and the protein of interest, bioinformaticians choose an appropriate peak-calling algorithm and normalization method [71]. ENCODE has used several algorithms including ChIP-seq processing pipeline [72], peak sequencing [73], and model-based analysis of ChIP-seq [74]. They have reported that when using standard peak-calling thresholds, successful experiments generally identify thousands of peaks for most transcription factors in mammalian genomes. Nonetheless, it is strongly recommended that mapped reads from an appropriate control experiment (e.g., from input DNA) are used in peak-calling [66]. Following peak calling, it is crucial to assess reproducibility. This is achieved by performing at least two biological replicates of each ChIP-seq experiments and to examine the reproducibility of each [71]. Peak annotation and motif analysis are also recommended to be determined. Peak annotation aims to associate ChIP-seq peaks with functionally relevant genomic regions such as gene promoters and enhancers. Motif analysis is useful not only to identify the DNA-binding motif seen in the peaks, but also when the motif is known to further validate the success of the ChIP-seq experiment [71]. Motif analysis is also useful to identify the DNA-binding motifs of other proteins that may bind in the same complex as the ChIP-ed protein, providing important clues into the mechanisms of transcriptional regulation. Thus, the importance of DNA-binding proteins and their biological functions has motivated the development of the diverse experimental and computational methods of ChIP-seq. These methods are transforming our understanding of the complex and dynamic networks that underlie transcription, impact translation, and all biological processes.

6.4 ChIP-Seq to Identify Biomarkers and Therapeutic Targets in Disease

Results from ChIP-seq experiments not only improve our understanding of normal cell biology but also provide essential information that will help determine the causes and consequences of abnormal cellular state linked with disease. One of the first papers to report ChIP-seq mapped genome wide the chromatin state of pluripotent and lineage-committed cells. The study determined that lineage commitment of pluripotent cells is accompanied by chromatin changes at promoters leading to changes in gene expression. The study discussed that chromatin-state maps from ChIP-seq should also be determined from situations of abnormal development such as cancer cells, as they are frequently associated with epigenetic defects [65].

As altered transcriptional programs are a hallmark of disease, numerous studies have tried to understand how these programs are established using ChIP-seq. Examples include the mapping of the transcription factor estrogen receptor (ER) or the pioneer transcriptional factor PBX1 in breast cancer. ER drives cell proliferation in over 70% of all breast cancers, where it acts as a transcriptional factor. ER+ breast cancers are generally treated with endocrine therapies such as tamoxifen, although resistance emerges in a significant number of women. Defining the basis of this drug resistance requires understanding of the genomic activity of ER. Our knowledge of ER activity in breast cancer has substantially increased in recent years as numerous studies have mapped ER binding through ChIP-seq. We can now appreciate the multitude of factors that augment or inhibit ER activity and better understand the *cis*-regulatory elements and enhancers of ER target genes in breast cancer [75–77]. The ER mapping studies above have not only been restricted to breast cancer cell lines. A more recent study used primary breast cancer from patients with different clinical outcomes and in distant ER-positive metastases to map genome-wide ER binding events by ChIP-seq [78]. By establishing transcription factor mapping in primary samples, the study concluded that there is plasticity in ER binding with distinct ER-binding profiles associated with clinical outcome [78]. Another study has used ChIP-seq experiments to elucidate the role of the pioneer transcriptional factor PBX1 in breast cancer. They determined PBX1, to be a novel factor defining aggressive ER-positive breast tumors, as it guides ER genomic activity and promoting a transcriptional program favorable to breast cancer progression [79]. This study introduced PBX1 as a potential biomarker and clinical tool with additive prognostic value to ER [79].

Moreover, ChIP-seq has also been used to provide critical information on other chromatin modifiers, such as histone marks and the enzymes that modify these marks in diseases such as cancer. Examples include work from Armstrong and colleagues who have used ChIP-seq to discover that mixed lineage leukemia (MLL)-rearranged leukemias are dependent on aberrant H3K79 methylation by the methyltransferase DOT1L [80]. These leukemias bear translocations involving the *MLL* gene. The researchers demonstrated an aberrant H3K79 methylation pattern, and the specific requirement of H3K79 methylation for the maintenance of the MLL translocation-associated oncogenic program [80]. Since DOT1L is responsible for this methylation pattern, this has profound therapeutic implications and DOT1L might represent a critical therapeutic target. Taken together, these few examples reveal how altered transcriptional programs are a hallmark of disease and ChIP-seq experiments are useful to explain how these programs are established.

7. CONCLUSIONS

Epigenetic modifications across the genome represent a fine-tuning mechanism for modulating the transcriptional output of the genetic code and to ultimately determine the identity of the cell. Identifying the aberrant changes associated with human disease and the factors promoting such alterations gives the potential to provide new targets for therapeutic intervention, as well as biomarkers for clinical management of the disease.

The implementation of available protocols and the development of new techniques to ascertain the epigenomic alterations underlying pathological conditions on diverse scenarios have already uncovered many reliable biomarkers. DNA methylation studies in cancer have provided with a plethora of biomarkers identifying different cancer subtypes, predicting risk of relapse, or response to therapy. The last years have witnessed an explosion of genome-wide studies exposing how the DNA methylation landscape gets disrupted in human cancer. The goal now is to translate such findings to a clinical setting by validating these clinical epigenetic biomarkers on a prospective manner and to further exploit epigenetic therapy to increase treatment success in patients. In addition to that, the histone code is also being progressively deciphered as we gain insights on the role of histone modifications in regulating gene transcription and function in normal and disease states. ChIP-seq experiments are revolutionizing our understanding of the complex associated with chromatin dynamics. Ongoing advances such as nano-ChIP-seq will allow ChIP-seq to be analyzed from far fewer cells necessary for embryology and development studies. The emergence of ChIP-exo that digests the ends of DNA fragments not bound to protein is quite promising. As research moves forward, we increase our understanding of chromatin-modifying enzymes that are aberrantly expressed in malignancies. Clinical trials have demonstrated clinical efficacy of epigenetic-directed therapies with histone deacetylase and DNA methyltransferase inhibitors. Beyond the FDA-approved epigenetics-based therapies, there are currently over 100 clinical trials with an epigenetics focus in diverse cancer types. Moving forward, as data from complementary assays accumulate, it remains critical to integrate all information to provide a more complete understanding of transcriptional networks and cellular processes.

LIST OF ABBREVIATIONS

AIMS Amplification of intermethylated sites
ASO Allele-specific oligo
BS Sodium bisulfite modification
CHARM Comprehensive high-throughput array for relative methylation
ChIP Chromatin immunoprecipitation
ChIP-seq Chromatin immunoprecipitation with high-throughput sequencing
DMH Differential methylation hybridization
DNMT1/3b DNA methyltransferases 1 and 3b
ENCODE Encyclopedia of DNA Elements
ER Estrogen receptor
hMeDIP-seq 5hmC immunoprecipitation with high-throughput sequencing
LSO Locus-specific oligo
MBD Methyl-CpG-binding domain proteins
MeDIP 5mC immunoprecipitation
MeDIP-chip 5mC immunoprecipitation with hybridization with array
MeDIP-seq 5mC immunoprecipitation with high-throughput sequencing
MIRA Methylated CpG Island Recovery Assay
MLL Mixed lineage leukemia
oxBS Oxidative bisulfite sequencing
RLGS Restriction landmark genomic scanning
RRBS Reduced representation bisulfite sequencing
TET Ten-eleven translocation enzymes
WGBS Whole-genome bisulfite sequencing

Acknowledgments

We would like to thank Natasha Morse for critical reading of the manuscript.

References

[1] Jones PA. Functions of DNA methylation: islands, start sites, gene bodies and beyond. Nat Rev Genet May 2012;13(7):484–92.
[2] Irizarry RA, Ladd-Acosta C, Carvalho B, et al. Comprehensive high-throughput arrays for relative methylation (CHARM). Genome Res 2008;18(5):780–90.
[3] Yang X, Han H, De Carvalho DD, Lay FD, Jones PA, Liang G. Gene body methylation can alter gene expression and is a therapeutic target in cancer. Cancer Cell 2014;26(4):577–90.
[4] Kulis M, Heath S, Bibikova M, et al. Epigenomic analysis detects widespread gene-body DNA hypomethylation in chronic lymphocytic leukemia. Nat Genet 2012;44(11):1236–42.
[5] Wu H, Zhang Y. Reversing DNA methylation: mechanisms, genomics, and biological functions. Cell 2014;156(1–2):45–68.
[6] Pastor WA, Aravind L, Rao A. TETonic shift: biological roles of TET proteins in DNA demethylation and transcription. Nat Rev Mol Cell Biol 2013;14(6):341–56.
[7] Plongthongkum N, Diep DH, Zhang K. Advances in the profiling of DNA modifications: cytosine methylation and beyond. Nat Rev Genet 2014;15(10):647–61.
[8] Esteller M. Cancer epigenomics: DNA methylomes and histone-modification maps. Nat Rev Genet 2007;8(4):286–98.
[9] Akama TO, Okazaki Y, Ito M, et al. Restriction landmark genomic scanning (RLGS-M)-based genome-wide scanning of mouse liver tumors for alterations in DNA methylation status. Cancer Res 1997;57(15):3294–9.

[10] Kawai J, Hirotsune S, Hirose K, Fushiki S, Watanabe S, Hayashizaki Y. Methylation profiles of genomic DNA of mouse developmental brain detected by restriction landmark genomic scanning (RLGS) method. Nucleid Acids Res 1993;21(24):5604–8.

[11] Huang TH, Perry MR, Laux DE. Methylation profiling of CpG islands in human breast cancer cells. Hum Mol Genet 1999;8(3):459–70.

[12] Frigola J, Ribas M, Risques RA, Peinado MA. Methylome profiling of cancer cells by amplification of inter-methylated sites (AIMS). Nucleic Acids Res 2002;30(7):e28.

[13] Paz MF, Wei S, Cigudosa JC, Rodriguez-Perales S, Peinado MA, Huang TH, et al. Genetic unmasking of epigenetically silenced tumor suppressor genes in colon cancer cells deficient in DNA methyltransferases. Hum Mol Genet 2003;12(17):2209–19.

[14] Rhee I, Bachman KE, Park BH, Jair KW, Yen RW, Schuebel KE, et al. DNMT1 and DNMT3b cooperate to silence genes in human cancer cells. Nature 2002;416(6880):552–6.

[15] Frommer M, McDonald LE, Millar DS, et al. A genomic sequencing protocol that yields a positive display of 5-methylcytosine residues in individual DNA strands. Proc Natl Acad Sci USA 1992;89(5):1827–31.

[16] Bibikova M, Fan JB. GoldenGate assay for DNA methylation profiling. Methods Mol Biol 2009;507:149–63.

[17] Martinez R, Martin-Subero JI, Rohde V, et al. A microarray-based DNA methylation study of glioblastoma multiforme. Epigenetics 2009;4(4):255–64.

[18] Bibikova M, Lin Z, Zhou L. High-throughput DNA methylation profiling using universal bead arrays. Genome Res 2006;16(3):383–93.

[19] Carmona FJ, Villanueva A, Vidal A, et al. Epigenetic disruption of cadherin-11 in human cancer metastasis. J Pathol 2012;228(2):230–40.

[20] Fernandez AF, Assenov Y, Martin-Subero JI, et al. A DNA methylation fingerprint of 1628 human samples. Genome Res 2012;22(2):407–19.

[21] Carmona FJ, Azuara D, Berenguer-Llergo A, et al. DNA methylation biomarkers for noninvasive diagnosis of colorectal cancer. Cancer Prev Res (Phila) 2012;6(7):656–65.

[22] Sanchez-Mut JV, Aso E, Panayotis N, et al. DNA methylation map of mouse and human brain identifies target genes in Alzheimer's disease. Brain 2013;136(Pt 10):3018–27.

[23] Bell CG, Teschendorff AE, Rakyan VK, Maxwell AP, Beck S, Savage DA. Genome-wide DNA methylation analysis for diabetic nephropathy in type 1 diabetes mellitus. BMC Med Genomics 2010;3:33.

[24] Irizarry RA, Ladd-Acosta C, Wen B, et al. The human colon cancer methylome shows similar hypo- and hypermethylation at conserved tissue-specific CpG island shores. Nat Genet 2009;41(2):178–86.

[25] Bibikova M, Barnes B, Tsan C, et al. High density DNA methylation array with single CpG site resolution. Genomics 2011;98(4):288–95.

[26] Sandoval J, Heyn H, Moran S, et al. Validation of a DNA methylation microarray for 450,000 CpG sites in the human genome. Epigenetics 2011;6(6):692–702.

[27] Moran S, Vizoso M, Martinez-Cardús A, et al. Validation of DNA methylation profiling in formalin-fixed paraffin-embedded samples using the Infinium HumanMethylation450 Microarray. Epigenetics 2014;9(6):829–33.

[28] Sandoval J, Mendez-Gonzalez J, Nadal E, et al. A prognostic DNA methylation signature for stage I non-small-cell lung cancer. J Clin Oncol 2013;31(32):4140–7.

[29] Ammerpohl O, Martín-Subero JI, Richter J, Vater I, Siebert R. Hunting for the 5th base: techniques for analyzing DNA methylation. Biochim Biophys Acta 2009;1790(9):847–62.

[30] Laird PW. Principles and challenges of genome-wide DNA methylation analysis. Nat Rev Genet 2010;11(3):191–203.

[31] Nestor C, Ruzov A, Meehan R, Dunican D. Enzymatic approaches and bisulfite sequencing cannot distinguish between 5-methylcytosine and 5-hydroxymethylcytosine in DNA. Biotechniques 2010;48(4):317–9.

[32] Huang Y, Pastor WA, Shen Y, Tahiliani M, Liu DR, Rao A. The behaviour of 5-hydroxymethylcytosine in bisulfite sequencing. PLoS One 2010;5(1):e8888.

[33] Lister R, Mukamel EA, Nery JR, Urich M, Puddifoot CA, Johnson ND, et al. Global epigenomic reconfiguration during mammalian brain development. Science 2012;341(6146):1237905.

[34] Yu M, Hon GC, Szulwach KE, Song CX, Jin P, Ren B, et al. Tet-assisted bisulfite sequencing of 5-hydroxymethylcytosine. Nat Protoc 2012;7(12):2159–70.

[35] Mariani CJ, Madzo J, Moen EL, Yesilkanal A, Godley LA. Alterations of 5-hydroxymethylcytosine in human cancers. Cancers (Basel) June 25, 2013;5(3):786–814.

[36] Chopra P, Papale LA, White AT, Hatch A, Brown RM, Garthwaite MA, et al. Array-based assay detects genome-wide 5-mC and 5-hmC in the brains of humans, non-human primates, and mice. BMC Genomics 2014;15:131.

[37] Zhubi A, Chen Y, Dong E, Cook EH, Guidotti A, Grayson DR. Increased binding of MeCP2 to the GAD1 and RELN promoters may be mediated by an enrichment of 5-hmC in autism spectrum disorder (ASD) cerebellum. Transl Psychiatry 2014;4:e349.

[38] Jacinto FV, Ballestar E, Ropero S, Esteller M. Discovery of epigenetically silenced genes by methylated DNA immunoprecipitation in colon cancer cells. Cancer Res 2007;67(24):11481–6.

[39] Weber M, Davies JJ, Wittig D, et al. Chromosome-wide and promoter-specific analyses identify sites of differential DNA methylation in normal and transformed human cells. Nat Genet 2005;37(8):853–62.

[40] Keshet I, Schlesinger Y, Farkash S, et al. Evidence for an instructive mechanism of de novo methylation in cancer cells. Nat Genet 2006;38(2):149–53.
[41] Down TA, Rakyan VK, Turner DJ, et al. A Bayesian deconvolution strategy for immunoprecipitation-based DNA methylome analysis. Nat Biotechnol 2008;26(7):779–85.
[42] Simmer F, Brinkman AB, Assenov Y, et al. Comparative genome-wide DNA methylation analysis of colorectal tumor and matched normal tissues. Epigenetics 2012;7(12):1355–67.
[43] Maunakea AK, Nagarajan RP, Bilenky M, et al. Conserved role of intragenic DNA methylation in regulating alternative promoters. Nature 2010;466(7303):253–7.
[44] Pomraning KR, Smith KM, Freitag M. Genome-wide high throughput analysis of DNA methylation in eukaryotes. Methods 2009;47(3):142–50.
[45] Bock C, Tomazou EM, Brinkman AB, et al. Quantitative comparison of genome-wide DNA methylation mapping technologies. Nat Biotechnol 2010;28(10):1106–14.
[46] Meissner A, Gnirke A, Bell GW, Ramsahoye B, Lander ES, Jaenisch R. Reduced representation bisulfite sequencing for comparative high-resolution DNA methylation analysis. Nucleic Acids Res 2005; 33(18):5868–77.
[47] Lee YK, Jin S, Duan S, Lim YC, Ng DP, Lin XM, et al. Improved reduced representation bisulfite sequencing for epigenomic profiling of clinical samples. Biol Proced Online 2014;16(1):1.
[48] Gu H, Bock C, Mikkelsen TS, et al. Genome-scale DNA methylation mapping of clinical samples at single-nucleotide resolution. Nat Methods 2010;7(2):133–6.
[49] Lister R, O'Malley RC, Tonti-Filippini J, et al. Highly integrated single-base resolution maps of the epigenome in *Arabidopsis*. Cell 2008;133(3):523–36.
[50] Cokus SJ, Feng S, Zhang X, et al. Shotgun bisulphite sequencing of the *Arabidopsis* genome reveals DNA methylation patterning. Nature 2008;452(7184):215–9.
[51] Carmona FJ, Davalos V, Vidal E, et al. A comprehensive DNA methylation profile of epithelial-to-mesenchymal transition. Cancer Res 2014;74(19):5608–19.
[52] Berman BP, Weisenberger DJ, Aman JF, et al. Regions of focal DNA hypermethylation and long-range hypomethylation in colorectal cancer coincide with nuclear lamina-associated domains. Nat Genet 2011; 44(1):40–6.
[53] Jjingo D, Conley AB, Yi SV, Lunyak VV, Jordan IK. On the presence and role of human gene-body DNA methylation. Oncotarget 2012;3(4):462–74.
[54] Clarke J, Wu HC, Jayasinghe L, Patel A, Reid S, Bayley H. Continuous base identification for single-molecule nanopore DNA sequencing. Nat Nanotechnol April 2009;4(4):265–70.
[55] Krivanek OL, Chisholm MF, Nicolosi V, Pennycook TJ, Corbin GJ, Dellby N, et al. Atom-by-atom structural and chemical analysis by annular dark-field electron microscopy. Nature 2010;464(7288):571–4.
[56] Flusberg BA, Webster DR, Lee JH, et al. Direct detection of DNA methylation during single-molecule, real-time sequencing. Nat Methods 2010;7(6):461–5.
[57] Park PJ. ChIP-seq: advantages and challenges of a maturing technology. Nat Rev Genet 2009;10:669–80.
[58] Ren B, Robert F, Wyrick JJ, Aparicio O, Jennings EG, Simon I, et al. Genome-wide location and function of DNA binding proteins. Science 2000;290:2306–9.
[59] Iyer VR, Horak CE, Scafe CS, Botstein D, Snyder M, Brown PO. Genomic binding sites of the yeast cell-cycle transcription factors SBF and MBF. Nature 2001;409:533–8.
[60] Lieb JD, Liu X, Botstein D, Brown PO. Promoter-specific binding of Rap1 revealed by genome-wide maps of protein-DNA association. Nat Genet 2001;28:327–34.
[61] Weinmann AS, Yan PS, Oberley MJ, Huang TH, Farnham PJ. Isolating human transcription factor targets by coupling chromatin immunoprecipitation and CpG island microarray analysis. Genes Dev 2002;16:235–44.
[62] Johnson DS, Mortazavi A, Myers RM, Wold B. Genome-wide mapping of in vivo protein–DNA interactions. Science 2007;316:1497–502.
[63] Barski A, Cuddapha S, Cui K, Roh TY, Schones DE, Wang Z, et al. High-resolution profiling of histone methylations in the human genome. Cell 2007;129:823–37.
[64] Robertson G, Hirst M, Bainbridge M, Bilenky M, Zhao Y, Zeng T, et al. Genome-wide profiles of STAT1 DNA association using chromatin immunoprecipitation and massively parallel sequencing. Nat Methods 2007;4:651–7.
[65] Mikkelsen TS, Ku M, Jaffe DB, Issac B, Lieberman E, Giannoukos G, et al. Genome-wide maps of chromatin state in pluripotent and lineage-committed cells. Nature 2007;448:553–60.
[66] Landt SG, Marinov GK, Kundaje A, Kheradpour P, Pauli F, Batzoglou S, et al. ChIP-seq guidelines and practices of the ENCODE and modENCODE consortia. Genome Res 2012;9:1813–31.
[67] Kim TH, Barrera LO, Zheng M, Qu C, Singer MA, Richmond TA, et al. A high-resolution map of active promoters in the human genome. Nature 2005; 436:876–80.
[68] Adli M, Bernstein BE. Whole-genome chromatin profiling form limited number of cells using nano ChIP-seq. Nat Protoc 2011;6:1656–68.
[69] Furey ST. ChIP-seq and beyond: new and improved methodologies to detect and characterize protein-DNA interactions. Nat Rev Genet 2012;12:840–52.
[70] Pepke S, Wold B, Mortazavi A. Computation for ChIP-seq and RNA-seq studies. Nat Methods 2009;6:S22–32.

[71] Bailey T, Krajewski P, Ladunga I, Lefebvre C, Li Q, Liu T, et al. Practical guidelines for the comprehensive analysis of ChIP-seq data. PLoS Comp Biol 2013;9(11):e1003326.
[72] Kharchenko PV, Tolstorukov MY, Park PJ. Design and analysis of ChIP-seq experiments for DNA-binding proteins. Nat Biotechnol 2008;12:1351–9.
[73] Rozowsky J, Euskirchen G, Auerbach RK, Zhang ZD, Gibson T, Bjornson R, et al. PeakSeq enables systematic scoring of ChIP-seq experiments relative to controls. Nat Biotechnol 2009;1:66–75.
[74] Zhang Y, Liu T, Meyer CA, Eeckhoute J, Johnson DS, Bernstein BE, et al. Genome Biol 2008;9:R137.2.
[75] Fullwood MJ, Liu MH, Pan YF, Liu J, Xu H, Mohamed YB, et al. An oestrogen-receptor-alpha-bound human chromatin interactome. Nature 2009;462:58–64.
[76] Carroll JS, Meyer CA, Song J, Li W, Geistlinger TR, Eeckhoute J, et al. Genome-wide analysis of estrogen receptor binding sites. Nat Genet 2006;38:1289–97.
[77] Hurtado A, Holmes KA, Ross-Innes CS, Schmidt D, Carroll JS. FOXA1 is a key determinant of estrogen receptor function and endocrine response. Nat Genet 2011;1:27–33.
[78] Ross-Innes CS, Stark R, Teschendorff AE, Holmes KA, Ali HR, Dunning MJ, et al. Differential oestrogen receptor binding is associated with clinical outcome in breast cancer. Nature 2012;481:389–93.
[79] Magnani L, Ballantyne EB, Zhang X, Lupien M. PBX1 genomic pioneer function drives ERα signaling underlying progression in breast cancer. PLOS Genet 2011;11:e1002368.
[80] Bernt KM, Zhu N, Sinha AU, Vempati S, Faber J, Krivtsov AV, et al. *MLL*-rearranged leukemia is dependent on aberrant H3K79 methylation by DOT1L. Cancer Cell 2011;20:66–78.

Further Reading

[1] Stewart SK, Morris TJ, Guilhamon P, Bulstrode H, Bachman M, Balasubramanian S, et al. oxBS-450K: a method for analyzing hydroxymethylation using 450K BeadChips. Methods 2015;15(72):9–15.
[2] Myers RM, Stamatoyannopoulos J, Snyder M, Dunham I, Hardison RC, Bernstein BE, et al. A user's guide to the Encyclopedia of DNA Elements (ENCODE). PLoS Biol 2011;9:e1001046. http://dx.doi.org/10.1371/journal.pbio.1001046.

CHAPTER

7

Sequenom MassARRAY Technology for the Analysis of DNA Methylation: Clinical Applications

Enrique J. Busó[1], *Marisa Iborra*[2]

[1]Unidad Central de Investigación, University of Valencia, Valencia, Spain; [2]Gastroenterology Department, Hospital Universitari i Politècnic La Fe, Valencia, Spain

OUTLINE

Epigenetic Biomarkers and Diagnostics
http://dx.doi.org/10.1016/B978-0-12-801899-6.00007-3

1. INTRODUCTION

Regulation of gene expression is a complex process and currently most human diseases are in some way related to either loss or gain of gene functions. Most of these effects are hereditary; however, the effects of external environment on genes can also influence these changes. Recently, epigenetic regulation of gene expression has emerged as a potentially important contributor in several normal cellular processes and in the development of certain pathologies. In biology, epigenetics is defined as the study of heritable changes that are not caused by changes in DNA sequence. The term was first used in 1942 by Conrad Waddington to describe gene–environment interactions that lead to manifestations of various phenotypes during development [1].

Epigenetic mechanisms influence gene expression and function without modification of the base sequence of DNA and may be reversible, heritable, and influenced by the environment [2]. Epigenetic mechanisms are often involved in the regulation of genes that produce permanent changes associated with the differentiation of various cell types. Currently, the study of human diseases has focused on epigenetics and the development of diagnostic tools which might themselves reveal the mechanisms involved in pathologies such as cancer, inflammation, aging, or degenerative disease. Moreover, the development of *epigenetic therapies* has shown promising effects in some diseases with great social impact [3].

A variety of processes have been included into epigenetics; these include DNA methylation, covalent histone modifications, chromatin folding, several types of regulatory RNAs (microRNAs (miRNAs), noncoding RNA, small RNAs), and polycomb group complexes [4]. In this chapter, we focus on DNA methylation.

2. DNA METHYLATION

The DNA methylation is perhaps the most extensively studied epigenetic modification in mammals. It is a biochemical process where a methyl (CH_3) group is added to cytosine or guanine nucleotides. It provides a stable gene silencing mechanism that plays an important role in regulating gene expression and chromatin architecture, in association with histone modifications and other chromatin-associated proteins [5].

DNA methylation primarily occurs in cytosine–phosphate–guanine (CpG) sites where cytosine is directly followed by guanine in the DNA sequence. This reaction is catalyzed by the enzyme DNA methyltransferase [6]. CpG dinucleotides are spread all over the genome. Human DNA has about 80–90% of CpG sites methylated, but there are other areas, called CpG islands, that remain unmethylated during development and in differentiated tissues. Methylation adds information not encoded in the DNA sequence, but it does not interfere with the Watson–Crick pairing of DNA. The pattern of methylation controls protein binding to target sites on DNA, affecting changes in gene expression and in chromatin organization, often silencing genes, which physiologically organizes processes like differentiation, and pathologically leads to cancer [7].

3. METHODS FOR DNA METHYLATION ANALYSIS

DNA methylation patterns cannot currently be analyzed by routine direct sequencing or by hybridization-based methods and are erased by polymerase chain reaction (PCR) or cloning of the DNA. Recently, various technologies for the analysis of DNA methylation have been developed [8–10]. Traditionally, DNA methylation has been studied in a gene- or region-specific manner. However, advances in global profiling technologies have expanded studies to the entire methylome. To explore DNA methylation profiles at genome scale, a wide range of approaches have been developed. Most of the methods were originally used for detecting methylation changes at the single gene level; but by coupling them with extensive cloning and sequencing work, or combining them with microarray platforms, genome-wide analysis has become a feasible and cost-efficient approach.

These technologies can be categorized into several well-characterized groups based on the approach or principle used to discriminate the methylated and unmethylated cytosine.

3.1 Methylation-Sensitive Restriction Enzyme Digestion

Several methods are based on restriction endonucleases that possess altered sensitivity toward methylated cytosine residues present in the cleavage site. The vast majority of methylation-sensitive restriction enzymes, such as HpaII and SmaI, are inactive on methylated CpG sites, but a unique methylation-sensitive restriction enzyme, McrBC, is inactive on unmethylated CpG sites. Thus, the genomic DNA can be digested by methylation-sensitive restriction enzymes like HpaII and McrBC to discriminate and/or enrich methylated or unmethylated DNA. The resulting restriction endonuclease digestion pattern depends on the methylation status of the cleavage sites and ultimately reflects methylation profiles of the given chromosomal region. However, the methods based on this principle can only provide methylation data at the restriction enzyme-recognition sites or adjacent regions, clearly limiting their usefulness.

3.2 Affinity Purification

These methods are based on the ability of certain antibodies or proteins to bind to methylated DNA. Thus, by using antibodies against methylated cytosine, methyl–CpG binding domains or other protein domains of the methylated or unmethylated fractions of a sample DNA can be immunoprecipitated [11–15]. The main limitation of this approach is that the exact methylation state of individual CpG sites cannot be determined.

3.3 Conversion of DNA Using Bisulfite

Most of the methods used for DNA methylation analysis rely on a chemical reaction called "bisulfite treatment," in which sodium bisulfite causes changes in the DNA sequence through selective chemical conversion of nonmethylated cytosine (C) to uracil (U) that depends on the methylation state of each CpG position. After treatment, all nonmethylated cytosine bases are converted to uracil in a specific manner, whereas all methylated cytosine bases remain as cytosine. These methylation-dependent C-to-U changes are later amplified as thymine during PCR. Afterward, the sequence changes introduced by this treatment can be analyzed using traditional or next-generation sequencing [16,17]. A key aspect of the methods involving sodium bisulfite is that the efficiency of bisulfite conversion is critical for the accuracy and the reliability of the results. The incomplete conversion of unmethylated cytosine to uracil or undesired conversion of methylcytosine to thymine can lead to a bias in the estimation of the methylation level [18,19].

Out of the three aforementioned approaches the next-generation sequencing of bisulfite-converted and PCR-amplified DNA molecules provides the most reliable and quantitative information on the methylation level for every single CpG site and is therefore currently regarded as the "gold standard" of DNA methylation analysis. In addition, it provides specific information about the methylation level of different haplotypes of DNA molecules in a reproducible and quantitative manner. Furthermore, in the case of genome-wide DNA methylation studies, bisulfite conversion coupled with high-throughput sequencing is currently considered the best choice among the available methods, mainly because bisulfite conversion can be performed at a genomic scale without the need of a restriction enzyme's recognition site or a high CpG density.

However, common high-throughput sequencing strategies rely on using large amounts of bisulfite-treated DNA. Different next-generation protocols recommend the use of 100–500ng of bisulfite-treated DNA per sequencing run [20–23]. In addition, quantitative DNA methylation assessment, unlike genotypic assessment, requires technical replicates to ensure accuracy. Consequently, the needed DNA amounts for this kind of studies are very difficult to get for most researchers, especially in the case of human samples, as DNA quantity usually is severely limited.

Another problem in the use of next-generation sequencing for methylation analysis arises from the experience of recent large-scale genetic association studies. These genome-wide experiments suggest that very large sample sizes are needed in order to detect the small differences in methylation of certain CpG sites that have a key role in the development of highly complex disorders or physiological processes [24]. The cost of such large-scale research projects remains a huge obstacle to many researchers.

3.4 Sequenom EpiTYPER System

Though the precise quantification of DNA methylation can greatly improve our knowledge about the origin and etiology of many complex disorders and can provide a tool for the early detection of several pathologies, current methods remain inefficient for profiling the large sample numbers needed to detect pathogenic epimutations, especially for complex disorders like cancer or settings in which several tissue types need to be analyzed.

In recent years, a novel technique for methylation analysis has been developed by Sequenom. The Sequenom EpiTYPER system (Sequenom Inc., CA, USA) technology employs sodium bisulfite, which converts unmethylated cytosine into uracil, while methylated cytosine remains unchanged. These sequence changes are preserved during subsequent PCR, with conversion to thymidine at unmethylated (but not methylated) cytosine positions. The resulting DNA is transcribed in vitro into a single-stranded RNA molecule and, subsequently, cleaved base specifically by an endoribonuclease. The exact weight of the fragments produced will depend upon bisulfite-treatment-induced variations in the DNA sequence. Samples are then analyzed by matrix-assisted laser desorption/ionization time-of-flight mass spectrometry (MALDI-TOF MS) to assess the size ratio of the cleaved products. The abundance of each fragment (signal-to-noise level in the spectrum) is indicative of the amount of DNA methylation in the interrogated sequence, which provides quantitative methylation estimation for CpG sites within a target region [25].

This method permits the high-throughput quantification of methylation levels at single or multiple CpG positions in a cost-effective manner using as little as 10ng of DNA per sample and assay and, therefore, has certain advantages over many other currently used methods.

In this section, each of the individual steps of the EpiTYPER assay will be described and their accuracy, precision, and sensitivity will be evaluated. Furthermore, we critically address the success rate and determine the limitations of this technique. Figure 1 shows an overview of EpiTYPER assay.

3.4.1 *Bisulfite Conversion and PCR Amplification*

Bisulfite treatment of the sample DNA can be performed by any of the available commercial methods. The bisulfite conversion protocol is susceptible to processing errors, and small deviation from the protocol can result in failure of the treatment or cause an over- or underestimation of the methylation levels. This chemical treatment introduces various DNA strand breaks and leads to a severely fragmented single-stranded DNA. Consequently, DNA degradation decreases the number of DNA molecules which are effectively available for PCR amplification. It has been

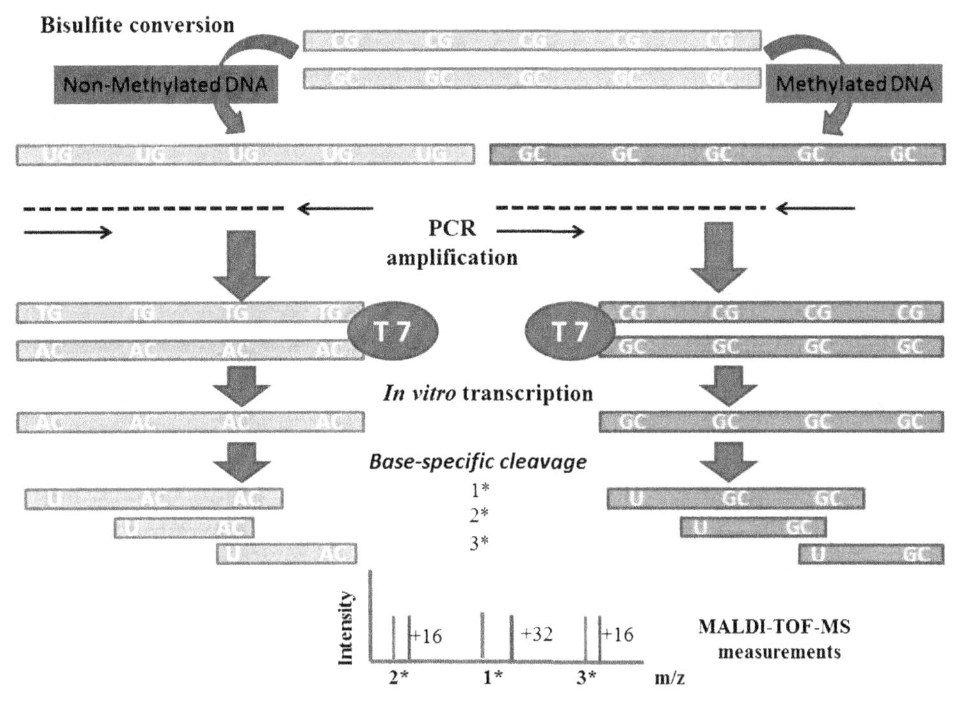

FIGURE 1 Overview of EpiTYPER® assay.

demonstrated that around 90% of DNA suffers severe degradation during this conversion, with depurination being the main cause for this loss of quality DNA [18,26]. Aggressive bisulfite treatment protocols (long incubation, high temperatures, and high molarity of bisulfite) result in the complete conversion of cytosine to uracil, but the treated DNA can be so damaged so as to render further PCR amplification very difficult. On the other hand, less aggressive treatments can lead to the overestimation of methylation levels due to detection of nonconverted cytosine. In conclusion, choosing the appropriate method to perform bisulfite conversion of DNA samples is an important aspect that may require some tests.

The next step in the process involves the amplification of converted DNA by PCR. PCR primers for the amplification of different regions of interest must be designed by using EpiDesigner software from Sequenom (www.epidesigner.com). The designer software introduces several small sequences that do not hybridize with the actual sequence of the regions of interest but which play an important role in the process. These include for each reverse primer, a T7 promoter tag for in vitro transcription and an 8-base-pair (bp) insert for prevention of abortive cycling and constant 5′- fragment for RNase A reaction. Meanwhile, the forward primer is designed to include a 10-mer tag to adjust for melting-temperature differences.

The recommended size range for PCR amplicons is 200–600 bp. Longer amplicons can also be analyzed but the DNA degradation caused by the bisulfite treatment limits the success of the PCR. The protocol establishes an initial sum of 10 ng of converted DNA per assay, which equals about 6500 copies of genomic DNA. Considering the aforementioned average 90% DNA degradation during bisulfite treatment, only relatively few molecules are left for PCR amplification. Moreover, the number of available molecules is also influenced by the length of the target amplicon. Longer amplicons are less likely to amplify, simply because the likelihood of finding a single intact starting template decreases. This is especially important when a quantitative methylation value is desired as it is in EpiTYPER. When only a few molecules are used as a starting template, statistical effect can have a dramatic consequence on the quantitative result and can lead to incorrect estimation of the methylation level of certain CpG sites. If no intact DNA fragments are available for the amplification of the region of interest, the PCR reaction will obviously fail. In addition, when the number of available molecules is drastically reduced due to DNA fragmentation, the results are no longer quantitative and show large variability when measured repeatedly.

The PCRs are carried out in a 5-μL format on 384-well microplates (view Table 1), with 10 ng bisulfite-treated DNA, 0.2 units of *Taq* DNA polymerase (Sequenom), 1X supplied *Taq* buffer, and 200 nM PCR primers. Amplification for the PCR will be as follows: preactivation of 95 °C for 4 min, 45 cycles of 94 °C denaturation for 20 s, annealing of 56 °C for 30 s, and a 72 °C extension for 1 min, finishing with a 72 °C incubation for 3 min. It is highly recommended to include a negative control, which consists of running the PCR reaction without any bisulfite-treated DNA. Doing this verifies if there is any PCR contamination and indicates excessive formation of primer dimers that lead to specific transcripts and respective cleavage products. Furthermore, additional signals that may appear later on the MALDI-TOF spectrum can also be cross-correlated with those from the negative control.

After amplification it is also helpful to run 0.5 μL of PCR product on 1.5% agarose gel to confirm successful PCR amplification and amplification specificity. This analysis of the PCR amplification quality can be replaced with a quicker SYBR Green PCR melt curve analysis, allowing for a high-throughput screen for PCR amplification without any further sample manipulations, which is essential in any high-throughput assay.

3.4.2 *Dephosphorylation of Unincorporated Deoxynucleotides (dNTPs) and Base-Specific Cleavage*

Before the base-specific cleavage is performed, it is necessary to dephosphorylate any remaining, unincorporated dNTPs in the amplification

TABLE 1 PCR Amplification

PCR	Reactive	Amount
	Bisulfite-treated DNA	10 ng
	Taq DNA polymerase (Sequenom)	0.2 units
	Taq buffer	1X
	PCR primers	200 nM
	Temperature	**Time**
1. Preactivation	95 °C	4 min
2. 45 cycles:	94 °C	20 s
	56 °C	30 s
	72 °C	1 min
3. Incubation	72 °C	3 min

products in order to render them unavailable for future polymerase reactions. This is carried away by Shrimp alkaline phosphatase (SAP) (Sequenom) which cleaves phosphates of unincorporated dNTPs. The protocol, summarized in Table 2, establishes that to each 5 μL of PCR-amplified product, 2 μL of H_2O containing 0.51 units of SAP must be added, incubating at 37 °C for 40 min, and then for 5 min at 85 °C to deactivate the enzyme.

At this point the in vitro transcription and base-specific cleavage of the PCR products can be performed in the same reaction. This is achieved by adding 2 μL of this SAP-treated PCR mixture to 5 μL of transcription/cleavage reaction and incubating at 37 °C for 3 h (summary in Table 3). The transcription reaction contains 27 units of T7 RNA and DNA polymerase (Sequenom), 0.64X of T7 RNA and DNA polymerase buffer, and 3.14 mM dithiothreitol (DTT). In the same step of the in vitro transcription, 0.22 μL cleavage mix (T or C mix according to the desired type of cleavage) (Sequenom) and 0.09 mg/mL RNase A (Sequenom) are added to cleave the RNA molecules.

Lastly, the mixture is further diluted with H_2O to a final volume of 27 mL. Conditioning of the phosphate backbone prior to MALDI-TOF MS was achieved by the addition of 6 mg CLEAN Resin (Sequenom Inc., San Diego, CA) and rotating the microplate with the resin for 10 min. This conditioning step is important to optimize MS analysis by reducing the presence of salt adducts in the reaction products. More detailed information about this protocol can be found elsewhere [27].

TABLE 2 Shrimp Alkaline Phosphatase (SAP) Incubation

	Reactive	**Amount**
	PCR-amplified product	5 μL
	H_2O	1.7 μL
	SAP	0.3 μL
	Temperature	**Time**
Incubation	37 °C	40 min
Enzyme deactivation	85 °C	5 min

TABLE 3 Cleavage and Transcription Cocktail

	Amount	**Mixed**
SAP-treated	2 μL	
Transcription reaction and cleavage mix	5 μL	• 27 units of T7 RNA and DNA polymerase • 0.64X of T7 RNA and DNA polymerase buffer • 3.14 mM DTT • 0.22 μL cleavage mix • 0.09 mg/mL RNase A
	Temperature	**Time**
Incubation	37 °C	3 h

SAP, shrimp alkaline phosphatase; DTT, dithiothreitol.

3.4.3 MALDI-TOF MS Measurements

Twenty-five nanoliters of the cleavage reactions are robotically dispensed onto silicon chips preloaded with matrix (SpectroCHIPs II; Sequenom, San Diego). Mass spectra are collected using a MassARRAY mass spectrometer (Bruker–Sequenom). Spectra are analyzed using proprietary peak picking and signal-to-noise calculations and spectra's methylation ratios are generated by the EpiTYPER software (Sequenom).

A critical aspect to obtain good data from the spectra obtained following this protocol involves the mathematical formula that the software uses to calculate the abundance of each fragment (signal-to-noise level in the spectrum) and consequently the methylation level of each CpG site.

The first formula used by the EpiTYPER software leads to biases when comparing fragments with different numbers of CpG sites. In the case of fragments with a single CpG site, the formula originally used calculated the level of methylation at that specific CpG. However, when analyzing a fragment with more than one CpG site, the formula determined the ratio of fragments containing one or more methylated positions. In consequence, the higher the number of CpG positions that a certain fragment has, the higher the methylation level will be as they have more chances of containing one or more methylated sites, thus introducing a bias [28].

Later updates of the EpiTYPER software modified the original formula to account for that problem. A few years back, some authors proposed a new method for calculating the methylation level that is not biased by the number of sites within the fragment and that incorporates the average methylation of each CpG site within a fragment [28]:

$$\text{Average methylation of fragment with } n\ CpG\ sites = \frac{\left(\frac{1}{n}\right) SNR_1 + \left(\frac{2}{n}\right) + \cdots + \left(\frac{n}{n}\right) SNR_n}{SNR_{NOME} + SNR_1 + SNR_2 + \cdots + SNR_n}$$

where SNR is the signal-to-noise ratio of either the unmethylated peak or one of the methylated peaks (1, 2,..., n) associated with this fragment.

3.4.4 Limitations of the EpiTYPER Protocol

Though the EpiTYPER protocol has several advantages for quantitative methylation detection over other common technologies, a number of limitations must be considered. These limitations affect the precision of this technology and are present in many of the aforementioned steps.

Bisulfite conversion remains the main source of process variability in the analysis of DNA methylation and in the case of EpiTYPER this is not an exception. Each of the reaction parameters of this step such as quality of the template DNA, pH, incubation time, and temperature and concentration of urea or hydroquinone can critically influence DNA quality during the reaction. In consequence, great attention must be paid to optimize those parameters in order to reduce DNA degradation. The number of available molecules for PCR amplification is essential for precise determination of methylation levels and is determined by the amount of starting genomic DNA, the level of fragmentation it experiences during this step, and amplicon length. If few high-quality molecules are present, the EpiTYPER protocol can still be performed successfully, but the reduced statistical population greatly diminishes the confidence level of the quantitative results. Recently, a quality control assay of bisulfite-treated DNA has been proposed. The assay includes four amplicons of increasing length which are located in the *IGF2/H19* region. It combines measuring amplification success with calculation of variance methylation values in order to give an estimate of DNA quality [29]. Including this quality control assay in the EpiTYPER protocol could greatly enhance the accuracy of the results. In addition, in this work it is suggested that even though the EpiTYPER protocol allows for an amplicon length of up to 600bp, depending on the temperature conditions of the bisulfite conversion used, it may be better to reduce the maximum length of the designed amplicons to 450–500bp.

To reduce the variability caused by the bisulfite conversion and posterior PCR amplification, it is also recommended that the PCR step be replicated at least three times. In cases where very large cohorts need to be analyzed, and the inclusion of this many replicates is not an option, it is recommended to at least pool multiple PCR reactions prior to downstream methylation analysis to minimize this PCR-induced variability [28]. These authors also suggest increasing the minimum mass of the expected fragments when designing the amplicons from the default 1500–1700Da. They found that for fragments below 1700Da, the calibration is not optimal, as peak position is not always called correctly. This is aggravated by the fact that fragments below this mass commonly contain only a small number of nucleotides (<6 nucleotides), thus increasing the chance of overlapping fragments with the same composition.

Other important limitation of the EpiTYPER protocol is caused by the presence of polymorphisms in the sequence of interest. The presence of single-nucleotide polymorphisms (SNPs) in the target region may cause a mass shift in the spectrum or even a different fragmentation pattern, which can lead to an incorrect interpretation of the obtained peaks and consequently to an over- or underestimation of the methylation values. The EpiTYPER software cannot account for the presence of SNPs and can only analyze the fragmentation pattern of a single sequence entered per spectrum. However, in this situation an experienced user can analyze those spectra that include altered peak patterns for SNP discovery, a high-throughput analysis of the allelic distribution of this SNP in the sample population and even be exploited for the analysis of the allele-specific methylation status [28].

Another source of variability when using MALDI-TOF MS to determine methylation levels comes from contaminant peaks that may appear in the spectrum. The MassCLEAVE® chemistry and the ionization process of MALDI-TOF introduce many contaminant peaks, some with intensities as high as 15% of the analyte peaks. Most of the contaminant peaks are far away from the analyte peaks, but some do fall close or even right on top of the informative peaks, a situation that could lead to an incorrect quantification of the methylation level. These additional peaks can originate from sources such as abortive amplification cycles during PCR, the presence of doubly charged particles, depurination of G (which creates an additional peak at −153Da of the original peak), etc., but the bulk of them can be grouped as adduct peaks, that is to say, the conjugation of analytes with salts present in the reaction well or material from the Spectro CHIP used in the MS. Their masses are more or less predictable within a spectrum and the EpiTYPER software can detect and correct them if they overlap with the analyte peaks. However, a manual inspection of the spectra is highly recommended to confirm that some were not overlooked.

The commonly observed adducts are sodium, with a 22Da higher mass than the expected peak, potassium (38Da), manganese (54Da), copper or carboxyl (62Da), and matrix adducts (95, 138, and 189Da). The sodium and potassium adducts are the most problematic because they are the most frequent and very close to peak separations between C-to-A (24Da) and G-to-T (40Da) changes. Using the Sequenom Spectro CHIP I, adducts are normally observed at 10–20% of parent peak for sodium, 5–15% for potassium, and 5–15% for copper and matrix adducts. With the new second-generation Spectro CHIPs, adduct intensities are greatly reduced mostly within 10% of the parent peaks, making analysis much easier. If that does not eliminate the adduct peaks a solution can be to increase the conditioning of the cleavage products with the CLEAN Resin to 20 or 30min.

4. DNA METHYLATION STUDIES BY SEQUENOM

Several studies focus on the use of MassARRAY assay to detect methylated CpG sites of candidate genes. In this chapter, we analyze the studies involving some of the most interesting pathologies (Table 4).

4.1 Cancer

The disruption of epigenetic mechanisms can alter the normal patterns of gene expression in various tissues, thereby inducing malignant cellular transformation. Because methylated genes are typically turned off, loss of DNA methylation can cause abnormally high gene activation by altering the arrangement of chromatin. On the other hand, too much methylation can undo the work of protective tumor suppressor genes. Moreover, the reversible nature of epigenetic aberrations has led to the emergence of the promising field of epigenetic therapy and the development of epigenetic drugs designed for cancer treatment [30]. In this sense, methylated genes may be interesting targets for chemotherapy allowing for the use of DNA methylation inhibitors in patients with cancer, as well as helping to clarify the importance of epigenetics in tumorigenesis [31].

The first human disease to be linked to epigenetics was cancer, in 1983. Researchers found that diseased tissue from patients with colorectal cancer had less DNA methylation than normal tissue from the same patients [32]. Since this publication, several studies have focused on the study of colorectal cancer by Sequenom analysis. Wang et al. reported that black raspberry powder has cancer-inhibitory effects, due to the decrease in promoter methylation of tumor suppressor genes caused by anthocyanins in patients with colorectal cancer [33]. MassARRAY assay was used to detect methylated CpG sites of candidate genes, CDKN2A, SFRP2, SFRP5, and WIF1, in HCT116, Caco2, and SW480 cells. Results showed different responses of human colon cancer cells to anthocyanin-induced demethylation when the cells were treated with anthocyanins. Different methylation levels were detected and the demethylation was found to be cell line dependent [33].

Moreover, there are studies in cancer where the authors have used MassARRAY to validate results [34]. In this sense, a study in neoplastic kidney specimens from pediatric patients analyzed the general methylation status of the tumor group using Illumina. The authors carried out a DNA methylation analysis to identify genes that are differentially methylated in tissue samples of pediatric tumors including rhabdoid tumor of the kidney, Ewing's sarcoma, clear cell sarcoma of the kidney, congenital mesoblastic nephroma, and Wilms' tumor. They showed that pediatric renal sarcomas possess a distinct DNA methylation profile in a tumor-type-specific manner. Then, by employing MassARRAY, the authors analyzed the surrounding CpG of some specific probes to validate the results of Infinium assay (Illumina Infinium HumanMethylation27). The Sequenom EpiTYPER assay was performed following the manufacturer's recommended protocol. The authors discovered that a combination of genes was sufficient for the DNA methylation profile-based differentiation of different types of renal tumors by clustering analysis. The DNA methylation status of the THBS1 CpG site was sufficient for the distinction of clear cell sarcoma of the kidney from other pediatric renal tumors, including Wilms' tumor and congenital mesoblastic nephroma. They concluded that the DNA methylation profile could be useful for the differential diagnosis of pediatric renal tumors and their observations shed light on the significance of the epigenetic diversity of these tumors on their biological features and mechanisms of pathogenesis [34].

MiRNAs play an important role in the regulation of gene expression that is frequently deregulated in cancer. In literature, MassARRAY technology has been used in carcinogenesis to analyze the methylation of miRNAs [35].

TABLE 4 Examples of Some Pathologies Studied Using MassARRAY Assay

Cancer/oncology	Breast cancer
	Renal cancer
	Colorectal cancer
	Gastric carcinoma
	Lymphoma and leukemia
Metabolic and endocrine diseases	Obesity
	Diabetes
	Circadian clock system
	Dietary and hormonal factors
Pregnancy	Preeclampsy
	Down syndrome placenta
	Fetal diseases
Neurological and neurodegenerative diseases	Alzheimer
	Rett syndrome
	Autism
	Friedreich's ataxia
	Hereditary osteodystrophy
Psychiatric disorders	Depression
	Schizophrenia
	Posttraumatic stress
Chronic inflammatory diseases	Inflammatory bowel disease (ulcerative colitis and Crohn's disease)
	Chronic gastritis
Infectious diseases	Human papiloma virus (HPV)
	Human immunodeficiency virus (HIV)
	Hepatitis C virus (HCV)
	Helicobacter pylori
	Malaria (*Plasmodium falciparum*)

Authors found a great portion of analyzed miRNA promoters to be hypermethylated in breast cancer cell lines, and the same promoters were targeted, although less frequently so, in breast tumor tissues. The hypermethylation of selected miRNA promoters was verified using MassARRAY technology [35]. MassARRAY DNA methylation data revealed that in 6 miRNA genes analyzed in more detail (mir-31, mir-130a, let-7a-3/let-7b, mir-155, mir-137, and mir-34b/mir-34c genes), the hypermethylation of miRNA promoters was linked to decreased levels of miRNAs expression. The study concluded that the proportion of miRNA promoters found aberrantly methylated in breast cancer suggests an important role of DNA methylation in miRNA in human breast carcinogenesis [35].

4.2 Obesity

Obesity has become an alarming epidemic in worldwide population because it increases the risk of hypertension, type II diabetes, and cardiovascular morbidity, which leads to increased morbidity and mortality. The main molecular mechanism involved in obesity is not yet well established. However, its multifactorial nature is well known and involves a complex interaction between both genetic predisposition and environmental factors (diet, physical exercise) regulated by epigenetic mechanisms such as DNA methylation [36]. In recent years, epigenetic mechanisms linked to dietary and hormonal factors have been investigated as potentially important contributors to excessive weight gain. The identification of changes in methylation profiles of specific genes has proved advantageous for predicting the development of obesity and preventing its complications.

Out of all the available studies most focus on obesity, with only a few using Sequenom MassARRAY technology. There is epidemiological evidence of perinatal factors that could later be involved in the increase of adiposity and risk of metabolic disease. These studies suggest that adult disease risk is associated with adverse environmental conditions early in development. Although the mechanisms behind these relationships are uncertain, a contribution of epigenetic dysregulation has been postulated [37]. The investigation in individuals conceived during Dutch Hunger Winter in 1944 revealed that the exposure to prenatal famine resulted in the persistent difference in DNA methylation of insulin-like growth factor 2 (IGF2) and other imprinted genes [38]. The methylation levels of IGF2 were reduced compared with their unexposed same-sex siblings. The differentially methylated region (DMR) was affected by folic acid intake before or during pregnancy, and depression in pregnancy was associated with obesity in the adult offspring [37]. The periconceptional exposure and the early-life environmental conditions can cause epigenetic marks in humans that persist throughout life [38].

Two studies using MassARRAY technology have suggested that adverse in utero environments are associated with obesity-related diseases and linked with altered DNA methylation of IGF2 [39,40]. Huang et al. investigated whether the levels of IGF2/H19 imprinting control region (ICR) methylation was associated with the anthropometry and adiposity. They found a positive association of more IGF2/H19 ICR methylation with greater subcutaneous adiposity [39]. Additionally, He et al. discovered that decreased methylation levels of IGF2 and GNAS DMRs in infants born to preeclamptic pregnancies was associated with high risk for metabolic diseases in the later life of the infants [40].

The circadian clock system controls gene expression in almost all cells, including adipocytes and is mainly regulated by epigenetic mechanisms. A study investigated the influence of obesity and metabolic syndrome features in clock gene methylation (CLOCK, BMAL1, and PER2 genes) [41]. White blood cells were obtained from normal, overweight, and obese women and DNA methylation levels at different CpG sites were analyzed using Sequenom MassARRAY. The methylation pattern of CpG sites of three genes showed statistical differences and an association was found between methylation status of CpG sites located in clock genes (CLOCK, BMAL1, and PER2) and obesity, metabolic syndrome, and weight loss [41].

Finally, a study in two types of mutant obese rats (WNIN) has measured the methylation levels of leptin, one of the most clinically relevant hormones associated with obesity [42]. Quantitative methylation analysis of the leptin gene promoter was performed with the MassARRAY

Compact system. This study demonstrated that the leptin gene expression was in line with leptin promoter DNA methylation patterns in WNIN obese mutant rats suggesting that genetic differences seemed to have an effect on methylation percentages and subsequently on the regulation of gene expression [42].

4.3 Alzheimer's Disease

The human adult brain is characterized by its flexibility and capacity to respond to external stimuli. The brain receives and processes signals from the outside world, transforming these signals into memories, which allow us to learn from past experiences. These processes are mediated at the cellular level by quick changes in gene expression, which are in turn regulated by epigenetic mechanisms such as histone modification and particularly DNA methylation. Recent work in humans has shown that epigenetic patterns remain "plastic" throughout all periods of brain development and aging and that ongoing dynamic regulation occurs even in neurons and other postmitotic constituents of the brain [43].

In addition, epigenetic changes play key roles in several neurodevelopmental syndromes of early childhood such as Rett syndrome [44], many adult-onset hereditary neurodegenerative disorders like Alzheimer's disease (AD) [45], and in psychiatric disorders like depression and schizophrenia [46,47]. The field of epigenetics is reshaping the current thinking about neurological and neurodegenerative processes and is opening the door to the development of chromatin-modifying drugs with therapeutic potential for a wide range of degenerative and functional disorders of the nervous system. This may lead to the development of new and more effective treatments for brain diseases.

Among the epigenetic changes that have a role in the aforementioned diseases, DNA methylation has received special attention. In recent years Sequenom's MS technique has been used in several projects to analyze the methylation level of certain key CpG sites involved in the development of neurodegenerative pathologies. An example is AD.

AD is a neurodegenerative disorder and the major cause of cognitive decline among the elderly population with an estimated prevalence of 1:100 in the population over 65 years of age in Western countries. It is characterized by a progressive impairment of episodic memory that becomes more severe and, eventually, incapacitating. Other common symptoms include delusions, confusion, poor judgment and decision-making, agitation as well as language disturbances. Neuropathologically, AD is characterized by extracellular deposits of β-amyloid and the intracellular accumulation of hyperphosphorylated tau protein [48]. Significant atrophy of some brain regions is observed in the early stages of the disease, mainly in the hippocampal formation and entorhinal cortex. Several studies suggest that alterations in DNA methylation affect specific genes in AD, like the promoter regions of APP, β-amyloid precursor protein, PSEN1, which enhances β-amyloid formation, BACE1, which codes for β-secretase, or APOE, the gene coding for apolipoprotein E that facilitates amyloid plaque formation [49].

In a recent study the authors wanted to test whether DNA methylation patterns in postmortem brains and lymphocytes from late-onset AD (LOAD) patients are different from patterns found in healthy individuals and if age affects the distribution of these profiles. With that objective, they analyzed DNA methylation patterns across candidate genes for which a priori evidence for a role in the etiology of AD existed using Sequenom's MassCLEAVE technology [50]. They found that a subset of the analyzed gene promoters (i.e., NCSTN, TFAM, SIN3A, or HTATIP) displayed abnormal methylation patterns observed only in the LOAD brains. In addition they confirmed that the largest interindividual variance in DNA methylation was observed in the PSEN1 and APOE promoters, the two genes that are genetically associated with LOAD.

This finding may be important because hypomethylation could induce an overexpression of PSEN1, which could result in an imbalance in β-amyloid production. That variability may contribute to different susceptibility to disease later in life. Lastly, since the main characteristic of LOAD is the late age of onset, the authors looked for age effects in the DNA methylation profiles and found out that the strongest age effects were detected for the NCSTN and TFAM genes in neurons. This suggests that the accumulation of epimutations in those genes could lead to the development of the disease.

However, detecting differences in the methylation levels of some of the aforementioned genes implied in AD between brain cells of controls and those of AD patients is not an easy task as several papers have shown. An example of this is a study in which the authors determined the methylation of certain regions of the MAPT, which encodes for the tau protein, and the previously commented PSEN1 and APP genes among others, in the frontal cortex and hippocampus of patients and controls. They were unable to find differences in the percentage of CpG methylation of any of the analyzed regions. The authors point out the robustness and reproducibility of Sequenom's MassCLEAVE technique for these kind of studies and interpret their findings as the result of the masking effect caused by working with homogenates of different cell populations. Thus, the small changes in methylation status of vulnerable cells would be lost among total homogenates [51].

This masking effect caused by using cell homogenates or simply the fact that methylation is not the only epigenetic mechanism for gene regulation in AD, may be the reason for not being able to detect differences in CpG methylation between healthy and affected patients. In this line of thought, several works show that genes like SNAP25 (synaptosomal-associated protein 25 kDa), CNP (2,3-cyclic nucleotide 3-phosphodiesterase), and DPYSL2 (dihydropyrimidinase-like 2) do exhibit the same methylation values between both groups. SNAP25 codifies the synaptosomal-associated protein 25 kDa, CNP has been used as an index of myelin and DPYSL2 and is involved in neural differentiation, neurotransmitter release, and stabilization of microtubules. As their protein levels have been described as decreased in AD patients, the authors conclude that promoter methylation is not the principal epigenetic mechanism responsible for their mRNA regulation and that there might be other mechanisms involved in this regulation, such as noncoding RNA silencing or histone modifications [52,53]. The same has been reported for the heat shock proteins HSPA8 and HSPA9, which show altered expression in AD patients but no different levels of methylation [53].

The importance of PSEN1, BACE1, and APP in the development of AD has been further confirmed in another work in which the authors treated human neuroblastoma cells by using an activator of stress-related MAPKs, anisomycin, to activate stress-related MAPKs and measured the expression of those genes and their methylation levels by Sequenom's MS. The activated stress-related signaling pathways resulted in the enhanced transcription of APP, BACE1, and PSEN1 genes through DNMT-dependent hypomethylation, which consequently led to an increase of β-amyloid [54].

5. CONCLUSIONS

In this chapter, we provide an overview of the different technologies available for the analysis of DNA methylation with emphasis on Sequenom MassARRAY technology and its applications in clinical practice.

DNA methylation is one of several epigenetic mechanisms that cells use to control gene expression and to develop different diseases. The enzyme involved in this process is DNA methyltransferase, which catalyzes the transfer of a methyl group from S-adenosylmethionine to cytosine residues to form 5-methylcytosine, a modified base that is found mostly at CpG sites in the genome. The presence of methylated CpG islands in the promoter region of genes can suppress their expression.

In recent years the Sequenom EpiTYPER system (Sequenom Inc., CA, USA) technology, a novel technique for methylation analysis, has been developed. This technique has an important place among the classical techniques with great performance and several advantages. A lot of studies in different research fields have been published using the Sequenom technology.

Cancer was the first human disease to be associated to epigenetics. Sequenom MassARRAY technology has been used in several cancer studies to identify genes differentially methylated or to analyze the methylation of miRNAs. These studies represent a major advance in understanding of the tumorigenesis and in the development of new drugs for cancer treatment.

Epigenetics could play a fundamental role in the development of obesity. Moreover, epigenetic diet-induced changes during the perinatal period have metabolic effects that may lead to an increase in the susceptibility to gain weight and to develop diseases related to obesity.

Currently, Sequenom's MS technique has been used in several projects to analyze the methylation level of certain key CpG sites involved in the onset of Alzheimer's disease, neurodegenerative pathology with great impact in the society. All studies concluded that abnormal methylation patterns could lead to the development of this disease.

Although research in epigenetics is just beginning, the number of studies is increasing every year. The coming years will undoubtedly show flourishing research in related fields and important advances in this topic.

Glossary

DNA methylation A biochemical process where a methyl group is added to the cytosine or adenine DNA nucleotides.

Epigenetic Heritable changes in gene expression that do not involve changes to the underlying DNA sequence; a change in phenotype without a change in genotype.

EpiTYPER assay A tool for the discovery and quantitative analysis of DNA methylation that uses base-specific cleavage and matrix-assisted laser desorption/ionization time-of-flight mass spectrometry (MALDI-TOF MS).

MassARRAY® system An ideal method for discovery of methylation, for discrimination between methylated and nonmethylated samples, and for quantifying the methylation levels of DNA.

MicroRNA (miRNA) A small noncoding RNA molecule (containing about 22 nucleotides) found in plants, animals, and some viruses, which functions in RNA silencing and posttranscriptional regulation of gene expression.

LIST OF ABBREVIATIONS

A Adenosine
AD Alzheimer's disease
Bp Base pair
C Cytosine
CNP 2,3-Cyclic nucleotide 3 phosphodiesterase
CpG Cytosine–phosphate–guanine
Da Dalton
dNTP Dinucleotide triphosphate
DPYSL2 Dihydropyrimidinase-like 2
G Guanine
IGF2 Insulin-like growth factor 2
LOAD Late-onset Alzheimer's disease
MALDI-TOF MS Matrix-assisted laser desorption/ionization time-of-flight mass spectrometry
miRNA MicroRNAs
PCR Polymerase chain reaction
SAP Shrimp alkaline phosphatase
SNAP25 Synaptosomal-associated protein 25 kDa
SNP Single-nucleotide polymorphism
T Thymine

References

[1] Liu L, Li Y, Tollefsbol TO. Gene-environment interactions and epigenetic basis of human diseases. Curr Issues Mol Biol 2008;10(1–2):25–36. Epub 2008/06/06.

[2] Chong S, Whitelaw E. Epigenetic germline inheritance. Curr Opin Genet Dev 2004;14(6):692–6. Epub 2004/11/09.

[3] Egger G, Liang G, Aparicio A, Jones PA. Epigenetics in human disease and prospects for epigenetic therapy. Nature 2004;429(6990):457–63. Epub 2004/05/28.

[4] Mansego ML, Milagro FI, Campion J, Martinez JA. Techniques of DNA methylation analysis with nutritional applications. J Nutrigenet Nutrigenomics 2013;6(2):83–96. Epub 2013/05/22.
[5] Keller TE, Yi SV. DNA methylation and evolution of duplicate genes. Proc Natl Acad Sci USA 2014;111(16):5932–7. Epub 2014/04/09.
[6] Bestor TH. The DNA methyltransferases of mammals. Hum Mol Genet 2000;9(16):2395–402. Epub 2000/09/27.
[7] Bird A. DNA methylation patterns and epigenetic memory. Genes Dev 2002;16(1):6–21. Epub 2002/01/10.
[8] Suzuki MM, Bird A. DNA methylation landscapes: provocative insights from epigenomics. Nat Rev Genet 2008;9(6):465–76. Epub 2008/05/09.
[9] Laird PW. Principles and challenges of genome-wide DNA methylation analysis. Nat Rev Genet 2010;11(3):191–203. Epub 2010/02/04.
[10] Beck S, Rakyan VK. The methylome: approaches for global DNA methylation profiling. Trends Genet 2008;24(5):231–7. Epub 2008/03/08.
[11] Keshet I, Schlesinger Y, Farkash S, Rand E, Hecht M, Segal E, et al. Evidence for an instructive mechanism of de novo methylation in cancer cells. Nat Genet 2006;38(2):149–53. Epub 2006/01/31.
[12] Rakyan VK, Down TA, Thorne NP, Flicek P, Kulesha E, Graf S, et al. An integrated resource for genome-wide identification and analysis of human tissue-specific differentially methylated regions (tDMRs). Genome Res 2008;18(9):1518–29. Epub 2008/06/26.
[13] Weber M, Davies JJ, Wittig D, Oakeley EJ, Haase M, Lam WL, et al. Chromosome-wide and promoter-specific analyses identify sites of differential DNA methylation in normal and transformed human cells. Nat Genet 2005;37(8):853–62. Epub 2005/07/12.
[14] Weber M, Hellmann I, Stadler MB, Ramos L, Paabo S, Rebhan M, et al. Distribution, silencing potential and evolutionary impact of promoter DNA methylation in the human genome. Nat Genet 2007;39(4):457–66. Epub 2007/03/06.
[15] Illingworth R, Kerr A, Desousa D, Jorgensen H, Ellis P, Stalker J, et al. A novel CpG island set identifies tissue-specific methylation at developmental gene loci. PLoS Biol 2008;6(1):e22. Epub 2008/02/01.
[16] Clark SJ, Harrison J, Paul CL, Frommer M. High sensitivity mapping of methylated cytosines. Nucleic Acids Res 1994;22(15):2990–7. Epub 1994/08/11.
[17] Frommer M, McDonald LE, Millar DS, Collis CM, Watt F, Grigg GW, et al. A genomic sequencing protocol that yields a positive display of 5-methylcytosine residues in individual DNA strands. Proc Natl Acad Sci USA 1992;89(5):1827–31. Epub 1992/03/01.
[18] Grunau C, Clark SJ, Rosenthal A. Bisulfite genomic sequencing: systematic investigation of critical experimental parameters. Nucleic Acids Res 2001;29(13):E65–5. Epub 2001/07/04.
[19] Genereux DP, Johnson WC, Burden AF, Stoger R, Laird CD. Errors in the bisulfite conversion of DNA: modulating inappropriate- and failed-conversion frequencies. Nucleic Acids Res 2008;36(22):e150. Epub 2008/11/06.
[20] Laurent L, Wong E, Li G, Huynh T, Tsirigos A, Ong CT, et al. Dynamic changes in the human methylome during differentiation. Genome Res 2010;20(3):320–31. Epub 2010/02/06.
[21] Taylor KH, Kramer RS, Davis JW, Guo J, Duff DJ, Xu D, et al. Ultradeep bisulfite sequencing analysis of DNA methylation patterns in multiple gene promoters by 454 sequencing. Cancer Res 2007;67(18):8511–8. Epub 2007/09/19.
[22] Lister R, Ecker JR. Finding the fifth base: genome-wide sequencing of cytosine methylation. Genome Res 2009;19(6):959–66. Epub 2009/03/11.
[23] Meissner A, Mikkelsen TS, Gu H, Wernig M, Hanna J, Sivachenko A, et al. Genome-scale DNA methylation maps of pluripotent and differentiated cells. Nature 2008;454(7205):766–70. Epub 2008/07/05.
[24] Genome-wide association study of 14,000 cases of seven common diseases and 3,000 shared controls. Nature 2007;447(7145):661–78. Epub 2007/06/08.
[25] Ehrich M, Nelson MR, Stanssens P, Zabeau M, Liloglou T, Xinarianos G, et al. Quantitative high-throughput analysis of DNA methylation patterns by base-specific cleavage and mass spectrometry. Proc Natl Acad Sci USA 2005;102(44):15785–90. Epub 2005/10/26.
[26] Raizis AM, Schmitt F, Jost JP. A bisulfite method of 5-methylcytosine mapping that minimizes template degradation. Anal Biochem 1995;226(1):161–6. Epub 1995/03/20.
[27] Hartmer R, Storm N, Boecker S, Rodi CP, Hillenkamp F, Jurinke C, et al. RNase T1 mediated base-specific cleavage and MALDI-TOF MS for high-throughput comparative sequence analysis. Nucleic Acids Res 2003;31(9):e47. Epub 2003/04/25.
[28] Coolen MW, Statham AL, Gardiner-Garden M, Clark SJ. Genomic profiling of CpG methylation and allelic specificity using quantitative high-throughput mass spectrometry: critical evaluation and improvements. Nucleic Acids Res 2007;35(18):e119. Epub 2007/09/15.
[29] Ehrich M, Zoll S, Sur S, van den Boom D. A new method for accurate assessment of DNA quality after bisulfite treatment. Nucleic Acids Res 2007;35(5):e29. Epub 2007/01/30.
[30] Sharma S, Kelly TK, Jones PA. Epigenetics in cancer. Carcinogenesis 2010;31(1):27–36. Epub 2009/09/16.
[31] Momparler RL, Bovenzi V. DNA methylation and cancer. J Cell Physiol 2000;183(2):145–54. Epub 2000/03/29.
[32] Feinberg AP, Vogelstein B. Hypomethylation distinguishes genes of some human cancers from their normal counterparts. Nature 1983;301(5895):89–92. Epub 1983/01/06.

[33] Wang LS, Kuo CT, Cho SJ, Seguin C, Siddiqui J, Stoner K, et al. Black raspberry-derived anthocyanins demethylate tumor suppressor genes through the inhibition of DNMT1 and DNMT3B in colon cancer cells. Nutr Cancer 2013;65(1):118–25. Epub 2013/02/02.
[34] Ueno H, Okita H, Akimoto S, Kobayashi K, Nakabayashi K, Hata K, et al. DNA methylation profile distinguishes clear cell sarcoma of the kidney from other pediatric renal tumors. PloS One 2013;8(4):e62233. Epub 2013/05/03.
[35] Vrba L, Munoz-Rodriguez JL, Stampfer MR, Futscher BW. miRNA gene promoters are frequent targets of aberrant DNA methylation in human breast cancer. PloS One 2013;8(1):e54398. Epub 2013/01/24.
[36] Campion J, Milagro FI, Martinez JA. Individuality and epigenetics in obesity. Obes Rev 2009;10(4):383–92. Epub 2009/05/06.
[37] Godfrey KM, Sheppard A, Gluckman PD, Lillycrop KA, Burdge GC, McLean C, et al. Epigenetic gene promoter methylation at birth is associated with child's later adiposity. Diabetes 2011;60(5):1528–34. Epub 2011/04/08.
[38] Heijmans BT, Tobi EW, Stein AD, Putter H, Blauw GJ, Susser ES, et al. Persistent epigenetic differences associated with prenatal exposure to famine in humans. Proc Natl Acad Sci USA 2008;105(44):17046–9. Epub 2008/10/29.
[39] Huang RC, Galati JC, Burrows S, Beilin LJ, Li X, Pennell CE, et al. DNA methylation of the IGF2/H19 imprinting control region and adiposity distribution in young adults. Clin Epigenet 2012;4(1):21. Epub 2012/11/15.
[40] He J, Zhang A, Fang M, Fang R, Ge J, Jiang Y, et al. Methylation levels at IGF2 and GNAS DMRs in infants born to preeclamptic pregnancies. BMC Genomics 2013;14:472. Epub 2013/07/13.
[41] Milagro FI, Gomez-Abellan P, Campion J, Martinez JA, Ordovas JM, Garaulet M. CLOCK, PER2 and BMAL1 DNA methylation: association with obesity and metabolic syndrome characteristics and monounsaturated fat intake. Chronobiol Int 2012;29(9):1180–94. Epub 2012/09/26.
[42] Kalashikam RR, Inagadapa PJ, Thomas AE, Jeyapal S, Giridharan NV, Raghunath M. Leptin gene promoter DNA methylation in WNIN obese mutant rats. Lipids Health Dis 2014;13:25. Epub 2014/02/06.
[43] Siegmund KD, Connor CM, Campan M, Long TI, Weisenberger DJ, Biniszkiewicz D, et al. DNA methylation in the human cerebral cortex is dynamically regulated throughout the life span and involves differentiated neurons. PloS One 2007;2(9):e895. Epub 2007/09/20.
[44] Chahrour M, Jung SY, Shaw C, Zhou X, Wong ST, Qin J, et al. MeCP2, a key contributor to neurological disease, activates and represses transcription. Science 2008;320(5880):1224–9. Epub 2008/05/31.
[45] Migliore L, Coppede F. Genetics, environmental factors and the emerging role of epigenetics in neurodegenerative diseases. Mutat Res 2009;667(1–2):82–97. Epub 2008/11/26.
[46] Poulter MO, Du L, Weaver IC, Palkovits M, Faludi G, Merali Z, et al. GABAA receptor promoter hypermethylation in suicide brain: implications for the involvement of epigenetic processes. Biol Psychiatry 2008;64(8):645–52. Epub 2008/07/22.
[47] Costa E, Chen Y, Dong E, Grayson DR, Kundakovic M, Maloku E, et al. GABAergic promoter hypermethylation as a model to study the neurochemistry of schizophrenia vulnerability. Expert Rev Neurother 2009;9(1):87–98. Epub 2008/12/24.
[48] Bird TD. Genetic aspects of Alzheimer disease. Genet Med 2008;10(4):231–9. Epub 2008/04/17.
[49] Scarpa S, Fuso A, D'Anselmi F, Cavallaro RA. Presenilin 1 gene silencing by S-adenosylmethionine: a treatment for Alzheimer disease?. FEBS Lett 2003;541(1–3):145–8. Epub 2003/04/23.
[50] Wang SC, Oelze B, Schumacher A. Age-specific epigenetic drift in late-onset Alzheimer's disease. PloS One 2008;3(7):e2698. Epub 2008/07/17.
[51] Barrachina M, Ferrer I. DNA methylation of Alzheimer disease and tauopathy-related genes in postmortem brain. J Neuropathol Exp Neurol 2009;68(8):880–91. Epub 2009/07/17.
[52] Furuya TK, Silva PN, Payao SL, Bertolucci PH, Rasmussen LT, De Labio RW, et al. Analysis of SNAP25 mRNA expression and promoter DNA methylation in brain areas of Alzheimer's Disease patients. Neuroscience 2012;220:41–6. Epub 2012/06/27.
[53] Silva PN, Furuya TK, Sampaio Braga I, Rasmussen LT, de Labio RW, Bertolucci PH, et al. CNP and DPYSL2 mRNA expression and promoter methylation levels in brain of Alzheimer's disease patients. J Alzheimers Dis 2013;33(2):349–55. Epub 2012/09/08.
[54] Guo X, Wu X, Ren L, Liu G, Li L. Epigenetic mechanisms of amyloid-beta production in anisomycin-treated SH-SY5Y cells. Neuroscience 2011;194:272–81. Epub 2011/08/17.

CHAPTER

8

The Role of Methylation-Specific PCR and Associated Techniques in Clinical Diagnostics

Fang Zhao[1,2], *Bharati Bapat*[1,2,3]

[1]Lunenfeld Tanenbaum Research Institute, Mount Sinai Hospital, Toronto, ON, Canada;
[2]Department of Laboratory Medicine and Pathobiology, University of Toronto, Toronto, ON, Canada;
[3]Department of Pathology, University Health Network, Toronto, ON, Canada

OUTLINE

Epigenetic Biomarkers and Diagnostics
http://dx.doi.org/10.1016/B978-0-12-801899-6.00008-5

1. OVERVIEW OF DNA METHYLATION AS AN EPIGENETIC MARKER

Epigenetics was originally defined as "The causal interactions between genes and their products, which bring the phenotype into being" by Conrad Waddington in 1942 [1]. The definition of epigenetics has changed over time as it has been found to be implicated in a wide variety of biological processes. The current definition for epigenetics could be summarized as "the study of heritable changes in gene expression that occur independent of changes in the primary DNA sequence" [2]. Epigenetics includes a variety of different phenomena such as histone modifications, RNA interference through microRNAs (miRNAs) and long noncoding RNAs (lncRNAs), as well as DNA methylation [2].

Among these, histone modifications are quite diverse in range including acetylation, methylation, and ubiquitination. These modifications play key roles in a variety of functional pathways including transcriptional activation/repression, DNA repair, genomic structural change between euchromatin and heterochromatin, and apoptosis [3,4].

MicroRNAs primarily serve to regulate gene translation through the Drosha/Dicer/RISC pathway [5]. lncRNAs may inhibit gene transcription through blocking access points for transcription factors or facilitate transcription through recruiting transcription factors [6].

On the other hand, DNA methylation is heritable, causes changes to gene expression, and is involved in cell differentiation. DNA methylation is a post-replication modification. It is predominantly but not exclusively found in the fifth carbon of the pyrimidine ring of cytosine nucleotides occurring usually in the palindromic sequence of 5′-CpG-3′ (CpG). CpG-rich regions, known as CpG islands, are located in the promoter regions of around 60% of genes. CpG islands are usually unmethylated in normal cells except for certain genes involved in the inactivation of the paternal X-chromosome [7], testis-specific genes in somatic tissues [8], genomic imprinting [9], and cell-type specific expression [10].

DNA methylation is involved in the control of gene expression. Unmethylated promoter CpG islands are usually associated with euchromatin structure and activate gene transcription. Conversely, promoter region CpG island methylation is associated with heterochromatin structure and gene silencing. Contrary to promoter methylation, gene body methylation is commonly associated with active gene expression. Although, the functional roles of gene body methylation are still poorly understood, a recent study described a global remethylation event following treatment with the demethylating agent 5-aza-2′-deoxycytidine (5-aza-CdR) that resulted in increased CpG methylation of the gene body. It was found that CpG methylation in the gene body directly increases gene expression [11]. These findings demonstrate the potential impact of demethylating drugs in a clinical setting. This will be discussed in a later section.

Aberrant methylation of CpG islands has been found to be associated with many disorders including cancer. These well-characterized epigenetic features form the foundation of MSP (methylation-specific polymerase chain reaction (PCR)) and its associated techniques.

2. DNA METHYLATION IN DIAGNOSTICS

In many cancers, a phenomenon of global hypomethylation combined with hypermethylation of the promoter regions of tumor suppressor genes is often observed [12]. Usually several genes are silenced through promoter hypermethylation in proliferating tumors cells. Tumor-specific DNA methylation changes are used in cancer diagnostics for both disease detection and classification through various DNA methylation analysis techniques. DNA hypermethylation can be measured in tissue samples such as surgically resected fresh or archival tumors and biopsies, or from body fluids collected noninvasively (or less invasively) such as saliva, urine, and serum.

Methylation biomarkers have already been integrated into clinical applications for the diagnosis of certain cancers. One such example is the detection of Septin 9 (*SEPT9*) methylation as a diagnostic marker in colorectal cancer (CRC, MIM #114500). *SEPT9* methylation in serum samples is detected with high sensitivity (90%) and specificity (88%) for all stages of CRC [13]. Compared with the current standard screening modalities for CRC detection including fetal-occult blood test (FOBT) and internal imaging through colonoscopy, circulating DNA methylation-based test is an approach both more sensitive than FOBT and far less invasive than colonoscopy for CRC diagnosis [14]. In addition to *SEPT9*, hypermethylation of several gene clusters, termed as the CpG island methylator phenotype (CIMP), is able to differentiate subgroups of CRC characterized by pathological, clinical, and molecular features. CIMP positive CRC tumors show distinct association with survival and chemotherapeutic response [14]. Hypermethylation of *CDKN2A*, *MLH1*, and *TMEFF2* genes are representative examples of CRC CIMPs. However, a consensus panel of markers has not yet been applied in clinical setting [15,16].

In prostate cancer (PCa, MIM #176807), a major limitation for accurate diagnosis is that up to 25% of all prostate biopsies are false negatives, due to the fact that less than 1% of the prostate gland is biopsied and occult cancer may be missed simply due to random chance. It has been observed that many cancers, including prostate tumors, display a "field effect" (sometimes also called "field cancerization effect") such that tumor-adjacent tissue, although normal in appearance, shows a methylation profile similar to that of the tumor. Thus, despite histologically normal biopsy cores, if an occult tumor were present, the "adjacent normal" core will show aberrant DNA methylation [17]. Based on this field effect, a three-gene panel consisting of adenomatous polyposis coli (*APC*), glutathione S-transferase P (*GSTP1*), and Ras-association domain family 1 isoform A (*RASSF1A*) methylation has been approved as a diagnostic biomarker panel to discriminate false-negative biopsies in PCa. Such a methylation detection-based test may reduce the rate of false-negative biopsies and improve the overall efficiency of diagnosis.

DNA methylation of *GSTP1* and *APC* is associated with aggressive breast cancer (MIM #114480) development and can serve as a screening test to identify high-risk breast cancers. These markers can be detected in circulating tumor cells isolated from blood samples which may offer a noninvasive prognostic marker for breast cancer in the near future [18].

For non-cancer diseases, some genes are epigenetically silenced through DNA methylation in a "parent-of-origin" manner, known as genomic imprinting. Mammalian imprinted genes remain methylated/silenced throughout development [19]. For example, insulin-like growth factor 2 (*IGF-2*) is only expressed from the paternal allele and the *H19* gene is only expressed from the maternal allele. Improper imprinting could result in two active copies or two inactive copies of the gene, and consequently lead to severe developmental disabilities, cancer, or other disorders [19]. Prader–Willi (PWS; MIM #176270, ORPHA739) and Angelman syndromes (AS; MIM #105830, ORPHA72) are examples of such disorders caused by improper imprinting on chromosome 15.

In this chapter, DNA methylation detection using MSP and its associated techniques will be discussed in relation to diagnosis and prognosis of cancers and other diseases.

3. RESTRICTION ENZYME DIGESTION FOR ANALYZING DNA METHYLATION

Restriction enzymes that are sensitive to methylation status of cytosines at the enzyme recognition sequence and are unable to digest methylated DNA can be used to detect distinct DNA methylation patterns. Enzyme pairs such as HpaII–MspI can be used to discriminate methylation status of CpG-containing sequences. For example, both HpaII and MspI target the same (CCGG) sequence during digestion. However, HpaII is methylation sensitive and digestion is inhibited if the cytosine nucleotides are methylated. While MspI, an isoschizomer of HpaII, can cleave the CCGG site regardless of methylation status. This difference in methylation sensitivity for the pair of enzymes allows for methylated and unmethylated CpGs to be distinguished in downstream applications such as methylation-sensitive multiplex ligation-dependent probe analysis (MS-MLPA), as described later.

4. METHYLATION-SPECIFIC PCR AND ASSOCIATED TECHNIQUES

MSP relies on PCR to amplify specifically methylated or unmethylated alleles. Since conventional PCR is unable to differentiate between methylated and unmethylated cytosines, prior to MSP, template DNA must first undergo sodium bisulfite treatment in a process known as bisulfite conversion or bisulfite modification to differentiate methylated cytosines [20].

5. SODIUM BISULFITE CONVERSION

Bisulfite conversion deaminates unmethylated cytosines to uracil. Methylated cytosines are resistant to the deamination process and are thus protected during bisulfite conversion. This unique characteristic of bisulfite-converted DNA, which was first discovered in 1970 by Shapiro et al. [21], provides the basis to distinguish between methylated and unmethylated CpG sites. Bisulfite conversion allows for downstream applications of a variety of techniques to determine the methylation status at selected/specific CpG sites and/or general methylation pattern of a particular region of the genome.

However, bisulfite conversion is a harsh process and known to cause fragmentation of DNA which may be a limitation. Furthermore, since bisulfite conversion is assumed to convert all unmethylated cytosines to uracil, it is necessary to have a methylation-independent control post-bisulfite conversion, to validate the success of the process [22]. One such example is *ALU-C4* (*ALU*) repeats, which do not harbor any CpGs, and therefore do not incur any methylation, and are ideal to be used as internal control [22]. Following bisulfite conversion, DNA samples are analyzed using locus-specific or genome-wide techniques.

6. METHYLATION-SPECIFIC PCR

In MSP, bisulfite-converted DNA is amplified using PCR with primers specifically designed to target either methylated or unmethylated CpGs. Following MSP, amplicons/bands are visualized on agarose, boric acid, or nondenaturing polyacrylamide gels that are specific to either methylated or unmethylated alleles. These bands are then assessed through densitometry analysis. Due to the nature of this technique, standard MSP is unable to quantify the number of methylated alleles if both methylated and unmethylated alleles are present in the DNA sample. Depending on the primer location, both methylated and unmethylated DNA-specific primers would produce bands of different sizes and can be easily distinguished by gel electrophoresis. DNA extracted from low-quality and/or limited quantity tissues such as from formalin-fixed paraffin-embedded (FFPE) samples following bisulfite conversion tends to be damaged and/or fragmented, which may limit MSP amplification restricted to short amplicons (Table 1).

TABLE 1 The Advantages, Limitations, and Clinical Applications of Methylation-Specific PCR and Associated Assays

Assay	Advantages	Limitations	Clinical applications	References
Methylation-specific PCR (standard MSP)	Methylation specific	Not quantitative	Analysis of X chromosomal inactivation and disorders linked to X-chromosomal inactivation such as fragile X syndrome, PWS, and AS	[20,24,26–30]
	Detection down to 0.1% of methylated alleles	Cannot determine methylation pattern if methylated and unmethylated alleles are present		
		Bisulfite conversion dependent		
		Cannot detect damaged/fragmented DNA		
Nested MSP	Rescues damaged/fragmented DNA for MSP analysis	Same as MSP other than damaged/fragmented DNA		[20]
Methylation-specific dot blot assay	Methylation specific	Takes longer than more advanced techniques	Detection of methylation of genes in prostate cancer (PCa)	[31,32]
	More sensitive than standard MSP			
	Semi-quantitative			
	Does not require high-tech equipment			
Methylation-specific qPCR (MethyLight (singleplex))	Methylation specific	Not apt for screening of large number of CpG sites	Assessment of methylation of specific CpG sites. May assess up to 8–12 CpG sites in the same assay. Promoter methylation of genes associated with PCa, bladder cancer, breast cancer	[33,37–43]
	10 times more sensitive than standard MSP	Primer/probe must contain 8–12 CpG sites		
	Quantitative	Bisulfite conversion dependent		
Multiplex MethyLight	Methylation specific	Gene combinations and target CpG sites are limited by primer probe combination	Detection of methylation of genes in CRC, PCa, and cervical cancer	[36,44,45]
	Can detect methylation in as little as 370 pg of input DNA	Bisulfite conversion dependent		
	Quantitative			
	High throughput compared with singleplex MethyLight			

Continued

TABLE 1 The Advantages, Limitations, and Clinical Applications of Methylation-Specific PCR and Associated Assays—cont'd

Assay	Advantages	Limitations	Clinical applications	References
Methylation-sensitive multiplex ligation-dependent probe amplification (MS-MLPA)	Detect changes in copy number of a gene	Can only detect copy number variations in regions targeted by probes	Diagnosis of PWS and AS, Lynch syndrome; tumor diagnosis and prognosis when used in combination of other methods	[46–50]
	DNA methylation and point mutations	Mutations or polymorphism in the target sequence may be detected as copy number variation leading to false negatives		
	Up to 50 different loci could be analyzed in the same MLPA reaction	Relies on enzyme digestion: restricted to CGCG sites		
	Bisulfite conversion independent	Cannot be used for unknown SNPs		
	Can be used on fragmented DNA 150bp long			
	Detection in as small as 20ng of input DNA			
Pyrosequencing	Fast compared to other techniques	Bisulfite conversion dependent for methylation detection	Diagnosis of PCa, CRC, lung cancer prognosis; cervical cancer prognosis; bladder cancer recurrence screening; soft tissue sarcoma prognosis; bladder cancer	[51–58]
	Can detect SNPs, insertions/deletions, copy number variation, DNA methylation	Only short stretches of DNA can be analyzed		
	Quantitative	Not suitable for discovery of new markers		
	High throughput			
Methylation-specific high-resolution melting analysis	Quantitative	Limited by available dyes	Diagnosis of PWS and AS, nasopharyngeal carcinoma	[59–63]
	Can quantitate methylation status of all CpGs in a sample	Not locus specific		
Digital MethyLight	Quantitative	Requires time-consuming optimization	CRC detection	[64–66]
	Up to three times more sensitive than MethyLight	Relatively more expensive		
	Able to measure methylation from a single molecule of DNA template			
	Low noise			

AS, Angelman syndrome; CRC, colorectal cancer; PCa, prostate cancer; PCR, polymerase chain reaction; PWS, Prader-Willi syndrome; qPCR, quantitative PCR.

Nested MSP (MN-MSP) [20] was developed to overcome limitations of standard MSP described above. MN-MSP involves an initial round of amplification after bisulfite conversion, using primers that are independent of the methylation status for the gene of interest (GOI) followed by amplification using methylation-specific primers. The advantage of this two-step process is that the first round of amplification may rescue fragmented input DNA. Although with modern and improved bisulfite conversion kits, this process may not be necessary. Current bisulfite conversion kits are able to convert >99% of all unmethylated cytosines and result in a DNA recovery of >80% [23].

Applications of standard MSP technique include assessment of X-chromosomal inactivation, diagnosis of fragile X syndrome (FXS), and diagnosis of PWS and AS.

X-chromosomal inactivation acts through DNA methylation and heterochromatin-dependent gene silencing of one of the two X-chromosomes randomly in females. Analysis of X-chromosomal inactivation patterns is widely used as a diagnostic tool for assessment of X-linked disorders in clinical settings. Traditionally, X-chromosomal inactivation was evaluated by methylation-sensitive restriction enzyme-based digestion analysis to assess differential methylation of genes between the active and the inactive X-chromosomes. MSP can also be used to determine the methylation levels of X-inactivation which could be used for DNA samples that are unsuitable for restriction enzyme digestion analysis [24].

FXS (MIM #300624, ORPHA908), also known as Martin–Bell syndrome, is a genetic disorder that is the most common cause of autism by a single gene [25]. It is associated with a lack of expression of the fragile X mental retardation protein (FMRP). Initially, expansion of trinucleotide CGG repeats in the *FMR-1* gene, which codes for FMRP, was thought to cause FXS. Subsequently, abnormal DNA methylation of a CpG island upstream of the CGG repeats was also found to be involved in the disease pathogenesis. Patients with these two alterations show silencing of *FMR-1* and FXS [26]. Historically, screening of FXS was performed by Southern blot analysis of DNA digested with methylation-sensitive restriction enzymes. However, Southern blotting is a slow, labor-intensive procedure to detect mutations of *FMR-1*. MSP is now routinely used as a diagnostic tool for FXS for the detection of both CGG repeat expansion and *FMR-1* methylation.

The MSP test for FXS takes advantage of the length of CGG repeats and methylation status of promoter and also CGG repeats, between normal (unmethylated promoter and unmethylated CGG repeats, CGG repeats less than 52), premutation (unmethylated promoter and unmethylated CGG repeats, CGG repeats 52–200), and full mutation of *FMR-1* alleles (de novo methylated promoter and de novo methylated CGG repeats, CGG repeats greater than 200) [27]. Separate sets of primers are designed to be specific for either methylated or unmethylated promoter sequences as well as methylated and unmethylated CGG repeat sequences. The PCR products are visualized on agarose gel postamplification. Comparing the methylation status of the promoter and the size of the CGG repeat amplicon, this semi-quantitative assay can determine the 11 distinct patterns encountered in nonaffected, carrier, and affected individuals [26]. The most current diagnosis strategy for fragile X includes methylation-specific fluorescent PCR which incorporates fluorescent primers complementary to bisulfite-converted methylated or unmethylated DNA. The PCR products are analyzed by capillary electrophoresis. The mutation status of *FMR1* is determined by fluorescent peak sizes and patterns on the GeneScan electropherogram and can accurately predict the CGG repeat lengths of all normal and premutation samples [28].

MSP-based diagnosis test of FXS is quicker and more cost-effective compared to Southern blot-based tests and may be the only method used in the near future [29].

PWS and AS are neurological disorders with distinct symptoms but share a common cause rooted in genomic imprinting defects of chromosome 15q11-q13 through either the loss of maternally inherited genes with paternal imprinting in AS or the loss of paternally inherited genes with maternal imprinting in PWS. Traditionally both PWS and AS are diagnosed using fluorescence in situ hybridization (FISH) DNA polymorphism analysis for the detection of chromosome 15q11-q13 deletion and DNA analysis with microsatellite markers for the detection of uniparental disomy (UPD). However, there are limitations to both methods as FISH cannot discover UPD, and microsatellite analysis requires parental blood samples. Furthermore, neither method will be able to detect rare imprinting mutations. MSP analysis of DNA methylation of the small nuclear ribonucleoprotein-associated polypeptide N gene (*SNRPN*) or at locus PW71 can detect PWS that is concordant with FISH and DNA microsatellite analysis, indicating specificity for chromosomal deletion in 15q11-q13 and UPD as well as PWS caused by imprinting mutations. Therefore MSP analysis of SNRPN is an efficient screening test for PWS [30].

Due to the limitations of standard MSP, diagnosis and prognosis of cancers are usually assessed through more advanced techniques such as MethyLight, pyrosequencing, and methylation-specific high-resolution melting analysis.

7. METHYLATION-SENSITIVE DOT BLOT ASSAY

Methylation-sensitive dot blot assays (MS-DBAs) were developed in 2005 in order to overcome limitations of standard MSP. MS-DBA is a semiquantitative technique and decreases the false-positive rate of MSP.

In MS-DBA, PCR primers are designed to flank CpG sites of interest but do not contain CpGs themselves and consequently are methylation independent. After PCR amplification, the PCR product is dotted and cross-linked to a nylon membrane. Internal probes labeled with digoxigenin-11-dUTP (DIG) are designed to target either methylated alleles (complementary to CpGs) or unmethylated alleles (complementary to UpGs). Probes are allowed to bind to the target templates and any excess/unbound amounts are washed away. Anti-DIG antibody is applied followed by chemiluminescent substrate application. Data can be analyzed visually or by densitometry, yielding a semiquantitative result [31]. MS-DBA can be completed without the need for high-tech equipment and is practical for use in developing countries with limited resources.

A recent publication has shown that MS-DBA performed following MSP assay can accurately detect methylation of *GSTP1* and *RASSF1A* in paraffin-embedded radical prostatectomy samples that is more sensitive than standard MSP [32].

Despite the advantages of MS-DBA over standard MSP, it requires optimization and can be time-consuming compared to more advanced techniques.

8. METHYLIGHT

One such technique is MethyLight or quantitative real-time MSP. MethyLight was first developed in 1999 [33]. MethyLight is based on standard MSP but is able to provide quantitative analysis using fluorescence-based quantitative PCR (qPCR) technology due to methylation-specific primers and TaqMan™ fluorescent probes used in the amplification of bisulfite-converted DNA. Since the distinguishing method is through qPCR amplification, high sensitivity, at up to a single-nucleotide resolution, may be achieved while maintaining high specificity. MethyLight is at least 10-fold more sensitive than standard MSP [33] and is capable of detecting very low frequencies of hypermethylated alleles. In MethyLight, *SssI* (a CpG-specific methylase) treated DNA also known as universal methylated DNA, in which every CpG in the entire genome is methylated, is used as a positive control to produce the percentage of methylated reference (PMR). PMRs are used to assess methylation levels of a specific gene when comparing against universal methylated DNA. Singleplex MethyLight is capable of detecting 1 methylated allele among 10,000 unmethylated alleles or a detection frequency of 0.01%. Due to its high sensitivity and specificity, MethyLight is ideal for the detection of DNA methylation biomarkers for disease diagnosis and prognosis.

Similar to standard MSP, primers are designed to specifically anneal with either methylated or unmethylated CpGs. With addition of the fluorescent probe, another layer of specificity is added which allows improved detection of methylated alleles. The ratio of methylated alleles among total number of alleles is then assessed.

9. TECHNICAL SPECIFICATIONS FOR METHYLIGHT

The target sequences for primers and probes for MethyLight assay should be in a CpG island. The primers should each contain two CpG dinucleotides minimum. The 3′ end should end in a G or CG to promote clamping. The melting temperature (Tm) of the primers should be between 60° and 65° and ±2° of each other. The primers should be ≥20 base pairs (bp) to ensure specificity. The probe should contain at least 3 CpG dinucleotides that are biased toward the center with no CpGs within 3 bp of either end of the probe to allow for cleavage of the quencher. The Tm of the probe should be 10° higher than the primers. The probe should be ≤30 bp. The probe and primers should be able to cover 8–12 CpGs. The amplicon length should be ≤130 bp to prevent secondary structures. There are many online tools that are able to aid in the design of primers and probes for MethyLight. We have successfully used Primer3 [34], OligoCalc [35], and online DNA tools from companies such as IDT DNA and Life Technologies in our MethyLight studies.

In order to quantify methylated alleles, a reference gene is used to calculate the PMR. As mentioned previously, *ALU* is an ideal reference gene due to its genome-wide distribution and methylation-independent characteristics. Universal methylated DNA is used both as positive control and for generating the standard curve for each assay. For singleplex MethyLight, the standard curve is prepared using 5 ng of universal methylated DNA which is serially diluted 1:25 for *ALU* (4 points) and 1:5 for GOI (5 points). For multiplex MethyLight, described in the next section, the dilution ratio for the standard curve depends on the detection level of your GOI and can range from 1:3 to 1:5 serial dilutions starting with 5 ng of universal methylated DNA. It is important to determine the lowest detection level of all GOI included in the multiplex panel. Positive control consists of 0.5 ng of universal methylated DNA [36]. PMR is calculated according to the formula of Eads et al. [33] as follows: [(GOI/ALU)sample/(GOI/ALU)universal methylated DNA] × 100.

Standard curves and samples should be run in duplicate, positive control in triplicate. Samples with high deviation between duplicates should be rerun in independent assays.

10. APPLICATIONS OF METHYLIGHT TECHNIQUE

MethyLight has been used to assess the diagnostic, prognostic, and predictive potential of biomarkers for various cancers. For example, analysis of methylation status of repetitive elements such as long interspersed element (*LINE-1*) and *Sat2* by MethyLight showed that *LINE-1* hypermethylation in circulating white blood cells was significantly associated with early diagnosis of pancreatic cancer (MIM #260350) [37].

Our group has extensive experience with MethyLight technology. Using DNA extracted from archival FFPE radical prostatectomy tissues samples of two large PCa patient cohorts consisting of several hundred patients, we have discovered and validated promising tumor-specific methylation biomarkers (*APC*, *TGFβ2*, *HOXD3*, *KLK10*, and *RASSF1A*) associated with PCa progression and/or biochemical recurrence [38–41].

In addition to tissue samples, MethyLight has also been used to analyze DNA methylation of noninvasive body fluids such as blood and urine. Representative examples include circulating cell-free DNA assessed by MethyLight, which showed that *RASSF1A* and/or *APC* methylation in pretherapeutic serum samples of breast cancer patients was linked with poor prognosis for breast cancer [42]. A panel of six DNA methylation biomarkers (*EOMES*, *HOXA9*, *POU4F2*, *TWIST1*, *VIM*, and *ZHF154*) obtained from urinary sediment DNA were found to be significantly correlated with recurrence of bladder cancer (Table 2) [43].

11. MULTIPLEX METHYLIGHT

Multiplex MethyLight can simultaneously analyze methylation of multiple genes in the same sample. The number of genes constituting the panel is only limited by the number of emission detectors of the qPCR machine.

Multiplex MethyLight was the technique first published by He Q. et al. in 2010 in the investigation of a panel of biomarkers (*ALX4*, *SEPT9*, and *TMEFF2*) for screening of CRC from cell-free DNA in peripheral blood samples or fresh tissue samples [44]. The specificity for detection of the selected panel of biomarkers was 87% for primary tissues and 90% for peripheral blood samples. However, the reference gene used in this study was *ACTB* which is a less stable and/or less reproducible measure of bisulfite-converted template DNA, especially in the context of analyzing tumor DNA where amplification or deletions of single genes may occur. To address this issue, standard MethyLight assays now include short interspersed *ALU* element as an internal control as previously mentioned, since it is methylation independent and also less sensitive to copy number variations due to its genome-wide distribution [22].

We recently developed a multiplex MethyLight assay using *ALU* as the reference gene with a sensitivity limit for methylation detection of *APC*, *HOXD3*, and *TGFβ2* at an estimated 370 pg of universal methylated DNA and an estimated 1.6 pg for *ALU*. In addition, multiplex MethyLight assay was able to discriminate between fully methylated alleles and unmethylated alleles with 100% specificity [36]. Multiplex MethyLight is highly accurate and reproducible when compared with singleplex MethyLight. Furthermore, in our study, multiplex MethyLight has successfully been applied to measure DNA methylation in FFPE tissue, fresh frozen tissue, tissue biopsy, and urine samples.

Multiplex MethyLight further reduces the requirement for quantity of initial input DNA when analyzing multiple genes, making it ideal for samples with low DNA yields such as saliva, urine, hair, or valuable clinical samples with limited quantity such as punch core biopsies. Since multiplex MethyLight is efficient and cost-effective, it is very likely that in the near future this particular technique will become more prevalent for locus-specific DNA methylation analysis.

Multiplex MethyLight has been used to detect methylation in PCa, CRC, and cervical cancer (MIM #603956) [36,44,45].

TABLE 2 Methylation Biomarkers of Various Cancers

Disease	Genes analyzed	Biomaterials analyzed	Specificity and sensitivity	Techniques used	References
Colorectal cancer (CRC)	*SEPT9*	Plasma	90% sensitivity and 88% specificity	MethyLight	[14]
	CDKN2A/p14	Tissue	Adenoma from FAP-Pts (41%, 13/32), multiple adenoma Pts (69%, 20/29), MSI-H CRC Pts (86%, 12/14), MSS/MSI-L CRC Pts (88%, 14/16)	MSP	[15]
		Tissue	MSS CRC (12%, 3/24), MSI CRC (39%, 17/28)	MSP	[15]
		Tissue	CRC (32%, 61/188)	MSP, MethyLight	[15]
	CDKN2A/P16	Tissue	Adenomas (34%, 14/41)	MSP	[15]
		Stool	Pts with adenomas (31%, 9/29), HD (16%, 3/19)	MSP	[15]
		Serum	CRC pts (71%, 12/17), HD (0%, 0/10)	MSP	[15]
	MLH1	Tissue	MSS (0%, 0/25) CRC, MSI CRC (39%, 11/28)	MSP	[15]
		Peripheral blood	CRC Pts (13.4%, 35/262)	MSP	[15]
		Serum	CRC Pts (39%, 19/49), HD (2%, 1/41)	MethyLight	[15]
	TMEFF2 (TPEF/HPP1)	Plasma	CRC Pts (65%, 87/133), HD (31%, 56/179)	MethyLight	[15]
	ALX4, SEPT9, and TMEFF2	Tissue	84% sensitivity and 87% specificity	Multiplex MethyLight	[44]
		Peripheral blood	81% sensitivity and 90% specificity	Multiplex MethyLight	[44]
	IGFBP3, miR137	Tissue	95.5% sensitivity and 90.5% specificity	Pyrosequencing	[54]
	THBD-M	Plasma	71% sensitivity and 80% specificity	Digital MethyLight	[66]
Breast cancer	*APC*	Serum	29% sensitivity	MethyLight	[18]
	GSTP1	Serum	18% sensitivity	MethyLight	[18]

Continued

TABLE 2 Methylation Biomarkers of Various Cancers—cont'd

Disease	Genes analyzed	Biomaterials analyzed	Specificity and sensitivity	Techniques used	References
Prostate cancer	*HOXD3*	Tissue	61.8% sensitivity and 86.9% specificity	MethyLight	[38,40]
	APC	Tissue	82.4% sensitivity and 95.2% specificity	MethyLight	[40]
	TGFb2	Tissue	42.4% sensitivity and 95.5% specificity	MethyLight	[40]
	RASSF1A	Tissue	92% sensitivity and 85.6% specificity	MethyLight	[40]
	RARβ2	Ejaculate	100% sensitivity and 15% specificity	Pyrosequencing	[53]
Bladder cancer	*EOMES*	Urine	94% sensitivity and 39% specificity	MethyLight	[43]
	HOXA9	Urine	92% sensitivity and 38% specificity	MethyLight	[43]
	POU4F2	Urine	87% sensitivity and 47% specificity	MethyLight	[43]
	TWIST1	Urine	89% sensitivity and 28% specificity	MethyLight	[43]
	VIM	Urine	90% sensitivity and 43% specificity	MethyLight	[43]
	ZHF154	Urine	93% sensitivity and 47% specificity	MethyLight	[43]
	SOX1, IRAK3, L1-MET	Urine	93% sensitivity and 94% specificity	Pyrosequencing	[58]
Nasopharyngeal carcinoma	*RASSF1A, WIF1, DAPK1, RARβ2*	Tissue	95.8% sensitivity and 67.4% specificity	MS-HRM	[63]
Glioblastoma	*MGMT*	Tissue	80% sensitivity and 67% specificity	MS-MLPA	[49]
Lynch syndrome	*MLH1*	Tissue	96% sensitivity and 66% specificity	MS-MLPA	[50]
Urothelial cancer	*SOX1, TJP2, MYOD, HOXA9_1, HOXA9_2, VAMP8, CASP8, SPP1, IFNG, CAPG, HLADPA1, RIPK3*	Tissue	94.3% sensitivity and 97.8% specificity	Pyrosequencing	[55]

MS-HRM, methylation-specific high-resolution melting; MS-MLPA, methylation-sensitive multiplex ligation-dependent probe amplification; MSP, methylation-specific polymerase chain reaction.

12. METHYLATION-SENSITIVE MULTIPLEX LIGATION-DEPENDENT PROBE AMPLIFICATION

MS-MLPA is a multiplex PCR-based technique that can detect changes in gene copy number status, DNA methylation, and point mutations, simultaneously.

Standard MLPA technique requires two hemiprobes which target adjacent sequences in the same stretch of the template DNA. Both probes contain a PCR primer-binding region. One of the probes also contains a nonhybridizing "stuffer sequence" at the 3′ end to give the PCR product of desired length. The MLPA probe mix is allowed to hybridize with the denatured template DNA overnight. The annealed hemiprobes are then enzymatically ligated. The ligation product is amplified by PCR using the primer pair specific to the primer-binding regions of the two probes. Due to differences in length of the "stuffer sequences," the amplification products can be separated, identified, and quantified by capillary electrophoresis. Up to 50 different probes can be amplified in a single MLPA reaction for 50 different loci [46].

MS-MLPA uses the endonuclease HhaI to digest unmethylated CpGs after probe hybridization. The MLPA probes target sequences that are 50–100 nucleotides in length which means it can detect methylation in highly fragmented DNA samples such as bisulfite-converted DNA or DNA extracted from FFPE tissue samples. MLPA is a highly sensitive technique with input DNA being as low as 20 ng and is capable of detecting small copy number variations such as heterozygous deletion (2 copies to 1 copy per cell) or duplication (2–3 copies per cell) in a sample. MS-MLPA does not require bisulfite conversion of template DNA and thus does not have the limitations associated with the bisulfite conversion process such as DNA fragmentation and/or incomplete conversion. However, MLPA is limited in detecting only copy number changes of the regions targeted by MLPA probes. In addition, a mutation or polymorphism in the target sequence may be detected as a change in copy number and potentially lead to a false-negative result. Furthermore, MS-MLPA relies on HhaI digestion which means only CpGs in the target sequence of HhaI (CGCG) will be identified. Thus, MLPA is ideal for the detection of deletions/insertions and known point mutations but is not suitable for detection of unknown single nucleotide polymorphisms (SNPs) [47].

MS-MLPA can be used to diagnose PWS and AS for the three major classes of genetic defects (See previous sections: 6. Methylation-Specific PCR; Table 1) [48].

In addition, MS-MLPA is used for tumor diagnosis and prognosis either alone or in combination with other techniques such as MSP. For example, MS-MLPA targeting *MGMT* methylation is able to detect glioblastoma (MIM #137800) progression after chemotherapy with 80% diagnostic accuracy [49], while assessment of *MLH1* hypermethylation by MS-MLPA showed 96% sensitivity and 66% specificity in ruling out Lynch syndrome (MIM #120435) in individuals with a family history of CRC [50].

13. PYROSEQUENCING

Pyrosequencing is a reliable, fast, and high-throughput technique that can analyze up to 96 bisulfite-converted DNA samples in approximately 4 h. It is based on sequential addition and incorporation of nucleotides that can be quantitated through conversion of naturally released pyrophosphate into a light signal in real time [51]. The released pyrophosphate is used in a sulfurylase reaction releasing ATP that is subsequently used by luciferase to produce light. Pyrosequencing has a wide range of applications including detection of SNPs, insertion/deletions, gene copy number, and DNA methylation.

Bisulfite-converted pyrosequencing, first developed in 2003 by Colella et al. [52], has been successfully applied in the detection of DNA methylation biomarkers for diagnosis and prognosis of cancer. For example, a recent report described a noninvasive test for PCa diagnosis using bisulfite-converted pyrosequencing to detect DNA methylation in body fluids such as ejaculate to predict PCa with *RARβ2* promoter methylation [53]. Quantitative bisulfite pyrosequencing was used to identify methylation of *IGFBP3* and *MIR137* to stratify CRC patients with 95.5% sensitivity and 90.5% specificity [54].

A limitation of pyrosequencing is that only short stretches of DNA can be analyzed. It is primarily used as a validation step for DNA methylation biomarkers and is not recommended for discovery of new biomarkers. For example, quantitative pyrosequencing was used to validate 12 differentially methylated loci from urothelial bladder cancer patients (MIM #109800). The 12 loci combined showed 94.3% sensitivity and 97.8% specificity as diagnostic markers [55]. Also, in a recent study by Sandoval et al. pyrosequencing was used to validate a panel of DNA methylation biomarkers selected from 10,000 CpGs obtained through DNA methylation microarray in the prognosis of non-small cell lung cancer (MIM #211980). It was found that *p16INK4a*, *CDH13*, *RASSF1A*, and *APC* hypermethylation is associated with early recurrence of stage I lung cancer [56].

Other examples of applications of pyrosequencing include the identification of a panel of 32 genes to be associated with prognosis of cervical squamous cell carcinoma. In particular, *VIM* gene methylation in cervical squamous cell carcinoma was shown to predict a favorable prognosis [57]. Also, a three-marker panel (*SOX1*, *IRAK3*, and *L1-MET*) was validated using pyrosequencing that can predict between bladder cancer recurrence with sensitivity of 93% and specificity of 94% [58]. This technique is described in detail in the following section.

14. HIGH-RESOLUTION MELTING ANALYSIS

High-resolution melting analysis (HRM) is a qPCR-based technique based on the Tm difference between purine and pyrimidine bases that can quantify the methylation status of all CpG sites in a sample. Following bisulfite conversion, unmethylated cytosines are converted to uracils. Due to differences in Tm between cytosines and thymines, a melting profile can be compiled with slowly increasing temperatures and by the incorporation of an intercalating fluorescent dye. The melting profiles of unmethylated and methylated variants of the same area of amplification will be significantly different due to the differences in GC content. This difference in the melting profile is used to assess the methylation status of the template DNA by comparing against the native DNA that has not been bisulfite converted. HRM will assess the entire amplified region rather than only a few selected CpG sites.

The melting curve analysis of DNA methylation was first developed in 2001 [59]. At that time, the sensitivity of the technique was limited by the availability of intercalating dyes. HRM for DNA methylation was developed in 2003 using the LCGreen dye [60]. LCGreen PCR HRM can discriminate between heterozygous and homozygous sequence variants. Methylation-specific HRM (MS-HRM) was developed in 2008 using methylated- and unmethylated-specific primers to amplify bisulfite-converted DNA in a PCR [61]. Although HRM is not locus specific, it has practical applications for diagnosis of diseases such as PWS and AS [62], and nasopharyngeal carcinoma (MIM #607107) [63] based on screening of methylated markers.

15. EMERGING APPLICATIONS

15.1 Digital MethyLight

Bisulfite-converted DNA samples analyzed using standard MSP techniques are limited by their throughput capacity, sensitivity of methylation detection, and resolving power. To address these issues, digital PCR technology is applied in MethyLight assays. Digital PCR employs distribution/dilution of DNA samples into multiple reaction chambers which can amplify signals from a single molecule of template DNA and, at the same time, decreases nonspecific noise [64]. Due to distribution of a sample over multiple PCR reaction wells to concentrations below that of a single-template molecule per well, specific and sensitive amplification is achievable for single-template molecules. Digital bisulfite genomic sequencing is able to omit the cloning step in bisulfite sequencing. Using SYBR Green, positive wells are visualized and positive PCR products can be sequenced directly after amplification. In addition, if the sample contains heterogeneous populations of DNA methylation patterns, templates may be sequenced separately to obtain individual DNA methylation patterns of the entire population.

Digital MethyLight is considerably more sensitive than traditional MethyLight assays with up to three times the sensitivity [65]. Although digital MethyLight was established in 2008, there have been very few publications using this method. Some potential disadvantages of digital MethyLight may be that optimization is required to find the most effective dilution. Also it may be too expensive to sequence all positive wells for an individual sample.

One study published in 2012 used digital MethyLight in a genome-wide discovery and validation of blood-based biomarkers for CRC detection. Stringent multi-step filtering criteria were achieved with digital MethyLight performed following MethyLight and array profiling of CRC tumors. Two candidate methylation biomarkers (*THBD* and *C9orf50*) were identified and validated in as little as 50 μL of sera/plasma compared with conventional *SEPT9* assays that use 4–5 mL of plasma [66].

15.2 A Novel Methylation Marker: 5-Hydroxymethylation

DNA hydroxymethylation is a recently discovered modification on CpG dinucleotides and involves the addition of a hydroxyl group on 5-methylcytosines (5mCs) to produce 5-hydroxymethylcytosine (5hmC). Although its biological role has not been fully elucidated, it has been speculated that 5hmC is an intermediate during the process of demethylation mediated through the TET family of proteins [67].

Conventional techniques using bisulfite-converted DNA cannot distinguish between CpG methylation and hydroxymethylation since both are protected from bisulfite conversion. Emerging MS-PCR applications are focusing on selective enrichment of 5hmC for epigenetic analysis. Therefore, future developments will provide new epigenetic biomarkers based on hydroxymethylation of specific genes.

16. FUTURE OF DNA METHYLATION IN CLINICAL APPLICATIONS

The future of DNA methylation-dependent diagnosis will involve panels of multiple biomarkers for screening and prognosis of specific diseases including cancers. Many recent publications describe development of biomarker screening panels for DNA methylation as well as other epigenetic markers, including miRNA, lncRNA, and histone modifications [68–70]. The panels of epigenetic biomarkers may offer more sensitive detection and accurate prognosis of diseases, as well as the discovery of potential therapeutic targets.

Since DNA methylation is a reversible phenomenon, it provides a potentially valuable target for the treatment of methylation-dependent diseases. Currently there are few studies investigating epigenetic modifications as potential therapeutic targets. Re-expression of methylation-silenced tumor suppressor genes by inhibiting DNA methyltransferases (DNMT1, DNMT3A, and DNMT3B) has emerged as an effective strategy against cancer [71]. In fact 5-aza-CdR, a drug which produces irreversible inactivation of DNA methyltransferases, has shown interesting antineoplastic activity in patients with leukemia, myelodysplastic syndrome, and non-small cell lung cancer [72]. Importantly, MSP can be used as a monitoring strategy for alterations in methylation levels following demethylation treatments.

In the past decade, it has become very apparent that epigenetic alterations are heavily involved in disease onset and progression. Specifically, aberrant DNA methylation of genes serves as biomarkers for screening of epigenetic diseases and many types of cancers through potentially noninvasive means such as cheek swabs, urine, and blood.

Emerging initiatives are now focused on not only discovery and validation of novel biomarkers, but also elucidation of the underlying mechanism of these diagnostic and prognostic epigenetic changes as well as exploration of their potential for therapeutic targets. DNA methylation investigation/assessment through MSP and its associated techniques have played a key role in this regard with the discovery of many clinically relevant diagnostic, prognostic, and predictive biomarkers. With continued advancement in technology, DNA methylation as both biomarkers and therapeutic targets will become even more important in the future for health care.

LIST OF ABBREVIATIONS

5-aza-CdR 5-Aza-2′-deoxycytidine
5hmC 5-Hydroxymethylcytosine
5mC 5-Methylcytosine
AS Angelman syndrome
Bp Base pairs
CIMP CpG island methylator phenotype
CpG 5′-Cytosine-guanine-3′
CRC Colorectal cancer
FFPE Formalin-fixed paraffin-embedded
FISH Fluorescence in situ hybridization
FMRP Fragile X mental retardation protein
FOBT Fetal-occult blood test
FXS Fragile X syndrome
GOI Gene of interest
HRM High-resolution melting analysis
lncRNAs Long noncoding RNA
miRNA MicroRNA
MN-MSP Nested MSP
MS-DBA Methylation-sensitive dot blot assay
MS-MLPA Methylation-sensitive multiplex ligation-dependent probe amplification
MSP Methylation-specific PCR
PCa Prostate cancer
PCR Polymerase chain reaction
PMR Percentage of methylated reference
PWS Prader–Willi syndrome
qPCR Quantitative PCR
RISC RNA-induced silencing complex
SNP Single-nucleotide polymorphism
Tm Melting temperature

Acknowledgment and Funding Support

The authors would like to acknowledge support from Ontario Institute of Cancer Research (OICR) Personalized Medicine Research Fund #10Nov-412 and Prostate Cancer Canada #2011-700; PCC TAG #2014-01 1417 (to B. Bapat). F. Zhao is supported by Ontario Student Opportunity Trust Funds award.

We would like to thank Andrea Savio for critical reading of the manuscript.

References

[1] Waddington CH. The epigenotype. 1942. Int J Epidemiol February 2012;41(1):10–3.
[2] Sharma S, Kelly TK, Jones PA. Epigenetics in cancer. Carcinogenesis January 2010;31(1):27–36.
[3] Kimura A, Horikoshi M. Tip60 acetylates six lysines of a specific class in core histones in vitro. Genes Cells December 1998;3(12):789–800.
[4] Clarke AS, Lowell JE, Jacobson SJ, Pillus L. Esa1p is an essential histone acetyltransferase required for cell cycle progression. Mol Cell Biol April 1999;19(4):2515–26.
[5] Fabbri M, Calin GA. Epigenetics and miRNAs in human cancer. Adv Genet 2010;70:87–99.
[6] Mercer TR, Mattick JS. Structure and function of long noncoding RNAs in epigenetic regulation. Nat Struct Mol Biol March 2013;20(3):300–7.

[7] Heard E, Clerc P, Avner P. X-chromosome inactivation in mammals. Annu Rev Genet 1997;31:571–610.

[8] Zendman AJ, Ruiter DJ, Van Muijen GN. Cancer/testis-associated genes: identification, expression profile, and putative function. J Cell Physiol March 2003;194(3):272–88.

[9] Paulsen M, Ferguson-Smith AC. DNA methylation in genomic imprinting, development, and disease. J Pathol September 2001;195(1):97–110.

[10] Futscher BW, Oshiro MM, Wozniak RJ, Holtan N, Hanigan CL, Duan H, et al. Role for DNA methylation in the control of cell type specific maspin expression. Nat Genet June 2002;31(2):175–9.

[11] Yang X, Han H, De Carvalho DD, Lay FD, Jones PA, Liang G. Gene body methylation can alter gene expression and is a therapeutic target in cancer. Cancer Cell October 2014;26(4):577–90.

[12] Feinberg AP, Tycko B. The history of cancer epigenetics. Nat Rev Cancer February 2004;4(2):143–53.

[13] Warren JD, Xiong W, Bunker AM, Vaughn CP, Furtado LV, Roberts WL, et al. Septin 9 methylated DNA is a sensitive and specific blood test for colorectal cancer. BMC Med 2011;9:133.

[14] Rawson JB, Bapat B. Epigenetic biomarkers in colorectal cancer diagnostics. Expert Rev Mol Diagn June 2012;12(5):499–509.

[15] Kim MS, Lee J, Sidransky D. DNA methylation markers in colorectal cancer. Cancer Metastasis Rev March 2010;29(1):181–206.

[16] Issa JP. CpG island methylator phenotype in cancer. Nat Rev Cancer December 2004;4(12):988–93.

[17] Wojno KJ, Costa FJ, Cornell RJ, Small JD, Pasin E, Van Criekinge W, et al. Reduced rate of repeated prostate biopsies observed in ConfirmMDx clinical utility field study. Am Health Drug Benefits May 2014;7(3):129–34.

[18] Matuschek C, Bölke E, Lammering G, Gerber PA, Peiper M, Budach W, et al. Methylated APC and GSTP1 genes in serum DNA correlate with the presence of circulating blood tumor cells and are associated with a more aggressive and advanced breast cancer disease. Eur J Med Res 2010;15:277–86.

[19] Reik W, Walter J. Genomic imprinting: parental influence on the genome. Nat Rev Genet January 2001;2(1):21–32.

[20] Licchesi JD, Herman JG. Methylation-specific PCR. Methods Mol Biol 2009;507:305–23.

[21] Shapiro R, Servis RE, Welcher M. Reactions of uracil and cytosine derivatives with sodium bisulfite. J Am Chem Soc 1970;92(2):422–4.

[22] Campan M, Weisenberger DJ, Trinh B, Laird PW. MethyLight. Methods Mol Biol 2009;507:325–37.

[23] Holmes EE, Jung M, Meller S, Leisse A, Sailer V, Zech J, et al. Performance evaluation of kits for bisulfite-conversion of DNA from tissues, cell lines, FFPE tissues, aspirates, lavages, effusions, plasma, serum, and urine. PLoS One 2014;9(4):e93933.

[24] Kubota T, Nonoyama S, Tonoki H, Masuno M, Imaizumi K, Kojima M, et al. A new assay for the analysis of X-chromosome inactivation based on methylation specific PCR. Hum Genet January 1999;104(1):49–55.

[25] Kidd SA, Lachiewicz A, Barbouth D, Blitz RK, Delahunty C, McBrien D, et al. Fragile X syndrome: a review of associated medical problems. Pediatrics November 2014;134(5):995–1005.

[26] Weinhäusel A, Haas OA. Evaluation of the fragile X (FRAXA) syndrome with methylation-sensitive PCR. Hum Genet June 2001;108(6):450–8.

[27] Carrel L, Willard HF. An assay for X inactivation based on differential methylation at the fragile X locus, FMR1. Am J Med Genet July 1996;64(1):27–30.

[28] Zhou Y, Lum JM, Yeo GH, Kiing J, Tay SK, Chong SS. Simplified molecular diagnosis of fragile X syndrome by fluorescent methylation-specific PCR and GeneScan analysis. Clin Chem August 2006;52(8):1492–500.

[29] Foundation Nfx. Testing-National Fragile X Foundation. 2014. [cited; Available from: https://fragilex.org/fragile-x-associated-disorders/testing.

[30] Kubota T, Sutcliffe JS, Aradhya S, Gillessen-Kaesbach G, Christian SL, Horsthemke B, et al. Validation studies of SNRPN methylation as a diagnostic test for Prader-Willi syndrome. Am J Med Genet December 1996;66(1):77–80.

[31] Clément G, Benhattar J. A methylation sensitive dot blot assay (MS-DBA) for the quantitative analysis of DNA methylation in clinical samples. J Clin Pathol February 2005;58(2):155–8.

[32] Lan VT, Trang NT, Van DT, Thuan TB, Van To T, Linh VD, et al. A methylation-specific dot blot assay for improving specificity and sensitivity of methylation-specific PCR on DNA methylation analysis. Int J Clin Oncol January 2015.

[33] Eads CA, Danenberg KD, Kawakami K, Saltz LB, Blake C, Shibata D, et al. MethyLight: a high-throughput assay to measure DNA methylation. Nucleic Acids Res April 2000;28(8):E32.

[34] Untergasser A, Cutcutache I, Koressaar T, Ye J, Faircloth BC, Remm M, et al. Primer3–new capabilities and interfaces. Nucleic Acids Res August 2012;40(15):e115.

[35] Kibbe WA. OligoCalc: an online oligonucleotide properties calculator. Nucleic Acids Res July 2007;35(Web Server issue):W43–6.

[36] Olkhov-Mitsel E, Zdravic D, Kron K, van der Kwast T, Fleshner N, Bapat B. Novel multiplex MethyLight protocol for detection of DNA methylation in patient tissues and bodily fluids. Sci Rep 2014;4:4432.

[37] Neale RE, Clark PJ, Fawcett J, Fritschi L, Nagler BN, Risch HA, et al. Association between hypermethylation of DNA repetitive elements in white blood cell DNA and pancreatic cancer. Cancer Epidemiol October 2014;38(5):576–82.

[38] Kron KJ, Liu L, Pethe VV, Demetrashvili N, Nesbitt ME, Trachtenberg J, et al. DNA methylation of HOXD3 as a marker of prostate cancer progression. Lab Invest July 2010;90(7):1060–7.

[39] Kron K, Liu L, Trudel D, Pethe V, Trachtenberg J, Fleshner N, et al. Correlation of ERG expression and DNA methylation biomarkers with adverse clinicopathologic features of prostate cancer. Clin Cancer Res May 2012;18(10):2896–904.

[40] Liu L, Kron KJ, Pethe VV, Demetrashvili N, Nesbitt ME, Trachtenberg J, et al. Association of tissue promoter methylation levels of APC, TGFβ2, HOXD3 and RASSF1A with prostate cancer progression. Int J Cancer November 2011;129(10):2454–62.

[41] Olkhov-Mitsel E, Van der Kwast T, Kron KJ, Ozcelik H, Briollais L, Massey C, et al. Quantitative DNA methylation analysis of genes coding for kallikrein-related peptidases 6 and 10 as biomarkers for prostate cancer. Epigenetics September 2012;7(9):1037–45.

[42] Müller HM, Widschwendter A, Fiegl H, Ivarsson L, Goebel G, Perkmann E, et al. DNA methylation in serum of breast cancer patients: an independent prognostic marker. Cancer Res November 2003;63(22):7641–5.

[43] Reinert T, Borre M, Christiansen A, Hermann GG, Ørntoft TF, Dyrskjøt L. Diagnosis of bladder cancer recurrence based on urinary levels of EOMES, HOXA9, POU4F2, TWIST1, VIM, and ZNF154 hypermethylation. PLoS One 2012;7(10):e46297.

[44] He Q, Chen HY, Bai EQ, Luo YX, Fu RJ, He YS, et al. Development of a multiplex MethyLight assay for the detection of multigene methylation in human colorectal cancer. Cancer Genet Cytogenet October 2010;202(1):1–10.

[45] Sohrabi A, Mirab-Samiee S, Rahnamaye-Farzami M, Rafizadeh M, Akhavan S, Hashemi-Bahremani M, et al. C13orf18 and C1orf166 (MULAN) DNA genes methylation are not associated with cervical cancer and precancerous lesions of human papillomavirus genotypes in Iranian women. Asian Pac J Cancer Prev 2014;15(16):6745–8.

[46] Hömig-Hölzel C, Savola S. Multiplex ligation-dependent probe amplification (MLPA) in tumor diagnostics and prognostics. Diagn Mol Pathol December 2012;21(4):189–206.

[47] MRC-Holland. MLPA: cost-effective and sensitive genomic and methylation profiling. 2014. cited; Available from: https://mlpa.com/WebForms/WebFormDBData.aspx?Tag=_zjCZBtdOUyAt3KF3EwRZhGvpJtUw-l0URKGIlLAToek.

[48] Procter M, Chou LS, Tang W, Jama M, Mao R. Molecular diagnosis of Prader-Willi and Angelman syndromes by methylation-specific melting analysis and methylation-specific multiplex ligation-dependent probe amplification. Clin Chem July 2006;52(7):1276–83.

[49] Park CK, Kim J, Yim SY, Lee AR, Han JH, Kim CY, et al. Usefulness of MS-MLPA for detection of MGMT promoter methylation in the evaluation of pseudoprogression in glioblastoma patients. Neuro Oncol February 2011;13(2):195–202.

[50] Gausachs M, Mur P, Corral J, Pineda M, González S, Benito L, et al. MLH1 promoter hypermethylation in the analytical algorithm of Lynch syndrome: a cost-effectiveness study. Eur J Hum Genet July 2012;20(7):762–8.

[51] Tost J, Gut IG. DNA methylation analysis by pyrosequencing. Nat Protoc 2007;2(9):2265–75.

[52] Colella S, Shen L, Baggerly KA, Issa JP, Krahe R. Sensitive and quantitative universal Pyrosequencing methylation analysis of CpG sites. Biotechniques July 2003;35(1):146–50.

[53] Zhang T, Zhang L, Yuan Q, Wang X, Zhang Y, Wang J. The noninvasive detection of RARβ2 promoter methylation for the diagnosis of prostate cancer. Cell Biochem Biophys March 2015;71(2):925–30.

[54] Perez-Carbonell L, Balaguer F, Toiyama Y, Egoavil C, Rojas E, Guarinos C, et al. IGFBP3 methylation is a novel diagnostic and predictive biomarker in colorectal cancer. PLoS One 2014;9(8):e104285.

[55] Chihara Y, Kanai Y, Fujimoto H, Sugano K, Kawashima K, Liang G, et al. Diagnostic markers of urothelial cancer based on DNA methylation analysis. BMC Cancer 2013;13:275.

[56] Sandoval J, Mendez-Gonzalez J, Nadal E, Chen G, Carmona FJ, Sayols S, et al. A prognostic DNA methylation signature for stage I non-small-cell lung cancer. J Clin Oncol November 2013;31(32):4140–7.

[57] Lee MK, Jeong EM, Kim JH, Rho SB, Lee EJ. Aberrant methylation of the VIM promoter in uterine cervical squamous cell carcinoma. Oncology 2014;86(5–6):359–68.

[58] Su SF, de Castro Abreu AL, Chihara Y, Tsai Y, Andreu-Vieyra C, Daneshmand S, et al. A panel of three markers hyper- and hypomethylated in urine sediments accurately predicts bladder cancer recurrence. Clin Cancer Res April 2014;20(7):1978–89.

[59] Worm J, Aggerholm A, Guldberg P. In-tube DNA methylation profiling by fluorescence melting curve analysis. Clin Chem 2001;47(7):1183–9.

[60] Wittwer CT, Reed GH, Gundry CN, Vandersteen JG, Pryor RJ. High-resolution genotyping by amplicon melting analysis using LCGreen. Clin Chem June 2003;49(6 Pt 1):853–60.

[61] Wojdacz TK, Dobrovic A, Hansen LL. Methylation-sensitive high-resolution melting. Nat Protoc 2008;3(12):1903–8.

[62] White HE, Hall VJ, Cross NC. Methylation-sensitive high-resolution melting-curve analysis of the SNRPN gene as a diagnostic screen for Prader-Willi and Angelman syndromes. Clin Chem November 2007;53(11):1960–2.

[63] Yang X, Dai W, Kwong DL, Szeto CY, Wong EH, Ng WT, et al. Epigenetic markers for noninvasive early detection of nasopharyngeal carcinoma by methylation-sensitive high resolution melting. Int J Cancer February 2015;136(4):E127–35.

[64] Vogelstein B, Kinzler KW, Digital PCR. Proc Natl Acad Sci USA August 1999;96(16):9236–41.

[65] Weisenberger DJ, Trinh BN, Campan M, Sharma S, Long TI, Ananthnarayan S, et al. DNA methylation analysis by digital bisulfite genomic sequencing and digital MethyLight. Nucleic Acids Res August 2008;36(14):4689–98.

[66] Lange CP, Campan M, Hinoue T, Schmitz RF, van der Meulen-de Jong AE, Slingerland H, et al. Genome-scale discovery of DNA-methylation biomarkers for blood-based detection of colorectal cancer. PLoS One 2012;7(11):e50266.

[67] Delatte B, Fuks F. TET proteins: on the frenetic hunt for new cytosine modifications. Brief Funct Genomics May 2013;12(3):191–204.

[68] Chiang JH, Cheng WS, Hood L, Tian Q. An epigenetic biomarker panel for glioblastoma multiforme personalized medicine through DNA methylation analysis of human embryonic stem cell-like signature. OMICS May 2014;18(5):310–23.

[69] Andresen K, Boberg KM, Vedeld HM, Honne H, Hektoen M, Wadsworth CA, et al. Novel target genes and a valid biomarker panel identified for cholangiocarcinoma. Epigenetics November 2012;7(11):1249–57.

[70] Lind GE, Danielsen SA, Ahlquist T, Merok MA, Andresen K, Skotheim RI, et al. Identification of an epigenetic biomarker panel with high sensitivity and specificity for colorectal cancer and adenomas. Mol Cancer 2011;10:85.

[71] Singh V, Sharma P, Capalash N. DNA methyltransferase-1 inhibitors as epigenetic therapy for cancer. Curr Cancer Drug Targets May 2013;13(4):379–99.

[72] Momparler RL. Epigenetic therapy of cancer with 5-aza-2′-deoxycytidine(decitabine).SeminOncolOctober2005;32(5):443–51.

CHAPTER

9

Pyrosequencing and Its Application in Epigenetic Clinical Diagnostics

Ana-Maria Florea

Department of Neuropathology, Heinrich-Heine-University Düsseldorf, Dusseldorf, Germany

OUTLINE

1. INTRODUCTION

Pyrosequencing was described in 1985 [1] and is currently used for quantitative analysis of nucleic acids. It is a method where the sequencing takes place during the deoxyribonucleic acid (DNA) synthesis, using the so-called "sequencing-by-synthesis" principle. It involves the synthesis of single-stranded DNA resulting in quick and accurate analysis of nucleotide sequences. In this method, since the added nucleotide order is known, the template sequence can be determined [2]. Thus, this method can be used not only for confirmatory sequencing but also for "de novo" sequencing (for review see Refs [1,3–5]). However, since pyrosequencing has only a short read length (into a pyrosequencing reaction a 45–60 nucleotide sequence can

Epigenetic Biomarkers and Diagnostics
http://dx.doi.org/10.1016/B978-0-12-801899-6.00009-7

be analyzed) [4], it was not used for genome sequencing but for genotyping [3].

With the use of pyrosequencing, the analyzed nucleic acid molecule might be RNA or DNA. However, since DNA polymerases "work better" than RNA polymerases (better catalytic activity), the primed DNA templates are usually used for pyrosequencing (for review see Refs [3,1]). For analysis of DNA methylation, a prior bisulfite modification pyrosequencing is required. This step converts the unmethylated CG to TG, while the methylated CG will not be affected ([6]; for practical example, Ref. [5]). In order to be able to perform pyrosequencing, labeled primers, labeled nucleotides, and agarose gel electrophoresis are required [3].

During the execution of pyrosequencing method, the DNA synthesis is monitored in "real time" with the help of 4-enzyme-dependent nucleotide incorporation that results in the bioluminescence release which is recorded by the pyrosequencer (for review see Refs [3,5,6]). The reaction starts with a nucleic acid polymerization in which inorganic pyrophosphate gets released due to nucleotide incorporation that is regulated with the help polymerase. Furthermore, with the help of adenosine triphosphate (ATP) sulfurylase, the pyrophosphate is then converted to ATP, that in turn provides the power to luciferase to oxidize luciferin and generate light (for review see Refs [1,3]). Thus, the pyrosequencer measures the pyrophosphate release due to dNTP (nucleotide triphosphates containing deoxyribose) incorporation in the DNA strand that is recorded in a form of peaks, the so-called "pyrogram," an example is shown in Figure 1 [4]. For the quality control of a pyrosequencing reaction a methylation positive and negative control is recommended to be used; an example is shown in Figure 2. A schematic overview of the steps required for pyrosequencing is shown in Figure 3.

To complete the picture and summarize, I would like to cite the work of Ahmadian et al. [6] which described pyrosequencing as following: "The technique relies on measurement of light signals (bioluminescence) that are generated by pyrophosphate release during DNA synthesis along a sequencing template with an annealed primer using specific enzymes (Klenow fragment of DNA polymerase I, ATP sulfurylase, luciferase, apyrase) and enzyme substrates (adenosine phosphosulfate, D-luciferin). The four nucleotides (C, T, G, A) are iteratively added in a cyclic reaction and the light signals emitted upon their incorporation into the newly synthesized complementary strand along the DNA template are recorded by a camera system. The nucleotide sequence of the DNA template can be deduced from the newly synthesized DNA sequence based on the Watson-Crick base pairing rule, according to which the nucleotide G will always pair with C and T will always pair with A" [6].

Several pyrosequencing platforms are available today such as PyroMark Q24, PyroMark Q24 Advanced, or PyroMark Q96 ID from Qiagen or the 454 Sequencing from Roche which is a high-throughput next-generation sequencing platform, e.g., GS FLX+ System. A comparison of these platforms is presented in Table 1. For clinical and diagnostic routine, when only a limited number of biomarkers are analyzed at once, the Pyromark platforms are more recommended since the acquisition of the instrument and the sample preparation is cheaper and it does not require a lot of lab space. However, for biomarker discovery, the 454 Sequencing from Roche is more recommended due to the high-throughput analysis (nevertheless, it has high acquisition costs and requires a lot of lab space) or microarray approach as published in Ref. [7].

For Pyromark platforms, PyroMark CpG Assays are available, which are ready designed assays for pyrosequencing analysis for human, mouse, and rat genomes (purchasable at Qiagen, Germany; or EpigenDX, USA) (for review see Ref. [5]). These assays are created using algorithms that provide specific quantification of CpG methylation. Furthermore, it is possible to create self-designed assays using commercial

Neuroblastoma SH-SY5Y
S100A6

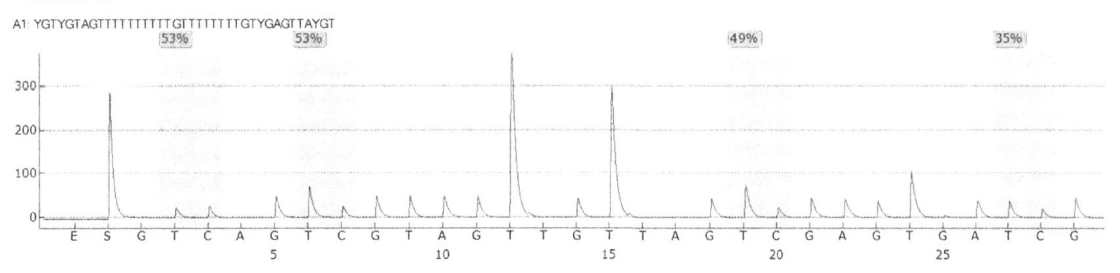

FIGURE 1 Pyrogram resulting from pyrosequencing of S100A6 gene in human neuroblastoma cell line SH-SY5Y using a Pyromark assay provided from Qiagen, Germany. The bisulfite conversion of the DNA was performed with the EZ DNA Methylation-GoldTM Kit (Zymo Research, USA) following the provider's recommendations. One microgram DNA was used for each reaction. Each PCR mix contained the 1xPCR buffer, 1.5μmmol/L of $MgCl_2$ (final concentration) (Qiagen), 0.2 mol/L of each dNTP (Biobudget, Germany), 1 μL of forward and reverse primer (10 pmol/L of each PCR primer), 2 U of HotStar Taq Polymerase (Qiagen), and 1 μL of sodium bisulfite-treated DNA in a total volume of 25 μL. PCR conditions were as follows: 95 °C→15 min; 45 cycles of 95 °C→20 s; 50 °C→20 s; 72 °C→20 s; and 72 °C→5 min. The PCR product quality was checked using 1% agarose gel electrophoresis. DNA pyrosequencing was performed with the PyroMark Q24 System and the PyroMark Gold Q24 Reagents Kit (Qiagen, Germany) as recommended. The biotinylated PCR product was purified and made single stranded to act as a template in a pyrosequencing reaction as recommended by the manufacturer using the pyrosequencing vacuum prep tool (Qiagen, Germany). A 25 μL of biotinylated PCR product was adjusted with MilliQ® water to a volume of 38 μL, mixed with 40 μL binding buffer (Qiagen, Germany) and 2 μL sepharose beads/reaction (GE Healthcare Bio-Sciences AB, Sweden), then shaken for 5 min at room temperature in order to get the biotinylated PCR product immobilized on streptavidin-coated sepharose beads. After that, the PCR product was washed in 70% ethanol, denatured using a 0.2 M NaOH solution, and washed again in washing buffer (Qiagen, Germany). Then, 0.3 μM pyrosequencing primer (30 μL) was annealed to the purified single-stranded PCR product (the PCR product—sequencing primer mix was incubated at 80 °C for 2 min) and then the pyrosequencing was performed by PyroMark Q24 (Qiagen, Germany) [5]. Nucleotides are added successively in the pyrosequencing reaction following the sequence to analyze (in this case: YGTYGTAGTTTTTTTTTTGTTTTTTTTGTYGAGT-TAYGT): if an adequate nucleotide is included in the growing DNA double strand, pyrophosphate is released and converted to ATP which catalyzes a luciferase reaction and generates a detectable light signal as a peak in the pyrogram. In the presented pyrogram, one can see the addition of the enzyme (E) followed by substrate (S), control nucleotide guanine (G) that is not present in the sequence to analyze, then the first CpG site marked with Y (in the pyrogram thymine T and cytosine C), and so on. Any excess nucleotides are degraded when the enzyme apyrase is injected in the reaction (PyroMark® Gold Q24 Reagents Handbook, Qiagen). The peak heights in the resulting pyrogram reflect the ratio of cytosine to thymine at each analyzed CpG site, which reflects the proportion of methylated DNA (in this case 53%, 53%, 49%, 35%).

software (PSQ assay design, Assay Design SW 2.0 or PyroMark CpG SW 1.0) or freely available software (Primer3).

2. DETECTION OF EPIGENETIC CHANGES BY PYROSEQUENCING: DNA METHYLATION

Epigenetically regulated processes are an important target in biomedical research. Epigenetic factors such as DNA methylation, histone modifications, and microRNAs (miRNAs) play a key role in the regulation of gene expression and but also of genomic stability (for review see Ref. [5]). The pattern of DNA methylation is maintained in the cells with the help of enzymatic activity: DNA methyltransferase (DNMT) family controls de novo and maintenance DNA methylation. Therefore, it is possible to modify the DNA methylation status of genes using agents that hinder the function of DNA methylation enzymes (for review see Ref. [5]).

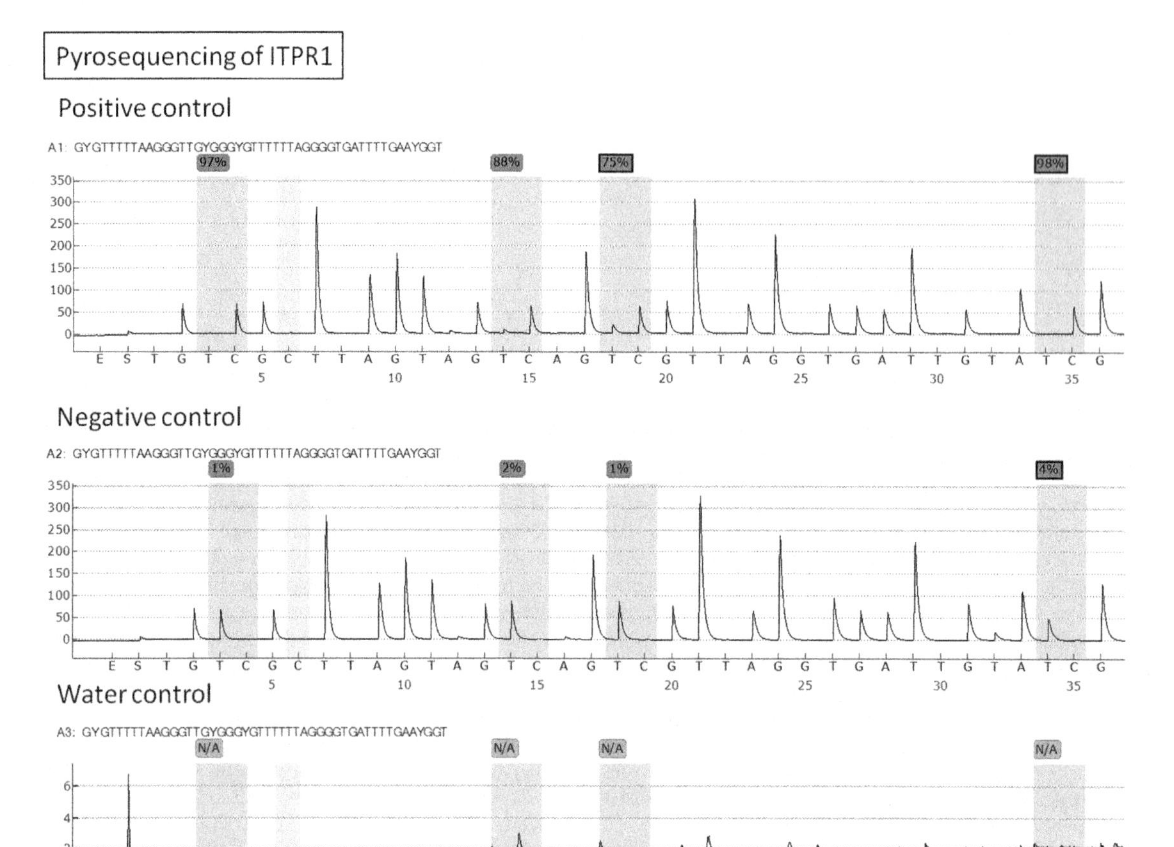

FIGURE 2 Pyrogram resulting from pyrosequencing of methylation positive, negative, and water control (Universal Methylated Human Standard from Zymo Research, Germany) using the Pyromark Assay specific for ITPR1 gene (sequence to analyze GYGTTTTTAAGGGTTGYGGGYGTTTTTTAGGGGTGATTTTGAAYGGT) provided from Qiagen, Germany, same experimental conditions were used as described in Figure 1. However, in the assay setup and additional cytosine (C) was introduced in order to check the success of the natrim-bisulfite conversion) (marked in the pyrogram) with the yellow bar. In addition to observe that the level of methylation for each of the CpG site of the positive methylation control is fully methylated while in the methylation negative control the CpG sites are completely demethylated. The water control did not give an analyzable sequence.

From these three epigenetic factors, the DNA methylation was previously the most studied, while pyrosequencing, as a method, is becoming more and more important to study the changes in the status of DNA methylation. Previous studies analyzed the methylation of the promoter region of genes involved in physiological and pathological processes, as well as in CpG islands. CG dinucleotides are found to be associated in CGs reach cluster regions: CpG islands, while DNA methylation could lead to transcriptional silencing of genes. Therefore,

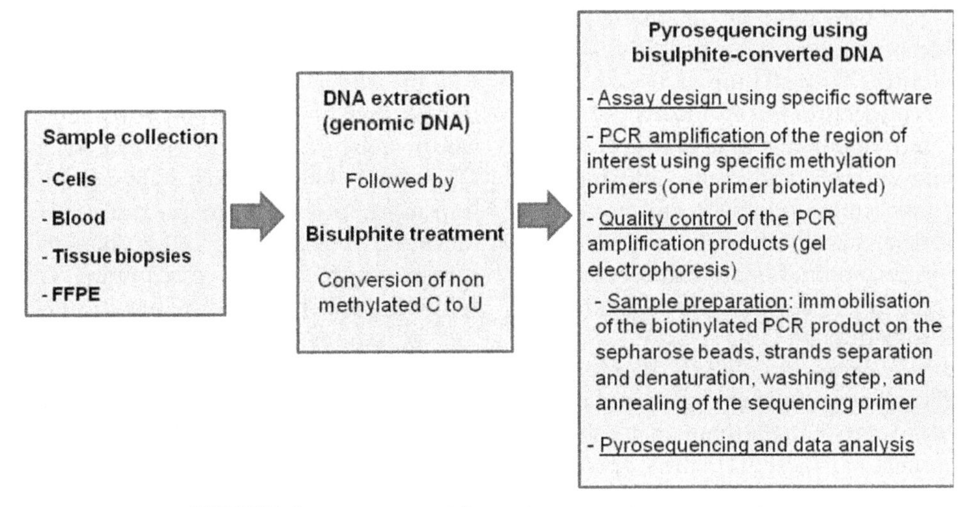

FIGURE 3 Steps required for performance of pyrosequencing.

TABLE 1 Description of the Different Pyrosequencing-Based Platforms

	Qiagen			Roche GS FLX+ System	
	PyroMark Q24 Advanced	PyroMark Q24	PyroMark Q96 ID		
Reagents/ Sequencing kits	PyroMark Q24 Advanced Reagents; PyroMark Q24 Advanced CpG Reagents	PyroMark Q24 Gold Reagents	PyroMark Gold Q96 Reagents	New! GS FLX Titanium XL+	GS FLX Titanium XLR70
Read length	10–140bp	10–80bp	10–80bp	Up to 1000bp	Up to 600bp
Throughput	1–24 samples	1–24 samples	1–96 samples	700Mb	450Mb
Applications	Complex mutation analysis; epigenetics (CpG and CpN analysis); resistance typing; and microbial ID	Mutation analysis; resistance typing	Mutation analysis; epigenetics; resistance typing; and microbial ID	Whole genome, transcriptome, amplicon sequencing; sequence capture; metagenomics	

Information obtained from the Qiagen and Roche Web site, and can be accessed directly under following links: https://www.qiagen.com/de/resources/technologies/pyrosequencing-resource-center/technology-overview/) and Roche http://454.com/products/gs-flx-system/.

low methylation (hypomethylation) of the gene promoter-associated CpG islands is related to active genes, while high methylation (hypermethylation) is associated with low-transcribed genes and with the repetitive elements such as long interspersed elements (LINEs) and short interspersed elements (SINEs) (retrotransposons) which represent important biomarkers that are often used in the evaluation of DNA methylation by pyrosequencing (for review see Ref. [5]). In particular, DNA methylation measured in LINE1 sequences has been considered a surrogate marker for global genome methylation [8,5].

Active retrotransposable elements (retrotransposons) include LINEs and SINEs while approximately 0.27% of all human disease mutations are in connection to retrotransposons [9]. These are gene elements that "move" within a gene using a "copy and paste" mechanism, therefore contributing to genetic variation or to disease-causing mutations such as insertional mutagenesis, recombination, retrotransposition-mediated and gene conversion-mediated deletion, and 3′ transduction [10].

Since gene expression is tightly regulated in normal cells any epigenetic change that disturbs this regulation might be a mechanism involved in the development of cancer [11]. Thus, epigenetic alterations, such as abnormal DNA methylation, are associated with human cancers; differences in methylation patterns between neoplastic and normal cells have diagnostic potential in cancer [12]. In particular, CpG island hypermethylation of gene promoters and regulatory regions is a well-known mechanism of epigenetic silencing of tumor suppressors and is directly linked to carcinogenesis [13]. Methylation of Alu and LINE1 is a well-established measure of DNA methylation often used in epidemiologic studies [10] and might be used as a biomarker in cancer.

2.1 Application of Pyrosequencing in Epigenetic Biomedical Research

Pyrosequencing is a molecular method with large array of applications. Applications of pyrosequencing range from microbial forensics (e.g., identification of microbes), molecular epidemiology [14] or clinical/environmental toxicology [5], multidrug-resistant microbes [15] to biomarker discovery [16] and prognostic indicators [17], epigenetic modifications in form of aberrant DNA methylation in infectious diseases [18] and cancers [16], but also analysis of short DNA sequences exemplified by single nucleotide polymorphism analysis and genotyping [19,6]. Furthermore, pyrosequencing can be used for mutation detection and single nucleotide polymorphism [6]. Additionally, the pyrosequencing method can be used for the analysis of cell-free circulating DNA isolated from the plasma of individuals with cancer, which can be related to cancer-associated changes in DNA methylation. Thus, pyrosequencing is an important tool for biomarker discovery but the detection of DNA methylation changes might be sometimes, technically challenging [20]. In the following subchapters, the use of pyrosequencing in various fields of medical research and clinical applications will be shown.

2.1.1 *Diabetes and Metabolic Syndrome Research*

Recent work in the field of diabetes and metabolic syndrome research points out to the possibility of using pyrosequencing not only for basic research in this field but also for biomarker discovery. In this regard, new findings have suggested the potential involvement of epigenetic mechanisms in type 2 diabetes (MIM #222100), as a crucial interface between the effects of genetic predisposition and environmental influences. Therefore methylation markers might be used for analysis of molecular changes related to diabetes. As a premier example, Martín-Núñez et al. [8] evaluated (using pyrosequencing) whether the change in global DNA methylation is able to predict the increase in risk for acquiring carbohydrate metabolism disorders. The authors used in their study a common biomarker applicable for the evaluation of global DNA methylation: the LINE1 element. The authors found that lower levels of global DNA methylation were associated with the subjects that did not improve their carbohydrate metabolism status and practiced less intense physical activity [8].

In a different study, using DNA pyrosequencing, Remely et al. [21] provide evidence that a different composition of gut microbiota in obesity and type 2 diabetes is able to affect the epigenetic regulation of FFAR3 gene. However, the marker of global methylation LINE1, failed to prove a significant difference between obese and type 2 diabetic patients compared to a lean control group, although methylation of type 2 diabetic patients had the tendency to increase over time [21]. In an additional study, Remely et al. [22] studied the methylation of four CpGs in the TLR4 gene and they found significantly lower methylation in obese individuals, but no significant difference between type 2 diabetics and lean controls was found. However, the methylation of seven CpGs in the promoter region of TLR2 was significantly lower in type 2 diabetics compared to obese subjects and lean controls [22].

2.1.2 *Infertility and Prenatal Diagnostic*

A new area of application for pyrosequencing is represented by the area of infertility and prenatal diagnostic. In this regard, recent studies have shown associations of aberrant DNA methylation in spermatozoa with idiopathic infertility. Therefore, the DNA methylation analysis of specific genes can be a diagnostic marker in clinical andrology. In this context, Kläver et al. [23] analyzed with the help of pyrosequencing the DNA methylation of the maternally imprinted gene *MEST*. The authors found that *MEST* DNA methylation was significantly associated with oligozoospermia, decreased bitesticular volume, and increased FSH levels. Thus, the authors concluded that *MEST* DNA methylation might be used as biomarker in the routine diagnosis [23].

In addition, DNA methylation might be a molecular mechanism through which fetal growth might be affected, such as the case of phthalate exposure. Zhao et al. [24] examined associations between prenatal phthalate exposure, infant growth, and global DNA methylation of LINE1 element in human placenta samples. The authors found that placental LINE1 methylation was positively associated with fetal birth weight, and negatively associated with urinary phthalate metabolites concentrations and thus, concluded that the birth weight might be mediated through LINE1 methylation. The authors also concluded that changes in placental LINE1 methylation might be part of the underlying biological pathway between prenatal phthalate exposure and adverse fetal growth [24].

Into a different study, Calvello et al. [25] showed that pyrosequencing might be used to analyze the Beckwith–Wiedemann syndrome (BWS, MIM #130650; ORPHA116), which is a rare disorder characterized by overgrowth and predisposition to embryonal tumors, caused due to epigenetic and/or genetic alterations that dysregulate the imprinted genes on chromosome region 11p15.5. In this study [25], the authors established a reliable pyrosequencing assay which evaluates the methylation profiles of *ICR1* and *ICR2* [25]. In this context, the supportive study of Lee et al. [26] concludes that methylation-specific pyrosequencing enhanced the detection rate of molecular defects in BWS and the Silver–Russell syndrome (SRS, MIM #180860) while the methylation status at 11p15.5 might have an important role in fetal growth [26].

Furthermore, Houde et al. [27] hypothesized that in utero environmental perturbations have been associated with epigenetic changes in the offspring and a lifelong susceptibility to cardiovascular diseases. Therefore, the authors analyzed, using pyrosequencing, whether DNA methylation at the ATP-binding cassette transporter A1 (*ABCA1*) gene is associated with cardiovascular diseases and if these epigenetic marks respond to changes in the maternal environment. In conclusion, the authors observed DNA methylation changes in placenta and cord blood that are likely to contribute to an optimal maternofetal cholesterol transfer that might potentially trigger the long-term susceptibility of the newborn to dyslipidemia and cardiovascular diseases [27]. Nevertheless, Desgagné et al. [28] determined by DNA pyrosequencing whether placental *IGF1R*, *IGFBP3*, *INSR*, and *IGF1* DNA methylation and mRNA levels were deregulated when exposed to maternal impaired glucose tolerance and investigated whether the epigenetic profile is associated with fetoplacental developmental markers. The authors found that the results support the growth-promoting role of the impaired glucose tolerance system in placental/fetal development and suggest that the *IGF1R* and *IGFBP3* DNA methylation profiles are deregulated in the impaired glucose tolerance, potentially affecting the fetal metabolic programming [28]. Nevertheless, the perspective of using the pyrosequencing method in this area of medical application is quite clear, however, it is very necessary to find robust biomarkers that can be used for the molecular detection of changes related to infertility and prenatal diagnostic.

2.1.3 *Environmental and Toxicological Biomarkers*

Epigenetic changes such as DNA methylation may be a molecular mechanism through which environmental exposures affect health [10]. The methylation of repetitive sequences, such as Alu and LINE1, is a well-established measure of DNA methylation that often used in epidemiologic studies [10]. This parameter might be used in the evaluation of environmental and clinical toxicology as well, e.g., the in vitro evaluation of global and focal (CpG site specific) DNA methylation upon occupational or chronic exposure to chemicals/drugs; but also in "in vitro" studies where cells are treated with chemicals/drugs (for review see Ref. [5]).

In this context, the study by Florea [5] described with the help of bisulfite genomic sequencing of LINE1 element that "the CpG sites within the LINE1 element are methylated at different levels. Furthermore, the in vitro cell cultures show a methylation level ranging from 56% to 49% while the cultures of drug resistant tumor cells show significant hypomethylation as compared with the originating nonresistant tumor cells. The in vitro testing of epigenetically active chemicals (5-methyl-2′-deoxycytidine and trichostatin A) revealed a significant change of LINE1 methylation status upon treatment, while specific CpG sites were more prone to demethylation than others (focal methylation). In conclusion, DNA methylation using pyrosequencing might be used not only for testing epigenetic toxins/drugs but also in risk assessment of drugs, food, and environmental relevant pollutants" (for review see Ref. [5]).

Modification of the epigenome may be a mechanism underlying toxicity and disease following chemical exposure [29,10]. For instance, Huen et al. [10] measured Alu and LINE1 methylation by pyrosequencing and characterized the relationship of age, sex, and prenatal exposure to persistent organic pollutants, dichlorodiphenyl trichloroethane (DDT), dichlorodiphenyldichloroethylene (DDE), and polybrominated diphenyl ethers (PBDEs) using bisulfite-treated DNA isolated from whole blood samples collected from newborns and 9-year-old children. The persistent organic pollutants were measured in maternal serum during late pregnancy. The authors found that the levels of DNA methylation were lower in 9-year-olds compared to newborns and were higher in boys compared to girls. Nevertheless, higher prenatal DDT/E exposure was associated with lower Alu methylation at birth. Associations of persistent organic pollutants with LINE1 methylation were only identified after examining the coexposure of DDT/E with PBDEs simultaneously. The authors concluded that the repeat element methylation can be an informative marker of epigenetic differences [10].

Additional research from Ref. [29] shows that other environmental toxicants such as mercury (Hg) impact DNA methylation. The authors hypothesized that methylmercury and inorganic Hg exposures from fish consumption and dental amalgams, respectively, may be associated with altered DNA methylation of LINE1 and candidate genes related to epigenetic processes (DNMT1) and protection against Hg toxicity (SEPW1, SEPP1) quantified via pyrosequencing of bisulfite-converted DNA isolated from buccal mucosa. The authors found in males a significant trend of SEPP1 hypomethylation with increasing hair Hg levels [29].

Furthermore, Shenker et al. [30] showed that bisulfite pyrosequencing might be used in a test as a predictive model of smoking status. A set of four genomic loci (AHRR, 6p21, and two at 2q37) had differential DNA methylation levels in peripheral blood DNA which was dependent on tobacco exposure. Combining four gene loci into a single methylation index it was possible to provide high positive predictive and sensitivity values for predicting former smoking status. This study provides a direct molecular measure of prior exposure to tobacco that can be performed using the quantitative approach of bisulfite pyrosequencing [30]. Therefore, epigenetic changes of specific biomarkers detectable in blood samples might be used as molecular biomarkers for other types of exposure and thus, can be used in epidemiological studies.

Altered DNA methylation of LINE1 can be used as biomarker for other types of chemical exposure like, for instance, environmental or occupational exposure. Evidence has been provided by the study of Tajuddin et al. [31] that investigates the association between levels of methylation in leukocyte DNA at LINE1 and genetic and nongenetic characteristics of the participants from the Spanish Bladder Cancer/EPICURO study. Also, in this case, the LINE1 methylation level was analyzed by pyrosequencing. The data showed that women had lower levels of LINE1 methylation than men while blond tobacco smokers showed lower methylation than nonsmokers. Furthermore, arsenic toenail concentration was inversely associated with LINE1 methylation while iron and nickel were positively associated with LINE1 methylation [31].

Similarly, the study conducted by Hossain et al. [32] investigated cadmium, as a common food pollutant, that alters DNA methylation in vitro. Epigenetic effects might therefore partly explain cadmium's toxicity, including its carcinogenicity. Methylation in CpG islands of LINE1 and promoter regions of *p16* (cyclin-dependent kinase inhibitor 2A (*CDKN2A*)) and *MLH1* (mutL homolog 1) in peripheral blood were measured by bisulfite polymerase chain reaction pyrosequencing. Cadmium exposure was low while urinary cadmium (natural log transformed) was inversely associated with LINE1 methylation but not with *p16* or *MLH1* methylation. Thus, environmental cadmium exposure was associated with DNA hypomethylation in peripheral blood [32].

Other indications that pyrosequencing can be used in this field of investigation hint to the possibility of using the serotonin transporter gene-linked polymorphic region (*5-HTTLPR*) that might be involved in moderating vulnerability to stress-related psychopathology upon exposure to environmental adversity. Working on this hypothesis, Alexander et al. [33] analyzed in healthy adults exposed to a laboratory stressor, the relationship between the cortisol response patterns as a function of 5-HTTLPR and the DNA methylation profiles in SLC6A4, using pyrosequencing. The results suggested that SLC6A4 methylation levels significantly moderate the association of 5-HTTLPR and cortisol stress reactivity and concluded that epigenetic changes may compensate for genotype-dependent differences in stress sensitivity [33].

2.1.4 Biomarker Pyrosequencing and the Cancer Diagnostics

It is now widely accepted in the scientific community that the changes in the methylation pattern, especially in the promoter regions of genes, as well as the mutations in the genetic sequence of genes play an important role in carcinogenesis. Thus, pyrosequencing can provide molecular analysis of the epigenetic and genetic modifications of cancer-related genes involved in the pathology of tumors [34]. Furthermore, better understanding of early epigenetic changes related to cancer will help finding new biomarkers that could be used in diagnosis, prognosis, and therapy of cancer (for review see Refs [35,5]) where pyrosequencing can be used for a large array of molecular investigations.

Many working groups are focusing their research on the changes in DNA methylation pattern in tumors. In this context, pyrosequencing method might have applicability to analyze the DNA methylation of specific genes and to screen for novel disease markers. Thus, pyrosequencing might have application in clinical studies and have potential in a diagnostic setting. In this regard, patients can be screened using selected DNA methylation "signatures" not only for cancers, response to chemotherapy, increased risk identification but also for prophylaxis, in order to reduce the incidence of cancer (for review see Refs [35,5]). Thus, pyrosequencing testing of DNA methylation has the potential to improve current noninvasive testing in diagnostics and it can be used as marker for treatment response [11].

A number of tumor suppressor genes have been shown to be inactivated by DNA promoter hypermethylation in tumors, including *MGMT*, *SCGB3A1*, *RASSF1A*, *HIC1*, and *PRSS21* ([11]; for review see Ref. [5]) and some of them (e.g., *MGMT*) are already clinically used. For instance, Kiss et al. [36] proposed that the alterations in methylation promoter of tumor suppressor genes and LINE1 repeat might be used in the diagnostic and prognostic applications in cancer. Kiss et al. [36] investigated by bisulfite pyrosequencing the CpG methylation abnormalities in a tumor panel and assessed possible relationships between metastatic disease and mutation status. Increased methylation was observed for *DCR2* (TNFRSF10D), *CDH1*, *P16* (CDKN2A), retinoic acid receptor beta (*RARB*), and *RASSF1A* [36]. Overall, methylation levels of tumor suppressor genes, and individual tumor suppressor genes, but not LINE1, correlated with metastatic disease, paraganglioma, disease predisposition, or the outcome [36]. In line with previous studies, some methylated genes have been proposed for cervical cancer detection by Chen et al. [37]. Using bisulfite pyrosequencing the authors found 14 genes, including *ADRA1D*, *AJAP1*, *COL6A2*, *EDN3*, *EPO*, *HS3ST2*, *MAGI2*, *POU4F3*, *PTGDR*, *SOX8*, *SOX17*, *ST6GAL2*, *SYT9*, and *ZNF614*, that were significantly hypermethylated in CIN3+ lesions [37].

2.1.4.1 GLIOBLASTOMA (MIM #137800; 613029)

Molecular markers play a key role in the diagnostics of gliomas and are very important for histopathological diagnosis and their molecular classification (see Table 2) [38]. A widely used molecular marker is the *MGMT* gene that encodes a DNA repair protein: O(6)-methylguanine-DNMT. The hypermethylation in the promoter region of *MGMT* gene is not only an important prognostic factor for glioblastoma patients but also a predictor for the outcome of the treatment to alkylating agents (e.g., temozolomide) [17]. MGMT status is determined in most clinical trials and frequently requested in routine diagnostics of glioblastoma [17] and is relevant for clinical decision-making and research applications [39].

Various techniques are available for *MGMT* promoter methylation analysis [17] but pyrosequencing might represent an important tool for determining MGMT status in patients. Furthermore, pyrosequencing proved to be a good method for assessing the degree of *MGMT* methylation in formalin-fixed paraffin-embedded glioma samples [38]. Nevertheless, *MGMT* testing kits are offered by the companies for sale, but only for research purpose, although the pyrosequencer (sold originally by Biotage and now by Qiagen) is approved for use in diagnostic applications. In this regard, Preusser et al. [39] concluded from their own study that "pyrosequencing-based assessment of *MGMT* promoter methylation status in glioblastoma meets the criteria of high analytical test performance and can be recommended for clinical application, provided that strict quality control is performed and provide practical instructions and open issues for *MGMT* promoter methylation testing in glioblastoma using pyrosequencing" [39].

TABLE 2 Overview of Possible Biomarkers for Pyrosequencing in Cancers

Cancer type	Type of biomaterial	Genes	Medical diagnostic	References
Glioblastoma (MIM #137800; 613029)	FFPE tissue, frozen brain tumor tissue samples	MGMT methylation	Recommended for clinical application	[38]
Glioblastoma (MIM #137800; 613029)	FFPE tissue, DNA tissue, DNA blood	MGMT methylation	Compliance with EU IVD Directive 98/79/EC	MGMT Pyro Kit, Qiagen
Glioblastoma (MIM #137800; 613029)	FFPE and RCLPE samples	MGMT methylation	Recommended for clinical application	[39]
Glioblastoma (MIM #137800; 613029)	FFPE and RCLPE samples	MGMT methylation	Prognostic and predictive value of the MGMT promoter methylation status in elderly glioblastoma	[39]
Glioblastoma (MIM #137800; 613029)	FFPE samples	MGMT methylation	Predicting progression-free survival	[17]
Glioblastoma (MIM #137800; 613029)	Snap-frozen and FFPE samples	MGMT methylation	Survival of glioblastoma patients treated with alkylating agents	[40]
Glioblastoma (MIM #137800; 613029)	Archival samples	MGMT methylation	Prognostically significant in glioblastomas given chemoradiotherapy, use in the routine clinic	[41]
Glioblastoma (MIM #137800; 613029)	Archival samples	MGMT methylation	Prognostic stratification	[41]
Glioblastoma (MIM #137800; 613029)	Archival samples	PTPN6/SHP1	Poor survival for anaplastic glioma patients	[42]
Glioblastoma (MIM #137800; 613029)	Cell culture	PTPN6/SHP2	Increased drug resistance	[42]
Prostate cancer (PCa, MIM #176808)	Human prostate tissues	APC	Associated with prostate cancer	[12]
Prostate cancer (PCa, MIM #176808)	Human prostate tissues	APC	Associated with prostate cancer	[12]
Prostate cancer (PCa, MIM #176808)	Human prostate tissues, cell lines	RARbeta	Hypermethylated and silenced in prostate cancer tissues and prostate cancer cell lines	[43]
Prostate cancer (PCa, MIM #176808)	Human prostate tissues, cell lines	PDLIM4	Hypermethylated and silenced in prostate cancer tissues and prostate cancer cell lines	[43]
Hepatocellular carcinoma (MIM #114550)	Diverse liver tissues	RASSF1A	Aberrantly and strongly methylated in early hepatocellular carcinoma	[44]

Continued

TABLE 2 Overview of Possible Biomarkers for Pyrosequencing in Cancers—cont'd

Cancer type	Type of biomaterial	Genes	Medical diagnostic	References
Hepatocellular carcinoma (MIM #114551)	Diverse liver tissues	CCND2	Aberrantly and strongly methylated in early hepatocellular carcinoma	[44]
Hepatocellular carcinoma (MIM #114552)	Diverse liver tissues	SPINT2	Aberrantly and strongly methylated in early hepatocellular carcinoma	[44]
Hepatocellular carcinoma (MIM #114553)	Diverse liver tissues	RUNX3	Aberrantly and strongly methylated in early hepatocellular carcinoma	[44]
Hepatocellular carcinoma (MIM #114554)	Diverse liver tissues	GSTP1	Aberrantly and strongly methylated in early hepatocellular carcinoma	[44]
Hepatocellular carcinoma (MIM #114555)	Diverse liver tissues	APC	Aberrantly and strongly methylated in early hepatocellular carcinoma	[44]
Hepatocellular carcinoma (MIM #114556)	Diverse liver tissues	CFTR	Aberrantly and strongly methylated in early hepatocellular carcinoma	[44]
CRC (MIM #114500)	Human preneoplastic tissue	PYCARD and NR1H4	Significant methylation differences in normal mucosa between colorectal cancer patients and controls	[46]
Pancreatic cancer (MIM #260350)	Pancreatic endocrine tumors	RASSF1	Cannot be considered a marker for this neoplasm	[47]
Bladder cancer (MIM #109800)	Human bladder cancer samples	ITIH5	Associated with progressive bladder cancers	[16]
Bile duct cancer	Mid/distal bile duct cancer samples	P16, DAPK, and RASSF-1	Methylation correlated with perineural invasion, tumor depth, and age	[48]
Bile duct cancer	Mid/distal bile duct cancer samples	P16	Presence of lymph node metastasis	[48]
AML (MIM #601626)	Blood	DNMT3A	Useful for risk stratification or choice of therapeutic regimen	[50]
Nonsmall cell lung cancer (MIM #211980)	Formalin-fixed tumor tissue	p16	Associated with a worse outcome in patients with age at diagnosis of 60 years or younger	[51]
Nonsmall cell lung cancer (MIM #211980)	Neoplastic and normal lung tissue	WT1	Higher promoter methylation islands of WT1 in the neoplastic tissues of the nonsmall-cell lung cancer patients	[13]

CRC, colorectal cancer; DNA, deoxyribonucleic acid; MGMT, O-6-methylguanine-DNA methyltransferase; AML, acute myeloid leukemia; FFPE, formalin-fixed, paraffin-embedded *tissue* sections; RCLPE, RCL2-fixed and paraffin-embedded tissue.

The MGMT potential as biomarker has been shown in several studies. For instance, Tuononen et al. [38] characterized the methylation pattern of *MGMT* in 51 gliomas by pyrosequencing. MGMT hypermethylation was observed in 100% of oligoastrocytomas, 93% of oligodendrogliomas, and 47% of glioblastomas while the deletions on 9p and 10q, and gain of 7p were associated with the unmethylated MGMT phenotype, whereas deletion of 19q and oligodendroglial morphology was associated with *MGMT* hypermethylation. The authors concluded that *MGMT* promoter methylation, analyzed by pyrosequencing, is a frequent event in oligodendroglial tumors, and it correlates with *IDH1* mutation and 19q loss in gliomas [38]. Furthermore, Christians et al. [17] aimed to determine a diagnostic method that provides the most accurate prediction of progression-free survival and they found out that pyrosequencing provides a significant improvement in predicting progression-free survival compared with established clinical prognostic factors alone. Furthermore, the authors recommend pyrosequencing for analyses of *MGMT* promoter methylation in high-throughput settings [17].

Another application of *MGMT* gene is for prediction of the outcome of cancer therapy and therefore an important biomarker when chemotherapy resistance takes place [40]. It has been shown that hypermethylation of the *MGMT* gene alters DNA repair and is associated with longer survival of glioblastoma patients treated with alkylating agents. Therefore, *MGMT* promoter methylation plays an important role as a predictive biomarker for chemotherapy resistance. In this context, Mikeska et al. [40] established an optimized pyrosequencing assay that provided a sensitive, robust, and easy-to-use method for quantitative assessment of *MGMT* methylation (in snap-frozen and paraffin-embedded specimens) [40].

Since promoter methylation of *MGMT* is associated with improved survival in glioblastoma-treated patients with alkylating agents, Dunn et al. [41] investigated MGMT promoter methylation. Using pyrosequencing of glioblastoma individual CpG sites, the authors found that pyrosequencing data were reproducible while a variation in methylation patterns of discrete CpG sites and intratumoral methylation heterogeneity were observed. The authors concluded that *MGMT* methylation is prognostically significant in glioblastomas given chemoradiotherapy in the routine clinic; furthermore, the extent of methylation may be used to provide additional prognostic stratification [41].

Into a newer study, Sooman et al. [42] analyzed the prognostic value of the expression of the protein tyrosine phosphatase nonreceptor type 6 (*PTPN6*, also referred to as SHP1) in high-grade glioma patients, where the DNA methylation was analyzed by pyrosequencing. The PTPN6 expression correlated to poor survival for anaplastic glioma patients while in glioma-derived cell lines, overexpression of PTPN6 caused increase resistance to the chemotherapeutic drugs bortezomib, cisplatin, and melphalan. PTPN6 expression may be a factor contributing to poor survival for anaplastic glioma patients, and in glioma-derived cells, its expression is epigenetically regulated and influences the response to chemotherapy [42].

2.1.4.2 PROSTATE CANCER (PCA, MIM #176808)

In prostate cancer, hypermethylation of CpG islands is a common epigenetic alteration [43], thus, epigenetic mechanisms together with alterations in the gene sequence have an impact on the carcinogenesis of prostate cancer (see Table 2). To date, only a few studies outlining methylation in prostate cancer and pyrosequencing have been published, but, nevertheless, better understanding of these factors might give new diagnostic or prognostic approaches [34]. In this regard, Yoon et al. [12] checked with quantitative pyrosequencing whether the detection of adenomatous polyposis coli hypermethylation will help in discrimination between normal and prostate cancer cells and for predicting tumor behaviors. The authors found that the target investigated was significantly higher methylated in prostate cancer specimens and is not only associated with prostate cancer and but it can predict its aggressive tumor features [12].

Other genes might represent possible biomarkers for prostate cancer. For instance, tumor suppressor genes *RARB* and *PDLIM4* are hypermethylated and silenced in prostate cancer tissues and prostate cancer cell lines compared to normal prostate cells [43]. To test the applicability of the two genes, He et al. [43] used in their study a benign prostate epithelial cell line RWPE1 to study the epigenetic regulation of Myc on the retinoic acid receptor beta and *PDLIM4* promoters. The authors found that forced Myc overexpression inhibited the retinoic acid receptor beta and *PDLIM4* expression. Furthermore, the pyrosequencing showed that Myc overexpression increased methylation in several CpG sites of both promoters while 5-aza-2′-deoxycytidine reversed the epigenetic alteration effect of Myc [43].

2.1.4.3 LIVER CANCER (HEPATOCELLULAR CARCINOMA, MIM #114550)

In the study of Moribe et al. [44], it was shown that DNA pyrosequencing can be used to find biomarkers with use in diagnostic, more precisely, in the diagnosis of early hepatocellular carcinoma (HCC) (see Table 2). As material, the authors used in their study, diverse liver tissues with characteristics of carcinoma and non-HCC. As method, the authors proceeded with DNA pyrosequencing of seven genes: *RASSF1A*, *CCND2*, *SPINT2*, *RUNX3*, *GSTP1*, *APC*, and *CFTR*. It was found that these genes were aberrantly methylated in stages I and II HCC: the pyrosequencing analysis confirmed that the seven genes were aberrantly and strongly methylated in early HCC, but not in any of the corresponding nontumor liver tissues [44].

2.1.4.4 COLORECTAL CANCER (CRC, MIM #114500)

In concordance with the facts described in previous sections, it is now accepted that aberrant DNA methylation plays an important role in genesis of colorectal cancer (CRC) as well (see Table 2) [45]. In a recent study by Leclerc et al. [46], it has been studied whether in human preneoplastic tissue there are changes in DNA methylation status of five orthologous genes *PDK4*, *SPRR1A*, *SPRR2A*, *NR1H4*, and *PYCARD*. These genes showed in previous analysis the clear changes in murine gene expression profiles. With the help of bisulfite pyrosequencing the authors identified 14 CpGs that show significant methylation differences in normal mucosa between CRC patients and controls. From all genes analyzed, the *PYCARD* and *NR1H4* genes seemed to be the most promising markers for the presence of polyps in controls [46].

2.1.4.5 PANCREATIC CANCER (MIM #260350)

Malpeli et al. [47] studied the methylation status of the *RASSF1* CpG islands using pyrosequencing since *RASSF1A* gene silencing by DNA methylation has been suggested as a major event in pancreatic endocrine tumor (see Table 2). The authors quantified the methylation of 51 CpGs that showed variable distribution and levels of methylation within and among samples with pancreatic endocrine tumor, while the average methylation is higher than normal samples. Furthermore, *RASSF1A* methylation was found to be inversely correlated with its expression. Nevertheless, *RASSF1A* gene methylation in pancreatic endocrine tumor is higher than normal pancreas in no more than 75% of cases thus, it cannot be considered a marker for this neoplasm. *RASSF1A* is always expressed in pancreatic endocrine tumor and normal pancreas and its levels are inversely correlated with gene methylation [47].

2.1.4.6 BLADDER CANCER (MIM #109800)

In a recent study of Rose et al. [16], the role of *ITIH5* methylation in bladder cancers has been analyzed (see Table 2). Here, the analysis on *ITIH5* promoter hypermethylation was performed using pyrosequencing. The authors found that the hypermethylation of the *ITIH5* promoter was closely associated with progressive bladder cancers while the aberrant DNA hypermethylation could be the reason of invasive phenotypes in human bladder cancer. Therefore, *ITIH5* has the potential to become a prognostic biomarker for relapse risk stratification in high-grade urothelial cancer patients. This hypothesis was sustained by the finding that reexpression of *ITIH5* led to both suppression of cell migration and inhibition of colony spreading of human RT112 bladder cancer cells [16].

2.1.4.7 BILE DUCT CANCER

Bile duct cancer has very poor prognosis thus, finding diagnostic and prognostic markers will be an advantage (see Table 2). Park et al. [48] evaluated the hypermethylation status of genes for the power to predict overall survival following curative resection of mid/distal bile duct cancer. Pyrosequencing hypermethylation status showed significant methylation frequencies (methylation >5%) were obtained for five genes: *P16*, *DAPK*, E-cadherin, *RASSF1*, and *MLH1*. Methylation status of *P16*, *DAPK*, and *RASSF1* was correlated with perineural invasion, tumor depth, and age while for the overall survival, the presence of lymph node metastasis and *P16* methylation status were identified as independent prognostic factors. Thus, the authors concluded that classification of mid/distal bile duct cancer by both genetic and epigenetic profiles may improve the accuracy in predicting outcome and the effectiveness of tailored therapy in these diseases [48].

2.1.4.8 LEUKEMIA (CLL, MIM #151400; AML, MIM #601626)

Queirós et al. [49] investigated in chronic lymphocytic leukemia (CLL) new biomarkers for investigation of the progression that could be used in clinical management. Thus, the authors developed a clinically applicable method to identify the molecular subgroups of CLL and to study their clinical relevance using the DNA methylation by bisulfite pyrosequencing assays [49]. Furthermore, Jost et al. [50] found that aberrant DNA hypermethylation within the *DNMT3A* gene, in analogy to *DNMT3A* mutations, is frequently observed in acute myeloid leukemia and both modifications seem to be useful for risk stratification or choice of therapeutic regimen (see Table 2) [50].

2.1.4.9 LUNG CANCER (MIM #211980)

Bradly et al. [51] assessed in nonsmall cell lung cancer the promoter methylation of selected tumor suppressor genes on formalin-fixed tumor tissue. Methylation was quantified in *p16*, *MGMT*, *DAPK*, *RASSF1*, *CDH1*, *LET7-3-a*, *NORE1(RASSF5)*, and *PTEN* promoters by pyrosequencing (see Table 2). The authors found that promoter methylation was higher in patients older than 60 years of age; the p16 promoter showed age-related differences while the p16 promoter hypermethylation was associated with a worse outcome in patients with age at diagnosis of 60 years or younger [51]. Furthermore, Bruno et al. [13] compared the methylation profile of *WT1* gene in samples of neoplastic and nonneoplastic lung tissue taken from the same patients with nonsmall cell lung cancer. The methylation status of 29 CpG islands in the 5′ region of *WT1* was determined by pyrosequencing and it was found higher promoter methylation islands of *WT1* in the neoplastic tissues of the nonsmall cell lung cancer patients as compared with the normal lung tissue from the same patients [13].

3. PYROSEQUENCING AND THE ASSESSMENT TO miRNAs

Previous studies have demonstrated that miRNA expression is altered in human cancer. However, the molecular mechanisms underlying these changes in miRNA expression remain still not fully unclear [52]. miRNAs belong to the heterogeneous class of noncoding RNAs (ncRNAs) that regulate the translation and degradation of target mRNAs, and control approximately 30% of human genes. miRNA genes might be silenced in human tumors (oncomiRs/miRNAs that are associated with cancer) due to aberrant hypermethylation of CpG islands that are adjacent to miRNA genes and/or by histone modifications [53].

Epigenetic inactivation by aberrant DNA methylation has been reported for many miRNA genes in various human malignancies [54]. In this regard, Kunej et al. [53] performed literature search for research articles describing epigenetically regulated miRNAs in cancer. The study revealed 122 miRNAs that were reported to be epigenetically regulated in 23 cancer types. Compared to protein-coding genes, human oncomiRs showed higher methylation frequency while about 45% epigenetically regulated miRNAs were associated with different cancer types. Interestingly, 55% of the miRNAs were present in only one type of cancer thus having cancer-specific biomarker potential [53]. Thus, analysis of miR genes using pyrosequencing might represent an important tool for diagnosis and prognosis of cancers but also of the biological processes that might be disturbed during carcinogenesis.

Scientific evidence is demonstrating the potential use of miRNA gene methylation in clinical applications. For instance, Wang et al. [52] investigated the epigenetic modification of miR-124 genes and the potential function of miR-124 in pancreatic cancer. Using pyrosequencing analysis, the authors found that miR-124 genes (including miR-124-1, miR-124-2, and miR-124-3) are highly methylated in pancreatic cancer tissues compared within noncancerous tissues. Hypermethylation mediated the silencing of miR-124, which was a frequent event in pancreatic duct adenocarcinoma while the miR-124 downregulation was significantly associated with worse survival of pancreatic duct adenocarcinoma patients. The authors also show that Rac1 as a direct target of miR-124. Thus, miR-124 is a tumor suppressor miRNA that is epigenetically silenced in pancreatic cancer [52].

Also, in bladder cancer cell lines (T24 and UM-UC-3) treated with 5-aza-2′-deoxycytidine (5-aza-dC) and 4-phenylbutyric acid (PBA), Shimizu et al. [55] found that the miRNA expression profiles have been changed as compared with untreated controls. The bisulfite pyrosequencing was used to assess miRNA gene methylation in cancer cell lines, primary tumors, and preoperative and postoperative urine samples. The authors found that 146 miRNAs were upregulated by 5-aza-dC plus PBA. Interestingly, CpG islands were identified in the proximal upstream of 23 miRNA genes, while 12 of those were hypermethylated in cell lines such as miR-137, miR-124-2, miR-124-3, and miR-9-3 were frequently and tumor-specifically methylated in primary cancers. These miRNAs enabled in urine specimens the bladder cancer detection with 81% sensitivity and 89% specificity; thus, the methylation of miRNA genes could be a useful biomarker for cancer detection [55].

Not at last, Anwar et al. [54] studied the miRNA gene methylation in HCC. The authors checked 39 intergenic CpG island-associated miRNA genes and found aberrant hypermethylation and downregulation of miRNA genes as a frequent event in human HCC: hsa-miR-9-2, hsa-miR-9-3, hsa-miR-124-1, hsa-miR-124-2, hsa-miR-124-3, hsa-miR-129-2, hsa-miR-596, and hsa-miR-1247 impacting about 90% of the HCC specimens. Furthermore, it was found that concomitant hypermethylation of three or more miRNA genes is a highly specific marker for the detection of HCC and for poor prognosis [54].

4. CONCLUSIONS

To conclude, pyrosequencing is a nucleic acid sequencing method that can be used in clinical and research applications. In particular, the pyrosequencing-based techniques for DNA methylation analysis will be adopted in clinical practice for diagnostic or prognostic. Since the advantages of this method are flexibility, parallel processing, and possible automatization [3,56], these procedures are suitable to be used in clinical laboratories. Pyrosequencing technique can be easily adapted for clinical routine when specific biomarkers are analyzed, e.g., MGMT, LINE1. In these cases, kits are available on the market for purchase. There are, however, some obstacles that hinder the use of pyrosequencing for DNA methylation analysis in clinical diagnostics such as (1) an international agreement/consensus regarding the quantitative methylation data (e.g., what level of hipo/hyper methylation is significant and in which region of the gene, etc.); (2) cheap equipment and chemicals; (3) fast standard operation procedure (SOP); (4) easy to establish in the diagnostic routine due to available SOPs; and (5) possibility of using the patient material that does not have a high quality. Thus, the role of pyrosequencing-based techniques for DNA methylation analysis in biomedical research, in basic research investigation, and in clinical diagnostic is increasing over time.

Furthermore, the pyrosequencing might be used to analyze the effects of environmental influences on stress response-related processes. Since, exposure of cells and organisms to stressors might result in epigenetic changes, investigation of DNA methylation using pyrosequencing might represent in the future, an important tool to evaluate in vitro and in vivo toxicity of epigenetic regulating chemicals and drugs. Overall, the use of pyrosequencing in the biomedical research is just at the beginning; more effort must be invested in order to find robust biomarkers that can be used in clinical applications but also to understand the role of epigenetics in the regulation of human diseases in public health issues.

Glossary

CpG islands CG dinucleotides reach cluster regions, they might play a role in transcriptional silencing of genes.

FSH levels The follicle-stimulating hormone (FSH) blood test measures the level of FSH in blood.

miRNA/microRNA Noncoding RNAs that regulate the translation and degradation of target mRNAs.

Pyrosequencing Method used for quantitative analysis of nucleic acids, where the sequencing takes place during the DNA synthesis, using the sequencing-by-synthesis principle.

LIST OF ABBREVIATIONS

5-aza-dC 5-Aza-2′-deoxycytidine
5-HTTLPR Serotonin transporter gene-linked polymorphic region
ABCA1 ATP-binding cassette transporter A1
ADRA1D Alpha-1D adrenergic receptor (α1D adrenoreceptor),
AHRR Aryl-hydrocarbon receptor repressor
AJAP1 Adherens junctions-associated protein 1
APC Adenomatous polyposis coli
ATP Adenosine triphosphate
CCND2 G1/S-specific cyclin-D2
CDH1 Cadherin-1
CFTR Cystic fibrosis transmembrane conductance regulator
CIN3+ Cervical intraepithelial neoplasia
COL6A2 Collagen alpha-2(VI) chain
DAPK Death-associated protein kinase 1
DCR2 (TNFRSF10D) Decoy receptor 2 (tumor necrosis factor receptor superfamily, member 10d, decoy with truncated death domain)
DDT Dichlorodiphenyl trichloroethane
DDE Dichlorodiphenyldichloroethylene
DNA Deoxyribonucleic acid
DNMT DNA methyltransferase family, controls de novo and maintains DNA methylation
EDN3 Endothelin 3
EPO Erythropoietin
FFAR Free fatty acid receptor
GSTP1 Glutathione S-transferase P
HS3ST2 Heparan sulfate glucosamine 3-O-sulfotransferase 2
HIC1 Hypermethylated in cancer 1
ICR Imprinting control regions
IGF1 Insulin like growth factor 1
IGFBP3 Insulin like growth factor binding protein 3
IGF1R Insulin like growth factor 1
INSR Insulin receptor
ITIH5 Inter-alpha-trypsin inhibitor heavy chain family, member 5
LINE Long interspersed elements
MAGI2 Membrane-associated guanylate kinase, WW and PDZ domain containing 2
MGMT O-6-Methylguanine-DNA methyltransferase
miRNA MicroRNAs

MLH1 MutL homolog 1, colon cancer, nonpolyposis type 2
Myc v-myc avian myelocytomatosis viral oncogene homolog
ncRNAs Noncoding RNAs
NR1H4 Nuclear receptor subfamily 1, group H, member 4
p16 Cyclin-dependent kinase inhibitor 2A (CDKN2A)
PBA 4-Phenylbutyric acid
PBDE Polybrominated diphenyl ethers
PDLIM4 PDZ and LIM domain 4
PDK4 Pyruvate dehydrogenase lipoamide kinase isozyme 4
POU4F3 POU class 4 homeobox 3
PRSS21 Protease, serine, 21 (testisin)
PTGDR Prostaglandin D2 receptor (DP)
PTPN6/SHP1 Protein tyrosine phosphatase nonreceptor type 6
PTEN Phosphatase and tensin homolog
PYCARD PYD and CARD domain containing
RAC1 Ras-related C3 botulinum toxin substrate 1 (rho family, small GTP-binding protein Rac1)
RARB Retinoic acid receptor beta
RASSF1A Ras association domain-containing protein 1a
RNA Ribonucleic acid
RUNX3 Runt-related transcription factor 3
SCGB3A1 Secretoglobin, family 3A, member 1
SINE Short interspersed elements
SOP Standard operation procedure
SOX SRY (sex-determining region Y)-box
SPINT2 Serine peptidase inhibitor, Kunitz type, 2
SPRR Small proline-rich protein
ST6GAL2 ST6 beta-galactosamide alpha-2,6-sialyltranferase 2
SYT9 Synaptotagmin IX
TLR Toll-like receptor
WT1 Wilms tumor protein
ZNF614 Zinc finger protein 614

References

[1] Melamede RJ. Automatable process for sequencing nucleotide. 1985. US Patent no. US4863849.

[2] Diggle MA, Clarke SC. Genotypic characterization of *Neisseria meningitidis* using pyrosequencing. Mol Biotechnol 2004;28:139–45.

[3] Ronaghi M. Pyrosequencing sheds light on DNA sequencing. Genome Res 2001;11:3–11.

[4] Deyde VM, Gubareva LV. Influenza genome analysis using pyrosequencing method: current applications for a moving target. Expert Rev Mol Diagn 2009;9: 493–509.

[5] Florea A-M. DNA methylation pyrosequencing assay is applicable for the assessment of epigenetic active environmental or clinical relevant chemicals. Biomed Res Int 2013;2013:486072.

[6] Ahmadian A, Ehn M, Hober S. Pyrosequencing: history, biochemistry and future. Clin Chim Acta 2006;363:83–94.

[7] Sandoval J, Mendez-Gonzalez J, Nadal E, Chen G, Carmona FJ, Sayols S, et al. A prognostic DNA methylation signature for stage I non-small-cell lung cancer. J Clin Oncol 2013;31:4140–7.

[8] María Martín-Núñez G, Rubio-Martín E, Cabrera-Mulero R, Rojo-Martínez G, Mikeska T, Bock C, et al. Optimization of quantitative MGMT promoter methylation analysis using pyrosequencing and combined bisulfite restriction analysis. J Mol Diagn 2007;9:368–81.

[9] Callinan PA, Batzer MA. Retrotransposable elements and human disease. Genome Dyn 2006;1:104–15.

[10] Huen K, Yousefi P, Bradman A, Yan L, Harley KG, Kogut K, et al. Effects of age, sex, and persistent organic pollutants on DNA methylation in children. Environ Mol Mutagen 2014;55:209–22.

[11] Lind GE, Skotheim RI, Lothe RA. The epigenome of testicular germ cell tumors. APMIS 2007;115:1147–60.

[12] Yoon HY, Kim YW, Kang HW, Kim WT, Yun SJ, Lee SC, et al. Pyrosequencing analysis of APC methylation level in human prostate tissues: a molecular marker for prostate cancer. Korean J Urol March 2013;54(3):194–8. http://dx.doi.org/10.4111/kju.2013.54.3.194.

[13] Bruno P, Gentile G, Mancini R, De Vitis C, Esposito MC, Scozzi D, et al. WT1 CpG islands methylation in human lung cancer: a pilot study. Biochem Biophys Res Commun 2012;426:306–9.

[14] Cebula TA, Brown EW, Jackson SA, Mammel MK, Mukherjee A, LeClerc JE. Molecular applications for identifying microbial pathogens in the post-9/11 era. Expert Rev Mol Diagn 2005;5:431–45.

[15] Tenover FC. Rapid detection and identification of bacterial pathogens using novel molecular technologies: infection control and beyond. Clin Infect Dis 2007;44:418–23.

[16] Rose M, Gaisa NT, Antony P, Fiedler D, Heidenreich A, Otto W, et al. Epigenetic inactivation of ITIH5 promotes bladder cancer progression and predicts early relapse of pT1 high-grade urothelial tumours. Carcinogenesis 2014;35:727–36.

[17] Christians A, Hartmann C, Benner A, Meyer J, von Deimling A, Weller M, et al. Prognostic value of three different methods of MGMT promoter methylation analysis in a prospective trial on newly diagnosed glioblastoma. PLoS One 2012;7:e33449.

[18] Okamoto Y, Shinjo K, Shimizu Y, Sano T, Yamao K, Gao W, et al. Hepatitis virus infection affects DNA methylation in mice with humanized livers. Gastroenterology 2014;146:562–72.

[19] Wu YY, Csako G. Rapid and/or high-throughput genotyping for human red blood cell, platelet and leukocyte antigens, and forensic applications. Clin Chim Acta 2006;363:165–76.

[20] Vaissière T, Cuenin C, Paliwal A, Vineis P, Hoek G, Krzyzanowski M, et al. Quantitative analysis of DNA methylation after whole bisulfitome amplification of a minute amount of DNA from body fluids. Epigenetics 2009;4:221–30.
[21] Remely M, Aumueller E, Jahn D, Hippe B, Brath H, Haslberger AG. Microbiota and epigenetic regulation of inflammatory mediators in type 2 diabetes and obesity. Benef Microbes 2014;5:33–43.
[22] Remely M, Aumueller E, Merold C, Dworzak S, Hippe B, Zanner J, et al. Effects of short chain fatty acid producing bacteria on epigenetic regulation of FFAR3 in type 2 diabetes and obesity. Gene March 1, 2014;537(1):85–92. http://dx.doi.org/10.1016/j.gene.2013.11.081.
[23] Kläver R, Tüttelmann F, Bleiziffer A, Haaf T, Kliesch S, Gromoll J. DNA methylation in spermatozoa as a prospective marker in andrology. Andrology 2013;1: 731–40.
[24] Zhao Y, Shi HJ, Xie CM, Chen J, Laue H, Zhang YH. Prenatal phthalate exposure, infant growth, and global DNA methylation of human placenta. Environ Mol Mutagen 2014. http://dx.doi.org/10.1002/em.21916.
[25] Calvello M, Tabano S, Colapietro P, Maitz S, Pansa A, Augello C, et al. Quantitative DNA methylation analysis improves epigenotype-phenotype correlations in Beckwith-Wiedemann syndrome. Epigenetics 2013;8:1053–60.
[26] Lee BH, Kim GH, Oh TJ, Kim JH, Lee JJ, Choi SH, et al. Quantitative analysis of methylation status at 11p15 and 7q21 for the genetic diagnosis of Beckwith-Wiedemann syndrome and Silver-Russell syndrome. J Hum Genet 2013;58:604–10.
[27] Houde AA, Guay SP, Desgagné V, Hivert MF, Baillargeon JP, St-Pierre J, et al. Adaptations of placental and cord blood ABCA1 DNA methylation profile to maternal metabolic status. Epigenetics 2013;8:1289–302.
[28] Desgagné V, Hivert MF, St-Pierre J, Guay SP, Baillargeon JP, Perron P, et al. Epigenetic dysregulation of the IGF system in placenta of newborns exposed to maternal impaired glucose tolerance. Epigenomics 2014;6:193–207.
[29] Goodrich JM, Basu N, Franzblau A, Dolinoy DC. Mercury biomarkers and DNA methylation among Michigan dental professionals. Environ Mol Mutagen 2013;54:195–203.
[30] Shenker NS, Ueland PM, Polidoro S, van Veldhoven K, Ricceri F, Brown R, et al. DNA methylation as a long-term biomarker of exposure to tobacco smoke. Epidemiology 2013;24:712–6.
[31] Tajuddin SM, Amaral AF, Fernández AF, Rodríguez-Rodero S, Rodríguez RM, Moore LE, et al. Genetic and non-genetic predictors of LINE-1 methylation in leukocyte DNA. Environ Health Perspect 2013;121:650–6.
[32] Hossain MB, Vahter M, Concha G, Broberg K. Low-level environmental cadmium exposure is associated with DNA hypomethylation in Argentinean women. Environ Health Perspect 2012;120:879–84.
[33] Alexander N, Wankerl M, Hennig J, Miller R, Zänkert S, Steudte-Schmiedgen S, et al. DNA methylation profiles within the serotonin transporter gene moderate the association of 5-HTTLPR and cortisol stress reactivity. Transl Psychiatry 2014;16(4):e443.
[34] Stadler TC, Jung A, Schlenker B, Nuhn P, Ellinger J, Kirchner T, et al. Pyrosequencing in uro-oncology: applications in prostate cancer. Urol A 2010;49:1356–64.
[35] Shames DS, Minna JD, Gazdar AF. Methods for detecting DNA methylation in tumors: from bench to bedside. Cancer Lett 2007;251:187–98.
[36] Kiss NB, Muth A, Andreasson A, Juhlin CC, Geli J, Bäckdahl M, et al. Acquired hypermethylation of the P16INK4A promoter in abdominal paraganglioma: relation to adverse tumor phenotype and predisposing mutation. Endocr Relat Cancer 2013;20:65–78.
[37] Chen YC, Huang RL, Huang YK, Liao YP, Su PH, Wang HC, et al. Methylomics analysis identifies epigenetically silenced genes and implies an activation of β-catenin signaling in cervical cancer. Int J Cancer 2014;135:117–27.
[38] Tuononen K, Tynninen O, Sarhadi VK, Tyybäkinoja A, Lindlöf M, Antikainen M, et al. The hypermethylation of the O6-methylguanine-DNA methyltransferase gene promoter in gliomas–correlation with array comparative genome hybridization results and IDH1 mutation. Genes Chromosom Cancer 2012;51:20–9.
[39] Preusser M, Berghoff AS, Manzl C, Filipits M, Weinhäusel A, Pulverer W, et al. Clinical Neuropathology practice news 1-2014: pyrosequencing meets clinical and analytical performance criteria for routine testing of MGMT promoter methylation status in glioblastoma. Clin Neuropathol January–February 2014;33(1):6–14.
[40] Mikeska T, Bock C, El-Maarri O, Hübner A, Ehrentraut D, Schramm J, et al. Optimization of quantitative MGMT promoter methylation analysis using pyrosequencing and combined bisulfite restriction analysis. J Mol Diagn 2007;9:368–81.
[41] Dunn J, Baborie A, Alam F, Joyce K, Moxham M, Sibson R, et al. Extent of MGMT promoter methylation correlates with outcome in glioblastomas given temozolomide and radiotherapy. Br J Cancer 2009;101:124–31.
[42] Sooman L, Ekman S, Tsakonas G, Jaiswal A, Navani S, Edqvist PH, et al. PTPN6 expression is epigenetically regulated and influences survival and response to chemotherapy in high-grade gliomas. Tumour Biol 2014;35:4479–88.
[43] He M, Vanaja DK, Karnes RJ, Young CY. Epigenetic regulation of Myc on retinoic acid receptor beta and PDLIM4 in RWPE1 cells. Prostate November 1, 2009;69(15): 1643–50. http://dx.doi.org/10.1002/pros.21013.

[44] Moribe T, Iizuka N, Miura T, Kimura N, Tamatsukuri S, Ishitsuka H, et al. Methylation of multiple genes as molecular markers for diagnosis of a small, well-differentiated hepatocellular carcinoma. Int J Cancer 2009;125:388–97.

[45] Sakai E, Ohata K, Chiba H, Matsuhashi N, Doi N, Fukushima J, et al. Methylation epigenotypes and genetic features in colorectal laterally spreading tumors. Int J Cancer 2014:1586–95.

[46] Leclerc D, Lévesque N, Cao Y, Deng L, Wu Q, Powell J, et al. Genes with aberrant expression in murine preneoplastic intestine show epigenetic and expression changes in normal mucosa of colon cancer patients. Cancer Prev Res (Phila) 2013;6:1171–81.

[47] Malpeli G, Amato E, Dandrea M, Fumagalli C, Debattisti V, Boninsegna L, et al. Methylation-associated down-regulation of RASSF1A and up-regulation of RASSF1C in pancreatic endocrine tumors. BMC Cancer 2011;11:351.

[48] Park JS, Park YN, Lee KY, Kim JK, Yoon DS. P16 hypermethylation predicts surgical outcome following curative resection of mid/distal bile duct cancer. Ann Surg Oncol August 2013;20(8):2511–7. http://dx.doi.org/10.1245/s10434-013-2908-7.

[49] Queirós AC, Villamor N, Clot G, Martinez-Trillos A, Kulis M, Navarro A, et al. A B-cell epigenetic signature defines three biologic subgroups of chronic lymphocytic leukemia with clinical impact. Leukemia 2014. http://dx.doi.org/10.1038/leu.2014.252.

[50] Jost E, Lin Q, Weidner CI, Wilop S, Hoffmann M, Walenda T, et al. Epimutations mimic genomic mutations of DNMT3A in acute myeloid leukemia. Leukemia 2014;28:1227–34.

[51] Bradly DP, Gattuso P, Pool M, Basu S, Liptay M, Bonomi P, et al. CDKN2A (p16) promoter hypermethylation influences the outcome in young lung cancer patients. Diagn Mol Pathol 2012;21:207–13.

[52] Wang P, Chen L, Zhang J, Chen H, Fan J, Wang K, et al. Methylation-mediated silencing of the miR-124 genes facilitates pancreatic cancer progression and metastasis by targeting Rac1. Oncogene 2014;33:514–24.

[53] Kunej T, Godnic I, Ferdin J, Horvat S, Dovc P, Calin GA. Epigenetic regulation of microRNAs in cancer: an integrated review of literature. Mutat Res 2011;717:77–84.

[54] Anwar SL, Albat C, Krech T, Hasemeier B, Schipper E, Schweitzer N, et al. Concordant hypermethylation of intergenic microRNA genes in human hepatocellular carcinoma as new diagnostic and prognostic marker. Int J Cancer 2013;133:660–70.

[55] Shimizu T, Suzuki H, Nojima M, Kitamura H, Yamamoto E, Maruyama R, et al. Methylation of a panel of microRNA genes is a novel biomarker for detection of bladder cancer. Eur Urol 2013;6:1091–100.

[56] Bushman FD, Hoffmann C, Ronen K, Malani N, Minkah N, Rose HM, et al. Massively parallel pyrosequencing in HIV research. AIDS 2008;22:1411–5.

CHAPTER

10

Mass Spectrometry for the Identification of Posttranslational Modifications in Histones and Its Application in Clinical Epigenetics

Roberta Noberini[1], *Alessandro Cuomo*[2], *Tiziana Bonaldi*[2]

[1]Center of Genomic Science, Istituto Italiano di Tecnologia, Milano, Italy; [2]Department of Experimental Oncology, European Institute of Oncology, Milano, Italy

OUTLINE

Epigenetic Biomarkers and Diagnostics
http://dx.doi.org/10.1016/B978-0-12-801899-6.00010-3

1. INTRODUCTION

In eukaryotes, DNA compaction and gene regulation are mediated by chromatin, a macromolecular complex whose basic unit is the nucleosome. Histones represent the protein component of the nucleosome and are sites for a variety of dynamic and reversible posttranslational modifications (PTMs), which together with DNA methylation and nucleosome positioning constitute the epigenetic signatures that regulate gene expression. According to the histone code hypothesis, the type, location, and combination of histone PTMs (hPTMs) generate a language that determines the functional state of the underlying genes by directly influencing chromatin configuration and accessibility and by generating binding platforms for the recruitment of proteins and enzyme complexes that mediate downstream events [1,2].

Given their involvement in the regulation of critical cellular processes, it is not surprising that alterations in histone modification patterns have been linked with various pathologies, including cancer. Tumor cells show a global decrease of acetylated histones H3 and H4, in particular monoacetylated H4K16 [3], a decrease of H3K4me3 [4] and H4K20me3 [5], and an increase of H3K9me [6] and H3K27me3 [7]. In prostate cancer, different acetylation and methylation patterns in histones H3 and H4 correlate with disease progression and clinical outcome [8] and the list of malignancies associated to aberrant hPTM patterns is continuously expanding. Alteration of acetylation and methylation marks is mainly caused by alterations of histone deacetylases activity and the aberrant expression of histone methyltransferases and demethylases [9] that are found in different tumor types. Targeting these enzymes for therapeutical use has gained much attention and various small molecule inhibitors of histone deacetylases are currently being evaluated in clinical trials for the treatment of different forms of cancer [10]. A role for histone modifications in noncancer diseases has also been described [11]. For instance, histone hypoacetylation is often observed in neurological disorders, including ALS, Parkinson's and Huntington's diseases, and Friedreich's ataxia [12]. Furthermore, the increase of H3K9me2 in genes associated with autoimmunity and inflammation in lymphocytes from patients with type 1 diabetes suggests a connection between altered histone modifications and autoimmune diseases [13]. However, despite the accumulating evidence linking aberrant histone modification patterns and human pathologies, the underlying epigenetic mechanisms have not yet been fully elucidated. It is therefore crucial to accurately detect and quantify hPTMs associated with diseased states to shed light on these mechanisms and to exploit them for prognostic, diagnostic, and therapeutic purposes.

Mass spectrometry (MS) has emerged as a powerful tool for the identification and quantification of hPTMs and has gained an increasingly important role in epigenetic research. MS provides several advantages over the traditional antibody-based methods used for hPTMs detection, such as immunofluorescence and immunoblotting. First, MS analysis is unbiased, since it does not require a priori knowledge of the modification site, as antibodies do. Thanks to this ability, in recent years MS has significantly expanded the list of existing hPTMs, revealing various previously unknown histone modifications, such as serine O-GlcNAcylation [14] and lysine propionylation [15,16], butyrylation [15,16], formylation [17], succinylation [18], malonylation [18], and crotonylation [19]. Second, while antibodies allow the detection of a limited number of modifications, MS offers a comprehensive view of complete histone modification patterns, also providing some information regarding the combinatorial aspect of the histone code. Third, antibody-based assays are limited by possible cross-reactivity with histones or nonhistone proteins with similar modifications or sequences and by their poor efficiency when used to detect a modification when another is present nearby. This effect, known as epitope masking, is particularly relevant in heavily modified portions of the histone sequence.

In this chapter, we will first provide an overview of the basic principles of MS and its applications to the identification, quantification, and dissection of combinations of hPTMs, following an ideal workflow that covers hPTM analysis from sample preparation to quantification (Figure 1). We will then focus on the clinical applications of the abovementioned techniques, with particular emphasis on the exciting but still underexploited perspectives that they offer, providing practical examples and results from the experience we gained from the analysis of breast cancer clinical samples.

2. MS ANALYSIS OF hPTMs

2.1 Histone Isolation and Sample Preparation for MS Analysis

Typical protocols for histone isolation take advantage of their hydrophilic nature and solubility in strong acids such as HCl or H_2SO_4 [20]. Cell nuclei are usually purified, for instance, through incubation in a buffer containing low salt concentrations and a sucrose cushion [21], and then subjected to acidic extraction to isolate histone proteins, which are then dialyzed and lyophilized or precipitated using trichloroacetic acid (Figure 1, blue panel). Alternative methods that overcome possible problems related to the use of strong acids, such as loss of labile modifications and generation of insoluble precipitates, are based on high salt-based extraction [22] or hydroxyapatite chromatography [23]. Once isolated, histones can be further purified and separated by SDS-PAGE electrophoretic gels or liquid chromatography, such as reversed-phase high-performance liquid chromatography (RP-HPLC) or hydrophilic interaction liquid chromatography (HILIC) [24]. Histones are usually enzymatically digested into peptides prior to MS analysis (Figure 1, orange panel). While trypsin is considered the gold standard for protein digestion in global proteomics studies [25], it produces peptides that are too small to be easily detected by liquid chromatography–mass spectrometry (LC-MS) when digesting histones, which are rich in basic residues. On the contrary, endopeptidase Arg-C, which cleaves with good specificity at the C-terminal of arginine residues, produces histone peptides of optimal length for LC-MS analysis, thus representing a better alternative [21]. Arg-C-like digestions can be obtained by derivatizing unmodified and monomethylated lysines with propionic or deuterated acetic anhydride [26,27], or other novel and less commonly used derivatizing agents, such as propionic acid N-hydroxysuccinimide ester [28], prior to trypsin digestion, thus preventing cleavage at these sites, with the advantage of being applicable to in-gel digestion [29]. Digested peptides can be desalted and concentrated prior to MS analysis through a combination of reversed-phase C18/carbon and ion exchange chromatography (SCX), which maximize the retention of very short as well as particularly hydrophilic peptides [21]. Glu-C or Asp-N, which cleave at less frequently occurring residues within histone sequences, can also be used and is particularly useful to obtain longer peptides for middle-down MS analysis (see below) [30].

Typically, digested peptides are chromatographically separated prior to MS analysis (Figure 1, purple panel). This step is particularly important when analyzing hypermodified proteins such as histones, where the identification and quantification of isobaric peptides often depends on the ability to chromatographically separate them [31,32]. RP-HPLC, which employs a hydrophobic stationary phase and polar mobile phase, is normally employed to separate Arg-C-digested histone peptides, using gradients that have been adapted to the highly hydrophilic nature of histones [33]. In modern instruments, flow rates are reduced to nanoliter per minute and HPLC systems are automatized,

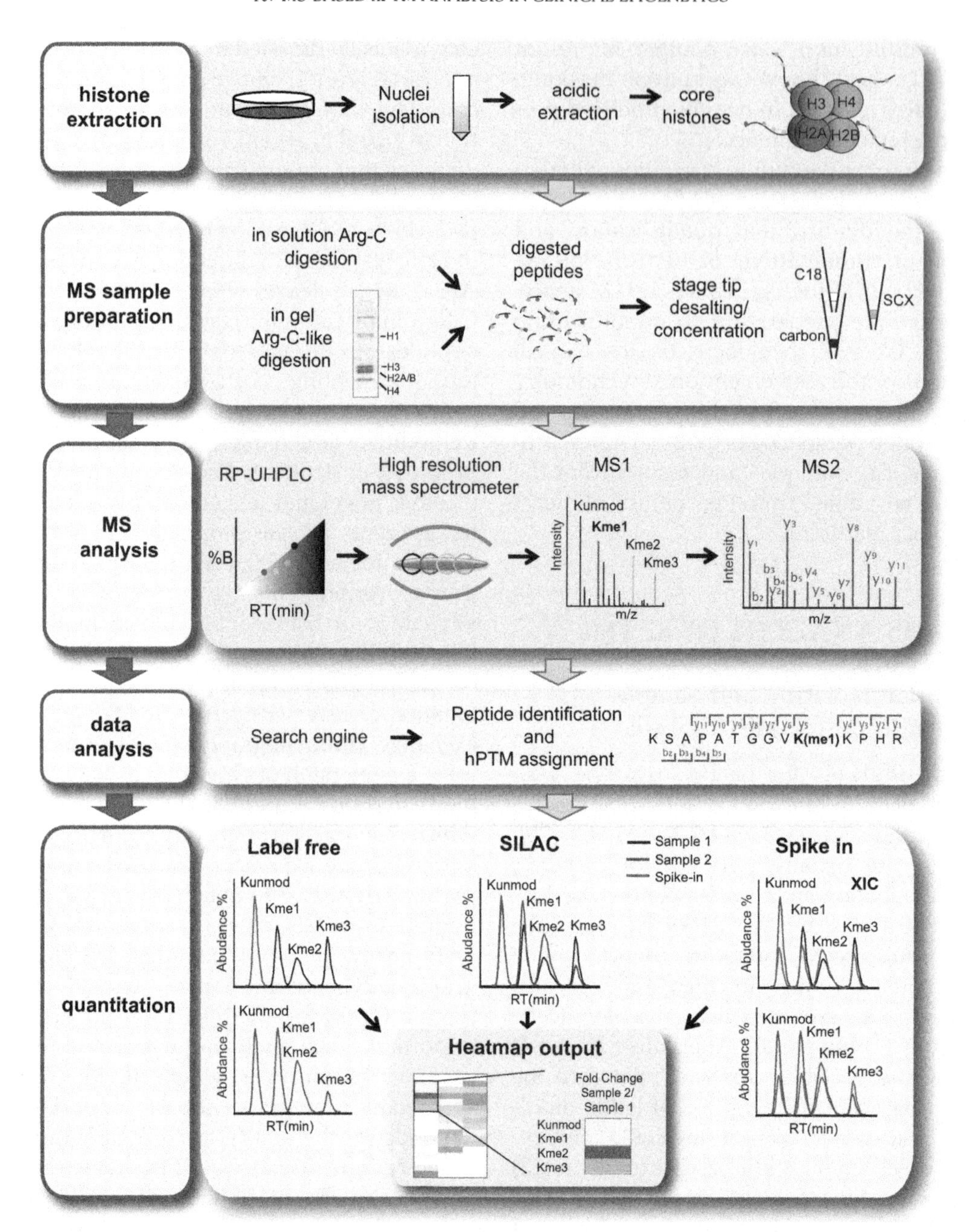

FIGURE 1 Representative workflow for histone posttranslational modification (hPTM) mass spectrometry (MS)-based analysis using a bottom-up approach. Histones are initially purified through nuclei isolation on a sucrose gradient and acidic extraction (blue panel) and then subjected to in-solution Arg-C digestion or in-gel Arg-C-like digestion, producing peptides that can be purified through a combination of C18/carbon and ion exchange chromatography on stage tips (orange panel). The peptides are further separated by high-performance liquid chromatography prior to MS and tandem mass spectra analyses (purple panel) and the MS data are analyzed using a search engine to determine protein identification and posttranslational modification assignment (green panel). Various methods (label-free, stable isotope labeling by amino acids in cell culture, and spike-in strategies are shown in the yellow panel) can be applied for hPTM quantitation, generating a heatmap that shows relative changes in modification abundances among samples.

which permits high-throughput sample analysis [34,35]. In addition, in ultra-HPLC systems (UHPLC), long columns packed with small particle size (<2 μm) greatly improve the loading capacity, leading to better separation of histone hypermodified peptides, as shown in our recent study [32]. Furthermore, UHPLC setups often allow to control the column temperature, which increases chromatography reproducibility [36]. Specific and reproducible retention times (RTs) can thus be associated to distinct modified histone peptides and serve as diagnostic parameters to increase the confidence of peptide identification and to transfer peptide identifications across multiple samples. Overall this boosts the identification of low-abundance modified peptides that are not consistently identified across multiple MS analyses [37,38]. Finally, the annotation of accurate and reproducible RTs can provide a priori knowledge to design targeted MS methods (see Ref. [39] and Section 2.4 in this chapter).

The longer, more hydrophilic, and multiply charged peptides that are generated by Glu-C or Asp-N endopeptidases typically require different separation methods, such as weak-cation exchange (WCX) in combination with HILIC. In this setup, modified histone peptides are separated orthogonally in two dimensions based on their charge state (acetylation degree) and hydrophilic content (the position and degree of methylation). This approach is therefore particularly useful to detect long distance co-occurrences and distinguish near-isobaric histone-modified peptides [40].

Given the low stoichiometry of many hPTMs, their enrichment is sometimes advisable to improve their detection. A common enrichment strategy that can be applied at both the protein and peptide levels involves the use of antibodies raised against specific modifications, such as tyrosine phosphorylation or lysine acetylation [41,42]. However, while the enrichment of above-mentioned modifications is relatively straightforward, also thanks to the availability of efficient reagents, the analysis of protein methylation is more challenging. This is due to the chemical properties of the methyl group and the complexity of methylation states, which involve different residues (arginine and lysine) and various degrees of modification (mono-, di-, and trimethylation). During the last few years, our group and others applied immunoaffinity purification of methylated residues at the protein or peptide level in global proteomic studies using various in-house or commercial antibodies [43–45]. The most encouraging results were obtained recently using a set of commercially available antibodies against monomethyl arginine and symmetric and asymmetric dimethyl arginine, which could likely be exploited also for hPTM analysis [46].

hPTMs can also be enriched through antibody-independent methods that take advantage of the physicochemical properties of different modified residues. One example is represented by titanium dioxide or Ti^{4+} enrichment for phosphorylated peptides [47,48], which dramatically improves the sensitivity and specificity of the detection of phosphorylation sites, also on histones [49]. Ubiquitinated peptides can be enriched through affinity purification with ubiquitin-binding proteins [50], while SUMO proteins can be purified by tandem affinity purification in cell lines expressing a tagged version of SUMO [51].

2.2 MS Methods for hPTM Analysis

Typically, mass spectrometers measure the mass-to-charge ratio (m/z) of ionized molecules in the gas phase, which can fly with different rates or circulate with different frequencies under electric and/or magnetic fields in the mass analyzer. Proteins and peptides are non-volatile and polar compounds, therefore requiring special methods to be transferred into the gas phase without extensive degradation. The most common ionization method for histone analysis is electrospray ionization (ESI), where a high voltage is used to create an electrically charged spray (electrospray) that triggers the desolvation of peptides/proteins from solvent droplets into the gas phase [52]. ESI sources are typically combined with RP-HPLC systems, which allow the separation of very complex peptide mixtures prior to MS analysis (LC-MS) (Figure 1, purple panel). LC-MS analysis of peptides and proteins provides information on the molecular weight and the elemental composition of the analyte, if sufficient mass resolution is achieved. In addition, further fragmentation of the ions (precursor ions) in tandem mass spectrometry (MS/MS) enables extrapolating the primary sequence of the peptide of interest and information on the presence and position of PTMs. Various fragmentation techniques can be employed. In the collision-induced dissociation (CID) and higher-energy collision dissociation (HCD) methods gas-phase protonated peptides/proteins are subjected to multiple collisions with rare gas atoms, resulting in the breakage of the peptide backbone at —CO—NH— bonds [53]. This type of fragmentation increases the frequency of relevant modification-specific fragmentation events in high-resolution (HR) MS/MS spectra and is therefore suitable for the analysis of modifications in bottom-up experiments (see below) [54]. Electron capture dissociation (ECD) and electron transfer dissociation (ETD) induce fragmentation of the peptide backbone based on gas-phase reactions using either thermal electrons or formation of radical ions, respectively, and are particularly suitable for sequencing longer peptides (>2 kDa), or intact proteins [55].

Three MS strategies can be used to characterize hPTMs: bottom-up, middle-down, and top-down approaches. The bottom-up approach, for which a typical workflow is shown in Figure 1, is most widely used in proteomic studies for global investigations of PTMs and combines high sensitivity of detection at the MS level with high efficiency of MS/MS fragmentation via CID or HCD [56]. Proteins are enzymatically digested into peptides prior to MS analysis, making this a "peptide-centric" approach. As an alternative, intact histones can be directly analyzed by top-down MC approaches [57], which provide information about the relative abundance and stoichiometry of modifications and a "protein-centric" view on the complete panel of histone isoforms present in a sample. Nonergodic fragmentation such as ETD and ECD must be used in this case, because of the high charge state of histones under the acidic conditions used [58–60]. This approach is less sensitive than bottom-up, and data analysis is more challenging due the higher complexity of both MS and MS/MS spectra obtained. Middle-down approaches combine the advantages from the global view of hPTMs gained through top-down and the sensitivity of bottom-up methods, analyzing long histone peptides (>2 kDa) and using either ETD or ECD as fragmentation methods. However, similarly to intact protein analysis, the much wider charge-state distributions of long peptides as compared to bottom-up peptides reduce the overall signal of each charged state. Prefractionating the enzymatic digestion products partially solves this sensitivity issue.

Matrix-assisted laser desorption ionization (MALDI) [61] represents an ionization method alternative to ESI. In a MALDI source, peptide mixtures are cocrystallized onto a metal plate with a solid-phase matrix, typically consisting of acidic organic compounds that facilitate the thermal desorption of protonated peptides when irradiated by laser pulses. A MALDI source can be coupled with different types of mass analyzers, but the most common is the time-of-flight analyzer [62], which separates the ions generated in the source based on the time that peptides of different masses (and more precisely of different mass/charge ratios) take to reach the detector, after being accelerated in the source region. A recent application of MALDI, namely MALDI imaging, combines the MS ability to analyze complex mixtures with the capability of obtaining a spatial distribution of proteins within biological samples through the analysis of thin sections of tissue that have been spray-coated or microspotted with a MALDI matrix [63]. By plotting ion intensities obtained at discrete locations within the section as a function of their x and y coordinates, unique multidimensional spatial expression maps can be generated for any given ion. Although several analytical challenges are associated with this method, such as biases toward lower molecular weight components and a resolution insufficient to observe organelles, in recent years MALDI-IMS has emerged as a promising technique to study biomolecules, including histones, in intact tissue specimens [64,65].

2.3 Analysis of MS Data for hPTM Characterization

Several database search engines have been implemented for the fast and accurate analysis of large proteomics data set (Figure 1, green panel). Among the most popular are Mascot, SEQUEST, pFind, OMSSA, and Andromeda [66–70]. However, in most cases the identification rate of MS/MS spectra is still limited compared with the number of spectra generated by modern MS instruments [71]. This is caused by different factors, including complex MS/MS spectra generated from coeluting precursors, unspecific protease cleavage, isolation of the incorrect monoisotopic m/z ratio or charge state, and incapability of the search engine to correctly assign similar sequences belonging to protein isoforms. In this scenario, the assignment of histone modifications at specific sites is particularly challenging due to the many coexisting modifications found in hypermodified regions, which increase the complexity of the database search and the incidence of false positives. HR mass analyzers can resolve and identify peptides bearing modifications with very similar delta-mass values as well as multiply charged ions in the MS/MS spectra [72], thus facilitating the analysis. Particularly useful is the ability of specialized software to perform data-dependent searches using HR-MS data. In this approach, PTMs are identified using an iterative database search. First, a "blind" database search is performed without specifying any PTMs. Then, all unassigned MS/MS spectra are subjected to a second search to match the difference between their molecular masses and those of the unmodified peptides to a list of delta-mass values corresponding to known modifications [73]. Because PTMs are identified through database search without specifying a priori any variable modifications, this unbiased approach can expand the search to hundreds of modifications at a time. However, most of the available methods are still suboptimal for the analysis of the very complex MS/MS spectra corresponding to the long peptide sequences and intact proteins deriving from top-down or middle-down analyses.

Although various types of software can support the analysis of hPTM by MS, this task still relies strongly on the visual inspection of the raw data by the operator and is therefore largely carried out manually. In this scenario, histone-specific data repositories developed within the currently available proteomic databases, such as the PRIDE repository [74], could represent useful resources for researchers, not only to facilitate data dissemination and exchange among different research groups but also to support data analysis. In fact, the possibility of accessing original information concerning the chromatographic distribution of the peaks and the raw MS/MS spectra, including modification-specific diagnostic peaks [75], would be extremely helpful to increase the confidence in the characterization of modified histone peptides.

2.4 hPTMs Quantitation

An accurate quantification of the various modified forms of a peptide is often essential for the characterization of a biological sample as their identification. Various strategies can be used for this purpose (Figure 1, yellow panel).

Label-free analysis of histones involves the direct comparison of unlabeled samples and relies on ion intensity-based quantitation, where the raw abundance of each modified peptide form is quantified using MS-extracted ion chromatograms. This method depends on the ability to chromatographically separate isobaric peptides, which has been dramatically improved by advances in HPLC separation. For instance, in a recent work we combined different ad hoc sample preparation strategies and UHPLC to separate and quantify histone modifications in hypermodified regions, exemplified by the H3 [27–40] and H4 [4–17] peptides [32]. Strategies to quantify isobaric histone isoforms when peptides cannot be separated using reversed-phase chromatography have also been reported. Isobaric peptides undistinguishable in full MS can be discerned based on the positional selectivity of ion fragmentation in MS/MS, thus allowing a relative quantitation based on the relative ratios of fragment ions [76,77].

Histones can also be quantified using in vivo metabolic labeling, for instance stable isotope labeling by amino acids in cell culture (SILAC), where two- or three-cell populations in different functional states (e.g., different drug treatments) are grown in media containing natural or isotope-encoded amino acids that can be distinguished by MS when mixed [78]. The different cell populations can be combined at a very early stage of the MS proteomics workflow, significantly reducing the impact of experimental variation during sample preparation and improving quantitation accuracy. SILAC was employed in several studies to quantify hPTMs [79–82]. When more than three samples need to be compared, an adapted version of the SILAC approach, involving a spike-in reference to which all the others are compared, can be used ([83] and unpublished results, see below). Another variation of SILAC, known as heavy methyl SILAC, can be employed for high confidence identification of methylations at lysine and arginine residues. In this approach, cells are grown in medium containing heavy methionine, which is converted into heavy *S*-adenosyl methionine, the sole methyl donor in enzymatic methylation reactions. Methylated histone and nonhistone proteins are enzymatically heavy methyl labeled and can be confidently identified and quantified based on the presence of light/heavy peak pairs corresponding to the methylated peptide. This method allows discriminating true methylations from those occurring during sample preparation due to various factors, including methanol-containing buffers. This strategy, first proposed by Ong et al. [41], has been applied in different studies addressing hPTM changes in different functional states [84–86].

Another labeling strategy that can be exploited for hPTMs quantitation is based on chemical derivatization using deuterated acetic or propionic anhydride. This method relies on the comparison of PTM abundances between labeled and unlabeled samples mixed in a 1:1 ratio and has been used to quantify histones in human and mouse melanoma cancer samples [87], as well as in other organisms [88–91]. An interesting application of a quantitative SILAC approach combined with chemical labeling of histone and nonhistone proteins has been recently proposed to quantify acetylation stoichiometry. This approach is based on controlled partial chemical acetylation of lysine residues with acetyl phosphate, which causes a defined increase in the intensity of the precursor ions of acetylated peptides, thus enabling reliable quantification also of low stoichiometry sites [92].

Finally, MS-targeted approaches, which differ from discovery methods because they require a priori knowledge of the peptide of interest, can also be used for hPTM quantification [93]. Targeted methods are particularly useful to analyze in a reproducible and quantitative fashion a predefined set of peptides, for instance candidate biomarkers, across multiple samples. Targeted experiments are usually performed using single or multiple reaction monitoring (SRM/MRM) MS, in triple quadrupole instruments, where target peptides are selected in the first quadrupole and fragmented by

CID in the second, while one or several transitions (fragment ions uniquely derived from the targeted peptide) are measured in the third quadrupole. Peptide transitions are usually acquired only during a scheduled time window, defined based on the known RT of the target peptides [39]. When a synthetic isotope-labeled peptide corresponding to the peptide of interest is used as internal standard, SRM and MRM can be employed for both relative and absolute quantitation of histones and their PTMs. The peptide ID is determined based on the presence of the transition and its quantification derives from the transition intensity relative to that of the standard [93]. This type of approach has been successfully employed to analyze H2B ubiquitination [94] and to absolutely quantify 42 histone peptides by MRM analysis [95]. Recently, a library of 93 synthetic Protein-Aqua™ peptides representing different modified forms of histone H3, H4, and H2A peptides was designed to analyze hPTM detection efficiencies. This study highlighted how modifications may affect the MS-based detection of peptides, thus biasing the calculation of their relative intensity, and provided a useful resource to correct for detection biases [96]. The application of SRM/MRM on an HR-MS spectrometer has been recently described and defined as "parallel reaction monitoring" (PRM) [97]. In one recent study, PRM has been applied to the analysis and label-free relative quantitation of histone modifications [98].

2.5 MS-Based Characterization of Histone Variants

With the exception of histone H4, each canonical histone forming the nucleosome has distinct variants, which synergize with PTMs to enforce the epigenetic regulation of gene expression [99]. Histone variants show differences in their expression patterns, in their enrichment at specific genomic regions, and in their PTM signatures, affecting different chromatin-related processes. Aberrations in the expression or deposition of histone variants at specific chromatin regions are linked with various diseases, including cancer [100,101]. Hence, while the importance of being able to distinguish among different variants is apparent, their often-small sequence differences makes their analysis using conventional antibody-based methods very challenging. Bottom-up MS has been used to characterize and even discover new histone variants from different organisms [102–106]; however, this approach is not suitable for those variants that display very limited sequence difference, such as H3.1 and H3.3. Top-down and middle-down MS approaches, which are usually employed on bulk histones previously LC-separated, have been used to characterize all canonical histones, comprising some variants [57,107]. For instance, histone variants were analyzed using intact mass and fragmentation spectra in top-down MS, leading to the characterization of 12 individual H2A sequences [108,109]. H3 variants from human, rat, and yeast were also analyzed by top- and middle-down MS techniques and found to have significant differences in their PTM profiles [57].

3. TOWARD CLINICAL APPLICATIONS OF MS ANALYSIS OF hPTMs

3.1 MS-Based Epigenetic Biomarker Discovery

Proteomic-based approaches for biomarker discovery can be exploited in various areas of medicine: for prognostic purposes (to identify individuals who are at a high risk of developing a disease), for therapy choices (to predict the therapeutical response and incidence of side effects in patients subjected to specific treatments), and finally for diagnostic purposes (to identify and classify diseases). MS has become one of the elective methods to quantify proteins from complex mixtures including plasma and has been increasingly used for biomarker discovery [110]. Recently, advances in MS-based hPTM analysis, together with increasing evidence regarding an epigenetic involvement in various diseases, has prompted the application of MS strategies to epigenetic biomarkers discovery, as reported in several studies.

In a seminal study, bottom-up MS was employed in combination with other techniques to compare hPTM patterns in normal lymphocytes and leukemia cells, showing a decrease of acetylation at H4K16 and trimethylation at H4K20 in cancer cells [3]. Recently, bottom-up MS has also been used to qualitatively identify H3 and H4 histone marks associated with cigarette smoking in mouse lungs and human bronchial epithelial cells, finding modification patterns that may play a role in the pathogenesis of smoking-induced chronic lung diseases, such as lung cancer and chronic obstructive pulmonary disease and, if validated, may represent biomarkers for these pathological states [111]. Other MS studies showed a decrease of the H3R8me2 mark in a mouse model of diet-induced hyperhomocysteinemia, which has been extensively associated with neurological and cardiovascular diseases [112] and a differential expression of the H4K79me2 mark in two types of esophageal squamous cell carcinoma with different invasiveness [113].

A SILAC approach was used by Garcia and colleagues to study histone H3 and H4 hypoacetylation in splenocytes from a mouse model of lupus compared with control, identifying differentially expressed modifications such as H3K18 methylation, H4K31 methylation, and H4K31 acetylation [114]. Histone H3 and H4 hypoacetylation in diseased mice was corrected by in vivo administration of the histone deacetylase inhibitor trichostatin A, with improvement of disease phenotype, thus establishing an association between aberrant histone modification patterns and the pathogenesis of lupus and also proving the usefulness of MS in detecting hPTM differences in an animal disease model. More recently, the same group profiled histone PTMs in 24 commonly used normal and cancerous cell lines, providing a proteomic atlas of histone modification signatures in different cancers, which could be potentially used for a novel cancer cell classification based on differential histone modification landscapes [115]. In particular, they observed an enrichment of peptides containing H3K27me in breast cancer cell lines and accordingly found that knockdown of the predominant H3K27 methyltransferase (EZH2) in a mouse mammary xenograft model significantly reduced tumor growth. These studies demonstrated the predictive power of proteomic techniques and support the possibility to exploit MS-derived epigenetic biomarkers to design epigenetic therapies.

We also employed stable isotope labeling, using a spike-in SILAC approach [83], to compare modifications occurring in breast cancer relative to normal cells, identifying hPTMs changes already known to occur in cancer, such as the reduction of H3K9me3, H4K20me3, and H4K16ac [3,116], as well as previously unknown PTMs variations, such as the increase of H3K4me1, H3K27me2/3, and H3K9me2. In addition, we identified hPTM patterns specific to distinct types of breast cancer cell lines associated with different aggressiveness or prognosis. The spike-in strategy overcomes the intrinsic impossibility of applying stable metabolic labeling strategies to the analysis of clinical samples. Recently, we have developed further this idea by generating a "super-SILAC" histone mixture to be used as an internal standard, in a histone-focused version of the previously described "super-SILAC" approach [117]. A super-SILAC mixture generated by combining heavy-labeled histones from different breast cancer cell lines is spiked into clinical samples of different nature, generating a universal reference for quantitation (unpublished results). A mixture of cell lines instead of a single one is expected to better represent the range of all the possible modifications found in a tissue and can thus be used for a more comprehensive quantification. In order to generate an appropriate super-SILAC mix, we first analyzed hPTM patterns in five breast cancer cell lines from different subtypes and grade of aggressiveness, using the spike-in approach previously developed [83], and then chose the ones to be included in the super-SILAC mix based on their divergent hPTM patterns. We applied this strategy to the analysis of a panel of breast cancer cells that are either sensitive or resistant to four HDAC inhibitors with different specificities, in order to correlate hPTM patterns with drug responsiveness. This analysis revealed hPTM signatures associated with the cellular response to the drugs and pinpointed the marks affected by HDAC inhibitors targeting different members of the HDAC family (unpublished results). The global and comprehensive hPTM profiles acquired for these breast cancer cell lines could also serve as resource for prospective studies which aim at finding biomarkers for response to any other epigenetic drug, such as histone demethylases or bromodomains inhibitors. We have also successfully applied this super-SILAC strategy to the analysis of other breast cancer primary samples, as described in more detail below, proving its potential for its application in clinical epigenetics.

As an alternative to LC-MS-MS analysis, MALDI imaging has gained much attention for biomarker discovery [118]. Recently, whole-cell MALDI MS biotyping, which is the direct and reagent-free analysis of whole cells homogenized in a solvent/MALDI matrix mixture, and MALDI imaging were adapted to monitor histone acetylation changes induced by HDAC inhibitors in cells and mouse gastrointestinal cancer tissue, respectively [65]. In addition, protein patterns obtained by MALDI imaging were combined with LC-MS/MS information to compare the tissue proteomes of hepatocellular carcinoma with and without microvascular invasion, which represents a major risk factor in postoperative mortality and tumor recurrence in hepatocellular carcinoma [64]. Among the proteins expressed in samples with microvascular invasion were identified modified forms of histone H4 bearing H4K16ac and H4K20me, which could represent new tissue biomarkers for diagnostic and prognostic purposes.

3.2 MS-Based hPTMs Analysis in Clinical Samples

To date, the majority of hPTM studies have been conducted on cultured cell lines or fresh animal tissues, with the exception of a study where human tissues were used to compare fetal brain histone modifications in normal human fetuses with fetuses with low folate levels and neural tube defects by Zhang et al. [119] who showed lack of H3K79me2 and H2bK5me1. Therefore, adaptations of the available protocols are needed for the analysis of clinical samples, such as frozen tissue biopsies, formalin-fixed paraffin-embedded (FFPE) tissues, and primary cells, whose amount and availability are limited. Below, we provide a concise overview of protocols and results recently obtained in our group from the analysis of primary breast cancer samples.

3.2.1 *Frozen Tissues*

Fresh-frozen specimens are considered the gold standard for clinical proteomics, as freezing avoids molecular degradation and biases that can be caused by other preservation techniques. We extracted histones from fresh-frozen breast specimens using a simple protocol that involves homogenization of the sample in the presence of a low percentage of a nonionic buffer, which allows nuclei purification following centrifugation. After lysis in a high-detergent buffer, histones are highly enriched in the sample and suitable for MS analysis following Arg-C-like in-gel digestion (Figure 2). All the major histone modifications can be identified and quantified from this kind of sample, as well as less characterized modifications (unpublished results). However, the heterogeneous nature of tumors represents a significant problem

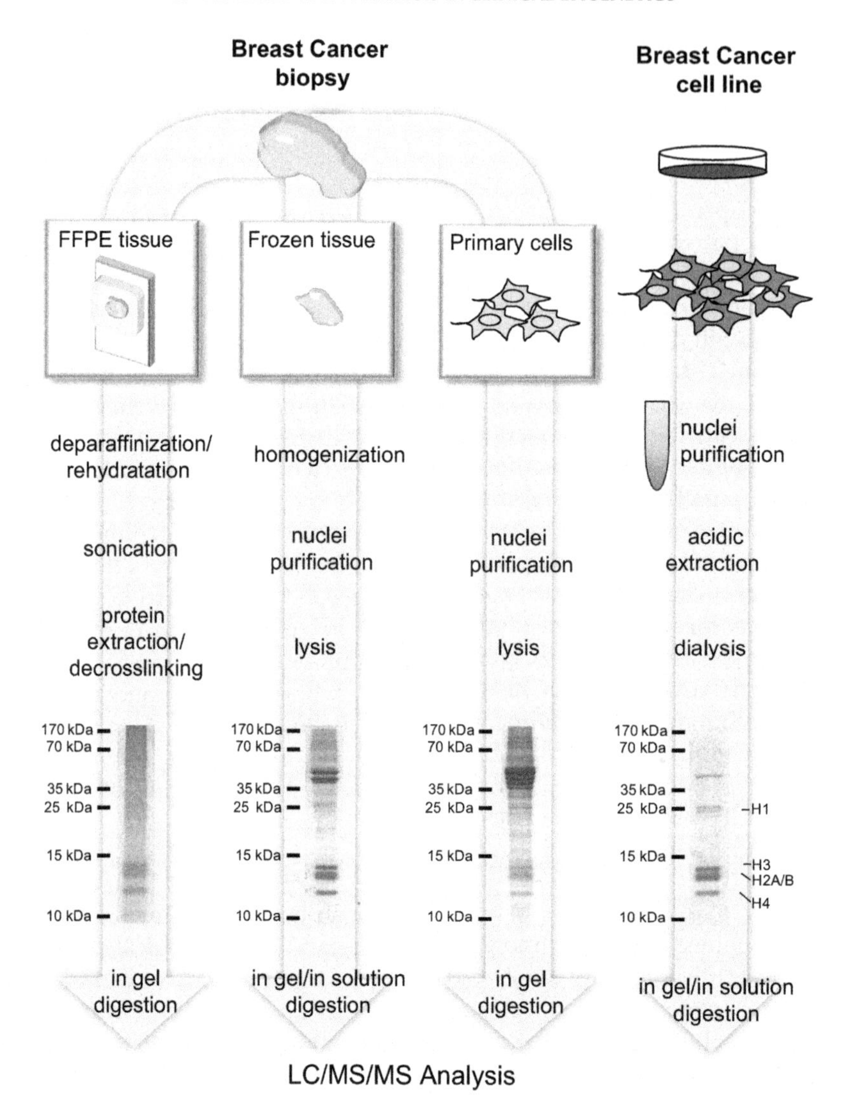

FIGURE 2 Isolation of histones from primary breast cancer tissues for mass spectrometry analysis. Scheme showing the isolation steps and the SDS-PAGE analysis of histones obtained from formalin-fixed paraffin-embedded tissue, fresh-frozen tissue, and primary cells, in comparison with a standard protocol used for histone isolation from cell lines. Based on the purity of the obtained histone preparation, samples can be digested in gel and/or in solution.

toward the identification of clinical biomarkers. Methods to overcome this limitation have been established, such as coupling MS with laser capture microdissection, which involves dissection of homogeneous histological cell types from thin fresh-frozen tissue sections to profile homogeneous, physically restricted, cell populations from solid tumors [120]. While this has only been applied so far to global proteome analysis, it appears as a very interesting application also for hPTM analysis, provided that enough starting material can be obtained.

3.2.2 FFPE Tissues

Because FFPE samples are routinely prepared for processing and storage of pathology specimens and their analysis is required for the diagnosis and clinical management of the vast majority of diseases, large archives of paraffin-embedded tissues are available in most hospitals and are linked to patient history, thus representing an invaluable source of clinical samples and retrospective information. However, analysis of archival FFPE tissues by proteomic methods has been limited by the extensive protein cross-linking generated by formaldehyde fixation. Lately, this problem has been overcome in global proteomic studies using extraction protocols largely based upon heat-induced antigen retrieval techniques borrowed from immunohistochemistry [121]. We have recently adapted these protocols for hPTM analysis (Figure 2), using a straightforward procedure (PAT-H-MS: PAThology tissue analysis of Histones by Mass Spectrometry) that involves initial deparaffinization and rehydration of the tissue, followed by protein extraction and de-cross-linking in a high-detergent buffer. In our experience, no further histone isolation steps are required, as histones represent the majority of the sample obtained from different tissues. Gel bands corresponding to the histones of interest can then be cut, in-gel digested, and MS analyzed, as non-FFPE samples would be. In a pairwise comparison with frozen tissues, histones from FFPE mouse spleen showed remarkably similar histone modification patterns on four histone H3 peptides carrying functionally relevant post-translational modifications, suggesting that paraffinization and cross-linking, as well as their reverse steps, do not affect hPTMs. Histone modifications could be efficiently detected in FFPE samples up to 6 years old and in various types of tissues, supporting the wide applicability of this technique. When combined with the super-SILAC approach, the analysis of either FFPE or fresh-frozen tissues revealed very similar pattern changes among three breast cancer human biopsies, with only minor differences which could be ascribed to variations in the tissue portion that was frozen or fixed. These results highlight the applicability of FFPE-obtained histones for hPTM analysis and pave the way for large retrospective studies that take advantage of the extended clinical information associated with large pathology FFPE archives. As an initial application of the PAT-H-MS approach, we analyzed breast cancer samples belonging to different subtypes, revealing significant changes in histone H3 methylation patterns among Luminal A-like and Triple Negative samples. As for fresh-frozen tissues, a limitation of this method is related to intrinsic FFPE tissue heterogeneity, for instance, the presence of normal cells in a cancer specimen, which impairs the possibility of analyzing individual cell populations. An interesting development of the technique would involve procedures able to increase the purity and homogeneity of cell populations, such as core needle biopsies and laser microdissection. These techniques have proven to be successful in the implementation of the pathology tissue chromatin immunoprecipitation (PAT-ChIP) [122,123], an approach for the extraction and high-throughput genomic analysis of chromatin derived from FFPE samples, which represent the genomic counterpart of our FFPE extraction protocol for MS-based proteomics.

3.2.3 Primary Cells

Primary cells can be used for various purposes, of which one of the most interesting is testing drug resistance. However, histone analysis of patient-derived primary cells poses several challenges, first the small number of cells that can be obtained from clinical samples. Typical protocols for histone preparations for MS require amounts of starting material (several millions) that are difficult if not impossible to obtain from primary cultured cells, given their limited life span. We have recently experienced that isolating nuclei with a mild detergent lysis without acidic extraction yields a good amount of histones from 0.5 million cells (Figure 2). Although less pure than the histones obtained with standard protocols involving acidic extraction, histones are sufficiently enriched in these samples for MS analysis following in-gel digestion. Despite their utility, the use of primary cells suffers from several limitations, including the challenges related to their ability to grow in two-dimensional (2D) culture conditions and the appearance of molecular alterations due to the cells' adaptation to such conditions, where they lack their physiological substrate, heterotypic cell–cell interactions, exposure to components of the extracellular matrix, and the three-dimensional (3D) tissue architecture found in vivo [124]. One interesting but technically challenging possibility to overcome the problems linked to 2D cultures is using 3D growing cultures, which better mimic in vivo cell growth conditions.

4. CONCLUSIONS AND PERSPECTIVES

MS holds great promise for clinical epigenetics and epigenetic biomarker discovery. On the one hand, MS represents the elective technique for dissecting hPTM combinations; on the other hand, it is emerging as a primary tool for biomarker discovery. Great advances have been accomplished in both areas, but their combination for the implementation of MS-based analysis of hPTMs in clinical practice is still at its infancy and will likely benefit from technological advances in several critical areas.

One important point for analyzing clinical samples is represented by the ability to comprehensively define modification patterns in a single analysis, including not only the most abundant modifications but also low stoichiometry and less frequent ones that may be linked with diseased states. Such ability would be important both to evaluate the most complete panel of modifications during the biomarker discovery phase and to reproducibly measure less abundant but relevant marks in patient samples. Most current data acquisition routines are based on data-dependent acquisition methods, which are designed to pick the ion precursors with the highest relative abundance in the full MS scan for subsequent fragmentation. As a consequence of this intensity bias, while the canonical and more abundant histone modifications from purified samples can be easily measured, the less abundant PTMs are less efficiently and reliably detected without specific enrichment/fractionation steps preceding MS. Alternative methods—such as targeted MS or data-independent acquisition approaches—which have been recently implemented in global proteome analysis, might represent a solution to gain sensitivity in the analysis of low-abundant histone modifications.

Detecting the combinatorial aspect of the histone code is crucial to define modification signatures of pathological states, which may improve the predictive power as compared with single biomarkers. While offering efficient amino acid sequencing and superior throughput for complex samples, bottom-up approaches only provide a partial view of the complex cross talks among different, sometimes distant, hPTMs. Recent advances in middle-down and top-down approaches greatly improved the detection of combinatorial histone PTMs, but the implementation of straightforward analytical workflows compatible with large-scale analyses of primary specimens and the development of bioinformatics tools to effectively handle the enormous complexity of MS/MS deriving from such analyses are still lagging behind the current needs. Middle-down and top-down MS could also facilitate the discrimination among histone variants [57,107], which is limited in bottom-up studies because of their small sequence differences.

Although histone MS analyses are usually performed from bulk isolated histones, the ability to analyze histones at specific genomic locations could provide valuable information regarding the hPTM patterns at the level of disease-associated genes. Therefore, the MS-based characterization of hPTM at single genomic loci represents an interesting application. Proteomic analysis of hPTMs has been combined with different biochemical methods to purify specific chromatic regions that employ either proteins or DNA as baits for ChIP. These methods include ChAP-MS (chromatin affinity purification with mass spectrometry), PICH (proteomics of isolated chromatin segments), MChIP (modified chromatin immunopurification), and ChIP-MS (chromatin interacting protein-mass spectrometry) (reviewed in Ref. [125]). Our group also developed an approach that couples standard ChIP with MS-SILAC-based quantitative proteomics to comprehensively investigate hPTMs at specific genomic regions [126]. Adaptations of the recently developed CRISPR/Cas9 system, which can induce precise cleavages at defined genomic loci [127], may represent another useful means to enrich specific genomic regions for proteomics analysis. Although detecting small amounts of histone hPTM at subgenome-wide regions has proven to be feasible in several studies, it still remains a daunting task. Lowering the limit of hPTM through further improvements in the sensitivity and dynamic range of MS instruments will be needed to streamline these types of analyses.

Altogether, future developments of MS-based analysis of hPTMs are expected to make this technique a valuable diagnostic and prognostic tool, providing epigenetic biomarkers that could lead to targeted epigenetic therapies for improved treatments and reduced side effects.

LIST OF ABBREVIATIONS

CID Collision-induced dissociation
ECD Electron capture dissociation
ESI Electrospray ionization
ETD Electron transfer dissociation
FFPE Formalin-fixed paraffin-embedded
HILIC Hydrophilic interaction liquid chromatography
HPLC High-performance liquid chromatography
hPTM Histone posttranslational modification
HR High resolution
LC-MS Liquid chromatography–mass spectrometry
MS Mass spectrometry
MS/MS Tandem mass spectra
RP Reversed phase
SCX Ion exchange chromatography
SILAC Stable isotope labeling by amino acids in cell culture
MALDI Matrix-assisted laser desorption ionization
WCX Weak-cation exchange

Acknowledgments

Tiziana Bonaldi's research group is supported by grants from the Giovanni Armenise-Harvard Foundation Career Development Program, the Italian Association for Cancer Research (AIRC), the Italian Ministry of Health, and CNR-EPIGEN flagship project.

References

[1] Jenuwein T, Allis CD. Translating the histone code. Science 2001;293(5532):1074–80.
[2] Bannister AJ, Kouzarides T. Regulation of chromatin by histone modifications. Cell Res 2011;21(3):381–95.
[3] Fraga MF, Ballestar E, Villar-Garea A, Boix-Chornet M, Espada J, Schotta G, et al. Loss of acetylation at Lys16 and trimethylation at Lys20 of histone H4 is a common hallmark of human cancer. Nat Genet 2005; 37(4):391–400.
[4] Hamamoto R, Furukawa Y, Morita M, Iimura Y, Silva FP, Li M, et al. SMYD3 encodes a histone methyltransferase involved in the proliferation of cancer cells. Nat Cell Biol 2004;6(8):731–40.
[5] Dalgliesh GL, Furge K, Greenman C, Chen L, Bignell G, Butler A, et al. Systematic sequencing of renal carcinoma reveals inactivation of histone modifying genes. Nature 2010;463(7279):360–3.
[6] Kondo Y, Shen L, Suzuki S, Kurokawa T, Masuko K, Tanaka Y, et al. Alterations of DNA methylation and histone modifications contribute to gene silencing in hepatocellular carcinomas. Hepatol Res 2007;37(11):974–83.
[7] Vire E, Brenner C, Deplus R, Blanchon L, Fraga M, Didelot C, et al. The polycomb group protein EZH2 directly controls DNA methylation. Nature 2006;439(7078):871–4.
[8] Seligson DB, Horvath S, Shi T, Yu H, Tze S, Grunstein M, et al. Global histone modification patterns predict risk of prostate cancer recurrence. Nature 2005; 435(7046):1262–6.

[9] Chi P, Allis CD, Wang GG. Covalent histone modifications–miswritten, misinterpreted and mis-erased in human cancers. Nat Rev Cancer 2010;10(7):457–69.
[10] Minucci S, Pelicci PG. Histone deacetylase inhibitors and the promise of epigenetic (and more) treatments for cancer. Nat Rev Cancer 2006;6(1):38–51.
[11] Portela A, Esteller M. Epigenetic modifications and human disease. Nat Biotechnol 2010;28(10):1057–68.
[12] Urdinguio RG, Sanchez-Mut JV, Esteller M. Epigenetic mechanisms in neurological diseases: genes, syndromes, and therapies. Lancet Neurol 2009;8(11):1056–72.
[13] Miao F, Smith DD, Zhang L, Min A, Feng W, Natarajan R. Lymphocytes from patients with type 1 diabetes display a distinct profile of chromatin histone H3 lysine 9 dimethylation: an epigenetic study in diabetes. Diabetes 2008;57(12):3189–98.
[14] Sakabe K, Wang Z, Hart GW. Beta-N-acetylglucosamine (O-GlcNAc) is part of the histone code. Proc Natl Acad Sci USA 2010;107(46):19915–20.
[15] Zhang K, Chen Y, Zhang Z, Zhao Y. Identification and verification of lysine propionylation and butyrylation in yeast core histones using PTMap software. J Proteome Res 2009;8(2):900–6.
[16] Chen Y, Sprung R, Tang Y, Ball H, Sangras B, Kim SC, et al. Lysine propionylation and butyrylation are novel post-translational modifications in histones. Mol Cell Proteomics 2007;6(5):812–9.
[17] Wisniewski JR, Zougman A, Mann M. Nepsilon-formylation of lysine is a widespread post-translational modification of nuclear proteins occurring at residues involved in regulation of chromatin function. Nucleic Acids Res 2008;36(2):570–7.
[18] Xie Z, Dai J, Dai L, Tan M, Cheng Z, Wu Y, et al. Lysine succinylation and lysine malonylation in histones. Mol Cell Proteomics 2012;11(5):100–7.
[19] Tan M, Luo H, Lee S, Jin F, Yang JS, Montellier E, et al. Identification of 67 histone marks and histone lysine crotonylation as a new type of histone modification. Cell 2011;146(6):1016–28.
[20] Shechter D, Dormann HL, Allis CD, Hake SB. Extraction, purification and analysis of histones. Nat Protoc 2007;2(6):1445–57.
[21] Bonaldi T, Imhof A, Regula JT. A combination of different mass spectroscopic techniques for the analysis of dynamic changes of histone modifications. Proteomics 2004;4(5):1382–96.
[22] von Holt C, Brandt WF, Greyling HJ, Lindsey GG, Retief JD, Rodrigues JD, et al. Isolation and characterization of histones. Methods Enzymol 1989;170:431–523.
[23] Simon RH, Felsenfeld G. A new procedure for purifying histone pairs H2A + H2B and H3 + H4 from chromatin using hydroxylapatite. Nucleic Acids Res 1979;6(2):689–96.
[24] Boersema PJ, Mohammed S, Heck AJ. Hydrophilic interaction liquid chromatography (HILIC) in proteomics. Anal Bioanal Chem 2008;391(1):151–9.
[25] Olsen JV, Ong SE, Mann M. Trypsin cleaves exclusively C-terminal to arginine and lysine residues. Mol Cell Proteomics 2004;3(6):608–14.
[26] Smith CM, Haimberger ZW, Johnson CO, Wolf AJ, Gafken PR, Zhang Z, et al. Heritable chromatin structure: mapping "memory" in histones H3 and H4. Proc Natl Acad Sci USA 2002;99(Suppl. 4):16454–61.
[27] Garcia BA, Mollah S, Ueberheide BM, Busby SA, Muratore TL, Shabanowitz J, et al. Chemical derivatization of histones for facilitated analysis by mass spectrometry. Nat Protoc 2007;2(4):933–8.
[28] Liao R, Wu H, Deng H, Yu Y, Hu M, Zhai H, et al. Specific and efficient N-propionylation of histones with propionic acid N-hydroxysuccinimide ester for histone marks characterization by LC-MS. Anal Chem 2013;85(4):2253–9.
[29] Bonaldi T, Regula JT, Imhof A. The use of mass spectrometry for the analysis of histone modifications. Methods Enzymol 2004;377:111–30.
[30] Taverna SD, Ueberheide BM, Liu Y, Tackett AJ, Diaz RL, Shabanowitz J, et al. Long-distance combinatorial linkage between methylation and acetylation on histone H3 N termini. Proc Natl Acad Sci USA 2007;104(7):2086–91.
[31] Jung HR, Pasini D, Helin K, Jensen ON. Quantitative mass spectrometry of histones H3.2 and H3.3 in Suz12-deficient mouse embryonic stem cells reveals distinct, dynamic post-translational modifications at Lys-27 and Lys-36. Mol Cell Proteomics 2010;9(5):838–50.
[32] Soldi M, Cuomo A, Bonaldi T. Improved bottom-up strategy to efficiently separate hypermodified histone peptides through ultra-HPLC separation on a bench top Orbitrap instrument. Proteomics 2014.
[33] Soldi M, Cuomo A, Bremang M, Bonaldi T. Mass spectrometry-based proteomics for the analysis of chromatin structure and dynamics. Int J Mol Sci 2013;14(3):5402–31.
[34] Wilm M, Mann M. Analytical properties of the nanoelectrospray ion source. Anal Chem 1996;68(1):1–8.
[35] Shen Y, Zhao R, Berger SJ, Anderson GA, Rodriguez N, Smith RD. High-efficiency nanoscale liquid chromatography coupled on-line with mass spectrometry using nanoelectrospray ionization for proteomics. Anal Chem 2002;74(16):4235–49.
[36] Thakur SS, Geiger T, Chatterjee B, Bandilla P, Frohlich F, Cox J, et al. Deep and highly sensitive proteome coverage by LC-MS/MS without prefractionation. Mol Cell Proteomics 2011;10(8). M110 003699.
[37] Tyanova S, Mann M, Cox J. MaxQuant for in-depth analysis of large SILAC datasets. Methods Mol Biol 2014;1188:351–64.

[38] Cox J, Hein MY, Luber CA, Paron I, Nagaraj N, Mann M. Accurate proteome-wide label-free quantification by delayed normalization and maximal peptide ratio extraction, termed MaxLFQ. Mol Cell Proteomics 2014;13(9):2513–26.

[39] Lesur A, Domon B. Advances in high-resolution accurate mass spectrometry application to targeted proteomics. Proteomics 2014.

[40] Young NL, DiMaggio PA, Plazas-Mayorca MD, Baliban RC, Floudas CA, Garcia BA. High throughput characterization of combinatorial histone codes. Mol Cell Proteomics 2009;8(10):2266–84.

[41] Ong SE, Mittler G, Mann M. Identifying and quantifying in vivo methylation sites by heavy methyl SILAC. Nat Methods 2004;1(2):119–26.

[42] Choudhary C, Kumar C, Gnad F, Nielsen ML, Rehman M, Walther TC, et al. Lysine acetylation targets protein complexes and co-regulates major cellular functions. Science 2009;325(5942):834–40.

[43] Cao XJ, Arnaudo AM, Garcia BA. Large-scale global identification of protein lysine methylation in vivo. Epigenetics 2013;8(5):477–85.

[44] Bremang M, Cuomo A, Agresta AM, Stugiewicz M, Spadotto V, Bonaldi T. Mass spectrometry-based identification and characterisation of lysine and arginine methylation in the human proteome. Mol Biosyst 2013;9(9):2231–47.

[45] Sylvestersen KB, Horn H, Jungmichel S, Jensen LJ, Nielsen ML. Proteomic analysis of arginine methylation sites in human cells reveals dynamic regulation during transcriptional arrest. Mol Cell Proteomics 2014;13(8):2072–88.

[46] Guo A, Gu H, Zhou J, Mulhern D, Wang Y, Lee KA, et al. Immunoaffinity enrichment and mass spectrometry analysis of protein methylation. Mol Cell Proteomics 2014;13(1):372–87.

[47] Zhou H, Ye M, Dong J, Han G, Jiang X, Wu R, et al. Specific phosphopeptide enrichment with immobilized titanium ion affinity chromatography adsorbent for phosphoproteome analysis. J Proteome Res 2008;7(9):3957–67.

[48] Rosenqvist H, Ye J, Jensen ON. Analytical strategies in mass spectrometry-based phosphoproteomics. Methods Mol Biol 2011;753:183–213.

[49] Garcia BA, Joshi S, Thomas CE, Chitta RK, Diaz RL, Busby SA, et al. Comprehensive phosphoprotein analysis of linker histone H1 from *Tetrahymena thermophila*. Mol Cell Proteomics 2006;5(9):1593–609.

[50] Kirkpatrick DS, Denison C, Gygi SP. Weighing in on ubiquitin: the expanding role of mass-spectrometry-based proteomics. Nat Cell Biol 2005;7(8):750–7.

[51] Rosas-Acosta G, Russell WK, Deyrieux A, Russell DH, Wilson VG. A universal strategy for proteomic studies of SUMO and other ubiquitin-like modifiers. Mol Cell Proteomics 2005;4(1):56–72.

[52] Fenn JB, Mann M, Meng CK, Wong SF, Whitehouse CM. Electrospray ionization for mass spectrometry of large biomolecules. Science 1989;246(4926):64–71.

[53] Olsen JV, Macek B, Lange O, Makarov A, Horning S, Mann M. Higher-energy C-trap dissociation for peptide modification analysis. Nat Methods 2007;4(9):709–12.

[54] Nagaraj N, D'Souza RC, Cox J, Olsen JV, Mann M. Feasibility of large-scale phosphoproteomics with higher energy collisional dissociation fragmentation. J Proteome Res 2010;9(12):6786–94.

[55] Zubarev RA, Horn DM, Fridriksson EK, Kelleher NL, Kruger NA, Lewis MA, et al. Electron capture dissociation for structural characterization of multiply charged protein cations. Anal Chem 2000;72(3):563–73.

[56] Zhang Y, Fonslow BR, Shan B, Baek MC, Yates 3rd JR. Protein analysis by shotgun/bottom-up proteomics. Chem Rev 2013;113(4):2343–94.

[57] Thomas CE, Kelleher NL, Mizzen CA. Mass spectrometric characterization of human histone H3: a bird's eye view. J Proteome Res 2006;5(2):240–7.

[58] Tian Z, Zhao R, Tolic N, Moore RJ, Stenoien DL, Robinson EW, et al. Two-dimensional liquid chromatography system for online top-down mass spectrometry. Proteomics 2010;10(20):3610–20.

[59] Eliuk SM, Maltby D, Panning B, Burlingame AL. High resolution electron transfer dissociation studies of unfractionated intact histones from murine embryonic stem cells using on-line capillary LC separation: determination of abundant histone isoforms and post-translational modifications. Mol Cell Proteomics 2010;9(5):824–37.

[60] Garcia BA. What does the future hold for top down mass spectrometry? J Am Soc Mass Spectrom 2010;21(2):193–202.

[61] Hillenkamp F, Karas M. Mass spectrometry of peptides and proteins by matrix-assisted ultraviolet laser desorption/ionization. Methods Enzymol 1990;193:280–95.

[62] Lagarrigue M, Lavigne R, Guevel B, Com E, Chaurand P, Pineau C. Matrix-assisted laser desorption/ionization imaging mass spectrometry: a promising technique for reproductive research. Biol Reprod 2012;86(3):74.

[63] Cornett DS, Reyzer ML, Chaurand P, Caprioli RM. MALDI imaging mass spectrometry: molecular snapshots of biochemical systems. Nat Methods 2007;4(10):828–33.

[64] Pote N, Alexandrov T, Le Faouder J, Laouirem S, Leger T, Mebarki M, et al. Imaging mass spectrometry reveals modified forms of histone H4 as new biomarkers of microvascular invasion in hepatocellular carcinomas. Hepatology 2013;58(3):983–94.

[65] Munteanu B, Meyer B, von Reitzenstein C, Burgermeister E, Bog S, Pahl A, et al. Label-free in situ monitoring of histone deacetylase drug target engagement by matrix-assisted laser desorption ionization-mass spectrometry biotyping and imaging. Anal Chem 2014;86(10):4642–7.

[66] Eng JK, McCormack AL, Yates JR. An approach to correlate tandem mass spectral data of peptides with amino acid sequences in a protein database. J Am Soc Mass Spectrom 1994;5(11):976–89.

[67] Perkins DN, Pappin DJ, Creasy DM, Cottrell JS. Probability-based protein identification by searching sequence databases using mass spectrometry data. Electrophoresis 1999;20(18):3551–67.

[68] Geer LY, Markey SP, Kowalak JA, Wagner L, Xu M, Maynard DM, et al. Open mass spectrometry search algorithm. J Proteome Res 2004;3(5):958–64.

[69] Wang LH, Li DQ, Fu Y, Wang HP, Zhang JF, Yuan ZF, et al. pFind 2.0: a software package for peptide and protein identification via tandem mass spectrometry. Rapid Commun Mass Spectrom 2007;21(18):2985–91.

[70] Cox J, Neuhauser N, Michalski A, Scheltema RA, Olsen JV, Mann M. Andromeda: a peptide search engine integrated into the MaxQuant environment. J Proteome Res 2011;10(4):1794–805.

[71] Michalski A, Cox J, Mann M. More than 100,000 detectable peptide species elute in single shotgun proteomics runs but the majority is inaccessible to data-dependent LC-MS/MS. J Proteome Res 2011;10(4):1785–93.

[72] Marshall AG, Hendrickson CL. High-resolution mass spectrometers. Annu Rev Anal Chem (Palo Alto Calif) 2008;1:579–99.

[73] Savitski MM, Nielsen ML, Zubarev RA. ModifiComb, a new proteomic tool for mapping substoichiometric post-translational modifications, finding novel types of modifications, and fingerprinting complex protein mixtures. Mol Cell Proteomics 2006;5(5):935–48.

[74] Riffle M, Eng JK. Proteomics data repositories. Proteomics 2009;9(20):4653–63.

[75] Kelstrup CD, Frese C, Heck AJ, Olsen JV, Nielsen ML. Analytical utility of mass spectral binning in proteomic experiments by SPectral Immonium Ion Detection (SPIID). Mol Cell Proteomics 2014;13(8):1914–24.

[76] Pesavento JJ, Mizzen CA, Kelleher NL. Quantitative analysis of modified proteins and their positional isomers by tandem mass spectrometry: human histone H4. Anal Chem 2006;78(13):4271–80.

[77] Garcia BA, Pesavento JJ, Mizzen CA, Kelleher NL. Pervasive combinatorial modification of histone H3 in human cells. Nat Methods 2007;4(6):487–9.

[78] Ong SE, Blagoev B, Kratchmarova I, Kristensen DB, Steen H, Pandey A, et al. Stable isotope labeling by amino acids in cell culture, SILAC, as a simple and accurate approach to expression proteomics. Mol Cell Proteomics 2002;1(5):376–86.

[79] Olsen JV, Blagoev B, Gnad F, Macek B, Kumar C, Mortensen P, et al. Global, in vivo, and site-specific phosphorylation dynamics in signaling networks. Cell 2006;127(3):635–48.

[80] Stunnenberg HG, Vermeulen M. Towards cracking the epigenetic code using a combination of high-throughput epigenomics and quantitative mass spectrometry-based proteomics. Bioessays 2011;33(7):547–51.

[81] Bonenfant D, Towbin H, Coulot M, Schindler P, Mueller DR, van Oostrum J. Analysis of dynamic changes in post-translational modifications of human histones during cell cycle by mass spectrometry. Mol Cell Proteomics 2007;6(11):1917–32.

[82] Pimienta G, Chaerkady R, Pandey A. SILAC for global phosphoproteomic analysis. Methods Mol Biol 2009;527:107–16. x.

[83] Cuomo A, Moretti S, Minucci S, Bonaldi T. SILAC-based proteomic analysis to dissect the "histone modification signature" of human breast cancer cells. Amino Acids 2011;41(2):387–99.

[84] Fodor BD, Kubicek S, Yonezawa M, O'Sullivan RJ, Sengupta R, Perez-Burgos L, et al. Jmjd2b antagonizes H3K9 trimethylation at pericentric heterochromatin in mammalian cells. Genes Dev 2006;20(12):1557–62.

[85] Zee BM, Levin RS, Xu B, LeRoy G, Wingreen NS, Garcia BA. In vivo residue-specific histone methylation dynamics. J Biol Chem 2010;285(5):3341–50.

[86] Sweet SM, Li M, Thomas PM, Durbin KR, Kelleher NL. Kinetics of re-establishing H3K79 methylation marks in global human chromatin. J Biol Chem 2010; 285(43):32778–86.

[87] Kapoor A, Goldberg MS, Cumberland LK, Ratnakumar K, Segura MF, Emanuel PO, et al. The histone variant macroH2A suppresses melanoma progression through regulation of CDK8. Nature 2010;468(7327): 1105–1109.

[88] Mandava V, Fernandez JP, Deng H, Janzen CJ, Hake SB, Cross GA. Histone modifications in *Trypanosoma brucei*. Mol Biochem Parasitol 2007;156(1):41–50.

[89] Plazas-Mayorca MD, Zee BM, Young NL, Fingerman IM, LeRoy G, Briggs SD, et al. One-pot shotgun quantitative mass spectrometry characterization of histones. J Proteome Res 2009;8(11):5367–74.

[90] Ouvry-Patat SA, Schey KL. Characterization of antimicrobial histone sequences and posttranslational modifications by mass spectrometry. J Mass Spectrom 2007;42(5):664–74.

[91] Voigt P, LeRoy G, Drury 3rd WJ, Zee BM, Son J, Beck DB, et al. Asymmetrically modified nucleosomes. Cell 2012;151(1):181–93.

[92] Weinert BT, Iesmantavicius V, Moustafa T, Scholz C, Wagner SA, Magnes C, et al. Acetylation dynamics and stoichiometry in *Saccharomyces cerevisiae*. Mol Syst Biol 2014;10:716.

[93] Picotti P, Aebersold R. Selected reaction monitoring-based proteomics: workflows, potential, pitfalls and future directions. Nat Methods 2012;9(6):555–66.
[94] Darwanto A, Curtis MP, Schrag M, Kirsch W, Liu P, Xu G, et al. A modified "cross-talk" between histone H2B Lys-120 ubiquitination and H3 Lys-79 methylation. J Biol Chem 2010;285(28):21868–76.
[95] Gao J, Liao R, Yu Y, Zhai H, Wang Y, Sack R, et al. Absolute quantification of histone PTM marks by MRM-based LC-MS/MS. Anal Chem 2014.
[96] Lin S, Wein S, Gonzales-Cope M, Otte GL, Yuan ZF, Afjehi-Sadat L, et al. Stable-isotope-labeled histone peptide library for histone post-translational modification and variant quantification by mass spectrometry. Mol Cell Proteomics 2014;13(9):2450–66.
[97] Gallien S, Bourmaud A, Kim SY, Domon B. Technical considerations for large-scale parallel reaction monitoring analysis. J Proteomics 2014;100:147–59.
[98] Tang H, Fang H, Yin E, Brasier AR, Sowers LC, Zhang K. Multiplexed parallel reaction monitoring targeting histone modifications on the QExactive mass spectrometer. Anal Chem 2014;86(11):5526–34.
[99] Yuan G, Zhu B. Histone variants and epigenetic inheritance. Biochim Biophys Acta 2012;1819(3–4):222–9.
[100] Skene PJ, Henikoff S. Histone variants in pluripotency and disease. Development 2013;140(12):2513–24.
[101] Schwartzentruber J, Korshunov A, Liu XY, Jones DT, Pfaff E, Jacob K, et al. Driver mutations in histone H3.3 and chromatin remodelling genes in paediatric glioblastoma. Nature 2012;482(7384):226–31.
[102] Bergmuller E, Gehrig PM, Gruissem W. Characterization of post-translational modifications of histone H2B-variants isolated from *Arabidopsis thaliana*. J Proteome Res 2007;6(9):3655–68.
[103] Bonenfant D, Coulot M, Towbin H, Schindler P, van Oostrum J. Characterization of histone H2A and H2B variants and their post-translational modifications by mass spectrometry. Mol Cell Proteomics 2006;5(3):541–52.
[104] Wiedemann SM, Mildner SN, Bonisch C, Israel L, Maiser A, Matheisl S, et al. Identification and characterization of two novel primate-specific histone H3 variants, H3.X and H3.Y. J Cell Biol 2010;190(5):777–91.
[105] Chu F, Nusinow DA, Chalkley RJ, Plath K, Panning B, Burlingame AL. Mapping post-translational modifications of the histone variant MacroH2A1 using tandem mass spectrometry. Mol Cell Proteomics 2006;5(1):194–203.
[106] Wisniewski JR, Zougman A, Kruger S, Mann M. Mass spectrometric mapping of linker histone H1 variants reveals multiple acetylations, methylations, and phosphorylation as well as differences between cell culture and tissue. Mol Cell Proteomics 2007;6(1):72–87.
[107] Jiang L, Smith JN, Anderson SL, Ma P, Mizzen CA, Kelleher NL. Global assessment of combinatorial post-translational modification of core histones in yeast using contemporary mass spectrometry. LYS4 trimethylation correlates with degree of acetylation on the same H3 tail. J Biol Chem 2007;282(38):27923–27934.
[108] Siuti N, Roth MJ, Mizzen CA, Kelleher NL, Pesavento JJ. Gene-specific characterization of human histone H2B by electron capture dissociation. J Proteome Res 2006;5(2):233–9.
[109] Boyne 2nd MT, Pesavento JJ, Mizzen CA, Kelleher NL. Precise characterization of human histones in the H2A gene family by top down mass spectrometry. J Proteome Res 2006;5(2):248–53.
[110] Hawkridge AM, Muddiman DC. Mass spectrometry-based biomarker discovery: toward a global proteome index of individuality. Annu Rev Anal Chem (Palo Alto Calif) 2009;2:265–77.
[111] Sundar IK, Nevid MZ, Friedman AE, Rahman I. Cigarette smoke induces distinct histone modifications in lung cells: implications for the pathogenesis of COPD and lung cancer. J Proteome Res 2014;13(2):982–96.
[112] Esse R, Florindo C, Imbard A, Rocha MS, de Vriese AS, Smulders YM, et al. Global protein and histone arginine methylation are affected in a tissue-specific manner in a rat model of diet-induced hyperhomocysteinemia. Biochim Biophys Acta 2013;1832(10):1708–14.
[113] Zhang K, Li L, Zhu M, Wang G, Xie J, Zhao Y, et al. Comparative analysis of histone H3 and H4 post-translational modifications of esophageal squamous cell carcinoma with different invasive capabilities. J Proteomics 2014.
[114] Garcia BA, Busby SA, Shabanowitz J, Hunt DF, Mishra N. Resetting the epigenetic histone code in the MRL-lpr/lpr mouse model of lupus by histone deacetylase inhibition. J Proteome Res 2005;4(6):2032–42.
[115] Leroy G, Dimaggio PA, Chan EY, Zee BM, Blanco MA, Bryant B, et al. A quantitative atlas of histone modification signatures from human cancer cells. Epigenetics Chromatin 2013;6(1):20.
[116] Tryndyak VP, Kovalchuk O, Pogribny IP. Loss of DNA methylation and histone H4 lysine 20 trimethylation in human breast cancer cells is associated with aberrant expression of DNA methyltransferase 1, Suv4-20h2 histone methyltransferase and methyl-binding proteins. Cancer Biol Ther 2006;5(1):65–70.
[117] Geiger T, Cox J, Ostasiewicz P, Wisniewski JR, Mann M. Super-SILAC mix for quantitative proteomics of human tumor tissue. Nat Methods 2010;7(5):383–5.
[118] Cazares LH, Troyer DA, Wang B, Drake RR, Semmes OJ. MALDI tissue imaging: from biomarker discovery to clinical applications. Anal Bioanal Chem 2011;401(1):17–27.

[119] Zhang Q, Xue P, Li HL, Bao YH, Wu LH, Chang SY, et al. Histone modification mapping in human brain reveals aberrant expression of histone H3 lysine 79 dimethylation in neural tube defects. Neurobiol Dis 2013;54:404–13.

[120] Mukherjee S, Rodriguez-Canales J, Hanson J, Emmert-Buck MR, Tangrea MA, Prieto DA, et al. Proteomic analysis of frozen tissue samples using laser capture microdissection. Methods Mol Biol 2013;1002: 71–83.

[121] Fowler CB, O'Leary TJ, Mason JT. Protein mass spectrometry applications on FFPE tissue sections. Methods Mol Biol 2011;724:281–95.

[122] Fanelli M, Amatori S, Barozzi I, Soncini M, Dal Zuffo R, Bucci G, et al. Pathology tissue-chromatin immunoprecipitation, coupled with high-throughput sequencing, allows the epigenetic profiling of patient samples. Proc Natl Acad Sci USA 2010;107(50): 21535–40.

[123] Amatori S, Ballarini M, Faversani A, Belloni E, Fusar F, Bosari S, et al. PAT-ChIP coupled with laser microdissection allows the study of chromatin in selected cell populations from paraffin-embedded patient samples. Epigenetics Chromatin 2014;7:18.

[124] Bissell MJ, Radisky DC, Rizki A, Weaver VM, Petersen OW. The organizing principle: microenvironmental influences in the normal and malignant breast. Differentiation 2002;70(9–10):537–46.

[125] Soldi M, Bremang M, Bonaldi T. Biochemical systems approaches for the analysis of histone modification readout. Biochim Biophys Acta 2014;1839(8):657–68.

[126] Soldi M, Bonaldi T. The proteomic investigation of chromatin functional domains reveals novel synergisms among distinct heterochromatin components. Mol Cell Proteomics 2013;12(3):764–80.

[127] Cong L, Ran FA, Cox D, Lin SL, Barretto R, Habib N, et al. Multiplex genome engineering using CRISPR/Cas systems. Science 2013;339(6121):819–23.

CHAPTER

11

High-Throughput Analysis of Noncoding RNAs: Implications in Clinical Epigenetics

Valerio Costa[1], *Maria R. Matarazzo*[1], *Miriam Gagliardi*[1], *Roberta Esposito*[1], *Alfredo Ciccodicola*[1,2]

[1]Institute of Genetics and Biophysics "Adriano Buzzati-Traverso", CNR, Naples, Italy; [2]Department of Science and Technology, University Parthenope of Naples, Italy

OUTLINE

Epigenetic Biomarkers and Diagnostics
http://dx.doi.org/10.1016/B978-0-12-801899-6.00011-5

1. INTRODUCTION

In recent years, the traditional view of RNA as a simple bridge between DNA and proteins has been challenged, and convincing evidence has revealed that the vast majority of the human genome encodes functional—even though noncoding—RNAs [1]. Indeed, advances in high-throughput sequencing have transformed the shared view on the complexity of transcriptome, highlighting that the protein-coding portion of the genome corresponds only to less than 2%. Conversely, tens of thousands of transcriptionally active regions produce transcripts with little-to-none protein-coding potential. The evolution of processes involved in the control of complex organisms is mainly due to the sophistication in regulatory potential of the noncoding portions of the genome. Surprisingly, the recent revolution in scientific knowledge has revealed that the vast majority of the genome is transcribed both in sense and antisense directions and that transcription is finely regulated in a cell context- and developmental stage-specific way. As an added level of complexity, a rigorous distinction between protein-coding and noncoding transcripts is misleading, since some noncoding RNAs (ncRNAs) can be translated and other mRNA transcripts may participate to regulatory and functional processes, rather than simply serving as templates for protein translation.

ncRNAs are divided into two major classes based on their size: (1) small ncRNAs (sncRNAs, 20–30nt) which are critical posttranscriptional regulators of target RNAs via RNA interference (RNAi), and/or able to modify other RNAs, including microRNAs (miRNAs), PIWI-interacting RNAs (piRNAs), and small nucleolar RNAs (snoRNAs); and (2) the heterogeneous group of long ncRNAs (lncRNAs, >200nt), such as transcribed ultraconserved regions (T-UCRs) and long intergenic ncRNAs (lincRNAs).

Despite the growing appreciation of the importance of ncRNAs in normal physiology and disease, our comprehension of clinical-related ncRNAs is still limited. Moreover, while most of the so far performed studies focus on cancer, only few of them have investigated the relevance of ncRNAs in other human disease.

In this chapter, we make an effort to provide a better understanding of how high-throughput sequencing platforms, and associated in silico tools led the scientific community to identify and characterize novel ncRNAs, allowing to predict their molecular and functional properties. In addition, we focus on how the epigenome regulates—and is regulated by—ncRNAs and what clinical significance these interplays may have in human pathologies.

2. ncRNAs: DEFINITIONS AND INNOVATIVE TECHNOLOGIES FOR THEIR ANALYSIS

2.1 General Concepts and Definitions

Among the above-mentioned ncRNAs, the most extensively studied are undoubtedly miRNAs. miRNAs are highly conserved 20–22 bp RNAs found in almost all eukaryotic cells. They have a role in posttranscriptional gene regulation mainly (but not exclusively) through the base pairing with the 3′ untranslated region (UTR) of the target transcript. This interaction leads to degradation of the target mRNA or to inhibition of its translation. siRNAs and piRNAs are sncRNAs generated through the processing of LINE-1 and other retrotransposons. Through the binding to PIWI proteins of the subfamily of Argonaute, these ncRNAs drive the silencing of repetitive elements and transposons [2].

Several evidences have shown that miRNAs are involved virtually in all biological processes, and it is likely that their alteration can contribute to—or in some cases determine—disease onset. Therefore, systematic analysis is needed to understand the biological processes in which miRNAs are involved and accurate quantification is crucial to assess their clinical significance in human diseases (discussed in the next paragraphs). However, several considerations can be done, especially when dealing with very sensitive and (relatively) expensive technologies.

Notably, the discovery of small RNAs has revealed only the tip of the iceberg [3]. Indeed, another widely studied class of ncRNAs comprises the lncRNAs. In the late 1990s at Yale University, John Rinn observed the presence of transcribed regions arising from genomic sites without any known protein-coding capability. In this pioneer study, focused on chromosome 22 transcriptional landscape, the authors discovered a steady stream of long interspersed transcribed regions without open reading frame (ORF) and other characteristics needed for protein translation [4]. Further independent large-scale studies in the mid-2000s definitely demonstrated that mammalian genomes produce thousands of lncRNAs of unknown function [5–8]. Generally, these transcripts are indicated as lncRNAs. This is "an umbrella term" used to define a class of transcripts longer than 200 bp not associated with protein-coding gene *loci* [5,7,9–12]. In particular, according to the GENCODE Consortium [12], lncRNAs can be classified with respect to protein-coding genes into "antisense" (if they intersect protein-coding loci on the opposite strand), "lincRNA" (when transcribed from intergenic regions), "sense overlapping" (that overlap intron and exon of a coding gene on the same strand), "sense intronic" (within the intron of a coding gene on the same strand) and "processed transcript" (without ORF and not classified in the other categories because of their complexity). Currently, the GENCODE Consortium has released a custom array, based on Gencode v15, to analyze the entire human lncRNA landscape (2 probes per lncRNA for 22,001 lncRNA transcripts; details of the entire array design can be found at http://www.gencodegenes.org/lncrna_microarray.html). Similar to RNA-encoding proteins, these transcripts can be spliced and polyadenylated, whereas they lack a protein-coding ORF. They have roles in transcriptional and epigenetic regulation by recruiting transcription factors and chromatin-modifying complexes to specific nuclear and genomic sites [8]. They are also involved in alternative splicing and other post-transcriptional RNA modifications through the assembly of nuclear domains containing RNA-processing factors [13,14]. The emerging structural and functional activities of lncRNAs have been categorized within a well-designed framework as molecules of signals, decoys, guides, and scaffolds [13].

Despite the large interest for the noncoding portion of the genome (i.e., the "dark matter" [15]), the most extensively studied class of long noncoding transcripts are the lincRNA. These have peculiar regulatory functions, even though only few of them have been biologically characterized. The first, and most convincing, examples were *H19*, *XIST*, *TSIX*, *HOTAIR*, and *AIR* (described below). Several works have definitely proven their crucial role in X-chromosome inactivation [16], *trans*-acting gene regulation [17], and imprinting [18].

In the past decade, large-scale international genomic projects have employed shotgun sequencing and microarray technology to analyze thousands of noncoding transcripts in different mammalian cells [5,7,8,19]. However, the first attempt to systematically identify a large catalog of lincRNAs has been performed by Guttman and colleagues in 2009 [7]. Using chromatin-state maps—created by chromatin immunoprecipitation followed by massively parallel sequencing (ChIP-Seq)—they discovered about 1600 new functional lincRNAs in mouse genome. The authors observed that genes actively transcribed by RNA Pol II have a typical "K4-K36 domain," i.e., the trimethylation of lysine 4 of histone H3 (H3K4me3) at their promoter and the trimethylation of lysine 36 of histone H3 (H3K36me3) along the gene body [20]. Thus, after the creation of a "K4-K36 domain" map they first identified—and then validated by microarray—transcriptionally active regions (of at least 5kb) outside protein-coding *loci* and not overlapping with annotated miRNAs or ribosomal RNAs (rRNAs). They also created a custom lncRNA expression array design with probes targeting the Gencode v15 human lncRNA annotation.

2.2 Purification Procedures for ncRNAs

miRNAs represent only a very small fraction (~0.01%) of the total RNA in a cell, and the lack of poly(A) makes it more difficult to efficiently enrich these RNAs from the "noisy" background of cellular RNA. In addition, the short length of mature miRNAs is not sufficient to use standard reverse transcription approach, based on oligonucleotide annealing and polymerase chain reaction assay. The presence of several miRNA families whose members differ by only one nucleotide, as well as the discovery of isomiRs variants, makes their analysis even more challenging [21]. Although unlike mRNAs, miRNAs are not polyadenylated, standard Trizol preparation (followed by ethanol precipitation) and commercially available column-based kits are useful approaches to isolate the entire small RNA fraction that mainly contains miRNAs. In addition, size selection by PAGE or AGO2 immunoprecipitation (AGO2-IP) are commonly used approaches (Figure 1(A)).

The introduction of innovative platforms for next-generation sequencing (NGS), and particularly one of its main applications, the RNA-Sequencing (RNA-Seq) [22], has significantly improved miRNA analysis. Indeed, one of the major advancements has been the possibility to simultaneously detect known miRNAs and to discover new ones, in a single experiment. Moreover, since it is a sequencing-based, rather than hybridization-based, procedure it allows to reliably distinguish between miRNAs within the same family, as well as to discriminate among isomiRs of variable length. A schematic workflow of the main experimental steps for miRNA sequencing on NGS platforms is provided in Figure 1(A).

As for miRNAs, the first fundamental step for lincRNA analysis, and above all for RNA-Seq-based experiments, is the isolation of RNA samples. In this regard, there are two main approaches that strictly depend on which RNA classes need to be analyzed: (1) ribodepletion (i.e., removal of abundant 5S, 5.8S, 18S, and 28S rRNA by magnetic beads or probes) or rRNA degradation (through exonuclease that specifically degrades 5′ uncapped RNA molecules) useful for the analysis of the whole transcriptome and (2) the poly(A) enrichment, needed to capture all the transcripts bearing a poly(A) tail (Figure 1(A)). However, a combination of both methods can also be used to increase the purity of poly(A) samples. Interestingly, although polyadenylation was thought to occur only at the 3′ end of mRNAs, the first RNA-Seq-based studies [11,19] demonstrated that a large fraction of lincRNAs are polyadenylated. In particular, Guttman and colleagues [19] using RNA-Seq reconstructed a large catalog of cell-specific conserved multiexonic lincRNAs in mouse embryonic stem cells, neuronal precursor cells, and lung fibroblasts [19]. Subsequently, John Rinn's group—applying RNA-Seq on the polyadenylated fraction of 24 tissues and cell types—developed a very comprehensive catalog of human lincRNAs [11], confirming the presence of the poly(A) tail in mature lincRNA transcripts.

2.3 New Technologies and Bioinformatics Tools for ncRNA Analysis

In the last years, NGS diffusion has revolutionized genetics and genomics studies [23]. The parallel development of new computational pipelines and analytical methodologies [22] has allowed to simultaneously analyze expressed ncRNAs in a single experiment. For instance, several independent studies based on RNA-Seq have allowed identifying new, often low-expressed and cell-specific, long noncoding

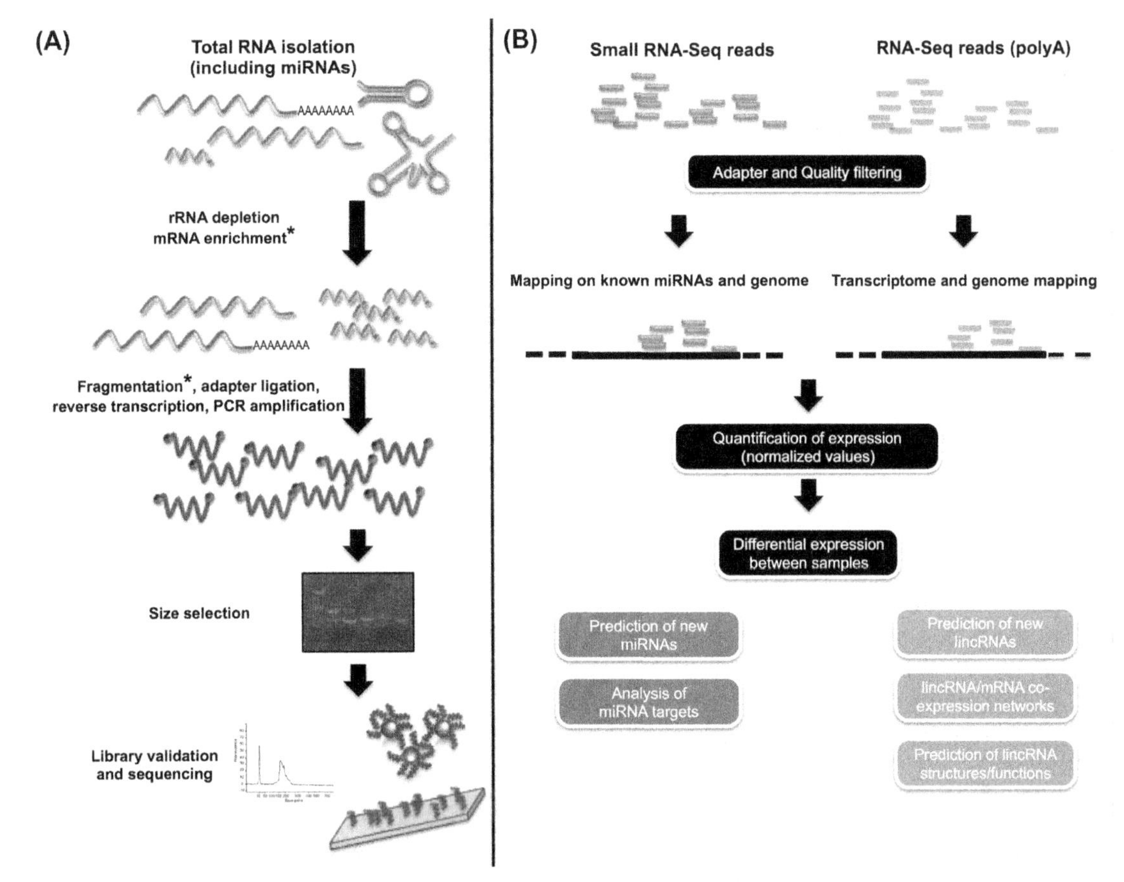

FIGURE 1 **Schematic workflow of the main experimental and computational steps in microRNA (miRNA) and long intergenic noncoding RNA (lincRNA) analysis.** (A) General scheme of the main "wet" phases needed for miRNA and lincRNA analysis. As depicted, most of currently used RNA-Sequencing (RNA-Seq) protocols start from total RNA and adopt different procedures to remove abundant ribosomal RNA (rRNA) molecules. Asterisks indicate experimental steps that are needed only for poly(A) transcripts (mRNA and lincRNA). Size selection is generally performed only for small RNA (miRNA) sequencing, even though some protocols (and platforms) still use this procedure to select complementary DNA libraries within a desired size range. For schematic purposes, solid-phase- (Illumina) and emulsion-based (Roche, ABI) sequencing supports are shown. (B) Simplified pipeline describing NGS data analysis. Black boxes indicate computational steps that are common to small RNA (miRNAs) and poly(A) (mRNA and lincRNA) sequencing, even though different parameters/tools are used for miRNA and lincRNA filtering and quantification. For instance, normalization of lincRNA expression takes into account also transcript length (Reads, or fragment, Per Kilobase of transcript per Million of mapped reads, RPKM or FPKM, described in the text). Blue (dark gray in print versions) and green (light in print versions) boxes indicate data analysis steps that are strictly specific of small RNAs and lincRNAs, respectively. PCR, polymerase chain reaction.

transcripts. Some of these, currently under clinical evaluation as potential cancer biomarkers, are discussed below. However, despite the pros in the use of NGS, there are some crucial points to take into account in the analysis of both miRNAs and lincRNAs.

Regardless of the platform used for high-throughput sequencing, the output of a typical NGS-based sequencing experiment consists in millions of short reads. In Figure 1(B), the main computational steps in RNA-Seq analysis are schematized. Briefly, after reads' quality evaluation (and filtering of low-quality reads), the first processing step consists in end trimming of the reads. This step, based on reads' quality in mRNAs experiments, is fundamental for miRNAs, as usually NGS platforms sequence more nucleotides than the exact length of mature miRNAs (normally up to 24 nt). Thus, reads will contain part of the 3′ adapters that have to be removed to avoid the bias in the next mapping procedure.

Cutadapt (http://journal.embnet.org/index.php/embnetjournal/article/view/200/458), Trim Galore [24], or Trimommatic [25] are some of the tools that are commonly used for this purpose. Subsequently, reads that have been filtered and trimmed can be aligned to the reference genome usually through short read aligners like GSNAP [26], Maq [27], SOAP [28], ELAND, or BWA [29]. For the analysis of known miRNAs, RNA-Seq reads can be aligned against miRNAs annotated in reference databases (i.e., miRBase; http://www.mirbase.org). Reads not corresponding to any previously annotated miRNA, but mapping elsewhere on the reference genome, can reliably represent a new source of miRNAs. Specific tools can be used in order to predict these new miRNAs. However, especially in this case, functional validations are needed to verify the presence of new miRNAs. A typical miRNA sequencing study is meant to assess their differential expression between "healthy" and "disease" samples, or among different experimental conditions. Differential miRNA analysis takes into account normalized expression values, computed by read counts (*rpm*, reads per million of mapped reads). However, even normalized read counts cannot be used for absolute miRNA quantification as fragment composition of each sample can be significantly altered by the different methods used for RNA extraction and library preparation [30,31]. Thus, miRNA analysis is generally limited to relative comparisons between samples. The most obvious next step in miRNA analysis is to understand miRNA–mRNA interaction network. To this purpose, it is essential to correlate miRNAs to gene expression data within the same biological system/sample. Currently, there are many freely available tools to address almost every aspect of miRNA analysis, including the exploration of validated and computationally predicted miRNA targets [32]. Some useful computational tools and algorithms are DIANA Tools (http://www.microrna.gr/), TargetScan (http://www.targetscan.org), miTalos (http://mips.helmholtz-muenchen.de/mitalos/), and StarBase (http://starbase.sysu.edu.cn). These software/algorithms allow (1) the identification of miRNAs' target and (2) of molecular pathways potentially altered by the miRNAs under evaluation, and (3) the reconstruction of entire competing endogenous RNAs (ceRNAs) networks.

One of the major advantages of RNA-Sequencing is the ability to detect even low-expressed transcripts, and to avoid background and cross-hybridization issues, often reported in microarray analyses [33,34,35]. Taking advantage of RNA-Seq, the first large-scale studies performed so far on ncRNAs, and particularly the works of Cabili and colleagues [11] and of the GENCODE Consortium [12], identified—and reliably quantified—even low-expressed lincRNAs. Moreover, since they used a sequencing-based approach they could also annotate for the first time their nucleotide sequences in public repositories, allowing the entire scientific community to study this new class of ncRNAs.

Since lncRNAs are longer than 200 nt, and most of them are polyadenylated, they can be studied with protein-coding genes in a single RNA-Seq experiment. In light of these considerations, a typical pipeline that researchers can use for lncRNAs data analysis is very similar to the ones proposed for protein-coding mRNAs. In detail, high-quality reads are mapped against the most updated release of the reference transcriptome and/or of the genome with a mapping algorithm, such as TopHat2 [36] and STAR [37], to cite few. Mapped reads not corresponding to any annotated transcript in existing databases can reliably represent novel unannotated transcriptional units, both coding and noncoding. To completely reconstruct the bona fide sequence of entire novel transcripts, ab initio assembly, performed by different bioinformatics software, provides an unbiased method for gene discovery. This approach has been successfully applied to detect novel lncRNAs. To narrow the analysis on lncRNAs, novel putative identified transcripts can, for instance, be filtered for their length, retaining only transcripts longer than 200 nt. Moreover, a fundamental step in the analysis of new lncRNAs is the assessment of protein-coding potential. Softwares such as CPC [38], CPAT [39], and GeneID [40] are some of the tools that are commonly used to determine the lack of potential ability for new lncRNAs.

Quantification of gene expression and the analysis of differentially expressed genes can be performed with the same tools available for a typical RNA-Seq experiment for coding mRNAs. For instance, tools like Cufflinks [41] and HTSeq [42] can be used to measure gene expression levels through reads' count, and Cuffdiff [41], or EdgeR [43] and DeSeq [44], implemented in R language, can be used to identify differentially expressed genes between different tissues and/or conditions. These considerations hold true both for the analysis of new data sets and for the reanalysis (or for large meta-analyses) of raw RNA-Seq data freely available in public genomic repositories such as Short Reads Archive (http://www.ncbi.nlm.nih.gov/sra), Gene Expression Omnibus (http://www.ncbi.nlm.nih.gov/geo/), and Expression Atlas (https://www.ebi.ac.uk/gxa/home).

However, to date, there is still a relatively low number—if compared to microarray data sets—of freely available RNA-Seq data sets in public repositories. Such consideration holds true particularly for clinically oriented studies. Moreover, as previously mentioned, sequencing coverage is a crucial aspect for RNA-Seq-based studies. Unfortunately, since NGS has been only recently employed to profile lincRNAs, there is a limited amount of public data sets with enough sequencing coverage to allow their comprehensive analysis. These considerations—discussed in detail in the next paragraphs—highlight some limitations of NGS to usefully identify clinically relevant lincRNAs [45].

3. HOW THE EPIGENOME CONTROLS ncRNAs AND HOW ncRNAs CONTROL THE EPIGENOME

3.1 Epigenome and miRNAs Interactions

Epigenetics is the study of heritable changes in gene expression that are not encoded in the nucleotide sequence. Rather they depend on the alteration of other properties such as DNA methylation and histone modification patterns, which are inherited through cell divisions and across generations. How cells and tissues with radically different phenotypic aspects can arise, although they all bring common genetic material, is explained by cascades of gene expression changes that are guided by epigenetic processes [46].

Epigenetic modifications and the control of multiple genes by single ncRNAs are the center of the same regulatory networks and are critical to guide the cellular differentiation in development and the regulation of gene expression in somatic tissues. Indeed, epigenetic chromatin remodeling mediated by DNA hypermethylation, histone modifications, or antisense RNA interactions can transcriptionally silence coding genes. On the other hand, miRNA molecules can posttranscriptionally regulate the expression of hundreds of genes, having broad direct and indirect effects. Therefore, there may potentially be a multitude of interactions at the interface of miRNAs and epigenetics. For instance, epigenetic mechanisms involving the regulatory regions of miRNA-encoding genes (i.e., methylation of their promoters) can inactivate their transcription.

An integrative study has calculated that DNA methylation is implicated in the epigenetic regulation of more than 11% of miRNA genes [47]. By looking at the distribution of miRNA genes related to CpG islands and host genes, it has been reported that more than half of the miRNA genes are at intergenic loci and are transcribed at their own transcription start sites. Such miRNA genes share all the features of protein-coding genes, being often the CpG island the only region regulating their transcription. Accordingly, primary transcripts of the majority of intergenic miRNAs are 3–4 kb in length and contain poly(A)-tail. Among the miRNAs transcribed as single units, miRNA gene clusters include 37% of the total miRNA in the human genome [48].

Up to 40% of miRNA genes are localized at intragenic position within host genes. Among intronic miRNA genes, less than 30% are transcribed from intrinsic promoters, and they may be controlled both by their own CpG island and the CpG island belonging to the host gene. Most intronic miRNA genes are expressed together with a protein-coding host gene under the control of its promoter and CpG island [49].

miRNAs can conversely be the mediators of transcriptional gene silencing. Indeed, by targeting complementary promoters, or silencing the expression of epigenetic modifier genes, such as DNA methyltransferases (DNMTs), histone deacetylases (HDACs), histone acetylases (HAT), histone demethylases (HDMs), and chromatin remodelers (SWI/SNF), miRNAs can lead to global changes in the epigenome (Figure 2).

3.2 Epigenome and lncRNA

Differently from miRNAs, lncRNAs have a wide variety of regulatory roles involving sensory, guiding, scaffolding, and allosteric capacities, which derive from the folding of their typical modular domains. In this diverse functional repertoire, their capability to function as epigenetic modulators has been characterized in detail [50,51]. Several lincRNAs physically interact with chromatin-modifying proteins and guide their catalytic activity to specific *loci* in the genome, thereby modulating chromatin states and affecting gene expression. Figure 3 summarizes the known and possible network linking epigenetic mechanisms and lincRNAs' activity.

This regulatory potential in combination with the huge number of lincRNAs suggests that lincRNAs may be part of a wide epigenetic regulatory network. Indeed, in a recent review Lee et al. highlighted that lncRNAs are involved in almost every epigenetic regulation event [52].

Furthermore, as it occurs for miRNAs, the transcription itself of lincRNAs may be silenced by DNA hypermethylation and histone modifications. In spite of the growing view that lincRNAs are modulators of cancer, the meaning of DNA methylation patterns of lincRNAs in cancer remains partial. In a latest study, lincRNAs were found to include the majority of the aberrantly methylated lncRNA promoters in the breast cancer (MIM#114480) samples and it was shown that aberrant methylation patterns occur not only on CGIs, but also on 5′ and 3′ CGI shores. Moreover, most hypermethylated lncRNAs were associated with decreased H3K4me3 density, and that the hypomethylated lncRNAs were associated with elevated H3K4me3 density. These results indicate that DNA methylation and histone modifications are two mechanisms regulating lncRNA expression [53].

LincRNAs can guide chromatin-modifying proteins and their catalytic activity in trans to multiple sites spread through the genome. For instance, the lincRNA *HOTAIR* is transcribed from within the HOXC cluster, and it is able to repress genes in the HOXD cluster, by binding to the polycomb repressive complex (PRC2) and recruiting it to the locus [17]. PRC2 is a methyltransferase that trimethylates H3K27 to silence the transcription of specific genes [54,55]. Many lncRNAs bring about local functions in *cis*, engaging chromatin-modifying proteins to modify their proximal epigenetic region. For example, *loci* with parent of origin-specific

miRNAs and the epigenetic machinery

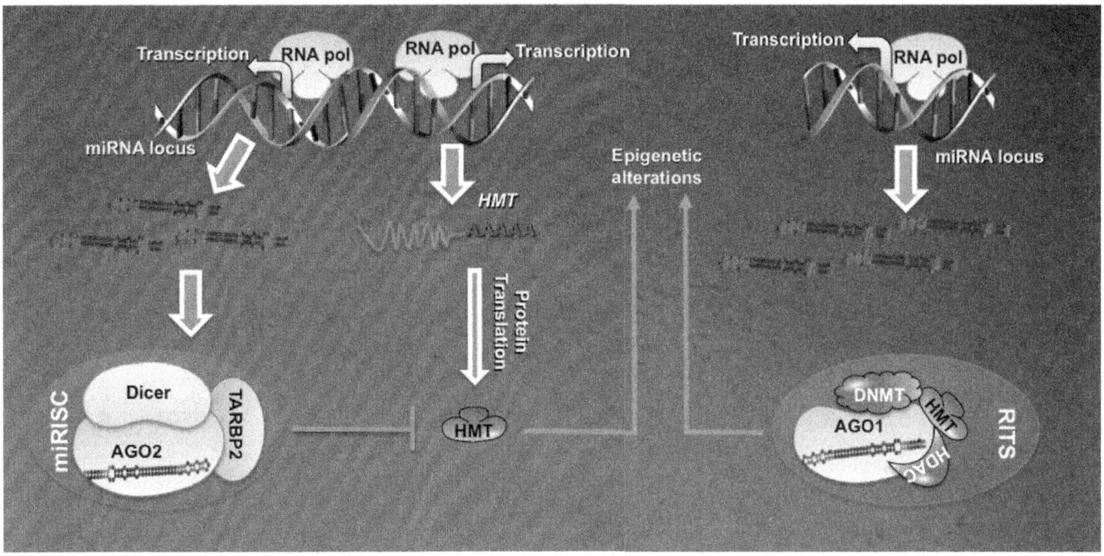

FIGURE 2 MicroRNAs (miRNAs) are able to modulate the epigenetic machinery. Mature miRNAs can regulate the expression of genes encoding epigenetic modifiers (histone methyltransferases or HMTs, in orange (light gray in print version)) by posttranscriptional gene silencing (on the left). Mature miRNA is loaded on the miRNA-RNA-induced silencing complex (miRISC) and reaches the cytoplasm to bind—and repress—its specific target (HMT-encoding mRNA). Silencing of HMT leads to global alterations in the epigenome. Mature miRNAs can also induce transcriptional gene silencing of complementary target promoters (on the right) if they are loaded into AGO1. Indeed, this protein is the core of the RNA-induced transcriptional silencing (RITS) complex (containing histone deacetylases, HDACs, in green (light gray in print version)). Mature miRNA functions as a guide for the RISC that induces the formation of heterochromatin as well as DNA methylation at the target promoter, driven by DNA methyltransferases (DNMTs, in green).

expression, named imprinted regions, often contain many lncRNAs [56]. These RNAs can silence the expression of neighboring genes on the parental allele from which they are expressed. In this context, the antisense transcript *AIR* is transcribed by the paternal allele, where it recruits the chromatin-modifying complex G9a, an H3K9me2 methyltransferase [18], thereby silencing the expression of neighboring paternal *Igf2r*, *Slc22a2*, and *Slc22a3* genes. As an additional example, the Kcnq1ot1 transcript is able to impart H3K4 trimethylation and H3K9 methylation, respectively, through the binding of both G9a and PRC2 [57].

Some studies have demonstrated that few lncRNAs bind chromatin proteins, which exert their function by inserting histone modifications permissive for transcription (e.g., trithorax) in mouse ES cells [58]. Local epigenetic changes initiated by lncRNAs can originate larger epigenetic effects, as *Xist* for the inactive X chromosome in female cells. After transcription, *Xist* accumulates at many sites across the X chromosome that subsequently results in the progression of facultative heterochromatin formation [16,59]. LncRNAs similarly allow for the spreading of heterochromatin at pericentric satellites [60]. These highly repeated regions, localized around the centromere, undergo bidirectional transcription during early development, and the strand-specific lncRNAs can guide heterochromatin protein 1 (HP1) to the satellite repeats to establish a precise modified chromatin profile [61].

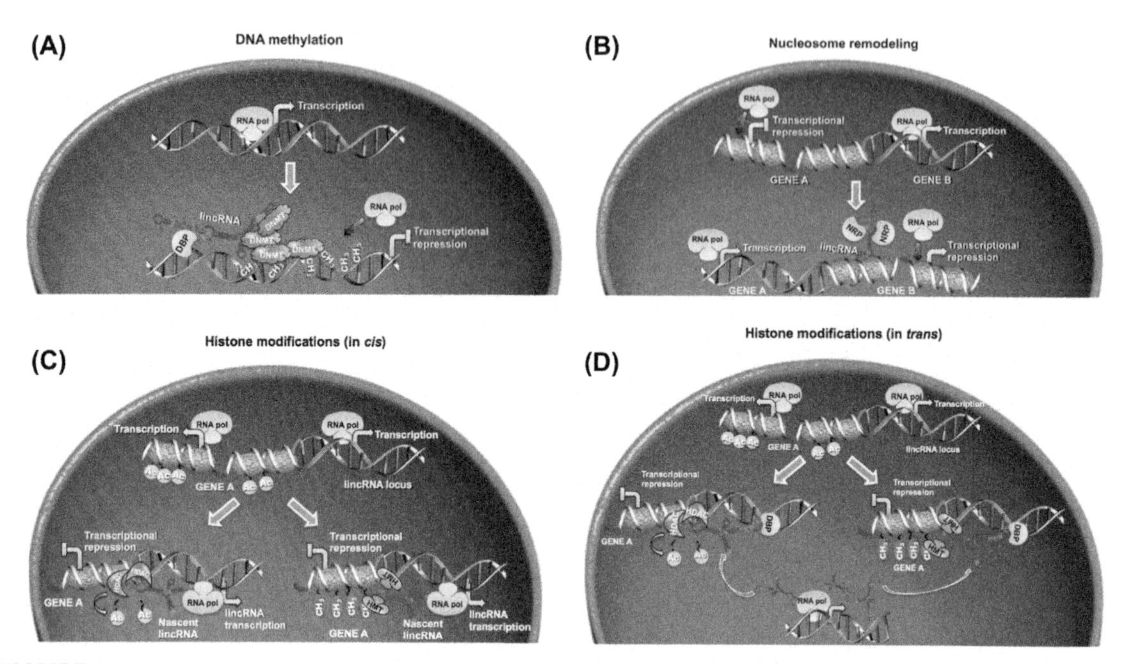

FIGURE 3 Potential epigenetic roles played by long intergenic noncoding RNAs (lincRNAs). (A) Schematic model of lincRNAs involved in DNA methylation. In this cartoon, the lincRNA (in red (dark gray in print versions)) directly interacts with DNA methyltransferases (DNMTs, in green (light gray in print versions)) and DNA-binding protein/s (DBP, in orange (gray in print versions)), functioning as scaffold to drive DNMTs to specific promoters. It leads to the methylation of the promoters of target genes and their transcriptional repression. (B) Model of lincRNAs involved in nucleosome remodeling. In the model, the lincRNA interacts with nucleosome remodeling proteins (NRPs, in violet (white in print versions)) determining the repositioning of the nucleosomes (blue (gray in print versions) ovals) from gene A (upper part) to gene B (lower part). Increased nucleosome compaction around gene B determines its repression. (C–D) Models of *cis*- and *trans*-acting lncRNAs. (C) RNA Polymerase II transcribes the lincRNA (shown in red (dark gray in print versions)) that is tethered in proximity of its transcriptional site. Here it recruits (on the left) histone deacetylases (HDACs, in green (gray in print versions)) that remove acetyl from histones, or recruits (on the right) histone methyltransferases (HMTs, in orange (light gray in print versions)) that lead to specific methylation (CH_3 groups in violet (white in print versions)) of histones. Both conditions lead to gene silencing. (D) Model of lincRNAs acting in *trans* on gene expression. The lincRNA (in red (dark gray in print versions))—transcribed from a distant *locus*—recruits histone deacetylases (HDACs, on the left) or histone methyltransferases (HMTs, on the right). In both cases the *trans*-acting lincRNA functions as scaffold to recruit proteins that induce gene silencing to a different, distant *locus*. On the opposite (models not shown), lincRNAs can act both in *cis*- and *trans*-recruiting demethylases and/or acetyltransferases to specific gene promoters determining their transcriptional activation.

4. ncRNAs AND EPIGENETICS: CLINICAL RELEVANCE

NcRNAs and epigenetics represent essential players in the regulation of biological processes. NcRNAs are in fact capable of modifying and being modified by the epigenetic machinery, highlighting the importance of this network in human physiology and pathology. Therefore, it is expected that several ncRNAs may be implicated in various pathological contexts. Indeed, a wide range of connections between the noncoding RNome and the epigenome, as well as the impact of their interaction in human pathogenesis, early diagnosis, and disease prognosis is emerging.

While the studies of aberrant epigenetic regulation patterns in miRNA and lncRNA genes at a global scale are still limited, the integration of whole-genome profiling of epigenetic modifications, e.g., DNA methylation, and transcriptional data sets promises to be a powerful tool to better understand the cancer biology. Remarkable is the latest finding that aberrant DNA methylation of either lncRNA or miRNA genes in breast cancer can perturb specific common pathways, such as the cell cycle, and MAPK signaling pathways, indicating that ncRNAs mediate the deregulation of these pathways in a coordinated manner [53].

Cell cycle perturbation can result in abnormal proliferation and inability of a cell to undergo differentiation and/or apoptosis, which contribute to the development of cancer. Specifically, the breast tumor-derived cell lines exhibited lower *CDK6* activity level as a consequence of promoter hypermethylation. Moreover, the gene was also found targeted by two hypomethylated miRNAs, hsa-miR-21 and hsa-miR-29b, which may further repress the activity of *CDK6*, implying complementary effects of DNA methylation and miRNA-mediated regulation. As an additional example, the authors show that the hypomethylated miRNA hsa-miR-16 regulates the hypermethylated cyclin D2 (*CCND2*) gene. This cyclin functions as a regulatory subunit of *CDK4* or *CDK6*, the activity of which is required for the G1/S transition in the cell cycle. Moreover, a similar aberrant methylation pattern to the *CCND2* gene was observed at two hypermethylated lncRNAs, suggesting that these two lncRNAs play key roles in cell cycle regulation.

Another pathway enriched among the dysregulated functions in breast cancer cell lines was the MAPK signaling pathway, which controls important cellular processes, such as growth, proliferation, differentiation, migration, and apoptosis. In breast cancer, most of genes belonging to this pathway were shown aberrantly methylated. For instance, *KRAS* gene was hypomethylated and was regulated by two miRNAs (hsa-miR-16 and has-miR-98). Moreover, these two miRNAs, which are implicated in the breast cancer development, were aberrantly methylated. Therefore, these integrated functional analyses indicated that aberrant methylation of ncRNAs widely disturbs processes associated with the development and progression of breast cancer [53].

In addition to unravel the cancer pathogenesis, there is no doubt that the RNome/epigenome interplay represents a new opportunity for future studies aimed to develop new therapeutic strategies [62]. However, only few studies have recently taken into account the role of ncRNAs from a clinical–pathological perspective, and most of them have focused on miRNAs.

4.1 miRNAs in Cancer

Over the last decade, many findings have contributed to clarify the role of epigenetics in miRNA expression. Moreover, ever since the recognition of DNA methylation as a promising biomarker in cancer, many studies have anticipated that epigenetically regulated miRNAs could have novel roles in prognosis, early detection, and predictive values in cancer [63,64]. Few examples of them are listed in Table 1.

For instance, the promoter of the gene encoding miR-129-2 has been frequently reported as methylated in hepatocellular carcinoma cells (HCC, MIM#114550) but not in normal liver cells and tissues [65]. In addition, promoter methylation of miR-129-2 gene has been detected in the plasma of HCC patients and not in healthy individuals or patients with cirrhosis, indicating that such methylated DNA region may be used as a specific, potential marker for early HCC diagnosis.

In acute lymphoblastic leukemia (ALL, MIM#613065), increased methylation of miR-124a encoding gene promoter has been detected and it serves as an independent prognostic factor for overall and disease-free survival and associated with higher relapse and mortality rates. Epigenetic regulation of this miRNA has been shown to mediate increased expression of

TABLE 1 List of Disease-Associated miRNAs and lncRNAs Linked to Epigenetic Regulation

Name	Class	Disease	References
miR-129-2	miRNA	HCC	[51]
miR-124-3	miRNA	Renal carcinoma	[52]
miR-124a	miRNA	ALL	[53]
miR-148a	miRNA	Colon cancer; melanoma	[54,55]
miR-34b/c	miRNA	Colon and breast cancer; melanoma	[54,55]
miR-9	miRNA	Colon and breast cancer; melanoma	[55]
miR-212	miRNA	Lung cancer	[56]
miRNA-29	miRNA	Lung cancer; Rett syndrome	[61–93]
miRNA-152	miRNA	Hepatobiliary cancer; ALL	[62,63]
miR-148	miRNA	Breast and gastric cancer; MM (MIM#254500); HCC	[64]
miR-449a	miRNA	Prostatic cancer	[65]
miR-144	miRNA	Bladder cancer	[66]
miR-101	miRNA	Bladder cancer; prostate and breast cancer	[67,68]
miR-30a/d	miRNA	Rett syndrome	[92]
miR-381	miRNA	Rett syndrome	[92]
miR-495	miRNA	Rett syndrome	[92]
miR-146	miRNA	Rett syndrome	[93]
miR-338	miRNA	ICF syndrome (Type1)	[95]
miR-196a	miRNA	ICF syndrome (Type1)	[95,96]
miR-483-5p	miRNA	BWS	[99]
ANRIL	lncRNA	Prostatic cancer	[59,60]
HOTAIR	lncRNA	Laryngeal carcinoma; primary breast and metastasis tumors	[73,74]
Xist	lncRNA	Hematologic cancer	[76]
PCAT1	lncRNA	Prostatic cancer	[77]
MALAT1	lncRNA	HCC; lung cancer; renal carcinoma	[78–82]
AANRASSF1	lncRNA	Breast and prostatic cancer	[83]
lincRNA-BC4	lncRNA	Breast cancer	[84,85]
lincRNA-BC5	lncRNA	Breast cancer	[84,85]
lincRNA-BC8	lncRNA	Breast cancer	[84,85]
linc-DMRT2	lncRNA	Cardio/metabolic diseases	[86]
linc-TP53I13	lncRNA	Cardio/metabolic diseases	[86]
AK081227	lncRNA	Rett syndrome	[94]
TERRAs	lncRNA	ICF syndrome (Type1)	[98]
KCNQ1OT1	lncRNA	BWS	[100–102]

Abbreviations: MM, multiple myeloma; HCC, hepatocellular carcinoma; ALL, acute lymphoblastic leukemia; BWS, Beckwith–Wiedemann syndrome; ICF, immunodeficiency, centromeric instability, and facial anomalies syndrome; miRNA, microRNA; lncRNA, long noncoding RNA.

CDK6, which contribute to abnormal proliferation of ALL cells in vitro and in vivo [66].

Similarly, hypermethylation of the CpG islands proximal to miR-124-3 encoding gene in renal carcinomas (RCC, MIM#144700) has been linked with adverse disease state, such as tumor metastasis formation. This hypermethylation appears tumor-specific and related to poor recurrence-free survival, indicating that it could be involved in cancer progression. It suggests miR-124-3 as a putative biomarker for patients that can take advantage from targeted therapy based on small molecule mediators [67].

Furthermore, Lujambio and colleagues [68] found specific DNA methylation pattern within the promoters of miRNAs associated with the metastatic stage of tumors, such as for lymph node metastasis derived from colon cancer (CRC, MIM#114500), melanoma (MIM#155600), and head and neck cancer [68] (HNSCC, MIM#275355). Epigenetic silencing of miR-148a and miR-34b/c genes has been shown to mediate the activation of oncogenes and metastasis-associated genes such as c-MYC, *E2F3*, *CDK6*, and *TGIF2* [68]. Also, CpG islands hypermethylation in genes encoding miR-34b, miR-34c, miR-148a, and miR-9 has been associated with metastasis in many different tumor types such as colon, melanoma, breast (summarized in Esteller et al. [69]).

Histone modifications, rather than DNA methylation, can induce miRNA suppression. Indeed, the silencing of miR-212 in lung cancer (MIM#211980)—associated to later stages of the disease—is due to altered histone methylation rather than DNA methylation, even though miRNA's promoter is located in a CpG island [70].

4.2 lncRNAs in Cancer

Epigenetic modifications may also occur in the promoters of lncRNAs-enconding genes, thereby controlling their spatial/temporal expression. Indeed, lncRNAs show more pronounced tissue-specific expression patterns than protein-coding genes, offering a variety of benefits in clinical applications [71] both for early diagnosis and disease prognosis.

Remarkably, it has been reported that CpG island hypermethylation-associated silencing of ncRNAs transcribed from ultraconserved regions (T-UCRs) in human cancer. This mechanism has been reported for several T-UCRs in cancer type-specific manner [72]. These findings suggest that T-UCRs may contribute to cancer development and progression acting as tumor suppressor genes.

ANRIL, also known as CDKN2b antisense or *CDKN2B-AS1* and mapping to the INK4b-ARF-INK4a *locus*, is an lncRNA regulated by the polycomb repressive complex (PRC). This *locus* is one of the most frequently altered in tumors and, in a subset of them; it is silenced by epigenetic modifications [73]. This finding indicates this lncRNA as potential diagnostic and prognostic target [74]. Notably, as already discussed, lncRNAs are also capable of modifying the epigenetic signature, targeting—directly or indirectly—the components of the epigenetic machinery via a regulatory loop (Figure 3). A relevant observation in this direction comes from the finding that miR-29 family negatively modulates DNMT3A and DNMT3B enzymes in lung cancer, reducing survival rates in patients with higher levels of DNMT3A [75]. DNMTs are also targets of miR-152 in hepatobiliary cancer [76] and in ALL [77], and targets of miR-148 in various cancer cell lines [78]. Other miRNAs regulate the epigenetic machinery by targeting crucial histone marks enzymes. For instance, miR-449a targets HDAC in prostatic cancer (MIM#176807) [79], miR-144 targets *EZH2* in bladder cancer (MIM#109800) [80], and miR-101 targets *EZH2* in bladder [81], prostate, and breast cancer [82].

As previously mentioned, different studies have reported lncRNA signatures associated to specific cancer and other human diseases [83–86]. It indicates the potential clinical relevance of all ncRNAs and their underestimated impact on human health. In this regard, a growing number of studies are proposing both miRNAs and lncRNAs as candidate disease biomarkers and as key factors to understand still uncharacterized aspects of tumor biology.

Among the lincRNAs, the well-known *HOTAIR* (homeobox transcript antisense RNA), is overexpressed in laryngeal squamous cell carcinomas when compared to adjacent healthy tissues, and is associated with poorly differentiated tumors, lymph node metastasis, and advanced disease stages, indicating it as a marker of poor prognosis [87]. *HOTAIR* is predicted to produce oncogenic phenotype through PTEN methylation. It is also capable to affect histone H3 methylation in a PRC2 regulation-dependent fashion. Moreover, its overexpression in both primary breast and metastatic tumors has been associated to higher incidence of metastasis and to reduced overall survival [88].

Although RNA-Sequencing has been used to study gene expression changes in different human diseases (reviewed in Ref. [89]), few studies have reported the alteration of other lincRNAs involved in the epigenetic machinery in the context of human health. The first—and probably well-studied lincRNA—*Xist* has been reported as potent tumor suppressor in hematologic malignancies in mice, suggesting a role beyond X inactivation and indicating it as a clinically relevant ncRNA [90].

Using innovative ab initio assembly of RNA-Sequencing reads on 102 prostate tissues and cells lines, Presner and colleagues [91] have recently identified the lincRNA *PCAT1* as a prostate-specific regulator of cell proliferation, target of the PRC2. ChIP (chromatin immunoprecipitation) revealed that the PRC2 complex directly binds to the promoter of *PCAT1* repressing it. Moreover, RIP (RNA immunoprecipitation) showed that this lincRNA reciprocally binds PRC2 creating a feedback inhibitory loop [91].

Another example of clinically relevant lincRNA is *MALAT1*, overexpressed in many cancer types [92]. Through Kaplan–Meier analyses, *MALAT1* has been identified as prognostic factor for patient survival in stage I non-small-cell lung cancer [93]. The same lincRNA has been also discovered in a rare subtype of renal cell carcinoma as fusion partner of *TFEB* gene in an oncogenic chimeric transcript [94,95]. The finding that *MALAT1* translocation activates *TFEB* expression suggests a new potential pathogenic mechanism of lincRNAs in cancer etiology, indicating *MALAT1* gene fusions as clinically relevant biomarkers of disease.

Interestingly, *MALAT1* binds—in cooperation with another lincRNA, *TUG*—a PRC1 subunit component (i.e., PC2). Yang and colleagues [96] demonstrated that methylation/demethylation of PC2 determines the relocation of growth control genes between polycomb bodies (PcGs) and interchromatin granules (ICGs) in response to growth signals. *TUG1* and *MALAT1*, located in PcGs and ICGs, respectively, differentially bond to methylated and unmethylated PC2 inducing this movement. Thus, through the assembly of multiple corepressor/coactivator complexes, *MALAT1* is able to alter histone marks, which are read by PC2 [96].

Another RNA-Seq-based study reported that the unspliced lncRNA *AANRASSF1*—transcribed antisense to the tumor suppressor gene *RASSF1*—is aberrantly expressed in breast and prostate tumor cell lines [97]. Its ectopic overexpression increases PRC2 occupancy and histone H3K27me3 repressive marks at the promoter of RAS-association domain family protein 1 isoform A (RASSF1A) through the formation of RNA/DNA hybrid. Such event reduces *RASSF1A* expression, leading to increased cell proliferation.

RNA-Sequencing has been recently used in a large panel of human cancers to identify aberrantly expressed lincRNAs [98,99]. However, despite the clinical relevance of specific lincRNAs, such as lincRNA-BC4, BC5, and BC8 (correlated with breast cancer stage and estrogen receptor's expression), a putative role in epigenetic mechanisms has not yet been proved for these lincRNAs. The same consideration holds true also in other clinical contexts where

RNA-Seq has been used to profile lincRNAs. Indeed, in a recent work, Liu et al. [100] have reported inflammatory lincRNAs induced in presence of endotoxemia. Two lincRNAs, linc-DMRT2 and linc-TP53I13, were completely suppressed, in presence of endotoxinemia, in the adipose tissue of obese individuals. In this work, the authors also provided the first relevant example of LPS-modulated lincRNAs that overlap single-nucleotide polymorphisms associated (by genome-wide association studies, GWAS) with cardio/metabolic traits in humans [100]. This finding is particularly relevant as more than 85% of the SNPs currently associated to complex human diseases/traits by GWAS are not located in protein-coding genes and may overlap ncRNAs, such as described in Liu et al. [100]. However, despite the identification of these two tissue-specific lincRNAs during endotoxinemia, a correlation with epigenetic mechanisms has not been hypothesized. Similarly, we recently reported the differential expression of 45 lincRNAs in endothelial progenitor cells in the context of Down syndrome (MIM#190685, ORPHA870) [101]. We believe that further efforts toward this direction would significantly help to assess the clinical relevance of these new potential biomarkers.

4.3 ncRNAs and Epigenetics: Their Relevance in Human Diseases

Looking at the general impact of ncRNA disruption in human diseases, we have only just lifted the lid of the Pandora's box. Like in cancer, also the so-called chromatin disease shows aberrant miRNA transcription signatures, and insights into the function of other ncRNAs in these human diseases are just beginning to emerge [102]. Indeed, latest findings just linked ncRNAs to the molecular and phenotypical defects of human disorders with deficiencies in (1) epigenetic modifiers, like MeCP2 (methyl-CpG-binding-protein-2) for the Rett syndrome (RTT, MIM#312750, ORPH778) and DNMT3B (DNA-methyltrasferase-3B) for the immunodeficiency, centromeric instability, and facial anomalies syndrome 1 (ICF1, MIM#242860, ORPHA2268) or (2) in the establishment of DNA methylation signature, as for the imprinting disorders [103,104].

In 2010, two large-scale analyses highlighted global dysregulation of miRNAs in mouse models of RTT. Wu and colleagues identified several up- and downregulated miRNAs in cerebella of MeCP2-deficient mice [105]. Remarkably, MeCP2 silences miR-30a/d, miR-381, and miR-495, which in turn represses brain-derived neurotrophic factor (BDNF), the downregulated target gene in MeCP2-null mice [106]; this suggests a multilayered MeCP2-mediated transcriptional regulation of BDNF [105]. Similarly, the analysis of MeCP2-null brains revealed alterations of several miRNAs. Noteworthy, deregulated miR-29 and miR-146 are known to have roles in neural and glial cells and its association with neurological disorders is reported [107]. Latest findings indicate that lncRNAs' deregulation in an MeCP2-deficient context may contribute to the RTT pathophysiology and underline a relationship between lncRNAs and protein-coding RNAs, proving that disrupted *cis*-regulated mechanisms are associated to the disorder [108].

While MeCP2 reads and interprets the DNA methylation epigenetic marks, the de novo DNMT3B is the epigenetic player establishing it during development. Mutations interfering with DNMT3B catalytic activity are reported to affect the transcriptional profile of several hundred protein-coding genes in ICF patients' derived cells, with enriched functional classes including immune function, development, and neurogenesis, which are relevant for the ICF clinical phenotype. Interestingly, a significant proportion of these genes are targets of miRNAs, which are inversely deregulated in ICF-derived cells, suggesting that miRNAs are integrated in the DNMT3B-mediated regulatory circuitry modulating cell-specific gene expression program [109]. miR-338 is the most upregulated miRNA with brain-specific functions, while among the downregulated miRNAs, miR-196a is transcribed from intergenic regions within Hox genes clusters in vertebrates. As for the Hox genes, miR-196a acts as regulators of the nervous system development in embryos [110]. Intriguingly, *LHX2*, which is crucial for the proper development of cerebral cortex in mouse embryo, is an miR-196a target gene, and results overexpressed in DNMT3B-deficient cells [109].

An important process in which the epigenetic players, MeCP2 and DNMT3B, interact with ncRNAs is the maintenance of the integrity at telomeric heterochromatin to prevent the telomere shortening [111]. Telomeric-repeat-containing RNAs (TERRAs) are lncRNAs transcribed from telomeres. TERRAs are involved in the formation of telomeric heterochromatin through a negative-feedback looping mechanism and interact with several heterochromatin-associated proteins, including MeCP2, HP1, and H3K9 methyltransferases. This interaction may be sufficient to nucleate heterochromatin complexes; the recruitment of DNMT3B and the establishment of CpG methylation at subtelomeres likely mediate this process. Indeed, in ICF-derived cells, TERRA levels are abnormally increased and telomeres are abnormally shortened [112].

NcRNAs have been found playing an important role in imprinting disorders, as, for instance, the Beckwith–Wiedemann syndrome (BWS, MIM#130650, ORPHA116). BWS results from increased expression of the paternally expressed growth promoter IGF2 and/or reduced expression (or loss of function) of the maternal growth suppressor CDKN1C. An intragenic miRNA of the imprinted IGF2 (miR-483-5p) regulates MeCP2 levels through a human-specific binding site in the MeCP2 long 3′-UTR. There is an inverse correlation of miR-483-5p and MeCP2 levels in developing human brains and fibroblasts from BWS patients. Importantly, the expression of miR-483-5p rescues abnormal dendritic spine phenotype of neurons overexpressing MeCP2 [113].

lncRNAs appear to play a key function in the BWS phenotype. KCNQ1OT1 is paternally transcribed from the imprinted domain 2 of chromosome 11p15.5 [114]. In the majority of BWS, loss of maternal methylation at IC2 is associated with biallelic transcription of the KCNQ1OT1 maternal allele [115] and the biallelic silencing of the imprinted domain genes, including the cell growth inhibitor CDKN1C [116]. Notably, CDKN1C loss-of-function causes the typical overgrowth and increased risk to develop embryonic tumor in BWS patients.

4.4 ncRNAs: From the Basic Knowledge to the Clinical Practice

Coping with ncRNAs in the clinical practice is still an open challenge. Indeed, whereas our knowledge about miRNAs in human diseases is more consolidated, for other ncRNAs (i.e., lnCRNAs) it is still in its infancy. Moreover, despite miRNA profiling panels for diagnostic use are already in the market, they have not yet become a widespread methodology for most of clinicians. Earliest clinical uses of miRNA-based expression profiles have been in the context of tumors; as in many cancers, miRNA levels are significantly altered. Changes in miRNA levels, both in tissue biopsies and in bloodstream, can be useful diagnostics in cancer. miRNA profiling arrays, or similar NGS-based approaches, can be used by physicians to profile a metastatic lesion and to determine with high accuracy the nature of the primary tumor site. In addition, as distinct miRNA signatures have been associated with better or worse prognosis, cancer panels that include miRNA signatures can be used to predict prognosis or some clinical outcomes (such as the response to a specific therapy). It is predictable that, in the near future, miRNA-based diagnostic kits will be released on the market to help clinicians in the characterization of hystological subtypes of cancer.

From a therapeutic point of view, there are human clinical trials with molecules that target overexpressed miRNAs, and it is arguable that antisense molecules targeting oncogenic miRNAs, or conversely miRNA mimics, will be available for clinical use.

Such considerations hold true especially for miRNAs, as our knowledge about the role of other ncRNAs (including lncRNAs) in human diseases is still at its infancy. In particular, despite the growing awareness that lncRNAs can regulate (and in some cases affect) crucial cellular processes in some cancers, our clinical understanding remains still limited. However, despite the open debate about lncRNAs' functionality, these molecules can reliably be considered for prognostic applications in human diseases, particularly in cancer. Indeed, a significant number of these ncRNAs have been associated with clinical outcomes and patients' survival in solid tumors [117]. Toward this direction, an interesting experimental procedure has been recently proposed by Garzon and colleagues for acute myeloid leukemia (AML, MIM#601626). By means of univariable Cox analysis, the authors identified 48 lncRNAs associated with event-free survival and then derived an outcome score. Applying such lncRNA-based scoring system to AML patients they were able to subdivide the patients in two groups, prospectively identifying those who are likely to have high complete response rates with standard therapy. On the opposite, the remaining patients were treated with novel targeted therapies to improve their complete response rates [117].

Despite lncRNA analysis has not yet been used in any clinical context, the road toward their clinical use is near, at least for prognostic/diagnostic purposes. Indeed, their potential therapeutic use is still far, as the biological role has not yet been assessed for most of these ncRNAs. Here, we highlight a general workflow that is valid for many human diseases, cancer included, containing some key points to be addressed before lncRNAs can significantly be used for clinical purposes. The first step is the identification of the molecular signatures of ncRNAs associated to the disease. To this aim, a very interesting approach has been recently proposed in the work of Li et al. [53]. The authors, integrating different omics data sets (RNA-Seq, miRNA-Seq, and MBDCap-Seq), found significant methylation of ncRNA gene promoters associated with transcriptional changes in breast cancer. Such approach, that combines different omic data sets to identify expression and/or methylation signatures, is very helpful even in small discovery cohorts of well-characterized patients to define new biomarkers associated to clinical outcomes. Regarding the oncological field, the approach could be also extended to find specific lncRNA signatures associated to the tumor staging, to patients' survival and to the metastatic potential of a given tumor. We believe that the integration of whole-genome profiling of epigenetic modifications (e.g., DNA methylation) and transcriptional data sets promises to be a powerful tool, first to better understand cancer biology, then to assess their potential clinical relevance. Indeed, in the absence of complete knowledge, through similar basic research studies, no clinical studies can be successfully carried out. However, before going into the clinical practice, newly identified biomarkers need to be validated in distinct (and larger) validation cohorts by independent multicenter studies.

5. CONCLUSIONS AND FUTURE PERSPECTIVES

Increasing reports highlighting aberrant expression of different classes of ncRNAs in cancer and other human diseases open new perspectives for the comprehension of the pathogenic mechanism of these disorders. Notably, although RNA-Seq proves to be a very powerful and innovative tool to profile small RNAs and lincRNAs, freely available RNA-Seq data sets in public repositories of "disease" versus "control" samples are relatively limited—particularly for cancer—if compared to the huge number of microarray-based studies. In addition, since NGS has been only recently employed to systematically profile lincRNAs, there is still a limited amount of RNA-Seq data sets with enough sequencing coverage (and with adequate sample numbers to reach the statistical power) needed to reliably discover clinically relevant long noncoding transcripts [45]. Conversely, there is a huge amount of data available from independent array-based gene expression studies across of tumor samples. In addition, as such array-based profiles of cancer specimen are often accompanied by complete matched clinical annotation, these are particularly useful for association studies [45].

The increasing number of studies based on RNA-Seq will surely provide a more detailed picture of the whole human ncRNAs transcriptome. However, since ncRNAs exert their biological functions by folding into complex secondary structures, it is evident that sequence-based alignments alone cannot provide sufficient information to fully understand their function. Such consideration holds true particularly for lncRNAs. Indeed, despite the development of new algorithms to predict RNA secondary structures (RNAfold, Pfold, and FOLDALIGN), to date we are not able to assign a precise biological function to the thousands of annotated lncRNAs. In addition, the comprehension of if—and especially of how—these ncRNAs exert a biological role is further complicated by the lack of knowledge about functional and repetitive motifs (or domains) in these molecules.

Therefore, we envision that RNA-Seq advancements should be supported by the parallel development of sophisticated computational methods for the systematic characterization of functional domains in ncRNA. It would represent a solid basis to predict the function of thousands of ncRNAs.

Obviously, only experimental approaches will confirm the biological role of these ncRNAs. However, in general, to infer the functionality of ncRNAs is quite challenging, particularly in relation to human disease. The generation of mouse models for single ncRNAs and the restricted spatiotemporal expression of many ncRNAs may be useful experimental approaches to reveal their function in the pathological context. Alternatively, gain- and loss-of-function experimental strategies focused on robust cellular models have begun to provide functional clues for some ncRNAs. The emerging findings about the regulation of ncRNAs by the epigenetic machinery and, vice versa, their role in the regulation of epigenetic processes are paving the way for the identification of accurate predictive and prognostic factors in cancer field.

Moreover, being the ncRNAs' druggable targets, the expectations in oncologic therapy have been progressively increased. Indeed, initial clinical trials using ncRNA-based molecules are ongoing and novel approaches are forthcoming, such as small molecule-based treatments that target the miRNA machinery. However, much has to be done, and the future challenge will be the generation of molecules able to overcome a key problem that is tackled by traditional cancer therapy: the development of resistance. Although being in its infancy, exciting and promising findings are expected in this area.

Glossary

AGO-IP Immunoprecipitation of Argonaute proteins.

ChIP-sequencing This is a method used to analyze in vivo interactions between protein and DNA. ChIP-seq combines chromatin immunoprecipitation (ChIP) with massively parallel DNA sequencing to identify the binding sites of DNA-associated proteins.

CpG island CpG island or CG sites are short interspersed DNA sequences that deviate significantly from the average genomic pattern by being GC-rich and CpG-rich.

DNA methylation This is a biochemical process where a methyl group is added to the cytosine or adenine DNA nucleotides.

Epigenetics This is the study of heritable changes in gene expression that are not encoded in the nucleotide sequence.

Genome-wide association study (GWA study, or GWAS) is defined as "any study of genetic variation across the entire human genome that is designed to identify genetic associations with observable traits or the presence of a disease or condition."

isomiRs These are miRNAs that exhibit variation from their "reference" sequences.
Long noncoding RNAs lncRNAs are non-protein-coding transcripts longer than 200 nucleotides.
MicroRNAs These are small noncoding RNA molecules (containing about 22 nucleotides) that function in RNA silencing and posttranscriptional regulation of gene expression.
Next-generation sequencing (NGS) Also known as high-throughput sequencing, is the catch-all term used to describe second-generation technologies to sequence DNA and RNA much more quickly and cheaply than the previously used Sanger sequencing. Millions or billions of DNA/RNA fragments can be sequenced in parallel, yielding substantially more throughput and minimizing the need for the fragment-cloning methods that are often used in Sanger sequencing of genomes.
Polycomb repressive complexes PRCs are two classes of polycomb-group proteins (PcG)—PRC1 and PRC2—that establish epigenetic patterns for maintaining gene repression.
Ribodepletion This indicates the removal of abundant 5S, 5.8S, 18S, and 28S ribosomal RNAs by magnetic beads or probes.
RNA-Sequencing Also called whole-transcriptome shotgun sequencing, it is an application of NGS technologies to profile the entire transcriptome in a cell/tissue.
Untranslated region UTR refers to either of two regions, one on each side of a coding sequence on a strand of mRNA. If it is found on the 5′ side, it is called the 5′ UTR, or if it is found on the 3′ side, it is called the 3′ UTR.

LIST OF ABBREVIATIONS

AGO2-IP AGO2 immunoprecipitation
ALL Acute lymphoblastic leukemia
AML Acute myeloid leukemia
BDNF Brain-derived neurotrophic factor
BWS Beckwith–Wiedemann syndrome
ceRNA Competing endogenous RNA
ChIP-seq Chromatin immunoprecipitation followed by massively parallel sequencing
ChIP Chromatin immunoprecipitation
DNMT DNA methyltransferase
GWAS Genome-wide association studies
H3K36me3 Trimethylation of lysine 36 of histone H3
H3K4me3 Trimethylation of lysine 4 of histone H3
HAT Histone acetyltransferase
HCC Hepatocellular carcinoma cells
HDM Histone demethylase
HMT Histone methyltransferase
HP1 Heterochromatin protein 1
ICF Immunodeficiency, centromere instability, and facial anomalies syndrome
ICGs Interchromatin granules
lincRNAs Long intergenic noncoding RNAs
lncRNAs Long noncoding RNAs
miRNAs MicroRNAs
ncRNAs Noncoding RNAs
NGS Next-generation sequencing
ORF Open reading frame
PcGs Polycomb bodies
piRNAs PIWI-interacting RNAs
PRC Polycomb repressive complex
RIP RNA immunoprecipitation
RNA-Seq RNA-Sequencing
RNAi RNA interference
Rpm Reads per million of mapped reads
rRNA Ribosomal RNA
RTT Rett syndrome
siRNA Small interfering RNA
sncRNA Small noncoding RNA
snoRNA Small nucleolar RNA
SWI/SNF Switch/sucrose nonfermentable
T-UCR Transcribed ultraconserved region
TERRAs Telomeric-repeat-containing RNAs
UTR Untranslated region

References

[1] Taft RJ, Pang KC, Mercer TR, Dinger M, Mattick JS. Non-coding RNAs: regulators of disease. J Pathol January 2010;220(2):126–39.
[2] Saxena A, Tang D, Carninci P. piRNAs warrant investigation in Rett Syndrome: an omics perspective. Dis Markers 2012;33(5):261–75.
[3] Jacquier A. The complex eukaryotic transcriptome: unexpected pervasive transcription and novel small RNAs. Nat Rev Genet December 2009;10(12):833–44.
[4] Rinn JL, Euskirchen G, Bertone P, Martone R, Luscombe NM, Hartman S, et al. The transcriptional activity of human chromosome 22. Genes Dev February 15, 2003;17(4):529–40.
[5] Carninci P, Kasukawa T, Katayama S, Gough J, Frith MC, Maeda N, et al. The transcriptional landscape of the mammalian genome. Science September 2, 2005;309(5740):1559–63.
[6] Consortium EP, Birney E, Stamatoyannopoulos JA, Dutta A, Guigo R, Gingeras TR, et al. Identification and analysis of functional elements in 1% of the human genome by the ENCODE pilot project. Nature June 14, 2007;447(7146):799–816.
[7] Guttman M, Amit I, Garber M, French C, Lin MF, Feldser D, et al. Chromatin signature reveals over a thousand highly conserved large non-coding RNAs in mammals. Nature March 12, 2009;458(7235):223–7.

[8] Khalil AM, Guttman M, Huarte M, Garber M, Raj A, Rivea Morales D, et al. Many human large intergenic noncoding RNAs associate with chromatin-modifying complexes and affect gene expression. Proc Natl Acad Sci USA July 14, 2009;106(28):11667–72.
[9] Kapranov P, Willingham AT, Gingeras TR. Genome-wide transcription and the implications for genomic organization. Nat Rev Genet June 2007;8(6):413–23.
[10] Clark MB, Mattick JS. Long noncoding RNAs in cell biology. Semin Cell Dev Biol June 2011;22(4):366–76.
[11] Cabili MN, Trapnell C, Goff L, Koziol M, Tazon-Vega B, Regev A, et al. Integrative annotation of human large intergenic noncoding RNAs reveals global properties and specific subclasses. Genes Dev September 15, 2011;25(18):1915–27.
[12] Derrien T, Johnson R, Bussotti G, Tanzer A, Djebali S, Tilgner H, et al. The GENCODE v7 catalog of human long noncoding RNAs: analysis of their gene structure, evolution, and expression. Genome Res September 2012;22(9):1775–89.
[13] Wang KC, Chang HY. Molecular mechanisms of long noncoding RNAs. Mol Cell September 16, 2011;43(6):904–14.
[14] Caudron-Herger M, Rippe K. Nuclear architecture by RNA. Curr Opin Genet Dev April 2012;22(2):179–87.
[15] Ponting CP, Belgard TG. Transcribed dark matter: meaning or myth? Hum Mol Genet October 15, 2010;19(R2):R162–8.
[16] Wutz A. Gene silencing in X-chromosome inactivation: advances in understanding facultative heterochromatin formation. Nat Rev Genet August 2011;12(8):542–53.
[17] Rinn JL, Kertesz M, Wang JK, Squazzo SL, Xu X, Brugmann SA, et al. Functional demarcation of active and silent chromatin domains in human HOX loci by noncoding RNAs. Cell June 29, 2007;129(7):1311–23.
[18] Nagano T, Mitchell JA, Sanz LA, Pauler FM, Ferguson-Smith AC, Feil R, et al. The Air noncoding RNA epigenetically silences transcription by targeting G9a to chromatin. Science December 12, 2008;322(5908):1717–20.
[19] Guttman M, Garber M, Levin JZ, Donaghey J, Robinson J, Adiconis X, et al. Ab initio reconstruction of cell type-specific transcriptomes in mouse reveals the conserved multi-exonic structure of lincRNAs. Nat Biotechnol May 2010;28(5):503–10.
[20] Smith AG, Box NF, Marks LH, Chen W, Smit DJ, Wyeth JR, et al. The human melanocortin-1 receptor locus: analysis of transcription unit, locus polymorphism and haplotype evolution. Gene December 27, 2001;281(1–2):81–94.
[21] Pritchard CC, Cheng HH, Tewari M. MicroRNA profiling: approaches and considerations. Nat Rev Genet May 2012;13(5):358–69.
[22] Costa V, Angelini C, De Feis I, Ciccodicola A. Uncovering the complexity of transcriptomes with RNA-Seq. J Biomed Biotechnol 2010;2010:853916.
[23] Metzker ML. Sequencing in real time. Nat Biotechnol February 2009;27(2):150–1.
[24] Krueger F, Kreck B, Franke A, Andrews SR. DNA methylome analysis using short bisulfite sequencing data. Nat Methods February 2012;9(2):145–51.
[25] Bolger AM, Lohse M, Usadel B. Trimmomatic: a flexible trimmer for Illumina sequence data. Bioinformatics August 1, 2014;30(15):2114–20.
[26] Wu TD, Nacu S. Fast and SNP-tolerant detection of complex variants and splicing in short reads. Bioinformatics April 1, 2010;26(7):873–81.
[27] Li H, Ruan J, Durbin R. Mapping short DNA sequencing reads and calling variants using mapping quality scores. Genome Res November 2008;18(11):1851–8.
[28] Li R, Yu C, Li Y, Lam TW, Yiu SM, Kristiansen K, et al. SOAP2: an improved ultrafast tool for short read alignment. Bioinformatics August 1, 2009;25(15):1966–7.
[29] Li H, Durbin R. Fast and accurate short read alignment with Burrows-Wheeler transform. Bioinformatics July 15, 2009;25(14):1754–60.
[30] Motameny S, Wolters S, Nurnberg P, Schumacher B. Next generation sequencing of miRNAs – strategies, resources and methods. Genes 2010;1(1):70–84.
[31] Linsen SE, de Wit E, Janssens G, Heater S, Chapman L, Parkin RK, et al. Limitations and possibilities of small RNA digital gene expression profiling. Nat Methods July 2009;6(7):474–6.
[32] Vlachos IS, Hatzigeorgiou AG. Online resources for miRNA analysis. Clin Biochem July 2013;46(10–11):879–900.
[33] Wang Z, Gerstein M, Snyder M. RNA-Seq: a revolutionary tool for transcriptomics. Nat Rev Genet January 2009;10(1):57–63.
[34] t Hoen PA, Ariyurek Y, Thygesen HH, Vreugdenhil E, Vossen RH, de Menezes RX, et al. Deep sequencing-based expression analysis shows major advances in robustness, resolution and inter-lab portability over five microarray platforms. Nucleic Acids Res December 2008;36(21):e141.
[35] Bloom JS, Khan Z, Kruglyak L, Singh M, Caudy AA. Measuring differential gene expression by short read sequencing: quantitative comparison to 2-channel gene expression microarrays. BMC Genomics 2009;10:221.
[36] Kim D, Pertea G, Trapnell C, Pimentel H, Kelley R, Salzberg SL. TopHat2: accurate alignment of transcriptomes in the presence of insertions, deletions and gene fusions. Genome Biol 2013;14(4):R36.

[37] Dobin A, Davis CA, Schlesinger F, Drenkow J, Zaleski C, Jha S, et al. STAR: ultrafast universal RNA-seq aligner. Bioinformatics January 1, 2013;29(1):15–21.

[38] Kong L, Zhang Y, Ye ZQ, Liu XQ, Zhao SQ, Wei L, et al. CPC: assess the protein-coding potential of transcripts using sequence features and support vector machine. Nucleic Acids Res July 2007;35(Web Server issue):W345–9.

[39] Wang L, Park HJ, Dasari S, Wang S, Kocher JP, Li W. CPAT: coding-potential assessment tool using an alignment-free logistic regression model. Nucleic Acids Res April 1, 2013;41(6):e74.

[40] Blanco E, Parra G, Guigo R. Using geneid to identify genes. Curr Protoc Bioinformatics June 2007;Chapter 4:Unit 4.3.

[41] Trapnell C, Williams BA, Pertea G, Mortazavi A, Kwan G, van Baren MJ, et al. Transcript assembly and quantification by RNA-Seq reveals unannotated transcripts and isoform switching during cell differentiation. Nat Biotechnol May 2010;28(5):511–5.

[42] Anders S, Pyl PT, Huber W. HTSeq–a Python framework to work with high-throughput sequencing data. Bioinformatics January 15, 2015;31(2):166–9.

[43] Robinson MD, McCarthy DJ, Smyth GK. edgeR: a Bioconductor package for differential expression analysis of digital gene expression data. Bioinformatics January 1, 2010;26(1):139–40.

[44] Anders S, Huber W. Differential expression analysis for sequence count data. Genome Biol 2010;11(10):R106.

[45] Du Z, Fei T, Verhaak RG, Su Z, Zhang Y, Brown M, et al. Integrative genomic analyses reveal clinically relevant long noncoding RNAs in human cancer. Nat Struct Mol Biol July 2013;20(7):908–13.

[46] Romanoski CE, Glass CK, Stunnenberg HG, Wilson L, Almouzni G. Epigenomics: roadmap for regulation. Nature February 19, 2015;518(7539):314–6.

[47] Kunej T, Godnic I, Ferdin J, Horvat S, Dovc P, Calin GA. Epigenetic regulation of microRNAs in cancer: an integrated review of literature. Mutat Res December 1, 2011;717(1–2):77–84.

[48] Altuvia Y, Landgraf P, Lithwick G, Elefant N, Pfeffer S, Aravin A, et al. Clustering and conservation patterns of human microRNAs. Nucleic Acids Res 2005;33(8):2697–706.

[49] Loginov VI, Rykov SV, Fridman MV, Braga EA. Methylation of miRNA genes and oncogenesis. Biochem Biokhimiia February 2015;80(2):145–62.

[50] Saxena A, Carninci P. Long non-coding RNA modifies chromatin: epigenetic silencing by long non-coding RNAs. Bioessays November 2011;33(11):830–9.

[51] Mercer TR, Mattick JS. Structure and function of long noncoding RNAs in epigenetic regulation. Nat Struct Mol Biol March 2013;20(3):300–7.

[52] Lee JT. Epigenetic regulation by long noncoding RNAs. Science December 14, 2012;338(6113): 1435–9.

[53] Li Y, Zhang Y, Li S, Lu J, Chen J, Wang Y, et al. Genome-wide DNA methylome analysis reveals epigenetically dysregulated non-coding RNAs in human breast cancer. Sci Rep 2015;5:8790.

[54] Bracken AP, Dietrich N, Pasini D, Hansen KH, Helin K. Genome-wide mapping of Polycomb target genes unravels their roles in cell fate transitions. Genes Dev May 1, 2006;20(9):1123–36.

[55] Ku M, Koche RP, Rheinbay E, Mendenhall EM, Endoh M, Mikkelsen TS, et al. Genomewide analysis of PRC1 and PRC2 occupancy identifies two classes of bivalent domains. PLoS Genet October 2008;4(10):e1000242.

[56] Mohammad F, Mondal T, Kanduri C. Epigenetics of imprinted long noncoding RNAs. Epigenetics July 1, 2009;4(5):277–86.

[57] Pandey RR, Mondal T, Mohammad F, Enroth S, Redrup L, Komorowski J, et al. Kcnq1ot1 antisense noncoding RNA mediates lineage-specific transcriptional silencing through chromatin-level regulation. Mol Cell October 24, 2008;32(2):232–46.

[58] Dinger ME, Amaral PP, Mercer TR, Pang KC, Bruce SJ, Gardiner BB, et al. Long noncoding RNAs in mouse embryonic stem cell pluripotency and differentiation. Genome Res September 2008;18(9):1433–45.

[59] Pinter SF, Sadreyev RI, Yildirim E, Jeon Y, Ohsumi TK, Borowsky M, et al. Spreading of X chromosome inactivation via a hierarchy of defined Polycomb stations. Genome Res October 2012;22(10):1864–76.

[60] Almouzni G, Probst AV. Heterochromatin maintenance and establishment: lessons from the mouse pericentromere. Nucleus September–October 2011;2(5):332–8.

[61] Maison C, Bailly D, Roche D, Montes de Oca R, Probst AV, Vassias I, et al. SUMOylation promotes de novo targeting of HP1alpha to pericentric heterochromatin. Nat Genet March 2011;43(3):220–7.

[62] Maia BM, Rocha RM, Calin GA. Clinical significance of the interaction between non-coding RNAs and the epigenetics machinery: challenges and opportunities in oncology. Epigenetics January 2014;9(1):75–80.

[63] Calin GA, Ferracin M, Cimmino A, Di Leva G, Shimizu M, Wojcik SE, et al. A MicroRNA signature associated with prognosis and progression in chronic lymphocytic leukemia. N Engl J Med October 27, 2005;353(17):1793–801.

[64] Calin GA, Trapasso F, Shimizu M, Dumitru CD, Yendamuri S, Godwin AK, et al. Familial cancer associated with a polymorphism in ARLTS1. N Engl J Med April 21, 2005;352(16):1667–76.

[65] Lu CY, Lin KY, Tien MT, Wu CT, Uen YH, Tseng TL. Frequent DNA methylation of MiR-129-2 and its potential clinical implication in hepatocellular carcinoma. Genes Chromosomes Cancer July 2013;52(7):636–43.

[66] Agirre X, Vilas-Zornoza A, Jimenez-Velasco A, Martin-Subero JI, Cordeu L, Garate L, et al. Epigenetic silencing of the tumor suppressor microRNA Hsa-miR-124a regulates CDK6 expression and confers a poor prognosis in acute lymphoblastic leukemia. Cancer Res May 15, 2009;69(10):4443–53.

[67] Gebauer K, Peters I, Dubrowinskaja N, Hennenlotter J, Abbas M, Scherer R, et al. Hsa-mir-124-3 CpG island methylation is associated with advanced tumours and disease recurrence of patients with clear cell renal cell carcinoma. Br J Cancer January 15, 2013;108(1):131–8.

[68] Lujambio A, Calin GA, Villanueva A, Ropero S, Sanchez-Cespedes M, Blanco D, et al. A microRNA DNA methylation signature for human cancer metastasis. Proc Natl Acad Sci USA September 9, 2008;105(36):13556–61.

[69] Esteller M. Non-coding RNAs in human disease. Nat Rev Genet December 2011;12(12):861–74.

[70] Incoronato M, Urso L, Portela A, Laukkanen MO, Soini Y, Quintavalle C, et al. Epigenetic regulation of miR-212 expression in lung cancer. PLoS One 2011;6(11):e27722.

[71] Beckedorff FC, Amaral MS, Deocesano-Pereira C, Verjovski-Almeida S. Long non-coding RNAs and their implications in cancer epigenetics. Biosci Rep 2013;33(4).

[72] Lujambio A, Portela A, Liz J, Melo SA, Rossi S, Spizzo R, et al. CpG island hypermethylation-associated silencing of non-coding RNAs transcribed from ultraconserved regions in human cancer. Oncogene December 2, 2010;29(48):6390–401.

[73] Yap KL, Li S, Munoz-Cabello AM, Raguz S, Zeng L, Mujtaba S, et al. Molecular interplay of the noncoding RNA ANRIL and methylated histone H3 lysine 27 by polycomb CBX7 in transcriptional silencing of INK4a. Mol Cell June 11, 2010;38(5):662–74.

[74] Popov N, Gil J. Epigenetic regulation of the INK4b-ARF-INK4a locus: in sickness and in health. Epigenetics November–December 2010;5(8):685–90.

[75] Fabbri M, Garzon R, Cimmino A, Liu Z, Zanesi N, Callegari E, et al. MicroRNA-29 family reverts aberrant methylation in lung cancer by targeting DNA methyltransferases 3A and 3B. Proc Natl Acad Sci USA October 2, 2007;104(40):15805–10.

[76] Braconi C, Huang N, Patel T. MicroRNA-dependent regulation of DNA methyltransferase-1 and tumor suppressor gene expression by interleukin-6 in human malignant cholangiocytes. Hepatology March 2010;51(3):881–90.

[77] Stumpel DJ, Schotte D, Lange-Turenhout EA, Schneider P, Seslija L, de Menezes RX, et al. Hypermethylation of specific microRNA genes in MLL-rearranged infant acute lymphoblastic leukemia: major matters at a micro scale. Leukemia March 2011;25(3):429–39.

[78] Huang JJ, Yu J, Li JY, Liu YT, Zhong RQ. Circulating microRNA expression is associated with genetic subtype and survival of multiple myeloma. Med Oncol December 2012;29(4):2402–8.

[79] Noonan EJ, Place RF, Pookot D, Basak S, Whitson JM, Hirata H, et al. miR-449a targets HDAC-1 and induces growth arrest in prostate cancer. Oncogene April 9, 2009;28(14):1714–24.

[80] Guo Y, Ying L, Tian Y, Yang P, Zhu Y, Wang Z, et al. miR-144 downregulation increases bladder cancer cell proliferation by targeting EZH2 and regulating Wnt signaling. FEBS J September 2013;280(18):4531–8.

[81] Friedman JM, Jones PA, Liang G. The tumor suppressor microRNA-101 becomes an epigenetic player by targeting the polycomb group protein EZH2 in cancer. Cell Cycle August 2009;8(15):2313–4.

[82] Ren G, Baritaki S, Marathe H, Feng J, Park S, Beach S, et al. Polycomb protein EZH2 regulates tumor invasion via the transcriptional repression of the metastasis suppressor RKIP in breast and prostate cancer. Cancer Res June 15, 2012;72(12):3091–104.

[83] Gutschner T, Diederichs S. The hallmarks of cancer: a long non-coding RNA point of view. RNA Biol June 2012;9(6):703–19.

[84] Reis EM, Verjovski-Almeida S. Perspectives of long non-coding RNAs in Cancer diagnostics. Front Genet 2012;3:32.

[85] Spizzo R, Almeida MI, Colombatti A, Calin GA. Long non-coding RNAs and cancer: a new frontier of translational research? Oncogene October 25, 2012;31(43):4577–87.

[86] Prensner JR, Chinnaiyan AM. The emergence of lncRNAs in cancer biology. Cancer Discov October 2011;1(5):391–407.

[87] Li D, Feng J, Wu T, Wang Y, Sun Y, Ren J, et al. Long intergenic noncoding RNA HOTAIR is overexpressed and regulates PTEN methylation in laryngeal squamous cell carcinoma. Am J Pathol January 2013;182(1):64–70.

[88] Gupta RA, Shah N, Wang KC, Kim J, Horlings HM, Wong DJ, et al. Long non-coding RNA HOTAIR reprograms chromatin state to promote cancer metastasis. Nature April 15, 2010;464(7291):1071–6.

[89] Costa V, Aprile M, Esposito R, Ciccodicola A. RNA-Seq and human complex diseases: recent accomplishments and future perspectives. Eur J Hum Genet February 2013;21(2):134–42.

[90] Yildirim E, Kirby JE, Brown DE, Mercier FE, Sadreyev RI, Scadden DT, et al. Xist RNA is a potent suppressor of hematologic cancer in mice. Cell February 14, 2013;152(4):727–42.
[91] Prensner JR, Iyer MK, Balbin OA, Dhanasekaran SM, Cao Q, Brenner JC, et al. Transcriptome sequencing across a prostate cancer cohort identifies PCAT-1, an unannotated lincRNA implicated in disease progression. Nat Biotechnol August 2011;29(8):742–9.
[92] Lin R, Maeda S, Liu C, Karin M, Edgington TS. A large noncoding RNA is a marker for murine hepatocellular carcinomas and a spectrum of human carcinomas. Oncogene February 8, 2007;26(6):851–8.
[93] Ji P, Diederichs S, Wang W, Boing S, Metzger R, Schneider PM, et al. MALAT-1, a novel noncoding RNA, and thymosin beta4 predict metastasis and survival in early-stage non-small cell lung cancer. Oncogene September 11, 2003;22(39):8031–41.
[94] Argani P, Yonescu R, Morsberger L, Morris K, Netto GJ, Smith N, et al. Molecular confirmation of t(6;11)(p21;q12) renal cell carcinoma in archival paraffin-embedded material using a break-apart TFEB FISH assay expands its clinicopathologic spectrum. Am J Surg Pathol October 2012;36(10):1516–26.
[95] Davis IJ, Hsi BL, Arroyo JD, Vargas SO, Yeh YA, Motyckova G, et al. Cloning of an Alpha-TFEB fusion in renal tumors harboring the t(6;11)(p21;q13) chromosome translocation. Proc Natl Acad Sci USA May 13, 2003;100(10):6051–6.
[96] Yang L, Lin C, Liu W, Zhang J, Ohgi KA, Grinstein JD, et al. ncRNA- and Pc2 methylation-dependent gene relocation between nuclear structures mediates gene activation programs. Cell November 11, 2011;147(4):773–88.
[97] Beckedorff FC, Ayupe AC, Crocci-Souza R, Amaral MS, Nakaya HI, Soltys DT, et al. The intronic long noncoding RNA ANRASSF1 recruits PRC2 to the RASSF1A promoter, reducing the expression of RASSF1A and increasing cell proliferation. PLoS Genet 2013;9(8):e1003705.
[98] Ding X, Zhu L, Ji T, Zhang X, Wang F, Gan S, et al. Long intergenic non-coding RNAs (LincRNAs) identified by RNA-seq in breast cancer. PLoS One 2014;9(8):e103270.
[99] Brunner AL, Beck AH, Edris B, Sweeney RT, Zhu SX, Li R, et al. Transcriptional profiling of long non-coding RNAs and novel transcribed regions across a diverse panel of archived human cancers. Genome Biol 2012;13(8):R75.
[100] Liu Y, Ferguson JF, Xue C, Ballantyne RL, Silverman IM, Gosai SJ, et al. Tissue-specific RNA-Seq in human evoked inflammation identifies blood and adipose LincRNA signatures of cardiometabolic diseases. Arterioscler Thromb Vasc Biol April 2014;34(4):902–12.
[101] Costa V, Angelini C, D'Apice L, Mutarelli M, Casamassimi A, Sommese L, et al. Massive-scale RNA-Seq analysis of non ribosomal transcriptome in human trisomy 21. PLoS One 2011;6(4):e18493.
[102] Della Ragione F, Gagliardi M, D'Esposito M, Matarazzo MR. Non-coding RNAs in chromatin disease involving neurological defects. Front Cell Neurosci 2014;8:54.
[103] Matarazzo MR, De Bonis ML, Vacca M, Della Ragione F, D'Esposito M. Lessons from two human chromatin diseases, ICF syndrome and Rett syndrome. Int J Biochem Cell Biol January 2009;41(1):117–26.
[104] Walter J, Paulsen M. Imprinting and disease. Semin Cell Dev Biol February 2003;14(1):101–10.
[105] Wu H, Tao J, Chen PJ, Shahab A, Ge W, Hart RP, et al. Genome-wide analysis reveals methyl-CpG-binding protein 2-dependent regulation of microRNAs in a mouse model of Rett syndrome. Proc Natl Acad Sci USA October 19, 2010;107(42):18161–6.
[106] Wang H, Chan SA, Ogier M, Hellard D, Wang Q, Smith C, et al. Dysregulation of brain-derived neurotrophic factor expression and neurosecretory function in Mecp2 null mice. J Neurosci October 18, 2006;26(42):10911–5.
[107] Urdinguio RG, Fernandez AF, Lopez-Nieva P, Rossi S, Huertas D, Kulis M, et al. Disrupted microRNA expression caused by Mecp2 loss in a mouse model of Rett syndrome. Epigenetics October 1, 2010;5(7):656–63.
[108] Petazzi P, Sandoval J, Szczesna K, Jorge OC, Roa L, Sayols S, et al. Dysregulation of the long non-coding RNA transcriptome in a Rett syndrome mouse model. RNA Biol July 2013;10(7):1197–203.
[109] Gatto S, Della Ragione F, Cimmino A, Strazzullo M, Fabbri M, Mutarelli M, et al. Epigenetic alteration of microRNAs in DNMT3B-mutated patients of ICF syndrome. Epigenetics July 1, 2010;5(5):427–43.
[110] Kosik KS. The neuronal microRNA system. Nat Rev Neurosci December 2006;7(12):911–20.
[111] Deng Z, Campbell AE, Lieberman PM. TERRA, CpG methylation and telomere heterochromatin: lessons from ICF syndrome cells. Cell Cycle January 1, 2010;9(1):69–74.
[112] Yehezkel S, Segev Y, Viegas-Pequignot E, Skorecki K, Selig S. Hypomethylation of subtelomeric regions in ICF syndrome is associated with abnormally short telomeres and enhanced transcription from telomeric regions. Hum Mol Genet September 15, 2008;17(18):2776–89.
[113] Han K, Gennarino VA, Lee Y, Pang K, Hashimoto-Torii K, Choufani S, et al. Human-specific regulation of MeCP2 levels in fetal brains by microRNA miR-483-5p. Genes Dev March 1, 2013;27(5):485–90.

[114] Weksberg R, Nishikawa J, Caluseriu O, Fei YL, Shuman C, Wei C, et al. Tumor development in the Beckwith-Wiedemann syndrome is associated with a variety of constitutional molecular 11p15 alterations including imprinting defects of KCNQ1OT1. Hum Mol Genet December 15, 2001;10(26):2989–3000.

[115] Choufani S, Shuman C, Weksberg R. Molecular findings in Beckwith-Wiedemann syndrome. Am J Med Genet Part C Semin Med Genet May 2013; 163C(2):131–40.

[116] Diaz-Meyer N, Day CD, Khatod K, Maher ER, Cooper W, Reik W, et al. Silencing of CDKN1C (p57KIP2) is associated with hypomethylation at KvDMR1 in Beckwith-Wiedemann syndrome. J Med Genet November 2003;40(11):797–801.

[117] Garzon R, Volinia S, Papaioannou D, Nicolet D, Kohlschmidt J, Yan PS, et al. Expression and prognostic impact of lncRNAs in acute myeloid leukemia. Proc Natl Acad Sci USA December 30, 2014;111(52):18679–84.

CHAPTER

12

Circulating Noncoding RNAs as Clinical Biomarkers

Francesco Russo[1,2,a], *Flavia Scoyni*[3,a], *Alessandro Fatica*[3], *Marco Pellegrini*[1], *Alfredo Ferro*[4], *Alfredo Pulvirenti*[4,a], *Rosalba Giugno*[4,a]

[1]Laboratory of Integrative Systems Medicine (LISM), Institute of Informatics and Telematics (IIT) and Institute of Clinical Physiology (IFC), National Research Council (CNR), Pisa, Italy; [2]Department of Computer Science, University of Pisa, Pisa, Italy; [3]Department of Biology and Biotechnology Charles Darwin, Sapienza University of Rome, Rome, Italy; [4]Department of Clinical and Experimental Medicine, University of Catania, Catania, Italy

OUTLINE

[a]These authors contributed equally to this work.

Epigenetic Biomarkers and Diagnostics
http://dx.doi.org/10.1016/B978-0-12-801899-6.00012-7

1. INTRODUCTION

For decades, it was thought that the major players in gene expression regulation were only proteins. RNA was considered to play a minor role in gene expression by converting genetic information from DNA into functional proteins or to serve as structural molecules for translation or RNA maturation. Many noncoding RNAs (ncRNAs) can act as biologically crucial regulatory molecules [1,2]. These RNA molecules are generally referred as noncoding because they are not translated into a protein product but nevertheless they have emerged as key factors in maintaining normal cellular function and therefore play several roles in human diseases, including cancer.

The human genome consists of a vast collection of ncRNAs, which can be grouped according to their sizes and functions. MicroRNAs (miRNAs) are the best-characterized class of small (around 22 nucleotides) ncRNAs, with 2588 mature human transcripts listed in miRBase v21 (http://www.mirbase.org/) [3], the database of published miRNA sequences and annotation. miRNAs function as posttranscriptional regulators of gene expression [4] by targeting specific mRNAs, leading to translational repression or promoting degradation [5] of mRNA. miRNAs have been proposed as tissue biomarkers and more recently as promising noninvasive biomarkers because they circulate in the bloodstream in highly stable extracellular forms [6].

The human genome also contains long ncRNAs (lncRNAs, molecules longer than 200 nucleotides), which are structurally similar to protein-coding genes with proximal promoter sequences and consist of exons and introns but possess no open reading frames (ORFs). lncRNAs constitute the majority of the transcriptome but little is known about their biological functions and their potential role as novel noninvasive biomarkers.

In order to establish specific extracellular RNA molecules as clinically relevant biomarkers, both reproducible recovery from biological samples and reliable measurements of isolated RNA are key instrumental variables.

The main challenge associated with circulating miRNA diagnostics is a lower efficacy of miRNA isolation from body fluids compared to miRNA isolation from cells and tissues. This limits the sensitivity of this approach for clinical purposes [7]. Furthermore, extracellular circulating miRNA exists in both free- and microvesicle-associated states (such as exosomes) [6,8]. Therefore, the method chosen for blood plasma processing can significantly affect the results of extracellular miRNA analysis [9] and many methods and commercial kits for RNA extraction from body fluids exist (see Table 1). Some RNA isolation methods could be superior to others for the recovery of RNA from biological fluids. Furthermore, addition of a carrier molecule, such as glycogen, could be beneficial for some isolation methods.

Exosome isolation protocols vary depending on the biological fluid of origin, but generally involve serial centrifugation at low speed, followed by ultracentrifugation at 100,000 g [10,11]. Alternatively, exosomes can be isolated by immunocapture or size-exclusion methods [10,12]. Recently, a method for exosome isolation called ExoQuick (System Biosciences, Mountain View, California, USA) has been made commercially available [13].

Real-time polymerase chain reaction (PCR) is the most commonly used approach for miRNA quantification. The sensitivity and specificity of real-time PCR and its flexibility, makes it a rather attractive approach to measuring miRNA abundance. The data stored in databases such as miRandola (the database of circulating miRNAs) [14,15], clearly show that next-generation sequencing (NGS) is not widely used yet, perhaps because a larger amount of RNA is needed [16]. Nevertheless, some groups have started to use this technology for profiling circulating miRNAs [17,18]. Yang and colleagues [17] for the first time

TABLE 1 Some Methods and Commercial Kits for RNA Extraction from Body Fluids

Kit	Company	Description
MaxRecovery BiooPure RNA Isolation Reagent	BiooScientific	A single-phase reagent for extraction of total RNA or enriched small RNA (including miRNA) from solid tissues, cultured cells, and cell-free fluids such as serum and plasma.
mirVana miRNA Isolation Kit	Ambion	Uses a rapid procedure to isolate small RNAs using an efficient glass fiber filter (GFF)-based method. The method isolates total RNA ranging in size from kilobases down to 10-mers.
TRI Reagent RT	Molecular Research Center	TRI Reagent RT is used for RNA isolation from tissues, pelleted cells, and cells grown in monolayer.
TRI Reagent RT-Blood	Molecular Research Center	TRI Reagent RT-Blood is specifically designed to isolate RNA from whole blood, plasma, or serum samples.
TRI Reagent RT-Liquid Samples	Molecular Research Center	TRI Reagent RT-Liquid Samples is used for cell suspensions and other liquid samples.
RNAzol RT	Molecular Research Center	RNAzol RT is used to isolate RNA from tissues, cells, liquid samples, or blood. One milliliter is sufficient to process up to 100 mg tissue yielding 50–700 μg of large RNA (>200 bases) and 8–120 μg of small RNA (200–10 bases).
MiRNeasy Serum/Plasma Kit	Qiagen	The miRNeasy Serum/Plasma Kit is designed for purification of cell-free total RNA—primarily miRNA and other small RNA—from small volumes of serum and plasma.
PureLink microRNA Isolation Kit	Ambion	The Ambion PureLink miRNA Isolation Kit provides a column-based method for isolating high-quality total miRNA from animal and plant cells, as well as from bacteria and yeast samples.
mirPremier	Sigma	Provides a method for purifying and enriching miRNA along with other small RNAs, allowing researchers to obtain miRNA directly from cells or tissues.
miRCURY RNA Isolation Kits—Exosome Isolation	Exiqon	The miRCURY Isolation Kits for exosome isolation are optimized for isolation of exosomes from various biofluids and for integration with the miRCURY RNA Isolation Kits. Exosomes can be isolated from various biofluids such as cell-conditioned media, urine, serum, or plasma.
QIAamp Circulating Nucleic Acid Kit	Qiagen	For isolation of free-circulating DNA and RNA from human plasma or serum.

analyzed the expression profiles of circulating miRNAs in the serum of four pregnant women with preeclampsia (PE, MIM #189800) and one healthy pregnant woman as normal control, using the SOLiD sequencer. Their results showed that miRNAs in serum of pregnant women could be detected more comprehensively by NGS.

In another recent study, small RNA fraction from cerebrospinal fluid (CSF) was profiled through NGS (HiSeq 2000) for the first time by Burgos and colleagues [18]. The aim of their study was to maximize RNA isolation from RNA-limited samples and apply these methods to profile miRNA in human CSF by small RNA deep sequencing. More recently, one study [19],

TABLE 2 Classes of Circulating ncRNAs

ncRNA	Length (nt)	Function	Extracellular form
MicroRNA (miRNA)	21–23	Posttranscriptional regulation	Ago2, vesicles, HDL
Piwi-interacting RNA (piRNA)	24–30	Genome stabilization	Vesicles
Long noncoding RNA (lncRNA)	>200	Transcription, splicing, transport regulation	Vesicles
Ribosomal RNA (rRNA)	120–4700	Translation	Vesicles
Transfer RNA (tRNA)	70–100	Translation	Vesicles
Small nuclear RNA (snRNA)	70–350	Splicing, mRNA processing	Vesicles
Small nucleolar RNA (snoRNA)	70–300	RNA modification, rRNA processing	Vesicles

aimed at the discovery and characterization of plasma-derived exosomal RNAs through NGS techniques, demonstrated that a wide variety of RNA species are embedded in circulating vesicles. Authors detected significant fractions of miRNAs (the most abundant) and other RNA species including ribosomal RNA (9.16% of all mappable counts), lncRNA (3.36%), Piwi-interacting RNA (1.31%), transfer RNA (1.24%), small nuclear RNA (0.18%), and small nucleolar RNA (0.01%) (see Table 2). Fragments of coding sequence (1.36%), 5′ untranslated region (0.21%), and 3′ untranslated region (0.54%) were also present.

Circulating nucleic acids in plasma and serum also include DNA [20]. In 1948, Mandel and Metais reported the existence of circulating extracellular nucleic acids in human blood [21]. Circulating DNAs are small fragments of genomic DNA present in the plasma or serum [22]. Many researchers proposed the application of circulating cell-free DNA as a noninvasive biomarker for early detection and prognosis of cancer. For example, the detection of point mutations, microsatellite alterations, DNA hypermethylations, and losses of heterozygosity in circulating cell-free DNA have been characterized in esophageal cancer (MIM #133239) [23].

In this chapter we present a brief description of miRNAs, lncRNAs, and the recently discovered class of circular RNAs (circRNAs). Then, we discuss the application of these ncRNAs as noninvasive biomarkers and, finally, we present the major online resources for biomarker annotation and identification. These tools are fundamental resources for extracellular nucleic acids giving researchers a comprehensive overview about noninvasive biomarkers.

2. MicroRNAs

2.1 miRNA Biogenesis and Function

miRNA genes are widely distributed in animals, plants, and viruses and constitute one of the most abundant gene families [24]. In animals, miRNAs represent the dominating class of small RNAs in most somatic tissues.

In humans, the majority of miRNAs reside in introns of their host genes. They share their regulatory elements and primary transcript, and have a similar expression profile. Other miRNA genes are transcribed from their own promoters, but few primary transcripts have been fully identified. miRNAs can be also encoded by exonic regions. Often, several miRNA loci constitute a polycistronic transcription unit [25]. miRNAs in the same cluster are generally co-transcribed, but the individual miRNAs can be additionally regulated at the posttranscriptional level.

miRNA transcription is carried out by RNA Pol II and is controlled by RNA Pol II-associated transcription factors and epigenetic regulators [4,26–29]. After the transcription phase, the primary miRNA (pri-miRNA) undergoes several steps of maturation [25]. The pri-miRNA contains a stem-loop structure, in which mature miRNA sequences are embedded. The nuclear RNase III Drosha initiates the maturation process by cropping the stem loop to release a small hairpin-shaped RNA [30]. Together with its essential cofactor DiGeorge syndrome chromosomal (or critical) region 8 (DGCR8), Drosha forms a complex called microprocessor [31–34]. Drosha cleaves the hairpin [35,36], and the resulting pre-miRNA is exported into the cytoplasm, where maturation can be completed. Upon export to the cytoplasm, pre-miRNA is cleaved near the terminal loop by the RNAse III enzyme Dicer. The resulting small RNA duplex [37–41] is subsequently loaded onto an Argonaute (AGO) protein to form an effector complex called RNA-induced silencing complex (RISC) [42–44]. Only one of the two strands of the original pre-miRNA stem remains bound to the RISC (the guide strand) as mature miRNA, whereas the other strand (passenger strand) may be eliminated.

In RNA silencing, miRNA functions as a guide by base pairing with its target mRNAs, whereas AGO proteins function as effectors by recruiting factors that induce translational repression, mRNA deadenylation, and mRNA decay.

miRNA-binding sites are usually located in the 3′ untranslated region of mRNAs [45]. The sequence of mature miRNAs at the 5′ end that spans from nucleotide position 2 to 7 is crucial for target recognition and has been termed the "miRNA seed."

miRNAs are released from cells in membrane-bound vesicles (e.g., exosomes and microvesicles), which protect them from blood RNase activity. Exosomes are 50- to 90-nm vesicles arising from multivesicular bodies and released by exocytosis [46]. They consist of a limiting lipid bilayer, transmembrane proteins, and a hydrophilic core containing proteins, mRNAs, and miRNAs. Exosomes are present in a wide range of biological fluids, including blood and urine. They are recognized by their size and the fact that they are formed intracellularly within multivesicular endosomes, while microvesicles (100–1000 nm in diameter) are shed from the plasma membrane surface directly [47]. Key mechanisms by which exosomes may exert their biological functions on cells include (1) direct contact between surface molecules of vesicles and cells, (2) endocytosis of vesicles, and (3) vesicle–cell membrane fusion [48]. Exosomes may horizontally transfer RNAs, including miRNAs that have been shown to be functional after exosome-mediated delivery [49]. Moreover, recent evidence indicates that viruses can export and deliver functional miRNAs through exosomes [50,51].

A significant portion of circulating miRNAs in human plasma and serum is associated with the Argonaute2 (Ago2) protein [6,8]. Ago2 is the effector component of the miRNA-induced silencing complex that directly binds miRNAs and mediates mRNA repression in cells [52], [53]. It has been hypothesized that most extracellular miRNAs might be by-products of dead cells that remain in extracellular space due to the high stability of the Ago2 protein and Ago2-miRNA complex [6].

2.2 miRNAs as Diagnostic Biomarkers

miRNAs have been used to identify cancer tissue of unknown primary origin [54], demonstrating the effectiveness of miRNAs as biomarkers. Moreover, it has been reported that miRNA expression accurately separated carcinomas from benign samples such as benign prostatic hyperplasia (BPH, MIM #600082) and also further classified the carcinoma tumors according to their androgen dependence, indicating the potential of miRNAs as a novel diagnostic and prognostic tool for prostate cancer (MIM #176807) [55].

Although there are only about 2000 different miRNAs found in human cells (see miRBase, http://www.mirbase.org/), current estimates indicate that more than one-third of the cellular transcriptome is regulated by miRNAs [45].

miRNAs have been considered as a promising next generation of diagnostic biomarkers because of the strong correlation between their expression patterns and disease progression. Recently, miRNAs have been found in extracellular human body fluids including plasma, serum, urine, and saliva [56–59]. Some circulating miRNAs in the blood have been successfully proven as biomarkers for several diseases, including cardiovascular diseases [59] and cancer [56,60]. It remains unclear, however, if this approach can be extended to all reported circulating miRNAs, since variations in isolation, detection, and quantification protocols make it hard to compare the significance of the results.

Recently, it has been shown that breast cancer cells secrete exosomes with specific capacity for cell-independent miRNA biogenesis, while normal cell-derived exosomes lack this ability [61]. In the same study, the authors found that exosomes derived from cancer cells and serum from patients with breast cancer (MIM #114480) contain the RISC-loading complex proteins (Dicer, TRBP, and AGO2). Moreover, cancer exosomes alter the transcriptome of target cells in a Dicer-dependent manner, which stimulates nontumorigenic epithelial cells to form tumors. The authors concluded that the presence of Dicer in exosomes may serve as biomarker for detection of cancer.

One of the most frequent circulating miRNAs is miR-21. It has been successfully characterized as a promising biomarker for several diseases including hepatocellular carcinoma [60], non-small-cell lung cancer (NSCLC, MIM #211980) [62] and other solid cancers [63], as well as cardiovascular diseases, such as myocardial fibrosis [64].

Recently, it has been shown that serum miR-21 is a potential biomarker of epigenetic therapy in myelodysplastic syndrome (MDS, MIM #614286) patients [65]. Furthermore, Song et al. [66] explore regulation of miR-21 expression by epigenetic change and its impact on chemoresistance and malignant properties of pancreatic cancer showing that the miR-21 upregulation induced by histone acetylation in the promoter zone is associated with chemoresistance to gemcitabine and enhanced malignant potential in pancreatic cancer cells.

Another promising circulating miRNA is miR-141. Serum levels of miR-141 (an miRNA expressed in prostate cancer) can distinguish patients with prostate cancer from healthy controls [56]. Moreover, plasma miR-141 may represent a novel biomarker in detecting colon cancer (MIM #114500) with distant metastasis and high levels of miR-141 in plasma are associated with poor prognosis [67].

So far, most approaches for the identification of circulating diagnostic RNAs have focused on quantifying circulating RNAs that are overexpressed or depleted in the cancer cells they may have originated from [56,68],69]. This approach has only revealed some of the highly abundant cellular miRNAs and other ncRNAs, suggesting that only a subset of cellular RNAs is released into the extracellular environment [70]. This release does not appear to be a mere mechanical consequence of cellular decay, and in a recent study, Pigati and colleagues [71] have found that nearly 30% of the released miRNAs in vitro and in vivo do not reflect cellular profile, suggesting that some miRNAs are retained or released selectively. Some selectively released miRNAs were enriched in body fluids conditioned by mammary cells, including mammary fluids, blood, and milk. This subset of miRNAs may have value in breast cancer diagnosis and biology.

2.3 Effects of Drugs and Diet on Circulating miRNAs

Circulating miRNAs appear to be affected by various parameters, including drugs and diet. For instance, De Boer and colleagues [72] have shown that aspirin intake should be accounted when considering circulating miR-126 as diagnostic biomarker for cardiovascular diseases, or, more generally, when studying the possible role of miRNAs as mediators of cardiovascular disease. Table 3 summarizes some recent reports on the effects of drugs on circulating miRNAs.

Likewise, diet-derived exogenous miRNAs (or "xenomiRs") should also be taken into account because they affect total miRNA profiles as part of a circulating miRNA homeostasis that is altered in many diseases. Because miRNAs are found in both animals and plants, almost all fresh foods contain small RNAs that could contribute to the circulating miRNA population. Even processed food (e.g., cooked rice, potatoes, cabbage [73], and baby milk) [74] contain miRNAs, albeit at reduced concentrations.

The fact that these miRNAs could enter the bloodstream is supported by oral delivery of pharmacological preparations of small interfering RNA (siRNA) [75–77]. When protected by lipids, proteins, or polysaccharides, from the acidic and enzymatic environment of the digestive tract, ingested small RNA molecules may enter into circulation through the gut. Protection is achieved by artificial shells for therapeutic siRNAs, but multiple protective means are available for food-acquired miRNAs, including natural lipid vesicles [78] and protein complexes [73].

Recently, Vickers and colleagues [79] reported evidence that the high-density lipoprotein (HDL, a delivery vehicle for the return of excess cellular cholesterol to the liver for excretion) transports endogenous miRNAs and delivers them to recipient cells with functional targeting capabilities. Many of the genes significantly altered in response to atherosclerotic HDL-miRNAs delivery have a role in lipid metabolism, inflammation, and atherosclerosis (NDST1 [80], NR1D2 [81,82], BMPR2 [82], VEGFA, and FLT1 [83]). The exact process of how HDL is loaded with miRNAs and which proteins, if any, facilitate this association is not known. However, the importance of this study is that HDL-miRNAs could potentially serve as novel diagnostic markers in much the same way exosomal miRNAs have.

Circulating miRNA resources can be used to estimate the potential of reported miRNAs and thus help prioritizing their systematic clinical evaluation. Table 4 shows some data retrieved from the miRandola database [14,15] (see the following sections) about potential biomarkers based on extracellular miRNAs.

3. LONG NONCODING RNAs

Over the last decade, development of new technologies for genome-wide analyses of the eukaryotic trascriptome has revealed that around

TABLE 3 The Effects of Drugs on Circulating miRNAs

miRBase ID	Expr.	Drug	Sample	Disease
Hsa-miR-134	Down	Lithium	Plasma	Bipolar disorder
Hsa-miR-21-5p	Up	Platinum	Plasma	Non-small-cell lung cancer
Hsa-miR-210	Up	Trastuzumab	Plasma	Breast cancer
Hsa-miR-126-3p	Down	Aspirin	Plasma	Type 2 diabetes
Hsa-miR-152	Up	Docetaxel	Malignant effusions	Lung cancer
Hsa-miR-122-5p	Up	Acetaminophen	Serum	Acetaminophen-induced acute liver injury
Hsa-miR-192-5p	Up	Acetaminophen	Serum	Acetaminophen-induced acute liver injury

TABLE 4 Some Data Retrieved from miRandola and PubMed about Potential Biomarkers Based on Extracellular miRNAs and Long Noncoding RNAs

ID	RNA type	Sample	Disease or cell line	Expr.
Hsa-miR-21-5p	miRNA	Plasma, serum	Esophageal cancer, aortic stenosis, breast cancer, colorectal cancer, gastric cancer, glioblastoma, hepatocellular carcinoma, lung cancer, non-small-cell lung cancer, pediatric Crohn disease, solitary pulmonary nodules	Up
Hsa-miR-210	miRNA	Plasma, serum	Breast cancer, conventional renal cell cancer, preeclampsia, solitary pulmonary nodules	Up
Hsa-miR-29a-3p	miRNA	Plasma, serum	Advanced colorectal neoplasia (carcinomas and advanced adenomas), colorectal cancer, colorectal liver metastasis , preeclampsia	Up
Hsa-miR-499a-5p	miRNA	Plasma	Acute heart failure, acute myocardial infarction, acute viral myocarditis, ST elevation myocardial infarction, troponin-positive acute coronary syndrome	Up
Hsa-miR-208b	miRNA	Plasma	Acute myocardial infarction, acute viral myocarditis, ST elevation myocardial infarction	Up
Hsa-miR-208a	miRNA	Plasma	Acute myocardial infarction, troponin-positive acute coronary syndrome	Up
Hsa-miR-200b-3p	miRNA	Plasma, serum	Biliary atresia, metastatic breast cancer, serous epithelial ovarian cancer	Up
Hsa-miR-1	miRNA	Plasma	Acute myocardial infarction, ST elevation myocardial infarction	Up
Hsa-miR-122-5p	miRNA	Serum	Acetaminophen-induced acute liver injury, chronic hepatitis C infection, endometriosis	Up
Hsa-miR-126-3p	miRNA	Plasma,urine	Endurance exercise, urothelial bladder cancer	Up
Hsa-miR-133a	miRNA	Plasma, serum	ST elevation myocardial infarction, troponin-positive acute coronary syndrome	Up
Hsa-miR-141-3p	miRNA	Plasma, serum	Colorectal cancer, pregnancy, prostate cancer	Up
Hsa-miR-20a-5p	miRNA	Plasma, serum	Chronic hepatitis c infection, multiple myeloma, pediatric Crohn disease	Up
Hsa-miR-92a-3p	miRNA	Plasma, serum	Advanced colorectal neoplasia (carcinomas and advanced adenomas), chronic hepatitis c infection	Up
Hsa-miR-122-5p	miRNA	Serum	Liver injury	Down
Hsa-miR-126-3p	miRNA	Plasma	Acute myocardial infarction, type 2 diabetes	Down
Hsa-miR-133a	miRNA	Serum	Malignant astrocytomas	Down
Hsa-miR-141-3p	miRNA	Plasma	Pregnancy	Down
Hsa-miR-20a-5p	miRNA	Plasma	Endometriosis	Down
Hsa-miR-92a-3p	miRNA	Serum	Breast cancer	Down

TABLE 4 Some Data Retrieved from miRandola and PubMed about Potential Biomarkers Based on Extracellular miRNAs and Long Noncoding RNAs—cont'd

ID	RNA type	Sample	Disease or cell line	Expr.
POU3F3	lncRNA	Plasma	Esophageal squamous cell carcinoma	Up
TapSAKI	lncRNA	Plasma	Acute kidney injury	Up
LIPCAR	lncRNA	Plasma	Cardiac remodeling after myocardial infarction	Up
LIPCAR	lncRNA	Plasma	Future cardiovascular death after cardiac failure	Down
H19	lncRNA	Plasma	Gastric cancer	Up
PCA3	lncRNA	Urine	Prostate cancer	Up
HULC	lncRNA	Blood	Hepatocellular carcinoma	Up
TUC399	lncRNA	Cell medium	Hep3B, HepG2, PLC/PRF/5	Up
linc-RoR	lncRNA	Cell medium	HepG2 and PLC-PRF5	Up

two-thirds of the human genome is pervasively transcribed. This extensive transcription generates a huge array of RNA molecules of which less than 2% are protein-coding RNAs [84–87].

A large proportion of the human transcriptome produces the heterogeneous class of lncRNAs, which are generally referred to as molecules longer than 200nt, often polyadenilated and lacking of evident ORFs [88–90].

lncRNAs are produced from RNA polymerase II and can be antisense, interleaved, or overlapping with protein-coding genes or produced from region devoid of coding genes (termed lincRNAs, long intergenic or intervening ncRNAs) [84–87]. In addition, some of them are produced from enhancer regions (termed eRNAs) or transcription start site. Defining lncRNAs simply on the basis of size and lack of protein-coding capability is intellectually far from satisfying for these noncoding molecules, which are characterized by heterogeneous functions. lncRNAs for their intrinsic nature are able to mediate base pairing with other nucleic acids in a sequence-specific manner and at the same time are able to form complex structures that allow interaction with proteins. In particular, their modular and flexible characteristic leads them to act as scaffold for partners that normally would not interact with a protein–protein interaction and, in addition, their nucleic acids nature make them able to localize the scaffold-based complex in a sequence-specific manner [91,88,90]. Thus, in many cases, the secondary structure is more important than nucleotide sequence. This is reflected in the lack of sequence conservation for the majority of lncRNAs between different species [92–95]. The production of such flexible and multifunctional RNA molecules led on to an important evolutionary advantage for the selection of regulatory molecules [88,90,91]. Indeed, the increase in the noncoding content of genome with organismal complexity supports the idea that the noncoding regions were fundamental to the genetic programming of complex eukaryotes [96].

The function of lncRNAs is often related to their localization inside the cell. Most nuclear lncRNAs function by recruiting chromatin-modifying machineries to specific genomic loci [88,90,91,97]. In particular, they can act as a guide for DNA methyltransferase 3 and histone modifier as polycomb repressive complex PRC2 [98,99], and histone H3 lysine 9 (H3K9) methyltransferases [100,101]. These modifications correlate with the leading of repressive heterochromatin and the resultant transcriptional repression. lncRNAs have also been shown to lead transcriptional activation by recruiting chromatin-modifying complexes as the histone H3K4 methyltransferase MLL1 complex [102,103]. In addition, the lncRNA transcription itself can negatively affect gene expression [104,105]. Based on the target sites, it is possible to distinguish between *cis*- and *trans*-acting lncRNAs. *Cis*-acting lncRNAs control the expression of genes located near their transcription sites, instead *trans*-acting can regulate gene expression at independent loci [88,90,91]. lncRNAs-mediated mechanisms of gene regulation have been also identified in the cytoplasm [96]. The regulation promoted by cytoplasmic lncRNAs is often based on sequence complementary with protein-coding transcripts and upon recognition of the target by base paring, they can modulate gene expression at the post-transcriptional level by regulating RNA translation and/or stability. Such examples include the positive regulation by the ubiquitin carboxyl-terminal hydrolase L1 antisense RNA 1 (Uchl1-as1) [106] and the negative regulation by the tumor protein p53 pathway corepressor 1 (Trp-53cor1; also known as lincRNA-p21) [107].

In the cytoplasm, lncRNAs can also act as competing endogenous RNAs (ceRNAs) [108]. This process is based on binding to and sequestering of specific miRNAs, acting as "miRNAs sponges" to lead to the derepression of the target mRNA. This phenomenon represents a new regulatory loop in which different types of RNAs (both coding and noncoding) can compete with each other for shared miRNAs binding [109].

Due to the great potential of lncRNA to modulate gene expression, there is an increasing interest in the likely involvement of these RNAs in disease etiology. Indeed, different studies have already shown that changes in lncRNA expression may contribute to the development of human diseases. Many studies demonstrate a correlation between lncRNAs misregulation, and so gene expression alteration, and occurrence of pathology, but now what is emerging is a potential role in gene regulation outside the cell and between different cells.

Recent publications have shown that in human fluids it is possible to detect the presence of lncRNAs (see Table 4), and this discovery is important in the development of new diagnostic and prognostic tools, thanks to the possible association between circulating lncRNA concentration and pathology development and evolution. A systematic analysis of human plasma-derived exosomal RNAs by deep-sequencing technique detected a significant fraction of other RNA species in addition to well known and abundant miRNAs among which lncRNAs represent around 3% of the mapped counts [19].

Cell-free RNA stability analysis demonstrated that the majority of circulating RNA is fragmented and unstable in plasma [110]. Nevertheless, at present, with the well-known instability of RNA species, detection of circulating RNA was perhaps rather surprising and probably associated to an RNA-packaging process to avoid nucleases degradation as apoptotic bodies. At present, few lncRNAs have been identified and characterized as potential biomarkers in human fluids. In particular, by examining the expression of lncRNAs in peripheral blood of major depressive disorder (MDD, MIM #608516) patients, it has been shown that there is a substantial difference in the expression of lncRNAs and mRNAs compared to the control group [111]. Among these differentially expressed mRNAs, 17 genes were documented as depression-related genes in previous studies and it has been speculated that the differentially expressed lncRNAs might contribute to MDD by regulating these coding genes. The detection of circulating lncRNAs in peripheral blood not only represents a new layer of complexity in the molecular architecture of MDD, but it also reveals the potential to use them as diagnostic markers and therapeutics.

Another example is provided by the measurement of lncRNA PCA3 (prostate cancer antigen 3) in patient urine sample, which it has been shown to allow more sensitive and specific diagnosis of prostate cancer than the widely used prostate-specific antigen (PSA) serum level [112–114]. Comparing the performance characteristic of PCA3 assay in a cohort of over 300 patients showed an increased specificity in cancer detection with respect to PSA assay and minimized unnecessary biopsies [115]. The lncRNA HULC (highly upregulated in liver cancer) is highly expressed in hepatocarcinoma (HCC, MIM #114550) patients and can be detected in the HCC patient blood in tissue samples [116]. This study highlights the potential role of HULC as novel biomarker for HCC.

Another study identified a previous unrecognized role of lincRNAs regulator of reprogramming (linc-ROR) as a mediator of cell-to-cell communication through the transfer of extracellular vesicle. This mechanism is involved in stress cell-to-cell signaling in order to activate stress responses such as activation of survival pathways. Cellular cross-talk pathway may then result in chemoresistance within the tissue and may contribute to loss of therapeutic effect of therapeutic drugs, such as sorafenib [117].

These studies demonstrate that circulating lncRNAs are present in human fluids and may play an important regulative role in cell-to-cell signaling and modulation of gene expression. Even if the mechanisms by which this could happen are not understood at all, cell-to-cell exchange of lncRNAs could lead drastic changes due to the wide landscape of function that these RNAs are able to dispatch both at physiological and pathological levels.

For the time being, circulating lncRNA studies represent a very promising field that in the next future will provide great advantages in the development of prognostic and diagnostic noninvasive tool for human diseases.

4. CIRCULAR RNAs

circRNAs are recent and new-growing components of the lncRNAs class. Although biologists identified the existence of circular transcripts more than 20 years ago [118], these circular molecules were long considered artifacts of aberrant splicing reactions [119] or prerogative of few viral pathogens [120,121].

However, recent works, by using specific computational pipeline for the identification of circular molecule, have revealed that in many cells the production of circRNAs is not as rare as once thought [122–124]. Evidence from expression data in Archaea and Mammalia indicated that circRNAs are abundant, conserved, and stably accumulated within the cell [125,118,119,126–128,122].

CircRNAs can be generated from exons (exonic circRNA) or introns (intronic circRNAs) through independent ways: direct ligation of 5′ and 3′ ends of linear RNAs, as intermediates in RNA processing reaction, or by "backsplicing," wherein a downstream 5′ splice site (splice donor) is joined to an upstream 3′ splice site (splice acceptor) [129].

Circular molecules present many advantages compared to linear molecule that could explain the evolutionary conservation from Archaea to Mammalia [125,124]. RNA circles can be much more stable than linear RNAs due to the inaccessibility to degradative enzymes [130–132], and in addition circRNAs can create constrained folded molecule, which may be useful for interacting with specific protein or other RNA molecules.

The functions of circRNAs remain to be fully understood. However, one recent study demonstrates that some intronic circRNAs enhance the transcription of the genes from which they are produced. Antisense oligonucleotides against a circRNA intron reduce expression of the resident gene, while oligonucleotides against other introns, or the region between the branchpoint and the 3′ splice site of the circRNA intron, which would be lacking in the stable circRNA intron, do not have any effect [133]. Moreover, intronic circRNAs can associate with RNA polymerase II [133]. These observations suggest that circRNAs might modulate RNA polymerase II in cis and thereby alter the expression of their hosting gene. Consistent with this idea, some intronic circRNAs accumulate in the nucleus and, where examined, can be localized to their sites of transcription [134,135,133]. One of the first circRNA identified in mammals is the one-antisense to the mRNA transcribed from the sex-determining region Y locus, which is highly expressed in testes of adult mouse and is essential for sex determination [136]. More interestingly, recent studies identified a new example of ceRNA [108] belonging to the circRNA class [128,137,138]. Whereas the linear ceRNAs have a short half-life that allows a rapid control of sponge activity, circRNAs have improved stability, and their turnover can be controlled by the presence of a perfectly matched miRNA [128,137,138]. In particular, in human neuronal cells a circRNA running antisense to the cerebellar degeneration-related protein 1 (CDR1) locus [128] has been identified, referred to as CDR1as (antisense) or CiRS-7 (circular RNA sponge for miR-7), which show about 70 conserved matches to the miR-7 seed [138], [137]. The striking presence of such abundant number of sites for miR-7 suggested a possible function as miRNA sponge, inhibiting an endogenous miRNA and leading to miR-targets derepression. So circRNAs may serve as transcription regulators [133] or as sponges for small RNA regulators [137,138]. CircRNAs could also act as RNA-binding proteins "sponge" or recruiter and bind RNAs besides miRNAs to form RNA–protein complexes. In addition, some circRNAs could be easily translated producing protein since internal ribosome entry site (IRES)-mediated translation can in principle occur on circRNAs [139]. Evidence that circular intronic RNAs can get passed on to offspring in *Xenopus* oocytes hints at their role in RNA-mediated inheritance and epigenetics [140]. It has also been demonstrated that they are differentially expressed in different cell types and conditions [124].

Due to their emerging role as gene expression regulators, circRNAs are very likely to become important players in cancer development [127] and pathologies as other type of ncRNAs (see Sections 2 and 3). Studies about emerging circular structured RNA molecules have just started and are a completely new field of study that will contribute to better understand the articulated regulatory networks occurring in complex organisms. Furthermore, the striking stability of these molecules suggests that soon or late circRNAs could be used as diagnostic marker and in addition, they could be easily identified in human fluids.

5. RESOURCES FOR CIRCULATING RNAs

Here, we present four online resources recently developed, collecting noninvasive biomarker data based on literature mining. miRandola and Human miRNA Disease Database (HMDD) are specific for miRNAs, while ExoCarta and Vesiclepedia are databases of membrane-bound vesicles. These resources provide a variety of information including diseases, RNA isolation methods, and the experimental protocols, essential data for researchers. Nowadays, these resources are mainly focused on miRNAs.

5.1 miRandola: Extracellular Circulating miRNAs Database

miRandola (http://atlas.dmi.unict.it/mirandola/) is the first comprehensive database of extracellular circulating miRNAs [14,15]. It is manually curated and miRNAs are classified into different categories, based on their main extracellular forms: solely complexed with Ago2 protein [6,8] encapsulated within exosomes [49] or bound to HDL [79]. The last database update is based on the compilation of 139 papers, resulting in 2366 entries,

599 unique mature miRNAs, and 23 samples. The database provides users a variety of information including diseases, samples description, potential biomarker role of miRNAs, the isolation method, and the experimental protocol.

By using the search section of miRandola (http://atlas.dmi.unict.it/mirandola/browse.php), users can mine information choosing between miRNA-Ago2, miRNA-exosome, miRNA-HDL, and miRNA-circulating. The latter, which constitutes the largest group, is used when authors of the corresponding paper do not distinguish between Ago2, exosome, or HDL. The miRNA-exosome type has 862 entries retrieved from papers mined from PubMed (http://www.ncbi.nlm.nih.gov/pubmed/) and ExoCarta [141], a database of exosomal proteins, RNA, and lipids (more details in the next sections).

In miRandola there are 173 miRNA-Ago2-type entries [6,8]. Extracellular miRNA in human blood plasma can be also immunoprecipitated with anti-Ago1 antibody; furthermore Ago1- and Ago2-associated miRNA profiles are significantly different [142]. This indicates that many other tissues and organs are probably contributing to the extracellular miRNA content of biological fluids (provided that the relative expression of AGO1 and AGO2 is cell-type specific).

The database also contains 20 entries retrieved from a work of Vickers and colleagues [79]. Altogether, they make a total of 16 unique mature miRNAs derived from HDL isolated from plasma. One of the most abundant miRNAs that was found in Vickers' work is miR-223, a highly conserved miRNA shown to be highly correlated with atherosclerosis [143]. Interestingly, miR-223 is one of the most frequent miRNAs in the database (together with miR-21 and miR-16).

The existence of HDL-associated miRNAs in the blood circulation have been recently confirmed by an independent research group, however the analyzed HDL-miRNAs (including miR-223) constituted only a minor proportion of the total circulating miRNAs [144].

The miRNA-circulating type constitutes the largest group. The classification of this biotype is used whenever authors report an extracellular miRNA without further characterization. miRandola contains 1311 entries of this kind.

The vast number of entries for the miRNA-circulating type, probably depends on the fact that the extracellular miRNA research topic is very new, yet it is growing very fast.

5.2 ExoCarta: The Exosome Database

ExoCarta (http://www.exocarta.org) is a manually curated database of exosomal proteins, RNA, and lipids [141]. It collects information from both published and unpublished exosomal studies, the mode of exosomal purification, and characterization and the biophysical and molecular properties. Currently, ExoCarta contains 13,333 protein entries (4563 proteins), 2375 mRNA entries (1639 mRNAs), 764 miRNAs, and 194 lipid entries retrieved from 146 studies.

The database has two modules. The Query module(http://exocarta.org/query.html)allows users to search for gene symbols and protein names while for more options to search data by tissue/cell-type or organism users can select the Browse module (http://exocarta.org/browse).

ExoCarta collects data from five species (*Homo sapiens*, *Rattus norvegicus*, *Mus musculus*, *Ovis aries*, and *Cavia porcellus*) and 52 samples.

All the data can be downloaded from the download web page of the database.

5.3 Vesiclepedia: The Extracellular Vesicles Database

Vesiclepedia is a manually curated compendium of molecular data (lipid, RNA, and protein) identified in different classes of extracellular vesicles (EVs), membraneous vesicles released by a variety of cells into the extracellular microenvironment [145]. Based on the mode of biogenesis, EVs can be classified into three classes: (1) ectosomes or shedding microvesicles (large EVs ranging between 50 and 1000 nm in diameter) (2) exosomes, and (3) apoptotic bodies (they are released from fragmented apoptotic cells and are 50–5000 nm in diameter).

Currently, Vesiclepedia comprises 92,897 protein entries, 27,642 mRNA entries, 4934 miRNA entries, 584 lipid entries, and 33 species collected from 538 studies.

The importance of this database is that the authors have initiated a continuous community annotation project with the active involvement of EV researchers setting a gold standard in data sharing for the field.

5.4 HMDD: The Human miRNA Disease Database

The HMDD (http://cmbi.bjmu.edu.cn/hmdd) is a collection of experimentally supported human miRNA and disease associations [146]. Currently, HMDD collected 10,368 entries that include 572 miRNA genes, 378 diseases from 3511 papers.

Users can browse the HMDD by miRNA names or disease names by clicking one miRNA or disease in the Browse page. Moreover, the authors provided a fuzzy search function for the entries by the full or partial names of miRNAs or diseases in the Search page. All data in the database can be freely downloaded and users can also submit novel data into the database.

HMDD collects specific classes of entries, for example, entries whose experimental evidence is from genetics, epigenetics, circulating miRNAs, and miRNA–target interactions. The information provided for the circulating miRNA data concerns the miRNA name, the associated disease, the PubMed ID, and a short description.

Interestingly, by analyzing the publication time regarding the miRNA–disease associations in the past decade included in HMDD, the authors show that this simple analysis could predict the tendency of miRNA–disease relationship investigations. Especially, entries from circulating miRNAs increase most dramatically in recent years, and it is expected that more data will be generated in the coming years suggesting that the identification of circulating miRNAs as biomarkers is one of the hottest topics in miRNA research.

6. CONCLUSION AND FUTURE PERSPECTIVE

An ideal biomarker should be accessible by applying noninvasive protocols, being inexpensive to quantify, specific to the disease of interest, translatable from model systems to humans, and a reliable early indication of disease before clinical symptoms appear [147]. Extracellular RNAs are promising candidate biomarkers, but optimal and standardized conditions for processing blood specimens for RNA measurement remain to be established.

The establishment of a standardized acquisition, processing, storage procedures, as well as the development of assays that are accurate, precise, specific, and robust, with regard to quantitation of miRNAs, are very much needed. Plasma and serum processing [148,149], choice of anticoagulant [150], and hemolysis [148,151] have been reported to affect miRNA measurement.

Given the increasing amount of data available about extracellular nucleic acids, databases are useful reference tools for anyone who wishes to investigate the role and function of ncRNAs as noninvasive biomarkers. The online resources presented in this chapter are mainly manually curated and the updation of data needs many biocurators and time. For this reason, in the spirit of modern collaborative research projects, scientists can collaborate through direct data submissions, in order to expand the knowledge base and give their contribution to the research topic. This will facilitate a better understanding of the role of circulating ncRNAs, while stimulating the discussion on the current knowledge and the controversial topics mentioned in this chapter.

LIST OF ABBREVIATIONS

BPH Benign prostatic hyperplasia
ceRNA Competing endogenous RNA
circRNAs Circular RNAs
CNA Circulating nucleic acids
HDL High-density lipoprotein
lncRNA Long noncoding RNA
MDD Major depressive disorder
MDS Myelodysplastic syndromes
miRNA MicroRNA
miRISC miRNA-induced silencing complex
ncRNAs Noncoding RNAs
NGS Next-generation sequencing
NSCLC Nonsmall cell lung cancer
PSA Prostate-specific antigen

Acknowledgments

Francesco Russo has been supported by a fellowship sponsored by Progetto Istituto Toscano Tumori-Grant 2012 Prot. A00GRT.

References

[1] Birney E, Stamatoyannopoulos JA, Dutta A, Guigó R, Gingeras TR, Margulies EH, et al. Identification and analysis of functional elements in 1% of the human genome by the encode pilot project. Nature 2007;447(7146):799–816.
[2] Kapranov P, Cheng J, Dike S, Nix DA, Duttagupta R, Willingham AT, et al. RNA maps reveal new RNA classes and a possible function for pervasive transcription. Science 2007;316(5830):1484–8.
[3] Kozomara A, Griffiths-Jones S. miRBase: annotating high confidence microRNAs using deep sequencing data. Nucleic Acids Res 2013;42:D68–73. gkt1181.
[4] Krol J, Loedige I, Filipowicz W. The widespread regulation of microRNA biogenesis, function and decay. Nat Rev Genet 2010;11(9):597–610.
[5] Bartel DP. Micrornas: genomics, biogenesis, mechanism, and function. Cell 2004;116(2):281–97.
[6] Turchinovich A, Weiz L, Langheinz A, Burwinkel B. Characterization of extracellular circulating microRNA. Nucleic Acids Res 2011;39(16):7223–33. gkr254.
[7] Kroh EM, Parkin RK, Mitchell PS, Tewari M. Analysis of circulating microRNA biomarkers in plasma and serum using quantitative reverse transcription-PCR (qRT-PCR). Methods 2010;50(4):298–301.
[8] Arroyo JD, Chevillet JR, Kroh EM, Ruf IK, Pritchard CC, Gibson DF, et al. Argonaute2 complexes carry a population of circulating microRNAs independent of vesicles in human plasma. Proc Natl Acad Sci 2011;108(12):5003–8.
[9] Turchinovich A, Weiz L, Burwinkel B. Isolation of circulating microRNA associated with RNA-binding protein. In: Circulating microRNAs. Springer; 2013. p. 97–107.
[10] Théry C, Amigorena S, Raposo G, Clayton A. Isolation and characterization of exosomes from cell culture supernatants and biological fluids. Curr Protoc Cell Biol 2006. 3.22.
[11] Rani S, O'Brien K, Kelleher FC, Corcoran C, Germano S, Radomski MW, et al. Isolation of exosomes for subsequent mRNA, microRNA, and protein profiling. In: Gene expression profiling. Springer; 2011. p. 181–95.
[12] Clayton A, Court J, Navabi H, Adams M, Mason MD, Hobot JA, et al. Analysis of antigen presenting cell derived exosomes, based on immuno-magnetic isolation and flow cytometry. J Immunol Methods 2001;247(1):163–74.
[13] Taylor DD, Zacharias W, Gercel-Taylor C. Exosome isolation for proteomic analyses and RNA profiling. In: Serum/plasma proteomics. Springer; 2011. p. 235–46.
[14] Russo F, Di Bella S, Nigita G, Macca V, Lagana A, Giugno R, et al. miRandola: extracellular circulating microRNAs database. PLoS One 2012;7(10): e47786.
[15] Russo F, Di Bella S, Bonnici V, Laganà A, Rainaldi G, Pellegrini M, et al. A knowledge base for the discovery of function, diagnostic potential and drug effects on cellular and extracellular miRNAs. BMC Genomics 2014;15(3):1–7.
[16] Chen X, Ba Y, Ma L, Cai X, Yin Y, Wang K, et al. Characterization of microRNAs in serum: a novel class of biomarkers for diagnosis of cancer and other diseases. Cell Res 2008;18(10):997–1006.
[17] Yang Q, Lu J, Wang S, Li H, Ge Q, Lu Z. Application of next-generation sequencing technology to profile the circulating microRNAs in the serum of preeclampsia versus normal pregnant women. Clin Chim Acta 2011;412(23):2167–73.
[18] Burgos KL, Javaherian A, Bomprezzi R, Ghaffari L, Rhodes S, Courtright A, et al. Identification of extracellular miRNA in human cerebrospinal fluid by next-generation sequencing. RNA 2013;19(5): 712–22.
[19] Huang X, Yuan T, Tschannen M, Sun Z, Jacob H, Du M, et al. Characterization of human plasma-derived exosomal RNAs by deep sequencing. BMC Genomics 2013;14(1):319.
[20] Peters DL, Pretorius PJ. Origin, translocation and destination of extracellular occurring DNA–a new paradigm in genetic behaviour. Clin Chim Acta 2011;412(11):806–11.
[21] Mandel P, Metais P. Les acides nucleiques du plasma sanguin chez l'homme. C R Seances Soc Biol Fil 1948;142(3–4):241–3.

[22] van der Vaart M, Pretorius PJ. Is the role of circulating DNA as a biomarker of cancer being prematurely overrated? Clin Biochem 2010;43(1):26–36.
[23] Ghorbian S, Ardekani AM. Non-invasive detection of esophageal cancer using genetic changes in circulating cell-free DNA. Avicenna J Med Biotechnol 2012;4(1):3.
[24] Griffiths-Jones S, Saini HK, van Dongen S, Enright AJ. miRBase: tools for microRNA genomics. Nucleic Acids Res 2008;36(Suppl. 1):D154–8.
[25] Lee Y, Jeon K, Lee J-T, Kim S, Kim VN. MicroRNA maturation: stepwise processing and subcellular localization. EMBO J 2002;21(17):4663–70.
[26] Cai X, Hagedorn CH, Cullen BR. Human microRNAs are processed from capped, polyadenylated transcripts that can also function as mRNAs. RNA 2004;10(12):1957–66.
[27] Lee Y, Kim M, Han J, Yeom K-H, Lee S, Baek SH, et al. MicroRNA genes are transcribed by RNA polymerase II. EMBO J 2004;23(20):4051–60.
[28] Davis-Dusenbery BN, Hata A. Mechanisms of control of microRNA biogenesis. J Biochem 2010;148(4):381–92.
[29] Kim VN, Han J, Siomi MC. Biogenesis of small RNAs in animals. Nat Rev Mol Cell Biol 2009;10(2):126–39.
[30] Lee Y, Ahn C, Han J, Choi H, Kim J, Yim J, et al. The nuclear RNase III Drosha initiates microRNA processing. Nature 2003;425(6956):415–9.
[31] Denli AM, Tops BB, Plasterk RH, Ketting RF, Hannon GJ. Processing of primary microRNAs by the microprocessor complex. Nature 2004;432(7014):231–5.
[32] Gregory RI, Yan K-P, Amuthan G, Chendrimada T, Doratotaj B, Cooch N, et al. The microprocessor complex mediates the genesis of microRNAs. Nature 2004;432(7014):235–40.
[33] Han J, Lee Y, Yeom K-H, Kim Y-K, Jin H, Kim VN. The Drosha-DGCR8 complex in primary microRNA processing. Genes Dev 2004;18(24):3016–27.
[34] Landthaler M, Yalcin A, Tuschl T. The human DiGeorge syndrome critical region gene 8 and its *D. melanogaster* homolog are required for miRNA biogenesis. Curr Biol 2004;14(23):2162–7.
[35] Han J, Lee Y, Yeom K-H, Nam J-W, Heo I, Rhee J-K, et al. Molecular basis for the recognition of primary microRNAs by the Drosha-DGCR8 complex. Cell 2006;125(5):887–901.
[36] Zeng Y, Yi R, Cullen BR. Recognition and cleavage of primary microRNA precursors by the nuclear processing enzyme Drosha. EMBO J 2005;24(1):138–48.
[37] Bernstein E, Caudy AA, Hammond SM, Hannon GJ. Role for a bidentate ribonuclease in the initiation step of RNA interference. Nature 2001;409(6818):363–6.
[38] Grishok A, Pasquinelli AE, Conte D, Li N, Parrish S, Ha I, et al. Genes and mechanisms related to RNA interference regulate expression of the small temporal RNAs that control *C. elegans* developmental timing. Cell 2001;106(1):23–34.
[39] Hutvágner G, McLachlan J, Pasquinelli AE, Bálint É, Tuschl T, Zamore PD. A cellular function for the RNA-interference enzyme dicer in the maturation of the let-7 small temporal RNA. Science 2001;293(5531):834–8.
[40] Ketting RF, Fischer SE, Bernstein E, Sijen T, Hannon GJ, Plasterk RH. Dicer functions in RNA interference and in synthesis of small RNA involved in developmental timing in C. elegans. Genes Dev 2001;15(20):2654–9.
[41] Knight SW, Bass BL. A role for the RNase III enzyme DCR-1 in RNA interference and germ line development in *Caenorhabditis elegans*. Science 2001;293(5538):2269–71.
[42] Hammond SM, Boettcher S, Caudy AA, Kobayashi R, Hannon GJ. Argonaute2, a link between genetic and biochemical analyses of RNAi. Science 2001; 293(5532):1146–50.
[43] Mourelatos Z, Dostie J, Paushkin S, Sharma A, Charroux B, Abel L, et al. miRNPs: a novel class of ribonucleoproteins containing numerous microRNAs. Genes Dev 2002;16(6):720–8.
[44] Tabara H, Sarkissian M, Kelly WG, Fleenor J, Grishok A, Timmons L, et al. The rde-1 gene, RNA interference, and transposon silencing in *C. elegans*. Cell 1999;99(2):123–32.
[45] Bartel DP. MicroRNAs: target recognition and regulatory functions. Cell 2009;136(2):215–33.
[46] Fevrier B, Raposo G. Exosomes: endosomal-derived vesicles shipping extracellular messages. Curr Opin Cell Biol 2004;16(4):415–21.
[47] Heijnen HF, Schiel AE, Fijnheer R, Geuze HJ, Sixma JJ. Activated platelets release two types of membrane vesicles: microvesicles by surface shedding and exosomes derived from exocytosis of multivesicular bodies and alpha-granules. Blood 1999;94(11):3791–9.
[48] György B, Szabó TG, Pásztói M, Pál Z, Misják P, Aradi B, et al. Membrane vesicles, current state-of-the-art: emerging role of extracellular vesicles. Cell Mol Life Sci 2011;68(16):2667–88.
[49] Valadi H, Ekström K, Bossios A, Sjöstrand M, Lee JJ, Lötvall JO. Exosome-mediated transfer of mRNAs and microRNAs is a novel mechanism of genetic exchange between cells. Nat Cell Biol 2007;9(6):654–9.
[50] Pegtel DM, Cosmopoulos K, Thorley-Lawson DA, van Eijndhoven MA, Hopmans ES, Lindenberg JL, et al. Functional delivery of viral miRNAs via exosomes. Proc Natl Acad Sci 2010;107(14):6328–33.
[51] Laganà A, Russo F, Veneziano D, Di Bella S, Pulvirenti A, Giugno R, et al. Extracellular circulating viral microRNAs: current knowledge and perspectives. Front Genet 2013;24(4):120.
[52] Song J-J, Liu J, Tolia NH, Schneiderman J, Smith SK, Martienssen RA, et al. The crystal structure of the argonaute2 PAZ domain reveals an RNA binding motif in RNAi effector complexes. Nat Struct Mol Biol 2003;10(12):1026–32.

[53] Ma J-B, Ye K, Patel DJ. Structural basis for overhang-specific small interfering RNA recognition by the PAZ domain. Nature 2004;429(6989):318–22.

[54] Rosenfeld N, Aharonov R, Meiri E, Rosenwald S, Spector Y, Zepeniuk M, et al. MicroRNAs accurately identify cancer tissue origin. Nat Biotechnol 2008;26(4):462–9.

[55] Porkka KP, Pfeiffer MJ, Waltering KK, Vessella RL, Tammela TL, Visakorpi T. MicroRNA expression profiling in prostate cancer. Cancer Res 2007;67(13):6130–5.

[56] Mitchell PS, Parkin RK, Kroh EM, Fritz BR, Wyman SK, Pogosova-Agadjanyan EL, et al. Circulating microRNAs as stable blood-based markers for cancer detection. Proc Natl Acad Sci 2008;105(30):10513–8.

[57] Park NJ, Zhou H, Elashoff D, Henson BS, Kastratovic DA, Abemayor E, et al. Salivary microRNA: discovery, characterization, and clinical utility for oral cancer detection. Clin Cancer Res 2009;15(17):5473–7.

[58] Zubakov D, Boersma AW, Choi Y, van Kuijk PF, Wiemer EA, Kayser M. MicroRNA markers for forensic body fluid identification obtained from microarray screening and quantitative RT-PCR confirmation. Int J Legal Med 2010;124(3):217–26.

[59] Gupta SK, Bang C, Thum T. Circulating microRNAs as biomarkers and potential paracrine mediators of cardiovascular disease. Circ Cardiovasc Genet 2010;3(5):484–8.

[60] Tomimaru Y, Eguchi H, Nagano H, Wada H, Kobayashi S, Marubashi S, et al. Circulating microRNA-21 as a novel biomarker for hepatocellular carcinoma. J Hepatol 2012;56(1):167–75.

[61] Melo SA, Sugimoto H, O'Connell JT, Kato N, Villanueva A, Vidal A, et al. Cancer exosomes perform cell-independent microRNA biogenesis and promote tumorigenesis. Cancer Cell 2014;26(5):707–21.

[62] Wei J, Gao W, Zhu C-J, Liu Y-Q, Mei Z, Cheng T, et al. Identification of plasma microRNA-21 as a biomarker for early detection and chemosensitivity of non–small cell lung cancer. Chin J Cancer 2011;30(6):407.

[63] Wang B, Zhang Q. The expression and clinical significance of circulating microRNA-21 in serum of five solid tumors. J Cancer Res Clin Oncol 2012;138(10):1659–66.

[64] Villar AV, García R, Merino D, Llano M, Cobo M, Montalvo C, et al. Myocardial and circulating levels of microRNA-21 reflect left ventricular fibrosis in aortic stenosis patients. Int J Cardiol 2013;167(6):2875–81.

[65] Kim Y, Cheong J-W, Kim Y-K, Eom J-I, Jeung H-K, Kim SJ, et al. Serum microRNA-21 as a potential biomarker for response to hypomethylating agents in myelodysplastic syndromes. PloS One 2014;9(2):e86933.

[66] Song W-F, Wang L, Huang W-Y, Cai X, Cui J-J, Wang L-W. MiR-21 upregulation induced by promoter zone histone acetylation is associated with chemoresistance to gemcitabine and enhanced malignancy of pancreatic cancer cells. Asian Pac J Cancer Prev 2012;14(12):7529–36.

[67] Cheng H, Zhang L, Cogdell DE, Zheng H, Schetter AJ, Nykter M, et al. Circulating plasma MiR-141 is a novel biomarker for metastatic colon cancer and predicts poor prognosis. PloS One 2011;6(3):e17745.

[68] Huang Z, Huang D, Ni S, Peng Z, Sheng W, Du X. Plasma microRNAs are promising novel biomarkers for early detection of colorectal cancer. Int J Cancer 2010;127(1):118–26.

[69] Skog J, Würdinger T, van Rijn S, Meijer DH, Gainche L, Curry WT, et al. Glioblastoma microvesicles transport RNA and proteins that promote tumour growth and provide diagnostic biomarkers. Nat Cell Biol 2008;10(12):1470–6.

[70] Chen TS, Lai RC, Lee MM, Choo ABH, Lee CN, Lim SK. Mesenchymal stem cell secretes microparticles enriched in pre-microRNAs. Nucleic Acids Res 2010;38(1):215–24.

[71] Pigati L, Yaddanapudi SC, Iyengar R, Kim D-J, Hearn SA, Danforth D, et al. Selective release of microRNA species from normal and malignant mammary epithelial cells. PloS One 2010;5(10):e13515.

[72] de Boer HC, van Solingen C, Prins J, Duijs JM, Huisman MV, Rabelink TJ, et al. Aspirin treatment hampers the use of plasma microRNA-126 as a biomarker for the progression of vascular disease. Eur Heart J 2013;34(44):3451–7. eht007.

[73] Zhang L, Hou D, Chen X, Li D, Zhu L, Zhang Y, et al. Exogenous plant MIR168a specifically targets mammalian LDLRAP1: evidence of cross-kingdom regulation by microRNA. Cell Res 2011;22(1): 107–26.

[74] Chen X, Gao C, Li H, Huang L, Sun Q, Dong Y, et al. Identification and characterization of microRNAs in raw milk during different periods of lactation, commercial fluid, and powdered milk products. Cell Res 2010;20(10):1128–37.

[75] Akhtar S. Oral delivery of siRNA and antisense oligonucleotides. J Drug Target 2009;17(7):491–5.

[76] Aouadi M, Tesz GJ, Nicoloro SM, Wang M, Soto E, Ostroff GR, et al. Orally delivered siRNA targeting macrophage Map4k4 suppresses systemic inflammation. Nature 2009;458(7242):1180–4.

[77] Xu J, Ganesh S, Amiji M. Non-condensing polymeric nanoparticles for targeted gene and siRNA delivery. Int J Pharm 2012;427(1):21–34.

[78] Hata T, Murakami K, Nakatani H, Yamamoto Y, Matsuda T, Aoki N. Isolation of bovine milk-derived microvesicles carrying mRNAs and microRNAs. Biochem Biophys Res Commun 2010;396(2): 528–33.

[79] Vickers KC, Palmisano BT, Shoucri BM, Shamburek RD, Remaley AT. MicroRNAs are transported in plasma and delivered to recipient cells by high-density lipoproteins. Nat Cell Biol 2011;13(4): 423–33.

[80] MacArthur JM, Bishop JR, Stanford KI, Wang L, Bensadoun A, Witztum JL, et al. Liver heparan sulfate proteoglycans mediate clearance of triglyceride-rich lipoproteins independently of LDL receptor family members. J Clin Invest 2007;117(1):153–64.
[81] Ramakrishnan S, Lau P, Burke L, Muscat G. Rev-erbbeta regulates the expression of genes involved in lipid absorption in skeletal muscle cells: evidence for cross-talk between orphan nuclear receptors and myokines. J Biol Chem 2005;280(10):8651–9.
[82] Yao Y, Shao ES, Jumabay M, Shahbazian A, Ji S, Boström KI. High-density lipoproteins affect endothelial BMP-signaling by modulating expression of the activin-like kinase receptor 1 and 2. Arterioscler Thromb Vasc Biol 2008;28(12):2266–74.
[83] Moreno PR, Purushothaman K-R, Sirol M, Levy AP, Fuster V. Neovascularization in human atherosclerosis. Circulation 2006;113(18):2245–52.
[84] Carninci P, Kasukawa T, Katayama S, Gough J, Frith M, Maeda N, et al. The transcriptional landscape of the mammalian genome. Science 2005;309(5740):1559–63.
[85] Kapranov P, Drenkow J, Cheng J, Long J, Helt G, Dike S, et al. Examples of the complex architecture of the human transcriptome revealed by race and high-density tiling arrays. Genome Res 2005;15(7):987–97.
[86] Mattick JS, Makunin IV. Non-coding RNA. Hum Mol Genet 2006;15(Suppl. 1):R17–29.
[87] Kapranov P, Willingham AT, Gingeras TR. Genome-wide transcription and the implications for genomic organization. Nat Rev Genet 2007;8(6):413–23.
[88] Rinn JL, Chang HY. Genome regulation by long noncoding RNAs. Annu Rev Biochem 2012;81:145–66.
[89] Derrien T, Johnson R, Bussotti G, Tanzer A, Djebali S, Tilgner H, et al. The GENCODE v7 catalog of human long noncoding RNAs: analysis of their gene structure, evolution, and expression. Genome Res 2012; 22(9):1775–89.
[90] Batista PJ, Chang HY. Long noncoding RNAs: cellular address codes in development and disease. Cell 2013;152(6):1298–307.
[91] Guttman M, Rinn JL. Modular regulatory principles of large non-coding RNAs. Nature 2012;482(7385): 339–46.
[92] Pang KC, Frith MC, Mattick JS. Rapid evolution of noncoding RNAs: lack of conservation does not mean lack of function. Trends Genet 2006;22(1):1–5.
[93] Ulitsky I, Shkumatava A, Jan CH, Sive H, Bartel DP. Conserved function of lincRNAs in vertebrate embryonic development despite rapid sequence evolution. Cell 2011;147(7):1537–50.
[94] Smith MA, Gesell T, Stadler PF, Mattick JS. Widespread purifying selection on RNA structure in mammals. Nucleic Acids Res 2013;41(17):8220–36.
[95] Johnson R, Guigó R. The RIDL hypothesis: transposable elements as functional domains of long noncoding RNAs. RNA 2014;20(7):959–76.
[96] Taft RJ, Pheasant M, Mattick JS. The relationship between non-protein-coding DNA and eukaryotic complexity. Bioessays 2007;29(3):288–99.
[97] Khalil AM, Guttman M, Huarte M, Garber M, Raj A, Morales DR, et al. Many human large intergenic noncoding RNAs associate with chromatin-modifying complexes and affect gene expression. Proc Natl Acad Sci 2009;106(28):11667–72.
[98] Rinn JL, Kertesz M, Wang JK, Squazzo SL, Xu X, Brugmann SA, et al. Functional demarcation of active and silent chromatin domains in human HOX loci by noncoding RNAs. Cell 2007;129(7):1311–23.
[99] Zhao J, Sun BK, Erwin JA, Song J-J, Lee JT. Polycomb proteins targeted by a short repeat RNA to the mouse X chromosome. Science 2008;322(5902):750–6.
[100] Nagano T, Mitchell JA, Sanz LA, Pauler FM, Ferguson-Smith AC, Feil R, et al. The air noncoding RNA epigenetically silences transcription by targeting G9a to chromatin. Science 2008;322(5908):1717–20.
[101] Pandey RR, Mondal T, Mohammad F, Enroth S, Redrup L, Komorowski J, et al. Kcnq1ot1 antisense noncoding RNA mediates lineage-specific transcriptional silencing through chromatin-level regulation. Mol Cell 2008;32(2):232–46.
[102] Wang KC, Yang YW, Liu B, Sanyal A, Corces-Zimmerman R, Chen Y, et al. A long noncoding RNA maintains active chromatin to coordinate homeotic gene expression. Nature 2011;472(7341):120–4.
[103] Bertani S, Sauer S, Bolotin E, Sauer F. The noncoding RNA Mistral activates Hoxa6 and Hoxa7 expression and stem cell differentiation by recruiting MLL1 to chromatin. Mol Cell 2011;43(6):1040–6.
[104] Martianov I, Ramadass A, Barros AS, Chow N, Akoulitchev A. Repression of the human dihydrofolate reductase gene by a non-coding interfering transcript. Nature 2007;445(7128):666–70.
[105] Latos PA, Pauler FM, Koerner MV, Senergin HB, Hudson QJ, Stocsits RR, et al. Airn transcriptional overlap, but not its lncRNA products, induces imprinted Igf2r silencing. Science 2012;338(6113):1469–72.
[106] Carrieri C, Cimatti L, Biagioli M, Beugnet A, Zucchelli S, Fedele S, et al. Long non-coding antisense RNA controls Uchl1 translation through an embedded SINEB2 repeat. Nature 2012;491(7424):454–7.
[107] Yoon J-H, Abdelmohsen K, Srikantan S, Yang X, Martindale JL, De S, et al. LincRNA-p21 suppresses target mRNA translation. Mol Cell 2012;47(4):648–55.
[108] Salmena L, Poliseno L, Tay Y, Kats L, Pandolfi PP. A cerna hypothesis: the rosetta stone of a hidden RNA language? Cell 2011;146(3):353–8.

[109] Tay Y, Rinn J, Pandolfi PP. The multilayered complexity of ceRNA crosstalk and competition. Nature 2014;505(7483):344–52.

[110] Wong BC, Chan KA, Chan AT, Leung S-F, Chan LY, Chow KC, et al. Reduced plasma RNA integrity in nasopharyngeal carcinoma patients. Clin Cancer Res 2006;12(8):2512–6.

[111] Liu Z, Li X, Sun N, Xu Y, Meng Y, Yang C, et al. Microarray profiling and co-expression network analysis of circulating lncRNAs and mRNAs associated with major depressive disorder. PloS One 2014;9(3):e93388.

[112] Fradet Y, Saad F, Aprikian A, Dessureault J, Elhilali M, Trudel C, et al. uPM3, a new molecular urine test for the detection of prostate cancer. Urology 2004;64(2):311–5.

[113] Tinzl M, Marberger M, Horvath S, Chypre C. DD3PCA3 RNA analysis in urine–a new perspective for detecting prostate cancer. Eur Urol 2004;46(2):182–7.

[114] Shappell SB, Fulmer J, Arguello D, Wright BS, Oppenheimer JR, Putzi MJ. PCA3 urine mRNA testing for prostate carcinoma: patterns of use by community urologists and assay performance in reference laboratory setting. Urology 2009;73(2):363–8.

[115] Lee GL, Dobi A, Srivastava S. Prostate cancer: diagnostic performance of the PCA3 urine test. Nat Rev Urol 2011;8(3):123–4.

[116] Panzitt K, Tschernatsch MM, Guelly C, Moustafa T, Stradner M, Strohmaier HM, et al. Characterization of HULC, a novel gene with striking up-regulation in hepatocellular carcinoma, as noncoding RNA. Gastroenterology 2007;132(1):330–42.

[117] Takahashi K, Yan IK, Kogure T, Haga H, Patel T. Extracellular vesicle-mediated transfer of long non-coding RNA ROR modulates chemosensitivity in human hepatocellular cancer. FEBS Open Bio 2014;4:458–67.

[118] Nigro JM, Cho KR, Fearon ER, Kern SE, Ruppert JM, Oliner JD, et al. Scrambled exons. Cell 1991;64(3):607–13.

[119] Cocquerelle C, Mascrez B, Hetuin D, Bailleul B. Mis-splicing yields circular RNA molecules. FASEB J 1993;7(1):155–60.

[120] Kos A, Dijkema R, Arnberg A, Van der Meide P, Schellekens H. The hepatitis delta (d) virus possesses a circular RNA. Nature 1986;323(6088):558–60.

[121] Sanger HL, Klotz G, Riesner D, Gross HJ, Kleinschmidt AK. Viroids are single-stranded covalently closed circular RNA molecules existing as highly base-paired rod-like structures. Proc Natl Acad Sci 1976;73(11):3852–6.

[122] Salzman J, Gawad C, Wang PL, Lacayo N, Brown PO. Circular RNAs are the predominant transcript isoform from hundreds of human genes in diverse cell types. PloS One 2012;7(2):e30733.

[123] Wu Q, Wang Y, Cao M, Pantaleo V, Burgyan J, Li W-X, et al. Homology-independent discovery of replicating pathogenic circular RNAs by deep sequencing and a new computational algorithm. Proc Natl Acad Sci 2012;109(10):3938–43.

[124] Jeck WR, Sorrentino JA, Wang K, Slevin MK, Burd CE, Liu J, et al. Circular RNAs are abundant, conserved, and associated with ALU repeats. RNA 2013;19(2):141–57.

[125] Danan M, Schwartz S, Edelheit S, Sorek R. Transcriptome-wide discovery of circular RNAs in archaea. Nucleic Acids Res 2011;40(7):3131–42. gkr1009.

[126] Chao C-C, Aust AE. Photochemical reduction of ferric iron by chelators results in DNA strand breaks. Arch Biochem Biophys 1993;300(2):544–50.

[127] Burd CE, Jeck WR, Liu Y, Sanoff HK, Wang Z, Sharpless NE. Expression of linear and novel circular forms of an ink4/arf-associated non-coding RNA correlates with atherosclerosis risk. PLoS Genet 2010;6(12):e1001233.

[128] Hansen TB, Wiklund ED, Bramsen JB, Villadsen SB, Statham AL, Clark SJ, et al. miRNA-dependent gene silencing involving Ago2-mediated cleavage of a circular antisense RNA. EMBO J 2011;30(21):4414–22.

[129] Lasda E, Parker R. Circular RNAs: diversity of form and function. RNA 2014;20(12):1829–42.

[130] Suzuki H, Zuo Y, Wang J, Zhang MQ, Malhotra A, Mayeda A. Characterization of RNase r-digested cellular RNA source that consists of lariat and circular RNAs from pre-mRNA splicing. Nucleic Acids Res 2006;34(8):e63.

[131] Vincent HA, Deutscher MP. Substrate recognition and catalysis by the exoribonuclease RNase r. J Biol Chem 2006;281(40):29769–75.

[132] Suzuki H, Tsukahara T. A view of pre-mRNA splicing from RNase R resistant RNAs. Int J Mol Sci 2014;15(6):9331–42.

[133] Zhang Y, Zhang X-O, Chen T, Xiang J-F, Yin Q-F, Xing Y-H, et al. Circular intronic long noncoding RNAs. Mol Cell 2013;51(6):792–806.

[134] Kopczynski CC, Muskavitch M. Introns excised from the delta primary transcript are localized near sites of delta transcription. J Cell Biol 1992;119(3):503–12.

[135] Qian L, Vu MN, Carter M, Wilkinson MF. A spliced intron accumulates as a lariat in the nucleus of T cells. Nucleic Acids Res 1992;20(20):5345–50.

[136] Capel B, Swain A, Nicolis S, Hacker A, Walter M, Koopman P, et al. Circular transcripts of the testis-determining gene Sry in adult mouse testis. Cell 1993;73(5):1019–30.

[137] Hansen TB, Jensen TI, Clausen BH, Bramsen JB, Finsen B, Damgaard CK, et al. Natural RNA circles function as efficient microRNA sponges. Nature 2013;495(7441):384–8.

[138] Memczak S, Jens M, Elefsinioti A, Torti F, Krueger J, Rybak A, et al. Circular RNAs are a large class of animal RNAs with regulatory potency. Nature 2013;495(7441):333–8.

[139] Chen C-y, Sarnow P. Initiation of protein synthesis by the eukaryotic translational apparatus on circular RNAs. Science 1995;268(5209):415–7.

[140] Talhouarne GJ, Gall JG. Lariat intronic RNAs in the cytoplasm of *Xenopus tropicalis* oocytes. RNA 2014;20(9):1476–87.

[141] Mathivanan S, Fahner CJ, Reid GE, Simpson RJ. ExoCarta 2012: database of exosomal proteins, RNA and lipids. Nucleic Acids Res 2012;40(D1):D1241–4.

[142] Turchinovich A, Burwinkel B. Distinct AGO1 and AGO2 associated miRNA profiles in human cells and blood plasma. RNA Biol 2012;9(8):1066–75.

[143] Fukao T, Fukuda Y, Kiga K, Sharif J, Hino K, Enomoto Y, et al. An evolutionarily conserved mechanism for microRNA-223 expression revealed by microRNA gene profiling. Cell 2007;129(3):617–31.

[144] Wagner J, Riwanto M, Besler C, Knau A, Fichtlscherer S, Röxe T, et al. Characterization of levels and cellular transfer of circulating lipoprotein-bound microRNAs. Arterioscler Thromb Vasc Biol 2013;33(6):1392–400.

[145] Kalra H, Simpson RJ, Ji H, Aikawa E, Altevogt P, Askenase P, et al. Vesiclepedia: a compendium for extracellular vesicles with continuous community annotation. PLoS Biol 2012;10(12):e1001450.

[146] Li Y, Qiu C, Tu J, Geng B, Yang J, Jiang T, et al. HMDD v2. 0: a database for experimentally supported human microRNA and disease associations. Nucleic Acids Res (1 January 2014);42(1):D1070–4.

[147] Weber JA, Baxter DH, Zhang S, Huang DY, Huang KH, Lee MJ, et al. The microRNA spectrum in 12 body fluids. Clin Chem 2010;56(11):1733–41.

[148] McDonald JS, Milosevic D, Reddi HV, Grebe SK, Algeciras-Schimnich A. Analysis of circulating microRNA: preanalytical and analytical challenges. Clin Chem 2011;57(6):833–40.

[149] Duttagupta R, Jiang R, Gollub J, Getts RC, Jones KW. Impact of cellular miRNAs on circulating miRNA biomarker signatures. PloS One 2011;6(6):e20769.

[150] Kim D-J, Linnstaedt S, Palma J, Park JC, Ntrivalas E, Kwak-Kim JY, et al. Plasma components affect accuracy of circulating cancer-related microRNA quantitation. J Mol Diagn 2012;14(1):71–80.

[151] Pritchard CC, Kroh E, Wood B, Arroyo JD, Dougherty KJ, Miyaji MM, et al. Blood cell origin of circulating microRNAs: a cautionary note for cancer biomarker studies. Cancer Prev Res 2012;5(3):492–7.

CHAPTER

13

DNA Methylation Biomarkers in Lung Cancer

Juan Sandoval, Paula Lopez Serra

Epigenetic and Cancer Biology Program (PEBC), Bellvitge Biomedical Research Institute (IDIBELL), Barcelona, Spain

OUTLINE

1. INTRODUCTION

The influence of the environment on cell biology and disease development is well known. In this sense, epigenetics has been defined as the link between the cells and their environment. Life style, stress, tobacco, alcohol, pathological situations, or pharmacological uptake may contribute to modify the DNA epigenome of the individuals. Classically, epigenetics has been

Epigenetic Biomarkers and Diagnostics
http://dx.doi.org/10.1016/B978-0-12-801899-6.00013-9

described as the branch of science that studies the heritable changes in gene expression that are not determined by changes in the DNA sequence. However, Adrian Bird recently redefined epigenetics as the study of all the mechanisms involved in the regulation of chromatin function, in other words, the study of the structural changes in chromosomes to register, modify, or perpetrate the gene expression activity [1].

1.1 Epigenetics in Physiological Conditions

Epigenetic regulation is essential for normal development of pluricellular organism. All cells are genetically identical but structurally and functionally heterogeneous. The temporal order of gene expression is a precise and critical feature to ensure a specific cell lineage and the final organogenesis. In mammals, epigenetic regulation is also involved in gene silencing of imprinted genes, X-chromosome inactivation in females as a method for dose compensation [2], and silencing of endoparasitic sequences as the transposable elements in ensuring chromosome stability.

The complexity of the epigenetic control over gene expression is represented by the number of different mechanisms and enzymatic reactions that are involved in the gene expression process: DNA methylation, posttranslational histone modifications, noncoding RNAs, and nucleosome remodeling. These mechanisms are not independent; they are related and coordinated to ensure a tight control over gene expression, DNA replication, and DNA repair [3]. In the next sections, we will briefly define the main epigenetic mechanisms.

2. METHYLATION

2.1 DNA Methylation

Methylation is a chemical modification of the DNA where a methyl group (CH_3) is covalently bound to the 5′ carbon of the cytosine in a cytosine–phosphate–guanine (CpG) dinucleotide context [4]. Recent studies suggest that non-CpG methylation is present in oocytes, pluripotent embryonic stem cells, and mature neurons [5–7], but the function of this methylation pattern remains unclear.

In the human genome, the CpG dinucleotide is lower than statistically expected, and it is not randomly distributed. They are allocated in the body of repetitive sequences and concentrated around the CpG islands (CGIs), CpG-rich regions in the promoter of approximately 70% of the genes [8]. In this sense, CGIs have been identified as interspersed with vast low-density CpG regions across the human genome. Different criteria have been reported to define a CGI. One of the accepted definitions considers a CGI to be a DNA sequence (>200 bp window) with a guanine–cytosine (GC) content greater than 50% and an observed-to-expected CpG ratio of more than 0.6 [9,10]. In normal conditions in somatic cells, CpG methylation is mainly found in the repetitive sequences, maintaining chromosome stability and preventing translocations derived from parasitic sequences. However, CGIs do not present 5′-methylcytosines (5mC) [11], they appear only in the imprinted genes and in silenced embryonic developmental genes [12].

Methylation patterns are established during embryonic development. Before the implantation, the zygote suffers a passive DNA demethylation which erases part of the inheriting parental methylation patterns. After implantation, the embryo undergoes a wave of de novo DNA methylation that will determine the methylation pattern of the new individual [13].

DNA methylation is driven by the DNA methyltransferases (DNMTs), the enzymes that catalyze the transference of a methyl group from the S-adenosil-L-metionine (SAM) to the position 5 in the cytosine. DNMT family is composed of five protein members: DNMT1, DNMT2, DNMT3a, DNMT3b, and DNMT3L

(DNMT3-like), although only DNMT1, 3a, and 3b present methyltransferase activity [14]. Every member of the family responds for a specific function.

DNMT1 is mainly involved in the DNA methylation maintenance after mitosis to ensure an identical methylation pattern between the two daughter cells. The recognition of the hemimethylated CpGs is driven by the DNMT1 partner UHRF1 (ubiquitin-like plant homeodomain and ring finger domain 1) [15].

DNMT3a and DNMT3b are mainly involved in de novo methylation during embryonic development, playing a critical role in the establishment of the genomic DNA methylation pattern [16]. This process occurs in the early stages of human development, being critical for reproductive health and neuronal maturation.

DNMT2 acts as an RNA methyltransferase transferring a methyl group to the C5 position in tRNA for aspartic acid [17]. DNMT2 shows a weak DNA methyltransferase activity in vitro [18]. However, its elimination does not affect the global DNA methylation level in embryonic cells. Therefore, although DNMT2 was initially assigned as a DNA methyltransferase, the results described above suggest that it is not involved in DNA methylation pattern determination [19].

DNMT3L does not show methyltransferase activity but it is related to gene repression, as it recruits histone deacetylases to methylated promoters [20]. Moreover, DNMT3L induces the methyltransferase activity of the DNMT3a by recognizing DNMT3a and the N-terminal end of the histone H3 when it is not methylated in the lysine 4, promoting de novo methylation in those regions [21].

2.2 DNA Demethylation

Global DNA demethylation is important for the maintenance of pluripotent states in early embryos and parental patterns erasing. DNA demethylation can be an active or a passive process [22]. In the passive mechanism, successive cycles of DNA replication in the absence of functional DNMT1/UHRF1 dilute the global content of 5mC. It can be due to the presence of cofactors that sequester the enzymes or by the presence of histone marks, like acetylation, that do not allow the enzyme to join the DNA. Active demethylation involves different enzymes that act in different ways: (1) Nucleotide excision repair is the excision of short genomic regions that contain 5mC. The Gadd45 (growth arrest and DNA-damage-inducible protein 45) protein family stimulate active DNA demethylation via this process, however, null mice showed no global alteration in DNA methylation levels, suggesting this pathway is still unknown [23]. (2) Direct base repair of 5mC by DNA glycosidases. The enzymes TDG and MBD4 (methyl-CpG-binding domain-containing protein 4) show 5mC excision activity, however, their activity over 5mC is much lower than over other nucleotides. (3) Deamination and repair of 5mC base generates thymine and the resulting T:G mismatch, which will be repaired by DNA glycosidases [24].

Another active mechanism for DNA demethylation involves the TET (ten-eleven translocation) proteins that catalyze the iterative oxidation of 5mC to 5-hydroximethylcytosine, 5-formylcytosine, and 5-carboxylcytosine. This mechanism mediated by TET proteins is a potential driver of the epigenetic reprogramming by DNA demethylation [25].

3. EPIGENETIC-ASSOCIATED REGULATORY ELEMENTS

Functional alterations in epigenetic factors do not occur randomly throughout the genome in human diseases, but are overlapping with regulatory important regions, such as transcription start site (TSS)/promoter, enhancers, or even intergenic regions. In the following sections, we overview the current knowledge regarding the epigenetic-associated regulatory elements participating in the regulation of gene expression, such as promoters, enhancers, and intergenic regions [26].

3.1 DNA Methylation in Promoters

The core promoter is generally defined to be the sequence that directs the initiation of transcription overlapping the TSSs and may contain many different sequence motifs. There are two main types of core promoters with single (focused) or diverse and complex transcriptional module (dispersed). Focused promoters contain either a single TSS or a distinct cluster of start sites over several nucleotides, while dispersed promoters contain several start sites over 50–100 nucleotides. They are typically found in CGIs in vertebrates [27].

Evidence has been accumulated on suitability of DNA promoter hypermethylation as a biomarker for diagnostics and therapy [28]. It has been proved that DNA hypermethylation, defined as the gain of methylation at specific sites that are unmethylated under normal conditions, occurs mainly in the regulatory regions, such as CGI promoter. In this sense, CGI promoter hypermethylation in tumor suppressor and the resulting gene silencing has been the most frequent epigenetic marker associated with cancer [29]. However in 2009, a study reported that the majority of DNA methylation changes involved in gene regulation are located at the promoter CGI shores [30]. CpG shores, defined as regions next to CGIs, up to 2kb long with comparatively low GC density, are now a main focus in cancer biology [31].

Several large-scale studies have led to significant progress in characterizing global levels of histone modifications in mammals. Interesting insights into the complex relationship between gene expression and histone modifications have been revealed. Usually, elevated levels of histone acetylation H3K9ac, H3K14ac, and histone methylation (H3K4methylation, H3K9me1, and H2A.Z) are detected in promoter regions of active genes, whereas elevated levels of H3K27me3 correlate with gene repression [32,33]. Interestingly, it has been recently reported that human genes with CGI promoters have a distinct transcription-associated chromatin organization [34].

3.2 DNA Methylation in Enhancers and Super-enhancers

Enhancers are distal *cis*-regulatory elements that orchestrate the regulation of gene expression patterns of developmental genes. These elements are composed of transcription factor binding site clusters (6- to 20-bp motifs) where combinations of *trans*-activating and repressive factors bind in sequence-specific manner. They can be located in intergenic regions, introns, and exons, tens to hundreds of kilobases from their target genes [35]. Specifically, enhancer elements, located in open chromatin, often exhibit cell type-specific localization patterns. Specific patterns of histone H3 acetylation and high levels of H3K4me1 combined with H3k4me2 define enhancers [32,36–38]. It has been reported that only a fraction that combines H3K4me1- and H3K27ac-marked elements are active enhancers [39,40]. The activation of distal regulatory elements is often accompanied by demethylation of these residues [41]. Recent publications have discovered cancer type-specific DMRs, that not only were located on promoters but also for enhancer elements that impact cancer gene expression in endometrial carcinoma and multiple myeloma [42,43].

Recently, a new subset of enhancers has been defined and called super-enhancers. They consist of enhancer clusters that are densely occupied by five key embryonic stem cell transcription factors and the mediator coactivator [44]. Super-enhancers are associated with key genes that control cell state. Aberrant DNA changes in these regions may be associated with specific diseases, including cancer. From all of the marks that are associated with enhancers, the histone H3K27ac modification was found enriched in super-enhancers [45].

3.3 Intergenic Hypomethylated Regions (HMRs)

Unequal distribution of DNA methylation levels throughout the genome have been previously defined, including regions with unique features acting as potential regulators. Although the main proportions of CpGs are highly methylated throughout the genome, distinct regions display low levels of methylation for high number of consecutive CpG sites. Hypomethylated regions (HMRs) possibly mark *trans*-acting regulatory regions by favoring accessibility of the transcriptional machinery and associated with promoting factors. Functionally, DNA hypomethylation occurs in many gene-poor genomic areas, including repetitive elements, retrotransposons, and introns, contributing to genomic instability [28]. In repetitive sequences, this is achieved by a higher rate of chromosomal rearrangements and, in retrotransposons, by a higher probability of translocation to other genomic regions [46]. Previous studies have shown how intergenic HMRs were enriched for factor binding sites. These intergenic HMRs participate in genome organization and regulation of gene expression during lineage-specific blood cell differentiation [47].

4. DNA METHYLATION IN CANCER

Tumor cells display rather lower DNA methylation level when compared to their normal counterparts. The loss of methylation is mainly due to the hypomethylation of the repetitive sequences, promoting next chromosomal instability and reactivation of transposable elements. Hypomethylation of DNA can also favor mitotic recombination, leading to deletions, translocations, and chromosomal rearrangements [48].

It is well established that the hypermethylation of the CGI in the promoter region of tumor suppressor genes is a characteristic feature for tumor cells. After analyzing, by methylation-sensitive restriction enzymes, the tumor cells of 21 retinoblastoma patients, Greger and collaborators described the first report of hypermethylation in the promoter region of a tumor suppressor gene, the retinoblastoma suppression gene (Rb) [49].

Hypermethylation of the CGI promoters affects tumor suppressor genes involved in all the stages of cancer progression. After the description of DNA methylation of the Rb gene, many other studies described the silencing of tumor suppressor genes by DNA hypermethylation of the promoter region in all cancer types[28] (Figure 1).

5. DNA METHYLATION IN LUNG CANCER

Cancer is the leading cause of death worldwide. According to the World Health Organization data from 2012, 8.2 million deaths are related to cancer, including here the most common cancer type, lung cancer with 1.59 million deaths (World Health Organization in 2011). Tobacco is the most important risk factor causing about 70% of global lung cancer-related deaths.

There are two major subtypes of lung cancer. The more frequent is non-small-cell lung cancer (NSCLC), representing 85–90% of all diagnosed cases. NSCLC is divided into three major groups: (1) Lung adenocarcinoma (accounting for approximately 40% of lung cancers) is the most frequent lung cancer in never-smokers. The cells that form these tumors are those that normally secrete mucus. (2) Squamous cell carcinoma accounts for 25–30% of the lung cancers. It has an epidermoid origin from the cells that line the airways in the lungs and is often linked

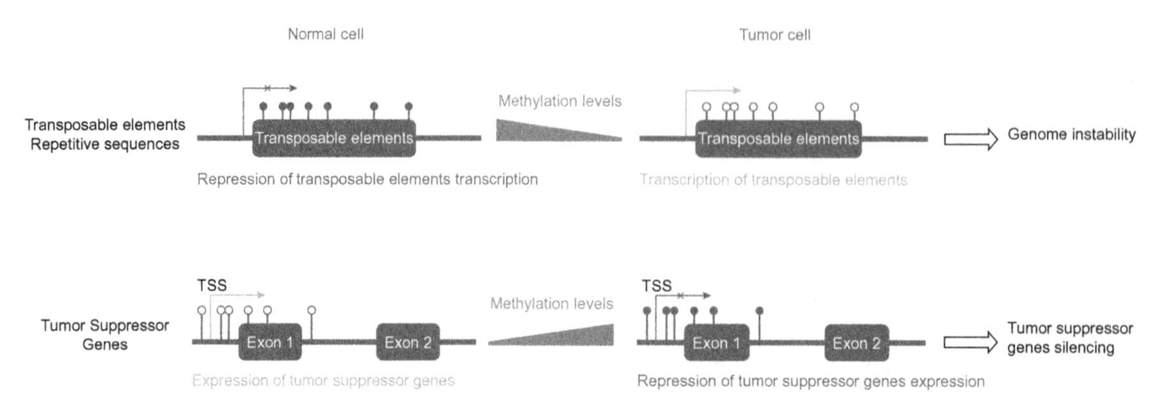

FIGURE 1 **Changes in the DNA methylation pattern in cancer cells.** Cancer cells suffer a global DNA hypomethylation of their genome (upper panel), however, tumor suppressor genes undergo DNA promoter hypermethylation, silencing their transcription (lower panel).

to smoking habits. (3) Large cell carcinoma represents about 10–15% of lung cancers. It is an undifferentiated tumor that appears anywhere in the lungs. It tends to grow and spread rapidly: reflection in complication due to its treatment [50].

The other major lung cancer subtype is called small cell lung cancer (SCLC) group. It represents about 10–15% of the lung cancer cases and represents the most aggressive subset of lung cancer. It is almost mostly linked to smoking habits. It is an undifferentiated neoplasm that is originated mainly by neuroendocrine cells [51].

Like other cancer types, lung cancer is affected by the global DNA hypomethylation and the regional hypermethylation in the promoter of tumor suppressor genes. Aberrant DNA methylation is an early feature in lung cancer development. Many studies have described the promoter hypermethylation of cancer-related genes in lung cancer. One of those examples is related to the reduction of expression of the well-known tumor suppressor gene PTEN, reported to be downregulated in about 70% of NSCLC tumors [52]. Its loss of expression has been attributed to allelic loss or promoter hypermethylation [53]. Other examples are including the silencing of genes involved in cell growth, apoptosis evasion, invasion of adjacent and distal tissues, recovering all the stages of tumor growth [54].

Aberrant DNA methylation in lung cancer can be detected in circulating DNA in the blood or isolated from the sputum. It is becoming a powerful tool for lung cancer diagnosis, patient prognosis, and response upon treatment [55].

5.1 Interplay between DNA Methylation and microRNAs

DNA methylation-associated silencing of microRNA (miRNA) expression has been related to several tumor types. For example, miR-34b/c promoter hypermethylation and subsequently reduced mature miRNA levels are associated with higher rates of tumor recurrence and thus decrease overall survival in NSCLC patients [56]. Promoter hypermethylation of miR-200 family and miR-205 has been related to epithelial-to-mesenchymal transition induction in early stages of lung carcinogenesis [57]. Moreover, miR-148a promoter methylation is associated to lymph node metastasis [58]. Although tumor suppressor miRNAs reported CGI promoter hypermethylation in tumor samples, other studies have also shown promoter hypomethylation in miRNAs promoters. Let-7 is mainly recognized as a tumor-suppressive member of the family of miRNAs, however, the promoter of let-7a-3 was hypermethylated in normal human tissues whereas it appears hypomethylated in a set of lung adenocarcinomas. Importantly, let-7a-3 exhibits oncogenic activity in lung cancer as it promotes anchorage-independent cell growth and positively regulates the oncogenes CDK6 and CXCL5, among others [59].

On the other hand, miRNAs can regulate the expression of the de novo DNMTs, the enzymes that drive the DNA methylation. miRNAs are frequently overexpressed in lung cancer and are associated with poor prognosis. The expression of the miR-29 family is inversely correlated to the expression levels of DNMT3A and DNMT3B in lung cancer. miR-29 members downregulate the expression levels of the two enzymes, promoting the reactivation of tumor suppressor genes regulated by promoter DNA methylation [60].

5.2 DNA Methylation Biomarkers with Diagnostic Value for Lung Cancer

Early diagnosis of cancer is the main factor to enhance patient survival. Nowadays, NSCLC patients diagnosed in stage I have a 5-year overall survival of almost 50%, dropping down to 1% if the cancer reaches the stage IV. The ratios are even worse for SCLC, the most aggressive clinical course of pulmonary tumor, with an overall survival at 5 years at about 5–10% (data obtained from World Health Organization, 2011). Therefore, markers for early detection and proper classification of the tumor the patient suffers are extremely needed to improve life expectancy.

Diagnostic methods for lung cancer detection include noninvasive methods (sputum cytology and imaging tests) and invasive (lung biopsy and bronchoscopy analysis).

Many studies have tried to find DNA methylation biomarkers to diagnose lung cancer both in invasive and noninvasive samples. Regarding noninvasive methods, a study presented by Hubers and collaborators compares DNA hypermethylation in sputum samples to sputum cytology for lung cancer diagnosis. To date, the cytological examination of sputum has a limited value with a clinical sensitivity ranging from 42 to 97% [61]. By using two sets of samples, one composed of 98 patients and 90 healthy controls and another composed of 60 patients and 445 controls, the authors described RASSF1A hypermethylation as a better diagnostic factor than the cytology analysis. They found no preference of hypermethylation for disease stage, indicating a sensitivity which does not depend on clinical tumor size or progression [62] (Table 1). Another study described a profile of DNA methylation in bronchial washes to get a better diagnosis of lung cancer. Nikolaidis and collaborators analyze 665 patients' samples to delineate a panel of DNA methylation biomarkers in bronchial washes that showed a better sensitivity of lung cancer detection. The panel included the analysis of the promoter methylation status of four genes (TERT, WT1, p16, and RASFF1) for correct prediction of lung cancer, highlighting the detection of small tumors in early stages of cancer development, a group of tumors that regular cytological analyses are not able to detect [63]. These two studies provide promising results. However, an independent validation using large cohorts of patients is needed to demonstrate the predictive sensitivity of these assays.

Another study by Xiao and collaborators describes the analysis of the methylation pattern of p16 in exhaled breath condensate for NSCLC diagnosis. The authors compare the methylation status of p16 among tumor tissue, blood, and exhaled breath condensate of 60 individuals, 30 NSCLC patients, and 30 healthy controls. Aberrant DNA methylation of p16 promoter was found in about 87% of the tumor tissues, in 50% of the blood plasma, and 40% of the exhaled breath condensates. None of the healthy samples or adjacent normal tissues showed p16 DNA methylation. The authors suggest the usability of p16 hypermethylation in exhaled breath condensates as a noninvasive method for NSCLC

TABLE 1 Epigenetic Biomarkers in Lung Cancer

Epigenetic biomarker	Lung cancer subtype	Biological sample	Diagnostic value	References
RASFF1A	NSCLC/SCLC	Sputum	45% Sensitivity 94–100% Specificity	[62]
TERT/WT1/p16/RASFF1	NSCLC/SCLC	Bronchial washings	82% Sensitivity 91% Specificity	[63]
p16	NSCLC	Tumor tissue Serum Exhaled breath condensate	87% Sensitivity 50% Sensitivity 40% Sensitivity Specificity ND	[64]
RASFF1/p16/RARβ	MPM NSCLC/SCLC	Pleural fluid	87% Sensitivity 79.5% Specificity	[67]
miR-34b/c	MPM	Serum	67% Sensitivity 77% Specificity	[76]
Epigenetic biomarker	**Lung cancer subtype**	**Biological material**	**Prognostic value**	**References**
p16/CDH13/APC/RASSF1A	NSCLC	Tumor tissue and nodes	Tumor recurrence	[68]
HIST1H4F/PCDHGB6/NBPWR1/ALX1/HOXA9	NSCLC	Tumor tissue	Stage I tumor risk	[69]
ADCY5/EVX1/GRFRA1/PDE9A/TBX20	Adenocarcinoma	Precancerous lung tissue	Tumor recurrence	[70]
miR-127	NSCLC	Tumor tissue	Overall survival	[71]
miR-34b	NSCLC	Tumor tissue	Invasive phenotype	[73]
miR-886-3p	SCLC	Tumor tissue	Clinical outcome	[77]
Epigenetic biomarker	**Lung cancer subtype**	**Biological material**	**Drug resistance**	**References**
RUNX3	Adenocarcinoma	Cell lines/tumor tissue	Docetaxel	[79]
IGFBP-3	NSCLC	Cell lines/tumor tissue	Cisplatin	[82]

NSCLC, non-small-cell lung cancer; SCLC, small cell lung cancer; MPM, malignant pleural mesothelioma.

diagnosis [64]. Other authors have identified circulating hypermethylated DNA in blood of lung cancer patients, in the tumor suppressor gene SHOX2 [65]. Specifically, the validated SHOX2 methylation biomarker has recently been certificated by the "Conformité Européenne *In vitro* Diagnostic" (CE IVD) in bronchial aspirates [66]. Unfortunately, the sensitivity and specificity for SHOX2 methylation was higher in bronchial aspirates than in blood samples. Thus, there is a real need to find good epigenetic biomarkers in noninvasive samples with higher sensitivity and specificity for implementing diagnostic tools.

DNA methylation profiling of the pleural liquid has been described as a feasible tool for mesothelioma diagnosis. Mesothelioma is a rare and very aggressive lung cancer type affecting the membranes lining the lungs and abdomen.

Mesothelioma diagnosis is particularly difficult as this tumor often presents symptoms that mimic other diseases. In this regard, Fujii and collaborators described a DNA methylation profile in pleural fluid for differential diagnosis of malignant pleural mesothelioma (MPM) from lung cancer. The authors analyze the methylation profile of the promoter region of five genes (RASFF1A, p16, RAR beta, MGMT, and DAPK) in the pleural fluid of 140 individuals: 46 diagnosed with lung cancer, 39 with MPM, 25 with benign pleurisy, and 39 with other pulmonary diseases (Table 1). The methylation ratio for the genes RASSF1A, p16, and RAR beta was significantly higher in lung cancer than in mesothelioma, being a potential marker to differentiate both pathologies [67].

The use of high-throughput technologies for the discovery of new epigenetic biomarkers in invasive samples has generated high number of cancer-specific candidates. In our laboratory, we will release a database generated using genome-wide DNA methylation fingerprints data from different tumors. This database will facilitate the identification of cancers of unknown primary origin, including those whose origin is the lung tissue.

5.3 DNA Methylation Biomarkers with Prognostic Value for Lung Cancer

Unfortunately, lung cancer outlook is not very promising. The prognosis depends on the stage of cancer when it is diagnosed. Overall for all lung cancer types, about 32% of the patients will live for at least 1 year, 10% will live for 5 years, and only 5% will live for at least 10 years (data obtained from the World Health Organization, 2011).

Main concern in NSCLC treatment is the recurrence of tumors, even if they are diagnosed in early stages and removed by surgery, the standard care of stage I tumors. However, 30–40% of stage I NSCLC patients die of recurrent disease. A study identified new biomarkers to predict the recurrence of stage I NSCLC patients by comparing, by MSP (methyl-specific PCR), the methylation pattern of 7 genes in 40 patients that suffered recurrence within 40 months after resection with 116 patients that underwent curative resection. They reported that the methylation pattern of four genes (p16, CDH13, APC, and RASSF1A) in the primary tumor could predict the risk of recurrence in early stages of lung cancer [68].

Using a high-throughput approach, we have recently described a prognostic DNA methylation signature for stage I NSCLC patients by using a 450-K CpG methylation array. We analyzed a cohort of 444 patients with NSCLC, including 237 cases of stage I tumors, to find a 5-gene methylation signature associated with a shorter relapse-free survival in stage I NSCLC. The presence of methylation in two or more of these five genes (HIST1H4F, PCDHGB6, NPBWR1, ALX1, and HOXA9) is associated with a higher risk of recurrence. The signature was validated in an independent cohort of 143 stage I NSCLC patients [69]. The next step would be to check if these high-risk low-stage patients would be eligible to receive standard chemotherapy, since current accepted treatment for these stage I patients is lung lobectomy.

Another study carried on a 27K DNA methylation array showed that DNA methylation profiles associated with recurrence of lung adenocarcinoma are established at precancerous stages. Hypermethylation of DNA promoter of the ADCY5, EVX1, GRFRA1, PDE9A, and TBX20 genes in precancerous lung tissues from adenocarcinoma patients was significantly correlated with tumor aggressiveness [70].

DNA hypermethylation in miRNA promoters has therapeutic value as it can define the clinical outcome of cancer patients. High DNA methylation levels in the promoter region of the miR-127 are associated with a reduced overall survival in NSCLC [71]. miR-127 targets BCL-6, a proto-oncogene that modulates DNA-damage-induced apoptosis, and the silencing of this miR-127 by DNA methylation contributes to BCL-6 activation in renal carcinoma [72]. Watanabe et al. have described the association between miR-34b methylation and lymphatic invasion in NSCLC tumors, proposing the DNA methylation status of this miRNA as a biomarker for the invasive phenotype in lung cancer [73]. Hypermethylation of miR-34b promoter has been found in 90% of the pleural mesotheliomas [74], an aggressive tumor with a median overall survival of 12 months [75]. Analysis of the DNA methylation status of miR-34b promoter in serum-circulating DNA of pleural mesothelioma patients has been proposed as a noninvasive method for diagnosis [76]. Another example of the prognostic value of DNA methylation analysis of miRNA promoters is the silencing of miR-886-3p by DNA hypermethylation in SCLC. The level of DNA methylation has potential readout in the patient outcome [77]. However, prior to the utilization of DNA methylation of miRNA in molecular diagnostics, the identification of appropriate controls and standardization protocols must be accomplished.

5.4 DNA Methylation Biomarkers for Drug Response Prediction in Lung Cancer

Tumor resection when possible, targeted therapy and chemotherapy compose the standard treatment in both primary and palliative care of patients with lung cancer. However, some tumor patients do not respond to the treatment or they respond initially but then gradually relapse. Drug resistance may be intrinsic in the tumor cells or acquired upon continued administration of the compound or continued treatment. DNA methylation alterations have been shown to be essential for drug resistance development by leading the transcriptional silencing of the genes involved in the response [78].

Docetaxel is one of the first-line chemotherapeutic agents used to treat advance NSCLC; however, many patients are nonresponsive or develop acquired resistance. Trying to identify the mechanisms by which the tumor cells avoid the drug effect, Zheng and collaborators analyzed the DNA methylation profile with the infinium 27K platform of two cell lines of lung adenocarcinoma and their corresponding resistance counterparts after continuous exposure to docetaxel. They have found that downregulation by DNA hypermethylation of RUNX3 is linked to the acquired resistance on the cell line model by an indirect activation of the AKT pathway. In normal conditions, RUNX3 is a repressor of AKT transcription. Moreover, low levels of RUNX3 in lung adenocarcinoma tissues is correlated with AKT overexpression, reduced sensitivity to docetaxel, and poor prognosis of the patients [79]. Accumulating evidence suggests that AKT pathway promotes acquired resistance to targeted therapy, chemotherapy, and radiation [80].

Another drug for advance NSCLC treatment is cisplatin; 70% of the patients are nonresponsive and those that do develop resistance upon treatment. In a similar study, the DNA methylation profiles of sensitive NSCLC cell line and its resistance derivate are compared. Three hundred and seventy two genes appear to be hypermethylated in the resistance cells and epigenetic treatment to restore the normal methylation status resensitizes the cells to cisplatin [81].

Another study links the activation of the AKT pathway with acquired resistance to cisplatin. In this case, the silencing of the insulin-like growth factor-binding protein-3 (IGFBP-3) by DNA hypermethylation is involved in acquired cisplatin resistance of NSCLC patients. IGFBP-3 downregulation activates the AKT pathway by inducing AKT phosphorylation and, thus, promoting cisplatin resistance (Table 1). In addition, in a proof-of-concept approach, the study confirms that IGFBP-3 promoter methylation only occurs in cisplatin-resistant patients [82].

5.5 Epigenetic Therapy in Lung Cancer

As we highlighted in the previous sections, altered DNA methylation of the tumor cells plays a key role in tumorigenesis from early stages of the disease. New approaches trying to improve the current treatment of lung cancer include the epigenetic therapy.

To date, only four drugs targeting epigenetic changes are approved by the US Food and Drug Administration, including the demethylating agents decitabine (for the treatment of myelodysplastic syndrome and acute myeloid leukemia) and 5-azacytidine (indicated for the treatment of myelodysplastic syndrome) the histone deacetylases vorinostat and romidepsin (both administrated for T-cell lymphoma).

In lung cancer, a phase I/II clinical trial has tested the combination of epigenetics in refractory patients with recurrent metastatic NSCLC. The observations suggest that reversal of epigenetic tumor suppressor genes silencing by administration of azacytidine (DNA methylation inhibitor) and entinostat (histone deacetylase inhibitor) could improve the clinical outcome in advanced lung cancer [83].

Combination therapy of epigenetic drugs plus chemotherapy has also been tried in clinical trials for NSCLC treatment. A combination of carboplatin and paclitaxel with either vorinostat (histone deacetylase inhibitor) or placebos has been tried in a phase II clinical trial. The study concludes that vorinostat treatment enhances the effect of carboplatin and paclitaxel in advanced NSCLC patients [84].

In vitro, the blockade of DNA methylation by 5-aza-2′-deoxycytidine sensitizes NSCLC cells to the therapeutic effect of gefitinib, a specific inhibitor of the epidermal growth factor receptor (EGFR). The treatment with EGFR tyrosine kinase inhibitors, such as gefitinib, of NSCLC patients carrying activating mutations is efficient but, unfortunately, most primary tumor cells are resistant to the treatment. EGFR gene is silenced by DNA methylation in many cell lines; its reexpression with the addition of 5-aza-2′-deoxy cytidine resensitizes the cells to gefitinib [85]. In a previous study, it was observed that stage IV NCSLC patients who received five cycles of 5-aza-2′-deoxycytidine treatment showed a prolonged survival. The authors suggest that one of the mechanisms underlying the effect is the reexpression of silenced tumor suppressor genes [86].

6. CONCLUSIONS

Personalized medicine is turning into reality for cancer treatment. Cancer is a pathology where multiple different diseases are included, being all of them being characterized by the uncontrolled growth of transformed cells. Finding the individual characteristics of every patient will make possible to apply the treatment that works efficiently. DNA methylation plays a critical role in tumorigenesis. Every year, new studies highlight the importance of DNA analysis to proper diagnosis and treatment for individual cancer patient. Therapy to restore tumor suppressor genes expression by demethylating their promoter region is already in use in acute myeloid leukemia and myelodysplastic syndrome. However, despite initial successes, major challenges remain that must be addressed by future studies and trials. These issues involve expanding the therapeutic reach beyond hematologic cancers, such as solid tumors including the most frequent as lung cancer. Furthermore, future testing for the synergistic potential of combinatory therapy (standard and epiChemotherapy) for better patient outcomes will be needed. Understanding the mechanistic basis of the pharmacological agents will be essential to narrow drug resistance and more precisely target aberrant marks. Moreover, since researchers have already described a considerable number of biomarkers with clinical value, scientific community should make an effort for cross-validating those already identified, most promising markers. Finally, it is clear that epigenetic therapy will impact the management of cancer patients in near future.

LIST OF ABBREVIATIONS

5mC 5-Methylcytosine
CGI CpG islands
CUPs Cancers of unknown primary origin
DNMTs DNA methyltransferases
EGFR Epidermal growth factor receptor
Gadd45 Growth arrest and DNA-damage-inducible protein 45
HMRs Intergenic hypomethylated regions
MSP Methyl-specific PCR
NSCLC Non-small-cell lung cancer
Rb Retinoblastoma suppression gene
SAM S-adenosil-L-metionine
TET Ten-eleven translocation
TSSs Transcription start sites
UHRF1 Ubiquitin-like plant homeodomain and ring finger domain 1

References

[1] Bird A. Perceptions of epigenetics. Nature 2007;447(7143):396–8.
[2] Lund AH, van Lohuizen M. Epigenetics and cancer. Genes Dev 2004;18(19):2315–35.
[3] Levine M, Tijan R. Transcription regulation and animal diversity. Nature 2003;424:147–51.
[4] DeAngelis JT, Farrington WJ, Tollefsbol TO. An overview of epigenetic assays. Mol Biotechnol 2008;38(2):179–83.
[5] Shirane K, Toh H, Kobayashi H, Miura F, Chiba H, Ito T, et al. Mouse oocyte methylomes at base resolution reveal genome-wide accumulation of non-CpG methylation and role of DNA methyltransferases. PLoS Genet 2013;9(4):e1003439.
[6] Xie W, Schultz MD, Lister R, Hou Z, Rajagopal N, Hou Z, et al. Epigenomic analysis of multilineage differentiation of human embryonic stem cells. Cell 2013;153(5):1134–48.
[7] Lister R, Mukamel EA, Nery JR, Urich M, Puddifoot CA, Johnson ND, et al. Global epigenomic reconfiguration during mammalian brain development. Science 2013;341(6146):1237905.
[8] Deaton AM, Bird A. CpG islands and the regulation of transcription. Genes Dev 2011;25(10):1010–22.
[9] Takai D, Jones PA. Comprehensive analysis of CpG islands in human chromosomes 21 and 22. Proc Natl Acad Sci USA 2002;99(6):3740–5.
[10] Wang Y, Leung FC. An evaluation of new criteria for CpG islands in the human genome as gene markers. Bioinformatics 2004;20(7):1170–7.
[11] Bird A. DNA methylation patterns and epigenetic memory. Genes Dev 2002;16(1):6–21.
[12] Hattori N, Imao Y, Nishino K, Hattori N, Ohgane J, Yagi S, et al. Epigenetic regulation of nanog gene in embryonic stem and trophoblast stem cells. Genes Cells 2007;12(3):522–31.
[13] Okano M, Bell DW, Haber DA, Li E. DNA methyltransferases Dnmt3a and Dnmt3b are essential for de novo methylation and mammalian development. Cell 1999;99(3):247–57.
[14] Jin B, Robertson KD. DNA methyltransferases, DNA damage repair, and cancer. Adv Exp Med Biol 2013;754:3–29.
[15] Bostick M, Kim JK, Estève PO, Clark A, Pradhan S, Jacobsen SE. UHRF1 plays a role in maintaining DNA methylation in mammalian cells. Science 2007;317(5845):1760–4.
[16] Leppert S, Matarazzo MR. De novo DNMTs and DNA methylation: novel insights into disease pathogenesis and therapy from epigenomics. Curr Pharm Des 2014;20(11):1812–8.
[17] Goll MG, Kirpekar F, Maggert KA, Yoder JA, Hsieh CL, Zhang X, et al. Methylation of tRNAAsp by the DNA methyltransferase homolog Dnmt2. Science 2006;311(5759):395–8.
[18] Herman JG, Baylin SB. Gene silencing in cancer in association with promoter hypermethylation. N Engl J Med 2003;349(21):2042–54.
[19] Okano M, Xie S, Li E. Dnmt2 is not required for de novo and maintenance methylation of viral DNA in embryonic stem cells. Nucleic Acids Res 1998;26(11):2536–40.
[20] Aapola U, Liiv I, Peterson P. Imprinting regulator DNMT3L is a transcriptional repressor associated with histone deacetylase activity. Nucleic Acids Res 2002;30(16):3602–8.
[21] Ooi SK, Qiu C, Bernstein E, Li K, Jia D, Yang Z, et al. DNMT3L connects unmethylated lysine 4 of histone H3 to de novo methylation of DNA. Nature 2007;448(7154):714–7.
[22] Messerschmidt DM, Knowles BB, Solter D. DNA methylation dynamics during epigenetic reprogramming in the germline and preimplantation embryos. Genes Dev 2014;28(8):812–28.
[23] Engel N, Tront JS, Erinle T, Nguyen N, Latham KE, Sapienza C, et al. Conserved DNA methylation in Gadd45a(-/-) mice. Epigenetics 2009;4(2):98–9.
[24] Piccolo FM, Fisher AG. Getting rid of DNA methylation. Trends Cell Biol 2014;24(2):136–43.
[25] Hill PWS, Amouroux R, Hajkova P. DNA demethylation, Tet proteins and 5-hydroxymethylcytosine in epigenetic reprogramming: an emerging complex story. Genomics 2014;104(5):324–33. S0888-7543(14)00154-2.
[26] Harmston N, Lenhard B. Chromatin and epigenetic features of long-range gene regulation. Nucleic Acids Res 2013;41(15):7185–99.

[27] Juven-Gershon T, Hsu JY, Theisen JW, Kadonaga JT. The RNA polymerase II core promoter – the gateway to transcription. Curr Opin Cell Biol 2008;20(3):253–9.
[28] Esteller M. Epigenetics in cancer. N Engl J Med 2008;358(11):1148–59.
[29] Baylin SB, Jones PA. A decade of exploring the cancer epigenome – biological and translational implications. Nat Rev Cancer 2011;11(10):726–34.
[30] Irizarry RA, Ladd-Acosta C, Wen B, Wu Z, Montano C, Onyango P, et al. The human colon cancer methylome shows similar hypo- and hypermethylation at conserved tissue-specific CpG island shores. Nat Genet 2009;41(2):178–86.
[31] Hansen KD, Timp W, Bravo HC, Sabunciyan S, Langmead B, McDonald OG, et al. Increased methylation variation in epigenetic domains across cancer types. Nat Genet 2011;43(8):768–75.
[32] Barski A, Cuddapah S, Cui K, Roh TY, Schones DE, Wang Z, et al. High-resolution profiling of histone methylations in the human genome. Cell 2007;129(4):823–37.
[33] Ernst J, Kheradpour P, Mikkelsen TS, Shoresh N, Ward LD, Epstein CB, et al. Mapping and analysis of chromatin state dynamics in nine human cell types. Nature 2011;473(7345):43–9.
[34] Vavouri T, Lehner B. Human genes with CpG island promoters have a distinct transcription-associated chromatin organization. Genome Biol 2012;13(11):R110.
[35] Kleinjan DA, van Heyningens V. Long-range control of gene expression: emerging mechanisms and disruption in disease. Am J Hum Genet 2005;76(1):8–32.
[36] Heintzman ND, Ren B. The gateway to transcription: identifying, characterizing and understanding promoters in the eukaryotic genome. Cell Mol Life Sci 2007;64(4):386–400.
[37] Heintzman ND, Hon GC, Hawkins RD, Kheradpour P, Stark A, Harp LF, et al. Histone modifications at human enhancers reflect global cell-type-specific gene expression. Nature 2009;459(7243):108–12.
[38] Schnetz MP, Handoko L, Akhtar-Zaidi B, Bartels CF, Pereira CF, Fisher AG, et al. CHD7 targets active gene enhancer elements to modulate ES cell-specific gene expression. PLoS Genet 2010;6(7):e1001023.
[39] Creyghton MP, Cheng AW, Welstead GG, Kooistra T, Carey BW, Steine EJ, et al. Histone H3K27ac separates active from poised enhancers and predicts developmental state. Proc Natl Acad Sci USA 2010;107(50):21931–6.
[40] Rada-Iglesias A, Bajpai R, Swigut T, Brugmann SA, Flynn RA, Wysocka J. A unique chromatin signature uncovers early developmental enhancers in humans. Nature 2011;470(7333):279–83.
[41] Serandour AA, Avner S, Percevault F, Demay F, Bizot M, Lucchetti-Miganeh C, et al. Epigenetic switch involved in activation of pioneer factor FOXA1-dependent enhancers. Genome Res 2011;21(4):555–65.
[42] Zhang B, Xing X, Li J, Lowdon RF, Zhou Y, Lin N, et al. Comparative DNA methylome analysis of endometrial carcinoma reveals complex and distinct deregulation of cancer promoters and enhancers. BMC Genomics 2014;15:868.
[43] Agirre X, Castellano G, Pascual M, Heath S, Kulis M, Segura V, et al. Whole-epigenome analysis in multiple myeloma reveals DNA hypermethylation of B cell-specific enhancers. Genome Res 2015. http://dx.doi.org/10.1101/gr.180240.114.
[44] Whyte WA, Orlando DA, Hnisz D, Abraham BJ, Lin CY, Kagey MH, et al. Master transcription factors and mediator establish super-enhancers at key cell identity genes. Cell 2013;153(2):307–19.
[45] Hnisz D, Abraham BJ, Lee TI, Lau A, Saint-Andre V, Sigova AA, et al. Super-enhancers in the control of cell identity and disease. Cell 2013;155(4):934–47.
[46] Eden A, Gaudet F, Waghmare A, Jaenisch R. Chromosomal instability and tumors promoted by DNA hypomethylation. Science 2003;300(5618):455.
[47] Hodges E, Molaro A, Dos Santos CO, Thekkat P, Song Q, Uren PJ, et al. Directional DNA methylation changes and complex intermediate states accompany lineage specificity in the adult hematopoietic compartment. Mol Cell 2011;44(1):17–28.
[48] Sandoval J, Esteller M. Cancer epigenomics: beyond genomics. Curr Opin Genet Dev 2012;22(1):50–5.
[49] Greger V, Passarge E, Höpping W, Messmer E, Horsthemke B. Epigenetic changes may contribute to the formation and spontaneous regression of retinoblastoma. Hum Genet 1989;83(2):155–8.
[50] Goldstraw P, Ball D, Jett JR, Le Chevalier T, Lim E, Nicholson AG, et al. Non-small-cell lung cancer. Lancet 2011;378(9804):1727–40.
[51] Sutherland KD, Proost N, Brouns I, Adriaensen D, Song JY, Berns A. Cell of origin of small cell lung cancer: inactivation of Trp53 and Rb1 in distinct cell types of adult mouse lung. Cancer Cell 2011;19(6):754–64.
[52] Soria JC, Lee HY, Lee JI, Wang L, Issa JP, Kemp BL, et al. Lack of PTEN expression in non-small cell lung cancer could be related to promoter methylation. Clin Cancer Res 2002;8(5):1178–84.
[53] Marsit CJ, Zheng S, Aldape K, Hinds PW, Nelson HH, Wiencke JK, et al. PTEN expression in non-small-cell lung cancer: evaluating its relation to tumor characteristics, allelic loss, and epigenetic alteration. Hum Pathol 2005;36:768–76.
[54] Berdasco M, Esteller M. Aberrant epigenetic landscape in cancer: how cellular identity goes awry. Dev Cell 2010;19(5):698–711.
[55] Sandoval J, Peiró-Chova L, Pallardó FV, García-Giménez JL. Epigenetic biomarkers in laboratory diagnostics: emerging approaches and opportunities. Expert Rev Mol Diagn 2013;13(5):457–71.

[56] Wang Z, Chen Z, Gao Y, Li N, Li B, et al. DNA hypermethylation of microRNA-34b/c has prognostic value for stage I non-small cell lung cancer. Cancer Biol Ther 2011;11:490–6.

[57] Tellez CS, Juri DE, Do K, Bernauer AM, Thomas CL, Damiani LA, et al. EMT and stem cell-like properties associated with miR-205 and miR-200 epigenetic silencing are early manifestations during carcinogen-induced transformation of human lung epithelial cells. Cancer Res 2011;71(8):3087–97.

[58] Lujambio A, Calin GA, Villanueva A, Ropero S, Sanchez-Cespedes M, Blanco D, et al. A microRNA DNA methylation signature for human cancer metastasis. Proc Natl Acad Sci USA 2008;105:13556–61.

[59] Brueckner B, Stresemann C, Kuner R, Mund C, Musch T, Meister M, et al. The human let-7a-3 locus contains an epigenetically regulated microRNA with oncogenic function. Cancer Res 2007;67:1419–23.

[60] Fabbri M, Garzon R, Cimmino A, Liu Z, Zanesi N, Callegari E, et al. MicroRNA-29 family reverts aberrant methylation in lung cancer by targeting DNA methyltransferases 3A and 3B. Proc Natl Acad Sci USA 2007;104(40):15805–10.

[61] Rivera MP, Mehta A,C, Wahidi MM. Establishing the diagnosis of lung cancer: diagnosis and management of lung cancer, 3rd ed: American College of Chest Physicians evidence-based clinical practice guidelines. Chest 2013;143:e142s–165S.

[62] Hubers AJ, van der Drift MA, Prinsen CF, Witte BI, Wang Y, Shivapurkar N, et al. Methylation analysis in spontaneous sputum for lung cancer diagnosis. Lung Cancer 2014;84(2):127–33.

[63] Nikolaidis G, Raji OY, Markopoulou S, Gosney JR, Bryan J, Warburton C, et al. DNA methylation biomarkers offer improved diagnostic efficiency in lung cancer. Cancer Res 2013;72(22):5692–701.

[64] Xiao P, Chen JR, Zhou F, Lu CX, Yang Q, Tao GH, et al. Methylation of P16 in exhaled breath condensate for diagnosis of non-small cell lung cancer. Lung Cancer 2014;83(1):56–60.

[65] Kneip C, Schmidt B, Seegebarth A, Weickmann S, Fleischhacker M, Liebenberg V, et al. SHOX2 DNA methylation is a biomarker for the diagnosis of lung cancer in plasma. J Thorac Oncol 2011;6(10):1632–8.

[66] Dietrich D, Kneip C, Raji O, Liloglou T, Seegebarth A, Schlegel T, et al. Performance evaluation of the DNA methylation biomarker SHOX2 for the aid in diagnosis of lung cancer based on the analysis of bronchial aspirates. Int J Oncol 2012;40(3):825–32.

[67] Fujii M, Fujimoto N, Hiraki A, Gemba K, Aoe K, Umemura S, et al. Aberrant DNA methylation profile in pleural fluid for differential diagnosis of malignant pleural mesothelioma. Cancer Sci 2012;103(3):510–4.

[68] Brock MV, Hooker CM, Ota-Machida E, Han Y, Guo M, Ames S, et al. DNA methylation markers and early recurrence in stage I lung cancer. N Engl J Med 2008;358(11):1118–28.

[69] Sandoval J, Mendez-Gonzalez J, Nadal E, Chen G, Carmona FJ, Sayols S, et al. A prognostic DNA methylation signature for stage I non-small-cell lung cancer. J Clin Oncol 2013;31(32):4140–7.

[70] Sato T, Arai E, Kohno T, Tsuta K, Watanabe S, Soejima K, et al. DNA methylation profiles at precancerous stages associated with recurrence of lung adenocarcinoma. PLoS One 2013;8(3):e59444.

[71] Tan W, Gu J, Huang M, Wu X, Hildebrandt MAT. Epigenetic analysis of microRNA genes in tumors from surgically resected lung cancer patients and association with survival. Mol Carcinog 2014;54(Suppl. 1):E45–51 (Epub ahead of print).

[72] Saito Y, Liang G, Egger G, Friedman JM, Chuang JC, Coetzee GA, et al. Specific activation of microRNA-127 with downregulation of the proto-oncogene BCL6 by chromatin-modifying drugs in human cancer cells. Cancer Cell 2006;9(6):435–43.

[73] Watanabe K, Emoto N, Hamano E, Sunohara M, Kawakami M, Kage H, et al. Genome structure-based screening identified epigenetically silenced microRNA associated with invasiveness in non-small-cell lung cancer. Int J Cancer 2012;130(11):2580–90.

[74] Kubo T, Toyooka S, Tsukuda K, Sakaguchi M, Fukazawa T, et al. Epigenetic silencing of MicroRNA-34b/c plays an important role in the pathogenesis of malignant pleural mesothelioma. Clin cancer Res 2011;17:4965–74.

[75] Robinson BW, Lake RA. Advances in malignant mesothelioma. N Engl J Med 2005;353:1591–603.

[76] Muraoka T, Soh J, Toyooka S, Aoe K, Fujimoto N, et al. The degree of microRNA-34b/c methylation in serum-circulating DNA is associated with malignant pleural mesothelioma. Lung Cancer 2013;82(3):485–90.

[77] Cao J, Song Y, Bi N, Shen J, Liu W, et al. DNA methylation-mediated repression of miR-886-3p predicts poor outcome of human small cell lung cancer. Cancer Res 2013;73:3326–35.

[78] Wilting RH, Dannenberg JH. Epigenetic mechanisms in tumorigenesis, tumor cell heterogeneity and drug resistance. Drug Resist Updat 2012;15(1–2):21–38.

[79] Zheng Y, Wang R, Song HZ, Pan BZ, Zhang YW, Chen LB. Epigenetic downregulation of RUNX3 by DNA methylation induces docetaxel chemoresistance in human lung adenocarcinoma cells by activation of the AKT pathway. Int J Biochem Cell Biol 2013;45(11):2369–78.

[80] Huang WC, Hung MC. Induction of Akt activity by chemotherapy confers acquired resistance. J Formos Med Assoc 2009;108:180–94.

[81] Zhang YW, Zheng Y, Wang JZ, Lu XX, Wang Z, Chen LB, et al. Integrated analysis of DNA methylation and mRNA expression profiling reveals candidate genes associated with cisplatin resistance in non-small cell lung cancer. Epigenetics 2014;9(6):896–909.
[82] Cortés-Sempere M, de Miguel MP, Pernía O, Rodriguez C, de Castro Carpeño J, Nistal M, et al. IGFBP-3 methylation-derived deficiency mediates the resistance to cisplatin through the activation of the IGFIR/Akt pathway in non-small cell lung cancer. Oncogene 2013;32(10):1274–83.
[83] Juergens RA, Wrangle J, Vendetti FP, Murphy SC, Zhao M, Coleman B, et al. Combination epigenetic therapy has efficacy in patients with refractory advanced non-small cell lung cancer. Cancer Discov 2011;1(7):598–607.
[84] Ramalingam SS, Maitland ML, Frankel P, Argiris AE, Koczywas M, Gitlitz B, et al. Carboplatin and paclitaxel in combination with either vorinostat or placebo for first-line therapy of advanced non-small-cell lung cancer. J Clin Oncol 2010;28(1):56–62.
[85] Li XY, Wu JZ, Cao HX, Ma R, Wu JQ, Zhong YJ, et al. Blockade of DNA methylation enhances the therapeutic effect of gefitinib in non-small cell lung cancer. Oncol Rep 2013;29(5):1975–82.
[86] Momparler RL, Ayoub J. Potential of 5-aza-2′-deoxycytidine (Decitabine) a potent inhibitor of DNA methylation for therapy of advanced non-small cell lung cancer. Lung Cancer 2001;34.

CHAPTER

14

DNA Methylation Alterations as Biomarkers for Prostate Cancer

João Ramalho-Carvalho[1], *Rui Henrique*[1,2,3,*], *Carmen Jerónimo*[1,3,*]

[1]Cancer Biology & Epigenetics Group, IPO-Porto Research Center (CI-IPOP), Portuguese Oncology Institute, Porto, Portugal; [2]Department of Pathology, Portuguese Oncology Institute, Porto, Portugal; [3]Department of Pathology and Molecular Immunology, Institute of Biomedical Sciences Abel Salazar – University of Porto (ICBAS-UP), Porto, Portugal

OUTLINE

1. INTRODUCTION

Prostate cancer (PCa) is the most commonly diagnosed cancer in men, with an estimated 250,000 new cases in the United States diagnosed each year, ranking second in cancer-related mortality. The risk factors for PCa include aging, race (e.g., Afro-American men are more prone to develop PCa whereas native Japanese are at low risk for developing PCa), and family history of PCa (susceptibility genes may contribute up to 5–10% of all detected PCa cases) [1].

*Joint senior authors.

Epigenetic Biomarkers and Diagnostics
http://dx.doi.org/10.1016/B978-0-12-801899-6.00014-0

Presently, most patients are diagnosed after detection of elevated serum prostate-specific antigen (PSA) levels or abnormal digital rectal examination, which entail diagnostic prostate biopsy. Clinically localized PCa, which is potentially curable, is usually treated with radical prostatectomy or radiation, although patients with low-risk disease or short life expectancy may be managed expectantly (e.g., through periodic serum PSA measurements and repeat biopsies, if required, to assess disease progression). Conversely, patients with locally advanced or metastatic (i.e., mostly incurable) disease are initially treated with androgen-deprivation therapy (ADT). However, almost all advanced PCa cases, after a period of ADT, progress to castration-resistant disease, an aggressive and highly lethal form of PCa [1,2].

Prognostication of PCa behavior mostly relies in histological grading, staging, and baseline serum PSA levels. PCa histological grading is based on the Gleason grading system, which combines 5 simple grades (from grade 1 (most differentiated) to 5 (least differentiated)) into 9 combined grades, the so-called Gleason score (GS) or sum (ranging from 1+1 to 5+5), a feature that incorporates information from the frequent morphological heterogeneity of PCa [3]. Among putative precursor lesions, the most widely acknowledged is high-grade prostatic intraepithelial neoplasia (PIN) which consists of malignant-appearing cells still confined to prostate acini, with at least partial preservation of basal layer cells [4]. Interestingly, it is hypothesized that PCa may originate from either luminal or basal epithelial cells, frequently arising as multiple disease foci, thus contributing to wide diversity and heterogeneity at the molecular, cellular, and morphological levels [2].

Prostate carcinogenesis seems to require the acquisition of large-scale genomic rearrangements and copy number alterations involving multiple chromosomes. Indeed, oncogenic fusions, such as *TMPRSS2-ERG*, are found in ~50% of all PCa, as well as in smaller proportion of high-grade PIN [2]. Moreover, loss of tumor-suppressor genes *PTEN*, *NKX3.1*, *TP53*, and *CDKN1B* are often identified in PCa and next-generation sequencing is providing further evidence for molecular subclassification of PCa based on *CDH1* alterations, *SPINK1* overexpression, ERG rearrangements, and *SPOP* mutations [2,5–7].

Given the lack of specific therapeutic targets, high prevalence, and the prolonged latency period of PCa, the role of chemoprevention has been emphasized. Currently, the most encouraging agents are 5-α reductase inhibitors, which prevent the conversion of testosterone to dihydrotestosterone, the most active prostatic androgen. However, early development of resistance to ADT with consequent clinical progression due to the acquisition of castration-resistant phenotype is a main concern [2]. The pathways involved in castration resistance are not fully elucidated, but accepted mechanisms include (1) intratumoral androgen biosynthesis, (2) androgen receptor (AR) pathway hypersensitivity via AR gene amplification, (3) expression of variant AR isoforms that are ligand independent, (4) selection of preexisting castration-resistant epithelial stem cells, (5) growth factor-mediated increase in AR transcription activity, and (6) activation of the PI3K-AKT-mTOR pathway [7–9]. Treatment of castration-resistant PCa (CRPCa) is mostly restricted to taxane-based chemotherapy and palliative care. Nevertheless, new drugs that seem to tackle CRPCa with proven survival benefits, such as abiraterone (blocks androgen production) and enzalutamide (inhibition of androgen binding to AR), are able to prolong survival in men with metastatic CRPCa after docetaxel treatment [7].

Owing to the importance of early diagnosis of PCa (which is clinically silent at its earliest stages) to increase disease survival, there has been an intensive search for specific PCa biomarkers. Serum PCa remains the most widely used biomarker but its usefulness has been recently questioned due to its lack of specificity and inability to accurately identify aggressive forms of PCa, causing overdiagnosis and overtreatment [10,11]. Moreover, currently used prognostic parameters, such as the GS, are limited in their ability to predict disease behavior, in particular for patients with GS 7, which constitutes most of diagnosed PCa cases at present. Thus, the discovery of novel biomarkers that may identify clinically significant PCa, discriminating these from indolent tumors, is a major challenge, which can be met through the study of PCa epigenetics.

The term "epigenetics" was first used to explain why genetic variations occasionally did not lead to phenotypic deviations and how genes might interact with their environment to generate a phenotype [12]. Currently, epigenetics refers particularly to the study of mitotically and/or meiotically heritable changes in gene expression not caused by alterations in the DNA sequence [13]. In general, epigenetic mechanisms comprise DNA methylation, covalent histone modifications, and noncoding RNA (ncRNA) regulation. Epigenetic homeostasis is fundamental to a multitude of biological processes such as transcription, DNA replication, and repair. The disruption of these regulatory mechanisms affects an array of shared and specific cellular processes, influencing the genomic output and triggering several diseases, including cancer [14,15].

Chemical modifications of DNA have been annotated as major players for maintenance of the cellular homeostasis and memory. Such DNA modifications include 5-methylcytosine (5mC), 5-hydroxymethylcytosine (5hmC), and the less common 5-formylcytosine (5fC) and 5-carboxycytosine (5caC) [16]. The addition of a methyl group to a carbon 5 position (5mC) is a common epigenetic mark in many eukaryotes and it is regularly found in the sequence context of CpG (cytosine nucleotide-phosphate-guanine nucleotide) or CpHpG (H=A/T/C) [17]. This is, by far, the most widely studied epigenetic modification in humans [15]. In mammals, this *cis*-regulatory alteration is primarily restricted to symmetrical CpG context. DNA methylation is dynamic and heritable: methyl groups can be added or removed and can remain stable throughout multiple cell divisions. The CpG dinucleotides are predisposed to organize in domains called CpG islands [18]. These are defined as regions of more than 200 bp in length, with a GC content of at least 50% and a ratio of observed to statistically expected CpG frequencies of at least 0.6. CpG dinucleotides are relatively uncommon in mammalian genomes (~1%). There are approximately 28 million CpGs in the human genome, 60–80% of which are generally methylated. Only less than 10% of CpGs are found in CpG islands. High levels of 5mC in CpG-rich promoters are strongly associated with transcriptional repression, whereas CpG-poor regions exhibit a more intricate and context-dependent relationship between DNA methylation and transcriptional activity.

Of all human gene promoters about 60% are associated with CpG islands and these are often unmethylated in normal cells, although some of them (~6%) progress to a methylated state in a tissue-specific manner through early development or in differentiated tissues [19]. Likewise, CpG nucleotides located in repetitive sequences, inserted viral sequences, and retrotransposons are also methylated in normal cells [18], to avoid transcription of these elements and maintain genomic integrity by obstructing recombination events that may lead to gene disruption, oncogene activation, translocations, and chromosomal instability during development and differentiation [20]. Thus, epigenomic states are established in normal cells as a result of development [18]. This is due to the marks present in DNA and chromatin structure, and the cell state is maintained through mitosis [18].

Chemically, 5mC is the result of the addition of a methyl group, donated by S-adenosylmethionine (SAM), to the fifth carbon of the cytosine residue ring, mediated by DNA methyltransferase (DNMT) [21,22]. There are five main

DNMTs [23]: *DNMT1* acts on hemimethylated DNA substrates created during DNA synthesis and maintains the existing methylation patterns after DNA replication and mitosis. *DNMT3A* and *DNMT3B* target previously unmethylated CpGs [23] and are thought to be responsible for the establishment of methylation patterns during embryogenesis and are also able to add methyl groups at non-CpG sites. *DNMT2* shows sequence and structural characteristics of the above-mentioned DNMTs (*DNMT1*, *DNMT3A* and *DNMT3B*) except for a putative nucleic acid binding cleft that cannot easily accommodate duplex DNA. It is responsible for the methylation of aspartic acid transfer RNA ($tRNA^{ASP}$), specifically at the cytosine-38 residue in the anticodon loop, beside its residual DNA methyltansferase activity [24]. *DNMT3L* is structurally similar to DNMTs but does not contain the catalytic domain necessary for methyltransferase activity [23]. However, it has the ability to recognize DNA or chromatin by specific domains and interact with the unmodified H3K4 [23].

Given the dual and rather unspecific activity of DNMTs (they are able to deposit methyl groups at non-CpG sites), whole-genome maps of 5mC have revealed interesting patterns such as cell state-dependent occurrences of 5mC in contexts other than canonical CpGs and in partially methylated domains (PMDs), and conserved regions depleted of 5mCs across mouse and human [16].

Generation and maintenance of non-CpG methylation seem to be strictly regulated, as such modifications are enriched in specific cell types (e.g., pluripotent cells and neural progenitors in adolescent and adult cortex tissues). It is still unclear if cytosine in non-CpG methylation has any functional relevance in normal development and cancer [25].

PMDs have been mostly found in nonpluripotent cells and noncortex tissue types. These PMDs seem to associate with low transcription rates, lamina-associated domains, and late-replicating domains. Different classes of methylation-depleted regions entitled unmethylated regions (UMRs), DNA methylation valleys (DMVs), and DNA methylation canyons (DMCs) have been defined, as well [26,27]. These regions tend to be conserved across cell types and across mouse and human species. Both methylation valleys and canyons tend to be marked with H3K4me3 or H3K27me3 or both, and each can lead to active, inactive, or poised transcriptional states, respectively. Interestingly, these regions cover most genes important for embryonic development [16,25,27].

Nonetheless, there is growing evidence that CpG DNA methylation is augmented at gene bodies of actively transcribed genes in mammals [28]. It has been proposed that it might be related to elongation efficiency and prevention of erroneous initiations of transcription [29].

The transcriptional repression by DNA methylation is thought to follow two different routes: a direct mode in which promoter methylation obstructs the binding of transcriptional activators to the target promoter region [30] or an indirect mode through the recruitment of methyl-CpG-binding domain proteins (MBD) [31]. MBDs can recruit chromatin remodeling complexes to the methylated sequence, resulting in chromatin conformation changes that also inhibit gene transcription [32].

DNA methylation does not occur exclusively at CpG islands. Recently, the concept of "CpG island shores" has been defined, referring to regions of lower CpG density that stretch out in close proximity (~2kb) of CpG islands [33]. Equally, the methylation of these CpG island shores is tightly associated with transcriptional inactivation. Indeed, the majority of the tissue-specific DNA methylation appears at CpG island shores instead at CpG islands [33,34]. Differentially methylated CpG island shores are sufficient to distinguish between specific tissues and are conserved between humans and mouse. Furthermore, 70% of the differentially methylated regions in reprogramming are associated with CpG island shores [34,35].

Alterations in DNA methylation equilibrium are hallmarks of human cancer, including PCa. Discrimination of driver cancer alterations from those that are merely passengers and, thus, not involved in malignant transformation, is a major challenge. During oncogenic transformation, dynamic alterations in hypermethylation and hypomethylation of DNA arise. Hypomethylation appears to be a global phenomenon, but some data indicate that it may take place at specific gene promoters, eventually causing protooncogene activation [15,20]. Hypermethylation mostly occurs in a gene promoter context and is one of the main mechanisms associated with gene expression disruption due to transcriptional repression [15,20,36].

Conventionally, DNA methylation has been associated with protein-coding genes. However, advances in the postgenomic era revealed a large number of RNA families which were thought to be nonfunctional elements of the genome. These include a plethora of long noncoding RNAs and small regulatory RNAs [37]. These are all putative targets for DNA methylation, and, among them, microRNAs (miRNAs) have often been considered to be regulated by DNA methylation [38]. miRNAs are ~22-nucleotide-long RNA molecules that mediate posttranscriptional gene silencing by guiding Argonaute (AGO) proteins to RNA targets [39]. Following transcription, miRNAs are processed in the nucleus by Drosha and then exported to cytoplasm by Exportin-5. In the cytoplasm, DICER processes pre-miRNA before being loaded into the miRNA-induced silencing complex, which contains AGO proteins. This complex is involved in translational repression, mRNA decay, or both [39]. DNA methylation disruption of miRNA transcription impairs miRNA processing and contributes to tumorigenesis [38]. miRNAs control the expression of tumor suppressors, and oncogenes known to be involved in tumorigenesis, angiogenesis, and metastasis. Given the tissue-specific expression signatures of miRNAs, they are tightly regulated and DNA methylation is a key effector in this process.

2. THE ABERRANT DNA METHYLATION LANDSCAPE OF PCa

Altered epigenetic patterns may serve as useful PCa biomarkers for detection, diagnosis, prognosis, and posttreatment surveillance, advancing on the limitations of currently available tools [40]. Over the last years, the role of aberrant DNA methylation in the development and progression in PCa has been increasingly recognized. Indeed, epigenetic alterations are early events in carcinogenesis that may even precede the acquisition of well-defined genetic alterations, and persist through invasion, metastasis, and life-threatening malignant progression. Because they do not directly affect the genome sequence, therapeutic modulation through drugs that can target epigenetic modifications opens new avenues for PCa prevention and treatment.

In theory, prostate cells with anomalous DNA methylation marks arise in nonmalignant lesions, and due to a "DNA methylation catastrophe," those cells with epigenetically silenced genes undergo clonal selection to form a tumor [41]. Conversely, during PCa progression, there is also a global loss of DNA methylation, probably as a consequence of reduced DNA methylation maintenance fidelity, contributing to distinctive cell-to-cell and lesion-to-lesion phenotypic heterogeneity [41]. Thus, both locus-specific hypermethylation and global hypomethylation are involved in neoplastic transformation and contribute to tumor progression.

Understanding how epigenetics contributes to the clonal evolution of PCa is also under scrutiny. Recent reports suggest that the clonal architecture of aggressive PCa is mediated by intratumor DNA methylation heterogeneity given the extensive epigenetic (and genetic) heterogeneity observed among different regions of the same tumor [42]. This intratumoral DNA methylation heterogeneity predominantly occurs at prostate-specific gene regulatory elements, with AR enhancer domains constituting a good example of that intratumoral variation in methylation patterns. Thus, not only DNA methylation plays a role in the regulatory activity of tumor subclones but it may also be a key element explaining the convergent tumorigenic processes and the diversity of metastatic origin [42]. The clues to understand the lethality of PCa may in fact arise from somatic DNA alterations. These aberrations, despite showing marked interindividual heterogeneity among patients with lethal metastatic PCa, seem to be maintained across all metastases within the same individual [43]. Regions that are frequently hypermethylated across individuals are markedly enriched in cancer- and development/differentiation-related genes [43]. Hence, DNA methylation alterations have the potential to generate selectable driver events in carcinogenesis and disease progression [43].

As previously mentioned, age, diet, and environmental factors are thought to be involved in prostate tumorigenesis [40,44], and its effects might be mediated by changes in the epigenetic homeostasis. Indeed, some studies pointed out that age, the most important risk factor for the development of PCa [44], is positively correlated with increased aberrant promoter methylation of various genes [40], of which the age-dependent methylation of estrogen receptor alpha (*ESR1*) may represent a mechanism linking aging and PCa [45]. Given the refined balance between stability and plasticity of DNA methylation patterns, it has been proposed that DNA methylation may provide a lifetime record of environmental exposures and might be used as a potential source of PCa biomarkers.

3. PROMOTER HYPERMETHYLATION IN PCa

The best characterized epigenetic alteration in PCa is promoter hypermethylation [36], and both protein-coding and noncoding genes are targeted [40]. This epigenetic modification is associated with silencing of classic tumor-suppressor genes as well as genes involved in different cellular pathways such as cell cycle, hormone response, DNA repair and damage prevention, signal transduction, tumor invasion and architecture, and apoptosis [15]. The hypermethylation of those genes in PCa may correlate with pathological grade, clinical stage, and castration resistance.

Prostate is an endocrine gland, in which normal cells maintain an appropriate balance between sex hormones (androgen, estrogen, and progesterone) and their specific receptors [2]. Several data show that DNA methylation participates in the transcriptional regulation of hormone receptors [40]. Androgen activity is essential for the development of both normal and PCa cells. From its earliest stages, PCa is androgen-dependent, a status that is kept during disease progression, until, eventually, tumors are enriched in cells that can grow independently of androgens, mostly as a result of ADT [44]. The transition to this state of androgen-independent growth has been associated with genetic alterations including mutations and amplifications of AR locus [46]. Such alterations alter the sensitivity of AR to androgens and are thought to play a role in the development of CRPCa [46]. Not surprisingly, CRPCa displays a heterogeneous loss of AR that is associated with aberrant methylation of its promoter [47]. Up to 28% of CRPCa cases show AR methylation [48], whereas about 20% of primary PCa display this alteration [49], probably at lower levels that do not significantly impair AR expression. Thus, paradoxically, DNA methylation might be an alternative mechanism involved in castration resistance in a subset of patients, through abrogation of AR expression, forcing neoplastic cells to develop alternative signaling pathways, increasing their biological aggressiveness. This may explain the observation that pharmacological reversion of AR promoter methylation, using 5,6-dihydro-5-azacytidine, in CRPCa cell lines restores AR activity, turning cells (again) sensitive to ADT [50], and attenuating the more aggressive phenotype.

The hormonal environment of the prostate is also dependent on estrogens, and both receptors—*ESR1* and *ESR2*—display low or diminished expression in PCa [51,52]. Although an association between *ESR1* methylation and tumor progression has been hypothesized [53], *ESR2* methylation is the main inactivation mechanism accounting for loss of *ESR2* expression in primary PCa [54]. However, during PCa progression, especially in metastatic PCa, cancer cells can reexpress *ESR2* [55] which is accompanied by promoter demethylation [56]. This dynamic reversion of *ESR2* expression is epigenetic in nature and consequently reversible.

Cell cycle is often deregulated in PCa and endows neoplastic cells with increased proliferative capabilities. Cell cycle is very strictly regulated, with multiple checkpoints, and all genes involved are putative targets for silencing through aberrant DNA methylation. Tumor-suppressor genes from the cyclin-dependent kinases family are frequently altered in PCa due to several mechanisms. *CDKN2A* has been shown to be downregulated due to DNA methylation [46], although methylation at exon 2 is more frequent than methylation at promoter region [57]. However, exonic methylation of *CDKN2A* is not associated with loss of gene expression [57], and, thus, further studies are needed to clarify how this may impact in prostate carcinogenesis.

D-type cyclins play a critical role in cell cycle regulation and their abnormal expression has been associated with several human malignancies. One of the best studied cyclins is cyclin D2. Promoter methylation levels of *CCND2* are significantly higher in PCa compared to high-grade PIN and nontumorous prostate tissues ($p < 0.01$), correlating with tumor stage and GS [58]. Moreover, *CCND2* mRNA levels are significantly lower in PCa, inversely correlating with promoter methylation, and demethylating treatment induces a substantial increase in *CCND2* mRNA, in LNCaP cells [58].

Methylation of *SFN* promoter is also a frequent event in malignant prostate lesions, but it also affects nontumorous tissue [59]. *SFN* is a putative tumor-suppressor gene involved in cell cycle regulation and apoptosis following DNA damage [59]. Since there is a progressive accumulation of neoplastic cells with *SFN* methylation from HGPIN to PCa, it is suggested that this epigenetic event might be relevant for prostate carcinogenesis [59].

Genes involved in signal transduction pathways are also affected by aberrant promoter methylation in PCa. Two of the best examples are endothelin receptor type B (*EDNRB*) [60] and the RAS-association domain family protein 1 isoform A (*RASSF1A*) [61]. The protein encoded by *EDNRB* is a G protein-coupled receptor which activates a phosphatidylinositol-calcium second messenger system [60]. Methylation of *EDNRB* is a frequent event in PCa and is associated with decreased mRNA expression [60,62], being restored after treatment with 5-azacytidine [60]. *EDNRB* silencing diminishes the capacity of PCa cells to clean and even block the expression the vasoconstrictor ET1, which accompanies PCa progression in vivo. Promoter methylation of *RASSF1A*, a well-known tumor suppressor, is also a common event in PCa as well as in high-grade PIN lesions [61,63], associating with more advanced tumor stage [61]. These data suggest that *RASSF1A* promoter methylation occurs early in prostate carcinogenesis and increases as PCa progresses, as methylation levels are higher in locally invasive tumors compared to those organ-confined [61].

The maintenance and regulation of normal prostate tissue architecture is based on cadherin–catenin adhesion systems. In PCa, loss of expression of the cell adhesion molecule E-cadherin, encoded by *CDH1*, has been reported in association with promoter methylation of *CDH1*, which increases with disease progression, suggesting a role in the metastatic potential of PCa [64]. Moreover, the promoters of genes involved in the cadherin–catenin axis are also frequently methylated (*APC*, *CAV1*, *LAMA3*, *TIMP3*) [40], theoretically contributing to tumor invasion and metastization.

One of the most acclaimed biomarkers in PCa is *GSTP1*, encoding the intracellular detoxification enzyme glutathione S-transferase π. *GSTP1* promoter methylation occurs at the earliest stages of tumor development and is associated with transcriptional silencing [65]. Aberrant methylation of *GSTP1* is considered a hallmark of PCa, present in about 90% of all PCa and in 75% of high-grade PIN lesions [66,67].

Disruption of apoptotic pathways is key to PCa development and progression and promoter methylation of apoptosis-related genes is common in PCa [68]. Aberrant promoter hypermethylation of *TMS1*, a proapoptotic tumor-suppressor gene, has been reported as an early event in prostate carcinogenesis, correlating with higher GS [69]. Treatment with a demethylating agent restored *TMS1* expression in LNCaP cells [69,70]. Paradoxically, the oncogene *BCL2* (which is antiapoptotic) is also a common target for gene silencing through DNA methylation in PCa [71], a finding that may support a role for apoptosis-inducing therapy in PCa. Because promoter methylation prevents the antiapoptotic role of *BCL2*, PCa cells could be more prone to endure apoptosis upon exposure to proapoptotic drugs.

miRNAs are critical regulators of many pathways and are often deregulated in carcinogenesis [72]. miRNAs are differentially expressed in PCa and may be targeted for downregulation by promoter methylation [73]. For instance, the ability to escape apoptosis is an important carcinogenic event facilitated by numerous miRNAs in PCa [72]. Other common pathways disrupted owing to miRNA deregulation include the cell cycle, intracellular signaling, DNA repair, and adhesion and migration [74]. Table 1 summarizes the most relevant information about miRNAs downregulated by promoter hypermethylation in PCa. One of the most interesting miRNAs in PCa is miR-132 that controls cellular adhesion and directly targets *HBEGF* and *TALIN2*, which have been found to be methylated in 42% of PCa, with higher methylation levels associating with higher GS and more advanced tumor stage [75]. Interestingly, reexpression of miR-132 in PC3 cells induced cell detachment followed by cell death (anoikis) [75]. Furthermore, miR-145 was found to be significantly downregulated in PCa compared to normal prostate tissues and 5-aza-2′-deoxycytidine treatment dramatically restored miR-145 expression [76] (Table 2).

miRNAs may be putative targets for epigenetic therapy in PCa since they have a role in key signaling pathways and are associated with multiple layers of gene regulation. Remarkably, it has been demonstrated that about one-third of downregulated miRNA loci show a matched pattern of DNA methylation and H3K9 acetylation [73]. This underlines the cooperation between the different levels of epigenetic regulation to accomplish transcriptional block of miRNAs.

4. HYPOMETHYLATION IN PCa

Although DNA hypomethylation was the first epigenetic alteration described in cancer, there are few reports on this aberration in PCa. Global DNA hypomethylation occurs at both early and late stages of PCa, and it might serve as biomarker for early detection and prognostication [77–79]. Global hypomethylation expands the chromatin predisposing to genomic instability and promoting deleterious mutations [80,81]. Loss of DNA methylation is

TABLE 1 Common microRNA Inactivated by DNA Methylation in Prostate Cancer

miRNA	Putative target mRNA	Phenotypic effect	References
miR-1	*FN1*, *LASP1* and *XPO6*	Inhibits cell proliferation and motility, represses mitosis, invasion and filipodia formation	Hudson et al. [122]
miR-31	*E2F1*, *E2F2*, *FOXM1*, *MCM2*	Alters androgen receptor homeostasis and controls cell cycle	Lin et al. [131]
miR-34a	*SIRT1*, *CDK6*	Tumor suppressive properties, mediator of apoptosis, cell cycle arrest, and senescence	Lodygin et al. [123]
miR-126	*EGFL7*, *VEGF-A*	May play a role in angiogenesis and cell proliferation	Saito et al. [124]
miR-132	*HBEGF*, *TALIN2*	Stimulates cell death by anoikis, and impedes cell migration and invasion	Formosa et al. [75]
miR-141	*BRD3*, *UBAP1*	Inhibit epithelial to mesenchymal transition	Vrba et al. [125]
miR-145	*IRS1*, *c-Myc*, *MUC1*, *Fli1*, *BNIP3*	Downregulation of miR-145 is associated with aggressive phenotype and poor prognosis in prostate cancer	Suh et al. [76], Zaman et al. [126]
miR-193b	*CCND1*, *ETS1*	Involved in prostate cancer cell proliferation and anchorage-independent growth	Ruahala et al. [127]
miR-196b	*c-myc*, *HOXB8*	Putative repressor of cell migration and metastasis	Hulf et al. [73]
miR-200c	*ZEB1*, *ZEB2*	Inhibit epithelial to mesenchymal transition	Vrba et al. [125]
miR-205	*BCL2L2*, *MED1*	Promote chemotherapeutic agents-induced apoptosis in prostate cancer cells; counteracts epithelial-to-mesenchymal transition and cell migration/invasion	Bhatnagar et al. [128]

associated with tumor progression and chromosome instability [80], and global lower levels of methylation are more evident in metastatic PCa [82]. Genome-wide hypomethylation may trigger inappropriate transcription of proviral and retrotransposon sequences, leading to disruption of neighboring genes [83,84]. The *LINE-1* retrotransposon is hypomethylated in 50% of PCa samples, especially in cases with lymph node metastases [80]. In addition, genome-wide hypomethylation is associated with gains or losses of sequences on chromosome 8, in localized PCa [80]. The progressive genome hypomethylation seems to be linked with deficient DNA methylation maintenance fidelity during DNA replication and it probably contributes to generate cell- and lesion-specific phenotypic heterogeneity [41]. This theory is supported by autopsy studies which showed low levels of DNA methylation and high frequency of copy number alterations in lethal PCa [78].

Gene-specific hypomethylation in PCa has been documented for *CAGE* [85], *CYP1B1* [86], *HPSE* [87], *PLAU* [88], as well as *CRIP1*, *S100P*, and *WNT5A* [89]. Increased expression of *HPSE*, encoding the extracellular matrix degradation protein heparanase which degrades heparan sulfate, has been implicated in tumor invasion and metastasis [87]. Interestingly, *PLAU* (urokinase-type plasminogen activator) is associated with the acquisition of castration resistance and increases tumorigenesis in both in vitro and in vivo models [88].

TABLE 2 DNA Methylation Alterations as Biomakers for Prostate Cancer Management

Gene(s)/miR(s)	Biomarker	Sample type	Method	Sensitivity/ Specificity	References
Hypermethylated					
GSTP1	Early detection, diagnosis, prognosis	Urine Plasma Serum	MSP qMSP Restriction endonuclease qPCR,	58%/na [100]	Jerónimo et al. [99] Cairns et al. [98] Gonzalgo et al. [100] Bastian et al. [112]
GSTP1/APC/MDR1	Diagnosis, prognosis	Prostatectomy	MSP	75.9%/84.1% [111] 72%/67.8% [111]	Enokida et al. [111]
p16/ARF/MGMT/GSTP1	Early detection	Urine	qMSP	87%/100% [101]	Hoque et al. [101]
GSTP1/APC/RARB2/ RASSF1A	Early detection	Urine	qMSP	86%/na [102]	Roupret et al. [102]
GSTP1/APC/RARB2	Early detection, prognosis	Prostate biopsy	qMSP	60%/80% [104]	Baden et al. [104]
GSTP1/PTGS2/RPRM/TIG1	Prognosis	Serum	Restriction endonuclease MSP	47%/92% [103]	Ellinger et al. [103]
GSTP1/APC/PTGS2/MDR1	Prognosis, diagnosis	Prostatectomy	qMSP,	100%/92% [110]	Yegnasubramanian et al. [110]
APC	Prognosis	Prostate biopsy	qMSP	na	Henrique et al. [117]
PTGS2/CD44	Prognosis	Prostatectomy	qMSP	na	Woodsoon et al. [113]
GPR7/ABHD9/Chr3-EST	Prognosis	Prostatectomy	qMSP		Cottrell et al. [114]
PITX2	Prognosis	Prostatectomy	EpiChip microarray	na	Banez et al. [129]
GABRE/miR-452/miR-224	Diagnosis, prognosis	Prostatectomy	MethyLight	95.5%/94.3%	Kristensen et al. [118]
miR-205	Prognosis	Prostatectomy	Sequenom MassArray	na	Hulf et al. [120]
Hypomethylated					
IGF2	Early detection	Prostatectomy	Pyrosequencing	na	Bhusari et al. [130]

MSP, methylation-specific PCR; qMSP, quantitative MSP; qPCR, quantitative polymerase chain reaction.

Loss of imprinting is also associated with aberrant biallelic expression of some genes in PCa, such as *IGF2* [90]. Thus, local and tissue-specific patterns of gene expression might prompt neoplastic transformation over a long period of time [40]. Moreover, *IGF2* hypomethylation may be a biomarker for early detection of PCa.

The impact of DNA hypomethylation in overexpression of oncogenic miRNAs in PCa is very limited. Thus far, only miR-615 has been reported as epigenetically activated in PCa cells due to DNA hypomethylation [73]. However, hypomethylation seems to play a role in deregulation of the ncRNA transcriptome. *XIST*, a single-copy gene heavily methylated in morphologically normal prostate cells, is hypomethylated in PCa, a feature that is associated with *LINE-1* hypomethylation, suggesting a global hypomethylation rather than a promoter-specific loss of methylation [91]. Moreover, *XIST* hypomethylation is increased in more aggressive tumors [91]. The melanoma antigen gene protein-A11 (*MAGE-11*) is a coregulator of the AR signaling. During PCa progression and ADT there is increased expression of *MAGE-11* mediated by promoter hypomethylation, providing an alternative mechanism for increased AR signaling in CRPCa [92].

The available data highlight that hypomethylation changes may not strongly correlate with functional gene sets or with *cis* activation of gene expression. Consequently, in case that DNA hypomethylation plays a driver role in PCa, it would most likely be through promotion of genomic instability (e.g., through promotion of retrotransposition [93]), rather than through direct *cis* regulation of specific genes.

5. DNA METHYLATION-BASED MARKERS FOR PCa DETECTION, MANAGEMENT, AND RISK ESTIMATION

In Western countries, PCa screening is carried out mainly by serum PSA testing. However, serum PSA is limited in the ability to specifically detect PCa and it does not discriminate between clinically aggressive and clinically indolent PCa. Thus novel and more powerful biomarkers are warranted. Because epigenetic alterations are highly prevalent and arise early in prostate tumorigenesis, DNA methylation-based biomarkers constitute promising biomarkers for PCa detection, diagnosis, assessment of prognosis, and prediction of response to therapy [40,94].

6. CANCER DETECTION AND DIAGNOSIS

GSTP1 is the best characterized epigenetic biomarker for PCa. Somatic DNA methylation of *GSTP1* is nearly universally present in almost all PCa cells but is absent or low in normal cells [95]. Indeed, more than 90% of PCa cases show aberrant promoter methylation of *GSTP1* [65] and it might be specifically detected using MSP-based approaches in a wide range of tissue samples and bodily fluids, mainly blood and urine. Testing for *GSTP1* could be used for screening or stratification for the need of prostate biopsy [95]. *GSTP1* performance methylation displays high specificity (86.8–100%) but low sensitivity, both in urine (18.8–38.9%) and serum/plasma (13.0–75.5%) [96–101]. This might be overcome by a multigene promoter methylation testing, and several different gene panels have been proposed, including *GSTP1/ARF/CDNK2A/MGMT* [101] and *GSTP1/APC/RARB2/RASSF1A* [102] in urine and *GSTP1/PTGS2/RPRM/TIG1* [103] in serum. As a result, the detection rate increased significantly to 86% in urine and 42–47% in serum, retaining high specificity: 89–100% for urine and 92% for serum [101–103]. The results gathered allowed for the design of a urine-based diagnostic test—the prostate cancer methylation (ProCaM) assay [104]—that interrogates the methylation levels of *GSTP1*, *APC*, and *RARB2*. The assay displayed 60% sensitivity and 80% specificity, with 97% informative rate. ProCaM has been validated in a multicentre prospective study testing samples of men with serum PSA levels of 2.0–10.0 ng/mL, in which its performance was compared with existing methods based on clinical workup and serum PSA levels [104]. The ProCaM predictive accuracy was higher than that of serum PSA or any of its related parameters (area under the curve (AUC) = 0.73, p = 0.038). Importantly, a positive result correlated not only with positive biopsy, but it also associated with increased risk for detecting high-grade PCa (GS≥7) with a substantial predictive accuracy (AUC = 0.79, p = 0.001). Indeed, men with positive ProCaM result were seven times more likely to be diagnosed high-grade PCa [104].

The EGF-containing fibulin-like extracellular matrix protein 1 (*EFEMP1*) is a tumor-suppressor gene epigenetically deregulated in PCa. *EFEMP1* methylation seems to be PCa-specific and accurately discriminates PCa from nonmalignant prostate tissues (AUC = 0.98; p < 0.001), as well as from bladder and renal tumors (AUC = 0.986, 96% sensitivity, and 98% specificity) [105]. The high accuracy (96%) of *EFEMP1* methylation test shows promise for its use as an ancillary tool in diagnostically challenging lesions. Moreover, when compared with other frequently methylated genes in PCa (e.g., *GSTP1*, *APC*, and *RARbeta*), the *EFEMP1* assay displays similar performance, even when compared with multigene panels [105].

Using prostate core biopsies, Paziewska et al. [106] compared the performance of expression and methylation markers to distinguishing cancerous from noncancerous prostate tissues. Although *HOXC6*, *AMACR*, and *PCA3* expression displayed the best discrimination between PCa and BPH (AUC = 0.94; 0.92; 0.955), they were not sensitive and specific enough to be considered PCa diagnostic biomarkers. However, DNA promoter methylation levels of *APC*, *TACC2*, *RARB*, *DGKZ*, and *HES5* identified PCa with high sensitivity and specificity (AUCs ranging between 0.95 and 1.0) [106]. Some overlap was observed for DNA methylation levels of PCa-positive and PCa-negative needle biopsies, but with minor impact. Combination of methylation levels of *RARB*, *HES5*, and *C5Orf4* displayed the highest performance (AUC = 0.909), detecting over 50% of cancer samples with 100% specificity [106]. Taking into account that problematic prostate core biopsies usually contain just a small amount of neoplastic cells, these biomarkers might be of help to pathologists, especially in patients with suspected PCa following a negative initial biopsy.

Fast and accurate DNA methylation analysis of multiple loci in clinical samples with limited DNA quantities would provide a clear advantage to patient management. A MethyLight multiplex assay combining *APC*, *HOXD3*, and *TGFB2* has been evaluated in patient samples, and its sensitivity was sufficient for detection methylation of those genes in formalin-fixed paraffin-embedded tissue and urine samples. These encouraging results, however, require validation in larger patient cohorts [107].

Limitations of sampling during prostatic biopsy raise the concern that cancer might not be sampled in high-risk men. This leads to a high frequency of follow-up procedures that are needed to ultimately confirm the absence of disease. Partin and collaborators developed an assay (DOCUMENT, Detection Of Cancer Using Methylated Events in Negative Tissue) aimed to screen, among patients with a negative prostate biopsy, those who are at low risk of harboring cancer which was not detected through biopsy due to inaccuracy and might avoid unnecessary repeat [108]. The assay is based on the quantification of methylation levels of three genes commonly methylated in PCa—*GSTP1*, *APC*, and *RASSF1*—using multiplex methylation-specific PCR [108]. In a multicenter study, this epigenetic assay was an independent predictor of PCa detection in a repeat biopsy 30 months after the initial negative results. The performance of the assay displayed 64% specificity, 88% negative predictive value (NPV), and 18% false-negative rate, based on the adjusted cancer prevalence in repeat biopsies. In multivariate models corrected for factors with diagnostic potential, the assay proved to be the most significant independent predictor of patient outcome (odds ratio (OR) = 2.69, 95% confidence interval (CI): 1.60–4.51). Owing to the high NPV, the DOCUMENT assay combined with other known risk factors may assist in clinical management of PCa suspects, reducing the rate of repeat biopsies [108].

7. PROGNOSIS AND PREDICTION OF RESPONSE TO THERAPY

Current prognostic markers for PCa are suboptimal, contributing to overtreatment of indolent PCa patients. A DNA methylation signature that predicts biochemical recurrence after radical prostatectomy has been uncovered. Hypermethylation of *AOX1*, *C1orf114*, *GAS6*, *HAPLN3*, *KLF8*, and *MOB3B* was shown to be highly PCa-specific (AUC ranging from 0.89 to 0.98) and high *C1orf114* methylation was significantly associated with biochemical recurrence after radical prostatectomy in multivariate analysis, both in the testing and validation sets (hazard ratio (HR) = 3.10; 95% CI: 1.89–5.09; HR = 3.27; 95% CI: 1.17–9.12, respectively) [109]. In this multicohort approach, a significant three-gene prognostic methylation signature (*AOX1-C1orf114-HAPLN3*) was used to classify patients into low- and high-methylation subgroups (HR = 1.91; 95% CI: 1.26–2.90; HR = 2.33; 95% CI: 1.31–4.13, for cohorts 1 and 2, respectively). *AOX1-C1Orf114-HAPLN3* and *C1Orf114-HAPLN3* panels were evaluated and these models successfully predicted biochemical recurrence after radical prostatectomy in multivariate analysis including standard clinicopathological variables and in two different cohorts, indicating that these methylation-based markers hold independent prognostic value. Further testing is needed to evaluate the prognostic potential of these markers in prostatic biopsies.

Hypermethylation of *PTGS2* in localized PCa predicted PCa recurrence after radical prostatectomy independently of tumor stage and GS [110]. The univariate Cox proportional hazard models revealed that high *PTGS2* methylation levels significantly increased the risk of recurrence (HR = 2.82; 95% CI: 1.07–7.44). Moreover, in the multivariate Cox proportional hazards model, only high *PTGS2* hypermethylation predicted for increased risk of recurrence independently of GS and pathological stage (HR = 4.26; 95% CI: 1.36–13.36).

A hypermethylation score derived from *GSTP1*, *APC*, and *MDR1* discriminated organ confined from locally advanced disease with 72% sensitivity and 67.8% specificity [111]. Moreover, considering patients with PSA levels <10 ng/mL, the M score had a sensitivity of 67.1% with 85.7% specificity. The circulating cell-free DNA carrying *GSTP1* hypermethylation was detected in 12% of men with clinically localized disease and 28% of men with metastatic cancer. Additionally, 8 men (15%) who developed PSA recurrence were positive for serum *GSTP1* hypermethylation, whereas patients who were disease-free tested negative. In multivariate analysis, serum *GSTP1* hypermethylation was the most significant predictor of PSA recurrence (HR = 4.4; 95% CI: 2.2–8.8) [112].

Furthermore, hypermethylation of *CD44* and *PTGS2* is also predictive of PSA recurrence after radical prostatectomy (HR = 8.87, 95% CI: 1.85–42.56). Kaplan–Meyer analysis showed that combined hypermethylation of *CD44* and *PTGS2* associated with shorter time to biochemical recurrence compared to absence of gene methylation [113]. Similar findings were reported for promoter hypermethylation of *ABHD9* [114]. *PITX2* hypermethylation is a strong marker for biochemical recurrence either in a univariate or in a multivariate analysis (HR = 3.4; 95%

CI: 1.9–6.0; HR=2.1; 95% CI: 1.2–3.9, respectively) [115]. In survival analysis, the estimated 8-year probability of biochemical recurrence-free was 79% in the group with high *PITX2* methylation as opposed to 94% in the low-methylation group. Interestingly, *PITX2* hypermethylation seems to be associated with biochemical recurrence in PCa with intermediate GS [115]. On the other hand, high methylation levels of *APC* and *CCND2* are strong predictors of short time to postradical prostatectomy recurrence in PCa with a GS 3+4=7 (multivariate model, HR=4.33; 95% CI: 1.52–12.33) [116], whereas in a prospective study, high *APC* promoter methylation levels predicted poor prognosis in prostate biopsy specimens, irrespective of GS (OR=3.5; 95% CI: 1.23–9.96) [117].

Promoter methylation levels of *GABRE/miR-452/miR-224* not only discriminate PCa from nonmalignant tissues (AUC=0.98; 95.5% sensitivity and 94.3% specificity), but also high methylation levels independently predicted early biochemical recurrence after radical prostatectomy in two independent cohorts (HR=1.75; 95% CI: 1.37–2.23 and HR=2.99; 95% CI: 1.71–5.21) [118]. Interestingly, high methylation was significantly associated with the standard clinicopathological parameters, including serum PSA, pathological stage, GS, and surgical margin status. In a large cohort (407 patients), high *HOXD3* promoter methylation levels were strongly associated with shorter recurrence-free survival (high methylation=43.1% vs low methylation=34.5%) [119]. Moreover, *HOXD3* methylation levels were associated with higher GS and more advanced tumor stage (HR=5.23; 95% CI: 1.31–20.96) [119].

Because miR-205 is epigenetically downregulated in PCa it may have a prognostic biomarker potential [120]. Indeed, hypermethylation of miR-205 locus is significantly associated with biochemical recurrence in patients with localized PCa and low preoperative PSA levels (HR=2.005; 95% CI: 1.109–3.625). Furthermore, on multivariate analysis, low miR-205 methylation was a significant predictor of biochemical relapse (HR=2.2; 95% CI: 0.99–5.0) [120].

DNA methylation-based biomarkers might also be useful for prediction of response to therapy. In a recent study, detectable baseline-methylated *GSTP1* in serum was associated with poorer overall survival in men with CRPCa (HR=4.2; 95% CI: 2.1–8.2) and a decrease of methylation levels after cycle 1 of chemotherapy was associated with PSA response [121]. These preliminary findings were further validated in another cohort and the results were similar (HR=2.4; 95% CI: 1.0–5.6), thus confirming that assessment of *GSTP1* methylation levels in plasma is a promising predictive biomarker for a subgroup of advanced PCa patients undergoing chemotherapy.

8. CONCLUSIONS AND PERSPECTIVES

Alterations in DNA methylation patterns are frequent and early events in prostate carcinogenesis, enabling its use as tumor biomarkers. These may accurately detect PCa at its earliest stages, by means of noninvasive techniques, increasing the likelihood of curative treatment. Moreover, the diagnostic performance of several methylated genes in PCa makes them potential ancillary tools of histopathological assessment of prostate biopsies. However, discrimination of clinically aggressive from indolent PCa is mandatory, and several methylation-based biomarkers have demonstrated to be of prognostic value. Finally, the predictive value of methylation markers is just starting to be unraveled in PCa and is likely to provide novel tools from clinical management (Figure 1).

Figure 2 provides an overview of the most promising methylation-based biomarkers at different stages of the workup of PCa, from risk assessment to disease follow-up. Those

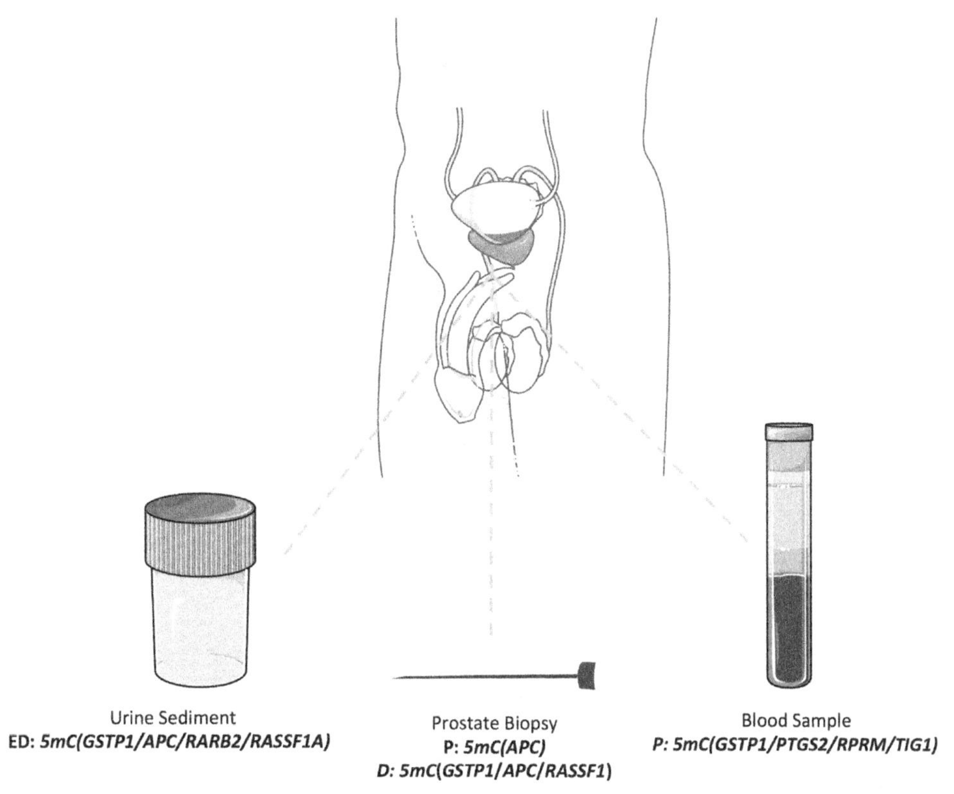

FIGURE 1 Clinical applications of DNA methylation analysis in prostate cancer (PCa). Urine biomarker testing employs a noninvasive approach to detect PCa by collecting voided urine samples, isolating DNA from cells in the urine sediment, and quantifying DNA methylation levels. The DNA methylation is also an ancillary tool in samples with low cellular content such as prostate biopsy. Alternatively, isolation of circulating PCa cells/DNA from patient serum and testing for DNA methylation levels might provide predictive information for PCa. 5mC (5-methylcytosine): quantitative detection of DNA methylation levels using quantitative methylation-specific PCR (qMSP). Most accepted panels are suggested for testing in prostate samples. ED, early detection; D, diagnosis; P, prognosis.

biomarkers were selected on the basis of the performance reported in the original and subsequent studies. Interestingly, at each stage of the workup the set of biomarkers differs whereas others remain across the algorithm. This may reflect, to the former, the biological stage at which a particular epigenetic alteration becomes more relevant to carcinogenesis, whereas for the latter its persistence or increase along tumor progression endows a particular relevance for the neoplastic process, from transformation to invasion and dissemination. Importantly, some of the biomarkers have been already tested in a clinical-level assay (e.g., ProCaM and DOCUMENT), meaning that its translation to routine use might be foreseen for the near future. However, it must be acknowledged that a long path lies still ahead for most of the DNA methylation biomarkers, requiring standardization at methodological level and adequate clinical trial design to ascertain whether it represents a significant step forward in the management of PCa.

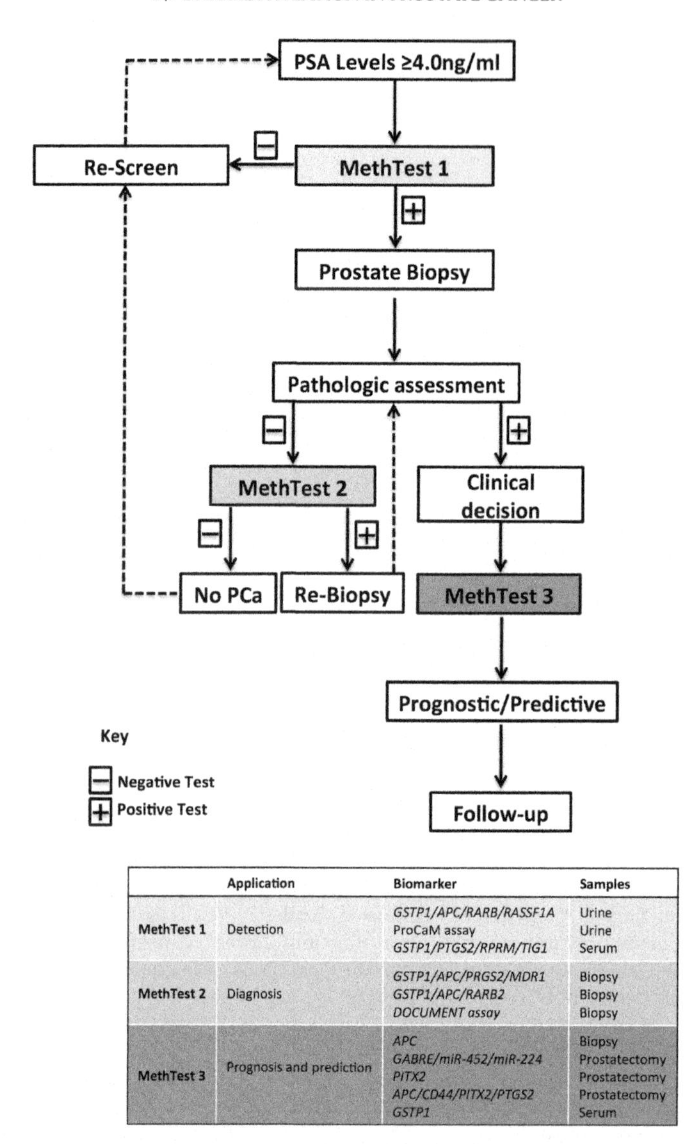

	Application	Biomarker	Samples
MethTest 1	Detection	*GSTP1/APC/RARB/RASSF1A* ProCaM assay *GSTP1/PTGS2/RPRM/TIG1*	Urine Urine Serum
MethTest 2	Diagnosis	*GSTP1/APC/PRGS2/MDR1* *GSTP1/APC/RARB2* *DOCUMENT assay*	Biopsy Biopsy Biopsy
MethTest 3	Prognosis and prediction	*APC* *GABRE/miR-452/miR-224* *PITX2* *APC/CD44/PITX2/PTGS2* *GSTP1*	Biopsy Prostatectomy Prostatectomy Prostatectomy Serum

FIGURE 2 **DNA methylation biomarkers in prostate cancer (PCa) management.** Specific panels of methylated genes may be used in the clinical management of PCa patients, not only in cancer detection (MethTest 1), but also are putative ancillary tools after pathologic assessment (MethTest 2). Additionally, these biomarkers may provide prognostic/predictive information of recurrence and progression following clinical decision (treatment or surveillance). MethTest, DNA methylation analysis; –, negative test; +, positive test; PSA, prostate-specific antigen.

Glossary

Biochemical relapse When, following radical prostatectomy, PSA levels are higher that 0.2 ng/mL, or when PSA consecutively rises in from the postoperative baseline. It is indicative of disease recurrence.

Castration-resistant PCa Highly aggressive and lethal form of PCa characterized by disease progression despite androgen-deprivation therapy (castrate serum levels of testosterone), presenting as one or any combination of a continuous rise in serum PSA levels, progression of preexisting disease or appearance of new metastases.

CpG island Regions of more than 200bp in length, with a GC content of at least 50% and a ratio of observed to statistically expected CpG frequencies of at least 0.6. CpG dinucleotides are relatively uncommon in mammalian genomes (~1%) but are frequently found at promoters. Those CpG islands at promoters are predominantly unmethylated across cell types and are very prevalent at transcription start sites of housekeeping and developmental regulatory genes.

DNA methylation canyon (DMC) Regions of low methylation covering conserved domains that frequently contain transcription factors and are distinct from CpG islands and shores. Approximately half of the genes in these canyons are marked with repressive histone marks.

DNA methylation valleys (DMV) Large genomic domains (≥5kb long) that are methylation devoid. Moreover, DMV genes are poised with bivalent states (H3K4me3 and H3K27me3). DMVs are uniquely enriched for transcription factors and developmental regulatory genes. Developmental regulatory genes are preferentially located in DMVs. Strikingly genes with DMVs tend to be hypermethylated in cancer.

Gleason score (GS) Prostate Cancer's microscopic grading system ranging from 1 to 5 based on the glandular architecture and patterns. The combined GS (2–10) is the result of the two most prevalent Gleason grade patterns per tumor. GS is a significant prognostic indicator: high GS (≥8) is associated with a poorer outcome than an intermediate GS (7), and low GS (≤6) is globally associated with a better prognosis.

Partially methylated domains (PMDs) Large contiguous regions of the genome (mean size ~153kb) that display intermediate methylation levels (average <70%).

Prostate-specific antigen Serine protease produced by prostate epithelial cells implicated in seminal fluid liquefaction. Serum PSA levels are used for detection and monitoring of prostate cancer.

Unmethylated regions (UMRs) Regions constitutively unmethylated independ of whether they are defined or not as CpG islands. Almost all UMRs are associated with known or predicted transcription start sites, and very often packaged with nucleosomes containing H3K4me3.

LIST OF ABBREVIATIONS

5caC 5-Carboxycytosine
5fC 5-Formylcytosine
5hmC 5-Hydroxymethylcytosine
5mC 5-Methylcytosine
ADT Androgen-deprivation therapy
AR Androgen receptor
CRPCa Castration-resistant PCa
DMC DNA methylation canyon
DMV DNA methylation valley
DNMT DNA methyltransferase
HDAC Histone deacetylase
MBP Methylcytosine-binding protein
miRNA MicroRNA
MSP Methylation-specific PCR
NPV Negative predictive value
PCa Prostate cancer
PMD Partially methylated domain
PSA Prostate-specific antigen
ProCaM Prostate cancer methylation
qMSP Quantitative methylation-specific PCR
SAM S-adenosylmethionine
UMR Unmethylated regions

References

[1] Hoffman RM. Clinical practice. Screening for prostate cancer. N Engl J Med November 24, 2011;365(21):2013–9.

[2] Shen MM, Abate-Shen C. Molecular genetics of prostate cancer: new prospects for old challenges. Genes Dev September 15, 2010;24(18):1967–2000.

[3] Epstein JI, Allsbrook Jr WC, Amin MB, Egevad LL, Committee IG. The 2005 International Society of Urological Pathology (ISUP) consensus conference on Gleason grading of prostatic carcinoma. Am J Surg Pathol September 2005;29(9):1228–42.

[4] Epstein JI. Precursor lesions to prostatic adenocarcinoma. Virchows Archiv January 2009;454(1):1–16.

[5] Baca SC, Prandi D, Lawrence MS, Mosquera JM, Romanel A, Drier Y, et al. Punctuated evolution of prostate cancer genomes. Cell April 25, 2013;153(3):666–77.

[6] Beltran H, Rubin MA. New strategies in prostate cancer: translating genomics into the clinic. Clin Cancer Res February 1, 2013;19(3):517–23.

[7] Barlow LJ, Shen MM. SnapShot: prostate cancer. Cancer Cell September 9, 2013;24(3):400.e1.

[8] Karantanos T, Corn PG, Thompson TC. Prostate cancer progression after androgen deprivation therapy: mechanisms of castrate resistance and novel therapeutic approaches. Oncogene December 5, 2013;32(49):5501–11.

[9] Saraon P, Jarvi K, Diamandis EP. Molecular alterations during progression of prostate cancer to androgen independence. Clin Chem October 2011;57(10):1366–75.

[10] Moyer VA, Force USPST. Screening for prostate cancer: U.S. Preventive Services Task Force recommendation statement. Ann Intern Med July 17, 2012;157(2):120–34.

[11] Catalona WJ, D'Amico AV, Fitzgibbons WF, Kosoko-Lasaki O, Leslie SW, Lynch HT, et al. What the U.S. Preventive Services Task Force missed in its prostate cancer screening recommendation. Ann Intern Med July 17, 2012;157(2):137–8.

[12] Waddington CH. The epigenotype. 1942. Int J Epidemiol February 2012;41(1):10–3.
[13] Berger SL, Kouzarides T, Shiekhattar R, Shilatifard A. An operational definition of epigenetics. Genes Dev April 1, 2009;23(7):781–3.
[14] Jones PA, Baylin SB. The epigenomics of cancer. Cell February 23, 2007;128(4):683–92.
[15] Esteller M. Epigenetics in cancer. N Engl J Med March 13, 2008;358(11):1148–59.
[16] Plongthongkum N, Diep DH, Zhang K. Advances in the profiling of DNA modifications: cytosine methylation and beyond. Nat Rev Genet October 2014; 15(10):647–61.
[17] Laird PW. Principles and challenges of genomewide DNA methylation analysis. Nat Rev Genet March 2010;11(3):191–203.
[18] Portela A, Esteller M. Epigenetic modifications and human disease. Nat Biotechnol October 2010; 28(10):1057–68.
[19] Straussman R, Nejman D, Roberts D, Steinfeld I, Blum B, Benvenisty N, et al. Developmental programming of CpG island methylation profiles in the human genome. Nat Struct Mol Biol May 2009;16(5):564–71.
[20] Esteller M. Cancer epigenomics: DNA methylomes and histone-modification maps. Nat Rev Genet April 2007;8(4):286–98.
[21] Lopez-Serra L, Esteller M. Proteins that bind methylated DNA and human cancer: reading the wrong words. Br J Cancer June 17, 2008;98(12):1881–5.
[22] Goldberg AD, Allis CD, Bernstein E. Epigenetics: a landscape takes shape. Cell February 23, 2007;128(4):635–8.
[23] Denis H, Ndlovu MN, Fuks F. Regulation of mammalian DNA methyltransferases: a route to new mechanisms. EMBO Rep July 2011;12(7):647–56.
[24] Goll MG, Kirpekar F, Maggert KA, Yoder JA, Hsieh CL, Zhang X, et al. Methylation of tRNAAsp by the DNA methyltransferase homolog Dnmt2. Science January 20, 2006;311(5759):395–8.
[25] Valdes-Mora F, Clark SJ. Prostate cancer epigenetic biomarkers: next-generation technologies. Oncogene May 19, 2014.
[26] Jeong M, Sun D, Luo M, Huang Y, Challen GA, Rodriguez B, et al. Large conserved domains of low DNA methylation maintained by Dnmt3a. Nat Genet January 2014;46(1):17–23.
[27] Xie W, Schultz MD, Lister R, Hou Z, Rajagopal N, Ray P, et al. Epigenomic analysis of multilineage differentiation of human embryonic stem cells. Cell May 23, 2013;153(5):1134–48.
[28] Shenker N, Flanagan JM. Intragenic DNA methylation: implications of this epigenetic mechanism for cancer research. Br J Cancer January 17, 2012;106(2):248–53.
[29] Zilberman D, Gehring M, Tran RK, Ballinger T, Henikoff S. Genome-wide analysis of *Arabidopsis thaliana* DNA methylation uncovers an interdependence between methylation and transcription. Nat Genet January 2007;39(1):61–9.
[30] Kuroda A, Rauch TA, Todorov I, Ku HT, Al-Abdullah IH, Kandeel F, et al. Insulin gene expression is regulated by DNA methylation. PLoS One 2009;4(9):e6953.
[31] Thomson JP, Skene PJ, Selfridge J, Clouaire T, Guy J, Webb S, et al. CpG islands influence chromatin structure via the CpG-binding protein Cfp1. Nature April 15, 2010;464(7291):1082–6.
[32] Vaissiere T, Sawan C, Herceg Z. Epigenetic interplay between histone modifications and DNA methylation in gene silencing. Mutat Res July–August 2008;659(1–2):40–8.
[33] Irizarry RA, Ladd-Acosta C, Wen B, Wu Z, Montano C, Onyango P, et al. The human colon cancer methylome shows similar hypo- and hypermethylation at conserved tissue-specific CpG island shores. Nat Genet February 2009;41(2):178–86.
[34] Doi A, Park IH, Wen B, Murakami P, Aryee MJ, Irizarry R, et al. Differential methylation of tissue- and cancer-specific CpG island shores distinguishes human induced pluripotent stem cells, embryonic stem cells and fibroblasts. Nat Genet December 2009;41(12):1350–3.
[35] Ji H, Ehrlich LI, Seita J, Murakami P, Doi A, Lindau P, et al. Comprehensive methylome map of lineage commitment from haematopoietic progenitors. Nature September 16, 2010;467(7313):338–42.
[36] Perry AS, Watson RW, Lawler M, Hollywood D. The epigenome as a therapeutic target in prostate cancer. Nat Rev Urol December 2010;7(12):668–80.
[37] Morris KV, Mattick JS. The rise of regulatory RNA. Nat Rev Genet June 2014;15(6):423–37.
[38] Lopez-Serra P, Esteller M. DNA methylation-associated silencing of tumor-suppressor microRNAs in cancer. Oncogene March 29, 2012;31(13):1609–22.
[39] Ha M, Kim VN. Regulation of microRNA biogenesis. Nat Rev Mol Cell Biol August 2014;15(8):509–24.
[40] Jeronimo C, Bastian PJ, Bjartell A, Carbone GM, Catto JW, Clark SJ, et al. Epigenetics in prostate cancer: biologic and clinical relevance. Eur Urol October 2011;60(4):753–66.
[41] Nelson WG, De Marzo AM, Yegnasubramanian S. Epigenetic alterations in human prostate cancers. Endocrinology September 2009;150(9):3991–4002.
[42] Brocks D, Assenov Y, Minner S, Bogatyrova O, Simon R, Koop C, et al. Intratumor DNA methylation heterogeneity reflects clonal evolution in aggressive prostate cancer. Cell Rep August 7, 2014;8(3):798–806.

[43] Aryee MJ, Liu W, Engelmann JC, Nuhn P, Gurel M, Haffner MC, et al. DNA methylation alterations exhibit intraindividual stability and interindividual heterogeneity in prostate cancer metastases. Sci Transl Med January 23, 2013;5(169):169ra10.

[44] Gronberg H. Prostate cancer epidemiology. Lancet March 8, 2003;361(9360):859–64.

[45] Li LC, Shiina H, Deguchi M, Zhao H, Okino ST, Kane CJ, et al. Age-dependent methylation of ESR1 gene in prostate cancer. Biochem Biophys Res Commun August 20, 2004;321(2):455–61.

[46] Li LC, Carroll PR, Dahiya R. Epigenetic changes in prostate cancer: implication for diagnosis and treatment. J Natl Cancer Inst January 19, 2005;97(2):103–15.

[47] Jarrard DF, Kinoshita H, Shi Y, Sandefur C, Hoff D, Meisner LF, et al. Methylation of the androgen receptor promoter CpG island is associated with loss of androgen receptor expression in prostate cancer cells. Cancer Res December 1, 1998;58(23):5310–4.

[48] Kinoshita H, Shi Y, Sandefur C, Meisner LF, Chang C, Choon A, et al. Methylation of the androgen receptor minimal promoter silences transcription in human prostate cancer. Cancer Res July 1, 2000;60(13):3623–30.

[49] Nakayama T, Watanabe M, Suzuki H, Toyota M, Sekita N, Hirokawa Y, et al. Epigenetic regulation of androgen receptor gene expression in human prostate cancers. Lab Invest December 2000;80(12):1789–96.

[50] Izbicka E, MacDonald JR, Davidson K, Lawrence RA, Gomez L, Von Hoff DD. 5,6 Dihydro-5′-azacytidine (DHAC) restores androgen responsiveness in androgen [illegible] prostate cancer cells. Anticancer Res 1999 Mar–Apr;19(2A):1285–91.

[51] Hobisch A, Hittmair A, Daxenbichler G, Wille S, Radmayr C, Hobisch-Hagen P, et al. Metastatic lesions from prostate cancer do not express oestrogen and progesterone receptors. J Pathol July 1997;182(3):356–61.

[52] Horvath LG, Henshall SM, Lee CS, Head DR, Quinn DI, Makela S, et al. Frequent loss of estrogen receptor-beta expression in prostate cancer. Cancer Res July 15, 2001;61(14):5331–5.

[53] Li LC, Chui R, Nakajima K, Oh BR, Au HC, Dahiya R. Frequent methylation of estrogen receptor in prostate cancer: correlation with tumor progression. Cancer Res February 1, 2000;60(3):702–6.

[54] Lau KM, LaSpina M, Long J, Ho SM. Expression of estrogen receptor (ER)-alpha and ER-beta in normal and malignant prostatic epithelial cells: regulation by methylation and involvement in growth regulation. Cancer Res June 15, 2000;60(12):3175–82.

[55] Leav I, Lau KM, Adams JY, McNeal JE, Taplin ME, Wang J, et al. Comparative studies of the estrogen receptors beta and alpha and the androgen receptor in normal human prostate glands, dysplasia, and in primary and metastatic carcinoma. Am J Pathol July 2001;159(1):79–92.

[56] Zhu X, Leav I, Leung YK, Wu M, Liu Q, Gao Y, et al. Dynamic regulation of estrogen receptor-beta expression by DNA methylation during prostate cancer development and metastasis. Am J Pathol June 2004;164(6):2003–12.

[57] Nguyen TT, Nguyen CT, Gonzales FA, Nichols PW, Yu MC, Jones PA. Analysis of cyclin-dependent kinase inhibitor expression and methylation patterns in human prostate cancers. Prostate May 15, 2000;43(3):233–42.

[58] Henrique R, Costa VL, Cerveira N, Carvalho AL, Hoque MO, Ribeiro FR, et al. Hypermethylation of cyclin D2 is associated with loss of mRNA expression and tumor development in prostate cancer. J Mol Med Berl November 2006;84(11):911–8.

[59] Henrique R, Jeronimo C, Hoque MO, Carvalho AL, Oliveira J, Teixeira MR, et al. Frequent 14-3-3 sigma promoter methylation in benign and malignant prostate lesions. DNA Cell Biol April 2005;24(4):264–9.

[60] Nelson JB, Lee WH, Nguyen SH, Jarrard DF, Brooks JD, Magnuson SR, et al. Methylation of the 5′ CpG island of the endothelin B receptor gene is common in human prostate cancer. Cancer Res January 1, 1997;57(1):35–7.

[61] Liu L, Yoon JH, Dammann R, Pfeifer GP. Frequent hypermethylation of the RASSF1A gene in prostate cancer. Oncogene October 3, 2002;21(44):6835–40.

[62] Jeronimo C, Henrique R, Campos PF, Oliveira J, Caballero OL, Lopes C, et al. Endothelin B receptor gene hypermethylation in prostate adenocarcinoma. J Clin Pathol January 2003;56(1):52–5.

[63] Jeronimo C, Henrique R, Hoque MO, Mambo E, Ribeiro FR, Varzim G, et al. A quantitative promoter methylation profile of prostate cancer. Clin Cancer Res December 15, 2004;10(24):8472–8.

[64] Li LC, Zhao H, Nakajima K, Oh BR, Ribeiro Filho LA, Carroll P, et al. Methylation of the E-cadherin gene promoter correlates with progression of prostate cancer. J Urol August 2001;166(2):705–9.

[65] Jeronimo C, Usadel H, Henrique R, Oliveira J, Lopes C, Nelson WG, et al. Quantitation of GSTP1 methylation in non-neoplastic prostatic tissue and organ-confined prostate adenocarcinoma. J Natl Cancer Inst November 21, 2001;93(22):1747–52.

[66] Kang GH, Lee S, Lee HJ, Hwang KS. Aberrant CpG island hypermethylation of multiple genes in prostate cancer and prostatic intraepithelial neoplasia. J Pathol February 2004;202(2):233–40.

[67] Henrique R, Jeronimo C, Teixeira MR, Hoque MO, Carvalho AL, Pais I, et al. Epigenetic heterogeneity of high-grade prostatic intraepithelial neoplasia: clues for clonal progression in prostate carcinogenesis. Mol Cancer Res January 2006;4(1):1–8.

[68] Murphy TM, Perry AS, Lawler M. The emergence of DNA methylation as a key modulator of aberrant cell death in prostate cancer. Endocr Relat Cancer March 2008;15(1):11–25.
[69] Das PM, Ramachandran K, Vanwert J, Ferdinand L, Gopisetty G, Reis IM, et al. Methylation mediated silencing of TMS1/ASC gene in prostate cancer. Mol Cancer 2006;5:28.
[70] Suzuki M, Shigematsu H, Shivapurkar N, Reddy J, Miyajima K, Takahashi T, et al. Methylation of apoptosis related genes in the pathogenesis and prognosis of prostate cancer. Cancer Lett October 28, 2006;242(2):222–30.
[71] Carvalho JR, Filipe L, Costa VL, Ribeiro FR, Martins AT, Teixeira MR, et al. Detailed analysis of expression and promoter methylation status of apoptosis-related genes in prostate cancer. Apoptosis August 2010;15(8):956–65.
[72] Catto JW, Alcaraz A, Bjartell AS, De Vere White R, Evans CP, Fussel S, et al. MicroRNA in prostate, bladder, and kidney cancer: a systematic review. Eur Urol May 2011;59(5):671–81.
[73] Hulf T, Sibbritt T, Wiklund ED, Bert S, Strbenac D, Statham AL, et al. Discovery pipeline for epigenetically deregulated miRNAs in cancer: integration of primary miRNA transcription. BMC Genomics 2011;12:54.
[74] Gandellini P, Folini M, Zaffaroni N. Emerging role of microRNAs in prostate cancer: implications for personalized medicine. Discov Med March 2010;9(46): 212–8.
[75] Formosa A, Lena AM, Markert EK, Cortelli S, Miano R, Mauriello A, et al. DNA methylation silences miR-132 in prostate cancer. Oncogene February 6, 2013;32(1):127–34.
[76] Suh SO, Chen Y, Zaman MS, Hirata H, Yamamura S, Shahryari V, et al. MicroRNA-145 is regulated by DNA methylation and p53 gene mutation in prostate cancer. Carcinogenesis May 2011;32(5):772–8.
[77] Cho NY, Kim JH, Moon KC, Kang GH. Genomic hypomethylation and CpG island hypermethylation in prostatic intraepithelial neoplasm. Virchows Archiv January 2009;454(1):17–23.
[78] Yegnasubramanian S, Haffner MC, Zhang Y, Gurel B, Cornish TC, Wu Z, et al. DNA hypomethylation arises later in prostate cancer progression than CpG island hypermethylation and contributes to metastatic tumor heterogeneity. Cancer Res November 1, 2008;68(21):8954–67.
[79] Brothman AR, Swanson G, Maxwell TM, Cui J, Murphy KJ, Herrick J, et al. Global hypomethylation is common in prostate cancer cells: a quantitative predictor for clinical outcome? Cancer Genet Cytogenet January 1, 2005;156(1):31–6.
[80] Schulz WA, Elo JP, Florl AR, Pennanen S, Santourlidis S, Engers R, et al. Genomewide DNA hypomethylation is associated with alterations on chromosome 8 in prostate carcinoma. Genes Chromosomes Cancer September 2002;35(1):58–65.
[81] Eden A, Gaudet F, Waghmare A, Jaenisch R. Chromosomal instability and tumors promoted by DNA hypomethylation. Science April 18, 2003;300(5618):455.
[82] Bedford MT, van Helden PD. Hypomethylation of DNA in pathological conditions of the human prostate. Cancer Res October 15, 1987;47(20):5274–6.
[83] Chalitchagorn K, Shuangshoti S, Hourpai N, Kongruttanachok N, Tangkijvanich P, Thong-ngam D, et al. Distinctive pattern of LINE-1 methylation level in normal tissues and the association with carcinogenesis. Oncogene November 18, 2004;23(54):8841–6.
[84] Roman-Gomez J, Jimenez-Velasco A, Agirre X, Cervantes F, Sanchez J, Garate L, et al. Promoter hypomethylation of the LINE-1 retrotransposable elements activates sense/antisense transcription and marks the progression of chronic myeloid leukemia. Oncogene November 3, 2005;24(48):7213–23.
[85] Cho B, Lee H, Jeong S, Bang YJ, Lee HJ, Hwang KS, et al. Promoter hypomethylation of a novel cancer/testis antigen gene CAGE is correlated with its aberrant expression and is seen in premalignant stage of gastric carcinoma. Biochem Biophys Res Commun July 18, 2003;307(1):52–63.
[86] Tokizane T, Shiina H, Igawa M, Enokida H, Urakami S, Kawakami T, et al. Cytochrome P450 1B1 is overexpressed and regulated by hypomethylation in prostate cancer. Clin Cancer Res August 15, 2005;11(16):5793–801.
[87] Ogishima T, Shiina H, Breault JE, Tabatabai L, Bassett WW, Enokida H, et al. Increased heparanase expression is caused by promoter hypomethylation and up-regulation of transcriptional factor early growth response-1 in human prostate cancer. Clin Cancer Res February 1, 2005;11(3):1028–36.
[88] Pakneshan P, Xing RH, Rabbani SA. Methylation status of uPA promoter as a molecular mechanism regulating prostate cancer invasion and growth in vitro and in vivo. Faseb J June 2003;17(9):1081–8.
[89] Wang Q, Williamson M, Bott S, Brookman-Amissah N, Freeman A, Nariculam J, et al. Hypomethylation of WNT5A, CRIP1 and S100P in prostate cancer. Oncogene October 4, 2007;26(45):6560–5.
[90] Jarrard DF, Bussemakers MJ, Bova GS, Isaacs WB. Regional loss of imprinting of the insulin-like growth factor II gene occurs in human prostate tissues. Clin Cancer Res December 1995;1(12):1471–8.
[91] Laner T, Schulz WA, Engers R, Muller M, Florl AR. Hypomethylation of the XIST gene promoter in prostate cancer. Oncol Res 2005;15(5):257–64.

[92] Karpf AR, Bai S, James SR, Mohler JL, Wilson EM. Increased expression of androgen receptor coregulator MAGE-11 in prostate cancer by DNA hypomethylation and cyclic AMP. Mol Cancer Res April 2009;7(4):523–35.

[93] Lee E, Iskow R, Yang L, Gokcumen O, Haseley P, Luquette 3rd LJ, et al. Landscape of somatic retrotransposition in human cancers. Science August 24, 2012;337(6097):967–71.

[94] Jeronimo C, Henrique R. Epigenetic biomarkers in urological tumors: a systematic review. Cancer Lett January 28, 2014;342(2):264–74.

[95] Henrique R, Jeronimo C. Molecular detection of prostate cancer: a role for GSTP1 hypermethylation. Eur Urol November 2004;46(5):660–9. discussion 9.

[96] Goessl C, Krause H, Muller M, Heicappell R, Schrader M, Sachsinger J, et al. Fluorescent methylation-specific polymerase chain reaction for DNA-based detection of prostate cancer in bodily fluids. Cancer Res November 1, 2000;60(21):5941–5.

[97] Goessl C, Muller M, Heicappell R, Krause H, Straub B, Schrader M, et al. DNA-based detection of prostate cancer in urine after prostatic massage. Urology September 2001;58(3):335–8.

[98] Cairns P, Esteller M, Herman JG, Schoenberg M, Jeronimo C, Sanchez-Cespedes M, et al. Molecular detection of prostate cancer in urine by GSTP1 hypermethylation. Clin Cancer Res September 2001;7(9):2727–30.

[99] Jerónimo C, Usadel H, Henrique R, Silva C, Oliveira J, Lopes C, et al. Quantitative GSTP1 hypermethylation in bodily fluids of patients with prostate cancer. Urology December 2002;60(6):1131–5.

[100] Gonzalgo ML, Pavlovich CP, Lee SM, Nelson WG. Prostate cancer detection by GSTP1 methylation analysis of postbiopsy urine specimens. Clin Cancer Res July 2003;9(7):2673–7.

[101] Hoque MO, Topaloglu O, Begum S, Henrique R, Rosenbaum E, Van Criekinge W, et al. Quantitative methylation-specific polymerase chain reaction gene patterns in urine sediment distinguish prostate cancer patients from control subjects. J Clin Oncol September 20, 2005;23(27):6569–75.

[102] Roupret M, Hupertan V, Yates DR, Catto JW, Rehman I, Meuth M, et al. Molecular detection of localized prostate cancer using quantitative methylation-specific PCR on urinary cells obtained following prostate massage. Clin Cancer Res March 15, 2007;13(6):1720–5.

[103] Ellinger J, Haan K, Heukamp LC, Kahl P, Buttner R, Muller SC, et al. CpG island hypermethylation in cell-free serum DNA identifies patients with localized prostate cancer. Prostate January 1, 2008;68(1):42–9.

[104] Baden J, Adams S, Astacio T, Jones J, Markiewicz J, Painter J, et al. Predicting prostate biopsy result in men with prostate specific antigen 2.0 to 10.0 ng/ml using an investigational prostate cancer methylation assay. J Urol November 2011;186(5):2101–6.

[105] Almeida M, Costa VL, Costa NR, Ramalho-Carvalho J, Baptista T, Ribeiro FR, et al. Epigenetic regulation of EFEMP1 in prostate cancer: biological relevance and clinical potential. J Cell Mol Med September 11, 2014;18(11):2287–97.

[106] Paziewska A, Dabrowska M, Goryca K, Antoniewicz A, Dobruch J, Mikula M, et al. DNA methylation status is more reliable than gene expression at detecting cancer in prostate biopsy. Br J Cancer August 12, 2014;111(4):781–9.

[107] Olkhov-Mitsel E, Zdravic D, Kron K, van der Kwast T, Fleshner N, Bapat B. Novel multiplex MethyLight protocol for detection of DNA methylation in patient tissues and bodily fluids. Sci Rep 2014;4:4432.

[108] Partin AW, Van Neste L, Klein EA, Marks LS, Gee JR, Troyer DA, et al. Clinical validation of an epigenetic assay to predict negative histopathological results in repeat prostate biopsies. J Urol October 2014;192(4):1081–7.

[109] Haldrup C, Mundbjerg K, Vestergaard EM, Lamy P, Wild P, Schulz WA, et al. DNA methylation signatures for prediction of biochemical recurrence after radical prostatectomy of clinically localized prostate cancer. J Clin Oncol September 10, 2013;31(26):3250–8.

[110] Yegnasubramanian S, Kowalski J, Gonzalgo ML, Zahurak M, Piantadosi S, Walsh PC, et al. Hypermethylation of CpG islands in primary and metastatic human prostate cancer. Cancer Res March 15, 2004;64(6):1975–86.

[111] Enokida H, Shiina H, Urakami S, Igawa M, Ogishima T, Li LC, et al. Multigene methylation analysis for detection and staging of prostate cancer. Clin Cancer Res September 15, 2005;11(18):6582–8.

[112] Bastian PJ, Palapattu GS, Lin X, Yegnasubramanian S, Mangold LA, Trock B, et al. Preoperative serum DNA GSTP1 CpG island hypermethylation and the risk of early prostate-specific antigen recurrence following radical prostatectomy. Clin Cancer Res June 1, 2005;11(11):4037–43.

[113] Woodson K, O'Reilly KJ, Ward DE, Walter J, Hanson J, Walk EL, et al. CD44 and PTGS2 methylation are independent prognostic markers for biochemical recurrence among prostate cancer patients with clinically localized disease. Epigenetics October–December 2006;1(4):183–6.

[114] Cottrell S, Jung K, Kristiansen G, Eltze E, Semjonow A, Ittmann M, et al. Discovery and validation of 3 novel DNA methylation markers of prostate cancer prognosis. J Urol May 2007;177(5):1753–8.

[115] Weiss G, Cottrell S, Distler J, Schatz P, Kristiansen G, Ittmann M, et al. DNA methylation of the PITX2 gene promoter region is a strong independent prognostic marker of biochemical recurrence in patients with prostate cancer after radical prostatectomy. J Urol April 2009;181(4):1678–85.
[116] Rosenbaum E, Hoque MO, Cohen Y, Zahurak M, Eisenberger MA, Epstein JI, et al. Promoter hypermethylation as an independent prognostic factor for relapse in patients with prostate cancer following radical prostatectomy. Clin Cancer Res December 1, 2005;11(23):8321–5.
[117] Henrique R, Ribeiro FR, Fonseca D, Hoque MO, Carvalho AL, Costa VL, et al. High promoter methylation levels of APC predict poor prognosis in sextant biopsies from prostate cancer patients. Clin Cancer Res October 15, 2007;13(20):6122–9.
[118] Kristensen H, Haldrup C, Strand S, Mundbjerg K, Mortensen MM, Thorsen K, et al. Hypermethylation of the GABRE~miR-452~miR-224 promoter in prostate cancer predicts biochemical recurrence after radical prostatectomy. Clin Cancer Res April 15, 2014;20(8):2169–81.
[119] Kron KJ, Liu L, Pethe VV, Demetrashvili N, Nesbitt ME, Trachtenberg J, et al. DNA methylation of HOXD3 as a marker of prostate cancer progression. Lab Invest July 2010;90(7):1060–7.
[120] Hulf T, Sibbritt T, Wiklund ED, Patterson K, Song JZ, Stirzaker C, et al. Epigenetic-induced repression of microRNA-205 is associated with MED1 activation and a poorer prognosis in localized prostate cancer. Oncogene June 6, 2013;32(23):2891–9.
[121] Mahon KL, Qu W, Devaney J, Paul C, Castillo L, Wykes RJ, et al. Methylated Glutathione S-transferase 1 (mGSTP1) is a potential plasma free DNA epigenetic marker of prognosis and response to chemotherapy in castrate-resistant prostate cancer. Br J Cancer August 21, 2014;111(9):1802–9.
[122] Hudson RS, Yi M, Esposito D, Watkins SK, Hurwitz AA, Yfantis HG, et al. MicroRNA-1 is a candidate tumor suppressor and prognostic marker in human prostate cancer. Nucleic Acids Res April 1, 2012;40(8):3689–703.
[123] Lodygin D, Tarasov V, Epanchintsev A, Berking C, Knyazeva T, Korner H, et al. Inactivation of miR-34a by aberrant CpG methylation in multiple types of cancer. Cell Cycle August 15, 2008;7(16):2591–600.
[124] Saito Y, Friedman JM, Chihara Y, Egger G, Chuang JC, Liang G. Epigenetic therapy upregulates the tumor suppressor microRNA-126 and its host gene EGFL7 in human cancer cells. Biochem Biophys Res Commun February 13, 2009;379(3):726–31.
[125] Vrba L, Jensen TJ, Garbe JC, Heimark RL, Cress AE, Dickinson S, et al. Role for DNA methylation in the regulation of miR-200c and miR-141 expression in normal and cancer cells. PLoS One 2010;5(1):e8697.
[126] Zaman MS, Chen Y, Deng G, Shahryari V, Suh SO, Saini S, et al. The functional significance of microRNA-145 in prostate cancer. Br J Cancer July 13, 2010;103(2):256–64.
[127] Rauhala HE, Jalava SE, Isotalo J, Bracken H, Lehmusvaara S, Tammela TL, et al. miR-193b is an epigenetically regulated putative tumor suppressor in prostate cancer. Int J Cancer September 1, 2010;127(6):1363–72.
[128] Bhatnagar N, Li X, Padi SK, Zhang Q, Tang MS, Guo B. Downregulation of miR-205 and miR-31 confers resistance to chemotherapy-induced apoptosis in prostate cancer cells. Cell Death Dis 2010;1:e105.
[129] Banez LL, Sun L, van Leenders GJ, Wheeler TM, Bangma CH, Freedland SJ, et al. Multicenter clinical validation of PITX2 methylation as a prostate specific antigen recurrence predictor in patients with post-radical prostatectomy prostate cancer. J Urol July 2010;184(1):149–56.
[130] Bhusari S, Yang B, Kueck J, Huang W, Jarrard DF. Insulin-like growth factor-2 (IGF2) loss of imprinting marks a field defect within human prostates containing cancer. Prostate November 2011;71(15): 1621–30.
[131] Lin PC, Chiu YL, Banerjee S, Park K, Mosquera JM, Giannopoulou E, et al. Epigenetic repression of miR-31 disrupts androgen receptor homeostasis and contributes to prostate cancer progression. Cancer Res 2013;73(3):1232–44.

CHAPTER

15

DNA Methylation in Breast Cancer

Raimundo Cervera, Alberto Ramos, Ana Lluch, Joan Climent

Hematology and Oncology Unit, Biomedical Research Institute INCLIVA, Valencia, Spain

OUTLINE

1. INTRODUCTION

Breast cancer (OMIM #114480) represents a complex and heterogeneous disease at the histopathological, molecular, and genetic levels, with a distinct clinical outcome, and it is one of the most common human malignancies of all cancers in women. The worldwide cancer incidence in 2012, according to the World Health Organization, is about 14 million cases per year, and breast cancer is the second most common type of cancer after lung cancer, 12% of the total (http://www.who.int/mediacentre/factsheets/fs297/en/). In Europe, there were an estimated 71.08 cases of breast cancer per 100,000 adults, whereas in the United States, there were an estimated 92.93 cases of breast cancer per 100,000 adults. Not only has breast cancer incidence been increasing over the last years, having a higher incidence in developed countries [1,2], but it is also the most commonly diagnosed cancer. Recently, interest has grown

Epigenetic Biomarkers and Diagnostics
http://dx.doi.org/10.1016/B978-0-12-801899-6.00015-2

in the role of epigenetics in breast cancer development and progression. Conrad Waddington described epigenetics as "the branch of biology that studies the causal interaction between genes and their products which bring the phenotype into being" [3], and this term has been used since then to describe epigenetic heritable changes in the profile of genetic expression that does not affect the genetic sequence. Epigenetic alterations include histone modifications, nucleosome remodeling, and DNA methylation, although this chapter will focus only on the effects of hypo- and hypermethylation in breast cancer cells.

2. MAJOR MILESTONES IN BREAST CANCER AND DNA METHYLATION RESEARCH

The oldest recorded case of breast cancer in the world was in ancient Egypt in 1500 BC, and it was reported that no treatment was possible. From those ancient times until the mid-nineteenth century there was almost no progress in breast cancer treatment, with surgery being the only available option, and for almost the next 100 years radical mastectomy became the key treatment. It was about the mid-twentieth century when the first chemotherapeutic agent was used, and systemic treatment became an option for patients (Figure 1). That twentieth century was a learning period in the conduct of clinical trials, especially with regard to experimental design. Interestingly, the recognition of 5-methylcytosine in nucleic acids [4], the DNA double helix structure [5], and the DNA methylation discovery occurred then [6]. It was in the 1980s, however, when the first observations of epigenetic abnormalities in cancer were made, showing that DNA methylation changes can occur to both oncogenes and tumor suppressors [7,8]. More direct evidence linking DNA methylation and cancer came several years later, supporting the idea that tumor suppressor genes are silenced in cancer by promoter hypermethylation and their function can be restored by inhibiting DNA methylation [9]. Simultaneously, breast cancer became the first type of solid tumor to be successfully treated with oncogene-targeted therapy, thanks to the discovery in 1985 of ErbB-2/HER2 [10] (for human epidermal growth factor 2). Years later, it would identify a specific subtype of breast tumors accounting for the 15–20% of all breast cancer cases.

The 1990s was the period when the first breast cancer susceptibility gene (*BRCA1*) was identified. To be more precise, the conclusive step was the identification of truncating mutations in the coding sequence of *BRCA1* in families with multiple cases of breast cancer [11]. Only a few years later research revealed the hypermethylation of the promoter region on the *BRCA1* gene in tumor DNA from patients with sporadic breast cancer without somatic gene mutations, but with the BRCA1 activity markedly decreased [12]. In parallel with the discovery of *BRCA1* methylation, there was the first published report on DNA methyltransferases (DNMTs) describing the role of the DNMT1 as that responsible for the wave of de novo methylation that occurs in early embryonic cells [13]. With the arrival of modern medicine and new diagnostic methods, by 1995 the development of novel therapies for breast cancer was remarkably noticeable, including hormone treatments, surgery, and biological therapies. Breast cancer became the first type of solid tumor cancer to be successfully treated with molecular-targeting therapy. In 1998, Herceptin™ was approved for the treatment of metastatic breast cancer. By this time, less than 10% of breast cancer-inflicted women had mastectomies.

The twenty-first century started a new era for breast cancer molecular classification, diagnosis, and treatment assignment. By introducing the complementary deoxyribonucleic acid (cDNA) microarrays in breast cancer research, Charles Perou showed that the phenotypic diversity of breast tumors was accompanied by

FIGURE 1 Breast cancer and DNA methylation time line milestones (Refs [1–25]).

Web links sources:

- http://am.asco.org/50-years-breast-cancer-dramatic-progress-treatment-based-improved-understanding-biology
- http://www.nature.com/milestones/milecancer/full/milecancer19.html
- http://epigeneticsunraveled.wix.com/epigenetics#!timeline/ckl8
- http://www.cancerprogress.net/timeline/breast
- http://www.news-medical.net/health/History-of-Breast-Cancer.aspx
- http://www.nature.com/milestones/milecancer/timeline.html

a corresponding diversity in gene expression patterns, and several subtypes of breast cancer were identified based on their molecular profile [14]. Five groups were identified and named: luminal A, luminal B, basal-like, Her-2 enriched, and normal-like subgroups. All of them have shown to be different in terms of tumor biology or patient clinical outcome [15,16]. Just 2 years later, Laura van't Veer and Marc J. van de Vijver described a 70 gene profile by gene expression analysis. It predicted the clinical outcome of breast cancer patients [17,18]. This gene expression signature was commercialized as the gene test MammaPrint (Agendia, Irvine, CA, www.agendia.com), and it was the first test to be a powerful predictor of outcome for patients with breast cancer, including the likelihood of cancer recurrence and the potential of the patient to benefit from more aggressive treatment. The second gene test for breast cancer outcome prediction was the Oncotype DX Breast Cancer Assay (Genomic Health Inc, Redwood City, CA, www.oncotypedx.com) that came out in 2005 as a 21-gene prediction test. It predicts chemotherapy benefits and the likelihood of distant breast cancer recurrence. In 2009, Parker et al. [19] reported on a risk prediction model for breast cancer developed from the expression data of 50 genes known as PAM50 that classify the five "intrinsic" subtypes of breast cancer. Currently it has been commonly employed to identify those clinically relevant molecular subtypes of breast cancer [20–22].

Today there are at least five different multi-gene expression-based prognostic tests offered as a laboratory test, each of them claiming to have additional prognostic information [17]. Currently, there are no commercial or lab tests for breast cancer DNA methylation signatures that predict clinical outcome. Today, as in 2014, several groups are working and publishing the results of tissue-specific DNA methylation patterns in normal and tumor human breast tissue, and DNA methylation signatures that could predict breast cancer patient outcomes [23–26], providing evidence in support of bringing DNA methylation-based biomarkers into the clinical setting.

3. DNA METHYLATION

Methylation is a process characterized by an enzyme-driven chemical modification of DNA based mostly on the addition of a methyl (CH3) group with a covalent link at the 5′ position of the cytosine pyrimidine ring in a CpG dinucleotide. Usually known as a CpG island, it has profound effects on the mammalian genome [27]. The CpG islands are defined as regions of DNA with at least 200 base pairs with a guanine–cytosine (GC) percentage that is greater than 50% [28]. In mammals, 70–80% of CpG cytosines are methylated [29].

DNA methylation is one of the most frequently occurring epigenetic events in the mammalian genome, and unlike other genetic alterations, it does not involve changes in the nucleotide sequence of DNA. It has been shown that DNA methylation is essential for normal development [30], and the methylation pattern is determined during embryogenesis and passed over to differentiating cells and tissues. It can, however, cause an increase in the mutation rate and genomic errors in the human genome. These errors or the disruption of the normal methylation patterns will trigger a wide variety of pathologies, including breast cancer.

3.1 DNA Methyltransferases

The methylation process is catalyzed by DNMTs, and it is the predominant epigenetic modification in mammals that causes gene silencing and noncoding genomic regions. Currently, the mammalian DNMT family consists of five members: DNMT1, DNMT2, DNMT3A, DNMT3B, DNMT3L, although DNMT2 and DNMT3L have not been shown to have a relation with DNA methylation [31,32]. There are two models of DNA methylation: de novo and for maintenance. In de novo methylation, the DNMT methylates previously unmethylated CpG islands, whereas in for maintenance, the DNMT methylates the paternal strand methylation pattern to the newly synthesized daughter strand [33]. DNMT1 is known as maintenance methyltransferase and is responsible for copying the methylation patterns after DNA replication. DNMT3A and DNMT3B are known as de novo methyltransferases and are responsible for the methylation pattern during the processes of embryogenesis and soft tissue development [34]. These three main DNMTs are responsible for coordinating the gene expression levels in normal tissue, and it has been shown that they cooperate in establishing and maintaining specific patterns of DNA methylation [35]. Variations in the expression level of these DNMTs have been reported in normal human tissues, whereas in tumors, the DNMT expression levels have been found to be elevated [30,33–37]. In breast cancer, some reports show the overexpression of *DNMT3B* as the only methyltransferase with a role in tumorigenesis, both in breast cancer patients associated with a higher histological grade, as well as in breast tumor progression [38] and breast cancer cell lines leading to the concurrent aberrant hypermethylation of numerous genes [36,39]. Recently, however, it has been shown that the overexpression of *DNMT1* and *DNMT3a* is associated with lymph node metastasis, advanced clinical stages, and a poor prognosis for breast cancer patients [37].

3.2 Breast Cancer DNA Methylation Patterns

DNA methylation patterns may differ largely between normal tissues and transformed tumorigenic cells. Numerous studies have supported the important role of DNA methylation in carcinogenesis, and the list of candidate genes with an aberrant methylation pattern has grown exponentially [26,40–43]. Although several studies on these aberrant patterns in breast cancer have been reported, acceptance of the methylation signatures as a tool for breast cancer diagnosis remains limited. Nevertheless, it has already been shown that DNA methylation can act as a disease biomarker [44–46]. Recent emerging data point to exceptional developments in the comprehension of cancer cell DNA methylation profiles, especially with the advent of the microarray and next-generation sequencing technologies. These have provided the research community with the possibility of analyzing the genome-wide DNA methylation profile, although in some cases the highly associated sequencing costs continue to limit the widespread application of the latter [47–49].

The methylation pattern of normal tissue, as indicated previously, is established during early development, in part due to the action of the de novo DNMT3A and DNMT3B. In normal cells, though, the CpG islands preceding gene promoters are generally unmethylated [50,51], and the "normal" CpG methylation profile is often inverted in cells that become tumorigenic [52].

In tumor tissue, however, DNA hypomethylation has been associated with gene overexpression leading to the oncogene activation and genome instability that may promote mitotic recombination and chromosome rearrangement. Withal, DNA hypermethylation is usually associated with gene repression due to the silencing of DNA repair, tumor suppressor, and other cancer-related genes involved in apoptosis, cell cycle regulation, cell adhesion, and invasion, among other biological functions [50,53,54]. In general, breast cancer is characterized by global hypomethylation with a more than 50% decreased methylation status when compared with normal breast tissue and, in consequence, it induces the overexpression of oncogenes and chromosomal instability in breast cancer cells [53,55,56]. In contrast, the promoter of tumor suppressor genes is hypermethylated, leading to the epigenetic silencing of these target genes.

Although DNA global hypomethylation and promoter hypermethylation have been reported occurring simultaneously as independent processes in breast carcinogenesis and at different stages of breast cancer, it has been shown that global hypomethylation may be a mechanism of late stages since the degree of hypomethylation of genomic DNA increases as the lesion progresses. Promoter hypermethylation, to the contrary, would be an early event in breast cancer directly linked to tumor development [57,58].

3.2.1 *Hormone Receptor Methylation*

Estrogen is essential for the development and proper function of the mammary gland, and it exerts its function by binding to the estrogen receptor (ER). Although there are two isoforms of ER (α and β), the ERα isoform is the one with higher expression in breast tissue and a predominant role in breast cells [59]. About 70% of all breast tumors express ER (ER-positive breast cancer), and despite its implication in breast tumorigenesis and progression, it is generally a good prognostic marker. Actually, *ESR1* (ER gene 1) expression is a good marker of endocrine therapy response; however, 30% of tumors do not express *ESR1*. These ER-negative breast tumors are more aggressive and invasive than ER-positive ones and in consequence, since antiestrogen therapies are inefficient, breast cancer patients with ER-negative tumors have a worse prognosis [60].

It is well established that DNA methylation plays an important role in hormone-dependent cancers. In breast cancer, DNA methylation of *ESR1* and progesterone receptor (PR) promoter and, in consequence, loss of gene expression has been proposed as a mechanism of development of receptor-negative tumor cells in both cell lines, as well as in primary tumors [61,62]. In fact, loss of ER expression due to DNA methylation has been reported to be increased in those tumors lacking the expression of ER, PR, and HER2 receptors, known as triple-negative breast tumors (TNBC) [39,63], however, the role of *ESR1* methylation in TNBC is heavily under debate. Increased levels of methyltransferase activity [64] and a higher expression of methyltransferases DNMT3B and DNMT1 have been observed compared with those for the ER-positive cells [39]. Interestingly, studies with demethylating agents in ER-negative breast cancer cell lines have shown that the inhibition of methylation could restore *ESR1* expression [65].

3.2.2 *BRCA1 Methylation Status*

Hereditary breast cancer is caused by inherited germ line mutations in genes of "high-penetrance" (i.e., BRCA1), "moderate-penetrance," and "low-penetrance" susceptibility [66]. Families carrying these germ lines apparently show a dominant inheritance pattern of breast cancer at an earlier age than patients developing sporadic breast tumors [67]. *BRCA1* encodes an 1863 amino acid protein that is important for normal embryonic development, and it is a typical breast tumor suppressor gene mutated in about 40–50% of hereditary breast cancers [68], with an 85% risk of developing breast tumors for *BRCA1* mutation carriers [25]; however, it accounts for less than 10% of all breast cancers. In hereditary breast cancer, the loss of one *BRCA1* allele is required for tumorigenesis, and it has been reported to be involved in the regulation of cell proliferation, participation in DNA repair/recombination processes related to the maintenance of genomic integrity, induction of apoptosis in damaged cells, and regulation of gene transcription [69,70].

In contrast to heritable breast tumors, mutations of *BRCA1* in somatic breast cancer are very rare events. Nevertheless, methylation of the *BRCA1* promoter has been reported in about 5–31% of breast cancers [71,72]. Some studies have shown that the gene expression profile of BRCA1-mutated breast cancers is similar to the somatic breast tumors where the inactivation of BRCA1 is due to methylation of the promoter [73,74], suggesting that this epigenetic event may also be considered a hereditary factor in breast cancer tumorigenesis. Additionally, sporadic breast tumors with BRCA1 methylation status have a similar profile of DNA copy number changes as the one observed in breast cancer BRCA1 mutation carriers [75]. In fact, a recent published work has indicated that *BRCA1* promoter methylation can be observed in normal breast tissue from patients with BRCA1 methylated tumors, but not in that breast tissue where BRCA1 was unmethylated in the tumor cells [76].

3.2.3 *Methylation and Breast Cancer Molecular Subtypes*

Breast cancer is a very heterogeneous disease characterized by diverse pathological features with highly different therapeutic responses. Breast cancer has been traditionally classified based on morphology, histopathological criteria, and other clinical parameters such as tumor size, level of invasiveness, and immunochemical characterization of the hormonal receptors (ER and PR) and HER2. The knowledge of breast cancer at the molecular level has allowed the identification of five different subtypes of breast tumors according to their gene expression pattern. This classification has had a major impact on the advances in breast cancer research [14,16], and the subgroups identified are known as luminal A, luminal B, basal-like, HER2 positive, and normal-like tumors. Later on these subgroups were extended to include a sixth one known as the "claudin-low" subtype, based on the low expression levels of the claudin genes and by its highly mesenchymal and invasive properties [77]. Together, the basal-like and claudin-low tumors represent the subset of tumors known as "TNBCs" because they lack both the expression of ER and PR receptors (ER-/PR-), as well as the expression and amplification of HER2 (HER2-) [78].

Currently, primary breast tumors can be diagnosed and assigned to these intrinsic subtypes using a 50-gene transcriptional signature known as PAM50 [19,20,22]. All these molecular subtypes are classified based on different levels of gene expression, and in breast cancer a significant alteration of gene expression patterns has proven to result from aberrant DNA methylation [24,39,40,72]. Indeed, since DNA methylation changes are reasonably critical components of the molecular mechanism involved in gene expression regulation of breast cancer cells, the specific subtypes of breast cancer might be expected to show distinct DNA methylation stages. In a recent study, Roll et al. classified about 950 primary breast tumors based on DNA hypermethylation status, identifying a strong association between aberrant DNA hypermethylation patterns and the basal-like and claudin-low (triple-negative) breast cancer subtypes [39].

The results of a previously reported analysis of the methylation status of the five molecular subtypes showed that only luminal A, luminal B, and basal-like tumors presented different methylation patterns, whereas tumors with the normal-like and HER2-amplified molecular subtypes did not present any distinct methylation profile [79]. Moreover, out of all of them, the luminal B tumors were the ones with the highest frequency of genes with a specific methylation pattern associated to the molecular subtype. That notwithstanding, little is known about the origins of the methylation patterns in the different molecular subtypes, even though it has been suggested that they reflect the different cellular origins [26,79]. Recently, Stefansson et al. defined two DNA methylation-based subtypes, associated with unfavorable clinical factors and worse survival, called Epi-LumB and Epi-Basal, and that few selected proxy markers can be used to detect the distinct DNA methylation-based subtypes [80]. In summary, similar to the gene expression profile tumor classification, DNA methylation signatures involving multiple variations in gene expression will be able to distinguish different subtypes of breast cancer and their prognosis with precision.

3.2.4 DNA Methylation Biomarkers in Breast Cancer

DNA methylation is an emerging field of biomarkers in many different tumors and specifically in breast cancer. There are a large number of genes that have been found inactivated in breast cancer due to promoter methylation. These genes regulate and control many different biological processes such as cell cycle regulation, apoptosis, tissue invasion and metastasis, DNA repair, and hormone receptor signaling, among others [24,39,50,57]. In addition, many of them are used to generate methylation profiles of breast tumors that are significantly associated with known prognostic factors and biomarkers of possible therapeutic response [41,81]. Due to the exponential increase of these methylated genes in breast cancer, a select list of the most frequently methylated ones is shown in Table 1. These genes are mainly tumor suppressor and other cancer-related genes. They have been found hypo- or hypermethylated in primary tumors, and some of them have been described previously as mutated in the germ lines of patients with inheritable cancers (e.g., *CDH1, p16INK4A/CDKN2A, RB, BRCA1*) [12,69,82,83].

The use of easily accessible body fluids like serum or plasma, the so-called "liquid biopsies," is a new and rapidly emerging field in biomarker discovery and cancer prediction, where novel methylated biomarkers are being described in breast cancer [84–87]. The main principle in this diagnostic system for early breast cancer is based on evidence that there is circulating-free DNA (cfDNA) released into the bloodstream from primary and metastatic tumors due to necrosis or apoptosis of the tumor cells [88]. Indeed, several biomarkers already described as methylated in breast tumors have been evaluated alone or in combination in serum. Nevertheless, although several potential DNA methylation biomarkers have been reported, and the number of groups working in this emerging field is growing exponentially, as of today none of these described biomarkers has reached clinical practice [87].

DNA methylation in breast cancer influences complex gene networks, rather than single genes, affecting distinct biological processes [24]. One example of gene deregulation is in the phosphatidylinositol 3-kinase (PI3K) pathway. Gene mutations of the components of this pathway occur in almost 70% of breast cancers [89]. The PI3K pathway is important in controlling several normal cellular functions that are also critical for tumorigenesis, including cellular proliferation, growth, survival, and mobility. One of these functions is autophagy, which plays an important role in cell differentiation, apoptosis, and the maintenance of cellular homeostasis [90]. In some studies, it has been shown that *PIK3C3* is hypermethylated in breast cancer cells, and the aberrant inhibition of PIK3C3 suppresses the correct autophagy of breast tissues inducing tumorigenesis [53].

Although previously explained, aberrant DNA methylation may affect normal cell pathways and regulation inducing different mechanisms of tumorigenesis. Moreover, methylation is implicated in the epithelial-to-mesenchymal transition (EMT). This transition is the most important factor for breast tumor cells in order to invade surrounding tissues in the body producing secondary tumors and metastasis. Some studies have shown that 90% of breast cancer deaths from solid tumors are the result of metastases, and not primary tumors [91]. Epithelial cells have a high level of plasticity, an important factor in cancer metastasis. The EMT starts with the loss of cell adhesion properties, cell–cell contact, remodeling the cell-matrix adhesion and epithelial markers, and acquiring invasive skills. In this way, they achieve the mesenchymal phenotype. These cells cross matrix and vascular tissue spread along the body to different organs and colonize new hospitable body regions [92].

Although much has been written about the EMT process in promoting tumor invasion and

TABLE 1 Methylated Genes in Breast Cancer

Gene	Description	Function	Methylation status
CDH13	Cadherin 13	Cell invasion/metastasis	Hyper
CST6	Cystatin E/M		Hyper
SYK	Spleen tyrosine kinase		Hyper
BCSG1	Breast cancer-specific gene 1 protein		Hypo
CAV1	Caveolin 1		Hypo
CDH1	Cadherin 1		Hypo
UPA	Plasminogen activator, urokinase		Hypo
CDH3	Cadherin 3		Hypo
BRCA1	Breast cancer 1, early onset	DNA repair	Hyper
MGMT	O-6-methylguanine-DNA methyltransferase		Hyper
MLH1	MutL Homolog1		Hyper
RAD9	RAD9 homolog A (*Schizosaccharomyces pombe*)		Hyper
AK5	Adenylate kinase 5	Cell cycle regulation	Hyper
SFN	Stratifin		Hyper
FOX2A			Hyper
RAR	Retinoic acid receptor		Hyper
CCND2	Cyclin D2		Hyper
RUNX3	Runt-related transcription factor 3		Hyper
ER	Estrogen receptor		Hyper
SFRP1	Secreted frizzles-related protein 1		Hyper
RASSF1A	RAS-association domain family member 1		Hyper
CDKN1C	Cyclin-dependent kinase inhibitor 1C		Hyper
CDKN2A	Cyclin-dependent kinase inhibitor 2A		Hyper
WIF1	WNT inhibitor factor 1		Hyper
PR	Progesterone receptor		Hyper
WRN	Werner syndrome. RecQ helicase-like		Hyper
APC	Adenomatous polyposis coli	Apoptosis	Hyper
BCL2	B cell CL/lymphoma 2		Hyper
DAPK	Death-associated protein kinase		Hyper

Continued

TABLE 1 Methylated Genes in Breast Cancer—cont'd

Gene	Description	Function	Methylation status
DCC	DCC netrin 1 receptor—Deleted in colorectal carcinoma		Hyper
HIC1	Hypermethylated in cancer 1		Hyper
HOXA5	Homeobox A5		Hyper
TMS1			Hyper
TWIST	Twist family bHLH transcription factor 1		Hyper
LDLRAP1	Low-density lipoprotein receptor adaptor protein 1	Cellular homeostasis	Hyper
GPC3	Glypican 3		Hyper
HOXD11	Homeobox D11		Hyper
LAMA3	Laminin, alpha 3		Hyper
LAMB3	Laminin, beta 3		Hyper
ROBO1	Roundabout, axon guidance receptor, homolog 1 (*Drosophila*)		Hyper
LAMC2	Laminin, gamma 2		Hyper

metastatic dissemination, less has been published about the opposite process, mesenchymal-to-epithelial transition (MET), which is as important to colonizing and developing new distant metastases as is EMT [93,94]. A combination of genetic and epigenetic aberrations leads to the activation of the tightly regulated cellular networks of EMT and MET in neoplastic cells, producing the manifestation of metastatic processes. Deregulation of heritable, fine-tuned epigenetic patterns inhibits and activates different factors that will eventually develop into tumors [95].

There are many examples of aberrant methylation patterns of genes implicated in EMT and MET processes, and the downregulation of E-cadherin by promoter hypermethylation is one of the most significant in breast cancer cells [39]. Another previously mentioned, well-known example is the loss of ER expression, an important regulator of proliferation and cell differentiation in mammary epithelial cells [61]. The loss of ER is related to a poorer prognosis due to the dedifferentiation of breast tumors. The main cause of ER downregulation, an increase in aggressiveness and proliferation in breast cancer, is the *ESR1* promoter hypermethylation [50,64,65]. Another example of an EMT aberrantly methylated gene is the high-temperature requirement A serine peptidase 1 (HtrA1). The downregulation of this gene is due to a combination of several epigenetic factors like DNA hypermethylation and histone deacetylation. The decrease in the expression of HrtA1 is related to higher tumor grade and indicates a poor prognosis in patients with breast cancer [96].

4. DNA METHYLATION AS A THERAPEUTIC TARGET

An increasing number of treatments are currently under research to unravel the problems of aberrant methylation. The investigation of the different groups has proven that the aberrant DNA hypermethylation observed in cell lines is actually identifiable among primary tumors and their metastasis [36,39], whereas other groups have already proven that breast cancer cell lines can be used as good models for evaluating the response of therapeutic agents [97]. Regulatory methylation enzymes are important targets for cancer therapy, and furthermore, different drugs with epigenetic activity, like DNA methylation, are already approved for the treatment of cancer patients. In fact, demethylating agents and DNMT inhibitors have been under clinical research for several decades since Jones and Taylor described in 1980 (see time line in Figure 1) that the nucleoside analog 5-azacytidine (5-aza-CR) induced marked changes in the differentiated state of cultured mouse embryo cells, and it also inhibited the methylation of newly synthesized DNA [98]. Nucleoside analogs are transformed into deoxynucleotide triphosphates and are incorporated into the DNA sequence replacing the cytosine in the process of replication, inhibiting the DNMTs and targeting them for degradation. Low doses of these DNA methylation inhibitors do not affect cell proliferation.

Currently, some of these inhibitors have been approved by the US Food and Drug Administration (FDA) for the treatment of different cancer types. azacitidine (Vidaza) and decitabine (Dacogen) are used in myelodysplastic diseases and leukemia, and vorinostat and romidepsin, two pan-HDAC inhibitors, are used for cutaneous T-cell lymphomas [92,99]. Additionally, DNMT inhibitors have been shown to sensitize breast cancer cell lines to the antibiotic doxorubicin, the most common chemotherapeutic agent used in breast cancer treatment. In fact, Sandhu et al. showed that breast cancer cell lines with aberrant DNA hypermethylation will respond better to chemotherapy after the inhibition of DNMTs through 5-aza-2′deoxycytidine, suggesting that combined epigenetic and cytotoxic treatment improves the efficacy of breast cancer chemotherapy [100]. In concordance with these results, recent studies have shown that the methylation status of cell cycle genes correlates with the response of treatment with doxorubicin in patients with locally advanced breast tumors. This suggests that methylation patterns of cell cycle genes may be potential predictive markers of anthracycline sensitivity [41].

The problem of DNMT inhibitors is the effect that these drugs may have in normal cells. The lack of methylation in normal cells causes hypomethylation, the activation of silenced genes like oncogenes, and genomic instability, and in the long term, it may have carcinogenic properties. As previously described, some studies have described *ESR1* as downregulated due to promoter hypermethylation. Treating breast cancer cells with a demethylating agent (aza-2-deoxycytidine) shows a partial reactivation of the ER expression [65]. By treating patients with downregulated E-cadherin by promoter hypermethylation with the same agent, a partial restoration of E-cadherin expression was observed, decreasing the effect of the EMT process, and in consequence, reducing the invasive properties of breast tumors and their capacity to metastasize to other tissues [101,102].

Combinations of different drugs have been used as a strategy to overcome chemoresistance inhibiting DNA methylation and histone deacetylase, respectively [103]. Moreover, treatments specific to reactivating silenced ER in the subgroup of TNBC is used to make tumor cells susceptible to responding to anti-ER therapies like tamoxifen by using a combination of demethylating inhibitors and histone deacetylase inhibitors [92]. New epigenetic drugs targeting DNA methylation are in development for the treatment of breast cancer by trying to target tumor-specific genes. Second-generation drugs integrate the knowledge of cellular control and metabolic pathways [92].

5. CONCLUDING REMARKS AND FUTURE PROSPECTS

Understanding the molecular consequences of normal and aberrant DNA methylation patterns in breast cancer remains a fundamental area of interest, especially because hypomethylating agents are one of the few epigenetic therapies with FDA approval that are currently used clinically. Forthcoming genome-wide epigenetic works will allow the discovery of changes in breast cancer-related genes identifying DNA methylation signatures. By integrating all these emerging information with the current and future data from genomics and transcriptomic fields, the understanding of breast tumorigenesis will expand considerably, permitting the use of that information as a better diagnostic or predictive marker and therapeutic response predictor in breast cancer research and treatment.

LIST OF ABBREVIATIONS

BC Breast cancer
BRCA1 Breast cancer susceptibility gene 1
CDH1 Cadherin-1 gene, also known as E-cadherin
cDNA Complementary deoxyribonucleic acid
cfDNA Circulating-free deoxyribonucleic acid
CH3 Methyl group
CpG (islands) Regions of DNA where cytosine occurs next to a guanine nucleotide and which are separated by only one phosphate.
DNA Deoxyribonucleic acid
DNMT DNA methyltransferase
EMT Epithelial-to-mesenchymal transition
ER Estrogen receptor
ESR1 Estrogen receptor gene 1
FDA Food and Drug Administration
GC Guanine–cytosine
HDAC Histone deacetylase
HER2/Her-2 (ErbB-2) Human epidermal growth factor 2 receptor
HtrA1 High-temperature requirement A serine peptidase 1
MET Mesenchymal-to-epithelial transition
OMIM *Online Mendelian Inheritance in Man*
p16INK4A/CDKN2A Cyclin-dependent kinase inhibitor 2A gene
PAM50 Gene signature that measures the expression levels of 50 genes in a surgically resected breast cancer sample to classify a tumor as one of five intrinsic subtypes (luminal A, luminal B, HER2-enriched, basal-like, and normal-like)
PI3K Phosphatidylinositol 3-kinase
PIK3C3 Phosphatidylinositol 3-kinase, catalytic subunit type 3
PR Progesterone receptor
RB Retinoblastoma gene
TNBC Triple-negative breast cancer

References

[1] Forouzanfar MH, Foreman KJ, Delossantos AM, Lozano R, Lopez AD, Murray CJ, et al. Breast and cervical cancer in 187 countries between 1980 and 2010: a systematic analysis. Lancet October 22, 2011;378(9801):1461–84.
[2] Wang F, Liu J, Liu L, Ma Z, Gao D, Zhang Q, et al. The status and correlates of depression and anxiety among breast-cancer survivors in Eastern China: a population-based, cross-sectional case-control study. BMC Public Health April 8, 2014;14:326.
[3] Waddington CH. The epigenotype. 1942. Int J Epidemiol February 2012;41(1):10–3.
[4] Wyatt GR. Recognition and estimation of 5-methylcytosine in nucleic acids. Biochem J May 1951;48(5):581–4.
[5] Watson JD, Crick FH. Molecular structure of nucleic acids; a structure for deoxyribose nucleic acid. Nature April 25, 1953;171(4356):737–8.
[6] Gold M, Gefter M, Hausmann R, Hurwitz J. Methylation of DNA. J Gen Physiol July 1966;49(6):5–28.
[7] Feinberg AP, Vogelstein B. Hypomethylation of ras oncogenes in primary human cancers. Biochem Biophys Res Commun February 28, 1983;111(1):47–54.
[8] Feinberg AP, Vogelstein B. Hypomethylation distinguishes genes of some human cancers from their normal counterparts. Nature January 6, 1983;301(5895): 89–92.
[9] Feinberg AP, Tycko B. The history of cancer epigenetics. Nat Rev Cancer February 2004;4(2):143–53.
[10] Semba K, Kamata N, Toyoshima K, Yamamoto T. A v-erbB-related protooncogene, c-erbB-2, is distinct from the c-erbB-1/epidermal growth factor-receptor gene and is amplified in a human salivary gland adenocarcinoma. Proc Natl Acad Sci USA October 1985;82(19):6497–501.
[11] Miki Y, Swensen J, Shattuck-Eidens D, Futreal PA, Harshman K, Tavtigian S, et al. A strong candidate for the breast and ovarian cancer susceptibility gene BRCA1. Science October 7, 1994;266(5182):66–71.
[12] Dobrovic A, Simpfendorfer D. Methylation of the BRCA1 gene in sporadic breast cancer. Cancer Res August 15, 1997;57(16):3347–50.
[13] Yoder JA, Soman NS, Verdine GL, Bestor TH. DNA (cytosine-5)-methyltransferases in mouse cells and tissues. Studies with a mechanism-based probe. J Mol Biol July 18, 1997;270(3):385–95.
[14] Perou CM, Sorlie T, Eisen MB, van de Rijn M, Jeffrey SS, Rees CA, et al. Molecular portraits of human breast tumours. Nature August 17, 2000;406(6797):747–52.
[15] Sorlie T, Tibshirani R, Parker J, Hastie T, Marron JS, Nobel A, et al. Repeated observation of breast tumor subtypes in independent gene expression data sets. Proc Natl Acad Sci USA July 8, 2003;100(14):8418–23.

[16] Sorlie T. Molecular portraits of breast cancer: tumour subtypes as distinct disease entities. Eur J Cancer December 2004;40(18):2667–75.

[17] van 't Veer LJ, Dai H, van de Vijver MJ, He YD, Hart AA, Mao M, et al. Gene expression profiling predicts clinical outcome of breast cancer. Nature January 31, 2002;415(6871):530–6.

[18] van de Vijver MJ, He YD, van't Veer LJ, Dai H, Hart AA, Voskuil DW, et al. A gene-expression signature as a predictor of survival in breast cancer. N Engl J Med December 19, 2002;347(25):1999–2009.

[19] Parker JS, Mullins M, Cheang MC, Leung S, Voduc D, Vickery T, et al. Supervised risk predictor of breast cancer based on intrinsic subtypes. J Clin Oncol March 10, 2009;27(8):1160–7.

[20] Bastien RR, Rodriguez-Lescure A, Ebbert MT, Prat A, Munarriz B, Rowe L, et al. PAM50 breast cancer subtyping by RT-qPCR and concordance with standard clinical molecular markers. BMC Med Genomics 2012;5:44.

[21] Prat A, Bianchini G, Thomas M, Belousov A, Cheang MC, Koehler A, et al. Research-based PAM50 subtype predictor identifies higher responses and improved survival outcomes in HER2-positive breast cancer in the NOAH study. Clin Cancer Res January 15, 2015;20(2):511–21.

[22] Prat A, Parker JS, Fan C, Perou CM. PAM50 assay and the three-gene model for identifying the major and clinically relevant molecular subtypes of breast cancer. Breast Cancer Res Treat August 2012;135(1):301–6.

[23] El Helou R, Wicinski J, Guille A, Adelaide J, Finetti P, Bertucci F, et al. Brief reports: a distinct DNA methylation signature defines breast cancer stem cells and predicts cancer outcome. Stem Cells November 2014;32(11):3031–6.

[24] Szyf M. DNA methylation signatures for breast cancer classification and prognosis. Genome Med 2012;4(3):26.

[25] Anjum S, Fourkala EO, Zikan M, Wong A, Gentry-Maharaj A, Jones A, et al. A BRCA1-mutation associated DNA methylation signature in blood cells predicts sporadic breast cancer incidence and survival. Genome Med 2014;6(6):47.

[26] Fleischer T, Frigessi A, Johnson KC, Edvardsen H, Touleimat N, Klajic J, et al. Genome-wide DNA methylation profiles in progression to in situ and invasive carcinoma of the breast with impact on gene transcription and prognosis. Genome Biol 2014;15(8):435.

[27] Bird A. DNA methylation patterns and epigenetic memory. Genes Dev January 1, 2002;16(1):6–21.

[28] Saxonov S, Berg P, Brutlag DL. A genome-wide analysis of CpG dinucleotides in the human genome distinguishes two distinct classes of promoters. Proc Natl Acad Sci USA January 31, 2006;103(5):1412–7.

[29] Jabbari K, Bernardi G. Cytosine methylation and CpG, TpG (CpA) and TpA frequencies. Gene May 26, 2004;333:143–9.

[30] Li E, Bestor TH, Jaenisch R. Targeted mutation of the DNA methyltransferase gene results in embryonic lethality. Cell June 12, 1992;69(6):915–26.

[31] Goll MG, Kirpekar F, Maggert KA, Yoder JA, Hsieh CL, Zhang X, et al. Methylation of tRNAAsp by the DNA methyltransferase homolog Dnmt2. Science January 20, 2006;311(5759):395–8.

[32] Bourc'his D, Xu GL, Lin CS, Bollman B, Bestor TH. Dnmt3L and the establishment of maternal genomic imprints. Science December 21, 2001;294(5551):2536–9.

[33] Bestor TH. The DNA methyltransferases of mammals. Hum Mol Genet October 2000;9(16):2395–402.

[34] Okano M, Bell DW, Haber DA, Li E. DNA methyltransferases Dnmt3a and Dnmt3b are essential for de novo methylation and mammalian development. Cell October 29, 1999;99(3):247–57.

[35] Liang G, Chan MF, Tomigahara Y, Tsai YC, Gonzales FA, Li E, et al. Cooperativity between DNA methyltransferases in the maintenance methylation of repetitive elements. Mol Cell Biol January 2002;22(2):480–91.

[36] Roll JD, Rivenbark AG, Jones WD, Coleman WB. DNMT3b overexpression contributes to a hypermethylator phenotype in human breast cancer cell lines. Mol Cancer 2008;7:15.

[37] Yu Z, Xiao Q, Zhao L, Ren J, Bai X, Sun M, et al. DNA methyltransferase 1/3a overexpression in sporadic breast cancer is associated with reduced expression of estrogen receptor-alpha/breast cancer susceptibility gene 1 and poor prognosis. Mol Carcinog January 25, 2014;54(9):707–19.

[38] Girault I, Tozlu S, Lidereau R, Bieche I. Expression analysis of DNA methyltransferases 1, 3A, and 3B in sporadic breast carcinomas. Clin Cancer Res October 1, 2003;9(12):4415–22.

[39] Roll JD, Rivenbark AG, Sandhu R, Parker JS, Jones WD, Carey LA, et al. Dysregulation of the epigenome in triple-negative breast cancers: basal-like and claudin-low breast cancers express aberrant DNA hypermethylation. Exp Mol Pathol December 2013;95(3):276–87.

[40] Baylin SB, Esteller M, Rountree MR, Bachman KE, Schuebel K, Herman JG. Aberrant patterns of DNA methylation, chromatin formation and gene expression in cancer. Hum Mol Genet April 2001;10(7):687–92.

[41] Klajic J, Busato F, Edvardsen H, Touleimat N, Fleischer T, Bukholm IR, et al. DNA methylation status of key cell cycle regulators such as CDKNA2/p16 and CCNA1 correlates with treatment response to doxorubicin and 5-fluorouracil in locally advanced breast tumors. Clin Cancer Res October 7, 2014;20(24):6357–66.

[42] Klajic J, Fleischer T, Dejeux E, Edvardsen H, Warnberg F, Bukholm I, et al. Quantitative DNA methylation analyses reveal stage dependent DNA methylation and association to clinico-pathological factors in breast tumors. BMC Cancer 2013;13:456.
[43] Ronneberg JA, Fleischer T, Solvang HK, Nordgard SH, Edvardsen H, Potapenko I, et al. Methylation profiling with a panel of cancer related genes: association with estrogen receptor, TP53 mutation status and expression subtypes in sporadic breast cancer. Mol Oncol February 2011;5(1):61–76.
[44] Sandoval J, Esteller M. Cancer epigenomics: beyond genomics. Curr Opin Genet Dev February 2012;22(1): 50–5.
[45] Jovanovic J, Ronneberg JA, Tost J, Kristensen V. The epigenetics of breast cancer. Mol Oncol June 2010;4(3):242–54.
[46] Huang Y, Nayak S, Jankowitz R, Davidson NE, Oesterreich S. Epigenetics in breast cancer: what's new? Breast Cancer Res 2011;13(6):225.
[47] Ziller MJ, Hansen KD, Meissner A, Aryee MJ. Coverage recommendations for methylation analysis by whole-genome bisulfite sequencing. Nat Methods November 2, 2014;12(3):230–2.
[48] Hsu YW, Huang RL, Lai HC. MeDIP-on-Chip for methylation profiling. Methods Mol Biol 2015;1249:281–90.
[49] Rhee JK, Kim K, Chae H, Evans J, Yan P, Zhang BT, et al. Integrated analysis of genome-wide DNA methylation and gene expression profiles in molecular subtypes of breast cancer. Nucleic Acids Res October 2013;41(18):8464–74.
[50] Jovanovic J, Ronneberg JA, Tost J, Kristensen V. The epigenetics of breast cancer. Mol Oncol June 2010;4(3):242–54.
[51] Straussman R, Nejman D, Roberts D, Steinfeld I, Blum B, Benvenisty N, et al. Developmental programming of CpG island methylation profiles in the human genome. Nat Struct Mol Biol May 2009;16(5):564–71.
[52] Esteller M. Cancer epigenomics: DNA methylomes and histone-modification maps. Nat Rev Genet April 2007;8(4):286–98.
[53] Wang F, Yang Y, Fu Z, Xu N, Chen F, Yin H, et al. Differential DNA methylation status between breast carcinomatous and normal tissues. Biomed Pharmacother July 2014;68(6):699–707.
[54] Esteller M. Aberrant DNA methylation as a cancer-inducing mechanism. Annu Rev Pharmacol Toxicol 2005;45:629–56.
[55] Schmutte C, Fishel R. Genomic instability: first step to carcinogenesis. Anticancer Res November–December 1999;19(6A):4665–96.
[56] Veeck J, Esteller M. Breast cancer epigenetics: from DNA methylation to microRNAs. J Mammary Gland Biol Neoplasia March 2010;15(1):5–17.
[57] Esteller M. Epigenetics in cancer. N Engl J Med March 13, 2008;358(11):1148–59.
[58] Tan J, Gu Y, Zhang X, You S, Lu X, Chen S, et al. Hypermethylation of CpG islands is more prevalent than hypomethylation across the entire genome in breast carcinogenesis. Clin Exp Med February 2013;13(1):1–9.
[59] Kuiper GG, Carlsson B, Grandien K, Enmark E, Haggblad J, Nilsson S, et al. Comparison of the ligand binding specificity and transcript tissue distribution of estrogen receptors alpha and beta. Endocrinology March 1997;138(3):863–70.
[60] Chia S, Bryce C, Gelmon K. The 2000 EBCTCG overview: a widening gap. Lancet May 14–20, 2005;365(9472): 1665–6.
[61] Kerdivel G, Flouriot G, Pakdel F. Modulation of estrogen receptor alpha activity and expression during breast cancer progression. Vitam Horm 2013;93:135–60.
[62] Hervouet E, Cartron PF, Jouvenot M, Delage-Mourroux R. Epigenetic regulation of estrogen signaling in breast cancer. Epigenetics March 2013;8(3):237–45.
[63] Martinez-Galan J, Torres-Torres B, Nunez MI, Lopez-Penalver J, Del Moral R, Ruiz De Almodovar JM, et al. ESR1 gene promoter region methylation in free circulating DNA and its correlation with estrogen receptor protein expression in tumor tissue in breast cancer patients. BMC Cancer 2014;14:59.
[64] Ottaviano YL, Issa JP, Parl FF, Smith HS, Baylin SB, Davidson NE. Methylation of the estrogen receptor gene CpG island marks loss of estrogen receptor expression in human breast cancer cells. Cancer Res May 15, 1994;54(10):2552–5.
[65] Ferguson AT, Lapidus RG, Baylin SB, Davidson NE. Demethylation of the estrogen receptor gene in estrogen receptor-negative breast cancer cells can reactivate estrogen receptor gene expression. Cancer Res June 1, 1995;55(11):2279–83.
[66] Rizzolo P, Silvestri V, Falchetti M, Ottini L. Inherited and acquired alterations in development of breast cancer. Appl Clin Genet 2011;4:145–58.
[67] Larsen MJ, Kruse TA, Tan Q, Laenkholm AV, Bak M, Lykkesfeldt AE, et al. Classifications within molecular subtypes enables identification of BRCA1/BRCA2 mutation carriers by RNA tumor profiling. PLoS One May 21, 2013;8(5):e64268.
[68] Easton DF, Bishop DT, Ford D, Crockford GP. Genetic linkage analysis in familial breast and ovarian cancer: results from 214 families. The Breast Cancer Linkage Consortium. Am J Hum Genet April 1993;52(4):678–701.
[69] Werner H, Bruchim I. IGF-1 and BRCA1 signalling pathways in familial cancer. Lancet Oncol December 2012;13(12):e537–44.
[70] Silver DP, Livingston DM. Mechanisms of BRCA1 tumor suppression. Cancer Discov August 2012;2(8):679–84.

[71] Bal A, Verma S, Joshi K, Singla A, Thakur R, Arora S, et al. BRCA1-methylated sporadic breast cancers are BRCA-like in showing a basal phenotype and absence of ER expression. Virchows Arch September 2012;461(3):305–12.

[72] Buyru N, Altinisik J, Ozdemir F, Demokan S, Dalay N. Methylation profiles in breast cancer. Cancer Invest March 2009;27(3):307–12.

[73] Esteller M, Silva JM, Dominguez G, Bonilla F, Matias-Guiu X, Lerma E, et al. Promoter hypermethylation and BRCA1 inactivation in sporadic breast and ovarian tumors. J Natl Cancer Inst April 5, 2000;92(7): 564–9.

[74] Wei M, Grushko TA, Dignam J, Hagos F, Nanda R, Sveen L, et al. BRCA1 promoter methylation in sporadic breast cancer is associated with reduced BRCA1 copy number and chromosome 17 aneusomy. Cancer Res December 1, 2005;65(23):10692–9.

[75] Stefansson OA, Jonasson JG, Johannsson OT, Olafsdottir K, Steinarsdottir M, Valgeirsdottir S, et al. Genomic profiling of breast tumours in relation to BRCA abnormalities and phenotypes. Breast Cancer Res 2009;11(4):R47.

[76] Otani Y, Miyake T, Kagara N, Shimoda M, Naoi Y, Maruyama N, et al. BRCA1 promoter methylation of normal breast epithelial cells as a possible precursor for BRCA1-methylated breast cancer. Cancer Sci October 2014;105(10):1369–76.

[77] Prat A, Parker JS, Karginova O, Fan C, Livasy C, Herschkowitz JI, et al. Phenotypic and molecular characterization of the claudin-low intrinsic subtype of breast cancer. Breast Cancer Res 2010;12(5):R68.

[78] Bosch A, Eroles P, Zaragoza R, Vina JR, Lluch A. Triple-negative breast cancer: molecular features, pathogenesis, treatment and current lines of research. Cancer Treat Rev May 2010;36(3):206–15.

[79] Holm K, Hegardt C, Staaf J, Vallon-Christersson J, Jonsson G, Olsson H, et al. Molecular subtypes of breast cancer are associated with characteristic DNA methylation patterns. Breast Cancer Res 2010;12(3):R36.

[80] Stefansson OA, Moran S, Gomez A, Sayols S, Arribas-Jorba C, Sandoval J, et al. A DNA methylation-based definition of biologically distinct breast cancer subtypes. Mol Oncol November 5, 2014;9(3): 555–68.

[81] Shinozaki M, Hoon DS, Giuliano AE, Hansen NM, Wang HJ, Turner R, et al. Distinct hypermethylation profile of primary breast cancer is associated with sentinel lymph node metastasis. Clin Cancer Res March 15, 2005;11(6):2156–62.

[82] Grady WM, Willis J, Guilford PJ, Dunbier AK, Toro TT, Lynch H, et al. Methylation of the CDH1 promoter as the second genetic hit in hereditary diffuse gastric cancer. Nat Genet September 2000;26(1):16–7.

[83] Berman H, Zhang J, Crawford YG, Gauthier ML, Fordyce CA, McDermott KM, et al. Genetic and epigenetic changes in mammary epithelial cells identify a subpopulation of cells involved in early carcinogenesis. Cold Spring Harb Symp Quant Biol 2005;70:317–27.

[84] Kloten V, Becker B, Winner K, Schrauder MG, Fasching PA, Anzeneder T, et al. Promoter hypermethylation of the tumor-suppressor genes ITIH5, DKK3, and RASSF1A as novel biomarkers for blood-based breast cancer screening. Breast Cancer Res 2013;15(1):R4.

[85] Chimonidou M, Tzitzira A, Strati A, Sotiropoulou G, Sfikas C, Malamos N, et al. CST6 promoter methylation in circulating cell-free DNA of breast cancer patients. Clin Biochem February 2013;46(3):235–40.

[86] Fackler MJ, Lopez Bujanda Z, Umbricht C, Teo WW, Cho S, Zhang Z, et al. Novel methylated biomarkers and a robust assay to detect circulating tumor DNA in metastatic breast cancer. Cancer Res April 15, 2014;74(8):2160–70.

[87] Wittenberger T, Sleigh S, Reisel D, Zikan M, Wahl B, Alunni-Fabbroni M, et al. DNA methylation markers for early detection of women's cancer: promise and challenges. Epigenomics June 2014;6(3):311–27.

[88] Jahr S, Hentze H, Englisch S, Hardt D, Fackelmayer FO, Hesch RD, et al. DNA fragments in the blood plasma of cancer patients: quantitations and evidence for their origin from apoptotic and necrotic cells. Cancer Res February 15, 2001;61(4):1659–65.

[89] Morgensztern D, McLeod HL. PI3K/Akt/mTOR pathway as a target for cancer therapy. Anticancer Drugs September 2005;16(8):797–803.

[90] Klionsky DJ, Emr SD. Autophagy as a regulated pathway of cellular degradation. Science December 1, 2000;290(5497):1717–21.

[91] Nickel A, Stadler SC. Role of epigenetic mechanisms in epithelial-to-mesenchymal transition of breast cancer cells. Transl Res April 12, 2014;165(1):126–42.

[92] Nowsheen S, Aziz K, Tran PT, Gorgoulis VG, Yang ES, Georgakilas AG. Epigenetic inactivation of DNA repair in breast cancer. Cancer Lett January 28, 2014;342(2):213–22.

[93] Polyak K, Weinberg RA. Transitions between epithelial and mesenchymal states: acquisition of malignant and stem cell traits. Nat Rev Cancer April 2009;9(4):265–73.

[94] Tsai JH, Donaher JL, Murphy DA, Chau S, Yang J. Spatiotemporal regulation of epithelial-mesenchymal transition is essential for squamous cell carcinoma metastasis. Cancer Cell December 11, 2012;22(6):725–36.

[95] Jones PA, Baylin SB. The epigenomics of cancer. Cell February 23, 2007;128(4):683–92.

[96] Lehner A, Magdolen V, Schuster T, Kotzsch M, Kiechle M, Meindl A, et al. Downregulation of serine protease HTRA1 is associated with poor survival in breast cancer. PLoS One 2013;8(4):e60359.

[97] Daemen A, Griffith OL, Heiser LM, Wang NJ, Enache OM, Sanborn Z, et al. Modeling precision treatment of breast cancer. Genome Biol 2013;14(10):R110.

[98] Jones PA, Taylor SM. Cellular differentiation, cytidine analogs and DNA methylation. Cell May 1980; 20(1):85–93.

[99] Olsen EA, Kim YH, Kuzel TM, Pacheco TR, Foss FM, Parker S, et al. Phase IIb multicenter trial of vorinostat in patients with persistent, progressive, or treatment refractory cutaneous T-cell lymphoma. J Clin Oncol July 20, 2007;25(21):3109–15.

[100] Sandhu R, Rivenbark AG, Coleman WB. Enhancement of chemotherapeutic efficacy in hypermethylator breast cancer cells through targeted and pharmacologic inhibition of DNMT3b. Breast Cancer Res Treat January 2011;131(2):385–99.

[101] Graff JR, Gabrielson E, Fujii H, Baylin SB, Herman JG. Methylation patterns of the E-cadherin 5′ CpG island are unstable and reflect the dynamic, heterogeneous loss of E-cadherin expression during metastatic progression. J Biol Chem January 28, 2000;275(4):2727–32.

[102] Graff JR, Herman JG, Lapidus RG, Chopra H, Xu R, Jarrard DF, et al. E-cadherin expression is silenced by DNA hypermethylation in human breast and prostate carcinomas. Cancer Res November 15, 1995;55(22): 5195–9.

[103] Candelaria M, Gallardo-Rincon D, Arce C, Cetina L, Aguilar-Ponce JL, Arrieta O, et al. A phase II study of epigenetic therapy with hydralazine and magnesium valproate to overcome chemotherapy resistance in refractory solid tumors. Ann Oncol September 2007;18(9):1529–38.

CHAPTER

16

DNA Methylation in Obesity and Associated Diseases

Ana B. Crujeiras[1,2], *Angel Diaz-Lagares*[3]

[1]Laboratory of Molecular and Cellular Endocrinology, Instituto de Investigación Sanitaria (IDIS), Complejo Hospitalario Universitario de Santiago (CHUS) and Santiago de Compostela University (USC), Santiago de Compostela, Spain; [2]CIBER Fisiopatología de la Obesidad y la Nutrición (CIBERobn), Madrid, Spain; [3]Cancer Epigenetics and Biology Program (PEBC), Bellvitge Biomedical Research Institute (IDIBELL), Barcelona, Spain

OUTLINE

1. INTRODUCTION

Obesity (MIM #601665) is a multifactorial chronic disease whose etiology is an imbalance between the energy ingested with food and the energy expended. This imbalance is promoted by complex interactions between inadequate dietary habits, diminished physical exercise, and genetic background [1]. The excess of energy is stored in fat cells that enlarge and/or increase in number. This process results in an adipose tissue dysfunction that leads to a state of chronic low-grade

Epigenetic Biomarkers and Diagnostics
http://dx.doi.org/10.1016/B978-0-12-801899-6.00016-4

inflammation that promotes oxidative stress and the alterations of adipocyte-derived hormone secretion and cytokine synthesis [2]. This condition was proposed as the major player in the pathological consequence of obesity, as well as other metabolic disorders such as alterations in insulin sensitivity, blood pressure, and the plasma lipid profile. Together, these risk factors define the metabolic syndrome [3].

The prevalence of obesity and its associated codisorders is rapidly increasing worldwide and are considered nowadays as pandemic noncommunicable diseases, representing a major challenge for the health-care system [4]. The worldwide prevalence of obesity has nearly doubled in the last three decades. According to data from the National Health and Nutrition Examination Survey (NHANES), 35.7% of American adults are obese together with 17% of children and adolescents aged 2–19 years [5]. In Europe, over 50% of the population is overweight, and approximately 23% is obese. Despite Mediterranean diet, Spain presented in 2006/2007 15.5% of men and 15.2% of women with obesity. This prevalence is currently increasing and reached a 22.9% in 2010 [6]. Besides, abdominal obesity is strongly associated with cardiovascular disease and type 2 diabetes, even among patients with low body mass index (BMI, kg/m^2), as demonstrated by data that were generated during the International Day for the Evaluation of Abdominal obesity (IDEA) study [7]. In consequence, the risk for metabolic diseases such as type 2 diabetes and cardiovascular diseases is increased by obesity and overweight.

Many of the metabolic alterations are due to interplay between environmental, lifestyle, and genetic factors [8]. It is well known that physical inactivity and unhealthy dietary patterns are associated with a major impact on metabolic syndrome, diabetes, and obesity [9]. Unfortunately, despite intensive genetic research on these alterations, the underlying mechanisms and the pathogenesis are still poorly understood. In recent years, it has been widely accepted that epigenetic alterations may underlie metabolic alterations. In particular, there has been an increasing interest in the possible role of these modifications in the field of metabolic syndrome, diabetes mellitus, and obesity and in how lifestyle through diet and physical activity could be related to these changes [9].

In this chapter, we provide an overview of the recent findings in the research area of epigenetics and obesity and its codisorders, as well as the impact of lifestyle interventions on DNA methylation profiles to mitigate obesity, diabetes mellitus, and metabolic syndrome.

2. AN OVERVIEW OF EPIGENETICS: FOCUS ON DNA METHYLATION

The biological regulatory system through which the organism responds to environmental pressures is mediated by epigenetic modifications of the genome [10]. Epigenetics was first proposed by Waddington in 1942 and this term has been traditionally used to explain the phenotypic events that cannot be explained by genetic mechanisms [11]. Nowadays, epigenetics refers to mitotically and/or meiotically heritable changes in gene expression that occur without altering the DNA sequence [12]. This factor plays an important role in regulating the gene expression of many biological processes that take place throughout a person's lifetime, such as during embryogenesis and the phenotypic variations of genetically identical individuals [10]. Epigenetics can illustrate the reason why an organism produces many different cell types during its development, despite the fact that most of the cells in a multicellular organism share the same genetic information. The epigenetic machinery presents several levels of regulation (Figure 1): DNA methylation, posttranslational histone modifications, nucleosome positioning, and noncoding RNAs (ncRNAs) including microRNAs (miRNAs) and long noncoding RNAs

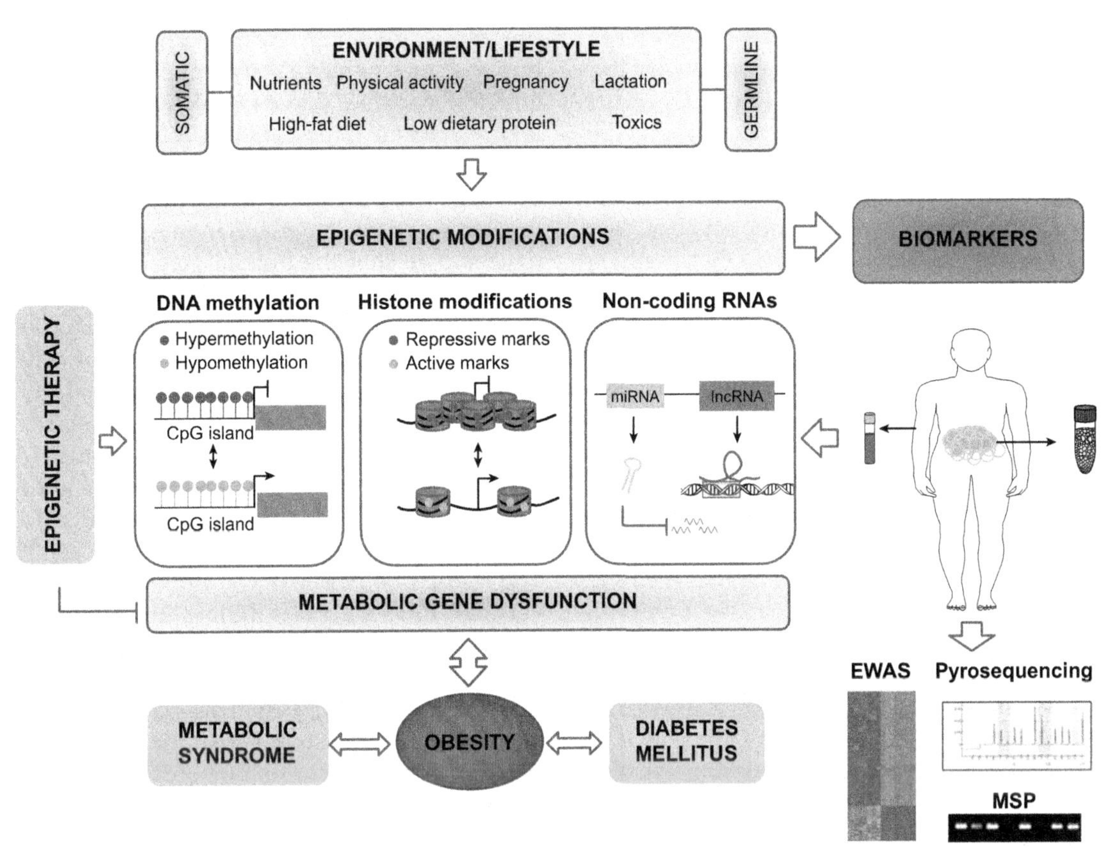

FIGURE 1 **Summary of the interplay between environment/lifestyle factors and metabolic diseases by means of the epigenetic machinery and its utility as epigenetic biomarkers.** Environment and lifestyle factors may contribute to metabolic disorders by means of the dysregulation of the epigenetic machinery. The alteration of the epigenetic mechanisms can modify the activity of metabolic-related genes triggering metabolic disorders such as obesity, diabetes mellitus, and metabolic syndrome. These epigenetic modifications can be used as clinical biomarkers for metabolic diseases. The samples can be obtained from patients' body fluids such as blood or urine and from adipose tissue biopsies. Epigenetic biomarkers could be analyzed by means of epigenome-wide association studies (EWAS) or using methylation-specific PCR (MSP) or pyrosequencing methods in candidate genes. The epigenetic modifications can be reverted by epigenetic drugs or functional foods, which could restore the right metabolic gene status in patients with metabolic diseases. miRNA, microRNA; lncRNA, long noncoding RNA; CpG, cytosine–phosphate–guanine.

(lncRNAs) [13,14]. Among these mechanisms, DNA methylation is the best-known epigenetic modification [15].

DNA methylation usually occurs in certain areas of the genome with a high density of cytosine–phosphate–guanine (CpG) dinucleotides named CpG islands (CGIs) leading to the silencing of both coding and noncoding genes [16,17]. Moreover, DNA methylation occurs in other genomic regions to maintain the conformation and integrity of chromosomes as well as to avoid potential damage by mobile genetic elements [10]. Recent array-based genome-wide methylation analyses have shown that DNA methylation can also affect areas of the genome with low CpG density increasingly far from CGIs: CGI shores (up to 2 kb from CGIs), shelves (2–4 kb from CGIs), and open sea (>4 kb from CGIs) [18].

These low-CpG-density regions may be especially important for gene regulation as it is reflected by the maintenance of small CpG clusters, despite the high mutation rates observed in CpG sites [19,20]. In addition, DNA methylation can affect not only intragenic but also genome intergenic regions increasing the level of complexity of the interplay between DNA methylation and gene expression regulation [21]. There are strong evidences that methylated CGIs at transcription start sites of promoters cannot initiate transcription after the DNA has been assembled into nucleosomes [22–24]. However, the issue of whether DNA methylation may not drive changes in gene expression but may be acting as a secondary stabilizing mechanism has long been a discussion in the field. Experiments by Lock et al. [25] clearly showed that methylation of the HPRT gene on the inactive X chromosome occurred after the chromosome had been inactivated. Therefore, methylation appeared to serve as a "lock" to reinforce a previously silenced state of X-linked genes [26]. This is an important concept when considering whether changes in DNA methylation are causal factors or a consequence of the disease.

The function of DNA methylation is intrinsically linked to the mechanisms for establishing, maintaining, and removing the methyl group. This process of DNA methylation is controlled by DNA methyltransferase enzymes (DNMTs). DNMT3A and DNMT3B are essential for de novo methylation, while DNMT1 maintains methylation patterns during cell division [27,28]. Establishment and maintenance of cell type-specific DNA methylation patterns are dependent on both methylation and demethylation. DNA demethylation is the process of the removal of a methyl group from nucleotides in DNA that can be a passive or active mechanism [29]. It has been generally accepted that passive DNA demethylation occurs by a reduction in activity or absence of DNMTs, whereas the mechanism of active DNA demethylation has been controversial during decades. Recently, it has been proposed that the DNA demethylation process is mediated by ten-eleven translocation enzymes, which catalyze the conversion of 5-methylcytosine (5mC) of DNA to 5-hydroxymethylcytosine (5hmC) [30].

On the other hand, it has been also shown that ncRNAs play an important role in controlling gene expression [31]. Genome-wide studies have shown that the human genome produces different types of regulatory ncRNAs [32], with miRNAs being the most widely studied. The miRNAs bind to specific regions of target mRNAs and mediate posttranscriptional gene silencing by blocking or degrading these transcripts [33]. Through this mechanism, miRNAs are involved in important processes including development, proliferation, cell differentiation, regulation of cell cycle, and apoptosis [31,34]. The number of ncRNAs identified is increasing very fast in recent years. Interestingly, lncRNAs represent a new emerged class of ncRNAs defined as transcripts longer than 200bp that lack protein coding [35]. This type of ncRNA presents important regulatory functions in the cell by means of chromatin organization and regulation of gene expression [36].

3. ENVIRONMENTAL AND LIFESTYLE FACTORS INFLUENCING DNA METHYLATION

All levels of epigenetic regulation appear to have wide-ranging effects on health. Aberrant epigenetic regulation has been described in many human diseases, including cancer [37], inflammatory processes [38], obesity [39], diabetes [40], and metabolic syndrome [41]. Alterations in epigenetic processes may induce long-term changes in gene function and metabolism that persist throughout the life course. Increasing evidences show that environmental and lifestyle factors including nutrition and physical activity may potentially alter the extent of epigenetic regulation. Therefore, epigenetics might explain the link between lifestyle and risk of disease [42]. These lifestyle factors may influence the regulation of the epigenetic machinery by means of different processes [43]. Most of the studies conducted so far have been centered on DNA methylation, whereas only a few investigations have studied lifestyle factors in relation to histone modifications and ncRNAs.

Altered DNA methylation has been shown to affect healthy aging and also to promote age-related health problems [44]. There are increasing evidences that lifestyle changes, including weight loss, can have an impact on DNA methylation and consequently in gene expression [45]. DNA methylation changes produce modifications in chromatin structure that are behind metabolic alterations leading to the phenomenon termed metabolic memory. The concept of metabolic memory has been introduced based on the observation that prolonged exposure to high glucose in diabetes deeply changes the human epigenome to the extent that alterations remain stable even in the absence of a hyperglycemic environment [46,47]. The underlying mechanisms of metabolic memory are still unclear, but DNA methylation has been suggested as one of the mechanisms that may maintain the reminiscence of the metabolic state and to contribute to the disease phenotype. Interestingly, nutrition in early life can alter the epigenome producing different phenotypes and alter disease susceptibilities, especially for impaired glucose metabolism [4]. It has been recently shown that lifestyle intervention resulted in a reduction in the risk of developing diabetes mellitus. This effect could be mediated by lifestyle changes as well as to the metabolic memory effect in glycemia. This benefit could also be explained by modulation of DNA methylation on glycemic control [46]. Therefore, DNA methylation represents an important epigenetic mechanism that underlies the metabolic alterations and it is influenced by environmental and lifestyle factors. However, not only these factors but also genetic background may have an impact on DNA methylation. Thus, while DNA methylation states are tightly maintained between genetically identical and related individuals, there remains a considerable epigenetic variation that may contribute to disease susceptibility. This fact indicates that maintenance of the methylation is influenced by the underlying DNA sequence [48,49].

Despite the inherent stability of epigenetic marks, all levels of epigenetic modifications are reversible [13,14]. In particular, DNA methylation may change from the state of hypermethylation to hypomethylation, and vice verse, induced by some different factors such as epigenetic drugs (e.g., the hypomethylating agent 5-aza-2′-deoxycytidine (5-aza-dC)) and healthy lifestyle habits [45,50]. In this line, developmental metabolic programming has shown to be potentially reversible. One major mechanism is that early developmental nutrition, such as pregnancy and lactation, can change the expression of some genes related to cell differentiation and glucose metabolism by means of DNA methylation modifications [51]. These periods of growth are very sensitive to nutrition changes. Therefore, the relationship between early-life nutrition and glucose metabolism alterations in later life, such as diabetes, could be explained by alterations in DNA methylation levels. In view of the reversibility of epigenetic mechanisms, it could be useful to discover novel epigenetic drugs for the prevention and intervention of glucose metabolism from the early stage of life to reduce the risk of these disorders in the future. In addition, lifestyle intervention based on functional foods and exercise may modify DNA methylation at a genome-wide level together with gene expression that could improve the phenotype associated with the disease [52].

4. DNA METHYLATION PATTERN AND OBESITY SUSCEPTIBILITY

There is a growing body of evidence that shows a relevant role of epigenetic marks with obesity and its codiseases susceptibility [39,52–55].

The specific epigenetic profile associated with obesity can be induced by environmental factors on somatic cells but also a number of studies demonstrate that ancestral environmental exposure can promote the epigenetic transgenerational inheritance of obesity through germ line epimutations (Figure 1), i.e., the germ line (sperm or ovocyte) transgenerational transmission of epigenetic marks that influence physiological parameters and disease, in the absence of direct environmental exposures [56]. For example, a recent study demonstrated the presence of differentially DNA methylated regions in the sperm of the F3 generation of male rodents with ancestral exposure to the insecticide dichlorodiphenyltrichloroethane (DDT) and a number of previously identified obesity-associated genes correlated with the epimutations identified [57]. This germ line transmission of epimutations associated with obesity in rodents was also observed after the exposure to plastics-derived endocrine disruptors (bisphenol-A, diethylhexyl phthalate, and dibutyl phthalate) [58] or jet fuel hydrocarbons [59]. In this line, there are studies that have assessed differences in DNA methylation of candidate genes in children in relation to maternal/paternal characteristics [55]. The intake of particular nutrient, such as methyl donors and vitamins, high-fat diet, or low dietary protein during the gestation can regulate the epigenetic mechanism involved in the pathogenesis of obesity and metabolic syndrome in the offsprings [52]. Duration of breastfeeding was also associated with lower DNA methylation levels of leptin (*LEP*), a nonimprinted gene implicated in appetite regulation and fat metabolism [60]. Differences in obesity characteristics and in DNA methylation profiles for genes involved in the regulation of glucose homeostasis and immune function were found between siblings born before and after maternal weight loss surgery [61].

Strikingly, therapeutic strategies for counteracting excess body weight are able to remodel DNA methylation profiles concomitant with the reduction of body weight. The DNA methylation and expression levels of several genes, which are related to metabolic processes and mitochondrial functions (e.g., peroxisome proliferator-activated receptor gamma, coactivator 1 alpha (*PGC-1α*) and pyruvate dehydrogenase kinase, isozyme 4 (*PDK4*)), are altered in the skeletal muscle of obese people and after Roux-en-Y gastric bypass (RYGB), a type of weight loss surgery, were normalized to levels observed in normal-weight, healthy controls [62]. A 6-month intervention of exercise can induce changes in the genome-wide DNA methylation patterns of human adipose tissue that potentially affect adipocyte metabolism [63]. Similar results have been observed in the muscle of patients with type 2 diabetes. In these patients, exercise altered the DNA methylation status of genes involved in retinol metabolism and calcium signaling pathways that are known to have functions in the muscle and type 2 diabetes [64]. The DNA methylation levels of specific genes in human leukocytes [65] and adipose tissue can also be altered by caloric restriction interventions [66]. Additionally, it has been demonstrated that responses to weight loss treatments can be influenced and predicted by DNA methylation status prior to beginning treatment; i.e., differences in the DNA methylation patterns of specific genes have been found between low and high responders to weight loss therapy [65,66] as well as between patients who are prone to regain lost weight and those who are able to maintain weight loss during a free-living period after dieting [67].

5. EPIGENETICS AS A LINK BETWEEN OBESITY AND ITS CODISEASES

Obesity is strongly associated with changes in the physiological function of adipose tissue that lead to its dysregulation, which in turn results in increased systemic levels of proinflammatory cytokines such as tumor necrosis factor α (TNFα), interleukin-6 (IL-6), C-reactive protein, and matrix metalloprotienases [68]. The chronic inflammation that is produced by adipocyte dysfunction induces increases in the release and accumulation of reactive oxygen species (ROS). Additionally, obesity alone induces an excessive generation of ROS due to inefficient energy metabolism [69]. Such obesity-related inflammatory and oxidative stress has been hypothesized to be a link between obesity and its comorbidities [70].

In this context, we hypothesize that one of the plausible mechanisms by which the obesity state may induce higher susceptibility to metabolic diseases is by contributing to DNA alterations that result in the regulation of metabolic function (Figure 1). One major factor that contributes to this metabolic regulation is oxidative stress throughout DNA damage induction (e.g., base modifications, deletions, strand breakages, and chromosomal rearrangements) that reduces the ability of DNA to be methylated by DNMTs resulting in global hypomethylation [71,72]. Moreover, oxidative damage has also been implicated in the regulation of histone modifications and miRNA expression [73,74]. Inflammation also induces epigenetic alterations in tissues that are associated with disease manifestations, as revealed by recent therapeutic interventions utilizing histone deacetylase and DNMT inhibitors. Certain anti-inflammatory dietary elements have effects on DNA methylation and chromatin remodeling, and the actions of several inflammatory-related transcription factors such as nuclear factor kappa B (NF-κB) can be interfered [52].

Several enzymes involved in the epigenetic modifications utilize cofactors or substrates that are crucial metabolites in core pathways of intermediary metabolism such as acetyl-CoA, glucose, α-ketoglutarate (α-KG), nicotinamide adenine dinucleotide (NAD+), flavin adenine dinucleotide (FAD), adenosine triphosphate (ATP), or S-adenosylmethionine (SAM) [75]. Histone acetylation primarily depends on glucose-derived, cytosolic pools of acetyl-CoA. This chromatin modification allows a feed-forward control mechanism for the selective expression of genes that regulate cellular function [75]. Therefore, in obesity, might also lead to stable epigenetic changes that are maintained through the germ line or they can occur in adult tissues and may affect the health of the organism, as was recently reviewed [75].

The obesity-related factors can induce epigenetic alterations in adult target tissues and cells but these factors can induce an epigenetic phenotype in germ line cells as well, that can be transmittable to the next generation. Even though the notion that epigenetic marks are transmitted across generations is still controversial because it is uncertain whether and to what extent such epigenetic inheritance exists [76], over the past few years, evidence has been accumulating that epigenetic transgenerational inheritance occurs in mammals [19,20,77]. Relatedly, a recent study observed that diet-induced paternal obesity modulates sperm miRNA content and germ cell methylation status (which are potential signals that program the health of the offspring) and impairs the metabolic health of future generations [78]. Similarly, maternal obesity adversely affects oocyte quality, embryo development, and the health of the offspring. The DNA methylation status of several imprinted genes and metabolism-related genes appears to be the underlying mechanism responsible for the adverse effects of maternal obesity on oocyte quality and the embryo development of the offspring. These findings have recently been corroborated in oocytes from an obesity mouse model and in oocytes and liver from their offspring [79]. The results of this study revealed that the DNA methylation patterns of several metabolism-related genes are not only altered in the oocytes of the obese mice but also in the oocytes and liver of their offspring [79].

6. DNA METHYLATION AS BIOMARKER OF DISEASE

Biomarkers are naturally occurring characteristics by which a particular pathological process or disease can be identified or monitored. They can reflect past environmental exposures, predict disease onset or course, or determine a patient's response to therapy. Epigenetic changes present these characteristics, with most epigenetic biomarkers discovered to date based on DNA methylation. Many tissue types are suitable for the identification of DNA methylation biomarkers including cell-based samples, such as blood cells and tissue material, and cell-free DNA samples, such as plasma [80]. In the search for biomarkers, multiple studies have used a candidate gene approach, analyzing methylation sites associated to genes with known relevant roles in a particular disease. In this type of studies, DNA methylation can be analyzed by qualitative or quantitative bisulfite conversion-based methods, including methylation-specific PCR (MSP) and pyrosequencing. MSP is a qualitative PCR method widely applied for DNA methylation analyses in patients since it is fast, highly sensitive, and specific [15]. On the other hand, pyrosequencing is a method of DNA sequencing, based on the "sequencing by synthesis" principle, that presents the advantage of being able to quantify the level of DNA methylation in a particular set of CpG sites [63]. This methodology is useful when the aim is to target a certain gene which is already known in terms of its biological function in that particular tissue. In several cases, the choice of genes has been based on prior analysis of gene expression differences. This candidate gene methylation approach has focused on a wide range of genes implicated in obesity, diabetes, and metabolic syndrome analyzing also their relationship with a variety of other well-known markers [55]. On the other hand, array-based genome-wide methylation analyses have the advantage of genome-wide testing, and the candidate gene approach could be underpowered in this sense [21]. However, the candidate gene approach analysis is useful to confirm the gene candidate's result of the genome-wide methylation analyses. Nowadays, methylation analysis is increasing in complexity with the methylation study at base nucleotide resolution by means of a next-generation genome-wide sequencing approach [81].

DNA methylation biomarkers with diagnostic, prognostic, and predictive power are already in clinical trials or in a clinical setting for some diseases as cancer. The strong evidences that complex diseases, such as metabolic disorders, are under the influence of epigenetic modifications even in early life are opening up exciting new avenues for the identification of DNA methylation biomarkers associated to these disorders and for estimation of future disease risk [80]. Therefore, epigenetic marks might explain the link between lifestyle and the risk for disease and have been proposed to be sensitive biomarkers of disease and potential therapeutic targets for disease management that could contribute to personalized medicine throughout life [42].

Regarding obesity susceptibility, several studies have investigated methylation sites in known candidate genes providing evidence that obesity is associated with altered epigenetic regulation of a number of metabolically important genes such as *TNFα*, *LEP*, proopiomelanocortin (*POMC*), melanin-concentrating hormone receptor 1 (*MCHR1*), or *IGF2/H19* imprinting region as was previously reviewed [52,55]. Moreover, with the recent development of genome-wide methods for quantifying site-specific DNA methylation, studies investigating associations across a large number of genes and CpGs are being driven [55]. These approaches identify obesity-associated differentially methylated sites enriched in obesity candidate genes and in genes with a wide diversity of other functions or even unknown properties related to obesity or adipose tissue functioning [55]. Thus, specific patterns of epigenetic factors, including DNA methylation, have been found to be associated with obesity itself in a genome-wide DNA methylation analysis in leucocytes and adipose tissue [82–84] and in analyses of specific genes, such as the circadian clock genes (e.g., *CLOCK*, clock circadian regulator; *BMAL1*, aryl hydrocarbon receptor nuclear translocator-like; and *PER2*, period circadian 2), whose methylation status in human leucocytes is associated with obesity [45].

As blood is easily accessible and it is routinely sampled in clinical and large-scale studies, peripheral blood cells are the most frequently used source of DNA for epigenetic studies. However, epigenetic changes may be more tissue specific and the blood cell methylation profile may not necessarily report the epigenetic state in other tissues [55]. Nowadays, a huge effort is being carried out to get a better insight in tissue-specific epigenetic signatures and their role in disease development. Importantly, for diagnostic purposes in a clinical setting, such as young children or to study inaccessible tissues such as brain tissue [67], epigenetic marks should be detectable in easily accessible samples, such as peripheral blood. In fact, a recent study comparing methylomes across 30 human tissues and cell types showed that there is a large "common methylation profile" across tissues and only a small fraction of CpGs show dynamic regulation during development [55]. In this regard, very recently, Dick and colleagues [84] examined the correlation between DNA methylation and *HIF3A* expression in adipose tissue, reporting a significant inverse correlation and drawing attention to the potential functional relevance of epigenetic variation at the identified locus. The importance of this finding relies on the fact that the assessment of DNA methylation in whole blood can identify robust and biologically relevant epigenetic variation related to BMI [85]. Dick and colleagues described the first systematic analysis of the association between variation in DNA methylation and BMI [84]. They discovered that for every 10% increase in methylation of a probe targeting specific CpG sites within intron 1 of *HIF3A*, BMI increased by 3.6% (95% confidence interval (CI): 2.4–4.9). This significant association between methylation and BMI was reported in a discovery cohort and subsequently confirmed in two independent cohorts. *HIF3A* encodes a component of the hypoxia inducible transcription factor that mediates the cellular response to hypoxia by regulating expression of many downstream genes [86]. The findings suggest that increased CpG methylation at three different sites in the *HIF3A* locus occurs as a consequence of increased BMI and might have a role in the development of obesity-associated metabolic dysfunction [85,87].

As other examples of reports recently published (Table 1), DNA methylation was also discovered recently as a potential molecular link between obesity and age-related diseases since, in an epigenome-wide association study (EWAS), 135 genomic sites corresponding to 10 obesity-susceptible genes (*RNH1, ADCY1, NNAT, CXADR, KCSN2, LMX1B, FNDC4, NAT8L, AQPEP*, and *FBLIM1*) were found in peripheral blood. These genes are subject to methylation changes during aging and that this process is influenced by BMI [88]. Potentially altered biological pathways were identified in visceral adipose tissue of severely obese men with metabolic syndrome compared with severely obese men without this disorder by genome-wide DNA methylation analysis [89]. Thus, the most overrepresented pathways differentially methylated were related to structural components of the cell membrane, inflammation and immunity, and cell cycle regulation and methylation levels of three overmethylated (*GSTM5, LRP1B*, and *RPTOR*) and three undermethylated (*DGKZ, PEAK1, ZNF234*) genes were validated [89]. Studying the DNA methylation levels of adipose tissue from monozygotic twin pairs discordant for type 2 diabetes and independent case-control cohorts, it was found that genes including *PPARG, KCNQ1, TCF7L2*, and *IRS1* were differentially methylated in adipose tissue from unrelated subjects with type 2 diabetes compared with control subjects [90]. Global methylation levels in subcutaneous adipose tissue are associated with measures of fat distribution and glucose homeostasis as well as the *GLUT4* methylation at one CpG locus within intron 1 [91]. In line, a differential influence of BMI on global DNA methylation was identified in healthy Korean women, indicating that BMI-related changes in Alu methylation might play a complex role in the etiology and pathogenesis

TABLE 1 Examples of More Recently Published Evidences on DNA Methylation Biomarkers in Obesity and Its Codiseases

Genes and reference	Tissue type	Approach	Epigenetic association
HIF3A [84]	SAT and PBL	EWAS	Increased CpG methylation at the HIF3A locus associates with BMI and might have a role in the development of obesity-related metabolic dysfunction
RNH1, ADCY1, NNAT, CXADR, KCSN2, LMX1B, FNDC4, NAT8L, AQPEP, and *FBLIM1* [88]	PBL	EWAS	DNA methylation as a potential molecular link between obesity and age-related diseases
Overmethylated: *GSTM5, LRP1B*, and *RPTOR*. Undermethylated: *DGKZ, PEAK1, ZNF234* [89]	VAT	EWAS	Differentially methylated pathways in severely obese men with MetS compared with severely obese men without MetS
PPARG, KCNQ1, TCF7L2, and *IRS1* [90]	SAT	EWAS	Differentially methylated genes in subjects with T2D compared with control subjects
Global methylation and *GLUT4* [91]	SAT	LUMA	Association with measures of fat distribution and glucose homeostasis
Global methylation [92]	PBL	Pyrosequencing	Alu methylation might play a complex role in the etiology and pathogenesis of obesity
ADRB3 [93]	VAT and PBL	Pyrosequencing	Associated with blood lipid levels, blood pressure, body fat distribution, and obesity as well as with common ADRB3 gene polymorphisms
HTR2A [94]	PBL	EWAS	Higher mean methylation levels associated with higher waist circumference and insulin levels and worst response to a 6-month weight loss treatment.

EWAS, epigenome-wide association study; LUMA, luminometric methylation assay; PBL, peripheral blood leukocytes; SAT, subcutaneous adipose tissue; VAT, visceral adipose tissue; CpG, cytosine–phosphate–guanine.

of obesity [92]. Regarding specific-gene methylation levels in association with anthropometric and metabolic traits, the methylation of the β3 adrenergic receptor (*ADRB3*) gene promoter in blood and visceral adipose tissue was associated with blood lipid levels, blood pressure, body fat distribution, and obesity as well as with common *ADRB3* gene polymorphisms in Caucasian men [93]. The methylation levels of the serotonin 2A receptor encoded by the 5-hydroxytryptamine receptor 2A (*HTR2A*) gene was also evaluated in white blood cells of subjects with metabolic syndrome enrolled in a behavioral weight loss program [94]. The results revealed that upper category of mean *HTR2A* methylation levels had a higher waist circumference and insulin levels. Interestingly, after the 6-month weight loss treatment those participants with a greater response to the dietary treatment had lower *HTR2A* gene promoter methylation levels which would suggest that *HTR2A* gene methylation in white blood cells could serve as a useful biomarker to predict weight loss response in subjects with metabolic syndrome [94].

Overall these results revealed that the epigenetic changes associated with obesity and its related comorbidities will provide new insights about the pathophysiological processes of these diseases leading to develop better prevention, management, and treatment strategies.

7. EPIGENETIC DRUGS AND FUNCTIONAL FOODS FOR OVERCOMING METABOLIC DISEASES

Epigenetic changes have been demonstrated in response to a wide range of foods and nutrients and drugs. Since the implication of DNA methylation in tumoral processes, different molecules are being studied in cancer therapy research. In this context, DNMT inhibitors, such as azacitidine, decitabine, were developed and approved by the US Food and Drug Administration. These compounds showed good antiproliferative effects in a number of cancer cell lines including leukemia, breast cancer and prostate cancer, and recent work has identified a number of even more potent analogs although none of them has progressed to the clinic due to toxicity and off-target effects [95]. Relatedly metabolic diseases, the reports that have analyzed differences in DNA methylation profile in relation to obesity and metabolic diseases reported variations of about 10–20% at the most which could be difficult to modulate by epigenetic drug [52]. To date, only one study has been performed to investigate whether inhibiting DNA methylation pharmacologically using 5-aza-dC ameliorates atherosclerosis in LDL receptor knockout (*Ldlr*$^{-/-}$) mice [96]. 5-aza-dC is a nucleoside-based DNMT inhibitor that induces demethylation and is employed as an anticancer drug. The results showed that 5-aza-dC treatment significantly diminished atherosclerosis development in Ldlr$^{-/-}$ mice without altering lipid and cholesterol metabolism suppressing macrophage inflammation and infiltration into atherosclerotic plaques, reduced ER stress signals, and reduced apoptosis. This effect was associated with decreased DNA methylation status at the promoters of liver X receptor-α (*LXR α*) and peroxisome proliferator-activated receptor γ1 (*PPARγ1*), important transcriptional factors that play key roles during the development of atherosclerosis through regulating macrophage inflammation and/or cholesterol homeostasis γγ[96].

On the other hand, several nutrients and food compounds were identified to be able to slightly modify the epigenetic patterns of different cell lines and tissues that might help to overcoming metabolic diseases [52]. Numerous studies have demonstrated effects on DNA methylation of alcohol, the B vitamins, proteins, micronutrients, functional food components, and general nutritional status as it was recently reviewed [97]. Dietary change may act directly on the epigenetic processes that result in health/disease but it can also program metabolism and the future response to nutrition itself [97]. The epigenetic diet has been proposed as an important mechanism that modulates and potentially slows the progression of age-related diseases such as cardiovascular disease, cancer, and obesity because it introduces bioactive medicinal chemistry compounds, such as sulforaphane (SFN), curcumin (CCM), epigallocatechin gallate (EGCG), and resveratrol (RSV), that are thought to aid in extending the human lifespan [98].

As an example of the effect on nutritional epigenetic modifiers from dietary components (Table 2), SFN found in broccoli, has been shown to suppress the proinflammatory response through transcription factor activation and therefore they could be potential epigenetic therapeutics for the treatment of cardiovascular disease [99]. Essential nutrient supplementation prevents heritable metabolic disease in multigenerational intrauterine growth-restricted rats in association with epigenomic modifications of *IGF-1* in a rodent model of multigenerational metabolic syndrome [100]. High vitamin A in postweaning diets was reported to be able to reduce postweaning weight gain and food intake and modify gene expression in food intake and reward pathways in association of high DNA methylation of *POMC* in the hypothalamus of male rats born to dams fed a high multivitamin diet [101].

TABLE 2 Examples of Nutritional and Drug Factors Targeting Obesity and Its Codiseases through DNA Methylation Modification

Nutritional Factor and Reference	Population	Metabolic Effect
Vitamin B12, folate, and methionine [103]	Periconceptional mature female sheep	Prevents insulin resistance, obesity, and hypertension
Genistein [104]	Avy mouse offspring	Body weight
Soy isoflavones [105]	Cynomolgus monkey	Body weight, insulin sensitivity
Curcumin [106]	HFD fed C57BL/6J mice	Prevents high-fat-diet-induced insulin resistance and obesity
Essential nutrient supplementation [100]	Rodent model of multigenerational MetS	Prevents heritable metabolic disease in association with epigenomic modifications of IGF-1
Vitamin A [101]	Rats born to dams fed a high-multivitamin diet	Reduce postweaning weight gain and food intake
5-aza-2′-deoxycytidine [96]	LDL receptor knockout (*Ldlr*$^{-/-}$) mice	Diminishes atherosclerosis development

8. CONCLUSION

Emerging evidence suggests that epigenetics is one of the links between the environmental factors and the higher predisposition to develop obesity and its codiseases. Epigenetic marks can be induced during the perinatal events and persist until the adult age. Therefore, these modifications could be used as early prognostic markers of disease to identify those individuals with more risk to develop a metabolic disorder before the phenotype develops and to design more personalized therapeutic strategies for prevention and management of the condition [52,102]. To achieve this aim, microarray-based or sequencing approaches are contributing to obtain relevant information about genome methylation. Since the epigenetic marks can be reversible or modifiable, an effort in research for counteracting metabolic diseases is being focused on drugs and food components as promising epigenetic therapeutic agents able to rescue unfavorable epigenomic profiles associated to metabolic dysfunction. However, it is still unclear whether the best therapeutic strategy is to target DNA methylation of genes using demethylation agents, for example, or to target the genes by means of other mechanisms that modulate the expression. On the other hand, the source of such biomarkers may be removed from a biological source more easily accessible and routinely sampled in clinic such as leukocytes which often may be of little relevance in the disease development. Consequently, in this case the biomarkers are useful as biomarkers of disease but have not usually value as a therapeutic target. Numerous evidences exist currently on the association between obesity-related genes and differential methylation profile such as *TNFα*, *LEP*, *POMC*, *MCHR1*, or *IGF2/H19* imprinting region, genes *CLOCK* or that more recently published *HIF3A*, *PPARG*, *GLUT4*, *ADRB3*, or *HTR2A*. Among them, only *HIF3A* could have a potential value for diagnosis and prognosis of obesity because it was demonstrated that the BMI increases with the increase in the methylation of specific CpG sites within intron 1 of this gene in two independent cohorts. However, in spite of the number of reports demonstrating an association between obesity and methylation profile of specific genes, the study of epigenetic marks in obesity is very recent and until today there is no enough evidence supporting the potential of these epigenetic biomarkers for clinical applications. In addition, due to the complexity of the interplay between DNA methylation and gene expression regulation, it would be interesting to identify the DNA methylation changes that are causal factors or a consequence of the obesity.

Altogether, this review highlights the need for further scientific research to elucidate the role of the epigenetic regulation as a potential molecular mechanism involved in obesity and its codiseases. Due to the fact that epigenetic modifications are dynamic and reversible and change in response to dietary patterns, physical activity, and weight loss, the epigenetic markers related to obesity may constitute therapeutic targets for the prevention of obesity-related disorders including type 2 diabetes, cardiovascular disease, and metabolic syndrome. Future efforts will therefore be needed to carry out genome-wide analysis that provides information on metabolic disorders-specific epigenetic marks, and longitudinal studies will be required to evaluate the effect of drugs or dietary components on reprogramming these disease-specific epigenetic modifications. Understanding the influence of the obesity-related microenvironment on the epigenetic regulation of the metabolic function machinery will provide new tools to improve the management as well as the prevention of the metabolic disorders.

LIST OF ABBREVIATIONS

5-aza-dC 5-Aza-2′-deoxycytidine
5hmC 5-Hydroxymethylcytosine
5mC 5-Methylcytosine
ATP Adenosine triphosphate
BMI Body mass index
CGIs CpG islands
CI Confidence interval
CpG Cytosine–phosphate–guanine
DNMTs DNA methyltransferase enzymes
EWAS Epigenome-wide association study
FAD Flavin adenine dinucleotide
kb Kilobase
lncRNAs Long noncoding RNAs
LUMA Luminometric methylation assay
miRNAs MicroRNAs
MSP Methylation-specific PCR
NAD+ Nicotinamide adenine dinucleotide
ncRNAs Noncoding RNAs
PBL Peripheral blood leukocytes
ROS Reactive oxygen species
SAM S-adenosylmethionine
SAT Subcutaneous adipose tissue
VAT Visceral adipose tissue

References

[1] Marti A, Martinez-Gonzalez MA, Martinez JA. Interaction between genes and lifestyle factors on obesity. Proc Nutr Soc February 2008;67(1):1–8.
[2] Zou C, Shao J. Role of adipocytokines in obesity-associated insulin resistance. J Nutr Biochem May 2008;19(5):277–86.
[3] Eckel RH, Grundy SM, Zimmet PZ. The metabolic syndrome. Lancet April 16–22, 2005;365(9468):1415–28.
[4] Apovian CM, Aronne LJ, Bessesen DH, McDonnell ME, Murad MH, Pagotto U, Ryan DH, Still CD. Endocrine society, pharmacological management of obesity: an endocrine society clinical practice guideline. J Clin Endocrinol Metab February 2015;100(2):342–62.
[5] Ogden CL, Carroll MD, Kit BK, Flegal KM. Prevalence of childhood and adult obesity in the United States, 2011–2012. JAMA February 26, 2014;311(8):806–14.
[6] Gutierrez-Fisac JL, Guallar-Castillon P, Leon-Munoz LM, Graciani A, Banegas JR, Rodriguez-Artalejo F. Prevalence of general and abdominal obesity in the adult population of Spain, 2008–2010: the ENRICA study. Obes Rev April 2011;13(4):388–92.
[7] Casanueva FF, Moreno B, Rodriguez-Azeredo R, Massien C, Conthe P, Formiguera X, et al. Relationship of abdominal obesity with cardiovascular disease, diabetes and hyperlipidaemia in Spain. Clin Endocrinol (Oxf) July 2010;73(1):35–40.
[8] de Mello VD, Pulkkinen L, Lalli M, Kolehmainen M, Pihlajamaki J, Uusitupa M. DNA methylation in obesity and type 2 diabetes. Ann Med May 2014;46(3):103–13.
[9] Leech RM, McNaughton SA, Timperio A. The clustering of diet, physical activity and sedentary behavior in children and adolescents: a review. Int J Behav Nutr Phys Act 2014;11:4.
[10] Herceg Z, Vaissiere T. Epigenetic mechanisms and cancer: an interface between the environment and the genome. Epigenetics July 2011;6(7):804–19.
[11] Waddington CH. The epigenotype. 1942. Int J Epidemiol February 2012;41(1):10–3.
[12] Berger SL, Kouzarides T, Shiekhattar R, Shilatifard A. An operational definition of epigenetics. Genes Dev April 1, 2009;23(7):781–3.

[13] Portela A, Esteller M. Epigenetic modifications and human disease. Nat Biotechnol October 2010;28(10):1057–68.
[14] Rodriguez-Paredes M, Esteller M. Cancer epigenetics reaches mainstream oncology. Nat Med March 2011;17(3):330–9.
[15] Esteller M. Epigenetics in cancer. N Engl J Med March 13, 2008;358(11):1148–59.
[16] Quintavalle C, Mangani D, Roscigno G, Romano G, Diaz-Lagares A, Iaboni M, et al. MiR-221/222 target the DNA methyltransferase MGMT in glioma cells. PLoS One 2013;8(9):e74466.
[17] Lujambio A, Ropero S, Ballestar E, Fraga MF, Cerrato C, Setien F, et al. Genetic unmasking of an epigenetically silenced microRNA in human cancer cells. Cancer Res February 15, 2007;67(4):1424–9.
[18] Qu Y, Lennartsson A, Gaidzik VI, Deneberg S, Karimi M, Bengtzen S, et al. Differential methylation in CN-AML preferentially targets non-CGI regions and is dictated by DNMT3A mutational status and associated with predominant hypomethylation of HOX genes. Epigenetics August 2014;9(8):1108–19.
[19] Guerrero-Bosagna C, Savenkova M, Haque MM, Nilsson E, Skinner MK. Environmentally induced epigenetic transgenerational inheritance of altered Sertoli cell transcriptome and epigenome: molecular etiology of male infertility. PLoS One 2013;8(3):e59922.
[20] Skinner MK, Guerrero-Bosagna C. Role of CpG deserts in the epigenetic transgenerational inheritance of differential DNA methylation regions. BMC Genomics 2014;15(1):692.
[21] Sandoval J, Heyn H, Moran S, Serra-Musach J, Pujana MA, Bibikova M, et al. Validation of a DNA methylation microarray for 450,000 CpG sites in the human genome. Epigenetics June 2011;6(6):692–702.
[22] Hashimshony T, Zhang J, Keshet I, Bustin M, Cedar H. The role of DNA methylation in setting up chromatin structure during development. Nat Genet June 2003;34(2):187–92.
[23] Kass SU, Landsberger N, Wolffe AP. DNA methylation directs a time-dependent repression of transcription initiation. Curr Biol March 1, 1997;7(3):157–65.
[24] Venolia L, Gartler SM. Comparison of transformation efficiency of human active and inactive X-chromosomal DNA. Nature March 3, 1983;302(5903):82–3.
[25] Lock LF, Takagi N, Martin GR. Methylation of the *Hprt* gene on the inactive X occurs after chromosome inactivation. Cell January 16, 1987;48(1):39–46.
[26] Jones PA. Functions of DNA methylation: islands, start sites, gene bodies and beyond. Nat Rev Genet July 2012;13(7):484–92.
[27] Gowher H, Jeltsch A. Molecular enzymology of the catalytic domains of the Dnmt3a and Dnmt3b DNA methyltransferases. J Biol Chem June 7, 2002;277(23):20409–14.
[28] Jaenisch R, Bird A. Epigenetic regulation of gene expression: how the genome integrates intrinsic and environmental signals. Nat Genet March 2003;(33 Suppl.):245–54.
[29] Bhutani N, Burns DM, Blau HM. DNA demethylation dynamics. Cell September 16, 2011;146(6):866–72.
[30] Ito S, D'Alessio AC, Taranova OV, Hong K, Sowers LC, Zhang Y. Role of Tet proteins in 5mC to 5hmC conversion, ES-cell self-renewal and inner cell mass specification. Nature August 26, 2010;466(7310):1129–33.
[31] Esteller M. Non-coding RNAs in human disease. Nat Rev Genet December 2011;12(12):861–74.
[32] Taft RJ, Pang KC, Mercer TR, Dinger M, Mattick JS. Non-coding RNAs: regulators of disease. J Pathol January 2010;220(2):126–39.
[33] Lee SK, Calin GA. Non-coding RNAs and cancer: new paradigms in oncology. Discov Med March 2011;11(58):245–54.
[34] Garzon R, Calin GA, Croce CM. MicroRNAs in cancer. Annu Rev Med 2009;60:167–79.
[35] Fatica A, Bozzoni I. Long non-coding RNAs: new players in cell differentiation and development. Nat Rev Genet January 2014;15(1):7–21.
[36] Rinn JL, Chang HY. Genome regulation by long non-coding RNAs. Annu Rev Biochem 2012;81:145–66.
[37] Berdasco M, Esteller M. Aberrant epigenetic landscape in cancer: how cellular identity goes awry. Dev Cell November 16, 2010;19(5):698–711.
[38] Ballestar E. Epigenetic alterations in autoimmune rheumatic diseases. Nat Rev Rheumatol May 2011;7(5):263–71.
[39] Campion J, Milagro F, Martinez JA. Epigenetics and obesity. Prog Mol Biol Transl Sci 2010;94:291–347.
[40] Toperoff G, Aran D, Kark JD, Rosenberg M, Dubnikov T, Nissan B, et al. Genome-wide survey reveals predisposing diabetes type 2-related DNA methylation variations in human peripheral blood. Hum Mol Genet January 15, 2012;21(2):371–83.
[41] DelCurto H, Wu G, Satterfield MC. Nutrition and reproduction: links to epigenetics and metabolic syndrome in offspring. Curr Opin Clin Nutr Metab Care July 2013;16(4):385–91.
[42] Hamilton JP. Epigenetics: principles and practice. Dig Dis 2011;29(2):130–5.
[43] Ziech D, Franco R, Pappa A, Panayiotidis MI. Reactive oxygen species (ROS)–induced genetic and epigenetic alterations in human carcinogenesis. Mutat Res June 3, 2012;711(1–2):167–73.
[44] Heyn H, Li N, Ferreira HJ, Moran S, Pisano DG, Gomez A, et al. Distinct DNA methylomes of newborns and centenarians. Proc Natl Acad Sci USA June 26, 2012;109(26):10522–7.
[45] Milagro FI, Gomez-Abellan P, Campion J, Martinez JA, Ordovas JM, Garaulet M. CLOCK, PER2 and BMAL1 DNA methylation: association with obesity and metabolic syndrome characteristics and monounsaturated fat intake. Chronobiol Int November 2012;29(9):1180–94.

[46] Brasacchio D, Okabe J, Tikellis C, Balcerczyk A, George P, Baker EK, et al. Hyperglycemia induces a dynamic cooperativity of histone methylase and demethylase enzymes associated with gene-activating epigenetic marks that coexist on the lysine tail. Diabetes May 2009;58(5):1229–36.

[47] Pirola L, Balcerczyk A, Tothill RW, Haviv I, Kaspi A, Lunke S, et al. Genome-wide analysis distinguishes hyperglycemia regulated epigenetic signatures of primary vascular cells. Genome Res October 2011;21(10):1601–15.

[48] Coolen MW, Statham AL, Qu W, Campbell MJ, Henders AK, Montgomery GW, et al. Impact of the genome on the epigenome is manifested in DNA methylation patterns of imprinted regions in monozygotic and dizygotic twins. PLoS One 2011;6(10): e25590.

[49] Fraga MF, Ballestar E, Paz MF, Ropero S, Setien F, Ballestar ML, et al. Epigenetic differences arise during the lifetime of monozygotic twins. Proc Natl Acad Sci USA July 26, 2005;102(30):10604–9.

[50] Estey EH. Epigenetics in clinical practice: the examples of azacitidine and decitabine in myelodysplasia and acute myeloid leukemia. Leukemia September 2013;27(9):1803–12.

[51] Langie SA, Achterfeldt S, Gorniak JP, Halley-Hogg KJ, Oxley D, van Schooten FJ, et al. Maternal folate depletion and high-fat feeding from weaning affects DNA methylation and DNA repair in brain of adult offspring. Faseb J August 2013;27(8):3323–34.

[52] Milagro FI, Mansego ML, De Miguel C, Martinez JA. Dietary factors, epigenetic modifications and obesity outcomes: progresses and perspectives. Mol Asp Med July–August 2013;34(4):782–812.

[53] Burgio E, Lopomo A, Migliore L. Obesity and diabetes: from genetics to epigenetics. Mol Biol Rep September 25, 2014;42(4):799–818.

[54] Drummond EM, Gibney ER. Epigenetic regulation in obesity. Curr Opin Clin Nutr Metab Care July 2013;16(4):392–7.

[55] van Dijk SJ, Molloy PL, Varinli H, Morrison JL, Muhlhausler BS. Epigenetics and human obesity. Int J Obes (Lond) February 25, 2014;39(1):85–97.

[56] Skinner MK, Manikkam M, Guerrero-Bosagna C. Epigenetic transgenerational actions of environmental factors in disease etiology. Trends Endocrinol Metab April 2010;21(4):214–22.

[57] Skinner MK, Manikkam M, Tracey R, Guerrero-Bosagna C, Haque M, Nilsson EE. Ancestral dichlorodiphenyltrichloroethane (DDT) exposure promotes epigenetic transgenerational inheritance of obesity. BMC Med 2013;11:228.

[58] Manikkam M, Tracey R, Guerrero-Bosagna C, Skinner MK. Plastics derived endocrine disruptors (BPA, DEHP and DBP) induce epigenetic transgenerational inheritance of obesity, reproductive disease and sperm epimutations. PLoS One 2013;8(1):e55387.

[59] Tracey R, Manikkam M, Guerrero-Bosagna C, Skinner MK. Hydrocarbons (jet fuel JP-8) induce epigenetic transgenerational inheritance of obesity, reproductive disease and sperm epimutations. Reprod Toxicol April 2013;36:104–16.

[60] Obermann-Borst SA, Eilers PH, Tobi EW, de Jong FH, Slagboom PE, Heijmans BT, et al. Duration of breastfeeding and gender are associated with methylation of the LEPTIN gene in very young children. Pediatr Res September 2013;74(3):344–9.

[61] Guenard F, Deshaies Y, Cianflone K, Kral JG, Marceau P, Vohl MC. Differential methylation in glucoregulatory genes of offspring born before vs. after maternal gastrointestinal bypass surgery. Proc Natl Acad Sci USA July 9, 2013;110(28):11439–44.

[62] Barres R, Kirchner H, Rasmussen M, Yan J, Kantor FR, Krook A, et al. Weight loss after gastric bypass surgery in human obesity remodels promoter methylation. Cell Rep April 25, 2013;3(4):1020–7.

[63] Ronn T, Volkov P, Davegardh C, Dayeh T, Hall E, Olsson AH, et al. A six months exercise intervention influences the genome-wide DNA methylation pattern in human adipose tissue. PLoS Genet June 2013;9(6):e1003572.

[64] Nitert MD, Dayeh T, Volkov P, Elgzyri T, Hall E, Nilsson E, et al. Impact of an exercise intervention on DNA methylation in skeletal muscle from first-degree relatives of patients with type 2 diabetes. Diabetes December 2012;61(12):3322–32.

[65] Milagro FI, Campion J, Cordero P, Goyenechea E, Gomez-Uriz AM, Abete I, et al. A dual epigenomic approach for the search of obesity biomarkers: DNA methylation in relation to diet-induced weight loss. Faseb J April 2011;25(4):1378–89.

[66] Bouchard L, Rabasa-Lhoret R, Faraj M, Lavoie ME, Mill J, Perusse L, et al. Differential epigenomic and transcriptomic responses in subcutaneous adipose tissue between low and high responders to caloric restriction. Am J Clin Nutr February 2010;91(2): 309–20.

[67] Crujeiras AB, Campion J, Diaz-Lagares A, Milagro FI, Goyenechea E, Abete I, et al. Association of weight regain with specific methylation levels in the NPY and POMC promoters in leukocytes of obese men: a translational study. Regul Pept September 10, 2013;186: 1–6.

[68] Dizdar O, Alyamac E. Obesity: an endocrine tumor? Med Hypotheses 2004;63(5):790–2.

[69] Crujeiras AB, Parra D, Goyenechea E, Abete I, Gonzalez-Muniesa P, Martinez JA. Energy restriction in obese subjects impact differently two mitochondrial function markers. J Physiol Biochem September 2008; 64(3):211–9.

[70] Vincent HK, Taylor AG. Biomarkers and potential mechanisms of obesity-induced oxidant stress in humans. Int J Obes (Lond) March 2006;30(3):400–18.
[71] Franco R, Schoneveld O, Georgakilas AG, Panayiotidis MI. Oxidative stress, DNA methylation and carcinogenesis. Cancer Lett July 18, 2008;266(1):6–11.
[72] Lim SO, Gu JM, Kim MS, Kim HS, Park YN, Park CK, et al. Epigenetic changes induced by reactive oxygen species in hepatocellular carcinoma: methylation of the E-cadherin promoter. Gastroenterology December 2008;135(6):2128–40. 2140.e1–8.
[73] Mateescu B, Batista L, Cardon M, Gruosso T, de Feraudy Y, Mariani O, et al. miR-141 and miR-200a act on ovarian tumorigenesis by controlling oxidative stress response. Nat Med December 2011;17(12):1627–35.
[74] Rajendran R, Garva R, Krstic-Demonacos M, Demonacos C. Sirtuins: molecular traffic lights in the crossroad of oxidative stress, chromatin remodeling, and transcription. J Biomed Biotechnol 2011;2011:368276.
[75] Gut P, Verdin E. The nexus of chromatin regulation and intermediary metabolism. Nature October 24, 2013;502(7472):489–98.
[76] Heard E, Martienssen RA. Transgenerational epigenetic inheritance: myths and mechanisms. Cell March 27, 2014;157(1):95–109.
[77] Pembrey M, Saffery R, Bygren LO. Human transgenerational responses to early-life experience: potential impact on development, health and biomedical research. J Med Genet September 2014;51(9):563–72.
[78] Fullston T, Ohlsson Teague EM, Palmer NO, DeBlasio MJ, Mitchell M, Corbett M, et al. Paternal obesity initiates metabolic disturbances in two generations of mice with incomplete penetrance to the F2 generation and alters the transcriptional profile of testis and sperm microRNA content. Faseb J October 2013;27(10):4226–43.
[79] Ge ZJ, Luo SM, Lin F, Liang QX, Huang L, Wei YC, et al. DNA methylation in oocytes and liver of Female mice and their offspring: effects of high-fat-diet-induced obesity. Environ Health Perspect December 6, 2013;122(2):159–64.
[80] Mikeska T, Craig JM. DNA methylation biomarkers: cancer and beyond. Genes (Basel) 2014;5(3):821–64.
[81] Ku CS, Naidoo N, Wu M, Soong R. Studying the epigenome using next generation sequencing. J Med Genet November 2011;48(11):721–30.
[82] Carless MA, Kulkarni H, Kos MZ, Charlesworth J, Peralta JM, Goring HH, et al. Genetic effects on DNA methylation and its potential relevance for obesity in Mexican Americans. PLoS One 2013;8(9): e73950.
[83] Xu X, Su S, Barnes VA, De Miguel C, Pollock J, Ownby D, et al. A genome-wide methylation study on obesity: differential variability and differential methylation. Epigenetics May 2013;8(5):522–33.
[84] Dick KJ, Nelson CP, Tsaprouni L, Sandling JK, Aissi D, Wahl S, et al. DNA methylation and body-mass index: a genome-wide analysis. Lancet June 7, 2014;383(9933): 1990–8.
[85] Murphy TM, Mill J. Epigenetics in health and disease: heralding the EWAS era. Lancet June 7, 2014;383(9933): 1952–4.
[86] Greer SN, Metcalf JL, Wang Y, Ohh M. The updated biology of hypoxia-inducible factor. Embo J May 30, 2012;31(11):2448–60.
[87] Osorio J. Obesity. Looking at the epigenetic link between obesity and its consequences–the promise of EWAS. Nat Rev Endocrinol May 2014;10(5):249.
[88] Almen MS, Nilsson EK, Jacobsson JA, Kalnina I, Klovins J, Fredriksson R, et al. Genome-wide analysis reveals DNA methylation markers that vary with both age and obesity. Gene September 10, 2014;548(1):61–7.
[89] Guenard F, Tchernof A, Deshaies Y, Perusse L, Biron S, Lescelleur O, et al. Differential methylation in visceral adipose tissue of obese men discordant for metabolic disturbances. Physiol Genomics March 15, 2014;46(6):216–22.
[90] Nilsson E, Jansson PA, Perfilyev A, Volkov P, Pedersen M, Svensson MK, et al. Altered DNA methylation and differential expression of genes influencing metabolism and inflammation in adipose tissue from subjects with type 2 diabetes. Diabetes September 2014;63(9):2962–76.
[91] Keller M, Kralisch S, Rohde K, Schleinitz D, Dietrich A, Schon MR, et al. Global DNA methylation levels in human adipose tissue are related to fat distribution and glucose homeostasis. Diabetologia November 2014;57(11):2374–83.
[92] Na YK, Hong HS, Lee DH, Lee WK, Kim DS. Effect of body mass index on global DNA methylation in healthy Korean women. Mol Cells June 2014;37(6):467–72.
[93] Guay SP, Brisson D, Lamarche B, Biron S, Lescelleur O, Biertho L, et al. ADRB3 gene promoter DNA methylation in blood and visceral adipose tissue is associated with metabolic disturbances in men. Epigenomics February 2014;6(1):33–43.
[94] Perez-Cornago A, Mansego ML, Zulet MA, Martinez JA. DNA hypermethylation of the serotonin receptor type-2A gene is associated with a worse response to a weight loss intervention in subjects with metabolic syndrome. Nutrients June 2014;6(6):2387–403.
[95] Dhanak D, Jackson P. Development and classes of epigenetic drugs for cancer. Biochem Biophys Res Commun July 10, 2014;455(1–2):58–69.
[96] Cao Q, Wang X, Jia L, Mondal AK, Diallo A, Hawkins GA, et al. Inhibiting DNA methylation by 5-aza-2′-deoxycytidine ameliorates atherosclerosis through suppressing macrophage inflammation. Endocrinology September 24, 2014;155(12):4925–38. http://dx.doi.org/10.1210/en.2014-1595.

[97] Haggarty P. Epigenetic consequences of a changing human diet. Proc Nutr Soc November 2013;72(4):363–71.
[98] Martin SL, Hardy TM, Tollefsbol TO. Medicinal chemistry of the epigenetic diet and caloric restriction. Curr Med Chem 2013;20(32):4050–9.
[99] Byrne MM, Murphy RT, Ryan AW. Epigenetic modulation in the treatment of atherosclerotic disease. Front Genet 2014;5:364.
[100] Goodspeed D, Seferovic MD, Holland W, McKnight RA, Summers SA, Branch DW, et al. Essential nutrient supplementation prevents heritable metabolic disease in multigenerational intrauterine growth-restricted rats. Faseb J November 13, 2014;29(3):807–19.
[101] Sanchez-Hernandez D, Cho CE, Kubant R, Reza-Lopez SA, Poon AN, Wang J, et al. Increasing vitamin A in post-weaning diets reduces food intake and body weight and modifies gene expression in brains of male rats born to dams fed a high multivitamin diet. J Nutr Biochem October 2014;25(10):991–6.
[102] Martinez JA, Milagro FI, Claycombe KJ, Schalinske KL. Epigenetics in adipose tissue, obesity, weight loss, and diabetes. Adv Nutr January 2014;5(1):71–81.
[103] Sinclair KD, Allegrucci C, Singh R, Gardner DS, Sebastian S, Bispham J, et al. DNA methylation, insulin resistance, and blood pressure in offspring determined by maternal periconceptional B vitamin and methionine status. Proc Natl Acad Sci USA December 4, 2007;104(49):19351–6.
[104] Dolinoy DC, Weidman JR, Waterland RA, Jirtle RL. Maternal genistein alters coat color and protects Avy mouse offspring from obesity by modifying the fetal epigenome. Environ Health Perspect April 2006;114(4):567–72.
[105] Howard TD, Ho SM, Zhang L, Chen J, Cui W, Slager R, et al. Epigenetic changes with dietary soy in cynomolgus monkeys. PLoS One 2011;6(10):e26791.
[106] Shao W, Yu Z, Chiang Y, Yang Y, Chai T, Foltz W, et al. Curcumin prevents high fat diet induced insulin resistance and obesity via attenuating lipogenesis in liver and inflammatory pathway in adipocytes. PLoS One 2012;7(1):e28784.

CHAPTER

17

DNA Methylation Biomarkers in Asthma and Allergy

Avery DeVries, Donata Vercelli

Department of Cellular and Molecular Medicine, Arizona Respiratory Center and Arizona Center for the Biology of Complex Diseases, University of Arizona, Tucson, AZ, USA

OUTLINE

1. INTRODUCTION

The role of epigenetic processes in allergic diseases has received increasing attention over the last few years. Indeed, the number of publications has been rising steadily and seems to be far from reaching its plateau (Figure 1). Remarkably, as we ourselves recently remarked [1] and Figure 1 clearly shows, in certain years the reviews about asthma/allergy epigenetics have even outnumbered the primary papers, a pattern that suggests an interest with few precedents and an urgency not easy to explain.

With this somewhat paradoxical scenario in mind, in this chapter we will first briefly discuss the main phenotypic and genetic characteristics of allergic diseases, and we will then explore three main questions: Why is epigenetics expected to be (so) critical in asthma and allergy? What have we learned from the studies performed so far? and Where is the field going next?

Epigenetic Biomarkers and Diagnostics
http://dx.doi.org/10.1016/B978-0-12-801899-6.00017-6

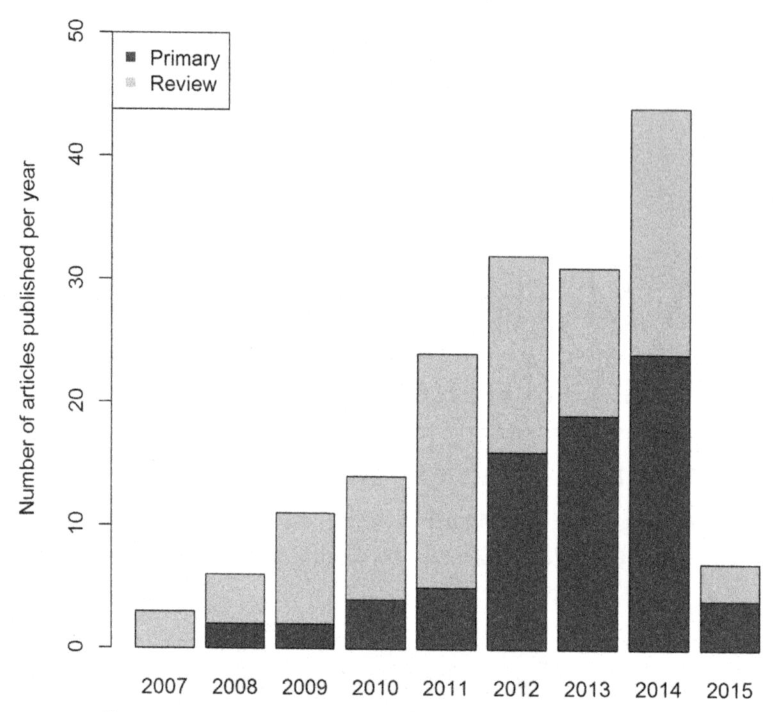

FIGURE 1 Literature on epigenetics and allergic disease, 2007–2015. The y-axis shows the number of journal articles published each year. Light gray bars indicate review articles, while dark gray bars indicate primary research articles. The literature search was performed entering asthma, allergy, DNA methylation, and epigenetics.

2. ALLERGIC DISEASES: PHENOTYPES AND GENES

Allergic diseases—primarily asthma, allergic rhinitis, and atopic dermatitis—are among the most common chronic complex diseases both in childhood and among adults. These diseases are quite heterogeneous in their clinical phenotypes and primary target organs. Indeed, asthma is characterized by recurrent episodes of airway obstruction, which reverse either spontaneously or after use of medication, and is usually associated with bronchial hyperresponsiveness and evidence of chronic airway inflammation. The latter is often, but not invariably, associated with increased expression of Th2 cytokines (IL-4, IL-13, IL-5) by T cells and type-2 innate lymphoid cells and with altered epithelial responses to environmental and infectious stimuli, with abnormal release of epithelial cytokines (IL-33, IL-25, TSLP) that activate Th2 and Th2-like responses in the respiratory mucosa [2].

Atopic dermatitis is the most common chronic inflammatory skin disease and often precedes the development of food allergy and asthma. Recent insights into the pathogenesis of atopic dermatitis reveal abnormalities in terminal differentiation of the epidermal epithelium leading to a defective stratum corneum, which allows enhanced allergen penetration and systemic immunoglobulin E (IgE) sensitization. Atopic skin is also predisposed to colonization or infection by pathogenic microbes, most notably *Staphylococcus aureus* and herpes simplex virus. Causes of this abnormal skin barrier are complex and driven by a combination of genetic, environmental, and immunologic factors. These factors likely account for the heterogeneity of atopic dermatitis onset and the severity and natural history of this skin disease [3].

Despite their biological, phenotypic, and clinical differences, allergic diseases are typically marked by the propensity to mount vigorous IgE responses to otherwise innocuous environmental antigens (allergens). IgE in turn plays a critical role in allergic disease pathogenesis by engaging a complex network of receptors expressed on cells from multiple lineages (mast cells, eosinophils, macrophages, dendritic cells, Langerhans cells), which upon allergen-mediated cross-linking of receptor-bound IgE release an array of mediators and cytokines that induce, maintain, and amplify inflammation in the target organs, decisively contributing to morbidity [2].

Because of their high prevalence, elusive pathogenesis, and strong family history, allergic diseases were aggressively investigated using candidate-gene and later genome-wide association studies (GWAS) aimed at discovering the genetic basis of their susceptibility and severity. GWAS have been particularly successful in that a number of plausible genetic pathways (e.g., those orchestrated by IL-33, TSLP, and Th2 cytokines in asthma) were identified as determinants of risk, and other previously unknown players (ORMDL3/GSDM first and foremost) came on stage [4]. Moreover, the hypothesis-generating nature of GWAS has provided unexpected insights about potential mechanistic similarities between selected allergy phenotypes (e.g., childhood asthma) and other disorders (e.g., inflammatory bowel disease) ostensibly unrelated to allergy [5]. Conversely, diseases such as atopic dermatitis and psoriasis, which have mutually exclusive clinical phenotypes and opposing immune mechanisms, have been shown to have distinct genetic mechanisms with opposing effects in shared pathways [6].

3. WHY HAS EPIGENETICS BECOME A BUZZ WORD IN ALLERGIC DISEASE?

In spite of these successes, GWAS have failed to account for more than a modest proportion of the phenotypic variance that marks allergic disease [7]. The current studies focused on rare variants (as opposed to the common ones that GWAS typically interrogate) will probably improve the situation but only marginally. The realization that for all its technological prowess, genetics cannot "explain" allergic disease (and the same has been true of other complex diseases, such as diabetes, hypertension, and obesity, to mention a few) has rekindled interest in other potential sources of phenotypic variance, primarily the environment and development. The fact that allergic diseases have a strong environmental component has been known for several decades and was clearly illustrated by the seminal epidemiologic studies that revealed major differences in allergic disease prevalence among countries with more or less Westernized lifestyles [2]. The point was reinforced by the discovery of asthma/allergy protection in the farming communities of Alpine Europe [8], not to mention the association between allergy susceptibility and exposure to smoking and respiratory viruses in early life [9]. The developmental trajectory of allergic diseases, particularly asthma, which typically begins in childhood with subtle respiratory and immune alterations even when overt symptoms leading to a firm diagnosis appear only much later in life [2] is also well established. Thus, much effort currently focuses on assessing the extent to which integrating environmental and developmental factors into studies of allergy susceptibility and severity can further account for the phenotypes of interest.

It is against this backdrop of renewed interest for the environmental and developmental components of allergic diseases, and more generally of human complex non-communicable diseases that the current burning interest in their epigenetic determinants can be best understood. Indeed, epigenetic mechanisms are exquisitely designed to faithfully and sensitively transduce environmental signals and preside over the time-dependent unfolding of complex developmental differentiation programs. On the other hand, interrogating the epigenome, particularly the methylome, to highlight its ability to influence complex disease phenotypes has also become much more feasible after the recent advent of technologies that allows to simultaneously sample relatively large numbers of potential DNA methylation sites [CpG (cytosine–phosphate–guanine) dinucleotides] throughout the genome using straightforward assays and streamlined analytical pipelines. Add to this that the DNA required for these methylation studies is adequate even if it was collected years earlier for whatever purpose, was stored under standard conditions, and is available in relatively modest (nanograms–low micrograms) amounts. These easy-to-meet technical requirements are a major reason why DNA methylation studies, unlike the more technically challenging analyses of post-translational histone modifications, are flourishing and currently represent the totality of the epigenetic studies performed in human populations with allergic diseases.

4. WHAT HAVE WE LEARNED SO FAR FROM ALLERGIC DISEASE EPIGENETICS?

The results obtained by genome-wide and candidate-gene DNA methylation studies performed in human allergic disease are summarized in Tables 1 and 2. Because we recently extensively reviewed candidate gene studies [1], we will focus the current discussion on the genome-wide ones. At first glance, the scenario emerging from these data is not too encouraging. Indeed, the regions where differential methylation was detected are spread throughout the genome, without any discernible pattern and with no internal consistency. If we borrow from the genetics world and consider result replication, strict or even loose, as the main criterion for robustness, we find that only two regions (MRI1 and FAM181A) replicated in two distinct studies, and neither of them harbors genes seemingly related to allergy pathogenesis. Moreover, even when significant phenotype-related differences in DNA methylation were detected, these differences were often modest, of the order of a few percent, raising questions about their biological significance and consequences.

There are, however, attenuating circumstances that put these results in a more favorable perspective. First, even though the studies in question examined "asthma" or "atopic dermatitis," in reality the phenotypic heterogeneity among study populations is so striking that very few if any studies can be directly compared to one another. Moreover, the relatively simple technical requirements of genome-wide DNA methylation analyses have allowed investigators to test samples collected for other purposes. In other words, oftentimes DNA methylation analysis was a by-product of existing data collections and was performed because it was feasible, not because it was necessary or perhaps even relevant. As a result, the questions (and the answers) often appear contrived. The numbers of cases and controls in each study also vary greatly, reflecting the lack of firm criteria to define population sizes adequate to generate robust results, but these numbers are overall relatively small, often below 50–60 subjects.

The tissues/cells on which these studies were performed also deserve a comment. Because of availability and ease of access, many studies relied on DNA isolated from peripheral blood leukocytes (granulocytes, lymphocytes, and monocytes) or peripheral blood mononuclear cells (lymphocytes and monocytes), most often unfractionated. Obviously, such an approach can create problems if epigenetic marks are tissue-/cell type-specific and the cells bearing the mark

TABLE 1 Top Hits from Genome-Wide DNA Methylation Studies in Asthma and Allergy

Chr	Gene symbol[a]	Phenotypes/Exposures	Sample size	Tissue	Direction of methylation difference	References[b]
1	AK5	Neutrophilic, eosinophilic, and paucigranulocytic asthma vs controls	n=62	CD14+ monocytes	↑ in all three asthma phenotypes vs controls	[17]
	CLK2	Childhood asthma and levels of air pollution	n=200	Whole blood	↓ methylation in highly polluted Ostrava region (in asthmatics only)	[18]
	EFNA3	Subjects with and without asthma and high total IgE levels	n=664	Peripheral blood leukocytes	↓ in asthmatic subjects with high IgE levels	[11]
	IL-10	Childhood asthma and levels of air pollution	n=200	Whole blood	↓ methylation in highly polluted Ostrava region (in asthmatics only)	[18]
	KCNN3	IgE-mediated food allergy	n=24	CD4+ T cells	↓ in allergics vs nonallergics at birth and age 12 months	[14]
	KCNQ4	Neutrophilic, eosinophilic, and paucigranulocytic asthma vs controls	n=62	CD14+ monocytes	↑ in all three asthma phenotypes vs controls	[17]
	LRRC8C	Atopic dermatitis (AD, lesional or nonlesional tissue) vs healthy controls	n=57	Epidermis	↑ in AD lesional vs healthy controls	[19]
	PM20D1	Fetal exposure to maternal asthma	n=40	Whole blood	↓ in children of asthmatic mothers	[20]
	S100A5	AD (lesional or nonlesional tissue) vs healthy controls	n=57	Epidermis	↑ in AD lesional vs healthy controls	[19]
	SERPINC1	Subjects with and without asthma and high total IgE levels	n=664	Peripheral blood leukocytes	↓ in asthmatic subjects with high IgE levels	[11]
	SH2D2A	AD (lesional or nonlesional tissue) vs healthy controls	n=57	Epidermis	↑ in AD lesional vs healthy controls	[19]
	SLC25A33	Subjects with and without asthma and high total IgE levels	n=664	Peripheral blood leukocytes	↓ in asthmatic subjects with high IgE levels	[11]
2	ACSL3	Childhood asthma and maternal airborne PAH exposure	n=56	Umbilical cord white blood cells	↑ methylation associated with maternal PAH exposure and asthma	[21]
	B3GALT1	Atopic vs nonatopic asthmatics vs normal controls	n=24	Bronchial mucosal tissue	↓ in atopic vs nonatopic asthmatics	[22]
	CFLAR	AD (lesional or nonlesional tissue) vs healthy controls	n=57	Epidermis	↑ in AD lesional vs healthy controls	[19]

Continued

TABLE 1 Top Hits from Genome-Wide DNA Methylation Studies in Asthma and Allergy—cont'd

Chr	Gene symbol[a]	Phenotypes/Exposures	Sample size	Tissue	Direction of methylation difference	References[b]
	CTLA4	Childhood asthma and levels of air pollution	n=200	Whole blood	↓ methylation in highly polluted Ostrava region (in asthmatics and controls)	[18]
	CYBRD1	Childhood asthma and levels of air pollution	n=200	Whole blood	↓ methylation in highly polluted Ostrava region (in asthmatics only)	[18]
	GPR55	AD (lesional or nonlesional tissue) vs healthy controls	n=57	Epidermis	↑ in AD lesional vs healthy controls	[19]
	IL-1R2	Childhood asthma and levels of air pollution	n=200	Whole blood	↓ methylation in highly polluted Ostrava region (in asthmatics only)	[18]
	MFSD6	Subjects with and without asthma and high total IgE levels	n=664	Peripheral blood leukocytes	↓ in asthmatic subjects with high IgE levels	[11]
3	FAM19A4	Neutrophilic, eosinophilic, and paucigranulocytic asthma vs controls	n=62	CD14+ monocytes	↑ in all three asthma phenotypes vs controls	[17]
	IL-5RA	Subjects with and without asthma and high total IgE levels	n=664	Peripheral blood leukocytes	↓ in asthmatic subjects with high IgE levels	[11]
	PIK3CB	Subjects with and without asthma and high total IgE levels	n=664	Peripheral blood leukocytes	↓ in asthmatic subjects with high IgE levels	[11]
	RBP1	Neutrophilic, eosinophilic, and paucigranulocytic asthma vs controls	n=62	CD14+ monocytes	↑ in all three inflammatory phenotypes vs controls	[17]
	SLMAP	Subjects with and without asthma and high total IgE levels	n=664	Peripheral blood leukocytes	↓ in asthmatic subjects with high IgE levels	[11]
	TMEM41A	Subjects with and without asthma and high total IgE levels	n=664	Peripheral blood leukocytes	↓ in asthmatic subjects with high IgE levels	[11]
	ZNF445	Childhood asthma and levels of air pollution	n=200	Whole blood	↓ methylation in highly polluted Ostrava region (in asthmatics and controls)	[18]
4	NAP1L5	Fetal exposure to maternal asthma	n=40	Whole blood	↓ in children of asthmatic mothers	[20]
	SLC7A11	Subjects with and without asthma and high total IgE levels	n=664	Peripheral blood leukocytes	↓ in asthmatic subjects with high IgE levels	[11]
	ZNF718	Severe asthmatics vs healthy controls	n=10	Whole blood	↑ in severe asthmatics	[23]

5	IL-4	Subjects with and without asthma and high total IgE levels	n=664	Peripheral blood leukocytes	↓ in asthmatic subjects with high IgE levels	[11]
	TSLP	AD and prenatal smoking exposure	n=150	Umbilical cord white blood cells	↓ in AD vs non-AD, associated with high-dose cotinine exposure	[24]
6	ACAT2	Fetal exposure to maternal asthma	n=40	Whole blood	↓ in children of asthmatic mothers	[20]
	HLA-DMA	Patients with seasonal allergic rhinitis (SAR) vs healthy controls	n=40	CD4+ T cells	↑ in patients with SAR	[25]
	MAP3K5	Atopic vs nonatopic asthmatics vs normal controls	n=24	Bronchial mucosal tissue	↑ in atopic vs nonatopic asthmatics	[22]
	MAP3K7	Childhood asthma and levels of air pollution	n=200	Whole blood	↓ in methylation in highly polluted Ostrava region (in asthmatics only)	[18]
	ME1	Neutrophilic, eosinophilic, and paucigranulocytic asthma vs controls	n=62	CD14+ monocytes	↑ in all three asthma phenotypes vs controls	[17]
	PKHD1	Atopic vs nonatopic asthmatics vs normal controls	n=24	Bronchial mucosal tissue	↓ in atopic vs nonatopic asthmatics	[22]
	RPP21	Patients with SAR vs healthy controls	n=40	CD4+ T cells	↑ in patients with SAR	[25]
	SLC17A4	Subjects with and without asthma and high total IgE levels	n=664	Peripheral blood leukocytes	↓ in asthmatic subjects with high IgE levels	[11]
7	KEL	Subjects with and without asthma and high total IgE levels	n=664	Peripheral blood leukocytes	↓ in asthmatic subjects with high IgE levels	[11]
	NAPRT1	IgE-mediated food allergy	n=24	CD4+ T cells	↑ in allergics vs nonallergics at age 12 months	[14]
	PIK3CG	Childhood asthma and levels of air pollution	n=200	Whole blood	↓ in methylation in highly polluted Ostrava region (in asthmatics only)	[18]
8	DEFA1	Fetal exposure to maternal asthma	n=40	Whole blood	↑ in children of asthmatic mothers	[20]
	EPB49	AD (lesional or nonlesional tissue) vs healthy controls	n=57	Epidermis	↑ in AD lesional vs healthy controls	[19]
	NRG1	Neutrophilic, eosinophilic, and paucigranulocytic asthma vs controls	n=62	CD14+ monocytes	↑ in all three asthma phenotypes vs controls	[17]

Continued

TABLE 1 Top Hits from Genome-Wide DNA Methylation Studies in Asthma and Allergy—cont'd

Chr	Gene symbol[a]	Phenotypes/Exposures	Sample size	Tissue	Direction of methylation difference	References[b]
	PHF20L1	Childhood asthma and levels of air pollution	n=200	Whole blood	↓ in methylation in highly polluted Ostrava region (in asthmatics only)	[18]
	TRPS1	Childhood asthma and levels of air pollution	n=200	Whole blood	↓ methylation in highly polluted Ostrava region (in asthmatics only)	[18]
9	CACNA1B	IgE-mediated food allergy	n=24	CD4+ T cells	↑ in allergics vs nonallergics at birth and age 12 mo	[14]
	CEL	Subjects with and without asthma and high total IgE levels	n=664	Peripheral blood leukocytes	↓ in asthmatic subjects with high IgE levels	[11]
	COL15A1	Subjects with and without asthma and high total IgE levels	n=664	Peripheral blood leukocytes	↓ in asthmatic subjects with high IgE levels	[11]
	PTGDS	Aspirin-intolerant (AIA) vs aspirin-tolerant asthmatics	n=9	Nasal polyps	↓ in AIA	[26]
	PTGES	AIA vs aspirin-tolerant asthmatics	n=9	Nasal polyps	↑ in AIA	[26]
	SPINK4	Subjects with and without asthma and high total IgE levels	n=664	Peripheral blood leukocytes	↓ in asthmatic subjects with high IgE levels	[11]
10	CYP26A1	House dust mite allergic asthmatics or aspirin-intolerant asthmatics vs controls	n=43	CD19+ B cells	↑ in allergic asthmatics vs nonallergic group (AIA and controls)	[27]
	IL-2RA	Obese vs nonobese asthmatics and controls	n=32	Peripheral blood mononuclear cells	↓ in obese vs nonobese asthmatics	[28]
	SNORA12	Response to in vitro HRV infection of nasal cells from asthmatics vs healthy controls	n=58	Nasal epithelial cells	↑ in healthy controls after in vitro HRV infection	[29]
	ZNF22	Subjects with and without asthma and high total IgE levels	n=664	Peripheral blood leukocytes	↓ in asthmatic subjects with high IgE levels	[11]
11	ALDH3B2	Subjects with and without asthma and high total IgE levels	n=664	Peripheral blood leukocytes	↓ in asthmatic subjects with high IgE levels	[11]
	C11orf47	Atopic vs nonatopic asthmatics vs normal controls	n=24	Bronchial mucosal tissue	↓ in atopic vs nonatopic asthmatics	[22]
	MMP7	AD (lesional or nonlesional tissue) vs healthy controls	n=57	Epidermis	↑ in AD lesional vs healthy controls	[19]

	PRG2	Subjects with and without asthma and high total IgE levels	n=664	Peripheral blood leukocytes	↓ in asthmatic subjects with high IgE levels	[11]
	PRG3	Subjects with and without asthma and high total IgE levels	n=664	Peripheral blood leukocytes	↓ in asthmatic subjects with high IgE levels	[11]
	SLC43A3	Subjects with and without asthma and high total IgE levels	n=664	Peripheral blood leukocytes	↓ in asthmatic subjects with high IgE levels	[11]
12	CHFR	Fetal exposure to maternal asthma	n=40	Whole blood	↑ in children of asthmatic mothers	[20]
	ERP27	AD (lesional or nonlesional tissue) vs healthy controls	n=57	Epidermis	↑ in AD lesional vs healthy controls	[19]
	PDE6H	Subjects with and without asthma and high total IgE levels	n=664	Peripheral blood leukocytes	↓ in asthmatic subjects with high IgE levels	[11]
	PIWIL1	Fetal exposure to maternal asthma	n=40	Whole blood	↑ in children of asthmatic mothers	[20]
	TBX5	Neutrophilic, eosinophilic, and paucigranulocytic asthma vs controls	n=62	CD14+ monocytes	↑ in all three asthma phenotypes vs controls	[17]
	TMEM52B	Subjects with and without asthma and high total IgE levels	n=664	Peripheral blood leukocytes	↓ in asthmatic subjects with high IgE levels	[11]
13	ALOX5AP	AIA vs aspirin-tolerant asthmatics	n=9	Nasal polyps	↓ in AIA	[26]
	LOC283487	AD (lesional or nonlesional tissue) vs healthy controls	n=57	Epidermis	↑ in AD lesional vs healthy controls	[19]
	RB1	Subjects with and without asthma and high total IgE levels	n=664	Peripheral blood leukocytes	↓ in asthmatic subjects with high IgE levels	[11]
14	CRIP1	Atopics vs asthmatics vs healthy controls	n=25	Airway epithelial cells	↓ in atopics vs asthmatics	[30]
	FAM181A	Severe asthmatics vs healthy controls	n=10	Whole blood	↓ in severe asthmatics	[23]
	FAM181A	Fetal exposure to maternal asthma	n=40	Whole blood	↑ in children of asthmatic mothers	[20]
	L2HGDH	Subjects with and without asthma and high total IgE levels	n=664	Peripheral blood leukocytes	↓ in asthmatic subjects with high IgE levels	[11]
	LTB4R	AIA vs aspirin-tolerant asthmatics	n=9	Nasal polyps	↓ in AIA	[26]
15	SYNM	Neutrophilic, eosinophilic, and paucigranulocytic asthma vs controls	n=62	CD14+ monocytes	↑ in all three asthma phenotypes vs controls	[17]

Continued

TABLE 1 Top Hits from Genome-Wide DNA Methylation Studies in Asthma and Allergy—cont'd

Chr	Gene symbol[a]	Phenotypes/Exposures	Sample size	Tissue	Direction of methylation difference	References[b]
16	CDH1	Atopic vs nonatopic asthmatics vs normal controls	n=24	Bronchial mucosal tissue	↓ in atopic vs nonatopic asthmatics	[22]
	LPCAT2	Subjects with and without asthma and high total IgE levels	n=664	Peripheral blood leukocytes	↓ in asthmatic subjects with high IgE levels	[11]
	MRPL28	Fetal exposure to maternal asthma	n=40	Whole blood	↑ in children of asthmatic mothers	[20]
	PIEZO1	Patients with SAR vs healthy controls	n=40	CD4+ T cells	↓ in patients with SAR	[25]
	SEPT12	Subjects with and without asthma and high total IgE levels.	n=664	Peripheral blood leukocytes	↓ in asthmatic subjects with high IgE levels	[11]
17	ALOX12	Asthma-related persistent wheeze and prenatal exposure to dichlorodiphenyldichloroethylene	n=358	Whole blood	↓ in persistent vs never wheezers	[31]
	CCL5	Obese vs nonobese asthmatics and controls	n=32	Peripheral blood mononuclear cells	↓ in obese asthmatics vs obese nonasthmatics	[28]
	ITGA2B	Subjects with and without asthma and high total IgE levels	n=664	Peripheral blood leukocytes	↓ in asthmatic subjects with high IgE levels	[11]
	MAPK8IP3	Fetal exposure to maternal asthma	n=40	Whole blood	↓ in children of asthmatic mothers	[20]
	PYY2	Neutrophilic, eosinophilic, and paucigranulocytic asthma vs controls	n=62	CD14+ monocytes	↑ in all three asthma phenotypes vs controls	[17]
	STAT5A	Atopics vs asthmatics vs healthy controls	n=25	Airway epithelial cells	↓ in atopics vs asthmatics	[30]
	TBX21	Obese vs nonobese asthmatics and controls	n=32	Peripheral blood mononuclear cells	↓ in obese asthmatics vs healthy controls	[28]
19	CLC	Subjects with and without asthma and high total IgE levels	n=664	Peripheral blood leukocytes	↓ in asthmatic subjects with high IgE levels	[11]
	FCER2	Obese vs nonobese asthmatics and controls	n=32	Peripheral blood mononuclear cells	↑ in obese vs nonobese asthmatics	[28]
	KLF1	Subjects with and without asthma and high total IgE levels	n=664	Peripheral blood leukocytes	↓ in asthmatic subjects with high IgE levels	[11]
	MRI1	Fetal exposure to maternal asthma	n=40	Whole blood	↑ in children of asthmatic mothers	[20]
	MRI1	Severe asthmatics vs healthy controls	n=10	Whole blood	↑ in severe asthmatics	[23]

	TGFB1	Obese vs nonobese asthmatics and controls	n=32	Peripheral blood mononuclear cells	↑ in obese asthmatics vs healthy controls	[28]
	TMEM86B	Subjects with and without asthma and high total IgE levels	n=664	Peripheral blood leukocytes	↓ in asthmatic subjects with high IgE levels	[11]
20	AURKA	Fetal exposure to maternal asthma	n=40	Whole blood	↑ in children of asthmatic mothers	[20]
	FAM112A	Subjects with and without asthma and high total IgE levels	n=664	Peripheral blood leukocytes	↓ in asthmatic subjects with high IgE levels	[11]
21	ADARB1	Subjects with and without asthma and high total IgE levels	n=664	Peripheral blood leukocytes	↓ in asthmatic subjects with high IgE levels	[11]
	TFF1	Subjects with and without asthma and high total IgE levels	n=664	Peripheral blood leukocytes	↓ in asthmatic subjects with high IgE levels	[11]
X	CXorf40A	AD or psoriasis vs healthy controls	n=37	Naïve CD4+ T cells	↑ in subjects with psoriasis	[32]
	EMD	AD or psoriasis vs healthy controls	n=37	Naïve CD4+ T cells	↑ in subjects with psoriasis	[32]
	GATA1	Subjects with and without asthma and high total IgE levels	n=664	Peripheral blood leukocytes	↓ in asthmatic subjects with high IgE levels	[11]
	HDAC6	AD or psoriasis vs healthy controls	n=37	Naïve CD4+ T cells	↑ in subjects with psoriasis	[32]
	IKBKG	AD or psoriasis vs healthy controls	n=37	Naïve CD4+ T cells	↑ in subjects with psoriasis	[32]
	SLITRK4	AD or psoriasis vs healthy controls	n=37	Naïve CD4+ T cells	↑ in subjects with psoriasis	[32]
	ZIC3	AD or psoriasis vs healthy controls	n=37	Naïve CD4+ T cells	↑ in subjects with psoriasis	[32]

PAH, polycyclic aromatic hydrocarbon; ↑, increase; ↓, decrease; HRV, human rhinovirus
[a] *Nonredundant gene names are listed from each individual study.*
[b] *96 loci from Martino et al. [13] are not included in the table.*

TABLE 2 Candidate Gene Studies: DNA Methylation in Asthma and Allergy

Chr	Gene Symbol	Phenotypes/Exposures	Sample Size	Tissue	Direction of methylation difference	References
1	FCER1G	Atopic dermatitis (AD) vs healthy controls	n=20	CD14+ monocytes	↓ in AD	[33]
	ORMDL1	Childhood asthma, farm exposure, and age	n=46	Whole blood	↓ in farm children; differentially methylated with age	[34]
4	IL-2	Childhood asthma	n=303	Cord blood mononuclear cells	↑ in methylation was associated with ↑ in likelihood of severe asthma outcomes	[35]
5	ADRB2	Mild vs severe childhood asthma and NO_2 exposure	n=182	Cord blood	↑ in severe asthmatics who were exposed to NO_2	[36]
	ADRB2	Childhood asthma	n=177	Whole blood and saliva	↑ in methylation was associated with ↓ in asthma severity	[37]
	CD14	Childhood asthma and allergic sensitization and age	n=157	Whole blood	↑ methylation from 2 to 10 years of age.	[38]
	CD14	Pet keeping and tobacco smoke exposure (TSE) at ages 2 and 10 years	n=157	Whole blood	↑ in methylation in children without pets vs children with pets; ↑ in methylation in TSE-unexposed children	[39]
	CD14	Prenatal farm exposure	n=94	Placentas	↓ in placentas of mothers living on a farm	[40]
	IL-13	Childhood asthma, farm exposure, and age	n=46	Whole blood	↑ in farm children; differentially methylated with age	[34]
	IL-13	Asthma-related lung function, prenatal maternal smoking and IL-13 SNPs	n=245	Peripheral blood leukocytes	↑ methylation was associated with an interaction between maternal prenatal smoking and rs20541. ↑ in methylation interacts with rs1800925, which significantly ↓ lung function	[41]
	IL-13	Association between genetic variation and total serum IgE and asthma risk	n=134	Whole blood	↑ in methylation was associated with the T allele in rs2240032 and was associated with total serum IgE levels	[42]
	IL-4	Childhood asthma, farm exposure, and age	n=46	Whole blood	Differentially methylated with age	[34]
	IL-4	Diisocyanate-induced occupational asthmatics (DA+) vs symptomatic exposed (DA-) and asymptomatic exposed workers (AW)	n=131	Whole blood	↓ in DA+ compared with AW among nonsmoking workers.	[43]

	IL-4	Asthma risk, genetic variation, and age	n=245	Peripheral blood leukocytes	↓ cg23943829 methylation between ages 10 and 18	[44]
	IL-4	In vitro allergen or PHA-stimulated vs nonstimulated cells from allergic asthmatics and controls	n=6	CD4+ T cells	↓ in methylation in asthmatics after simulation with Dp or Df	[45]
	RAD50	Childhood asthma, farm exposure, and age	n=46	Whole blood	↑ in farm children; differentially methylated with age	[34]
	TSLP	AD vs healthy controls	n=20	Epidermis	↓ in skin lesions from AD patients	[46]
6	ARG1	Childhood asthma and FeNO	n=940	Buccal cells	↑ methylation was associated with ↓ in FeNO (stronger among asthmatics than nonasthmatics)	[47]
7	ADCYAP1R1	Risk for childhood asthma	n=516	White blood cells	↑ in asthmatics vs control (more prevalent in males than females)	[48]
	IL-6	Childhood asthma, FeNO, FEV1, and wheezing	n=35	Nasal cells	↓ methylation was associated with ↑ FeNO	[49]
	NOS3	Childhood respiratory disease and air pollution	n=940	Buccal cells	↑ PM2.5 was associated with ↑ methylation	[50]
	NPSR1	Asthma and different environmental exposures	n=171	Whole blood	↓ in child and adult asthmatics; methylation was associated with parental smoking and sampling season in children, and with current and former smoking in adults	[51]
10	GATA3	Asthma risk, genetic variation, and age	n=245	Peripheral blood leukocytes	At age 10, ↓ asthma risk was associated with ↑ methylation	[44]
12	IFNG	In vitro allergen or PHA-stimulated vs nonstimulated cells from allergic asthmatics and controls	n=6	CD4+ T cells	↓ in methylation in the control group after PHA stimulation	[45]
	IFNG	Impact of sex, age, and airway vs systemic tissue on allergic asthmatics	n=74	Buccal cells and CD4+ T cells	↑ in the CD4+ lymphocytes and buccal cells of children vs adults; ↑ in males vs females in CD4+ lymphocytes but not buccal cells	[52]
	IFNG	Diisocyanate-induced occupational asthmatics (DA+) vs symptomatic exposed (DA-) and asymptomatic exposed workers (AW)	n=131	Whole blood	↑ in DA+ compared to other groups	[43]

Continued

TABLE 2 Candidate Gene Studies: DNA Methylation in Asthma and Allergy—cont'd

Chr	Gene Symbol	Phenotypes/Exposures	Sample Size	Tissue	Direction of methylation difference	References
	IFNG	Adult twins discordant for asthma and second-hand smoke exposure	n=42	Effector T cells	↑ in methylation in Teff from asthmatic twins	[53]
	IFNG	Maternal PAH exposure	n=53	Umbilical cord buffy coat	↑ IFNG was associated with maternal PAH exposure	[54]
	STAT6	Childhood asthma, farm exposure, and age	n=46	Whole blood	↓ in farm children	[34]
13	PCDH20	Asthmatic vs nonasthmatic smokers (non-COPD)	n=695	Sputum	↑ methylation was associated with asthma	[55]
14	ARG2	Childhood asthma and FeNO	n=940	Buccal cells	↑ methylation associated with ↓ in FeNO (stronger among asthmatics than nonasthmatics)	[47]
	PTGDR	Allergic asthmatics vs controls	n=637	CD19+ B cells	↓ in allergic individuals	[56]
16	IL-4R	Risk for asthma at age 18 and IL-4R SNPs	n=245	Whole blood	↑ in risk for asthma was associated with an interaction between rs3024685 and ↑ in methylation	[57]
	IL-4R	Asthma risk, genetic variation, and age	n=245	Peripheral blood leukocytes	IL-4R cg26937798 methylation decreases between ages 10 and 18. At age 18, ↑ in cg26937798 methylation among subjects with rs3024685 genotype AA or rs8832 genotype GG was associated with a ↓ in asthma risk	[44]
17	IKZF3	Childhood asthma risk (healthy controls vs controlled persistent asthma vs severe asthma) and asthma-associated SNPs	n=2079	Whole blood and nine purified blood cell populations	↑ in children with controlled persistent asthma vs healthy controls and children with severe asthma	[58]
	iNOS (NOS2)	Childhood asthma, FeNO, FEV1, and wheezing	n=35	Nasal cells	↓ methylation was associated with ↑ FeNO	[49]
	iNOS (NOS2)	Ambient air pollutants, genetic variation, and FeNO levels in children	n=940	Buccal cells	↑ in short-term PM2.5 exposure was associated with ↓ in methylation	[59]
	NOS2A	Childhood respiratory disease and air pollution	n=940	Buccal cells	↑ PM2.5 was associated with ↓ methylation	[50]

	ORMDL3	Childhood asthma risk (healthy controls vs controlled persistent asthma vs severe asthma) and asthma-associated SNPs	n=2079	Whole blood and nine purified blood cell populations	↑ in children with controlled persistent asthma vs healthy controls and children with severe asthma	[58]
	ORMDL3	Childhood asthma, farm exposure, and age	n=46	Whole blood	↑ in asthmatics; differentially methylated with age	[34]
	ZPBP2	Sex- and age-specific patterns of genetic association to childhood asthma	n=80	Peripheral blood mononuclear cells	↓ in methylation in males vs females; ↑ in adult males vs boys.	[60]
X	FOXP3	Childhood wheeze and/or asthma and air pollution	n=92	Saliva	↑ in methylation was associated with ↑ risk of wheeze. Children with ↑ methylation were more likely to have asthma	[61]
	FOXP3	Atopic children (asthma and/or allergic rhinitis) and ambient PAH exposure	n=256	Regulatory T cells	↑ in methylation associated with ↑ in average PAH exposure, conditional on atopic status	[62]
	FOXP3	Risk for AD	n=346	Whole blood	↑ in methylation at birth was associated with ↑ risk to develop AD and food allergy during the first year of life	[63]
	FOXP3	Childhood asthma and farm exposure	n=298	Whole blood	↓ in farm milk exposed children	[64]
	FOXP3	Asthmatics vs nonasthmatics and levels of ambient air pollution	n=181	Regulatory T cells	↓ in children exposed to high ambient air pollution	[65]
	FOXP3	Asthma: monozygotic twin (adult) pairs discordant for disease	n=42	Regulatory T cells	↑ in methylation in Tregs from asthmatic twins	[53]
	FOXP3	Prenatal farm exposure	n=82	Cord blood mononuclear cells	↓ in methylation in children of mothers with farm milk exposure	[66]
	FOXP3	Dual sublingual immunotherapy (SLIT)	n=30	Regulatory T cells	↓ in methylation associated with dual SLIT treatment vs placebo	[67]

PHA, phytohemagglutinin; FEV1, forced expiratory volume in 1 s;

of interest are present in different proportions among cases and controls. Conversely, if certain epigenetic regulatory processes are tissue specific, detection of the relevant marks might be hindered when analyzing complex cellular mixtures. On the other hand, allergic diseases have a major immunologic component, and in this perspective focusing on DNA methylation signatures of peripheral immune blood cells may be acceptable, at least for an initial screening phase. Moreover, methods have been recently developed to infer the proportion of immune cell populations in peripheral blood from the DNA methylation data themselves [10]. Such methods allow adjusting for differences in these proportions, leading to more robust results, and large effect sizes in specific cells may also allow detecting associations in cell mixtures, especially with large sample sizes. For instance, a recent genome-wide study of the association between DNA methylation and total serum IgE concentrations in peripheral blood leukocytes from 95 nuclear pedigrees identified 36 IgE-associated loci in genes that encode eosinophil products. Methylation at these loci differed significantly in isolated eosinophils from subjects with and without asthma and high IgE levels [11].

On balance, we would argue that the search for epigenetic biomarkers of allergic disease is still in its infancy, and the studies performed so far provide food for thought in terms of fundamental variables such as study design, population size, depth of phenotypic assessment, and choice of DNA source. Some studies have reported hits in plausible genes (e.g., TSLP, an epithelial cytokine that acts on dendritic cells to promote Th2 cell differentiation; IL2RA, a key regulator of the immune response; TBX21/Tbet, the master transcriptional regulator of Th1 responses; FCER2, the low-affinity IgE receptor; TGFB1, the central molecule in the development of both T-regulatory and Th17 cells: Table 1), but the vast majority of results point to regions that hold no clear mechanistic relationship to allergy pathogenesis and will require further functional explorations. Moreover, interestingly enough, none of the hits generated in peripheral blood replicated in more "physiologic" airway tissues (for asthma) or skin (for atopic dermatitis), and vice versa. This discrepancy may reflect the inability of peripheral blood to elucidate tissue-based disease mechanisms, but may also point to distinct components of disease pathogenesis. Finally, it is interesting, and possibly informative, that overall DNA methylation signatures did not replicate not only across studies of the same disease (with the caveats about phenotypic heterogeneity mentioned above) but also when comparing the three main disease phenotypes typically associated with allergy (asthma, allergic rhinitis, and atopic dermatitis). This situation resembles the results generated by GWAS, when to the surprise of many, different genes were found to be associated with asthma and IgE levels [12] despite the striking clinical overlap between these two phenotypes. This pattern may indicate that distinct pathways underpin these processes.

So far, most if not all DNA methylation studies in allergy and asthma were aimed at the identification of epigenetic *determinants* of allergic diseases, but new efforts are also ongoing to discover epigenetic disease *biomarkers*. In particular, a recent Australian study explored the possibility of using DNA methylation data as clinical biomarkers for food allergy [13]. The diagnosis of true food allergy is often challenging because approximately half of food-sensitized patients are asymptomatic. Food allergen-specific IgE are excellent markers of sensitization but poor predictors of clinical reactivity. Thus, oral food challenges remain the golden standard to determine a patient's risk of reactivity, but they are fraught with risks of anaphylactic reactions. The goal of this study was to assess whether DNA methylation profiles in an easily accessible tissue, peripheral blood mononuclear cells, could predict food challenge outcomes. Genome-wide DNA methylation was profiled in 58 food-sensitized children (aged 11–15 months), half of whom were clinically reactive. These children had undergone oral food challenges, concurrent skin prick tests, and specific IgE tests.

Thirteen non-allergic subjects were also included as controls. A supervised learning approach identified a 96 CpG site DNA methylation signature that predicted clinical outcomes. Diagnostic scores derived from these 96 sites outperformed allergen-specific IgE and skin prick tests for predicting oral food challenge outcomes, and food allergy status was correctly predicted in a replication cohort with an accuracy of 79.2%. Albeit preliminary, this study suggests that in time, DNA methylation profiles may become clinically valuable biomarkers [13]. On the other hand, it is noteworthy that these results are difficult to interpret because the only canonical pathway enriched in the list of genes associated with the CpGs that best predict food challenge outcome is the MAP kinase pathway (with MAPK8IP1, MAPK8IP2, MAP3K1, RPS6KA2, RASGRP2, NF1, ZAK, and FGF12). An association between clinical food allergy and profiles of DNA methylation in MAP kinase signaling genes had been previously reported in another Australian cohort [14], but different MAP kinase genes were implicated. Despite these difficulties, the attempt to turn epigenetic information into clinical disease biomarkers should be considered a move in the right direction.

5. CONCLUSION: WHERE IS THE FIELD GOING?

This takes us to the last and perhaps the most important question about the future of allergy epigenetics. There is undoubtedly much need (and room) for improvement in this field. One unavoidable hurdle of the initial genome-wide studies is that even the most exhaustive ones relied on platforms that provide limited genomic coverage and were often designed primarily to address cancer-related differential DNA methylation. In GWAS, patterns of linkage disequilibrium across the genome allow tagging groups of variants and genotyping a single one to get dependable information about the profile of the entire group [15]. Unfortunately, similar information is not available for the methylome, and tagging strategies might work only to a limited extent (e.g., in one tissue but not in another) or even not at all. Therefore, collection of more robust information will require the development of more inclusive array platforms and/or the coming of age of next-generation sequencing-based methods—not unthinkable goals, considering how rapidly the relevant technologies have been advancing.

Another major hurdle for the current genome-wide DNA methylation analyses is that, with few exceptions, they leveraged existing populations designed for other purposes. For epigenetics to make a difference in the discovery of allergy biomarkers, the design of future studies will need to incorporate the notion that epigenetics explores *mechanisms* potentially influencing allergy susceptibility and severity, and that such mechanisms are typically triggered in response to environmental signals and developmental programs. Therefore, well-phenotyped, *longitudinal* mother–child cohorts with cord blood and early infancy samples and in-depth assessments of environmental exposures (to chemicals, diet, and also microbiota) might provide epigenetic studies with an opportunity to unveil determinants of disease susceptibility. While obtaining samples other than cord and peripheral blood may be impossible in such populations for obvious logistical reasons, such cohorts will allow for the analysis of *epigenetic trajectories over time*, a theme that appears to be exquisitely relevant to the developmental processes epigenetic mechanisms primarily regulate. Such birth cohorts will also allow asking intriguing questions about the relationships between maternal and fetal methylome, fetal methylome and maternal exposures (for instance, to smoking or to specific environmental microbial profiles), and fetal methylome and early (and later) life allergic disease outcomes. Several such birth cohorts [e.g., the Tucson Infant Immune Study (IIS) [16], the Wisconsin Childhood Origins of ASThma (COAST, http://www.medicine.wisc.edu/coast/family) study, the Danish Copenhagen Prospective Study of Asthma in Childhood (COPSAC, http://www.copsac.com/

content/asthma), the UK Manchester Asthma and Allergy Study (MAAS, http://www.maas.org.uk/) exist in the United States and Europe, and research along these lines is already beginning. Because the unique time- and environment-dependent nature of epigenetic marks appear much better suited for the analysis of prenatal and early postnatal trajectories to disease than for conventional, static assessments of disease risk, we expect that this second generation of epigenetic studies will avoid most of the pitfalls the initial round unavoidably fell into and will return exciting results highlighting the potential of epigenetic biomarkers to foster a better understanding of allergic disease pathogenesis.

LIST OF ABBREVIATIONS

AD Atopic dermatitis
AIA Aspirin-intolerant asthma
DA Diisocyanate-induced occupational asthma
DEP Diesel exhaust pollution
Df *Dermatophagoides farina*
Dp *Dermatophagoides pteronyssinus*
FeNO Fractional exhaled nitric oxide
FEV1 Forced expiratory volume in 1 s
GWAS Genome-wide association studies
NO_2 Nitrogen dioxide
PAH Polycyclic aromatic hydrocarbon
PHA Phytohemagglutinin
PM2.5 Particulate matter ≤2.5 μm aerodynamic diameter
PM10 Particulate matter ≤10 μm aerodynamic diameter

References

[1] DeVries A, Vercelli D. The epigenetics of human asthma and allergy: promises to keep. Asian Pac J Allergy Immunol 2013;31(3):183–9.

[2] Martinez FD, Vercelli D. Asthma. Lancet October 19, 2013;382(9901):1360–72.

[3] Leung DY, Guttman-Yassky E. Deciphering the complexities of atopic dermatitis: shifting paradigms in treatment approaches. J Allergy Clin Immunol October 2014;134(4):769–79.

[4] Meyers DA, Bleecker ER, Holloway JW, Holgate ST. Asthma genetics and personalised medicine. Lancet Respir Med May 2014;2(5):405–15.

[5] Manolio TA. Genomewide association studies and assessment of the risk of disease. N Engl J Med July 8, 2010;363(2):166–76.

[6] Baurecht H, Hotze M, Brand S, Buning C, Cormican P, Corvin A, et al. Genome-wide comparative analysis of atopic dermatitis and psoriasis gives insight into opposing genetic mechanisms. Am J Hum Genet January 8, 2015;96(1):104–20.

[7] Manolio TA, Collins FS, Cox NJ, Goldstein DB, Hindorff LA, Hunter DJ, et al. Finding the missing heritability of complex diseases. Nature October 8, 2009;461(7265):747–53.

[8] von Mutius E, Vercelli D. Farm living: effects on childhood asthma and allergy. Nat Rev Immunol December 2010;10(12):861–8.

[9] Ober C, Vercelli D. Gene-environment interactions in human disease: nuisance or opportunity? Trends Genet 2011;27:107–14.

[10] Houseman EA, Accomando WP, Koestler DC, Christensen BC, Marsit CJ, Nelson HH, et al. DNA methylation arrays as surrogate measures of cell mixture distribution. BMC Bioinforma 2012;13:86.

[11] Liang L, Willis-Owen SA, Laprise C, Wong KC, Davies GA, Hudson TJ, et al. An epigenome-wide association study of total serum immunoglobulin E concentration. Nature February 18, 2015.

[12] Moffatt MF, Gut IG, Demenais F, Strachan DP, Bouzigon E, Heath S, et al. A large-scale, consortium-based genomewide association study of asthma. N Engl J Med September 23, 2010;363(13):1211–21.

[13] Martino D, Dang T, Sexton-Oates A, Prescott S, Tang ML, Dharmage S, et al. Blood DNA methylation biomarkers predict clinical reactivity in food-sensitized infants. J Allergy Clin Immunol February 10, 2015.

[14] Martino D, Joo JE, Sexton-Oates A, Dang T, Allen K, Saffery R, et al. Epigenome-wide association study reveals longitudinally stable DNA methylation differences in CD4+ T cells from children with IgE-mediated food allergy. Epigenetics July 2014;9(7):998–1006.

[15] Carlson CS, Eberle MA, Kruglyak L, Nickerson DA. Mapping complex disease loci in whole-genome association studies. Nature 2004;429:446–52.

[16] Rothers J, Stern DA, Spangenberg A, Lohman IC, Halonen M, Wright AL. Influence of early day-care exposure on total IgE levels through age 3 years. J Allergy Clin Immunol November 2007;120(5):1201–7.

[17] Gunawardhana LP, Gibson PG, Simpson JL, Benton MC, Lea RA, Baines KJ. Characteristic DNA methylation profiles in peripheral blood monocytes are associated with inflammatory phenotypes of asthma. Epigenetics September 2014;9(9):1302–16.

[18] Rossnerova A, Tulupova E, Tabashidze N, Schmuczerova J, Dostal M, Rossner Jr P, et al. Factors affecting the 27K DNA methylation pattern in asthmatic and healthy children from locations with various environments. Mutat Res January–February 2013;741–742:18–26.

[19] Rodriguez E, Baurecht H, Wahn AF, Kretschmer A, Hotze M, Zeilinger S, et al. An integrated epigenetic and transcriptomic analysis reveals distinct tissue-specific patterns of DNA methylation associated with atopic dermatitis. J Invest Dermatol July 2014;134(7):1873–83.
[20] Gunawardhana LP, Baines KJ, Mattes J, Murphy VE, Simpson JL, Gibson PG. Differential DNA methylation profiles of infants exposed to maternal asthma during pregnancy. Pediatr Pulmonol September 2014;49(9): 852–62.
[21] Perera F, Tang WY, Herbstman J, Tang D, Levin L, Miller R, et al. Relation of DNA methylation of 5′-CpG island of ACSL3 to transplacental exposure to airborne polycyclic aromatic hydrocarbons and childhood asthma. PLoS One 2009;4(2):e4488.
[22] Kim YJ, Park SW, Kim TH, Park JS, Cheong HS, Shin HD, et al. Genome-wide methylation profiling of the bronchial mucosa of asthmatics: relationship to atopy. BMC Med Genet 2013;14:39.
[23] Wysocki K, Conley Y, Wenzel S. Epigenome variation in severe asthma. Biol Res Nurs October 6, 2014.
[24] Wang IJ, Chen SL, Lu TP, Chuang EY, Chen PC. Prenatal smoke exposure, DNA methylation, and childhood atopic dermatitis. Clin Exp Allergy May 2013;43(5):535–43.
[25] Nestor CE, Barrenas F, Wang H, Lentini A, Zhang H, Bruhn S, et al. DNA methylation changes separate allergic patients from healthy controls and may reflect altered CD4+ T-cell population structure. PLoS Genet January 2014;10(1):e1004059.
[26] Cheong HS, Park SM, Kim MO, Park JS, Lee JY, Byun JY, et al. Genome-wide methylation profile of nasal polyps: relation to aspirin hypersensitivity in asthmatics. Allergy May 2011;66(5):637–44.
[27] Pascual M, Suzuki M, Isidoro-Garcia M, Padron J, Turner T, Lorente F, et al. Epigenetic changes in B lymphocytes associated with house dust mite allergic asthma. Epigenetics 2011;6(9):1131–7.
[28] Rastogi D, Suzuki M, Greally JM. Differential epigenome-wide DNA methylation patterns in childhood obesity-associated asthma. Sci Rep 2013;3:2164.
[29] McErlean P, Favoreto Jr S, Costa FF, Shen J, Quraishi J, Biyasheva A, et al. Human rhinovirus infection causes different DNA methylation changes in nasal epithelial cells from healthy and asthmatic subjects. BMC Med Genomics 2014;7:37.
[30] Stefanowicz D, Hackett TL, Garmaroudi FS, Gunther OP, Neumann S, Sutanto EN, et al. DNA methylation profiles of airway epithelial cells and PBMCs from healthy, atopic and asthmatic children. PLoS One 2012;7(9):e44213.
[31] Morales E, Bustamante M, Vilahur N, Escaramis G, Montfort M, de Cid R, et al. DNA hypomethylation at ALOX12 is associated with persistent wheezing in childhood. Am J Respir Crit Care Med May 1, 2012;185(9):937–43.
[32] Han J, Park SG, Bae JB, Choi J, Lyu JM, Park SH, et al. The characteristics of genome-wide DNA methylation in naive CD4+ T cells of patients with psoriasis or atopic dermatitis. Biochem Biophys Res Commun May 25, 2012;422(1):157–63.
[33] Liang Y, Wang P, Zhao M, Liang G, Yin H, Zhang G, et al. Demethylation of the FCER1G promoter leads to FcepsilonRI overexpression on monocytes of patients with atopic dermatitis. Allergy March 2012;67(3):424–30.
[34] Michel S, Busato F, Genuneit J, Pekkanen J, Dalphin JC, Riedler J, et al. Farm exposure and time trends in early childhood may influence DNA methylation in genes related to asthma and allergy. Allergy March 2013;68(3):355–64.
[35] Curtin JA, Simpson A, Belgrave D, Semic-Jusufagic A, Custovic A, Martinez FD. Methylation of IL-2 promoter at birth alters the risk of asthma exacerbations during childhood. Clin Exp Allergy March 2013;43(3):304–11.
[36] Fu A, Leaderer BP, Gent JF, Leaderer D, Zhu Y. An environmental epigenetic study of ADRB2 5′-UTR methylation and childhood asthma severity. Clin Exp Allergy November 2012;42(11):1575–81.
[37] Gaffin JM, Raby BA, Petty CR, Hoffman EB, Baccarelli AA, Gold DR, et al. beta-2 adrenergic receptor gene methylation is associated with decreased asthma severity in inner-city schoolchildren: asthma and rhinitis. Clin Exp Allergy 2014;44(5):681–9.
[38] Munthe-Kaas MC, Torjussen TM, Gervin K, Lodrup Carlsen KC, Carlsen KH, Granum B, et al. CD14 polymorphisms and serum CD14 levels through childhood: a role for gene methylation? J Allergy Clin Immunol June 2010;125(6):1361–8.
[39] Munthe-Kaas MC, Bertelsen RJ, Torjussen TM, Hjorthaug HS, Undlien DE, Lyle R, et al. Pet keeping and tobacco exposure influence CD14 methylation in childhood. Pediatr Allergy Immunol December 2012;23(8):747–54.
[40] Slaats GG, Reinius LE, Alm J, Kere J, Scheynius A, Joerink M. DNA methylation levels within the CD14 promoter region are lower in placentas of mothers living on a farm. Allergy July 2012;67(7):895–903.
[41] Patil VK, Holloway JW, Zhang H, Soto-Ramirez N, Ewart S, Arshad SH, et al. Interaction of prenatal maternal smoking, interleukin 13 genetic variants and DNA methylation influencing airflow and airway reactivity. Clin Epigenet 2013;5(1):22.
[42] Schieck M, Sharma V, Michel S, Toncheva AA, Worth L, Potaczek DP, et al. A polymorphism in the TH 2 locus control region is associated with changes in DNA methylation and gene expression. Allergy September 2014;69(9):1171–80.
[43] Ouyang B, Bernstein DI, Lummus ZL, Ying J, Boulet LP, Cartier A, et al. Interferon-gamma promoter is hypermethylated in blood DNA from workers with confirmed diisocyanate asthma. Toxicol Sci June 2013;133(2):218–24.

[44] Zhang H, Tong X, Holloway JW, Rezwan FI, Lockett GA, Patil V, et al. The interplay of DNA methylation over time with Th2 pathway genetic variants on asthma risk and temporal asthma transition. Clin Epigenet 2014;6(1):8.

[45] Kwon NH, Kim JS, Lee JY, Oh MJ, Choi DC. DNA methylation and the expression of IL-4 and IFN-gamma promoter genes in patients with bronchial asthma. J Clin Immunol March 2008;28(2):139–46.

[46] Luo Y, Zhou B, Zhao M, Tang J, Lu Q. Promoter demethylation contributes to TSLP overexpression in skin lesions of patients with atopic dermatitis. Clin Exp Dermatol January 2014;39(1):48–53.

[47] Breton CV, Byun HM, Wang X, Salam MT, Siegmund K, Gilliland FD. DNA methylation in the arginase-nitric oxide synthase pathway is associated with exhaled nitric oxide in children with asthma. Am J Respir Crit Care Med July 15, 2011;184(2):191–7.

[48] Chen W, Boutaoui N, Brehm JM, Han YY, Schmitz C, Cressley A, et al. ADCYAP1R1 and asthma in Puerto Rican children. Am J Respir Crit Care Med March 15, 2013;187(6):584–8.

[49] Baccarelli A, Rusconi F, Bollati V, Catelan D, Accetta G, Hou L, et al. Nasal cell DNA methylation, inflammation, lung function and wheezing in children with asthma. Epigenomics February 2012;4(1):91–100.

[50] Breton CV, Salam MT, Wang X, Byun HM, Siegmund KD, Gilliland FD. Particulate matter, DNA methylation in nitric oxide synthase, and childhood respiratory disease. Environ Health Perspect September 2012;120(9):1320–6.

[51] Reinius LE, Gref A, Saaf A, Acevedo N, Joerink M, Kupczyk M, et al. DNA methylation in the Neuropeptide S Receptor 1 (NPSR1) promoter in relation to asthma and environmental factors. PLoS One 2013;8(1):e53877.

[52] Lovinsky-Desir S, Ridder R, Torrone D, Maher C, Narula S, Scheuerman M, et al. DNA methylation of the allergy regulatory gene interferon gamma varies by age, sex, and tissue type in asthmatics. Clin Epigenet 2014;6(1):9.

[53] Runyon RS, Cachola LM, Rajeshuni N, Hunter T, Garcia M, Ahn R, et al. Asthma discordance in twins is linked to epigenetic modifications of T cells. PLoS One 2012;7(11):e48796.

[54] Tang WY, Levin L, Talaska G, Cheung YY, Herbstman J, Tang D, et al. Maternal exposure to polycyclic aromatic hydrocarbons and 5′-CpG methylation of interferon-gamma in cord white blood cells. Environ Health Perspect August 2012;120(8):1195–200.

[55] Sood A, Petersen H, Blanchette CM, Meek P, Picchi MA, Belinsky SA, et al. Methylated genes in sputum among older smokers with asthma. Chest August 2012;142(2):425–31.

[56] Isidoro-Garcia M, Sanz C, Garcia-Solaesa V, Pascual M, Pescador DB, Lorente F, et al. PTGDR gene in asthma: a functional, genetic, and epigenetic study. Allergy December 2011;66(12):1553–62.

[57] Soto-Ramírez N, Arshad SH, Holloway JW, Zhang H, Schauberger E, Ewart S, et al. The interaction of genetic variants and DNA methylation of the interleukin-4 receptor gene increase the risk of asthma at age 18 years. Clin Epigenet 2013;5(1):1.

[58] Acevedo N, Reinius LE, Greco D, Gref A, Orsmark-Pietras C, Persson H, et al. Risk of childhood asthma is associated with CpG-site polymorphisms, regional DNA methylation and mRNA levels at the GSDMB/ORMDL3 locus. Hum Mol Genet September 25, 2014.

[59] Salam MT, Byun HM, Lurmann F, Breton CV, Wang X, Eckel SP, et al. Genetic and epigenetic variations in inducible nitric oxide synthase promoter, particulate pollution, and exhaled nitric oxide levels in children. J allergy Clin Immunol January 2012;129(1). 232–239.e1–7.

[60] Naumova AK, Al Tuwaijri A, Morin A, Vaillancourt VT, Madore AM, Berlivet S, et al. Sex- and age-dependent DNA methylation at the 17q12-q21 locus associated with childhood asthma. Hum Genet July 2013;132(7):811–22.

[61] Brunst KJ, Leung YK, Ryan PH, Khurana Hershey GK, Levin L, Ji H, et al. Forkhead box protein 3 (FOXP3) hypermethylation is associated with diesel exhaust exposure and risk for childhood asthma. J Allergy Clin Immunol February 2013;131(2). 592–594.e1–3.

[62] Hew KM, Walker AI, Kohli A, Garcia M, Syed A, McDonald-Hyman C, et al. Childhood exposure to ambient polycyclic aromatic hydrocarbons is linked to epigenetic modifications and impaired systemic immunity in T cells. Clin Exp Allergy July 22, 2014.

[63] Hinz D, Bauer M, Roder S, Olek S, Huehn J, Sack U, et al. Cord blood Tregs with stable FOXP3 expression are influenced by prenatal environment and associated with atopic dermatitis at the age of one year. Allergy March 2012;67(3):380–9.

[64] Lluis A, Depner M, Gaugler B, Saas P, Casaca VI, Raedler D, et al. Increased regulatory T-cell numbers are associated with farm milk exposure and lower atopic sensitization and asthma in childhood. J Allergy Clin Immunol August 28, 2014;133(2):551–9.

[65] Nadeau K, McDonald-Hyman C, Noth EM, Pratt B, Hammond SK, Balmes J, et al. Ambient air pollution impairs regulatory T-cell function in asthma. J Allergy Clin Immunol October 2010;126(4). 845–852.e10.

[66] Schaub B, Liu J, Hoppler S, Schleich I, Huehn J, Olek S, et al. Maternal farm exposure modulates neonatal immune mechanisms through regulatory T cells. J Allergy Clin Immunol April 2009;123(4):774–82.

[67] Swamy RS, Reshamwala N, Hunter T, Vissamsetti S, Santos CB, Baroody FM, et al. Epigenetic modifications and improved regulatory T-cell function in subjects undergoing dual sublingual immunotherapy. J allergy Clin Immunol July 2012;130(1). 215–224.e7.

CHAPTER

18

DNA Methylation for Prediction of Adverse Pregnancy Outcomes

Deepali Sundrani, Vinita Khot, Sadhana Joshi

Department of Nutritional Medicine, Interactive Research School for Health Affairs, Bharati Vidyapeeth University, Pune, Maharashtra, India

OUTLINE

Epigenetic Biomarkers and Diagnostics
http://dx.doi.org/10.1016/B978-0-12-801899-6.00018-8

1. INTRODUCTION

Epidemiological studies have shown that alterations in key developmental processes in utero can predispose the offspring to postnatal noncommunicable diseases (NCDs) and are referred to as "Developmental Origins of Health and Disease" (DOHaD). It is well known that suboptimal maternal nutrition during critical periods of development (prenatal and early postnatal) "programs" the offspring to an increased risk of developing diseases in adult life [1]. During pregnancy, the placenta responds to perturbations in the maternal compartment, such as altered nutrition and reduced uteroplacental blood flow, and plays a key role in programming the fetus to diseases. These programmed changes are likely to be a result of changes in the gene expression patterns which are influenced by epigenetic modifications like DNA methylation, histone modification, and RNA regulation [2]. They are involved in processes such as cell differentiation and morphogenesis and may be affected by nutritional, environmental, or lifestyle factors [3].

Various animal and human studies have demonstrated that maternal nutrition during pregnancy can alter the DNA methylation patterns of specific genes leading to permanent phenotypic changes [4,5]. Micronutrients (folate and vitamin B_{12}) involved in the one-carbon cycle are known to play an important role in DNA methylation processes since they influence the biochemical pathways of methylation processes by the supply of methyl groups [6]. Animal studies have shown that alterations in the one-carbon cycle during pregnancy and periconceptional period influence gene expression through altered DNA methylation patterns [7,8]. In addition to the micronutrients, it has been suggested that polyunsaturated fatty acids (PUFA) can also influence the DNA methylation patterns, thereby modifying the epigenome [9]. Dietary components can therefore selectively activate or inactivate gene expression patterns via the epigenetic mechanisms.

It is suggested that in utero insults can disturb placental epigenetics and thereby placental development and function, leading to maternal morbidity and increasing disease susceptibility in later life [10]. This is of relevance since children born to mothers with pregnancy complications are known to be at increased risk of developing diseases in adult life. It is likely that during pregnancy complications the fetus is most vulnerable since maximum epigenetic changes taking place during this critical period will influence vital genes involved in key mechanisms like angiogenesis, glucose metabolism, lipid metabolism, and neurodevelopment. In view of this, the current chapter focuses on the role of epigenetic mechanisms, especially DNA methylation in pregnancy complications. Identification of various epigenetic biomarkers for detection of chromosomal abnormalities and pregnancy complications is also discussed.

2. EPIGENETICS

In the 1940s, Conrad Waddington coined the term epigenetics as a blend of two words, genetics and epigenesis. He provided the first definition of epigenetics as "the causal interactions between genes and their products, which bring the phenotype into being" [11]. It was observed that environmental factors can alter the expected phenotype to produce an entirely new one from the same genotype. Therefore, epigenetics was used as a term that defined ways in which the environment influences phenotype (the way that an organism looks) [12]. Today, epigenetics is defined as the study of heritable changes in phenotype or gene expression by mechanisms other than changes in the underlying DNA sequence [13].

Epigenetic events control the gene expression patterns and are involved in many physiological and pathological conditions [14]. Further, aberrant epigenetic regulation has been found to be associated with a number of human diseases such as cancer, neurodegenerative disorders, and mental retardation [15].

3. EPIGENETIC MECHANISMS

The molecular targets of epigenetic changes include DNA, histones, and noncoding RNAs (ncRNAs). In order to understand the molecular mechanisms of epigenetic regulation, it is necessary to understand the chromatin structure which is made up of DNA and histone molecules [16]. In eukaryotes, DNA is packaged into a chromatin, which is a tight complex of DNA and histone molecules in the nucleus.

The chromatin is composed of repeated structural units called "nucleosomes" that influence gene expression. The nucleosome consists of a histone octamer and DNA that wraps around it [17]. Histones are globular basic proteins with a flexible aminoterminal tail, which protrudes from the nucleosome. The histone octamer consists of two copies of four core histones (H2A, H2B, H3, and H4 histones) surrounded by DNA [18].

There are three epigenetic mechanisms of gene regulation, which collectively make up the epigenome namely DNA methylation, histone modification, and regulation of ncRNAs. These modifications can affect the accessibility of the gene for regulatory proteins such as methyl-CpG (cytosine–phosphate–guanine)-binding proteins, transcription factors, RNA polymerase II, and other components of the transcriptional machinery, thereby altering gene expression patterns [19].

3.1 DNA Methylation

DNA methylation is one of the most well-studied epigenetic mechanisms. It occurs via covalent modification of cytosines by adding a methyl group to a 5′ carbon of the cytosine ring located within cytosine–guanine (CpG) dinucleotides [3]. This reaction is catalyzed by DNA methyltransferases (DNMTs) [20]. Around 80% of CpG dinucleotides across the genome are heavily hypermethylated and the remaining 20% are unmethylated and clustered in CpG-rich sequences, termed "CpG islands" within the promoter region of gene [21]. Unmethylated CpG islands result in the transcription of a particular gene in presence of corresponding gene transcription factors [22]; while, the methylation of CpG islands in the promoter region silences gene expression [23]. DNA methylation profile can be analyzed globally throughout the genome or specifically at a particular gene promoter region.

DNA methylation patterns are established and maintained by a family of enzymes called DNMTs which use the methyl donor, S-adenosyl methionine (SAM). The human genome contains three active enzymes: DNMT1, to maintain the DNA methylation marks in somatic cells after each replication cycle [24] and DNMT3a and DNMT3b, to establish methylation patterns on the genome following embryonic fertilization [25].

3.2 Histone Modifications

The epigenetic histone modifications include methylation, acetylation, ubiquitination, and SUMOylation of lysine residues; methylation of arginine residues; and phosphorylation of serines [17]; among others. These posttranslational modifications (PTMs) of histones occur primarily at specific positions in the aminoterminal tails of histones [13] and are catalyzed by histone-modifying enzymes that add and remove these modifications. The histone acetyltransferases and methyltransferases add the acetyl and methyl groups to histone residues respectively, while the histone deacetylases and demethylases remove these modifications [26]. In Chapter 21 of this volume, Vaquero et al. describe a number of histone PTMs and the function of histone-modifying enzymes.

3.3 RNA Regulation through ncRNAs

Epigenetic regulation through RNA is more complex and is the least-known epigenetic mechanism. ncRNA-mediated epigenetic regulation can be driven by long and small ncRNAs. Some of the long ncRNAs have been shown to be involved in X-chromosome inactivation [27] and genomic imprinting [28]. Small ncRNAs are suggested to regulate gene expression through different mechanisms including complementary binding to the target RNAs, RNA degradation, and RNA splicing [29].

4. REGULATION OF GENE EXPRESSION BY EPIGENETIC MECHANISMS

As mentioned in the above sections, epigenetic mechanisms determine the phenotype without changing the genotype through chemical modifications of DNA and histone proteins leading to long-term changes in gene expression [13]. In mammalian cells, most of the chromatin exists as heterochromatin which is highly condensed and transcriptionally silent. In contrast, euchromatin is less condensed and contains most of the actively transcribed genes. Epigenetic changes in the DNA and histones alter the electrostatic nature of the chromatin or alter the binding of chromatin proteins affecting thereby the chromatin structure [30].

DNA methylation can affect gene expression by directly preventing the binding of transcription factors to the CpG-containing DNA-binding elements and/or by promoting the binding of methyl-DNA-binding proteins. This hampers the access of transcription factors to DNA [31]. These DNA-binding proteins influence enzymes that catalyze PTMs of histones and thereby lead to transcription repression [32]. Histone modifications can therefore affect chromatin structure directly or indirectly by various chromatin remodeling complexes which interact with the methylated histones and thereby affect transcription [33]. Thus, the mechanism for CpG island-associated gene silencing involves methylated DNA-binding proteins and their interactions with transcription factors, DNA methyltransferases, histone-modifying enzymes, and chromatin remodeling factors [34]. These protein–DNA interactions alter the DNA conformation and influence gene regulation [35]. Epigenetic mechanisms operate at the transcriptional and posttranscriptional level of gene activity as well as at the level of protein translation and posttranslational modifications.

5. EPIGENETIC CHANGES DURING THE MAMMALIAN LIFE CYCLE

Several types of epigenetic modifications facilitate the complex pattern required for proper mammalian development, and DNA methylation is one of the important epigenetic mechanisms [36]. During the development of multicellular organisms, different cells and tissues acquire different programs of gene expression that are regulated by epigenetic modifications such as DNA methylation and histone modifications. Thus, each cell type in the body has its own epigenetic signature and these epigenetic marks become fixed in most of the differentiated cells once they exit the cell cycle [37]. During mammalian development, these epigenetic marks in differentiated cells confer stability to gene expression, as DNA methylation patterns are replicated with each mitotic division [38]. This is because the "hemimethylated" DNA so formed is the template for the maintenance DNMT1 to add a methyl group to the cytosine on the nascent daughter strand opposite the methylated cytosine on the parental strand [39].

Genomic methylation patterns in somatic differentiated cells are generally stable and heritable. However, in mammals there are at least two developmental time periods—in germ cells and in developing embryos, where genome-wide epigenetic reprogramming occurs resulting in removal and reestablishment of epigenetic patterns [40]. The early embryonic development is the time period of greatest epigenetic change, when epigenetic marks across the entire genome (termed the epigenome) undergo programmed events in which nearly all epigenetic marks are removed and reestablished.

6. DNA METHYLATION DURING EMBRYONIC DEVELOPMENT

DNA methylation patterns are established during embryogenesis. Genome-wide demethylation occurs after fertilization and before embryo implantation, resulting into a state of hypomethylation through three different mechanism: (1) passive replication-dependent dilution consisting in the loss of 5-methylcytosine (5mC) in mitotic cells by downregulation of DNMT1, (2) active demethylation by the activation of DNA repair mechanisms such as base excision repair, and (3) 5-hydroxymethylcytosine (5hmC) ten–eleven translocation (TET)-mediated DNA demethylation. De novo global remethylation then occurs following embryo implantation in response to organ development. This further leads to development of gene-specific methylation patterns, which determine tissue-specific transcription [2]. Epigenetic regulation, thus, evolves before implantation and then across gestation (Figure 1).

At every stage of fetal development de novo methylation and chromatin remodeling takes place which influences the placental structure and function by switching on and off

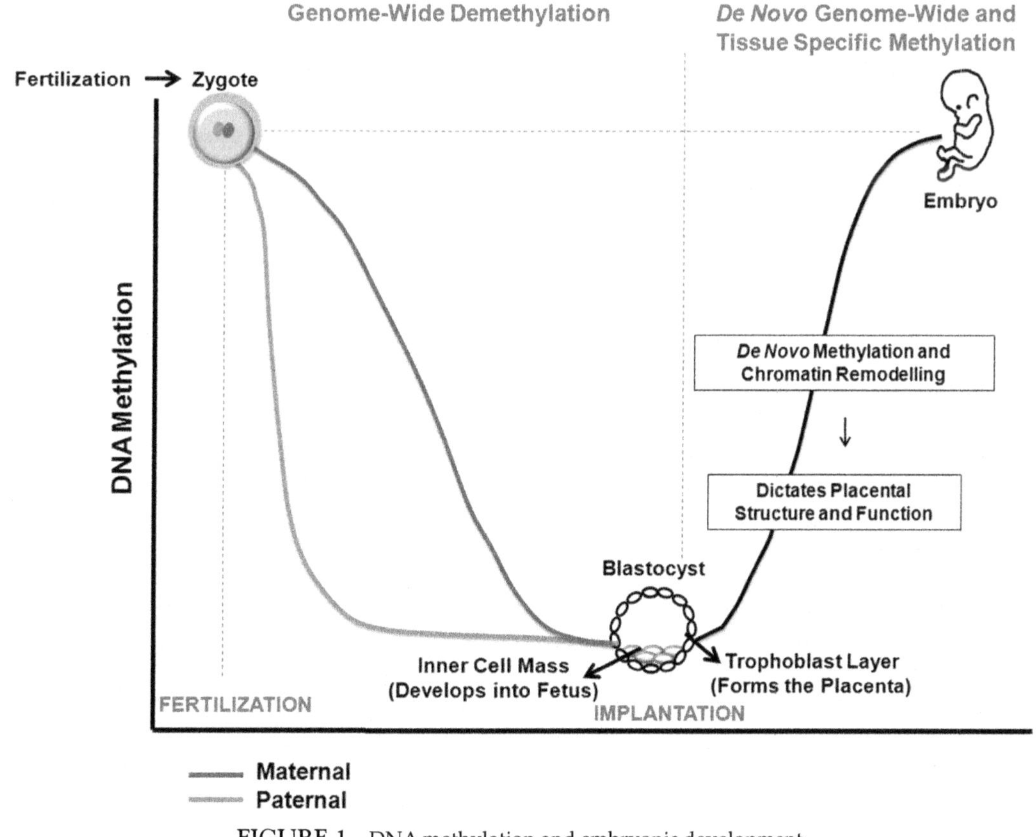

FIGURE 1 DNA methylation and embryonic development.

various genes. These programs must be completed within the critical spatiotemporal windows during pregnancy, and failure to complete these programs in time may lead to long-term consequences [41]. It is therefore, necessary to understand the methylation patterns of various vital genes involved in placental development. Alterations in the methylation pattern of these genes may affect the gene expression patterns for different cell types leading to abnormal placentation, thereby affecting birth outcome. Furthermore, maternal diet has been shown to have a significant impact on epigenetic processes like DNA methylation, and the next section discusses the role of nutrition in epigenetic regulation.

7. MATERNAL NUTRITION, DOHaD, AND EPIGENETICS

Studies have well established the association between maternal nutrition and birth weight and risk for NCDs, coronary heart disease, and type II diabetes mellitus (MIM #125853) in the offspring at adult age [42,43]. This concept was initially put forth by Late Professor David Barker, who proposed that the origins of diseases in adults begin in utero [44].

It is suggested that alterations in maternal nutrition may result in developmental adaptations that permanently change the structure, physiology, and metabolism of the offspring; thereby predisposing the infant to cardiovascular, metabolic, and endocrine diseases in adult life [45]. Maternal nutrition can "program" gene expression patterns in the embryo that persist into adulthood and contribute to the metabolic diseases [46]. There is growing interest in the field of DOHaD; however, the mechanisms that lead to adverse programmed changes in growth, development, and metabolism are not clear [47]. These programmed changes are likely to be a result of stable changes in the gene expression patterns which are influenced by epigenetic modifications, as a result of a number of external and internal environmental factors [2]. Epigenetic mechanisms may therefore, provide a possible explanation for how environmental influences in early life cause long-term changes in later life.

Evidence suggests that maternal nutritional exposure in utero can change the stable expression of genes in the offspring through epigenetic modification [48]. Under- and overnutrition during pregnancy has been linked to development of chronic diseases in later life. Nutrients can reverse or change the epigenetic patterns by inhibiting enzymes that catalyze DNA methylation and histone modifications or by altering the availability of substrates, thereby modifying the expression of critical genes associated with physiological and pathological processes, including embryonic development [49].

A number of animal studies have shown that maternal nutrition during pregnancy can alter the DNA methylation patterns of specific genes in the offspring leading to permanent phenotypic changes such as coat color, body weight, and blood pressure [4,50]. Reports indicate that individuals who were exposed to famine in utero during the Dutch Hunger had different methylation patterns in genes involved in growth and metabolic disease compared with controls [5]. Dietary factors that are involved in the one-carbon cycle play an important role in DNA methylation processes since they control the supply of methyl groups.

7.1 Nutritional Regulation of DNA Methylation through One-Carbon Cycle

The one-carbon metabolism is used to describe a set of reactions that involve the continuous transfer of a single carbon unit from folate, the key methyl donor, to various methyl acceptors and recycling of the by-product homocysteine. Briefly, methionine is converted to SAM by the action of methionine adenosyltransferase (MAT). SAM, the major methyl donor, is converted to S-adenosyl homocysteine (SAH) on donating the methyl group to various methyl acceptors. Homocysteine obtained by the hydrolysis of SAH is then remethylated to methionine in a reaction catalyzed by methionine tetrahydrofolate reductase (MTR). This reaction requires vitamin B_{12} as a cofactor and folate in the form of 5-methyltetrahydrofolate (5-MTHF) obtained from dietary folates by the action of methylene tetrahydrofolate reductase (MTHFR). The alternative folate-independent pathway is catalyzed by the enzyme betaine–homocysteine methyltransferase (BHMT) and uses betaine direct from the diet or from choline [51]. Homocysteine can alternatively be utilized for cysteine synthesis catalyzed by cystathionine b-synthase (CBS) which requires vitamin B_6 as an essential cofactor.

The one-carbon metabolism is regulated by its substrates (methionine, cysteine), cofactors (folate, vitamin B_{12}, and vitamin B_6), and intermediates (SAM, SAH, and homocysteine). It thus entails pathways which conserve methionine to assure methyl group availability and removes unwanted homocysteine out of the system [52]. The one carbon metabolism has been extensively studied and discussed, especially in the context of its role in DNA methylation and fetal programming [53]. It has been suggested that the transsulfuration pathway is more active during early gestation while higher rate of remethylation pathway exists during late gestation in normal pregnancy in humans [54].

Hypomethylation of various regulatory genes like *c-myc*, *c-fos*, *H-ras*, and *p53* has been observed when animals were fed with methyl-deficient diet suggesting that dietary factors influence gene expression through altered gene-specific DNA methylation patterns [8]. Studies in sheep provide the first evidence that, during the periconceptional period, reduced dietary inputs (vitamin B_{12}, folate, and methionine) to the one-carbon cycle can lead to widespread epigenetic alterations to DNA methylation in offspring with long-term implications for adult health [7]. A number of studies in humans and animals have suggested that altered micronutrients (folic acid and vitamin B_{12}) along with reduced long-chain PUFA (LCPUFA) in pregnancy affect the placental global methylation levels [55–59]. Nutrients during pregnancy directly affect the DNA methylation patterns possibly via changes in the expression of genes involved in the one-carbon cycle and result in adverse pregnancy outcomes. The next section discusses in detail the role of maternal nutrition (macronutrients, micronutrients, and LCPUFA) in DNA methylation (Figure 2).

7.2 Macronutrients and DNA Methylation

Macronutrients are energy-yielding nutrients (carbohydrate, fat, and protein) required in relatively larger amounts. Studies have shown that feeding rats with protein-restricted diet during pregnancy resulted in permanent changes in the DNA methylation and expression patterns of the glucocorticoid receptor (*GR*) and peroxisome proliferator-activated receptor-α (*PPAR-α*) genes in the liver and heart of the offspring [60–62]. Altered methylation patterns of hepatic *PPAR-α* and *GR* promoters have been shown to carry forward to the F2 offspring even though the F1 offspring were fed a normal diet. This suggests that epigenetic marks can persist for at least two generations [63]. Consumption of maternal high-fat diet is reported to result in significant hyperacetylation of histone H3K14 in fetal hepatic tissues which resulted in an obese offspring [64].

Global hypomethylation and gene-specific promoter DNA hypomethylation of genes related to dopamine and opioids have been reported in the brain of offspring from dams that consumed a high-fat diet [65]. The same group further reports that supplementation of maternal diet with methyl donors attenuates global hypomethylation in the brain, suggesting the importance of balanced methylation status during pregnancy [66].

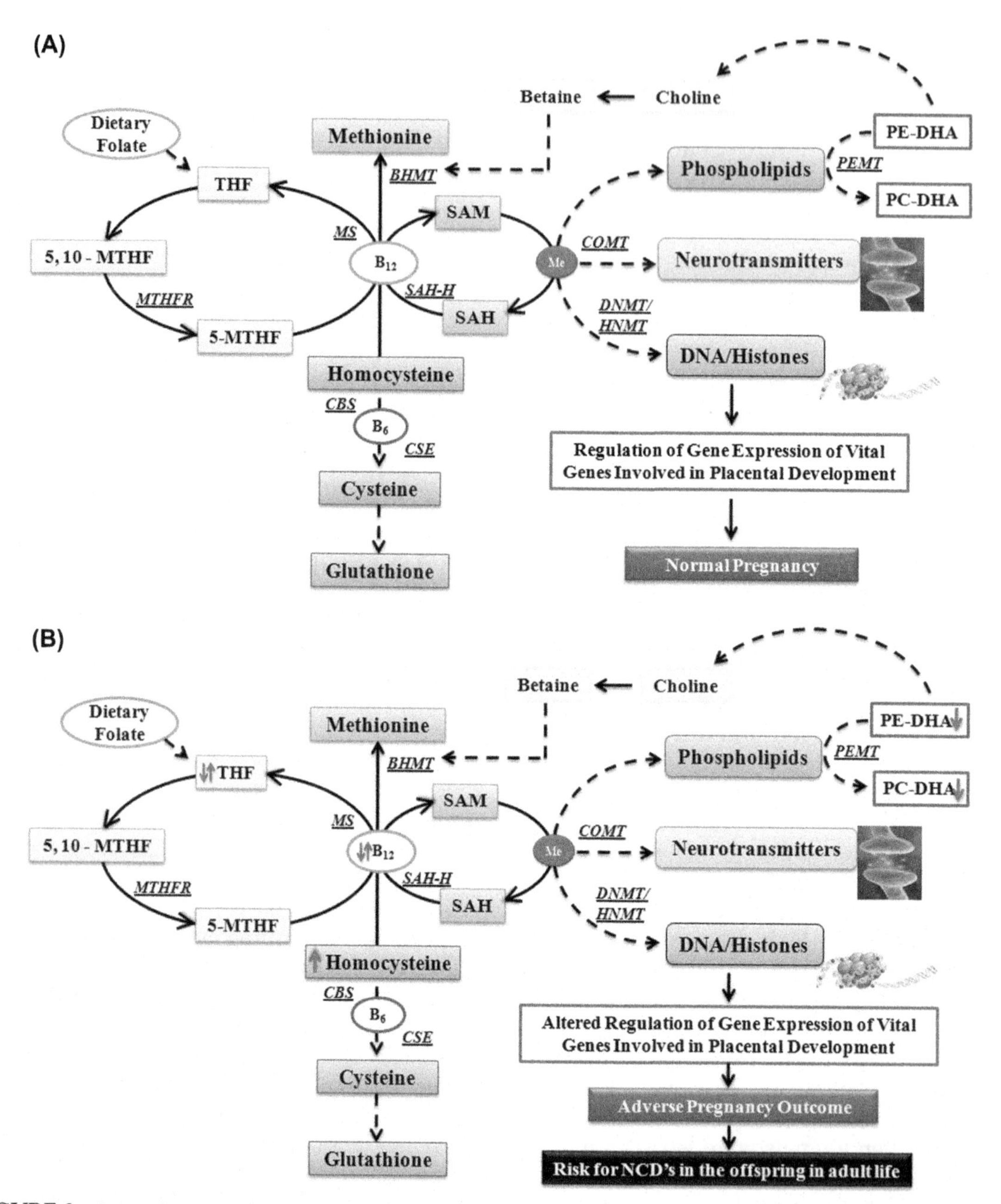

FIGURE 2 One-carbon cycle highlighting the role of micronutrients (folate and vitamin B_{12}) and fatty acids in gene regulation. (A) Normal pregnancy: The methyl groups generated in the one-carbon cycle, where folic acid and vitamin B_{12} act as essential cofactors are utilized by the various methyl acceptors like phospholipids, neurotransmitters, and DNA. The DNA methylation patterns regulate the expression levels of vital genes involved in placental development important for a normal pregnancy outcome. (B) Adverse pregnancy outcome: the imbalance in maternal micronutrients and LCPUFA may alter the one-carbon cycle which affects the methyl group transfer for DNA methylation. This leads to changes in the expression levels of vital genes involved in placental development resulting in an adverse pregnancy outcome and also increasing the risk for noncommunicable diseases in the offspring at later life. THF, tetrahydrofolate; 5,10-MTHF, 5,10-methylenetetrahydrofolate; 5-MTHF, 5-methyltetrahydrofolate; SAM, S-adenosyl methionine; SAH, S-adenosyl homocysteine; B_{12}, vitamin B_{12}; B_6, vitamin B_6; MTHFR, methyltetrahydrofolate reductase; MS, methionine synthase; MAT, methionine adenosyltransferase; SAH-H, S-adenosyl homocysteine hydrolase; CBS, cystathionine beta synthase; CSE, cystathionine gamma-lyase; PEMT, phosphatidylethanolamine N–methyltransferase; COMT, catechol-O-methyltransferase; DNMT, DNA methyltransferase; HNMT, histone methyltransferase; PE, phosphatidylethanolamine; PC, phosphatidylcholine; Me, methyl group; NCDs, noncommunicable diseases.

7.3 Micronutrients and DNA Methylation

Micronutrients are required in small quantities during pregnancy and include vitamins and minerals. Micronutrient deficiencies alter placental development and the expression of key signaling molecules involved in growth and regulation of the fetoplacental unit [67]. It is suggested that micronutrients like folate and vitamin B_{12} play a significant role in altering DNA methylation patterns by regulating the transfer of methyl groups through the one-carbon cycle [68]. Methyl groups needed for DNA methylation are derived in part from foods that contain methionine, folate, and choline [69]. Reports suggest that dietary deficiency in methyl donors, especially during pregnancy, may affect DNA methylation [68]. Studies show that animals fed diets deficient in methyl donors result in hypomethylated DNA [70,71]. These changes occur not only in global methylation [72] but also in the methylation of specific genes [73]. Studies in pregnant rats show that methyl-deficient diets low in folic acid, choline, and methionine increase homocysteine concentrations both in the maternal and fetal plasma [74]. It is suggested that folate is essential for DNA methylation reprogramming during the early embryonic period [49]. Periconceptional folic acid supplementation is known to be associated with imprinting status of insulin growth factor 2 (*IGF2*) in the child, affecting intrauterine programming of growth and development with consequences for health and disease throughout life [75].

Reports indicate that variance in dietary methyl nutrients like folate, vitamin B_{12}, and methionine during the periconceptional period can change DNA methylation patterns in the offspring and may result in modified adult phenotypes [7]. Studies report altered placental global methylation patterns in pregnancy complications like preeclampsia (MIM #189800) [56] and in preterm pregnancies (MIM #610504) [57], wherein these women had altered levels of maternal micronutrients like folic acid, vitamin B_{12}, and omega-3 fatty acids and increased homocysteine [58,59]. Other studies by the same group in Wistar rats also reports global hypomethylation in the placenta and in the offspring brain at birth as a consequence of altered maternal micronutrients like folic acid and vitamin B_{12} [55,76]. A maternal diet imbalanced in micronutrients, particularly vitamin B_{12} deficiency in the presence of excess folic acid levels, is known to affect the expression of key genes encoding enzymes in the one-carbon cycle [77]. Similarly, the effect of omega-3 fatty acids on these critical enzyme genes of the homocysteine metabolism has been documented [78]. Other studies in humans and animals have also demonstrated a link between micronutrients (folate and vitamin B_{12}) and LCPUFA in the one-carbon cycle that influences DNA methylation [55,76,79]. The next section therefore discusses the role of LCPUFA in DNA methylation.

7.4 LCPUFA and DNA Methylation

LCPUFA are crucial for the growth and development of the placenta and fetus [80]. There are two main families of LCPUFA namely omega-3 and omega-6 fatty acids. Alpha-linolenic acid (ALA) (omega-3) and linoleic acid (LA) (omega-6) are the dietary essential fatty acids and precursors of more highly unsaturated members of their family, docosahexaenoic acid (DHA) and arachidonic acid (AA), respectively [81].

In the one-carbon cycle, methyl groups from SAM are transferred to major acceptors like phospholipids, DNA, histones, and neurotransmitters. The enzyme phosphatidylethanolamine N–methyltransferase (PEMT) utilizes three SAM molecules for interconversion of phosphatidylethanolamine (PE) to phosphatidylcholine (PC). PC synthesized by the PEMT enzyme is critical for the delivery of important LCPUFA such as DHA from the liver to the plasma and distribution to peripheral tissues. It has been reported that dietary folate deficiency diminishes choline and PC levels in the liver [82].

A recent review summarizes emerging findings that support the suggestion that fatty acids, in particular, PUFA, can modify the epigenome [9]. Another recent review highlights the role of the phospholipids as major methyl acceptors from the one-carbon cycle [83]. In an altered one-carbon scenario if the phospholipid conversion by PEMT is low, as indicated by low DHA levels [84], then the methyl groups will be directed to the DNMTs for DNA methylation leading to altered methylation of various genes [83].

Studies indicate that prenatal omega-3 fatty acid supplementation to pregnant dams receiving an imbalanced maternal micronutrient diet (folate and vitamin B_{12}) normalizes the global DNA methylation levels to that of the control group in the placenta and in the brain of the offspring at 3 months of age indicating that omega-3 fatty acids may be involved in reversing the methylation patterns [55,76].

A number of studies show the link between poor nutrition with persistent changes in offspring phenotype specifically acting through epigenetic processes. A recent randomized trial in pregnant women indicates that omega-3 fatty acid supplementation influences the promoter methylation levels of inflammatory markers in cord blood mononuclear cells [85]. A study in rats indicates that maternal fat exposure influences epigenetic regulation of *Fads2* gene (which encodes Δ6 desaturase, the rate-limiting enzyme in polyunsaturated fatty acid synthesis) thereby resulting in altered hepatic LCPUFA status in the offspring [86]. Similarly, a study has also demonstrated an association between alterations in the *Fads2* DNA methylation and maternal ALA availability [87].

8. EPIGENETIC REGULATION IN THE PLACENTA AND ADVERSE PREGNANCY COMPLICATIONS

The previous section discusses the role of maternal nutrition in inducing alterations in gene expression through epigenetic modifications like DNA methylation. Epigenetic events therefore constitute an important mechanism by which dietary components can selectively activate or inactivate gene expression. Several environmental factors including maternal nutrition are known to influence placental epigenetic profile and any disturbances in DNA methylation patterns can lead to altered placental development and function.

Placenta is considered as the key programming agent of adult health and disease. During pregnancy, the placenta anchors the conceptus, prevents its rejection by the maternal immune system, and transports nutrients and wastes between the mother and the fetus. It is suggested that any alterations in the placental growth and development in response to maternal diet affects the supply of nutrients to the fetus resulting in an adverse pregnancy outcome thereby predisposing the offspring to adult diseases [88]. Thus, the placenta plays a critical structural and functional role during pregnancy and is a key link in the chain of events that leads to intrauterine programming of adult health. These functions are performed through multiple specialized cell types derived from lineage-committed precursors that either proliferate or differentiate. As in other organs, cell proliferation and differentiation in the placenta involves different epigenetic processes which, if altered, may affect the gene expression patterns for different cell types [89,90].

It is known that the placenta displays unique epigenetic profile [2], and a number of studies are examining the role of epigenetic regulation in placental development both at genome-wide and gene-specific levels. A study by Novakovic et al. reports an increase in overall DNA methylation levels in the human placenta across gestation at three different time points (first, second, and third trimester). In addition, they also report the existence of CpG sites that show interindividual variability in third trimester compared to first and second and suggest that such variability could be attributed to cumulative differences in environmental exposure [91]. Similarly, another study reports 20,893 differentially methylated positions (DMPs) (most of them being hypomethylated) in placental DNA of growth-restricted neonates and appropriately grown neonates which were associated with gestational age [92]. This study also observed that umbilical cord blood DNA of fetal growth-restricted (FGR) offspring born at term had 839 DMPs as compared with appropriately grown (AGA) offspring. Out of the 839 DMPs, 53 DMPs had a beta methylation value difference of >10% and were comethylated at several CpG sites on the genes (*FOXP1*, *RIOK3*, and *TAF5*) which control the transcription factors. The authors therefore suggest that these DMPs in the placenta and cord blood DNA are involved in gene regulation and transcription pathways related to organ development and metabolic function [92]. It has also been reported that there is a significant difference in loss of imprinting (LOI) (i.e., biallelic expression) in 14 genes (*CD44*, *EPS15*, *PEG3*, *DLK1*, *MEG3*, *TP73*, *PEG10*, *PHLDA2*, *SNRPN*, *H19*, *PLAGL1*, *SLC22A18*, *MEST*, and *IGF2*) but no difference in gene expression of most of the genes in first-trimester placentas as compared to term placentas of uncomplicated pregnancies suggesting that there is no association between gene expression and LOI. The authors also mention that genomic imprinting in placenta is a dynamic, maturational process and these epigenetic imprints may continue to evolve across gestation [93].

Studies also report difference in DNA methylation patterns among different cell types within tissues like breast tissue [94] and embryonic and adult stem cells [95]. Cell-specific methylation differences in cytotrophoblasts and fibroblasts of the human placenta are reported and suggest that CpG methylation of the placental tissue is more representative of cytotrophoblast methylation than that of fibroblast methylation [96]. Genome-wide DNA methylation studies have also identified differentially methylated regions between maternal blood and placental DNA as fetal DNA methylation markers. This has resulted in development of efficient noninvasive method for detecting or diagnosis of fetal chromosomal disease, aneuploidies, and placental pathologies like preeclampsia [97,98].

A study by Schroeder et al. reported that 37% of the human placental genome contains partially methylated domains (PMDs) which are developmentally important regions of the placental methylome and can be used as epigenetic biomarkers [99]. PMDs are large regions over 100 kb in length with lower DNA methylation and surrounded by regions with high DNA methylation. These PMDs are stable throughout gestation and cover tissue-specific repressed genes in the placenta [100]. Any alterations in the PMD regions can result in placental dysfunction leading to implantation failure or miscarriage (MIM #614389) [101].

Evidence suggests that epigenetics is linked to adverse pregnancy outcomes including early pregnancy loss (MIM #614389), IUGR (MIM #614389), preeclampsia, and preterm birth [102] and thereby plays an important role in the developmental origin of adult disease [47]. Studies have examined the role of epigenetic modifications in human placental-related pathologies like preeclampsia, gestational diabetes (MIM #125851), and preterm birth and are discussed below.

8.1 Preeclampsia

Preeclampsia, a hypertensive disorder of pregnancy, characterized by the new onset of hypertension and proteinuria after 20 weeks of gestation, is hypothesized to be associated with abnormal placental function. Many studies have investigated changes in gene expression patterns in placenta from women with preeclampsia; despite which the role of epigenetics in preeclampsia-associated placental dysfunction remains unclear.

A study by Rahat et al. reported higher DNA methylation and higher H3K27me3 levels at the c-myc promoter region along with lower gene expression of c-myc in preeclamptic placental tissue. Similarly, the authors also observed higher DNA methylation levels of c-myc promoter region in the maternal plasma suggesting c-myc as a possible fetal epigenetic marker for preeclampsia [103]. Hypermethylation at CpG sites within the exon 1 of the *H19* gene has been reported in women with preeclampsia indicating that they may be involved in trophoblast dysfunction and pathogenesis of preeclampsia [104]. Aberrant methylation of genes associated with cell adhesion (neural cell adhesion 1 (*NCAM1*), cadherin 11 (*CDH11*), collagen type V, alpha 1 (*COL5A1*), and tumor necrosis factor (*TNF*)) has been reported in first-trimester trophoblasts as well as in placentas from preterm preeclampsia as compared to controls. The epigenetic regulation of these genes involved in cell adhesion is suggested to contribute to altered migration and invasion of placental trophoblast cells leading to abnormal placental development in preeclampsia [105].

A report indicates that the placental promoter region of serine protease inhibitor-3 (*SERPINA3*) has hypomethylated CpGs situated in close vicinity of binding sites of transcription factors associated with hypoxia and inflammation along with increased mRNA expression in placental pathologies like preeclampsia and IUGR. Thus, the authors suggest that increased SERPINA3 may affect the process of implantation in preeclamptic placenta by inhibiting matrix disintegration. Further, it is also suggested that this protein may be present in the maternal blood and may provide markers for early diagnosis [106]. A recent study in women with preeclampsia reported that hypermethylation of key placental genes, such as 1α-hydroxylase (*CYP27B1*), vitamin D receptor (*VDR*), and retinoid X receptor (*RXR*), which are involved in vitamin D metabolism, may affect placental development and availability of vitamin D at the maternal–fetal interface [107]. Other report provides first evidence of altered DNA methylation and gene expression patterns of angiogenic genes like vascular endothelial growth factor (*VEGF*), fms-like tyrosine kinase-1 (*FLT-1*), and kinase domain receptor (*KDR*) in placentas from women with preeclampsia suggesting a role of altered DNA methylation in placental angiogenesis. Further, in this study, although the mean methylation of the *FLT-1* and *KDR* promoters was comparable between groups, the mRNA levels of *FLT-1* and *KDR* were significantly higher in the women with preeclampsia delivering preterm as compared to controls. Thus, the authors state that changes in *FLT-1* and *KDR* gene expression may not be mediated through DNA methylation but possibly through other factors affecting gene expression such as transcription factors, mRNA stability, and histone modifications [108]. Other recent study reported that few CpG sites of leptin promoter that were in the vicinity of the transcription start site (TSS) including the binding sites of transcription factors Sp1, LP1, and CEBPa were demethylated in placentas from women with preeclampsia when compared to controls [109]. This is consistent with another study by Hogg et al. which also reported hypomethylation of the leptin gene promoter in early onset preeclampsia placentas but not in late-onset preeclampsia placentas [110]. These studies suggest that hypomethylation in *LEP* promoter region is the epigenetic cause of the increased placental leptin expression and circulating maternal leptin levels which are observed in women with preeclampsia. Increased leptin levels in preeclampsia have been proposed as a compensatory response to decreased placental perfusion thereby increasing the nutrient delivery to the fetus [111].

Promoter hypomethylation of tissue inhibitor of matrix metalloproteinase-3 (*TIMP-3*) which is associated with a number of processes, such as angiogenesis, tissue remodeling, invasion, cell growth, and apoptosis, has been reported in placentas affected by preeclampsia in the Canadian [112] and Chinese populations [113]. These studies suggest that *TIMP-3* may be a potential marker for early diagnosis of preeclampsia. Promoter demethylation of matrix metalloproteinase-9 (*MMP-9*) gene along with increased *MMP-9* expression in placentas complicated with preeclampsia as compared to controls has also been reported [114]. Thus, higher expression of *MMP-9* may contribute to shallow placentation observed in preeclamptic placenta.

Now, it is known that hypoxia alters the methylation profile of functionally relevant genes involved in the differentiation of villous cytotrophoblast to syncytiotrophoblast [115]. Studies report reduced placental expression of syncytin-1 protein involved in the terminal differentiation of placental trophoblast linage along with hypermethylated syncytin-1 promoter in preeclamptic placentas [116] and cytotrophoblasts [117] as compared to controls. Studies also reported higher DNMT1 and DNMT3B activity in placentas from women with preeclampsia that may contribute to the hypermethylation of syncytin-1 in preeclampsia. These studies suggest that epigenetic changes play an important role in the pathophysiology of preeclampsia [116].

A study by Yuen et al. reported 34 differentially methylated CpG sites (hypomethylated) in early-onset (<34 weeks gestation), but not in late-onset (>34 weeks gestation), preeclampsia placental tissues with their gestational age-matched tissue samples using the Illumina microarray. Hypomethylation of four loci (*CAPG, GLI2, KRT13,* and *TIMP-3*) in early-onset preeclampsia was further confirmed by bisulfite pyrosequencing [112]. Further, the same group confirmed the above results on a larger sample size and number of analyzed loci. They reported 282 differentially methylated CpG sites (74.5% hypomethylated and 25.5% hypermethylated) in the early-onset preeclampsia samples as compared to gestational age-matched controls indicating that it may be associated with changes in placental function [118]. Therefore these genes can be potential biomarkers for DNA methylation-based noninvasive prenatal diagnosis of risk for pregnancy disorders.

A recent study demonstrates that only a single CpG site melanocortin 1 receptor (*MC1R*) is differentially methylated between early-onset and late-onset preeclampsia samples, indicating that there is minimal impact of gestational age at delivery on DNA methylation. This study also reports fetal gender-dependent changes in placental DNA methylation with a large number of differentially methylated loci in samples from pregnancies in which the fetal gender was female [119]. More genes have appeared aberrantly methylated during this pathological process. Therefore, no clear candidates have been validated yet. In this regard, changes in DNA methylation levels of cortisol-signaling genes like GR (*NR3C1*), corticotropin-releasing hormone (*CRH*), and corticotropin releasing hormone-binding protein (*CRHBP*) in early-onset preeclampsia-associated placentas as compared to controls have also been reported [120].

Genome-wide methylation patterns in maternal leukocyte DNA at the time of delivery demonstrated DNA hypermethylation in women with preeclampsia compared to normotensive women with most of the differentially methylated genes implicated in seizure disorders [121]. The changes in global DNA methylation patterns in preeclampsia placentas have also been shown to be associated with blood pressure [56]. In contrast, Nomura et al. reported lower global DNA methylation levels in placentas of women with preeclampsia [122]. Global DNA methylation levels in early-onset preeclampsia placentas have also been shown to be hypomethylated as compared to normal controls [123]. It is suggested that genome-wide methylation studies both in normal and pathological pregnancies are important so as to identify novel genes involved in these processes [121].

Altered global DNA methylation levels in maternal blood vessels of women with preeclampsia with most of the differentially methylated genes related to inflammation and immune response have been reported [124]. Another report by the same group suggests hypomethylation of matrix metalloproteinases (*MMP-1* and *MMP-8*) that are involved in collagen metabolism in blood vessels of women with preeclampsia [125]. Hypermethylation is mostly associated with gene silencing; however, a few studies mentioned above report that methylation at promoter-distal sites or gene body is positively associated with gene expression [105,109].

Studies have also demonstrated that preeclampsia induces reduced methylation levels at DMR of the imprinted gene *IGF2* in the umbilical cord lymphocytes of neonates born to mothers with preeclampsia [126]. These studies suggest these epigenetic changes may underlie the associations between intrauterine exposure to preeclampsia and high risk for metabolic diseases in later life in the infants. Thus, maternal preeclampsia may epigenetically program the placental tissue that may adversely influence fetal growth. Further, it is well known that maternal complications like preeclampsia or eclampsia are the common reasons for indicated preterm birth [127]. Therefore the next section provides an overview of DNA methylation studies of preterm birth.

8.2 Preterm Birth

Preterm birth is an adverse pregnancy outcome and a leading cause for maternal and perinatal morbidity and mortality [128]. It is defined as delivery of a baby before 37 completed weeks of gestation and accounts for 9.6% of all births worldwide [129]. Abnormal placentation is associated with pregnancy complications and can affect the birth outcome. Despite extensive research in medical care, preterm birth remains a major public health problem.

Gene–environment interactions have been suggested to be associated with the risk of delivering preterm, possibly through epigenetic regulation [130]. It is suggested that epigenetic modifications such as DNA methylation in response to various environmental factors may influence the risk of preterm birth or induce changes in the fetal epigenome thereby predisposing the child to adult-onset diseases [131].

A number of studies using various samples like maternal blood and placenta report epigenetic differences to be associated with gestational age. An inverse association of global methylation levels in cervical swabs with gestational age suggests that cervical DNA methylation is associated with the length of gestation [132]. In contrast, another study reports a positive association between placental global DNA methylation levels and gestational age in normotensive women [57]. During the course of pregnancy, placental development involves spatiotemporally programmed epigenetic processes. These results therefore suggest that children born preterm impede the normal spatiotemporal pattern of gene expression resulting in fetal programming of adult diseases.

Study by Maccani and colleagues reports that placental methylation pattern at one of the loci of *RUNX3* (runt-related transcription factor 3) gene involved in normal immune system development is associated with decreased gestational age [133]. It has been reported that increased DNA methylation levels of the imprinted gene, *PLAGL1*, in the cord blood of preterm infants affect the growth and development at birth and also increase the susceptibility to adult-onset diseases [134]. It is reported that differential epigenetic changes in umbilical cord blood leukocytes exist in fetuses born at different gestational ages particularly at CpG sites of genes that play an important role in embryonic development (*HDAC4*, *DNMT1*, *DNMT3A*, *DNMT3B*, and *TET1*) and extracellular matrix degradation (*MMP-9* and *TIMP-2*) [135].

A study by Kantake et al. reports higher promoter methylation of glucocorticoid gene in the peripheral blood of preterm infants than the term infants at postnatal day 4 suggesting that postnatal environment influences epigenetic programming of *GR* expression resulting in relative glucocorticoid insufficiency during the postnatal period [136]. Another study reports hypermethylation of prostaglandin D2 receptor (*PTGDR*) gene involved in myometrial contractions in women delivering preterm compared to women delivering at term. This study also demonstrated that differential DNA methylation of myometrial contraction-associated genes occurs at sites outside the CpG island of the promoter region indicating a need to examine DNA methylation changes across the genome [137].

Reports also indicate that widespread DNA methylation differences between extreme preterm and term infants at birth are largely resolved by 18 years of age, suggesting that DNA methylation differences at birth are mainly driven by factors relating to gestational age. In addition, the authors also identified 10 probes with mean methylation difference of more than 5% at both time points (at birth and 18 years) in preterm individuals suggesting a possibility of long-term epigenetic inheritance of preterm birth [138].

Preterm babies are known to be at increased risk of neurodevelopment and metabolic disorders in later life. It is therefore necessary to understand the methylation patterns of relevant genes in women delivering preterm. In view of these reports, future studies examining DNA methylation patterns may help in identification of a biomarker for the preterm birth.

8.3 Gestational Diabetes

Epidemiological studies have shown that gestational diabetes mellitus (GDM) is associated with large for gestational age baby at birth. Evidence suggests that the relationship between maternal gestational diabetes and adverse short- and long-term fetal outcomes is possibly via the epigenetic modifications. Study by Chen et al. suggests that GDM-induced cord blood hypermethylation in CpG sites of the imprinted gene, guanine nucleotide-binding protein alpha subunit (*GNAS*) in GDM women could be a mechanism underlying the increased risk of metabolic diseases in the offspring in later life [139].

It is suggested that the maternal metabolic status before and during pregnancy can alter placental DNA methylation profile at birth and subsequently contribute to metabolic programming of obesity and related conditions in later life [140]. A study by Ruchat et al. reports differential placental and cord blood DNA methylation levels between samples exposed or not exposed to GDM, and the ingenuity pathway analysis (IPA) identified most of differentially methylated genes to be involved in the metabolic pathways [141]. The same group recently reports lower DNA methylation levels of insulin like growth factor 1 receptor (*IGF1R*) and IGF-binding protein 3 (*IGFBP3*) genes in placentas exposed to impaired glucose tolerance (*IGT*) as compared to the normal placenta. This suggests that the epigenetic dysregulation of *IGF1R* and *IGFBP3* (two key genes involved in growth and metabolism) increases the risk of developing metabolic diseases in later life [142]. An earlier study by the same group also reported that placental DNA methylation at the *IGF2/H19* locus correlated with the newborns' weight although there was no difference based on maternal glucose tolerance status (normal vs impaired glucose tolerance) [143].

Further, a number of studies also indicate that altered maternal glucose metabolism affects the DNA methylation patterns of genes involved in energy and glucose metabolism like leptin [140,144] or adiponectin gene [145], and imprinted genes like *MEST* [146], which may contribute to the development of obesity in the offspring. A recent review suggests that future randomized trials are needed to find out if early intervention could decrease the risk for gestational diabetes and prevent long-term adverse outcome [147].

A list of epigenetic biomarkers for adverse pregnancy outcomes, which are identified by different methylation techniques, has been summarized in Table 1.

9. CONCLUSION AND FUTURE PROSPECTS

Recent data from World Health Organization indicate that approximately 800 women die as a result of complications during pregnancy and childbirth every day, especially in developing countries [148]. Similarly, three

TABLE 1 List of Epigenetic Biomarkers for Adverse Pregnancy Outcomes Identified by Different Methylation Techniques

Pathology	Epigenetic biomarker	Differential methylation	Biomaterial	Technique	References
Preeclampsia	c-myc	Hypermethylated	Placenta	Methylation-sensitive high-resolution melting (MS-HRM)	[102]
Preeclampsia	c-myc	Hypermethylated	Maternal plasma	Methylation-sensitive high-resolution melting (MS-HRM)	[102]
Preterm preeclampsia	CDH11, COL5A1, and TNF	Hypermethylated	Placenta	Methylation 450 array and pyrosequencing	[104]
Preterm preeclampsia	NCAM1	Hypomethylated	Placenta	Methylation 450 array and pyrosequencing	[104]
Preeclampsia	SERPINA3	Hypomethylated	Placenta	Pyrosequencing	[105]
Preeclampsia	CYP27B1, VDR, and RXR	Hypermethylated	Placenta	Human DNA methylation 2.1 M microarrays (NimbleGen)	[106]
Preterm preeclampsia	VEGF	Hypomethylated	Placenta	MassARRAY EpiTYPER	[107]
Preterm and term preeclampsia	FLT-1	Hypomethylated	Placenta	MassARRAY EpiTYPER	[107]
Preterm and term preeclampsia	KDR	Hypermethylated	Placenta	MassARRAY EpiTYPER	[107]
Preeclampsia	Leptin	Hypomethylated	Placenta	MassARRAY EpiTYPER	[108]
Early-onset preeclampsia	Leptin	Hypomethylated	Placenta	Pyrosequencing	[109]
Early-onset preeclampsia	TIMP-3	Hypomethylated	Placenta	Pyrosequencing	[110]
Preeclampsia	TIMP-3	Hypomethylated	Placenta	MassARRAY EpiTYPER	[111]
Preeclampsia	MMP-9	Hypomethylated	Placenta	Methylation-sensitive restriction enzymes—PCR	[112]
Preeclampsia	Syncytin-1	Hypermethylated	Placenta	COBRA, methylation-specific PCR, and DNA sequencing	[114]
Preeclampsia	Syncytin-1	Hypermethylated	Placenta	Pyrosequencing	[115]
Early onset preeclampsia	NR3C1 and CRHBP	Hypermethylated	Placenta	Pyrosequencing	[118]

Early-onset preeclampsia	CRH, CYP11A1, HSD3B1, TEAD3, and CYP19	Hypomethylated	Placenta	Pyrosequencing	[118]
Preeclampsia	MMP-1 and MMP-8	Hypomethylated	Omental arteries	Illumina Infinium HumanMethylation27 BeadChip assay	[123]
Preeclampsia	TIMP-3, COL8A1, COL9A3, COL19A1	Hypermethylated	Omental arteries	Illumina Infinium HumanMethylation27 BeadChip assay	[123]
Preeclampsia	IGF2	Hypomethylated	Umbilical cord blood	MassARRAY EpiTYPER	[124]
Preterm	RUNX3	Hypermethylated	Placenta	Pyrosequencing	[131]
Preterm and infection (chorioamnionitis or funisitis)	PLAGL1	Hypermethylated	Umbilical cord blood	Pyrosequencing	[132]
Preterm	GSK3B, HDAC4	Hypermethylated	Umbilical cord blood	Illumina Infinium HumanMethylation27 BeadChip assay	[133]
Preterm	MAML1, MMP-9	Hypomethylated	Umbilical cord blood	Illumina Infinium HumanMethylation27 BeadChip assay	[133]
Preterm infants	GR	Hypermethylated	Peripheral blood (postnatal day 4)	MQuant method	[134]
Preterm	PTGDR	Hypermethylated	Myometrium	Methylated DNA immunoprecipitation (MeDIP)-chip analysis	[135]
Gestational diabetes	GNAS	Hypermethylated	Umbilical cord blood	MassARRAY EpiTYPER	[137]
Gestational diabetes	IGF1R and IGFBP3	Hypomethylated	Placenta	Pyrosequencing	[140]
Gestational diabetes	Leptin	Hypermethylated	Placenta	Pyrosequencing	[138]
Gestational diabetes	Leptin	Hypermethylated	Placenta	MassARRAY EpiTYPER	[142]
Gestational diabetes	Adiponectin	Hypomethylated	Placenta	Pyrosequencing	[143]
Gestational diabetes	MEST and NR3C1	Hypomethylated	Umbilical cord blood and placenta	Pyrosequencing	[144]
Obese adults	MEST	Hypomethylated	Blood	Pyrosequencing	[144]

million newborn babies die every year, and an additional 2.6 million babies are stillborn [149]. It is known that children born to mothers with an adverse pregnancy outcome are at increased risk of diseases in later life. An appropriate epigenetic regulation of various genes is crucial for placental development and any alterations may affect birth outcome and disease susceptibility in later life.

In view of this, it is necessary to undertake studies for identification of epigenetic biomarkers which may be useful in the prediction of adverse pregnancy outcomes like preeclampsia, gestational diabetes, fetal growth restriction, and preterm birth. Epigenetic approaches, which include screening of fetal DNA methylation markers, can be used for noninvasive or semi-invasive prenatal testing for diagnosis of fetal chromosomal disease and placental pathologies. However, this needs to be confirmed on large populations and validated before its use in clinical practice. Further, longitudinal studies (across pregnancy as well as follow-up of children from birth to adulthood) are required to describe changes in DNA methylation levels of various biomarkers. Identification of epigenetic biomarkers like the methylation of leptin, *VEGF*, *TIMP-3*, *RUNX3*, *GR*, *MEST* genes, and others during the progression of gestation may be useful in the prediction of pregnancy complications (Figure 3).

Currently, methylation analyses use techniques like methylated DNA immunoprecipitation–sequencing (MeDIP-Seq), Infinium Human Methylation450 BeadChip assays, methylation-sensitive high-resolution melting (MS-HRM), microarrays, pyrosequencing, and MassARRAY EpiTYPER. However, for routine clinical practices, there is a need to identify

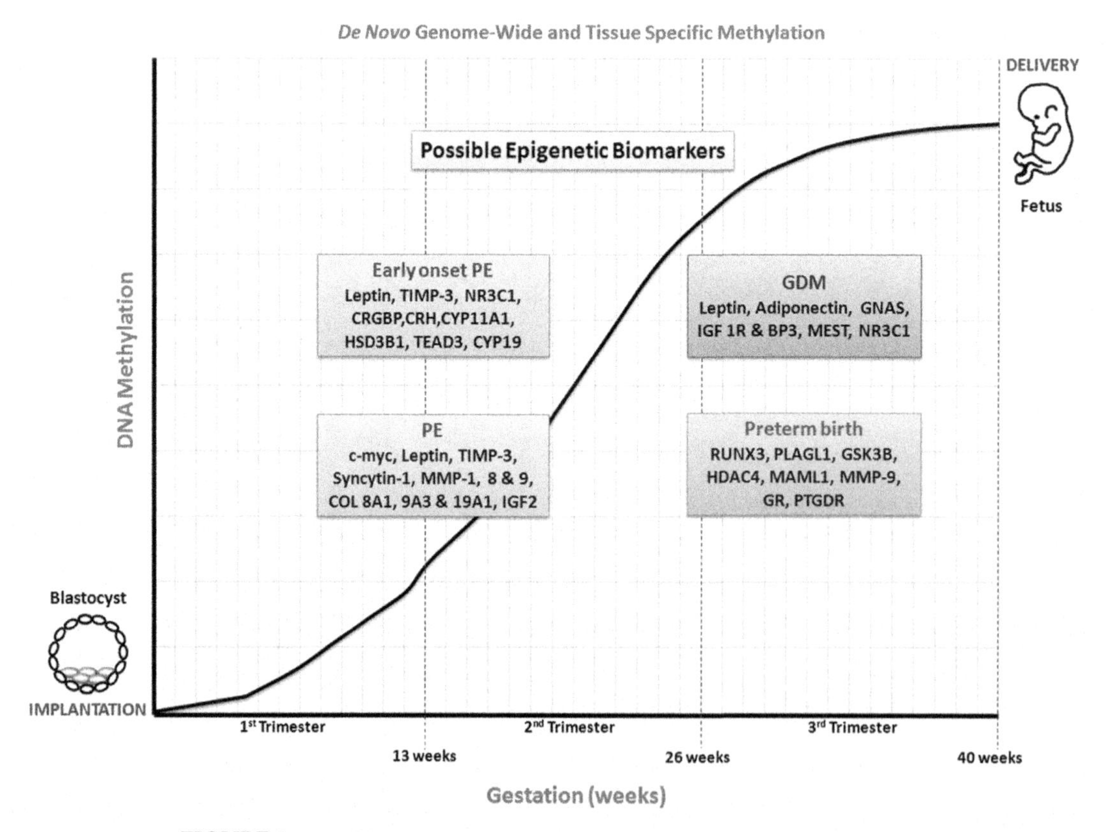

FIGURE 3 Possible epigenetic biomarkers during the progression of gestation.

cost-effective epigenetic techniques. A proposed schematic diagram describing the procedure for the study of epigenetic biomarkers in pregnancy complications that could be implemented in a clinical laboratory has been illustrated in Figure 4.

Future studies also need to examine the effect of maternal diet on epigenetic changes in various pregnancy complications through which it may be possible to ameliorate the severity of such varied complications during pregnancy.

Glossary

Adiponectin Adiponectin is a protein hormone encoded by the *ADIPOQ* gene in humans and involved in regulating glucose levels as well as fatty acid breakdown.

Angiogenesis Angiogenesis is the physiological process involving the growth of new blood vessels from preexisting vessels.

Cervix It is the lower part of the uterus in the human female reproductive system that connects the uterine cavity and the lumen of the vagina.

Chromatin Chromatin is a complex of macromolecules found in cells, consisting of DNA, protein, and RNA that forms chromosomes within the nucleus of eukaryotic cells.

CpG site The CpG sites are regions of DNA where a cytosine nucleotide occurs next to a guanine nucleotide in the linear sequence of bases along its length.

Dopamine Dopamine is a hormone and a neurotransmitter that plays an important role in the human brain and body.

Epidemiology Epidemiology is the science that studies the patterns, causes, and effects of health and disease conditions in defined populations.

Epigenetic/developmental programming The relationship between the periconceptual, fetal, and early infant phases of life and the subsequent development of

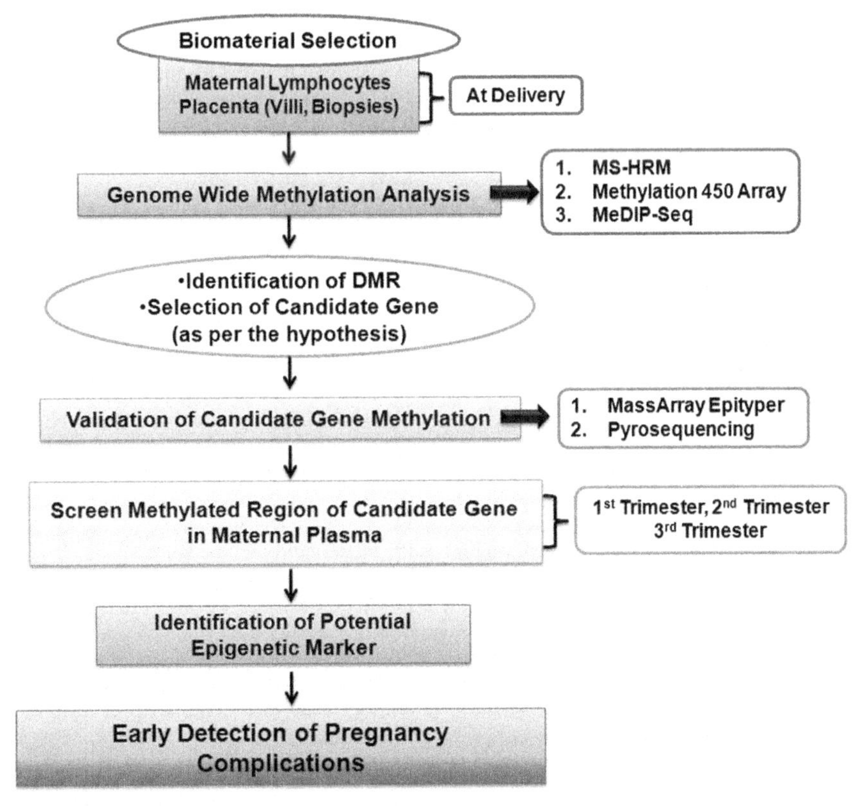

FIGURE 4 Schematic diagram describing the procedure for the study of epigenetic biomarkers in pregnancy complications. DMR, differentially methylated regions; MS-HRM, methylation-sensitive high-resolution melting; MeDIP-Seq, methylated DNA immunoprecipitation-sequencing.

adult noncommunicable disorders through epigenetic mechanisms is termed as epigenetic/developmental programming.

Epigenetics It is the study of heritable changes that determine the phenotype of an organism by factors other than the underlying DNA sequences.

Epigenome Epigenome literally means "above" the genome and consists of chemical compounds (methyl groups) that modify the genome.

Gene-specific methylation/promoter methylation It is defined as the methylation over a specific gene promoter which regulates the gene expression.

Genome The genetic material of an organism comprising of the complete set of DNA in a cell carrying instructions that make each living creature unique is termed as genome.

Genomic imprinting Genomic imprinting is an epigenetic process that can involve DNA methylation and histone modulation by which certain genes can be expressed in a parent-of-origin-specific manner without altering the genetic sequence.

Gestational diabetes mellitus (GDM) It is one of the most common metabolic disorders complicating pregnancy, defined as any degree of glucose intolerance with onset or first recognition during pregnancy.

Global DNA methylation Global DNA methylation is a widely used term to denote methylation of CpG dinucleotides over the entire genome.

Glucocorticoids It is a class of steroid hormones that bind to their specific receptors and involved in the regulation of the glucose metabolism.

Leptin Leptin is a hormone made by fat cells and regulates the amount of fat stored in the body.

Myometrium It is the middle layer of the uterine wall consisting mainly of uterine smooth muscle cells that induces contractions during labor.

Opioid An opioid is any chemical that resembles morphine or other opiates in its pharmacological effects of relieving pain in the body by reducing the intensity of pain signals reaching the brain.

Periconceptional period The period from before conception to early pregnancy, which is considered as an important time for interventions, is defined as periconceptional period.

Phenotype The observable characteristics of an organism that result from the interaction of its genotype (total genetic inheritance) with the environment are termed as phenotype.

Placenta The placenta is an organ that connects the developing fetus to the uterine wall of the mother for the nutrient uptake, waste elimination, gas exchange, and carry out other important functions to support pregnancy.

LIST OF ABBREVIATIONS

5-MTHF 5-Methyltetrahydrofolate
AA Arachidonic acid
AGA Appropriate for gestational age
ALA Alpha-linolenic acid
BHMT Betaine–homocysteine methyltransferase
CBS Cystathionine b-synthase
CpG Cytosine–guanine dinucleotides
CRH Corticotropin-releasing hormone
CRHBP Corticotropin-releasing hormone-binding protein
DHA Docosahexaenoic acid
DMPs Differentially methylated positions
DMRs Differentially methylated regions
DNMTs DNA methyltransferases
DOHaD Developmental Origins of Health and Disease
FGR Fetal growth restricted
FLT-1 Fms-like tyrosine kinase-1
GDM Gestational diabetes mellitus
GNAS Guanine nucleotide-binding protein alpha subunit
GR Glucocorticoid receptor
ICD International classification of diseases
IGF1R Insulin-like growth factor 1 receptor
IGFBP3 IGF-binding protein 3
IGT Impaired glucose tolerance
IPA Ingenuity pathway analysis
IUGR Intrauterine growth restriction
KDR Kinase domain receptor
LA Linoleic acid
LBW Low birth weight
LCPUFA Long-chain polyunsaturated fatty acids
MAT Methionine adenosyltransferase
MC1R Melanocortin 1 receptor
MeDIP-Seq Methylated DNA immunoprecipitation–sequencing
MEST Mesoderm-specific transcript
MIM Mendelian inheritance in man
MMP-9 Matrix metalloproteinase-9
MTHFR Methylene tetrahydrofolate reductase
MTR Methionine tetrahydrofolate reductase
NCDs Noncommunicable diseases
ncRNAs Noncoding RNAs
PC Phosphatidylcholine
PE Phosphatidylethanolamine
PEMT Phosphatidylethanolamine N–methyltransferase
PLAGL1 Pleiomorphic adenoma gene-like 1
PMDs Partially methylated domains
PPAR-α Peroxisomal proliferator-activated receptor-alpha
PTGDR Prostaglandin D2 receptor
RUNX3 Runt-related transcription factor 3
RXR Retinoid X receptor
SAH S-adenosyl homocysteine

SAM S-adenosyl methionine
SERPINA3 Serine protease inhibitor-3
TIMP-3 Tissue inhibitor of matrix metalloproteinase-3
TSS Transcription start site
VDR Vitamin D receptor
VEGF Vascular endothelial growth factor

References

[1] Thornburg KL, Shannon J, Thuillier P, Turker MS. In utero life and epigenetic predisposition for disease. Adv Genet 2010;71:57–78.
[2] Novakovic B, Saffery R. The ever growing complexity of placental epigenetics–role in adverse pregnancy outcomes and fetal programming. Placenta 2012;33:959–70.
[3] Halusková J. Epigenetic studies in human diseases. Folia Biol (Praha) 2010;56:83–96.
[4] Waterland RA, Jirtle RL. Transposable elements: targets for early nutritional effects on epigenetic gene regulation. Mol Cell Biol 2003;23:5293–300.
[5] Heijmans BT, Tobi EW, Stein AD, Putter H, Blauw GJ, Susser ES, et al. Persistent epigenetic differences associated with prenatal exposure to famine in humans. Proc Natl Acad Sci USA 2008;105:17046–9.
[6] Selhub J. Folate, vitamin B12 and vitamin B6 and one carbon metabolism. J Nutr Health Aging 2002;6:39–42.
[7] Sinclair KD, Allegrucci C, Singh R, Gardner DS, Sebastian S, Bispham J, et al. DNA methylation, insulin resistance, and blood pressure in offspring determined by maternal periconceptional B vitamin and methionine status. Proc Natl Acad Sci USA 2007;104:19351–6.
[8] Ross SA, Poirier L. Proceedings of the Trans-HHS Workshop: diet, DNA methylation processes and health. J Nutr 2002;132:2329S–32S.
[9] Burdge GC, Lillycrop KA. Fatty acids and epigenetics. Curr Opin Clin Nutr Metab Care 2014;17:156–61.
[10] Nelissen EC, van Montfoort AP, Dumoulin JC, Evers JL. Epigenetics and the placenta. Hum Reprod Update 2011;17:397–417.
[11] Ordovás JM, Smith CE. Epigenetics and cardiovascular disease. Nat Rev Cardiol 2010;7:510–9.
[12] Godfrey KM, Lillycrop KA, Burdge GC, Gluckman PD, Hanson MA. Epigenetic mechanisms and the mismatch concept of the developmental origins of health and disease. Pediatr Res 2007;61:5R–10R.
[13] Goldberg AD, Allis CD, Bernstein E. Epigenetics: a landscape takes shape. Cell 2007;128:635–8.
[14] Baccarelli A, Rienstra M, Benjamin EJ. Cardiovascular epigenetics: basic concepts and results from animal and human studies. Circ Cardiovasc Genet 2010;3:567–73.
[15] Santos-Rebouças CB, Pimentel MM. Implication of abnormal epigenetic patterns for human diseases. Eur J Hum Genet 2007;15:10–7.
[16] Choudhuri S, Cui Y, Klaassen CD. Molecular targets of epigenetic regulation and effectors of environmental influences. Toxicol Appl Pharmacol 2010;245:378–93.
[17] Turunen MP, Ylä-Herttuala S. Epigenetic regulation of key vascular genes and growth factors. Cardiovasc Res 2011;90:441–6.
[18] Cosgrove MS, Wolberger C. How does the histone code work? Biochem Cell Biol 2005;83:468–76.
[19] Ling C, Groop L. Epigenetics: a molecular link between environmental factors and type 2 diabetes. Diabetes 2009;58:2718–25.
[20] Bashir Q, William BM, Garcia-Manero G, de Lima M. Epigenetic therapy in allogeneic hematopoietic stem cell transplantation. Rev Bras Hematol Hemoter 2013;35:126–33.
[21] Bird A. DNA methylation patterns and epigenetic memory. Genes Dev 2002;16:6–21.
[22] Esteller M, Herman JG. Cancer as an epigenetic disease: DNA methylation and chromatin alterations in human tumours. J Pathol 2002;196:1–7.
[23] Davis CD, Uthus EO. DNA methylation, cancer susceptibility, and nutrient interactions. Exp Biol Med (Maywood) 2004;229:988–95.
[24] Dhe-Paganon S, Syeda F, Park L. DNA methyltransferase 1: regulatory mechanisms and implications in health and disease. Int J Biochem Mol Biol 2011;2:58–66.
[25] Xie S, Wang Z, Okano M, Nogami M, Li Y, He WW, et al. Cloning, expression and chromosome locations of the human DNMT3 gene family. Gene 1999;236:87–95.
[26] Kouzarides T. Chromatin modifications and their function. Cell 2007;128:693–705.
[27] Memili E, Hong YK, Kim DH, Ontiveros SD, Strauss WM. Murine Xist RNA isoforms are different at their 3′ ends: a role for differential polyadenylation. Gene 2001;266:131–7.
[28] Sleutels F, Zwart R, Barlow DP. The non-coding Air RNA is required for silencing autosomal imprinted genes. Nature 2002;415:810–3.
[29] Pozharny Y, Lambertini L, Clunie G, Ferrara L, Lee MJ. Epigenetics in women's health care. Mt Sinai J Med 2010;77:225–35.
[30] Donkena KV, Young CY, Tindall DJ. Oxidative stress and DNA methylation in prostate cancer. Obstet Gynecol Int 2010;2010:302051.
[31] Baylin SB, Herman JG, Graff JR, Vertino PM, Issa JP. Alterations in DNA methylation: a fundamental aspect of neoplasia. Adv Cancer Res 1998;72:141–96.
[32] Miranda TB, Jones PA. DNA methylation: the nuts and bolts of repression. J Cell Physiol 2007;213:384–90.

[33] Greer EL, Shi Y. Histone methylation: a dynamic mark in health, disease and inheritance. Nat Rev Genet 2012;13:343–57.
[34] Brown R, Strathdee G. Epigenomics and epigenetic therapy of cancer. Trends Mol Med 2002;8:S43–8.
[35] Santini V, Kantarjian HM, Issa JP. Changes in DNA methylation in neoplasia: pathophysiology and therapeutic implications. Ann Intern Med 2001;134: 573–86.
[36] Bogdanović O, Gómez-Skarmeta JL. Embryonic DNA methylation: insights from the genomics era. Brief Funct Genomics 2014;13:121–30.
[37] Morgan HD, Santos F, Green K, Dean W, Reik W. Epigenetic reprogramming in mammals. Hum Mol Genet 2005;14(1):R47–58.
[38] Reik W, Dean W, Walter J. Epigenetic reprogramming in mammalian development. Science 2001;293:1089–93.
[39] Tomizawa S, Nowacka-Woszuk J, Kelsey G. DNA methylation establishment during oocyte growth: mechanisms and significance. Int J Dev Biol 2012;56: 867–75.
[40] Inbar-Feigenberg M, Choufani S, Butcher DT, Roifman M, Weksberg R. Basic concepts of epigenetics. Fertil Steril 2013;99:607–15.
[41] Hales CN, Barker DJ. The thrifty phenotype hypothesis. Br Med Bull 2001;60:5–20.
[42] Fall CH. Fetal programming and the risk of noncommunicable disease. Indian J Pediatr 2013;80:S13–20.
[43] Pasternak Y, Aviram A, Poraz I, Hod M. Maternal nutrition and offspring's adulthood NCD's: a review. J Matern Fetal Neonatal Med 2013;26:439–44.
[44] Barker DJ. The fetal and infant origins of adult disease. BMJ 1990;301:1111.
[45] Godfrey KM, Barker DJ. Fetal nutrition and adult disease. Am J Clin Nutr 2000;71:1344S–52S.
[46] Stover JP, Caudill MA. Genetic and epigenetic contributions to human nutrition and health: managing genome-diet interactions. J Am Diet Assoc 2008;108:1480–7.
[47] Cutfield WS, Hofman PL, Mitchell M, Morison IM. Could epigenetics play a role in the developmental origins of health and disease? Pediatr Res 2007;61:68R–75R.
[48] Simmons R. Epigenetics and maternal nutrition: nature v. nurture. Proc Nutr Soc 2011;70:73–81.
[49] Choi SW, Friso S. Epigenetics: a new bridge between nutrition and health. Adv Nutr 2010;1:8–16.
[50] Bogdarina I, Welham S, King PJ, Burns SP, Clark AJ. Epigenetic modification of the renin-angiotensin system in the fetal programming of hypertension. Circ Res 2007;100:520–6.
[51] Pellanda H. Betaine homocysteine methyltransferase (BHMT)-dependent remethylation pathway in human healthy and tumoral liver. Clin Chem Lab Med 2013;51:617–21.
[52] Dominguez-Salas P, Moore SE, Cole D, da Costa KA, Cox SE, Dyer RA, et al. DNA methylation potential: dietary intake and blood concentrations of one-carbon metabolites and cofactors in rural African women. Am J Clin Nutr 2013;97:1217–27.
[53] Kalhan SC. One-carbon metabolism, fetal growth and long-term consequences. Nestle Nutr Inst Workshop Ser 2013;74:127–38.
[54] Dasarathy J, Gruca LL, Bennett C, Parimi PS, Duenas C, Marczewski S, et al. Methionine metabolism in human pregnancy. Am J Clin Nutr 2010;91: 357–65.
[55] Kulkarni A, Dangat K, Kale A, Sable P, Chavan-Gautam P, Joshi S. Effects of altered maternal folic acid, vitamin B12 and docosahexaenoic acid on placental global DNA methylation patterns in Wistar rats. PLoS One 2011;6:e17706.
[56] Kulkarni A, Chavan-Gautam P, Mehendale S, Yadav H, Joshi S. Global DNA methylation patterns in placenta and its association with maternal hypertension in pre-eclampsia. DNA Cell Biol 2011;30: 79–84.
[57] Chavan-Gautam P, Sundrani D, Pisal H, Nimbargi V, Mehendale S, Joshi S. Gestation-dependent changes in human placental global DNA methylation levels. Mol Reprod Dev 2011;78:150.
[58] Dhobale M, Chavan P, Kulkarni A, Mehendale S, Pisal H, Joshi S. Reduced folate, increased vitamin B(12) and homocysteine concentrations in women delivering preterm. Ann Nutr Metab 2012;61:7–14.
[59] Dhobale MV, Wadhwani N, Mehendale SS, Pisal HR, Joshi SR. Reduced levels of placental long chain polyunsaturated fatty acids in preterm deliveries. Prostagl Leukot Essent Fat Acids 2011;85:149–53.
[60] Lillycrop KA, Phillips ES, Jackson AA, Hanson MA, Burdge GC. Dietary protein restriction of pregnant rats induces and folic acid supplementation prevents epigenetic modification of hepatic gene expression in the offspring. J Nutr 2005;135: 1382–6.
[61] Lillycrop KA, Slater-Jefferies JL, Hanson MA, Godfrey KM, Jackson AA, Burdge GC. Induction of altered epigenetic regulation of the hepatic glucocorticoid receptor in the offspring of rats fed a protein-restricted diet during pregnancy suggests that reduced DNA methyltransferase-1 expression is involved in impaired DNA methylation and changes in histone modifications. Br J Nutr 2007;97:1064–73.
[62] Slater-Jefferies JL, Lillycrop KA, Townsend PA, Torrens C, Hoile SP, Hanson MA, et al. Feeding a protein-restricted diet during pregnancy induces altered epigenetic regulation of peroxisomal proliferator-activated receptor-α in the heart of the offspring. J Dev Orig Health Dis 2011;2:250–5.

[63] Burdge GC, Hanson MA, Slater-Jefferies JL, Lillycrop KA. Epigenetic regulation of transcription: a mechanism for inducing variations in phenotype (fetal programming) by differences in nutrition during early life? Br J Nutr 2007;97:1036–46.
[64] Aagaard-Tillery KM, Grove K, Bishop J, Ke X, Fu Q, McKnight R, et al. Developmental origins of disease and determinants of chromatin structure: maternal diet modifies the primate fetal epigenome. J Mol Endocrinol 2008;41:91–102.
[65] Vucetic Z, Kimmel J, Totoki K, Hollenbeck E, Reyes TM. Maternal high-fat diet alters methylation and gene expression of dopamine and opioid-related genes. Endocrinology 2010;151:4756–64.
[66] Carlin J, George R, Reyes TM. Methyl donor supplementation blocks the adverse effects of maternal high fat diet on offspring physiology. PLoS One 2013;8:e63549.
[67] Ashworth CJ, Antipatis C. Micronutrient programming of development throughout gestation. Reproduction 2001;122:527–35.
[68] Crider KS, Yang TP, Berry RJ, Bailey LB. Folate and DNA methylation: a review of molecular mechanisms and the evidence for folate's role. Adv Nutr 2012;3:21–38.
[69] Niculescu MD, Zeisel SH. Diet, methyl donors and DNA methylation: interactions between dietary folate, methionine and choline. J Nutr 2002;132: 2333S–5S.
[70] Tryndyak VP, Han T, Muskhelishvili L, Fuscoe JC, Ross SA, Beland FA, et al. Coupling global methylation and gene expression profiles reveal key pathophysiological events in liver injury induced by a methyl-deficient diet. Mol Nutr Food Res 2011;55:411–8.
[71] Tsujiuchi T, Tsutsumi M, Sasaki Y, Takahama M, Konishi Y. Hypomethylation of CpG sites and c-myc gene overexpression in hepatocellular carcinomas, but not hyperplastic nodules, induced by a choline-deficient L-amino acid-defined diet in rats. Jpn J Cancer Res 1999;90: 909–13.
[72] Pogribny IP, Karpf AR, James SR, Melnyk S, Han T, Tryndyak VP. Epigenetic alterations in the brains of Fisher 344 rats induced by long-term administration of folate/methyl-deficient diet. Brain Res 2008;1237:25–34.
[73] Steinmetz KL, Pogribny IP, James SJ, Pitot HC. Hypomethylation of the rat glutathione S-transferase pi (GSTP) promoter region isolated from methyl-deficient livers and GSTP-positive liver neoplasms. Carcinogenesis 1998;19:1487–94.
[74] Maloney CA, Hay SM, Rees WD. Folate deficiency during pregnancy impacts on methyl metabolism without affecting global DNA methylation in the rat fetus. Br J Nutr 2007;97:1090–8.
[75] Steegers-Theunissen RP, Obermann-Borst SA, Kremer D, Lindemans J, Siebel C, Steegers EA, et al. Periconceptional maternal folic acid use of 400 microg per day is related to increased methylation of the IGF2 gene in the very young child. PLoS One 2009;4:e7845.
[76] Sablet P, Randhir K, Kale A, Chavan-Gautam P, Joshi S. Maternal micronutrients and brain global methylation patterns in the offspring. Nutr Neurosci 2015;18:30–6.
[77] Khot V, Kale A, Joshi A, Chavan-Gautam P, Joshi S. Expression of genes encoding enzymes involved in the one carbon cycle in rat placenta is determined by maternal micronutrients (folic acid, vitamin B12) and omega-3 fatty acids. Biomed Res Int 2014;2014:613078.
[78] Huang T, Wahlqvist ML, Li D. Effect of n-3 polyunsaturated fatty acid on gene expression of the critical enzymes involved in homocysteine metabolism. Nutr J 2012;11:6.
[79] Kale A, Naphade N, Sapkale S, Kamaraju M, Pillai A, Joshi S, et al. Reduced folic acid, vitamin B12 and docosahexaenoic acid and increased homocysteine and cortisol in never-medicated schizophrenia patients: implications for altered one-carbon metabolism. Psychiatry Res 2010;175:47–53.
[80] Haggarty P. Fatty acid supply to the human fetus. Annu Rev Nutr 2010;21:237–55.
[81] Benatti P, Peluso G, Nicolai R, Calvani M. Polyunsaturated fatty acids: biochemical, nutritional and epigenetic properties. J Am Coll Nutr 2004;23:281–302.
[82] da Silva RP, Kelly KB, Al Rajabi A, Jacobs RL. Novel insights on interactions between folate and lipid metabolism. Biofactors 2014;40:277–83.
[83] Khot V, Chavan-Gautam P, Joshi S. Proposing interactions between maternal phospholipids and the one carbon cycle: a novel mechanism influencing the risk for cardiovascular diseases in the offspring in later life. Life Sci 2015;129:16–21.
[84] da Costa KA, Rai KS, Craciunescu CN, Parikh K, Mehedint MG, Sanders LM, et al. Dietary docosahexaenoic acid supplementation modulates hippocampal development in the Pemt-/- mouse. J Biol Chem 2010;285:1008–15.
[85] Lee HS, Barraza-Villarreal A, Biessy C, Duarte-Salles T, Sly PD, Ramakrishnan U, et al. Dietary supplementation with polyunsaturated fatty acid during pregnancy modulates DNA methylation at IGF2/H19 imprinted genes and growth of infants. Physiol Genomics 2014. http://dx.doi.org/10.1152/physiolgenomics.00061.
[86] Hoile SP, Irvine NA, Kelsall CJ, Sibbons C, Feunteun A, Collister A, et al. Maternal fat intake in rats alters 20:4n-6 and 22:6n-3 status and the epigenetic regulation of Fads2 in offspring liver. J Nutr Biochem 2013;24:1213–20.

[87] Niculescu MD, Lupu DS, Craciunescu CN. Perinatal manipulation of α-linolenic acid intake induces epigenetic changes in maternal and offspring livers. FASEB J 2013;27:350–8.
[88] Tarrade A, Panchenko P, Junien C, Gabory A. Placental contribution to nutritional programming of health and diseases: epigenetics and sexual dimorphism. J Exper Biol 2015;2018(Pt1):50–8.
[89] Szyf M. The early life environment and the epigenome. Biochim Biophys Acta 2009;1790:878–85.
[90] Dolinoy DC, Jirtle RL. Environmental epigenomics in human health and disease. Environ Mol Mutagen 2008;49:4–8.
[91] Novakovic B, Yuen RK, Gordon L, Penaherrera MS, Sharkey A, Moffett A, et al. Evidence for widespread changes in promoter methylation profile in human placenta in response to increasing gestational age and environmental/stochastic factors. BMC Genomics 2011;12:529.
[92] Hillman SL, Finer S, Smart MC, Mathews C, Lowe R, Rakyan VK, et al. Novel DNA methylation profiles associated with key gene regulation and transcription pathways in blood and placenta of growth-restricted neonates. Epigenetics 2015;10:50–61.
[93] Pozharny Y, Lambertini L, Ma Y, Ferrara L, Litton CG, Diplas A, et al. Genomic loss of imprinting in first-trimester human placenta. Am J Obstet Gynecol 2010;202:391.e1–8.
[94] Bloushtain-Qimron N, Yao J, Snyder EL, Shipitsin M, Campbell LL, Mani SA, et al. Cell type-specific DNA methylation patterns in the human breast. Proc Natl Acad Sci USA 2008;105:14076–81.
[95] Bloushtain-Qimron N, Yao J, Shipitsin M, Maruyama R, Polyak K. Epigenetic patterns of embryonic and adult stem cells. Cell Cycle 2009;8:809–17.
[96] Grigoriu A, Ferreira JC, Choufani S, Baczyk D, Kingdom J, Weksberg R. Cell specific patterns of methylation in the human placenta. Epigenetics 2011;6:368–79.
[97] Xiang Y, Zhang J, Li Q, Zhou X, Wang T, Xu M, et al. DNA methylome profiling of maternal peripheral blood and placentas reveal potential fetal DNA markers for non-invasive prenatal testing. Mol Hum Reprod 2014;20:875–84.
[98] Tsui DW, Chiu RW, Lo YD. Epigenetic approaches for the detection of fetal DNA in maternal plasma. Chimerism 2010;1:30–5.
[99] Schroeder DI, Blair JD, Lott P, Yu HO, Hong D, Crary F, et al. The human placenta methylome. Proc Natl Acad Sci USA 2013;110:6037–42.
[100] Schroeder DI, LaSalle JM. How has the study of the human placenta aided our understanding of partially methylated genes? Epigenomics 2013;5:645–54.
[101] Robinson WP, Price EM. The human placental methylome. Cold Spring Harb Perspect Med 2015. pii: a023044.
[102] Robins JC, Marsit CJ, Padbury JF, Sharma SS. Endocrine disruptors, environmental oxygen, epigenetics and pregnancy. Front Biosci (Elite Ed) 2011;3:690–700.
[103] Rahat B, Hamid A, Ahmad Najar R, Bagga R, Kaur J. Epigenetic mechanisms regulate placental c-myc and hTERT in normal and pathological pregnancies; c-myc as a novel fetal DNA epigenetic marker for pre-eclampsia. Mol Hum Reprod 2014;20:1026–40.
[104] Lu L, Hou Z, Li L, Yang Y, Wang X, Zhang B, et al. Methylation pattern of H19 exon 1 is closely related to preeclampsia and trophoblast abnormalities. Int J Mol Med 2014;34:765–71.
[105] Anton L, Brown AG, Bartolomei MS, Elovitz MA. Differential methylation of genes associated with cell adhesion in preeclamptic placentas. PLoS One 2014;9:e100148.
[106] Chelbi ST, Mondon F, Jammes H, Buffat C, Mignot TM, Tost J, et al. Expressional and epigenetic alterations of placental serine protease inhibitors: SERPINA3 is a potential marker of preeclampsia. Hypertension 2007;49:76–83.
[107] Anderson CM, Ralph JL, Johnson L, Scheett A, Wright ML, Taylor JY, et al. First trimester vitamin D status and placental epigenomics in preeclampsia among Northern Plains primiparas. Life Sci 2015; 129:10–5.
[108] Sundrani DP, Reddy US, Joshi AA, Mehendale SS, Chavan-Gautam PM, Hardikar AA, et al. Differential placental methylation and expression of VEGF, FLT-1 and KDR genes in human term and preterm preeclampsia. Clin Epigenetics 2013;5:6.
[109] Xiang Y, Cheng Y, Li X, Li Q, Xu J, Zhang J, et al. Up-regulated expression and aberrant DNA methylation of LEP and SH3PXD2A in pre-eclampsia. PLoS One 2013;8:e59753.
[110] Hogg K, Blair JD, von Dadelszen P, Robinson WP. Hypomethylation of the LEP gene in placenta and elevated maternal leptin concentration in early onset pre-eclampsia. Mol Cell Endocrinol 2013;367:64–73.
[111] Miehle K, Stepan H, Fasshauer M. Leptin, adiponectin and other adipokines in gestational diabetes mellitus and preeclampsia. Clin Endocrinol(Oxf) 2012;76:2–11.
[112] Yuen RK, Peñaherrera MS, von Dadelszen P, McFadden DE, Robinson WP. DNA methylation profiling of human placentas reveals promoter hypomethylation of multiple genes in early-onset preeclampsia. Eur J Hum Genet 2010;18:1006–12.
[113] Xiang Y, Zhang X, Li Q, Xu J, Zhou X, Wang T, et al. Promoter hypomethylation of TIMP3 is associated with pre-eclampsia in a Chinese population. Mol Hum Reprod 2013;19:153–9.

[114] Wang Z, Lu S, Liu C, Zhao B, Pei K, Tian L, et al. Expressional and epigenetic alterations of placental matrix metalloproteinase 9 in preeclampsia. Gynecol Endocrinol 2010;26:96–102.

[115] Yuen RK, Chen B, Blair JD, Robinson WP, Nelson DM. Hypoxia alters the epigenetic profile in cultured human placental trophoblasts. Epigenetics 2013;8:192–202.

[116] Zhuang XW, Li J, Brost BC, Xia XY, Chen HB, Wang CX, et al. Decreased expression and altered methylation of syncytin-1 gene in human placentas associated with preeclampsia. Curr Pharm Des 2014;20:1796–802.

[117] Ruebner M, Strissel PL, Ekici AB, Stiegler E, Dammer U, Goecke TW, et al. Reduced syncytin-1 expression levels in placental syndromes correlates with epigenetic hypermethylation of the ERVW-1 promoter region. PLoS One 2013;8:e56145.

[118] Blair JD, Yuen RK, Lim BK, McFadden DE, von Dadelszen P, Robinson WP. Widespread DNA hypomethylation at gene enhancer regions in placentas associated with early-onset pre-eclampsia. Mol Hum Reprod 2013;19:697–708.

[119] Chu T, Bunce K, Shaw P, Shridhar V, Althouse A, Hubel C, et al. Comprehensive analysis of preeclampsia-associated DNA methylation in the placenta. PLoS One 2014;9:e107318.

[120] Hogg K, Blair JD, McFadden DE, von Dadelszen P, Robinson WP. Early onset pre-eclampsia is associated with altered DNA methylation of cortisol-signalling and steroidogenic genes in the placenta. PLoS One 2013;8:e62969.

[121] White WM, Brost B, Sun Z, Rose C, Craici I, Wagner SJ, et al. Genome-wide methylation profiling demonstrates hypermethylation in maternal leukocyte DNA in preeclamptic compared to normotensive pregnancies. Hypertens Pregnancy 2013;32:257–69.

[122] Nomura Y, Lambertini L, Rialdi A, Lee M, Mystal EY, Grabie M, et al. Global methylation in the placenta and umbilical cord blood from pregnancies with maternal gestational diabetes, preeclampsia, and obesity. Reprod Sci 2014;21:131–7.

[123] Gao WL, Li D, Xiao ZX, Liao QP, Yang HX, Li YX, et al. Detection of global DNA methylation and paternally imprinted H19 genemethylation in preeclamptic placentas. Hypertens Res 2011;34:655–61.

[124] Mousa AA, Archer KJ, Cappello R, Estrada-Gutierrez G, Isaacs CR, Strauss 3rd JF, et al. DNA methylation is altered in maternal blood vessels of women with preeclampsia. Reprod Sci 2012;19:1332–42.

[125] Mousa AA, Cappello RE, Estrada-Gutierrez G, Shukla J, Romero R, Strauss 3rd JF, et al. Preeclampsia is associated with alterations in DNA methylation of genes involved in collagen metabolism. Am J Pathol 2012;181:1455–63.

[126] He J, Zhang A, Fang M, Fang R, Ge J, Jiang Y, et al. Methylation levels at IGF2 and GNAS DMRs in infants born to preeclamptic pregnancies. BMC Genomics 2013;14:472.

[127] Goldenberg RL, Culhane JF, Iams JD, Romero R. Epidemiology and causes of preterm birth. Lancet 2008;371:75–84.

[128] Saigal S, Doyle LW. An overview of mortality and sequelae of preterm birth from infancy to adulthood. Lancet 2008;371:261–9.

[129] Beck S, Wojdyla D, Say L, Betran AP, Merialdi M, Requejo JH, et al. The worldwide incidence of preterm birth: a systematic review of maternal mortality and morbidity. Bull World Health Organ 2010;88:31–8.

[130] Burris HH, Collins Jr JW. Race and preterm birth–the case for epigenetic inquiry. Ethn Dis 2010;20:296–9.

[131] Menon R, Conneely KN, Smith AK. DNA methylation: an epigenetic risk factor in preterm birth. Reprod Sci 2012;19:6–13.

[132] Burris HH, Baccarelli AA, Motta V, Byun HM, Just AC, Mercado-Garcia A, et al. Association between length of gestation and cervical DNA methylation of PTGER2 and LINE 1-HS. Epigenetics 2014;9:1083–91.

[133] Maccani JZ, Koestler DC, Houseman EA, Marsit CJ, Kelsey KT. Placental DNA methylation alterations associated with maternal tobacco smoking at the RUNX3 gene are also associated with gestational age. Epigenomics 2013;5:619–30.

[134] Liu Y, Hoyo C, Murphy S, Huang Z, Overcash F, Thompson J, et al. DNA methylation at imprint regulatory regions in preterm birth and infection. Am J Obstet Gynecol 2013;208:395.e1–7.

[135] Parets SE, Conneely KN, Kilaru V, Fortunato SJ, Syed TA, Saade G, et al. Fetal DNA methylation associates with early spontaneous preterm birth and gestational age. PLoS One 2013;8:e67489.

[136] Kantake M, Yoshitake H, Ishikawa H, Araki Y, Shimizu T. Postnatal epigenetic modification of glucocorticoid receptor gene in preterm infants: a prospective cohort study. BMJ Open 2014;4:e005318.

[137] Mitsuya K, Singh N, Sooranna SR, Johnson MR, Myatt L. Epigenetics of human myometrium: DNA methylation of genes encoding contraction-associated proteins in term and preterm labor. Biol Reprod 2014;90:98.

[138] Cruickshank MN, Oshlack A, Theda C, Davis PG, Martino D, Sheehan P, et al. Analysis of epigenetic changes in survivors of preterm birth reveals the effect of gestational age and evidence for a long term legacy. Genome Med 2013;5:96.

[139] Chen D, Zhang A, Fang M, Fang R, Ge J, Jiang Y, et al. Increased methylation at differentially methylated region of GNAS in infants born to gestational diabetes. BMC Med Genet 2014;15:108.

[140] Lesseur C, Armstrong DA, Paquette AG, Li Z, Padbury JF, Marsit CJ. Maternal obesity and gestational diabetes are associated with placental leptin DNA methylation. Am J Obstet Gynecol 2014;211(654):e1–9.
[141] Ruchat SM, Houde AA, Voisin G, St-Pierre J, Perron P, Baillargeon JP, et al. Gestational diabetes mellitus epigenetically affects genes predominantly involved in metabolic diseases. Epigenetics 2013;8:935–43.
[142] Desgagné V, Hivert MF, St-Pierre J, Guay SP, Baillargeon JP, Perron P, et al. Epigenetic dysregulation of the IGF system in placenta of newborns exposed to maternal impaired glucose tolerance. Epigenomics 2014;6:193–207.
[143] St-Pierre J, Hivert MF, Perron P, Poirier P, Guay SP, Brisson D, et al. IGF2 DNA methylation is a modulator of newborn's fetal growth and development. Epigenetics 2012;7:1125–32.
[144] Bouchard L, Thibault S, Guay SP, Santure M, Monpetit A, St-Pierre J, et al. Leptin gene epigenetic adaptation to impaired glucose metabolism during pregnancy. Diabetes Care 2010;33:2436–41.
[145] Bouchard L, Hivert MF, Guay SP, St-Pierre J, Perron P, Brisson D. Placental adiponectin gene DNA methylation levels are associated with mothers' blood glucose concentration. Diabetes 2012;61:1272–80.
[146] El Hajj N, Pliushch G, Schneider E, Dittrich M, Müller T, Korenkov M, et al. Metabolic programming of MEST DNA methylation by intrauterine exposure to gestational diabetes mellitus. Diabetes 2013;62:1320–8.
[147] Hiersch L, Yogev Y. Impact of gestational hyperglycemia on maternal and child health. Curr Opin Clin Nutr Metab Care 2014;17:255–60.
[148] Say L, Chou D, Gemmill A, Tunçalp Ö, Moller AB, Daniels J, et al. Global causes of maternal death: a WHO systematic analysis. Lancet Glob Health 2014;2:e323–33.
[149] Cousens S, Blencowe H, Stanton C, Chou D, Ahmed S, Steinhardt L, et al. National, regional, and worldwide estimates of stillbirth rates in 2009 with trends since 1995: a systematic analysis. Lancet 2011;377:1319–30.

CHAPTER

19

Epigenetics in Infectious Diseases

Genevieve Syn, Jenefer M. Blackwell, Sarra E. Jamieson
Telethon Kids Institute, The University of Western Australia, Subiaco, WA, Australia

OUTLINE

1. INTRODUCTION

Pathogens (such as viruses, bacteria, and parasites) have been coevolving with their hosts for a long time. They have developed strategies to be able to successfully invade and colonize their host. It has now been widely published that infectious agents use epigenetics as part of their life cycle to modulate host processes, creating the optimal environment for their survival [1]. Epigenetics is a dynamic process in which changes to gene expression occur without any alterations to the underlying genomic sequence [2]. The mechanisms behind these changes in gene expression include methylation of DNA [3], histone modifications [4], and gene silencing by noncoding RNAs such as microRNA (miRNA) [5]. Therefore, the host epigenome serves as a superb platform for pathogens to be able to control their host processes, such as the downregulation of the immune system, enabling persistent infection. Modulation of the host epigenome by pathogens can leave unique marks on the host's DNA and histones, illustrated by *Helicobacter pylori* as an example in Figure 1.

We can use these marks in the host epigenetic landscape as potential epigenetic biomarkers to

Epigenetic Biomarkers and Diagnostics
http://dx.doi.org/10.1016/B978-0-12-801899-6.00019-X

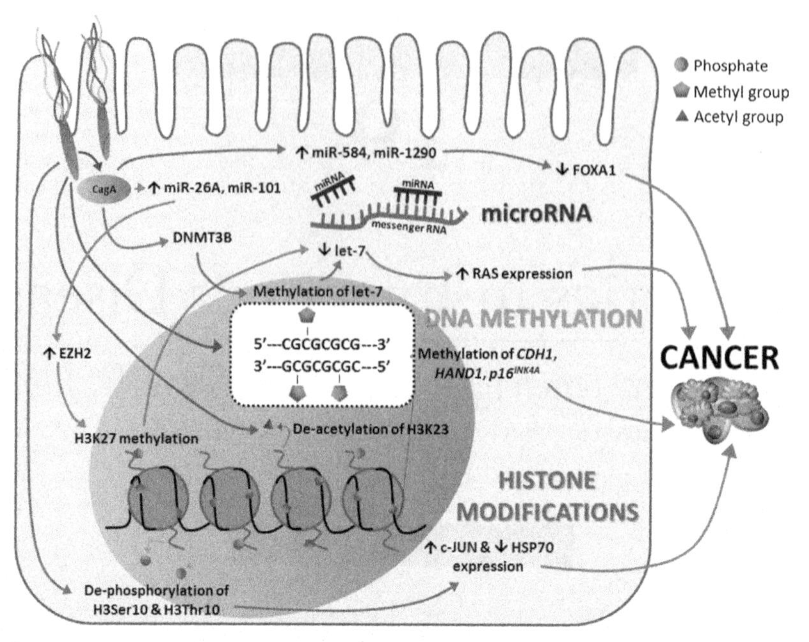

FIGURE 1 An illustration of how gram-negative bacterium *Helicobacter pylori* is able to modify its host's DNA methylation, histones, and microRNA (miRNAs) in a gastric epithelium cell contributing to the pathogenesis of *H. pylori*-induced gastric cancer. *Helicobacter pylori* secretes a protein CagA, which increases the levels of miR-584 and miR-1290. These two miRNAs target the gene forkhead box A1 (*FOXA1*) and downregulate its expression. Suppression of *FOXA1* promotes epithelial–mesenchymal transition and dysregulation of stem cell differentiation, which potentially leads to oncogenesis. CagA can also increase levels of miR-26a and miR-101, which increase enhancer of zeste homolog 2 (EZH2), a histone lysine methyltransferase. EZH2 then methylates lysine residue 27 on histone H3 in the miRNA let-7 promoter, reducing the expression of let-7. The reduction of let-7 is also contributed by CagA-mediated increase in DNA methyltransferase 3 (DNMT3). DNMT3 methylates the promoter of let-7, preventing its transcription. Let-7 regulates the *RAS* oncogenic genes, and decreased cytoplasmic let-7 levels cause the upregulation of *RAS* contributing to carcinogenesis. *Helicobacter pylori* also increases DNA methylation of tumor suppressor genes E-cadherin (*CDH1*), heart and neural crest derivatives expressed 1 (*HAND1*), and cyclin-dependent kinase inhibitor 2A, multiple tumor suppressor 1 ($p16^{INK4A}$), downregulating their expression allowing oncogenesis. Lastly, *H. pylori* is able to modify its host epigenome in a CagA-independent manner. It deacetylates lysine residue 23 and de-phosphorylates serine residue 10 and threonine residue 10 on histone H3. The actual mechanisms of these histone modifications are still unknown. Dephosphorylation of these residues is linked to an increase in expression of JUN proto-oncogene (*c-JUN*) and heat shock protein 70 (*HSP70*), which are associated with tumor formation. These mechanisms orchestrated by *H. pylori* deregulate the host epigenome and aid in the processes leading to *H. pylori*-induced gastric cancers.

provide physicians with information regarding the downstream effects of the infection, such as the onset of cancer. Ideally, an effective epigenetic biomarker should be easily detected in high enough levels in tissue samples. These samples should preferably come from noninvasive sources such as serum, plasma, saliva, urine, or stool. However, there are caveats to bear in mind when obtaining measurements of epigenetic biomarkers in tissue samples not relevant to the disease. For example, measuring the levels of biomarkers in blood to detect for the presence of a solid tumor is heavily reliant on the rationale that solid tumors may shed epigenetically altered DNA into the bloodstream or that cells circulating in the blood undergo epigenetic changes representative of those present in the tumor [6]. In this chapter, we detail all published results for pathogen-induced epigenetic changes in their hosts (see Table 1).

TABLE 1 Summary of Published Literature of Viral-, Bacterial-, and Parasite-Mediated Changes to the Host Epigenome

Microorganism	Epigenetic mechanism	Consequence in host	Cell type	PubMed ID
Viruses				
Cytomegalovirus	Histone modification	General increase in H3K9 methylation and histone tail acetylation affecting all four classes of histones and H3K4 methylation; methylation of H3K4 preferentially associated with postreplicative CMV chromatin; decreased de novo acetylation of H4 at early phase of the infection; and CMV IE1 protein interacts directly with HDAC3	Human fetal diploid lung fibroblasts (MRC-5)	[83,84]
	microRNA	Downregulation of miR-21, miR-99, miR-100, miR-101, miR-155, miR-181, miR-213, miR-222, miR-223, and miR-320; upregulation of miR-17, miR-20, miR-106, and miR-219. Downregulation of mouse-miR-27a; downregulation of 39 genes and upregulation of 10 genes during CMV latency	MRC-5 lung cells; HeLa cells; HEK293T cells NIH-3T3 fibroblast cells; SVEC4-10 endothelial cells; TCMK-1 and C127 epithelial cells; primary bone marrow-derived macrophages; THP-1 cells	[85–87]
Epstein–Barr virus	DNA methylation[a]	Upregulation of DNMT1, DNMT3A, and DNMT3B leads to hypermethylation of E-cadherin promoter, hypermethylation of viral genes *BZLF1* and *BRLF1*, and host gene *Bim* during latent phase of the infection; global differentially methylated CpG islands; hypermethylation of *DAPK, RASSF1A, p16, WIF1, CHFR, RIZ1,RARβ2,* and *CDH1*	NPCTW02 cell line; B cells; oral keratinocytes; nasopharyngeal brushings; nasopharyngeal carcinoma tissue; gastric tumors	[37,88–93]
	Histone modification	Decreased acetylation of histones and increased methylation of H3K27 at site of *Bim* promoter; increased methylation of H3K27 and H4K20; and decreased histone acetylation and methylation H3K4 at viral gene *BZLF1*	B cells; Raji cells	[94]

Continued

TABLE 1 Summary of Published Literature of Viral-, Bacterial-, and Parasite-Mediated Changes to the Host Epigenome—cont'd

Microorganism	Epigenetic mechanism	Consequence in host	Cell type	PubMed ID
Hepatitis B virus	microRNA[a]	Differential expression of miR-375, miR-92a, miR-10a, miR-223, miR-423, miR-23b, miR-23a, miR-342-3p, miR-99a, miR-122a, miR-125b, miR-150, and let-7c between HBV-negative and -positive patients; differential expression of miR-122, miR-192, miR-21, miR-233, miR-26a, miR-27a, and miR-801 between patients with HBV-induced HCC; inhibition of miR-15a/16 complex; downregulation of miR-152, miR-148a, and let-7 family; upregulation of miR-181a, miR-181b, miR-200b, miR-501, and miR-146a	Serum, hepatocytes	[10,11,95–99]
Hepatitis C virus	microRNA[a]	Upregulation of miR-122, miR-141, miR-200a, miR-200b, and miR-200c; downregulation of miR-100, miR-10a, miR-198, and miR-145 in HCC samples	Liver nodules; serum; primary hepatocytes	[15–17]
Herpes simplex virus	Histone modification	Increased acetylation of H3K9 and H3K14 at the *LAT* genes and decreased acetylation of E and IE genes during the latent phase of the infection	Dorsal root ganglia	[59]
Human adenovirus	Histone modification	Blocks IFN-γ-induced monoubiquitination of H2BK120; causes H3K4 and H3K79 trimethylation at IFN-stimulated genes; and decreases global H3K19 acetylation	A549 lung adenocarcinoma; IMR90 primary fibroblasts	[100]
Human papillomavirus	DNA methylation[a]	Differential methylation of promoters at multiple genes in oropharyngeal squamous cell carcinoma, anal squamous cell carcinoma, and cervical cancer	OPSCC tumors, anal mucosa samples, and cervical	[31,32,34,35]

TABLE 1 Summary of Published Literature of Viral-, Bacterial-, and Parasite-Mediated Changes to the Host Epigenome—cont'd

Microorganism	Epigenetic mechanism	Consequence in host	Cell type	PubMed ID
	microRNA	Upregulation of miR-15a, -146a, -223, -15b, -16, -17, -20a, -20b, -93, -106a, -155,- 224, -21, -182, -183, -210, -124, -135b, -141, -301b, -449a, -449b, -517a, -517c, -545, -10a, -132, -148a, -196a, -302b, -9, -127, -199b, -199s, and -214; and downregulation of miR-23b, -34a, -101, -143, -145, -218,-42b, let-7b, -10b, -29a, -125b, -126, -375, -424, let-7a-c, -196b, -195, -368, -497, -433, -26a, -99a, -203, -513, -149 in HPV-positive cervical cancer cells	Cervical cancer cells including HPV-16 and HPV-18 positive cells	[101–106]
Human polyomavirus BKV	DNA methylation	Strongly activates *DNMT1* expression	Primary HPTE cells	[107]
Influenza A virus	microRNA	Increased expression of miR-7, miR-132, miR-146-a, miR-187, miR-200c, and miR-1275 which regulate genes of the immune system including *IRAK1* and *MAPK3*	Human lung cells	[108]
	DNA methylation	Increased methylation at promoters of *CXCL14, CCL25, CXCL6,* and *IL4R*; decreased methylation at promoters of *IL13* and *IL17C*	Human lung epithelial cells (A549)	[109]
Kaposi's sarcoma-associated herpesvirus	DNA methylation	Recruits DNMT3A and DNMT3B to hypermethylate the promoter of H-cadherin (*CDH13*)	TIME cells	[110]
Simian virus 40	DNA methylation	Induced DNMT3b that may contribute to oncogenic transformation. Increased methylation of *HPP1, RASSF1A,* cyclin P2, and *RRAD*	Normal human bronchial epithelial	[111]
	Histone modification	Increases histone lysine methyltransferase activity for methylating H3K9 and H3K27	Normal human bronchial epithelial	[112]
Varicella zoster virus	miRNA	Increased expression of miR-21	Human malignant melanoma, human embryonic lung fibroblast	[113]

Continued

TABLE 1 Summary of Published Literature of Viral-, Bacterial-, and Parasite-Mediated Changes to the Host Epigenome—cont'd

Microorganism	Epigenetic mechanism	Consequence in host	Cell type	PubMed ID
Bacteria				
Anaplasma phagocytophilum	Histone modification	Decreased acetylation of histone H3 at *CYBB, MPO, DEFA1, DEFA4, DEFA5, DEFA6, BPI, LYZ*, and *AZU1*; increased methylation of histone H3 at *CYBB, MPO, DEFA1, DEFA4, DEFA5, DEFA6, BPI, LYZ, AZU1, GNLY*, and *DCD*	Monocytes	[114]
Bacillus anthracis	Histone modification	Methylation of histone H1	Macrophages	[115]
Campylobacter rectus	DNA methylation	Increased methylation of *Igf2* promoter	Placenta	[39]
Chlamydia trachomatis	Histone modification	Methylation of histone H2B, H3, and H4	HeLa cells/3T3 cells	[116]
Clostridium perfringens	Histone modification	Dephosphorylation of H3Ser10	HeLa cells	[117]
Escherichia coli	DNA methylation	Increased methylation at *CDKN2A* exon 1	Uroepithelial	[118]
Helicobacter pylori	DNA methylation	Increased methylation of *LOX, HAND1, THBD, CDH1*, $p16^{INK4A}$ and *p41ARC*	Gastric mucosae	[119]
	Histone modification	Dephosphorylation of H3Ser10 and H3Thr3; deacetylation of H3K23	Gastric epithelial	[120,121]
	microRNA	Downregulation of let-7a, let-7b, let-7d, let-7e, let-7f, miR-1, miR-31, miR-32, miR-34b, miR-34c, miR-101, miR-103, miR-106b, miR-125a, miR-130a, miR-133, miR-141, miR-200a, miR-200b, miR-200c, miR-203, miR-204, miR-214, miR-218, miR-320, miT-372, miR-373, miR-375, miR-377, miR-379, miR-429, miR-449, miR-455, miR-491-5p, miR-500, miR-532, and miR-652. Upregulation of miR-17, miR-20a,, miR-146a, miR-155, miR-222, miR-223, miR-584, miR-1290	Gastric mucosa	[21–23, 122–124]
Legionella pneumophila	Histone modification	Acetylation of H3K14 and phosphorylation of H3Ser10 at *IL-8* gene promoter; trimethylation and deacetylation of H3K14 on another 4870 gene promotes	Lung epithelial; monocytes	[64]

TABLE 1 Summary of Published Literature of Viral-, Bacterial-, and Parasite-Mediated Changes to the Host Epigenome—cont'd

Microorganism	Epigenetic mechanism	Consequence in host	Cell type	PubMed ID
Listeria monocytogenes	Histone modification	Acetylation of H4K8 and phosphorylation and acetylation of H3Ser10 and H3K14	Umbilical vein endothelium	[125]
	microRNA	Upregulation of mmu-miR-146a, mmu-miR-146b, mmu-miR-16, mmu-let-7a1 and mmu-miR-155, mmu-miR-147, mmu-miR-191, mmu-miR-125a-5p, mmu-mir-132, mmu-miR-497, mmu-miR-125a-3p, mmu-miR-455, mmu-miR-149, and mmu-miR-29b; downregulation of mmu-miR-145	Epithelial cells; macrophages	[126,127]
Moraxella catarrhalis	Histone modification	Increased global acetylation at histone H3 and H4, and at the promoter of *IL-8*	Bronchial epithelial	[128]
Mycobacterium tuberculosis	Histone modification	Inhibits IFN-γ-induced acetylation of histone H3 and H4 at CIITA promoter IV	Macrophages	[129]
	microRNA	Differential expression of miR-30a, miR-30e, miR-155, miR-1275, miR-3665, miR-3178, miR-4484, miR-4668-5p, and miR-4497 when compared between cells infected with virulent and avirulent strains	Macrophages	[129]
Salmonella enterica	microRNA	Downregulation of mmu-miR-let-7a, mmu-miR-let-7c, mmu-miR-let-7d, mmu-miR-let-7f, mmu-miR-let-7g, mmu-miR-let-7i, and mmu-miR-98	Monocytes	[130]
Shigella flexneri	Histone modification	Inhibits MAPK-induced phosphorylation of H3Ser10	HeLa cells	[131]
Streptococcus pneumonia	Histone modification	Dephosphorylation of H3Ser10	HeLa cells	[117]
Streptococcus pyogenes	Histone modification	Dephosphorylation of histone H1	Pharyngeal cells	[132]
Parasites—Protozoa				
Leishmania donovani	DNA methylation	Differential methylation of 443 CpG sites upon infection affecting genes that play a role in host defense	Macrophages	[42]

Continued

TABLE 1 Summary of Published Literature of Viral-, Bacterial-, and Parasite-Mediated Changes to the Host Epigenome—cont'd

Microorganism	Epigenetic mechanism	Consequence in host	Cell type	PubMed ID
Leishmania major	microRNA	Consistent deregulation of 64 miRNA affecting genes belonging to pro- and antiapoptotic, proliferative, and immune pathways	Human PBMC	[133]
Plasmodium chabaudi	DNA methylation	Organ-specific DNA methylation of gene promoters. *Pigr*, *Ncf1*, *Klkb1*, *Emr1*, *Ndufb11*, and *Tlr6* in the liver and *Apol6* in the spleen were differentially methylated	Liver, spleen	[134]
Toxoplasma gondii	Histone modification	Inhibits lipopolysaccharide-induced phosphorylation of H3S10 and acetylation of H3K9 and H3K14 at the promoter of *IL-10*; renders host cell unresponsive to IFN-γ which can be rescued by histone deacetylases	Macrophages	[67,68]
Parasites—Metazoa				
Brugia malayi	microRNA	Differential expression of miR-125-5p, miR-146-5p, miR-199b-5p, and miR-378-3p	Macrophages	[135]
Clonorchis sinensis	Histone modification	Increased in acetylation at histone H3 and H4 at the promoter of *Mcm7*	Human cholangiocarcinoma cells	[72]
Opisthorchis viverrini	DNA methylation	*OPCML*, *SFRP1*, *HIC1*, *PTEN*, and *DcR* were highly methylated and *MINT25*, *p16*, *RASSF1A*, and *BLU* were moderately methylated in *O. viverrini*-induced cholangiocarcinoma	Human cholangiocarcinoma cells	[45]
Schistosoma japonicum	microRNA	Elevated levels of miR-223 and differential expression of more than 130 miRNAs during the development of schistosomal hepatopathy	Serum, liver	[27,28]

CMV, cytomegalovirus; HBV, hepatitis B virus; HCV, hepatitis C virus; HCC, hepatocellular carcinoma; HPV, human papillomavirus; LAT, latency-associated transcript; HSV, herpes simplex virus; E, early HSV-1 gene; IE, immediate-early HSV-1 gene; miRNA, microRNA.

[a] *Refer to Table 2 for sensitivity and specificity information.*

TABLE 2 Summary of Known Sensitivity and Specificity for Epigenetic Biomarker Panels

Pathogen	To diagnose	Specimen	Diagnostic panel	Sensitivity	Specificity	PubMed ID
microRNA						
Hepatitis B virus (HBV)	HBV infection	Serum	miR-375, miR-10a, miR-233, miR-423	99.3%	98.8%	[9]
	HBV-induced HCC against HBV infection only	Serum	miR-10a, miR-125b	98.5%	98.5%	[9]
	HBV-related HCC against no HBV infection	Serum	miR-23b, mir-423, miR-375, miR-23a, miR-342-3p	96.9%	99.4%	[9]
	HBV-specific HCC against non-HBV-related HCC	Serum	miR-375, miR-25, let-7f	97.9%	99.1%	[9]
	HBV-induced HCC against no HBV infection	Plasma	miR-122, miR-192, miR-21, miR-233, miR-269, miR-27a, miR-801	83.2%	93.9%	[10]
	HBV-induced HCC against chronic hepatitis B			79.1%	76.4%	[10]
	HBV-induced HCC against liver cirrhosis			75.0%	91.1%	[10]
Hepatitis C virus (HCV)	Differentiate between HBV and HCV infection	Serum	miR-92a, miR-423	97.9%	99.4%	[9]
DNA Methylation						
Epstein–Barr virus	Nasopharyngeal carcinoma	Nasopharyngeal brushings	*RASSF1A*, *p16*, *WIF1*, *CHFR* and *RIZ1*	98%	96%	[92]
Human papillomavirus	Cervical cancer	Cervical scrapings/cervical lesions	*CADM1*, *MAL*	70%	78%	[35]
		Cervical scrapings	*DLX-1*, *ITGA4*, *RXFP3*, *SOX17*, *ZNF671*	96.2% (in women above 30 years old)	76.6% (in women above 30 years old)	[34]

We will also highlight some pathogens of particular interest capable of infecting humans from each of the taxonomic groups (viruses, bacteria, parasites) and explore the possibility of using changes in host's miRNA profiles, DNA methylation patterns, and histone marks occurring during infection as potential epigenetic biomarkers. In addition, we have summarized any sensitivity and specificity information for these epigenetic biomarker panels when available (see Table 2).

2. miRNA PROFILES

miRNAs are short, noncoding RNA molecules used in the posttranslational regulation of gene expression, and aberrant expression of miRNA can lead to the progression of disease. Upon microbial infection, host miRNA profiles have been shown to be altered. Recently, the suggestion of using circulating miRNA in the plasma or serum as epigenetic biomarkers for solid cancers has emerged as miRNA profiles are proving to be consistent within species [6]. They are also very stable under harsh conditions, such as high or low temperature, and can even be harvested from formalin-fixed paraffin-embedded (FFPE) tissue samples [7]. We now describe some pathogens where host miRNA profiles have been shown to be altered upon infection.

2.1 Viruses

Hepatitis B Virus (HBV) is from the Hepadnaviridae family of viruses and is the cause of the inflammatory liver disease hepatitis B (OMIM #610424). The acute phase of the infection is characterized by the swelling of the liver and jaundice but rarely leads to death [8]. Chronic infection of HBV is associated with liver cirrhosis (OMIM #215600) and hepatocellular carcinoma (HCC) (OMIM #114550). Comparison of miRNA expression levels in the serum of patients infected with HBV against patients negative for HBV antibodies yielded a panel of 13 miRNAs (miR-375, miR-92a, miR-10a, miR-223, miR-423, miR-23b, miR-23a, miR-342-3p, miR-99a, miR-122a, miR-125b, miR-150, and let-7c) which were differentially expressed [9]. The origins of these serum-derived miRNAs are still unknown, but the authors speculate that they may have been shed by the tissues affected by the disease [9]. This miRNA panel is also successful in distinguishing patients infected with HBV from the patients infected with hepatitis C virus (HCV—See below). Using 2 of the 13 miRNAs (miR-375 and miR-92a), the authors determined that it is possible to identify HBV-induced HCC from HCC caused by other etiologies. An independent group screened three cohorts consisting of patients with chronic hepatitis B, liver cirrhosis, and HCC [10]. This resulted in a seven plasma-derived miRNA panel (miR-122, miR-192, miR-21, miR-233, miR-26a, miR-27a, and miR-801) which not only provides an accurate diagnosis of HCC but also is able to differentiate it from chronic hepatitis and liver cirrhosis [10]. HBV is also capable of directly interfering with its host endogenous miRNA expression. It is able to directly bind and inhibit its host miR-15a/16 complex with its own miR-15a/16 complementary site leading to the downregulation of the miRNA complex. This leads to the upregulation of antiapoptotic protein B-cell lymphoma 2 (BCL-2) [11], proposing a mechanism for inhibiting apoptosis in HBV-infected hepatocellular cells. These studies all suggest that serum-derived miRNAs represent potential noninvasive epigenetic biomarkers that could be used in the identification of HBV-induced HCC and its intermediate phenotypes.

HCV is an RNA virus from the family Flaviviridae and it is the cause of hepatitis C (OMIM #609532) in humans [12]. It is the leading cause of liver transplants in developing countries and may also lead to HCC [13]. Unlike HBV, there are currently no vaccines available for HCV, though progress is being made on this front [14]. Profiling of miRNA expression of HCV-associated HCC showed activation of 10 miRNAs and repression of 19 miRNAs when comparing HCV-positive HCC FFPE liver tissue to normal liver parenchyma across the different grades of HCC [15]. Out of the 29 differentially expressed miRNAs, five miRNAs (miR-122, miR-100, miR-10a, miR-198, and miR-145) showed consistent deregulated expression when further examined in 52 liver nodules arising from HCV infection [15]. miR-122 was strongly induced in the malignant liver nodules and had been previously shown to be imperative for HCV replication

[16]. It is of note that HBV also causes miR-122 to be dysregulated during HCC which could suggest that miR-122 may not be a virus-specific marker but possibly a biomarker specific to HCC or liver damage. Of interest, expression levels of miR-198 and miR-145 were not only downregulated in these HCV-induced liver nodules but also correlated with the severity of the disease, ranging from liver cirrhosis to HCC [15]. This suggests that these miRNAs are not only potential epigenetic biomarkers in predicting HCV-related HCC but could also be used in distinguishing the different stages leading to HCC. A recent study also identified increased expression of miR-141, miR-200a, miR-200b, and miR-200c upon HCV infection of primary human hepatocytes [17]. The expression of miR-141 correlated inversely with the expression of tumor suppressor gene, deleted in liver cancer-1 (*DLC-1*), suggesting that miR-141 is responsible for the downregulation of *DLC-1*. Genetic polymorphisms and epigenetic modifications leading to reduced expression of *DLC-1* have previously been identified in HCC [18]. HCV-infected hepatocytes had increased cell proliferation, which was countered by overexpressing *DLC-1* [17]. These findings are consistent with HCV-induced HCC. Together, these miRNAs could be used as epigenetic biomarkers for the propensity of carcinogenesis following HCV infection.

2.2 Bacteria

Bacteria are also capable of inducing changes in the host miRNA expression. *Helicobacter pylori* is a gram-negative bacterium that is able to survive the low pH of gastric acid and colonize the stomach mucosa lining [19]. Chronic infection of *H. pylori* leads to gastritis (OMIM #219721) and is one of the strongest risk factors for stomach cancer (OMIM #613659) [20]. Figure 1 demonstrated some of the different mechanisms *H. pylori* use to modify the host's epigenome promoting gastric cancer. We now discuss in more detail the change in host miRNA profiles upon infection with *H. pylori* in relation to their possibilities as epigenetic biomarkers.

Comparison of the miRNA profiles of gastric endoscopy specimens between *H. pylori* positive and *H. pylori* negative patients resulted in 55 differentially expressed miRNA [21]. Reverse transcription polymerase chain reaction (RT-PCR) of miRNA expression in an independent set of *H. pylori* positive versus *H. pylori* negative patients confirmed that 30 miRNAs were downregulated while only one miRNA was upregulated [21]. Following successful elimination of *H. pylori* infection, 14 (let-7a, let-7b, let-7d, let-7e, miR-106b, miR-130a, miR-141, miR-200a, miR-200b, miR-200c, miR-31, miR-500, miR-532, and miR-652) of the 31 differentially expressed miRNAs had significant changes [21]. These 14 miRNAs may be harnessed as a potential epigenetic biomarker panel to allow doctors to monitor the success of *H. pylori* eradication following chemotherapeutic interventions but not as epigenetic biomarkers of gastric carcinogenesis, with the authors unable to find any significant correlation with intestinal metaplasia [21].

However, other studies have identified several different miRNAs, which could be used for the prediction of carcinogenesis following *H. pylori* infection. miR-222 was found to be over expressed in *H. pylori* positive gastric cancers. It exerts its oncogenic properties by targeting and inhibiting the gene reversion-inducing-cysteine-rich protein with kazal motifs (*RECK*) leading to increased cell proliferation [22]. Increased expressions of miR-584 and miR-1290, induced by the *H. pylori*-secreted protein CagA, were also associated with the risks of *H. pylori*-related carcinoma [23]. Upregulation of these two miRNAs in a knockin mouse model led to metastasis of gastric epithelia cells [23]. These three miRNAs may thus be used in the diagnosing of *H. pylori* specific gastric cancer. A full list of differentially expressed miRNAs during an *H. pylori* infection and their potential in gastric carcinogenesis has recently been reviewed by Noto and Peek (2012).

2.3 Parasites

In addition to viral- and bacterial-induced modification of human miRNA expression profiles, parasites are also able to modulate the miRNA profiles of its host. *Schistosoma japonicum* is a metazoan blood liver fluke that is endemic in China and parts of Southeast Asia [24,25]. It is spread by direct contact of freshwater contaminated with *S. japonicum* released by infected freshwater snails [26]. The parasites are then able to penetrate the skin and enter the bloodstream of its mammalian host [26]. Infection of *S. japonicum* leads to the chronic disease schistosomiasis (OMIM #181460), which may progress to liver fibrosis. During *S. japonicum* infection, the levels of serum miR-223 were found to be elevated in mice, humans, buffalos, and rabbits [27]. These levels correlated with the number of *S. japonicum* eggs in the liver and the extent of liver damage following infection. miR-223 levels were also elevated in Kupffer, hepatocytes, and hepatic stellate cells, suggesting these elevated miR-223 levels measured in the serum were derived from the liver [27]. After administering the antischistosome drug praziquantel, the levels of serum miR-223 decreased and returned to baseline [27]. Thus, miR-223 can both serve as an epigenetic biomarker for disease diagnosis and track patient response to chemotherapy.

This observation of induced miR-223 expression was further supported by another study profiling miRNA expression in the mouse following *S. japonicum* infection [28]. The authors reported that miR-223 had the greatest increase in expression at 45 days postinfection and suggested that this upregulation prevents the activation of granulocytes and diminishes the immune response in the liver [28]. They also identified more than 130 additional miRNAs differentially expressed during the development of schistosomal hepatopathy [28] that can be further studied for a prospective miRNA biomarker panel relevant for schistosomiasis.

The results of all of these studies suggest that the profiling of the host-miRNA expression levels following infection could assist with diagnosis, identification of the pathogen, as well as contributing to prediction of disease prognosis. Moreover, miRNAs could be used to allow the levels of infection to be monitored following treatment. As such, profiling of host miRNAs during infection could lead to the elucidation of diagnostically useful biomarker panels. There is another possible epigenetic biomarker not discussed in this chapter, namely the use of microbial miRNAs found in the bloodstream of the patients, which could also be used in the diagnosis of microbe-specific infections. However, these may not be as useful in the monitoring of disease progression. Although miRNAs make ideal epigenetic biomarkers due to their robust nature, their relationship to their target genes and their downstream effects must first be well characterized. Furthermore, each miRNA can target multiple genes and these associations can be cell specific. Therefore, care must be taken in interpreting miRNA dysregulation during infection before determining its specificity as an infection-related epigenetic biomarker.

3. DNA METHYLATION PATTERNS

DNA methylation is, generally, the covalent addition of a methyl (CH_3) group to a cytosine–phosphate–guanine (CpG), defined as a cytosine residue that occurs directly before a guanine residue bonded by phosphate [29]. This process is mediated by DNA methyltransferases (DNMT). Methylation occurring at the CpGs in a promoter will prevent transcription factors from directly binding to the promoter, leading to repression of gene expression [3]. Pathogenic infections have been shown to lead to aberrant host DNA methylation patterns and we now discuss some of these methylome changes in the host and their propensity to be epigenetic biomarkers.

3.1 Viruses

The human papillomavirus (HPV) is a DNA virus of the Papillomavirus family capable of infecting humans. It is a sexually transmitted virus and has been associated with several cancers including cervical, anogenital, skin, and head and neck carcinomas [30]. A five-gene methylation signature has been identified in HPV-driven tumors (oropharyngeal squamous cell carcinomas (OPSCC) (OMIM #275355)), which predict its clinical outcome. These five genes (aldehyde dehydrogenase 1 family member A2 (*ALDH1A2*); odd-skip related transcription factor 2 (*OSR2*); GATA binding protein 4 (*GATA4*); glutamate receptor, ionotropic, AMPA3 (*GRIA3*); and iroquois homeobox 4 (*IRX4*)) were found to be differentially methylated at their promoters between HPV-driven tumors and non-HPV-driven tumors and their transcript levels correlated inversely with their methylation status [31]. This methylation signature, consisting of the hypomethylation of *ALDH1A2* and *OSR2* and the hypermethylation of *GATA4, GRIA3,* and *IRX4* promoters, was indicative of a positive outcome and increased survival in HPV-induced OPSCC [31].

In HPV-associated anal squamous cell carcinoma (SCC) (OMIM #105580), 20 differentially methylated genes were identified in FFPE anal mucosa samples from patients with HPV-induced invasive SCC when compared to patients with HPV-induced preinvasive SCC [32]. These differentially methylated genes fell into five main categories: genes associated with growth regulation and cell-cycle control (transforming growth factor beta 3 (*TGFβ3*), fyn-related kinase (*FRK*), peptidyl arginine deaminase type IV (*PADI4*), and inhibitor of DNA binding 1, dominant negative helix-loop-helix protein (*ID1*)), apoptosis (tumor necrosis factor receptor superfamily member 10b (*TNFRSF10B*), death-associated protein kinase 1 (*DAPK1*), homeobox A5 (*HOXA5*), BCL2-related protein A1 (*BCL2A1*), and Sema domain, immunoglobulin domain (Ig), short basic domain, secreted, (semaphoring) 3B (*SEMA3B*)), differentiation (keratin 1 (*KRT1*), keratin 5 (*KRT5*), and protease serine 8 (*PRSS8*)), angiogenesis (FMS-related tyrosine kinase 1 (*FLT1*) and kinase domain receptor (a type III receptor kinase domain receptor) (*KDR*)), and others (chemokine (C–C motif) ligand 3 (*CCL3*), purinergic receptor, P2X, ligand-gated ion channel 7 (*P2RX7*), CD9 molecule (*CD9*), deiodinase iodothyronine type III *(DIO3)*, and gamma-aminobutyric acid (GABA) A Receptor, alpha 5(*GABRA5*)) [32]. In particular, increasing levels of methylation at *DAPK1* was associated with the progression of anal carcinogenesis [32]. Hence, the methylation status of *DAPK1* could be used to track the stages of HPV-associated anal SCC.

Cervical cancers (OMIM #603596) are nearly synonymous with HPV infections, as HPV have been found in virtually all cervical cancer cases [33]. Two groups have tested the efficacy of using the methylation signatures of different genes to detect malignant cervical lesions in HPV-infected women [34,35]. Quantitative methylation-specific PCR (q-MSP) of the promoters of two genes cell adhesion molecule 1 (*CADM1*) and Mal, T cell differentiation protein *(MAL)* was carried out in cervical tissue specimens ranging from cervical intraepithelial neoplasia (CIN) grade 1 (mild abnormal cell growth) to CIN3+ (cervical cancer). A correlation was found between increasing levels of methylation and increasing severity of the cervical lesions where almost all CIN3 lesions and cervical carcinomas had methylated *CADM1* and *MAL* promoter regions [35]. The same methylation pattern was also found in cervical scrapings obtained via pap smears in a cross-sectional study done on 79 women to evaluate the efficacy of the *CADM1/MAL* methylation biomarker panel [35]. This resulted in a specificity of 78% and sensitivity of 70%. Another study identified a five-gene methylation panel from cervical scrapes of HPV-positive cases to be specific for CIN3 and cervical cancer [34]. This methylation signature consists of distal-less homeobox 1 (*DLX-1*), integrin alpha 4 (*ITGA4*), relaxin/insulin-like family peptide receptor 3 (*RXFP3*), SRY (sex determining region Y)-box17 (*SOX17*), and zinc finger protein 671 (*ZNF671*). The sensitivity and specificity were extremely high (96.2% and 76.6%, respectively) in detecting CIN3+ lesions in women over the age of 30. Though these biomarker panels have already been proven to have high sensitivity and specificity, they have yet to be translated into commercial assays for diagnostics.

In addition to the host DNA methylome, the methylation patterns of the viral genome integrated into the host DNA may also be harnessed as a marker to detect progression of the disease. Epstein–Barr virus (EBV), also known as glandular fever [36], is one of the most common infections in man and the cause of infectious mononucleosis (OMIM #308240). It is also associated with several cancers such as Burkitt's lymphoma (OMIM #113970) [36] and Hodgkin's lymphoma (OMIM #236000) [36]. Upon infection, EBV may enter its lytic cycle and replicate. This process is quickly halted in immunocompetent patients, resulting in a latent infection. Two EBV proteins, BamHI Z fragment leftward open reading frame 1 (BZLF1) and BamHI R fragment leftward open reading frame 1 (BRLF1), are responsible for the switch from the latent to lytic cycle [37]. Methylation profiling of the viral promoters regulating expression of these genes in EBV positive tumors revealed that the promoters were hypermethylated during the latent phase [37]. This is surprising because the EBV viral DNA lacks any DNA methylation when it first enters its host cell, as it does not encode a viral cytosine methyltransferase [38]. When the promoters of *BZLF1* and *BRLF1* were demethylated using the drug azacitidine, EBV was induced to undergo the lytic cascade in these cells [37]. Therefore, the methylation patterns of these viral promoters can be used as a biomarker to detect the current status of infection in EBV-infected patients.

3.2 Bacteria

Infections acquired during pregnancy can also lead to changes in the host's DNA methylation profiles, potentially impacting on the unborn fetus. In the murine model, maternal oral infection with the bacterial pathogen *Campylobacter rectus* has been shown to induce hypermethylation at the promoter region of the imprinted insulin-like growth factor 2 (*Igf2*) gene in the placenta [39]. This is accompanied by the downregulation of placental *Igf2* expression following infection with *C. rectus* [39]. The paternally expressed *Igf2* gene plays an essential role during gestation, directly controlling the supply of maternal nutrients to the fetus [40]. Knock out of this gene in a mouse model resulted in reduced placental and fetal growth [40], suggesting that the hypermethylation of *Igf2* gene induced by *C. rectus* could mediate restriction of fetal growth in utero [41]. Thus, levels of DNA methylation at the promoter of *Igf2* could potentially be indicative of fetal growth restriction and future health outcomes.

3.3 Parasites

Changes in host DNA methylation patterns may also arise from parasitic infections. Though modifications to the host DNA methylome has been observed during protozoan parasite infections such as *Leishmania donovani* [42], we focus here on changes in host DNA methylation caused by metazoan parasites where there are more immediate opportunities for translation. *Opisthorchis viverrini* is a metazoan liver fluke endemic to Thailand and Laos and humans may get infected by eating raw or undercooked freshwater fish contaminated with the cysts of *O. viverrini* [43]. It resides in the biliary ducts of its mammalian host where it attaches to the mucosa. Prolonged infection of *O. viverrini* may lead to the metastasis of the epithelial cells in the bile duct, also known as cholangiocarcinoma (CCA) (OMIM 615619) [43]. The prognosis of CCA is poor with less than 5% of patients surviving more than 5 years after initial diagnosis of the disease [44]. This is attributed to the difficulty in diagnosing CCA before it presents at a later clinical stage [45]. When comparing the methylation profiles of *O. viverrini*-related CCA tumor samples to normal tissue samples, nine genes were differentially methylated [45]. Five genes were highly methylated: opioid-binding protein/cell adhesion molecule-like (*OPCML*), secreted frizzled-related protein 1 (*SFRP1*), hypermethylated in cancer 1 (*HIC1*), phosphatase and tensin homolog (*PTEN*), and Dicer 1 ribonuclease III (*DCR1*), while four genes were moderately methylated: Msx2-interacting nuclear target (*MINT25*), cyclin-dependent kinase inhibitor 2A (*p16*), Ras association (RalGDS/AF-6) domain family member 1 (*RASSF1A*), and zinc finger, MYND-type containing 10 (*ZMYND10*, previously known as *BLU*) [45]. Of interest, patients with methylated *DCR1* showed a longer overall survival from CCA than those without *DCR1* methylation [45]. Also, methylation of the tumor suppressor gene *OPCML* was found more frequently in less differentiated subtypes of CCA compared to those which were well differentiated [45]. These two hypermethylated genes in CCA tumor samples may thus be used as methylation biomarkers of CCA to predict disease outcomes.

As DNA methylation marks are more robust than an RNA profile, it makes them very appealing as an epigenetic biomarker in a clinical setting, since the samples do not require special storage conditions such as cryopreservation. It can also be readily detected in a great variety of samples collected noninvasively such as blood, saliva, stool, urine, and plasma [46]. DNA methylation is also easy to amplify and identify using PCR-based assays even if the sample consists of only a few cells [47]. However, there are several issues to consider when using DNA methylation profiles to detect for the presence of infection and its downstream effects. Pathogens often have a preferred area of colonization within the human body as such DNA methylation changes may be tissue specific. However, measuring DNA methylation profiles of infected tissues may prove difficult as there are some tissues, such as the brain, eye, and heart, only obtainable surgically (i.e., during biopsy) or at postmortem. In addition, if there is heterogeneity in DNA methylation profiles at the target gene promoter in a healthy control population, it may reduce the sensitivity and specificity of the assay. Currently, there are no assays used in the clinical settings measuring DNA methylation profiles as a biomarker for infection. Nonetheless, this is a promising field as DNA methylation is already being used as an epigenetic biomarker in other clinical applications such as the diagnosis of cancer [46,48].

4. HISTONE MARKS

Histones are highly conserved, positive-charged proteins found in the nucleus of the cell [49]. They are responsible for packaging DNA into nucleosomes. A nucleosome is made up of an octamer of core histones (two each of histone H2A, H2B, H3, and H4) and the DNA strand is wrapped around these histones, enabling DNA to be highly compact. The affinity of DNA binding to histone proteins can be manipulated to allow for transcriptional activation or repression by posttranslational modifications (PTMs) on the tails found on histone H3 and H4. This includes acetylation of lysine residues; methylation of either arginine or lysine residues; phosphorylation of serine, threonine, and tyrosine residues; and ubiquitination of lysine residues [49]. Histone modification is a highly dynamic process that reflects the condition of the cell at a particular point in time. We will now highlight several pathogenic agents capable of modifying the PTMs of host intracellular histones and evaluate their potential as biomarkers for infectious disease.

4.1 Viruses

In viruses, the PTMs of host intracellular histones can be used to discriminate between a latent infection and a lytic infection. The herpes simplex virus type 1 (HSV-1) does not seem to contain any histones within its capsid [50], yet it is able to hijack the host histones and form nucleosomes within the host cell [51]. HSV-1, a double-stranded DNA virus of the herpes virus family, is able to infect humans causing cold sores at the orofacial and genital areas [52]. It is able to escape the immune system by infecting the peripheral nervous system and establishing a latent infection for life [53]. Although the primary and recurrent infection of HSV is usually self-limiting, it can cause debilitating diseases such as herpes simplex encephalitis (OMIM #610551) [54] and neonatal herpes simplex infection [55]. When the HSV-1-infected host is under stress, the virus can be reactivated from its latent phase causing sores near or at the initial site of infection [56]. This switch between the latent and lytic cycle of HSV-1 infection has been associated with the PTMs of host histones on which the viral DNA is wrapped around [57]. During latency, HSV latency-associated transcript (LAT) is the only viral gene product generated by HSV-1 in the sensory neurons where it stays quiescent [58]. Chromatin immunoprecipitation (ChIP) analysis of latent HSV-1 DNA showed that a portion of LAT is associated with the increased acetylation of lysine residue 9 and 14 in histone H3, which is consistent with an open chromatin state [59]. ChIP analysis of early (E) and immediate-early (IE) HSV-1 genes in murine dorsal root ganglions during the latent phase shows that these genes were not acetylated at the same lysine residues, indicating they were transcriptionally repressed [59]. When E and IE HSV-1 genes are expressed, the virus undergoes replication and the infected cell is destined for lysis [60]. During the lytic phase, the opposite was found to occur where there was increased acetylation of histone H3 lysine 9 and 14 at the E and IE genes [61] in an in vitro study of neuronal cells. In addition to acetylation, histone H3 lysine 4 was found to have increased methylation at the E and IE genes during the lytic cycle [62], which is a mark of active transcription. Therefore using a combination of both acetylation and methylation statuses on the lysine residues of histone H3 at the LAT, E and IE genes would be useful in determining the status (latent or lytic) of an HSV-1 infection at any given time. However, these HSV-1-mediated histone modifications have only been observed in the nervous system to date and it may prove difficult to obtain these tissues for sampling.

4.2 Bacteria

Bacteria too do not contain histones within their genome but have developed the capabilities of modifying their host's histone marks. *Legionella pneumophila* has been shown to induce changes in global histone modifications upon infection of its host cell [63,64]. *L. pneumophila* is a pathogenic gram-negative bacillus capable of causing a type of pneumonia called Legionnaire's disease (OMIM #608556) in humans. The infection is spread by inhalation of aerosols or water droplets contaminated with *L. pneumophila* species, but not everyone infected with *L. pneumophila* will develop Legionnaire's disease [65]. Diagnosis of Legionnaire's disease involves a range of assays such as culturing and staining of the bacteria and testing urine, serum, sputum, and throat swabs for *L. pneumophila* DNA [65]. However there are some concerns about the reliability of these current assays [65], signifying the need for more efficient ways of diagnosis.

Upon *L. pneumophila* infection in alveolar epithelial cells (A549), genome-wide histone modifications were induced including acetylation of lysine residue 14 and phosphorylation of serine residue 10 at histone H3 at the interleukin-8 (*IL-8*) gene promoter [63]. Moreover, an *L. pneumophila*-excreted protein RomA (regulator of methylation A) has been found to directly target specific PTMs of the host histones [64]. It causes trimethylation and deacetylation of lysine residue 14 on histone H3 on 4870 gene promoter regions, including those of the innate immune system [64]. Methylation and deacetylation of lysine residue 14 on histone H3 were found to be mutually exclusive [64]. This unique pattern of host histone PTMs combined with the fact that it was induced solely by excreted *L. pneumophila* protein could act as epigenetic biomarkers aiding in a more accurate diagnosis of Legionnaire's disease.

4.3 Parasites

As eukaryotic organisms, parasite genomes contain histones and various histone-modifying enzymes [66]. Therefore, it is not surprising that they can alter the histone marks of their host. Protozoan parasites such as *Toxoplasma gondii* have been shown to change the PTMs of their host histones [67,68]. Similarly, metazoan parasites have also been shown to induce histone modifications in their hosts. We now focus on one such metazoan parasite *Clonorchis sinensis* whose induced histone modifications have opportunities for translation. *Clonorchis sinensis*, also known as the Chinese liver fluke, is able to infect humans and is the most prevalent liver fluke infection in Asia [69,70]. It resides in the liver feeding on the host's bile and has been proven to be one of the causative agents of CCA [71]. Humans are the definitive host for *C. sinensis* where they undergo sexual reproduction producing eggs. The infection is acquired when raw or undercooked infected fish is consumed. Exposure to the excretory–secretory products of *C. sinensis* in vitro has revealed a global increase in acetylation at histone H3 [72]. As the gene minichromosome maintenance complex component 7 (*MCM7*) was previously shown to be upregulated in a *C. sinensis* infection [73], ChIP analysis was performed to determine the status of histone acetylation at its promoter in human CCA cells. A significant 21-fold increase in acetylation was seen in histone H3 and a threefold increase was seen in histone H4 at the promoter of *MCM7* [72], correlating with the transactivation of *MCM7* previously documented [73]. *MCM7* is crucial for DNA replication and cell cycle control and has been associated with several cancers due to its oncogenic properties [74]. Therefore, *C. sinensis*-induced hyperacetylation of histone H3 and H4 at the promoter of *MCM7* in CCA tissues could help physicians detect the progression or presence of *C. sinensis*-related CCAs. This allows for more accurate treatment plans leading to a better prognosis of the disease.

The potential of using histone PTMs as a biomarker of disease has already been explored extensively in the field of cancer research [75]. Relative to DNA methylation and miRNA, the methodologies used to study histone PTMs such as ChIP have a longer turnaround time and can be expensive when interrogating more than one histone PTM target. Furthermore, storage and processing of patient samples may not be as straightforward due to the dynamic nature of histone modification. Nonetheless, histone PTMs still remains an untapped resource of biomarkers for infectious disease. In addition to the effects of PTMs found on host intracellular histones, there is increasing evidence that host extracellular histones may too have an important role in the human disease [76]. These extracellular histones can function as damage-associated molecular pattern (DAMP) molecules [77], which are signals of tissue damage leading to an inflammatory response [78]. These histones are released by neutrophils during an innate immune response to an infection [79]. Infection of gram-negative bacillus *Escherichia coli* in the bloodstream prompts circulating neutrophils to release histones leading to sepsis [80]. Though one group has shown that PTMs on extracellular histones have a role to play in inducing the autoimmune disease systemic lupus erythematosus (OMIM #152700) [81], there are no studies done to date on the PTMs of these extracellular histones released during an infection. This is an underresearched area and has great potential to be an accessible source of histone biomarkers for infection.

5. CONCLUDING REMARKS

As researchers continue to understand the interactions between infectious agents and their hosts, there is an increasing interest in unraveling modifications occurring in the host's epigenome following infection. The epigenetic signatures of the infected host cell may be harnessed as an epigenetic biomarker for more accurate diagnoses of the infection or predictor of disease outcomes. It is also useful to determine if the infection is in the active or quiescent phase and can be an indicator for the host's susceptibility to pathogen-induced carcinogenesis. Overall, we have listed (Table 1) all pathogens whose infections lead to modifications of their host epigenomes. We have also discussed the alterations in the host's miRNA profiles, DNA methylation patterns, and histone marks with regard to their propensity to be translated into biomarkers. The pathogens we have reviewed in detail were specifically chosen for their potential contributions to existing diagnostic and prognostic tests. To our knowledge, none of the epigenetic biomarkers discussed are currently used in practice, though the sensitivity and specificity of some of these biomarker panels have been measured (see Table 2). The validated epigenetic biomarkers seen in HBV, HCV, EBV, and HPV infections show promising diagnostic efficacy and are candidates to be adopted promptly into the clinical setting to aid in the early detection of pathogen-induced oncogenesis.

Although the development of epigenetic biomarkers has been focused mainly on DNA methylation [82], histone modifications and miRNA may complement DNA methylation biomarkers in the future. To create the most accurate biomarker panels for infection, we believe it is important to determine whether the change in the host epigenome is host specific (i.e., epigenetic modifications induced by the host as a general response to an infection) or pathogen specific (i.e., epigenetic modifications induced by the pathogen to create a replicative niche for itself). The latter could be optimal and more accurate as an epigenetic biomarker as the modification would be unique to the pathogen. One such method of investigating if the epigenetic modifications are pathogen specific would be to determine if the pathogens are secreting any proteins into the host cells capable of directly modifying the host's epigenome. This has already been shown in certain microorganisms such as *H. pylori*-secreted protein CagA inducing a change in the host's miRNA [23] and *L. pneumophila*-secreted protein RomA causing trimethylation and deacetylation of lysine residues on host's histone tails [64]. Thus, using epigenetic biomarkers would definitely be beneficial in the area of infectious disease, as it would help with early detection of pathogen-induced diseases and cancers leading to prompt treatment and better prognosis of the patients.

Glossary

Biomarkers A biological marker that can be measured and evaluated as an indicator of normal biological processes, pathogenic processes, or a response to a therapeutic intervention.

Chromatin A complex made up of DNA, RNA, and proteins (mainly histones) which functions to compact and reinforce DNA for mitosis and control gene expression.

Chromatin immunoprecipitation (ChIP) An experimental technique used to investigate the interaction between specific proteins and specific genomic sites; for example, to determine specific locations in the genome that histone modifications are associated with.

Differentially expressed genes Genes with increased or decreased expression when compared between the control and treated samples.

Differentially methylated genes Genes with different percentages of methylated cytosine nucleotides when compared between the control and treated samples.

DNA methylation The addition of a methyl group (CH_3) to a cytosine in a CpG dinucleotide context. Non-CpG DNA methylation has also been shown to occur in specific cell types, such as embryonic stem cells.

Epigenetics The study of heritable changes in gene expression that occur without change in the DNA sequence.

Histone posttranslational modifications Covalent modifications of histones including acetylation, methylation, ubiquitination, sumoylation, and phosphorylation.

Metazoa Multicellular eukaryotic organisms.

MicroRNA A class of naturally occurring, small noncoding RNA about 22 nucleotides long found in plants, animals, and viruses which function to downregulate gene expression by translational repression, messenger RNA cleavage, and deadenylation.

Nucleosome The basic structural and functional repeating unit of chromatin made up of a segment of DNA wound around eight histone proteins.

Protozoa Single-cell eukaryotic organisms

Imprinted genes Genes whose expression is determined by the parent who contributed it, where one of the alleles contributed is silenced by epigenetic mechanisms.

LIST OF ACRONYMS AND ABBREVIATIONS

BRLF1 BamHI R fragment leftward open reading frame 1
BZLF1 BamHI Z fragment leftward open reading frame 1
CCA Cholangiocarcinoma
ChIP Chromatin immunoprecipitation
CIN Cervical intraepithelial neoplasia
DAMP Damage-associated molecular pattern
DNA Deoxyribonucleic acid
DNMT DNA methyltransferase
E Early
EBV Epstein–Barr virus
ESP Excretory–secretory products
FFPE Formalin-fixed paraffin-embedded
HBV Hepatitis B virus
HCC Hepatocellular carcinoma
HCV Hepatitis C virus
HPV Human papillomavirus
HSV-1 Herpes simplex virus type 1
IE Immediate-early
LAT Latent-associated transcript
miRNA micro ribonucleic acid
OPSCC Oropharyngeal squamous cell carcinomas
PCR Polymerase chain reaction
PTMs Posttranslational modifications
q-MSP Quantitative methylation-specific polymerase chain reaction
RNA Ribonucleic acid
SCC Squamous cell carcinomas
SLE Systemic lupus erythematosus

References

[1] Gómez-Díaz E, Jordà M, Peinado MA, Rivero A. Epigenetics of host–pathogen interactions: the road ahead and the road behind. PLoS Pathog 2012;8(11):e1003007.

[2] Bird A. Perceptions of epigenetics. Nature 2007; 447(7143):396–8.

[3] Phillips T. The role of methylation in gene expression. Nat Educ 2008;1(1):116.

[4] Lennartsson A, Ekwall K. Histone modification patterns and epigenetic codes. Biochimica Biophysica Acta (BBA) 2009;1790(9):863–8.

[5] Chuang JC, Jones PA. Epigenetics and microRNAs. Pediatr Res 2007;61(5 Part 2):24R–9R.

[6] Redova M, Sana J, Slaby O. Circulating miRNAs as new blood-based biomarkers for solid cancers. Future Oncol 2013;9(3):387–402.

[7] Liu A, Xu X. MicroRNA isolation from formalin-fixed, paraffin-embedded tissues. Methods Mol Biol 2011;724:259–67.

[8] Aspinall EJ, Hawkins G, Fraser A, Hutchinson SJ, Goldberg D. Hepatitis B prevention, diagnosis, treatment and care: a review. Occup Med 2011;61(8): 531–40.

[9] Li L-M, Hu Z-B, Zhou Z-X, et al. Serum microRNA profiles serve as novel biomarkers for HBV infection and diagnosis of HBV-positive hepatocarcinoma. Cancer Res 2010;70(23):9798–807.

[10] Zhou J, Yu L, Gao X, et al. Plasma microRNA panel to diagnose hepatitis B virus-related hepatocellular carcinoma. J Clin Oncol 2011;29(36):4781–8.

[11] Liu N, Zhang J, Jiao T, et al. Hepatitis B virus inhibits apoptosis of hepatoma cells by sponging the microRNA 15a/16 cluster. J Virol 2013;87(24):13370–8.

[12] Choo QL, Kuo G, Weiner AJ, Overby LR, Bradley DW, Houghton M. Isolation of a cDNA clone derived from a blood-borne non-A, non-B viral hepatitis genome. Science 1989;244(4902):359–62 (New York, NY).

[13] Shepard CW, Finelli L, Alter MJ. Global epidemiology of hepatitis C virus infection. Lancet Infect Dis 2005;5(9):558–67.

[14] Man John Law L, Landi A, Magee WC, Lorne Tyrrell D, Houghton M. Progress towards a hepatitis C virus vaccine. Emerg Microbes Infect 2013;2:e79.

[15] Varnholt H, Drebber U, Schulze F, et al. MicroRNA gene expression profile of hepatitis C virus–associated hepatocellular carcinoma. Hepatology 2008;47(4): 1223–32.

[16] Jopling CL, Yi M, Lancaster AM, Lemon SM, Sarnow P. Modulation of hepatitis C virus RNA abundance by a liver-specific microRNA. Science 2005;309(5740): 1577–81 (New York, NY).

[17] Banaudha K, Kaliszewski M, Korolnek T, et al. MicroRNA silencing of tumor suppressor DLC-1 promotes efficient hepatitis C virus replication in primary human hepatocytes. Hepatology 2011;53(1):53–61.

[18] Wong CM, Lee JM, Ching YP, Jin DY, Ng IO. Genetic and epigenetic alterations of DLC-1 gene in hepatocellular carcinoma. Cancer Res 2003;63(22):7646–51.

[19] Robin Warren J, Marshall B. Unidentified curved bacilli on gastric epithelium in active chronic gastritis. Lancet 1983;321(8336):1273–5.

[20] Uemura N, Okamoto S, Yamamoto S, et al. *Helicobacter pylori* infection and the development of gastric cancer. N Engl J Med 2001;345(11):784–9.

[21] Matsushima K, Isomoto H, Inoue N, et al. MicroRNA signatures in *Helicobacter pylori*-infected gastric mucosa. Int J Cancer 2011;128(2):361–70.

[22] Li N, Tang B, Zhu E-D, et al. Increased miR-222 in *H. pylori*-associated gastric cancer correlated with tumor progression by promoting cancer cell proliferation and targeting RECK. FEBS Lett 2012;586(6):722–8.

[23] Zhu Y, Jiang Q, Lou X, et al. MicroRNAs up-regulated by CagA of *Helicobacter pylori* induce intestinal metaplasia of gastric epithelial cells. PLoS One 2012;7(4):e35147.

[24] Li YS, Sleigh AC, Ross AG, Williams GM, Tanner M, McManus DP. Epidemiology of *Schistosoma japonicum* in China: morbidity and strategies for control in the Dongting Lake region. Int J Parasitol 2000;30(3):273–81.

[25] Leonardo LR, Acosta LP, Olveda RM, Aligui GDL. Difficulties and strategies in the control of schistosomiasis in the Philippines. Acta Trop 2002;82(2):295–9.

[26] Mitreva M. The genome of a blood fluke associated with human cancer. Nat Genet 2012;44(2):116–8.

[27] He X, Sai X, Chen C, et al. Host serum miR-223 is a potential new biomarker for *Schistosoma japonicum* infection and the response to chemotherapy. Parasit Vectors 2013;6(1):272.

[28] Cai P, Piao X, Liu S, Hou N, Wang H, Chen Q. MicroRNA-gene expression network in murine liver during *Schistosoma japonicum* infection. PLoS One 2013;8(6):e67037.

[29] Lim DHK, Maher ER. DNA methylation: a form of epigenetic control of gene expression. Obstet Gynaecol 2010;12(1):37–42.

[30] zur Hausen H. Papillomaviruses in the causation of human cancers–a brief historical account. Virology 2009;384(2):260–5.

[31] Kostareli E, Holzinger D, Bogatyrova O, et al. HPV-related methylation signature predicts survival in oropharyngeal squamous cell carcinomas. J Clin Invest 2013;123(6):2488–501.

[32] Hernandez JM, Siegel EM, Riggs B, et al. DNA methylation profiling across the spectrum of HPV-associated anal squamous neoplasia. PLoS One 2012;7(11):e50533.

[33] Bosch FX, Lorincz A, Munoz N, Meijer CJ, Shah KV. The causal relation between human papillomavirus and cervical cancer. J Clin Pathol 2002;55(4):244–65.

[34] Hansel A, Steinbach D, Greinke C, et al. A promising DNA methylation signature for the triage of high-risk human papillomavirus DNA-positive women. PLoS One 2014;9(3):e91905.

[35] Overmeer RM, Louwers JA, Meijer CJLM, et al. Combined CADM1 and MAL promoter methylation analysis to detect (pre-)malignant cervical lesions in high-risk HPV-positive women. Int J Cancer 2011;129(9):2218–25.

[36] Okano M, Gross TG. Acute or chronic life-threatening diseases associated with Epstein-Barr virus infection. Am J Med Sci 2012;343(6):483–9.

[37] Li L, Su X, Choi G, Cao Y, Ambinder R, Tao Q. Methylation profiling of Epstein-Barr virus immediate-early gene promoters, BZLF1 and BRLF1 in tumors of epithelial, NK- and B-cell origins. BMC Cancer 2012;12(1):125.

[38] Woellmer A, Hammerschmidt W. Epstein–Barr virus and host cell methylation: regulation of latency, replication and virus reactivation. Curr Opin Virol 2013;3(3):260–5.

[39] Bobetsis YA, Barros SP, Lin DM, et al. Bacterial infection promotes DNA hypermethylation. J Dent Res 2007;86(2):169–74.

[40] Constancia M, Hemberger M, Hughes J, et al. Placental-specific IGF-II is a major modulator of placental and fetal growth. Nature 2002;417(6892):945–8.

[41] Yeo A, Smith MA, Lin D, et al. *Campylobacter rectus* mediates growth restriction in pregnant mice. J Periodontol 2005;76(4):551–7.

[42] Marr AK, MacIsaac JL, Jiang R, Airo AM, Kobor MS, McMaster WR. *Leishmania donovani* infection causes distinct epigenetic DNA methylation changes in host macrophages. PLoS Pathog 2014;10(10):e1004419.

[43] Kaewpitoon N, Kaewpitoon SJ, Pengsaa P, Sripa B. *Opisthorchis viverrini*: the carcinogenic human liver fluke. World J Gastroenterol 2008;14(5):666–74.

[44] Shaib Y, El-Serag HB. The epidemiology of cholangiocarcinoma. Seminars Liver Dis 2004;24(2):115–25.

[45] Sriraksa R, Zeller C, El-Bahrawy MA, et al. CpG-island methylation study of liver fluke-related cholangiocarcinoma. Br J Cancer 2011;104(8):1313–8.

[46] Paluszczak J, Baer-Dubowska W. Epigenetic diagnostics of cancer–the application of DNA methylation markers. J Appl Genet 2006;47(4):365–75.

[47] Herman JG, Graff JR, Myohanen S, Nelkin BD, Baylin SB. Methylation-specific PCR: a novel PCR assay for methylation status of CpG islands. Proc Natl Acad Sci USA 1996;93(18):9821–6.

[48] Mikeska T, Craig J. DNA methylation biomarkers: cancer and beyond. Genes 2014;5(3):821–64.

[49] Bannister AJ, Kouzarides T. Regulation of chromatin by histone modifications. Cell Res 2011;21(3):381–95.

[50] Oh J, Fraser NW. Temporal association of the herpes simplex virus genome with histone proteins during a lytic infection. J Virol 2008;82(7):3530–7.

[51] Lacasse JJ, Schang LM. Herpes simplex virus 1 DNA is in unstable nucleosomes throughout the lytic infection cycle, and the instability of the nucleosomes is independent of DNA replication. J Virol 2012;86(20):11287–300.

[52] Sucato G, Wald A, Wakabayashi E, Vieira J, Corey L. Evidence of latency and reactivation of both herpes simplex virus (HSV)-1 and HSV-2 in the genital region. J Infect Dis 1998;177(4):1069–72.

[53] Decman V, Freeman ML, Kinchington PR, Hendricks RL. Immune control of HSV-1 latency. Viral Immunol 2005;18(3):466–73.

[54] Sabah M, Mulcahy J, Zeman A. Herpes simplex encephalitis. BMJ 2012. http://dx.doi.org/10.1136/bmj.e3166.

[55] Kimberlin DW. Neonatal herpes simplex infection. Clin Microbiol Rev 2004;17(1):1–13.

[56] Padgett DA, Sheridan JF, Dorne J, Berntson GG, Candelora J, Glaser R. Social stress and the reactivation of latent herpes simplex virus type 1. Proc Natl Acad Sci 1998;95(12):7231–5.

[57] Knipe DM, Cliffe A. Chromatin control of herpes simplex virus lytic and latent infection. Nat Rev Microbiol 2008;6(3):211–21.

[58] Umbach JL, Kramer MF, Jurak I, Karnowski HW, Coen DM, Cullen BR. MicroRNAs expressed by herpes simplex virus 1 during latent infection regulate viral mRNAs. Nature 2008;454(7205):780–3.

[59] Kubat NJ, Tran RK, McAnany P, Bloom DC. Specific histone tail modification and not DNA methylation is a determinant of herpes simplex virus type 1 latent gene expression. J Virol 2004;78(3):1139–49.

[60] Lehman IR, Boehmer PE. Replication of herpes simplex virus DNA. J Biol Chem 1999;274(40):28059–62.

[61] Kent JR, Zeng P-Y, Atanasiu D, Gardner J, Fraser NW, Berger SL. During lytic infection herpes simplex virus type 1 is associated with histones bearing modifications that correlate with active transcription. J Virol 2004;78(18):10178–86.

[62] Huang J, Kent JR, Placek B, et al. Trimethylation of histone H3 lysine 4 by Set1 in the lytic infection of human herpes simplex virus 1. J Virol 2006;80(12):5740–6.

[63] Schmeck B, Lorenz J, N'Guessan PD, et al. Histone acetylation and flagellin are essential for *Legionella pneumophila*-induced cytokine expression. J Immunol 2008;181(2):940–7.

[64] Rolando M, Sanulli S, Rusniok C, et al. *Legionella pneumophila* effector RomA uniquely modifies host chromatin to repress gene expression and promote intracellular bacterial replication. Cell Host Microbe 2013;13(4):395–405.

[65] Phin N, Parry-Ford F, Harrison T, et al. Epidemiology and clinical management of Legionnaires' disease. Lancet Infect Dis 2014;14(10):1011–21.

[66] Dalmasso MC, Sullivan Jr WJ, Angel SO. Canonical and variant histones of protozoan parasites. Front Biosci (Landmark Ed) 2011;16:2086–105.

[67] Lang C, Hildebrandt A, Brand F, Opitz L, Dihazi H, Lüder CGK. Impaired chromatin remodelling at STAT1-regulated promoters leads to global unresponsiveness of *Toxoplasma gondii*-infected macrophages to IFN-γ. PLoS Pathog 2012;8(1):e1002483.

[68] Leng J, Denkers EY. *Toxoplasma gondii* inhibits covalent modification of histone H3 at the IL-10 promoter in infected macrophages. PLoS One 2009;4(10):e7589.

[69] Qian M-B, Chen Y-D, Fang Y-Y, et al. Epidemiological profile of *Clonorchis sinensis* infection in one community, Guangdong, People's Republic of China. Parasit Vectors 2013;6(1):194.

[70] Cho SH, Sohn WM, Na BK, et al. Prevalence of *Clonorchis sinensis* metacercariae in freshwater fish from three latitudinal regions of the Korean Peninsula. Korean J Parasitol 2011;49(4):385–98.

[71] Choi BI, Han JK, Hong ST, Lee KH. Clonorchiasis and cholangiocarcinoma: etiologic relationship and imaging diagnosis. Clin Microbiol Rev 2004;17(3):540–52. Table of contents.

[72] Kim D-W, Kim J-Y, Moon JH, et al. Transcriptional induction of minichromosome maintenance protein 7 (Mcm7) in human cholangiocarcinoma cells treated with *Clonorchis sinensis* excretory–secretory products. Mol Biochem Parasitol 2010;173(1):10–6.

[73] Pak JH, Kim DW, Moon JH, et al. Differential gene expression profiling in human cholangiocarcinoma cells treated with *Clonorchis sinensis* excretory-secretory products. Parasitol Res 2009;104(5):1035–46.

[74] Luo JH. Oncogenic activity of MCM7 transforming cluster. World J Clin Oncol 2011;2(2):120–4.

[75] Kurdistani SK. Histone modifications as markers of cancer prognosis: a cellular view. Br J Cancer 2007;97(1):1–5.

[76] Chen R, Kang R, Fan XG, Tang D. Release and activity of histone in diseases. Cell Death Dis 2014;5:e1370.

[77] Huang H, Evankovich J, Yan W, et al. Endogenous histones function as alarmins in sterile inflammatory liver injury through Toll-like receptor 9 in mice. Hepatology 2011;54(3):999–1008.

[78] Jounai N, Kobiyama K, Takeshita F, Ishii KJ. Recognition of damage-associated, nucleic acid-related molecular patterns during inflammation and vaccination. Front Cell Infect Microbiol 2013;2.

[79] Brinkmann V, Reichard U, Goosmann C, et al. Neutrophil extracellular traps kill bacteria. Science 2004;303(5663):1532–5.

[80] Xu J, Zhang X, Pelayo R, et al. Extracellular histones are major mediators of death in sepsis. Nat Med 2009;15(11): 1318–21.
[81] Liu CL, Tangsombatvisit S, Rosenberg JM, et al. Specific post-translational histone modifications of neutrophil extracellular traps as immunogens and potential targets of lupus autoantibodies. Arthritis Res Ther 2012;14(1):R25.
[82] Bock C. Epigenetic biomarker development. Epigenomics 2009;1(1):99–110.
[83] Nitzsche A, Paulus C, Nevels M. Temporal dynamics of cytomegalovirus chromatin assembly in productively infected human cells. J Virol 2008;82(22):11167–80.
[84] Nevels M, Paulus C, Shenk T. Human cytomegalovirus immediate-early 1 protein facilitates viral replication by antagonizing histone deacetylation. Proc Natl Acad Sci USA 2004;101(49):17234–9.
[85] Wang FZ, Weber F, Croce C, Liu CG, Liao X, Pellett PE. Human cytomegalovirus infection alters the expression of cellular microRNA species that affect its replication. J Virol 2008;82(18):9065–74.
[86] Buck AH, Perot J, Chisholm MA, et al. Post-transcriptional regulation of miR-27 in murine cytomegalovirus infection. RNA 2010;16(2):307–15 (New York, NY).
[87] Fu M, Gao Y, Zhou Q, et al. Human cytomegalovirus latent infection alters the expression of cellular and viral microRNA. Gene 2014;536(2):272–8.
[88] Zazula M, Ferreira AM, Czopek JP, et al. CDH1 gene promoter hypermethylation in gastric cancer: relationship to Goseki grading, microsatellite instability status, and EBV invasion. Diagn Mol Pathol 2006; 15(1):24–9.
[89] Birdwell CE, Queen KJ, Kilgore PC, et al. Genome-wide DNA methylation as an epigenetic consequence of Epstein-Barr virus infection of immortalized keratinocytes. J Virol 2014;88(19):11442–58.
[90] Paschos K, Parker GA, Watanatanasup E, White RE, Allday MJ. BIM promoter directly targeted by EBNA3C in polycomb-mediated repression by EBV. Nucleic Acids Res 2012;40(15):7233–46.
[91] Tong JH, Tsang RK, Lo KW, et al. Quantitative Epstein-Barr virus DNA analysis and detection of gene promoter hypermethylation in nasopharyngeal (NP) brushing samples from patients with NP carcinoma. Clin Cancer Res 2002;8(8):2612–9.
[92] Hutajulu SH, Indrasari SR, Indrawati LP, et al. Epigenetic markers for early detection of nasopharyngeal carcinoma in a high risk population. Mol Cancer 2011;10:48.
[93] Yang X, Dai W, Kwong DL, et al. Epigenetic markers for noninvasive early detection of nasopharyngeal carcinoma by methylation-sensitive high resolution melting. Int J Cancer 2015;136(4):E127–35.
[94] Murata T, Kondo Y, Sugimoto A, et al. Epigenetic histone modification of Epstein-Barr virus BZLF1 promoter during latency and reactivation in Raji cells. J Virol 2012;86(9):4752–61.
[95] Huang J, Wang Y, Guo Y, Sun S. Down-regulated microRNA-152 induces aberrant DNA methylation in hepatitis B virus-related hepatocellular carcinoma by targeting DNA methyltransferase 1. Hepatology 2010;52(1):60–70.
[96] Xu X, Fan Z, Kang L, et al. Hepatitis B virus X protein represses miRNA-148a to enhance tumorigenesis. J Clin Invest 2013;123(2):630–45.
[97] Wang Y, Lu Y, Toh ST, et al. Lethal-7 is down-regulated by the hepatitis B virus x protein and targets signal transducer and activator of transcription 3. J Hepatol 2010;53(1):57–66.
[98] Liu Y, Zhao JJ, Wang CM, et al. Altered expression profiles of microRNAs in a stable hepatitis B virus-expressing cell line. Chin Med J 2009;122(1):10–4.
[99] Jin J, Tang S, Xia L, et al. MicroRNA-501 promotes HBV replication by targeting HBXIP. Biochem Biophys Res Commun 2013;430(4):1228–33.
[100] Fonseca GJ, Thillainadesan G, Yousef AF, et al. Adenovirus evasion of interferon-mediated innate immunity by direct antagonism of a cellular histone posttranslational modification. Cell Host Microbe 2012;11(6):597–606.
[101] Wang X, Tang S, Le SY, et al. Aberrant expression of oncogenic and tumor-suppressive microRNAs in cervical cancer is required for cancer cell growth. PLoS One 2008;3(7):e2557.
[102] Li Y, Wang F, Xu J, et al. Progressive miRNA expression profiles in cervical carcinogenesis and identification of HPV-related target genes for miR-29. J Pathol 2011;224(4):484–95.
[103] Lui WO, Pourmand N, Patterson BK, Fire A. Patterns of known and novel small RNAs in human cervical cancer. Cancer Res 2007;67(13):6031–43.
[104] Martinez I, Gardiner AS, Board KF, Monzon FA, Edwards RP, Khan SA. Human papillomavirus type 16 reduces the expression of microRNA-218 in cervical carcinoma cells. Oncogene 2008;27(18):2575–82.
[105] Pereira PM, Marques JP, Soares AR, Carreto L, Santos MA. MicroRNA expression variability in human cervical tissues. PLoS One 2010;5(7):e11780.
[106] Lee JS, Choi YD, Lee JH, et al. Expression of PTEN in the progression of cervical neoplasia and its relation to tumor behavior and angiogenesis in invasive squamous cell carcinoma. J Surg Oncol 2006;93(3):233–40.
[107] McCabe MT, Low JA, Imperiale MJ, Day ML. Human polyomavirus BKV transcriptionally activates DNA methyltransferase 1 through the pRb/E2F pathway. Oncogene 2006;25(19):2727–35.

[108] Buggele WA, Johnson KE, Horvath CM. Influenza A virus infection of human respiratory cells induces primary microRNA expression. J Biol Chem 2012;287(37):31027–40.

[109] Mukherjee S, Vipat VC, Chakrabarti AK. Infection with influenza A viruses causes changes in promoter DNA methylation of inflammatory genes. Influenza Other Respir Viruses 2013;7(6):979–86.

[110] Shamay M, Krithivas A, Zhang J, Hayward SD. Recruitment of the de novo DNA methyltransferase Dnmt3a by Kaposi's sarcoma-associated herpesvirus LANA. Proc Natl Acad Sci USA 2006;103(39):14554–9.

[111] Soejima K, Fang W, Rollins BJ. DNA methyltransferase 3b contributes to oncogenic transformation induced by SV40T antigen and activated Ras. Oncogene 2003;22(30):4723–33.

[112] Watanabe H, Soejima K, Yasuda H, et al. Deregulation of histone lysine methyltransferases contributes to oncogenic transformation of human bronchoepithelial cells. Cancer Cell Int 2008;8(1):15.

[113] Li Y, Wu R, Liu Z, Fan J, Yang H. Enforced expression of microRNA-21 influences the replication of varicella-zoster virus by triggering signal transducer and activator of transcription 3. Exp Ther Med 2014;7(5):1291–6.

[114] Garcia-Garcia JC, Barat NC, Trembley SJ, Dumler JS. Epigenetic silencing of host cell defense genes enhances intracellular survival of the Rickettsial pathogen *Anaplasma phagocytophilum*. PLoS Pathog 2009;5(6):e1000488.

[115] Mujtaba S, Winer BY, Jaganathan A, et al. Anthrax SET Protein: a potential virulence determinant that epigenetically represses NF-κB activation in infected macrophages. J Biol Chem 2013;288(32):23458–72.

[116] Pennini ME, Perrinet S, Dautry-Varsat A, Subtil A. Histone methylation by NUE, a novel nuclear effector of the intracellular pathogen *Chlamydia trachomatis*. PLoS Pathog 2010;6(7):e1000995.

[117] Hamon MA, Batsché E, Régnault B, et al. Histone modifications induced by a family of bacterial toxins. Proc Natl Acad Sci 2007;104(33):13467–72.

[118] Tolg C, Sabha N, Cortese R, et al. Uropathogenic *E. coli* infection provokes epigenetic downregulation of CDKN2A (p16INK4A) in uroepithelial cells. Lab Invest 2011;91(6):825–36.

[119] Maekita T, Nakazawa K, Mihara M, et al. High levels of aberrant DNA methylation in *Helicobacter pylori*-infected gastric mucosae and its possible association with gastric cancer risk. Clin Cancer Res 2006;12(3):989–95.

[120] Ding S-Z, Fischer W, Kaparakis-Liaskos M, et al. *Helicobacter pylori*-induced histone modification, associated gene expression in gastric epithelial cells, and its implication in pathogenesis. PLoS One 2010;5(4):e9875.

[121] Fehri LF, Rechner C, Janssen S, et al. *Helicobacter pylori*-induced modification of the histone H3 phosphorylation status in gastric epithelial cells reflects its impact on cell cycle regulation. Epigenetics 2009;4(8):577–86.

[122] Motoyama K, Inoue H, Nakamura Y, Uetake H, Sugihara K, Mori M. Clinical significance of high mobility group A2 in human gastric cancer and its relationship to let-7 microRNA family. Clin Cancer Res 2008;14(8):2334–40.

[123] Saito Y, Suzuki H, Tsugawa H, et al. Dysfunctional gastric emptying with down-regulation of muscle-specific microRNAs in *Helicobacter pylori*-infected mice. Gastroenterology 2011;140(1):189–98.

[124] Belair C, Baud J, Chabas S, et al. *Helicobacter pylori* interferes with an embryonic stem cell micro RNA cluster to block cell cycle progression. Silence 2011;2(1):7.

[125] Schmeck B, Beermann W, van Laak V, et al. Intracellular bacteria differentially regulated endothelial cytokine release by MAPK-dependent histone modification. J Immunol 2005;175(5):2843–50.

[126] Izar B, Mannala GK, Mraheil MA, Chakraborty T, Hain T. microRNA response to *Listeria monocytogenes* infection in epithelial cells. Int J Mol Sci 2012;13(1):1173–85.

[127] Schnitger AKD, Machova A, Mueller RU, et al. *Listeria monocytogenes* infection in macrophages induces vacuolar-dependent host miRNA response. PLoS One 2011;6(11):e27435.

[128] Slevogt H, Schmeck B, Jonatat C, et al. *Moraxella catarrhalis* induces inflammatory response of bronchial epithelial cells via MAPK and NF-kappaB activation and histone deacetylase activity reduction. Am J Physiol Lung Cell Mol Physiol 2006;290(5):L818–26.

[129] Pennini ME, Pai RK, Schultz DC, Boom WH, Harding CV. *Mycobacterium tuberculosis* 19-kDa lipoprotein inhibits IFN-gamma-induced chromatin remodeling of MHC2TA by TLR2 and MAPK signaling. J Immunol 2006;176(7):4323–30.

[130] Schulte LN, Eulalio A, Mollenkopf HJ, Reinhardt R, Vogel J. Analysis of the host microRNA response to Salmonella uncovers the control of major cytokines by the let–7 family. EMBO J 2011;30(10):1977–89.

[131] Arbibe L, Kim DW, Batsche E, et al. An injected bacterial effector targets chromatin access for transcription factor NFκB to alter transcription of host genes involved in immune responses. Nat Immunol 2007;8(1):47–56.

[132] Agarwal S, Agarwal S, Jin H, Pancholi P, Pancholi V. Serine/Threonine phosphatase (SP-STP), secreted from *Streptococcus pyogenes*, is a pro-apoptotic protein. J Biol Chem 2012;287(12):9147–67.

[133] Lemaire J, Mkannez G, Guerfali FZ, et al. MicroRNA expression profile in human macrophages in response to *Leishmania major* infection. PLoS Negl Trop Dis 2013;7(10):e2478.
[134] Al-Quraishy S, Dkhil MA, Abdel-Baki AA, Delic D, Santourlidis S, Wunderlich F. Genome-wide screening identifies *Plasmodium chabaudi*-induced modifications of DNA methylation status of Tlr1 and Tlr6 gene promoters in liver, but not spleen, of female C57BL/6 mice. Parasitol Res 2013;112(11):3757–70.
[135] Ruckerl D, Jenkins SJ, Laqtom NN, et al. Induction of IL-4Ralpha-dependent microRNAs identifies PI3K/Akt signaling as essential for IL-4-driven murine macrophage proliferation in vivo. Blood 2012;120(11):2307–16.

CHAPTER

20

DNA Methylation in Neurodegenerative Diseases

Sahar Al-Mahdawi[1,2], *Sara Anjomani Virmouni*[1,2], *Mark A. Pook*[1,2]

[1]Department of Life Sciences, College of Health & Life Sciences, Brunel University London, Uxbridge, UK; [2]Synthetic Biology Theme, Institute of Environment, Health & Societies, Brunel University London, Uxbridge, UK

OUTLINE

1. INTRODUCTION

Mammalian DNA methylation is an epigenetic modification that occurs predominantly due to DNA methyltransferase (DNMT)-catalyzed conversion of cytosine to 5-methylcytosine (5mC) primarily within the context of CpG dinucleotides, although non-CpG methylation also occurs [1]. The DNMT family of enzymes includes DNMT1, which preferentially methylates hemimethylated DNA and is involved in maintaining methylation after DNA replication [2], while DNMT3a and DNMT3b act equally on hemimethylated and nonmethylated DNA and

Epigenetic Biomarkers and Diagnostics
http://dx.doi.org/10.1016/B978-0-12-801899-6.00020-6

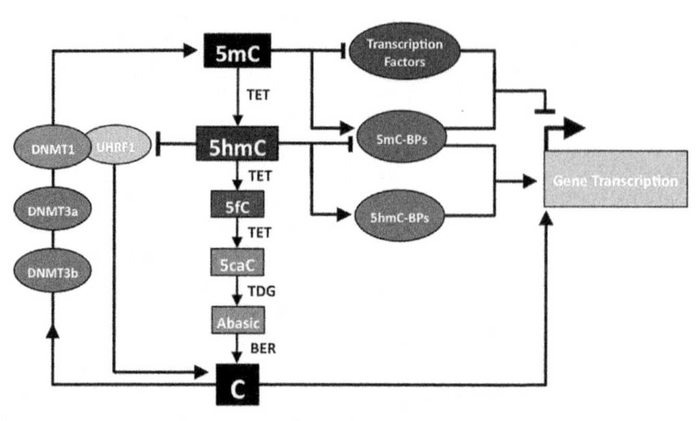

FIGURE 1 The role of DNA methylation in regulation of gene transcription. Unmethylated cytosine residues (C) at the gene promoter are associated with active gene transcription. 5-Methylcytosine (5mC) blocks the binding of transcription factors, such as Ets-1 and CCTC-binding factor, and binds 5mC-binding proteins (5mC-BPs), such as MeCP2, MBD1, MBD2, and MBD4, to block transcription initiation at the gene promoter. 5-Hydroxymethylcytosine (5hmC), which is generated by ten-eleven translocation (TET) oxidation of 5mC, selectively binds 5hmC-BPs rather than 5mC-BPs to activate gene transcription. At the same time 5hmC acts as an intermediate [1] in passive DNA demethylation due to poor binding between 5hmC and ubiquitin-like, containing PHD and RING finger domain 1 (UHRF1), the partner of DNMT1 [2]; in active DNA demethylation by acting as an intermediate in the TET, thymine DNA glycosylase (TDG), and base excision repair (BER)-induced conversion of 5mC to unmethylated cytosine.

so have been classified as de novo methyltransferases [3], although they are also involved with DNMT1 in maintenance methylation [4]. CpG dinucleotides are often clustered together as CpG islands (CGIs), which locate to distinct regions of the gene, and the 5mC profile of a gene has been shown to regulate its expression [5]. For example, there is a strong correlation between DNA methylation of CGIs at promoters and transcriptional silencing. This is caused by two mechanisms (Figure 1). Firstly, 5mC can directly interfere with the binding of transcription factors, such as Ets-1, or the boundary element factor CCTC-binding factor (CTCF). Secondly, specific 5mC-binding proteins, including MeCP2, MBD1, MBD2, and MBD4, repress transcription [6]. However, it has also been reported that intragenic CpG methylation can contribute to gene silencing [7], stimulation of transcription elongation, or gene splicing [8]. At the same time, CpG methylation can itself be influenced by the neighboring DNA sequence, as evidenced by single base-pair variants, designated "epimutations," that can result in increased promoter DNA methylation followed by reduced gene expression, and vice versa [9]. In addition to 5mC, recent studies have revealed alternative cytosine modifications, 5-hydroxymethylcytosine (5hmC), 5-formylcytosine (5fC), and 5-carboxylcytosine (5caC), which are formed by sequential oxidation of 5mC by ten-eleven translocation (TET) enzymes [10–12]. These DNA modifications may either be intermediates in the removal of 5mC by passive or active demethylation processes [13] or they may be epigenetic modifications in their own right, regulating gene transcription by influencing the ratio of 5mC- versus 5hmC-binding proteins [14] (Figure 1). In addition to its role in the regulation of gene expression, DNA methylation has been regarded as either a cause or a consequence of several other processes, including mammalian development, X inactivation, genomic imprinting, and the immobilization of transposons [5]. Furthermore, aberrant DNA methylation profiles are known to be associated with many different human diseases [15], including Rett syndrome [16];

immunodeficiency, centromeric region instability, and facial anomalies syndrome [17]; and cancer [18], where there are generally alterations of global DNA methylation patterns, and fragile X syndrome (FXS), where there is specific methylation of the CCG repeat mutation in the fragile X mental retardation-1 (*FMR1*) gene [19]. DNA methylation has proven to be particularly important in neurodevelopment and normal adult central nervous system (CNS) functioning [20]. When defective, this contributes either to neurodevelopmental disorders such as Rett syndrome, autism spectrum disorders, schizophrenia and psychotic disorders, and fetal alcohol spectrum disorder [21] or to neurodegenerative diseases, which are characterized by progressive loss of neurons in the CNS [22]. This review article will focus on our current understanding of the role of DNA methylation in such neurodegenerative diseases. We will begin by considering studies that have revealed the importance of DNA methylation in the CNS. We will then describe the progress of DNA methylation research in each of the major neurodegenerative diseases: Alzheimer's disease (AD), Parkinson's disease (PD), amyotrophic lateral sclerosis (ALS), Huntington's disease (HD), fragile X-associated tremor/ataxia syndrome (FXTAS), Friedreich ataxia (FRDA), and spinocerebellar ataxia type 7 (SCA7). We will discuss both global and locus-specific DNA methylation alterations that have been identified with regard to the potential cause or effect of each disease, together with consideration of DNA methylation changes as a potential biomarker or therapeutic target of disease.

2. DNA METHYLATION AND THE CNS

The first evidence that 5mC DNA methylation could be important for the CNS came from studies that identified mutations in the *MECP2* gene as the cause of Rett syndrome [16]. This is one of the most common causes of mental retardation in females, occurring sporadically once in every 10,000–22,000 female births [6]. It is characterized by normal development for about 1 year followed by rapid loss of acquired speech and motor skills, seizures, autism, and ataxia. However, examination of the brain has revealed no evidence of a progressive neurodegeneration, but rather there is a pronounced microcephaly caused by decreased neuronal size and packing [20]. The reason why Rett syndrome primarily affects females is because *MECP2* is X-linked and patients are heterozygous for the mutated allele. Following random X-chromosome inactivation, half of their cells will express abnormally functioning MeCP2 protein, resulting in a mosaic pattern of abnormal global gene expression. A second line of evidence that 5mC DNA methylation is important for the CNS comes from the finding that neurons in adult mouse brain express very high levels of *Dnmt1* [23]. It is plausible that high levels of this maintenance DNMT in postmitotic neurons indicate a high demand to remethylate cytosine residues produced by DNA damage repair following deamination of the original methylated cytosines to thymines [20]. This suggests that DNA methylation is essential for the survival and correct functioning of CNS neurons. More recent studies have also shown that the de novo DNMT Dnmt3a is expressed in postnatal neural stem cells (NSCs) and is required for neurogenesis. *Dnmt3a*-null mice exhibit impaired neurogenesis and genome-wide analysis of NSCs has shown that Dnmt3a occupies and methylates DNA at nonpromoter intragenic regions of regulators of neurogenesis [24]. The subsequent identification of 5hmC, 5fC, and 5caC, in addition to 5mC, has led to a more complex understanding of the dynamic roles of DNA methylation and demethylation in CNS development and function. The overall levels of 5hmC in the mammalian genome have been reported to be approximately 10% of 5mC levels [25]. However, higher levels have been detected in tissues of the CNS [26]. For example, 5hmC has been reported to be approximately 40% as abundant as 5mC in Purkinje cells of the cerebellum [10]. Recent studies of human and mouse brain from postnatal through to adult stages have revealed positive correlations between 5hmC levels and brain development [13,14,27,28], particularly in the cerebellum [29,30] and hippocampus [31], suggesting the involvement of active DNA demethylation processes. 5fC and 5caC levels have also been shown to accumulate throughout NSC differentiation, further supporting the role of active DNA demethylation in brain development [32]. Furthermore, recent comprehensive analysis of 5mC and 5hmC at single-base resolution in human and mouse frontal cortex from fetal to adult stages has revealed a conserved, genome-wide increase in non-CpG 5mC (i.e., methyl cytosine H (mCH), where H is adenine, cytosine, or thymine) throughout development, in addition to the previously described increase in 5hmC, suggesting that there is yet further complexity to be considered for the role of DNA methylation in the CNS. Recent studies have investigated the variations in the patterns of 5mC, 5hmC, 5fC, and 5caC between normal and neurodegenerative states of CNS tissues, and the results suggest an important role for DNA methylation dynamics in neurodegenerative diseases. Changes in 5mC, 5hmC, 5fC, or 5caC levels may occur at a specific locus and affect the transcription of only a single gene, or they may occur globally and affect the transcription of many genes, as we will now describe.

3. ALZHEIMER'S DISEASE

AD (OMIM: 104300) is the most common neurodegenerative disorder, characterized by progressive decline of cognitive functions, neuronal cell loss, and two hallmarks of pathology, extracellular amyloid beta plaques and intracellular neurofibrillary tangles composed of hyperphosphorylated tau protein [33]. AD accounts for over 50% of all cases of dementia, and it is estimated that it affects more than 24 million people worldwide [34]. The causes of AD are unknown, but evidence has emerged from the identification of abnormalities in genes, including overexpression of the beta-amyloid precursor protein gene, *APP*, and functional mutations of the presenilin genes, *PSEN1* and *PSEN2* [35]. Approximately 2% of AD cases are early onset, before the age of 60 years, with defined mutations in *APP*, *PSEN1*, or *PSEN2* genes. These findings have led to the hypothesis that increased amyloid beta aggregation inducing cellular toxicity is the major cause of AD. However, treatments aimed at reducing amyloid beta pathology have thus far proved unsuccessful. This observation, together with other evidence that hyperphosphorylated tau protein closely correlates with disease progression, points to tau hyperphosphorylation as the primary cause of AD. In addition, there may be alterations of epigenetic factors due to aging or in response to environmental stresses [36–38]. Several studies have recently investigated the global levels of DNA methylation-based enzymes, DNMT1 and TET1, and modified cytosine residues, 5mC, 5hmC, 5fC, and 5caC, in AD brain tissues using immunohistochemical analysis, with somewhat contradictory results (Table 1). Initial studies of human temporal lobe brain samples revealed evidence of decreased levels of 5mC and DNMT1 in neurons of AD patients [39]. Similar decreases in 5mC, together with decreased levels of 5hmC, were subsequently identified in the hippocampal region of AD brains [40]. However, other studies have identified the opposite effects, reporting increased levels of both 5mC and 5hmC in AD brains (Table 1). For example, increased levels of 5mC have been detected in frontal cortex of AD patients [41]. Also, increased levels of TET1, 5mC, and 5hmC, accompanied by decreases in the levels of 5fC and 5caC, have been detected in the hippocampus of AD patients, while no changes were detected in cerebellum tissues [42]. Finally, increased levels of both 5mC and 5hmC have been detected in the frontal and temporal cortex of AD patients [43].

TABLE 1 DNA Methylation Changes in Neurodegenerative Diseases

Disease	DNA methylation change	References
AD	Decreased global DNMT1 and 5mC in AD temporal cortex	[39]
	Decreased global 5mC and 5hmC in AD hippocampus	[40]
	Increased global 5mC in AD frontal cortex	[41]
	Increased global Tet1, 5mC and 5hmC, and decreased global 5fC and 5caC in AD hippocampus, but no change in cerebellum	[42]
	Increased global 5mC and 5hmC in AD frontal and temporal cortex	[43]
	Decreased 5mC levels in the *APP* promoter of AD patient cortex	[47]
	Decreased 5mC levels in the *PSEN1* promoter of AD brain	[38]
	Increased 5mC levels in the *SIRT1* gene of AD patient blood	[48]
	Decreased 5mC levels in the *APP* gene of AD patient blood	[48]
PD	Decreased 5mC of the *SNCA* gene and enhanced α-synuclein expression in PD patient brains	[51,52]
	α-Synuclein sequesters DNMT1 in PD brain	[53]
	Decreased 5mC of the *SNCA* gene in PD blood samples	[55]
ALS	Differential 5mC levels of genes in ALS brain	[61]
	Increased global 5mC and 5hmC in sporadic ALS spinal cord	[62]
	Increased 5mC in the 5′ region of the *C9ORF72* hexanucleotide repeat in familial ALS patient blood, brain, and spinal cord, together with downregulation of the *C9ORF72* mRNA	[66]
HD	Global DNA demethylation induces CAG repeat instability	[70,71]
	The HD mutation induces global 5mC changes in mouse striatal neurons	[72]
	Increased 5mC and decreased 5hmC in the 5′ UTR of the *ADORA2A* gene in HD putamen	[73]
	Decreased global 5hmC in YAC128 HD mouse model striatum and cortex	[75]
FXTAS	Decreased global 5hmC in rCGG FXTAS mouse model cerebellum	[80]
FRDA	Increased 5mC at the 5′ GAA repeat region and decreased 5mC at the 3′ GAA repeat region of the *FXN* gene in FRDA patient cells and FRDA patient and FRDA mouse model cerebellum and heart	[87–91]
	Increased 5hmC at the 5′ GAA repeat region in FRDA patient cerebellum and heart	[92]
SCA7	Global DNA demethylation induces CAG repeat instability	[70,71]
	5mC hypermethylation of CTCF-binding sites enhances CAG repeat instability in a SCA7 mouse model	[95]

AD, Alzheimer's disease; PD, Parkinson's disease; ALS, amyotrophic lateral sclerosis; HD, Huntington's disease; FXTAS, fragile X-associated tremor/ataxia syndrome; FRDA, Friedreich ataxia; SCA7, spinocerebellar ataxia type 7; 5mC, 5-methylcytosine; 5hmC, 5-hydroxymethylcytosine; 5fC, 5-formylcytosine; 5caC, 5-carboxylcytosine; UTR, untranslated region.

The reasons for the discrepancies between the different studies are not clear, but they may be due to the analysis of different regions of the brain using different immunohistochemical quantification techniques. The investigation of locus-specific DNA methylation in AD is also unclear. A large-scale study of the *APP* promoter methylation status in postmortem frontal cortex and hippocampus found no difference in 5mC levels in AD patients compared with unaffected controls [44]. Another study that investigated cortex and cerebellum similarly found no differences in the 5mC levels of the *APP* gene in AD patients compared with unaffected controls [45]. Furthermore, a study of the *APP* promoter identified complex tissue-specific patterns of DNA methylation that differed in different regions of the brain, indicating the difficulties that are involved in the analysis of this locus [46]. However, there are reports of decreased 5mC levels in the *APP* promoter of AD patient cortex [47] and blood leukocytes [48]. The latter report also identified hypermethylation of the *SIRT1* gene in blood leukocytes from AD patients, noting a correlation between DNA methylation levels and AD severity, suggesting the possible use of DNA methylation as a biomarker for AD [48] (Table 2). There is also a report of decreased 5mC levels in the *PSEN1* promoter of AD brain [38]. It would now be interesting to investigate levels of locus-specific 5hmC, 5fC, and 5caC at specific genes related to AD pathology, such as the *APP*, *PSEN1*, and *PSEN2* genes, which have previously shown inconclusive AD-related changes in 5mC levels [44–47].

4. PARKINSON'S DISEASE

PD (OMIM: 168600) is the second most common neurodegenerative disorder, affecting more than four million people worldwide. It is characterized by the progressive loss of substantia nigra dopaminergic neurons, together with aggregates of misfolded α-synuclein called Lewy bodies, resulting in muscle rigidity, bradykinesia, tremor, and instability [49]. Most cases of PD are sporadic and the causes are unknown. However, mutations in several genes have now been identified in rare familial forms of PD, including the genes *SNCA* (α-synuclein), *PARK2* (parkin), *PTEN*-induced putative kinase 1 (*PINK1*), *PARK7* (DJ-1), leucine-rich repeat kinase 2 (*LRRK2*), and *ATP13A2* [50]. Subsequent DNA methylation studies of PD have focused on the *SNCA* gene, particularly with regard to regulation of *SNCA* gene expression (Table 1). For example, two studies have revealed decreased levels of DNA methylation of the *SNCA* gene and enhanced α-synuclein expression in PD patient brains [51,52]. Furthermore, α-synuclein has been shown to sequester DNMT1 away from the nucleus in brain samples from PD patients and α-synuclein transgenic mice, suggesting a potential mechanism to produce the decreased levels of DNA methylation at the *SNCA* gene [53]. However, there is conflicting evidence when considering DNA methylation levels and α-synuclein expression in blood samples from PD patients. One study has failed to detect any corresponding hypomethylation at the *SNCA* gene in blood samples of PD patients [54]. However, another study has detected hypomethylation of the *SNCA* gene in PD patient blood samples, although this decrease in DNA methylation does not show any association with α-synuclein expression levels [55] (Table 2). More recently, 5hmC levels have been studied in striatal brain tissues of the 6-OHDA-induced rat model of PD, but while 5hmC content generally increased with age, no changes in 5hmC levels were detected compared with controls [56].

5. AMYOTROPHIC LATERAL SCLEROSIS

ALS (OMIM: 105400) is the third most common adult-onset neurodegenerative disease, with a global incidence of about 2 in 100,000 individuals [57]. The average age of onset of

TABLE 2 DNA Methylation as a Biomarker of Neurodegenerative Diseases

Disease	Sample	Method	Measured effect	Sensitivity/Specificity[a]	References
AD	Blood	Bisulfite cloning sequence + MSP	Increased DNA methylation in *SIRT1* gene	Low: 30–80% in AD patients 10–40% in controls	[48]
PD		Bisulfite cloning sequence	Decreased DNA methylation in *SNCA* gene	Low: 14.2% in PD patients 16.9% in controls	[55]
ALS		Direct bisulfite sequence	Increased DNA methylation in *C9ORF72* gene	High: 73% in ALS patients 4% in controls	[66]
FRDA		Bisulfite pyrosequence	Increased DNA methylation in *FXN* gene (5′ GAA region)	Medium: 75–95% in FRDA patients 20–70% in controls	[89]
		Bisulfite MASS array EpiTYPER	Increased DNA methylation in *FXN* gene (5′ GAA region)	Low: 85–95% in FRDA patients 60–90% in controls	[90]
			Decreased DNA methylation in *FXN* gene (3′ GAA region)	Low: 40–80% in FRDA patients 85–95% in controls	
	Buccal cells	Bisulfite MASS array EpiTYPER	Increased DNA methylation in *FXN* gene (5′ GAA region)	Low: 80–95% in FRDA patients 40–90% in controls	[90]
			Decreased DNA methylation in *FXN* gene (3′ GAA region)	Low: 35–75% in FRDA patients 80–95% in controls	

MSP, methylation-specific PCR; AD, Alzheimer's disease; PD, Parkinson's disease; ALS, amyotrophic lateral sclerosis; FRDA, Friedreich ataxia.
[a] *% = percentage of DNA methylation at CpGs.*

disease is 55–65 years, and it is a progressive disease characterized by selective loss of motor neurons within the brain and spinal cord [58]. The cause of ALS is largely unknown and it is typically identified as sporadic. However, about 10% of ALS cases are identified as familial due to defective genes, including those that encode superoxide dismutase 1 (*SOD1*), TAR DNA-binding protein (*TARDBP*), fused in sarcoma (*FUS*), ubiquilin2 (*UBQLN2*), and chromosome 9 open reading frame 72 (*C9ORF72*) [59]. For the larger percentage of sporadic ALS cases, environmental factors such as exposure to toxins or dietary factors may be driving global DNA methylation changes [60]. Therefore, it is interesting to note that one initial genome-wide study of brain DNA methylation identified differentially methylated genes in ALS cases compared with controls [61]. This was followed by a second study that reported global alterations in both 5mC and 5hmC levels in sporadic ALS spinal cord, but not in blood samples [62] (Table 1). The significance of such global DNA methylation changes in brain and spinal cord requires further investigation. However, the lack of altered global DNA methylation levels in blood samples suggests that neither global 5mC nor 5hmC levels would be suitable as biomarkers of ALS. At the locus-specific level, DNA methylation was initially investigated in the *SOD1* gene of sporadic ALS cases, but this study failed to identify any ALS-specific DNA methylation differences compared with controls [63]. Subsequently, the G_4C_2 hexanucleotide repeat expansion mutation in the *C9ORF72* gene has been identified as the most common known genetic cause of ALS [64,65], and recent studies have shown that the CGI in the 5′ region of the *C9ORF72* hexanucleotide repeat is hypermethylated in familial ALS patient blood, brain, and spinal cord samples, together with downregulation of the *C9ORF72* mRNA [66]. This report is very interesting because it supports the hypothesis that the *C9ORF72* hexanucleotide repeat induces gene silencing to cause loss of function and subsequent ALS disease, together with the possible use of DNA methylation as a biomarker for ALS (Table 2). Furthermore, CGI methylation levels were associated with more aggressive ALS disease, suggesting their potential use as a biomarker of disease phenotype.

6. HUNTINGTON'S DISEASE

HD (OMIM: 143100) is a late-onset autosomal dominant neurodegenerative disease, characterized by chorea, dystonia, and cognitive decline. It has a prevalence of about 1 in 10,000 individuals [22]. The main sites of pathology are the medium spiny neurons of the striatum, which comprises the caudate and putamen basal ganglia, located deep within the forebrain. HD is caused by CAG repeat expansion mutation within exon 1 of the *HTT* gene, leading to abnormal polyglutamine formation within the amino terminus of the huntingtin protein, HTT [67]. Unaffected individuals have 17–20 CAG repeats, while HD patients have 36 or more CAG repeats. In addition, the CAG repeat expansion mutation exhibits both intergenerational and somatic instability, which contribute to the development and progression of disease. The mechanism of HD is unknown, although there is evidence for polyglutamine or RNA toxic gains of function and haploinsufficiency or alternative splicing of the *HTT* gene [68,69]. To our knowledge, there are no reports of specific DNA methylation alterations at the *HTT* gene in HD, although genome-wide DNA demethylation has been reported to promote CAG repeat expansion as a more general phenomenon [70,71], and this may have some bearing on HD causality. However, several lines of evidence have recently been put forward to suggest alterations of DNA methylation as a consequence of HD molecular disease mechanisms (Table 1). Firstly, a genome-wide 5mC analysis of mouse striatal neurons expressing HTT, either with or without mutant polyglutamine, identified significant changes of

DNA methylation resulting from the HD mutation, and the authors further identified AP-1 and SOX2 as specific transcriptional regulators associated with the DNA methylation changes [72]. Secondly, increased levels of 5mC and corresponding decreased levels of 5hmC have been identified in the 5" untranslated region (UTR) of the *ADORA2A* gene in the striatum of HD patients compared with unaffected controls [73]. The *ADORA2A* gene encodes the adenosine A_{2A} receptor, a G-protein-coupled receptor that is normally highly expressed in the basal ganglia, but severely reduced in HD [74]. Finally, genome-wide loss of 5hmC has been reported in YAC128 HD mouse brain tissues compared with age-matched wild-type controls [75]. Of the 747 differentially hydroxymethylated regions that were identified in the striatum, 49 showed HD-related increases of 5hmC, enriched in gene bodies and positively correlated with gene transcription, while 698 showed HD-related decreases of 5hmC. The authors suggest that alteration of 5hmC is a novel dynamic DNA methylation feature of HD, involved in abnormal neurogenesis and neuronal function in HD brain. To the best of our knowledge, there are no reports that describe the potential use of DNA methylation as a biomarker for HD, so this would be an interesting area of research to pursue.

7. FRAGILE X-ASSOCIATED TREMOR/ATAXIA SYNDROME

FXTAS (OMIM: 300623) is a late-onset neurodegenerative disease characterized by severe tremor, ataxia, and progressive cognitive decline in the fifth decades of life. FXTAS is caused by CGG trinucleotide repeat expansion within the 5" UTR of the *FMR1* gene [76]. FXTAS patients carry 55–200 CGG repeats, regarded as premutation alleles, whereas individuals who carry over 200 CGG repeats, regarded as full mutations, develop FXS, which is the commonest form of inherited mental retardation [77]. In FXS, the CGG repeat expansion mutation becomes hypermethylated, as does the CGI within the *FMR1* promoter region, resulting in reduced expression of the *FMR1* gene [19]. However, in FXTAS the CGG repeats are unmethylated and there is increased expression of *FMR1*, resulting in a toxic RNA gain of function [78]. To our knowledge, there are no reports of global alterations of 5mC levels or locus-specific alterations of 5hmC, 5fC, or 5caC levels at the *FMR1* gene in either FXS or FXTAS. However, global levels of 5hmC have been investigated in the rCGG mouse model of FXTAS, which is characterized by overexpression of human CGG repeats within the 5" UTR of the *FMR1* gene in Purkinje cells, leading to cell death and mouse behavioral deficits [79]. A genome-wide decrease of 5hmC levels was identified in the cerebellum of rCGG mice compared to controls, mainly within gene bodies and CGIs [80]. However, there were also increases of 5hmC levels in repetitive elements and cerebellum-specific enhancers that correlated with neurodevelopmental genes and transcription factors (Table 1). These findings strongly suggest a role for 5hmC in FXTAS molecular disease mechanisms, and further studies are now required to investigate this as a potential biomarker.

8. FRIEDREICH ATAXIA

FRDA (OMIM: 229300) is a rare autosomal recessive multisystem neurodegenerative disorder, with a prevalence of about 1 in 50,000, characterized by early-onset progressive ataxia. FRDA is caused by GAA repeat expansion mutation within intron 1 of the *FXN* gene, leading to decreased expression of the essential mitochondrial protein frataxin [81]. The main sites of pathology are the large sensory neurons of the dorsal root ganglia and the dentate nucleus of the cerebellum [82], but there are also non-CNS pathologies, including hypertrophic cardiomyopathy [83] and diabetes [84]. To our knowledge, there are no reports of global DNA methylation studies for FRDA, but several studies have identified consistent locus-specific DNA methylation changes of the *FXN* gene (Table 1) [85,86]. For example, increased 5mC levels have been identified at specific CpG sites upstream of the GAA repeat in FRDA patient lymphoblasts [87], fibroblasts, induced pluripotent (iPS) cells and iPS-derived neuronal cells [88], blood and buccal cells [89,90], and FRDA patient and FRDA mouse model cerebellum and heart tissues [91]. Interestingly, decreased 5mC levels have also been reported in the downstream GAA repeat region of FRDA patient tissues [88,90,91]. In the studies of blood and buccal cells from FRDA patients, an inverse correlation has been reported between the 5mC level in the upstream GAA region and the level of *FXN* expression [90]. In addition, there is a direct correlation between the 5mC level in the upstream GAA repeat region and the length of the GAA repeats and an inverse correlation with the age of disease onset [89]. Therefore, there is some evidence that allows 5mC to be considered as a potential biomarker for FRDA (Table 2). However, more recent analysis of the upstream GAA repeat region in FRDA cerebellum and heart tissues, using methodology that distinguishes between 5hmC and 5mC, has shown that the majority of the hypermethylated DNA at one particular CpG residue comprises 5hmC rather than 5mC [92] (Table 1). Therefore, it will be interesting to further investigate the alterations of 5mC and 5hmC levels within blood and buccal cells, within other regions of the *FXN* gene, and within other genes that may be associated with FRDA.

9. SPINOCEREBELLAR ATAXIA 7

SCA7 (OMIM: 164500) is an autosomal dominant neurodegenerative disorder characterized by adult-onset of progressive cerebellar ataxia and macular dystrophy. It is caused by CAG trinucleotide repeat expansion mutation within the *ATXN7* gene. Normal *ATXN7* alleles have 4–35 CAG repeats, whereas pathologic alleles have from 37 to approximately 200 CAG repeats, which encode an abnormal polyglutamine tract within the ataxin-7 protein [93]. Mutated ataxin-7 protein forms nuclear inclusions and binds the nuclear cone-rod homeobox (CRX) transcription factor, resulting in reduced CRX activity and subsequent retinal degeneration [94]. Intergenerational and somatic CAG repeat instabilities have been identified and are thought to play important roles in SCA7 anticipation and disease progression, respectively. Therefore, global DNA methylation changes may impact upon CAG repeat instability and subsequent SCA7 disease in a similar manner to that proposed for the CAG repeats of HD and other polyglutamine disorders [70,71]. However, in studies of SCA7 transgenic mice, hypermethylation of CTCF-binding sites that flank the CAG repeat disrupts CTCF-binding and produces dramatic enhancement of somatic CAG repeat instability [95]. Therefore, a hypothetical reduction of DNA methylation at this specific locus may actually reduce CAG repeat expansion and ameliorate SCA7 disease progression (Table 1). Our understanding of such conflicting global and locus-specific DNA methylation effects in SCA7 will require further investigation, particularly if consideration is to be given to DNA methylation as a potential biomarker of disease.

10. CONCLUSION

DNA methylation clearly has significant involvement in a number of neurodegenerative diseases. However, further studies are needed to determine whether the observed alterations in DNA methylation are the causes or consequences of each primary disease mutation. Also, revisions of previous 5mC studies may be necessary to take into account the more recently identified 5hmC, 5fC, and 5caC DNA modifications. Thus, where a neurodegenerative disease-relevant gene has previously been shown to be hypermethylated in a specific brain tissue and associated with decreased expression of the gene in that tissue, it may now be necessary to investigate how much of the methylated DNA is actually 5mC, associated with gene silencing and how much is 5hmC, 5fC, or 5caC, associated with processes of demethylation or other functions as yet unknown. Of the studies reported to date, no common features of DNA methylation alterations are apparent for all neurodegenerative diseases. Both global increases and decreases of 5mC and 5hmC levels have been identified in different diseases, and for AD, both increases and decreases are seen within the same disease. It is possible that differences between selected tissues or cell populations together with poorly reproducible techniques may have contributed to this lack of clarity. Therefore, standardization of tissue cell samples and more accurate advanced technologies may provide better understanding in future.

A major impetus to perform the recent DNA methylation studies for neurodegenerative diseases has been to identify potential biomarkers of disease progression. Some insight has been obtained from locus-specific DNA methylation analysis of some diseases. For example, analysis of blood and buccal cells from FRDA patients has revealed a direct correlation between 5mC levels and GAA repeat lengths and inverse correlations between 5mC levels and both *FXN* expression and age of disease onset [89,90]. These studies suggest that DNA methylation in blood and buccal cells may be a useful biomarker of FRDA disease, although current data suggest that any tests for such a biomarker would lack both sensitivity and specificity (Table 2). Similarly for PD, one study has detected decreased 5mC levels of the *SNCA* gene in blood samples from sporadic PD patients compared with controls [55]. However, the average 5mC decrease is minimally significant and, therefore, such a biomarker would lack both sensitivity and specificity (Table 2). In addition, other studies of PD have revealed no correlation between decreased levels of 5mC at the *SNCA* gene in the brain and blood [54], indicating that further studies are required for clarification of DNA methylation as a potential biomarker for PD. A similar situation exists for AD, where one study has reported increased DNA methylation of the *SIRT1* gene and decreased DNA methylation of the *APP* gene in AD patient blood samples compared with controls [48]. However, sample numbers were low and there are significant overlaps between the levels of DNA methylation that were detected in AD patients and controls, indicating that such tests would lack sensitivity and selectivity as biomarkers (Table 2). In contrast, studies of familial ALS cases that are due to the *C9ORF72* hexanucleotide repeat have identified significant increases in the DNA methylation status of ALS patients compared with controls, associated with a more aggressive ALS disease [66]. This suggests that DNA methylation levels of the *C9ORF72* 5′ repeat region in blood samples may be a useful biomarker to pursue as an indicator of ALS disease severity, with tests being both sensitive and specific (Table 2). However, studies of sporadic ALS have revealed no correlation between global alterations of 5mC or 5hmC in spinal cord and blood [62], suggesting that such measurements of DNA methylation status would not be suitable as biomarkers for this neurodegenerative disease.

Another driver behind 5hmC studies of neurodegenerative disease is the potential to identify novel targets for therapy. Thus, if 5mC levels are globally increased in a neurodegenerative disease, it may be possible to consider treatment with DNMT inhibitors, such as 5-azacytidine (5-aza-CR or Vidaza), 5-aza-2′-deoxycytidine (5-aza-CdR or decitabine), and zebularine [96]. Also, if 5hmC proves to be a significant epigenetic mark of neurodegenerative disease, then it may be possible to develop drugs that modify the 5hmC status, either to decrease 5hmC levels by inhibiting TET activity [97] or to increase 5hmC levels by enhancing TET activity [98]. However, such approaches would have to proceed with great caution, because global hypomethylation of 5mC or 5hmC could induce severe adverse effects, such as cancer [99]. However, further investigations of 5mC and 5hmC, as well as 5fC, 5caC, and mCH, are first required for all neurodegenerative diseases, both at global and locus-specific levels. Such studies will be enhanced by the continued development of state-of-the-art technologies, including third-generation sequencing and array-based hybridization platforms [100].

LIST OF ABBREVIATIONS

5caC 5-Carboxylcytosine
5fC 5-Formylcytosine
5hmC 5-Hydroxymethylcytosine
5mC 5-Methylcytosine
AD Alzheimer's disease
ALS Amyotrophic lateral sclerosis
BER Base excision repair
CNS Central nervous system
CTCF CCTC-binding factor
DNMT DNA methyltransferase
FRDA Friedreich ataxia
FXS Fragile X syndrome
FXTAS Fragile X-associated tremor/ataxia syndrome
HD Huntington's disease
mCH Methyl cytosine H, where H is adenine, cytosine, or thymine
MeCP2 Methyl-CpG-binding proteins 2
MBD1, MBD2, and MBD4 Methyl-CpG-binding domain proteins 1, 2, and 4
PD Parkinson's disease
SCA7 Spinocerebellar ataxia type 7
TDG Thymine DNA glycosylase
TET Ten-eleven translocation
UHRF1 Ubiquitin-like, containing PHD and RING finger domain 1

References

[1] Robertson KD. DNA methylation, methyltransferases, and cancer. Oncogene 2001;20:3139–55.
[2] Pradhan S, Bacolla A, Wells RD, Roberts RJ. Recombinant human DNA (cytosine-5) methyltransferase. I. Expression, purification, and comparison of de novo and maintenance methylation. J Biol Chem 1999; 274:33002–10.
[3] Okano M, Bell DW, Haber DA, Li E. DNA methyltransferases Dnmt3a and Dnmt3b are essential for de novo methylation and mammalian development. Cell 1999;99:247–57.
[4] Jones PA, Liang G. Rethinking how DNA methylation patterns are maintained. Nat Rev Genet 2009;10:805–11.
[5] Bird AP, Wolffe AP. Methylation-induced repression–belts, braces, and chromatin. Cell 1999;99:451–4.
[6] Kriaucionis S, Bird A. DNA methylation and Rett syndrome. Hum Mol Genet 2003;12(Spec No 2):R221–7.
[7] Lorincz MC, Dickerson DR, Schmitt M, Groudine M. Intragenic DNA methylation alters chromatin structure and elongation efficiency in mammalian cells. Nat Struct Mol Biol 2004;11:1068–75.
[8] Jones PA. Functions of DNA methylation: islands, start sites, gene bodies and beyond. Nat Rev Genet 2012;13:484–92.
[9] Hitchins MP, Rapkins RW, Kwok CT, Srivastava S, Wong JJ, Khachigian LM, et al. Dominantly inherited constitutional epigenetic silencing of MLH1 in a cancer-affected family is linked to a single nucleotide variant within the 5′UTR. Cancer Cell 2011;20:200–13.
[10] Kriaucionis S, Heintz N. The nuclear DNA base 5-hydroxymethylcytosine is present in Purkinje neurons and the brain. Science 2009;324:929–30.
[11] Tahiliani M, Koh KP, Shen Y, Pastor WA, Bandukwala H, Brudno Y, et al. Conversion of 5-methylcytosine to 5-hydroxymethylcytosine in mammalian DNA by MLL partner TET1. Science 2009;324:930–5.
[12] Ito S, Shen L, Dai Q, Wu SC, Collins LB, Swenberg JA, et al. Tet proteins can convert 5-methylcytosine to 5-formylcytosine and 5-carboxylcytosine. Science 2011;333:1300–3.
[13] Guo JU, Su Y, Zhong C, Ming GL, Song H. Hydroxylation of 5-methylcytosine by TET1 promotes active DNA demethylation in the adult brain. Cell 2011;145: 423–34.
[14] Szulwach KE, Li X, Li Y, Song CX, Wu H, Dai Q, et al. 5-hmC-mediated epigenetic dynamics during postnatal neurodevelopment and aging. Nat Neurosci 2011;14:1607–16.
[15] Robertson KD. DNA methylation and human disease. Nat Rev Genet 2005;6:597–610.
[16] Amir RE, Van den Veyver IB, Wan M, Tran CQ, Francke U, Zoghbi HY. Rett syndrome is caused by mutations in X-linked MECP2, encoding methyl-CpG-binding protein 2. Nat Genet 1999;23:185–8.
[17] Ehrlich M. The ICF syndrome, a DNA methyltransferase 3B deficiency and immunodeficiency disease. Clin Immunol 2003;109:17–28.
[18] Kulis M, Esteller M. DNA methylation and cancer. Adv Genet 2010;70:27–56.

[19] Naumann A, Hochstein N, Weber S, Fanning E, Doerfler W. A distinct DNA-methylation boundary in the 5′-upstream sequence of the FMR1 promoter binds nuclear proteins and is lost in fragile X syndrome. Am J Hum Genet 2009;85:606–16.
[20] Tucker KL. Methylated cytosine and the brain: a new base for neuroscience. Neuron 2001;30:649–52.
[21] Cheng Y, Bernstein A, Chen D, Jin P. 5-Hydroxymethylcytosine: a new player in brain disorders? Exp Neurol 2014;268:3–9.
[22] Lu H, Liu X, Deng Y, Qing H. DNA methylation, a hand behind neurodegenerative diseases. Front Aging Neurosci 2013;5:85.
[23] Inano K, Suetake I, Ueda T, Miyake Y, Nakamura M, Okada M, et al. Maintenance-type DNA methyltransferase is highly expressed in post-mitotic neurons and localized in the cytoplasmic compartment. J Biochem 2000;128:315–21.
[24] Wu H, Coskun V, Tao J, Xie W, Ge W, Yoshikawa K, et al. Dnmt3a-dependent nonpromoter DNA methylation facilitates transcription of neurogenic genes. Science 2010;329:444–8.
[25] Branco MR, Ficz G, Reik W. Uncovering the role of 5-hydroxymethylcytosine in the epigenome. Nat Rev Genet 2011;13:7–13.
[26] Globisch D, Munzel M, Muller M, Michalakis S, Wagner M, Koch S, et al. Tissue distribution of 5-hydroxymethylcytosine and search for active demethylation intermediates. PLoS One 2010;5:e15367.
[27] Chen Y, Damayanti NP, Irudayaraj J, Dunn K, Zhou FC. Diversity of two forms of DNA methylation in the brain. Front Genet 2014;5:46.
[28] Lister R, Mukamel EA, Nery JR, Urich M, Puddifoot CA, Johnson ND, et al. Global epigenomic reconfiguration during mammalian brain development. Science 2013;341:1237905.
[29] Song CX, Szulwach KE, Fu Y, Dai Q, Yi C, Li X, et al. Selective chemical labeling reveals the genome-wide distribution of 5-hydroxymethylcytosine. Nat Biotechnol 2011;29:68–72.
[30] Wang T, Pan Q, Lin L, Szulwach KE, Song CX, He C, et al. Genome-wide DNA hydroxymethylation changes are associated with neurodevelopmental genes in the developing human cerebellum. Hum Mol Genet 2012; 21:5500–10.
[31] Chen H, Dzitoyeva S, Manev H. Effect of aging on 5-hydroxymethylcytosine in the mouse hippocampus. Restor Neurol Neurosci 2012;30:237–45.
[32] Wheldon LM, Abakir A, Ferjentsik Z, Dudnakova T, Strohbuecker S, Christie D, et al. Transient accumulation of 5-carboxylcytosine indicates involvement of active demethylation in lineage specification of neural stem cells. Cell Rep 2014;7:1353–61.
[33] Tanzi RE. The genetics of Alzheimer disease. Cold Spring Harb Perspect Med 2012;2.
[34] Bekris LM, Yu CE, Bird TD, Tsuang DW. Genetics of Alzheimer disease. J Geriatr Psychiatry Neurol 2010;23:213–27.
[35] Mastroeni D, Grover A, Delvaux E, Whiteside C, Coleman PD, Rogers J. Epigenetic mechanisms in Alzheimer's disease. Neurobiol Aging 2011;32:1161–80.
[36] Bihaqi SW, Schumacher A, Maloney B, Lahiri DK, Zawia NH. Do epigenetic pathways initiate late onset Alzheimer disease (LOAD): towards a new paradigm. Curr Alzheimer Res 2012;9:574–88.
[37] Coppieters N, Dragunow M. Epigenetics in Alzheimer's disease: a focus on DNA modifications. Curr Pharm Des 2011;17:3398–412.
[38] Wang SC, Oelze B, Schumacher A. Age-specific epigenetic drift in late-onset Alzheimer's disease. PLoS One 2008;3:e2698.
[39] Mastroeni D, Grover A, Delvaux E, Whiteside C, Coleman PD, Rogers J. Epigenetic changes in Alzheimer's disease: decrements in DNA methylation. Neurobiol Aging 2010;31:2025–37.
[40] Chouliaras L, Mastroeni D, Delvaux E, Grover A, Kenis G, Hof PR, et al. Consistent decrease in global DNA methylation and hydroxymethylation in the hippocampus of Alzheimer's disease patients. Neurobiol Aging 2013;34:2091–9.
[41] Rao JS, Keleshian VL, Klein S, Rapoport SI. Epigenetic modifications in frontal cortex from Alzheimer's disease and bipolar disorder patients. Transl Psychiatry 2012;2:e132.
[42] Bradley-Whitman MA, Lovell MA. Epigenetic changes in the progression of Alzheimer's disease. Mech Ageing Dev 2013;134:486–95.
[43] Coppieters N, Dieriks BV, Lill C, Faull RL, Curtis MA, Dragunow M. Global changes in DNA methylation and hydroxymethylation in Alzheimer's disease human brain. Neurobiol Aging 2014;35: 1334–44.
[44] Barrachina M, Ferrer I. DNA methylation of Alzheimer disease and tauopathy-related genes in postmortem brain. J Neuropathol Exp Neurol 2009; 68:880–91.
[45] Brohede J, Rinde M, Winblad B, Graff C. A DNA methylation study of the amyloid precursor protein gene in several brain regions from patients with familial Alzheimer disease. J Neurogenet 2010; 24:179–81.
[46] Rogaev EI, Lukiw WJ, Lavrushina O, Rogaeva EA, St George-Hyslop PH. The upstream promoter of the beta-amyloid precursor protein gene (APP) shows differential patterns of methylation in human brain. Genomics 1994;22:340–7.

[47] Tohgi H, Utsugisawa K, Nagane Y, Yoshimura M, Genda Y, Ukitsu M. Reduction with age in methylcytosine in the promoter region-224 approximately-101 of the amyloid precursor protein gene in autopsy human cortex. Brain Res Mol Brain Res 1999;70:288–92.

[48] Hou Y, Chen H, He Q, Jiang W, Luo T, Duan J, et al. Changes in methylation patterns of multiple genes from peripheral blood leucocytes of Alzheimer's disease patients. Acta Neuropsychiatr 2012;25:66–76.

[49] Obeso JA, Rodriguez-Oroz MC, Goetz CG, Marin C, Kordower JH, Rodriguez M, et al. Missing pieces in the Parkinson's disease puzzle. Nat Med 2010;16:653–61.

[50] Yang YX, Wood NW, Latchman DS. Molecular basis of Parkinson's disease. Neuroreport 2009;20:150–6.

[51] Jowaed A, Schmitt I, Kaut O, Wullner U. Methylation regulates alpha-synuclein expression and is decreased in Parkinson's disease patients' brains. J Neurosci 2010;30:6355–9.

[52] Matsumoto L, Takuma H, Tamaoka A, Kurisaki H, Date H, Tsuji S, et al. CpG demethylation enhances alpha-synuclein expression and affects the pathogenesis of Parkinson's disease. PLoS One 2010;5:e15522.

[53] Desplats P, Spencer B, Coffee E, Patel P, Michael S, Patrick C, et al. Alpha-synuclein sequesters Dnmt1 from the nucleus: a novel mechanism for epigenetic alterations in Lewy body diseases. J Biol Chem 2011;286:9031–7.

[54] Richter J, Appenzeller S, Ammerpohl O, Deuschl G, Paschen S, Bruggemann N, et al. No evidence for differential methylation of alpha-synuclein in leukocyte DNA of Parkinson's disease patients. Mov Disord 2012;27:590–1.

[55] Ai SX, Xu Q, Hu YC, Song CY, Guo JF, Shen L, et al. Hypomethylation of SNCA in blood of patients with sporadic Parkinson's disease. J Neurol Sci 2014;337:123–8.

[56] Zhang B, Lv Q, Pu J, Lei X, Mao Y. Distribution of 5-hydroxymethylcytosine in rat brain tissues and the role of 5-hydroxymethylcytosine in aging and vivo model of Parkinson's disease. Mov Disord 2013;28:852.

[57] Martin LJ, Wong M. Aberrant regulation of DNA methylation in amyotrophic lateral sclerosis: a new target of disease mechanisms. Neurotherapeutics 2013;10:722–33.

[58] Calvo AC, Manzano R, Mendonca DM, Munoz MJ, Zaragoza P, Osta R. Amyotrophic lateral sclerosis: a focus on disease progression. Biomed Res Int 2014;2014:925101.

[59] Chen S, Sayana P, Zhang X, Le W. Genetics of amyotrophic lateral sclerosis: an update. Mol Neurodegener 2013;8:28.

[60] Ahmed A, Wicklund MP. Amyotrophic lateral sclerosis: what role does environment play? Neurol Clin 2011;29:689–711.

[61] Morahan JM, Yu B, Trent RJ, Pamphlett R. A genome-wide analysis of brain DNA methylation identifies new candidate genes for sporadic amyotrophic lateral sclerosis. Amyotroph Lateral Scler 2009;10:418–29.

[62] Figueroa-Romero C, Hur J, Bender DE, Delaney CE, Cataldo MD, Smith AL, et al. Identification of epigenetically altered genes in sporadic amyotrophic lateral sclerosis. PLoS One 2012;7:e52672.

[63] Oates N, Pamphlett R. An epigenetic analysis of SOD1 and VEGF in ALS. Amyotroph Lateral Scler 2007;8:83–6.

[64] DeJesus-Hernandez M, Mackenzie IR, Boeve BF, Boxer AL, Baker M, Rutherford NJ, et al. Expanded GGGGCC hexanucleotide repeat in noncoding region of C9ORF72 causes chromosome 9p-linked FTD and ALS. Neuron 2011;72:245–56.

[65] Renton AE, Majounie E, Waite A, Simon-Sanchez J, Rollinson S, Gibbs JR, et al. A hexanucleotide repeat expansion in C9ORF72 is the cause of chromosome 9p21-linked ALS-FTD. Neuron 2011;72:257–68.

[66] Xi Z, Zinman L, Moreno D, Schymick J, Liang Y, Sato C, et al. Hypermethylation of the CpG island near the G4C2 repeat in ALS with a C9orf72 expansion. Am J Hum Genet 2013;92:981–9.

[67] MacDonald ME, Gines S, Gusella JF, Wheeler VC. Huntington's disease. Neuromol Med 2003;4:7–20.

[68] Landles C, Bates GP. Huntingtin and the molecular pathogenesis of Huntington's disease. Fourth in molecular medicine review series. EMBO Rep 2004;5:958–63.

[69] Sathasivam K, Neueder A, Gipson TA, Landles C, Benjamin AC, Bondulich MK, et al. Aberrant splicing of HTT generates the pathogenic exon 1 protein in Huntington disease. Proc Natl Acad Sci USA 2013;110:2366–70.

[70] Dion V, Lin Y, Price BA, Fyffe SL, Seluanov A, Gorbunova V, et al. Genome-wide demethylation promotes triplet repeat instability independently of homologous recombination. DNA Repair (Amst) 2008;7:313–20.

[71] Gorbunova V, Seluanov A, Mittelman D, Wilson JH. Genome-wide demethylation destabilizes CTG.CAG trinucleotide repeats in mammalian cells. Hum Mol Genet 2004;13:2979–89.

[72] Ng CW, Yildirim F, Yap YS, Dalin S, Matthews BJ, Velez PJ, et al. Extensive changes in DNA methylation are associated with expression of mutant huntingtin. Proc Natl Acad Sci USA 2013;110:2354–9.

[73] Villar-Menendez I, Blanch M, Tyebji S, Pereira-Veiga T, Albasanz JL, Martin M, et al. Increased 5-methylcytosine and decreased 5-hydroxymethylcytosine levels are associated with reduced striatal A2AR levels in Huntington's disease. Neuromol Med 2013;15:295–309.

[74] Glass M, Dragunow M, Faull RL. The pattern of neurodegeneration in Huntington's disease: a comparative study of cannabinoid, dopamine, adenosine and GABA(A) receptor alterations in the human basal ganglia in Huntington's disease. Neuroscience 2000;97:505–19.
[75] Wang F, Yang Y, Lin X, Wang JQ, Wu YS, Xie W, et al. Genome-wide loss of 5-hmC is a novel epigenetic feature of Huntington's disease. Hum Mol Genet 2013;22:3641–53.
[76] Hagerman RJ, Leehey M, Heinrichs W, Tassone F, Wilson R, Hills J, et al. Intention tremor, parkinsonism, and generalized brain atrophy in male carriers of fragile X. Neurology 2001;57:127–30.
[77] Brouwer JR, Willemsen R, Oostra BA. The FMR1 gene and fragile X-associated tremor/ataxia syndrome. Am J Med Genet B Neuropsychiatr Genet 2009;150B:782–98.
[78] Jacquemont S, Hagerman RJ, Leehey M, Grigsby J, Zhang L, Brunberg JA, et al. Fragile X premutation tremor/ataxia syndrome: molecular, clinical, and neuroimaging correlates. Am J Hum Genet 2003;72:869–78.
[79] Hashem V, Galloway JN, Mori M, Willemsen R, Oostra BA, Paylor R, et al. Ectopic expression of CGG containing mRNA is neurotoxic in mammals. Hum Mol Genet 2009;18:2443–51.
[80] Yao B, Lin L, Street RC, Zalewski ZA, Galloway JN, Wu H, et al. Genome-wide alteration of 5-hydroxymethylcytosine in a mouse model of fragile X-associated tremor/ataxia syndrome. Hum Mol Genet 2014;23:1095–107.
[81] Campuzano V, Montermini L, Molto MD, Pianese L, Cossee M, Cavalcanti F, et al. Friedreich's ataxia: autosomal recessive disease caused by an intronic GAA triplet repeat expansion. Science 1996;271:1423–7.
[82] Koeppen AH. Nikolaus Friedreich and degenerative atrophy of the dorsal columns of the spinal cord. J Neurochem 2013;126(Suppl. 1):4–10.
[83] Weidemann F, Stork S, Liu D, Hu K, Herrmann S, Ertl G, et al. Cardiomyopathy of Friedreich ataxia. J Neurochem 2013;126(Suppl. 1):88–93.
[84] Cnop M, Mulder H, Igoillo-Esteve M. Diabetes in Friedreich ataxia. J Neurochem 2013;126(Suppl. 1): 94–102.
[85] Evans-Galea MV, Hannan AJ, Carrodus N, Delatycki MB, Saffery R. Epigenetic modifications in trinucleotide repeat diseases. Trends Mol Med 2013;19:655–63.
[86] Sandi C, Sandi M, Anjomani Virmouni S, Al-Mahdawi S, Pook MA. Epigenetic-based therapies for Friedreich ataxia. Front Genet 2014;5:165.
[87] Greene E, Mahishi L, Entezam A, Kumari D, Usdin K. Repeat-induced epigenetic changes in intron 1 of the frataxin gene and its consequences in Friedreich ataxia. Nucleic Acids Res 2007;35:3383–90.
[88] Soragni E, Miao W, Iudicello M, Jacoby D, De Mercanti S, Clerico M, et al. Epigenetic therapy for Friedreich ataxia. Ann Neurol 2014;76:489–508.
[89] Castaldo I, Pinelli M, Monticelli A, Acquaviva F, Giacchetti M, Filla A, et al. DNA methylation in intron 1 of the frataxin gene is related to GAA repeat length and age of onset in Friedreich ataxia patients. J Med Genet 2008;45:808–12.
[90] Evans-Galea MV, Carrodus N, Rowley SM, Corben LA, Tai G, Saffery R, et al. FXN methylation predicts expression and clinical outcome in Friedreich ataxia. Ann Neurol 2012;71:487–97.
[91] Al-Mahdawi S, Pinto RM, Ismail O, Varshney D, Lymperi S, Sandi C, et al. The Friedreich ataxia GAA repeat expansion mutation induces comparable epigenetic changes in human and transgenic mouse brain and heart tissues. Hum Mol Genet 2008;17:735–46.
[92] Al-Mahdawi S, Sandi C, Mouro Pinto R, Pook MA. Friedreich ataxia patient tissues exhibit increased 5-hydroxymethylcytosine modification and decreased CTCF binding at the FXN locus. PLoS One 2013;8: e74956.
[93] David G, Abbas N, Stevanin G, Durr A, Yvert G, Cancel G, et al. Cloning of the SCA7 gene reveals a highly unstable CAG repeat expansion. Nat Genet 1997;17:65–70.
[94] La Spada AR, Fu YH, Sopher BL, Libby RT, Wang X, Li LY, et al. Polyglutamine-expanded ataxin-7 antagonizes CRX function and induces cone-rod dystrophy in a mouse model of SCA7. Neuron 2001;31:913–27.
[95] Libby RT, Hagerman KA, Pineda VV, Lau R, Cho DH, Baccam SL, et al. CTCF cis-regulates trinucleotide repeat instability in an epigenetic manner: a novel basis for mutational hot spot determination. PLoS Genet 2008;4:e1000257.
[96] Sandi C, Al-Mahdawi S, Pook MA. Epigenetics in Friedreich's ataxia: challenges and opportunities for therapy. Genet Res Int 2013;2013:12.
[97] Xiao M, Yang H, Xu W, Ma S, Lin H, Zhu H, et al. Inhibition of alpha-KG-dependent histone and DNA demethylases by fumarate and succinate that are accumulated in mutations of FH and SDH tumor suppressors. Genes Dev 2012;26:1326–38.
[98] Yin R, Mao SQ, Zhao B, Chong Z, Yang Y, Zhao C, et al. Ascorbic acid enhances Tet-mediated 5-methylcytosine oxidation and promotes DNA demethylation in mammals. J Am Chem Soc 2013;135:10396–403.
[99] Ehrlich M, Lacey M. DNA hypomethylation and hemimethylation in cancer. Adv Exp Med Biol 2013; 754:31–56.
[100] Plongthongkum N, Diep DH, Zhang K. Advances in the profiling of DNA modifications: cytosine methylation and beyond. Nat Rev Genet 2014;15:647–61.

CHAPTER

21

The Histone Code and Disease: Posttranslational Modifications as Potential Prognostic Factors for Clinical Diagnosis

Nicolas G. Simonet, George Rasti, Alejandro Vaquero

Chromatin Biology Laboratory, Cancer Epigenetics and Biology Program (PEBC), Bellvitge Biomedical Research Institute (IDIBELL), Barcelona, Spain

OUTLINE

Epigenetic Biomarkers and Diagnostics
http://dx.doi.org/10.1016/B978-0-12-801899-6.00021-8

1. INTRODUCTION

Eukaryotic genomes are packed with histones and accessory proteins that, when combined with DNA, form chromatin. The basic unit of chromatin, the *nucleosome*, comprises two copies of each core histone protein (H2A, H2B, H3, and H4) assembled into an octamer containing a total of 145–147 base pairs of DNA [1]. Electron micrographs have elucidated the basic organization of nucleosomal arrays, which adopt a "beads on a string" structure known as the *11-nm fiber* (i.e., a nucleosome with a diameter of 11 nm) [2]. However, the next level of chromatin compaction is less understood. In vitro studies suggest the existence of a compacted structure, tentatively named the *30-nm fiber*, which would be stabilized by the binding of the linker histone H1 and whose DNA would be condensed sixfold relative to the 11-nm fiber. However, to date, no clear evidence of the 30-nm fiber has been found in vivo.

Histones are among the most highly conserved proteins in eukaryotes. Each core histone of the histone octamer adopts a similar secondary structure in its globular domain: the *histone fold*, a three-helix motif that heterodimerizes to enable histone–histone interactions, by forming a handshake-like structure between H2A-H2B and H3-H4. Additionally, there are two small unstructured domains that protrude from the globular domain: an *N-terminal domain*, comprising 20–35 (lysine- and arginine-rich) residues and a *C-terminal domain* of histone H2A, which contains a ladle-shaped docking domain that is important for the interaction between H2A/H2B dimer and the H3/4 tetramer and that strongly contributes to the nucleosomal surface [1]. In addition to the *canonical histones* (i.e., the most frequent isoform of each histone), eukaryotic cells posses several histone variants that are implicated in the regulation of cellular processes such as DNA repair and transcriptional activity [3].

Interestingly, albeit most histone posttranslational modifications (PTMs) occur in the N- and C-terminal domains of histones, many occur in the globular domain. The countless possible PTMs of histones provide a complex regulatory program with features of a sophisticated language, also known as histone code that enables cross-talk between different modifications (Figure 1). This epigenetic network is fundamental in normal development and in pathogenesis.

1.1 Histone Modifications and Their Functional Significance

Histone modifications were first discovered 50 years ago by Vincent Allfrey and colleagues, who found acetylation and methylation of histones [4]. They already hypothesized that these modifications might have a functional influence on gene expression [4]. Since their pioneering work, remarkable efforts have been made to identify additional histone PTMs, study their functional relevance, and ascertain their implications for pathogenesis. Here, we summarize the best-known PTMs of histones, provide an overview of their respective histone-modifying enzymes and discuss their significance in chromatin structure and function.

1.2 Histone Acetylation

Histone acetylation occurs at specific lysine residues. Neutralization of the positive charge of the targeted lysine leads to destabilization of the nucleosome, resulting in an open chromatin architecture that is conventionally associated with active gene expression [5]. Histone acetylation can provide a binding site for regulatory proteins that are involved in gene activation. These proteins are members of the bromodomain-containing family of proteins that, upon recognizing (*reading*) modified lysine residues within histone and nonhistone proteins, lead to a cascade of additional modifications implicated in transcriptional coactivation [6].

There are two principal groups of enzymes involved in maintaining the balance between

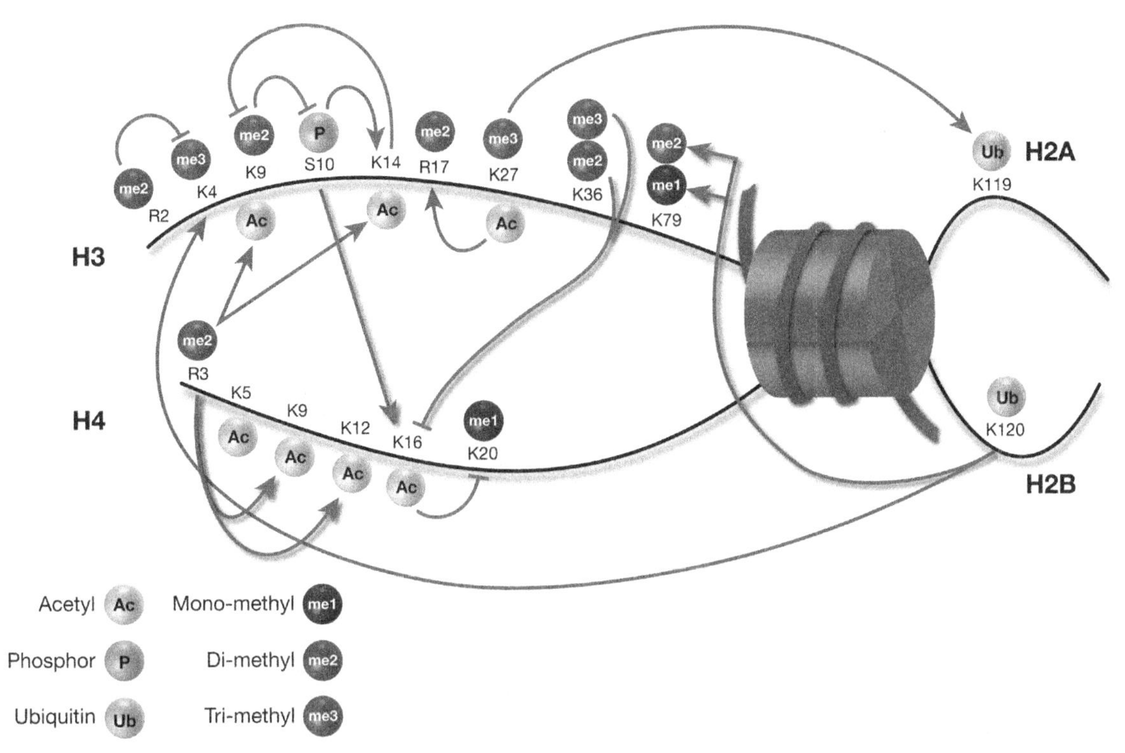

FIGURE 1 **Cross talk between histone modifications.** Positive and negative interplays described between PTMs. Such cross talk can take place on the same histone tail (cis cross talk) or between different histone tails (trans-tail cross talk). The arrow heads indicate positive effects and flat heads indicate negative effects.

protein acetylation and deacetylation: *histone acetyltransferases* (HATs), which transfer the acetyl group from acetyl CoA to the ε-NH_2 group of lysine residues, and *histone deacetylases* (HDACs), which remove the acetyl group [7] (Table 1).

HATs have been classified into two different ways. The earliest system divided HATs into two classes according to their cellular localization: *Type A HATs*, which are nuclear enzymes that have similar roles in transcription-related processes [8], and *Type B HATs*, which are mostly cytoplasmic and whose characteristic activity is acetylation of newly synthesized histones H4 (K5, K12) and H3 [9]. In the latter and (now) more widely accepted classification, HATs are divided into several groups based on phylogeny (namely, their homology with yeast proteins) and on their catalytic domain: *GNAT (Gcn5-related N-acetyltransferase)-related HATs* (Gcn5 and PCAF); *MYST (**M**OZ (monocytic leukemia zinc-finger protein), **Y**bf2 (yeast binding factor 2)/Sas3 (Something about silencing 3), **S**as2, **T**ip60 (Tat interactive protein-60))-related HATs*; and the *p300/CBP family* (cAMP response element-binding protein (CREB)-binding protein) [8].

Similarly to HATs, HDACs also have been classified by phylogeny into four groups: *Class I*, *Class II*, and *Class IV* HDACs are zinc (Zn)-dependent aminohydrolases, whereas *Class III* HDACs (the *sirtuins*) are NAD^+-dependent enzymes. Class I HDACs (HDAC1, 2, 3, and 8) are predominantly nuclear, exhibit homology to the yeast protein Rpd3, and are ubiquitously

TABLE 1 Histone Acetyltransferases (HATs) and Histone Deacetylases (HDACs)

Group	Enzyme	Remarks
HATs	HAT1	DNA replication and repair.
GNC5/PCAF family	Gcn5	Malfunction leads to diabetes, aging problems, and cancer.
	PCAF	PCAF knockout mice exhibit memory impairment, psychological anxiety, and defective stress control.
MYST family	MOF	Essential for embryonic stem cells and for oncogenesis.
	HBO1	Regulates prereplication complex assembly, transcription regulation, and cellular stress response; coactivator of TP53 transcription and represses AR-mediated transcription.
	MOZ	Coactivator of RUNX1 and RUNX2; control of p53 transcriptional activity.
	MORF	Required for RUNX2-dependent transcriptional activation; mutation causes genitopatellar syndrome.
	Tip60	Involved in cell signaling, apoptosis, cell cycle progression, and transcriptional regulation; essential protein for cellular survival and DNA damage.
P300/CBP family	P300	Implicated in cell proliferation, apoptosis, differentiation, cell cycle, and DNA repair.
	CBP	CBP and p300 together are essential for acetylation of H3K18 and of H3K27. Activate or repress transcription.
HDACs	HDAC1	DNA damage control, cell cycle, and differentiation.
Class I	HDAC2	Involved in neurodegeneration and Alzheimer's disease.
	HDAC3	Mitochondrial bioenergetics, lipid metabolism, and immune response.
		Repress c-Myc transcription.
	HDAC8	Knockdown inhibits proliferation of human lung, colon, and cervical cancer cell lines. Involved in Cornelia de Lange syndrome.
Class IIa	HDAC4	Accumulation is critical for ataxia-telangiectasia (A-T) neurodegeneration; blocks autophagy to trigger podocyte injury and might contribute to glomerular lesions and proteinuria.
	HDAC5	Knockdown could inhibit cancer cell proliferation in human hepatocellular carcinoma; negatively regulates sclerostin levels in osteocytes; associated with vascular endothelial dysfunction and cardiovascular diseases.
	HDAC7	Involved in Epstein–Barr virus infection; represses genes responsible for maintaining myeloid lineage potential.
	HDAC9	Associated with large-vessel stroke; its downregulation is associated with lung adenocarcinomas.
Class IIb	HDAC6	Central role in microtubule-dependent cell motility; critical in degradation of misfolded proteins.
	HDAC10	Contributes to transcriptional repression.

TABLE 1 Histone Acetyltransferases (HATs) and Histone Deacetylases (HDACs)—cont'd

Group	Enzyme	Remarks
Class III	SirT1	Tumor suppression, energy homeostasis, autophagy, DNA damage repair, Wnt signaling, PI3K/AKT pathway signaling, hypoxia, NF-κB signaling.
	SirT2	Tumor suppression, cell cycle, chromatin compaction, energy homeostasis.
	SirT3	Cellular energy metabolism.
	SirT4	Tumor suppressor, lipid homeostasis, and downregulates insulin secretion.
	SirT5	Regulator of lysine succinylation in mitochondria; responsible for growth and drug resistance in human nonsmall cell lung cancer.
	SirT6	Regulation of cell cycle, gene expression, glucose homeostasis, genome stability, and DNA damage repair.
	SirT7	Activation of rDNA transcription; contributes to tumorigenesis; regulator of hepatic lipid metabolism and mitochondrial homeostasis.
Class IV	HDAC11	Critical regulator of IL-10 gene expression and immune tolerance.

expressed in human cell lines and tissues [10]. Class II HDACs show homology to the yeast protein Hdac1p and their expression is tissue specific. They can shuttle between the nucleolus and the cytoplasm in response to various triggers, including their binding to MEF2 transcription factors and to cytoplasmic 14-3-3 chaperones. They are subdivided into *Class IIa* (HDAC4, 5, 7, and 9) and *Class IIb* (HDAC6 and 10) based on the double deacetylase domain of HDAC6 and HDAC10 [11]. Class IV is currently known to have only one member, HDAC11, which contains conserved residues in the catalytic region shared by Class I and Class II HDACs.

1.3 Histone Methylation

Histone methylation is a reversible modification of arginine and lysine residues. Unlike acetylation and phosphorylation, histone methylation does not alter the charge of these residues in the histone tail. However, methylation of the ε-amino group in these residues does increase their basicity and hydrophobicity, thereby facilitating selective recruitment of effector proteins (known as *readers*) and/or transcriptions factors to DNA [12].

Methylation in lysine residues can adopt three different states: monomethylation (me1), dimethylation (me2), or trimethylation (me3). In contrast, arginine residues can be monomethylated (me1) or dimethylated, and in turn dimethylation can be present symmetrically (me2s) or asymmetrically (me2a) on their guanidinyl group [12]. In mammals, methylation of histone lysines typically occurs at Lys4, -9, -27, -36, and -79 of histone H3 and at Lys20 of histone H4 [13]. In the case of arginine methylation, the most frequently methylated residues are Arg2, -8, -17, and -26 of histone H3 and Arg3 of histone H4 [14].

The enzymes involved in histone methylation are lysine methyltransferases (KMTs) and protein arginine methyltransferases (PRMTs). KMTs and PRMTs catalyze the transfer of a methyl group from S-adenosyl-L-methionine (SAM) to the ε-amino group of lysine or to the guanidine group of arginine.

KMTs have been classified into two families (Table 2): the *SET-domain family* [15], which encompasses the vast majority of KMTs, and *DOT1* [16]. The SET-domain family is in turn subdivided into eight subfamilies: *SUV39*, *SET1*, *SET2*, *EZ*, *RIZ*, *SMYD*, *SUV4-20*, and *the orphan*

TABLE 2 Histone Methyltransferases and Histone Demethylases

Type and family	Enzyme	Remarks
HISTONE METHYLTRANSFERASES		
SUV39	SUV39H1	Heterochromatin formation and gene transcription regulation.
	SUV39H2	Regulation of telomere length in mammalian cells.
	G9a	Essential role in cocaine-induced plasticity; represses developmental genes; essential for cardiac morphogenesis.
	GLP/EHMT1	Control of brown adipose cell fate and thermogenesis; represses developmental genes; essential for cardiac morphogenesis.
	SETDB1	Contributes to human lung tumorigenesis and melanoma.
	SETDB2	Chromosome condensation and segregation during mitosis.
	SUV420H1	Role in myogenesis.
	SUV420H2	Important for recruitment of cohesin to heterochromatin.
SET1	MLL1	Required for neurogenesis; important for DNA damage repair and disruption of its gene contributes to development of human leukemias.
	MLL2	Mutations of it are a major cause of Kabuki syndrome; coactivator of estrogen receptor.
	MLL3	Contributes to genome-scale circadian transcription; mutations of it are common in colorectal cancer.
	MLL4	Essential for enhancer activation during cell differentiation.
	MLL5	Promotes myogenic differentiation; enriched in euchromatic regions.
	hSET1A	Required for survival of embryonic, epiblast, and neural stem cells and for reprogramming of neural stem cells.
	hSET1B	Essential after gastrulation.
	ASH2l	Key regulator of open chromatin in embryonic stem cells.
SET2	SETD2	Key regulator during transcriptional elongation, splicing, and DNA-mismatch repair. Mutations in specific solid tumors (kidney, lung, and bladder) and in leukemia.
	NSD1	Mutations of its gene are the major cause of Sotos syndrome.
	SMYD1	Transcriptional repressor; required for cardiomyogenesis and sarcomere assembly in skeletal and cardiac muscles.
	SMYD2	Repression of p53 activity.
	SMYD3	Regulator of oncogenic Ras signaling; negative regulator of hepatitis C virus particle production.
EZ	EZH1	Important for development, cell homeostasis, and cancer.
	EZH2	Important for development, cell homeostasis, and cancer; implicated in neurodegeneration in ataxia-telangiectasia.

TABLE 2 Histone Methyltransferases and Histone Demethylases—cont'd

Type and family	Enzyme	Remarks
RIZ	RIZ1	Tumor suppressor.
DOT1	DOT1L	Activation and maintenance of gene transcription; essential for embryonic development and important for normal functions of the hematopoietic system, kidney, and heart.
PRMT	PRMT1	Regulator of ERα.
	PRMT2	Coactivator of the androgen receptor, inhibits NF-κB function and promotes apoptosis.
	PRMT3	Implicated in ribosomal biosynthesis and cancer; regulates hepatic lipogenesis.
	CARM1	Implicated in inflammation apoptosis and cancer.
	PRMT5	Role in various cancers (lung, leukemia, gastric and colorectal cancers, and lymphoma and melanoma).
	PRMT6	Role in cancer (lung and bladder); inhibits HIV-1 replication.
	PRMT7	Regulates cellular response to DNA damage; role in breast cancer.
	PRMT8	Plasma membrane-associated protein.
Others	PR-SET7	Important for genome stability.
	SET7/9	Regulates the activity of p53 and E2F1 upon genotoxic stress.
HISTONE DEMETHYLASES		
LSD demethylases	LSD1	Involved in development, cellular differentiation, embryonic pluripotency, and cancer.
	LSD2	Required for de novo DNA methylation of some imprinted genes in oocytes.
JMJC demethylases	JMJD5	Regulates cancer cell proliferation.
	JMJD6	Regulates macrophage cytokine responses, cell differentiation during embryogenesis, and hematopoietic differentiation.
	FBXL10	Represses the transcription of ribosomal RNA genes and cancer development; regulates ubiquitylation of H2A.
	FBXL11	Regulator of hepatic gluconeogenesis.
	JHDM1D	Tumor suppression via regulation of angiogenesis.
	PHF2	Promotes fat-cell differentiation, bone formation, and control of proinflammatory gene programs.
	PHF8	Cell cycle progression, cytoskeleton dynamics, and neuronal differentiation; causal factor for X-linked mental retardation.
	JMJD1A	Regulation of sex determination in mice; induces cell migration and invasion; has crucial role as tumor suppressor.
	UTX	UTX mutations are exclusively present in male T cell acute lymphoblastic leukemia. Cellular differentiation.

Continued

TABLE 2 Histone Methyltransferases and Histone Demethylases—cont'd

Type and family	Enzyme	Remarks
	JMJD3	Regulation of inflammation and senescence; tumor suppressor.
	JMJD2A	Promotes cellular transformation and is involved in human carcinogenesis; contributes to breast cancer progression.
	JMJD2B	Role in carcinogenesis.
	JMJD2C	Overexpression confers a progrowth effect on colon cancer cells; a HIF-1 coactivator required for breast cancer progression.
	JMJD2D	Stimulates cell proliferation and survival.
	RBP2	Links chronic inflammation to tumor development in gastric epithelial cells; promotes lung tumorigenesis and cancer metastasis.
	JARID1B	Controls mammary gland development; serves as a luminal lineage-driving oncogene in breast cancer.
	JARID1C	Related to X-linked intellectual disability, short stature, and speech delay.
	JARID1D	Role in spermatogenesis.
	JARID2	Regulates activation and differentiation of mouse epidermal stem cells.
	NO66	Myc-induced transcriptional activation; implicated in ribosome biogenesis; important for osteoblast and stem cell differentiation.

members, which comprise *SET7/9* and *SET8* (PR-SET7) [17].

PRMTs are divided into two main classes: *Type I PRMTs* (PRMT1, 2, 3, 4, 6, and 8), which catalyze the formation of both monomethylarginine and asymmetric dimethylarginine, and *Type II PRMTs* (PRMT5 and 7), which catalyze monomethylarginine and symmetric dimethylarginine modifications [18].

Histone demethylation is mediated by two evolutionarily conserved families of histone demethylases. The first family, the *lysine-specific histone demethylases* (LSDs), has two members: *LSD1* and *LSD2*. These factors use flavin adenine dinucleotide (FAD) as coenzyme to catalyze the demethylation. LSD1 demethylates specific mono- and dimethylated lysine residues in histone H3 as well as in nonhistone proteins such as p53, DNMT1, and E2F1. In contrast, LSD2 is highly specific: it only demethylates mono- and dimethyl Lys4 in histone H3 [19]. The second family of histone demethylases is the *Jumonji family*, which contains 30 members that share the catalytic domain Jumonji-C (JMJC). These enzymes catalyze a dioxygenase reaction that depends on Fe(II) and α-ketoglutarate for the demethylation of mono-, di-, and trimethylated residues [20].

Genome-wide studies have shown that histone methylation is enriched at gene promoters, insulators, enhancers, and transcribed regions. Unlike acetylation, methylation can occur in either transcriptional repression or activation, depending on the residue and the number of methyl groups added. For instance, both H3K4me3 and H3K36me3 are found in active genes. In the first case, high levels of H3K4me3 are found in promoters, whereas H3K36me3 is associated with elongating RNA polymerase II. High levels of H3K27me3 and H3K79me3 in promoter and gene-body regions are hallmarks of inactive genes. Interestingly, active promoters are rich in H3K9me1, H3K27me1, H3K36me1, H4K20me1, and H2BK5me1 [21], whereas H3K4me1 is often related to enhancer function [22].

1.4 Histone Phosphorylation

Phosphorylation of histones, like that of other proteins, occurs at serine (Ser), threonine (Thr), and tyrosine (Tyr) residues. It is characterized by ATP-dependent phosphorylation of the hydroxyl group of the target-residue side chain, which increases its negative charge and influences chromatin structure and function. Histone phosphorylation is involved in many cellular processes such as mitosis, meiosis, replication, transcription, and DNA repair [23,24].

Phosphorylation of specific histone residues is reversible: protein kinases (PKs) catalyze the addition of the phosphates groups, whereas protein phosphatases remove them. Activation or inactivation of these kinases is modulated by upstream signaling pathways that lead to cascades of protein phosphorylation, including phosphorylation of histones, and to regulation of nuclear transcription. An interesting example is phosphorylation of histone H2A.X (γ-H2A.X) at Ser139, an event that occurs in all phases of the cell cycle and is involved in DNA damage response mediated by the activation of phosphatidylinositol-3-OH kinases (PI3Ks) such as ATM, ATR, and DNA-PK. The presence of γ-H2A.X induced by upstream signaling pathways defines the boundaries of chromatin domains around the site of DNA breakage. Conversely, protein phosphatase 2 (PP2A) is the main player in dephosphorylation of γ-H2A.X after DNA repair [25]. Another noteworthy example is phosphorylation of histone H3 at Ser10 and Ser28 by aurora kinases (chiefly, aurora-B). The resulting phosphorylated residues associate to chromosome condensation and segregation and each has been commonly used as a reference mark of mitosis and meiosis [26].

1.5 Histone Ubiquitination

The 76-amino acid protein ubiquitin is covalently attached to lysine residues in three coupled sequential reactions catalyzed by three different enzymatic activities performed by three different enzymes: *activation* by an activating enzyme (E1); *conjugation* by a conjugating enzyme (E2); and *attachment of ubiquitin to the protein* by an isopeptide ligase (E3) [27]. Lysines can be mono- or polyubiquitinated, depending of the cellular context: monoubiquitination plays a role in cell-signaling transduction in myriad functional contexts [28], whereas polyubiquitination is associated principally with protein degradation via the 26S proteasome.

The functional consequences of histone ubiquitination and deubiquitination on gene expression have been far more extensively studied at histones H2A and H2B than at histones H3, H4, and H1 [29]. In fact, H2A was the first protein ever identified as being ubiquitinated. H2A that is monoubiquitinated at Lys119 is associated with inactive chromatin, including the inactive X chromosome in mammalian females and silenced developmental genes [30]. In contrast, H2B that is monoubiquitinated at Lys123 is important for transcriptional elongation and has been associated with either repressive or active DNA regions, depending on the positional context [31]. Interestingly, there appears to be important interplay between monoubiquitination of H2B and other marks. For instance, this modification is required for lysine methylation in histone H3K4, and conversely, this H3K4 methylation is inhibited by monoubiquitination of H2A [32].

The ubiquitin group can be removed from target residues by a class of thiol proteases known as *deubiquitinating enzymes* (DUBs) [33]. Although there are specific DUBs for H2A (e.g., USP16, 2A-DUB, USP21, BAP1) or H2B (e.g., USP3, USP7, USP12, and USP49 [34]), various DUBs exhibit dual specificity for the deubiquitination of both H2Aub and H2Bub. For example, USP22 (a component of SAGA complex) is required for proper cell cycle progression and is recruited to specific genes regulated by the oncoprotein Myc [35].

1.6 Histone SUMOylation

SUMOylation is a reversible PTM that entails covalent ligation of small ubiquitin-related modifier (SUMO) groups at specific lysine residues. The SUMO family members (SUMO-1, -2, -3, and -4) belong to the group of ubiquitin-like protein-modifying enzymes. Similarly to ubiquitination, SUMOylation is a multistep process that involves an activating heterodimer enzyme (E1: SAE1/SAE2), a conjugating enzyme (E2: Ubc9), and a SUMO ligase (E3) [36].

Growing evidence corroborates a link between protein SUMOylation and critical process, including cellular localization, chromatin organization, genome stability, signal transduction, protein–protein or protein-DNA interactions, and transcriptional regulation [36]. Early studies on histone SUMOylation showed that this modification on histone H4 associates with transcriptional repression through recruitment of HDAC and HP1 [37]. Later studies in *Saccharomyces cerevisiae* identified histone SUMOylation in all four core histones and associated the presence of SUMO with transcriptional repression. The importance of this PTM was underscored in a recent study that identified more than 4300 SUMOylation sites in more than 1600 proteins. Interestingly, many SUMOylated lysines can be subjected to methylation, acetylation, or ubiquitination—a finding that suggests cross-talk between SUMO and other PTMs [38].

1.7 Histone ADP-Ribosylation

ADP-ribosylation is a reversible modification resulting from transfer of an ADP-ribose moiety from NAD^+ to specific residues such as lysine, arginine, glutamate, aspartate, cysteine, phosphoserine, and asparagine [39]. Mono- and poly-ADP-ribosylation have been described. There are three families of ADP-ribosyltransferases. The first family is the *diphtheria toxinlike ADP-ribosyltransferases* (ARTDs, formerly known as *PARPs*), which can be both mono- and poly-ADP-ribosyltransferases. The other two families are the *clostridial toxin-like ADP-ribosyltransferases* (ARTCs) and the *Sir2 family of NAD^+-dependent protein deacetylases* (sirtuins); both of which are exclusively mono-ADP-ribosyltransferases [40].

Addition of a single ADP-ribose group implies addition of a negative charge to the modified protein; consequently, poly-ADP-ribosylation implies a large increase in negative charge in the substrate. Considering that ADP-ribosylation not only neutralizes the positive charge of an amino acid but also can actually reverse its net charge to negative, the functional consequences of ADP-ribosylation can be considered to be more drastic than those of other modifications [41].

Since ARTCs are expressed at the cell surface or are secreted into the extracellular environment, only ARTDs and sirtuins are capable of histone ADP-ribosylation. Among all nuclear ARTDs (ARTD1, ARTD2, ARTD3, ARTD5, and ARTD6), ARTD1 (also known as *PARP1*) accounts for up to 90% of the total cellular poly-ADP-ribosylation. Interestingly, the main target of PARP1—by far the most studied ARTD member—is PARP1 itself, although it also targets a wide variety of nuclear proteins and all five histone proteins [40]. This enzyme is involved in many cellular processes, including cell cycle regulation, genotoxic stress response, gene expression, and differentiation.

Interestingly, mono- or poly-ADP-ribosylation of histones seems to reduce their phosphorylation levels, an observation that corroborates the idea that ADP-ribose groups generally alter the binding of other factors (in this case, kinases) [42].

2. HISTONE MODIFICATIONS AND DISEASE

Aberrant distribution patterns of PTMs in the genome have been associated to some of the most frequent human pathologies. Below, we have included the most relevant associations (Figure 2).

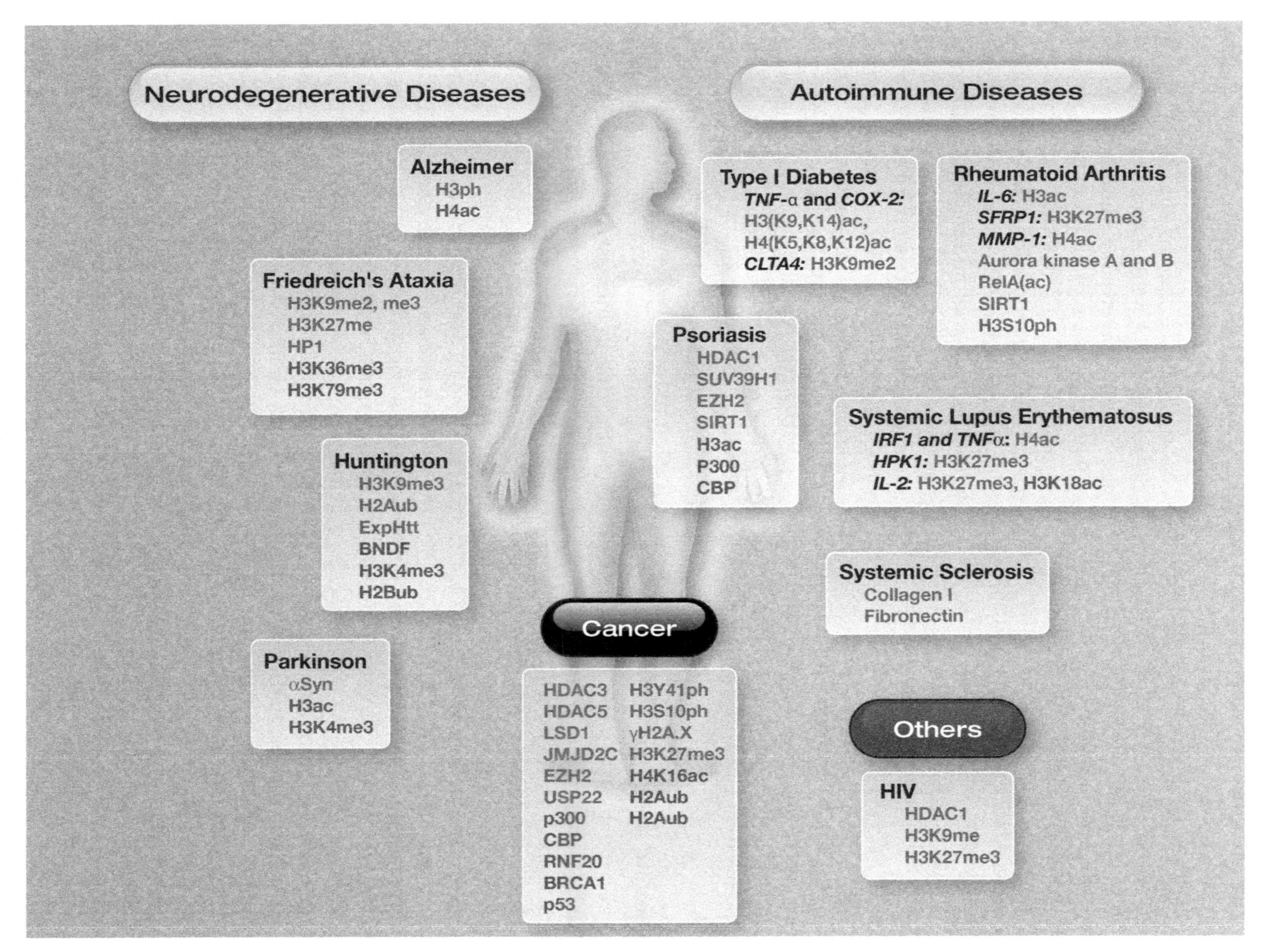

FIGURE 2 **A schematic representation of the best-known epigenetic alterations involved in neurodegenerative, autoimmune diseases, and cancer.** The histone posttranslational modifications and proteins up- or downregulated are depicted in blue (gray in print versions) or red (dark gray in print versions), respectively. In some cases, a single gene affected is indicated (italic bold) together with the described epigenetic mark alterations.

2.1 Autoimmune Diseases

Autoimmune diseases develop when the immune system wrongly attacks healthy cells and tissues in the body. They are generally considered to be complex diseases with two important components: *genetic predisposition* and *environmental influence*. Although genome-wide studies have identified numerous genetic risk variants associated with various autoimmune diseases, the concordance rate for these diseases between monozygotic twins tends to be less than 50% [43], suggesting that epigenetic mechanisms might be behind their pathogenesis. The study of the epigenetics of autoimmune diseases, particularly of histone PTMs, has helped elucidate the common bases shared by many of these diseases, including rheumatoid arthritis (RA), systemic lupus erythematosus (SLE), psoriasis, systemic sclerosis (SSc), and type 1 diabetes (Figure 2).

2.1.1 Rheumatoid Arthritis

Rheumatoid arthritis (RA; OMIM #180300; ORPHA284130) is a long-term autoimmune disease characterized by synovial inflammation, autoantibody production (rheumatoid factor and anticitrullinated protein antibody), and cartilage and bone destruction, all of which lead to development of systemic disorders. Synovial fibroblasts from patients with RA produce high levels of the inflammatory cytokine interleukin-6 (IL-6), which is involved in hematopoiesis, acute-phase response, and immune regulation—namely, by inducing T cell activation and differentiation, immunoglobulin production in B cells, platelet maturation, and production of acute-phase proteins [44].

Mounting evidence suggests that development of RA implies cooperation between genetic and environmental factors. Based on genetic studies done on twins, the estimated rate of heritability of RA is 65% [45] and more than 60 risk loci for RA have been identified [46]; however, the concordance rates between identical twins are surprisingly low (12–15%). Interestingly, specific environmental factors such as diet, cigarette smoke, and infection have been shown to contribute to RA pathogenesis [47].

Most of the epigenetic studies on RA have focused on DNA methylation or microRNAs; there have only been a few studies on PTMs, and on their molecular or physiological impact, in RA patients. In the first study on the functional role of histone methylation in RA synovial fibroblasts (RASFs), the authors reported overexpression of histone methyltransferase enhancer of zeste homolog 2 (EZH2) [48]. In mammals it is the chief enzyme responsible for H3K27 trimethylation (H3K27me3) [49], a mark associated with transcriptional silencing [50]. Stimulation of RASFs with the cytokine tumor necrosis factor alpha (TNF-α), a major inflammatory mediator in the pathogenesis of RA, induces EZH2 overexpression that does not affect global levels of H3K27me3. However, EZH2 overexpression reduces the levels of secreted frizzled-related protein 1 (SFRP1), an inhibitor of Wnt signaling, due to increased H3K27 methylation (H3K27me3) in its promoter [48].

There is growing evidence that histone PTMs are involved in the pathogenesis of RA. Chromatin-modifying enzymes, such as histones, kinases, acetyltransferases, deacetylases, methyltransferases, demethylases, and ubiquitin ligases, are encoded by a group of related genes with arthritis-specific expression. In one study, arthritic mice and treatment-naive RA patients both expressed high levels of aurora kinases A and B and exhibited increased phosphorylation of serine 10 in the tail of histone H3 [51]. Said mark is an epigenetic signal for the recruitment of nuclear factor-kappa B (NF-κB) during cytokine-triggered inflammatory responses [52].

An imbalance of histone and nonhistone acetylation/deacetylation clearly contributes to development of RA, as it does to that of other diseases. Specifically, it stems from a shift in the balance between HATs and HDACs toward histone hyperacetylation [53]. For example, in RASFs the distal region of the promoter of the matrix metalloproteinase 1 (MMP-1) gene is hyperacetylated in histone H4 [54]. MMP-1 is responsible for destroying cartilage, which it does by cleaving type II collagen, and its overexpression contributes significantly to degradation of the extracellular matrix in cartilage [55]. Another example of abnormal histone hyperacetylation occurs at the *IL-6* promoter, at which histone H3 is hyperacetylated in RASFs compared to in osteoarthritis synovial fibroblasts [56].

In addition to histone modifications, the acetylation levels of nonhistone proteins also influence RA development. A clear example corresponds to hyperacetylation of the RelA subunit of NF-κB by p300/CBP acetyltransferases. This modification controls several features of RelA, such as DNA binding, transcriptional activation, and assembly with its inhibitor IκBα [57]. Furthermore, recent work has shown that sirtuin 1 (SirT1) is overexpressed in synovial tissues from patients with RA, and its levels are further increased upon stimulation with TNF-α in RASFs and monocytes, to ultimately promote production of proinflammatory cytokines and resistance to apoptosis [58].

2.1.2 Systemic Lupus Erythematosus

SLE (OMIM #152700, ORPHA536) is a chronic inflammatory autoimmune disease associated with T cell overactivation and polyclonal B cell hyperactivity [59]. Its symptoms include rashes, photosensitivity, joint and muscle pain, pericarditis, and nephritis [59]. SLE patients exhibit multiple autoantibodies directed against the nucleosome, in which, in both spontaneous and drug-induced SLE, histone H2B is the principle target [60].

Among histone modifications, global histone H4 acetylation is significantly altered in monocytes from SLE patients. In one study, 63% of the H4-hyperacetylated genes were targets of interferon regulatory factor 1 (*IRF1*) [61]. In the same context, increased acetylation of histone H4 is observed in the *TNF-α* locus, which in turn correlates with higher levels of *TNF-α* transcripts in peripheral blood monocytes from SLE patients [62]. Contrariwise, T lymphocytes from SLE patients who fail to express IL-2 [63] had reduced H3K18 acetylation and increased H3K27 methylation on the *IL-2* gene [64]. Another study has shown that enrichment of H3K27me3 at the *hematopoietic progenitor kinase 1* (*HPK1*) promoter is higher in $CD4^+$ T cells from SLE patients than in those from healthy controls and is associated with loss of JMJD3 binding. Moreover, similar studies have reported lower expression levels of HPK1, a mammalian Ste20-related serine/threonine PK involved in multiple cellular events (e.g., MAPK, NF-κB, cellular proliferation, apoptosis, cytokine signaling, etc.) [65].

2.1.3 Systemic Sclerosis

SSc (OMIM #181750, ORPHA90291) is a multisystemic disease characterized by vasculopathy, immune activation, and extensive fibrosis of the skin and internal organs, such as the lungs, the kidneys, and the gastrointestinal tract [66]. The pathogenesis of SSc remains poorly understood; however, it is known to be strongly influenced by genetic, epigenetic, and environmental factors [67].

Experimental studies have shown that histone acetylation is important in SSc pathogenesis. The administration of HDAC inhibitor trichostatin A (TSA) attenuated the expression of collagen I (COL1A1) and fibronectin (FN1) in both normal and SSc skin fibroblasts [68,69]. The overexpression of both proteins, together with reduced expression of matrix metalloproteinases 1 and 3, leads to excessive buildup of matrix tissue fibrosis in SSc patients [68].

2.1.4 Type I Diabetes

Type I diabetes (T1D, OMIM #222100) is a chronic autoimmune disorder characterized by destruction of insulin-secreting beta cells in the islets of Langerhans [70]. Monocytes from patients with T1D exhibit significant enrichment in H3K9ac, which directly correlates with glycosylated hemoglobin (HbA_{1C}) levels [71]. Additionally, monocytes cultured in high glucose conditions that mimic diabetes show increased recruitment at *TNF-α* and *COX-2* promoters of HATs, such as CBP/p300 and p/CAF, leading to increased hyperacetylation at H3K9, H3K14, H4K5, H4K8, and H4K12 [72]. Furthermore, histone methylation patterns have been shown to be altered in T1D patients: lymphocytes isolated from these patients show significantly increased levels of H3K9me2 at the *CLTA4* gene, a known T1D susceptibility gene [73].

2.1.5 Psoriasis

Psoriasis (e.g., OMIM #603935, #610707, and #604316) is an inflammatory skin disorder elicited by chronic activation of cutaneous T cells and characterized by excessive keratinocyte proliferation and the presence of inflamed scaly patches of skin [74]. Genome-wide DNA methylation studies in $CD4^+$ T cells in psoriasis patients have revealed hypomethylation in pericentromeric regions of 10 chromosomes [75].

Studies with the HDAC inhibitor TSA in regulatory T cells from psoriasis patients have shown that histone/protein deacetylation is critical for transformation of these cells into potentially pathogenic effector T cells [76]. In fact, skin from psoriasis patients exhibits higher levels of *HDAC1* mRNA than does skin from healthy subjects [77]. Furthermore, SirT1 activation or inhibition with resveratrol or nicotinamide, respectively, has opposite effects on keratinocyte differentiation: whereas SirT1 upregulation induces calcium-dependent keratinocyte differentiation, SirT1 loss leads to the opposing consequence [78]. Additionally, peripheral blood mononuclear cells (PBMCs) from psoriasis patients show significantly lower levels of *p300* and *CBP* mRNA, and significantly higher levels of *HDAC1*, *SUV39H1*, and *EZH2* mRNA, than do PBMCs from healthy controls. In fact, a negative correlation has been established between the degree of global histone H4 acetylation and the severity of the disease (as measured by the Psoriasis Area Severity Index) [79].

2.1.6 Human Immunodeficiency Virus

Human immunodeficiency virus (HIV) is the retrovirus that causes acquired immunodeficiency syndrome (AIDS) [80]. The HIV genome preferentially integrates into intragenic portions of actively transcribed regions of the host DNA. Viral genome chromatin formation is subject to cellular epigenetic regulation in which proviral long terminal repeats (LTRs) seem to be important. Nucleosomes on the 5′ LTR of latent proviruses exhibit patterns of repressive heterochromatin structures that contain deacetylated and trimethylated histones [81]. It is noteworthy that knockdown of H3K27me3-specific EZH2 can induce up to 40% of latent HIV proviruses [81]. The H3K9me3-specific SUV39H1 and the H3K9me2-specific G9a also stabilize to a lesser extent the heterochromatinized structure of the viral genome through heterochromatin protein 1 gamma (HP1γ) recruitment [82,83]. Additionally, under basal conditions transcription factors such as Yin-Yang 1 (YY1), late SV40 factor (LSF), COUP-TF-interacting protein (CTIP2), c-promoter-binding factor-1 (CBF-1), NF-κB p50 homodimer, c-Myc, and Sp1, can recruit HDAC1 to the HIV LTR promoter to promote histone deacetylation and transcriptional silencing [84].

2.2 Neurodegenerative Diseases

More than 600 diseases are characterized by progressive nervous system dysfunction or neurodegeneration. Their frequency ranges from relatively high, for common diseases such as Alzheimer's disease (AD), Parkinson's disease (PD), and Huntington's disease (HD), to very low, for rare diseases such as Friedreich's ataxia (FA), amyotrophic lateral sclerosis (ALS), and Creutzfeldt–Jakob disease. Neurodegenerative diseases are generally associated with protein misfolding and with abnormal aggregation of native proteins—phenomena that have local and systemic consequences. In this section we describe aberrant histone PTMs that are common in neurodegenerative diseases (Figure 2). We also discuss the diagnostic and therapeutic potential of recent findings from this area.

2.2.1 *Alzheimer's Disease*

AD (OMIM #104300) is the most common age-dependent neurodegenerative disorder. It is clinically characterized by progressive memory loss and cognitive impairment. Age-related changes in the metabolism of amyloid precursor protein (APP) and τ (tau) protein produce intraneural filaments (called *neurofibrillary tangles*) and extracellular β-amyloid aggregates (also known as *senile plaques*) that strongly affect neuronal function, synaptic plasticity, and cell death, mainly in the hippocampus and in associated regions of the neocortex [85]. The molecular mechanisms behind AD pathogenesis are not completely defined. However, mounting evidence is imputing epigenetic mechanisms such as DNA methylation and histone modification in the altered synaptic function and memory loss associated with AD.

2.2.1.1 PTMs OF HISTONES IN AD

The first evidence of aberrant PTMs of histones in AD was that of increased levels of phosphorylation in histone H3 in hippocampal neurons of AD patients [86] (Figure 1). Additionally, an APP/PS1 mouse model of AD under fear-conditioning training showed lower levels of histone H4 acetylation than did wild-type littermates. Interestingly, pretraining treatment of the AD mice with TSA restored H4 acetylation back to wild-type levels [87]. A recent proteomics study revealed that global levels of histone acetylation in the temporal lobe are significantly lower in AD patients than in age-matched control subjects [88].

Among HDACs, those that seem to be involved in the pathogenesis of AD are HDAC6 and HDAC2: the former is related to increased levels of tau phosphorylation and the latter, in repression of genes implicated in learning and memory [89]. Despite all the aforementioned evidence, the role of histone acetylation in AD remains controversial, as some studies suggest that the link between histone hyperacetylation and AD might simply be coincidental [90]. The lack of concordance among researchers partially stems from the wide variation in observed histone acetylation levels according to cell type, animal model, and even the brain region studied.

2.2.1.2 EPIGENETIC INTERVENTIONS IN AD

Modulation of histone acetylation through the use of HDAC inhibitors significantly restores memory and alleviates cognitive deficits in animal models of AD [91]. In one model, inhibition of HDACs with valproic acid decreased the generation of β-APP and reduced senile plaque formation in the brain [92]. Similar beneficial effects have been observed with nicotinamide, a competitive inhibitor of sirtuins: specifically, a mouse model of AD that was treated with nicotinamide exhibited reduced levels of phosphorylated tau and improved cognition [93].

2.2.2 *Parkinson's Disease*

PD (OMIM #168600) is the second most common age-related neurodegenerative disorder. It is characterized by numerous motor and nonmotor symptoms including muscle rigidity, bradykinesia, resting tremor, and postural instability. PD mainly involves loss of dopaminergic neurons in the substantia nigra and deposition of α-synuclein (αSyn), an intraneural protein whose aggregation causes *Lewy pathology* (OMIM #602404), which is commonly found in PD patients [94].

2.2.2.1 PTMs OF HISTONES IN PD

Syn accumulation promotes hypoacetylation of histone H3, thereby promoting neurotoxicity in PD by masking histone proteins: specifically, by preventing histone acetylation, condensing chromatin, repressing gene expression, and ultimately, provoking cell death [95]. Interestingly, in murine as well as primate models of PD, a reduction in H3K4me3 has been associated with depletion of dopamine in striatal neurons [96].

2.2.2.2 EPIGENETIC INTERVENTIONS IN PD

Among epigenetic factors, SirT1, SirT2, and Class I and II HDACs seem to be directly involved in PD. Activation of SirT1 in in vivo and in vitro PD models slows down neuronal death as well as neurodegeneration [97]. Furthermore, activation of sirtuins with resveratrol protects neuroblastoma (SK-N-BE) cells from αSyn-induced toxicity associated with PD. However, the specific SirT1 inhibitor sirtinol has the opposite effect, suggesting that activation of SirT1 protects against αSyn-triggered toxicity [98]. Interestingly, SirT2 seems to play an opposite role to SirT1, as suggested by findings in a *Drosophila* model of PD, in which inhibition of SirT2 reduced the toxicity of αSyn in cultured cells and in dopaminergic neurons against the degenerative effects of this toxicity both in vitro and in a *Drosophila* model of PD [99]. Different animal models of PD have provided evidence imputing HDACs in the disease: treatment of animals with sodium butyrate, a classic inhibitor of Class I and II HDACs, reduced αSyn levels and alleviated cognitive deficits [100].

2.2.3 *Huntington's Disease*

HD (OMIM #143100, ORPHA399) is an autosomal dominant neurodegenerative disease associated with expanded CAG repeats in the gene encoding for the protein huntingtin (Htt). HD is characterized by the presence of a form of Htt that contains polyglutamine repeats, named *expanded polyQ-Huntingtin* (*ExpHtt*). These extra CAG repeats (60 or more) code for a series of glutamine residues that can affect both the conformation and aggregation of Htt. Htt aggregates increase oxidative stress, neurotoxicity, and mitochondrial and proteasome dysfunctions and lead to progressive loss of cortical neurons, especially in the cortex and striatum, thereby leading to cognitive dysfunction, dementia, and loss of muscle coordination. Wild-type Htt sequesters R element-1 silencing transcription factor (REST), a key transcriptional repressor of neuronal survival factor, inhibits the expression of brain-derived neurotrophic factor (BDNF), which is crucial for growth, maturation, and maintenance of neurons. Interestingly, HD patients, and mouse models of HD, exhibit decreased levels of BDNF, and an increase in BDNF expression levels alleviates the HD phenotype [101].

2.2.3.1 PTMs OF HISTONES IN HD

Biochemical evidence indicates that ExpHtt forms a complex with the different lysine acetyltransferases, including CBP and P/CAF [102]. Moreover, Htt interacts with transcription corepressors such as mSin3a and N-CoR, which in turn recruit HDACs, suggesting a functional link between Htt and HDACs in transcription repression [103]. HD patients and R6/2 transgenic mice (a mouse model of HD) show increased levels of the repressive heterochromatin mark H3K9me3, which is directly associated to overexpression of the H3K9-HMTs ESET and SETDB1. In contrast, the methylation marks connected with active genes, such as H3K4me3 [104], are decreased at the promoters of common downregulated genes associated with HD [105]. Htt protein has also been associated with polycomb repressive complex 2 (PRC2): in vitro, the presence of mutant Htt leads to a decrease in global levels of H3K27 methylation, whereas wild-type Htt stimulates the histone H3K27 methyltransferase activity of PRC2 [106].

Alterations in histone monoubiquitylation and SUMOylation also are crucial for transcriptional deregulation in HD. For instance, the brains of transgenic R6/2 mice have been reported to exhibit higher levels of ubiquitinated H2A and lower levels of ubiquitinated H2B than do the brains of control mice. These differences correlate to the respective levels of promoter binding and transcriptional regulation: having higher levels of ubiquitinated H2A reverses transcriptional repression and inhibits methylation of histone H3K9, whereas having lower levels of ubiquitinated H2B leads to transcriptional repression, by inhibiting methylation of histone H3K4 [107].

2.2.3.2 EPIGENETIC INTERVENTIONS IN HD

Recent studies have shown that treatment of a mouse model of HD with the HDAC inhibitor sodium butyrate ameliorates the neurodegenerative phenotype and motor deficits [102,108]. Interestingly, SirT1 and SirT2 seem to play opposite roles in HD, as they do in PD. In nematodes, SirT1 overexpression and treatment with resveratrol have each been shown to diminish polyglutamine cytotoxicity, whereas in mouse models of HD, treatment with the SirT2 inhibitor AK-7 has been reported to be neuroprotective [109]. Additionally, treatment of R6/2 transgenic mice with mithramycin, a DNA-binding drug that can prevent methylation of histone H3, not only prolongs their survival but also exerts beneficial effects on their behavior and their neuropathological phenotype [110,111].

2.2.4 *Friedreich's Ataxia*

FA (OMIM #601992, #229300, ORPHA95) is a rare neurodegenerative disease that is inherited in an autosomal recessive pattern. It chiefly affects the nervous system and is characterized primarily by progressive truncal ataxia, limb ataxia, muscle weakness, and dysarthria [112]. FA is caused by loss of expression of the *frataxin* gene (*FXN*), due to an expansion of GAA repeats in its first intron. Frataxin is a mitochondrial protein that appears to be crucial for regulation of iron homeostasis. Interestingly, lymphoblastoid cells from FA patients show higher levels of H3K9me2, H3K9me3, and H3K27me3—histone marks associated with inactive gene expression at upstream regions (GAATTC) of *FXN* gene [113]. These alterations are shown to be associated with increased levels of heterochromatin protein 1 (HP1) in FA lymphoblasts [114]. Other studies have reported lower levels of H3K36me3 and H3K79me3 at the upstream and downstream GAA repeat regions of the *FXN* gene in cells derived from FA patients. In the case of H3K4me3, the levels are only reduced at the upstream GAA repeat region, but not at the promoter region—a finding that corroborates a defect in the postinitiation and elongation stages of *FXN* gene expression [115].

2.2.4.1 EPIGENETIC INTERVENTIONS IN FA

HDAC inhibitors can partially reverse the silencing of *FXN* in cultured cells obtained from FA patients [113], suggesting that heterochromatin conformation might be important for transcriptional regulation of frataxin. In fact, histone H3K9 methylation and H3K9 acetylation are crucial signatures in *FXN* silencing and activation, respectively [116]. Importantly, treatment with the HDAC inhibitor 2-aminobenzamide has been reported to increase *FXN* mRNA and protein levels. In fact, a Phase I clinical trial on 2-aminobenzamide as a treatment for FA recently ended successfully: the average fold-induction of *FXN* mRNA expression observed was 1.5- to 1.6-fold (relative to untreated controls). Based on these encouraging results, the compound will soon enter a Phase II trial [116].

2.3 Cancer

Histone PTMs are at the core of cancer for many reasons: they are crucial not only in gene expression and in chromatin structure but also in virtually all DNA-associated functions. In particular, they lead to abnormalities that cause dysregulation of signaling networks that are required for physiological mechanisms, such as cell growth, apoptosis, and DNA repair (Figure 2). Consequently, they can lead to diverse cancers. However, among the most relevant contributions of epigenetic modifications to tumorigenesis is a hallmark of cancer: *genome instability*. In this section, we focus on the relationship between histone PTMs and tumorigenesis, and the possible mechanisms and pathways by which PTMs might contribute to the onset of cancer.

2.3.1 *Histone Acetylation and Cancer*

HATs and HDACs are frequently altered in most cancers, resulting in an unusual chromatin phenotype. Loss of heterozygosity of the HATs p300 or CBP has been reported in a significant proportion of cancer cell lines: 51% in the case of p300 and 35% in the case of CBP [117]. These observations suggest that both HATs are crucial tumor suppressors that can be lost due to loss of heterozygosity in several types of cancer. Among HATs, the MYST family appears to be essential for onset of acute myeloid leukemia (AML) (OMIM #601626) [118]. In several subtypes of AML (M4/M5), a stable and frequent translocation t(8;16)(p11;p13) leads to the fusion of MOZ and CBP, resulting in aberrant chromatin acetylation [119]. Similarly, in a subset of AML cases, following a t(8;22)(p11;q13) translocation, MOZ and p300 were observed to be fused [120].

The role of histone deacetylation in cancer is even clearer than that of histone acetylation: indeed, it has been shown to be an early step in carcinogenesis [121]. Analogously to HATs, HDACs target nonhistone proteins involved in carcinogenesis, such as p53, STAT3, and YY1 [122]. HDACs tend to be overexpressed in many tumor types, including breast [123], prostate [124], and colorectal tumors [125]. Furthermore, in several cancer cell lines, including primary lymphomas and colorectal adenomas, histones are hypoacetylated in comparison with normal cells, suggesting that histone deacetylation is prevalent in cancer [121]. For instance, early loss of histone H4K16 monoacetylation has been identified in a mouse model of multistage skin carcinogenesis [121]. Another clear example is HDAC3, which seems to be overexpressed in colon carcinogenesis and alters oncogenic pathways such as Wnt/B-catenin signaling and the growth-regulatory signal of 1,25-dihydroxyvitamin D3 [126].

To date, three HDAC inhibitors have been approved by the U.S. Food and Drug Administration (FDA). The first two, Zolinza® (vorinostat; Merck) and Istodax® (romidepsin; Celgene), are indicated for treatment of cutaneous T cell lymphoma (CTCL) [127]. The third one, Beleodaq® (belinostat; Spectrum Pharmaceuticals), has been approved for treatment of peripheral T cell lymphoma. It selectively inhibits two classes of the zinc-dependent HDACs (Class I and Class II) [128] and is active against multidrug-resistant tumors (including those that have developed resistance to platinum drugs, taxanes, and topoisomerase II inhibitors) [129].

2.3.2 *Histone Methylation and Cancer*

Like histone acetylation, histone methylation has been extensively linked to tumorigenesis. Albeit the functional effects of histone methylation dysregulation are not well understood, histone methyltransferases and demethylases are both known to be important in cancer development and therefore, represent novel therapeutic targets in oncology [130]. The histone methylation and acetylation marks known to be dysregulated in cancer are summarized in Table 3.

EZH2, the main H3K27me3-specific HMT in mammals, is essential for tissue-specific stem cell maintenance, cell identity and differentiation, and its overexpression has been correlated with tumor progression and poor prognosis in various cancers [131]. In normal cells, EZH2 controls DNA methylation by interacting with DNA methyltransferases [132]. Interestingly, in cancer cells H3K27me3 has been shown to repress gene expression independently of DNA methylation [133].

In contrast, other methylation marks are directly linked to transcriptional activation. For instance, mixed lineage leukemia (MLL) (OMIM +159555) is a H3K4me3-specific HMT. Chromosomal translocations of the *MLL* gene with other genes result in the formation of oncogenic MLL fusion proteins (MLL-FPs) with other partners. These FPs have been widely proven to be an important feature of leukemia.

TABLE 3 Alterations of Histone Modifications in Cancer

Posttranslational modification	Cancer type	References
H3K4me1	Prostate	[165]
	Kidney	[166]
H3K4me2	Lung	[167]
	Prostate	[168,169]
	Kidney	[166]
	Pancreas	[170]
H3K4me3	Kidney	[166]
	Liver	[171]
	Colon	[172]
H3K9ac	Lung	[167,173]
H3K9me1	Kidney	[174]
H3K9me2	Prostate	[165]
	Breast	[175]
	Pancreas	[170]
H3K9me3 H3K9me3k14ac	Lung	[173]
	Prostate	[165]
	Breast	[172,176]
	Leukemia	[172,177]
	Stomach	[178]
	Breast	[172]
H3K18ac H3K18acK23un	Lung	[179]
	Prostate	[168,169]
	Breast	[175]
	Esophagus	[180,181]
	Pancreas	[170]
	Lung	[172]
	Kidney	[172]
H3K27ac	Colon	[182]
H3K27me1	Kidney	[183]
H3K27me2	Kidney	[183]

Continued

TABLE 3 Alterations of Histone Modifications in Cancer—cont'd

Posttranslational modification	Cancer type	References
H3K27me3	Prostate	[184]
	Stomach	[185]
	Esophagus	[180]
	Liver	[186]
	Kidney	[183]
	Breast	[172]
	Osteosarcoma	[172,187]
	Colon	[182]
H3K36me1	Leukemia	[172,188]
H3K36me2	Leukemia	[172,188]
H4K5ac	Lung	[189]
H4K8ac	Lung	[189]
H4K12ac	Lung	[189]
	Breast	[175]
H4K16ac	Lung	[173,189]
	Breast	[172,175]
H4K20me1	Prostate	[190]
H4K20me2	Prostate	[172,190]
H4K20me3	Lung	[172,189]
	Breast	[175,176]
	Prostate	[121]
H4R3me2	Breast	[175]
	Esophagus	[180,181]

Histone demethylases also have been linked to oncogenesis. For instance, amplification of *JMJD2C* has been identified in many different types of cancer, such as breast and esophageal cancers (OMIM #114480 and #133239, respectively) [134,135]. Additionally, LSD1 has recently been shown to be overexpressed in bladder cancers (OMIM #109800) [136], estrogen receptor-negative breast cancer [137], and mesenchymal tumors [138].

The replacement of histone variants into chromatin and their respective mutations can lead to aberrant gene expression and global defects in chromatin structure. Particularly, the somatic missense mutations of the histone H3 variant, H3.3, are connected to some childhood and young-adult tumors, such as giant cell tumor of bone, chondroblastoma, and pediatric high-grade astrocytomas [139–141]. The H3.3 mutations Lys27Met (K27M) and Gly34Arg or

Gly34Val (G34R/V) lead to global histone modification changes, especially in H3K27me3 and H3K36me3, respectively. The mutation K27M inhibits the H3K27me3-specific methyltransferase EZH2, the catalytic component of the PRC2, resulting in global reduction of the mark H3K27me3 and upregulation of polycomb-repressed genes [142]. The mutation G34R/V notably results in redistribution of H3K36me3, probably by redirecting its enzyme SET domain-containing 2 (SETD2) promoting upregulation of genes such as *MYCN* [143].

2.3.3 Histone Phosphorylation and Cancer

Phosphorylation of histones is also important in cancer development: it has been directly linked to cell cycle progression, DNA damage response, chromosome stability, chromatin remodeling and transcription [144]. For instance, Janus kinase 2 (JAK2) is activated by chromosomal translocations or point mutations in blood-related cancers. The phosphorylation of H3Y41 (H3Y41ph) by JAK2 prevents binding of heterochromatin protein1α (HP1α) to H3 and consequently, induces gene expression of hematopoietic oncogene Imo2 [145]. Phosphorylation of H3 at S10 and S28 (leading to H3S10ph and H3S28ph, respectively) is crucial during chromosome condensation in mitosis and during transcriptional activation of immediate early genes. Interestingly, growth factors that induce Ras/MAPK and increase H3S10ph at transcriptionally active loci are implicated in aberrant gene expression, as observed in breast cancer progression [146]. Phosphorylation of histone H3 at specific residues is recognized by the 14-3-3 chaperones, which comprise seven isoforms in mammals. These proteins contain highly conserved phosphoserine-binding domains that they use to bind to H3S10ph and H3S28ph. Abnormal expression of certain isoforms of the 14-3-3 chaperones is implicated in various cancers; hence, these isoforms might prove useful as therapeutic targets in oncology [147].

Interestingly, a study aimed to determine the global-level signature of PTMs in 24 cancer cell lines showed that the phosphorylation status of H3S10 is strongly influenced by some methylation patterns, such as the combination of H3K9me3 and H3K14 unmethylation [148]. Based on a methyl-phospho switch model [15], in which H3S10ph blocks the binding of HP1 to H3K9me2 or H3K9me3, altered patterns of H3S10ph can contribute to aberrant gene expression and oncogenic transformation; hence, H3S10ph levels strongly (directly) correlate to tumor grade [149].

In order to maintain genomic stability to ensure cell survival, DNA damage is normally repaired by DNA repair response. Nevertheless, long-term activation of this response can bypass cell growth barriers created by it, ultimately causing malignancy [150]. Since double-strand breaks induce phosphorylation of H2AX (γH2AX), detection of this mark can be exploited diagnostically to detect premalignant lesions and early stages of cancer [151]. Interestingly, the presence of γH2AX can be used in the diagnosis of metastatic renal cell carcinoma [152].

2.3.4 Histone Ubiquitination and Cancer

A growing body of evidence supports the role for histone ubiquitination in cancer. For instance, a drastic global downregulation in monoubiquitination of histone H2A and of histone H2B is observed in prostate and breast tumors, respectively [153,154]. One explanation for this observation is that certain E3 ubiquitin ligases are downregulated during tumorigenesis. For instance, depletion of RNF20, the major H2B-specific E3 ubiquitin ligase in mammals, causes a significant reduction in cellular levels of monoubiquitinated H2B and induces expression of certain growth-related proto-oncogenes (e.g., c-Fos and c-Myc) [155]. Additionally, RNF20-depleted cells exhibit reduced expression of the tumor suppressor p53, which results in enhanced tumorigenesis, migration, and invasion. Also, tumor cells show DNA hypermethylation of the RNF20 promoter [155]. Altogether, these findings pinpoint RNF20 as a novel putative tumor suppressor.

Another important E3 ubiquitin-protein ligase involved in cancer is BRCA1, whose inactivation provokes breast and ovarian cancer (OMIM #167000) [156].

In vitro, this enzyme catalyzes the monoubiquitination of different substrates, including histones H2A, H2B, H3, and H4, conferring genomic stability at heterochromatin regions, especially via H2A-monoubiquitination [157,158].

The DUB USP22 is a ubiquitin hydrolase that removes ubiquitin groups from histones H2A and H2B [159,160]. Overexpression of USP22 has been associated with cancer stem cells, metastasis, chemotherapy resistance, and poor prognosis in malignant tumors such as colorectal and breast tumors [159,161]. USP22 is recruited to specific promoters [35] by Myc, and its depletion causes overexpression of p53 and p21, leading to cell-cycle arrest at the G1 phase and, consequently, inhibition of proliferation [35,162]. Loss of global H2B monoubiquitination (H2Bub1) has been reported in various cancers, including breast [154], colorectal (OMIM #114500) [163], lung (OMIM #211980), [163] and parathyroid (OMIM #608266) [164] cancers. For example, in the case of parathyroid cancer, a mechanistic explanation for loss of H2Bub1 is provided by the frequent occurrence of mutations in CDC73, a member of the tumor suppressor complex PAF1 [164]; said mutations disrupt the complex, whose proper function is required for regulation of H2Bub1.

3. CONCLUSIONS

During the past 15 years, the study of the histone PTMs and the underlying histone code has become a growing research topic in human health. The study of epigenetic alterations in human diseases provides a relevant tool not only to understand the molecular bases of these pathologies but also in clinical prognosis, diagnosis, and therapeutics. However, to be able to compare them and use them clinically, a standardization of the way we study these marks is urgently required. This should allow us in the future to establish the specificity, sensitivity, and reliability of the detected changes of each of these modifications. The full development of these procedures promises to make the study of these PTMs a key tool for biomedicine research in the near future.

Glossary

Ataxia Inability to coordinate movement.

Chromatin readers Protein domains that show high binding affinity for histone PTMs and function in downstream effects.

Chromatin remodeling An ATP-dependent enzymatic process that alters histone–DNA interactions or regulates nucleosomes positioning. In some cases, this process can also be ATP-independent, as in the case of the FAcilitates Chromatin Transcription (FACT) complex.

DNA methylation The addition of a methyl group to the 5′ carbon of cytosine, mostly in the context of CpG dinucleotides.

Euchromatin A form of chromatin that is fairly decondensed and is transcriptionally active.

Heterochromatin Highly condensed chromatin that is transcriptionally inactive.

Histone code Distinct histone PTMs in one or more histone tails that can act sequentially or in combination to form a "code" that can be read by reader proteins containing specific interacting domains.

Polycomb repressive complex (PRC) An epigenetic regulator of gene expression that silences target genes by establishing a repressive chromatin state. PRC2 trimethylates histone H3 at lysine 27. This repressive histone modification is recognized by PRC1, which has ubiquitylating activity. Since PRCs can maintain states of gene expression, they have key roles in cell fate maintenance and transitions during development.

LIST OF ABBREVIATIONS

AD Alzheimer's disease
AIDS Acquired immunodeficiency syndrome
ALS Amyotrophic lateral sclerosis
AML Acute myeloid leukemia
APP Amyloid precursor protein

ARTC Clostridial toxin-like ADP-ribosyltransferase
ARTD Diphtheria toxinlike ADP-ribosyltransferase
CREB cAMP response element-binding protein
CTCL Cutaneous T cell lymphoma
DUB Deubiquitinating enzyme
ExpHtt Expanded polyQ-Huntingtin
FA Friedreich's ataxia
FAD Flavin adenine dinucleotide
FDA Food and Drug Administration
FXN Frataxin
HD Huntington's disease
HDAC Histone deacetylase
HIV Human immunodeficiency virus
HP1 Heterochromatin protein 1
Htt Huntingtin
JMJC Jumonji-C
KMT Lysine methyltransferase
LSD Lysine-specific demethylases
LTR Long terminal repeat
MLL Mixed lineage leukemia
OMIM Online Mendelian Inheritance in Man
PBMC Peripheral blood mononuclear cells
PTM Posttranslational modification
RA Rheumatoid arthritis
RASF Rheumatoid arthritis synovial fibroblast
SAM S-adenosyl-L-methionine
SLE Systemic lupus erythematosus
SS Systemic sclerosis
SUMO Small ubiquitin-related modifier
T1D Type I diabetes
TSA Trichostatin A
Zn Zinc

Note: See tables or body text for additional gene and protein abbreviations.

References

[1] Luger K, Mäder AW, Richmond RK, Sargent DF, Richmond TJ. Crystal structure of the nucleosome core particle at 2.8 Å resolution. Nature 1997;389:251–60.
[2] Kornberg RD. Chromatin structure: a repeating unit of histones and DNA. Science 1974;184:868–71.
[3] Biterge B, Schneider R. Histone variants: key players of chromatin. Cell Tissue Res 2014;356:457–66.
[4] Allfrey VG, Faulkner R, Mirsky AE. Acetylation and methylation of histones and their possible role in the regulation of RNA synthesis. Proc Natl Acad Sci USA 1964;51:786–94.
[5] Bannister AJ, Kouzarides T. Regulation of chromatin by histone modifications. Cell Res 2011;21:381–95.
[6] Filippakopoulos P, Picaud S, Mangos M, et al. Histone recognition and large-scale structural analysis of the human bromodomain family. Cell 2012;149:214–31.
[7] Yang X-J, Seto E. HATs and HDACs: from structure, function and regulation to novel strategies for therapy and prevention. Oncogene 2007;26:5310–8.
[8] Lee KK, Workman JL. Histone acetyltransferase complexes: one size doesn't fit all. Nat Rev Mol Cell Biol 2007;8:284–95.
[9] Parthun MR. Hat1: the emerging cellular roles of a type B histone acetyltransferase. Oncogene 2007;26: 5319–28.
[10] Dell'Aversana C, Lepore I, Altucci L. HDAC modulation and cell death in the clinic. Exp Cell Res 2012;318:1229–44.
[11] Verdin E, Dequiedt F, Kasler HG. Class II histone deacetylases: versatile regulators. Trends Genet 2003; 19:286–93.
[12] Greer EL, Shi Y. Histone methylation: a dynamic mark in health, disease and inheritance. Nat Rev Genet 2012; 13:343–57.
[13] Jung HR, Pasini D, Helin K, Jensen ON. Quantitative mass spectrometry of histones H3.2 and H3.3 in Suz12-deficient mouse embryonic stem cells reveals distinct, dynamic post-translational modifications at Lys-27 and Lys-36. Mol Cell Proteomics 2010;9: 838–50.
[14] Zhang Y, Reinberg D. Transcription regulation by histone methylation: interplay between different covalent modifications of the core histone tails. Genes Dev 2001;15:2343–60.
[15] Rea S, Eisenhaber F, O'Carroll D, et al. Regulation of chromatin structure by site-specific histone H3 methyltransferases. Nature 2000;406:593–9.
[16] Feng Q, Wang H, Ng HH, et al. Methylation of H3-lysine 79 is mediated by a new family of HMTases without a SET domain. Curr Biol 2002;12: 1052–8.
[17] Dillon SC, Zhang X, Trievel RC, Cheng X. The SET-domain protein superfamily: protein lysine methyltransferases. Genome Biol 2005;6:227.
[18] Bedford MT, Richard S. Arginine methylation an emerging regulator of protein function. Mol Cell 2005; 18:263–72.
[19] Ciccone DN, Su H, Hevi S, et al. KDM1B is a histone H3K4 demethylase required to establish maternal genomic imprints. Nature 2009;461:415–8.
[20] Kooistra SM, Helin K. Molecular mechanisms and potential functions of histone demethylases. Nat Rev Mol Cell Biol 2012;13:297–311.
[21] Barski A, Cuddapah S, Cui K, et al. High-resolution profiling of histone methylations in the human genome. Cell 2007;129:823–37.
[22] Heintzman ND, Stuart RK, Hon G, et al. Distinct and predictive chromatin signatures of transcriptional promoters and enhancers in the human genome. Nat Genet 2007;39:311–8.

[23] Utley RT, Lacoste N, Jobin-Robitaille O, Allard S, Côté J. Regulation of NuA4 histone acetyltransferase activity in transcription and DNA repair by phosphorylation of histone H4. Mol Cell Biol 2005;25:8179–90.
[24] Zheng Y, John S, Pesavento JJ, et al. Histone H1 phosphorylation is associated with transcription by RNA polymerases I and II. J Cell Biol 2010;189:407–15.
[25] Chowdhury D, Keogh M-C, Ishii H, Peterson CL, Buratowski S, Lieberman J. γ-H2AX dephosphorylation by protein phosphatase 2A facilitates DNA double-strand break repair. Mol Cell 2005;20:801–9.
[26] Wei Y, Yu L, Bowen J, Gorovsky MA, Allis CD. Phosphorylation of histone H3 is required for proper chromosome condensation and segregation. Cell 1999; 97:99–109.
[27] Ye Y, Rape M. Building ubiquitin chains: E2 enzymes at work. Nat Rev Mol Cell Biol 2009;10:755–64.
[28] Miller J, Gordon C. The regulation of proteasome degradation by multi-ubiquitin chain binding proteins. FEBS Lett 2005;579:3224–30.
[29] Wright DE, Wang CY, Kao CF. Histone ubiquitylation and chromatin dynamics. Front Biosci (Landmark Ed) 2012;17:1051–78.
[30] Joo HY, Zhai L, Yang C, et al. Regulation of cell cycle progression and gene expression by H2A deubiquitination. Nature 2007;449:1068–72.
[31] Batta K, Zhang Z, Yen K, Goffman DB, Pugh BF. Genome-wide function of H2B ubiquitylation in promoter and genic regions. Genes Dev 2011;25:2254–65.
[32] Nakagawa T, Kajitani T, Togo S, et al. Deubiquitylation of histone H2A activates transcriptional initiation via trans-histone cross-talk with H3K4 di- and trimethylation. Genes Dev 2008;22:37–49.
[33] Reyes-Turcu FE, Ventii KH, Wilkinson KD. Regulation and cellular roles of ubiquitin-specific deubiquitinating enzymes. Annu Rev Biochem 2009;78:363–97.
[34] Zhang Z, Jones A, Joo HY, et al. USP49 deubiquitinates histone H2B and regulates cotranscriptional pre-mRNA splicing. Genes Dev 2013;27:1581–95.
[35] Zhang XY, Varthi M, Sykes SM, et al. The putative cancer stem cell marker USP22 is a subunit of the human SAGA complex required for activated transcription and cell-cycle progression. Mol Cell 2008;29:102–11.
[36] Flotho A, Melchior F. Sumoylation: a regulatory protein modification in health and disease. Annu Rev Biochem 2013;82:357–85.
[37] Shiio Y, Eisenman RN. Histone sumoylation is associated with transcriptional repression. Proc Natl Acad Sci USA 2003;100:13225–30.
[38] Hendriks IA, D'Souza RC, Yang B, Verlaan-de Vries M, Mann M, Vertegaal AC. Uncovering global SUMOylation signaling networks in a site-specific manner. Nat Struct Mol Biol 2014;21:927–36.
[39] Pearson CK. ADP-ribosylation reactions. Princ Med Biol 1995;4:305–22.
[40] Messner S, Hottiger MO. Histone ADP-ribosylation in DNA repair, replication and transcription. Trends Cell Biol 2011;21:534–42.
[41] Messner S, Altmeyer M, Zhao H, et al. PARP1 ADP-ribosylates lysine residues of the core histone tails. Nucleic Acids Res 2010;38:6350–62.
[42] Tanigawa Y, Tsuchiya M, Imai Y, Shimoyama M. ADP-ribosylation regulates the phosphorylation of histones by the catalytic subunit of cyclic AMP-dependent protein kinase. FEBS Lett 1983;160:217–20.
[43] Quintero-Ronderos P, Montoya-Ortiz G. Epigenetics and autoimmune diseases. Autoimmune Dis 2012; 2012:1–16.
[44] Ohshima S, Saeki Y, Mima T, et al. Interleukin 6 plays a key role in the development of antigen-induced arthritis. Proc Natl Acad Sci USA 1998;95:8222–6.
[45] MacGregor AJ, Snieder H, Rigby AS, et al. Characterizing the quantitative genetic contribution to rheumatoid arthritis using data from twins. Arthritis Rheum 2000;43:30–7.
[46] Kim K, Bang SY, Lee HS, et al. High-density genotyping of immune loci in Koreans and Europeans identifies eight new rheumatoid arthritis risk loci. Ann Rheumatic Dis 2015;74:e13.
[47] Grabiec AM, Reedquist KA. The ascent of acetylation in the epigenetics of rheumatoid arthritis. Nat Rev Rheumatol 2013;9:311–8.
[48] Trenkmann M, Brock M, Gay RE, et al. Expression and function of EZH2 in synovial fibroblasts: epigenetic repression of the Wnt inhibitor SFRP1 in rheumatoid arthritis. Ann Rheumatic Dis 2011;70:1482–8.
[49] Cao R, Wang L, Wang H, et al. Role of histone H3 lysine 27 methylation in Polycomb-group silencing. Science 2002;298:1039–43.
[50] Margueron R, Reinberg D. The Polycomb complex PRC2 and its mark in life. Nature 2011;469:343–9.
[51] Glant TT, Besenyei T, Kádár A, et al. Differentially expressed epigenome modifiers, including aurora kinases A and B, in immune cells in rheumatoid arthritis in humans and mouse models. Arthritis Rheum 2013;65:1725–35.
[52] Saccani S, Pantano S, Natoli G. p38-Dependent marking of inflammatory genes for increased NF-κ B recruitment. Nat Immunol 2002;3:69–75.
[53] Huber LC, Brock M, Hemmatazad H, et al. Histone deacetylase/acetylase activity in total synovial tissue derived from rheumatoid arthritis and osteoarthritis patients. Arthritis Rheum 2007;56:1087–93.
[54] Maciejewska-Rodrigues H, Karouzakis E, Strietholt S, et al. Epigenetics and rheumatoid arthritis: the role of SENP1 in the regulation of MMP-1 expression. J Autoimmun 2010;35:15–22.
[55] Murphy G, Lee MH. What are the roles of metalloproteinases in cartilage and bone damage? Ann Rheum Dis 2005;64:iv44–7.

[56] Wada TT, Araki Y, Sato K, et al. Aberrant histone acetylation contributes to elevated interleukin-6 production in rheumatoid arthritis synovial fibroblasts. Biochem Biophys Res Commun 2014;444:682–6.
[57] Chen LF, Greene WC. Assessing acetylation of NF-κB. Methods 2005;36:368–75.
[58] Niederer F, Ospelt C, Brentano F, et al. SIRT1 overexpression in the rheumatoid arthritis synovium contributes to proinflammatory cytokine production and apoptosis resistance. Ann Rheum Dis 2011;70:1866–73.
[59] Tsokos GC. Systemic lupus erythematosus. N Engl J Med 2011;365:2110–21.
[60] Burlingame RW, Rubin RL. Drug-induced antihistone autoantibodies display two patterns of reactivity with substructures of chromatin. J Clin Invest 1991;88:680–90.
[61] Zhang Z, Song L, Maurer K, Petri MA, Sullivan KE. Global H4 acetylation analysis by ChIP-chip in systemic lupus erythematosus monocytes. Genes Immun 2010;11:124–33.
[62] Sullivan KE, Suriano A, Dietzmann K, Lin J, Goldman D, Petri MA. The TNFα locus is altered in monocytes from patients with systemic lupus erythematosus. Clin Immunol 2007;123:74–81.
[63] Lieberman LA, Tsokos GC. The IL-2 defect in systemic lupus erythematosus disease has an expansive effect on host immunity. J Biomed Biotechnol 2010; 2010:1–6.
[64] Hedrich CM, Rauen T, Tsokos GC. Camp-responsive element modulator (CREM) protein signaling mediates epigenetic remodeling of the human Interleukin-2 gene: implications in systemic lupus erythematosus. J Biol Chem 2011;286:43429–36.
[65] Zhang Q, Long H, Liao J, et al. Inhibited expression of hematopoietic progenitor kinase 1 associated with loss of jumonji domain containing 3 promoter binding contributes to autoimmunity in systemic lupus erythematosus. J Autoimmun 2011;37:180–9.
[66] Varga J, Abraham D. Systemic sclerosis: a prototypic multisystem fibrotic disorder. J Clin Invest 2007;117:557–67.
[67] Broen JC, Radstake TR, Rossato M. The role of genetics and epigenetics in the pathogenesis of systemic sclerosis. Nat Rev Rheumatol 2014;10:671–81.
[68] Wang Y, Fan PS, Kahaleh B. Association between enhanced type I collagen expression and epigenetic repression of the FLI1 gene in scleroderma fibroblasts. Arthritis Rheum 2006;54:2271–9.
[69] Huber LC, Distler JH, Moritz F, et al. Trichostatin A prevents the accumulation of extracellular matrix in a mouse model of bleomycin-induced skin fibrosis. Arthritis Rheum 2007;56:2755–64.
[70] Bluestone JA, Herold K, Eisenbarth G. Genetics, pathogenesis and clinical interventions in type 1 diabetes. Nature 2010;464:1293–300.
[71] Miao F, Chen Z, Genuth S, et al. Evaluating the role of epigenetic histone modifications in the metabolic memory of type 1 diabetes. Diabetes 2014;63:1748–62.
[72] Miao F, Gonzalo IG, Lanting L, Natarajan R. In vivo chromatin remodeling events leading to inflammatory gene transcription under diabetic conditions. J Biol Chem 2004;279:18091–7.
[73] Miao F, Smith DD, Zhang L, Min A, Feng W, Natarajan R. Lymphocytes from patients with type 1 diabetes display a distinct profile of chromatin histone H3 lysine 9 dimethylation: an epigenetic study in diabetes. Diabetes 2008;57:3189–98.
[74] Soler DC, McCormick TS. The dark side of regulatory T cells in psoriasis. J Invest Dermatol 2011;131:1785–6.
[75] Han J, Park SG, Bae JB, et al. The characteristics of genome-wide DNA methylation in naïve CD4+ T cells of patients with psoriasis or atopic dermatitis. Biochem Biophys Res Commun 2012;422:157–63.
[76] Bovenschen HJ, van de Kerkhof PC, van Erp PE, et al. Foxp3+ regulatory T cells of psoriasis patients easily differentiate into IL-17A-producing cells and are found in lesional skin. J Invest Dermatol 2011;131:1853–60.
[77] Tovar-Castillo LE, Cancino-Díaz JC, García-Vázquez F, et al. Under-expression of VHL and over-expression of HDAC-1, HIF-1α, LL-37, and IAP-2 in affected skin biopsies of patients with psoriasis. Int J Dermatol 2007;46:239–46.
[78] Blander G, Bhimavarapu A, Mammone T, et al. SIRT1 promotes differentiation of normal human keratinocytes. J Invest Dermatol 2009;129:41–9.
[79] Zhang P, Su Y, Zhao M, Huang W, Lu Q. Abnormal histone modifications in PBMCs from patients with psoriasis vulgaris. Eur J Dermatol 2011;21:552–7.
[80] Blattner W, Gallo RC, Temin HM. HIV causes AIDS. Science 1988;241:515–6.
[81] Friedman J, Cho WK, Chu CK, et al. Epigenetic silencing of HIV-1 by the histone H3 lysine 27 methyltransferase enhancer of Zeste 2. J Virol 2011;85:9078–89.
[82] du Chéné I, Basyuk E, Lin YL, et al. Suv39H1 and HP1γ are responsible for chromatin-mediated HIV-1 transcriptional silencing and post-integration latency. EMBO J 2007;26:424–35.
[83] Imai K, Togami H, Okamoto T. Involvement of histone H3 lysine 9 (H3K9) methyltransferase G9a in the maintenance of HIV-1 latency and its reactivation by BIX 01294. J Biol Chem 2010;285:16538–45.
[84] Shirakawa K, Chavez L, Hakre S, Calvanese V, Verdin E. Reactivation of latent HIV by histone deacetylase inhibitors. Trends Microbiol 2013;21:277–85.
[85] Hardy J. A hundred years of Alzheimer's disease research. Neuron 2006;52:3–13.
[86] Ogawa O, Zhu X, Lee HG. Ectopic localization of phosphorylated histone H3 in Alzheimer's disease: a mitotic catastrophe? Acta Neuropathol 2003;105: 524–8.

[87] Francis YI, Fà M, Ashraf H, et al. Dysregulation of histone acetylation in the APP/PS1 mouse model of Alzheimer's disease. J Alzheimers Dis 2009;18:131–9.
[88] Zhang K, Schrag M, Crofton A, Trivedi R, Vinters H, Kirsch W. Targeted proteomics for quantification of histone acetylation in Alzheimer's disease. Proteomics 2012;12:1261–8.
[89] Gräff J, Rei D, Guan J-S, et al. An epigenetic blockade of cognitive functions in the neurodegenerating brain. Nature 2012;483:222–6.
[90] Marques SC, Lemos R, Ferreiro E, et al. Epigenetic regulation of BACE1 in Alzheimer's disease patients and in transgenic mice. Neuroscience 2012;220:256–66.
[91] Karagiannis TC, Ververis K. Potential of chromatin modifying compounds for the treatment of Alzheimer's disease. Pathobiol Aging Age Relat Dis 2012;2:14980.
[92] Qing H, He G, Ly PT, et al. Valproic acid inhibits Aβ production, neuritic plaque formation, and behavioral deficits in Alzheimer's disease mouse models. J Exp Med 2008;205:2781–9.
[93] Green KN, Steffan JS, Martinez-Coria H, et al. Nicotinamide restores cognition in Alzheimer's disease transgenic mice via a mechanism involving sirtuin inhibition and selective reduction of Thr231-phosphotau. J Neurosci 2008;28:11500–10.
[94] Olanow CW, Brundin P. Parkinson's disease and alpha synuclein: is Parkinson's disease a prion-like disorder? Mov Disord 2013;28:31–40.
[95] Kontopoulos E, Parvin JD, Feany MB. Alpha-synuclein acts in the nucleus to inhibit histone acetylation and promote neurotoxicity. Hum Mol Genet 2006;15:3012–23.
[96] Nicholas AP, Lubin FD, Hallett PJ, et al. Striatal histone modifications in models of levodopa-induced dyskinesia. J Neurochem 2008;106:486–94.
[97] Donmez G, Outeiro TF. SIRT1 and SIRT2: emerging targets in neurodegeneration. EMBO Mol Med 2013;5:344–52.
[98] Albani D, Polito L, Batelli S, et al. The SIRT1 activator resveratrol protects SK-N-BE cells from oxidative stress and against toxicity caused by alpha-synuclein or amyloid-β (1-42) peptide. J Neurochem 2009;110: 1445–56.
[99] Outeiro TF, Kontopoulos E, Altmann SM, et al. Sirtuin 2 inhibitors rescue α-synuclein-mediated toxicity in models of Parkinson's disease. Science 2007;317:516–9.
[100] St Laurent R, O'Brien LM, Ahmad ST. Sodium butyrate improves locomotor impairment and early mortality in a rotenone-induced Drosophila model of Parkinson's disease. Neuroscience 2013;246:382–90.
[101] Xie Y, Hayden MR, Xu B. BDNF overexpression in the forebrain rescues Huntington's disease phenotypes in YAC128 mice. J Neurosci 2010;30:14708–18.
[102] Steffan JS, Bodai L, Pallos J, et al. Histone deacetylase inhibitors arrest polyglutamine-dependent neurodegeneration in Drosophila. Nature 2001;413:739–43.
[103] Boutell JM, Thomas P, Neal JW, et al. Aberrant interactions of transcriptional repressor proteins with the Huntington's disease gene product, huntingtin. Hum Mol Genet 1999;8:1647–55.
[104] Parkel S, Lopez-Atalaya JP, Barco A. Histone H3 lysine methylation in cognition and intellectual disability disorders. Learn Mem 2013;20:570–9.
[105] Vashishtha M, Ng CW, Yildirim F, et al. Targeting H3K4 trimethylation in Huntington disease. Proc Natl Acad Sci 2013;110:E3027–36.
[106] Seong IS, Woda JM, Song JJ, et al. Huntingtin facilitates polycomb repressive complex 2. Hum Mol Genet 2010;19:573–83.
[107] Kim MO, Chawla P, Overland RP, Xia E, Sadri-Vakili G, Cha JH. Altered histone monoubiquitylation mediated by mutant huntingtin induces transcriptional dysregulation. J Neurosci 2008;28:3947–57.
[108] Ferrante RJ, Kubilus JK, Lee J, et al. Histone deacetylase inhibition by sodium butyrate chemotherapy ameliorates the neurodegenerative phenotype in Huntington's disease mice. J Neurosci 2003;23:9418–27.
[109] Chopra V, Quinti L, Kim J, et al. The sirtuin 2 inhibitor AK-7 is neuroprotective in Huntington's disease mouse models. Cell Rep 2012;2:1492–7.
[110] Ferrante RJ, Ryu H, Kubilus JK, et al. Chemotherapy for the brain: the antitumor antibiotic mithramycin prolongs survival in a mouse model of Huntington's disease. J Neurosci 2004;24:10335–42.
[111] Ryu H, Lee J, Hagerty SW, et al. ESET/SETDB1 gene expression and histone H3 (K9) trimethylation in Huntington's disease. Proc Natl Acad Sci USA 2006; 103:19176–81.
[112] Pandolfo M. Friedreich ataxia. Arch Neurol 2008;65: 1296–303.
[113] Herman D, Jenssen K, Burnett R, Soragni E, Perlman SL, Gottesfeld JM. Histone deacetylase inhibitors reverse gene silencing in Friedreich's ataxia. Nat Chem Biol 2006;2:551–8.
[114] Chan PK, Torres R, Yandim C, et al. Heterochromatinization induced by GAA-repeat hyperexpansion in Friedreich's ataxia can be reduced upon HDAC inhibition by vitamin B3. Hum Mol Genet 2013;22:2662–75.
[115] Kumari D, Biacsi RE, Usdin K. Repeat expansion affects both transcription initiation and elongation in Friedreich ataxia cells. J Biol Chem 2011;286:4209–15.
[116] Soragni E, Miao W, Iudicello M, et al. Epigenetic therapy for Friedreich ataxia. Ann Neurol 2014;76:489–508.
[117] Tillinghast GW, Partee J, Albert P. Analysis of genetic stability at the EP300 and CREBBP loci in a panel of cancer cell lines. Genes Chromosomes Cancer 2003;37:121–31.

[118] Yang XJ, Ullah M. MOZ and MORF, two large MYSTic HATs in normal and cancer stem cells. Oncogene 2007;26:5408–19.
[119] Borrow J, Stanton VP, Andresen JM, Becher R, Behm FG, Chaganti RS, et al. The translocation t(8;16)(p11;p13) of acute myeloid leukaemia fuses a putative acetyltransferase to the CREB-binding protein. Nat Genet 1996;14:33–41.
[120] Chaffanet M, Gressin L, Preudhomme C, Soenen-Cornu V, Birnbaum D, Pébusque MJ. MOZ is fused to p300 in an acute monocytic leukemia with t(8;22). Genes Chromosomes Cancer 2000;28:138–44.
[121] Fraga MF, Ballestar E, Villar-Garea A, Boix-Chornet M, Espada J, Schotta G, et al. Loss of acetylation at Lys16 and trimethylation at Lys20 of histone H4 is a common hallmark of human cancer. Nat Genet 2005;37:391–400.
[122] Glozak MA, Sengupta N, Zhang X, Seto E. Acetylation and deacetylation of non-histone proteins. Gene 2005;363:15–23.
[123] Krusche CA, Wülfing P, Kersting C, et al. Histone deacetylase-1 and -3 protein expression in human breast cancer: a tissue microarray analysis. Breast Cancer Res Treat 2005;90:15–23.
[124] Weichert W, Röske A, Niesporek S, et al. Class I histone deacetylase expression has independent prognostic impact in human colorectal cancer: specific role of class I histone deacetylases in vitro and in vivo. Clin Cancer Res 2008;14:1669–77.
[125] Weichert W, Röske A, Gekeler V, Beckers T, Stephan C, Jung K, et al. Histone deacetylases 1, 2 and 3 are highly expressed in prostate cancer and HDAC2 expression is associated with shorter PSA relapse time after radical prostatectomy. Br J Cancer 2008;98:604–10.
[126] Godman CA, Joshi R, Tierney BR, et al. HDAC3 impacts multiple oncogenic pathways in colon cancer cells with effects on Wnt and vitamin D signaling. Cancer Biol Ther 2008;7:1570–80.
[127] Tiffon C, Adams J, van der Fits L, Wen S, Townsend P, Ganesan A, et al. The histone deacetylase inhibitors vorinostat and romidepsin downmodulate IL-10 expression in cutaneous T-cell lymphoma cells. Br J Pharmacol 2011;162:1590–602.
[128] Ratner M. Small biotech steers HDAC inhibitor to clinic. Nat Biotechnol 2014;32:853–4.
[129] Goldenberg MM. Pharmaceutical approval update. Pharm Ther 2014;39:553–66.
[130] Hoffmann I, Roatsch M, Schmitt ML, et al. The role of histone demethylases in cancer therapy. Mol Oncol 2012;6:683–703.
[131] Sauvageau M, Sauvageau G. Polycomb group proteins: multi-faceted regulators of somatic stem cells and cancer. Cell Stem Cell 2010;7:299–313.
[132] Viré E, Brenner C, Deplus R, Blanchon L, Fraga M, Didelot C, et al. The Polycomb group protein EZH2 directly controls DNA methylation. Nature 2006;439:871–4.
[133] Kondo Y, Shen L, Cheng AS, Ahmed S, Boumber Y, Charo C, et al. Gene silencing in cancer by histone H3 lysine 27 trimethylation independent of promoter DNA methylation. Nat Genet 2008;40:741–50.
[134] Liu G, Bollig-Fischer A, Kreike B, van de Vijver MJ, Abrams J, Ethier SP, et al. Genomic amplification and oncogenic properties of the GASC1 histone demethylase gene in breast cancer. Oncogene 2009;28:4491–500.
[135] Yang ZQ, Imoto I, Fukuda Y, Pimkhaokham A, Shimada Y, Imamura M, et al. Identification of a novel gene, GASC1, within an amplicon at 9p23-24 frequently detected in esophageal cancer cell lines. Cancer Res 2000;60:4735–9.
[136] Hayami S, Kelly JD, Cho HS, Yoshimatsu M, Unoki M, Tsunoda T, et al. Overexpression of LSD1 contributes to human carcinogenesis through chromatin regulation in various cancers. Int J Cancer 2011;128:574–86.
[137] Lim S, Janzer A, Becker A, Zimmer A, Schüle R, Buettner R, et al. Lysine-specific demethylase 1 (LSD1) is highly expressed in ER-negative breast cancers and a biomarker predicting aggressive biology. Carcinogenesis 2010;31:512–20.
[138] Schildhaus HU, Riegel R, Hartmann W, Steiner S, Wardelmann E, et al. Lysine-specific demethylase 1 is highly expressed in solitary fibrous tumors, synovial sarcomas, rhabdomyosarcomas, desmoplastic small round cell tumors, and malignant peripheral nerve sheath tumors. Hum Pathol 2011;42:1667–75.
[139] Schwartzentruber J, Korshunov A, Liu XY, Jones DT, Pfaff E, Jacob K, et al. Driver mutations in histone H3.3 and chromatin remodelling genes in paediatric glioblastoma. Nature 2012;482:226–31.
[140] Behjati S, Tarpey PS, Presneau N, Scheipl S, Pillay N, Van Loo P, et al. Distinct H3F3A and H3F3B driver mutations define chondroblastoma and giant cell tumor of bone. Nat Genet 2013;45:1479–82.
[141] Wu G, Broniscer A, McEachron TA, Lu C, Paugh BS, Becksfort J, et al. Somatic histone H3 alterations in pediatric diffuse intrinsic pontine gliomas and non-brainstem glioblastomas. Nat Genet 2012;44:251–3.
[142] Chan KM, Fang D, Gan H, Hashizume R, Yu C, Schroeder M, et al. The histone H3.3K27M mutation in pediatric glioma reprograms H3K27 methylation and gene expression. Genes Dev 2013;27:985–90.
[143] Bjerke L, Mackay A, Nandhabalan M, Burford A, Jury A, Popov S, et al. Histone H3.3. mutations drive pediatric glioblastoma through upregulation of MYCN. Cancer Discov 2013;3:512–9.

[144] Rossetto D, Avvakumov N, Côté J. Histone phosphorylation: a chromatin modification involved in diverse nuclear events. Epigenetics 2012;7:1098–108.
[145] Dawson MA, Bannister AJ, Göttgens B, Foster SD, Bartke T, Green AR, et al. JAK2 phosphorylates histone H3Y41 and excludes HP1α from chromatin. Nature 2009;461:819–22.
[146] Espino PS, Li L, He S, Yu J, Davie JR. Chromatin modification of the trefoil factor 1 gene in human breast cancer cells by the Ras/mitogen-activated protein kinase pathway. Cancer Res 2006;66:4610–6.
[147] Yang X, Cao W, Zhang L, Zhang W, Zhang X, Lin H. Targeting 14-3-3ζ in cancer therapy. Cancer Gene Ther 2012;19:153–9.
[148] LeRoy G, Dimaggio PA, Chan EY, Zee BM, Blanco MA, Bryant B, et al. A quantitative atlas of histone modification signatures from human cancer cells. Epigenetics Chromatin 2013;6:20.
[149] Loddo M, Kingsbury SR, Rashid M, Proctor I, Holt C, Young J, et al. Cell-cycle-phase progression analysis identifies unique phenotypes of major prognostic and predictive significance in breast cancer. Br J Cancer 2009;100:959–70.
[150] Huang X, Halicka HD, Traganos F, Tanaka T, Kurose A, Darzynkiewicz Z. Cytometric assessment of DNA damage in relation to cell cycle phase and apoptosis. Cell Prolif 2005;38:223–43.
[151] Löbrich M, Rief N, Kühne M, Heckmann M, Fleckenstein J, Rübe C, et al. In vivo formation and repair of DNA double-strand breaks after computed tomography examinations. Proc Natl Acad Sci USA 2005;102:8984–9.
[152] Wasco MJ, Pu RT. Utility of antiphosphorylated H2AX antibody (γ-H2AX) in diagnosing metastatic renal cell carcinoma. Appl Immunohistochem Mol Morphol 2008;16:349–56.
[153] Zhu P, Zhou W, Wang J, Puc J, Ohgi KA, Erdjument-Bromage H, et al. A histone H2A deubiquitinase complex coordinating histone acetylation and H1 dissociation in transcriptional regulation. Mol Cell 2007;27:609–21.
[154] Prenzel T, Begus-Nahrmann Y, Kramer F, Hennion M, Hsu C, Gorsler T, et al. Estrogen-dependent gene transcription in human breast cancer cells relies upon proteasome-dependent monoubiquitination of histone H2B. Cancer Res 2011;71:5739–53.
[155] Shema E, Tirosh I, Aylon Y, Huang J, Ye C, Moskovits N, et al. The histone H2B-specific ubiquitin ligase RNF20/hBRE1 acts as a putative tumor suppressor through selective regulation of gene expression. Genes Dev 2008;22:2664–76.
[156] Ford D, Easton DF, Bishop DT, Narod SA, Goldgar DE. Risks of cancer in BRCA1-mutation carriers. Breast Cancer Linkage Consortium. Lancet 1994;343:692–5.
[157] Mallery DL, Vandenberg CJ, Hiom K. Activation of the E3 ligase function of the BRCA1/BARD1 complex by polyubiquitin chains. EMBO J 2002;21:6755–62.
[158] Zhu Q, Pao GM, Huynh AM, Suh H, Tonnu N, Nederlof PM, et al. BRCA1 tumour suppression occurs via heterochromatin-mediated silencing. Nature 2011;477:179–84.
[159] Zhang Y, Yao L, Zhang X, Ji H, Wang L, et al. Elevated expression of USP22 in correlation with poor prognosis in patients with invasive breast cancer. J Cancer Res Clin Oncol 2011;137:1245–53.
[160] Zhao Y, Lang G, Ito S, Bonnet J, Metzger E, Sawatsubashi S, et al. A TFTC/STAGA module mediates histone H2A and H2B deubiquitination, coactivates nuclear receptors, and counteracts heterochromatin silencing. Mol Cell 2008;29:92–101.
[161] Glinsky GV. Death-from-cancer signatures and stem cell contribution to metastatic cancer. Cell Cycle 2005;4:1171–5.
[162] Lv L, Xiao XY, Gu Z-H, Zeng FQ, Huang LQ, Jiang GS. Silencing USP22 by asymmetric structure of interfering RNA inhibits proliferation and induces cell cycle arrest in bladder cancer cells. Mol Cell Biochem 2011;346:11–21.
[163] Urasaki Y, Heath L, Xu CW. Coupling of glucose deprivation with impaired histone H2B monoubiquitination in tumors. PLoS One 2012;7:e36775.
[164] Hahn MA, Dickson KA, Jackson S, Clarkson A, Gill AJ, Marsh DJ. The tumor suppressor CDC73 interacts with the ring finger proteins RNF20 and RNF40 and is required for the maintenance of histone 2B monoubiquitination. Hum Mol Genet 2012;21:559–68.
[165] Ellinger J, Kahl P, Gathen von der J, Rogenhofer S, Heukamp LC, Gütgemann I, et al. Global levels of histone modifications predict prostate cancer recurrence. Prostate 2010;70:61–9.
[166] Ellinger J, Kahl P, Mertens C, Rogenhofer S, Hauser S, Hartmann W, et al. Prognostic relevance of global histone H3 lysine 4 (H3K4) methylation in renal cell carcinoma. Int J Cancer 2010;127:2360–6.
[167] Barlési F, Giaccone G, Gallegos-Ruiz MI, Loundou A, Span SW, Lefesvre P, et al. Global histone modifications predict prognosis of resected non small-cell lung cancer. J Clin Oncol 2007;25:4358–64.
[168] Seligson DB, Horvath S, Shi T, Yu H, Tze S, et al. Global histone modification patterns predict risk of prostate cancer recurrence. Nat Cell Biol 2005;435:1262–6.
[169] Bianco-Miotto T, Chiam K, Buchanan G, Jindal S, Day TK, Thomas M, et al. Global levels of specific histone modifications and an epigenetic gene signature predict prostate cancer progression and development. Cancer Epidemiol Biomarkers Prev 2010;19:2611–22.

[170] Manuyakorn A, Paulus R, Farrell J, Dawson NA, et al. Cellular histone modification patterns predict prognosis and treatment response in resectable pancreatic adenocarcinoma: results from RTOG 9704. J Clin Oncol 2010;28:1358–65.

[171] He C, Xu J, Zhang J, Xie D, Ye H, Xiao Z, et al. High expression of trimethylated histone H3 lysine 4 is associated with poor prognosis in hepatocellular carcinoma. Hum Pathol 2012;43:1425–35.

[172] LeRoy G, Dimaggio PA, Chan EY, Zee BM, Blanco MA, Bryant B, et al. A quantitative atlas of histone modification signatures from human cancer cells. Epigenetics Chromatin 2013;6:20.

[173] Song JS, Kim YS, Kim DK, Park SI, Jang SJ. Global histone modification pattern associated with recurrence and disease-free survival in non-small cell lung cancer patients. Pathol Int 2012;62:182–90.

[174] Rogenhofer S, Kahl P, Holzapfel S, Ruecker von A, Mueller SC, Ellinger J. Decreased levels of histone H3K9me1 indicate poor prognosis in patients with renal cell carcinoma. Anticancer Res 2012;32:879–86.

[175] Elsheikh SE, Green AR, Rakha EA, Powe DG, Ahmed RA, Collins HM, et al. Global histone modifications in breast cancer correlate with tumor phenotypes, prognostic factors, and patient outcome. Cancer Res 2009;69:3802–9.

[176] Leszinski G, Gezer U, Siegele B, Stoetzer O, Holdenrieder S. Relevance of histone marks H3K9me3 and H4K20me3 in cancer. Anticancer Res 2012;32:2199–205.

[177] Müller-Tidow C, Klein HU, Hascher A, Isken F, Tickenbrock L, Thoennissen N, et al. Profiling of histone H3 lysine 9 trimethylation levels predicts transcription factor activity and survival in acute myeloid leukemia. Blood 2010;116:3564–71.

[178] Park YS, Jin MY, Kim YJ, Yook JH, Kim BS, Jang SJ. The global histone modification pattern correlates with cancer recurrence and overall survival in gastric adenocarcinoma. Ann Surg Oncol 2008;15:1968–76.

[179] Seligson DB, Horvath S, McBrian MA, Mah V, Yu H, Tze S, et al. Global levels of histone modifications predict prognosis in different cancers. Am J Pathol 2009;174:1619–28.

[180] Tzao C, Tung HJ, Jin JS, Sun GH, Hsu HS, et al. Prognostic significance of global histone modifications in resected squamous cell carcinoma of the esophagus. Mod Pathol 2008;22:252–60.

[181] I H, Ko E, Kim Y, Cho EY, Han J, et al. Association of global levels of histone modifications with recurrence-free survival in stage IIB and III esophageal squamous cell carcinomas. Cancer Epidemiol Biomarkers Prev 2010;19:566–73.

[182] Karczmarski J, Rubel T, Paziewska A, et al. Histone H3 lysine 27 acetylation is altered in colon cancer. Clin Proteomics 2014;11:24.

[183] Rogenhofer S, Kahl P, Mertens C, Hauser S, et al. Global histone H3 lysine 27 (H3K27) methylation levels and their prognostic relevance in renal cell carcinoma. BJU Int 2012;109:459–65.

[184] Ellinger J, Kahl P, Gathen von der J, Heukamp LC, et al. Global histone H3K27 methylation levels are different in localized and metastatic prostate cancer. Cancer Invest 2012;30:92–7.

[185] Zhang L, Zhong K, Dai Y, Zhou H. Genome-wide analysis of histone H3 lysine 27 trimethylation by ChIP-chip in gastric cancer patients. J Gastroenterol 2009;44:305–12.

[186] Cai MY, Hou JH, Rao HL, Luo RZ, et al. High expression of H3K27me3 in human hepatocellular carcinomas correlates closely with vascular invasion and predicts worse prognosis in patients. Mol Med 2011;17:12–20.

[187] Sasaki H, Setoguchi T, Matsunoshita Y, Gao H, Hirotsu M, Komiya S. The knock-down of overexpressed EZH2 and BMI-1 does not prevent osteosarcoma growth. Oncol Rep 2010;23:677–84.

[188] Hamamoto R, Furukawa Y, Morita M, Iimura Y, Silva FP, Li M, et al. NUP98-NSD1 links H3K36 methylation to Hox-A gene activation and leukaemogenesis. Nat Cell Biol 2007;9:804–12.

[189] Van Den Broeck A, Brambilla E, Moro-Sibilot D, Lantuejoul S, et al. Loss of histone H4K20 trimethylation occurs in preneoplasia and influences prognosis of non-small cell lung cancer. Clin Cancer Res 2008;14:7237–45.

[190] Behbahani TE, Kahl P, Gathen von der J, Heukamp LC, et al. Alterations of global histone H4K20 methylation during prostate carcinogenesis. BMC Urol 2012;12:5.

CHAPTER

22

Histone Posttranslational Modifications and Chromatin Remodelers in Prostate Cancer

Filipa Quintela Vieira[1,2,a], *Diogo Almeida-Rios*[1,3,a], *Inês Graça*[1,2], *Rui Henrique*[1,3,4], *Carmen Jerónimo*[1,4]

[1]Cancer Biology and Epigenetics Group – Research Center, Portuguese Oncology Institute, Porto, Portugal; [2]School of Allied Health Sciences (ESTSP), Polytechnic of Porto, Porto, Portugal; [3]Department of Pathology, Portuguese Oncology Institute, Porto, Portugal; [4]Department of Pathology and Molecular Immunology, Institute of Biomedical Sciences Abel Salazar (ICBAS), University of Porto, Porto, Portugal

OUTLINE

[a]Joint first authors.

Epigenetic Biomarkers and Diagnostics
http://dx.doi.org/10.1016/B978-0-12-801899-6.00022-X

1. INTRODUCTION

1.1 Prostate Cancer

Prostate cancer (PCa; MIM #176807) is the second most frequently diagnosed cancer in men and fourth most incident cancer, globally. However, almost 75% of cases were diagnosed in developed countries, making PCa the most frequent malignant tumor among men in those regions [1,2]. Worldwide, incident rates vary more than 25-fold, due to differences in ethnicity, environment, and prevalence of prostate-specific antigen (PSA) screening for subsequent biopsy [3]. In contrast, mortality rates vary considerably less (5- to 10-fold) but, worldwide, PCa constitutes the fifth leading cause of death from cancer in men with an estimated 307,000 deaths in 2012 [2].

PCa is frequently asymptomatic at its early stages and, clinically, it ranges from indolent disease carrying good prognosis (the vast majority of cases) to extremely aggressive tumors leading to patients' death [4]. Prevention might have a slight impact on disease-related mortality, whereas diagnostic and prognostic evaluation is crucial both to earlier identification of clinically relevant tumors and to better adjust therapeutic options in accordance with disease aggressiveness [5,6]. Thus, a major challenge is to identify men with aggressive disease, for which suitable treatment is critical to attain a cure, and spare patients with clinically insignificant disease from unnecessary treatment morbidity.

Abnormalities in epigenetic mechanisms, namely histone modifications, contribute to cancer initiation and progression [7]. Because deregulation of histone modifiers is involved in tumorigenesis, these might be used as biomarkers to assist in diagnostics or for stratification of cancer patients into subgroups with different prognosis, aiming at appropriate selection for targeted therapies [8–11]. Additionally, understanding the role of these enzymes and consequently of histone modifications patterns in carcinogenesis and addressing the mechanisms underlying cancer cells, phenotypic plasticity may allow for the discovery of novel therapeutic targets.

1.2 Histone Modifications

Currently, epigenetics comprises a set of processes with long-term effects on gene expression programs but no interference with the DNA sequence [12]. These mechanisms consist of several heritable changes that are established early during embryonic development, responsible for initiation and maintenance of cellular differentiation, even within cell replication and division, which might explain different cell phenotypes for identical genetic information [13,14].

Epigenetics machinery is critical to chromatin structure and gene transcriptional activity, comprising three major mechanisms: DNA methylation, covalent modifications of histones, and noncoding RNAs [15]. Although part of cellular natural and normal physiology, epigenetic patterns might be modified in response to intrinsic and extrinsic stimuli that can lead to gene expression deregulation and, ultimately, to disease onset [12,16].

Histones are highly conserved proteins that are tightly bound to DNA, involved in regulation of gene expression. Together, histones and DNA form the so-called nucleosomes that consist of 146 bp of DNA coiled in sequence around a core of eight histone proteins, two of each H2A, H2B, H3, and H4, linked by H1 [17,18].

Histones maintain contact with DNA through their flexible globular domain and amino acid residue regions (lysine, serine, and arginine) that protrude from the nucleosome, the so-called histone "tails." The majority of post-translational modifications (PTMs) in histones occur at this region [19]. The pattern of histone modifications determines chromatin status (euchromatin vs heterochromatin), and accessibility of DNA to nuclear factors and subsequent gene transcription [20]. Heterochromatin has a highly packaged conformation comprising mostly inactive genes, while euchromatin is relatively uncondensed and represents *loci* being actively transcribed [18]. Posttranslational histone modifications encompass methylation, acetylation, phosphorylation, ubiquitylation, and sumoylation, and the first two are the most abundant and better characterized so far [21,22]. Acetylated histones are generally associated with less compact chromatin that facilitates access to transcriptional machinery and thereby leads to gene activation. In contrast, methylated histones may be associated either to gene repression or activation, depending on the amino acidic residue and the number of methyl groups added. Thus, the "histone code" comprises a combination of modifications on each histone and/or nucleosome that is strongly related with gene transcription status. The primary protein families involved in this process are the enzymes responsible for adding acetyl or methyl groups to histone tails, termed writers, including histone acetyltransferases (HATs) and histone methyltransferases (HMTs), as well as erasers, responsible for the removal of these marks, which include histone deacetylases (HDACs) and demethylases (HDMs) [11,23].

Histone modifications are thought to alter histones' electrostatic charge resulting in a structural change that results in higher or lower affinity to DNA binding. Additionally, these modifications may serve as binding sites for protein recognition modules, such as the bromodomains or chromodomains, which are termed readers, since they recognize acetylated lysines or methylated lysines, respectively.

1.2.1 Histone Acetylation

Protein acetylation is one of the most extensively characterized and critical PTM by which crucial steps of cell functions are regulated at a molecular level [24]. Acetylation of histones' N-terminal domains is a dynamic process controlled by the antagonistic actions of two large families of enzymes: the HATs and the HDACs, which maintain the equilibrium of acetyl groups added or removed from lysine residues, respectively [17]. HATs are considered as transcription coactivators since they induce the relaxation of chromatin structure, whereas HDACs promote chromatin compaction and act as corepressors. Therefore, the final balance between the interplay of these enzymes will contribute to gene expression regulation that can be crucial for the developmental of cancer [17]. Although HDACs were initially identified as involved in transcriptional regulation by targeting histones, it is known that they also target nonhistone proteins (e.g., transcription factors or cytoskeletal proteins) [24].

HATs catalyze the transfer of an acetyl group from acetyl-CoA to ε-amino group of a histone lysine residue, allowing transcriptional machinery to access DNA. HATs can be grouped into two classes concerning subcellular localization: Type A HATs are localized in the nucleus and are mainly responsible for acetylation of nucleosomal histones, whereas type B HATs are cytoplasmic and acetylate newly synthesized histones to promote their assembly into nucleosomes [25]. Based on the structural and functional similarity of their catalytic domains, HATs can be categorized into five families, some of them embracing several enzymes, namely Gcn5-related N-acetyltransferases and MYST families, as well as other small families like HAT1, p300/CBP, and Rtt109. Although other nuclear HAT subfamilies have been identified, their HAT activities have not been studied as extensively as the five major HAT classes, thus far [26].

To date, 18 HDACs have been identified in humans that differ in sequence similarity and cofactor dependency and therefore are categorized into two families and four classes. The classical zinc-dependent HDAC family has 11 members and includes class I, II, and IV, while NAD+-dependent sirtuin family is composed by class III (sirtuins 1–7). Class I contains

HDAC1, 2, 3, and 8; class II is divided into class IIa (HDAC4, 5, 7, and 9) and IIb (HDAC6 and 10); and class IV comprises as unique member HDAC11 [27,28].

1.2.2 *Histone Methylation*

The role of histone methylation in PCa has been increasingly elucidated over the past decades. In contrast to histone acetylation, histone methylation does not alter histone tail ionic charge, but instead it determines its basicity, hydrophobicity, and transcription factors' affinity toward DNA [29]. Histone methylation may occur at lysine residues, by lysine histone methyltransferases (KMT), or at arginine residues, by protein arginine methyltransferases (PRMT), and it may positively or negatively regulate gene transcription. The reversibility of histone methylation has been established through the discovery of histone lysine and arginine demethylases (HDMs), uncovering a new level of histone plasticity [30,31]. Whereas lysine residues might be modified into mono-, di-, or trimethyl states, arginine can only be modified to mono- or dimethyl states (symmetrically or asymmetrically) [32]. Thus, different degrees of methylation may be associated with distinct chromatin regions or transcriptional states [33].

Tri- and dimethylation of lysine 4 of histone H3 (H3K4me3 and H3K4me2, respectively) are highly enriched at transcriptionally competent or active gene promoters, but the monomethylated form of H3K4 is associated with gene enhancers. Trimethylation of lysine 27 of histone H3 (H3K27me3), on the other hand, is found at transcriptionally repressed promoters, whereas silent pericentric heterochromatin is marked by trimethylation of lysine 27 of histone H3 (H3K9me3) [34]. Recent data have enlightened the role of histone arginine methylation in transcription regulation: it promotes or antagonizes the interaction of nuclear factors with other nearby histone marks [35,36].

Thus far, more than 50 HMTs and 30 HDMs have been identified [23,37–39]. Concerning methylation, all HMTs use S-adenosylmethionine as a cosubstrate for methyl groups transfer, and it may be divided into three classes of histone-methylating enzymes: SET domain lysine methyltransferases, non-SET domain lysine methyltransferases, and arginine methyltransferases [40]. The specificity of these enzymes has become a controversial subject as they may target nonhistone proteins alongside with several histone residues [37]. The arginine methylases might be classified according to the type of methylation: type I enzymes catalyze symmetric arginine methylation, whereas type II enzymes are responsible for the asymmetric process [35]. Regarding HDMs, they are categorized into two different groups: Lys-specific demethylases (LSD) and Jumonji C (JMJC) HDMs. The JMJC family demethylates mono-, di-, and trimethylated lysines, whereas LSD family enzymes proved unable to catalyze demethylation of the trimethylated state [39,41].

1.3 Histone Modifications in PCa

Histone PTMs have been associated with prostate carcinogenesis but current knowledge is still limited and further investigation is required [13].

Concerning histone acetylases, little is known concerning their involvement in PCa. Recently, p300, a HAT, was reported to be overexpressed and associated with proliferation and progression of this malignancy [42].

Higher expression levels of HDAC1 [43–46], HDAC2 [46,47], and HDAC3 [46] have been reported in PCa, suggesting a role of these histone modifiers in cancer progression. Additionally, HDAC2 expression was found to be of prognostic value in PCa [46]. Among class III HDACs, SIRT6 proved to be associated with an increase of cell viability and its depletion enhanced chemotherapeutics sensitivity [48]. The role of SIRT1 in PCa is not clearly understood. Some authors support the status of SIRT1 as a tumor suppressor with lower expression levels in prostate tumors [49,50], while others demonstrated that silencing SIRT1 leads to a decrease of cellular growth, viability, and chemoresistance in PCa cell lines defending an oncogenic role of this enzyme [51]. Additionally, SIRT1 was shown to be associated with androgen-mediated transcription, and its activity was related with sensitivity of PCa cells to the transcriptional and proliferative activities of androgens [52].

Regarding histone methylation, dimethylation of arginine 3 of histone H4 (H4R3me2) and H3K4me2 levels enabled the distinction between two groups of low-grade PCa (Gleason score (GS) 6 or less) with different outcomes [10]. Moreover, some histone-modifying patterns, namely H3K4me2 and H3K4me1, were associated with increased risk of PCa recurrence [8,53]. Similarly, both methylation of H3K4 and H3K27 have been correlated with tumor grade and recurrence [8,54]. Nevertheless, it is noteworthy that the presence of multiple epigenetic marks should be interpreted carefully; indeed, the importance of the balance of PTMs it is currently recognized, especially those of bivalent marks such as H3K4/K27me3 [55].

Deregulation of some lysine histone methylases—MLL2, MLL3, NSD1, EZH2, or SMYD3—in PCa tissues has been also demonstrated [8,9]. However, due to inappropriate tissue sampling and/or the reduced number of samples tested, the reliability of most studies is rather limited. Nevertheless, EZH2, a member of the Polycomb complex components, was already proved to be a driver in prostate carcinogenesis [10,56]. EZH2 is an HMT that catalyzes repressive marks, such as H3K27me3 and, occasionally, H3K9me2, associated with heterochromatinization, gene repression, and ability to influence DNA methylation [57-61]. Furthermore, EZH2 was found overexpressed in metastatic PCa, being associated with high proliferation rate and tumor aggressiveness [62,63].

Since methylation of H3K9 is linked with repression of androgen receptor (AR)-regulated genes in PCa cells, silencing of its specific demethylases, namely LSD1, JHDM2A, or JMJD2C, leads to increased levels of those repressive marks on AR-targeted genes' regulatory regions and, thus, to a decrease in its expression [64,65]. In fact, LSD1 demethylates H3K4 and H3K9, and its upregulation is associated with aggressive and castration-resistant PCa as well as with higher risk of disease relapse [66–70].

Recent studies suggested that histone modifiers' deregulation might be crucial for cancer onset and progression. Indeed, in cancer cells, some of these enzymes may display altered expression and/or activity, and the mechanisms underlying this deregulation may include genetic alterations such as chromosomal translocations, gene mutations, or fusion proteins. Genetic alterations in histone modifiers may serve as biomarkers for patient stratification enabling treatment with specific inhibitors of those enzymes [71,72].

2. EPIGENETIC-BASED MARKERS FOR PCa DETECTION, MANAGEMENT, AND RISK ESTIMATION

Owing to biological and clinical heterogeneity of PCa, a major challenge for clinicians is to decide the best treatment option for each individual patient, balancing the survival benefits of therapy and its morbidity, with impact in patients' quality of life [73]. The lack of specific symptoms while PCa is organ confined leads to late diagnosis and compromises not only prognosis but also treatment success [74]. Thus, unraveling the mechanisms underlying prostate carcinogenesis may allow for the development of molecular markers with clinical application for early detection, diagnosis, and prognostication of PCa.

Currently, there is no single biomarker able to predict PCa progression at the time of diagnosis. Serum PSA is the only biomarker available for detection and monitoring of treatment efficacy in PCa patients but it meets with important limitations due to its lack of specificity as well as its inability to discriminate clinically significant from clinically insignificant PCa. Moreover, these limitations of PSA testing have been causally related with overdiagnosis and overtreatment [75].

Epigenetic alterations have been proposed as potential biomarkers for PCa diagnosis, prognosis, and response to treatment since they are highly prevalent in PCa, being involved in tumor initiation and progression, occurring more frequently than mutations [76]. In this section, the biomarker potential of aberrations in expression levels of both histone modifications and histone modifiers in PCa is discussed (summarized in Table 1).

2.1 Histone Covalent PTMs

Few studies have been performed to ascertain the usefulness of global patterns of histone

TABLE 1 Histone Modifications and Modifiers Biomarker Profile for Diagnosis and Prognosis of Prostate Cancer

Biomarker	Significance in PCa	Detection/prognostic
Histone Covalent Posttranslational Modifications		
Global levels of H3Ac and H3K9me2	Global levels were able to distinguish malignant from nonmalignant prostate tissues	Detection
Global levels of H3K4me2 and H3K18Ac	Correlated with biochemical recurrence in low-grade PCa; independent predictors of PCa progression, independent of tumor grade	Prognostic
Global levels of H3K4me1/H3K4me3	High levels of H3K4me1 correlated with biochemical recurrence in RP-treated patients; H3K4me3 levels were found to predict biochemical recurrence in low-grade tumors	Prognostic
Global levels of H3K27me3	Using IHC analysis, showed an overexpression in metastatic tumors compared to nonmalignant tissues; using ELISA to analyze plasma levels in PCa patients, decreased levels were found in metastatic tumors compared to localized PCa	Prognostic
Global levels of H4K20me1/H4K20me2 and H4K20me3	H4K20me1 levels were correlated with lymph node metastases; H4K20me2 levels were significantly correlated with Gleason score in patients with localized PCa; along with preoperative PSA levels, H4K20me3 levels could predict the risk of the biochemical recurrence after RP	Prognostic
Histone Modifiers		
EZH2	High expression associated with Gleason score, TNM stage, and disease progression; in localized PCa its overexpression correlated with risk of biochemical recurrence	Prognostic
LSD1	Higher expression levels were associated with shorter relapse-free survival in patients treated with RP	Prognostic
SMYD3	Overexpression in PCa correlated with shorter time to tumor recurrence in organ-confined PCa after RP	Prognostic
DAXX	Strong expression was correlated with early PSA recurrence	Prognostic

Pca, prostate cancer; RP, radical prostatectomy; IHC, immunohistochemistry; PSA, prostate-specific antigen.

modifications for PCa detection as well as its prognostic and predictive value. Ellinger and coworkers found a reduction of global levels of H3K4me1, H3K9me2, H3K9me3, acetylation of histone 3 (H3Ac), and acetylation of histone 4 (H4Ac) in PCa compared to nonneoplastic prostate tissues [53]. Remarkably, levels of H3Ac and H3K9me2 were able to distinguish between malignant and nonmalignant prostatic tissue samples with approximately 80% sensitivity and over 90% specificity [53].

Similarly, global levels of specific histone marks have been also associated with clinical behavior of PCa. Seligson et al., using immunohistochemistry (IHC), analyzed global levels of H3K9Ac, H3K18Ac, H4K12Ac, H4R3me2, and H3K4me2 in a series of 183 primary PCa samples [10]. They found that a subset of low-grade tumors (GS 6 or less) displaying intense immunostaining for H3K4me2 and H3K18Ac had a lower recurrence risk [10]. Subsequently, in a series of primary and metastatic PCa samples, Bianco-Miotto et al. reported that H3K4me2 and H3K18Ac were independent predictors of cancer progression, independently of tumor grade. In fact, high global levels of those marks were correlated with a threefold increased risk of PCa recurrence [8].

Global levels of H3K4me1 have been also associated with PCa clinical course as high levels of this histone mark were significantly correlated with biochemical recurrence after radical prostatectomy (RP) [53]. Moreover, H3K4me3 was found to predict biochemical recurrence after RP in patients with low-grade PCa (GS 6 or less) [77]. Together, these findings suggest that analysis of H3K4 methylation status might be of prognostic value.

Another histone mark with prognostic significance is H3K27me3. This repressive mark was found overexpressed in metastatic tumors compared with nonmalignant prostate tissue [78]. Contrarily, another study that evaluated the expression of this mark in PCa patients found decreased levels in metastatic compared to organ-confined PCa [79]. Differences observed in those studies might be a result of different methodologies. Whereas Ellinger and colleagues evaluated H3K27me3 in prostate tissues using IHC analysis, whereas Deligezer and coworkers performed enzyme-linked immunosorbent assay (ELISA) in plasma samples.

Finally, H4K20 methylation levels have been also reported to bear prognostic significance. Increased H4K20me1 levels were correlated with lymph node metastases, whereas H4K20me2 was significantly correlated with Gleason score, in patients with localized PCa [80]. In a different study, H4K20me3 levels, along with preoperative PSA levels, were shown to predict the risk of the biochemical recurrence after RP [77].

2.2 Histone Modifiers

In similarity to altered patterns of histone modification, enzymes that establish these marks were found deregulated in PCa, the best characterized of which is EZH2 [8,81]. EZH2 expression was able to predict tumor progression with a better performance than serum PSA and Gleason score, and in organ-confined PCa, its overexpression correlated with risk of biochemical recurrence [8]. Moreover, EZH2 expression levels in PCa tissue samples were able to predict biochemical recurrence after RP and overexpression of this methyltransferase was also correlated with Gleason score, TNM stage, and disease progression [82]. We also found that EZH2 transcript levels were associated with shorter time to biochemical recurrence, although only in univariate analysis [81].

LSD1 is another histone modifier deregulated in PCa with prognostic significance, as higher expression levels were associated with shorter relapse-free survival in patients treated with RP [68].

Recently, higher SMYD3 expression has been associated with shorter time to biochemical relapse after RP, retaining statistical significance as independent prognostic biomarker in multivariate analysis, in addition to Gleason score and pathological stage [81]. Nevertheless, to confirm SMYD3 levels as putative risk stratification tool in PCa, further validation is required.

Finally, even if in principle death-associated protein 6 (DAXX) may not be considered a histone modifier, this histone variant H3.3-specific chaperone was recently evaluated by IHC on a tissue microarray of PCa samples and strong expression of this enzyme was also correlated with early PSA recurrence [83].

Epigenetic alterations are promising biomarkers to improve PCa detection and management. However, the role of covalent histone PTMs in PCa is still poorly understood, hampering their potential value as cancer biomarkers. Indeed, few studies have evaluated the diagnostic or predictive potential of histone modifications in PCa. A major limitation lies in the methodologies currently used to assess histone modifications and modifiers, which vary from study to study. IHC has been the most widely used method to analyze global patterns of histone modifications. Nevertheless, variations in the technical procedure (e.g., different antigen retrieval or different antibodies) impair the reliability and reproducibility of the results. Therefore, improvements and standardization of techniques is mandatory, as well as the development of more accurate noninvasive methodologies to detect histone marks in body fluids. Finally, large multicentric studies including large cohorts of PCa patients with age-matched controls are imperative to ascertain the clinical utility of histone modifications as PCa biomarkers, comparing histone modifications in body fluids with currently available clinicopathological parameters used for PCa detection and management.

3. EPIGENETIC SILENCING AS A THERAPEUTIC TARGET IN PCa

Several promising agents that inhibit key enzymes involved in establishing/writers, removing/editors, and maintaining/readers the epigenetic profiles have been identified [14]. The majority of these inhibitors have already been tested in preclinical models of PCa and a few are already in clinical trials.

3.1 Histone Deacetylase Inhibitors

Because classical HDACs overexpression is common in cancer, HDAC inhibitors (HDACi) have emerged as promising therapeutic agents. Currently the best studied are HDACi target class I and class II HDACs. HDACi are chemically classified into different subgroups based on their structure: aliphatic acids (phenylbutyrate and valproic acid (VPA)), benzamides (entinostat), cyclic peptides (romidepsin), and hydroxamic acids (trichostatin A (TSA) and vorinostat/SAHA) [84,85].

Inhibition of HDAC activity frequently reduces cell proliferation and angiogenesis, and induces differentiation and apoptosis, which are highly desirable features for cancer therapy. However, HDAC targeting is quite complex, because they have multiple subclasses, some of which with yet unknown action mechanisms and functions [86]. Currently, it is widely acknowledged that enzymatic activity of HDACs is not restricted to histones, but it extends to several other proteins. Indeed, exposure of human cell lines to highly specific HDACi induced hyperacetylation of 1750 proteins, revealing that non-histone proteins are the main substrate of these compounds [87].

3.1.1 *Preclinical*

Several HDACi demonstrated encouraging results in preclinical phase studies, showing promise as candidates for future clinical trials.

Concerning the aliphatic acids family, exposure of PC-3cells to sodium butyrate increased differentiation and apoptosis [88]. VPA not only induced similar effects in in vitro models but also was able to reduce tumor growth in xenograft models [89]. Remarkably, this compound induced AR and E-cadherin expression in PCa cell lines [90].

Among hydroxamic acids, vorinostat inhibits PCa cell line proliferation and reduces tumor growth in vivo [91,92]. This compound enhanced radiation-induced apoptosis in DU145 cells [93] and demonstrated a synergistic effect with zoledronic acid, increasing LNCaP and PC-3 cell death [94]. A broader effect was achieved by panobinostat, which induces cell cycle arrest and DNA damage, and reduces PCa tumor growth in vivo [95]. Moreover, exposure to panobinostat not only decreased AR levels but also reversed resistance to hormone therapy in castration-resistant PCa cell lines [96]. When combined with radiotherapy (RT), panobinostat greatly improved the efficiency of cell death and induced persistent DNA double-strand breaks, compared to RT alone [97]. Therefore, panobinostat might increase radiosensitivity of PCa cells. Belinostat (PXD101) has pronounced antitumor effects in androgen-sensitive PCa cell lines increasing P21, P27, and P53 expression and leading to G2/M cell cycle arrest [98].

Remarkably, in mice inoculated with the 22Rv1 cell line, treatment with romidepsin not only reduced metastasis formation but also achieved a 61% increase in survival [99]. Moreover, the combination of this drug with docetaxel showed greater cytotoxic effects in castration-resistant PCa cell lines as well as a significant reduction of tumor growth in mice inoculated with PC-3 [100].

The exposure of PCa cell lines to MS-275, a benzamide derivative, resulted in an increase of H3 acetylation, p21 expression, growth arrest in LNCaP and PC-3, and apoptosis in DU145 cells. This drug also reduced tumor growth in xenograft mice [101], particularly when acting synergistically with radiation [102]. Moreover, MS-275 increased H3K4 methylation marks, inducing expression of tumor suppressor and cell differentiation genes [103].

Thus, available data suggest that HDACi constitute promising agents for PCa treatment, particularly in more differentiated PCa cells that express AR, and in combination with conventional therapy regimens.

3.1.2 Clinical Trials

HDACi is the most well-studied class of inhibitors of histone modifiers in PCa, as they have been tested in clinical trials, either as single agents or combined with conventional therapeutics (summarized in Table 2).

3.1.2.1 SINGLE AGENTS

A phase II clinical trial was conducted with vorinostat in metastatic castration-resistant prostate cancer (CRPC) patients with disease progression and previously exposed to chemotherapy [104]. Patients were daily treated with orally administrated 400 mg vorinostat. The 6-month progression rate was considered the primary end point. The best objective response was stable disease in 7% of the patients (2 out of the 27 enrolled in this trial). Median time to progression was 2.8 months, with a median overall survival of 11.7 months. Grade 3 or 4 toxicities (fatigue, nausea, vomiting, anorexia, diarrhea, and weight loss) were experienced by 48% of patients and 11 (41%) had to discontinue therapy due to toxicity. This trial revealed that vorinostat at this dose was minimally effective and associated with significant toxicities that might limit drug efficacy.

More recently, a phase II clinical trial evaluated the efficacy of panobinostat in CRPC patients that presented disease progression after chemotherapy [105]. The rate of progression-free survival (PFS) at 24 weeks was set as primary end point. Thirty-five patients received 20 mg/m^2 of panobinostat intravenously on days 1 and 8 of a 21-day cycle. There were no documented objective responses. Four patients (11.4%) did not present progression of disease at 24 weeks. All patients experienced adverse effects (toxicity grades 3 and 4). This study concluded that treatment with panobinostat alone was insufficient to achieve clinical efficacy.

A phase II study with romidepsin was conducted in 35 metastatic CRPC patients. Romidepsin was administrated intravenously at 13 mg/m^2 on days 1, 8, and 15 of a 28-day

TABLE 2 Histone-Modifying Drugs in Clinical Trials for PCa

Drug	Clinical Trial ID	Phase	Status	Protocol	Outcome
Panobinostat, docetaxel, and prednisone	NCT00663832	I	Completed	CRPC received panobinostat 10, 15, or 20 mg IV on days 1 and 8 in combination with docetaxel IV on day 1 and prednisone PO 5 mg bid every day of a 21-day cycle. n = 44	Determined the maximum tolerated dose (MTD) of IV panobinostat is 15 mg in combination with docetaxel and prednisone in patients with CRPC; 63% of patients showed >50% decline in PSA levels
Panobinostat, bicalutamide	NCT00878436	I/II	Recruiting	CRPC receiving panobinostat at either 60 or 120 mg per week for 2 consecutive weeks, with 1 week rest and bicalutamide 50 mg PO daily, continuously. Predicted enrollment: n = 78	Investigating safety and efficacy of combined treatment of panobinostat at 2 dose levels combined with bicalutamide for CRPC as measured by time to PSA progression and proportion of patients that achieve a >50% PSA decline by 9 months of therapy
Vorinostat	NCT00330161	II	Completed	mCRPC with disease progression and adequate organ function received 400 mg vorinostat orally each day. n = 27	7% patients achieved an objective response rate; no PSA decline >50% observed; significant toxicities reported
Vorinostat and androgen deprivation therapy (ADT)	NCT00589472	II	Active, not recruiting	Localized PCa patients received ADT and oral vorinostat administered for a minimum of 6 weeks and maximum of 8 weeks before radical prostatectomy. n = 38	Determine the rate of pathologic complete response in patients with localized PCa treated with ADT and vorinostat before radical prostatectomy measuring androgens in blood
Vorinostat and mTOR inhibitor temsirolimus	NCT01174199	I	Recruiting	mCRPC patients received oral vorinostat once daily on days 1–14 and temsirolimus intravenously on days 1, 8, and 15 of a 21-day cycle. n = 29	Determine the safety, tolerability, partial and complete objective response rates, progression-free survival and overall survival, and PSA response
Romidepsin	NCT00106418	II	Completed	mCRPC patients received romidepsin $13\,mg/m^2$ intravenously over 4 h on days 1, 8, and 15 every 4 weeks. n = 35	Two patients reached a confirmed radiological partial response of over 6 months, in addition to >50% PSA decline
SB939	NCT01075308	II	Active, not recruiting	Recurrent or mCRPC patients received SB939 orally on days 1, 3, and 5 for 3 consecutive weeks followed by 1 week off-dosing of a 28-day cycle. n = 32	To determine the efficacy, measured by PSA response and progression-free survival, of HDAC inhibitor SB939 in patients with recurrent or mCRPC

Valproic acid	NCT00670046	II	Not provided	Nonmetastatic with biochemical progression PCa patients received oral valproic acid twice daily for up to 1 year. n = 50	Percentage of patients exhibiting observed or predicted PSA doubling time >10 months after initiation of the study
Curcumin and radiotherapy	NCT01917890	Not provided	Recruiting	PCa patients undergo 74-Gy radiotherapy 5 times a week for 7–8 weeks and take 3 g of curcumin. n = 40	Assess biochemical or clinical progression-free survival by magnetic resonance spectroscopy and PSA rebound
Curcumin and taxotere	NCT02095717	II	Recruiting	mCRPC patients. n = 100	Assess time to progression of metastatic disease by tumor response rate, increase in PSA levels (≥25% and ≥2 ng/mL increase) or the appearance of new metastasis
Curcumin	NCT02064673	II	Recruiting	PCa patients with localized disease receive curcumin or placebo 500 mg PO twice a day for 6 months. n = 600	Determine recurrence-free survival as total PSA <0.2 ng/mL
Phenelzine	NCT02217709	II	Recruiting	Recurrent nonmetastatic PCa received phenelzine daily and orally during 12 months. n = 46	Determine biochemical recurrent prostate cancer by PSA decline to ≥50% following at least 12 weeks of treatment
Phenelzine and docetaxel	NCT01253642	II	Recruiting	PCa patients with progressive disease receive phenelzine orally once a day on days -7 to -4 and twice a day on days -3 to 21 and docetaxel on day 1. n = 40	Determine the proportion of patients who experience a PSA decline of at least 30% and duration of progression-free survival
OTX015	NCT02259114	IB	Not yet recruiting	Advanced solid tumors including CRPC. Patients divided into two regimens: (1) continuous, once daily for 21 consecutive days; (2) days 1 and 7 of a 21-day cycle (1 week on/2 weeks off). n = 98	Determine MTD and the efficacy of OTX015 in solid tumors

CRPC, castration-resistant prostate cancer; IV, intravenous; mCRPC, metastatic castration-resistant prostate cancer; PCa, prostate cancer; PO, by mouth; PSA, prostate-specific antigen.

cycle [106]. The primary end point was rate of disease control defined as no evidence of radiological progression at 6 months. Partial response was achieved in two patients confirmed by radiology and PSA decline. However, 11 patients discontinued therapy due to significant drug toxicity. With this drug schedule, romidepsin demonstrated minimal anticancer activity in metastatic CRPC patients.

Hence, HDACi as single agents in PCa did not show promising results. Their inability to significantly accumulate in solid tumors, associated with their fast excretion and off-target toxicity, might contribute to their failure as efficient drugs against PCa.

3.1.2.2 COMBINATION THERAPY

In a phase I trial, combined oral vorinostat (administered on days 1, 2, and 3 with a planned dose escalation of 600 mg given twice a day in two divided doses) and 20 mg/m^2 doxorubicin (topoisomerase II inhibitor, infused on third day, 4 h after the last vorinostat dose) achieved partial response in one of the two PCa patients enrolled [107]. A parallel, two-arm, open-label phase IA/IB study with orally panobinostat alone (20 mg administered on days 1, 3, and 5 for 2 consecutive weeks) or in combination with docetaxel and prednisone (15 mg of panobinostat administered in the same schedule and 75 mg/m^2 of docetaxel every 21 days) enrolled 16 CRPC patients. All individuals treated with panobinostat alone developed progressive disease, whereas the combined therapy resulted in 5 (63%) partial responses. Patients from both arms showed grade 3 toxicities [108].

3.2 Inhibitors of Histone Acetyltransferases

Inhibitors of histone acetyltransferases (HATi) have been less studied in the past, but they are now emerging due to promising results in preclinical models of solid tumors.

Exposure of PCa cells to curcumin, a HAT p300 inhibitor, increased apoptosis, inhibited proliferation, and downregulated several important metastasis-promoting genes, including *COX2*, *SPARC*, and *EFEMP*. Moreover, it reduced metastasis formation in vivo [109]. Curcumin also had the ability to demethylate and restore expression of *Neurog1* and decrease methyl CpG-binding protein 2 (MeCP2) binding to *Neurog1* promoter in LNCaP cells [110]. Furthermore, C646, another small-molecule inhibitor of p300/CBP, induced caspase-dependent apoptosis and decreased migration and invasion of PCa cells [111]. Therefore, HAT inhibitors might be helpful tools to inhibit PCa dissemination.

3.3 Histone Methyltransferase Inhibitor and Histone Demethylase Inhibitor

The development of histone methyltransferase inhibitor (HMTi) and histone demethylase inhibitor (HDMi) is less advanced than that of HDACi, but several ongoing studies are assessing their specificity for targeted epigenetic therapy [112]. These compounds are thought to be more attractive than HDACi because they can eliminate selective histone marks, which in turn might enable a better tailored therapy, minimizing undesirable side effects.

Among HMTi, DZNeP (3-dezaneplanocin-A) was first described as EZH2 inhibitor, decreasing H3K27 methylation, but is currently considered a global HMTi [113]. This compound has shown antitumoral activity in PCa cells, inducing cell cycle arrest in LNCaP and apoptosis in DU145 cells, as well as reducing PCa cell invasion. Furthermore, this drug decreased PCa stem cells' self-renewal and reduced tumor growth in mice [114].

Concerning HDMi, the most studied thus far are LSD1 inhibitors. Most studies were performed with nonselective amine oxidase (monoamine oxidase (MAO)) inhibitors. Pargyline inhibited growth of LNCaP cells, blocking H3K9 demethylation [68]. Furthermore, it induced cell cycle arrest at G1 and increased apoptosis of LNCaP cells [115]. In contrast, namoline, a novel and selective LSD1 inhibitor, has been recently proposed as potential therapeutic agent against hormone-sensitive PCa, since it induced silencing of AR-regulated genes and decreased PCa cells proliferation both in vitro and in vivo [116].

Currently, there are no clinical studies involving HMTi in PCa. Two clinical trials will be conducted with the nonspecific MAO inhibitor phenelzine, either alone or in combination with docetaxel.

3.4 Histone Readers

JQ1 and I-BET are novel compounds that inhibit bromodomain proteins interfering with their binding to histone acetylated lysine residues. Bromodomain proteins binding to acetylated histones increase proliferation and can lead to overexpression of several oncogenes such as *MYC* [117]. Both compounds had shown promising results in inhibiting PCa cell growth. I-BET762 decreased tumor cell proliferation of PCa cell lines and, in a patient-derived tumor model, it reduced tumor burden in vivo. It was suggested that these encouraging results might be due to MYC downregulation [118]. JQ1 also demonstrated a significant anticancer activity, especially in CRPC cell lines [119]. This compound acts downstream of AR, disrupting its recruitment to target gene loci and by negatively regulating the expression and oncogenic activity of *TMPRSS2-ETS* gene fusion products and *MYC*. These data suggest that BET bromodomain inhibitors might be therapeutically useful tools in PCa. However, much remains to be investigated concerning the molecular mechanisms that determine the activity of BET inhibitors upon *MYC* and *AR* regulation in PCa. A clinical trial with the BET inhibitor OTX015 in solid tumors, including CRPC is planned and might shed some light on the potential clinical usefulness of these compounds.

4. CONCLUSION

Disruption of normal patterns of expression/activity of histone-modifying enzymes, as well as of their corresponding histone marks, has been increasingly recognized as key events in prostatic tumorigenesis. However, the role of these covalent histone PTMs is still poorly understood, hampering their potential value as cancer biomarkers, as illustrated by the relative scarcity of studies on this issue. Indeed, there are two main issues that hamper the study of histone PTMs and histone modifiers in PCa. Both the high heterogeneity of this neoplasm and sampling procedures limit the assessment of the biological role of a specific histone modification in this tumor model. Microdissection techniques might overcome this limitation but are not routinely used in diagnostic pathology. Moreover, the methodology used in this type of studies needs to be updated since IHC has several limitations, as previously detailed. Chromatin immunoprecipitation (ChIP)-related technologies are the most commonly used approach to evaluate genome-wide changes in histone modifications levels and ChIP-chip is likely to be replaced in the near future by ChIP-seq. Interestingly, global histone modification level detection and measurement in cell-free body fluids (derived from mononucleosomes and oligonucleosomes originated from tumor cells) has been proposed. However, large validation studies are required to ascertain the clinical usefulness of histone PTMs as PCa biomarkers.

Additionally, the reversible character of histone modifications grants an opportunity for the development of molecular strategies that might attenuate the impact of abnormal histone modifications patterns. Indeed, promising results have been recently reported for several histone-modifying enzymes inhibitors, either in preclinical assays or in clinical trials for PCa treatment. HDACi have emerged as a powerful class of small-molecule therapeutics, however, their limitations should be considered, including ineffectively low concentrations in solid tumors and cardiac toxicity. The major limitation of targeting epigenetic enzymes, like HDACs, is lack of specificity. In fact, HDACi affects both histone-specific proteins as well as on nonhistone proteins. Therefore, the observed therapeutic response might not only be entirely due to inhibition of histone deacetylation but also due to the changes in acetylation status of these nonhistone proteins. In order to appropriately monitor drug response a better comprehension of HDACi effects in PCa is required.

Over the past few years, however, it became clear that single agents, targeting a single pathway, are not effective for PCa treatment (as well as for most solid malignancies). Thus, the most recent clinical trials are testing the combination of different drugs targeting the epigenome with conventional therapeutic agents. An effective use of these compounds requires a deeper understanding of its mechanisms of action as well as more profound knowledge of its impact on the epigenetic machinery.

Glossary

Biochemical relapse when, following radical prostatectomy, PSA levels are higher than 0.2 ng/mL, or when PSA consecutively rises in from the postoperative baseline. It is indicative of disease recurrence.

Castration-resistant PCa (CRPC) Advanced stage of PCa in which AR maintains functional signaling regardless of castrate levels of androgens.

Chromatin DNA coiled around histone proteins and compacted into highly ordered structures in the nucleus.

Gleason score (GS) Prostate cancer's microscopic grading system ranging from 1–5 based on the glandular architecture and patterns. The combined GS (2–10) is the result of the two most prevalent Gleason grade patterns per tumor. GS is a significant prognostic indicator: high GS (≥8) is associated with a poorer outcome than an intermediate GS (7), and low GS (≤6) is globally associated with a better prognosis.

Histone acetyltransferases (HATs) Class of enzymes responsible to add acetyl groups to histones' N-terminal domains associated with gene activation.

Histone deacetylases (HDACs) Class of enzymes responsible to remove acetyl groups from histones' N-terminal domains associated with gene repression.

Histone demethylases (HDMs) Class of enzymes responsible to remove methyl groups from histones' N-terminal domains that can lead to gene repression or activation depending on the residue and state of methylation (mono, di, or trimethylation).

Histone methyltransferases (HMTs) Class of enzymes responsible to add methyl groups to histones' N-terminal domains that can lead to gene repression or activation depending on the residue and state of methylation (mono, di, or trimethylation).

Posttranslational modifications (PTMs) Chemical alterations of histone tails, namely methylation, acetylation, phosphorylation, ubiquitylation, and sumoylation that can affect gene expression.

Prostate-specific antigen (PSA) Serine protease produced by prostate epithelial cells implicated in seminal fluid liquefaction. Serum PSA levels are used for detection and monitoring of prostate cancer.

LIST OF ABBREVIATIONS

AR Androgen receptor
BET Bromodomain and extraterminal
COX2 Cytochrome c oxidase subunit 2
CRPC Castration-resistant prostate cancer
DAXX Death-associated protein 6
DZNeP 3-Dezaneplanocin-A
EFEMP EGF-containing fibulin-like extracellular matrix protein
ELISA Enzyme-linked immunosorbent assay
EZH2 Enhancer of zeste 2
GS Gleason score
HATi Histone acetyltransferase inhibitor
HATs Histone acetyltransferases
HDACi Histone deacetylase inhibitor
HDACs Histone deacetylases
HDMi Histone demethylase inhibitor
HDMs Histone demethylases
HMTi Histone methyltransferase inhibitor
HMTs Histone methyltransferases
IHC Immunohistochemistry
JMJC Jumonji C
KMTs Lysine methyltransferases
LSD Lys-specific demethylase
MAO Monoamine oxidase
mCRPC Metastatic castration-resistant prostate cancer
MeCP2 Methyl CpG-binding protein 2
MYC V-myc avian myelocytomatosis viral oncogene homolog
PCa Prostate cancer
PFS Progression-free survival
PRMTs Protein arginine methyltransferases
PSA Prostate-specific antigen
PTMs Posttranslational modifications

RP Radical prostatectomy
RT Radiotherapy
SMYD3 SET and MYND domain-containing protein 3
SPARC Osteonectin
TMPRSS2-ETS Transmembrane protease, serine 2
TSA Trichostatin A
VPA Valproic acid

References

[1] Jemal A, Bray F, Center MM, Ferlay J, Ward E, Forman D. Global cancer statistics. CA Cancer J Clin 2011;61(2):69–90. Epub 2011/02/08.

[2] Ferlay J, Soerjomataram I, Dikshit R, Eser S, Mathers C, Rebelo M, et al. Cancer incidence and mortality worldwide: sources, methods and major patterns in GLOBOCAN 2012. Int J Cancer 2014;136(5):E359–86.

[3] Ferlay J, Steliarova-Foucher E, Lortet-Tieulent J, Rosso S, Coebergh JW, Comber H, et al. Cancer incidence and mortality patterns in Europe: estimates for 40 countries in 2012. Eur J Cancer 2013;49(6):1374–403.

[4] Moyer VA. Screening for prostate cancer: U.S. Preventive Services Task Force recommendation statement. Ann Intern Med 2012;157(2):120–34.

[5] Chen FZ, Zhao XK. Prostate cancer: current treatment and prevention strategies. Iran Red Crescent Med J 2013;15(4):279–84. Epub 2013/10/02.

[6] Ramon J, Denis L. Prostate cancer. Berlin: Springer; 2007.

[7] Chakrabarti SK, Francis J, Ziesmann SM, Garmey JC, Mirmira RG. Covalent histone modifications underlie the developmental regulation of insulin gene transcription in pancreatic beta cells. J Biol Chem 2003;278(26):23617–23.

[8] Bianco-Miotto T, Chiam K, Buchanan G, Jindal S, Day TK, Thomas M, et al. Global levels of specific histone modifications and an epigenetic gene signature predict prostate cancer progression and development. Cancer Epidemiol Biomarkers Prev 2010;19(10):2611–22. Epub 2010/09/16.

[9] Ke XS, Qu Y, Rostad K, Li WC, Lin B, Halvorsen OJ, et al. Genome-wide profiling of histone h3 lysine 4 and lysine 27 trimethylation reveals an epigenetic signature in prostate carcinogenesis. PLoS One 2009;4(3):e4687. Epub 2009/03/06.

[10] Seligson DB, Horvath S, Shi T, Yu H, Tze S, Grunstein M, et al. Global histone modification patterns predict risk of prostate cancer recurrence. Nature 2005;435(7046):1262–6. Epub 2005/07/01.

[11] Miremadi A, Oestergaard MZ, Pharoah PD, Caldas C. Cancer genetics of epigenetic genes. Hum Mol Genet 2007;16(Spec No. 1):R28–49. Epub 2007/09/26.

[12] Delcuve GP, Rastegar M, Davie JR. Epigenetic control. J Cell Physiol 2009;219(2):243–50. Epub 2009/01/08.

[13] Sharma S, Kelly TK, Jones PA. Epigenetics in cancer. Carcinogenesis 2010;31(1):27–36. Epub 2009/09/16.

[14] Rodriguez-Paredes M, Esteller M. Cancer epigenetics reaches mainstream oncology. Nat Med 2011;17(3):330–9. Epub 2011/03/10.

[15] Sandoval J, Esteller M. Cancer epigenomics: beyond genomics. Curr Opin Genet Dev 2012;22(1):50–5. Epub 2012/03/10.

[16] Esteller M. Epigenetics in cancer. N Engl J Med 2008;358(11):1148–59. Epub 2008/03/14.

[17] Jenuwein T, Allis CD. Translating the histone code. Science 2001;293(5532):1074–80.

[18] Dawson MA, Kouzarides T, Huntly BJ. Targeting epigenetic readers in cancer. N Engl J Med 2012;367(7):647–57. Epub 2012/08/17.

[19] Cosgrove MS, Boeke JD, Wolberger C. Regulated nucleosome mobility and the histone code. Nat Struct Mol Biol 2004;11(11):1037–43. Epub 2004/11/04.

[20] Lund AH, van Lohuizen M. Epigenetics and cancer. Genes Dev 2004;18(19):2315–35. Epub 2004/10/07.

[21] Kouzarides T. Chromatin modifications and their function. Cell 2007;128(4):693–705. Epub 2007/02/27.

[22] Zentner GE, Henikoff S. Regulation of nucleosome dynamics by histone modifications. Nat Struct Mol Biol 2013;20(3):259–66. Epub 2013/03/07.

[23] Arrowsmith CH, Bountra C, Fish PV, Lee K, Schapira M. Epigenetic protein families: a new frontier for drug discovery. Nat Rev Drug Discov 2012;11(5):384–400. Epub 2012/04/14.

[24] Glozak MA, Sengupta N, Zhang X, Seto E. Acetylation and deacetylation of non-histone proteins. Gene 2005;363:15–23.

[25] Brownell JE, Allis CD. Special HATs for special occasions: linking histone acetylation to chromatin assembly and gene activation. Curr Opin Genet Dev 1996;6(2):176–84. Epub 1996/04/01.

[26] Marmorstein R, Zhou MM. Writers and readers of histone acetylation: structure, mechanism, and inhibition. Cold Spring Harb Perspect Biol 2014;6(7):a018762. Epub 2014/07/06.

[27] de Ruijter AJ, van Gennip AH, Caron HN, Kemp S, van Kuilenburg AB. Histone deacetylases (HDACs): characterization of the classical HDAC family. Biochem J 2003;370(Pt 3):737–49.

[28] Gregoretti IV, Lee YM, Goodson HV. Molecular evolution of the histone deacetylase family: functional implications of phylogenetic analysis. J Mol Biol 2004;338(1):17–31.

[29] Zhang Y, Reinberg D. Transcription regulation by histone methylation: interplay between different covalent modifications of the core histone tails. Genes Dev 2001;15(18):2343–60.

[30] Shi Y, Lan F, Matson C, Mulligan P, Whetstine JR, Cole PA, et al. Histone demethylation mediated by the nuclear amine oxidase homolog LSD1. Cell 2004;119(7):941–53. Epub 2004/12/29.
[31] Chang B, Chen Y, Zhao Y, Bruick RK. JMJD6 is a histone arginine demethylase. Science 2007;318(5849):444–7. Epub 2007/10/20.
[32] Brame CJ, Moran MF, McBroom-Cerajewski LD. A mass spectrometry based method for distinguishing between symmetrically and asymmetrically dimethylated arginine residues. Rapid Commun Mass Spectrom 2004;18(8):877–81. Epub 2004/04/20.
[33] Lee DY, Northrop JP, Kuo MH, Stallcup MR. Histone H3 lysine 9 methyltransferase G9a is a transcriptional coactivator for nuclear receptors. J Biol Chem 2006;281(13):8476–85. Epub 2006/02/08.
[34] Varier RA, Timmers HT. Histone lysine methylation and demethylation pathways in cancer. Biochim Biophys Acta 2011;1815(1):75–89. Epub 2010/10/19.
[35] Di Lorenzo A, Bedford MT. Histone arginine methylation. FEBS Lett 2011;585(13):2024–31. Epub 2010/11/16.
[36] Litt M, Qiu Y, Huang S. Histone arginine methylations: their roles in chromatin dynamics and transcriptional regulation. Biosci Rep 2009;29(2):131–41. Epub 2009/02/18.
[37] Spannhoff A, Hauser AT, Heinke R, Sippl W, Jung M. The emerging therapeutic potential of histone methyltransferase and demethylase inhibitors. ChemMedChem 2009;4(10):1568–82. Epub 2009/09/10.
[38] Dambacher S, Hahn M, Schotta G. Epigenetic regulation of development by histone lysine methylation. Heredity 2010;105(1):24–37. Epub 2010/05/06.
[39] Cloos PA, Christensen J, Agger K, Helin K. Erasing the methyl mark: histone demethylases at the center of cellular differentiation and disease. Genes Dev 2008;22(9):1115–40. Epub 2008/05/03.
[40] Smith BC, Denu JM. Chemical mechanisms of histone lysine and arginine modifications. Biochim Biophys Acta 2009;1789(1):45–57. Epub 2008/07/08.
[41] Kooistra SM, Helin K. Molecular mechanisms and potential functions of histone demethylases. Nat Rev Mol Cell Biol 2012;13(5):297–311. Epub 2012/04/05.
[42] Zhong J, Ding L, Bohrer LR, Pan Y, Liu P, Zhang J, et al. p300 acetyltransferase regulates androgen receptor degradation and PTEN-deficient prostate tumorigenesis. Cancer Res 2014;74(6):1870–80.
[43] Patra SK, Patra A, Dahiya R. Histone deacetylase and DNA methyltransferase in human prostate cancer. Biochem Biophys Res Commun 2001;287(3):705–13.
[44] Waltregny D, North B, Van Mellaert F, de Leval J, Verdin E, Castronovo V. Screening of histone deacetylases (HDAC) expression in human prostate cancer reveals distinct class I HDAC profiles between epithelial and stromal cells. Eur J Histochem 2004;48(3):273–90.
[45] Halkidou K, Gaughan L, Cook S, Leung HY, Neal DE, Robson CN. Upregulation and nuclear recruitment of HDAC1 in hormone refractory prostate cancer. Prostate 2004;59(2):177–89.
[46] Weichert W, Roske A, Gekeler V, Beckers T, Stephan C, Jung K, et al. Histone deacetylases 1, 2 and 3 are highly expressed in prostate cancer and HDAC2 expression is associated with shorter PSA relapse time after radical prostatectomy. Br J Cancer 2008;98(3):604–10.
[47] Adams H, Fritzsche FR, Dirnhofer S, Kristiansen G, Tzankov A. Class I histone deacetylases 1, 2 and 3 are highly expressed in classical Hodgkin's lymphoma. Expert Opin Ther Targets 2010;14(6):577–84.
[48] Liu Y, Xie QR, Wang B, Shao J, Zhang T, Liu T, et al. Inhibition of SIRT6 in prostate cancer reduces cell viability and increases sensitivity to chemotherapeutics. Protein Cell 2013;4(9):702–10. Epub 2013/08/29.
[49] Di Sante G, Pestell TG, Casimiro MC, Bisetto S, Powell MJ, Lisanti MP, et al. Loss of Sirt1 promotes prostatic intraepithelial neoplasia, reduces mitophagy, and delays PARK2 translocation to mitochondria. Am J Pathol 2015;185(1):266–79. Epub 2014/12/23.
[50] Baptista T, Graca I, Sousa EJ, Oliveira AI, Costa NR, Costa-Pinheiro P, et al. Regulation of histone H2A.Z expression is mediated by sirtuin 1 in prostate cancer. Oncotarget 2013;4(10):1673–85. Epub 2013/10/16.
[51] Long Q, Xu J, Osunkoya AO, Sannigrahi S, Johnson BA, Zhou W, et al. Global transcriptome analysis of formalin-fixed prostate cancer specimens identifies biomarkers of disease recurrence. Cancer Res 2014;74(12):3228–37. Epub 2014/04/10.
[52] Dai Y, Ngo D, Forman LW, Qin DC, Jacob J, Faller DV. Sirtuin 1 is required for antagonist-induced transcriptional repression of androgen-responsive genes by the androgen receptor. Mol Endocrinol 2007;21(8):1807–21.
[53] Ellinger J, Kahl P, von der Gathen J, Rogenhofer S, Heukamp LC, Gutgemann I, et al. Global levels of histone modifications predict prostate cancer recurrence. Prostate 2010;70(1):61–9. Epub 2009/09/10.
[54] Seligson DB, Horvath S, McBrian MA, Mah V, Yu H, Tze S, et al. Global levels of histone modifications predict prognosis in different cancers. Am J Pathol 2009;174(5):1619–28. Epub 2009/04/08.
[55] Ke XS, Qu Y, Cheng Y, Li WC, Rotter V, Oyan AM, et al. Global profiling of histone and DNA methylation reveals epigenetic-based regulation of gene expression during epithelial to mesenchymal transition in prostate cells. BMC Genomics 2010;11:669. Epub 2010/11/27.
[56] Karanikolas BD, Figueiredo ML, Wu L. Comprehensive evaluation of the role of EZH2 in the growth, invasion, and aggression of a panel of prostate cancer cell lines. Prostate 2010;70(6):675–88. Epub 2010/01/21.

[57] Hoffmann MJ, Engers R, Florl AR, Otte AP, Muller M, Schulz WA. Expression changes in EZH2, but not in BMI-1, SIRT1, DNMT1 or DNMT3B are associated with DNA methylation changes in prostate cancer. Cancer Biol Ther 2007;6(9):1403–12. Epub 2008/07/22.

[58] Varambally S, Dhanasekaran SM, Zhou M, Barrette TR, Kumar-Sinha C, Sanda MG, et al. The polycomb group protein EZH2 is involved in progression of prostate cancer. Nature 2002;419(6907):624–9. Epub 2002/10/11.

[59] Gibbons RJ. Histone modifying and chromatin remodelling enzymes in cancer and dysplastic syndromes. Hum Mol Genet 2005;14(Spec No 1):R85–92. Epub 2005/04/06.

[60] Jeronimo C, Bastian PJ, Bjartell A, Carbone GM, Catto JW, Clark SJ, et al. Epigenetics in prostate cancer: biologic and clinical relevance. Eur Urol 2011;60(4):753–66. Epub 2011/07/02.

[61] Yang YA, Yu J. EZH2, an epigenetic driver of prostate cancer. Protein Cell 2013;4(5):331–41. Epub 2013/05/03.

[62] Bachmann IM, Halvorsen OJ, Collett K, Stefansson IM, Straume O, Haukaas SA, et al. EZH2 expression is associated with high proliferation rate and aggressive tumor subgroups in cutaneous melanoma and cancers of the endometrium, prostate, and breast. J Clin Oncol 2006;24(2):268–73. Epub 2005/12/07.

[63] Yu J, Rhodes DR, Tomlins SA, Cao X, Chen G, Mehra R, et al. A polycomb repression signature in metastatic prostate cancer predicts cancer outcome. Cancer Res 2007;67(22):10657–63. Epub 2007/11/17.

[64] Chin SP, Dickinson JL, Holloway AF. Epigenetic regulation of prostate cancer. Clin Epigenetics 2011;2(2):151–69. Epub 2012/06/19.

[65] Chen Z, Wang L, Wang Q, Li W. Histone modifications and chromatin organization in prostate cancer. Epigenomics 2010;2(4):551–60. Epub 2011/02/15.

[66] Opel M, Lando D, Bonilla C, Trewick SC, Boukaba A, Walfridsson J, et al. Genome-wide studies of histone demethylation catalysed by the fission yeast homologues of mammalian LSD1. PLoS One 2007;2(4):e386. Epub 2007/04/19.

[67] Metzger E, Wissmann M, Yin N, Muller JM, Schneider R, Peters AH, et al. LSD1 demethylates repressive histone marks to promote androgen-receptor-dependent transcription. Nature 2005;437(7057):436–9. Epub 2005/08/05.

[68] Kahl P, Gullotti L, Heukamp LC, Wolf S, Friedrichs N, Vorreuther R, et al. Androgen receptor coactivators lysine-specific histone demethylase 1 and four and a half LIM domain protein 2 predict risk of prostate cancer recurrence. Cancer Res 2006;66(23):11341–7.

[69] Scoumanne A, Chen X. The lysine-specific demethylase 1 is required for cell proliferation in both p53-dependent and -independent manners. J Biol Chem 2007;282(21):15471–5. Epub 2007/04/06.

[70] Hayami S, Kelly JD, Cho HS, Yoshimatsu M, Unoki M, Tsunoda T, et al. Overexpression of LSD1 contributes to human carcinogenesis through chromatin regulation in various cancers. Int J Cancer 2011;128(3):574–86. Epub 2010/03/25.

[71] Tian X, Zhang S, Liu HM, Zhang YB, Blair CA, Mercola D, et al. Histone lysine-specific methyltransferases and demethylases in carcinogenesis: new targets for cancer therapy and prevention. Curr Cancer Drug Targets 2013;13(5):558–79. Epub 2013/05/30.

[72] Chi P, Allis CD, Wang GG. Covalent histone modifications–miswritten, misinterpreted and mis-erased in human cancers. Nat Rev Cancer 2010;10(7):457–69.

[73] Roobol MJ, Carlsson SV. Risk stratification in prostate cancer screening. Nat Rev Urol 2012;10(1):38–48.

[74] Hammerich KH, Ayala GE, Wheeler TM. Anatomy of the prostate gland and surgical pathology of prostate cancer. Cambridge: Cambridge University; 2009. 1–10.

[75] Lilja H, Ulmert D, Vickers AJ. Prostate-specific antigen and prostate cancer: prediction, detection and monitoring. Nat Rev Cancer 2008;8(4):268–78.

[76] Chan TA, Glockner S, Yi JM, Chen W, Van Neste L, Cope L, et al. Convergence of mutation and epigenetic alterations identifies common genes in cancer that predict for poor prognosis. PLoS Med 2008;5(5):e114.

[77] Zhou L-X, Li T, Huang Y-R, Sha J-J, Sun P, Li D. Application of histone modification in the risk prediction of the biochemical recurrence after radical prostatectomy. Asian J Androl 2010;12(2):171–9.

[78] Ellinger J, Kahl P, von der Gathen J, Heukamp LC, Gütgemann I, Walter B, et al. Global histone H3K27 methylation levels are different in localized and metastatic prostate cancer. Cancer Invest 2012;30(2):92–7.

[79] Deligezer U, Yaman F, Darendeliler E, Dizdar Y, Holdenrieder S, Kovancilar M, et al. Post-treatment circulating plasma BMP6 mRNA and H3K27 methylation levels discriminate metastatic prostate cancer from localized disease. Clin Chim Acta 2010;411(19):1452–6.

[80] Behbahani TE, Kahl P, von der Gathen J, Heukamp LC, Baumann C, Gütgemann I, et al. Alterations of global histone H4K20 methylation during prostate carcinogenesis. BMC Urol 2012;12(1):5.

[81] Vieira FQ, Costa-Pinheiro P, Ramalho-Carvalho J, Pereira A, Menezes FD, Antunes L, et al. Deregulated expression of selected histone methylases and demethylases in prostate carcinoma. Endocr Relat Cancer 2014;21(1):51–61. Epub 2013/11/10.

[82] Li J, Fan Q, Fan X, Zhou W, Qiu Y, Qiu L. EZH2 expression in human prostate cancer and its clinicopathologic significance. Zhonghua Nan Ke Xue 2010;16(2):123–8.

[83] Tsourlakis MC, Schoop M, Plass C, Huland H, Graefen M, Steuber T, et al. Overexpression of the chromatin remodeler death-domain–associated protein in prostate cancer is an independent predictor of early prostate-specific antigen recurrence. Hum Pathol 2013;44(9):1789–96.

[84] Barneda-Zahonero B, Parra M. Histone deacetylases and cancer. Mol Oncol 2012;6(6):579–89. Epub 2012/09/12.

[85] Kozikowski A, Butler K, Kalin J. Hdac inhibitors and therapeutic methods using the same. WO Patent 2011011186; 2011.

[86] Gray SG, Ekstrom TJ. The human histone deacetylase family. Exp Cell Res 2001;262(2):75–83. Epub 2001/01/05.

[87] Choudhary C, Kumar C, Gnad F, Nielsen ML, Rehman M, Walther TC, et al. Lysine acetylation targets protein complexes and co-regulates major cellular functions. Science 2009;325(5942):834–40. Epub 2009/07/18.

[88] Huang H, Reed CP, Zhang JS, Shridhar V, Wang L, Smith DI. Carboxypeptidase A3 (CPA3): a novel gene highly induced by histone deacetylase inhibitors during differentiation of prostate epithelial cancer cells. Cancer Res 1999;59(12):2981–8. Epub 1999/06/26.

[89] Shabbeer S, Kortenhorst MS, Kachhap S, Galloway N, Rodriguez R, Carducci MA. Multiple molecular pathways explain the anti-proliferative effect of valproic acid on prostate cancer cells in vitro and in vivo. Prostate 2007;67(10):1099–110. Epub 2007/05/05.

[90] Chou Y-W, Chaturvedi NK, Ouyang S, Lin F-F, Kaushik D, Wang J, et al. Histone deacetylase inhibitor valproic acid suppresses the growth and increases the androgen responsiveness of prostate cancer cells. Cancer Lett 2011;311(2):177–86.

[91] Lakshmikanthan V, Kaddour-Djebbar I, Lewis RW, Kumar MV. SAHA-sensitized prostate cancer cells to TNFalpha-related apoptosis-inducing ligand (TRAIL): mechanisms leading to synergistic apoptosis. Int J Cancer 2006;119(1):221–8. Epub 2006/02/02.

[92] Butler LM, Agus DB, Scher HI, Higgins B, Rose A, Cordon-Cardo C, et al. Suberoylanilide hydroxamic acid, an inhibitor of histone deacetylase, suppresses the growth of prostate cancer cells in vitro and in vivo. Cancer Res 2000;60(18):5165–70. Epub 2000/10/04.

[93] Chinnaiyan P, Vallabhaneni G, Armstrong E, Huang S-M, Harari PM. Modulation of radiation response by histone deacetylase inhibition. Int J Radiat Oncol Biol Phys 2005;62(1):223–9.

[94] Sonnemann J, Bumbul B, Beck JF. Synergistic activity of the histone deacetylase inhibitor suberoylanilide hydroxamic acid and the bisphosphonate zoledronic acid against prostate cancer cells in vitro. Mol Cancer Ther 2007;6(11):2976–84.

[95] Pettazzoni P, Pizzimenti S, Toaldo C, Sotomayor P, Tagliavacca L, Liu S, et al. Induction of cell cycle arrest and DNA damage by the HDAC inhibitor panobinostat (LBH589) and the lipid peroxidation end product 4-hydroxynonenal in prostate cancer cells. Free Radic Biol Med 2011;50(2):313–22. Epub 2010/11/17.

[96] Liu X, Gomez-Pinillos A, Liu X, Johnson EM, Ferrari AC. Induction of bicalutamide sensitivity in prostate cancer cells by an epigenetic Purα-mediated decrease in androgen receptor levels. Prostate 2010;70(2):179–89.

[97] Xiao W, Graham PH, Hao J, Chang L, Ni J, Power CA, et al. Combination therapy with the histone deacetylase inhibitor LBH589 and radiation is an effective regimen for prostate Cancer cells. PLoS One 2013;8(8):e74253.

[98] Sante D. Differential effects of PXD101 (belinostat) on androgen-dependent and androgen-independent prostate cancer models. Int J Oncol 2012;40(3):711–20.

[99] Lai MT, Yang CC, Lin TY, Tsai FJ, Chen WC. Depsipeptide (FK228) inhibits growth of human prostate cancer cells. Urol Oncol 2008;26(2):182–9. Epub 2008/03/04.

[100] Zhang Z, Stanfield J, Frenkel E, Kabbani W, Hsieh JT. Enhanced therapeutic effect on androgen-independent prostate cancer by depsipeptide (FK228), a histone deacetylase inhibitor, in combination with docetaxel. Urology 2007;70(2):396–401. Epub 2007/09/11.

[101] Qian DZ, Wei YF, Wang X, Kato Y, Cheng L, Pili R. Antitumor activity of the histone deacetylase inhibitor MS-275 in prostate cancer models. Prostate 2007;67(11):1182–93. Epub 2007/05/24.

[102] Camphausen K, Scott T, Sproull M, Tofilon PJ. Enhancement of xenograft tumor radiosensitivity by the histone deacetylase inhibitor MS-275 and correlation with histone hyperacetylation. Clin Cancer Res 2004;10(18):6066–71.

[103] Huang P-H, Chen C-H, Chou C-C, Sargeant AM, Kulp SK, Teng C-M, et al. Histone deacetylase inhibitors stimulate histone H3 lysine 4 methylation in part via transcriptional repression of histone H3 lysine 4 demethylases. Mol Pharmacol 2011;79(1):197–206.

[104] Bradley D, Rathkopf D, Dunn R, Stadler WM, Liu G, Smith DC, et al. Vorinostat in advanced prostate cancer patients progressing on prior chemotherapy (National Cancer Institute Trial 6862). Cancer 2009;115(23):5541–9.

[105] Rathkopf DE, Picus J, Hussain A, Ellard S, Chi KN, Nydam T, et al. A phase 2 study of intravenous panobinostat in patients with castration-resistant prostate cancer. Cancer Chemother Pharmacol 2013;72(3):537–44.

[106] Molife L, Attard G, Fong P, Karavasilis V, Reid A, Patterson S, et al. Phase II, two-stage, single-arm trial of the histone deacetylase inhibitor (HDACi) romidepsin in metastatic castration-resistant prostate cancer (CRPC). Ann Oncol 2010;21(1):109–13.

[107] Munster PN, Marchion D, Thomas S, Egorin M, Minton S, Springett G, et al. Phase I trial of vorinostat and doxorubicin in solid tumours: histone deacetylase 2 expression as a predictive marker. Br J Cancer 2009;101(7):1044–50. Epub 2009/09/10.

[108] Rathkopf D, Wong BY, Ross RW, Anand A, Tanaka E, Woo MM, et al. A phase I study of oral panobinostat alone and in combination with docetaxel in patients with castration-resistant prostate cancer. Cancer Chemother Pharmacol 2010;66(1):181–9. Epub 2010/03/11.

[109] Killian PH, Kronski E, Michalik KM, Barbieri O, Astigiano S, Sommerhoff CP, et al. Curcumin inhibits prostate cancer metastasis in vivo by targeting the inflammatory cytokines CXCL1 and -2. Carcinogenesis 2012;33(12):2507–19. bgs312.

[110] Shu L, Khor TO, Lee J-H, Boyanapalli SS, Huang Y, Wu T-Y, et al. Epigenetic CpG demethylation of the promoter and reactivation of the expression of Neurog1 by curcumin in prostate LNCaP cells. AAPS J 2011;13(4):606–14.

[111] Santer FR, Höschele PP, Oh SJ, Erb HH, Bouchal J, Cavarretta IT, et al. Inhibition of the acetyltransferases p300 and CBP reveals a targetable function for p300 in the survival and invasion pathways of prostate cancer cell lines. Mol Cancer Ther 2011;10(9):1644–55.

[112] Rius M, Lyko F. Epigenetic cancer therapy: rationales, targets and drugs. Oncogene 2012;31(39):4257–65. Epub 2011/12/20.

[113] Miranda TB, Cortez CC, Yoo CB, Liang G, Abe M, Kelly TK, et al. DZNep is a global histone methylation inhibitor that reactivates developmental genes not silenced by DNA methylation. Mol Cancer Ther 2009;8(6):1579–88.

[114] Crea F, Hurt EM, Mathews LA, Cabarcas SM, Sun L, Marquez VE, et al. Pharmacologic disruption of polycomb repressive complex 2 inhibits tumorigenicity and tumor progression in prostate cancer. Mol Cancer 2011;10(1):40.

[115] Lee HT, Choi MR, Doh MS, Jung KH, Chai YG. Effects of the monoamine oxidase inhibitors pargyline and tranylcypromine on cellular proliferation in human prostate cancer cells. Oncol Rep 2013;30(4):1587–92.

[116] Willmann D, Lim S, Wetzel S, Metzger E, Jandausch A, Wilk W, et al. Impairment of prostate cancer cell growth by a selective and reversible lysine-specific demethylase 1 inhibitor. Int J Cancer 2012;131(11): 2704–9.

[117] Yang Z, He N, Zhou Q. Brd4 recruits P-TEFb to chromosomes at late mitosis to promote G1 gene expression and cell cycle progression. Mol Cell Biol 2008;28(3):967–76.

[118] Wyce A, Degenhardt Y, Bai Y, Le B, Korenchuk S, Crouthamel M-C, et al. Inhibition of BET bromodomain proteins as a therapeutic approach in prostate cancer. Oncotarget 2013;4(12):2419.

[119] Asangani IA, Dommeti VL, Wang X, Malik R, Cieslik M, Yang R, et al. Therapeutic targeting of BET bromodomain proteins in castration-resistant prostate cancer. Nature 2014;510(7504):278–82.

CHAPTER

23

Histone Posttranslational Modifications in Breast Cancer and Their Use in Clinical Diagnosis and Prognosis

Luca Magnani[1,a], *Annita Louloupi*[2], *Wilbert Zwart*[2,a]

[1]Department of Surgery and Cancer Imperial Centre for Translational and Experimental Medicine, Imperial College Hammersmith, London, UK; [2]Division of Molecular Pathology, The Netherlands Cancer Institute, Amsterdam, The Netherlands

OUTLINE

[a]Senior authorship

Epigenetic Biomarkers and Diagnostics
http://dx.doi.org/10.1016/B978-0-12-801899-6.00023-1

1. HISTONE MARKS AND THE EPIGENETIC LANDSCAPE

Cell-type-specific transcription is achieved by an orchestrated interplay between transcription factors and the chromatin state of the genomic regions where these transcription factors can potentially bind the DNA. Transcription factors have evolved according to their capacity to recognize specific DNA sequences [1] in response to specific signals. For example, the liganded estrogen receptor alpha (ERα) can recognize specific DNA consensus motif referred to as estrogen response elements (ERE). Importantly, binding seems to be dependent on a relaxed chromatin state surrounding the ERE-containing genomic regions [2]. In addition, functional ERE sequences are often found in regions marked by specific epigenetic marks [3].

Epigenetics is broadly defined as transmission of genetic information through mitosis without changes in the DNA sequence. Such epigenetic information is stored as chemical modifications on the DNA template (DNA methylation) or on the minimal chromatin unit, the nucleosome. Nucleosomes are multisubunit particles composed of four couples of histones (H2A, H2B, H3, and H4) [4]. The DNA is wrapped around the nucleosome in 147bp segments in what is generally described as the "beads-on-a-string model" [5]. Histone proteins can be modified by a variety of histone-modifying enzymes to include covalent chemical modifications such as lysine and arginine methylation, lysine acetylation, and serine phosphorylation [6]. Only recently, the field has begun to understand the potential mechanisms behind epigenetic inheritance (reviewed in [7,8]), and we will only discuss the mechanistic consequence of epigenetic bookmarking in relation to clinical outcome of breast cancer patients.

Recent technological advances allow for the investigation of active, inactive, or poised regulatory regions using combinatorial histone modifications as annotation tools. The histone code theory postulates that regulatory elements are defined by a combinatorial chemical alphabet written on histones. ChIP-seq (chromatin immunoprecipitation followed by massive parallel sequencing) is now used to quickly and accurately to map epigenetic modifications on a genome-wide scale [9]. Histone modifications such as the promoter-enriched active histone 3 lysine 4 trimethylation mark (H3K4me3) play crucial roles in the regulation of transcription [10,11]. In addition, active enhancers correlate with the presence of histone 3 lysine 27 acetylation mark (H3K27ac) and the histone 3 lysine 4 monomethylation mark (H3K4me1) [12]. Not all epigenetic modifications, however, facilitate transcription. H3K27me3, for example, a modification directly regulated by the EZH2 (enhancer of zeste homolog 2)-driven PRC2 (polycomb repressive complex 2), is associated with inactive chromatin and epigenetically silenced genes [13,14]. Interestingly, some promoters can contain both active (H3K4me3) as well as inactive histone marks (H3K27me3) and are often referred to as bivalent or poised promoters [15]. Analogous to all other tumor types, breast cancer is represented by aberrant transcription. Therefore, it is not surprising that histone marks have gained substantial attention from the scientific community as epigenetic defects may underlie activation of oncogenic pathways as well as the repression of proapoptotic signaling.

2. BREAST CANCER SUBTYPES, HORMONAL DEPENDENCE, AND EPIGENETIC REGULATION

Breast cancer (OMIM #114480) is the most commonly diagnosed malignancy in women with over 1.4million new cases and up to 450,000 deaths each year worldwide [16]. Over the past decade, it has become apparent that breast cancer is, above all, a heterogeneous disease that is represented by distinct molecular subtypes [17,18]. The classification of breast cancer is based on variations in gene expression profiles, tumor characteristics, and pathological features. The discovery of these subtypes of breast cancer was of the utmost clinical relevance, since they not only represent distinct groups of patients with differential prognosis [19] but also yield information on the most suitable and appropriate therapeutic option for that specific breast cancer subpopulation. Even though these subtypes have recently been refined into yet more sophisticated and smaller subgroups [20], still the most often-used and well-described subtypes of breast cancer are the HER2, basal, and luminal (A and B) subtypes. The largest subgroup of breast cancer patients is found in the luminal subtype, representing around 70% of all breast cancer cases [17,18]. These tumors are positive for ERα, and depend on the activity of this hormone-dependent transcription factor for cell proliferation and tumor growth. Such full-hormone dependence creates an ideal inroad for pharmaceutical interventions, and multiple drugs have been developed that prevent ERα activation and consequently block tumor cell proliferation (commonly referred to as endocrine therapies). The most-often prescribed endocrine therapies in breast cancer are tamoxifen, which blocks ERα action through competitive binding to the receptor [21], and aromatase inhibitors that act by blocking estrogen production [22].

Being a transcription factor, ERα binds promoter and enhancer regions throughout the genome upon hormone binding. Subsequently, ERα recruits its designated coregulators, ultimately forming the ERα transcription complex, which regulates gene expression programs involved in tumor cell proliferation [21,23,24]. Functional regulatory regions associated with estrogen-dependent transcription are often associated with ERα binding and a favorable epigenetic landscape to enable chromatin binding of this hormone-induced transcriptional regulator. Even though some ERα–chromatin binding is observed at promoter regions, the receptor preferably binds enhancer sites [25,26], demarcated by enhancer-specific epigenetic marks including H3K4me1/2 [27] and H3K27ac [28]. The link between the epigenome and ERα binding is thought to be mediated by a distinct family of transcription factors known as pioneer factors. These DNA-binding proteins associate to epigenetically defined regulatory regions to maintain accessible chromatin in the absence of ERα [3,29]. This pioneer factor family includes FOXA1 [26,30], PBX1 [31], GATA3 [32], and AP-2γ [33], even though additional ERα pioneer factors may be discovered in time.

ERα requires FOXA1 for binding the chromatin [26]. While FOXA1–chromatin interactions appear estrogen independent, the local epigenetic state dictates its activity [34] and consequently determines ERα binding. FOXA1 preferably binds hypomethylated DNA regions, and FOXA1–chromatin interactions subsequently facilitate H3K4 monomethylation [35]; a hallmark of enhancer regions [28]. As such, ERα's biological function and its DNA-binding capacity are partly dependent on histone modifications and a permissive epigenetic state (Figure 1).

3. EPIGENETIC FACTORS IN MAMMARY GLAND DEVELOPMENT AND CANCER

Recent studies support the hypothesis that longevity of normal breast stem cells may increase the possibility of accumulating genetic mutations, often associated with tumorigenesis. The finding that exposure to radiation early in life increases risk of breast cancer development further supports this hypothesis [36]. But next to genetic variations, also epigenetic programs are thought to play a key role in tumorigenesis. The maintenance of mammary stem cells as well as mammary gland development is under the control of a well-defined epigenetic program involving DNA methylation as well as histone modifications [10,37]. For example, the H3K27me3-mediating protein EZH2 is crucial

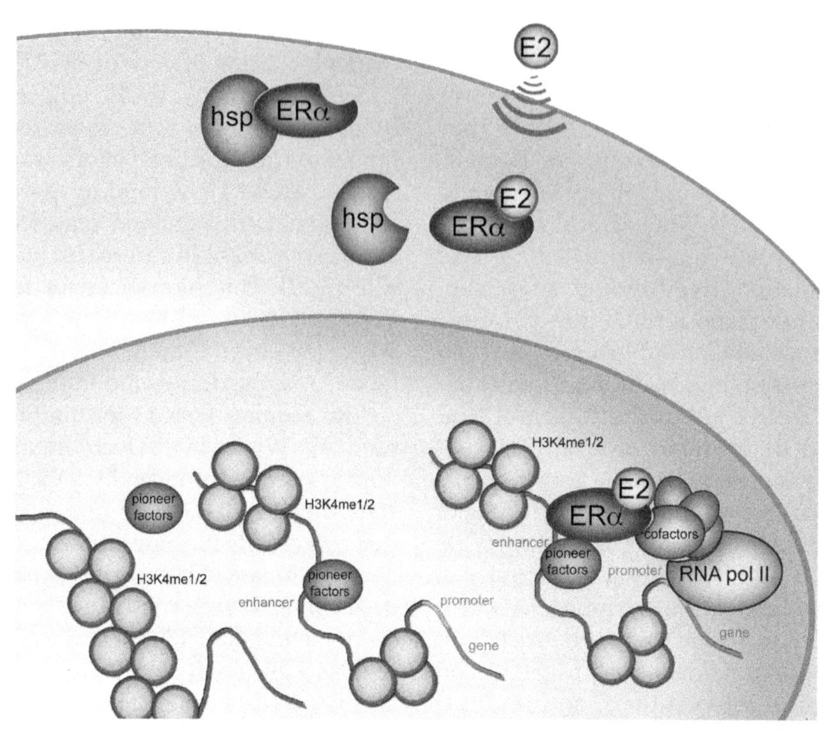

FIGURE 1 Estrogen receptor pathway in breast cancer and the role of epigenetics. Estradiol (E2) diffuses into the target cell, after which it binds the estrogen receptor alpha (ERα). Following ligand binding, ERα dissociates from heat shock protein 90 (hsp). For ERα to bind the chromatin, it requires the function of pioneer factors as well as a permissive epigenetic landscape (demarcated by H3K4me1 and H3K4me2) that jointly render the chromatin accessible at these sites. ERα is mainly found at enhancer regions, from where chromatin loops are formed toward promoter regions to control gene expression. Chromatin-bound ERα recruits cofactors and components of the RNA polymerase II complex to induce transcriptional programs that dictate cell proliferation in breast cancer cells.

for regulating mammary gland morphogenesis and maintenance of luminal progenitors [38]. In addition, epigenetic modifications are causally involved in large-scale chromatin conformational alterations and chromatin loops to stimulate or repress responsive gene transcription [10,39,40]. Deregulation of these epigenetic features may provide a basis for uncontrolled clonal proliferation of progenitor cells resulting in epithelial tumorigenesis [10,37].

Histone modifications can influence gene expression by establishing active or inactive chromatin states [41,42]. Genome-wide histone modification analysis of H3K4me3 and H3K27me3 in three mammary epithelial cell subpopulations (mammary stem cells, luminal progenitor, and mature luminal cells) illustrated that the number of epigenetically repressed genes, hallmarked by H3K27me3, is increased during luminal lineage differentiation in mice. Interestingly, hormonal deprivation by ovariectomy as well as pregnancy affected H3K27me3 levels, which was mediated by EZH2 [43]. Cell cycle regulators were among the genes with differential H3K27me3 signal after ovariectomy, which is in consistency with the role of steroid hormones in cell proliferation [43].

EZH2, the core enzymatic subunit of PRC2, can promote cell proliferation for ductal elongation and branching during puberty by repressing cell cycle inhibitor expression. In addition, EZH2 is directly involved in the maintenance and clonogenic potential of mammary stem cells, also in luminal and basal progenitors [38]. Interestingly, the expression levels of epigenetic modifiers are not uniformly distributed among all breast cancer subtypes, and EZH2 is found highly expressed in basal-type breast cancers, leading to transcription repression of tumor suppressor genes [44,45]. Paradoxically, while EZH2 is high in basal cancers, its downstream epigenetic modification, H3K27me3, is selectively enriched in luminal A type breast cancers [46]. These observations suggest that EZH2 expression levels cannot be extrapolated toward its biological activity, and regulation of EZH2 activation may play a determining role between different breast cancer subtypes.

4. EPIGENETIC COMPLEXES AS MUTATIONAL TARGETS IN LUMINAL BREAST CANCER

While epigenetic defects do not directly impinge on the genetic code, somatic mutations can have a direct impact on the epigenome. Over the past years, whole-exome sequencing data of thousands of ERα-positive breast cancer patients have been reported [47–50]. One key finding was that luminal cancers are not strongly characterized by common recurrent mutations, with the exception of PI3K mutations [51]. However, the histone methyltransferase *MLL3* was identified as the fifth most commonly mutated gene with over 8% of luminal patients carrying somatic alterations in this locus. MLL3 catalyzes the deposition of methyl groups on lysine residues including H3K4 [6]. Presently, we can only speculate about the functional consequence of *MLL3* mutations, but it is tempting to hypothesize that *MLL3* mutations may play a role in rearranging the breast cancer epigenetic landscape. Moreover, although identified at much lower frequency, multiple chromatin modifiers were found mutated in a whole-genome sequencing analysis of 46 aromatase inhibitor-resistant patients (histone methyltransferases: *MLL2*, *MLL3*, *MLL4*, and *MLL5*; histone demethyltranferases: *KDM6A*, *KDM4A*, *KDM5B*, and *KDM5C*; acetyltranferases: *MYST1*, *MYST3*, and *MYST4* [50]). These data are indirectly supported by many other genomic clues indicating an extensive cross-talk between the genome and the epigenome. One such example is provided by *GATA3*, another commonly mutated gene in ERα-positive breast cancer. GATA3 overlaps with ERα–chromatin binding sites where it affects ERα binding [32]. Equally to *MLL3*, the functional implications of *GATA3* mutations remain not fully understood. GATA3 mutations may regulate the response to estrogen signaling by stabilizing chromatin interactions of the complex [52]. Conversely, GATA3 mutations can also preclude GATA3 binding to the chromatin altogether [53], thus interfering with ERα access to the chromatin [32]. Considering the extensive epigenetic network underlying ERα signaling, we expect that few of these mutations could completely and directly reprogram ERα binding by indirect epigenetic reprogramming, favoring alternative binding. More importantly, they might contribute to activation of alternative survival pathways and could consequently stratify patients on outcome [54]. To what degree mutations in *MLL3* and *GATA3* affect tumor characteristics or response to specific therapeutics remains poorly understood and currently understudied.

5. EPIGENETIC MODIFICATIONS AS BIOMARKERS FOR TREATMENT SENSITIVITY AND OUTCOME

Loss of EZH2 in ERα-negative breast cancer cells leads to decreased tumor cell proliferation, both *in vitro* as well as *in vivo* [55]. In addition, elevated EZH2 levels in benign breast tissues serve as a prognostic marker for risk of breast cancer development [56]. High EZH2 expression levels determined by immunohistochemistry correlate in ERα-positive breast tumors with resistance to tamoxifen [57].

TABLE 1 Biomarker Studies on EZH2 and H3K27me3

Marker	Detection method	Population	High levels indicative for?	References
EZH2	IHC	Healthy women	Breast cancer risk	[56]
EZH2	IHC	Tamoxifen-treated breast cancer patients	Poor outcome	[57]
EZH2	IHC	Mixed subtypes, different treatments	Poor outcome	[61]
H3K27me3	IHC	Mixed subtypes, different treatments	Good outcome	[61]
H3K27me3	Distinct ChIP-seq profiles	Aromatase inhibitor-treated breast cancer patients	Good outcome/poor outcome	[62]

IHC, immunohistochemistry; ChIP-seq, chromatin immunoprecipitation followed by massive parallel sequencing.

Still, the tumorigenic potential of EZH2 is not necessarily linked with its epigenetic capacities and other biological roles of EZH2 may explain its involvement in tumorigenesis. EZH2 can regulate transcription of estrogen-responsive genes via non-PRC2-mediated mechanisms through interactions with repressor of estrogen receptor activity (REA), an ERα corepressor [58]. Alternatively, EZH2 has also been described to physically interact with ERα in an estrogen-dependent fashion, indicating that EZH2 may function as coactivator of estrogen signaling [59], parallel to mechanisms that have been reported for androgen receptor function in prostate [60].

While immunohistochemical analyses illustrated that high EZH2 levels are indicative of poor prognosis, high H3K27me3 levels are on the other hand associated with a good outcome [61]. Yet, since H3K27me3 is a histone modification with distinct genomic features and selective distributions throughout the genome, it remained to be discovered whether specific distributions of H3K27me3 would bear prognostic features as well. H3K27me3 ChIP-seq profiling in a small cohort of breast cancer patients identified a selective gain of H3K27me3 signal at distinct sites for patients with poor outcome after aromatase inhibitor treatment [62]. Gene ontology analysis on genes proximal to the differential H3K27me3 patterns between good and poor outcome patients showed enrichment for pathways involving developmental processes such as cell cycle regulation and cell adhesion. Interestingly, these biological processes strongly overlap with those identified by H3K27me3 ChIP-seq in developing mouse mammary gland [43]. These data position H3K27me3 and its upstream regulator, EZH2, in a quite unusual role in treatment resistance, where divergent expression levels and histone modification states ultimately give rise to distinct genomic programs and patient outcome. An overview of studies describing the biomarker potential of EZH2 and H3K27me3 is depicted in Table 1.

Distinct ERα-binding profiles in samples from patients with different outcomes were not correlated with altered epigenetic H3K4me3 and H3K27me3 marks [62]. These data suggest that the dynamic ERα binding behavior uses accessible chromatin regions that are already imprinted in a static epigenetic landscape. In agreement with these data, it was recently suggested that the chromatin landscape of endocrine therapy-resistant cells may disengage from ERα binding by remodeling ERE-containing loci [63]. Therefore, it is possible that resistant cells switch to other nuclear receptors or alternative transcription factor for their growth. Epigenetic profiling could then be used to detect epigenetic

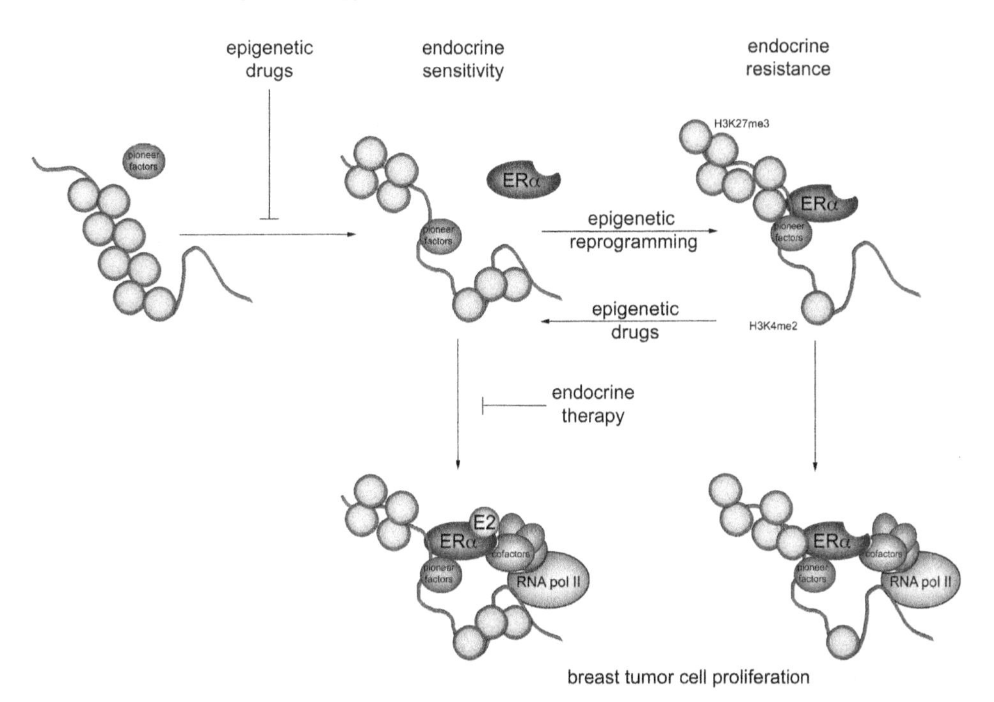

FIGURE 2 Epigenetic drugs in breast cancer treatment. Both ERα and its pioneer factors require a distinct set of epigenetic modifications that may be involved in facilitating chromatin interactions and functioning. In endocrine resistance, epigenetic reprogramming is observed, which renders the drug inactive, so that breast tumor cell proliferation can still occur in the presence of endocrine therapy. Epigenetic drugs can affect ERα functioning and endocrine resistance on multiple levels, which prevents hormonal action and may reverse treatment resistance. With this, epigenetic drugs and endocrine therapies can function independently but also in functional synergy to block breast tumor cell proliferation.

reprogramming to yield information on survival or response to therapy.

6. EPIGENETIC MODIFIERS AS POTENTIAL DRUG TARGETS IN CANCER

Epigenetic modifiers play crucial roles in mammary cell proliferation and oncogenic activity in breast cancer while acting as cofactors in hormone receptor regulation. This situation provides exciting opportunities for pharmaceutical intervention, since such therapeutics may be broadly applicable over multiple breast cancer subtypes and could function synergistically with currently available treatment regimens, including endocrine therapies (Figure 2).

The role of EZH2 in promoting breast cancer as either a transcriptional repressor or activator makes it an attractive target for antitumor drugs. High-throughput screens have led to the development of several EZH2 inhibitors which effectively decreased tumor cell proliferation in xenograft models [64].

In mouse models, estrogen deprivation directly affects EZH2 action and consequently alters H3K27me3 levels [43]. Since the analogous situation is clinically achieved through aromatase inhibitor therapies, EZH2 function may be affected in a similar fashion and give rise to an altered epigenetic landscape, ultimately contributing to aromatase inhibitor treatment resistance [65]. Therefore, combining conventional breast cancer therapies with EZH2 inhibitors could be significantly beneficial in blocking cancer therapy resistance. For the histone deacetylases (HDAC) inhibitor entinostat, such combinatorial therapy trials are currently running (NCT02115594 and NCT02115282). The rationale for such trials is provided by *in vitro* studies, where entinostat treatment sufficed in reactivation of ESR1 gene expression in ERα-negative breast tumors [66]. If successful in clinical trials, such novel epigenetic therapeutics in combination with endocrine therapies [67] may pave the way for other epigenetic drugs for assessment in clinical studies.

7. CONCLUSIVE REMARKS

With the broad-spectrum tissue distribution of epigenetic modifiers and general physiological effects of these proteins, one could argue whether epigenetic drugs could really be considered as a means of "targeted therapy." Yet, since epigenetic regulation represents a crucial component in breast cancer development and treatment resistance, pharmaceutical intervention based on epigenetic modifiers could provide a promising inroad in breast cancer treatment. EZH2 expression levels are increased in treatment-resistant breast cancers, and distinct profiles of H3K27me3 are indicative of aromatase inhibitor treatment resistance in patients. These findings would position EZH2 inhibition as a promising therapeutic target, with increased EZH2 expression levels or distinct H3K27me3 as an ideal companion diagnostic tool. Importantly, such epigenetic drugs may be prescribed as a combinational therapy, both targeting the ERα pathway and also as blocking the epigenetically-driven development of resistance to antiestrogen therapy.

The epigenetic profile of tumor cells is radically altered as compared to a normal cell [68], which suggests that selective epigenetic targeting of tumor cells may be achievable at some level. Rather than targeting directly the enzymes responsible for the deposition of histone modifications, future approaches may alternatively focus on tumor-specific enhancers to modulate expression of tumor drivers. Proof-of-principle studies on "epigenetic editing," based on the ability of targeting enzymes such as MLL3 and EZH2 on specific regulatory elements using zinc-finger-guided approaches [69], yielded promising results in the context of HER2 silencing [70]. In this proof-of-principle study using a number of breast and ovarian cancer cell lines, the epigenetic landscape at the *HER2* promoter was specifically edited to induce a repressive chromatin state at this locus. It remains, however, unknown how stable the transcriptional silencing in response to epigenetic editing will be. Nonetheless, this approach does still bear great promise, yielding the possibility to exploit individual epigenetic signatures to interfere with the proproliferative and proinvasive transcriptional programs of tumor cells.

EZH2 and H3K27me3 (expression and genomic profile) may represent interesting potential biomarkers to be of use for the clinical decision-making in the treatment of breast cancer patients. Yet, many issues need to be resolved before considering these factors truly as promising biomarkers for clinical implementation. Firstly, these factors need to be validated in additional cohorts. Secondly, sensitivity and specificity of these potential biomarkers need to be determined. Thirdly, applicability of these biomarkers needs to be assessed in randomized clinical trials. And lastly, a prospective clinical study would be needed to truly claim applicability of EZH2 and H3K27me3 as biomarkers for daily clinical practice. In summary, we are still far away from applying any of these findings in the clinic, and future research may elucidate whether epigenetic biomarkers are truly the "new black" or merely "a red herring."

Glossary

Breast cancer subtypes Distinct subgroups of breast cancer patients, hallmarked by clear differences in gene expression profiles, prognosis, and sensitivities to therapeutic intervention.

Enhancer Short DNA region that can be bound by proteins to regulate expression of genes.

Epigenetics/epigenome Alterations in genetic material, not represented as changes in the genetic code. In contrast to DNA sequence changes, epigenetics are intrinsically reversible.

Estrogen receptor alpha Transcription factor that is activated by the hormone estradiol. It represents the main driver of luminal breast cancers.

GATA3 Trans-acting T-cell-specific transcription factor, binding specifically to the "GATA" DNA sequence.

Luminal breast cancer The most frequently diagnosed breast cancer subtype. About 75% of all human breast cancers fall within this category. It is subdivided into luminal A and luminal B.

Pioneer factor Transcription factor that can bind condensed chromatin, making it accessible.

Promoter Short DNA region proximal to a gene that regulates its expression.

Transcription factor Protein that binds specific DNA sequences in order to regulate expression of a responsive gene.

LIST OF ABBREVIATIONS

AP-2γ Activating protein 2 gamma
ERα Estrogen receptor alpha
EZH2 Enhancer of zeste homolog 2
FOXA1 Forkhead box protein A1
H Histone
H3K4me1 Monomethylation on lysine 4 of histone 3
H3K4me2 Dimethylation on lysine 4 of histone 3
H3K4me3 Trimethylation on lysine 4 of histone 3
H3K27me3 Trimethylation on lysine 27 of histone 3
HDAC Histone deacetylases
KDM Lysine (K)-specific demethylase
MLL Myeloid/lymphoid leukemia
PBX1 Pre-B cell leukemia homeobox 1
PRC Polycomb repressive complex 2
REA Repressor of estrogen receptor activity

References

[1] Weirauch MT, Yang A, Albu M, Cote AG, Montenegro-Montero A, Drewe P, et al. Determination and inference of eukaryotic transcription factor sequence specificity. Cell September 11, 2014;158(6):1431–43.

[2] Magnani L, Lupien M. Chromatin and epigenetic determinants of estrogen receptor alpha (ESR1) signaling. Mol Cell Endocrinol January 25, 2014;382(1):633–41.

[3] Magnani L, Eeckhoute J, Lupien M. Pioneer factors: directing transcriptional regulators within the chromatin environment. Trends Genet November 2011;27(11):465–74.

[4] Luger K, Mader AW, Richmond RK, Sargent DF, Richmond TJ. Crystal structure of the nucleosome core particle at 2.8 A resolution. Nature September 18, 1997;389(6648):251–60.

[5] Fussner E, Ching RW, Bazett-Jones DP. Living without 30nm chromatin fibers. Trends Biochem Sci January 2011;36(1):1–6.

[6] Kouzarides T. Chromatin modifications and their function. Cell February 23, 2007;128(4):693–705.

[7] Zhu B, Reinberg D. Epigenetic inheritance: uncontested? Cell Res March 2011;21(3):435–41.

[8] Campos EI, Stafford JM, Reinberg D. Epigenetic inheritance: histone bookmarks across generations. Trends Cell Biol September 18, 2014. http://dx.doi.org/10.1016/j.tcb.2014.08.004.

[9] Magnani L, Carroll J, Zwart W, Palmieri C. ChIPing away at breast cancer. Lancet Oncol December 2012;13(12):1185–7.

[10] Huang TH, Esteller M. Chromatin remodeling in mammary gland differentiation and breast tumorigenesis. Cold Spring Harb Perspect Biol September 2010;2(9):a004515.

[11] Jaenisch R, Bird A. Epigenetic regulation of gene expression: how the genome integrates intrinsic and environmental signals. Nat Genet March 2003;(Suppl. 33):245–54.

[12] Ernst J, Kheradpour P, Mikkelsen TS, Shoresh N, Ward LD, Epstein CB, et al. Mapping and analysis of chromatin state dynamics in nine human cell types. Nature May 5, 2011;473(7345):43–9.

[13] Schuettengruber B, Cavalli G. Recruitment of polycomb group complexes and their role in the dynamic regulation of cell fate choice. Development November 2009;136(21):3531–42.

[14] Margueron R, Reinberg D. The polycomb complex PRC2 and its mark in life. Nature January 20, 2011;469(7330):343–9.

[15] Brown R, Curry E, Magnani L, Wilhelm-Benartzi CS, Borley J. Poised epigenetic states and acquired drug resistance in cancer. Nat Rev Cancer September 25, 2014. http://dx.doi.org/10.1038/nrc3819.

[16] Kamangar F, Dores GM, Anderson WF. Patterns of cancer incidence, mortality, and prevalence across five continents: defining priorities to reduce cancer disparities in different geographic regions of the world. J Clin Oncol May 10, 2006;24(14):2137–50.

[17] Sorlie T, Perou CM, Tibshirani R, Aas T, Geisler S, Johnsen H, et al. Gene expression patterns of breast carcinomas distinguish tumor subclasses with clinical implications. Proc Natl Acad Sci USA September 11, 2001;98(19):10869–74.

[18] Perou CM, Sorlie T, Eisen MB, van de Rijn M, Jeffrey SS, Rees CA, et al. Molecular portraits of human breast tumours. Nature August 17, 2000;406(6797):747–52.

[19] Parker JS, Mullins M, Cheang MC, Leung S, Voduc D, Vickery T, et al. Supervised risk predictor of breast cancer based on intrinsic subtypes. J Clin Oncol March 10, 2009;27(8):1160–7.
[20] Curtis C, Shah SP, Chin SF, Turashvili G, Rueda OM, Dunning MJ, et al. The genomic and transcriptomic architecture of 2,000 breast tumours reveals novel subgroups. Nature June 21, 2012;486(7403):346–52.
[21] Shiau AK, Barstad D, Loria PM, Cheng L, Kushner PJ, Agard DA, et al. The structural basis of estrogen receptor/coactivator recognition and the antagonism of this interaction by tamoxifen. Cell December 23, 1998;95(7):927–37.
[22] Osborne CK. Aromatase inhibitors in relation to other forms of endocrine therapy for breast cancer. Endocr Relat Cancer June 1999;6(2):271–6.
[23] Zwart W, Theodorou V, Kok M, Canisius S, Linn S, Carroll JS. Oestrogen receptor-co-factor-chromatin specificity in the transcriptional regulation of breast cancer. EMBO J November 30, 2011;30(23):4764–76.
[24] Zwart W, de Leeuw R, Rondaij M, Neefjes J, Mancini MA, Michalides R. The hinge region of the human estrogen receptor determines functional synergy between AF-1 and AF-2 in the quantitative response to estradiol and tamoxifen. J Cell Sci April 15, 2010;123(Pt 8):1253–61.
[25] Kittler R, Zhou J, Hua S, Ma L, Liu Y, Pendleton E, et al. A comprehensive nuclear receptor network for breast cancer cells. Cell Rep February 21, 2013;3(2):538–51.
[26] Carroll JS, Liu XS, Brodsky AS, Li W, Meyer CA, Szary AJ, et al. Chromosome-wide mapping of estrogen receptor binding reveals long-range regulation requiring the forkhead protein FoxA1. Cell July 15, 2005;122(1):33–43.
[27] Joseph R, Orlov YL, Huss M, Sun W, Kong SL, Ukil L, et al. Integrative model of genomic factors for determining binding site selection by estrogen receptor-alpha. Mol Syst Biol December 21, 2010;6:456.
[28] Hah N, Murakami S, Nagari A, Danko CG, Kraus WL. Enhancer transcripts mark active estrogen receptor binding sites. Genome Res August 2013;23(8):1210–23.
[29] Jozwik KM, Carroll JS. Pioneer factors in hormone-dependent cancers. Nat Rev Cancer June 2012;12(6):381–5.
[30] Hurtado A, Holmes KA, Ross-Innes CS, Schmidt D, Carroll JS. FOXA1 is a key determinant of estrogen receptor function and endocrine response. Nat Genet January 2011;43(1):27–33.
[31] Magnani L, Ballantyne EB, Zhang X, Lupien M. PBX1 genomic pioneer function drives ERα signaling underlying progression in breast cancer. PLoS Genet November 2011;7(11):e1002368.
[32] Theodorou V, Stark R, Menon S, Carroll JS. GATA3 acts upstream of FOXA1 in mediating ESR1 binding by shaping enhancer accessibility. Genome Res January 2013;23(1):12–22.
[33] Tan SK, Lin ZH, Chang CW, Varang V, Chng KR, Pan YF, et al. AP-2gamma regulates oestrogen receptor-mediated long-range chromatin interaction and gene transcription. EMBO J July 6, 2011;30(13):2569–81.
[34] Eeckhoute J, Lupien M, Meyer CA, Verzi MP, Shivdasani RA, Liu XS, et al. Cell-type selective chromatin remodeling defines the active subset of FOXA1-bound enhancers. Genome Res March 2009;19(3):372–80.
[35] Serandour AA, Avner S, Percevault F, Demay F, Bizot M, Lucchetti-Miganeh C, et al. Epigenetic switch involved in activation of pioneer factor FOXA1-dependent enhancers. Genome Res April 2011;21(4):555–65.
[36] Land CE, McGregor DH. Breast cancer incidence among atomic bomb survivors: implications for radiobiologic risk at low doses. J Natl Cancer Inst January 1979;62(1):17–21.
[37] Bloushtain-Qimron N, Yao J, Shipitsin M, Maruyama R, Polyak K. Epigenetic patterns of embryonic and adult stem cells. Cell Cycle March 15, 2009;8(6):809–17.
[38] Michalak EM, Nacerddine K, Pietersen A, Beuger V, Pawlitzky I, Cornelissen-Steijger P, et al. Polycomb group gene Ezh2 regulates mammary gland morphogenesis and maintains the luminal progenitor pool. Stem Cells September 2013;31(9):1910–20.
[39] Cairns BR. The logic of chromatin architecture and remodelling at promoters. Nature September 10, 2009;461(7261):193–8.
[40] Harmston N, Lenhard B. Chromatin and epigenetic features of long-range gene regulation. Nucleic Acids Res August 2013;41(15):7185–99.
[41] Bernstein BE, Humphrey EL, Erlich RL, Schneider R, Bouman P, Liu JS, et al. Methylation of histone H3 Lys 4 in coding regions of active genes. Proc Natl Acad Sci USA June 25, 2002;99(13):8695–700.
[42] Barski A, Cuddapah S, Cui K, Roh TY, Schones DE, Wang Z, et al. High-resolution profiling of histone methylations in the human genome. Cell May 18, 2007;129(4):823–37.
[43] Pal B, Bouras T, Shi W, Vaillant F, Sheridan JM, Fu N, et al. Global changes in the mammary epigenome are induced by hormonal cues and coordinated by Ezh2. Cell Rep February 21, 2013;3(2):411–26.
[44] Kleer CG, Cao Q, Varambally S, Shen R, Ota I, Tomlins SA, et al. EZH2 is a marker of aggressive breast cancer and promotes neoplastic transformation of breast epithelial cells. Proc Natl Acad Sci USA September 30, 2003;100(20):11606–11.
[45] Bracken AP, Pasini D, Capra M, Prosperini E, Colli E, Helin K. EZH2 is downstream of the pRB-E2F pathway, essential for proliferation and amplified in cancer. EMBO J October 15, 2003;22(20):5323–35.

[46] Healey MA, Hu R, Beck AH, Collins LC, Schnitt SJ, Tamimi RM, et al. Association of H3K9me3 and H3K27me3 repressive histone marks with breast cancer subtypes in the Nurses' Health Study. Breast Cancer Res Treat September 16, 2014. http://dx.doi.org/10.1007/s10549-014-3089-1.

[47] Cancer Genome Atlas N. Comprehensive molecular portraits of human breast tumours. Nature October 4, 2012;490(7418):61–70.

[48] Shah SP, Roth A, Goya R, Oloumi A, Ha G, Zhao Y, et al. The clonal and mutational evolution spectrum of primary triple-negative breast cancers. Nature June 21, 2012;486(7403):395–9.

[49] Nik-Zainal S, Alexandrov LB, Wedge DC, Van Loo P, Greenman CD, Raine K, et al. Mutational processes molding the genomes of 21 breast cancers. Cell May 25, 2012;149(5):979–93.

[50] Ellis MJ, Ding L, Shen D, Luo J, Suman VJ, Wallis JW, et al. Whole-genome analysis informs breast cancer response to aromatase inhibition. Nature June 21, 2012;486(7403):353–60.

[51] Polyak K, Metzger Filho O. SnapShot: breast cancer. Cancer Cell October 16, 2012;22(4):562-e1.

[52] Adomas AB, Grimm SA, Malone C, Takaku M, Sims JK, Wade PA. Breast tumor specific mutation in GATA3 affects physiological mechanisms regulating transcription factor turnover. BMC Cancer 2014;14:278.

[53] Yan W, Cao QJ, Arenas RB, Bentley B, Shao R. GATA3 inhibits breast cancer metastasis through the reversal of epithelial-mesenchymal transition. J Biol Chem April 30, 2010;285(18):14042–51.

[54] Jiang YZ, Yu KD, Zuo WJ, Peng WT, Shao ZM. GATA3 mutations define a unique subtype of luminal-like breast cancer with improved survival. Cancer May 1, 2014;120(9):1329–37.

[55] Gonzalez ME, Li X, Toy K, DuPrie M, Ventura AC, Banerjee M, et al. Downregulation of EZH2 decreases growth of estrogen receptor-negative invasive breast carcinoma and requires BRCA1. Oncogene February 12, 2009;28(6):843–53.

[56] Ding L, Erdmann C, Chinnaiyan AM, Merajver SD, Kleer CG. Identification of EZH2 as a molecular marker for a precancerous state in morphologically normal breast tissues. Cancer Res April 15, 2006;66(8):4095–9.

[57] Reijm EA, Timmermans AM, Look MP, Meijer-van Gelder ME, Stobbe CK, van Deurzen CH, et al. High protein expression of EZH2 is related to unfavorable outcome to tamoxifen in metastatic breast cancer. Ann Oncol September 5, 2014. http://dx.doi.org/10.1093/annonc/mdu391.

[58] Hwang C, Giri VN, Wilkinson JC, Wright CW, Wilkinson AS, Cooney KA, et al. EZH2 regulates the transcription of estrogen-responsive genes through association with REA, an estrogen receptor corepressor. Breast Cancer Res Treat January 2008;107(2):235–42.

[59] Shi B, Liang J, Yang X, Wang Y, Zhao Y, Wu H, et al. Integration of estrogen and Wnt signaling circuits by the polycomb group protein EZH2 in breast cancer cells. Mol Cell Biol July 2007;27(14):5105–19.

[60] Xu K, Wu ZJ, Groner AC, He HH, Cai C, Lis RT, et al. EZH2 oncogenic activity in castration-resistant prostate cancer cells is Polycomb-independent. Science December 14, 2012;338(6113):1465–9.

[61] Holm K, Grabau D, Lovgren K, Aradottir S, Gruvberger-Saal S, Howlin J, et al. Global H3K27 trimethylation and EZH2 abundance in breast tumor subtypes. Mol Oncol October 2012;6(5):494–506.

[62] Jansen MP, Knijnenburg T, Reijm EA, Simon I, Kerkhoven R, Droog M, et al. Hallmarks of aromatase inhibitor drug resistance revealed by epigenetic profiling in breast cancer. Cancer Res November 15, 2013;73(22):6632–41.

[63] Magnani L, Stoeck A, Zhang X, Lanczky A, Mirabella AC, Wang TL, et al. Genome-wide reprogramming of the chromatin landscape underlies endocrine therapy resistance in breast cancer. Proc Natl Acad Sci USA April 16, 2013;110(16):E1490–9.

[64] Tan JZ, Yan Y, Wang XX, Jiang Y, Xu HE. EZH2: biology, disease, and structure-based drug discovery. Acta Pharmacol Sin February 2014;35(2):161–74.

[65] Pathiraja TN, Nayak SR, Xi Y, Jiang S, Garee JP, Edwards DP, et al. Epigenetic reprogramming of HOXC10 in endocrine-resistant breast cancer. Sci Transl Med March 26, 2014;6(229):229ra41.

[66] Sabnis GJ, Goloubeva O, Chumsri S, Nguyen N, Sukumar S, Brodie AM. Functional activation of the estrogen receptor-α and aromatase by the HDAC inhibitor entinostat sensitizes ER-negative tumors to letrozole. Cancer Res March 1, 2011;71(5):1893–903.

[67] Entinostat plus exemestane has activity in ER+ advanced breast cancer. Cancer Discov July 2013;3(7):OF17.

[68] Akhtar-Zaidi B, Cowper-Sal-lari R, Corradin O, Saiakhova A, Bartels CF, Balasubramanian D, et al. Epigenomic enhancer profiling defines a signature of colon cancer. Science May 11, 2012;336(6082):736–9.

[69] Magnani L. Epigenetic engineering and the art of epigenetic manipulation. Genome Biol 2014;15(6):306.

[70] Falahi F, Huisman C, Kazemier HG, van der Vlies P, Kok K, Hospers GA, et al. Towards sustained silencing of HER2/neu in cancer by epigenetic editing. Mol Cancer Res September 2013;11(9):1029–39.

CHAPTER

24

Histone Variants and Posttranslational Modifications in Spermatogenesis and Infertility

Juan Ausio[1], *Yinan Zhang*[2], *Toyotaka Ishibashi*[2,3]

[1]Department of Biochemistry and Microbiology, University of Victoria, Victoria, BC, Canada; [2]Division of Life Science, Hong Kong University of Science and Technology, Kowloon, Hong Kong, HKSAR; [3]Department of Biomedical Engineer, Hong Kong University of Science and Technology, Kowloon, Hong Kong, HKSAR

OUTLINE

Epigenetic Biomarkers and Diagnostics
http://dx.doi.org/10.1016/B978-0-12-801899-6.00024-3

1. INTRODUCTION

Genomic DNA is organized into discrete chromatin particles known by the name of nucleosomes. They consist of DNA stretches of approximately 150 bp wrapped around a histone octamer (containing two copies each of canonical core histones H2A, H2B, H3, and H4). In the chromatin fiber, nucleosomes are linked by DNA regions that bind to linker histone H1 [1]. In addition to canonical histones, which are expressed mainly during the S phase, there are several histone variants that are expressed at different times during the cell cycle and confer different stabilities and/or plasticities to the nucleosome [2]. Each histone variant has a specific role within the cells; for instance, H2A.Z and H2A.X are involved in gene expression and DNA damage recognition, respectively. Some histone variants are germ line cell-specific. This is the case with H1t, HILS1 (histone linker spermatid-specific 1), H2AL.1, H2AL.2, H2A.Bbd, TH2A, TH2B, H2BL.1, H2BFW, and H3t in testes (see Figure 1 and Table 1 for a comprehensive list), and with H1foo, the only histone variant so far identified in oocytes [2].

During spermatogenesis, canonical histones (core histone H2A, H2B, H3, H4, and linker histone H1) and the somatic histone variants (H2A.Z, H2A.X, and H3.3) are replaced by the aforementioned testes-specific histone variants. The process involves the posttranslational modifications (PTMs) of several somatic histones. H4 becomes heavily acetylated at the N-terminal tail, and H2A.X phosphorylates its C-terminal tail [3,4] while H2A/H2B exhibits a significant extent of polyubiquitination [5] and ADP-ribosylation [6]. These histone PTMs play a critical role in the repair of DNA double-strand breaks which lead to the changes in chromatin topology associated with the nucleosome to nucleoprotamine transition [7]. PTMs also facilitate the protein replacements which take place during the structural transition. During spermiogenesis, in elongating spermatids, the majority of testes-specific histones and posttranslationally modified histones are gradually replaced by transition proteins (TNPs), which are subsequently replaced by protamines. This histone-to-protamine transition provides chromatin with the highly compact conformation found in mature sperm (Figure 2). Thus, the spermatogenic process involves a tight functional interconnection between histone variant expression and histone PTMs [8,9] (Figure 2). However, the molecular details of those processes involved in the histone-to-protamine transition still remain largely unknown. Several recent studies have shown that abnormalities in the testes-specific histone variants are implicated in infertility.

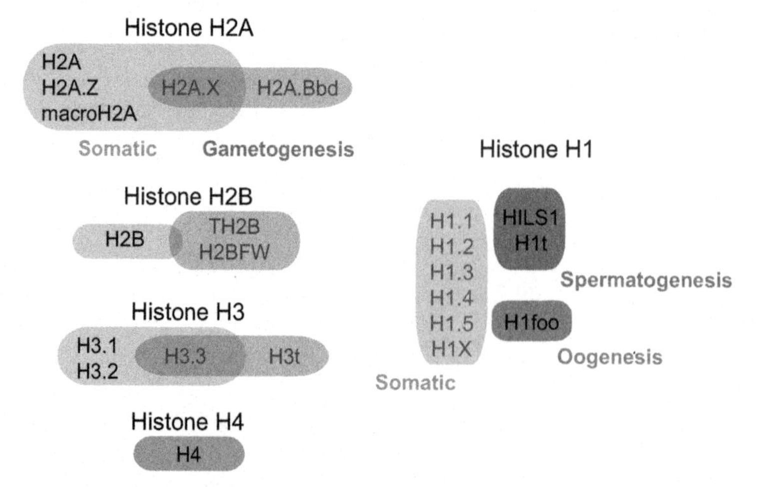

FIGURE 1 List of human somatic and gametogenesis histone variants.

TABLE 1 Lists of the Potential Functional Involvement of Histone Variants and Histone Posttranslational Modifications (PTMs) in Spermatogenesis and Oogenesis

Histone variant/PTM	Function in gametogenesis	Function in somatic cells
H2A.X	Participation in gametogenic meiosis [15,17]	DNA repair [3,14]/genome stability [16]
H2A.Bbd	Chromatin remodeling during spermiogenesis [54]	The role of the natively expressed form is unknown
TH2B	Chromatin remodeling during spermatogenesis [38,39] Protamine to histone reversal transition in the male pronucleus by the oocyte form [38]	Chromatin remodeling during early development [43]
H2BFW	Unknown but -9C > T SNP impairs fertility [46,47]	Unknown
H3.3	Involvement in the histone–protamine (chromatin remodeling) transition during spermiogenesis [29] Proper nucleosome assembly on the paternal DNA after fertilization [33]	Transcription memory [31]
H3t	Destabilization of nucleosome during spermatogenic chromatin remodeling [27]	N/A
H1t	Chromatin remodeling transitions during spermatogenesis [62–64]	N/A
HILS1	Chromatin remodeling during spermiogenesis [58]	N/A
H1foo	Chromatin flexibility and dynamics during oogenesis [11,67]	Chromatin remodeling during early development [11]
H4 acetylation	Protamine-to-histone transition during spermiogenesis [69]	Manifold roles involved in transcription activation
H2A.X phosphorylation	Repair and resealing of DSB during spermatogenesis [21]	DNA repair [3,14]
H1t phosphorylation	Chromatin remodeling transitions during spermatogenesis [62–64]	N/A

By the end of mammalian spermatogenesis, 2–15% of histones—depending on the species—still remain in mature sperm, and have been proposed to play a paternal epigenetic role after fertilization [10].

2. HISTONE VARIANTS AND THEIR ROLE IN INFERTILITY

In contrast to that of spermatogenesis, the literature regarding the roles of histone variants in mammalian oogenesis and postfertilization (with the exception of H1foo [11]) is almost nonexistent. Therefore, this chapter will focus mainly on the role of histone variants in spermatogenesis and their putative involvement in infertility.

2.1 H2A.X: A DNA Damage Response Histone Variant with an Important Fertility Role

H2A.X is one of the well-studied histone H2A variants [12,13]. It is specifically phosphorylated at the serine of the consensus SQEY sequence at its extreme C-terminal end, in response to DNA damage [3,14]. The phosphorylated form

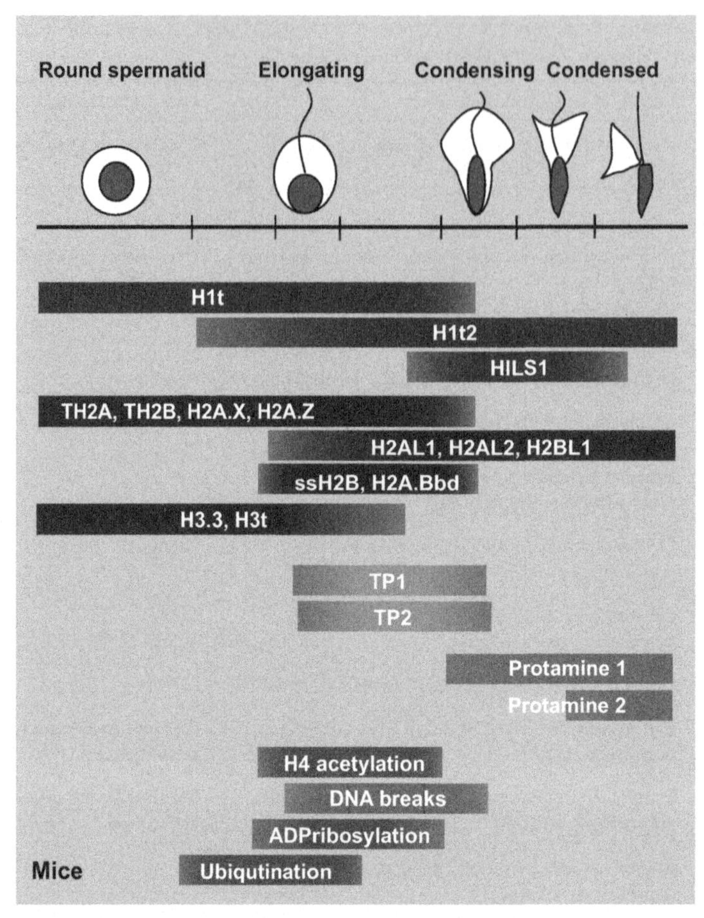

FIGURE 2 Scheme displaying the major histone variants and chromosomal proteins expressed during spermatogenesis in mice. *This figure is adapted from Ref. [102].*

of H2A.X is called γH2A.X, and it acts as a signal to recruit several DNA repair factors [3]. The histone variant H2A.X is maximally expressed in spermatogonia [15], and plays a critical role in the repair processes involving the DNA double-strand breaks that take place in the middle of the spermatid stage during spermatogenesis (Figure 2). Experiments with knockout H2A.X mice have shown that this variant is not essential for survival, but their growth is retarded and they do exhibit impaired embryo fibroblast growth [16]. Female H2A.X knockout mice are fertile, though the size of their litter is smaller than that of wild-type mice (Figure 3). However, male H2A.X knockout mice are infertile, and the sizes of their testes and seminiferous tubules are also smaller than those of the wild type [16]. In addition, the X and Y chromosomes of histone H2A.X-deficient spermatocytes failed to condense to form a sex body, and the spermatocytes were also severely defective in meiotic pairing [17]; this suggests that, in addition to double-strand break DNA repair, H2A.X is also critical to the chromatin remodeling occurring during mouse male meiosis.

Despite all the above, a thorough screening of 302 infertile men with azoospermia or severe oligospermia, and 198 normospermic

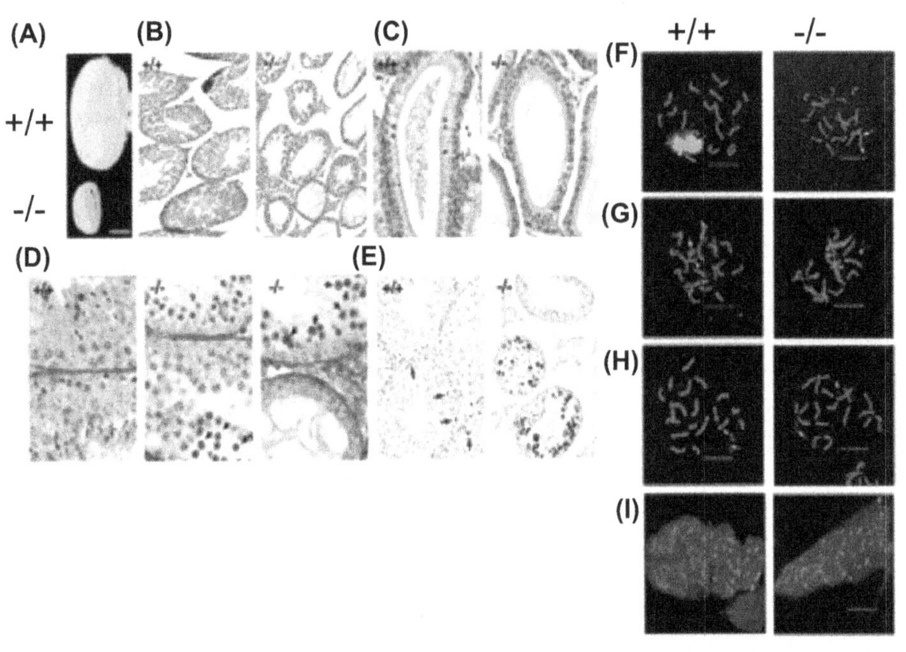

FIGURE 3 Defective spermatogenesis in H2AX−/− mice. (A) Comparison of testis size in 2-month-old H2AX wild-type (+/+) and knockout (−/−) mice. Bar, 2 mm. (B) Sections of seminiferous tubules of 7-week-old (+/+ and −/−) littermates stained with hematoxylin–eosin. Magnification, X10. (C) Hematoxylin–eosin-stained sections of epididymis from the same mice shown in (B). Magnification, X40. (D) High-magnification (X100) images of periodic acid—Schiff—stained paraffin sections of seminiferous tubules of 2-month-old wild-type (+/+) (left panel) and knockout (−/−) (two right panels) mice. Primary spermatocytes in early pachytene, late pachytene, and zygotene are indicated. Apoptotic nuclei with condensing chromatin are present in H2AX−/− tubules (arrows). (E) TUNEL (terminal deoxynucleotidyl transferase-mediated dUTP nick-end labeling) assay detects very few apoptotic cells in normal tubules (arrows, left panel), whereas H2AX-mutant tubules contain a large number of dying cells. Magnification, X40. (F to I) Indirect immunofluorescence of H2AX+/− (+/−; left panels) and H2AX−/− (−/−; right panels) pachytene spermatocytes. (F) Merged image of Scp3 (red, dark gray in print versions) and γ-H2AX (green, light gray in print versions). (G) Merged image of Scp3 (green, light gray in print versions) and Scp1 (red, dark gray in print versions). (H) Merged image of Scp1 (red, dark gray in print versions) and Mlh1 (green, light gray in print versions). (I) Diplotene (left) and diakinesis (right) H2AX+/− stages (two individual cells separated by the dotted line), and fragmented synaptonemal complex in H2AX−/− spermatocyte, visualized with antibody to Scp3 (red, dark gray in print versions) and counterstained with DAPI (blue, gray in print versions). For (F) to (H), the arrowhead indicates the Y chromosome, and the arrow shows the X chromosome. *Images and legend are from Ref. [16].*

controls, found no evidence for the presence of H2A.X mutations in infertile men [18]. This apparent contradiction could arise from either different roles of H2A.X in human and mice spermatogenesis, or from the putatively high deleterious effects of H2A.X in genome instability in humans [19]. As pointed out in Ref. [15], it would be of interest to obtain a mutation screening of H2A.X in infertile men of this population.

At the zygote level, H2A.X is the most abundant H2A variant present at the one- and two-cell stages of the zygote after fertilization [20]. The authors showed that the C-terminal 23 amino acids of H2A.X are critical for the role of this variant in the chromatin remodeling that takes place in the early stages immediately after fertilization [20]. It has also been shown that γH2A.X is present during the paternal chromatin remodeling process, to provide the signaling required for the repair of any DNA damage present at that time. However, neither H2A.X nor γH2A.X is directly involved in the replacement mechanisms that revert the nucleoprotamine to somatic chromatin organization that takes place within the first 2h after fertilization [20,21]. As in yeast, the purpose of the high abundance of H2A.X during the stages that immediately follow fertilization seems to provide a chromatin template with an enhanced dynamic plasticity [16]. This is in agreement with the structural data showing that yeast- and H2A.X-containing nucleosomes have a reduced stability [4,22].

2.2 The Important Role of H3.3 in Male Germ Line Development

The histone H3 family consists of two canonical histones (H3.1, H3.2), a centromere-specific histone variant CenH3 (CENP-A) [23], and the replication-independent histone variants H3.3 and testes-specific H3t. H3t is mainly expressed in the testes of both humans and mice [24–26]. Like H2A.X, H3t has been shown to decrease the stability of the nucleosome both *in vitro* and *in vivo* [27]; however, the role of H3t in spermatogenesis remains largely unknown.

H3.3 differs from H3.1 by only four amino acids. However, the histones exhibit important functional differences. For example, they are deposited onto the chromatin by different histone chaperones (HIRA for H3.3, CAF1 for H3.1) [28]. Also, unlike H3.1, which is deposited to chromatin in S phase, H3.3 can be deposited in DNA replication-independent manner. Several studies suggest that this variant is essential for male fertility, and plays an important role in the chromatin dynamics of the male germ line [29]. Deletion of the two copies of the H3.3 gene in *Drosophila* resulted in flies that were able to survive to adulthood and had an apparently normal morphology [30]. However, these flies had a reduced viability (semilethality), and were in all instances sterile in both male and female specimens [30–32]. Expression of only one copy of H3.3 rescues viability and fertility [30]. This result is in contrast to what it is observed in mammals, where disruption of the *H3f3b* gene results in a reduction of the H3.3 levels, leading to anomalous sperm morphology and male infertility [29].

It has been shown that histone H3.3 is enriched in PTMs associated with transcriptionally active chromatin, in particular H3.3K4 di- and tri-methylation [31]. Mutation to H3.3K4A was not able to rescue fertility, and hence it was concluded that this epigenetic mark was additionally important for fertility [32]. Interestingly, a mutant expressing H3.3K4R did not enhance viability, but it was able to rescue fertility [30]. This suggests that, whereas methylation of this residue may be important for viability, a positive charge at this position is critical for the ability of H3.3 to restore fertility [30].

The reduced levels of H3.3 using the knockout mouse model for the *H3f3b* gene resulted in altered germ cell chromatin reorganization and an incomplete replacement of histones by protamines in the late stages of spermatogenesis. This indicates that, in mammals, H3.3 plays a critical role in the dynamic transitions of chromatin during spermatogenesis [29]. Furthermore, depletion of H3.3 in mouse zygotes prevented de novo formation of nucleosomes onto paternal DNA, indicating that H3.3 is also essential for nucleosome formation in the early stage of development [33].

2.3 TH2B Plays a Key Role in Spermatogenesis and in the Early Developmental Stages of the Zygote

TH2B is one of the first described testes-specific histones [34]. While highly homologous to H2B, the N-terminal tail sequence of TH2B is unique [35] (Figure 4). Circular dichroism spectra analysis shows that TH2B has slightly higher alpha-helical content in this region. *In vitro*

```
HsH2B      ----------------------MPEP-AKSAPAPKKGS---------KKAVTKAQKKDGK 28
HsTH2B     ----------------------MPEVSSKGATISKKGF---------KKAVVKTQKKEGK 29
HsH2BFW    MLRTEVPRLPRSTTAIVWSCHLMATASAMAGPSSETTSEEQLITQEPKEANSTTSQKQSK 60
                                 *.   : ... .:.           *:*  .:.:*:.*

HsH2B      KRKRSRK-----------ESYSVYVYKVLKQVHPDTGISSKAMGIMNSFVNDIFERIAGE 77
HsTH2B     KRKRTRK-----------ESYSIYIYKVLKQVHPDTGISSKAMSIMNSFVTDIFERIASE 78
HsH2BFW    QRKRGRHGPRRCHSNCRGDSFATYFRRVLKQVHQGLSLSREAVSVMDSLVHDILDRIATE 120
           :*** *:           :*:: *. :****** . .:* :*:.:*:*:* **::*** *

HsH2B      ASRLAHYNKRSTITSREIQTAVRLLLPGELAKHAVSEGTKAVTK---YTSSK--- 126
HsTH2B     ASRLAHYSKRSTISSREIQTAVRLLLPGELAKHAVSEGTKAVTK---YTSSK--- 127
HsH2BFW    AGHLARSTKRQTITAWETRMAVRLLLPGQMGKLAESEGTKAVLRTSLYAIQQQRK 175
           *.:**: .**.**:: * : ********::.* * ******* :   *: .:
```

FIGURE 4 Amino acid sequence comparison among HsTH2B, HsH2BFW, and canonical HsH2B.

biochemical analysis shows that the histone octamer containing TH2B is less stable than the one consisting of canonical histone H2B [36]. However, when the TH2B-containing octamer is assembled into nucleosomes, the stability and dynamics of the resulting nucleosome is not affected [36].

TH2B starts being expressed in early spermatocytes, and gradually replaces canonical H2B. The majority of H2B is already replaced by TH2B in elongating spermatids [37,38] (Figure 2). These replacements take place genome-wide, except at regions near the transcription start sites [38]. These results suggest a global structural involvement of TH2B in spermatogenesis which might facilitate histone–protamine transition by contributing to change in the plasticity of the nucleosome. It remains unclear as to whether this H2B variant has any other additional function similar to H2A.Z and H2A.X, which are involved in transcription regulation and DNA repair, respectively.

A couple of papers highlight the relevance of TH2B to spermatogenesis and the subsequent implications for fertility [38,39]. Ramos et al. studied the sperm of oligoasthenoteratozoospermia (OAT) patients, who have a lower concentration and/or quality of sperm due to unknown causes. They were able to show that, compared to the sperm of normospermic males, only about half the amount of TH2B is present in the nuclei of OAT sperm [39]. They also measured the amount of nucleosomes remaining at the end of spermatogenesis by using a monoclonal antibody that recognizes H2A.H2B DNA in a nucleosomal context, and concluded that the sperm of OAT patients had more nucleosomes than did fertile sperm [39]. This result suggests that the insufficient amount of TH2B hinders the nucleosome to protamine transition in these OAT patients, and results in infertility. In fact, a recent paper from Khochbin's group provides direct evidence for the role of TH2B in orchestrating the replacement of histones by protamines [8,38]. They showed that mice expressing C-terminally tagged TH2B were able to proceed normally through the premeiotic spermatogenic stages; however, the tagging severely interfered with the histone-to-protamine transition, underscoring the role of TH2B in this process (Figure 5). Furthermore, they found that in TH2B knockdown mice, the lack of TH2B induces compensatory mechanisms which allow the histone-to-protamine replacement to take place through an increase of targeted histone PTMs. Specifically, H3K122 crotonylation, H4R36, R55, R67 methylation, H4K77 crotonylation, H2BR72 methylation, and decreased H2BK108 crotonylation were observed [38].

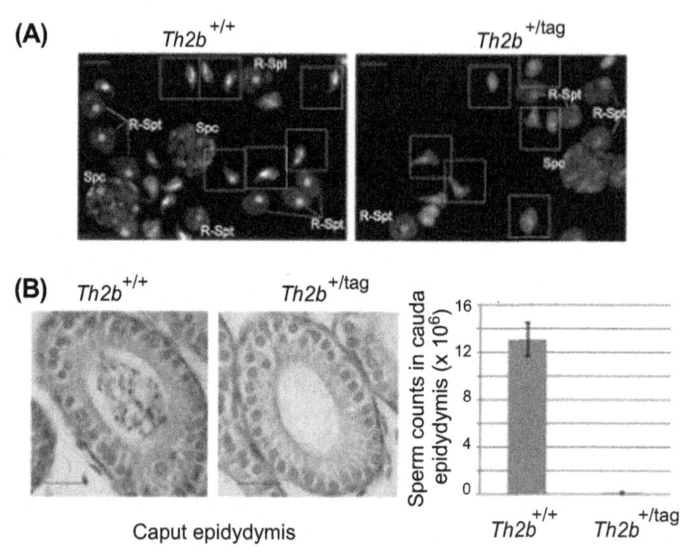

FIGURE 5 TH2B-tag induces late arrest of spermatogenesis. (A) Different cell types stained with DAPI from the indicated genotypes are shown. Bars, 5 μm. Spermatocyte (Spc) and round spermatids (R-Spt) are indicated. Square boxes represent elongating spermatids. (B) Cauda epididymis sections stained with hematoxylin are shown in the left panel, and spermatozoa counts from isolated cauda epididymis from mice with the indicated genotypes are presented as histograms in the right panel. Bars, 20 μm. Bars represent standard deviations of sperm counts from cauda epididymis of five mice of each genotype. *Images and legends are from Ref. [38]. (Note: In previous experiments, the authors have shown that the addition of a C-terminal tag to TH2B, namely the TH2B-tag genotype, has no effect on overall nucleosome dynamics or gene expression. The authors used a TH2B-tag to avoid knocking out* Th2a *that shares the same promoter as* Th2b.*)*

Like some of the canonical histones that undergo PTMs during spermatogenesis (such as the ones just described), TH2B is also post-translationally modified. Acetylated TH2B has been observed specially in spermatogonia, and phosphorylation of T116 and methylation of K117 were observed in spermatogonia, spermatocytes, and round spermatids [40]. A recent mass spectrometric mapping has provided additional information about the TH2B PTMs in tetraploid spermatocytes and haploid spermatids [41]. Evidence was provided for acetylation at K7, K14, K17, and K22 and methylation of K27 and K110 in both developmental stages, with a distinctive S5 phosphorylation in the tetraploid stage. A close inspection of the location occupied by some of these PTMs in the crystallographic structure of the nucleosome [42] suggested that they might be directly involved in structural chromatin changes [41]. However, at present the role of these TH2B PTMs is unknown.

In addition to testes, Montellier et al. [38] showed that TH2B is also present in mature oocytes and plays a major role in the protamine-to-histone reversal transition of the male pronucleus. This result suggests that, in addition to spermatogenesis, TH2B plays a critical role in some of the chromatin remodeling events that take place immediately after fertilization. Recently, Shinagawa et al. [43] reported that TH2B and TH2A (which has been less studied than TH2B) are highly expressed in oocytes. The *Th2a* and *Th2b* genes share a regulatory region, and are actively expressed in oocytes and in two- to eight-cell embryos—albeit at a lower level (one-tenth to one-third of the amount) of TH2B than is found in testes. TH2B and TH2A induce an open chromatin structure which is presumably critical for the remodeling of the chromatin structure during development [43]. Therefore, it is not surprising that alterations in the expression of the *Th2b* gene resulting in dosage alterations or mutations might have important consequences for fertility.

2.4 A Recently Described Histone Variant H2BFW May Also Be Involved in Sperm Fertilization

H2BFW is a recently identified testes-specific histone variant in humans. It is located in chromosome 22 and shares only about 45% identity and 70% amino acid sequence homology with canonical H2B [44]. The most prominent structural difference between H2BFW and H2B is the larger size and longer N-terminal tail of the former (Figure 4). In addition, H2BFW gene is interrupted by two introns, in contrast to the replication-dependent canonical histones genes that are intronless. GFP-tagged H2BFW colocalized with telomeric DNA in Chinese hamster V-79 cells [44], suggesting a telomeric functional association. Using salt dissociation analyses in acrylamide gels and fluorescence recovery after photobleaching (FRAP) assays, Dimitrov's group studied the stability of nucleosomes containing an N-terminal truncated version of H2BFW, and found it to be identical to that of canonical ones [45]. The authors also claimed that the N-terminal tail of H2BFW plays an important role in mitotic chromosome assembly [45]. Note that the authors used the 153 amino acid version of H2BFW protein for these analyses, which is 22 amino acids shorter at the N-terminal tail [45] (Figure 4).

In addition to the biochemical studies, Lee *et al.* [46] compared H2BFW transcripts from sperm samples of both fertile and infertile patients. They found that 36.2% contained a single nucleotide polymorphism (SNP) (-9C > T) (at position −9 of the 5′ UTR position). The percentile increased to 46.2% when only samples from infertile men were analyzed. The (-9C > T) SNP results in a reduction of the rate at which the histone gene is translated, without affecting the levels of transcription of the gene. The concentration of sperm and vitality was also much lower in the (-9C > T) SNP sperm samples [46]. All these suggest that the (-9C > T) SNP can be considered a susceptibility factor for male infertility. A similar result was obtained in a later study, in which two SNPs were analyzed for the same gene (-9C > T) and (368A > G) [47]. The frequency of these SNPs was also found to be considerably higher in infertile patients than in controls. The (-9C > T) SNP might also have an effect at the protein level [46]. The molecular mechanism by which (-9C > T) SNP reduces the level of H2BFWT—and how this impairs fertility—is still not clear.

2.5 H2A.Bbd Destabilizing Nucleosomes during Spermatogenesis

The histone variant H2A.Bbd is only present in mammals and humans, and three copies of this gene are present in the X chromosome [48,49]. These copies exhibit a sequence and identity that clearly distinguishes them from other histone H2A variants such as H2A.Z, H2A.X (see above), and macroH2A. H2A.Bbd is a very fast-evolving histone [49]. For example, human and mouse H2A.Bbd have only about 50% amino acid sequence identity. It has been shown that the H2A.Bbd-containing nucleosome core particle associates with 116 bp of DNA—compared to 147 bp of DNA in canonical nucleosome core particles, and is less stable [49] (Figure 6). Incorporation of H2A.Bbd variant into the nucleosome results in the abrogation of the so-called "acidic patch" domain. This highly charged region in the outer surface of the nucleosome arises from the clustering of a set of acidic amino acids (six on H2A and two on H2B) [42], and has been proposed to play a role in chromatin folding, mediated by its electrostatic interaction with the positively charged N-terminal tails

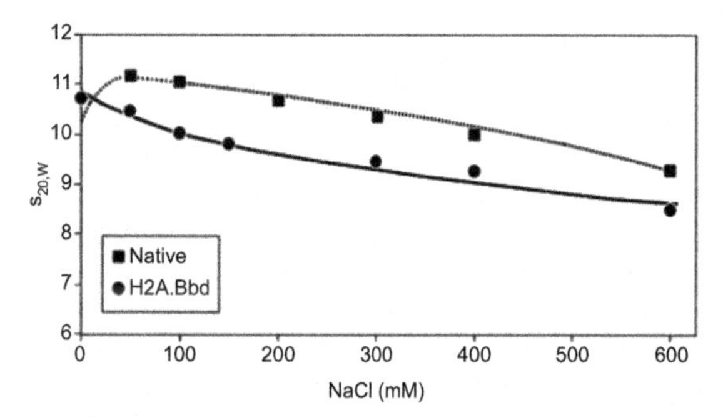

FIGURE 6 Sedimentation profile of nucleosome core particles reconstituted with H2A.Bbd or a full native histone complement under different ionic strengths (0–600 mM NaCl). The dotted line and squares are the predicted behavior obtained from native nucleosome core particles [103]. *Image and legend are from Ref. [48].*

of H4 in adjacent nucleosomes [42]. In the H2A. Bbd sequence, three contributing acidic amino acids are substituted with neutral residues [50]. It has therefore been suggested and experimentally shown that H2A.Bbd destabilizes the folding of nucleosome arrays [50].

In agreement with the in vitro evidence for the loss of nucleosome stability conferred by this variant, FRAP experiments indicate that H2A.Bbd has a fast nuclear turnover compared to canonical H2A in cells [48]. This is in tune with the results obtained with cells in which this histone variant is expressed ectopically. Using this approach, localization of this variant was observed in the more open euchromatic regions that are transcriptionally active in somatic cells where it colocalizes with acetylated H4 [51]. H2A.Bbd was also found to be enriched at sites of DNA synthesis [52], and it shows an association with high cell proliferation.

Of interest to this chapter, the first experimental evidence for the existence of this histone variant in a tissue was in mouse testes [53]. It was found to be present in the spermatogenic fraction with higher levels of histone H4 acetylation, at a time when the histone-to-protamine replacement starts to take place [53]. H2A.Bbd was also found to be present in human sperm [53,54]. Therefore, it is quite possible that this variant plays an important role in the histone-to-protamine transition during mammalian spermatogenesis, as well as in the early events that involve the chromatin remodeling of the male pronucleus after fertilization in humans [10]. It would be of interest to analyze the presence of this histone in infertile men and to further study its involvement in early development immediately after fertilization.

2.6 Testes-Specific Linker Histones and Infertility

The H1 family includes a group of highly microheterogeneous histones [55] that bind to the linker DNA regions of chromatin [56]. Their binding is extremely dynamic and it plays an important role in the folding transitions of the chromatin fiber. There are a large variety of linker histones; these include the somatic and many development-specific variants, including the testes-specific linker histones H1t, H1T2, and H1-like protein HILS1, and the oogenesis-specific linker histone H1foo [2].

Elongating and elongated spermatids express high levels of HILS1, and its expression time partially overlaps with the expression of TNPs and protamines (see Figure 2); yet they are absent from mature sperm [57]. So far, the detailed roles of HILS1 during spermatogenesis are unknown. However, alterations of HILS1 could have important consequences in spermatogenesis. *Drosophila* Mst77F, which shows significant similarity to mammalian HILS1 protein, is involved in sperm maturation, and its mutation leads to infertility [58].

Another important testes-specific linker histone variant is H1t. This variant is expressed during meiosis in midpachytene cells. Surprisingly, H1t knockouts do not have any major effects on spermatogenesis. However, it appears that its absence is only compensated for by somatic linker histones [33]. Because H1t binds more weakly and has a lower condensation

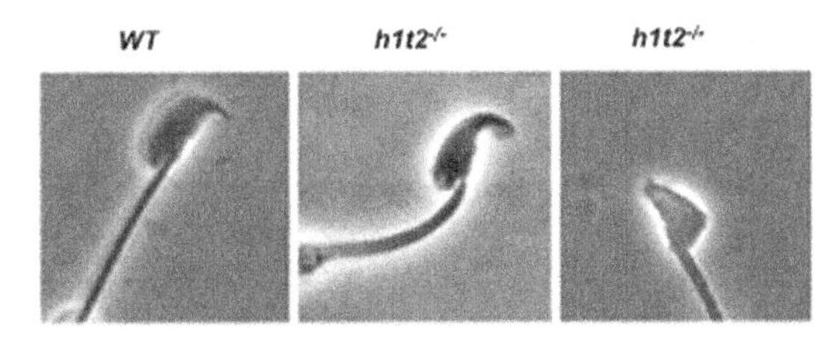

FIGURE 7 Abnormalities in sperm from cauda epididymis. Light microscope images (×5000) of representative individual spermatozoa from WT and H1t2-mutant cauda epididymis. *Image and legend are from Ref. [64].*

effect on the chromatin fiber [59], the knockout model suggests that H1t is functionally similar to H1-deficient chromatin [60]. H1t is the predominant linker histone in round spermatids, where it amounts to 55% of the total H1. In elongating mice spermatids, about 30% of H1t are phosphorylated [61]. However, no significant differences in the phosphorylation pattern were observed between H1t and the somatic linker histone H1 counterpart [62]. As H1t phosphorylation takes place in spermatids at an overlapping timing with H4 acetylation [63], all of this suggests that the role of H1t phosphorylation is there to further destabilize the interaction of this linker histone with chromatin, to facilitate the histone replacement by TNPs [61,62].

Another testes-specific linker histone H1T2 is specifically expressed after meiosis [64,65]. While female H1T2 knockout mice were fertile, the mutant males had decreased sperm motility, and were infertile [64,65] (Figure 7). While the analyses carried out so far indicate that H1T2 also plays a critical role in proper elongation and DNA condensation, the functional correlation—if any—among the different testes-specific linker histones remains unknown.

H1foo is a mammalian oocyte-specific linker histone with a maximal pattern of expression during oogenesis that is maintained until the two-cell embryo stage [11]. During nuclear transfer in mice, the somatic linker histones are rapidly replaced by H1foo—presumably resulting in a more flexible chromatin organization [66]. Knockdown of H1foo impaired the rate of meiotic maturation of germinal vesicle stage oocytes [66]. Further studies are required to elucidate the involvement of H1foo and other linker histones and histone variants (see Ref. [67] for a recent overview) in oogenesis.

3. THE EFFECTS OF HISTONE PTMs

As mentioned in previous sections, several histone PTMs (e.g., H1t phosphorylation, γH2A.X, H4 acetylation) are involved in spermatogenesis; they add an extra layer of complexity to the molecular events of the chromatin transitions during spermatogenesis. In this regard, histone H4 in particular occupies a prominent place. In contrast to all other histones, H4 is the only histone that does not have any significantly expressed histone variants. Instead, its acetylation has a historically well-recognized function in spermatogenesis [68], where it accumulates in elongated spermatids; its impairment may have important consequences for fertility. H4 hyperacetylation is thought to be essential in transitioning from histones to protamines during spermatogenesis. However, the effect of alterations of this histone's PTMs on infertility is still controversial. Sonnack et al. have reported a decrease in the amount of acetylated H4 present in the spermatids of azoospermic and oligospermic patients [69]. In contrast, Ramos *et al.* [39] showed that the sperm of OAT patients is more frequently labeled by an antibody against H4K8, K12, K16 acetylation, with preference for apoptotic nuclei. The difference could be attributed to the different role of H4 acetylation in spermatids, facilitating the histone-to-protamine transition or its use as an apoptotic marker for mature sperm. Alternatively, the discrepancy could simply be the result of differences in the specific antibodies used for each study.

Very recently, Dottermusch-Heidel *et al.* [70] have shown that H3K79 methylation, a histone PTM initially related to transcriptional activity [71], exhibits a strong correlation with the histone H4 hyperacetylation in elongating spermatids of flies and rats [70]. In rats, active RNA polymerase II is detected in spermatocytes and round spermatids, but not in elongating spermatids. Therefore, H3K79 methylation and H4 acetylation must be involved in opening the chromatin structure that is required by the histone-to-protamine transition [70].

Another highly conserved H4 PTM in metazoans which has a major involvement in spermatogenesis is H4S1 phosphorylation [72–74]. Its functional role appears to be that of chromatin compaction. However, its participation in the spermatogenesis process and its relation, if any, to the histone-to-protamine transition remain to be elucidated.

Other histone PTMs are associated with the different testes-specific histone variants, and play a role in the intricate process of chromatin remodeling during spermatogenesis, as has been described in the corresponding preceding sections. However, their functional involvement and potential implications for fertility are poorly understood.

4. BRIEF OVERVIEW OF THE ROLE OF SPERMIOGENESIS-SPECIFIC NUCLEAR BASIC PROTEINS: TNPs AND PROTAMINES IN INFERTILITY

During the process of spermiogenesis in mammals, testes-specific histones are replaced by TNPs that are subsequently replaced by protamines. Eutherian mammals have two copies of TNPs (TNP1 and TNP2) and two types of protamines (P1 and P2) [75]. P1 represents the ancestral protamine, and P2 is only found in certain species of eutherian mammals, for instance, human and mice. During the process of chromatin deposition, a protamine P2 precursor undergoes a sequential posttranslational cleavage, and three different species can be visualized (P2, P3, and P4), with P2 representing the most abundant component in the sperm of fertile men [76]. Protamines compact chromatin by a six- to sevenfold higher extent than do canonical histones. This tight packaging is important for the hydrodynamic properties of sperm, as well as for DNA protection against damaging agents during its journey along the male and female reproductive tracts, and for preventing the inheritance of damaged DNA. Indeed, Aoki *et al.* were able to show that patients with lower protamine levels were observed to have more DNA fragmentations, a result that could come from a lack of protection by protamines [77]. Of note, reactive oxygen species can have a detrimental effect, and induce sperm DNA fragmentation. Supplemental intake of antioxidant vitamins, especially vitamin C, has been used for the treatment of male infertility in certain patients, resulting in a decreased DNA fragmentation and increased sperm concentration [49,50]. Protamines also contribute to erasing most of the histone epigenetic components from the early spermatogenic cells while potentially providing some imprinting to defined regions of the sperm genome [76]. Incomplete protamination leads to altered level of retained histone as well as alterations in the P1/P2 ratio [78,79]. In mature sperm, both play an important role in the chromatin organization, and are very important fertility indicators (see Refs [76] and [78] for detailed reviews on this topic). The P1/P2 ratio is usually within 0.8–1.2 in fertile humans. Males with a higher P1/P2 ratio are reported to have reduced fertility. Incomplete P2 precursor processing is often responsible for the altered values, although an incomplete histone-to-protamine transition can also be involved [76]. Interestingly, external factors have the potential to alter these ratios. For instance, it has been shown that heavy cigarette smoking (>20 cigarettes/day) can affect the histone-to-protamine transition [80], increasing the H2B to protamine and the P1/P2 ratios in comparison to nonsmokers. Furthermore, smokers have a lower sperm count, and their sperm exhibits a decreased motility [81].

At the gene level, Tanaka *et al.* have described an SNP in the open reading frame (ORF) of the protamine 2 precursor gene (*PRM2*) [82]. For more information about the role of the protamine genes (*PRM1* and *PRM2*) in infertility, the reader is referred to Francis *et al.* [78].

At the level of protamine PTMs, phosphorylation has been shown to play an important role in proper protamine deposition during the protamine-to-histone transition, and it is gradually removed during sperm chromatin maturation [83]. Protein kinase CAMK4 is expressed in spermatids and it is responsible for P2 phosphorylation. CAMK4 knockout mice exhibit a decreased level of P2, increased levels of TNP2, impairment of spermatogenesis in late elongating spermatids, and infertility [84]. This result suggests that P2 phosphorylation is also necessary for the replacement of TNPs by protamines.

Compared to protamines, the potential involvement of TNPs in infertility still remains unclear. TNP1 knockout mice show no difference in testes weight and sperm count compared to wild type. However, TNP1 deletion results in a higher probability of blunted tips of the epididymal sperm heads, and a decrease in the sperm motility (Figure 8) [85]. TNP1 deletion mice have higher levels of TNP2 and an altered P1:P2 ratio. In addition, the P2 precursor, which is expressed in spermatids, can still be observed in mature sperm [85]. This result suggests that TNP1 has a role in P2 processing, and agrees with a study that found an increase in P2 precursors in infertile men [86].

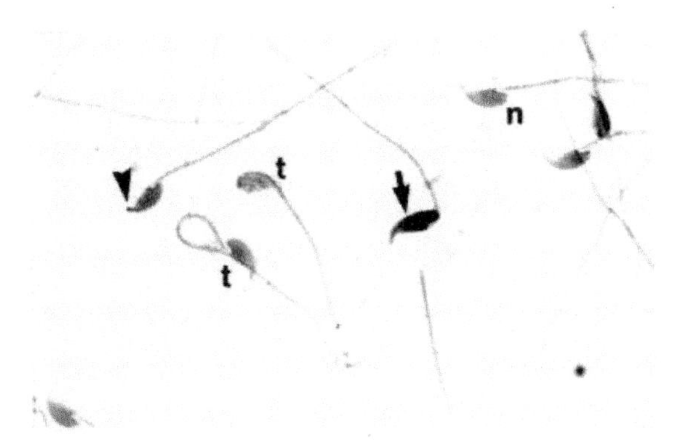

FIGURE 8 Spermatozoa from a *Tnp1−/−*mouse (B6 ×129 F2). **n**, normal sperm; arrowhead, sperm with blunted apex; arrow, dark-staining sperm; **t**, sperm with tails coiled around heads (×900). *Image and legend are from Ref. [85].*

5. CAN HISTONE VARIANTS BE USED AS BIOMARKERS FOR INFERTILITY?

Histone PTMs and histone variants both play important epigenetic roles [87–89]. Therefore, they both have the potential to play an important role in sperm fertility and implications for the developing embryo [90].

In recent years, with the help of genome-wide analysis (GWA) [91], a lot of attention has been paid to the histones that are retained in the mature spermatozoa of some mammals like human and mice [92,93]. Approximately, 15% histones are present in mature fertile human sperm [94] and 1% in mice [93,95]. A detailed list of the histones and their variants found in this fraction in the sperm of fertile patients has recently been provided using mass spectrometry [96]. Although these histones have been implied to play a critical role in the events that take place immediately after fertilization in the early development of the embryo [92], their role within a broader context in mammalian infertility is not straightforward. For instance, any other species of mammals including horses, cows, and dogs replace 100% of their histones during spermatogenesis and yet they are equally fertile.

Given the role in early development described for such histone fraction in human sperm [92], it is plausible to speculate that they might have a role in human infertility. However, whether the precise amount of histones retained in a given individual sperm cell is constant in all the sperm cells in semen or if the number 15% value represents an average is yet unknown. The extent of the potential variation in the amounts of each of the different histone variants in unfertile patients when compared to the fertile population is not clear either. An early attempt in this regard which monitored the variability of TH2B in the sperm of donors and subfertile patients using Western blot analysis, demonstrated a higher variability of this variant in the latter [97]. However, a direct correlation could not be established. These results are reminiscent of the data obtained through a much more expensive and involved GWA of a comparison between fertile men and several cohorts of infertile patients which systematically exhibited moderate decreases in the levels of histone H3 lysine 4 and lysine 27 methylation (H3K4me/H3K27me) at random loci [98]. All these uncertainties and ambiguity need to be addressed before histone variants, and their PTMs can be used as potential biomarkers for human infertility.

The current lack of clear-cut sperm-fertility biomarkers in terms of both histone variants and PTMs makes it premature to envisage how the information currently available and that described in the previous sections will transpire into the clinical practice for clinicians that primarily rely on semen analysis to predict male reproductive potential.

6. CLOSING REMARKS

Several testes-specific histone variants (H1t, HILS1, H2AL.1, H2AL.2, H2A.Bbd, TH2A, TH2B, H2BL.1, H2BFW, and H3t) and a set of well-defined histone PTMs (i.e., H4 hyperacetylation, γH2A.X, and H1t phosphorylation) are involved in the process of spermatogenesis, underscoring the complexity of the chromatin remodeling process involved. Alterations that directly or indirectly affect any of these components can potentially lead to infertility. In the mammalian species that retain specific histone variants in their sperm (i.e., mouse and human), such variants play an important role in the paternal chromatin remodeling in early zygote development. Hence, abnormalities in this histone complement can also have additional consequences for fertility. Nevertheless, a recurrent topic throughout most of the sections discussed in this chapter is the insufficient amount of knowledge currently available about the detailed mechanisms played by most of the different germ line-specific histone variants, as well as to how their PTMs contribute to all these processes. In particular, the roles of histone variants in oogenesis remain largely understudied.

Another important remaining question has to do with the potential involvement in infertility of the different histone variants (H2A.Z, H2A.X, H2A.Bbd, H1t) present in the sperm of unfertile patients [96]. At present, this remains an important and relatively unchartered territory that would involve an extensive analysis of different infertile populations. Furthermore, external factors such as air pollution and endocrine disruptors have been shown to have an environmental impact affecting global DNA hypermethylation and heritable epigenetic transmission [99,100]. Also, advanced paternal age has been shown to affect the sperm DNA damage and chromatin integrity [101]. It would also be of interest to analyze the effects of environmental and aging insult on the histone PTM and variant composition of sperm.

Glossary

Histone variants It is the name that is given to histones that are often expressed outside of the S phase and throughout the rest of the cell cycle in a nonreplication-dependent way. Their amino acid composition can differ significantly from that of the canonical (H2A, H2B, H3, and H4) core histones.

Protamines Arginine-rich and highly basic small chromosomal proteins that replace histones during spermiogenesis.

Azoospermia It is a medical infertility condition in which no sperm can be detected in semen.

Oligospermia Medical condition related to male infertility that arises from low sperm count in semen.

Teratozoospermia Medical condition in which sperm exhibits an abnormal morphology in the semen.

LIST OF ABBREVIATIONS

CENP-A Centromeric protein A
FRAP Fluorescence recovery after photobleaching
GWA Genome-wide analysis
OAT Oligoasthenoteratozoospermia
PTMs Posttranslational modifications
ORF Open reading frame
TNPs Transition proteins
SNP Single nucleotide polymorphism

Acknowledgments

This work was supported by a Natural Sciences and Engineering Research Council of Canada (NSERC) 46399-2012 to JA and by a grant from the Research Grants Council of the Hong Kong SAR (26100214) to TI.

References

[1] Van Holde K. Chromatin. New York: Springer; 1988.
[2] Ausió J. Histone variants–the structure behind the function. Brief Funct Genomic Proteomic 2006;5(3):228–43.
[3] Paull TT, Rogakou EP, Yamazaki V, Kirchgessner CU, Gellert M, Bonner WM. A critical role for histone H2AX in recruitment of repair factors to nuclear foci after DNA damage. Curr Biol 2000;10:886–95.
[4] Li A, Yu Y, Lee S-C, Ishibashi T, Lees-Miller SP, Ausió J. Phosphorylation of histone H2A.X by DNA-dependent protein kinase is not affected by core histone acetylation, but it alters nucleosome stability and histone H1 binding. J Biol Chem 2010;285(23):17778–88.
[5] Baarends WM, Van Der LR, Grootegoed JA. Specific aspects of the ubiquitin system in spermatogenesis. J Endocrinol Invest 2000;23:597–604.
[6] Faraone Mennella MR. Mammalian spermatogenesis, DNA repair, poly (ADP-ribose) turnover: the state of the art. In: Storici F, editor. DNA repair-on pathways fixing DNA damage errors. 2010. p. 235–54.
[7] Boissonneault G. Chromatin remodeling during spermiogenesis: a possible role for the transition proteins in DNA strand break repair. FEBS Lett 2002;514:111–4.
[8] Govin J, Khochbin S. Histone variants and sensing of chromatin functional states. Nucleus 2013;4(6):438–42.
[9] Lewis JD, Abbott DW, Ausió J. A haploid affair: core histone transitions during spermatogenesis. Biochem Cell Biol 2003;81(3):131–40.
[10] Hammoud SS, Nix DA, Zhang H, Purwar J, Carrell DT, Cairns BR. Distinctive chromatin in human sperm packages genes for embryo development. Nature 2009;460(7254):473–8.
[11] Tanaka M, Hennebold JD, Macfarlane J, Adashi EY. A mammalian oocyte-specific linker histone gene H1oo: homology with the genes for the oocyte-specific cleavage stage histone (cs-H1) of sea urchin and the B4/H1M histone of the frog. Development 2001;128(5):655–64.
[12] Ausio J, Abbott DW. New concepts. The many tales of a tail : carboxyl-terminal tail heterogeneity specializes histone H2A variants for defined chromatin function. Biochemistry 2002;41(19):5945–9.
[13] Li A, Eirín-López JM, Ausió J. H2AX: tailoring histone H2A for chromatin-dependent genomic integrity. Biochem Cell Biol 2005;83(4):505–15.
[14] Rogakou EP, Pilch DR, Orr AH, Ivanova VS, Bonner WM. DNA double-stranded breaks induce histone H2AX phosphorylation on serine 139. J Biol Chem 1998;273(10):5858–68.
[15] Meistrich ML, Bucci LR, Trostle-Weige PK, Brock WA. Histone variants in rat spermatogonia and primary spermatocytes. Dev Biol 1985;112(1):230–40.
[16] Celeste A, Petersen S, Romanienko PJ, Fernandez-Capetillo O, Chen HT, Sedelnikova OA, et al. Genomic instability in mice lacking histone H2AX. Science 2002;296(5569):922–7.
[17] Fernandez-Capetillo O, Mahadevaiah SK, Celeste A, Romanienko PJ, Camerini-Otero RD, Bonner WM, et al. H2AX is required for chromatin remodeling and inactivation of sex chromosomes in male mouse meiosis. Dev Cell 2003;4(4):497–508.
[18] Zhang W, Yang Y, Su D, Ma Y, Zhang S. Absence of the H2AX mutations in idiopathic infertile men with spermatogenic impairment. Syst Biol Reprod Med 2008;54(2):93–5.
[19] Fernandez-Capetillo O, Lee A, Nussenzweig M, Nussenzweig A. H2AX: the histone guardian of the genome. DNA Repair (Amst) 2004;3(8–9):959–67.
[20] Nashun B, Yukawa M, Liu H, Akiyama T, Aoki F. Changes in the nuclear deposition of histone H2A variants during pre-implantation development in mice. Development 2010;137(22):3785–94.
[21] Derijck AA, van der Heijden GW, Giele M, Philippens MEP, van Bavel CC, de Boer P. gammaH2AX signalling during sperm chromatin remodelling in the mouse zygote. DNA Repair (Amst) 2006;5(8):959–71.
[22] Pineiro M, Puerta C, Palacian E. Yeast nucleosomal particles: structural and transcriptional properties. Biochemistry 1991;28(1990):5805–10.
[23] Talbert PB, Ahmad K, Almouzni G, Ausió J, Berger F, Bhalla PL, et al. A unified phylogeny-based nomenclature for histone variants. Epigenetics Chromatin 2012;5(7):1–19.
[24] Witt O, Albig W, Doenecke D. Testis-specific expression of a novel human H3 histone gene. Exp Cell Res 1996;229(2):301–6.

[25] Govin J, Escoffier E, Rousseaux S, Kuhn L, Ferro M, Thévenon J, et al. Pericentric heterochromatin reprogramming by new histone variants during mouse spermiogenesis. J Cell Biol 2007;176(3):283–94.
[26] Govin J, Caron C, Rousseaux S, Khochbin S. Testes-specific histone H3 expression in somatic cells. Trends Biochem Sci 2005;30(7):357–9.
[27] Tachiwana H, Kagawa W, Osakabe A, Kawaguchi K, Shiga T, Hayashi-Takanaka Y, et al. Structural basis of instability of the nucleosome containing a testis-specific histone variant, human H3T. Proc Natl Acad Sci USA 2010;107(23):10454–9.
[28] Tagami H, Ray-Gallet D, Almouzni G, Nakatani Y. Histone H3.1 and H3.3 complexes mediate nucleosome assembly pathways dependent or independent of DNA synthesis. Cell 2004;116:51–61.
[29] Yuen BTK, Bush KM, Barrilleaux BL, Cotterman R, Knoepfler PS. Histone H3.3 regulates dynamic chromatin states during spermatogenesis. Development 2014;141(18):3483–94.
[30] Sakai A, Schwartz BE, Goldstein S, Ahmad K. Transcriptional and developmental functions of the H3.3 histone variant in *Drosophila*. Curr Biol 2009;19(21):1816–20.
[31] McKittrick E, Gafken PR, Ahmad K, Henikoff S. Histone H3.3 is enriched in covalent modifications associated with active chromatin. Proc Natl Acad Sci USA 2004;101(6):1525–30.
[32] Hödl M, Basler K. Transcription in the absence of histone H3.3. Curr Biol 2009;19(14):1221–6.
[33] Inoue A, Zhang Y. Nucleosome assembly is required for nuclear pore complex assembly in mouse zygotes. Nat Struct Mol Biol 2014;21(7):609–16.
[34] Shires A, Carpenter MP, Chalkley R. A cysteine-containing H2B-like histone found in mature mammalian testis. J Biol Chem 1976;251:4155–8.
[35] Zalensky AO, Siino JS, Gineitis AA, Zalenskaya IA, Tomilin NV, Yau P, et al. Human testis/sperm-specific histone H2B (hTSH2B). Molecular cloning and characterization. J Biol Chem 2002;277(45):43474–80.
[36] Li A, Maffey AH, Abbott WD, Conde e Silva N, Prunell A, Siino J, et al. Characterization of nucleosomes consisting of the human testis/sperm-specific histone H2B variant (hTSH2B). Biochemistry 2005;44(7):2529–35.
[37] Brock WA, Trostle PK, Meistrich ML. Meiotic synthesis of testis histones in the rat. Proc Natl Acad Sci USA 1980;77(1):371–5.
[38] Montellier E, Boussouar F, Rousseaux S, Zhang K, Buchou T, Fenaille F, et al. Chromatin-to-nucleoprotamine transition is controlled by the histone H2B variant TH2B. Genes Dev 2013;27(15):1680–92.
[39] Ramos L, van der Heijden GW, Derijck A, Berden JH, Kremer JAM, van der Vlag J, et al. Incomplete nuclear transformation of human spermatozoa in oligo-astheno-teratospermia: characterization by indirect immunofluorescence of chromatin and thiol status. Hum Reprod 2008;23(2):259–70.
[40] Lu S, Xie YM, Li X, Luo J, Shi XQ, Hong X, et al. Mass spectrometry analysis of dynamic post-translational modifications of TH2B during spermatogenesis. Mol Hum Reprod 2009;15(6):373–8.
[41] Pentakota SK, Sandhya S, Sikarwar AP, Chandra N, Rao MRS. Mapping post-translational modifications of mammalian testicular specific histone variant TH2B in tetraploid and haploid germ cells and their implications on the dynamics of nucleosome structure. J Proteome Res 2014;13(12):5603–17.
[42] Luger K, Mader AW, Richmond RK, Sargent DF, Richmond TJ. Crystal structure of the nucleosome core particle at 2.8 Å resolution. Nature 1997;389:251–60.
[43] Shinagawa T, Takagi T, Tsukamoto D, Tomaru C, Huynh LM, Sivaraman P, et al. Histone variants enriched in oocytes enhance reprogramming to induced pluripotent stem cells. Cell Stem Cell 2014;14(2):217–27.
[44] Churikov D, Siino J, Svetlova M, Zhang K, Gineitis A, Morton Bradbury E, et al. Novel human testis-specific histone H2B encoded by the interrupted gene on the X chromosome. Genomics 2004;84(4):745–56.
[45] Boulard M, Gautier T, Mbele GO, Gerson V, Hamiche A, Angelov D, et al. The NH 2 tail of the novel histone variant H2BFWT exhibits properties distinct from conventional H2B with respect to the assembly of mitotic chromosomes. Mol Cell Biol 2006;26(4):1518–26.
[46] Lee J, Park HS, Kim HH, Yun Y-J, Lee DR, Lee S. Functional polymorphism in H2BFWT-5′UTR is associated with susceptibility to male infertility. J Cell Mol Med 2009;13(8B):1942–51.
[47] Ying H, Scott MB, Zhou-Cun a. Relationship of SNP of H2BFWT gene to male infertility in a Chinese population with idiopathic spermatogenesis impairment. Biomarkers 2012;17(5):402–6.
[48] Gautier T, Abbott DW, Molla A, Verdel A, Ausio J, Dimitrov S. Histone variant H2ABbd confers lower stability to the nucleosome. EMBO Rep 2004;5(7):715–20.
[49] Eirín-López JM, Ishibashi T, Ausió J. H2A.Bbd: a quickly evolving hypervariable mammalian histone that destabilizes nucleosomes in an acetylation-independent way. FASEB J 2008;22(1):316–26.
[50] Zhou J, Fan JY, Rangasamy D, Tremethick DJ. The nucleosome surface regulates chromatin compaction and couples it with transcriptional repression. Nat Struct Mol Biol 2007;14(11):1070–6.

[51] Tolstorukov MY, Goldman JA, Gilbert C, Ogryzko V, Kingston RE, Park PJ. Histone variant H2A.Bbd is associated with active transcription and mRNA processing in human cells. Mol Cell 2012;47(4):596–607.
[52] Sansoni V, Casas-Delucchi CS, Rajan M, Schmidt A, Bönisch C, Thomae AW, et al. The histone variant H2A.Bbd is enriched at sites of DNA synthesis. Nucleic Acids Res 2014;42(10):6405–20.
[53] Ishibashi T, Li A, Eirín-López JM, Zhao M, Missiaen K, Abbott DW, et al. H2A.Bbd: an X-chromosome-encoded histone involved in mammalian spermiogenesis. Nucleic Acids Res 2010;38(6):1780–9.
[54] Castillo J, Amaral A, Oliva R. Sperm nuclear proteome and its epigenetic potential. Andrology 2014;2(3): 326–38.
[55] Cole RD. Microheterogeneity in H1 histones and its consequences. Int J Pept Protein Res 1987;30(4):433–49.
[56] Gonzalez-Romero R, Ausio J. dBigH1, a second histone H1 in Drosophila, and the consequences for histone fold nomenclature. Epigenetics 2014;9(6):791–7.
[57] Yan W, Ma L, Burns KH, Matzuk MM. HILS1 is a spermatid-specific linker histone H1-like protein implicated in chromatin remodeling during mammalian spermiogenesis. Proc Natl Acad Sci USA 2003;100(18): 10546–51.
[58] Jayaramaiah Raja S, Renkawitz-pohl R. Replacement by Drosophila melanogaster protamines and Mst77F of histones during chromatin condensation in late spermatids and role of sesame in the removal of these proteins from the male pronucleus. Mol Cell Biol 2005;25(14):6165–77.
[59] Khadake JR, Rao MRS. DNA- and chromatin-condensing properties of rat testes H1a and H1t compared to those of rat liver H1bdec: H1t is a poor condenser of chromatin. Biochemistry 1995;34:15792–801.
[60] Fantz DA, Hatfield WR, Horvath G, Kistler MK, Kistler WS. Mice with a targeted disruption of the H1t gene are fertile and undergo normal changes in structural chromosomal proteins during spermiogenesis. Biol Reprod 2001;64(2):425–31.
[61] Rose KL, Li A, Zalenskaya I, Zhang Y, Unni E, Hodgson KC, et al. C-Terminal phosphorylation of Murine testis-specific histone H1t in elongating spermatids. J Proteome Res 2008:4070–8.
[62] Sarg B, Chwatal S, Talasz H, Lindner HH. Testis-specific linker histone H1t is multiply phosphorylated during spermatogenesis. Identification of phosphorylation sites. J Biol Chem 2009;284(6):3610–8.
[63] Meistrich ML, Trostle-Weige PK, Van Beek ME. Separation of specific stages of spermatids from vitamin A-synchronized rat testes for assessment of nucleoprotein changes during spermiogenesis. Biol Reprod 1994;51(2):334–44.
[64] Martianov I, Brancorsini S, Catena R, Gansmuller A, Kotaja N, Parvinen M, et al. Polar nuclear localization of H1T2, a histone H1 variant, required for spermatid elongation and DNA condensation during spermiogenesis. Proc Natl Acad Sci USA 2005;102(8): 2808–13.
[65] Tanaka H, Iguchi N, Isotani A, Toyama Y, Matsuoka Y, Masai K, et al. Protein involved in nuclear formation and sperm fertility HANP1/H1T2, a novel histone H1-like protein involved in nuclear formation and sperm fertility. Mol Cell Biol 2005;25(16):7107–19.
[66] Furuya M, Tanaka M, Teranishi T, Matsumoto K, Hosoi Y, Saeki K, et al. H1foo is indispensable for meiotic maturation of the mouse oocyte. J Reprod Dev 2007;53(4):895–902.
[67] Yang P, Wu W, Macfarlan TS. Maternal histone variants and their chaperones promote paternal genome activation and boost somatic cell reprogramming. BioEssays 2015;37(1):52–9.
[68] Oliva R, Dixon G. Vertebrate protamine genes and the histone-to-protamine replacement reaction. Prog Nucleic Acid Res Mol Biol 1991;40:25–94.
[69] Sonnack V, Failing K, Bergmann M, Steger K. Expression of hyperacetylated histone H4 during normal and impaired human spermatogenesis. Andrologia 2002;34(6):384–90.
[70] Dottermusch-Heidel C, Klaus ES, Gonzalez NH, Bhushan S, Meinhardt A, Bergmann M, et al. H3K79 methylation directly precedes the histone-to-protamine transition in mammalian spermatids and is sensitive to bacterial infections. Andrology 2014;4:1–11.
[71] Steger DJ, Lefterova MI, Ying L, Stonestrom AJ, Schupp M, Zhuo D, et al. DOT1L/KMT4 recruitment and H3K79 methylation are ubiquitously coupled with gene transcription in mammalian cells. Mol Cell Biol 2008;28(8): 2825–39.
[72] Krishnamoorthy T, Chen X, Govin J, Cheung WL, Dorsey J, Schindler K, et al. Phosphorylation of histone H4 Ser1 regulates sporulation in yeast and is conserved in fly and mouse spermatogenesis. Genes Dev 2006;20(18):2580–92.
[73] Wendt KD, Shilatifard A. Packing for the germy: the role of histone H4 Ser1 phosphorylation in chromatin compaction and germ cell development. Genes Dev 2006;20(18):2487–91.
[74] Govin J, Dorsey J, Gaucher J, Rousseaux S, Khochbin S, Berger SL. Systematic screen reveals new functional dynamics of histones H3 and H4 during gametogenesis. Genes Dev 2010;24(16):1772–86.
[75] Ausió J, Eirín-López JM, Frehlicj LJ. Evolution of vertebrate chromosomal sperm proteins: implications for fertility and sperm competition. Soc Reprod Fertil Suppl 2007;65:63–79.

[76] Oliva R. Protamines and male infertility. Hum Reprod Update 2006;12(4):417–35.
[77] Aoki VW, Moskovtsev SI, Willis J, Liu L, Mullen JBM, Carrell DT. DNA integrity is compromised in protamine-deficient human sperm. J Androl 2004;26(6):741–8.
[78] Francis S, Yelumalai S, Jones C, Coward K. Aberrant protamine content in sperm and consequential implications for infertility treatment. Hum Fertil (Camb) 2014;17(2):80–9.
[79] Simon L, Liu L, Murphy K, Ge S, Hotaling J, Aston KI, et al. Comparative analysis of three sperm DNA damage assays and sperm nuclear protein content in couples undergoing assisted reproduction treatment. Hum Reprod 2014;29(5):904–17.
[80] Niederberger C. Re: cigarette smoking is associated with abnormal histone-to-protamine transition in human sperm. J Urol 2014;192(4):1193.
[81] Hamad MF, Shelko N, Kartarius S, Montenarh M, Hammadeh ME. Impact of cigarette smoking on histone (H2B) to protamine ratio in human spermatozoa and its relation to sperm parameters. Andrology 2014;2(5):666–77.
[82] Tanaka H, Miyagawa Y, Tsujimura A, Matsumiya K, Okuyama A, Nishimune Y. Single nucleotide polymorphisms in the protamine-1 and -2 genes of fertile and infertile human male populations. Mol Hum Reprod 2003;9(2):69–73.
[83] Lewis JD, Song Y, de Jong ME, Bagha SM, Ausió J. A walk though vertebrate and invertebrate protamines. Chromosoma 2003;111(8):473–82.
[84] Wu JY, Ribar TJ, Cummings DE, Burton KA, McKnight GS, Means AR. Spermiogenesis and exchange of basic nuclear proteins are impaired in male germ cells lacking Camk4. Nat Genet 2000;25(4):448–52.
[85] Yu YE, Zhang Y, Unni E, Shirley CR, Deng JM, Russell LD, et al. Abnormal spermatogenesis and reduced fertility in transition nuclear protein 1-deficient mice. Proc Natl Acad Sci USA 2000;97(9):4683–8.
[86] De Yebra L. Detection of P2 precursors in the sperm cells of infertile patients who have reduced protamine P2 levels. Fertil Steril 1998;69(4):755–9.
[87] Rothbart SB, Strahl BD. Interpreting the language of histone and DNA modifications. Biochim Biophys Acta - Gene Regul Mech 2014;1839:627–43.
[88] Maze I, Noh K-M, Soshnev AA, Allis CD. Every amino acid matters: essential contributions of histone variants to mammalian development and disease. Nat Rev Genet 2014;15(4):259–71.
[89] Henikoff S, Smith MM. Histone variants and epigenetics. Cold Spring Harb Perspect Biol 2015;7(1).
[90] Jenkins TG, Carrell DT. The sperm epigenome and potential implications for the developing embryo. Reproduction 2012;143:727–34.
[91] Hisano M, Erkek S, Dessus-Babus S, Ramos L, Stadler MB, Peterson AH. Genome-wide chromatin analysis in mature mouse and human spermatozoa. Nat Protoc 2013;8(12):2449–70.
[92] Hammoud SS, Nix DA, Zhang H, Purwar J, Carrel DT, Cairns BR. Distinctive chromatin in human sperm packages genes for embryo development. Nature 2009;460(7254):473–8.
[93] Erkek S, Hisano M, Liang C-Y, Gill M, Murr R, Dieker J, et al. Molecular determinants of nucleosome retention at CpG-rich sequences in mouse spermatozoa. Nat Struct Mol Biol 2013;20(7):868–75.
[94] Tanphaichitr N, Sobhon P, Taluppeth N, Chalermisarachap P. Basic nuclear proteins in testicular cells and ejaculated spermatozoa in man. Exp Cell Res 1978;117:347–56.
[95] Brykczynska U, Hisano M, Erkek S, Ramos L, Oakeley EJ, Roloff TC, et al. Repressive and active histone methylation mark distinct promoters in human and mouse spermatozoa. Nat Struct Mol Biol 2010;17(6):679–87.
[96] Castillo J, Amaral A, Vavouri T, Estanyol JM, Ballesca JL, Oliva R. Genomic and proteomic dissection and characterization of the human sperm chromatin. Mol Hum Reprod 2014;20(11):1041–53.
[97] Singleton S, Zalensky A, Doncel GF, Morshedi M, Zalenskaya IA. Testis/sperm-specific histone 2B in the sperm of donors and subfertile patients: variability and relation to chromatin packaging. Hum Reprod 2007;22:743–50.
[98] Hammoud SS, Nix DA, Hammoud AO, Gibson M, Cairns BR, Carrell DT. Genome-wide analysis identifies changes in histone retention and epigenetic modifications at developmental and imprinted gene loci in the sperm of infertile men. Hum Reprod 2011;26:2558–69.
[99] Yauk C, Polyzos A, Rowan-Carroll A, Somers CM, Godschalk RW, Van Schooten FJ, et al. Germ-line mutations, DNA damage, and global hypermethylation in mice exposed to particulate air pollution in an urban/industrial location. Proc Natl Acad Sci USA 2008;105:605–10.
[100] Anway M, Skinner M. Epigenetic programming of the germ line: effects of endocrine disruptors on the development of transgenerational disease. Reprod Biomed Online 2008;16(1):23–5.
[101] Wyrobek AJ, Eskenazi B, Young S, Arnheim N, Tiemann-Boege I, Jabs EW, et al. Advancing age has differential effects on DNA damage, chromatin integrity, gene mutations, and aneuploidies in sperm. Proc Natl Acad Sci USA 2006;103:9601–6.
[102] Rathke C, Baarends WM, Awe S, Renkawitz-Pohl R. Chromatin dynamics during spermiogenesis. Biochim Biophys Acta 2014;1839(3):155–68.
[103] Ausio J, van Holde K. Histone hyperacetylation: its effects on nucleosome conformation and stability. Biochemistry 1986;25(6):1421–8.

CHAPTER

25

Circulating Histones and Nucleosomes as Biomarkers in Sepsis and Septic Shock

José Luis García Giménez[1,2], *Carlos Romá Mateo*[1,2], *Marta Seco Cervera*[1,2], *José Santiago Ibañez Cabellos*[1,2], *Federico V. Pallardó*[1,2]

[1]Center for Biomedical Network Research on Rare Diseases, Medicine and Dentistry School, University of Valencia, Valencia, Spain; [2]Biomedical Research Institute INCLIVA, Valencia, Spain

OUTLINE

Epigenetic Biomarkers and Diagnostics
http://dx.doi.org/10.1016/B978-0-12-801899-6.00025-5

1. INTRODUCTION

1.1 General Aspects of Sepsis

Sepsis (ICD-10: A26.7) is one of the first mentioned and most elusive syndromes in medicine. In 1992 the international consensus defined sepsis as a systemic inflammatory response syndrome (SIRS) as a result of infection [1]. SIRS is established after the identification of the following features: temperature higher than 38 °C or lower than 36 °C, heart rate over 90 beats/min, respiratory rate over 20 breaths/min or $PaCO_2$ under 32 mmHg, and white blood cell count over 12,000/mm^3, under 4000/mm^3, or over 10% immature (band) forms. In the same international consensus the term severe sepsis (ICD-10: R65.2) was used to describe the condition in which sepsis is complicated by organ dysfunction or tissue hypoperfusion (except when the organ failure was already present 48 h before the onset of sepsis), while septic shock (ICD-10: R65.21) was defined as sepsis complicated by either hypotension that is refractory to fluid resuscitation or by hyperlactatemia [2,3].

However, many patients with sepsis die despite successful elimination of the pathogen or microorganisms that originated the infection. Therefore, the modern view of sepsis suggests that it is the host and its intrinsic nature, not the microorganism, which drives the pathogenesis of sepsis [4]. From this point of view, evolution of the septic process will largely depend on the genetic and epigenetic background of the patient. Risk factors for severe sepsis are related to the patient's predisposition to infection and to the likelihood of acute organ dysfunction if the infection progresses. In addition, other risk factors such as chronic diseases (e.g., acquired immunodeficiency syndrome, chronic obstructive pulmonary disease, acute respiratory distress syndrome, and many types of cancer) and the use of immunosuppressive agents contribute to severe sepsis and septic shock. Resistance to infection and its outcomes are influenced by a complex interplay between the host (genetics), the nature and particularities of the microorganism, and the environment (epigenetics). Over the past few years, a series of key players have been arising, as a result of the dissection of molecular pathways involving inflammatory response, oxidative stress, and necrotic and apoptotic processes, all of them related to different stages of the septic process. This has significantly widened the field of sepsis research, with a strong focus on novel molecules with the potential to become important targets for both biomarker and drug design. Among these, histones and histone-related proteins are some of the most important candidates in unveiling the particular steps of septic progression at the molecular level, hopefully leading to a better understanding of sepsis and septic shock.

1.2 Nuclear Components Circulating in Body Fluids

Tissue injury and trauma produces an increase in circulating levels of damage-associated molecular patterns like nucleic acids, mono- and oligonucleosomes, histones, histone-complexed DNA, high-mobility group box1 (HMGB1), and lamins, among others [5–7]. Small amounts of histones and nucleosomes can be found in plasma and serum of healthy people produced after physiological cell death [8] or secreted by lymphocytes or released to plasma during erythroblast differentiation [6,9]. Cellular apoptosis may contribute to the liberation of nuclear components to the body fluids. However, upon conditions of enhanced cell death (apoptosis or necrosis), inflammation, ischemia, trauma, and toxin-mediated diseases, the level of circulating nucleosomes in body fluids can increase considerably. The release of nucleosomes, histones, and histone-complexed DNA occurs between 24 and 48 h after apoptosis induction [10] and can remain in the bloodstream during hours, participating in cellular signaling as we describe in this chapter. Other important sources of chromatin components are neutrophils. Activated neutrophils expose chromatin and granule proteins, forming the neutrophil extracellular traps (NETs) composed of histones, free DNA, and neutrophil-derived proteases. These web-shape structures bind to gram-positive and gram-negative bacteria facilitating the destruction of bacteria [11] (we describe this process in more detail in the following sections). It is very relevant that nucleosome degradation occurs in few minutes in blood, although a small amount (10%) is eliminated from blood in a slower manner [12,13]. However, the process of histone elimination is saturated by the presence of high amounts of histones, as demonstrated in experiments in which nucleosomes are injected to animal models [14]. More information regarding nucleosomes metabolism is described by Holdenrieder and Stieber in Ref. [15]. In this chapter, we will also explain how histones can also be found as free proteins in plasma, mediating extensive cellular damage, and amplifying the inflammatory response, as occurs in trauma-associated lung injury [16] and in septic patients [17].

2. EPIDEMIOLOGY AND CLINICAL FEATURES OF SEPSIS, SEVERE SEPSIS, AND SEPTIC SHOCK

The number of cases of sepsis in the United States exceeds 750,000 per year [18] and an increase by 1.5% each year is expected, due to the aging population and the rise in antibiotic resistance, among others causes [19]. Furthermore, it has been estimated that there are up to 19 million cases worldwide per year [20]. In Europe, the Sepsis Occurrence in Acutely Ill Patients study (SOAP study) reported a high frequency of sepsis in European intensive care units (ICUs), with more than 35% of patients suffering sepsis during their ICU stay. This study also reported high mortality rates, reaching 27% in sepsis patients during ICU stay, and rising more than 50% in patients with septic shock [21]. Fortunately, a decrease in mortality during septic shock has been reported in countries such as Australia and New Zealand in the 2000–2012 period [22].

It is of special relevance that sepsis is a leading cause of in-hospital deaths in developed countries, as monetary resources for technology and training would indicate a far more optimistic outcome [23,24]. Therefore, morbidity and mortality from sepsis remain unacceptably high. In a study performed in France on 100,554 ICU admissions, however, Annane et al. found that the frequency of septic shock increased from 7.0% of hospital admissions in 1993 to 9.7% in 2000 [25].

The incidence of severe sepsis and septic shock depends on two main issues: the first being how acute organ dysfunction is defined; the second being whether or not that dysfunction is attributed to an infection. However, during sepsis, diverse physiopathological events take place that produce complications in the patient's outcome. For example, damage of the vascular endothelia as a result of inflammatory responses might itself accelerate the progression of severe sepsis, even though the original infection might already be eliminated. Disseminated intravascular coagulation (DIC) is a secondary feature underlying sepsis. DIC is found in 25–50% of septic patients and has a strong correlation with mortality [26]. DIC is characterized by widespread microvascular thrombosis and profuse bleeding as a consequence of the reduction of platelets and coagulation factors, such as thrombocytopenia, prolonged clotting time, and elevated fibrin-related markers. Intravascular thrombosis interrupts blood circulation to organs, resulting in organ failure. During sepsis, extensive lymphocyte apoptosis occurs in a number of organs [27–29] and in the circulatory system [30,31]. It is worth noting that sepsis causes long-term immunosuppression (also known as immunoparalysis), which contributes to multiple organ failure and death [24]. The most common organ dysfunctions produced as a result of sepsis are related to the respiratory and cardiovascular systems, but the brain and kidneys are also often affected. The balance between proinflammatory responses to infection and anti-inflammatory immunosuppressive responses, activated as a control measure against the former, is a key turning point in the development of the septic process; around this fragile equilibrium many feedback processes are activated and the outcome is difficult to assess. Epigenetic regulation of gene expression has shown to be a critical modifier of each patient's own particular genetic background, making the already complex inflammatory response still more unpredictable. At the root of this kind of tissue damage, lie molecular determinants related to the hypoxia resulting from respiratory impairment and which in turn leads to cell deoxygenation and the consequent overproduction of oxidative damage. However, the particular details of the interplay among different molecules regulating pro- or anti-inflammatory processes, oxidative stress responses, or apoptotic and necrotic regulators are still largely unknown. For instance, in human patients with sepsis, lower levels of circulating cytokines have been detected, as compared to those observed in animal models with sepsis. This partially explains why some of the treatments that function in animal models have failed in human patients [32]. Therefore, a better understanding of molecular processes, genetic profiles, and epigenetic changes that occur during postseptic immunosuppression would improve diagnostic tools and also contribute to the identification of novel targets for new therapies.

3. GENETICS, EPIGENETICS, AND MOLECULAR PATHWAYS UNDERLYING SEPSIS

There is great interest in understanding the genetic determinants of the host response to infection. From a genomic point of view, it is of special relevance to know which genes are associated with poor outcomes after an infection. It has been found that susceptibility and response to infectious diseases is heritable [33–36]. However, at the present time there exist scarce studies assessing the genetics of sepsis, and more studies are needed to increase the cohort of patients, to study the influence of ethnicity, and to validate these clinical studies by means of multicenter research. Potential associations between the clinical outcome of patients with sepsis and polymorphisms in cytokine, innate immunity, and coagulation cascade genes have been identified. For example, single nucleotide polymorphisms (SNPs) in genes such as tumor necrosis factor-α (*TNF-α*), interleukin-6 (*IL-6*), the innate immunity receptor for lipopolysaccharide, peptidoglycan, lipoteichoic acid CD14, and toll-like receptor-2 (*TLR-2*) have been found. All of them were related to differences in patient survival, increased risk of developing septic shock, and increased mortality (for more detailed information see the review by Sutherland and Walley [37]) (Figure 1). Genome-wide association studies using next-generation sequencing have been performed to genotype up to two million SNPs, allowing the analysis of allele frequencies in an unbiased selection of specific genetic regions across the whole genome. Therefore, once these kinds of studies are completed it will be possible to clarify the genetic determinants for sepsis or septic shock predisposition.

Unfortunately, despite the diverse studies performed during more than a decade of research and the associations identified between polymorphisms in the above mentioned genes and survival, the role of genetic determinants participating in sepsis, severe sepsis, and septic shock still remains unclear and, sometimes, conflicting results appear. Therefore, it is evident that infection susceptibility and patient outcome will be due to both genetic and epigenetic interactions. Proof of this is that studies involving human and animal models have demonstrated that immunosuppression is one of the consequences of severe sepsis, and that it is produced

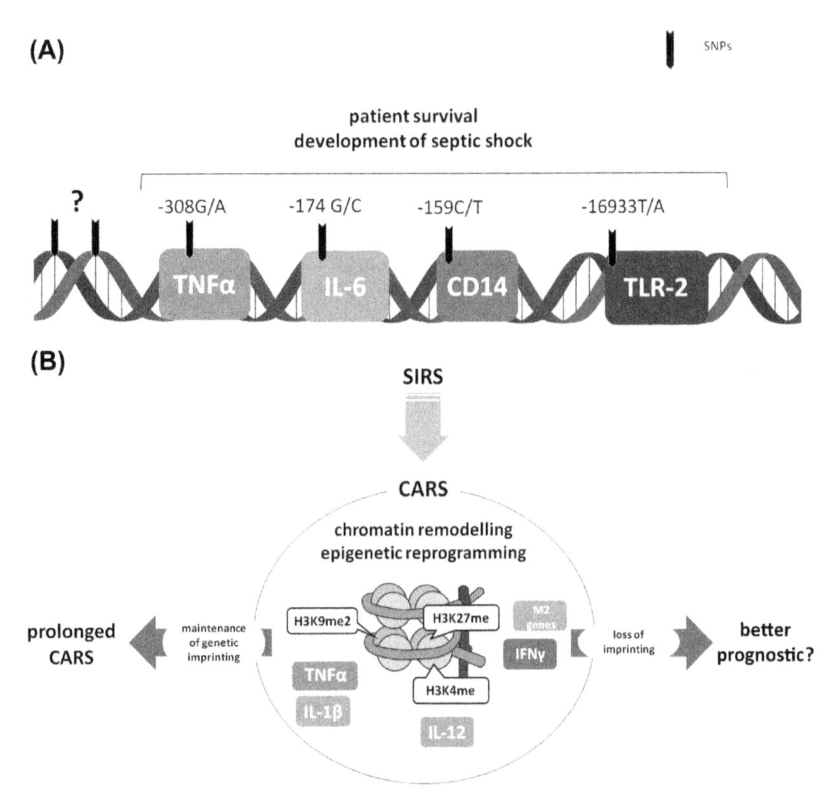

FIGURE 1 **Proposed genetic and epigenetic players in the prognostic of sepsis.** (A) Candidate gene single nucleotide polymorphisms (SNPs) associations in sepsis. Next-generation sequencing has provided candidate genes associated to adverse outcome of patients with sepsis. The proposed SNPs are the A allele of a G-to-A polymorphism at position −308 of the TNF-α gene, the C allele of a G-to-C polymorphism at position −174 of the IL-6 gene, C-to-T polymorphism at position −159 of the CD14 gene, and T to A at position −16,933 of the TLR-2 gene. (B) Chromatin remodeling and epigenetic reprogramming after SIRS determining the prognostic of patients. After systemic inflammatory response syndrome (SIRS), a compensatory anti-inflammatory response syndrome (CARS) sometimes occurs. During CARS, an epigenetic imprinting takes place, producing an epigenetic reprogramming that is retained in the progenitor cells of patients that survive SIRS. This epigenetic reprogramming at histone posttranslational level in proinflammatory genes can be studied in differentiated immune cells in order to obtain diagnostic epigenetic biomarkers (see text for details).

by chromatin remodeling and epigenetic reprogramming events. It is known that after SIRS, a compensatory anti-inflammatory response syndrome (CARS) also occurs (Figure 1). The pathophysiology of CARS persists in patients for an extended period of time, lasting anywhere from weeks to years. Epigenetic imprinting that occurs on mature immune cells may come from progenitor cells in the bone marrow during sepsis. This epigenetic reprogramming may be retained in the progenitor cells of patients that survive SIRS, allowing progenitor cells to perpetuate the epigenetic marks into differentiated cells, as proposed by Carson et al. [38]. Due to the existence of this phenomenon, studying the epigenetic reprogramming occurring in immune cells is advisable in order to obtain diagnostic tools, prognostic biomarkers, and therapies that aim to improve the long-term survival of affected patients. ChIP-seq experiments, whole methylome sequencing, and miRNA-seq may help to clarify this hypothesis.

As occurs in SIRS, epigenetics may be one of the driving processes involved in the long-term immunosuppression that often appears after severe sepsis. An example of this can be found in bacterial-induced SIRS responses to prokaryotic lipopolysaccharides (LPS) from the cell walls of gram-negative bacteria. LPS stimulation of human peripheral blood monocytes (PBMCs) induces chromatin remodeling pro-inflammatory gene promoters for chemokines and cytokines [39], a process that depends on the acetylation state of histones [40]. During primary responses to gram-negative bacteria by the host, the increase of proinflammatory factors (IL-1β and TNF-α) is critical to induce inflammation and to detoxify the microorganism. After these responses, an increase in the levels of repressive histone modification marks (H3K9me2) at the promoter regions of IL-1β and TNF-α has been detected in experiments using both human monocytic cell lines and PBMCs from septic patients [41,42] (Table 1). The epigenetic reprogramming and chromatin remodeling at proinflammatory genes can condition the patients to compensatory responses by the immune system. In fact, in macrophages, LPS stimulation has been shown to increase the expression of the histone demethylase KDM6B (JMJD3), responsible for the demethylation of H3K27me3 [43] at the promoters of M2 gene loci (e.g., IL-10, Arginase-1, Ym1, IL-1Rn, etc.) (Table 1) (Figure 1). Since KDM6B is responsible for removing repressive histone methylation marks of several anti-inflammatory genes, this enzyme could contribute to postseptic immunosuppression. Therefore, it would be of great interest to measure the expression of this demethylase after severe sepsis in order to evaluate the predisposition of patients to CARS.

One of the features of postseptic immunosuppression is the predisposition of survivors to opportunistic infections. This is explained, in part, by the fact that SIRS negatively affects the function of dendritic cells (DCs), which exert an important role during sepsis. As occurs with postseptic macrophages, postseptic DCs show reduced capacity to produce IL-12 in response to microbial stimulation. However, IL-12 expression requires chromatin remodeling at the promoter region of the gene by means of the elimination of the repressive methylation mark H3K27me3 and the addition of an activating methylation mark H3K4me (Figure 1). Experiments performed using mice models demonstrated that postseptic lung DCs exhibited increased recruitment of chromatin remodelers mediating H3K27me3, and decreased recruitment of H3K4me remodelers to the promoter regions of the *IL-12* gene [44], therefore silencing the *IL-12* gene (Table 1).

Regarding adaptive immunity, sepsis also exerts several effects on the lymphoid compartment. In fact, $CD4^+$ T-helper response is affected

TABLE 1 Epigenetic Regulation of the Immune Response of the Host to Infection

Disease	Type of cells	Epigenetic mark	References
Sepsis	Monocytes and PBMCs from septic patients	Increase in the levels of H3K9me2 at promoters of IL-1β and TNF-β genes	[41,42]
Septic shock (LPS-induced)	Mouse macrophages	Increased JMJD3 and demethylation of H3K27me3 at the promoters of M2 genes	[43]
Sepsis and SIRS	Mouse dendritic cells	Decreased demethylation of H3K27me3 at the promoters of IL-12	[44]
Sepsis	$CD4^+$-T patient cells	Increased H3K27me3 at the promoter of IFNγ gene	[48]

IFNγ, interferon γ; LPS, lipopolysaccharide; PBMCs, peripheral blood mononuclear cells; SIRS, systemic inflammatory response syndrome.

during the postseptic period. Reduced lymphocyte proliferation capacity, increased lymphocyte apoptosis, and cell death of $CD4^+$ T cells and T cell precursors have been described in the thymus [45], spleen [46], and blood [47] of both humans and animals during the acute phase of sepsis. In addition, interferon γ (IFNγ) production by $CD4\gamma^+$ T cells programmed to become T_H1 effectors is decreased in postseptic naïve $CD4^+$ T cells in mice models, an effect that was related to an increase in the heterochromatin mark H3K27me3 at the promoter region of the gene that codifies for IFNγ, which was corroborated by ChIP experiments [48] (Table 1). In this way, $CD4^+$ T cells are capable of increasing regulatory T cells (T_{reg}) [49], which are important suppressors of immune responses. Functional T_{reg} requires the expression of the transcription factor Foxp3 for the maintenance of the suppressing immune responses, which in turn depends on DNA demethylation at the promoter region and also on euchromatin marks at the histone level in the *Foxp3* locus [50]. It is important to note that Foxp3 expression in $CD4^+$ T cells increases after sepsis. Therefore, the plasticity of the epigenome of T cells makes it possible to increase repressive marks in histones, producing the deregulation of cytokine expression (i.e., IFNγ) and it also allows an increase in the transcription activation marks in other T lymphocyte subsets, increasing T_{reg}. Both of these increases contribute to immunosuppression after sepsis.

All these data provide a nice overview of the relevance of epigenetic regulation in the immune response of the host to infection, but also clear the path for understanding the long-term consequences of the immunosuppression process that encompasses so many cases of sepsis. It will be critical to finely dissect these pathways at the chromatin remodeling and histone modification levels to develop important diagnostic, prognostic, and therapeutic tools to modulate patient immune response, as we will see in the next sections.

4. EXTRACELLULAR NUCLEOSOMES AND HISTONES AND THEIR ROLE IN THE PHYSIOPATHOLOGY OF SEPSIS AND SEPTIC SHOCK

4.1 Sources of Extracellular Histones and Nucleosomes

Neutrophils may represent a unique defense mechanism against bacterial, fungal, and viral infections. These cells are able to destroy microorganisms by phagocytosis [51,52]. Moreover, neutrophils can extracellularly kill pathogens by releasing structures called NETs, occasioning a form of cell death called NETosis [11,53]. These structures include a web of fibers composed of stacked modified nucleosome components, granular components, and some cytoplasmic proteins. These include enzymes (lysozyme, proteases), antimicrobial peptides (bactericidal/permeability-increasing protein (BPI), defensins), ion chelators (calgranulin), and histones [11]. The release of NETs by the activated neutrophils needs severe morphological transformations (Figure 2). Immediately after activation of receptor/s, these transformations start with adhesion to the endothelium and assembly of NADPH oxidase enzyme complex or phagocytic oxidase (PHOX) at the cell membranes. PHOX reduces molecular oxygen into superoxide anions, which superoxide dismutase transforms into hydrogen peroxide. This hydrogen peroxide is a substrate for myeloperoxidase (MPO), the most abundant enzyme in the neutrophil granules that produces hypochlorous acid (HOCl). After that the granules, rich in MPO, protease neutrophil elastase (NE), and other proteases, disintegrate, with the succeeding migration of NE and MPO to the nucleus [54]. At the same time the structure of the nucleus is disassembled and chromatin decondensation takes place by virtue of different events: degradation of linker histone H1 and core histones by NE; enhancement of chromatin decondensation thanks to

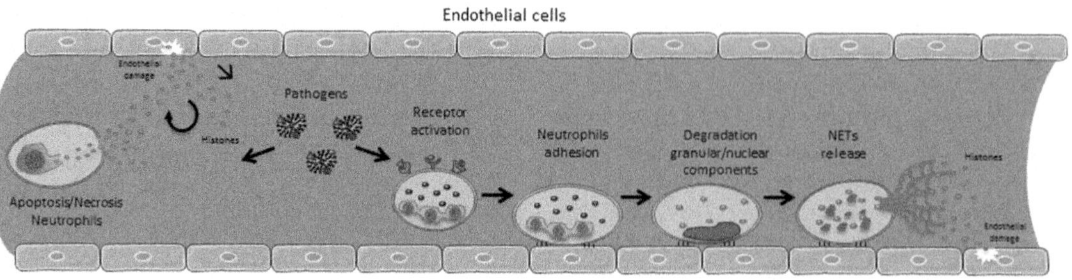

FIGURE 2 **Schematic representation of the different sources for the release of histones into the bloodstream after infection by pathogens.** Two possibilities are depicted: for entire elimination of pathogens, neutrophils can take an apoptotic/necrotic pathway (left). As a result of this process, histones can be released into bloodstream. Moreover, these free histones are able to generate endothelial damage and induce apoptosis or necrosis in these cells with the resulting release of more histones to the bloodstream. Other source for circulating histones is the neutrophil extracellular traps (NETs) (right). In this process, receptors are stimulated by different antigens. This triggers neutrophils attachment to the endothelium and mobilization of granular components (black points), the granular components disintegrate (red points), and nuclear components are disassembled (dark blue). Then, intracellular membranes disaggregate forming a homogenous mass. Finally, cell membrane is degraded, releasing into the bloodstream the cytoplasm and nucleoplasm mixture known as NETs. Histones present in NETs damage the endothelial cells.

hypochlorous acid produced by MPO [54]; and modifications of histones, such as citrullination by peptidylarginine deiminase 4 (PAD4) that converts arginine residues to citrulline in three out of the four core histones [55,56]. Once these changes end, the nucleoplasm merges with the cytoplasm, occupying a uniform area inside the neutrophil. Finally, the cell membrane ruptures, and the mixture mass is ejected outside the neutrophil, forming NETs [57]. Histones present in NETs act as a potent antimicrobial, more so than most other antimicrobials [58]. As described in the introduction section, these processes release histones to the bloodstream. In this regard, histones can act as cytotoxic agents due to their cationic charge, which may allow attachment to the microbial membrane. This binding could either destroy the microorganism or make the membrane more permeable to small particles, including histones. Inside the microbial cytoplasm, histones could bind to DNA [59] and block DNA gyrase activity, an essential bacterial enzyme, involved in DNA replication [60].

Histones from NETs could be a relevant source of histone presence on plasma in a sepsis context; another source could be cell death via apoptotic or necrotic pathways [61] of other kind of cells from vascular tissue, i.e., monocytes or damaged endothelial cells (Figure 2). NETs in sepsis are crucial for attaching and destroying pathogens, but their presence can also be dangerous for the host. This danger comes from the possibility that NETs might result in vessel clotting [62] and, moreover, liberated histones could kill endothelial cells [61] producing severe problems, even death, in sepsis patients. Furthermore, the posttranslational modifications in histones produced during NETosis offer us specific signatures that may serve as biomarkers.

4.2 Cytotoxic Effect of Circulating Histones

Recent evidence supports the idea that extracellular histones contribute to human disease. Circulating nucleosomes were recently associated with disease progression in patients with thrombotic microangiopathies [63]. Core and linker histones, and also whole nucleosomes, were observed in extranuclear spaces, in the outer mitochondrial membrane [64], in the cytoplasm [65,66], and also in the cell membrane surface [67,68]. The release of histones into the cytoplasm during apoptosis has been documented by several groups [69,70]. Taking into account the aforementioned role of histones as potent antimicrobial agents, extracellular histone H4 has been suggested as an additional antimicrobial component, which induces the death of microbes in human tissues [71]. However, the role of histones during infection might play a more sophisticated role than mere antimicrobial activity. In fact, it has also been proposed that circulating histone H4 can facilitate other histone release from tissues, thus exacerbating the antimicrobial action and probably spreading the signaling mediated by these proteins [61].

Histone variants also mediate several effects when they are released from the nucleus. For example, histone variant H1.2 triggers apoptotic mechanisms in different tumor cell lines [72]. Additionally, H1.2 released to the cytoplasm has been correlated with the presence of genetic abnormalities (e.g., deletions at 17p13 and 11q22) and with best clinical response to clinical treatments based on fludarabine, mitoxantrone, etoposide, and X-rays. The release of this histone variant seems to be involved in apoptosis induction in chronic lymphocytic leukemia, particularly in treatment-resistant cases with poor prognosis associated to p53-related drug resistance [73].

It is known since the 1940s that histones are cytotoxic to a variety of bacteria and mammalian cells [58,74,75]. In the latter case, although not completely understood, it has been proposed that the linker histone H1 disrupts the inner membrane potential and alters the mitochondrial NAD^+ pool [76], or also that it activates caspase-3 in neurons [77]. More recently, Abrams et al. [16] demonstrated by performing several experiments in vitro and in vivo that circulating histones can affect cell membranes after binding to phospholipids, producing the disruption of Ca^{+2} influx, and hence contributing to the release of other intracellular mediators that produce cell death.

Huang and colleagues have reported that histones can interact with TLR9, activating inflammation in mice [78]. In addition, extracellular histones contribute to thrombus formation by activating TLRs and the NLRP3 inflammasome. Extracellular histones can also propitiate the microvascular complications of sepsis, major trauma, and vasculitis in small vessels as well as acute liver, kidney, and lung injury [79].

To summarize, histones are not only cytotoxic for bacteria, but also for mammalian cells. Consequently, these observations provide histones with novel roles beyond their well-established involvement in structural chromatin organization and epigenetics.

4.3 Circulating Histones as Mediators of the Clinical Features of Sepsis

In 2009, a manuscript by Xu et al. demonstrated that extracellular histones mediate endothelial cell death and sepsis [61] and cause thrombocytopenia through platelet aggregation [80]. This research set the basis for exploring the role of histones in the pathogenesis of sepsis, and since then other manuscripts have been published with the aim to clarify the molecular mechanisms mediated by histones. In that regard, the histone role has been described as the histone-mediated apoptosis of lymphocytes in a time and dose-dependent manner [81] and it is known that lymphocyte apoptosis is one of the main reasons for immunoparalysis as a consequence of sepsis, as described above. It has also been reported that histone H4 induces platelet aggregation in a dose-dependent manner [82], which may explain DIC during the septic process. In fact, in vitro experiments demonstrated that 10 μg/mL of histone H4 induced platelet aggregation almost at the same level as that produced by 0.5 U/mL of thrombin [82].

Therefore, circulating histones present in plasma may induce critical damage in endothelial tissue, platelets, and PBMCs by activating several intracellular signaling pathways that produce cell death, contributing to the physiopathology of sepsis.

H4 has been proposed as the main histone-mediating lymphocyte apoptosis, so neutralization of this histone by means of immunotherapy has been proposed as a potential target in clinical interventions against sepsis [81]. Other studies have identified histone H3 and nucleosomes in plasma from septic patients [16,61,83]. Furthermore, histones mediate the apoptosis of cells in the lymphoid compartment observed in the thymus [45], spleen [46], and blood [47], compromising the adaptive immunity and contributing to CARS, as described above. Recent studies have shown elevated histone levels in blood from septic patients compared with other patients in the ICUs [17]. In fact, Ekaney et al. have shown higher levels of histone H4 for septic patients (0.35 ng/mL) compared to patients with multiple organ failure without infection (0.08 ng/mL) or to patients with minor trauma (0.11 ng/mL) [17] (Table 2).

These findings set the notion that increases in the concentration of extracellular histones in the bloodstream of patients with sepsis and septic shock is associated with prognosis and

TABLE 2 Circulating Nucleosomes and Histones as Biomarkers for Diagnostics and Prognostics of Sepsis

Biomarker	Type of sample	Experimental approach for detection	Levels	Diagnostic/Prognostic value	References
H3 and H4	Plasma	Western blot	–	–	[61][a]
H3	Plasma	ELISA	10–1600 ng/mL	Sepsis and DIC	[89]
H4	Plasma	Commercial ELISA	0.01–1.08 ng/mL 0.46–3 ng/mL 0.21 ng/mL versus 0.54 ng/mL	Sepsis Multiple trauma Survivors versus nonsurvivors[b]	[17]
Nucleosomes and histones	Serum	Commercial ELISA	Nucleosomes 100–350 AU Histones 10–230 μg/mL	Severe blunt trauma High levels of circulating histones were associated with the incidence of acute lung injury. Levels of circulating histones correlate with SOFA	[16]
Nucleosomes	Plasma EDTA	In-house made ELISA	38 (<35–285) units/mL 53 (<35–793) units/mL 269 (<35–1947) units/mL 814 (52–1979) units/mL	Fever SIRS Severe sepsis Septic shock[c]	[85,86]
Nucleosomes	Plasma EDTA	Commercial ELISA	AUC of 0.67 (95% CI 0.55–0.79) >2.09 AU (Sn 64%, Sp 76%)	Sepsis Patients experienced a more severe inflammatory response	[88]

AU, arbitrary units; AUC, area under the ROC curves; CI, confidence interval; DIC, disseminated intravascular coagulation; SOFA, sequential organ failure assessment; Sn, sensitivity; Sp, specificity; SIRS, systemic inflammatory response syndrome.

[a] In this study the authors did not quantify the levels of circulating nucleosomes.

[b] *With respect to ICU and 28-day mortality, on samples collected on day 1 from sepsis patients and ICU patients without sepsis.*

[c] *Levels of circulating nucleosomes correlate with SOFA.*

mortality. Thus, it is possible to use circulating histones as biomarkers of prognosis in these hyperinflammatory syndromes.

5. CIRCULATING HISTONES AS BIOMARKERS OF PROGNOSIS AND TREATMENT MONITORING

Early, accurate diagnosis and risk stratification of sepsis remains an important challenge in the ICU worldwide. Rapid diagnosis and treatment with antibiotics is critical in saving lives from this disease, but since there are currently no biomarkers in clinical use to enable fast diagnosis, it can take up to 2 or 3 days to diagnose sepsis, thus compromising the treatment of patients. Furthermore, as mentioned above, the pathophysiology of CARS and the epigenetic imprinting occurring during sepsis highlight epigenetics as a relevant tool in the diagnosis and prognosis of sepsis.

Since traditional biomarker strategies have not yielded a gold standard marker for sepsis, focus is shifting toward novel strategies that improve assessment capabilities. Epigenetics has already shown to be a promising tool in diagnosing sepsis. In fact, a study by Ma et al. [84] using next-generation sequencing identified microRNA candidates from blood samples of three groups of patients: those with sepsis, patients with other SIRS that do not respond to antibiotics, and healthy subjects. After validation of these candidates using qRT-PCR procedures, they found that microRNAs miR-150 and miR-4772-5p-iso were able to discriminate between SIRS patients and patients with sepsis. This finding was also validated in an independent cohort with an average diagnostic accuracy of 86%.

One of the most important findings opening a new frontier for the diagnostic and prognostic capabilities of sepsis was the identification of circulating histones. Pioneer investigations by Zeerleder et al. found circulating nucleosomes and histones in plasma from septic patients. Zeerleder et al. found increased nucleosome levels in systemic inflammation and septic shock [85,86] and studied how these levels correlated with the severity of the inflammatory response and mortality in children affected by meningococcal sepsis [86]. A recent work by Ekaney et al. demonstrated that with respect to ICU and 28-day mortality, on day 1 the levels of histone H4 detected in survivors who suffered sepsis (0.21 ng/mL) were lower than those observed for nonsurvivors (0.54 ng/mL) [17]. These results were more evident for 90-day mortality, and it was found that circulating H4 levels were about 0.19 ng/mL for survivors and about 0.48 ng/mL for nonsurvivors (Table 2).

Nonetheless, in order to make a proper biomarker out of circulating histones, one of the key points is the possibility of a quick, reproducible, and quantifiable technique that permits rapid and significant measures from patient blood samples. Several attempts in this direction have been previously made and are currently under development.

5.1 Experimental Approaches to Analyze Circulating Histones

Different approaches have been used to identify the presence of histones in plasma samples. Histones were identified in frozen archival plasma samples from baboon models with sepsis after infection with *Escherichia coli* [61,87] using Western blot (WB). This approach detected 15 μg/mL of histone H3 in plasma from baboons 8 h after infection with *E. coli*. However, the authors indicated that it was impossible to measure other histones by WB because the antibodies tested were unable to detect other nucleosomal histones [61]. The use of WB to detect histones can be applied in research investigations, but not in clinical practice, since WB is a semiquantitative analytical tool, and it requires relatively long times to obtain results, which is not always affordable in a clinical laboratory.

Other techniques used to measure the levels of histones in blood samples are based on ELISA procedures, which are more available in hospitals and are routinely used in clinical diagnostics laboratories. Chen's group used a commercial ELISA kit to measure circulating nucleosomes (Cell Death Detection ELISA^{PLUS} Kit, Roche, Manheim, Germany), a technique through which they found high reproducibility [88]. Zeerleder et al. [85] used an immunosorbent ELISA kit made in-house to evaluate circulating nucleosomes. This ELISA used a CLB-ANA/60 antibody, which recognizes histone H3, as a "catching" antibody, and for detection used the monoclonal antibody CLB-ANA/58 (CLB, Amsterdam, The Netherlands), which recognizes an exposed epitope formed by histones H2A, H2B, and dsDNA. Nakahara et al. also produced an ELISA assay made in-house using polystyrene microtiter plates which were coated with a commercial antihuman histone H3 polyclonal antibody (Abcam, Cambridge, United Kingdom) [89]. Ekaney et al. used a commercial ELISA kit to detect circulating histone H4 (USCN Life Science, China). Therefore, these commercial kits can be used to detect circulating nucleosomes or histones in body fluids for diagnostics or prognostics in a variety of pathological conditions including many cancers, autoimmune diseases, inflammatory conditions, stroke, among others [15].

5.2 Diagnostic and Prognostic Value

Research by Xu [61] and Abrams et al. [16] reinforces the possibility of the use of H3 and H4 as biomarkers of disease progression and also as therapeutic targets in sepsis, septic shock, and other inflammatory diseases.

Other studies performed by Zeerleder and collaborators tried to quantify circulating nucleosome levels comparing four groups of patients suffering fever, SIRS, severe sepsis, and septic shock [85]. It has been confirmed that circulating nucleosomes are low or even undetectable in healthy people [90,91], but interestingly they found elevated nucleosome levels in 64%, 60%, 94%, and 100% of the aforementioned groups of patients, respectively. Levels of circulating nucleosomes were significantly higher in plasma samples from patients suffering septic shock (814 (52–1979) units/mL) as compared with the other groups: fever, 38 (<35–285) units/mL; SIRS, 53 (<35–793) units/mL; and severe sepsis, 269 (<35–1947) units/mL. The values obtained for circulating histones correlated with cytokine plasma levels and with variables predictive of patient outcome (Table 2). In fact, when patients with sequential organ failure assessment (SOFA) scores of >12 were considered separately, nucleosome levels correlated significantly with a number of inflammatory and fibrinolytic parameters. Therefore, Zeerleder et al. suggested that nucleosomes might be involved in the pathogenesis of multiple organ dysfunction syndrome occurring in septic shock, and thus show the potential of nucleosomes as prognostic biomarkers for septic shock.

More recently, Chen et al. [88] prospectively investigated circulating nucleosomes in septic patients, aiming for the diagnosis and prognosis of sepsis. The study developed by these authors consisted of the measurement of circulating nucleosomes within 24h of admission in two different cohorts. The former cohort consisted of the recruitment of 74 admitted patients with a length of stay in the ICU of more than 48h, and the second cohort was set by 91 postsurgery patients. Chen et al. demonstrated that serum levels of circulating nucleosomes on ICU admission were significantly elevated in septic patients versus nonseptic patients, results that correlate with the Acute Physiology and Chronic Health Evaluation II (APACHE II) score. The authors also explored the relationship between the sensitivity and specificity of circulating nucleosomes and other biomarkers used for predicting sepsis, such as soluble triggering receptor expressed on myeloid cells, procalcitonin, and C-reactive protein [92], by means of the use of receiver operating characteristic (ROC) curves, in order to discriminate patients with sepsis from those without sepsis. The area under the ROC curve for circulating nucleosomes was 0.67 (95% CI 0.55–0.79). Furthermore, the authors established the best cutoff value of circulating nucleosomes for predicting sepsis at above 2.09 AU, with sensitivity (Sn) of 64% and specificity (Sp) of 76% (Table 2). Interestingly, the authors also observed that patients with circulating nucleosome levels above this value suffered a more severe inflammatory response.

In addition, Chen et al. used a multiple stepwise logistic regression model to evaluate the diagnostic value of circulating nucleosomes for sepsis, demonstrating that circulating nucleosomes are an independent predictor of sepsis [88]. They investigated the relationships between circulating nucleosomes and clinical features by means of the Spearman's correlation test, and observed that circulating nucleosome levels correlated with a greater immunosuppressive response (by apoptosis of T and B cells and the release of the anti-inflammatory cytokine IL-10, which contributes to immunosuppression) and organ dysfunction in septic patients, assessed by the SOFA score. However, the association between the admission levels of circulating nucleosomes and mortality rates in sepsis did not show significance, although a trend toward the admission levels of circulating nucleosomes in cases with fatal outcomes were higher than those in survivors (median 2.58 AU vs 1.97; $p=0.06$) (Table 2).

In another study, Abrams et al. [16] measured circulating histone levels in a cohort of 52 patients with severe blunt trauma. Among survivors they found that 76.5% of patients with circulating histone levels above 50 μg/mL developed respiratory failure, compared with 18.8% of patients with serum levels of circulating histones below this value. Through a longitudinal analysis of circulating nucleosomes and circulating histones, the authors showed that nucleosomes reached maximal levels within 4h and suffered a complete clearance after 24h, whereas histones were elevated at 4h after trauma, reached a maximum after about 24h, and were still detectable at 72h in many cases. Thus, histones proved to be better biomarkers for injury than nucleosomes, since total circulating histone levels ranged from 10 to 230 μg/mL within 4h in patients with severe trauma (Table 2). Abrams et al. found a correlation between circulating histone levels and SOFA scores, used for the evaluation of multiple organ dysfunction, and suggested that circulating histones may be a cause of lung injury and multiple organ failure. These results reinforce the notion that circulating histones, as occurs with circulating nucleosomes, might be proper biomarkers of sepsis and septic shock.

Circulating histone H3 has also been identified in patients with sepsis and DIC [89]. Whereas histone H3 was not detectable in plasma of healthy volunteers ($n=15$), histone H3 levels were higher in patients with sepsis and DIC ($n=26$). These levels were also higher in nonsurvivors than in survivors. The authors of this study then used this information as a starting point in order to work with mouse models injected with purified histones, reaching concentrations of about 1–10 μg/mL, which are comparable to those observed in these patients, as well as in previous studies [61,93]. These mice recapitulated the vascular effects and subsequent systemic failure observed in patients with DIC, and the results obtained may serve as a valuable tool in evaluating the cytotoxicity of circulating histones, and their potential value as a prognostic biomarker of thrombocytopenia and microvessel occlusion complications encompassing DIC and septic processes.

Ekaney et al. also studied the levels of circulating histone H4 in ICU and 28-day mortality, showing that nonsurvivors contained higher levels of circulating histone H4, with a median of 0.54 (0.34–0.45) ng/mL compared to survivors, which showed median values of 0.21 (0.06–0.4) ng/mL. The values for 28-day mortality in nonsurvivors were as high as 0.67 (0.25–0.75) ng/mL. The authors also found that histone levels were good predictors of the patient outcome at onset of sepsis [17] (Table 2). Ekaney et al. report lower concentrations in the nanogram range of circulating histones than other authors, in the microgram range [17]. The treatment of patients (i.e., heparin, the use of renal replacement therapy, etc.), and the technical approach to measure the levels of circulating histones, may be directly related with these discrepancies.

5.3 Determination of Circulating Histones for Treatment Monitoring

From the previous examples, it can be inferred that precisely determining the presence and identity of circulating and/or extracellular histones and nuclear components harbors the potential to serve as a powerful prognostic tool that will allow a close tracking of the physiopathological consequences of sepsis as they progress: the aforementioned immunoparalysis, CARS, and DIC. This also provides critical advantages for the monitoring of treatments, independently of whether or not their primary goal is the decrease of circulating histone levels.

Another example of how these molecular data might result in treatment monitoring can be applied to a recent study by Wildhagen and collaborators in which they proposed a nonanticoagulant derivative of heparin as an effective treatment against several histone-mediated cytotoxic processes involved in sepsis [94]. The therapeutic potential of this research is discussed in detail in the next section, but it should be noted that the authors used histone H3 quantification by WB as a means of defining their experimental animal models, setting a correlation between circulating histone H3 levels, sepsis development, and the effect of nonanticoagulant heparin as a therapeutic agent. Since their results are directly based on abrogation of histone-mediated cell damage, the need for a quick, reliable, and quantifiable method for determining circulating histones would largely improve the clinical prognostic value of this and similar approaches.

6. THERAPEUTIC APPROACHES TO ANTAGONIZE HISTONE-MEDIATED TOXICITY

Although the molecular mechanisms underlying histone-mediated toxicity are still not completely understood, there is a wide consensus in considering circulating histones as a pathogenic state with a pivotal role in the progress of sepsis. Besides this general notion, a good amount of data provides more detailed descriptions on how extracellular histones generate cell damage or promote the immunological alterations that unbalance the host's immune response to infection, creating a positive feedback toward systemic failure. In the above examples we have covered a series of cases in which extracellular presence of histones can be used as powerful tracking evidence of how sepsis progresses, and obviously, this permits a careful monitoring of any treatment applied to the septic patient. Nonetheless, it must be taken into account that circulating histones are biomarkers whose presence is directly related to the progress of the disease, and the numerous evidence of cytotoxic effects of histones highlight their potential as targets for treatment themselves. Here we will review, in the first place, the current approaches focused on diminishing or inactivating circulating histones, in an attempt to slow the feedback process and to avoid effects like damage to endothelial cells, microvascular occlusion, and other side effects directly related to the presence of extracellular histones. Secondly, we will review other recent approaches based on already known histone modifiers that might have an effect on histone-mediated toxicity.

6.1 Antagonizing Histone-Mediated Toxicity

Activated protein C (APC) is a serine protease encoded by the protein C, which is an inactivator of coagulation factors Va and VIIIa (PROC) genes. APC (drotrecogin alpha activated, marketed as Xigris) was approved in 2001 by the Food and Drug Administration (FDA) and in 2002 by the European Medicines Agency (EMA) for the treatment of severe sepsis, based on its APACHE II score. Although not completely known, its protective effect seems to be due to its anti-inflammatory activity, rather than due to its anticoagulant function [95–98]. Furthermore, it was observed in clinical practice that acquired severe protein C deficiency correlates with early death in septic patients [99]. However, the clinical use of APC has been the object of some controversy because its use does not demonstrate the improvement of survival rates [100] and in 2007 clinical trials failed to replicate favorable results from the pivotal recombinant human APC worldwide evaluation in severe sepsis (PROWESS) study [101]. Consequently, commercial preparations of recombinant human APC were removed from the market in 2011. In contrast, Xu et al. demonstrated that APC reduced the in vitro cytotoxicity caused by a mixture of histone H3 and H4 in endothelial cells because of its ability to cleave these histones in a dose-dependent manner [61]. They also demonstrate that APC is able to cleave histone H3 present in plasma from some septic patients and to increase the survival rates of mice in which APC (5 mg/kg) and histones (75 mg/kg) were coinjected, demonstrating the role of APC in regulating extracellular histone levels in vivo.

A clarifying set of results regarding APC and anticoagulatory features of sepsis was recently provided by Nakahara and collaborators [89] (see Section 6). They injected mice with histones H3 and H4 in concentrations similar to those found in patients with sepsis (1–10 μg/mL) and thoroughly examined the consequences of circulating histones, but focused on the intravascular coagulation processes. They found that histones mediated thrombocytopenia via platelet aggregation, and produced an outcome of microvessel obstruction; interestingly, platelet depletion prior to histone injection was protective for mice, but only in the first stages of sepsis. This points to histone-mediated microvessel occlusion as an early phase effect of circulating histones in mice, but also shows that other deleterious mechanisms mediated by histones might occur after long-term extracellular histone circulation. The study by Nakahara et al. [89] shows that histones H3 and H4 strongly attach to platelets, and that this attachment is impaired by the action of recombinant thrombomodulin (rTM), a drug which is now being used in Japan for the treatment of DIC. The authors postulate that the effect of rTM may be mediated, in part, by the specific activation of APC that results in an anticoagulation effect, but also that rTM exerts its effect by direct binding to histones. The lack of histone H3 degradation products in histone-injected mice after treatment with rTM further supports this hypothesis. Still, the mechanisms involving rTM-histone binding remain unclear and warrant further research aimed to develop rTM-based treatments that rely on the abrogation of extracellular histone-mediated vascular complications of sepsis. In any case, the possibility that rTM could diminish extracellular levels of histones by itself proves to be a powerful therapeutic approach and the use of circulating histones a good biomarker to evaluate the therapeutical effect mediated by the rTM. In humans, Ekaney et al. demonstrated a negative correlation between histone levels and endogenous APC levels in patients with sepsis, at onset of sepsis day 1, day 3, and day 5 of ICU stay [17], highlighting the role of this serine protease during sepsis.

More data point to a beneficial effect of anticoagulant proteins in histone-mediated cytotoxicity. Using nonanticoagulant derivatives of heparin on different animal models with sepsis, two independent groups [94,102] found a decrease in circulating histone levels correlating with a better survival rate, postulating that binding of heparin derivatives to histones promotes a sequestration effect that clears the bloodstream. The use of nonanticoagulant derivatives of heparin provides the important benefit of not overwhelming the already compromised coagulatory system of the septic patient, and thus might be a promising path for future clinical treatments.

In addition, histones can bind to and interact with a large variety of plasma proteins, including albumin [82], which decreases the deleterious effects of histones. As already known, albumin is the most abundant circulating protein in the bloodstream and, as Lam et al. demonstrated, it is probably one of the main natural mechanisms utilized by the organism to inhibit the pathological effects of circulating histones. The role of albumin forming a complex with histone H4 was evidenced by in vitro experiments with human serum albumin, and experiments using normal or albumin-depleted plasma demonstrated that only plasma containing albumin was able to inhibit histone H4-induced platelet aggregation [82]. Therefore, the addition of exogenous albumin may serve to capture circulating histones released from activated neutrophils during NETosis, as well as histones released from apoptotic and necrotic cells. The results by Lam et al. support previous results by Khan and collaborators demonstrating the beneficial effects of albumin reducing platelet activation and platelet–neutrophil aggregates in blood derived from patients with septic shock [103].

Another feasible tool in antagonizing the effect of histones is the use of specific antibodies. Xu et al. demonstrated that damage to human endothelial cells was blocked by the use of a monoclonal antibody against histone H4 [61]. Since circulating H4 has been proposed as a mediator for further histone release from cells and tissues, the use of H4 antibodies could be an interesting therapeutical approach for sepsis, as the results of Xu et al. suggest, in which coinfusion of H4 antibody rescued mice injected with LPS. Zhang and collaborators also used this strategy to antagonize the deleterious effect of histone H4 in their animal models in order to validate their experimental results [102].

6.2 The Use of HDAC Inhibitors as Therapeutic Agents against Sepsis

Histone deacetylase inhibitors (HDACIs) are being used in preclinical studies in inflammatory and immune diseases. They represent a class of therapeutic drugs in sepsis, given that HDACIs mediate anti-inflammatory responses and also have effects on the natural host defenses in patients [104]. In fact, the effects of sodium butyrate (NaB), suberoylanilide hydroxamic acid (SAHA), trichostatin A (TSA), and valproic acid (VPA) have been explored in preclinical studies in animal models of septic shock [105–107]. The studies performed by Roger et al. demonstrate that VPA reduced the cytokine burst induced by caecal ligation and puncture, and Pam_3CSK_4 lipopeptide, which mimics gram-positive bacteria lipopeptides [105]. The effect of HDACI may come from its function negatively regulating critical immune receptors and antimicrobial products by immune cells [105].

Despite the preliminary data that suggest using HDACIs as drug candidates to counteract the deleterious effect of histones in sepsis progression, it should be noted that histone deacetylation is a biologically relevant modification of histones involved in a plethora of processes. Therefore, it is noteworthy that the design of drugs that selectively inhibit HDAC to affect the expression of specific genes will surely be a bottleneck in their way toward a standard use as antagonists of sepsis progression. Several studies in this direction are addressing the specificity of the different kinds of HDACIs in terms of the signaling cascades and immune processes they mainly affect, although knowledge of the long-term effects of these therapies is still needed to assess the implications for the patient's immune system, chromatin remodeling in genes of the adaptive immune response, and CARS, especially after surviving infection.

7. CONCLUSIONS

The early, accurate diagnosis and risk stratification of sepsis, severe sepsis, and septic shock remains an important challenge in medicine, producing unacceptable mortality rates and expensive costs during hospitalization. Traditional biomarkers have not yielded a definitive candidate to improve triage decisions, the administration of appropriate therapy or to predict future immunosuppression in sepsis patients. Resistance to infection and its outcomes are influenced by a complex interplay between the host, the infective microorganism, and the environmental conditions that control the epigenome. Therefore, sepsis progression will depend on the genetics and epigenetics of the patient. Although there are no major studies of epigenetic modifications in sepsis, recent investigations have proposed miRNAs as biomarkers for sepsis. But one of the most important findings that paves the way for the diagnostic and prognostic handling of sepsis has been the identification of circulating histones as contributors to cytotoxicity and tissue damage. However, there exists no consensus on the concrete levels of circulating histones and nucleosomes that discriminate sepsis from severe sepsis and septic shock, or survivors from nonsurvivors, due in part to the diversity of procedures used in their determination. Furthermore, the use of antibodies against histones and HDACIs, which negatively regulate crucial immune receptors and antimicrobial products by immune cells, offers several therapeutic approaches against sepsis. In conclusion, the identification of circulating histones and nucleosomes as key players in the pathophysiology of sepsis has represented a promising starting point for setting new biomarkers in the diagnosis and prognosis of this devastating disease.

Glossary

APACHE II Acute Physiology and Chronic Health Evaluation II score. This score measures the severity of disease for adult patients admitted to ICUs by evaluating 12 physiologic variables (PaO_2; temperature; arterial pressure; arterial pH; heart and respiratory rate; sodium, potassium, and creatinine levels; hematocrit; white blood cell count, and Glasgow Coma Score; age; and chronic health evaluation status) during the first 24 h after admission.

SOFA Sequential organ failure assessment score (sepsis-related organ failure assessment). This score system is used to determine the extent of organ function and it is based on six different scores, corresponding to the respiratory, cardiovascular, hepatic, coagulation, renal, and neurological functions.

LIST OF ABBREVIATIONS

CARS Compensatory anti-inflammatory response syndrome
DC Dendritic cells
DIC Disseminated intravascular coagulation
ICU Intensive care unit
LPS Lipopolysaccharide
NETs Neutrophil extracellular traps
SIRS Systemic inflammatory response syndrome
T_{reg} Regulatory T cells

Acknowledgments

This work was supported by the INCLIVA Biomedical Research Institute and Generalitat Valenciana (GV/2014/132) for the grants to J.L.G–G, the Grand Challenges Canada to J.L.G–G. and F.V.P. and the Ministerio de Economía y Competitividad, Instituto de Salud Carlos III through CIBERer (Biomedical Network Research Center for Rare Diseases and INGENIO2010).

References

[1] Bone RC, Balk RA, Cerra FB, Dellinger RP, Fein AM, Knaus WA, et al. Definitions for sepsis and organ failure and guidelines for the use of innovative therapies in sepsis. The ACCP/SCCM Consensus Conference Committee. American College of Chest Physicians/Society of Critical Care Medicine. Chest June 1992;101(6):1644–55.

[2] Levy MM, Fink MP, Marshall JC, Abraham E, Angus D, Cook D, et al. 2001 SCCM/ESICM/ACCP/ATS/SIS International Sepsis Definitions Conference. Crit Care Med April 2003;31(4):1250–6.

[3] Dombrovskiy VY, Martin AA, Sunderram J, Paz HL. Rapid increase in hospitalization and mortality rates for severe sepsis in the United States: a trend analysis from 1993 to 2003. Crit Care Med May 2007;35(5):1244–50.
[4] Cerra FB. The systemic septic response: multiple systems organ failure. Crit Care Clin November 1985;1(3):591–607.
[5] Johansson PI, Windelov NA, Rasmussen LS, Sorensen AM, Ostrowski SR. Blood levels of histone-complexed DNA fragments are associated with coagulopathy, inflammation and endothelial damage early after trauma. J Emerg Trauma Shock July 2013;6(3):171–5.
[6] Rumore PM, Steinman CR. Endogenous circulating DNA in systemic lupus erythematosus. Occurrence as multimeric complexes bound to histone. J Clin Invest July 1990;86(1):69–74.
[7] Zeerleder S, Zwart B, te Velthuis H, Manoe R, Bulder I, Rensink I, et al. A plasma nucleosome releasing factor (NRF) with serine protease activity is instrumental in removal of nucleosomes from secondary necrotic cells. FEBS Lett November 27, 2007;581(28):5382–8.
[8] Bell DA, Morrison B. The spontaneous apoptotic cell death of normal human lymphocytes in vitro: the release of, and immunoproliferative response to, nucleosomes in vitro. Clin Immunol Immunopathol July 1991;60(1):13–26.
[9] Stroun M, Maurice P, Vasioukhin V, Lyautey J, Lederrey C, Lefort F, et al. The origin and mechanism of circulating DNA. Ann NY Acad Sci April 2000;906:161–8.
[10] van Nieuwenhuijze AE, van Lopik T, Smeenk RJ, Aarden LA. Time between onset of apoptosis and release of nucleosomes from apoptotic cells: putative implications for systemic lupus erythematosus. Ann Rheum Dis January 2003;62(1):10–4.
[11] Brinkmann V, Reichard U, Goosmann C, Fauler B, Uhlemann Y, Weiss DS, et al. Neutrophil extracellular traps kill bacteria. Science March 5, 2004;303(5663):1532–5.
[12] Rumore P, Muralidhar B, Lin M, Lai C, Steinman CR. Haemodialysis as a model for studying endogenous plasma DNA: oligonucleosome-like structure and clearance. Clin Exp Immunol October 1992;90(1):56–62.
[13] Burlingame RW, Volzer MA, Harris J, Du Clos TW. The effect of acute phase proteins on clearance of chromatin from the circulation of normal mice. J Immunol June 15, 1996;156(12):4783–8.
[14] Gauthier VJ, Tyler LN, Mannik M. Blood clearance kinetics and liver uptake of mononucleosomes in mice. J Immunol February 1, 1996;156(3):1151–6.
[15] Holdenrieder S, Stieber P. Clinical use of circulating nucleosomes. Crit Rev Clin Lab Sci 2009;46(1):1–24.
[16] Abrams ST, Zhang N, Manson J, Liu T, Dart C, Baluwa F, et al. Circulating histones are mediators of trauma-associated lung injury. Am J Respir Crit Care Med January 15, 2013;187(2):160–9.
[17] Ekaney ML, Otto GP, Sossdorf M, Sponholz C, Boehringer M, Loesche W, et al. Impact of plasma histones in human sepsis and their contribution to cellular injury and inflammation. Crit Care 2014;18(5):543.
[18] Angus DC, Linde-Zwirble WT, Lidicker J, Clermont G, Carcillo J, Pinsky MR. Epidemiology of severe sepsis in the United States: analysis of incidence, outcome, and associated costs of care. Crit Care Med July 2001;29(7):1303–10.
[19] Lagu T, Rothberg MB, Shieh MS, Pekow PS, Steingrub JS, Lindenauer PK. Hospitalizations, costs, and outcomes of severe sepsis in the United States 2003 to 2007. Crit Care Med March 2012;40(3):754–61.
[20] Adhikari NK, Fowler RA, Bhagwanjee S, Rubenfeld GD. Critical care and the global burden of critical illness in adults. Lancet October 16, 2010;376(9749):1339–46.
[21] Vincent JL, Sakr Y, Sprung CL, Ranieri VM, Reinhart K, Gerlach H, et al. Sepsis in European intensive care units: results of the SOAP study. Crit Care Med February 2006;34(2):344–53.
[22] Kaukonen KM, Bailey M, Suzuki S, Pilcher D, Bellomo R. Mortality related to severe sepsis and septic shock among critically ill patients in Australia and New Zealand, 2000–2012. JAMA April 2, 2014;311(13):1308–16.
[23] Hotchkiss RS, Karl IE. The pathophysiology and treatment of sepsis. N Engl J Med January 9, 2003;348(2):138–50.
[24] Stearns-Kurosawa DJ, Osuchowski MF, Valentine C, Kurosawa S, Remick DG. The pathogenesis of sepsis. Annu Rev Pathol 2011;6:19–48.
[25] Annane D, Aegerter P, Jars-Guincestre MC, Guidet B, CUB-Réa Network. Current epidemiology of septic shock: the CUB-Réa Network. Am J Respir Crit Care Med July 15, 2003;168(2):165–72.
[26] Zeerleder S, Hack CE, Wuillemin WA. Disseminated intravascular coagulation in sepsis. Chest October 2005;128(4):2864–75.
[27] Ayala A, Herdon CD, Lehman DL, DeMaso CM, Ayala CA, Chaudry IH. The induction of accelerated thymic programmed cell death during polymicrobial sepsis: control by corticosteroids but not tumor necrosis factor. Shock April 1995;3(4):259–67.
[28] Coopersmith CM, Stromberg PE, Dunne WM, Davis CG, Amiot 2nd DM, Buchman TG, et al. Inhibition of intestinal epithelial apoptosis and survival in a murine model of pneumonia-induced sepsis. JAMA April 3, 2002;287(13):1716–21.
[29] Hotchkiss RS, Swanson PE, Freeman BD, Tinsley KW, Cobb JP, Matuschak GM, et al. Apoptotic cell death in patients with sepsis, shock, and multiple organ dysfunction. Crit Care Med July 1999;27(7):1230–51.

[30] Hotchkiss RS, Osmon SB, Chang KC, Wagner TH, Coopersmith CM, Karl IE. Accelerated lymphocyte death in sepsis occurs by both the death receptor and mitochondrial pathways. J Immunol April 15, 2005;174(8):5110–8.
[31] Chang KC, Unsinger J, Davis CG, Schwulst SJ, Muenzer JT, Strasser A, et al. Multiple triggers of cell death in sepsis: death receptor and mitochondrial-mediated apoptosis. FASEB J March 2007;21(3):708–19.
[32] Marshall JC. Why have clinical trials in sepsis failed? Trends Mol Med April 2014;20(4):195–203.
[33] Sorensen TI, Nielsen GG, Andersen PK, Teasdale TW. Genetic and environmental influences on premature death in adult adoptees. N Engl J Med March 24, 1988;318(12):727–32.
[34] Burgner D, Levin M. Genetic susceptibility to infectious diseases. Pediatr Infect Dis J January 2003;22(1):1–6.
[35] Bellamy R, Hill AV. Genetic susceptibility to mycobacteria and other infectious pathogens in humans. Curr Opin Immunol August 1998;10(4):483–7.
[36] Choi EH, Zimmerman PA, Foster CB, Zhu S, Kumaraswami V, Nutman TB, et al. Genetic polymorphisms in molecules of innate immunity and susceptibility to infection with *Wuchereria bancrofti* in South India. Genes Immun August 2001;2(5):248–53.
[37] Sutherland AM, Walley KR. Bench-to-bedside review: association of genetic variation with sepsis. Crit Care 2009;13(2):210.
[38] Carson WF, Cavassani KA, Dou Y, Kunkel SL. Epigenetic regulation of immune cell functions during post-septic immunosuppression. Epigenetics March 2011;6(3):273–83.
[39] Brogdon JL, Xu Y, Szabo SJ, An S, Buxton F, Cohen D, et al. Histone deacetylase activities are required for innate immune cell control of Th1 but not Th2 effector cell function. Blood February 1, 2007;109(3):1123–30.
[40] Tsaprouni LG, Ito K, Adcock IM, Punchard N. Suppression of lipopolysaccharide- and tumour necrosis factor-alpha-induced interleukin (IL)-8 expression by glucocorticoids involves changes in IL-8 promoter acetylation. Clin Exp Immunol October 2007;150(1):151–7.
[41] Chan C, Li L, McCall CE, Yoza BK. Endotoxin tolerance disrupts chromatin remodeling and NF-kappaB transactivation at the IL-1beta promoter. J Immunol July 1, 2005;175(1):461–8.
[42] Lebre MC, Burwell T, Vieira PL, Lora J, Coyle AJ, Kapsenberg ML, et al. Differential expression of inflammatory chemokines by Th1- and Th2-cell promoting dendritic cells: a role for different mature dendritic cell populations in attracting appropriate effector cells to peripheral sites of inflammation. Immunol Cell Biol October 2005;83(5):525–35.
[43] De Santa F, Totaro MG, Prosperini E, Notarbartolo S, Testa G, Natoli G. The histone H3 lysine-27 demethylase Jmjd3 links inflammation to inhibition of polycomb-mediated gene silencing. Cell September 21, 2007;130(6):1083–94.
[44] Wen H, Dou Y, Hogaboam CM, Kunkel SL. Epigenetic regulation of dendritic cell-derived interleukin-12 facilitates immunosuppression after a severe innate immune response. Blood February 15, 2008;111(4):1797–804.
[45] Wang SD, Huang KJ, Lin YS, Lei HY. Sepsis-induced apoptosis of the thymocytes in mice. J Immunol May 15, 1994;152(10):5014–21.
[46] Ayala A, Chung CS, Xu YX, Evans TA, Redmond KM, Chaudry IH. Increased inducible apoptosis in CD4+ T lymphocytes during polymicrobial sepsis is mediated by Fas ligand and not endotoxin. Immunology May 1999;97(1):45–55.
[47] Le Tulzo Y, Pangault C, Gacouin A, Guilloux V, Tribut O, Amiot L, et al. Early circulating lymphocyte apoptosis in human septic shock is associated with poor outcome. Shock December 2002;18(6):487–94.
[48] Carson 4th WF, Cavassani KA, Ito T, Schaller M, Ishii M, Dou Y, et al. Impaired CD4+ T-cell proliferation and effector function correlates with repressive histone methylation events in a mouse model of severe sepsis. Eur J Immunol April 2010;40(4):998–1010.
[49] Monneret G, Debard AL, Venet F, Bohe J, Hequet O, Bienvenu J, et al. Marked elevation of human circulating CD4+CD25+ regulatory T cells in sepsis-induced immunoparalysis. Crit Care Med July 2003;31(7):2068–71.
[50] Lal G, Bromberg JS. Epigenetic mechanisms of regulation of Foxp3 expression. Blood October 29, 2009;114(18):3727–35.
[51] Metchnikoff E. Leçons sur la pathologie comparée de l'inflammation. Paris: G. Masson; 1892.
[52] Ehrlich P. Methodologische beitrage zur physiologie und pathologie der verschisdenen formen der leukocyten. Z Klin Med 1880;1:553–8.
[53] Steinberg BE, Grinstein S. Unconventional roles of the NADPH oxidase: signaling, ion homeostasis, and cell death. Sci STKE March 27, 2007;2007(379):pe11.
[54] Papayannopoulos V, Metzler KD, Hakkim A, Zychlinsky A. Neutrophil elastase and myeloperoxidase regulate the formation of neutrophil extracellular traps. J Cell Biol November 1, 2010;191(3):677–91.
[55] Neeli I, Dwivedi N, Khan S, Radic M. Regulation of extracellular chromatin release from neutrophils. J Innate Immun 2009;1(3):194–201.
[56] Wang Y, Li M, Stadler S, Correll S, Li P, Wang D, et al. Histone hypercitrullination mediates chromatin decondensation and neutrophil extracellular trap formation. J Cell Biol January 26, 2009;184(2):205–13.

[57] Fuchs TA, Abed U, Goosmann C, Hurwitz R, Schulze I, Wahn V, et al. Novel cell death program leads to neutrophil extracellular traps. J Cell Biol January 15, 2007;176(2):231–41.

[58] Miller BF, Abrams R, Dorfman A, Klein M. Antibacterial properties of protamine and histone. Science November 6, 1942;96(2497):428–30.

[59] Kawasaki H, Koyama T, Conlon JM, Yamakura F, Iwamuro S. Antimicrobial action of histone H2B in *Escherichia coli*: evidence for membrane translocation and DNA-binding of a histone H2B fragment after proteolytic cleavage by outer membrane proteinase T. Biochimie November–December 2008;90(11–12):1693–702.

[60] Lemaire S, Trinh TT, Le HT, Tang SC, Hincke M, Wellman-Labadie O, et al. Antimicrobial effects of H4-(86–100), histogranin and related compounds–possible involvement of DNA gyrase. FEBS J November 2008;275(21):5286–97.

[61] Xu J, Zhang X, Pelayo R, Monestier M, Ammollo CT, Semeraro F, et al. Extracellular histones are major mediators of death in sepsis. Nat Med November 2009;15(11):1318–21.

[62] Clark SR, Ma AC, Tavener SA, McDonald B, Goodarzi Z, Kelly MM, et al. Platelet TLR4 activates neutrophil extracellular traps to ensnare bacteria in septic blood. Nat Med April 2007;13(4):463–9.

[63] Fuchs TA, Kremer Hovinga JA, Schatzberg D, Wagner DD, Lammle B. Circulating DNA and myeloperoxidase indicate disease activity in patients with thrombotic microangiopathies. Blood August 9, 2012;120(6):1157–64.

[64] Choi YS, Hoon Jeong J, Min HK, Jung HJ, Hwang D, Lee SW, et al. Shot-gun proteomic analysis of mitochondrial D-loop DNA binding proteins: identification of mitochondrial histones. Mol Biosyst May 2011;7(5):1523–36.

[65] Gabler C, Blank N, Hieronymus T, Schiller M, Berden JH, Kalden JR, et al. Extranuclear detection of histones and nucleosomes in activated human lymphoblasts as an early event in apoptosis. Ann Rheum Dis September 2004;63(9):1135–44.

[66] Gabler C, Blank N, Winkler S, Kalden JR, Lorenz HM. Accumulation of histones in cell lysates precedes expression of apoptosis-related phagocytosis signals in human lymphoblasts. Ann NY Acad Sci December 2003;1010:221–4.

[67] Bolton SJ, Perry VH. Histone H1; a neuronal protein that binds bacterial lipopolysaccharide. J Neurocytol December 1997;26(12):823–31.

[68] Watson K, Gooderham NJ, Davies DS, Edwards RJ. Nucleosomes bind to cell surface proteoglycans. J Biol Chem July 30, 1999;274(31):21707–13.

[69] Wu D, Ingram A, Lahti JH, Mazza B, Grenet J, Kapoor A, et al. Apoptotic release of histones from nucleosomes. J Biol Chem April 5, 2002;277(14):12001–8.

[70] Nur EKA, Gross SR, Pan Z, Balklava Z, Ma J, Liu LF. Nuclear translocation of cytochrome c during apoptosis. J Biol Chem June 11, 2004;279(24):24911–4.

[71] Lee DY, Huang CM, Nakatsuji T, Thiboutot D, Kang SA, Monestier M, et al. Histone H4 is a major component of the antimicrobial action of human sebocytes. J Invest Dermatol October 2009;129(10):2489–96.

[72] Konishi A, Shimizu S, Hirota J, Takao T, Fan Y, Matsuoka Y, et al. Involvement of histone H1.2 in apoptosis induced by DNA double-strand breaks. Cell September 19, 2003;114(6):673–88.

[73] Gine E, Crespo M, Muntanola A, Calpe E, Baptista MJ, Villamor N, et al. Induction of histone H1.2 cytosolic release in chronic lymphocytic leukemia cells after genotoxic and non-genotoxic treatment. Haematologica January 2008;93(1):75–82.

[74] Alström L, von Euler H. Toxic action of histones and protamines from thymus. Ark Kemi Mineral och Geol A 1946:23.

[75] Hirsch JG. Bactericidal action of histone. J Exp Med December 1, 1958;108(6):925–44.

[76] Cascone A, Bruelle C, Lindholm D, Bernardi P, Eriksson O. Destabilization of the outer and inner mitochondrial membranes by core and linker histones. PloS One 2012;7(4):e35357.

[77] Gilthorpe JD, Oozeer F, Nash J, Calvo M, Bennett DL, Lumsden A, et al. Extracellular histone H1 is neurotoxic and drives a pro-inflammatory response in microglia. F1000Res 2013;2:148.

[78] Huang H, Evankovich J, Yan W, Nace G, Zhang L, Ross M, et al. Endogenous histones function as alarmins in sterile inflammatory liver injury through toll-like receptor 9 in mice. Hepatology September 2, 2011;54(3):999–1008.

[79] Allam R, Kumar SV, Darisipudi MN, Anders HJ. Extracellular histones in tissue injury and inflammation. J Mol Med May 2014;92(5):465–72.

[80] Fuchs TA, Bhandari AA, Wagner DD. Histones induce rapid and profound thrombocytopenia in mice. Blood September 29, 2011;118(13):3708–14.

[81] Liu ZG, Ni SY, Chen GM, Cai J, Guo ZH, Chang P, et al. Histones-mediated lymphocyte apoptosis during sepsis is dependent on p38 phosphorylation and mitochondrial permeability transition. PloS One 2013;8(10):e77131.

[82] Lam FW, Cruz MA, Leung HC, Parikh KS, Smith CW, Rumbaut RE. Histone induced platelet aggregation is inhibited by normal albumin. Thromb Res July 2013;132(1):69–76.

[83] Xu J, Ji Y, Zhang X, Drake M, Esmon CT. Endogenous activated protein C signaling is critical to protection of mice from lipopolysaccharide-induced septic shock. J Thromb Haemost May 2009;7(5):851–6.

[84] Ma Y, Vilanova D, Atalar K, Delfour O, Edgeworth J, Ostermann M, et al. Genome-wide sequencing of cellular microRNAs identifies a combinatorial expression signature diagnostic of sepsis. PloS One 2013;8(10):e75918.

[85] Zeerleder S, Zwart B, Wuillemin WA, Aarden LA, Groeneveld AB, Caliezi C, et al. Elevated nucleosome levels in systemic inflammation and sepsis. Crit Care Med July 2003;31(7):1947–51.
[86] Zeerleder S, Stephan F, Emonts M, de Kleijn ED, Esmon CT, Varadi K, et al. Circulating nucleosomes and severity of illness in children suffering from meningococcal sepsis treated with protein C. Crit Care Med December 2012;40(12):3224–9.
[87] Taylor Jr FB, Chang A, Esmon CT, D'Angelo A, Vigano-D'Angelo S, Blick KE. Protein C prevents the coagulopathic and lethal effects of *Escherichia coli* infusion in the baboon. J Clin Invest March 1987;79(3):918–25.
[88] Chen Q, Ye L, Jin Y, Zhang N, Lou T, Qiu Z, et al. Circulating nucleosomes as a predictor of sepsis and organ dysfunction in critically ill patients. Int J Infect Dis 2012;16(7):e558-e64.
[89] Nakahara M, Ito T, Kawahara K, Yamamoto M, Nagasato T, Shrestha B, et al. Recombinant thrombomodulin protects mice against histone-induced lethal thromboembolism. PloS One 2013;8(9):e75961.
[90] Holdenrieder S, Stieber P, Bodenmuller H, Fertig G, Furst H, Schmeller N, et al. Nucleosomes in serum as a marker for cell death. Clin Chem Lab Med July 2001;39(7):596–605.
[91] Kuroi K, Tanaka C, Toi M. Clinical significance of plasma nucleosome levels in cancer patients. Int J Oncol July 2001;19(1):143–8.
[92] Gibot S, Kolopp-Sarda MN, Bene MC, Cravoisy A, Levy B, Faure GC, et al. Plasma level of a triggering receptor expressed on myeloid cells-1: its diagnostic accuracy in patients with suspected sepsis. Ann Intern Med July 6, 2004;141(1):9–15.
[93] Xu H, Ye X, Steinberg H, Liu SF. Selective blockade of endothelial NF-kappaB pathway differentially affects systemic inflammation and multiple organ dysfunction and injury in septic mice. J Pathol March 2010;220(4):490–8.
[94] Wildhagen KC, Garcia de Frutos P, Reutelingsperger CP, Schrijver R, Areste C, Ortega-Gomez A, et al. Nonanticoagulant heparin prevents histone-mediated cytotoxicity in vitro and improves survival in sepsis. Blood February 13, 2014;123(7):1098–101.
[95] Bernard GR, Vincent JL, Laterre PF, LaRosa SP, Dhainaut JF, Lopez-Rodriguez A, et al. Efficacy and safety of recombinant human activated protein C for severe sepsis. N Engl J Med March 8, 2001;344(10):699–709.
[96] Russell JA. Management of sepsis. N Engl J Med October 19, 2006;355(16):1699–713.
[97] Kerschen EJ, Fernandez JA, Cooley BC, Yang XV, Sood R, Mosnier LO, et al. Endotoxemia and sepsis mortality reduction by non-anticoagulant activated protein C. J Exp Med October 1, 2007;204(10):2439–48.
[98] Mosnier LO, Zlokovic BV, Griffin JH. The cytoprotective protein C pathway. Blood April 15, 2007;109(8):3161–72.
[99] Macias WL, Vallet B, Bernard GR, Vincent JL, Laterre PF, Nelson DR, et al. Sources of variability on the estimate of treatment effect in the PROWESS trial: implications for the design and conduct of future studies in severe sepsis. Crit Care Med December 2004;32(12):2385–91.
[100] Marti-Carvajal AJ, Sola I, Agreda-Perez LH. Treatment for avascular necrosis of bone in people with sickle cell disease. Cochrane Database Syst Rev 2014;7:CD004344.
[101] Ranieri VM, Thompson BT, Barie PS, Dhainaut JF, Douglas IS, Finfer S, et al. Drotrecogin alfa (activated) in adults with septic shock. N Engl J Med May 31, 2012;366(22):2055–64.
[102] Zhang D, He J, Shen M, Wang R. CD16 inhibition increases host survival in a murine model of severe sepsis. J Surg Res April 2014;187(2):605–9.
[103] Khan R, Kirschenbaum LA, Larow C, Astiz ME. The effect of resuscitation fluids on neutrophil-endothelial cell interactions in septic shock. Shock November 2011;36(5):440–4.
[104] Ciarlo E, Savva A, Roger T. Epigenetics in sepsis: targeting histone deacetylases. Int J Antimicrob Agents June 2013;42(Suppl.):S8–12.
[105] Roger T, Lugrin J, Le Roy D, Goy G, Mombelli M, Koessler T, et al. Histone deacetylase inhibitors impair innate immune responses to toll-like receptor agonists and to infection. Blood January 27, 2011;117(4):1205–17.
[106] Li Y, Liu B, Zhao H, Sailhamer EA, Fukudome EY, Zhang X, et al. Protective effect of suberoylanilide hydroxamic acid against LPS-induced septic shock in rodents. Shock November 2009;32(5):517–23.
[107] Zhang L, Jin S, Wang C, Jiang R, Wan J. Histone deacetylase inhibitors attenuate acute lung injury during cecal ligation and puncture-induced polymicrobial sepsis. World J Surg July 2010;34(7):1676–83.

CHAPTER

26

Chromatin Landscape and Epigenetic Signatures in Neurological Disorders: Emerging Perspectives for Biological and Clinical Research

Pamela Milani, Ernest Fraenkel

Department of Biological Engineering, Massachusetts Institute of Technology, Cambridge, MA, USA

OUTLINE

Epigenetic Biomarkers and Diagnostics
http://dx.doi.org/10.1016/B978-0-12-801899-6.00026-7

1. INTRODUCTION

Epigenetic marks have become a major focus of investigation in the postgenomic era. These modifications, which include nucleosome positioning, posttranslational histone modifications, DNA methylation, and noncoding RNAs, cause changes in chromatin structure and influence the access of the transcriptional complex to the promoter of specific genes. Growing evidence suggests that epigenetic events are important regulatory factors in brain functions, exerting a vital role in gene expression regulation in the mammalian central nervous system (CNS) during development and in adulthood [1]. Moreover, it is known that environmental stressors exert prominent roles in shaping epigenetic patterns and that the epigenetic make up of an individual significantly affects the aging process of the CNS, thus potentially influencing the onset and progression of several neurodegenerative disorders. Therefore, a comprehensive understanding of the molecular mechanisms governing epigenetic events may shed new light on the basis of adult-onset neurodegeneration, in which the interplay between environmental factors and genetic susceptibility leads to the pathological state. Furthermore, epigenetic modifications might serve as potential biological markers for the identification and prognosis of neurological disorders and for the development and optimization of effective mechanism-based therapeutic options, since they could provide a valid instrument for measuring disease progression at molecular level and monitoring drug efficacy during clinical trials.

In this chapter, we provide a detailed description of the contribution of epigenetic events to the onset and progression of two neurodegenerative disorders, Huntington's disease (HD) and amyotrophic lateral sclerosis (ALS). We also summarize recent progress in the development of safe and effective therapeutic strategies based on the modulation of epigenetic components.

2. THE INVOLVEMENT OF EPIGENETIC MODIFICATIONS IN HD

2.1 Introduction to HD

HD (MIM #143100; ORPHA399) is a monogenic, autosomal-dominant neurodegenerative disorder that most often strikes individuals in the prime of life, with onset typically between the ages of 35 and 50 years. The disease relentlessly advances through progressive deterioration, lasting years before patients succumb to complications such as heart and respiratory failure and aspiration pneumonia [2]. HD is caused by a dynamic cytosine–adenine–guanine (CAG) triplet repeat expansion mutation in exon 1 of the *HTT* gene, located on chromosome 4p16.3, resulting in expanded polyglutamine stretch in the aminoterminal region of the corresponding protein product, huntingtin (Htt). In unaffected individuals, this gene contains fewer than 35 repeats, while incomplete penetrance and late onset is observed with 36–39 repeats. When the expansion reaches or exceeds 40 CAG repeats, the penetrance becomes complete and triplet length positively correlates with disease severity and inversely correlates with age of onset. The expanded trinucleotide repeat is unstable and continues to dynamically extend through successive generations, providing a molecular explanation for the genetic feature of anticipation, a phenomenon whereby symptoms manifest at earlier age in the next generation.

The disease is characterized clinically by cognitive decline, psychiatric symptoms, and prominent motor alterations. Cognitive deficits include executive dysfunction, episodic and procedural memory deficits, learning difficulties, and deterioration of reasoning and verbal fluency. Neuropsychiatric manifestations comprise a broad spectrum of behavioral signs, such as anxiety, depressed mood, social withdrawal, obsessions and compulsions, addictions, changes in personality, and psychosis. Motor symptoms classically present as involuntary choreic movements, and muscle rigidity, dystonia, and postural abnormalities may manifest during disease progression. The worldwide prevalence of HD is approximately 3 cases per 100,000 [3]. No effective therapy halts the progression of the pathology and existing treatments provide only symptom relief [4]. Although mutant Htt is ubiquitously expressed, selective neuronal cell types degenerate: overt tissue-specific degeneration is observed in HD patient brain, particularly of medium spiny neurons in the striatum, with atrophy of the cortex. The pathophysiology of HD has not been completely elucidated. Nevertheless, several pathological hallmarks have been described in HD, including impaired energy metabolism, oxidative stress, excitotoxicity, formation of toxic aggregates, and transcriptional dysregulation, which eventually lead to neuronal cell death [5].

Transcriptional dysregulation is one of the molecular phenotypes strongly implicated in HD: widespread alteration in gene regulatory network occurs in cortex and striatum during disease progression in both laboratory models and human disease and it is an early event in the pathological process of HD [6]. Repression of key neuronal transcripts is consistently observed and implicated in the disease pathogenesis. Genes repressed in HD mouse models and human brain tissues include neurotransmitter receptors, cytoskeletal proteins and intracellular signaling molecules. Particularly, aberrant expression has been reported for the dopamine receptor 2, preproenkephalin, the cannabinoid receptors, and brain-derived neurotrophic factor (BDNF), which is expressed in the cortex and provides trophic support to GABAergic medium spiny neurons [7]. Of these genes, BDNF has particular relevance for the pathology, as demonstrated by several studies consistently reporting downregulation of *BDNF* mRNA level in HD, which severely affects the corticostriatal circuitry and causes the degeneration of target striatal neuronal cells [7].

Widespread alteration of gene expression events detected in HD can be ascribed to several interdependent molecular mechanisms involving changes in transcriptional networks and variations at the epigenetic level. Both wild-type Htt and the mutant protein, which harbors an elongated polyglutamine stretch (expanded Htt), were shown to interact with numerous components of the transcriptional machinery, some of which showed increased affinity for the mutant form. For example, the global transcriptional regulator TATA-binding protein/TFIID and TAFII130, a coactivator of cAMP-responsive element-binding protein (CREB)-dependent transcriptional activation, preferentially interact with the expanded polyglutamine stretch [8]. Similarly, p53 and Sp1 abnormally interact with mutant Htt [9–11]. Transcriptional dysregulation in HD is also a consequence of mutation-dependent loss of physiological protein–protein interactions. Particularly, R element-1 silencing transcription factor (REST) is sequestered in the cytoplasmic compartment by wild-type Htt; by contrast, in the presence of mutant Htt, it is free to translocate to the nucleus where it represses the expression of several target genes, including BDNF [12]. Besides modifications in the activity of specific transcriptional factors, a mounting body of research suggests that dysregulation of epigenetic events represents one of the main pathognomonic molecular features of HD and a key process by which mutant Htt alters the patterns of gene expression.

2.2 Chromatin Remodeling and Histone Modifications in HD

Chromatin remodeling is a fine-tuned mechanism which involves dynamic, often synergistic interactions between DNA, enzymes, RNA, and nucleosomes in the nuclear compartment. It is controlled by the interplay between reversible posttranslational modifications (PTMs) of core histone proteins (H2A, H2B, H3, and H4), DNA methylation, and the activity of a vast array of remodeling enzymes, which modulate the plasticity of chromatin structures and lead to changes in nucleosome condensation and DNA accessibility [13,14]. Covalent modifications mostly occur on specific amino acids located in the basic N-terminal tails of histones and include methylation, acetylation, phosphorylation sumoylation, ADP-ribosylation, and ubiquitylation. PTMs act individually or in combination to define a "code" that is interpreted by an array of multiprotein enzymatic complexes which control downstream activities of chromatin through changes of DNA packaging. Indeed, the formation of higher-order chromatin structures affects the binding of transcription factors to enhancers and promoter regions of genes, with associated variations in gene expression [15]. The chromatin landscape is modulated by the activities of various enzymes, comprising histone acetyltransferases (HATs), histone deacetylases (HDACs), histone methyltransferases (HMTs), and histone demethylases (HDMs) [16].

Aberrant regulation of chromatin-modifying enzymes and widespread changes in histone modification patterns have been frequently reported in HD (as reviewed in Ref. [6] and shown in Figure 1). For example, chromatin immunoprecipitation (ChIP) experiments performed

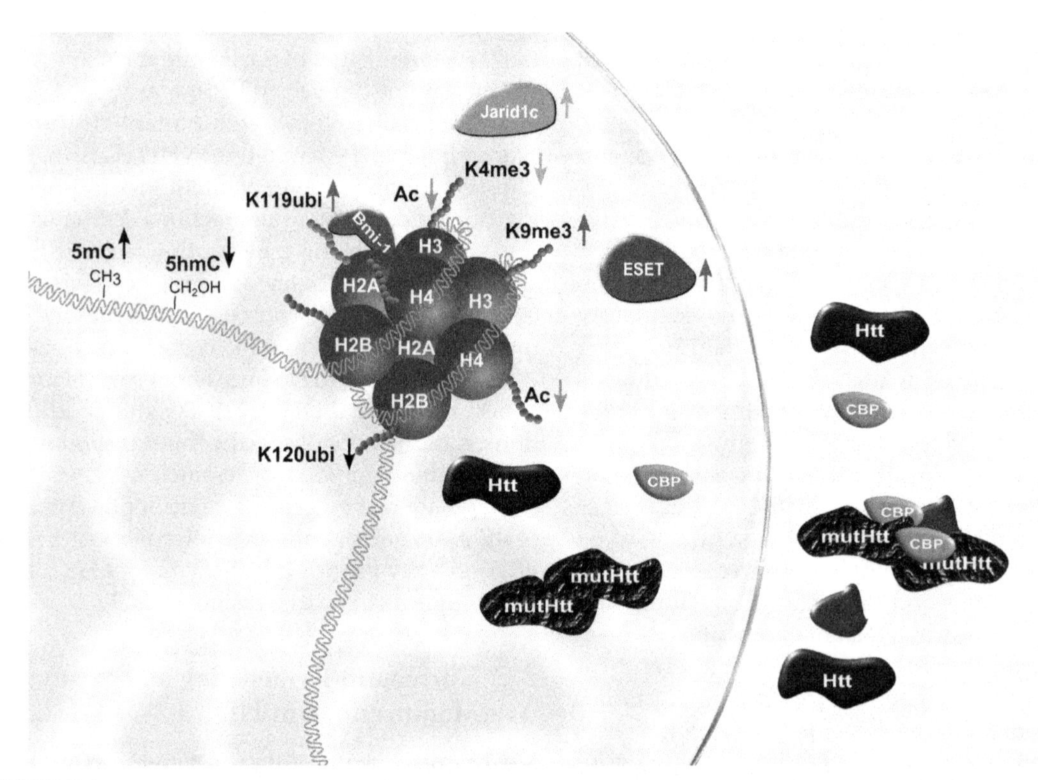

FIGURE 1 Schematic of epigenetic modifications detected in Huntington's disease (HD). In the presence of expanded Htt (mutHtt) CREB-binding protein (CBP) is sequestered in cytoplasmic aggregates with consequent hypoacetylation of histones H3 and H4 in the promoter region of specific genes; ERG-associated protein with SET domain (ESET) is upregulated, which triggers hypermethylation of H3K9 (H3K9me3); levels of histone demethylases such as Jarid1c are increased, which leads to decreased level of H3K4me3; increased Bmi-1 binding to histone H2A enhances monoubiquitylation of H2AK119; by contrast, monoubiquitylation of H2BK120 is reduced; increased levels of 5-methylcytosine (5mC) and decreased levels of 5-hydroxymethylcytosine (5hmC) reduce the expression of specific genes.

on brain tissues from R6/2 mice, a transgenic mouse model containing exon 1 of *HTT* with an expanded CAG stretch under the control of human *HTT* promoter, showed that histone H3 was hypoacetylated at promoters of downregulated genes [17], while global acetylated histone H3 levels were unchanged in transgenic mice compared to wild-type animals. A possible explanation accounting for decreased histone acetylation at promoters of repressed genes is that mutant Htt directly interferes with the activity of HATs. In this regard, mutant Htt was shown to interact with and inhibit the activity of CREB-binding protein (CBP) [11,18–22]. CBP is a ubiquitous nuclear protein that not only acts as a transcriptional coactivator able to recruit components of the basal transcriptional machinery but also possesses an intrinsic HAT activity, which enhances gene expression by reshaping chromatin topology [23,24]. Nucifora and coworkers [18] demonstrated that CBP specifically interacted with the mutant form of Htt and coaggregated with the polyglutamine stretch in cytoplasmatic insoluble inclusions in an in vitro cellular model. Notably, CBP was also detected in cytoplasmic aggregates in HD transgenic mice and patients' postmortem brain tissues. Furthermore, mutant Htt can affect CBP functions via additional mechanisms: Jiang and colleagues [19] demonstrated that, in the presence of expanded Htt, CBP expression was reduced due to enhanced ubiquitin-dependent proteasomal degradation. This finding was confirmed by Cong and coworkers [20] in neuronal pheochromocytoma rat PC12 cells, stably inducible for htt exon 1 with either normal or expanded glutamines. CBP depletion was accompanied by cellular toxicity as reported in both immortalized mouse hippocampal cell line (HT22) and in mouse primary cortical neurons expressing mutant Htt [21]. Importantly, the decreased level of CBP was associated with hypoacetylation of histones H3 and H4, while CBP overexpression was demonstrated to rescue physiological acetylated histone levels and improve cell viability.

Interestingly, CBP was also shown to exert a pivotal role in memory consolidation processes: diminished expression of CBP and decreased levels of histone H3 acetylation in the hippocampus accounted for long-term impairment of spatial and recognition memory in heterozygous HD knock-in mutant mice (Hdh(Q7/Q111)) [22]. Administration of trichostatin A (TSA), an inhibitor of HDAC (HDACi), rescued cognitive deficits in this HD mouse model.

Expanded Htt can also cause transcriptional dysregulation by affecting the activity of HMTs: altered expression levels of these enzymes have been reported in both animal mo5els of HD and tissues from patients [25–26]. Ryu and colleagues [26] showed that the levels of ERG-associated protein with SET domain (ESET), a histone H3K9-specific methyltransferase, were significantly upregulated in the striatum and neocortex from HD patients compared to age- and postmortem interval-matched control patients. Similar results were obtained in the brain from R6/2 mice where hypermethylation of H3K9 (H3K9me3), a mark of transcriptional repression, was also detected. H3K9me3-dependent heterochromatin condensation was demonstrated to directly affect transcript levels in HD models: genes highly enriched in H3K9me3 in their promoters presented decreased mRNA levels in HD knock-in striatal cell line expressing expanded Htt (ST[26]*Hdh*Q111/111) compared to cells expressing wild-type Htt (ST*Hdh*Q7/7) [25]. These genes were highly enriched in the functional categories of cell motility, neuronal differentiation, and synaptic transmission. Particularly, downregulation of levels of cholinergic receptor M1 (CHRM1) was shown to affect Ca^{2+}-dependent signal transduction, leading to neuronal dysfunction.

Besides changes in HMTs, aberrant expression of HDMs can also account for transcriptional dysregulation involved in HD pathogenesis: a recent paper published by Vashishtha and colleagues [27] reported increased expression of five HDMs, including Jarid1c, Jarid1b, Phf2, Mina, and Hspbap1, in brain tissues from R6/2 mice. Notably, decreased levels of trimethylation at lysine 4 of histone H3 (H3K4me3), a mark of transcription start sites (TSSs) and active chromatin [28–30], were observed at the promoters of specific downregulated genes, such as *Bdnf*, in both mouse model and human postmortem brain tissues.

Additionally, the authors were able to describe a unique pattern of the chromatin mark H3K4me3 at transcriptionally repressed promoters: by employing ChIP coupled with deep sequencing (ChIP-seq) and by performing global transcriptome analysis in brain tissues from R6/2 mice and wild-type littermates, they found that genes showing lower expression levels in mutant mice exhibited a characteristic peak of histone H3K4me3 downstream of the TSS in normal animals. In mutant animals, this peak was observed to be reduced in size as disease progressed. Genes in which this peak diminished in HD animals also had lower expression. These findings provided an important mechanistic clue to understanding transcriptional dysregulation in HD by revealing the vulnerability of promoters with a specific "signature" of H3K4me3 occupancy to transcriptional shutdown in HD. This observation could provide a path to the development of potential therapeutic options: as the author showed, Jarid1c knockdown in primary cortical neurons reversed the downregulation of key neuronal genes caused by mutant Htt. Moreover, the partial loss of the single Jarid1 homolog (*little imaginal disks—lid*) in a *Drosophila* model of HD was protective. Specifically, the number of mutant Htt flies surviving to adulthood was higher when they also had reduced *lid* expression due to a heterozygous loss-of-function mutation.

Mutant Htt was also demonstrated to alter the epigenome by affecting the monoubiquitylation of lysine residues of histone H2A and H2B [31]. Kim and coworkers demonstrated that monoubiquitylation levels at lysine 119 of histone H2A (H2AK119ubi) were upregulated in brain tissues from transgenic R6/2 mice compared to wild-type littermates. This was caused by increased binding of Bmi-1 protein, a component of polycomb-repressive complex 1-like (PRC1L) with E3-ubiquitin ligase activity, to the nucleosomes in the presence of expanded Htt. Several studies linked H2A monoubiquitylation to transcriptional repression [32,33], and in agreement with these findings, increased H2AK119ubi levels were detected at promoters of downregulated genes in both transgenic mice and cellular models of HD [31]. Small interfering RNA inhibition of Ring2, another component of PCR1L, reduced H2A monoubiquitylation and rescued transcriptional repression. Surprisingly, Ring2-inhibition also decreased methylation at H3K9, which suggests that cross talks between different histone modifications could modulate chromatin domain structures and affect gene expression outputs. By contrast to H2AK119ubi, monoubiquitylation at lysine 120 of histone H2B (H2BK120ubi), a mark preferentially located at transcriptionally active chromatin sites [34], was overall decreased in the cerebellum of R6/2 mice and its association with promoters of genes repressed in HD was reduced in cellular models of the pathology [31].

Figure 1 provides a summary of histone modifications occurring in HD.

2.3 DNA Methylation in HD

DNA methylation is the covalent addition of a methyl group to the C5 position of the pyrimidine ring of cytosine, which generates 5-methylcytosine (5mC). This reaction is catalyzed by DNA methyltransferase (DNMT) enzymes and mostly occurs at CpG dinucleotides. However, recent observations demonstrate that methylation in non-CpG context is present in human somatic tissues, particularly in brain [35]. Together with histone PTMs, this epigenetic modification exerts a key function in regulating chromatin packaging and modulating gene expression by recruiting several chromatin-remodeling complexes which usually lead to chromatin condensation and transcriptional silencing [36].

Changes in DNA methylation patterns have recently been detected in HD [37–39]. Ng and coworkers performed reduced representation bisulfite sequencing (RRBS), a technique used to analyze DNA methylation profiles genome-wide, and found large changes in DNA methylation in the presence of mutant Htt in a knock-in striatal cell culture model of HD [37]. Regions with low CpG content, which were shown to undergo methylation changes in response to neuronal activity, were disproportionately affected. By performing sequence motif analysis, the authors identified AP-1 and SOX2 as transcriptional regulators associated with DNA methylation changes, and confirmed these findings by ChIP-Seq. These results highlighted new mechanisms for the effects of expanded Htt. Furthermore, considering that there is increasing evidence of DNA methylation involvement in neuronal activity, memory, and learning [40,41], these findings raised important questions about the potential effects of changes in DNA methylation on neurogenesis and cognitive decline in patients with HD.

Another recent paper demonstrated that decreased expression of adenosine A2A receptor (A2AR) correlated with increased levels of 5mC in the 5′-untranslated region (UTR) of its gene in the striatum of HD transgenic mice and human patients [38]. By contrast, the levels of 5-hydroxymethylcytosine (5hmC), an oxidation product of 5mC, were decreased. 5hmC has been identified in mammalian genome in all tissues, but its content is particularly elevated in the brain, where it is enriched in the gene bodies of actively transcribed genes and distal regulatory elements [42]. Consistent with this result, another paper described decreased levels of the 5hmC signal in the striatum and cortex of transgenic HD mice when compared to age-matched wild-type littermates; pathway analysis of differentially hydroxymethylated sites showed that the impairment of pathways involved in neuronal differentiation and survival could be important for HD pathogenesis [43].

2.4 Modulating Epigenetic Features as Therapeutic Approach for HD

The findings described above regarding alterations in specific patterns of histone marks and DNA methylation provide very strong new support for the concept that therapeutic intervention directed toward modulating the epigenetic machinery of the cell can be beneficial in treating HD. Consistent with the observation of decreased levels of histone acetylation in HD, several data showed that HDACis, such as suberoylanilide hydroxamic acid (SAHA), sodium butyrate (NaB), and sodium phenylbutyrate (NaPB), HDACi 4b, and selisistat, ameliorated behavioral performance and decreased neuronal loss in numerous models of the disease [44–54].

Specifically, SAHA, also known as vorinostat, and NaB were shown to inhibit neurodegeneration in *Drosophila* models expressing expanded *htt* exon 1 [44] and to improve survival and motor impairment in R6/2 mouse model [45–47]. At the molecular level, SAHA administration increased histone H2B and H4 acetylation in the brain of R6/2 mice [46] and decreased HDAC2 and four protein levels, without affecting the expression of their respective mRNAs [47]. Prolonged SAHA treatment diminished insoluble protein aggregates and restored *Bdnf* cortical transcript levels [47].

Additionally, Gardian and coworkers showed that the HDACi NaPB extended survival and decreased brain atrophy in a transgenic mouse model expressing the first 171 amino acids of the human *HTT* gene carrying 82 trinucleotide repeats (N171-82Q) [48]. When compared to wild-type littermates, these transgenic mice presented increased levels of H3K9 methylation, which was markedly attenuated by the administration of NaPB. Increased histone H3 and H4 acetylation was also observed after drug treatment, which led to transcriptional rescue of prosurvival genes. Furthermore, a dose-ranging study was conducted to assess the safety and tolerability of NaPB in HD symptomatic subjects. The drug proved safe and well-tolerated at 12–15 g/day (clinical trial identifier: NCT00212316) [55]. Interestingly, Borovecki and coworkers identified 322 mRNAs with altered expression in blood samples from HD individuals compared to nonaffected subjects [56]. A subgroup of statistically significant upregulated mRNAs displayed concordant changes in human postmortem brain tissue from HD patients when compared to nonneurological control samples. Notably, the expression of these transcripts correlated with disease progression and their levels in blood cells were significantly reduced after treatment with NaPB. These observations suggested that epigenetic components likely exert a prominent role in determining gene expression alterations observed in tissues from HD cases and that peripheral cells may represent a valuable tool for detecting disease-specific molecular signatures and monitoring drug efficacy during clinical trials.

HDACi 4b, a new pimelic diphenylamide HDACi, was found to improve HD phenotype in R6/2 and N171-82Q mice [49–51]. Chronic oral administration of HDACi 4b improved motor performance and body weight of symptomatic mice and attenuated striatum atrophy. This was accompanied by reversal of mutant Htt-induced histone hypoacetylation and transcriptional abnormalities. However, a subsequent study failed to replicate some of these findings: no beneficial effects on body weight, survival, and motor performance were detected in both N171-82Q and R6/2 mice; striatal atrophy was decreased in 4b-treated R6/2 mice but with no significant improvement of global brain weight loss [57].

Beneficial effects were also observed with inhibitors of class III HDACs in several cellular and animal models of the disease [53,54]. Specifically, genetic and pharmacological reduction of Sir2 in a *Drosophila* model of HD suppressed pathogenetic features, such as loss of photoreceptor neurons [53]. The efficacy of pharmacological inhibition of sirtuins was also assessed in a recent paper published by Smith and colleagues [54]: selisistat (6-chloro-2,3,4,9-tetrahydro-*1H*-carbazole-1-carboxamide, SEN0014196, EX-527), a potent and selective Sir2/SirT1 inhibitor, reduced neuronal loss and extended lifespan in HD flies, decreased mutant Htt-dependent toxicity in PC12 mammalian cell cultures and postmitotic primary rat striatal neurons transfected with expanded Htt, and increased survival and locomotor performance in R6/2 mice. Notably, in a recent double-blind, randomized, placebo-controlled study, selisistat was found to be safe and well-tolerated by male and female healthy volunteers after single administration up to 600 mg and multiple doses up to 300 mg/day (clinical trial identifier: NCT01521832) [58]. Another double-blind, placebo-controlled trial proved the safety and tolerability of selisistat in individuals affected by early stage HD (clinical trial identifier: NCT01485952) [59].

It is worth noting that NaPB and selisistat have great potential to be translated into clinical practice since they were not only able to reduce pathological features in multiple animal and cellular models of the disease but also have proven safe and well-tolerated in clinical trials.

The list of epigenetic drugs relevant for the treatment of HD is reported in Table 1.

3. THE INVOLVEMENT OF EPIGENETIC MODIFICATIONS IN ALS

3.1 Introduction to ALS

ALS (ORPHA803) is the most common adult-onset motor neuron disease. It is characterized by rapidly progressive degeneration of both upper and lower motor neurons in spinal cord, brainstem, and motor cortex. Neuronal death is accompanied by the activation of glial cells and muscle weakness, wasting, and atrophy, which eventually leads to paralysis and death, generally due to respiratory insufficiency. ALS affects 1.5 to 2.7 individuals per 100,000 per year in

TABLE 1 Therapeutic Modulation of Epigenetic Mechanisms in HD

Drugs	Activities	Effects	References
SAHA	Inhibition of class I and II HDACs	Improved viability in expanded *htt* ex1 *Drosophila* models	[44]
SAHA	Inhibition of class I and II HDACs	Improved motor performance in R6/2 mouse model; increased histone acetylation in the brain	[46]
SAHA	Inhibition of class I and II HDACs	Improved motor performance in R6/2 mouse model; reduced expression of HDAC2 and 4 proteins and protein aggregates in cortex and brainstem	[47]
NaB	Inhibition of class I and II HDACs	Improved viability in *htt* ex1 expanded *Drosophila* models	[44]
NaB	Inhibition of class I and II HDACs	Increased body weight and motor performance; extended survival in R6/2 mouse model; improved oxidative phosphorylation and transcription	[45]
NaPB	Inhibition of class I and II HDACs	Extended survival and decreased brain atrophy in HD transgenic mice model (N171-82Q)	[48]
NaPB	Inhibition of class I and II HDACs	Safe and well-tolerated in human HD	[55]
HDACi 4b	Inhibition of class I HDACs	Improved motor performance, brain weight, and body weight of R6/2 mice	[49]
HDACi 4b	Inhibition of class I HDACs	Improved body weight, motor performance, and cognitive functions in N171-82Q mice; reduced aggregation of mutant htt	[50]
HDACi 4b	Inhibition of class I HDACs	Restored gene expression abnormalities in R6/2 mice; improved metabolic functions in immortalized striatal cells (ST*Hdh*Q111)	[51]
HDACi 4b	Inhibition of class I HDACs	No beneficial effects on body weight, survival, and motor performance in N171-82Q and R6/2 mice. Protection from striatal atrophy in R6/2 mice.	[57]
Trichostatin A	Inhibition of class I and II HDACs	Rescued cognitive deficits in heterozygous HD knock-in mutant mice (HdhQ7/Q111)	[22]
Selisistat	Inhibition of sirtuin 1	Neuroprotective in multiple HD cellular models; extended survival in HD flies; increased survival; and improved motor performance in R6/2 mice	[54]
Selisistat	Inhibition of sirtuin 1	Safe and well-tolerated in healthy volunteers (double-blind, randomized, placebo-controlled trial)	[58]
Selisistat	Inhibition of sirtuin 1	Safe and well-tolerated in HD patients (double-blind, randomized, placebo-controlled trial)	[59]

HD, Huntington's disease; HDACs, histone deacetylases; SAHA, suberoylanilide hydroxamic acid; NaB, sodium butyrate; NaPB, sodium phenylbutyrate.

Europe and North America [60]. Approximately 10% of ALS patients have a family history of the disease (FALS), and 75% of these familial cases have known mutations in genes with Mendelian transmission [61]. ALS-related genetic variants have been described in several genes, including *SOD1* (ALS1, MIM #105400), *TARDBP* (ALS10, MIM #612069), *FUS/TLS* (ALS6, MIM #608030), *C9ORF72* (FTDALS1, MIM #105550). The vast majority of ALS cases are sporadic (SALS), with no evident familial clustering. Nevertheless, approximately 10–15% of SALS patients also have mutations in FALS genes [61,62]. The etiology of SALS is still unknown, but is likely caused by the joint effect of several susceptibility genes and environmental risk factors. The clinical phenotypes of FALS and SALS are usually very similar and comprise limb weakness, fasciculation, hyperreflexia, cramps, spasticity, dysarthria, and dysphagia. Increasing evidence demonstrates that ALS is a multifactorial disorder characterized by the involvement of several, nonmutually exclusive pathogenic events including glutamate-dependent excitotoxicity, protein misfolding and aggregation, oxidative and inflammatory processes, mitochondrial damage and bioenergetics defects, cytoskeletal dysfunction and impaired axonal transport, changes in gene expression, and epigenetic abnormalities (as reviewed in Ref. [63]). In particular, both formation of insoluble inclusion bodies (IBs) and changes in RNA metabolism have been largely implicated in ALS [64–66]. A landmark breakthrough in ALS research was the discovery of mutation in the genes encoding for the DNA/RNA-binding proteins, TDP-43 and FUS. Since then many groups have focused their investigation on understanding RNA abnormalities associated to ALS pathogenesis and progression.

Mutations in TDP-43 and FUS cause their misfolding, mislocalization, and aggregation in the cytoplasm, which is accompanied by widespread changes in RNA metabolism. Interestingly, there are several lines of evidence that impaired RNA-processing events and toxic protein aggregates could represent general mechanisms in ALS. Furthermore, the discovery of hexanucleotide GGGGCC expansions in intron 1 and promoter region of the *C9ORF72* (*chromosome 9 open reading frame 72)* gene have recently been shown to be the most common cause of FALS and a frequent cause of SALS and frontotemporal dementia (FTD), a disease often described in comorbidity with ALS [67–69]. The mutated form of the *C9ORF72* transcript was shown to cause the formation of RNA foci in the nucleus, hinting to a potential toxic RNA gain-of-function mechanism [67]. Indeed, *C9ORF72*-positive foci can recruit RNA-binding proteins and change the cellular homeostatic protein balance [70]. It has been also suggested that the expansion may lead to decreased expression of C9ORF72 transcript and protein (haploinsufficiency) [67]. Notably, reduced expression of expanded C9ORF72 variants has been reported in motor cortex, frontal cortex, spinal cord, lymphoblast cell lines, and induced pluripotent stem cells from patients affected by ALS and FTD [71–75]. The physiological role of C9ORF72 protein is still unknown; however, recent evidence suggested that it may be involved in the regulation of endosomal trafficking and autophagy [76]. Furthermore, expanded *C9ORF72* transcripts could be processed by the non-ATG translation (RAN translation) mechanism and originate very short potentially toxic peptides [77].

Besides widespread changes in RNA metabolism and protein homeostasis, recent findings have demonstrated that epigenetic events may also play a key role in mediating the clinical manifestation of ALS, as described comprehensively in the next section.

3.2 Epigenetic Modifications in ALS

Some of the first evidence about epigenetic abnormalities in ALS was provided by Morahan and coworkers [78], who performed methylated DNA immunoprecipitation (MeDIP) in combination with tiling arrays in postmortem brain tissues. They identified significant changes in DNA methylation levels in the promoters of genes involved in calcium homeostasis, excitotoxicity, and oxidative stress in SALS brain samples compared to neurologically normal patients. More recently, it has been shown that alterations in DNMT levels mediate apoptosis in motoneuron-like hybridoma cell line NSC-34 [79]. Specifically, the overexpression of DNMT3a was found to exert cytopathic effects in this cellular model, as assessed by caspase-3 enzyme activity assay, while NSC-34 transfection with mutant catalytically inactive DNMT3a did not affect cell viability. DNMT3a was also shown to accumulate in spinal cord synapses and in mitochondria in vivo, with consequent changes in mtDNA methylation. This likely results in metabolic deficits, synaptic die-back and, eventually, muscle denervation. Furthermore, the authors detected increased expression of DNMT1 and DNMT3a in motor cortex and spinal cord motor neurons from ALS patients compared to control subjects by immunohistochemistry, and 5mC immunoreactivity was increased in motor cortical pyramidal neurons in SALS. Global changes in DNA methylation patterns have also been described by Figueroa-Romero and colleagues [80]: enzyme-linked immunosorbent assays (ELISAs) showed increased levels of global 5mC in SALS human spinal cord samples when compared to nonneurological control patients. High-throughput microarray-based profiling of both DNA methylation and gene expression detected that differentially methylated sites (hyper- or hypo-5mC) correlated with altered levels of 112 transcripts (down- and upregulation, respectively). These were significantly enriched in disease-relevant biological categories such as immune and defense response and neuron adhesion. Global DNA methylation was increased also in whole-blood from 96 ALS patients when compared to 87 healthy subjects [81].

A recent interesting study identified hypermethylation at the CpG islands located in the 5′ region of expanded *C9ORF72* gene in blood, brain, and spinal cord from ALS patients [82]. Notably, increased methylation was associated with decreased expression of the corresponding transcript and significantly correlated with an aggressive phenotype, consisting in a more rapid disease progression. This evidence provided compelling support to the hypothesis that the mutation can exert its deleterious effect through a loss-of-function mechanism which leads to downregulation of gene expression. Similar results have been reported in tissues from FTD patients harboring the expanded allele [83,84].

Decreased C9ORF72 expression in mutated ALS/FTD cases has also been ascribed to modifications in histone marks: ChIP experiments performed in brain tissues from 28 individuals showed increased association of trimethylation of histone H3 at lysines 9, 27, 79, and of histone H4 at lysine 20 to the regulatory region of the *C9ORF72* gene in expanded repeat carriers [73]. These modifications, which are associated with transcriptional repression [85], likely cause *C9ORF72* gene silencing. Consistent with this hypothesis, treatment of fibroblast cell lines derived from *C9ORF72*-mutated ALS/FTD patients with 5-aza-2-deoxycytidine, an epigenetic modifier acting as a demethylating agent [86], decreased histone trimethylation and rescued *C9ORF72* mRNA expression [73]. Notably, ChIP experiments coupled with quantitative PCR also detected hypermethylathion of H3K9 and H3K27 at the regulatory region of *C9ORF72* gene in peripheral blood mononuclear cells from mutated patients when compared to cells from subjects carrying normal *C9ORF72* alleles. This result suggests that peripheral cells may be useful in identifying and testing hypothesis on ALS pathomechanisms and might help the search for biomarkers for an early diagnosis or prognosis. Indeed, the validation of this finding in samples from a larger patient cohort may establish these epigenetic signatures as potential blood-detectable disease classifiers able to translate the experimental readouts in relevant outcomes for effective drug development programs and improved clinical practice.

Impairment in epigenetic mechanisms has also been associated with mutations in another ALS/FTD-related gene, *FUS/TLS*. FUS is a ubiquitously expressed heterogeneous nuclear ribonucleoprotein (hnRNP) with a prion-like domain; it acts as a multifunctional protein involved in transcription, RNA splicing and transport, mRNA/microRNA metabolism regulation, and stress granules formation [87]. Although it is predominantly localized in the nuclear compartment, some of its functions require nucleocytoplasmic shuttling, which is controlled by fine-tuned PTM events mediated by protein arginine methyltransferase 1 (PRMT1) [88]. *FUS/TLS* variants account for ~3–5% of FALS cases and, when mutated, the protein is aberrantly localized in the cytoplasm of motor neurons and glial cells, where it associates with components of stress granules and forms IBs [89,90]. Recent studies demonstrated that FUS exerts an important role in DNA damage response (DDR), which requires FUS direct interaction with histone deacetylase 1 (HDAC1) and involves its recruitment to DNA double-stranded break sites to preserve genomic stability and neuronal viability [91]. The authors showed that FALS-causing mutations impaired FUS interaction with HDAC1 and, consequently, DNA repair processes. Notably, increased markers of DNA damage were observed in patients carrying *FUS/TLS* mutations. A possible explanation for the impairment of DNA repair mechanisms is that the decreased binding of FUS to HDAC1 leads to alterations in histone acetylation patterns; this can affect the dynamics of DNA-damage-induced remodeling events which reshape chromatin topology around DNA lesions.

Another recent study linked FUS mutations to aberrant histone modifications: FUS depletion from the nuclear compartment caused a concomitant redistribution of PRMT1 in the cytoplasm of motor neurons of primary murine spinal cord cultures expressing either wild-type or mutant human FUS [92]. As a consequence of PRMT1 subcellular mislocalization, reduced methylation of its nuclear targets, including arginine 3 of histone 4 (H4R3), was detected in motor neurons with predominantly cytoplasmatic FUS. Decreased methylation of H4R3 has been associated with the formation of repressive heterochromatin through a mechanism involving a cross talk between histone modifications [93]. Consistent with this finding, Tibshirani and coworkers observed that in neurons with depletion of nuclear FUS, hypomethylation of H4R3 was accompanied by reduced acetylation at lysines 9 and 14 of histone 3 (H3K9/K14) [92]. Similar changes in histone PTMs occurred when motor neurons were treated with AMI-1, a cell-permeable inhibitor of PRMTs. H3K9/K14 acetylation was rescued by SAHA treatment in spinal cord cultures expressing mutant FUS. Furthermore, cells with predominantly cytoplasmatic FUS showed decreased transcriptional activities, as assessed by 5-bromouridine (BrU) incorporation assay, a technique which allows the detection of nascent RNAs. Treatment with AMI-1 repressed gene expression even in cells with nuclear FUS, demonstrating that PRMT1 activity is required for proper transcriptional regulation. It is particularly noteworthy that aberrant cytoplasmic accumulation of FUS has also been described in samples from nonmutated ALS cases [94]: this suggests that FUS mislocalization could be a mechanism potentially shared by FALS and SALS. However, it still remains to be determined whether FUS/PRMT1-related modification in histone marks and transcription can associate more generally with ALS pathomechanisms or whether this is a molecular event restricted to specific ALS-related genotypes.

3.3 Modulating Epigenetic Features as a Potential Therapeutic Approach for ALS

The growing number of evidence regarding the involvement of epigenetic mechanisms in ALS points to new druggable targets, with the potential to open up innovative avenues for the development of meaningful therapies. Considering that increased levels of DNMT and 5mC have been observed in ALS cellular and animal models as well as in biological specimens from patients [79–81], it seems plausible that compounds directed toward modulating DNA methylation can be beneficial in treating the disease. In this regard, promising candidates for ALS treatment are N-phthalyl-L-tryptophan (RG108) and procainamide. RG108 is nonnucleoside catalytic DNMT inhibitor which significantly decreases the methylation of genomic DNA without detectable toxicity [95]. This characteristic differentiates it from nucleoside-based inhibitors, such as 5-azacytidine, which showed dose-limiting toxicity and potential adverse side effects, including carcinogenesis. Procainamide is a drug employed for the treatment of cardiac arrhythmias and hypertension which was also found to act as a DNA methylation inhibitor [96]. Remarkably, both RG108 and procainamide were shown to prevent the increase in 5mC and apoptosis levels induced by sciatic nerve avulsion in motor neurons of adult mouse spinal cord [79]; this suggests that they might be effective in reducing motor neuronal degeneration in ALS.

Besides DNA methylation inhibitors, preclinical studies demonstrated that also HDACis, particularly NaPB [97], TSA [98], and valproic acid (VPA) [99], can be valuable epigenetic-based therapeutic options for ALS. Ryu and coworkers [97] showed that administration of NaPB prolonged the lifespan in a dose-dependent manner and ameliorated both clinical and neuropathological features of transgenic mice expressing a mutated form of SOD1 protein (SOD1 G93A). At the molecular level, this pharmacological treatment improved H2A, H2B, H3, and H4 hypoacetylation observed in mutant mice compared to wild-type littermates. Moreover, it increased the expression of NF-κB p50 and bcl-2, and reduced cytochrome c release, caspase activation, and SOD1-positive insoluble inclusions in spinal cord ventral horn neurons from SOD1 G93A mice. A phase 2 clinical study was undertaken to assess the tolerability, pharmacodynamics, and biological effects of escalating doses of NaPB in patients with ALS (clinical trial identifier: NCT00107770) [100]. NaPB proved to be safe in ALS participants and tolerability was comparable to that observed in a similar study conducted in HD patients [55]. Importantly, histone hypoacetylation detected in blood cells from ALS participants at the screening visit was improved by NaPB treatment. The largest increase in histone acetylation was detected at 9 g/day, 2 weeks after the beginning of the clinical trial. These results obtained with human subjects make NaPB a promising candidate for disease-modifying ALS treatment. Another HDACi, TSA, was found to slow ALS progression and extend survival in SOD1 G93A transgenic mice [98]. Specifically, intraperitoneal injection of TSA in symptomatic animals improved motor neuronal death, axonal loss, neuromuscular junction denervation, and skeletal muscle atrophy. Similarly, treatment of presymptomatic ALS mice with VPA significantly prolonged their lifespan [99]. The list of epigenetic drugs possibly relevant for the treatment of ALS is reported in Table 2.

4. CONCLUSIONS AND FUTURE PERSPECTIVES

Recent epigenetic discoveries have highlighted new challenges toward a more comprehensive understanding of the nervous system pathophysiology with important implications for translational research. The epigenetic components of neurodegenerative diseases like HD and ALS are being unraveled rapidly. This holds the promise to open up exciting avenues for drug development, biomarker discovery, molecular diagnostic, and personalized medicine. The advent of next-generation sequencing has provided detailed views of epigenetic landscapes, revealing high-resolution profiles

TABLE 2 Therapeutic Modulation of Epigenetic Mechanisms in ALS

Drugs	Activities	Effects	References (PMID)
RG108	Inhibition of DNMTs	Reduced 5mC levels and apoptosis of motor neurons in mice with unilateral SNA lesions	[79]
Procainamide	Inhibition of DNMT1	Reduced 5mC levels and apoptosis of motor neurons in mice with unilateral SNA lesions	[79]
Sodium phenylbutyrate	Inhibition of class I and II HDACs	Improved histone hypoacetylation, reduced apoptosis, and improved survival of mutant SOD1 (G93A) transgenic mice	[97]
Trichostatin A	Inhibition of class I and II HDACs	Reduced motor neuronal death, neuromuscular junction denervation, and skeletal muscle atrophy in mutant SOD1 (G93A) transgenic mice	[98]
Valproic acid	Inhibition of HDACs	Slowed disease progression in mutant SOD1 (G93A) transgenic mice	[99]

5mC, 5-methylcytosine; HDACs, histone deacetylases; DNMT, DNA methyltransferase; SNA, sciatic nerve avulsion.

of histone and DNA modifications genome-wide, identifying their interconnections and cross talk and defining the multilayered relationships with transcriptional mechanisms and association with resulting gene expression patterns. These findings, together with the parallel advancements in small molecule design and synthesis, are laying the foundation for new rational drug discovery.

Nevertheless, given the complexity of epigenetic mechanisms in the nervous system, much needs to be accomplished. First, it remains to be determined whether epigenetic anomalies play prominent roles as early and causal features of neurological disorders. Indeed, the knowledge of the molecular changes occurring prior to manifestation of symptom or during the earliest stages of adult-onset neurodegenerative diseases can be extremely important for designing effective pharmacological intervention aimed at halting disease progression. In this regard, studies centering on the prodromal disease stage represent an invaluable tool to assess the involvement of epigenetic alterations as primary triggers of disease onset. Second, it will be important to determine systematically the direct influence of epigenetic and transcriptional deregulation on clinical parameters. Third, investigating the impact of environmental constituents on chromatin structures will be essential to gain meaningful mechanistic insight into the causes of idiopathic neurodegenerative disorders with complex etiology such as sporadic ALS. The increasing interest in epigenetic mechanisms may lead to the identification of disease-specific signatures, which can be employed clinically for early diagnosis, more accurate prognosis and biomarker-driven, individualized treatment. Indeed, the ultimate goal would be to use chromatin signatures, either individually or in combination with other biochemical markers, to detect biologically distinct clinical categories, inform therapeutic decision-making, assess the effectiveness of specific treatments, and discriminate between clinical subtypes. However, epigenetic-based biomarker discovery of brain disorders is still in its infancy. As described above, concordant changes in postmortem brain tissues and more easily accessible biological specimens, such as peripheral blood cells, have been observed for some epigenetic marks in both ALS and

HD. These results open up promising perspectives for epigenetic-based biomarker discovery for neurological disorders; however, extensive studies should be undertaken to evaluate whether disease-associated changes in epigenetic machinery can be accurately detected with high sensitivity and reproducibility in larger patient cohorts.

Glossary

Amyotrophic lateral sclerosis A fatal, late-onset neurological disease characterized by progressive loss of motor neurons in motor cortex, brainstem, and spinal cord. Both sporadic and familial forms occur, and they are clinically similar. ALS symptoms include muscle weakness and atrophy, fasciculations, cramps, spasticity, dysarthria, and dysphagia. To date, there is no effective treatment for this disorder.

CpG islands Regions with more than 200 nucleotides, a G+C content greater than 50% and an observed-to-statistically expected CpG ratio of at least 0.6.

DNA methylation A biochemical mechanism consisting in the addition of a methyl group (-CH_3) to the fifth position of cytosine. This event typically occurs at CpG dinucleotides (cytosine and guanine separated by one phosphate) and is associated with reduction of gene expression.

Histone deacetylases Enzymes that remove acetyl groups from amino acids on histones. The HDAC family comprises 18 enzymes which can be subclassified into four principal groups: class I (HDAC1, 2, 3, and 8), class II (HDAC4, 5, 6, 7, 9, and 10), class III (SIRT1-7), and class IV (HDAC 11).

Histone modifications Covalent posttranslational modifications of residues of histone tails that dynamically modulate chromatin structures and functions. Modifications include acetylation, methylation, phosphorylation sumoylation, ADP-ribosylation, and ubiquitylation.

Huntington's disease A neurodegenerative disorder caused by a CAG expansion resulting in a polyglutamine (poly-Q) extension in the huntingtin protein (Htt) which leads to problems with metabolism, movement, cognition, behavioral function, and ultimately death.

Reduced representation bisulfite sequencing (RRBS) A high-throughput technique to study DNA methylation genome-wide with single-nucleotide resolution. This assay is based on DNA bisulfite treatment and digestion with the methylation-insensitive MspI enzyme. This restriction enzyme systematically cuts DNA to enrich for CpG dinucleotides.

LIST OF ABBREVIATIONS

5hmC 5-Hydroxymethylcytosine
5mC 5-Methylcytosine
A2AR Adenosine A2A receptor
ALS Amyotrophic lateral sclerosis
BDNF Brain-derived neurotrophic factor
CBP CREB-binding protein
ChIP-seq Chromatin immunoprecipitation coupled with deep sequencing
CHRM1 Cholinergic receptor M1
CNS Central nervous system
DDR DNA damage response
DNMT DNA methyltransferase
ELISA Enzyme-linked immunosorbent assay
ESET ERG-associated protein with SET domain
FALS Familial ALS
FTD Frontotemporal dementia
H2AK119ubi Histone 2A lysine 119 monoubiquitylation
H2BK120ubi Histone 2B lysine 120 monoubiquitylation
H3K27 Histone 3 lysine 27
H3K4 Histone 3 lysine 4
H3K4me3 histone 3 lysine 4 trimethylation
H3K9 Histone 3 lysine 9
H3K9/K14 Histone 3 lysines 9 and 14
H3K9me3 Histone 3 lysine 9 trimethylation
H4R3 Histone 4 arginine 3
HATs Histone acetyltransferases
HD Huntington's disease
HDACis HDAC inhibitors
HDMs Histone demethylases
HMTs Histone methyltransferases
hnRNPs Heterogeneous nuclear ribonucleoproteins
HTT Huntingtin
IBs Inclusion bodies
MeDIP Methylated DNA immunoprecipitation
NaB Sodium butyrate
NaPB Sodium phenylbutyrate
PRC1L Polycomb-repressive complex 1-like
PRMT Protein arginine methyltransferase 1
PTMs Posttranslational modifications
REST R element-1 silencing transcription factor
RRBS Reduced representation bisulfite sequencing
SAHA Suberoylanilide hydroxamic acid
SALS Sporadic ALS
TSA Trichostatin A
UTR Untranslated region
VPA Valproic acid

Acknowledgments

This work was supported by NIH U54-NS-091046, R01-GM089903, and R01-NS089076 grants.

References

[1] Jakovcevski M, Akbarian S. Epigenetic mechanisms in neurological disease. Nat Med August 2012;18(8): 1194–204.

[2] Heemskerk AW, Roos RA. Aspiration pneumonia and death in Huntington's disease. PLoS Curr 2012;4: Rrn1293.

[3] Pringsheim T, Wiltshire K, Day L, Dykeman J, Steeves T, Jette N. The incidence and prevalence of Huntington's disease: a systematic review and meta-analysis. Mov Disord August 2012;27(9):1083–91.

[4] Bonelli RM, Wenning GK, Kapfhammer HP. Huntington's disease: present treatments and future therapeutic modalities. Int Clin Psychopharmacol March 2004;19(2):51–62.

[5] Jones L, Hughes A. Pathogenic mechanisms in Huntington's disease. Int Rev Neurobiol 2011;98:373–418.

[6] Moumne L, Betuing S, Caboche J. Multiple aspects of gene dysregulation in Huntington's disease. Front Neurol 2013;4:127.

[7] Zuccato C, Cattaneo E. Role of brain-derived neurotrophic factor in Huntington's disease. Prog Neurobiol April 2007;81(5–6):294–330.

[8] Shimohata T, Nakajima T, Yamada M, Uchida C, Onodera O, Naruse S, et al. Expanded polyglutamine stretches interact with TAFII130, interfering with CREB-dependent transcription. Nat Genet September 2000;26(1):29–36.

[9] Boutell JM, Thomas P, Neal JW, Weston VJ, Duce J, Harper PS, et al. Aberrant interactions of transcriptional repressor proteins with the Huntington's disease gene product, huntingtin. Hum Mol Genet September 1999;8(9):1647–55.

[10] Dunah AW, Jeong H, Griffin A, Kim YM, Standaert DG, Hersch SM, et al. Sp1 and TAFII130 transcriptional activity disrupted in early Huntington's disease. Science (New York, NY) June 21, 2002;296(5576): 2238–43.

[11] Steffan JS, Kazantsev A, Spasic-Boskovic O, Greenwald M, Zhu YZ, Gohler H, et al. The Huntington's disease protein interacts with p53 and CREB-binding protein and represses transcription. Proc Natl Acad Sci USA June 6, 2000;97(12):6763–8.

[12] Buckley NJ, Johnson R, Zuccato C, Bithell A, Cattaneo E. The role of REST in transcriptional and epigenetic dysregulation in Huntington's disease. Neurobiol Dis July 2010;39(1):28–39.

[13] Rothbart SB, Strahl BD. Interpreting the language of histone and DNA modifications. Biochim Biophys Acta August 2014;1839(8):627–43.

[14] Swygert SG, Peterson CL. Chromatin dynamics: interplay between remodeling enzymes and histone modifications. Biochim Biophys Acta August 2014;1839(8):728–36.

[15] Petty E, Pillus L. Balancing chromatin remodeling and histone modifications in transcription. Trends Genet TIG November 2013;29(11):621–9.

[16] Lalonde ME, Cheng X, Cote J. Histone target selection within chromatin: an exemplary case of teamwork. Genes Dev May 15, 2014;28(10):1029–41.

[17] Sadri-Vakili G, Bouzou B, Benn CL, Kim MO, Chawla P, Overland RP, et al. Histones associated with downregulated genes are hypo-acetylated in Huntington's disease models. Hum Mol Genet June 1, 2007;16(11):1293–306.

[18] Nucifora Jr FC, Sasaki M, Peters MF, Huang H, Cooper JK, Yamada M, et al. Interference by huntingtin and atrophin-1 with cbp-mediated transcription leading to cellular toxicity. Science (New York, NY) March 23, 2001;291(5512):2423–8.

[19] Jiang H, Nucifora Jr FC, Ross CA, DeFranco DB. Cell death triggered by polyglutamine-expanded huntingtin in a neuronal cell line is associated with degradation of CREB-binding protein. Hum Mol Genet January 1, 2003;12(1):1–12.

[20] Cong SY, Pepers BA, Evert BO, Rubinsztein DC, Roos RA, van Ommen GJ, et al. Mutant huntingtin represses CBP, but not p300, by binding and protein degradation. Mol Cell Neurosci September 2005;30(1):12–23.

[21] Jiang H, Poirier MA, Liang Y, Pei Z, Weiskittel CE, Smith WW, et al. Depletion of CBP is directly linked with cellular toxicity caused by mutant huntingtin. Neurobiol Dis September 2006;23(3):543–51.

[22] Giralt A, Puigdellivol M, Carreton O, Paoletti P, Valero J, Parra-Damas A, et al. Long-term memory deficits in Huntington's disease are associated with reduced CBP histone acetylase activity. Hum Mol Genet March 15, 2012;21(6):1203–16.

[23] Bannister AJ, Kouzarides T. The CBP co-activator is a histone acetyltransferase. Nature December 19–26, 1996;384(6610):641–3.

[24] Martinez-Balbas MA, Bannister AJ, Martin K, Haus-Seuffert P, Meisterernst M, Kouzarides T. The acetyltransferase activity of CBP stimulates transcription. EMBO J May 15, 1998;17(10):2886–93.

[25] Lee J, Hwang YJ, Shin JY, Lee WC, Wie J, Kim KY, et al. Epigenetic regulation of cholinergic receptor M1 (CHRM1) by histone H3K9me3 impairs Ca(2+) signaling in Huntington's disease. Acta Neuropathol May 2013;125(5):727–39.

[26] Ryu H, Lee J, Hagerty SW, Soh BY, McAlpin SE, Cormier KA, et al. ESET/SETDB1 gene expression and histone H3 (K9) trimethylation in Huntington's disease. Proc Natl Acad Sci USA December 12, 2006;103(50):19176–81.

[27] Vashishtha M, Ng CW, Yildirim F, Gipson TA, Kratter IH, Bodai L, et al. Targeting H3K4 trimethylation in Huntington disease. Proc Natl Acad Sci USA August 6, 2013;110(32):E3027–36.

[28] Bernstein BE, Humphrey EL, Erlich RL, Schneider R, Bouman P, Liu JS, et al. Methylation of histone H3 Lys 4 in coding regions of active genes. Proc Natl Acad Sci USA June 25, 2002;99(13):8695–700.
[29] Kim TH, Barrera LO, Zheng M, Qu C, Singer MA, Richmond TA, et al. A high-resolution map of active promoters in the human genome. Nature August 11, 2005;436(7052):876–80.
[30] Santos-Rosa H, Schneider R, Bannister AJ, Sherriff J, Bernstein BE, Emre NC, et al. Active genes are tri-methylated at K4 of histone H3. Nature September 26, 2002;419(6905):407–11.
[31] Kim MO, Chawla P, Overland RP, Xia E, Sadri-Vakili G, Cha JH. Altered histone monoubiquitylation mediated by mutant huntingtin induces transcriptional dysregulation. J Neurosci April 9, 2008;28(15): 3947–57.
[32] de Napoles M, Mermoud JE, Wakao R, Tang YA, Endoh M, Appanah R, et al. Polycomb group proteins Ring1A/B link ubiquitylation of histone H2A to heritable gene silencing and X inactivation. Dev Cell November 2004;7(5):663–76.
[33] Wang H, Wang L, Erdjument-Bromage H, Vidal M, Tempst P, Jones RS, et al. Role of histone H2A ubiquitination in Polycomb silencing. Nature October 14, 2004;431(7010):873–8.
[34] Nickel BE, Allis CD, Davie JR. Ubiquitinated histone H2B is preferentially located in transcriptionally active chromatin. Biochemistry February 7, 1989;28(3):958–63.
[35] Lister R, Mukamel EA, Nery JR, Urich M, Puddifoot CA, Johnson ND, et al. Global epigenomic reconfiguration during mammalian brain development. Science (New York, NY) August 9, 2013;341(6146):1237905.
[36] Cheng X. Structural and functional coordination of DNA and histone methylation. Cold Spring Harbor Perspect Biol 2014;6(8).
[37] Ng CW, Yildirim F, Yap YS, Dalin S, Matthews BJ, Velez PJ, et al. Extensive changes in DNA methylation are associated with expression of mutant huntingtin. Proc Natl Acad Sci USA February 5, 2013;110(6):2354–9.
[38] Villar-Menendez I, Blanch M, Tyebji S, Pereira-Veiga T, Albasanz JL, Martin M, et al. Increased 5-methylcytosine and decreased 5-hydroxymethylcytosine levels are associated with reduced striatal A2AR levels in Huntington's disease. NeuroMol Med June 2013;15(2):295–309.
[39] Wood H. Neurodegenerative disease: altered DNA methylation and RNA splicing could be key mechanisms in Huntington disease. Nat Rev Neurol March 2013;9(3):119.
[40] Day JJ, Sweatt JD. DNA methylation and memory formation. Nat Neurosci November 2010;13(11):1319–23.
[41] Miller CA, Gavin CF, White JA, Parrish RR, Honasoge A, Yancey CR, et al. Cortical DNA methylation maintains remote memory. Nat Neurosci June 2010;13(6):664–6.
[42] Wen L, Tang F. Genomic distribution and possible functions of DNA hydroxymethylation in the brain. Genomics September 6, 2014.
[43] Wang F, Yang Y, Lin X, Wang JQ, Wu YS, Xie W, et al. Genome-wide loss of 5-hmC is a novel epigenetic feature of Huntington's disease. Hum Mol Genet September 15, 2013;22(18):3641–53.
[44] Steffan JS, Bodai L, Pallos J, Poelman M, McCampbell A, Apostol BL, et al. Histone deacetylase inhibitors arrest polyglutamine-dependent neurodegeneration in *Drosophila*. Nature October 18, 2001;413(6857):739–43.
[45] Ferrante RJ, Kubilus JK, Lee J, Ryu H, Beesen A, Zucker B, et al. Histone deacetylase inhibition by sodium butyrate chemotherapy ameliorates the neurodegenerative phenotype in Huntington's disease mice. J Neurosci October 15, 2003;23(28):9418–27.
[46] Hockly E, Richon VM, Woodman B, Smith DL, Zhou X, Rosa E, et al. Suberoylanilide hydroxamic acid, a histone deacetylase inhibitor, ameliorates motor deficits in a mouse model of Huntington's disease. Proc Natl Acad Sci USA February 18, 2003;100(4):2041–6.
[47] Mielcarek M, Benn CL, Franklin SA, Smith DL, Woodman B, Marks PA, et al. SAHA decreases HDAC 2 and 4 levels in vivo and improves molecular phenotypes in the R6/2 mouse model of Huntington's disease. PLoS One 2011;6(11):e27746.
[48] Gardian G, Browne SE, Choi DK, Klivenyi P, Gregorio J, Kubilus JK, et al. Neuroprotective effects of phenylbutyrate in the N171-82Q transgenic mouse model of Huntington's disease. J Biol Chem January 7, 2005;280(1):556–63.
[49] Thomas EA, Coppola G, Desplats PA, Tang B, Soragni E, Burnett R, et al. The HDAC inhibitor 4b ameliorates the disease phenotype and transcriptional abnormalities in Huntington's disease transgenic mice. Proc Natl Acad Sci USA October 7, 2008;105(40):15564–9.
[50] Jia H, Kast RJ, Steffan JS, Thomas EA. Selective histone deacetylase (HDAC) inhibition imparts beneficial effects in Huntington's disease mice: implications for the ubiquitin-proteasomal and autophagy systems. Hum Mol Genet December 15, 2012;21(24):5280–93.
[51] Jia H, Pallos J, Jacques V, Lau A, Tang B, Cooper A, et al. Histone deacetylase (HDAC) inhibitors targeting HDAC3 and HDAC1 ameliorate polyglutamine-elicited phenotypes in model systems of Huntington's disease. Neurobiol Dis May 2012;46(2):351–61.
[52] Kazantsev AG, Thompson LM. Therapeutic application of histone deacetylase inhibitors for central nervous system disorders. Nat Rev Drug Discov October 2008;7(10):854–68.
[53] Pallos J, Bodai L, Lukacsovich T, Purcell JM, Steffan JS, Thompson LM, et al. Inhibition of specific HDACs and sirtuins suppresses pathogenesis in a *Drosophila* model of Huntington's disease. Hum Mol Genet December 1, 2008;17(23):3767–75.

[54] Smith MR, Syed A, Lukacsovich T, Purcell J, Barbaro BA, Worthge SA, et al. A potent and selective Sirtuin 1 inhibitor alleviates pathology in multiple animal and cell models of Huntington's disease. Hum Mol Genet June 1, 2014;23(11):2995–3007.

[55] Hogarth P, Lovrecic L, Krainc D. Sodium phenylbutyrate in Huntington's disease: a dose-finding study. Mov Disord October 15, 2007;22(13):1962–4.

[56] Borovecki F, Lovrecic L, Zhou J, Jeong H, Then F, Rosas HD, et al. Genome-wide expression profiling of human blood reveals biomarkers for Huntington's disease. Proc Natl Acad Sci USA August 2, 2005;102(31):11023–8.

[57] Chen JY, Wang E, Galvan L, Huynh M, Joshi P, Cepeda C, et al. Effects of the pimelic diphenylamide histone deacetylase inhibitor HDACi 4b on the R6/2 and N171-82Q mouse models of Huntington's disease. PLoS Curr 2013;5.

[58] Westerberg G, Chiesa JA, Andersen CA, Diamanti D, Magnoni L, Pollio G, et al. Safety, pharmacokinetics, pharmacogenomics and QT concentration-effect modelling of the SirT1 inhibitor selisistat in healthy volunteers. Br J Clin Pharmacol March 2015;79(3):477–91.

[59] Sussmuth SD, Haider S, Landwehrmeyer GB, Farmer R, Frost C, Tripepi G, et al. An exploratory double-blind, randomized clinical trial with selisistat, a SirT1 inhibitor, in patients with Huntington's disease. Br J Clin Pharmacol March 2015;79(3):465–76.

[60] Wijesekera LC, Leigh PN. Amyotrophic lateral sclerosis. Orphanet J Rare Dis 2009;4:3.

[61] Al-Chalabi A, Jones A, Troakes C, King A, Al-Sarraj S, van den Berg LH. The genetics and neuropathology of amyotrophic lateral sclerosis. Acta Neuropathol September 2012;124(3):339–52.

[62] Andersen PM, Al-Chalabi A. Clinical genetics of amyotrophic lateral sclerosis: what do we really know? Nat Rev Neurol November 2011;7(11):603–15.

[63] Cozzolino M, Pesaresi MG, Gerbino V, Grosskreutz J, Carri MT. Amyotrophic lateral sclerosis: new insights into underlying molecular mechanisms and opportunities for therapeutic intervention. Antioxid Redox Signaling November 1, 2012;17(9):1277–330.

[64] Ling SC, Polymenidou M, Cleveland DW. Converging mechanisms in ALS and FTD: disrupted RNA and protein homeostasis. Neuron August 7, 2013;79(3):416–38.

[65] Polymenidou M, Lagier-Tourenne C, Hutt KR, Bennett CF, Cleveland DW, Yeo GW. Misregulated RNA processing in amyotrophic lateral sclerosis. Brain Res June 26, 2012;1462:3–15.

[66] Robberecht W, Philips T. The changing scene of amyotrophic lateral sclerosis. Nat Rev Neurosci April 2013;14(4):248–64.

[67] DeJesus-Hernandez M, Mackenzie IR, Boeve BF, Boxer AL, Baker M, Rutherford NJ, et al. Expanded GGGGCC hexanucleotide repeat in noncoding region of C9ORF72 causes chromosome 9p-linked FTD and ALS. Neuron October 20, 2011;72(2):245–56.

[68] Renton AE, Majounie E, Waite A, Simon-Sanchez J, Rollinson S, Gibbs JR, et al. A hexanucleotide repeat expansion in C9ORF72 is the cause of chromosome 9p21-linked ALS-FTD. Neuron October 20, 2011;72(2):257–68.

[69] Garcia-Redondo A, Dols-Icardo O, Rojas-Garcia R, Esteban-Perez J, Cordero-Vazquez P, Munoz-Blanco JL, et al. Analysis of the C9orf72 gene in patients with amyotrophic lateral sclerosis in Spain and different populations worldwide. Hum Mutat January 2013;34(1):79–82.

[70] Lee YB, Chen HJ, Peres JN, Gomez-Deza J, Attig J, Stalekar M, et al. Hexanucleotide repeats in ALS/FTD form length-dependent RNA foci, sequester RNA binding proteins, and are neurotoxic. Cell Rep December 12, 2013;5(5):1178–86.

[71] Gijselinck I, Van Langenhove T, van der Zee J, Sleegers K, Philtjens S, Kleinberger G, et al. A C9orf72 promoter repeat expansion in a Flanders-Belgian cohort with disorders of the frontotemporal lobar degeneration-amyotrophic lateral sclerosis spectrum: a gene identification study. Lancet Neurol January 2012;11(1):54–65.

[72] Almeida S, Gascon E, Tran H, Chou HJ, Gendron TF, Degroot S, et al. Modeling key pathological features of frontotemporal dementia with C9ORF72 repeat expansion in iPSC-derived human neurons. Acta Neuropathol September 2013;126(3):385–99.

[73] Belzil VV, Bauer PO, Prudencio M, Gendron TF, Stetler CT, Yan IK, et al. Reduced C9orf72 gene expression in c9FTD/ALS is caused by histone trimethylation, an epigenetic event detectable in blood. Acta Neuropathol December 2013;126(6):895–905.

[74] Ciura S, Lattante S, Le Ber I, Latouche M, Tostivint H, Brice A, et al. Loss of function of C9orf72 causes motor deficits in a zebrafish model of amyotrophic lateral sclerosis. Ann Neurol August 2013;74(2):180–7.

[75] Fratta P, Poulter M, Lashley T, Rohrer JD, Polke JM, Beck J, et al. Homozygosity for the C9orf72 GGGGCC repeat expansion in frontotemporal dementia. Acta Neuropathol September 2013;126(3):401–9.

[76] Farg MA, Sundaramoorthy V, Sultana JM, Yang S, Atkinson RA, Levina V, et al. C9ORF72, implicated in amyotrophic lateral sclerosis and frontotemporal dementia, regulates endosomal trafficking. Hum Mol Genet July 1, 2014;23(13):3579–95.

[77] Gendron TF, Belzil VV, Zhang YJ, Petrucelli L. Mechanisms of toxicity in C9FTLD/ALS. Acta Neuropathol March 2014;127(3):359–76.
[78] Morahan JM, Yu B, Trent RJ, Pamphlett R. A genome-wide analysis of brain DNA methylation identifies new candidate genes for sporadic amyotrophic lateral sclerosis. Amyotrophic Lateral Scler 2009 Oct-Dec;10(5–6):418–29.
[79] Chestnut BA, Chang Q, Price A, Lesuisse C, Wong M, Martin LJ. Epigenetic regulation of motor neuron cell death through DNA methylation. J Neurosci November 16, 2011;31(46):16619–36.
[80] Figueroa-Romero C, Hur J, Bender DE, Delaney CE, Cataldo MD, Smith AL, et al. Identification of epigenetically altered genes in sporadic amyotrophic lateral sclerosis. PloS One 2012;7(12):e52672.
[81] Tremolizzo L, Messina P, Conti E, Sala G, Cecchi M, Airoldi L, et al. Whole-blood global DNA methylation is increased in amyotrophic lateral sclerosis independently of age of onset. Amyotrophic Lateral Scler Frontotemporal Degener March 2014;15(1–2):98–105.
[82] Xi Z, Zinman L, Moreno D, Schymick J, Liang Y, Sato C, et al. Hypermethylation of the CpG island near the G4C2 repeat in ALS with a C9orf72 expansion. Am J Hum Genet June 6, 2013;92(6):981–9.
[83] Belzil VV, Bauer PO, Gendron TF, Murray ME, Dickson D, Petrucelli L. Characterization of DNA hypermethylation in the cerebellum of c9FTD/ALS patients. Brain Res October 10, 2014;1584:15–21.
[84] Xi Z, Rainero I, Rubino E, Pinessi L, Bruni AC, Maletta RG, et al. Hypermethylation of the CpG-island near the C9orf72 G4C2-repeat expansion in FTLD patients. Hum Mol Genet November 1, 2014;23(21):5630–7.
[85] Barski A, Cuddapah S, Cui K, Roh TY, Schones DE, Wang Z, et al. High-resolution profiling of histone methylations in the human genome. Cell May 18, 2007;129(4):823–37.
[86] Christman JK. 5-Azacytidine and 5-aza-2′-deoxycytidine as inhibitors of DNA methylation: mechanistic studies and their implications for cancer therapy. Oncogene August 12, 2002;21(35):5483–95.
[87] Lagier-Tourenne C, Cleveland DW. Rethinking ALS: the FUS about TDP-43. Cell March 20, 2009;136(6):1001–4.
[88] Yamaguchi A, Kitajo K. The effect of PRMT1-mediated arginine methylation on the subcellular localization, stress granules, and detergent-insoluble aggregates of FUS/TLS. PloS One 2012;7(11):e49267.
[89] Kwiatkowski Jr TJ, Bosco DA, Leclerc AL, Tamrazian E, Vanderburg CR, Russ C, et al. Mutations in the FUS/TLS gene on chromosome 16 cause familial amyotrophic lateral sclerosis. Science (New York, NY) February 27, 2009;323(5918):1205–8.
[90] Vance C, Rogelj B, Hortobagyi T, De Vos KJ, Nishimura AL, Sreedharan J, et al. Mutations in FUS, an RNA processing protein, cause familial amyotrophic lateral sclerosis type 6. Science (New York, NY) February 27, 2009;323(5918):1208–11.
[91] Wang WY, Pan L, Su SC, Quinn EJ, Sasaki M, Jimenez JC, et al. Interaction of FUS and HDAC1 regulates DNA damage response and repair in neurons. Nat Neurosci October 2013;16(10):1383–91.
[92] Tibshirani M, Tradewell ML, Mattina KR, Minotti S, Yang W, Zhou H, et al. Cytoplasmic sequestration of FUS/TLS associated with ALS alters histone marks through loss of nuclear protein arginine methyltransferase 1. Hum Mol Genet September 30, 2014.
[93] Huang S, Litt M, Felsenfeld G. Methylation of histone H4 by arginine methyltransferase PRMT1 is essential in vivo for many subsequent histone modifications. Genes Dev August 15, 2005;19(16):1885–93.
[94] Deng HX, Zhai H, Bigio EH, Yan J, Fecto F, Ajroud K, et al. FUS-immunoreactive inclusions are a common feature in sporadic and non-SOD1 familial amyotrophic lateral sclerosis. Ann Neurol June 2010;67(6):739–48.
[95] Brueckner B, Garcia Boy R, Siedlecki P, Musch T, Kliem HC, Zielenkiewicz P, et al. Epigenetic reactivation of tumor suppressor genes by a novel small-molecule inhibitor of human DNA methyltransferases. Cancer Res July 15, 2005;65(14):6305–11.
[96] Lee BH, Yegnasubramanian S, Lin X, Nelson WG. Procainamide is a specific inhibitor of DNA methyltransferase 1. J Biol Chem December 9, 2005;280(49):40749–56.
[97] Ryu H, Smith K, Camelo SI, Carreras I, Lee J, Iglesias AH, et al. Sodium phenylbutyrate prolongs survival and regulates expression of anti-apoptotic genes in transgenic amyotrophic lateral sclerosis mice. J Neurochem June 2005;93(5):1087–98.
[98] Yoo YE, Ko CP. Treatment with trichostatin A initiated after disease onset delays disease progression and increases survival in a mouse model of amyotrophic lateral sclerosis. Exp Neurol September 2011;231(1):147–59.
[99] Sugai F, Yamamoto Y, Miyaguchi K, Zhou Z, Sumi H, Hamasaki T, et al. Benefit of valproic acid in suppressing disease progression of ALS model mice. Eur J Neurosci December 2004;20(11):3179–83.
[100] Cudkowicz ME, Andres PL, Macdonald SA, Bedlack RS, Choudry R, Brown Jr RH, et al. Phase 2 study of sodium phenylbutyrate in ALS. Amyotrophic Lateral Scler April 2009;10(2):99–106.

CHAPTER

27

MicroRNA Deregulation in Lung Cancer and Their Use as Clinical Tools

Paula Lopez-Serra, Juan Sandoval

Epigenetic and Cancer Biology Program (PEBC), Bellvitge Biomedical Research Institute (IDIBELL), Barcelona, Spain

OUTLINE

Epigenetic Biomarkers and Diagnostics
http://dx.doi.org/10.1016/B978-0-12-801899-6.00027-9

1. INTRODUCTION TO NONCODING RNA IN HUMAN CELL PHYSIOLOGY

Noncoding RNA (ncRNA) has been considered a by-product of the transcriptional process driven by the RNA polymerase during the synthesis of functional RNA. However, more than 90% of the whole human genome is transcribed in a set of overlapping transcripts that do not codify any protein [1]. Different types of ncRNA have been described and the number keeps increasing. Among all the ncRNAs, microRNAs (miRNAs) are small molecules of ~22 nucleotides that regulate gene expression by inhibiting messenger RNA (mRNA) translation. They were discovered in *Caenorhabditis elegans* in 1993 [2] and are the most studied and well characterized ncRNAs. Other ncRNAs are PIWI-interacting RNAs (piRNAs), small nucleolar ribonucleoprotein complexes (snoRNPs), or long noncoding RNAs (lncRNAs).

piRNAs are 24–30-nucleotide-long molecules that bind the PIWI proteins to maintain genome stability in germ line cells; the mechanism of action consists of the repression of transposable elements transcription [3]. snoRNAs are 60–300-nucleotide-long molecules that form part of the snoRNPs, ribonucleoprotein complexes responsible for ribosomal RNA stability [4]. lncRNAs represent the main portion of ncRNA in a cell. They are defined as a heterogeneous group of transcripts with more than 200 nucleotides involved in many biological processes. lncRNAs are essential in X-chromosome inactivation in mammals, an essential process in dosage compensation. In this regard, the lncRNA X-inactivation-specific transcript recruits the silencing complex polycomb to the X chromosome from which it is transcribed, rendering the chromosome transcriptionally silent [5]. lncRNAs are also essential in key cell biological processes; the temporal and spatial expression of hundreds of lncRNAs from the human HOX loci during embryonic development regulates the chromatin accessibility to histone modification enzymes and RNA polymerases, contributing to cell identity maintenance [6]. Another classes of lncRNAs are the long intergenic noncoding RNAs (lincRNA), transcribed from the intergenic regions of the chromosomes and associated with chromatin signatures for active transcription regions [7]; and the transcribed ultraconserved RNAs (tUCR), ncRNAs transcribed from ultraconserved regions of the genome. Their function remains unknown but it has been reported their implication in miRNAs biogenesis [8].

There are other poorly defined ncRNAs, like transcriptional start site-associated RNAs, associated with the transcriptional start sites of genes (this group of ncRNAs includes the promoter-associated small RNAs [9], promoter upstream transcripts [10], and transcription initiation RNAs [11]). Seila and collaborators [12] suggest that these 20–200-nucleotide transcripts could help to maintain active the promoter regions of the surrounded genes.

ncRNAs play a critical role in the maintenance of telomeric length. During DNA replication, DNA polymerases cannot replicate the sequences present at the 3′ end of the chromosomes; this leads to the loss of genetic information in each round of cell division. Telomerase is the ribonucleoprotein responsible for the maintenance of telomeres' integrity. The transcription of telomeric repeat-containing RNAs from these regions regulates the telomerase activity [13] and the heterochromatinization of the telomeric regions [14].

Table 1 summarizes the principal characteristics of the ncRNAs described.

These few examples of ncRNA functions reflect their importance in cell biology. Alterations in ncRNA have been described in human pathologies like Alzheimer, cardiovascular disease, and cancer [15]. Here we will focus on miRNA disruption in lung cancer development.

TABLE 1 Resume of the Main Characteristic on the Noncoding RNA Described in Human Cells

Name	Abbreviation	Average size (nucleotides)	Function	References
microRNA	miRNA	19–24	Regulation of mRNA translation	[2]
PIWI RNA	piRNA	24–30	Genome stability	[3]
Small nucleolar RNA	snoRNA	60–300	Ribosomal RNA stability	[4]
Long noncoding RNA	lncRNA	>200	X-chromosome inactivation/development regulation	[5,6]
Long intergenic RNA	lincRNA		Chromatin scaffold	[7]
Transcribed ultraconserved RNA	T-UCR		miRNA biosynthesis	[8]
Transcriptional start site-associated RNA	TSSa-RNA		Gene transcription regulation	
Promoter-associated small RNAs	PASRs	22–200		[9]
Promoter upstream transcripts	PROMPTs	20–200		[10]
Transcription initiation RNAs	tiRNAs	17–18		[11]
Telomeric repeat-containing RNA	TERRA	Heterogeneous length	Telomere maintenance	[13,14]

2. miRNA BIOLOGY AND MECHANISM OF ACTION

2.1 miRNA Synthesis in Human Cells

miRNAs are small RNA transcripts ranging from 19 to 24 nucleotides in length that regulate gene expression as posttranscriptional repressors. It has been suggested that miRNAs play a role in the majority of biological cell processes, since more than 60% of the human coding genes contain at least 1 conserved miRNA-binding site [16]. miRNA synthesis is a tight regulated process. Briefly, miRNA loci are transcribed by the RNA polymerase II to produce a long (typically 1 kb), poly-(A)-tailed precursor molecule, named pri-miRNA, where several miRNA sequences are embedded [17]. This primary transcript is processed by the microprocessor complex, composed by RNase III Drosha and the DGCR8 RNA-binding protein (DiGeorge syndrome critical region gene 8), to produce a 70-nucleotide-long molecule called pre-miRNA [18]. The pre-miRNA is transferred to the cell cytoplasm by XPO5 (exportin 5) where miRNA maturation ends [19]. Once in the cytoplasm, the RNase III Dicer/TRPB (TAR RNA-binding protein 2) complex processes the pre-miRNA to generate the final functional ~22-nucleotide-long miRNA [20] that specifically binds the target mRNA.

The mature miRNA is loaded into a member of the Argonaute family of proteins, a set of proteins with endonuclease activity, to form the RNA-induced silencing complex (RISC). The recognition between the miRNA and its targeted mRNA is initiated through the specific base pairing complementarity between the seed region of the miRNA and the binding sites present in the mRNA, mostly in the 3′-untranslated region (UTR) [21]. The seed region in animals is formed by 6–8 nucleotides in the 5′ end of the miRNA [22]. Usually, the 3′-UTR of the mRNA is recognized by multiple miRNAs, amplifying the repressive effect.

2.2 Mechanisms of Action of miRNAs in Gene Silencing

The main mechanism of RNA silencing regulated by miRNAs is the disruption of translational initiation in the ribosome [23]. The cooperation of multiple RISCs provides the most efficient translational repression, explaining the presence of multiple miRNA-binding sites in the 3′-UTR of the mRNAs [24]. The RISC-mRNA complexes are accumulated in processing bodies (P-bodies) in the cytoplasm. P-bodies could be defined as cellular sites of silent mRNA storage where they can be degraded or returned to translation [25].

miRNAs have been associated to mRNA degradation. The entry of mRNAs in this degradative pathway depends on the degree of complementarity between the miRNA and the mRNA. Only a few examples of endogenous miRNAs entering this pathway of mRNA cleavage have been described in mammals [26,27]. mRNA cleavage is one of the mechanisms of RNA silencing driven by small ncRNAs. Small interference RNAs (siRNAs) or interfering RNAs regulate the degradation of mRNA of identical sequence; siRNAs are transcribed from segments included in the mRNA they match or come from an exogenous origin, like virus infection, directing an autosilencing process.

Other mechanisms of gene silencing have been described for miRNA regulation of gene expression. Direct targeting of ribonucleoproteins by miRNAs blocks the protein function in a process called decoy. Direct binding of miR-328 with the translational inhibitor hnRNP E2 rescues *CEBPA* mRNA translation in leukemic blasts, rescuing cell differentiation. miR-328 may compete with *CEBPA* for hnRNP E2 binding, releasing *CEBPA* mRNA and allowing its translation [28].

At a chromosomal level, gene silencing by heterochromatin formation mediated by miRNA was first described in yeast. Heterochromatin is a densely packed state of the DNA that does not allow access to the transcription factors, maintaining these loci silent. The RNA-induced transcriptional silencing complex recruits histone methyltransferases in a DNA locus determined by the RNA transcribed from these loci, orchestrating chromatin compaction and transcription inhibition [29]. In humans, a complex formed by the proteins hTERT (human telomerase reverse transcriptase), BRG1 (Brahma-related gene 1), and NS (nucleostemin) contributes to heterochromatin maintenance at the centromeric regions of the chromosomes and at the transposon elements. Therefore, it has been reported that this complex produces siRNAs that target chromatin remodelers to these sites in order to promote the heterochromatin assembly [30].

2.3 Mechanisms of Action of miRNAs in Gene Activation

Although the main mechanism of action for miRNA regulation of gene expression is gene silencing, function as gene expression activators has also been established. The miRNA let-7 switches from repression to activation function depending on the cell cycle: it upregulates its targeted mRNAs during cell cycle arrest, and it silences gene transcription in proliferative cells [31].

Another example of miRNA-regulated gene activation is the miRNA-10a. miRNA-10a targets the 5′-UTR of the ribosomal mRNA, activating their translation. This leads to the production of more ribosomes, promoting global protein synthesis [32].

3. miRNA ALTERATION IN LUNG CANCER PROGRESSION

3.1 Introduction to Lung Cancer

Cancer is a heterogeneous group of diseases that can be classified into different categories depending on the tissue of origin, such as epithelial, connective, blood. Lung cancer is the most common type of cancer and the leading cause of cancer-related deaths worldwide (1.59 million deaths in 2012 according to the World Health Organization, WHO). There are two major types of lung cancer: non-small-cell lung cancer (NSCLC, accounting for 85–90% of the cases) and small cell lung cancer (SCLC, accounting for 10–15% of the cases) [33].

NSCLC is defined by pathological characteristics and can be divided mainly into two major NSCLC histological types depending on the morphology and position of the tumor cells: adenocarcinomas (ADCs), that represent approximately 50% of the NSCLC cases, often show a glandular histology, and are usually located in distal airways region, whereas squamous cell lung cancers (SCCs), approximately ~40% of the NSCLC cases, are characterized by squamous differentiation, reminding the stratified epithelium in the trachea and upper airways, where they used to be located. There is the third NSCLC subtype, large cell carcinoma (~10% of the NSCLC cases), classified by exclusion if the tumor does not fall within the other two categories [34].

Although some neuroendocrine tumors are classified as NSCLC, the most common and aggressive forms of neuroendocrine tumors in the lungs are SCLC. SCLC primary tumor is often located in the bronchi near the chest. It is a very aggressive tumor that shows a very high mortality rate, with a 5-year survival rate much lower than NSCLC [35].

A major factor in the high mortality rate of lung cancer patients is the late diagnosis of the disease, independently of the cancer subtype, with presence of metastases in distal tissues in approximately 70% of the patients at the time of diagnosis [36].

There are other uncommon lung tumors like lung carcinoids (<5% of the cases), also from neuroendocrine origin, or pleural mesotheliomas, originated by the cells that form the inner layer of the lung.

3.2 miRNA Deregulation in Lung Cancer

Classically, cancer research has been focused on coding genes and their protein products, classifying them as oncogenes if they promote cell growth or tumor suppressor genes if they prevent it. With the discovery of the ncRNAs at the end of the twentieth century and their regulatory role in cell homeostasis, a new front in cancer research was opened.

miRNAs have emerged as crucial regulators of cell growth due to their pleiotropic functions; in other words, one single miRNA can regulate many different genes involved in many different cell processes. As coding genes, they can be classified as oncomiRs if their expression promotes tumorigenesis or tumor suppressor miRNAs if they inhibit it [37].

Hanahan and Weinberg [38] have described six hallmarks in cancer, as the capabilities acquired by the tumor cells during the multistep process of cancer development. Altered expression of miRNAs, together with changes in the synthesis machinery have been described in practically all the tumor types and all stages of cancer development. Here we show a few examples of altered miRNA in every stage of cancer progression. Figure 1 highlights the pivotal role of miRNA deregulation in all stages of cancer development.

Cell cycle is a tightly controlled process regulated by the activity of positive and negative growth factors; loss of checkpoints involved in cell cycle control may result in DNA damage, genetic mutations, or uncontrolled mitosis, contributing

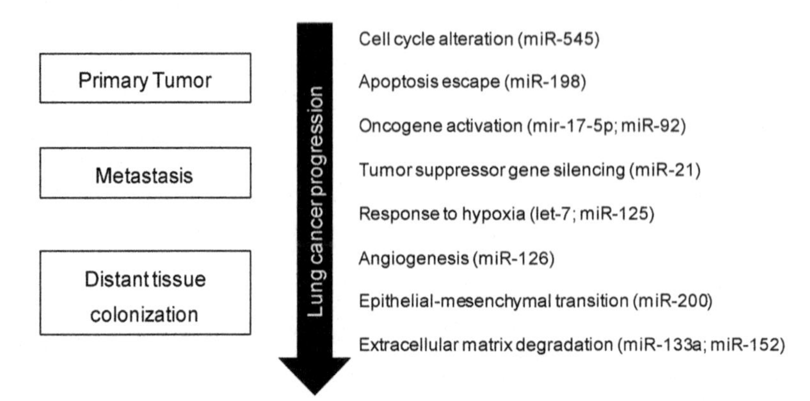

FIGURE 1 MicroRNA alteration is present in all the stages of lung cancer progression. Specific examples appear between brackets.

to carcinogenesis. miR-545 expression promotes cell cycle arrest and inhibition of lung cancer cells by targeting the cyclin D1 and *CDK4*, regulators of the G1/S-phase transition [39].

Failure to enter apoptosis, defined as the activation of the programmed cell death, represents one of the mechanisms by which cell transformation occurs, allowing tumor cells to proliferate. miR-198 expression promotes cell apoptosis by targeting *FGFR1* (fibroblast growth factor receptor 1) [40], a member of the fibroblast growth receptor family that acts as a tyrosine kinase for proliferative and survival signaling. FGFR1-altered activity derives from gene amplification, active mutations, and FGFR1 translocation [41]. miR-198 inhibits *FGFR1* translation in lung cancer cells and it is found to be strongly downregulated in NSCLC [42].

Angiogenesis, described as the process of the generation of new blood vessels, has a pivotal role in physiological embryonic development. In the adult, this process remains quiescent (except punctual physiological processes like wound healing and reproductive cycle in females). However, during tumor progression, the angiogenic process becomes active, resulting in new blood vessels that provide oxygen and nutrients to the tumor cells. A recent study in a mouse model of NSCLC demonstrates that a global depletion of miRNA activity by knocking out *Dicer 1* leads to poorly vascularized tumors despite the hypoxic conditions. Hypoxia activates HIF-1 (hypoxia-inducible factor 1), which promotes the transcription of *Vegf* (vascular endothelial growth factor), involved in the vascularization process. *Hif-1* is regulated by the factor inhibiting hypoxia-inducible factor 1 (*Fih1*), a gene highly regulated by miRNAs. FIH1 3′-UTR contains binding sites for multiple miRNA families like let-7 and miR-125. In the absence of miRNAs, *Fih1* translation is deregulated, disrupting the response to hypoxia [43].

Specific alterations of miRNAs controlling different pathways that trigger angiogenesis have been described. miR-126 was suggested to be an endothelial-specific miRNA that modulates the endothelial phenotype and the vessel integrity, being reported its overexpression in lung cancer to promote the new vascularization [44].

More than 90% of the cancer-related deaths are due to the migration of tumor cells in a process called metastasis. The metastatic process implies cell invasion capacity, intravasation from the primary tumor to the local bloodstream and lymphatic vessels, and extravasation and invasion of distant tissues. Epithelial–mesenchymal transition (EMT) is an early step in the acquisition of the invasive phenotype.

Epithelial cells change their shape and lose their attachment to other cells within the tissue and to the extracellular matrix, a characteristic that allows them to migrate from the primary tumor. The best characterized alterations in this process are loss of E-cadherin (CDH1), an essential molecule for cell–cell adhesion, loss of cell polarity, and the secretion of ECM-degrading proteins to invade the adjacent stroma and migrate toward the circulatory system [45]. miR-200 is a tumor suppressor miRNA family involved in EMT inhibition by targeting and inhibiting *ZEB1* and *ZEB2* translation, two transcriptional repressors of E-cadherin and the polarity genes, *CRB3* and *LGL2* [46]. This family of miRNA also regulates the effectors of extracellular matrix degradation, a key process for cell migration toward the blood and lymphatic vessels, and invadopodia formation to facilitate cell migration [47,48]; downregulation of this miRNA family has been described in many solid tumors, including lung ADC [49]. Other miRNAs have been related to extracellular matrix degradation: miR-133a specifically targets matrix metalloproteinase 14 (*MMP-14*) in lung cancer cell lines [50] and miR-152 targets the ADAM metallopeptidase domain 17 (*ADAM17*) in lung cancer cells, as shown by an inverse correlation between *ADAM17* and miR-152 in NSCLC patients [51].

4. miRNA-BASED BIOMARKERS FOR LUNG CANCER DIAGNOSIS, PROGNOSIS, AND TREATMENT

4.1 Introduction to Lung Cancer Diagnosis, Prognosis, and Treatment

Technological and research advances in lung cancer have enhanced the diagnosis and prognosis of patients. However, the mortality rates are still high as the primary tumor is difficult to detect at early stages when cancer is treatable. Unfortunately, less than 15% of the patients will survive 5 years after diagnosis. The prognosis of the patients is strongly correlated with the state of the disease at the time of diagnosis. Whereas patients diagnosed at stage I (tumor confined to the lung) have a 5-year survival rate of 60%, clinical stages II-IV (II/III the tumor is confined to the lung but presents lymph nodes invasion; IV the tumor has spread to distal tissues from the lung) have 5-year survival rates varying from 40% to less than 5%. Therefore, there is an urgent need of new diagnostic, prognostic, and drug-response biomarkers to improve the life expectancy for lung cancer patients.

Lung cancer diagnosis accomplishes a wide set of noninvasive and invasive methods. Noninvasive methods include computed tomography scan (CT-SCAN) [52], sputum cytology examination [53], or exhaled breath analysis [54]. Invasive methods imply surgery to obtain biopsies from the primary tumor and from the potentially invaded lymphatic nodes.

Regarding miRNA analysis, the development of high-throughput techniques, like next-generation sequencing or miRNA microarrays, has opened a new window in cancer monitoring by converting the detection of miRNAs in biological samples in feasible tools for lung cancer diagnosis, patient prognosis, and drug response [55].

Table 2 summarizes the best candidates described so far for lung cancer diagnosis, prognosis, and drug resistance.

4.2 miRNA Signatures in Lung Cancer Diagnosis

miRNA expression signatures are a promising tool for clinical purposes, ranging from diagnosis to personalized treatment. Many studies highlight the clinical potential of the miRNA expression profiles. miRNAs are an attractive option for disease biomarker discovery due to the latest advances on high-throughput methods for miRNA profiling and the higher stability of these molecules compared with mRNA.

TABLE 2 Selected microRNA for Lung Cancer Diagnosis, Prognosis, and Drug Resistance

Epigenetic biomarker	Lung cancer subtype	Biological material	Diagnostic value	References
miR-205/miR-99b/miR-203/ miR-202/miR-120/miR-204	NSCLC	Tumor sample	ND	[42]
miR-21	NSCLC	Sputum	ND	[57]
miR-205/miR-210/miR-708	SCC	Sputum	ND	[58]
miR-25/miR-223	NSCLC	Plasma/serum	ND	[59]
Epigenetic biomarker	**Lung cancer subtype**	**Biological material**	**Prognostic value**	**References**
High miR-155/low let-7a-2	ADC	Tumor sample	Overall survival	[42]
miR-150/miR-886-33p	SCLC	Tumor sample	Survival in early stage	[64,65]
let-7a/miR-221/miR-137/ miR-372/miR-182	NSCLC	Tumor sample	Survival and relapse	[66]
High miR-486/miR-30d+low miR-1/miR-499	NSCLC	Serum	Overall survival	[67]
miR-142-3p	ADC	Serum	Tumor recurrence	[68]
Epigenetic biomarker	**Lung cancer subtype**	**Biological material**	**Drug resistance**	**References**
miR-1290/miR-196b/miR-135a	ADC	Tumor sample	Platinum-based therapy	[74]
miR-149/miR-375	NSCLC	Tumor sample	Cisplatin-vinorelbine	[75]

NSCLC, non-small-cell lung cancer; SCC, squamous cell lung cancer; ADC, adenocarcinoma; SCLC, small cell lung cancer; ND, not determined.

miRNA microarray analysis is the most commonly used high-throughput technique to obtain cancer-specific expression of hundreds of miRNAs in a large number of samples. Using this approach, Volinia et al. [56] established cancer gene targets by defining an miRNA signature of solid human tumors. This study describes the miRNA expression profile of 363 human tumor samples representing six solid tumors (breast, colon, lung, pancreas, prostate, and stomach) and 177 normal tissue samples. They described a specific signature for each tumor type with 21 miRNAs upregulated in at least three tumor types. Surprisingly, miR-21 is upregulated in all the tumor types analyzed. Most of the 21 miRNAs deregulated in solid tumors targeted well-known genes involved in cancer like retinoblastoma (*RB1*), TGF-b2 receptor (*TGFBR2*), or *BCL-2* [41].

Regarding lung cancer, Yanaihara et al. [42] have defined a set of 43 differentially expressed miRNAs in a study with 104 matched pairs of primary tumors and its noncancerous lung tissue counterpart from NSCLC patients. They found 43 miRNAs deregulated in tumor samples, 17 of them exclusive for ADCs and 16 specific for SCC. Additionally, they found a correlation between miRNA expression profiles and the disease progression, being high miR-155 expression levels an unfavorable prognostic factor independent of other clinicopathological factors in addition to disease stage.

Most of the signatures described for lung cancer diagnosis have been obtained from the analysis of invasive primary tumor biopsies. Noninvasive approaches to diagnose lung cancer patients are required to improve early diagnosis optimizing the patient care. In this regard, in a proof-of-concept study with 23 lung cancer patients and 17 healthy individuals, Xie et al. [57] showed that the presence of miR-21 in sputum specimens was significantly higher in NSCLC patients than in healthy individuals, being even more sensitive than the sputum cytology analysis. A bigger sample size was used to define an miRNA signature in early SCC diagnosis, comparing sputum samples from 48 patients and 48 cancer-free individuals. Three miRNAs' (miR-205, miR-210, and miR-708) signature distinguished patients from healthy individuals. A validation cohort of 67 SCC patients and 55 healthy individuals was used to confirm the signature obtained [58].

Circulating miRNA in the plasma could be a suitable noninvasive tool for cancer diagnosis. The first study characterizing circulating miRNAs identified these molecules in the plasma and serum of healthy individuals and patients of different diseases. Chen et al. [59] describe a circulating miRNAs profile exclusive for NSCLC patients, not detected in healthy subjects. In this study, the authors propose miR-25 and miR-223 as the most promising circulating miRNAs for cancer diagnosis.

The majority of the published studies were restricted to the analysis of already diagnosed and symptomatic patient tumor samples, but a screening for asymptomatic high-risk patients is necessary, since early diagnosis is the main way to improve disease prognosis. CT population screening tests led to an overdiagnosis and overtreatment of benign tumors without clinical relevance [60]. Circulating miRNA may serve to develop new more accurate tests. For this purpose, Bianchi and collaborators [61] have developed a test for the detection of 34 circulating miRNAs in the serum that can distinguish with 80% accuracy asymptomatic NSCLC patients. This kind of screening test could be implemented in the clinics, as they have reduced cost, are easy to perform and minimally invasive.

4.3 miRNA Signatures in Lung Cancer Prognosis

There are four main stages in lung cancer progression, from I to IV, which define the anatomical extent of the disease with important prognostic information. The best predicted outcome is for stage I, when the tumor is confined to the tissue of origin and has not spread to any lymph node, making possible to completely remove this primary tumor. The 5-year survival rate for NSCLC patients diagnosed in stage I ranges from 35% to 50%. As the tumor progresses, the overall survival decreases, showing stage IV patients, with the tumor invading lymph nodes and distal tissues, a 5-year survival rate of 2–13%. SCLC accounts for ~15% of the lung cancer cases. The 5-year survival rate is lower than for NSCLC in all the four stages. Surgery is rarely used as it spreads around the tumor focal point and to the lymph nodes quickly, being the 5-year survival rate when diagnosed in stage IV of only 1%.

Although early-stage diagnosis predicts a better outcome, 53% of the NSCLC patients diagnosed in stage I and II with a complete surgical resection of the primary tumor will suffer tumor recurrence within 5 years after diagnosis [62]. Robust prognostic markers to identify patients with a high risk of relapse after surgery are needed to improve their overall survival. Similar to cancer diagnosis, miRNA signatures could be useful tools for this purpose based on their stability compared to other markers.

An analysis of the miRNA expression profile of 527 stage I NSCLC biopsy samples identified four different miRNA signatures associated with histological subtypes, brain metastasis, and recurrence/relapse-free survival. miRNA profiles were able to differentiate ADCs from squamous cell carcinoma, being 92 miRNAs overexpressed only in ADC and 53 only in SCC. Interestingly, 10 miRNAs were exclusively associated to brain metastasis, including the downregulation of miR-145, a tumor suppressor gene associated to cell invasion and metastasis inhibition by targeting the mucin 1 (*MUC1*). Moreover, miR-29c and miR-28 are involved in extracellular matrix remodeling in nasopharyngeal carcinomas and invasion inhibition in gastric cancer, respectively. In the same study, two signatures were able to predict the recurrence/relapse-free survival of the patients, one taking into account the two NSCLC subtypes and the other exclusive for ADCs. Both signatures were confirmed in a validation cohort of 170 stage I lung cancer patients [63].

Regarding SCLC, there are a few studies defining a prognostic signature, since surgery is often not suitable for the patients. A recent study defines an miRNA signature to predict survival in early stages of SCLC patients where surgery was applicable. Biopsies from 82 patients with focal SCLC and adjuvant therapy were analyzed to define a survival miRNA signature; the authors found that higher expression levels of miR-150 and miR-886-3p was correlated with a better overall survival and both miRNA expression levels were lower in SCLC samples than normal lung samples [64]. It was previously described that the decrease of miR-886-3p levels in SCLC patients predicts a poor outcome in two independent cohorts of 42 and 40 patients, being DNA hypermethylation of its promoter the responsible of its downregulation in tumors [65].

There is also a growing interest in defining an miRNA signature using noninvasive methods but with an accurate prognosis prediction. A study using 112 NSCLC sputums described a signature of five miRNAs to predict patient survival, being significantly associated with relapse-free survival of stage I patients and with overall survival for stage III patients. They also reported that the five miRNAs were needed for the predictive signature as any combination of just four does not correlate with either relapse-free or overall survival. These results were confirmed in an independent cohort with 62 patients [66].

Circulating miRNAs in the serum emerge as a powerful tool for clinicians as well as for prognosis. Hu et al. [67] describe a four miRNA signature from patient serum as a predictor for the overall survival in NSCLC. Analyzing serum from patients divided into long-survival group (more than 30 months of survival time) or short-survival group (less than 25 months of survival time), they found four miRNAs significantly different between the two groups. High serum expression levels of miR-486 and miR-30d and low expression levels of miR-1 and miR-499 were individually associated with unfavorable survival. Taking together these findings, they concluded that patients carrying two or more high-risk miRNAs present a shortened survival probability than those carrying none or just one high-risk miRNA.

Another study regarding tumor recurrence after surgery identifies an miRNA signature in the serum of the patients with early relapse in lung ADC. Circulating miR-142-3p was higher in the sera of early-stage ADC patients that suffered tumor recurrence 24 months after surgery [68].

4.4 miRNA Signatures as Predictors of Drug Resistance in Lung Cancer

Standard lung cancer care includes surgery when possible and adjuvant therapy, which typically involves radiation, chemotherapy, and targeted therapy, or combined treatments. Early-stage SCLC chemotherapy includes either cisplatin or carboplatin (both drugs interfere with DNA repair) in combination with etoposide (topoisomerase inhibitor) or gemcitabine (nucleoside analog); for recurrent tumors, chemotherapy is supplemented with agents to inhibit DNA integrity (cyclophosphamide, doxorubicin) and mitosis (vincristine) [69].

NSCLC chemotherapy includes a combination of cisplatin or carboplatin plus mitotic inhibitors (vinorelbine, paclitaxel, or docetaxel) and DNA synthesis inhibitors (doxorubicin, pemetrexed, or gemcitabine) [70].

Increasing knowledge of the molecular mechanisms underlying cancer development promotes the improvement of target-directed drugs to decrease the side effects of the chemotherapy. Personalized medicine became a major clinical concern as each patient should only receive the set of drugs specific for their tumor type and subtype. Targeted therapy refers to the blockage of cancer-specific genes, proteins, or microenvironment alterations that contribute to tumor growth. In lung cancer, targeted therapy is directed against activity-enhancing mutations in EGFR (erlotinib, gefitinib, and gilotrif) and ALK (crizotinib) [71].

A key challenge in cancer treatment is the emergence of inherent and acquired drug resistance by the tumor cells. Resistance to anticancer therapy agents is frequently observed in both SCLC and NSCLC patients. miRNA signatures have been studied to predict drug resistance to the cancer therapy. Lung ADC patients with *EGFR*, *BRAF*, *HER2*, *RET*, and/or *ROS1* mutations are treated with specific tyrosine kinase inhibitors. However, more than 60% of the European patients do not have any of the mutations listed [72]. In this case, a chemotherapeutic combination based on platinum (cisplatin or carboplatin) plus another agent including microtubule-targeted agents (paclitaxel, docetaxel, or vinorelbine) and DNA-damaging agents (gemcitabine and irinotecan) is preferred. All the combinations have a similar efficacy, with response rates from 30% to 40% [73]. Saito et al. [74] describe a signature of just three miRNAs for response prediction to platinum-based chemotherapy after primary tumor resection. miR-1290, miR-196b, and miR-135a were able to discriminate between responder and nonresponder patients independently of their genome abnormalities. In the same line, a study published by Berghmans and collaborators [75] found a signature of two miRNAs (miR-149 and miR-375) associated to drug response and progression-free survival. miR-149 is involved in cell apoptotic processes, epithelial–mesenchymal transition, control of cell proliferation, motility, and invasion [76]. miR-375 is altered in neuroendocrine lung cancer among other cancer types like hepatocellular carcinoma or gastric cancer [77].

5. INTERPLAY BETWEEN GENETICS AND EPIGENETICS IN miRNA REGULATION AND LUNG CANCER

As we highlighted in the previous section, alterations in the profile of miRNAs are common features in tumor progression. Similar to coding genes, multiple mechanisms can alter the miRNA profile in cancer cells.

5.1 Transcription Factors Interplay

The promoter region of miRNA loci contains CpG islands, TATA boxes, and histone modification patterns that regulate their transcription. Moreover, different transcription factors, including tumor suppressor genes and oncogenes, regulate the expression of the primary transcripts of miRNAs. For example, the protooncogene *MYC* and the tumor suppressor gene *p53* modulate miRNA transcription.

MYC, a transcription factor pathologically activated in many human malignancies, including lung ADC [78], promotes miRNA silencing by direct binding to the pri-miRNA promoter region where MYC antagonizes the activity of positive effectors of transcription [79]. However, the opposite effect is also described for the miRNAs, miR-17-5p and miR-92. It has been shown that Myc activates these miRNAs in human cell lines. miR-17-5p and miR-92 specifically silence the translation of *E2F1* transcription factor, a crucial regulator of cell cycle. On the other hand, MYC directly promotes *E2F1* gene transcription, adding a new layer of cell cycle control by *MYC* oncogene in a loop tightly regulated [80,81].

p53 is a transcriptional activator of genes involved in growth arrest, DNA repair, and apoptosis, playing a key role in cell homeostasis by inducing cell apoptosis upon DNA damage. p53 activity is downregulated in the majority of human cancers. miR-21 is an oncomiR highly expressed in different tumor types like breast, ovarian, colorectal, leukemia, and lung that contains a p53 binding site at its promoter region [82]. miR-21 targets and represses the translation of a set of tumor suppressor genes contributing to cell transformation like *PTEN* or negative regulators of the RAS signaling pathway [83,84].

5.2 Genetic Interplay

Variations in the DNA sequence like single nucleotide polymorphisms (SNPs) have been studied to predict disease susceptibility. The miRNA regulation of its mRNA target can be disrupted depending on the presence of a determined nucleotide in a specific position. Saunders et al. [85] highlight that 90% of human pre-miRNAs do not display sequence polymorphisms and most of the SNPs found are not in the seed region, suggesting a negative selection over sequence variability. However, although not in the seed region, polymorphisms in the pre-miRNA can alter the miRNA processing affecting the final levels of the miRNAs in a cell.

The miRNA-binding site in the target mRNA can also present sequence variability, altering the set of miRNA that can recognize it [86] or inhibiting the miRNA binding. Polymorphisms in both miRNA and mRNA are being thoroughly studied for cancer risk prediction. The G to T genetic variation in the rs3134615 SNP present in the 3′-UTR-binding site in *MYCL1* mRNA inhibits the binding capacity of the miR-1827, a negative effector of *MYCL1* translation; as MYC family of proteins act as oncogenes, the alteration of the binding site for miRNA regulation is another way to activate their oncogenic role in transformed cells [87]. Moreover, the presence of the SNP significantly increases the risk of SCLC [88]. An miR-SNP located in the 3′-UTR of the *XPO5* mRNA has been associated with the time to recurrence in resected NSCLC patients and overall survival in SCLC, being the genotype AA in the rs11077 position associated with a better survival in SCLC [89] but with a reduced time to recurrence in NSCLC [90]. The same study defines the genetic variation in the SNP rs3660 in the *KRT81* mRNA as a useful prognosis marker of the clinical outcome for time NSCLC patients, associating the genotype CC to a reduced time to recurrence after tumor resection.

5.3 Epigenetic Interplay

Histone marks are a set of posttranslational modifications of the histones that regulate gene expression and hence a wide range of cellular processes. miRNAs are also regulated by these mechanisms, for example, miR-212 in NSCLC. The silencing histone mark H3K27me3/H3K9Ac or H3K9me3/H3K9Ac downregulates miR-212 expression. miR-212 downregulation is associated with the severity of the disease, being its levels lower as the stage of the tumor advances [91].

Cancer cell genomes frequently show chromosome instability, genomic rearrangements, and aberrant mitotic recombination that lead to deletions and translocations. Tumor suppressor gene deletions and amplifications of oncogenes are very well documented in cancer development. Translocations that drive the formation of aberrant fusion proteins and the activation of oncogenes by active promoters are found in many cancer types. miRNAs are often located in these fragile sites commonly altered in cancer. Deletions and LOH (loss of heterogeneity) of tumor suppressor have been described in lung cancer, contributing to tumor progression [92].

The hypothesis of gene expression regulation by competitive endogenous RNA (ceRNA) has been recently proposed. It was first proposed by Poliseno et al. [93] when they discover that the pseudogene *PTENP1* actively regulates the protein levels of the tumor suppressor gene *PTEN* by its ability to compete for the miRNA present in the cell, acting as a sponge that binds the common miRNAs. This mechanism adds a new layer of gene expression regulation in normal and malignant cells as many genes involved in cell transformation, such as *KRAS*, have pseudogenes.

6. miRNA-BASED ANTICANCER THERAPY

As miRNAs are altered in cancer cells, they have become new target candidates for new anticancer therapies. Inhibition of oncomiRs and replacement of tumor suppressor miRNAs are being thoroughly studied to restore a nonpathogenic phenotype in cancer cells and to sensitize malignant cells to a particular chemotherapy. Oncogenic miRNAs inhibition is based on antisense oligonucleotides, also known as anti-miRs or antagomirs, that specifically bind the upregulated miRNA and block its downstream function. It was first tested in vivo by Krützfeldt et al. [94]; in this study they showed that an antagomir can effectively silence miR-122, a liver-specific miRNA related to cholesterol synthesis in the liver of the mice, by high-affinity binding to the targeted miRNA. The mechanism is specific to miR-122 as no alterations of the expression of off-target miRNAs were found. This strategy is the first miRNA-based treatment and is currently being tested in a clinical trial for hepatitis C [95]. Another strategy to inhibit the function of upregulated miRNAs is the introduction into the cells of artificial miRNA sponges that anchor the upregulated miRNAs, mimicking the ceRNA mechanism [96].

miRNA replacement is the exogenous introduction of the depleted miRNA in a cancer cell. Systemic restoration of the tumor suppressor miR-34a in an NSCLC mouse model blocks tumor growth, promoting the downregulation of its direct targets [97]. RNA interference-based therapy has been investigated to inhibit the translation of upregulated oncogenes since the siRNAs were first described in 1998 by Fire and collaborators [98]. siRNAs are small molecules that perfectly match the targeted sequence in the mRNA, leading to its degradation. The first siRNA-based therapy (CALAA-01) is being tested in a phase I clinical trial to treat solid human tumors [99]. CALAA-01 is a nanoparticle containing siRNA that specifically binds the M2 subunit of ribonucleotide reductase (RRM2), an established anticancer target [100]. Another miRNA replacement therapy, called MRX34, is being tested in a phase I clinical trial in primary liver cancer treatment. MRX34 mimics the miR-34 miRNA, a tumor suppressor miRNA downregulated not only in liver cancer but also in lung cancer, suggesting a potential use in lung cancer patients [101].

7. CONCLUDING REMARKS

Personalized medicine is becoming a reality for the treatment of cancer patients accordingly with their individual characteristics. It is important to find new tools to guide the decisions made by the clinicians regarding diagnosis and proper treatment of the disease. miRNAs have emerged as a feasible instrument for this purpose; their small size makes them stable molecules in blood, serum, sputum, and tumor samples. Therefore, specific signatures may contribute to better diagnose and treat cancer. Different miRNA profiles have been defined not only in tumor samples but also in body fluids, facilitating the noninvasive analysis of the specific patient signature.

miRNAs have also emerged as appropriate targets and tools for cancer treatment. Their use diverges from the targeting of overexpressed miRNAs in tumor cells by complementary sequences to the introduction of downregulated miRNAs in tumor cells to restore their function, opening a promising approach for the development of new therapies.

LIST OF ABBREVIATIONS

ADC Adenocarcinoma
CT Computed tomography
EMT Epithelial–mesenchymal transition
LOH Loss of heterogeneity
miRNA, miR MicroRNA
mRNA Messenger RNA
ncRNA Noncoding RNA
NSCLC Non-small-cell lung cancer
RISC RNA-induced silencing complex
SCC Squamous cell lung cancer
SCLC Small cell lung cancer
siRNA Small interference RNA
SNP Single nucleotide polymorphism
UTR Untranslated region
WHO World Health Organization

References

[1] ENCODE Project Consortium, Birney E, Stamatoyannopoulos JA, Dutta A, et al. Identification and analysis of functional elements in 1% of the human genome by the ENCODE pilot project. Nature 2007;447(7146):799–816.
[2] Lee RC, Feinbaum RL, Ambros V. The *C. elegans* heterochronic gene lin-4 encodes small RNAs with antisense complementarity to lin-14. Cell 1993;75(5):843–54.
[3] Aravin AA, Sachidanandam R, Girard A, Fejes-Toth K, Hannon GJ. Developmentally regulated piRNAs clusters implicate MILI in transposon control. Science 2007;316(5825):744–7.
[4] King TH, Liu B, McCully RR, Fournier MJ. Ribosome structure and activity are altered in cells lacking snoRNPs that form pseudouridines in the peptidyl transferase center. Mol Cell 2003;11(2):425–32.
[5] Plath K, Fang J, Mlynarczyk-Evans SK, et al. Role of histone H3 lysine 27 methylation in X inactivation. Science 2003;300(5616):131–5.
[6] Rinn JL, Kertesz M, Wang JK, et al. Functional demarcation of active and silent chromatin domains in human HOX loci by noncoding RNAs. Cell 2007;129(7):1311–23.
[7] Guttman M, Amit I, Garber M, et al. Chromatin signature reveals over a thousand highly conserved large non-coding RNAs in mammals. Nature 2009;458(7235):223–7.
[8] Liz J, Portela A, Soler M, et al. Regulation of pri-miRNA processing by a long noncoding RNA transcribed from an ultraconserved region. Mol Cell 2014;55(1):138–47.
[9] Kapranov P, Cheng J, Dike S, et al. RNA maps reveal new RNA classes and a possible function for pervasive transcription. Science 2007;316(5830):1484–8.
[10] Preker P, Nielsen J, Kammler S, et al. RNA exosome depletion reveals transcription upstream of active human promoters. Science 2008;322(5909):1851–4.
[11] Taft RJ, Glazov EA, Cloonan N, et al. Tiny RNAs associated with transcription start sites in animals. Nat Genet 2009;41(5):572–8.
[12] Seila AC, Calabrese JM, Levine SS, et al. Divergent transcription from active promoters. Science 2008;322(5909):1849–51.
[13] Luke B, Lingner J. TERRA: telomeric repeat-containing RNA. EMBO J 2009;28(17):2503–10.
[14] Deng Z, Norseen J, Wiedmer A, Riethman H, Lieberman PM. TERRA RNA binding to TRF2 facilitates heterochromatin formation and ORC recruitment at telomeres. Mol Cell 2009;35:403–13.
[15] Esteller M. Non-coding RNAs in human disease. Nat Rev Genet 2011;12(12):861–74.
[16] Friedman RC, Farh KK, Burge CB, Bartel DP. Most mammalian mRNAs are conserved targets of microRNAs. Genome Res 2009;19(1):92–105.
[17] Lee Y, Kim M, Han J, et al. MicroRNA genes are transcribed by the RNA polymerase II. EMBO J 2004; 23:4051–60.
[18] Han J, Lee Y, Yeom KH, et al. Molecular basis for the recognition of primary microRNAs by the Drosha-DGCR8 complex. Cell 2006;125:887–901.
[19] Yi R, Qin Y, Macara IG, Cullen BR. Exportin-5 mediates the nuclear export of pre-microRNAs and short hairpin RNAs. Genes Dev 2003;17:3011–6.
[20] Hutvágner G, McLachlan J, Pasquinelli AE, Bálint E, Tuschl T, Zamore PD. A cellular function for the RNA-interference enzyme Dicer in the maturation of the let-7 small temporal RNA. Science 2001;293:834–8.
[21] Grimson A, Farh KK, Johnston WK, Garrett-Engele P, Lim LP, Bartel DP. MicroRNA targeting specificity in mammals: determinants beyond the seed pairing. Mol Cell 2007;27(1):91–105.
[22] Lewis BP, Burge CB, Bartel DP. Conserved seed pairing, often flanked by adenosines, indicates that thousands of human genes are microRNA targets. Cell 2005;120(1):15–20.
[23] Pillai RS, Bhattacharyya SN, Artus CG, et al. Inhibition of translational initiation by let-7 microRNA in human cells. Science 2005;309(5740):1573–6.
[24] Doench JG, Petersen CP, Sharp PA. siRNAs can function as miRNAs. Genes Dev 2003;17(4):438–42.
[25] Parker R, Sheth U. P bodies and the control of mRNA translation and degradation. Mol Cell 2007;25(5):635–46.
[26] Yekta S, Shih IH, Bartel DP. MicroRNA-directed cleavage of HOXB8 mRNA. Science 2004;304(5670):594–6.
[27] Davis E, Caiment F, Tordoir X, et al. RNAi-mediated allelic trans-interaction at the imprinted Rtl1/Peg11 locus. Curr Biol 2005;15(8):743–9.

[28] Eiring AM, Harb JG, Neviani P, et al. miR-328 functions as an RNA decoy to modulate hnRNP E2 regulation of mRNA translation in leukemic blasts. Cell 2010;140(5):652–65.
[29] Verdel A, Jia S, Gerber S, et al. RNAi-mediated targeting of heterochromatin by the RITS complex. Science 2004;303:672–6.
[30] Maida Y, Yasukawa M, Okamoto N, et al. Involvement of telomerase reverse transcriptase in heterochromatin maintenance. Mol Cell Biol 2014;34(9):1576–93.
[31] Vasudevan S, Tong Y, Steitz JA. Switching from repression to activation: microRNAs can up-regulate translation. Science 2007;318:1931–4.
[32] Ørom UA, Nielsen FC, Lund AH. MicroRNA-10a binds the 5′UTR of ribosomal protein mRNAs and enhances their translation. Mol Cell 2008;30:460–71.
[33] Herbst RS, Heymach JV, Lippman SM. Lung cancer. N Engl J Med 2008;359(13):1367–80.
[34] Chen Z, Fillmore CM, Hammerman PS, Kim CF, Wong KK. Non-small-cell lung cancers: a heterogeneous set of disease. Nat Rev Cancer 2014;14(8):535–46.
[35] Park KS, Liang MC, Raiser DM, et al. Characterization of the cell of origin for small cell lung cancer. Cell Cycle 2011;10(16):2806–15.
[36] International Early Lung Cancer Action Program Investigators, Henschke CI, Yankelevitz DF, Libby DM, Pasmantier MW, Smith JP, et al. Survival of patients with stage I lung cancer detected on CT screening. N Engl J Med 2006;355(17):1763–71.
[37] Melo SA, Esteller M. Dysregulation of microRNAs in cancer: playing with fire. FEBS Lett 2011;585(13):2087–99.
[38] Hanahan D, Weinberg RA. Hallmarks of cancer: the next generation. Cell 2011;144(5):646–74.
[39] Du B, Wang Z, Zhang X, Feng S, Wang G, He J, et al. MicroRNA-545 suppresses cell proliferation by targeting cyclin D1 and CDK4 in lung cancer cells. PLoS One 2014;9(2):e88022.
[40] Yang J, Zhao H, Xin Y, Fan L. MicroRNA-198 inhibits proliferation and induces apoptosis of lung cancer cells via targeting FGFR1. J Cell Biochem 2014;115(5):987–95.
[41] Singh D, Chan JM, Zoppoli P, et al. Transforming fusions of FGFR1 and TACC genes in human glioblastoma. Science 2012;337(6099):1231–5.
[42] Yanaihara N, Caplen N, Bowman E, et al. Unique microRNA molecular profiles in lung cancer diagnosis and prognosis. Cancer Cell 2006;9(3):189–98.
[43] Chen S, Xue Y, Wu X, et al. Global microRNA depletion suppresses tumor angiogenesis. Genes Dev 2014;28:1054–67.
[44] Gallach S, Calabuig-Fariñas S, Jantus-Lewintre E, Camps C. MicroRNAs: promising new antiangiogenic targets in cancer. Biomed Res Int 2014;2014. 878450.
[45] Yang J, Weinberg A. Epithelial–mesenchymal transition: at the crossroads of development and tumor metastasis. Dev Cell 2008;14(6):818–29.
[46] Gibbons DL, Lin W, Creighton CJ, et al. Contextual extracellular cues promote tumor cell EMT and metastasis by regulating miR-200 family expression. Genes Dev 2009;23(18):2140–51.
[47] Schliekelman MJ, Gibbons DL, Faca VM, et al. Targets of the tumor suppressor miR-200 in regulation of the epithelial–mesenchymal transition in cancer. Cancer Res 2011;71(24):7670–82.
[48] Bracken CP, Li X, Wrigh JA, et al. Genome-wide identification of miR-200 targets reveals a regulatory network controlling cell invasion. EMBO J 2014;33(18):2040–56.
[49] Gill BJ, Gibbons DL, Roudsari LC, et al. A synthetic matrix with independently tunable biochemistry and mechanical properties to study epithelial morphogenesis and EMT in a lung adenocarcinoma model. Cancer Res 2012;72(22):6013–23.
[50] Xu M, Wang YZ. miR-133a suppresses cell proliferation, migration and invasion in human lung cancer by targeting MMP-14. Oncol Rep 2013;30(3):1398–404.
[51] Su Y, Wang Y, Zhou H, Lei L, Xu L. MicroRNA-152 targets ADAM17 to suppress NSCLC progression. FEBS Lett 2012;588(10):1983–8.
[52] Sone S, Takashima S, Li F, et al. Mass screening for lung cancer with mobile spiral computed tomography scanner. Lancet 1998;351(9111):1242–5.
[53] Thunnissen FB. Sputum examination for early detection of lung cancer. J Clin Pathol 2003;56(11):805–10.
[54] Bajtarevic A, Ager C, Pienz M, et al. Noninvasive detection of lung cancer by analysis of exhaled breath. BMC Cancer 2009;9:348.
[55] Uso M, Jantus-Lewintre E, Sirera R, Bremnes RM, Camps C. miRNA detection methods and clinical implications in lung cancer. Future Oncol 2014;10(14):2279–92.
[56] Volinia S, Calin GA, Liu C, et al. A microRNA expression signature of human solid tumors defines cancer gene targets. Proc Natl Acad Sci USA 2005;103(7):2257–61.
[57] Xie Y, Todd NW, Liu Z, et al. Altered miRNA expression in sputum for diagnosis of non-small cell lung cancer. Lung Cancer 2010;67(2):170–6.
[58] Xing L, Todd NW, Yu L, Fang H, Jiang F. Early detection of squamous cell lung cancer in sputum by a panel of microRNA markers. Mod Pathol 2010;23(8):1157–64.
[59] Chen X, Ba Y, Ma L, et al. Characterization of microRNAs in serum: a novel class of biomarkers for diagnosis of cancer and other diseases. Cell Res 2008;18(10):997–1006.
[60] Aberle DR, Berg CD, Black WC, et al. The national lung screening trial: overview and study design. Radiology 2011;258(1):243–53.

[61] Bianchi F, Nicassio F, Marzi M, et al. A serum circulating miRNA diagnostic test to identify asymptomatic high-risk individuals with early state lung cancer. EMBO Mol Med 2011;3(8):495–503.

[62] Groome PA, Bolejack V, Crowley JJ, et al. The IASLC lung cancer staging project: validation of the proposals for revision of the T, N, and M descriptors and consequent stage groupings in the forthcoming (seventh) edition of the TNM classification of malignant tumours. J Thorac Oncol 2007;2(8):694–705.

[63] Lu Y, Govindan R, Wang L, et al. MicroRNA profiling and prediction of recurrence/relapse-free survival in stage I lung cancer. Carcinogenesis 2012;33(5):1046–54.

[64] Bi N, Cao J, Song Y, et al. A microRNA signature predicts survival in early stage small-cell lung cancer treated with surgery and adjuvant chemotherapy. PLoS One 2014;9(3):e91388.

[65] Cao J, Song Y, Bi N, et al. DNA methylation-mediated repression of miR-886-3p predicts poor outcome of human small cell lung cancer. Cancer Res 2013;73(11):3326–35.

[66] Yu SL, Chen HY, Chang GC, et al. MicroRNA signature predicts survival and relapse in lung cancer. Cancer Cell 2008;13(1):48–57.

[67] Hu Z, Chen X, Zhao Y, et al. Serum microRNA signatures identified in a genome-wide serum microRNA expression profiling predict survival of non-small-cell lung cancer. J Clin Oncol 2010;28(10):1721–6.

[68] Kaduthanam S, Gade S, Meister M, et al. Serum miR-142-3p is associated with early relapse in operable lung adenocarcinoma patients. Lung Cancer 2013;80(2):223–7.

[69] Kalemkerian GP. Advances in pharmacotherapy of small-cell lung cancer. Expert Opin Parmacother 2014;15(16):2385–96.

[70] Pilkington G, Boland A, Brown T, Oyee J, Bagust A, Dickson R. A systematic review of the clinical effectiveness of first-line chemotherapy for adult patients with locally advanced or metastatic non-small cell lung cancer. Thorax 2015. PII:thoraxjnl-2014-205914.

[71] Rolfo C, Passiglia F, Ostrowski M, et al. Improvement in lung cancer outcomes with targeted therapies: an update for family physicians. J Am Board Fam Med 2015;28(1):124–33.

[72] Oxnard GR, Binder A, Janne PA. New targetable oncogenes in non-small cell lung cancer. J Clin Oncol 2013;31(8):1097–104.

[73] Scagliotti GV, De Marinis F, Rinaldi M, et al. Phase II randomized trial comparing three platinum-based doublets in advanced non-small-cell lung cancer. J Clin Oncol 2002;20(21):4285–91.

[74] Saito M, Shiraishi K, Matsumoto K, et al. A three-microRNA signature predicts responses to platinum-based doublet chemotherapy in patients with lung adenocarcinoma. Clin Cancer Res 2014;20(18):4784–93.

[75] Berghmans T, Ameye L, Willems L, et al. Identification of microRNA-based signatures for response and survival for non-small cell lung cancer treated with cisplatin-vinorelbine A ELCWP prospective study. Lung Cancer 2013;82(2):340–5.

[76] Jin L, Hu WL, Jiang CC, et al. MicroRNA-149*, a p53-responsive microRNA, functions as an oncogenic regulator in human melanoma. Proc Natl Acad Sci USA 2011;108(38):15840–5.

[77] Nishikawa E, Osada H, Okazaki Y, et al. miR-375 is activated by ASH1 and inhibits YAP1 in a lineage-dependent manner in lung cancer. Cancer Res 2011;71(19):6165–73.

[78] The Cancer Genome Atlas Research Network. Comprehensive molecular profiling of lung adenocarcinoma. Nature 2014;511(7511):543–50.

[79] Chang TC, Yu D, Lee YS, et al. Widespread microRNA repression by Myc contributes to tumorigenesis. Nat Genet 2008;40(1):43–50.

[80] O'Donnell KA, Wentzel EA, Zeller KI, Dang CV, Mendell JT. c-Myc-regulated microRNAs modulate E2F1 expression. Nature 2005;435(7043):839–43.

[81] Coller HA, Forman JJ, Legesse-Miller A. "Myc'ed messages": myc induces transcription of E2F1 while inhibiting its translation via a microRNA polycistron. PLoS Genet 2007;3:e146.

[82] Krichevsky AM, Gabriely G. miR-21: a small multi-faceted RNA. J Cell Mol Med 2009;13(1):39–53.

[83] Zhang JG, Wang JJ, Zhao F, Liu Q, Jiang K, Yang GH. MicroRNA-21 (miR-21) represses tumor suppressor PTEN and promotes growth and invasion in non-small cell lung cancer (NSCLC). Clin Chim Acta 2010;411(11–12):846–52.

[84] Hatley ME, Patrick DM, Garcia MR, et al. Modulation of K-Ras-dependent lung tumorigenesis by MicroRNA-21. Cancer Cell 2010;18(3):282–93.

[85] Saunders MA, Liang H, Wen-Hsiung L. Human polymorphism at microRNAs and microRNA target sites. Proc Natl Acad Sci USA 2007;104(9):3300–5.

[86] Chen K, Rajewsky N. Natural selection on human microRNA binding sites inferred from SNP data. Nat Genet 2006;38(1):1452–6.

[87] Xiong F, Wu C, Chang J, et al. Genetic variation in a miRNA-1827 binding site in MYCL1 alters susceptibility to small-cell lung cancer. Cancer Res 2011;71(15):5175–81.

[88] van Meerbeeck JP, Fennell DA, De Ruysscher DK. Small-cell lung cancer. Lancet 2011;378(9804):1741–55.

[89] Guo Z, Wang H, Li Y, Li B, Li C, Ding C. A microRNA-related nucleotide polymorphism of the XPO5 gene is associated with survival of small cell lung cancer patients. Biomed Rep 2013;1(4):545–8.

[90] Campayo M, Navarro A, Viñolas N, et al. A dual role for KRT81: a miR-SNP associated with recurrence in non-small-cell lung cancer and a novel marker of squamous cell lung carcinoma. PLoS One 2011;6(7):e22509.

[91] Incoronato M, Urso L, Portela A, et al. Epigenetic regulation of miR-212 expression in lung cáncer. PLoS One 2011;6:e27722.

[92] Calin GA, Sevignani C, Dumitru CD, et al. Human microRNA genes are frequently located at fragile sites and genomic regions involved in cancers. Proc Natl Acad Sci USA 2004;101(9):2999–3004.

[93] Poliseno L, Salmena L, Zhang J, Carver B, Haveman WJ, Pandolfi PP. A coding-independent function of gene and pseudogene mRNAs regulates tumor biology. Nature 2010;465(7301):1033–8.

[94] Krützfeldt J, Rajewsky N, Braich R, et al. Nature 2005;438(7068):685–9.

[95] Janssen HL, Reesink HW, Lawitz EJ, et al. Treatment of HCV infection by targeting microRNAs. N Engl J Med 2013;368(18):1685–94.

[96] Ebert MS, Sharp PA. MicroRNA sponges: progress and possibilities. RNA 2010;16(11):2043–50.

[97] Trang P, Wiggins JF, Daige CL, et al. Systemic delivery of tumor suppressor microRNA mimics using a neutral lipid emulsion inhibits lung tumors in mice. Mol Ther 2010;19(6):1116–22.

[98] Fire A, Xu S, Montgomery MK, Kostas SA, Driver SE, Mello CC. Potent and specific genetic interference by double-stranded RNA in *Caenorhabditis elegans*. Nature 1998;391(6669):806–11.

[99] Zuckerman JE, Gritli I, Tolcher A, et al. Correlating animal and human phase Ia/Ib clinical data with CALAA-01, a targeted, polymer-based nanoparticle containing siRNA. Proc Natl Acad Sci USA 2014;111(31):11449–54.

[100] Shao J, Zhou B, Chu B, Yen Y. Ribonucleotide reductase inhibitors and future drug design. Curr Cancer Drug Targets 2006;6(5):409–31.

[101] Fortunato O, Boeri M, Verri C, Moro M, Sozzi G. Therapeutic use of MicroRNAs in lung cancer. Biomed Res Int 2014:756975.

CHAPTER

28

Circulating Nucleic Acids as Prostate Cancer Biomarkers

Claire E. Fletcher, Ailsa Sita-Lumsden, Akifumi Shibakawa, Charlotte L. Bevan

Department of Surgery & Cancer, Imperial Centre for Translational & Experimental Medicine, Imperial College London, Hammersmith Hospital Campus, London, UK

OUTLINE

1. INTRODUCTION—THE URGENT REQUIREMENT FOR NEW PROSTATE CANCER BIOMARKERS

Prostate cancer (PCa, OMIM #176807) is the most frequently diagnosed cancer of men in the Western world. The median age at diagnosis is 72, and while diagnosis rates in older men are relatively high, PCa-specific mortality rates are comparatively much lower. For example, for men in the USA, the overall lifetime risk of developing PCa is 1 in 6, while the mortality risk—1 in 36—is considerably less [1]. Thus a considerable complication for clinicians treating PCa is that

Epigenetic Biomarkers and Diagnostics
http://dx.doi.org/10.1016/B978-0-12-801899-6.00028-0

some tumors remain relatively indolent, while others will become lethal and highly aggressive, and there is no accurate, noninvasive test to distinguish between such tumors. This yields two problems: first, overtreatment of indolent cases with little therapeutic benefit, but development of life-altering side effects that include impotence, bone demineralization, gynecomastia, and urinary problems. The second issue is in preventing growth, metastasis, and development of therapy resistance ("castration resistance") in individuals with aggressive disease. Thus, identification of new biomarkers that can accurately and noninvasively permit diagnosis, predict prognosis/treatment response, and/or stratify indolent from aggressive cancers remains the "holy grail" of PCa diagnosis and treatment.

Over the last two decades, deaths from PCa have decreased considerably, thanks in part to the introduction of prostate-specific antigen (PSA) testing. Twenty years since its introduction, PSA detection in serum remains the gold standard diagnostic biomarker and is also used to monitor treatment response to the extent that biochemical relapse is defined in terms of an elevation in PSA. Current diagnostic regimes involve initial digital rectal examination and serum PSA testing. Abnormal results indicate possible presence of PCa, at which point, a transrectal biopsy is indicated. However, PSA testing has a number of limitations, coming with a substantial risk of overdiagnosis and overtreatment. For example, elevated PSA is also found in men with other, nonmalignant, prostatic diseases including prostatitis and benign prostate hyperplasia (BPH). Thus, an unacceptably high false-positive rate results, as exemplified by the finding that less than 50% of men who have a biopsy after an elevated PSA test are subsequently diagnosed with PCa [2]. As prostate biopsy carries a substantial risk of infection, such a false-positive rate represents major clinical issue. Conversely to issues of false-positive detection, detection of very low circulating PSA does not entirely discount the presence of PCa. Indeed, the false-negative rate (where PCa is diagnosed despite a negative PSA test (0–4ng/mL)) is approximately 15% [3]. A number of alternatives to PSA testing have been proposed to improve sensitivity and specificity, including the use of age- or race-specific PSA reference ranges. While investigations into their clinical utility are ongoing, none of these alternatives have been demonstrated to decrease unnecessary biopsies or improve patient outcome [4]. The medical and financial costs of overdiagnosis are fueling a considerable drive toward discovery of novel PCa biomarkers, and already, a number of potential diagnostic tests are undergoing clinical trials. For example, increased levels of PCa antigen 3 (PCA3) have shown diagnostic biomarker potential [5]. Expression of the TMPRSS2:ERG fusion gene has also been assayed in cells isolated from urine of men following rectal examination, since this fusion event is rarely described in PCa-free individuals [6].

As for other diseases, the prospect of noninvasive "liquid biopsies" that can be used for accurate tests for PCa is an enticing one, and so circulating nucleic acids found in the blood are a major area of biomarker research. The possibility was first raised for microRNAs (miRNAs). Since the first publications describing the role of miRNAs in PCa development and progression in 2007, hundreds of articles have explored and highlighted their potential use as future diagnostic and prognostic tools. Other authors have investigated additional circulating nucleic acid species, including cell-free DNA, mRNA, and mitochondrial DNA (mtDNA) for their ability to predict PCa. It should be noted that PCa is a highly heterogeneous disease (as for other tumors, samples consist of different clonal subpopulations), also the epigenetic and nontranscriptional control processes that contribute to initiation or disease progression, or dictate therapy response, are heterogeneous and affect many pathways. To some extent, this may be mitigated by the use of "liquid biopsies" rather than tissue samples, however, the "noise" associated with heterogeneity could cause differences between subgroups to be missed. In response to this, novel methods of outlier detection have been developed (reviewed and compared in Ref. [7]). In this chapter, we aim to summarize recent research efforts in exploring circulating nucleic acids as novel PCa biomarkers.

2. CIRCULATING miRNAs AS BIOMARKERS

miRNAs are short noncoding RNAs (typically 21–23 nucleotides) that suppress gene expression at the posttranscriptional level. They suppress expression of target mRNAs through binding to complementary sequences, most often located in the 3′-untranslated region (3′-UTR), leading to transcriptional degradation or translational inhibition [8]. With respect to cancer, miRNAs can function as oncomiRs, promoting tumor growth or tumor suppressors, repressing it (reviewed in Ref. [9]). Interestingly, some miRNAs have been reported to function as both oncomiRs and tumor suppressors depending on the cancer subtype. For example, miR-125b is oncogenic in PCa, while it has been shown to act as a tumor suppressor in breast and ovarian cancer [10]. This apparent contradiction may be explained by the fact that a single miRNA has potentially thousands of target genes, and the function of a given miRNA depends on the relative availability of its target mRNAs that are likely to vary between cancer types.

In the context of biomarkers, circulating miRNAs have many favorable properties which include availability in biological fluids, stability, and ease of detection. Cell-free miRNAs have been readily detected in many biological fluids, including plasma, urine, saliva, seminal fluid, and even tears [10–12,42]. There are multiple sources for circulating miRNAs; tumor cells have been reported to release miRNAs in circulation in protective exosomes as well as in stable complexes by packaging them with specific proteins [9,13]. It is still unclear as to why disease causes changes in the levels of circulating miRNAs, but it is possible that cell lysis and/or an increase in exosome shedding may play a role. Circulating miRNAs are stable and can survive harsh handling conditions and can even be reliably quantified from archival samples. Finally, detection of circulating miRNAs is relatively easily achieved by quantitative polymerase chain reaction (qPCR), which is sensitive and specific [14].

2.1 Circulating miRNAs as Diagnostic Biomarkers

Diagnostic markers are particularly useful for early detection of disease, averting the morbidity and mortality associated with late-stage diagnosis. For PCa, PSA still remains the chief diagnostic marker. However, the actual benefit of using PSA as a marker of diagnosis has recently been questioned. Two large multinational randomized prospective control trials (European Randomized Study of Screening for Prostate Cancer (ERSPC) and Prostate, Lung, Colorectal, and Ovarian Cancer Screening Trial (PLCO)) have shown that PSA screening provided no substantial benefit in overall patient survival [2]. Subsequently, the US Preventive Services Task Force declared that there is no conclusive evidence to suggest the population benefit of PSA screening and they do not recommend it for men of any age [15]. There is, therefore, an urgent need for specific diagnostic markers that can differentiate PCa patients needing treatment and those with indolent PCa who do not require treatment.

The diagnostic utility of circulating miRNAs in PCa has been heavily investigated in recent years since in 2008, Mitchell et al. assessed the levels of 6 miRNAs by qPCR in the serum of 25 men with metastatic PCa and 25 healthy men. Of six miRNAs, serum levels of miR-141 were greatly elevated in men with metastatic PCa. In addition, serum levels of miR-141 and PSA showed significant correlation and could identify men with metastatic PCa with 60%

sensitivity and 100% specificity [11]. Although a great initial step forward in this area of research, in reality this low sensitivity would translate to 40 out of every 100 men with a raised serum miR-141 being falsely diagnosed with PCa.

In another study, Moltzahn et al. screened serum samples of 36 men with untreated PCa and 12 healthy men for the levels of 384 miRNAs by multiplex qPCR [16]. Receiver operating characteristic curves of individual miRNAs revealed that several miRNAs possessed significant diagnostic ability. Serum levels of miR-93 and miR-106a were consistently elevated in the high-risk group according to the Cancer of the Prostate Risk Assessment (CAPRA) score (CAPRA score >5 with lymph node metastasis) when compared with the control group.

In other studies, Bryant et al. assessed circulating levels of 742 miRNAs in 78 men with PCa and 20 healthy men and found significantly altered levels of 12 miRNAs with the highest fold change in miR-107 (>11-fold increase) [12], and Chen et al. reported that a panel of 5 miRNAs (upregulation of miR-622 and miR-1285 and downregulation of let-7c, let-7e, and miR-30c) generated an area under the curve of 86% in differentiating PCa patients from healthy controls [17]. Notably, miR-141 has been consistently identified as being at increased levels in circulation across multiple studies. Yaman Agaoglu et al. assessed the diagnostic ability of miR-21, miR-141, and miR-221 plasma levels in PCa patients. The authors found that plasma levels of miR-141 were significantly elevated in PCa patients with bone metastases compared with localized or locally advanced PCa. In addition, plasma levels of miR-21 and miR-221 were elevated in PCa patients when compared with healthy controls [18]. In the study by Mahn et al., serum levels of let-7i, miR-26a, and miR-195 were found to be significantly higher in patients with localized PCa when compared to patients with BPH. In addition, the importance of age-matched controls was highlighted by the study [19].

See Table 1 for a summary of circulating miRNAs with potential diagnostic usage.

2.2 Circulating miRNAs as Prognostic Biomarkers

Prognostic biomarkers are used to predict the disease progression in untreated individuals. Currently, there is no reliable method to distinguish the minority of patients with PCa that will progress to life-threatening aggressive disease from those whose cancers are more indolent. Approximately, 1 in 4 men treated for localized PCa is expected to relapse within 5 years [20]. The current method to identify patients likely to relapse and who therefore require adjuvant therapy relies on an evaluation of several biomarkers, including PSA, Gleason score, and histological score. The parameters are collated into a prognostic algorithm such as the CAPRA score or the D'Amico score [21], which aids in deciding treatment choices. However, like the PSA test, lack of specificity and sensitivity is a major issue. An ideal prognostic marker will identify the subset of patients with aggressive PCa who require adjuvant therapy, while preventing overtreatments in the majority of men whose cancers are indolent.

Prognostic circulating miRNA biomarkers have been investigated by two approaches. The first approach uses a panel of miRNAs that show altered levels in circulation of metastatic PCa patients. The levels of these miRNAs are assessed in circulation of localized PCa patients to determine whether they can be used to identify the subgroup of patients with poor prognosis. Brase et al. initially identified elevated levels of 69 miRNAs in circulation of metastatic PCa patients and subsequently assessed circulation levels of a subset from 69 miRNAs in localized PCa patients. Circulating levels of three miRNAs (miR-131, miR-200b, and miR-375) showed positive correlation with Gleason score and tumor stage [22]. Nguyen et al. used a similar approach and identified that levels of miR-141

TABLE 1 Circulating MicroRNA (miRNA) Markers of Prostate Cancer (PCa)

miRNAs	Observations	References
DIAGNOSTIC		
Let-7c	Decreased in PCa vs BPH and healthy controls	Chen et al. [17]
Let-7e	Decreased in PCa vs BPH and healthy controls	Chen et al. [17]
Let-7i	Increased in metastatic PCa vs BPH	Mahn et al. [19]
miR-21	Increased in PCa vs healthy controls	Yaman Agaoglu et al. [18]
miR-24	Increased in PCa vs healthy controls	Moltzahn et al. [16]
miR-26a	Increased in metastatic PCa vs BPH	Mahn et al. [19]
miR-30c	Decreased in PCa vs BPH and healthy controls	Chen et al. [17]
miR-93	Increased in PCa vs healthy controls	Moltzahn et al. [16]
miR-106a	Increased in PCa vs healthy controls	Moltzahn et al. [16]
miR-107	Increased in PCa vs healthy controls	Bryant et al. [12]
miR-141	Increased in metastatic PCa or CRPC vs localized disease	Mitchell et al. [11], Brase et al. [22], and Nguyen et al. [23]
miR-195	Increased in metastatic PCa vs BPH	Mahn et al. [19]
miR-221	Increased in PCa vs healthy controls	Yaman Agaoglu et al. [18]
miR-375	Increased in metastatic PCa or CRPC vs localized disease	Brase et al. [22] and Nguyen et al. [23]
miR-378	Increased in CRPC vs localized disease	Nguyen et al. [23]
miR-409-3p	Increased in CRPC vs localized disease	Nguyen et al. [23]
miR-622	Increased in PCa vs BPH and healthy controls	Chen et al. [17]
PROGNOSTIC		
miR-20c	Correlated to tumor stage and D'Amico score	Shen et al. [24]
miR-21	Correlated to CAPRA score and D'Amico score	Shen et al. [24]
miR-24	Decreased with increasing CAPRA score	Moltzahn et al. [16]
miR-106a	Correlated with increasing CAPRA score	Moltzahn et al. [16]
miR-93	Correlated with increasing CAPRA score	Moltzahn et al. [16]
miR-141	Correlated to Gleason score and lymph node status	Brase et al. [22]
miR-141	Correlated to CTC, PSA, and LDH	Gonzales et al. [100]
miR-145	Correlated to D'Amico score	Shen et al. [24]
miR-200b	Correlated to tumor stage and Gleason score	Brase et al. [22]
miR-221	Correlated to D'Amico score	Shen et al. [24]
miR-375	Correlated to Gleason score and lymph node status	Brase et al. [22]
miR-375	Increased in HRPC vs localized PCa	Nguyen et al. [23]
miR-378	Increased in HRPC vs localized PCa	Nguyen et al. [23]
miR-409-3p	Increased in HRPC vs localized PCa	Nguyen et al. [23]
miR-453	Correlated with increasing CAPRA score	Moltzahn et al. [16]

BPH, benign prostate hyperplasia; CRPC, castration-resistant prostate cancer; CAPRA, Cancer of the Prostate Risk Assessment; PSA, prostate-specific antigen; LDH, lactate dehydrogenase; HRPC, hormone-refractory prostate cancer; CTC, circulating tumor cells.

and miR-375 in circulation can differentiate men with metastatic PCa from those with low-risk localized PCa [23].

In the second approach, patients were divided into different risk groups according to their D'Amico and CAPRA score, and levels of circulating miRNAs are compared between the groups. The study of Moltzahn et al. mentioned in the previous section used such an approach to identify elevated levels of miR-93 and miR-106a in high-risk group. In a similar manner, Shen et al. reported an elevation in circulating levels of miR-20a and miR-21 in patients with high CAPRA score [24]. A longitudinal study of larger cohorts of patients is required to validate the prognostic ability of these candidate miRNAs. With a reliable prognostic marker, patients with undetectable micrometastases could potentially be identified and treated with systemic treatment at the time of diagnosis, conferring a mortality advantage.

See Table 1 for a summary of circulating miRNAs with potential prognostic usage.

2.3 Circulating miRNAs as Predictive Biomarkers

Predictive biomarkers are biological parameters that can be used to predict how patients will respond to a therapeutic intervention and thus can be used to stratify patients for treatment. Treatments for castration-resistant PCa (CRPC), such as docetaxel chemotherapy, are only effective (i.e., they alleviate symptoms or prolong survival) in a subset of patients and result in unnecessary toxicity with no substantial benefit in nonresponders. Hence reliable predictive biomarkers will improve quality of life for patients by reducing treatment-induced morbidity.

In one study, Zhang et al. assessed the serum levels of miR-21 in a small cohort of 10 CRPC patients prior to docetaxel-based chemotherapy. Serum levels of miR-21 were higher in CRPC patients who did not respond to chemotherapy than in those sensitive to chemotherapy [25]. In a recent study, Lin et al. assessed the plasma/serum levels of 46 miRNAs in a larger cohort of 97 CRPC patients prior to docetaxel treatment. Nonresponders generally had higher predocetaxel levels of two members of the miR-200 family (miR-200b and miR-200c) or lower levels of miR-146a [26]. In addition, higher prechemotherapy serum levels of miR-21 were associated with shorter survival. As well as predicting efficacy of a given therapy, the identification of predictive miRNA may contribute to understanding the underlying mechanisms of resistance to a given therapy.

2.4 HormomiRs: Circulatory miRNAs as Hormones

Evidence is accumulating that circulating miRNAs are not simply the result of lysis of dying cells: first, miRNAs are highly protected within the blood and are resistant to nucleases and extremes of pH, temperature, and long-term storage [11,27]. In addition, they are often contained within such lipid vesicles as microvesicles [28,29], exosomes [30–32], or associated with RNA-binding proteins such as AGO2 [33], nucleophosmin [34], and high-density lipoprotein (HDL) [35]. The presence of miRNAs in different lipid-containing entities hints at the potential to isolate tissue-specific vesicles in order to select miRNA populations derived from that tissue by exploiting vesicle surface proteins [36]. Such a process has been described for ovarian tumor-derived exosomes [13], where it was also reported that the tumor-derived miRNA repertoire significantly correlated with primary tumor miRNA expression patterns, and has also been reported in lung adenocarcinoma [37].

The "hormomiR" hypothesis, which is increasingly evidenced by in vivo and in vitro experiments, suggests that miRNAs can be passively or actively exported by donor cells and travel through the circulation in complex with accessory proteins and vesicles to regulate gene expression in distal target cells [10]. In this way, hormomiRs function as long- or short-range signals, similar to hormones, perhaps sustaining regular homeostasis or modulating disease pathogenesis. It has been hypothesized that cancer cells may actively release miRNAs into their surrounding microenvironment to promote their own growth and invasion or even "prime" potential metastatic sites for tumor cell invasion.

Evidence for distal signaling capacities of circulating miRNAs is provided chiefly by in vitro studies showing that vesicle/exosome/HDL-associated miRNAs can modulate expression of endogenous target genes and reporter constructs in recipient cells [29,31,32,38]. Initial evidence for such cell-to-cell communication came from a study by Valadi et al., who demonstrated the presence of miRNAs in mouse vesicles, which were transferable to human mast cells [31]. Additionally, Skog et al. demonstrated that exosomes released from glioblastoma cells (containing miRNAs, mRNAs, and angiogenic EGFRVIII protein) are taken up by brain microvascular endothelial cells [29] to promote tumor progression through induction of glial cell proliferation and tubule formation. However, such cell-autonomous miRNA functionality at distal sites has yet to be definitively demonstrated in vivo. It has also been established that exosomes containing miRNAs are found in other bodily fluids, including urine, saliva, seminal fluid, breast milk, amniotic fluid, bronchial lavage, cerebrospinal fluid, and tears [39–43].

The discovery that miRNAs are detectable in body fluids and show altered levels in many malignant tissues has generated considerable excitement about their putative use as predictive prognostic and diagnostic markers to monitor disease and inform treatment decisions. It is conceivable that in the near future, panels of biomarker miRNAs could be robustly, routinely, accurately, and inexpensively profiled to provide detailed information for patient stratification [44]. However, additional studies are required to ascertain how miRNAs are packaged for cell excretion, how this process is regulated and which proteins are involved, how exosomal pre-miRs escape the miRNA biogenesis pathway, how miRNAs are taken up by target cells, the influence of their packaging status on function at distal sites, whether excreted, tumor tissue-derived miRNAs in the blood of cancer patients originate purely from tumor cells or whether they may exist as a response of the surrounding tissues to tumor growth and invasion, and finally, the impact of circulating miRNAs on normal bodily processes and homeostasis remains to be fully revealed.

3. CIRCULATING mRNAs AS BIOMARKERS

At present, there is little evidence as to the clinical efficacy of cell-free circulating mRNA or DNA as diagnostic, predictive, and/or prognostic biomarkers in the management and treatment of PCa. Such markers have considerable potential for noninvasive detection and monitoring of disease progression, and are thus attractive to both clinicians and scientists. However, the clinical use of these circulating nucleic acids is hampered by issues of specificity, sensitivity, and nucleic acid stability. In particular, low specificity of qPCR assays and selection of mRNAs that are specific to the prostate, but not its cancer, present considerable technical challenges and preclude medical application. As an example, PSA mRNA is detectable in serum of healthy men as well as in PCa patients, following prostate biopsy [45,46]. Additionally, circulating mRNA demonstrates considerably lower stability than DNA, resulting in reduced amplicon abundance for qPCR procedures. Despite the numerous caveats, some circulating mRNAs have demonstrable potential for distinguishing patients with metastatic cancer from those with organ-confined PCa. For example, bone morphogenetic protein 6 (BMP6) mRNA levels are significantly higher in primary tumors of patients with metastatic disease than those with localized PCa tumors or benign prostate tissue. BMP6 is thought to enhance establishment of bone metastases by promoting the osteoblastic and invasive potential of PCa cells [47,48]. Further, it was shown that posttreatment circulating BMP6 mRNA levels were significantly higher in patients with metastatic PCa compared to those with localized or locally advanced disease, and plasma BMP6 mRNA levels were able to distinguish metastatic from localized/locally advanced PCa in 71% of cases [49]. These data thus provide evidence that, while plasma BMP6 mRNA lacks sufficient sensitivity and specificity to act only as a blood-based biomarker, it may be used in combination with PSA as a marker for treatment response and/or disease progression, or as one component of a biomarker panel test to identify PCa micrometastasis at the time of diagnosis.

Telomerase activity is increased in 85–100% of human cancers with reference to benign tissue [50]. It has additionally been shown that telomerase reverse transcriptase (hTERT) mRNA is detectable in plasma and serum of PCa patients [51,52], forming a precedent for investigations into the capacity of hTERT mRNA to function as a PCa biomarker. March-Villalba et al. assessed hTERT mRNA levels in plasma of 105 PCa patients with elevated PSA and 68 healthy controls to examine diagnostic accuracy and ability to predict recurrence and correlation with clinicopathological features. Interestingly, plasma hTERT mRNA showed increased sensitivity, specificity, positive predictive value, and negative predictive value in comparison to serum PSA, and was found to correlate significantly with clinical markers of poor prognosis and to independently predict diagnosis of PCa and biochemical recurrence in a highly significant manner, unlike PSA [53]. It should be noted, however, that only seven patients with biochemical recurrence were included in the study. Taken together, these suggest that plasma hTERT mRNA levels may offer increased prognostic and diagnostic value compared to PSA and may thus constitute a highly specific PCa biomarker, particularly when combined with other putative biomarkers. However, much larger trials are required to support the early promise of hTERT in this regard.

An additional mRNA with biomarker potential in PCa is anterior gradient 2 (AGR2). AGR2 protein is implicated in metastatic progression and migration of PCa cells, and it has been investigated in urine as a diagnostic biomarker [54,55]. Kani et al. used qPCR to demonstrate that AGR2 mRNA levels are significantly increased in plasma of patients with metastatic PCa, and interestingly, levels were highest for individuals whose tumors display pathological features of neuroendocrine-predominant CRPC (NP-CRPC) [56]. As PSA levels are often not elevated in such patients, its applications for treatment response and diagnosis are severely limited, and as such NP-CRPC-classified patients have very poor prognosis. Thus, the prospect of AGR2 mRNA as a noninvasive biomarker that can inform clinical management of this disease subtype is an exciting one. However, given that data from only three patients with NP-CRPC were used in this study, far larger investigations are warranted to assess the clinical utility of AGR2 mRNA as a biomarker for NP-CRPC.

The recent introduction of the novel antiandrogens, enzalutamide and abiraterone, to clinical practice has increased treatment options available to patients with advanced CRPC. However, resistance to these therapies is now being observed. One proposed mechanism of resistance is the prevalence of alternative splice forms of androgen receptor (AR). For example, AR transcript splice variant 7 (AR-V7) lacks the ligand-binding domain, yielding a constitutively active form of the AR protein upon translation that is thought to contribute to therapy resistance. Antonarakis et al. investigated the association of AR-V7 in circulating tumor cells with resistance to enzalutamide and abiraterone [57]. Within the cohort receiving enzalutamide, AR-V7-positive patients demonstrated lower PSA response and shorter progression-free survival than their AR-V7-negative counterparts [57]. Similarly, AR-V7-positive patients treated with abiraterone also had lower PSA response and shorter progression-free survival than AR-V7-negative individuals [57]. Far larger trials are required to rigorously assess the response/prognostic biomarker potential of this transcript variant in PCa. However, AR-V7 assays may prove valuable in clinical decision-making. For example, a CRPC patient with high AR-V7 levels may derive greater benefit from a non-AR targeting therapy than from either enzalutamide or abiraterone.

Circulating mRNAs, and also cell-free DNAs, with clinical potential are summarized in Table 2.

4. CELL-FREE DNAs AS BIOMARKERS

Circulating DNA-based tumor markers display increased stability and greater tumor specificity when compared to mRNA. This provides possibilities for accurate tumor grading/staging, treatment decision-making, and prognostic estimation, in order to improve clinical disease management. Surprisingly, the first evidence for the presence of cell-free DNA in blood plasma and serum was presented in 1948 [58], although its biomarker potential was only revealed in later studies, which showed that levels of circulating DNA are higher in cancer patients than healthy controls [59–61]. In support of these observations, recent evidence cites threefold higher levels of cell-free circulating DNA in blood of PCa patients compared to individuals with BPH [62]. With reference to PCa biomarker potential, cell-free circulating DNA is correlated with a number of clinicopathological features, including stage, Gleason score, extraprostatic invasion, and metastasis [61,63–65]. Unfortunately, reports vary considerably in their estimations of the proportion of circulating cell-free DNA that is directly of tumor origin, with figures between 3% and 93% stated [66]. This suggests that circulating DNA profiles may alter dramatically in response to various stimuli and environmental changes, bringing into question the utility of such a widely fluctuating marker. That said, alterations in circulating DNA have been shown to correlate with mutations present in primary cancers [67], asserting that at least a portion of cell-free DNA in the blood derives from tumor, released by processes including apoptosis, necrosis, and active export. Although levels of circulating DNA of tumor derivation are likely to be a function of tumor type, disease stage, metastatic status, treatment regimens, and even patient age, among other factors, it is widely held that the majority of circulating DNA derives from healthy tissue [66,68].

Three modes of DNA alteration have been investigated as blood-based biomarkers for PCa: mtDNA mutations [62,69–71], gene promoter hypermethylation [61,63,72–81], and microsatellite instability (MI) [65,72,73,82,83]. Such modifications represent potentially attractive tumor markers for a number of reasons: first, mtDNA, with 20–200 copies per cell, is far more abundant than genomic DNA (2 copies per cell) and thus theoretically much more readily detectable; second, methylation-specific PCR (MSP) is a highly sensitive application for the detection of gene promoter hypermethylation and requires only 0.1–0.001% of serum DNA to be of tumor origin [84]. The major advantage, however, is that these markers are mostly tumor-specific events and may thus reduce the risk of false-positive diagnosis.

4.1 Hypermethylation Events

Glutathione S-transferase P1 (GSTP1) is a tumor suppressor in PCa and has roles in detoxification. Hypermethylation of cytosine–phosphate–guanine islands within its gene promoter has been identified as a very early event in tumorigenesis and results in loss of gene transcription.

TABLE 2 Circulating mRNA and DNA Markers of Prostate Cancer (PCa)

Name	Type of biomarker	Observations	References
BMP6	Circulating mRNA	• Can distinguish metastatic from locally advanced or localized PCa	Deligezer et al. [49]
hTERT	Circulating mRNA	• Levels correlate with poor prognosis • Predicts diagnosis and recurrence with greater sensitivity and specificity than PSA • Very small patient numbers	March-Villalba et al. [53]
AGR2	Circulating mRNA	• Increased levels in patients with metastatic PCa • Potential as a treatment response biomarker for neuroendocrine-predominant CRPC, where PSA is often not elevated	Kani et al. [56]
Cell-free DNA	Circulating DNA	• Increased levels in blood of PCa patients vs BPH patients • Levels of circulating DNA correlate with stage, Gleason grade, invasion, and metastasis	Altimari et al. [61], Ellinger et al. [62], Bastian et al. [74], Jung et al. [64], and Schwarzenbach et al. [65]
GSTP1 promoter hypermethylation	Circulating DNA	• Absent in serum samples from BPH cases, but detectable in serum of 56% of T_{2-3} and 93% of T_4 PCa • Predicts recurrence after radical prostatectomy and correlates with Gleason score	Goessl et al. [84], Goessl et al. [89], Bastian et al. [90], and Catalona et al. [91]
RASSF1A hypermethylation	Circulating DNA	• Hypermethylation is an adverse prognostic marker in PCa	Kang et al. [93], Liu et al. [94], and Maruyama et al. [95]
RARB2 hypermethylation	Circulating DNA	• Correlated with poor prognosis, Gleason score, and serum PSA	Kang et al. [93], Maruyama et al. [95], and Nakayama et al. [96]
Hypermethylation of RF219 upstream region	Circulating DNA	• Distinguishes PCa from BPH with sensitivity of 61% and specificity of 71%	Cortese et al. [97]
Loss of repetitive pericentromeric DNA on chromosome 10	Circulating DNA	• Greater copy number loss observed from repetitive sequences in serum samples from stage III PCa patients vs stage II men and BPH controls	Cortese et al. [97]
Trimethylation of histone H3 at lysine 27	Circulating DNA	• Trimethylated by polycomb repressive complex 2 • H3K27me3 levels lower in plasma of metastatic PCa patients vs less aggressive disease • H3K27me3 can discriminate 74% of metastatic cases from localized or locally advanced PCa	Deligezer et al. [49]
Mitochondrial DNA (mtDNA)	mtDNA	• Correlates with levels of PSA and increased PCa-specific mortality when assessed in patient plasma • Increased levels of short mtDNA in patients with early postprostatectomy recurrence	Ellinger et al. [62] and Mehra et al. [70]

TABLE 2 Circulating mRNA and DNA Markers of Prostate Cancer (PCa)—cont'd

Name	Type of biomarker	Observations	References
Mutations of D-loop region, 16S rRNA, and complex I of mtDNA	mtDNA	• Mutations found in patient plasma were recapitulated in primary tumors • Rare events: identified in only 3 of 16 patients investigated	Jeronimo et al. [69]
Androgen receptor transcript splice variant 7 (AR-V7)	mRNA from circulating tumor cells	• AR-V7 positivity associated with lower PSA response and shorter progression-free survival in patients treated with enzalutamide or abiraterone	Antonarakis et al. [57]

BMP6, bone morphogenetic protein 6; hTERT, telomerase reverse transcriptase; AGR2, anterior gradient 2; PSA, prostate-specific antigen; CRPC, castration-resistant prostate cancer; BPH, benign prostate hyperplasia; GSTP1, glutathione S-transferase P1; RASSF1A, Ras association domain family member 1; RARB2, retinoic acid receptor beta variant 2.

Epigenetic modulation of the GSTP1 promoter is detectable in up to 100% of PCa tumors [85,86], providing robust justification for investigation of its cell-free circulating DNA as a noninvasive PCa biomarker, where promoter methylation was detected in blood of 90% of early-stage PCa patients [84]. This supports its use as a diagnostic biomarker for PCa, perhaps in combination with PSA. There is currently controversy over the extent of GSTP1 promoter hypermethylation (GPH) in cases of BPH and prostatic intraepithelial neoplasia (PIN). For example, in two studies by the same group, Jeronimo et al. detected GPH in 30% of BPH tissues in one [69], while describing GPH to be undetectable in benign prostate tissues, but present in 66% of high-grade PINs in a second [87]. A number of additional studies, however, failed to detect GPH in benign prostate tissue [84–86,88], and it has been shown that quantitative differences in GPH levels between BPH and prostate tumors are approximately 150-fold [87], thus supporting the robustness of GPH as a PCa biomarker if this ratio is maintained in serum/plasma analysis. Meticulous identification of a suitable threshold would permit the use of GPH as a blood-based biomarker in PCa.

MSP can be used for rapid, sensitive, and potentially automatable methylation-based biomarker detection. In one of the first studies to use MSP to investigate the diagnostic potential of GPH in PCa, Goessl et al. found GPH to be absent in 26 BPH cases, but detectable in 94% of 33 PCa tumor samples and 72% of serum samples from PCa patients [84]. They later identified this epigenetic event in 56% of plasma samples from patients with T_{2-3} PCa and 93% of plasma samples from men with lymph node-positive and/or metastatic T_4 PCa [89]. This suggests that plasma-based GPH profiling may be diagnostically informative, and may correlate with disease stage, prognosis, and/or outcome. In corroboration of this, Goessl et al. detected GPH in all plasma/serum samples from patients with locally advanced (T_4) or metastatic disease [78] and found specificity of the MSP approach to be 100%. In addition, GSP was found to be a highly specific predictor of recurrence after radical prostatectomy [90], and demonstrated stringent correlation with Gleason score and metastatic spread in patients with CRPC [81]. It is interesting to note that GPH provided a higher degree of diagnostic sensitivity in serum samples compared to other fluids, since only 50% of ejaculates and 36% of urine samples from PCa patients demonstrated GPH positivity [89]. However, it should be noted that patient numbers in this study were very small. Since GPH has been shown to be a tumor-specific event, and is found in pools of cell-free DNA of tumor origin from PCa patients, it may represent a very valuable, accurate PCa diagnosis biomarker, particularly in view of the fact that PCa is detectable in repeat biopsies of 30% of patients who were given a negative diagnosis upon first biopsy [91,92]. This provides potential to diagnose men with PCa who would otherwise be missed by standard biopsy procedures.

Hypermethylation of two further tumor suppressor genes, Ras association domain family member 1 (RASSF1A) and retinoic acid receptor beta variant 2 (RARB2) have been proposed as adverse prognostic biomarkers in PCa, since they are hypermethylated in primary prostate tumors [93–96] and correlate significantly with Gleason score and serum PSA levels [73].

In a more high-throughput approach, circulating DNA methylomes were analyzed from 19 PCa cases, 20 BPH patients, and 20 disease-free men by DNA modification-sensitive restriction digestion, followed by profiling on microarrays comprising over 12,000 GC-rich clones. Thirty-nine circulating DNA modifications were identified as significantly associated with PCa, and seven of these were validated in an independent patient cohort [97]. In particular, modification of the DNA region upstream of the ring finger protein 219 gene was able to distinguish PCa from BPH with relatively high sensitivity and specificity (61% and 71%, respectively) [97]. Additionally, significant loss of repetitive pericentromeric DNA on chromosome 10 was frequently observed in stage III PCa patients with respect to stage II patients and BPH controls [97]. Thus chromosome 10 may be subject to copy number loss within repetitive DNA sequences in case of advanced PCa, and identification of such events may be used as a prognostic tool to distinguish patients with more aggressive disease, or who are likely to progress quickly. Interestingly, the authors also utilized machine-based learning to develop a multilocus biomarker panel for PCa diagnosis. However, this panel was only able to distinguish cancer from BPH in 72% of cases, and thus will require further refinement if it is to provide any clinical benefit.

A further epigenetic event, trimethylation of histone H3 at lysine 27 (H3K27me3) by EZH2-containing polycomb repressive complex 2, leads to transcriptional silencing. Global loss or gain of H3K27me3 is associated with poor disease outcome and is suggested as a novel biomarker for a number of cancers, including PCa [98,99]. Upon qPCR- and enzyme-linked immunosorbent assay (ELISA)-based analysis of this event in plasma of PCa patients with localized, locally advanced, or metastatic disease, it was found that H3K27me3 levels were significantly lower in metastatic PCa patients compared to men with less aggressive disease, and that plasma H3K27me3 could discriminate 74% of metastatic PCa cases from localized or locally advanced PCa [49]. It is thus suggested that plasma H3K27me3 levels could be used in combination with PSA testing, pre- and posttreatment, in order to evaluate disease progression and/or drug response.

4.2 MI in Serum/Plasma

MI refers to the reduction in fidelity during replication of repetitive elements of DNA, often occurring in tumor cells and thought to be caused by strand slippage during DNA replication. This is often due to mutations in mismatch repair enzymes. Several studies have identified MI events in cell-free DNA from serum or plasma of PCa patients [65,72,73,83], although some allelic imbalances that result from MI have also been detected in the plasma or serum of BPH patients [72,83]. Sunami et al. examined a panel of 6 MI events leading to allelic imbalances in 83 PCa patients and 40 disease-free controls. It was found that the MI panel was able to distinguish PCa with a specificity of 100%, although the sensitivity of this biomarker set was much lower, identifying 1 or more of the 6 MI events in only 47% of serum samples from PCa patients [73]. In a similar way, 14 markers of MI detected only 45% of PCa from blood serum samples, albeit with specificity of 100%, in a further study [83]. Using a larger cohort of 230 patients and 43 controls, a biomarker panel of 13 MI events was able to distinguish 57% of PCa patients, but with a lower specificity of 70%, although metastatic PCa patients were found to have increased frequency of MI markers in their plasma [65]. The above data suggest that the use of MI events as biomarkers in isolation may not be of clinical relevance for diagnostic or prognostic testing. In corroboration of this, Sunami et al. demonstrated that combining MI event markers with promoter methylation assays as a biomarker platform was more informative and predictive than either marker individually, at least 1 marker of allelic imbalance, or promoter hypermethylation was detected in the circulating DNA pool of 63% of PCa patients, compared to detection of 16% of cancers using only an MI marker and 34% of cancers using promoter hypermethylation events [73]. Thus the authors propose that a combined panel of epigenetic and allelic imbalance event markers could be used in combination with standard PSA testing for diagnostic or disease-monitoring purposes. In support of this, the combined analysis of promoter methylation and MI events in PCa patient circulating DNA, alongside traditional PSA testing, detected 18% more cases than PSA analysis alone (89% and 71%, respectively), without increasing the rates of false-positive detection [73]. Additional markers of MI or DNA methylation may need to be included on any PCa biomarker panel in order to enhance diagnostic sensitivity, particularly given the high degree of tumor heterogeneity observed in PCa, and interpatient differences in circulating DNA turnover.

4.3 mtDNA in Plasma/Serum

It has previously been reported that levels of plasma/serum mtDNA are similar between patients with localized PCa and individuals with BPH, and a lack of correlation of mtDNA levels with clinicopathological indicators, such as PSA, Gleason score, and lymph node invasion, has been described in PCa [62,71]. However, a separate study has identified a correlation between levels of PSA and mtDNA and mtRNA in PCa patient plasma, demonstrating 2.5-fold higher levels of mtDNA in PCa patients compared to BPH controls [70]. It appears that mtDNA may be of greater prognostic than diagnostic utility, particularly in advanced PCa cases, since increased frequency of short mtDNA fragments was observed in PCa patients with early postprostatectomy PSA-assessed recurrence [62], and mtDNA was correlated with increased PCa-specific mortality [70]. Further, patients who survived for less than 2 years after initial presentation had 2.6-fold higher levels of circulating mtDNA at diagnosis than surviving patients, and 2-year survival was lower for patients with high levels of circulating mtDNA (35%) compared to those without increased circulating mtDNA (73%) [70]. In a different approach, the D-loop region, 16S rRNA, and complex I of mtDNA were subjected to sequencing in primary tumors, urine, and plasma from PCa patients to assess mutations of mtDNA in PCa [69]. Twenty such mutations were identified in the primary tumors, and where mtDNA mutations were described in plasma, this was recapitulated in the primary tumor from the same patient. However, it should be noted that mtDNA mutations were relatively rare events, identified in only 3 of 16 patients investigated [69], suggesting that the diagnostic potential of mtDNA mutations is limited.

5. CONCLUSIONS

PCa is in urgent need of new biomarkers, both to improve diagnosis of lethal forms of the disease and also to aid decision-making in the use of novel therapies, many of which are currently in clinical trials or have recently been approved. Due to their relative stability in serum and plasma, which can be obtained by noninvasive procedures, circulating nucleic acids currently represent the most promising source of such biomarkers.

However, in order to provide clinical benefit and avoid risks posed by detection of false positives and false negatives, putative biomarkers must satisfy a number of stringent criteria:

1. **High stability in blood**—even if samples are stored at room temperature for several hours prior to serum/plasma extraction and storage.
2. **PCa specific**—the major issue with the use of PSA as a biomarker is that it can be detected in blood of patients with BPH or PIN.
3. **Minimally influenced by metabolism, circadian rhythms, and environmental factors**—levels of nucleic acids can fluctuate widely during the course of a day and are often regulated by metabolic processes. Capturing of samples at the same time of day and, for example, following overnight fasting may counteract some of this variability.
4. **Simple, reproducible, inexpensive detection method(s)**—qPCR-based assays demonstrate very high specificity (can distinguish between single base pair differences in the case of miRs) at low levels of circulating nucleic acids and are potentially automatable. Several assays can be incorporated into a biomarker panel. One issue surrounding qPCR-based detection is that of normalization, since there is no "gold-standard" normalization control gene/miR. This could likely be avoided by employing absolute quantification methods for qPCR analysis.
5. **Greater sensitivity and specificity than current tests for diagnosis, prognosis, or response prediction**—a putative biomarker provides no clinical benefit if it cannot outperform PSA, the most widely used PCa biomarker. Other biomarker tests that have made it into clinical use are summarized in Table 3.
6. **Relatively abundant in patient blood**—low-abundance biomarkers increase the risk posed by assay contamination from other nucleic acid sources.
7. **Validated in a large, suitably powered patient cohort with appropriate controls and suitable patient follow-up**—in order to progress from lab bench to bedside, a potential biomarker must be able to reproduce its diagnostic, prognostic, or predictive utility in a large, independent cohort of many thousands of patients, including men with BPH or PIN, and healthy controls in several multicenter studies with access to all relevant clinical information.
8. **Biomarker potential identified in multiple studies**—a putative biomarker is more likely to represent a viable test if it has been examined by several research groups.

The ability of the putative biomarkers described in this chapter to fulfill these criteria is summarized in Table 4 [65]. It is clear that none to date satisfy all of these requirements and are unlikely to replace PSA testing in the very near future. For some, however, it is simply a matter of time. The major stumbling block is that promising small-scale biomarker trials have not been followed up with large cohort, multicenter trials, likely because such studies are extremely expensive and require coordination between large clinical, scientific, and management teams. As an example, GSTP1 gene promoter hypermethylation represents an extremely promising diagnostic and prognostic biomarker for PCa: it is a PCa-specific event that occurs very early in the pathogenesis of the disease and thus has considerable potential in aiding early PCa diagnosis, which represents a major hurdle to effective treatment. In addition, its detection by MSP is highly sensitive and specific, its levels are 150-fold higher in PCa patients compared to BPH individuals, it outperforms PSA in diagnosis and prognosis, and has been described in five different studies. The major stumbling to its clinical introduction is the lack of a large-scale study into its prognostic and/or diagnostic use. The same could be said to be true for the miR-141 and miR-21, both of which are highly stable in blood

TABLE 3 Biomarker Tests in Clinical Use for Prostate Cancer (PCa)

	Body fluid	Methodology	Biomarker type	Sensitivity	Specificity	References
Prostate-specific antigen (PSA), the current gold-standard biomarker	Blood	Measures level of PSA in the blood	Diagnostic	72%	93%	Ankherst et al. [101]
Progensa PCA3	Urine	Detects overexpression of PCA3 noncoding mRNA. Used in combination with PSA. FDA approved following negative prostate biopsy to add timing of rebiopsy	Diagnostic	52% (when combined with PSA)	87%	de la Taille et al. [102] and Crawford et al. [103]
TMPRSS2:ERG	Urine	Detects TMPRSS2:ERG gene fusion, a dominant variant in 40–80% of PCa	Prognostic	93% (when combined with PCA3 and PSA)	87%	Tomlins et al. [104], Salami et al. [105], and Leyten et al. [106]
ConfirmMDx	Negative biopsy tissue	Analyses epigenetic field effect around the cancer lesion to help determine the presence of occult cancers missed by standard biopsies	Diagnostic	74% (when combined with histological findings)	63%	Stewart et al. [107] and Aubry et al. [108]
Prostate Health Index (phi)	Blood	Level of p2PSA combined with PSA and free PSA phi = (−2) proPSA/ fPSA × PSA1/2; proPSA is a PSA subtype and fPSA is free PSA. FDA approved for when the PSA is in the 4–10 ng/mL range	Diagnostic	90%	31%	Stephan et al. [109]
Prolaris diagnostic test	Biopsy tissue	46-RNA expression predicts 10-year survival but identifying patients with an increase in cell cycle progression gene mutation signatures	Prognostic	69%	79%	Cuzick et al. [110]
Prostate Core Mitomic Test	Biopsy tissue	Identifies a large-scale depletion in mitochondrial DNA that indicates cellular change associated with undiagnosed PCa	Diagnostic	83%	79%	Robinson et al. [111]

PCA3, prostate cancer antigen 3; FDA, US Food and Drug Administration.

TABLE 4 A Summary of Putative Prostate Cancer Biomarkers and Their Ability to Meet Optimal Biomarker Requirements

			Optimum prostate cancer biomarker criteria[a]			
Name	**Category**	**Biomarker type**	**1. Highly stable in blood**	**2. Prostate tumor-specific event**	**3. Minimally influenced by circadian rhythms, metabolism, environmental factors**	**4. Simple, reproducible, inexpensive, rapid detection method**
miR-141	miR	Prognostic	Y	N		Y
miR-93	miR	Prognostic	Y	N		Y
miR-106a	miR	Prognostic	Y	N		Y
miR-107	miR	Diagnostic	Y	N		Y
5 miR panel (↑ 622, 1285; ↓ let-7c, let-7e, 30c)	miR panel	Diagnostic	Y	N		Y
miR-21	miR	Diagnostic	Y	N		Y
		Prognostic	Y	N		Y
		Predictive	Y	N		Y
miR-221	miR	Diagnostic	Y	N		Y
Let-7i	miR	Diagnostic	Y	N		Y
miR-26a	miR	Diagnostic	Y	N		Y
miR-195	miR	Diagnostic	Y	N		Y
miR-131	miR	Prognostic	Y	N		
miR-200b	miR	Prognostic	Y	N		
		Predictive	Y	N		Y

5. Greater sensitivity and specificity than PSA for diagnosis, prognosis or response prediction	6. Relatively abundant in biofluids	7. Validated in suitably powered patient cohort with appropriate controls and follow-up	8. Biomarker potential identified in multiple studies (number indicated)	Notes	References
	Y		3	Differentiates metastatic from localized PCa	Mitchell et al. [11], Yaman Agaoglu et al. [18], and Nguyen et al. [23]
	Y	N	1		Moltzahn et al. [16]
	Y	N	1		Moltzahn et al. [16]
	Y	N	1		Bryant et al. [12]
			1		Chen et al. [17]
	Y	N	4		Yaman Agaoglu et al. [18]
		N		High prechemotherapy levels associated with shorter survival	Shen et al. [24], Lin et al. [26]
	Y	N		Levels higher in chemotherapy nonresponders	Zhang et al. [25]
	Y		1		Yaman Agaoglu et al. [18]
			1		Mahn et al. [19]
			1		Mahn et al. [19]
			1		Mahn et al. [19]
			1		Brase et al. [22]
			2		Brase et al. [22]
	Y	N		Docetaxel nonresponders have higher predocetaxel levels of miR-200b than responders	Lin et al. [26]

Continued

TABLE 4 A Summary of Putative Prostate Cancer Biomarkers and Their Ability to Meet Optimal Biomarker Requirements—cont'd

			Optimum prostate cancer biomarker criteria[a]			
Name	**Category**	**Biomarker type**	**1. Highly stable in blood**	**2. Prostate tumor-specific event**	**3. Minimally influenced by circadian rhythms, metabolism, environmental factors**	**4. Simple, reproducible, inexpensive, rapid detection method**
miR-375	miR	Prognostic	Y	N		
miR-146a	miR	Predictive	Y	N		
BMP6	mRNA	Predictive	N	N		Y
hTERT	mRNA	Diagnostic	N	N	?Y	Y
		Prognostic	N	N	?Y	Y
AGR2	mRNA	Prognostic	N	N		Y
AR-V7	mRNA—AR splice variant in CTCs	Predictive for enzalutamide and abiraterone response	Y—in CTCs	?Y	?N	N—isolation of CTCs currently costly

5. Greater sensitivity and specificity than PSA for diagnosis, prognosis or response prediction	6. Relatively abundant in biofluids	7. Validated in suitably powered patient cohort with appropriate controls and follow-up	8. Biomarker potential identified in multiple studies (number indicated)	Notes	References
			2	Differentiates metastatic from localized PCa	Brase et al. [22] and Nguyen et al. [23]
			1	Docetaxel nonresponders have lower levels of miR-200b than responders	Lin et al. [26]
N	N	N	2	Circulating levels higher in patients with metastatic disease vs localized/locally advanced. Could be used in combination with PSA	Yuen et al. [48] and Deligezer et al. [49]
Y	Y	N	2	Independently predicts PCa diagnosis in highly significant manner, unlike PSA	Dasí et al. [52] and March-Villalba et al. [53]
Y	Y	N		Levels correlate with markers of poor prognosis, independently predicts recurrence	
N	Y	N	2	Levels higher in patients with metastatic disease and highest in patients with neuroendocrine-predominant CRPC, in which PSA levels not elevated. Could inform clinical management of these patients	Bu et al. [54] and Kani et al. [56]
	Y	N	1	AR-V7-positive patients taking enzalutamide or abiraterone had lower PSA response and shorter progression-free survival—may inform clinical decision-making	Antonarakis et al. [57]

Continued

TABLE 4 A Summary of Putative Prostate Cancer Biomarkers and Their Ability to Meet Optimal Biomarker Requirements—cont'd

			Optimum prostate cancer biomarker criteria[a]			
Name	**Category**	**Biomarker type**	**1. Highly stable in blood**	**2. Prostate tumor-specific event**	**3. Minimally influenced by circadian rhythms, metabolism, environmental factors**	**4. Simple, reproducible, inexpensive, rapid detection method**
Circulating cell-free DNA	DNA	Diagnostic	Y	N	N	Y
		Prognostic	Y	N	N	Y
GSTP1 promoter hypermethylation	DNA promoter methylation	Diagnostic	Y	Y		Y—extremely sensitive
		Prognostic	Y	Y		Y—extremely sensitive
Gene methylation panel (including RFP219 promoter methylation, loss of pericentromeric DNA on chromosome 10)	Promoter methylation, loss of pericentromeric DNA	Diagnostic	?Y			Y—multilocus panel developed
Trimethylayion of histone H3 at lysine 27	Histone methylation	Prognostic	?Y	N		Y—PCR and ELISA
Microsatellite instability (MI) panel—6, 13, or 14 events	MI in DNA	Diagnostic	Y	?	Y	Y

5. Greater sensitivity and specificity than PSA for diagnosis, prognosis or response prediction	6. Relatively abundant in biofluids	7. Validated in suitably powered patient cohort with appropriate controls and follow-up	8. Biomarker potential identified in multiple studies (number indicated)	Notes	References
N	Y	N	7	Higher in PCa patients than healthy controls	Delgado et al. [59], Leon et al. [60], Altimari et al. [61], Ellinger et al. [62], Bastian et al. [74], Jung et al. [64] and Schwarzenbach et al. [65]
N	Y	N		Associated with extraprostatic invasion and metastasis	
Y		N	5	Very early event in tumourigenesis—could detect cancers at very early stage. Detected in blood of 90% of PCa patients but not healthy controls/PIN. 150-fold differences in levels between BPH and PCa	Goessl et al. [84], Goessl et al. [89], Reibenwein et al. [81] and Bastian et al. [90]
Y		N		Correlates with disease stage and metastatic spread, and is a highly specific predictor of recurrence after prostatectomy	
N	Y	N	1	Methylation of upstream region of RFP219 increased in PCa vs healthy controls. Loss of pericentromeric DNA on chromosome 10 increased in stage III patients vs stage II and BPH controls	Cortese et al. [97]
?N	Y	N	2	H3K27me3 levels lower in metastatic patients vs localized disease	Deligezer et al. [49] and Ke et al. [99]
N—100% specificity but low sensitivity	N	N	3	Panels of 6, 13, or 14 events identify PCa patients with 100% specificity but low sensitivity	Schwarzenbach et al. [65], Sunami et al. [73] and Schwarzenbach et al. [83]

Continued

TABLE 4 A Summary of Putative Prostate Cancer Biomarkers and Their Ability to Meet Optimal Biomarker Requirements—cont'd

			Optimum prostate cancer biomarker criteria[a]			
Name	**Category**	**Biomarker type**	**1. Highly stable in blood**	**2. Prostate tumor-specific event**	**3. Minimally influenced by circadian rhythms, metabolism, environmental factors**	**4. Simple, reproducible, inexpensive, rapid detection method**
Combined analysis of promoter methylation and MI alongside PSA	DNA promoter methylation, MI, PSA	Diagnostic	Y	? (PSA= prostate specific)		Y—although difficult to combine into one panel
Short mitochondrial DNA (mtDNA) fragment abundance	mtDNA	Prognostic	Y	N		Y

PSA, prostate-specific antigen; BMP6, bone morphogenetic protein 6; hTERT, telomerase reverse transcriptase; AGR2, anterior gradient 2; CRPC, castration-resistant prostate cancer; AR-V7, androgen receptor transcript splice variant 7; GSTP1, glutathione S-transferase P1; BPH, benign prostate hyperplasia; PCR, polymerase chain reaction; ELISA, enzyme-linked immunosorbent assay; PIN, prostatic intraepithelial neoplasia; PCa, prostate cancer; CTC, circulating tumor cells.

[a] *Y = yes, N = no, black = not known.*

5. Greater sensitivity and specificity than PSA for diagnosis, prognosis or response prediction	6. Relatively abundant in biofluids	7. Validated in suitably powered patient cohort with appropriate controls and follow-up	8. Biomarker potential identified in multiple studies (number indicated)	Notes	References
Y—detected 18% more cases than PSA alone	Y	N	1	Combined analysis of epigenetic and allelic imbalance event markers in combination with PSA testing detected 18% more PCa cases than PSA alone (89% vs 71%) without increased false positives.	Sunami et al. [73],
	Y	N	2	Higher levels of short mtDNA fragments associated with early postprostatectomy recurrence and increased PCa-specific mortality	Ellinger et al. [62] and Mehra et al. [70]

and can be readily detected in a highly specific manner: miR-21 in particular shows promise in predicting response to chemotherapy. Robust analysis must also be performed as a matter of urgency to assess whether extraction of nucleic acids from tumor-derived exosomes provides any benefit in terms of prognostic or diagnostic accuracy over examination of miRs, mRNAs, or DNAs from total blood serum or plasma.

What is also increasingly apparent is that the most informative biomarkers are likely to be incorporated into multiplexed panels, which can assess abundance of, or epigenetic modification to, potentially dozens of promising nucleic acid markers in parallel (perhaps in conjunction with PSA) to constitute a test of the highest possible sensitivity and specificity. Patients would then be allocated a score based on the assay results and filtered into the most appropriate treatment stream. Such biomarker "libraries" represent the future of PCa management and are likely to revolutionize treatment while reducing the very significant mortality and morbidity associated with PCa and its current aggressive treatment strategies. Further, changes in circulating nucleic acids may inform as to disease processes and prove a valuable source of new therapeutic targets.

LIST OF ABBREVIATIONS

3′-UTR 3′-Untranslated region
AGR2 Anterior gradient 2
AR Androgen receptor
AR-V7 Androgen receptor transcript splice variant 7
BMP6 Bone morphogenetic protein 6
BPH Benign prostate hyperplasia
CAPRA Cancer of the Prostate Risk Assessment
CRPC Castration-resistant prostate cancer
ELISA Enzyme-linked immunosorbent assay
ERG ETS-related gene
ERSPC European Randomized Study of Screening for Prostate Cancer
FDA US Food and Drug Administration
GPH GSTP1 promoter hypermethylation
GSTP1 Glutathione S-transferase P1
H3K27me3 Trimethylation of histone H3 at lysine 27
HDL High-density lipoprotein
hTERT Telomerase reverse transcriptase
MI Microsatellite instability
miRNA MicroRNA
MSP Methylation-specific PCR
mtDNA Mitochondrial DNA
mtRNA Mitochondrial RNA
NP-CRPC Neuroendocrine-predominant castration-resistant prostate cancer
PCa Prostate cancer
PCA3 Prostate cancer antigen 3
PIN Prostatic intraepithelial neoplasia
PLCO Prostate, Lung, Colorectal, and Ovarian Cancer Screening Trial
PSA Prostate-specific antigen
qPCR Quantitative polymerase chain reaction
RARB2 Retinoic acid receptor beta variant 2
RASSF1A Ras association domain family member 1
rRNA Ribosomal RNA
TMPRSS2 Transmembrane serine protease 2

Acknowledgments

The authors are grateful for support from the Rosetrees Trust, Prostate Cancer UK, and Cancer Research UK during the writing of this chapter.

References

[1] Siegel R, Naishadham D, Jemal A. Cancer statistics, 2013. CA Cancer J Clin 2013;63(1):11–30.
[2] Schroder FH, Roobol MJ. Defining the optimal prostate-specific antigen threshold for the diagnosis of prostate cancer. Curr Opin Urol 2009;19(3):227–31.
[3] Thompson IM, Pauler DK, Goodman PJ, Tangen CM, Lucia MS, Parnes HL, et al. Prevalence of prostate cancer among men with a prostate-specific antigen level ≤4.0 ng per milliliter. N Engl J Med 2004;350(22):2239–46.
[4] Polascik TJ, Oesterling JE, Partin AW. Prostate specific antigen: a decade of discovery-what we have learned and where we are going. J Urol August 1999;162(2): 293–306.
[5] Bradley LA, Palomaki GE, Gutman S, Samson D, Aronson N. Comparative effectiveness review: prostate cancer antigen 3 testing for the diagnosis and management of prostate cancer. J Urol August 2013;190(2): 389–98.
[6] Nam RK, Sugar L, Wang Z, Yang W, Kitching R, Klotz LH, et al. Expression of TMPRSS2:ERG gene fusion in prostate cancer cells is an important prognostic factor for cancer progression. Cancer Biol Ther January 1, 2007;6(1):40–5.

[7] Tang Y, Yan W, Chen J, Luo C, Kaipia A, Shen B. Identification of novel microRNA regulatory pathways associated with heterogeneous prostate cancer. BMC Syst Biol 2013;7(Suppl. 3):S6.

[8] Lee RC, Feinbaum RL, Ambros V. The *C. elegans* heterochronic gene lin-4 encodes small RNAs with antisense complementarity to lin-14. Cell December 3, 1993;75(5):843–54.

[9] Heneghan HM, Miller N, Kerin MJ. MiRNAs as biomarkers and therapeutic targets in cancer. Curr Opin Pharmacol October 2010;10(5):543–50.

[10] Cortez MA, Bueso-Ramos C, Ferdin J, Lopez-Berestein G, Sood AK, Calin GA. MicroRNAs in body fluids – the mix of hormones and biomarkers. Nat Rev Clin Oncol August 2011;8(8):467–77.

[11] Mitchell PS, Parkin RK, Kroh EM, Fritz BR, Wyman SK, Pogosova-Agadjanyan EL, et al. Circulating microRNAs as stable blood-based markers for cancer detection. Proc Natl Acad Sci USA July 29, 2008; 105(30):10513–8.

[12] Bryant RJ, Pawlowski T, Catto JWF, Marsden G, Vessella RL, Rhees B, et al. Changes in circulating microRNA levels associated with prostate cancer. Br J Cancer February 14, 2012;106(4):768–74.

[13] Taylor DD, Gercel-Taylor C. MicroRNA signatures of tumor-derived exosomes as diagnostic biomarkers of ovarian cancer. Gynecol Oncol July 2008;110(1): 13–21.

[14] Hastings ML, Palma J, Duelli DM. Sensitive PCR-based quantitation of cell-free circulating microRNAs. Methods October 2012;58(2):144–50.

[15] Lin K, Lipsitz R, Miller T, Janakiraman S. Benefits and harms of prostate-specific antigen screening for prostate cancer: an evidence update for the U.S. Preventive Services Task Force. Ann Intern Med August 5, 2008;149(3):192–9.

[16] Moltzahn F, Olshen AB, Baehner L, Peek A, Fong L, Stöppler H, et al. Microfluidic-based multiplex qRT-PCR identifies diagnostic and prognostic microRNA signatures in the sera of prostate cancer patients. Cancer Res January 15, 2011;71(2):550–60.

[17] Chen ZH, Zhang GL, Li HR, Luo JD, Li ZX, Chen GM, et al. A panel of five circulating microRNAs as potential biomarkers for prostate cancer. Prostate February 1, 2012.

[18] Yaman Agaoglu F, Kovancilar M, Dizdar Y, Darendeliler E, Holdenrieder S, Dalay N, et al. Investigation of miR-21, miR-141, and miR-221 in blood circulation of patients with prostate cancer. Tumour Biol June 2011;32(3):583–8.

[19] Mahn R, Heukamp LC, Rogenhofer S, von Ruecker A, Muller SC, Ellinger J. Circulating microRNAs (miRNA) in serum of patients with prostate cancer. Urology May 2011;77(5):1265.e9-16.

[20] Greene KL, Meng MV, Elkin EP, Cooperberg MR, Pasta DJ, Kattan MW, et al. Validation of the Kattan preoperative nomogram for prostate cancer recurrence using a community based cohort: results from cancer of the prostate strategic urological research endeavor (capsure). J Urol June 2004;171(6 Pt 1):2255–9.

[21] Cooperberg MR, Moul JW, Carroll PR. The changing face of prostate cancer. J Clin Oncol November 10, 2005;23(32):8146–51.

[22] Brase JC, Johannes M, Schlomm T, Falth M, Haese A, Steuber T, et al. Circulating miRNAs are correlated with tumor progression in prostate cancer. Int J Cancer February 1, 2011;128(3):608–16.

[23] Nguyen HC, Xie W, Yang M, Hsieh CL, Drouin S, Lee GS, et al. Expression differences of circulating microRNAs in metastatic castration resistant prostate cancer and low-risk, localized prostate cancer. Prostate March 2013;73(4):346–54.

[24] Shen J, Hruby GW, McKiernan JM, Gurvich I, Lipsky MJ, Benson MC, et al. Dysregulation of circulating microRNAs and prediction of aggressive prostate cancer. Prostate February 1, 2012.

[25] Zhang HL, Yang LF, Zhu Y, Yao XD, Zhang SL, Dai B, et al. Serum miRNA-21: elevated levels in patients with metastatic hormone-refractory prostate cancer and potential predictive factor for the efficacy of docetaxel-based chemotherapy. Prostate February 15, 2011;71(3):326–31.

[26] Lin HM, Castillo L, Mahon KL, Chiam K, Lee BY, Nguyen Q, et al. Circulating microRNAs are associated with docetaxel chemotherapy outcome in castration-resistant prostate cancer. Br J Cancer May 13, 2014;110(10):2462–71.

[27] Chen X, Ba Y, Ma L, Cai X, Yin Y, Wang K, et al. Characterization of microRNAs in serum: a novel class of biomarkers for diagnosis of cancer and other diseases. Cell Res October 2008;18(10):997–1006.

[28] Zhang Y, Liu D, Chen X, Li J, Li L, Bian Z, et al. Secreted monocytic miR-150 enhances targeted endothelial cell migration. Mol Cell July 9, 2010;39(1):133–44.

[29] Skog J, Wurdinger T, van Rijn S, Meijer DH, Gainche L, Curry WT, et al. Glioblastoma microvesicles transport RNA and proteins that promote tumour growth and provide diagnostic biomarkers. Nat Cell Biol December 2008;10(12):1470–6.

[30] Kuwabara Y, Ono K, Horie T, Nishi H, Nagao K, Kinoshita M, et al. Increased microRNA-1 and microRNA-133a levels in serum of patients with cardiovascular disease indicate myocardial damage. Circ Cardiovasc Genet August 1, 2011;4(4):446–54.

[31] Valadi H, Ekstrom K, Bossios A, Sjostrand M, Lee JJ, Lotvall JO. Exosome-mediated transfer of mRNAs and microRNAs is a novel mechanism of genetic exchange between cells. Nat Cell Biol June 2007;9(6):654–9.

[32] Zernecke A, Bidzhekov K, Noels H, Shagdarsuren E, Gan L, Denecke B, et al. Delivery of microRNA-126 by apoptotic bodies induces CXCL12-dependent vascular protection. Sci Signal December 8, 2009;2(100):ra81.
[33] Arroyo JD, Chevillet JR, Kroh EM, Ruf IK, Pritchard CC, Gibson DF, et al. Argonaute2 complexes carry a population of circulating microRNAs independent of vesicles in human plasma. Proc Natl Acad Sci USA March 22, 2011;108(12):5003–8.
[34] Wang K, Zhang S, Weber J, Baxter D, Galas DJ. Export of microRNAs and microRNA-protective protein by mammalian cells. Nucleic Acids Res November 1, 2010;38(20):7248–59.
[35] Vickers KC, Palmisano BT, Shoucri BM, Shamburek RD, Remaley AT. MicroRNAs are transported in plasma and delivered to recipient cells by high-density lipoproteins. Nat Cell Biol April 2011;13(4):423–33.
[36] Taylor DD, Gercel-Taylor C. The origin, function and diagnostic potential of RNA within extracellular vesicles present in human biological fluids. Front Genet July 30, 2013;4.
[37] Rabinowits G, Gerçel-Taylor C, Day JM, Taylor DD, Kloecker GH. Exosomal microRNA: a diagnostic marker for lung cancer. Clin Lung Cancer January 2009;10(1):42–6.
[38] Yuan A, Farber EL, Rapoport AL, Tejada D, Deniskin R, Akhmedov NB, et al. Transfer of microRNAs by embryonic stem cell microvesicles. PLoS One 2009;4(3):e4722.
[39] Hunter MP, Ismail N, Zhang X, Aguda BD, Lee EJ, Yu L, et al. Detection of microRNA expression in human peripheral blood microvesicles. PLoS One 2008;3(11):e3694.
[40] Michael A, Bajracharya SD, Yuen PST, Zhou H, Star RA, Illei GG, et al. Exosomes from human saliva as a source of microRNA biomarkers. Oral Dis 2010;16(1):34–8.
[41] Hanke M, Hoefig K, Merz H, Feller AC, Kausch I, Jocham D, et al. A robust methodology to study urine microRNA as tumor marker: microRNA-126 and microRNA-182 are related to urinary bladder cancer. Urol Oncol November 2010;28(6):655–61.
[42] Park NJ, Zhou H, Elashoff D, Henson BS, Kastratovic DA, Abemayor E, et al. Salivary microRNA: discovery, characterization, and clinical utility for oral cancer detection. Clin Cancer Res September 1, 2009;15(17):5473–7.
[43] Weber JA, Baxter DH, Zhang S, Huang DY, How Huang K, Jen Lee M, et al. The microRNA spectrum in 12 body fluids. Clin Chem November 1, 2010;56(11):1733–41.
[44] Sita-Lumsden A, Fletcher CE, Dart DA, Brooke GN, Waxman J, Bevan CL. Circulating nucleic acids as biomarkers of prostate cancer. Biomark Med December 1, 2013;7(6):867–77.
[45] Moreno JG, O'Hara SM, Long JP, Veltri RW, Ning X, Alexander AA, et al. Transrectal ultrasound-guided biopsy causes hematogenous dissemination of prostate cells as determined by RT-PCR. Urology April 1997;49(4):515–20.
[46] Polascik T, Wang Z, Shue M, Di S, Gurganus R, Hortopan S, et al. Influence of sextant prostate needle biopsy or surgery on the detection and harvest of intact circulating prostate cancer cells. J Urol 1999;162:749–52.
[47] Dai J, Keller J, Zhang J, Lu Y, Yao Z, Keller ET. Bone morphogenetic protein-6 promotes osteoblastic prostate cancer bone metastases through a dual mechanism. Cancer Res September 15, 2005;65(18):8274–85.
[48] Yuen H-F, Chan Y-P, Cheung W-L, Wong Y-C, Wang X, Chan K-W. The prognostic significance of BMP-6 signaling in prostate cancer. Mod Pathol October 2008;21(12):1436–43.
[49] Deligezer U, Yaman F, Darendeliler E, Dizdar Y, Holdenrieder S, Kovancilar M, et al. Post-treatment circulating plasma BMP6 mRNA and H3K27 methylation levels discriminate metastatic prostate cancer from localized disease. Clin Chim Acta October 9, 2010;411(19–20):1452–6.
[50] Tricoli JV, Schoenfeldt M, Conley BA. Detection of prostate cancer and predicting progression: current and future diagnostic markers. Clin Cancer Res June 15, 2004;10(12):3943–53.
[51] Dalle Carbonare L, Gasparetto A, Donatelli L, Dellantonio A, Valenti MT. Telomerase mRNA detection in serum of patients with prostate cancer. Urol Oncol February 2013;31(2):205–10.
[52] Dasí F, Martínez-Rodes P, March JA, Santamaría J, Martínez-Javaloyas JM, Gil M, et al. Real-time quantification of human telomerase reverse transcriptase mRNA in the plasma of patients with prostate cancer. Ann N Y Acad Sci 2006;1075(1):204–10.
[53] March-Villalba JA, Martínez-Jabaloyas JM, Herrero MJ, Santamaria J, Aliño SF, Dasí F. Cell-free circulating plasma hTERT mRNA is a useful marker for prostate cancer diagnosis and is associated with poor prognosis tumor characteristics. PLoS One 2012;7(8):e43470.
[54] Bu H, Bormann S, Schäfer G, Horninger W, Massoner P, Neeb A, et al. The anterior gradient 2 (AGR2) gene is overexpressed in prostate cancer and may be useful as a urine sediment marker for prostate cancer detection. Prostate 2011;71(6):575–87.
[55] Zhang J-S, Gong A, Cheville JC, Smith DI, Young CYF. AGR2, an androgen-inducible secretory protein overexpressed in prostate cancer. Genes Chromosomes Cancer 2005;43(3):249–59.
[56] Kani K, Malihi PD, Jiang Y, Wang H, Wang Y, Ruderman DL, et al. Anterior gradient 2 (AGR2): blood-based biomarker elevated in metastatic prostate cancer associated with the neuroendocrine phenotype. Prostate 2013;73(3):306–15.

[57] Antonarakis ES, Lu C, Wang H, Luber B, Nakazawa M, Roeser JC, et al. AR-V7 and resistance to enzalutamide and abiraterone in prostate cancer. N Engl J Med 2014;371(11):1028–38.
[58] Mandel P, Metais P. Les acides nucleiques du plasma sanguin chez l'homme. CR Acad Sci Paris 1948;142:241–3.
[59] Delgado PO, Alves BC, Gehrke Fde S, Kuniyoshi RK, Wroclavski ML, Del Giglio A, et al. Characterization of cell-free circulating DNA in plasma in patients with prostate cancer. Tumour Biol April 1, 2013;34(2):983–6.
[60] Leon S, Shapiro B, Sklaroff D, Yaros M. Free DNA in the serum of cancer patients and the effect of therapy. Cancer Res 1977;37(3):646–50.
[61] Altimari A, Grigioni ADE, Benedettini E, Gabusi E, Schiavina R, Martinell A, et al. Diagnostic role of circulating free plasma DNA detection in patients with localized prostate cancer. Am J Clin Pathol May 1, 2008;129(5):756–62.
[62] Ellinger J, Müller SC, Wernert N, Von Ruecker A, Bastian PJ. Mitochondrial DNA in serum of patients with prostate cancer: a predictor of biochemical recurrence after prostatectomy. BJU Int August 5, 2008;102(5):628–32.
[63] Bastian PJ, Palapattu GS, Yegnasubramanian S, Lin X, Rogers CG, Mangold LA, et al. Prognostic value of preoperative serum cell-free circulating DNA in men with prostate cancer undergoing radical prostatectomy. Clin Cancer Res September 15, 2007;13(18):5361–7.
[64] Jung K, Stephan C, Lewandowski M, Klotzek S, Jung M, Kristiansen G, et al. Increased cell-free DNA in plasma of patients with metastatic spread in prostate cancer. Cancer Lett March 18, 2004;205(2):173–80.
[65] Schwarzenbach H, Alix-Panabières C, Müller I, Letang N, Vendrell J-P, Rebillard X, et al. Cell-free tumor DNA in blood plasma as a marker for circulating tumor cells in prostate cancer. Clin Cancer Res February 1, 2009;15(3):1032–8.
[66] Jahr S, Hentze H, Englisch S, Hardt D, Fackelmayer FO, Hesch R-D, et al. DNA fragments in the blood plasma of cancer patients: quantitations and evidence for their origin from apoptotic and necrotic cells. Cancer Res February 2, 2001;61(4):1659–65.
[67] Stroun M, Anker P, Maurice P, Lyautey J, Lederrey C, Beljanski M. Neoplastic characteristics of the DNA found in the plasma of cancer patients. Oncology 1989;46(5):318–22.
[68] Ellinger J, Müller SC, Stadler TC, Jung A, von Ruecker A, Bastian PJ. The role of cell-free circulating DNA in the diagnosis and prognosis of prostate cancer. Urol Oncol March 2011;29(2):124–9.
[69] Jeronimo C, Nomoto S, Caballero O, Usadel H, Henrique R, Varzim G, et al. Mitochondrial mutations in early stage prostate cancer and bodily fluids. Oncogene August 23, 2001;20(37):5195–8.
[70] Mehra N, Penning M, Maas J, van Daal N, Giles RH, Voest EE. Circulating mitochondrial nucleic acids have prognostic value for survival in patients with advanced prostate cancer. Clin Cancer Res January 15, 2007;13(2):421–6.
[71] Ellinger J, Müller DC, Müller SC, Hauser S, Heukamp LC, von Ruecker A, et al. Circulating mitochondrial DNA in serum: a universal diagnostic biomarker for patients with urological malignancies. Urol Oncol July 2012;30(4):509–15.
[72] Müllerü I, Urban K, Pantel K, Schwarzenbach H. Comparison of genetic alterations detected in circulating microsatellite DNA in blood plasma samples of patients with prostate cancer and benign prostatic hyperplasia. Ann N Y Acad Sci 2006;1075(1):222–9.
[73] Sunami E, Shinozaki M, Higano CS, Wollman R, Dorff TB, Tucker SJ, et al. Multimarker circulating DNA assay for assessing blood of prostate cancer patients. Clin Chem March 1, 2009;55(3):559–67.
[74] Bastian PJ, Palapattu GS, Yegnasubramanian S, Rogers CG, Lin X, Mangold LA, et al. CpG island hypermethylation profile in the serum of men with clinically localized and hormone refractory metastatic prostate cancer. J Urol February 2008;179(2):529–34. Discussion 34–5.
[75] Bryzgunova OE, Morozkin ES, Yarmoschuk SV, Vlassov VV, Laktionov PP. Methylation-specific sequencing of GSTP1 gene promoter in circulating/extracellular DNA from blood and urine of healthy donors and prostate cancer patients. Ann N Y Acad Sci 2008;1137(1):222–5.
[76] Chuang C-K, Chu D-C, Tzou R-D, Liou S-I, Chia J-H, Sun C-F. Hypermethylation of the CpG islands in the promoter region flanking GSTP1 gene is a potential plasma DNA biomarker for detecting prostate carcinoma. Cancer Detect Prev 2007;31(1):59–63.
[77] Ellinger J, Haan K, Heukamp LC, Kahl P, Büttner R, Müller SC, et al. CpG island hypermethylation in cell-free serum DNA identifies patients with localized prostate cancer. Prostate 2008;68(1):42–9.
[78] Goessl C, Muller M, Miller K. Methylation-specific PCR (MSP) for detection of tumour DNA in the blood plasma and serum of patients with prostate cancer. Prostate Cancer Prostatic Dis 2000;3(S1):S17.
[79] Jernimo C, Usadel H, Henrique R, Silva C, Oliveira J, Lopes C, et al. Quantitative GSTP1 hypermethylation in bodily fluids of patients with prostate cancer. Urology December 2002;60(6):1131–5.
[80] Papadopoulou E, Davilas E, Sotiriou V, Georgakopoulos E, Georgakopoulou S, Koliopanos A, et al. Cell-free DNA and RNA in plasma as a new molecular marker for prostate and breast cancer. Ann N Y Acad Sci 2006;1075(1):235–43.
[81] Reibenwein J, Pils D, Horak P, Tomicek B, Goldner G, Worel N, et al. Promoter hypermethylation of GSTP1, AR and 14-3-3sigma in serum of prostate cancer patients and its clinical relevance. Prostate March 1, 2007;67(4):427–32.

[82] Schwarzenbach H, Chun FKH, Lange I, Carpenter S, Gottberg M, Erbersdobler A, et al. Detection of tumor-specific DNA in blood and bone marrow plasma from patients with prostate cancer. Int J Cancer 2007; 120(7):1465–71.

[83] Schwarzenbach H, Chun FKH, Müller I, Seidel C, Urban K, Erbersdobler A, et al. Microsatellite analysis of allelic imbalance in tumour and blood from patients with prostate cancer. BJU Int 2008;102(2):253–8.

[84] Goessl C, Krause H, Müller M, Heicappell R, Schrader M, Sachsinger J, et al. Fluorescent methylation-specific polymerase chain reaction for DNA-based detection of prostate cancer in bodily fluids. Cancer Res November 1, 2000;60(21):5941–5.

[85] Lee WH, Isaacs WB, Bova GS, Nelson WG. CG island methylation changes near the GSTP1 gene in prostatic carcinoma cells detected using the polymerase chain reaction: a new prostate cancer biomarker. Cancer Epidemiol Biomark Prev June 1, 1997;6(6): 443–50.

[86] Lee WH, Morton RA, Epstein JI, Brooks JD, Campbell PA, Bova GS, et al. Cytidine methylation of regulatory sequences near the pi-class glutathione S-transferase gene accompanies human prostatic carcinogenesis. Proc Natl Acad Sci USA November 22, 1994;91(24):11733–7.

[87] Jerónimo C, Usadel H, Henrique R, Oliveira J, Lopes C, Nelson WG, et al. Quantitation of GSTP1 methylation in non-neoplastic prostatic tissue and organ-confined prostate adenocarcinoma. J Natl Cancer Inst November 21, 2001;93(22):1747–52.

[88] Nakayama M, Bennett CJ, Hicks JL, Epstein JI, Platz EA, Nelson WG, et al. Hypermethylation of the human glutathione S-transferase-π gene (GSTP1) CpG island is present in a subset of proliferative inflammatory atrophy lesions but not in normal or hyperplastic epithelium of the prostate: a detailed study using laser-capture microdissection. Am J Pathol September 2003;163(3):923–33.

[89] Goessl C, Müller M, Heicappell R, Krause H, Miller K. DNA-based detection of prostate cancer in blood, urine, and ejaculates. Ann N Y Acad Sci 2001;945(1):51–8.

[90] Bastian PJ, Palapattu GS, Lin X, Yegnasubramanian S, Mangold LA, Trock B, et al. Preoperative serum DNA GSTP1 CpG island hypermethylation and the risk of early prostate-specific antigen recurrence following radical prostatectomy. Clin Cancer Res June 1, 2005;11(11):4037–43.

[91] Catalona WJ, Beiser JA, Smith DS. Serum free prostate specific antigen and prostate specific antigen density measurements for predicting cancer in men with prior negative prostatic biopsies. J Urol December 1997;158(6):2162–7.

[92] Chon CH, Lai FC, McNeal JE, Presti Jr JC. Use of extended systematic sampling in patients with a prior negative prostate needle biopsy. J Urol June 2002;167(6):2457–60.

[93] Kang GH, Lee S, Lee HJ, Hwang KS. Aberrant CpG island hypermethylation of multiple genes in prostate cancer and prostatic intraepithelial neoplasia. J Pathol 2004;202(2):233–40.

[94] Liu L, Yoon J, Dammann R, Pfeifer G. Frequent hypermethylation of the RASSF1A gene in prostate cancer. Oncogene October 3, 2002;21(44):6835–40.

[95] Maruyama R, Toyooka S, Toyooka KO, Virmani AK, Zöchbauer-Müller S, Farinas AJ, et al. Aberrant promoter methylation profile of prostate cancers and its relationship to clinicopathological features. Clin Cancer Res February 1, 2002;8(2):514–9.

[96] Nakayama T, Watanabe M, Yamanaka M, Hirokawa Y, Suzuki H, Ito H, et al. The role of epigenetic modifications in retinoic acid receptor beta2 gene expression in human prostate cancers. Lab Invest 2001;81(7):1049–57.

[97] Cortese R, Kwan A, Lalonde E, Bryzgunova O, Bondar A, Wu Y, et al. Epigenetic markers of prostate cancer in plasma circulating DNA. Hum Mol Genet August 15, 2012;21(16):3619–31.

[98] He L-R, Liu M-Z, Li B-K, Rao H-L, Liao Y-J, Guan X-Y, et al. Prognostic impact of H3K27me3 expression on locoregional progression after chemoradiotherapy in esophageal squamous cell carcinoma. BMC Cancer 2009;9(1):461.

[99] Ke X-S, Qu Y, Rostad K, Li W-C, Lin B, Halvorsen OJ, et al. Genome-Wide profiling of histone H3 lysine 4 and lysine 27 trimethylation reveals an epigenetic signature in prostate carcinogenesis. PLoS One 2009;4(3):e4687.

[100] Gonzales JC, Fink LM, Goodman Jr OB, Symanowski JT, Vogelzang NJ, Ward DC. Comparison of circulating microRNA 141 to circulating tumor cells, lactate dehydrogenase, and prostate-specific antigen for determining treatment response in patients with metastatic prostate cancer. Clin Genitourin Cancer September 2011;9(1):39–45.

[101] Ankerst DP, Thompson IM. Sensitivity and specificity of prostate-specific antigen for prostate cancer detection with high rates of biopsy verification. Arch Ital Urol Androl December 2006;78(4):125–9.

[102] de la Taille A, Irani J, Graefen M, Chun F, de Reijke T, Kil P, et al. Clinical evaluation of the PCA3 assay in guiding initial biopsy decisions. J Urol June 2011;185(6):2119–25.

[103] Crawford ED, Rove KO, Trabulsi EJ, Qian J, Drewnowska KP, Kaminetsky JC, et al. Diagnostic performance of PCA3 to detect prostate cancer in men with increased prostate specific antigen: a prospective study of 1,962 cases. J Urol November 2012;188(5):1726–31.

[104] Tomlins SA, Aubin SM, Siddiqui J, Lonigro RJ, Sefton-Miller L, Miick S, et al. Urine TMPRSS2:ERG fusion transcript stratifies prostate cancer risk in men with elevated serum PSA. Sci Transl Med August 3, 2011;3(94):94ra72.

[105] Salami SS, Schmidt F, Laxman B, Regan MM, Rickman DS, Scherr D, et al. Combining urinary detection of TMPRSS2:ERG and PCA3 with serum PSA to predict diagnosis of prostate cancer. Urol Oncol July 2013;31(5):566–71.

[106] Leyten GH, Hessels D, Jannink SA, Smit FP, de Jong H, Cornel EB, et al. Prospective multicentre evaluation of PCA3 and TMPRSS2-ERG gene fusions as diagnostic and prognostic urinary biomarkers for prostate cancer. Eur Urol March 2014;65(3):534–42.

[107] Stewart GD, Van Neste L, Delvenne P, Delree P, Delga A, McNeill SA, et al. Clinical utility of an epigenetic assay to detect occult prostate cancer in histopathologically negative biopsies: results of the MATLOC study. J Urol March 2013;189(3):1110–6.

[108] Aubry W, Lieberthal R, Willis A, Bagley G, Willis 3rd SM, Layton A. Budget impact model: epigenetic assay can help avoid unnecessary repeated prostate biopsies and reduce healthcare spending. Am Health Drug Benefits January 2013;6(1):15–24.

[109] Stephan C, Vincendeau S, Houlgatte A, Cammann H, Jung K, Semjonow A. Multicenter evaluation of [−2] proprostate-specific antigen and the prostate health index for detecting prostate cancer. Clin Chem January 2013;59(1):306–14.

[110] Cuzick J, Swanson GP, Fisher G, Brothman AR, Berney DM, Reid JE, et al. Prognostic value of an RNA expression signature derived from cell cycle proliferation genes in patients with prostate cancer: a retrospective study. Lancet Oncol March 2011;12(3):245–55.

[111] Robinson K, Creed J, Reguly B, Powell C, Wittock R, Klein D, et al. Accurate prediction of repeat prostate biopsy outcomes by a mitochondrial DNA deletion assay. Prostate Cancer Prostatic Dis June 2010;13(2):126–31.

CHAPTER

29

MicroRNAs in Breast Cancer and Their Value as Biomarkers

Olafur Andri Stefansson

Cancer Research Laboratory, Faculty of Medicine, University of Iceland, Reykjavik, Iceland

OUTLINE

Epigenetic Biomarkers and Diagnostics
http://dx.doi.org/10.1016/B978-0-12-801899-6.00029-2

1. INTRODUCTION

In 1998, Andrew Z. Fire and Craig C. Mello described gene silencing by double-stranded RNA in *Caenorhabditis elegans* (the nematode worm) as an active catalytic system that can achieve transcriptional silencing of complementary RNA sequences: a system known as RNA interference (RNAi) [1]. This discovery provided an explanation for several previous observations made using antisense RNA molecules and later earned them the Nobel Prize in physiology or medicine in 2006. It is now known that RNAi is present in many different organisms, including humans, as a posttranscriptional regulatory mechanism. The type of RNA molecules central to RNAi are short interfering (siRNAs) and microRNAs (miRNAs) and, indeed, the two are intimately connected [2]. RNAi-induced transcriptional silencing appears to be maintained in daughter cells following cellular divisions and, for this reason, many researchers have proposed that the RNAi system represents an additional layer of epigenetic information—although this is still controversial. Nonetheless, recent research based on animal models has continued to support a link between RNAi and mechanisms of epigenetic regulation [3].

miRNAs are processed from longer RNA transcripts that can form stem-loop structures due to partial self-complementary base pairs [2] (Figure 1). These RNA molecules, referred to as pri-miRNAs (primary miRNA), can be transcribed as an independent transcript or as a polycistron from which multiple miRNAs are derived [4]. Furthermore, miRNAs are also found within introns of mRNA transcripts from which they are spliced for further processing. In most cases, the pri-miRNAs are processed by Drosha together with its partner called Pasha from which ~70-bp RNA molecules are released as precursor miRNAs (pre-miRNAs). Exceptions to this, wherein pri-miRNAs are able to bypass Drosha, are known. This includes miRNAs located within introns of mRNAs where the splicing sites define the ends of the miRNA, in which case it is named a mirtron [5]. These will then form a lariat that undergoes debranching to then fold and form a pre-miRNA without any processing by Drosha (Figure 1). A protein called Exportin enables the export of pre-miRNAs from the nucleus for further processing in the cytoplasm. Subsequently, once in the cytoplasm, pre-miRNAs are recognized by Dicer, which possesses endonuclease activity critical for further processing of pre-miRNAs to smaller RNA molecules of about ~20–30 bp in length [2]. These smaller RNA molecules then represent the final miRNA duplexes that are essentially equivalent to the aforementioned siRNAs, although it should be noted here that pre-miRNAs are not the only source of siRNAs, as the RNAi system is thought to be an important mechanism of defense against viruses and, possibly, also propagation of transposons [6]. Nucleotides 2–7 from the 5′end of the miRNA duplex are described as the seed sequence—the most important part in terms of target recognition. Only one of the two strands in the miRNA duplex, called the guiding strand, is then finally loaded into an Argonaute-containing complex called RNA-induced silencing complex (RISC). How this selection takes place, i.e., how the guiding strand is preferentially loaded into the RISC complex, is currently poorly understood [7]. In any case, the sequence of the guiding strand is critical in directing the activity of RISC toward specific targets based on base pair complementarity to other single-stranded RNA molecules, e.g., mRNAs. The outcome of this process often involves the destruction of the complementary RNAs, which is mediated by cleavage carried out by the Argonaute protein. Alternatively, it has been shown that in some cases, instead of cleaving, they inhibit their target mRNA transcripts from being translated to protein products. Interestingly, the role for miRNAs in promoting translation of target mRNA transcripts has been described [8].

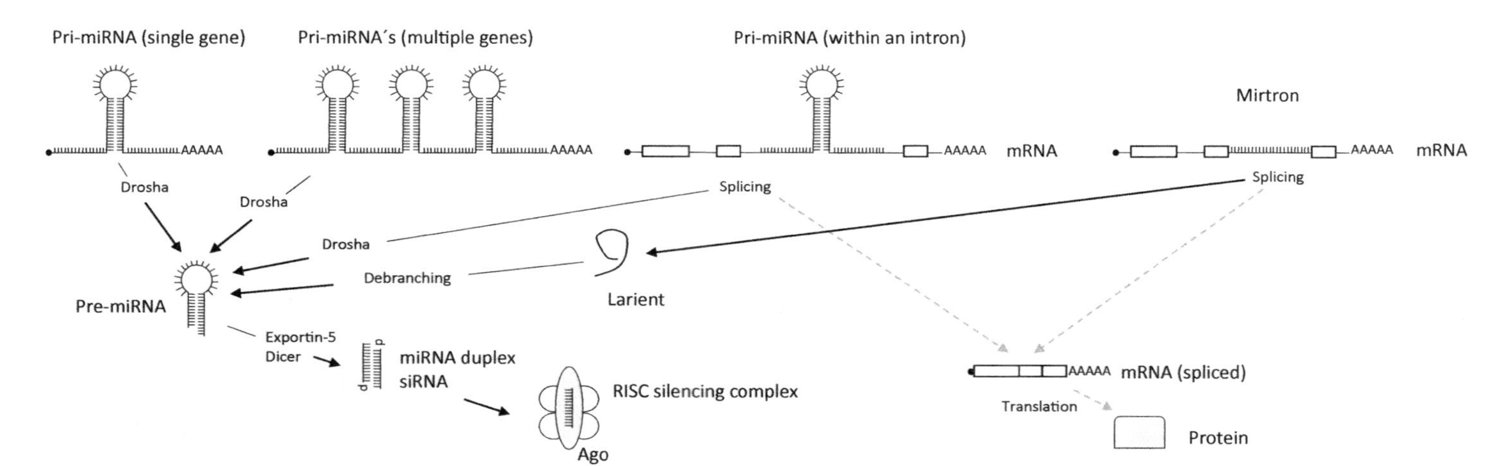

FIGURE 1 The transcription of miRNA genes leads to the formation of a primary miRNA (pri-miRNA) transcript. In a polycistron, many different miRNAs are found and considered as different genes, even though they are transcribed as one from the same locus. Introns of protein-coding genes can contain miRNA genes. After splicing, the intron is recognized by Drosha and is processed as a pri-miRNA transcript. A mirtron is an miRNA gene that constitutes an entire intron from a protein-coding gene wherein the splice site defines the miRNA. Following splicing, the mirtron will be released from the mRNA structure to form a lariat that undergoes debranching to then form a hairpin structure that is characteristic of pre-miRNAs. The key difference is that mirtrons effectively bypass processing by Drosha. Further processing will then take place in the cytosol. The export of pre-miRNAs from the nucleus required Exportin-5 and, once in the cytoplasm, the activity of Dicer will then lead to the formation of an miRNA duplex (essentially identical to short interfering RNAs, i.e., siRNAs). One of the two strands making up the miRNA duplexes will then be incorporated into the so-called RNA-induced silencing complex (RISC) complex. Here, the affinity of miRNA-containing RISC complex to mRNA transcripts depends on the base complementarity of the miRNA present in the RISC complex. The Ago protein is able to cleave mRNA transcripts recognized by complementary base pairing to the miRNA/siRNA found within the RISC complex.

A comprehensive catalog enlisting genomic coordinates for known and putative miRNA genes is maintained by the miRBase [9]. The underlying data used by the developers of the miRBase are based on ultradeep sequencing of RNA samples in different tissue types. According to these data, the human genome contains at least 1881 miRNA genes from which 2588 mature miRNA transcripts are processed. miRNAs are now known to be involved in many different biological processes and, accordingly, their expression levels vary widely across different tissues of the human body [10]. Moreover, the expression patterns of miRNAs are highly dynamic during development including the development of breast tissue—thereby highlighting the importance of miRNAs in the biology of the breast [11].

2. GLOBAL DEREGULATION OF miRNAs IN CANCER

By genome-wide miRNA expression profiling, a consistent pattern emerges that is seen across many different cancer types, i.e., a trend toward downregulation of multiple miRNA genes. It is now clear that, although downregulation is dominant, a relatively smaller subset of miRNAs is found overexpressed. Thus, it can be said that miRNAs are found deregulated, i.e., they can be either up- or downregulated. Nonetheless, the systematic effect toward downregulation of multiple miRNAs in cancer likely has an underlying causative component. Here, a key finding is that components of the RNAi machinery itself are found recurrently mutated in various types of human cancers. In particular, acquired mutations have been identified in Dicer as described in testicular germ cell tumors, embryonal rhabdomyosarcoma, Wilms' tumors, and nonepithelial ovarian cancers [12,13]. Moreover, germline mutations in Dicer have been described in relation to familial pleuropulmonary blastoma [14,15]. Other components of the RNAi machinery found defective in cancer include the TARBP2 gene (named TAR RNA-binding protein 2), an integral component of the Dicer-containing complex, which occurs through acquired mutations in colon and gastric cancers [16]. The nuclear export of pre-miRNAs is found defective in a subset of colon cancers due to acquired mutations in the Exportin-5 (XPO5) gene leading to reduced processing of miRNAs [17]. Additionally, it has recently been shown that Drosha is a target of acquired mutations in Wilms' tumors [18]. Although mutations in components of the RNAi machinery are not found in breast cancer, these findings are relevant in shedding light on the mechanisms by which miRNAs are perturbed in cancer while also firmly establishing miRNAs as important contributors to cancer development.

3. THE BIOLOGICAL SIGNIFICANCE OF miRNAs IN BREAST CANCER DEVELOPMENT

miRNAs are deregulated in various different cancer types including breast cancer (MIM #114480) [19]. A subset of these miRNAs have been shown to directly target and downregulate mRNA transcripts derived from some well-established oncogenes and tumor suppressor genes; including those known to be of critical importance in breast cancer development. For example, the BRCA1 gene, a central tumor suppressor in breast and ovarian cancers, is known to be targeted for downregulation by miR-182, miR-146a, and miR-146b-5p [20,21]. Of note here, in a recent study, the relevance of miR-146a in targeting BRCA1 for downregulation was not supported [22]. The BRCA2 gene has also been shown to be regulated by at least one miRNA known as miR-1245 [23]. Other examples include loss of miR-339-5p expression resulting in high expression of MDM2—a well-characterized oncogene [24] and poor disease prognosis [25]. Additionally, the downregulation of either miR-548d-3p or miR-559 has been described in relation to overexpression of the well-known ERBB2 gene (also known as HER2) found amplified in approximately 20% of all breast cancers [26].

Among the most consistently overexpressed miRNAs in breast cancer are miR-155 and miR-21 [19]. Here, miR-155 has been shown to positively regulate proliferation of breast cancer cells [27]. This was found to occur through direct interaction between miR-155 and SOCS1 mRNA transcripts leading to downregulation of SOCS1 (a JAK1 interacting partner) resulting in increased capacity for colony formation in vitro and enhanced proliferative rates when xenografted in nude mice [27]. The authors propose that miR-155 may act as a bridge between breast cancer and inflammation. This is partly supported by a subsequent genome-wide miRNA expression study in breast cancer by Dvinge et al. based on the use of freshly frozen primary specimens [28]. In this study, the miR-155 gene was highlighted among four other miRNAs as strongly correlated with immune responses. However, it still remains to be determined whether this observation reflects expression of miR-155 in infiltrating lymphocytes or breast cancer cells. However, studies in breast cancer cell lines (devoid of stroma) provide support for the latter and furthermore have revealed an important role for BRCA1 as an epigenetic regulator of the miR-155 promoter region [29,30]. Here, Chang et al. discovered that a known genetic variant in the BRCA1 gene (R1699Q) does not affect the ability of the BRCA1 variant protein to carry out DNA repair but still contributes to increased breast cancer risk through loss of its ability to interact with histone deacetylases and epigenetically suppress the miR-155 gene [29]. It is well established that BRCA1 abnormalities are associated with the development of basal-like breast cancers [31,32]. In line with this, the miR-155 gene shows a subtype-specific expression pattern wherein high levels of miR-155 are specifically seen in basal-like breast cancers [28]. Recent studies have shown miR-155 as a regulator of RAD51 [33] and TRF1 [34] of potential relevance for genetic instability in breast cancer. This finding is of particular relevance for breast cancer of the basal-like subtype wherein aneuploidy and copy number changes are highly prevalent [32]. Several other studies have supported a biologically relevant role for miR-155 in breast cancer development [35]. Thus, there is now a significant body of evidence to support the classification of miR-155 as a genuine oncogene, i.e., an oncomiR.

Another promising oncomiR candidate in breast cancer is miR-21 [19,36]. Here, data have emerged suggesting that a subset of breast cancers is highly dependent on the presence of miR-21 [37,38]. This growth dependency is thought to be mediated by the antiapoptotic roles played out by miR-21 through its direct interaction with the mRNA transcripts of BCL-2 [37] and PDCD4 [39,40]. Another study demonstrated a role for miR-21 in mediating anchorage independence, i.e., through targeting the mRNA transcripts of TPM1 [41]. In a pan-cancer study, the most prominent miRNA expression changes across many different cancer types involved the miR-21 gene based on analyses carried out using freshly frozen tumor specimens [42]. Importantly, this study further describes novel oncomiR candidates, including miR-17, miR-19, miR-130, miR-93, miR-18, miR-455, and miR-210, based on their coregulated expression patterns across multiple cancer types, including breast cancers, targeting tumor suppressor genes through the same seed motif, including PTEN, ZBTB4 (a p53-regulated gene), TGFBR2, and SMAD4 [42]. Although this is an exciting discovery, these results were based solely on bioinformatics analyses, and the experimental evidence demonstrating the claimed cotargeting by the aforementioned oncomiR "superfamily" is therefore still lacking.

The miR-15 and miR-16 genes were among the first miRNAs to be described as tumor suppressor genes [43]. This discovery, however, was made in chronic lymphocytic leukemia. In breast cancer, miR-16 is markedly downregulated and has been identified as a regulator of the Wip1 phosphatase of potential relevance to the DNA damage response (DDR)—thus, miR-16 could be of potential relevance as a tumor suppressor in breast cancer [44]. A large number of miRNAs have now been implicated as tumor suppressor genes in breast cancer, as well as, in other cancer types, whereas only very few of them have been validated as such. Nonetheless, there are a few notable examples such as miR-145, which has been shown to suppress the invasion phenotype of breast cancer cells [45,46]. Indeed, the most prominently downregulated miRNAs in breast cancer includes miR-145 together with two others, miR-10b and miR-125b [19]. Another compelling example includes the miR-200c gene, which was found to suppress the ability of normal mammary stem cells to differentiate, while also negatively affecting tumor formation capacity of breast cancer stem cells [47]. Shimono et al. described a role for miR-200c as a regulator of epithelial–mesenchymal transition (EMT), i.e., a process known to be linked to the metastatic capacity of cancer cells. Collectively, multiple different miRNAs have been described as candidate contributors to the development and/or progression of breast cancer, see Table 1.

4. GENETIC VARIANTS IN miRNAs AND THEIR BINDING SITES

Inherited variation in miRNA sequences has been linked to breast cancer susceptibility, see Table 1(c). For example, a common single nucleotide polymorphism (SNP) in the miR-146a gene was identified as a risk factor for breast and ovarian cancer [48] while additionally this miRNA was linked to increased risk for other cancer types as well, i.e., lung, cervical, and hepatocellular carcinomas [49–51]. Similarly, a common SNP in miR-196a-2 has been associated with decreased risk for developing breast cancer [52]. The miR-196a-2 variant, however, has not been firmly validated as a risk modifier [53,54]. Shen et al. have further studied the DNA sequences of miRNAs and identified rare variants in miR-17 and miR-30c-1 [55]. These variants were identified in a cohort of only 42 breast cancer patients with family history of the disease. Thus, although the miR-17 and miR-30c-1 variants can be considered as candidate risk variants, their contribution to breast cancer risk is currently unknown [55]. Common genetic variants in miR-27a, miR-423, and miR-499 have also been identified as risk modifiers in breast cancer [54,56,57]. Interestingly, genetic variation in the miR-423 gene has been described as a risk modifier in BRCA2 mutation carriers [56]. Recently, both miR-499a (rs3746444) and miR-146a (rs2910164) were validated as breast cancer susceptibility loci in a large study cohort involving approximately 8300 cases and 8500 controls [58]. Additionally, in 2013, a meta-analysis of more than 20,000 breast cancer cases and controls reported a risk-associated polymorphism proximal to the miR-1208 gene [59]. More recently, Rawlings-Goss et al. by next-generation sequencing focused on 1524 miRNAs, a rich harvest of novel germline variants was identified of which at least 5 were found within the seed region and thus predicted to have significant biological effects [60]. The contribution of these novel variants to cancer susceptibility remains to be determined.

Inherited genetic variants have also been described in putative miRNA binding sites of protein-coding genes and are implicated in breast cancer. A notable example includes a common variation found in the 3′UTR (untranslated region) of estrogen receptor (ESR1) as protective against the development of breast cancer [61]. This variation was found to modify the binding affinity between ESR1 transcripts and miR-453. As estrogen is a known risk factor, this finding is of potential relevance in terms of breast cancer susceptibility. An inherited variation in the 3′UTR of SET8 representing a binding site for miR-502 has been linked with the risk of developing breast cancer [62]. Interestingly, Song et al. showed that the risk-associated genotype was linked with reduced SET8 expression in breast cancer specimens. This same risk allele in

TABLE 1 List of miRNAs of Potential Value as Diagnostic Biomarkers in Breast Cancer that are Found Significantly (a) Upregulated and (b) Downregulated. (c) Genetic Predisposition Linked to miRNA Genes and 3′UTR Binding Sites of Target mRNA's

miRNA	mRNA target (selected examples)	Biological effects	Diagnostic value[a]	References
(A) UPREGULATED miRNAs IN BREAST CANCER				
miR-182, miR-146a, and miR-146b-5p	BRCA1	Loss of tumor suppressor gene BRCA1/genetic instability	Predictive for drug/targeted inhibitor (e.g., PARPi)/basal-like (miR-146a)	[20,21]
miR-1245	BRCA2	Loss of tumor suppressor gene BRCA2/genetic instability	Predictive for drug/targeted inhibitor (e.g., PARPi)	[23]
miR-548d-3p and miR-559	ERBB2	Oncogenic ERBB2 (HER2) signaling/proliferation stress	ERBB2/HER2-targeted therapy response	[26]
miR-155	SOCS1, RAD51, and TRF1	Proliferation stress/genetic instability	Prognostic/basal-like (miR-155)	[27,33,34, 154]
miR-21	BCL-2, PDCD4, and TPM1	Antiapoptotic/anchorage-independent growth	Prognostic/early detection in blood	[37,39,41, 153]
miR-17, miR-19, miR-130, miR-93, miR-18, miR-455, and miR-210	PTEN, ZBTB4, TGFBR2, and SMAD4	Loss of tumor suppressor genes	PI3K, TGFβ, and p53 pathway perturbations/basal-like (all except for miR-130)	[42]
miR-196a-2	HOXB2, HOXB3, HOXC13, and HOXB5	Loss of differentiation potential	Genetic risk assessment	[52]
miR-4728	MAPK1 and SOS1	Negative feedback loop for ERBB2 (HER2) signaling	ERBB2/HER2-targeted therapy response	[68]
miR-9-1	CDH1	Loss of adherence	Prognostic/basal-like (miR-9)	[70]
miR-126	Sdf1-a (CXCL12)	Inhibition of mesechymal stem cell infiltration and inflammatory monocytes	Prognostic/normal-like (miR-126)	[73]
miR-181a and miR-181b	ATM	Loss of tumor suppressor gene ATM/impaired DNA damage response	Predictive for drug/targeted inhibitor (e.g., PARPi)	[104]
miR-18a	ATM	Loss of tumor suppressor gene ATM/impaired DNA damage response	Predictive for drug/targeted inhibitor (e.g., PARPi)/early detection in blood/basal-like (miR-18a/b)	[106,152]
miR-107	RAD51	Impaired DNA damage response	Predictive for drug/targeted inhibitor (e.g., PARPi)/early detection in blood	[107]
miR-222	RAD51	Impaired DNA damage response	Predictive for drug/targeted inhibitor (e.g., PARPi)	[107]

Continued

TABLE 1 List of miRNAs of Potential Value as Diagnostic Biomarkers in Breast Cancer that are Found Significantly (a) Upregulated and (b) Downregulated. (c) Genetic Predisposition Linked to miRNA Genes and 3′UTR Binding Sites of Target mRNA's—cont'd

miRNA	mRNA target (selected examples)	Biological effects	Diagnostic value[a]	References
miR-1255b, miR-148b, or miR-193b	BRCA1, BRCA2, and RAD51	Loss of tumor suppressor genes/impaired DNA damage response	Predictive for drug/targeted inhibitor (e.g., PARPi)/luminal-B (miR-193b)	[108]
miR-373	RAD23B and RAD52	Impaired DNA damage response	Predictive for drug/targeted inhibitor (e.g., PARPi)	[109]
miRs 142-3p, 505, 1248, 181a-2, 25, and 340	Multiple	BRCAness molecular characteristics	Prognostic/predictive for drug/targeted inhibitors (e.g., PARPi) and/or BRCA mutation status	[115]
miR-345	BAG3 and ZEB2		Prognostic	[28]
miR-834 and miR-729	Various		Prognostic/iClust-4	[28]
miR-210		Invasion	Prognostic ductal carcinoma in situ/basal-like and HER2 (miR-210)	[120]
miR-200b	ZEB1, ZEB2, RAB21, RAB23, RAB18, and RAB3B	Invasion/migration/proliferation	Prognostic	[139]
miR-22	TET1, TET2	Loss of differentiation/EMT/migration	Prognostic	[143]
miR-199 and miR-31		Metastasis	Prognostic/luminal-A (miR-199a-3p, miR-199b-5p)	[144]
miR-15a, miR-18a, miR-107, miR-133a, miR-139-5p, miR-143, miR-145, miR-365, and miR-425	Multiple		Early detection in blood/basal-like (miR-18a), normal-like (miR-145, miR-143, and miR-139-5p), and luminal-B (miR-425)	[152]
miR-302 (high)/miR-203 (low)	Multiple	Loss of differentiation/metastasis	Prognostic	[150]
(B) DOWNREGULATED miRNAs IN BREAST CANCER				
miR-339-5p	MDM2	Oncogenic MDM2-mediating inhibition of p53	Prognostic	[24]
miR-16	WIP1	Impaired DNA damage response	Predictive of anticancer drug responses	[44]
miR-145	MUC1	Migration and metastasis	Prognostic	[45,46]
miR-10b	HOXD10 [137]	Metastasis	Prognostic/Loss in basal-like (miR-10b)	[19]
miR-125b	ETS1	Oncogenic/proliferation	Prognostic/Loss in luminal-B (miR-125b)	[19]

miR-200c	HMGB1	Loss of differentiation/EMT/migration/ Loss in normal-like (miR-200c)	Prognostic/loss in normal-like (miR-200c)	[47,141]
miR-195/miR-497	Raf-1, Cyclin-D1	Invasion/proliferation	Prognostic/loss in basal-like and HER2 (miR-195)	[71]
miR-148a	IGF1R, IRS1	Proliferation/migration	Prognostic	[72]
miR-29b	Multiple (VEGFA, PDGF, etc.)	EMT	Prognostic	[145]
miR-34a	BCL2, CDK4/6	Loss of p53-mediated checkpoint arrest	Prognostic	[149]
(C) RISK-ASSOCIATED GENETIC VARIATION AND miRNAs				
miR-146a	BRCA1	Loss of tumor suppressor BRCA1/ genetic instability	Genetic risk assessment (rs2910164)/ upregulated in basal-like and downregulated in luminal-B	[48,58]
miR-196a-2	HOXB2, HOXB3, HOXC13, and HOXB5	Loss of differentiation potential	Genetic risk assessment (rs11614913)/ upregulated in HER2 and downregulated in basal-like	[52]
miR-17 and miR-30c-1	PTEN (not validated)	PI3K pathway deregulation	Genetic risk assessment (novel rare variants)/upregulated in basal-like (miR-17)/downregulated in HER2 (miR-30c)	[55]
miR-423	Multiple		Genetic risk assessment (rs6505162)	[56]
miR-499	Multiple		Genetic risk assessment (rs3746444)	[54,58]
miR-27a	Multiple		Genetic risk assessment (rs895819)	[57]
miR-1208	Multiple		Genetic risk assessment (rs11780156)	[59]
ESR1 3′UTR in association with miR-453 binding site	ESR1	Estrogen receptor signaling	Genetic risk assessment (rs2747648)	[56]
SET8 3′UTR in association with miR-502 binding site	SET8	Aberrant histone methylation patterns (epigenetic maintenance)	Genetic risk assessment (rs16917496)	[62]
Drosha 3′UTR in association with miR-1298 binding site	DROSHA	miRNA processing	Genetic risk assessment (rs10719)	[66]
KRAS 3′UTR in association with let-7 binding site	KRAS, RAD52	MAPK signaling activation	Genetic risk assessment (rs61764370)	[65]

[a] *Information on association to breast cancer subtype was obtained from Ref. [28].*

SET8 had previously been reported as a cancer-associated variant thereby providing additional support for this finding [63]. Another protective variant is found in the 3′UTR region of the RAD52 gene likely reflecting modifications in a let-7 miRNA binding site [64]. A particularly striking example, based on the reported twofold increased risk estimate, is that of the 3′UTR of KRAS representing a binding site for let-7 identified in relation to triple-negative breast cancer susceptibility [65]. The authors carried out expression analyses in breast cancers arising in carriers of the KRAS variant and showed reduced levels of let-7 miRNA together with an activation of the MAPK activation signature. Thus, there were clear indications for alterations in the RAS pathway activity in breast cancers arising in carriers of the KRAS variant [65]. Future studies will undoubtedly address the relevance of the KRAS variant in terms of disease prognosis and response to MAPK inhibitors. This study clearly illustrates how genetic studies can contribute to the identification of biomarkers of potential relevance to the clinical management of breast cancer patients. Based on approximately 42,000 cases and 42,000 controls, genetic variations in the 3′ UTR of CASP8, HDDC3, DROSHA, MUSTN1, and MYCL1, representing putative miRNA binding sites, were identified in association with breast cancer susceptibility [66]. Of note here, Drosha has a central role in miRNA processing. The 3′UTR variant in Drosha (rs10719) interrupts the binding site for miR-1298 [66]—see further on Drosha in Section 1. More information as summary for miRNA-associated genetic predisposition to breast cancer is described in Table 1(c).

5. ACQUIRED GENOMIC AND EPIGENOMIC CHANGES AFFECTING miRNAs

DNA copy number changes are frequently observed in breast cancers [32]. Indeed, a subset of copy number gains or losses are known to affect the expression levels of some miRNA genes. In fact, miRNA genes preferentially reside in genomic regions affected by copy number changes in various types of cancers [67] including breast cancer [28]. Persson et al. described a new miRNA referred to as miR-4728 located within the ERBB2 (HER2) oncogene at chromosome 17q12 [68]. The amplification and overexpression of HER2 was found to co-occur with overexpression of miR-4728 suggesting that amplification of HER2 could have more widespread implications, i.e., posttranscriptional repression of genes targeted by miR-4728.

In a recent study, however, it was found that only a small proportion of copy number changes are associated with changes in expression levels of the corresponding miRNA genes [28]. Thus, even though Dvinge et al. confirmed the preference for copy number changes to occur proximal to miRNA genes, they find that most of these changes do not systematically affect miRNA expression levels [68]. Nonetheless, as alluded to by Persson et al., many miRNAs are expressed at very low levels in normal breast cells and could thus be more sensitive than other miRNAs to copy number changes, i.e., with even slightly elevated expression being of potential relevance to the development of breast cancer.

Epigenetics has been described as a mechanism leading to transcriptional silencing of miRNAs in cancer. CpG methylation events affecting the promoter region of miRNA genes are well described [69]. In this relation, by far most of the miRNAs detected as differentially methylated between normal breast tissue and cancer cells involve acquired promoter methylation, rather than hypomethylation, in the cancer cells. For example, miR-9-1 was identified as a target of CpG promoter methylation and transcriptional silencing in breast cancers [70]. Notably, this study highlights miR-9-1 promoter methylation as an early event during the development of breast cancer. In a study performed on the miR-195/miR-497 cluster (located <10kbp apart on chromosome 17q), Li et al. describe

CpG promoter methylation and downregulation of miR-195 and miR-497 in breast cancer [71]. The biological significance was further established by revealing Raf-1 and Cyclin D1 as direct targets of miR-195 and miR-497. Ectopic expression of either miR-195 or miR-497 was found to inhibit growth and invasion in vitro [71]. In another study, Xu et al. described CpG promoter methylation and downregulation of miR-148a [72]. In this study, ectopic expression of miR-148a inhibited cellular proliferation, colony formation, and angiogenesis. Further, the effects of miR-148a were shown to involve suppression of the PI3K/AKT and MAPK/ERK signaling pathways through targeting of both IGF1R and IRS1 [72]. An interesting study by Zhang et al. describes epigenetic regulation of miR-126 through CpG promoter methylation of its host gene called Egfl7 in relation to tumor cell invasion and metastasis [73]. A comprehensive analysis of miRNA gene promoter regions in terms of CpG promoter methylation events was recently carried out by Vrba et al. [74].

6. miRNAs AS REGULATORS OF THE DNA DAMAGE RESPONSE

The DDR is composed of a complex network of biological processes collectively aimed at maintaining genomic integrity [75]. These processes involve signaling for the presence of DNA damage, DNA repair, and activation of cell cycle checkpoints. In cancer, the DDR is often abolished by mutations in key components such as the TP53 gene. It can be said that defects in the DDR have been exploited for clinical management of cancer patients ever since anticancer drugs were introduced around the 1950s. For example, the effectiveness of platinum-based drugs (approved in 1978) is now known to involve the induction of DNA cross-links, which can be repaired by the homologous recombination (HR) machinery. Thus, response to platinum-based drugs in treatment of some cancers is now thought to be dependent on the inherent weaknesses found in some cancers involving defective DNA repair by HR. Evidence for this hypothesis has come from the study of cancers arising in carriers of either BRCA1 or BRCA2 mutations wherein the wild-type allele is frequently lost which leads to deficiency of BRCA1 or BRCA2, respectively. In fact, cancers arising in BRCA carriers are defective for DNA repair by HR. Indeed, breast and ovarian cancers arising in BRCA1 and BRCA2 mutation carriers are now known to be highly responsive to platinum drugs [75]. Further, patients with Fanconi anemia (FA; ORPHA84, MIM #227646) are homozygous carriers of mutations in genes that have been identified as key components in DNA repair of cross-links. Indeed, the BRCA2 gene itself is now known to represent a Fanconi complementation group referred to as FANCD1 (MIM #605724). The link to platinum drug sensitivity is seen in that normal cells derived from FA patients are hypersensitive to cross-linking agents. In fact, testing for FA in the clinic involves the use of a sensitivity assay for DNA cross-linking drugs [75,76].

The involvement of miRNAs in cellular responses to DNA damage is suggested by the observation that changes in miRNA expression are seen following exposure of human cells to genotoxic agents [77,78]. Further, the p53 gene has been identified as a transcription factor for the miR-34 family [79]. By experimental manipulation of human cells, miR-34 overexpression was found to be sufficient for inducing cell cycle arrest. In much the same way, miR-192 and miR-215 are transcriptionally regulated by p53 and shown to be biologically significant for inducing cell-cycle arrest following DNA damage [80,81]. In this way, by influencing the expression of just a few miRNAs a single DNA damage-responsive factor (such as p53) can influence a larger network of other genes. Other transcription factors involved in regulation of miRNAs following DNA damage likely remain to be discovered as already indicated by recent findings [82,83]. In addition, some miRNAs are responsive to p53 activation following DNA damage without being transcriptionally regulated by p53. The mechanism behind this relation was described in 2009 by Suzuki et al., wherein a previously established interaction between p53 and p68/p72 was found to be of significance in protein complex with Drosha for selectively inducing maturation of specific miRNAs involving miR-16-1, miR-143, and miR-145 [84]. These authors further demonstrated that mutations in the TP53 gene (p53) resulted in decreased miRNA processing activity. Other researchers have demonstrated a role for BRCA1 as an interacting partner of Drosha [85]. This interaction was shown to be of importance in positively regulating the expression of let-7a-1, miR-16-1, miR-145, and miR-34a. Remarkably, ATM is also involved in regulating miRNA biogenesis following DNA damage through direct interaction and phosphorylation of KSRP [86]. Further, ATM was shown to accelerate nuclear export of pre-miRNAs following DNA damage [87]. This involves ATM-dependent activation of AKT which then leads to enhanced interaction between Nup153 and XPO5 - proteins directly involved in nuclear export of pre-miRNAs. Taken together, these findings highlight a biologically relevant role for miRNAs in the DDR network.

It has long been known that p53-deficient cancers are inherently resistant to widely used anticancer drug treatments including doxorubicin and CMF-based treatments (cyclophosphamide, methotrexate, and 5-fluorouracil) [88,89]. The identification of miRNAs as key effector molecules of the p53 response to DNA damage raises the question of whether p53-responsive miRNAs can be used to predict if a patient is likely to respond (or not) to chemotherapy. Thus, moving towards defining a larger group of cancers with acquired defects in the p53 response pathway. Alternative mechanisms of p53 inactivation involve MDM2 amplification and overexpression in breast and several other cancer types [90] as well as loss of miR-339-5p [24,25]. Further, miR-125b, miR-25, and miR-30d have been identified as negative regulators of p53 expression levels through physical interactions with the mRNA product of the TP53 gene at the 3′UTR region [91,92]—although this has still not been reported as a mechanism of p53 inactivation in breast cancer. Patients having developed p53-deficient cancer can possibly be identified by expression profiling of a selected subset of miRNAs (those identified as p53 responsive). Supporting this hypothesis, in the breast cancer context, is the finding that miR-34a (a p53-regulated miRNA gene) has been described as downregulated in triple-negative (mostly p53 mutated) breast cancer cells [93]. The clinical relevance relates not only to decisions regarding other treatment options, i.e., whether to perform mastectomy or breast conservative surgery, but also to the selection of drugs and targeted inhibitors for overcoming chemotherapy resistance due to p53 deficiency. Although still in its infancy, it has been shown that p53 mutated cancer cells are dependent on functional ATM gene products for survival following chemotherapy treatment [94]. This suggests that ATM inhibition could be a useful strategy in the clinic for sensitizing p53-mutated cancers to chemotherapeutic drugs. Similar results have already been obtained from other researchers in different cancer types including breast cancer [95,96]. In this way, breast cancer expression profiling for miRNAs indicative of p53 abnormalities could be used to identify patients that will require targeted treatment, i.e., ATM inhibitors, together with chemotherapy due to inherent p53-associated drug resistance. Surely, other genetic dependencies remain to be discovered within the DDR network with exciting data having already emerged for PLK1 in the context of p53 mutations [97,98].

7. DNA REPAIR GENES AS TARGETS OF miRNAs: A BRIDGE TOWARD PERSONALIZED TREATMENT?

It is now well established that genes involved in DNA repair processes are themselves directly regulated by miRNAs [99]. The importance of this observation relates to emerging potentials in exploiting DNA repair deficiencies present in a subset of all cancers [75,76]. In this relation, targeted drug inhibitors such as poly ADP-ribose polymerase inhibitors (PARPi) are known to be effective against cancer cells with defects in components of the HR repair machinery [100]. This is particularly relevant for cancers arising in carriers of a BRCA1 or BRCA2 mutations. Cancer cells deficient in either BRCA1 or BRCA2 are extremely sensitive to inhibition of the PARP1 enzyme [76]. The mechanism underlying HR defects and PARPi sensitivity is thought to involve a functional role for the PARP1 enzyme in signaling the presence of single-stranded DNA breaks throughout the poly(ADP)-ribosylation of histones. By inhibiting the activity of PARP1 (through PARPi or knockdown of PARP1 by RNAi), these single-stranded breaks are left unrepaired and converted into double-stranded DNA breaks at replication forks. As double-stranded DNA breaks are highly lethal, their accumulation in cells lacking HR due to inhibition of PARP1 activity (e.g., using PARPi) is incompatible with cellular survival. The importance of PARPi in breast cancer relates not only to responses seen in patients with inherited mutations in BRCA1 or BRCA2 but also to a subset of sporadically arising triple-negative (basal-like) breast cancers. The responses seen in sporadic cases are likely due to acquired mutations or epigenetic silencing of the BRCA1 gene frequently observed in basal-like breast cancers [31,101]. PARPi represent one of the most exciting advances in breast cancer treatment over the last 10 years (as well as in other cancer types, e.g., ovarian cancer) and, for this reason, researchers are actively pursuing biomarkers for identifying patients that can derive benefits from treatment with PARPi.

As previously mentioned, different miRNAs have been found to directly regulate BRCA1 (miR-182, miR-146a, and miR-146b-5p) [20,21]. Here, breast cancer cells with miR-182 overexpression are sensitive to PARPi (e.g., olaparib) in vitro and in animal models [20]. In a recent study, Krishnan et al. show that miR-182 targets a large number of mRNA transcripts enriched for genes involved in the DDR [102]. This study reveals that compensation of miR-182 overexpressing cells with CHEK2 can partially rescue sensitivity for PARPi. Thus, the contribution of miR-182 overexpression to cellular sensitivity against PARPi is not only mediated by its repression of the BRCA1 gene [102]. The study by Garcia et al. similarly demonstrates miR-146a and miR-146b-5p as regulators of BRCA1 [21]. In both studies, miRNAs 182, 146a, and 146b-5p were shown to be overexpressed in breast cancers of the basal-like or triple-negative subtype. Indeed, the vast majority of breast cancers arising in BRCA1 mutation carriers are classified as basal-like. In this way, loss of BRCA1 through posttranscriptional repression by miRNAs represents a mechanism likely involved in the development of a subset of sporadically arising basal-like breast cancers.

The BRCA2 gene is posttranscriptionally repressed by miR-1245 [23,103]. Arbini et al. describe a combined effect of miR-1245 and Skp1 as negative regulators of BRCA2 and provide evidence to show that upregulation of these two factors following mitochondrial depletion effectively induces cellular sensitivity for PARPi [103]. Song et al. describe upregulation of miR-1245 by the c-Myc oncogene - known to be amplified in a substantial proportion of all breast cancers [92]. In another study, miRNAs involved in transcriptional repression of ATM were described, i.e., miR-181a and miR-181b, in breast cancer [104]. ATM is a core factor involved in the initial response to the formation of double-stranded breaks in DNA [76]. Among its substrates are BRCA1 and p53. Ectopic overexpression of miR-181a and miR-181b (miR-181a/b) was found to repress the expression of ATM resulting in lack of phosphorylated BRCA1 [104]. Importantly, loss of ATM through miR-181a/b overexpression was associated with sensitivity for PARPi. The authors describe high expression of miR-181a/b in high-grade and aggressive breast cancers thereby emphasizing the need for additional treatment for these patients wherein PARPi are clearly relevant [104]. ATM has also been found posttranscriptionally repressed by miR-18a [106]. In an miRNA "mimetic" screen, Neijenhuis et al. discovered miR-107 and miR-222 as predictors of PARPi response of potential relevance in breast and ovarian cancers [107]. Additionally, Choi et al. have recently described miRs 1255b, 148b, and 193b as suppressors of the HR pathway in the G1 phase of the cell cycle [108]. The identification of these miRNAs was based on their impact on PARPi sensitivity. Indeed, the expression of each of miR-1255b, miR-148b, or miR-193b leads to significantly reduced expression of BRCA1, BRCA2, and RAD51 [108]. These three miRNAs all lack canonical binding sites for their target genes, i.e., BRCA1/BRCA2/RAD51. Choi et al. propose that the interaction between the proposed target mRNA (i.e., the mRNA transcripts of BRCA1, BRCA2, and RAD51) and miR-1225b, miR-148b, and miR-193b is mediated by noncanonical miRNA response element binding, thereby highlighting weaknesses in currently available miRNA target prediction algorithms [108]. Other miRNAs of potential relevance in this context include miR-373 as a negative regulator of RAD23B and RAD52 [109], the recent discovery of miR-155 as a regulator of RAD51 [33], and still other miRNAs undoubtedly remain to be discovered as potential biomarkers of treatment responses in breast cancer [77,78,110–112].

Of considerable interest, in terms of identifying patients likely to derive benefits from treatment with PARPi is the identification of molecular characteristics for breast cancers arising in BRCA1 or BRCA2 mutation carriers. The hope is that BRCA-associated characteristics, collectively referred to as BRCAness [113], can then be used to identify breast cancers that are BRCA deficient [101]. Indeed, several miRNAs have now been identified as highly specific for breast cancers arising in either BRCA1 or BRCA2 mutation carriers [114,115]. Overexpression of miR-142-3p, miR-505, miR-1248, miR-181a-2, miR-25, and miR-340 are promising as molecular characteristics of both BRCA1- and BRCA2-mutated breast cancers [115]. Most patients attended by clinical practitioners have not undergone BRCA mutation testing and, for this reason, the BRCA-ness concept could possibly also be used to identify patients that should be referred to genetic counseling. Further, based on Tanic et al., miRNA expression patterns will help to clarify clinically relevant subtypes within the otherwise heterogeneous group of breast cancers arising in familial cases without involving BRCA1 or BRCA2 mutations [116].

8. miRNAs AS PROGNOSTIC BIOMARKERS

The progression from ductal carcinoma in situ (DCIS) is still poorly understood [117]. DCIS is prognostic for subsequent development of invasive breast cancer later in life, i.e., recurrence is seen in approximately 30% of all patients. Understanding the molecular characteristics of DCIS lesions associated with later onset of breast cancer could lead to a greater understanding of the invasive phenotype in cancer while, additionally, enabling personalized approach to monitor high-risk DCIS patients. Several studies have already addressed the role of miRNAs in DCIS [118–120]. Notably, Volinia et al. demonstrate miR-210 as an important player in the transition from premalignant to the invasive phenotype and further highlight its value as a significant prognostic biomarker [120]. In other studies, the expression of miR-21 has been found to be increased from normal breast to DCIS lesions [36]. This occurs in a gradual stepwise manner from low to high miR-21 expression in the transition from noninvasive lesions to the formation of invasive breast cancer, respectively [121]. In an analysis of DCIS compared with paired normal breast tissues from the same individuals, Hannafon et al. identified elevated expression of miR-21 in association with downregulation of NFIB, a predicted target for miR-21, while also identifying loss of miR-195 in association with upregulation of a predicted target, i.e., CCND1 (Cyclin D1). It remains to be seen whether variation in the degree of miR-21 upregulation, consistently observed in DCIS lesions, can be used as a prognostic biomarker, i.e., indicative of later onset of invasive disease. Li et al. demonstrated loss of miR-140 in DCIS lesions and further characterized the functional consequences involving upregulation of SOX9 and ALDH1 [82]. Restoration of miR-140 was found to reverse the phenotype and reduced tumor growth in vivo suggesting a promising preventative strategy.

An important problem in breast cancer treatment (as well as in treatment of other cancers) involves decisions regarding which patients should be offered chemotherapeutic treatment. In general, a large fraction of breast cancer patients do not require or will not benefit from chemotherapy. However, the clinical markers currently in use are not sufficiently reliable to accurately identify patients who will require chemotherapy, i.e., patients who are likely to progress toward early disease relapse without chemotherapy. As a result, many patients are being unnecessarily treated with chemotherapy. To a large extent, this problem reflects the need for specific and sensitive biomarkers that can reliably stratify patients into prognostic subgroups. Currently, routinely used clinical parameters for assessing patient prognosis are clinical staging, histological grade, and hormonal receptor status. Further improvements have emerged following the definition of biologically relevant subtypes. By genome-wide expression analysis, breast cancers have been classified into the following subtypes (often referred to as the "intrinsic" subtypes): luminal-A, luminal-B, HER2-enriched, and basal-like [122]. Other gene expression-based subtypes have been proposed, e.g., claudin-low, normal-like, and luminal-C, but these are less well characterized. Even though clinical data was not used to define these subtypes, they are nonetheless unique in terms of clinical features, prognosis, and response to therapy [123]. This greatly supports the notion that biologically relevant subtypes could be informative for the clinical management of patients. In this respect, the PAM50 method has gained considerable attention enabling classification of breast cancers, based on measuring the expression of 50 protein-coding genes, into the 4 so-called intrinsic subtypes, i.e., luminal-A, luminal-B, HER2-enriched, and basal-like. Recent data on more than 1000 patients have shown that the PAM50 method has superior prognostic value beyond other methods [124] and this is further supported by other researchers [125,126]. The most widely used gene expression signatures, however, for predicting outcome in patients are Oncotype DX [127] and MammaPrint [128].

Multiple studies have already indicated that a subset of miRNAs is highly correlated in expression with breast cancer subtypes [19,28,120,129]. It can therefore be hypothesized that incorporating miRNA expression data into the definition of breast cancer subtypes could provide better quality classification; thereby leading to improvements in evaluating patient prognosis. A notable example is miR-155, a well-characterized oncomiR in breast cancer, found overexpressed in hormone receptor-negative breast cancers [130]. A recent study showed overexpression of miR-155 specifically in breast cancers of the basal-like subtype [28]. As basal-like breast cancers are strongly associated with early disease relapse and less favorable outcomes, the value of miR-155 should be further explored as a biomarker for increased accuracy in classification of breast cancers into prognostic subgroups; especially in cases where other methods have provided uninformative or ambiguous results. A study focused on miR-155 in breast cancer showed overexpression in cases with lymph node metastases, high proliferation index, and advanced TNM stage [131]. The value of miR-155 has further been supported in a recent meta-analysis [132]. Indeed, the expression of miR-155 could simultaneously provide information on prognosis and sensitivity to PARPi as described in Section 7. A role for miR-155 in inducing metastasis following activation of TGF-beta signaling has been proposed [133]. Kong et al. show that miR-155 is a key factor in TGF-beta-mediated epithelial mesenchymal transformation (EMT) and cell migration/metastasis thus further highlighting a potential value for miR-155 expression as a prognostic biomarker [133]. The widely studied miR-21 oncomiR has frequently, although not consistently, been associated with breast cancers classified as hormone receptor positive [134]. In many studies, miR-21 has been identified as a prognostic biomarker with high expression associated with reduced time to death and clinical markers of advanced staging and grading [135].

In terms of downregulated miRNAs as an indicator of poor prognosis, the miR-125b gene was described by Iorio et al. as one of the most significantly downregulated miRNA in breast cancer [19]. Zhang et al. demonstrate downregulation of miR-125b in association with CpG promoter methylation of the miR-125b promoter and reduced time to patient death [136]. In this study, ETS1 was identified as a direct target of miR-125b which is of potential relevance for explaining aggressiveness associated with loss of miR-125b expression [136]. A further finding, derived from the study of Dvinge et al., is that downregulation of the miR-125b gene is significantly associated with the Luminal-B subtype which represents a well-known breast cancer subtype associated with poor prognosis [28]. In this way, loss of miR-125b expression could provide helpful information for classifying breast cancers to the luminal-B subtype, thereby of potential value as a biomarker of disease prognosis.

Beyond single miRNAs, the definition of subtype-specific miRNA expression signatures have been explored and found to be more informative in terms of disease prognosis. A comprehensive genome-wide analysis of miRNA expression on the largest breast cancer cohort to date (including approximately 1300 tumors) was carried out by Dvinge et al. [28]. These authors find that the prognostic value associated with miRNAs appears to be mostly confined to hormone receptor-positive breast cancers. In particular, Dvinge et al. demonstrate that the expression of miR-345 is a robust biomarker for poor disease outcome; a consistent finding across many different study cohorts [28]. However, an even more striking observation made by Dvinge et al., is that the expression of specific miRNAs appear to be strongly prognostic in a particular breast cancer subtype referred to as iClust-4 (previously defined by the same group). Moreover, the miRNA signature found to be of prognostic value within the iClust-4 subtype was also, although less strongly, prognostic across other subtypes as well [28]. The miRNAs found within the iClust-4 prognostic signature include previously undescribed putative miRNAs such as miR-834 and miR-729. This study clearly demonstrates the value of analyzing multiple miRNAs simultaneously in the context of biologically significant subtypes for the discovery of clinically relevant miRNA biomarkers.[28]

Other biologically potent and well-established miRNAs represent candidate prognostic biomarkers independently of whether or not they are linked to specific subtypes. For example, miR-10b was shown to be involved in metastasis by directly binding to and inhibiting translation of homeobox D10 mRNA transcripts thereby resulting in increased expression of RHOC—a well-known gene involved in metastasis [137]. However, the value of miR-10b as a clinical biomarker for predicting disease relapse subsequently proved discouraging [138]. Several other miRNA biomarkers have been implicated in disease progression, for example, Ye et al. described miR-200b as a prognostic factor in breast cancer and further identified the relevant mechanism to be mediated through direct targeting of miR-200b to mRNA transcripts derived from some of the RAB gene family, i.e., RAB21, RAB23, RAB18, and RAB3B [139]. As previously mentioned, Shimono et al. described a role for miR-200c as a regulator of epithelial mesenchymal transition (EMT), i.e., a process known to be linked to the metastatic capacity of cancer cells. The expression of miR-200c is therefore a likely candidate for use as a prognostic biomarker in breast cancer, i.e., for predicting the likely course of disease [140]. Other studies have already described the miR-200c gene as downregulated in breast cancer metastases thereby further supporting a role for miR-200c in disease progression [141]. The miR-200 family is located in two clusters on chromosomes 1p36 and 12p13. In general, the miR-200 family is considered to have tumor suppressor properties in many different cancer types [142]. A recent study has now shown how loss of miR-200 expression is mediated by overexpression of the miR-22 gene [143]. Song et al. show that this is mediated by miR-22-targeting mRNA transcripts of TET1 and TET2 subsequently leading to the inability of breast cancer cells to demethylate miR-200 promoter regions. Other miRNAs have been linked to the regulation of breast cancer metastasis including miR-199 and miR-31 [144]. GATA-3 was shown to regulate the expression of miR-29b, which, in turn, inhibits the EMT process required for breast cancer cells to metastasize [145]. Chou et al. show that loss of GATA-3, frequently observed in breast cancer, leads to metastasis through its role as regulator of miR-29b expression, i.e., loss of miR-29b is required for breast cancer cells to undergo EMT. Indeed, the prominent role of miRNAs implicated in the metastasis process across different cancer types has led to the definition of the "Metastamir" term to collectively encapsulate all functionally relevant metastasis-associated miRNAs [146]. The detection of changes in the expression of MetastamiRs in tumor samples could lead toward improvements in assessing disease prognosis. An important step in enabling the transition of miRNA analyses into the clinic is the development of in situ hybridization techniques. This is because RNA samples obtained from tumor tissue reflect only the "average" expression level across tumor cells present in a given sample. With in situ techniques, individual tumor cells can be assessed for changes in miRNA expression levels. This is particularly relevant for the assessment of MetastamiR expression levels as they are likely to show heterogeneous patterns of expression across the same tumor with only a few tumor cells, or clusters of tumor cells, displaying changes in MetastmiR expression levels. The use of in situ hybridization to detect and assess miRNA expression in breast cancer has been reported and already showed promising results [25,147]. Recently, Agarwal et al. used qISH (an in situ based method) to detect loss of miR-34a in a large number of breast cancer patients (>1500) and showed significant prognostic value in node-negative breast cancer patients (not in node-positive) [148]. As previously mentioned, miR-34a is transcriptionally regulated by the p53 tumor suppressor gene and selectively processed for maturation by protein–protein interactions between Drosha and BRCA1. The effects of miR-34 loss emerge in upregulation of its target genes involved in metastasis and proliferation including BCL2, Src, CD44, c-MET, E2F1, and CDK4/6 [149]. Additionally, miR-302 was investigated by in situ hybridization in a cohort of 318 breast cancers [150]. In this study, high expression of miR-302 together with low miR-203 was associated with embryonic/induced pluripotency characteristics and was associated with metastasis and shorter patient survival.

9. THE USE OF miRNAs IN SCREENING FOR EARLY BREAST CANCER DEVELOPMENT

Following the demonstration that miRNAs are stable in blood samples, the analysis of miRNAs has gained substantial attention for early cancer detection [151]. This is due to the potential relevance for early disease detection in otherwise apparently healthy individuals; thereby opening up the possibility for a blood-based screening approach. In terms of breast cancer, a recent report describes the use of serum samples to define a nine-miRNA gene expression signature able to detect the presence of early-stage breast cancer, i.e., assaying for miR-15a, miR-18a, miR-107, miR-133a, miR-139-5p, miR-143, miR-145, miR-365, and miR-425 [152]. Kodahl et al. were further able to validate this signature in a previously published dataset. Other miRNAs have been described in this relation, including the oncogenic miR-155, miR-181b, and miR-21, as potential biomarkers of breast cancer development [153,154]. Nonetheless, there exist contradictory results. For example, miR-155, miR-145, and miR-21 were not found to be elevated in serum obtained from a cohort of breast cancer patients [155]. These discrepancies are likely attributed to differences in factors such as cohort composition and different sample types (whole blood, plasma, or serum) and miRNA isolation procedures. For example, miR-155 is a basal-like associated miRNA and is therefore not expected to be detected in a large proportion of all patients or individuals with early breast cancer (as basal-like breast cancers represent less than 20% of all cases). Thus, it is understandable that some subtype-specific or rarely overexpressed miRNAs have escaped detection in smaller cohorts and thus erroneously excluded as promising diagnostic biomarkers. Nonetheless, irrespective of miR-155, based on the analysis of paired samples of normal breast tissue, primary tumor, and serum samples obtained from the same set of patients, it is clear that the pattern of miRNAs present in the primary tumor only partially reflect the observed pattern found in serum of the same patient [156]. Based on the findings presented by Chan et al., it can be hypothesized that miRNAs are selectively released into the extracellular fluid. In this context, it has been shown that miRNAs are, indeed, selectively released to the extracellular space based on observations in cell line model systems [157]. Nonetheless, it is known that tumor cells undergo apoptosis and necrosis at the site of origin due to hypoxia and thereby release their content to the microenvironment. The remains can then subsequently enter the bloodstream, including DNA and RNA molecules. Thus, regardless of whether or not miRNAs are released by tumor cells into the extracellular fluid, miRNAs found highly expressed in tumor should also be found in blood; although detecting them could be more difficult, i.e., detecting miRNAs that are not readily released by the tumor cells themselves but still highly expressed in the primary cancer cells will be more challenging as this relies on cancer cell death at the site of origin. For these purposes, novel techniques for the amplification of nucleic acids in blood plasma or serum are emerging. Notably, the application of the so-called "digital PCR" currently holds great promise in overcoming the hurdles that lie in detecting miRNAs present at low levels in biological samples [158]. In addition, improved miRNA isolation procedures from blood samples have been proposed that can be of added value in this regard [159].

10. CONCLUSIONS

Many different studies have supported the use of miRNAs as diagnostic biomarkers in breast cancer; see Table 1. In particular, the discovery of miRNAs as important regulators of the DDR network, and vice versa, highlights potentials with respect to targeted therapy directed against components of the DDR network. Notable examples include the identification of several miRNAs such as miR-182, miR-107, miR-1255b, miR-148b, and miR-193b as promising biomarkers for predicting responses to PARPi. Indeed, PARPi represent one of the most promising advances in breast cancer treatment over the last 10 years. Currently, PARPi have mostly been tested in selected patients focusing on BRCA1 and BRCA2 germline mutation carriers with exceptional responses already observed in many cases. Therefore, intense efforts have been initiated for the discovery of biomarkers that can be used to identify patients that can derive benefits from treatment with PARPi—regardless of whether or not they are BRCA mutation carriers. The clinical importance of these research activities relates to the definition of a much larger group of patients that should be treated with PARPi, i.e., a larger group than the subset of patients tested positive for germline mutations in either BRCA1 or BRCA2. The design of future clinical trials should give more attention to miRNAs as potential biomarkers for predicting responses to PARPi.

The advent of high-throughput methods enabling genome-wide profiling of miRNAs (e.g., by microarrays or next-generation sequencing) has now provided a comprehensive picture of miRNA expression patterns in breast cancer. This has not only provided useful information on which miRNAs are deregulated but has also led to novel insights regarding the value of miRNAs as prognostic biomarkers. In a landmark study, Dvinge et al. identified an miRNA expression signature, including miR-834 and miR-729, for stratifying breast cancer patients into prognostic subgroups—in particular, within the subtype referred to as iClust-4 [28]. Intriguingly, the iClust-4 subtype is characterized by lack of genomic instability emphasizing the biological significance of miRNAs in this particular breast cancer subtype. This study was only made possible by the advent of microarray technology for miRNA expression profiling together with extensive clinical data and pre-existing genetic and transcriptomic analyses in a large study cohort including more the 1000 breast cancer patients.

Other well-characterized miRNAs, including miR-155 and miR-21, have shown promising results with respect to prognosis—although there are still conflicting data. Further studies based on large cohorts including extensive clinical and pathological data will be required to clarify the relevance of these and other candidate miRNAs as prognostic biomarkers in breast cancer. The stability of miRNAs in blood indicates their use in early disease detection, i.e., screening for breast cancer in apparently healthy individuals, and for monitoring disease progression following treatment. The available studies published so far have revealed promising results, although still conflicting with respect to miR-155 and miR-21, i.e., two of the best characterized miRNAs in breast cancer. Improvements in this regard lie in making use of advanced technologies, such as the so-called "digital PCR", to provide high-quality data together with improved miRNA isolation procedures. Further, breast cancer is a heterogeneous disease and, therefore, the inclusion of multiple miRNAs is necessary to enable detection of breast cancer development. For an example, miR-155 is linked to the basal-like subtype. And, as this subtype accounts for approximately 15% of all breast cancer cases, it is not expected to show strong performance in detecting breast cancer in general—whereas it may well be an exceptionally good biomarker for detecting breast cancers of the basal-like subtype. Thus, miR-155 clearly warrants further investigation as a biomarker of early disease diagnosis in blood samples emphasizing a subtype-specific outcome measure for assessment of biomarker specificity and sensitivity.

Glossary

Epigenetics The study of mechanisms by which cells transmit information on expression potential and chromatin configuration to daughter cells without introducing any changes to the DNA sequence.

Fanconi anemia A rare inherited disorder mainly leading to bone marrow failure and cancer predisposition.

LIST OF ABBREVIATIONS

ATM Ataxia telangiectasia mutated
BRCA1 Breast cancer 1, early onset
BRCA2 Breast cancer 2, early onset
DCIS Ductal carcinoma in situ
DDR DNA damage response
EMT Epithelial–mesenchymal transition
ER Estrogen receptor 1
FA Fanconi anemia
HER2 Human epidermal growth factor receptor 2 (also known as ERBB2; erb-b2 receptor tyrosine kinase 2)
HR Homologous recombination
miRNA microRNA
PARP Poly ADP-ribose polymerase
PARPi Poly ADP ribose polymerase inhibitor
PCR Polymerase chain reaction
pre-miRNA Precursor microRNA
pri-miRNA Primary miRNA
RISC RNA-induced silencing complex
TGF Transforming growth factor
TP53 Tumor protein p53 (also known as p53)
UTR Untranslated region

Acknowledgments

The author would like to acknowledge the following funding bodies: Gongum Saman, Minningarsjodur Eggerts Briem and RANNIS (14193-051).

References

[1] Fire A, Xu SQ, Montgomery MK, Kostas SA, Driver SE, Mello CC. Potent and specific genetic interference by double-stranded RNA in *Caenorhabditis elegans*. Nature 1998;391:806–11.
[2] Carthew RW, Sontheimer EJ. Origins and mechanisms of miRNAs and siRNAs. Cell 2009;136(4):642–55.
[3] Gu SG, Pak J, Guang S, Maniar JM, Kennedy S, Fire A. Amplification of siRNA in *Caenorhabditis elegans* generates a transgenerational sequence-targeted histone H3 lysine 9 methylation footprint. Nat Genet January 8, 2012;44(2):157–64.
[4] He L, Thomson JM, Hemann MT, Hernando-Monge E, Mu D, Goodson S, et al. A microRNA polycistron as a potential human oncogene. Nature 2005;435(7043): 828–33.
[5] Berezikov E, Chung WJ, Willis J, Cuppen E, Lai EC. Mammalian mirtron genes. Mol Cell 2007;28(2):328–36.
[6] Muerdter F, Guzzardo PM, Gillis J, Luo Y, Yu Y, Chen C, et al. A genome-wide RNAi screen draws a genetic framework for transposon control and primary piRNA biogenesis in *Drosophila*. Mol Cell 2013;50(5):736–48.
[7] Meijer HA, Smith EM, Bushell M. Regulation of miRNA strand selection: follow the leader? Biochem Soc Trans 2014;42(4):1135–40.
[8] Vasudevan S, Tong Y, Steitz JA. Switching from repression to activation: microRNAs can up-regulate translation. Science 2007;318(5858):1931–4.
[9] Kozomara A, Griffiths-Jones S. miRBase: annotating high confidence microRNAs using deep sequencing data. Nucleic Acids Res 2014;42(Database issue): D68–73.
[10] Tsang JS, Ebert MS, van Oudenaarden A. Genome-wide dissection of microRNA functions and cotargeting networks using gene set signatures. Mol Cell 2010;38(1):140–53.
[11] Avril-Sassen S, Goldstein LD, Stingl J, Blenkiron C, Le Quesne J, Spiteri I, et al. Characterisation of microRNA expression in post-natal mouse mammary gland development. BMC Genomics 2009;10:548.
[12] Wu MK, Sabbaghian N, Xu B, Addidou-Kalucki S, Bernard C, Zou D, et al. Biallelic DICER1 mutations occur in Wilms tumours. J Pathol 2013;230(2):154–64.
[13] Heravi-Moussavi A, Anglesio MS, Cheng SW, Senz J, Yang W, Prentice L, et al. Recurrent somatic DICER1 mutations in nonepithelial ovarian cancers. N Engl J Med 2012;366(3):234–42.
[14] Hill DA, Ivanovich J, Priest JR, Gurnett CA, Dehner LP, Desruisseau D, et al. DICER1 mutations in familial pleuropulmonary blastoma. Science 2009;325(5943):965.
[15] Rio Frio T, Bahubeshi A, Kanellopoulou C, Hamel N, Niedziela M, Sabbaghian N, et al. DICER1 mutations in familial multinodular goiter with and without ovarian Sertoli-Leydig cell tumors. JAMA 2011;305(1):68–77.
[16] Melo SA, Ropero S, Moutinho C, Aaltonen LA, Yamamoto H, Calin GA, et al. A TARBP2 mutation in human cancer impairs microRNA processing and DICER1 function. Nat Genet 2009;41(3):365–70.
[17] Melo SA, Moutinho C, Ropero S, Calin GA, Rossi S, Spizzo R, et al. A genetic defect in exportin-5 traps precursor microRNAs in the nucleus of cancer cells. Cancer Cell 2010;18(4):303–15.
[18] Rakheja D, Chen KS, Liu Y, Shukla AA, Schmid V, Chang TC, et al. Somatic mutations in DROSHA and DICER1 impair microRNA biogenesis through distinct mechanisms in Wilms tumours. Nat Commun 2014;2:4802.

[19] Iorio MV, Ferracin M, Liu CG, Veronese A, Spizzo R, Sabbioni S, et al. MicroRNA gene expression deregulation in human breast cancer. Cancer Res 2005;65(16): 7065–70.

[20] Moskwa P, Buffa FM, Pan Y, Panchakshari R, Gottipati P, Muschel RJ, et al. miR-182-mediated downregulation of BRCA1 impacts DNA repair and sensitivity to PARP inhibitors. Mol Cell 2011;41(2):210–20.

[21] Garcia AI, Buisson M, Bertrand P, Rimokh R, Rouleau E, Lopez BS, et al. Down-regulation of BRCA1 expression by miR-146a and miR-146b-5p in triple negative sporadic breast cancers. EMBO Mol Med 2011;3(5):279–90.

[22] Sandhu R, Rein J, D'Arcy M, Herschkowitz JI, Hoadley KA, Troester MA. Overexpression of miR-146a in basal-like breast cancer cells confers enhanced tumorigenic potential in association with altered p53 status. Carcinogenesis 2014;35(11):2567–75.

[23] Song L, Dai T, Xie Y, Wang C, Lin C, Wu Z, et al. Up-regulation of miR-1245 by c-myc targets BRCA2 and impairs DNA repair. J Mol Cell Biol 2012;4(2):108–17.

[24] Jansson MD, Damas ND, Lees M, Jacobsen A, Lund AH. miR-339-5p regulates the p53 tumor-suppressor pathway by targeting MDM2. Oncogene 2014. http://dx.doi.org/10.1038/onc.2014.130.

[25] Wu ZS, Wu Q, Wang CQ, Wang XN, Wang Y, Zhao JJ, et al. MiR-339-5p inhibits breast cancer cell migration and invasion in vitro and may be a potential biomarker for breast cancer prognosis. BMC Cancer 2010;10:542.

[26] Chen H, Sun JG, Cao XW, Ma XG, Xu JP, Luo FK, et al. Preliminary validation of ERBB2 expression regulated by miR-548d-3p and miR-559. Biochem Biophys Res Commun 2009;385(4):596–600.

[27] Jiang S, Zhang HW, Lu MH, He XH, Li Y, Gu H, et al. MicroRNA-155 functions as an OncomiR in breast cancer by targeting the suppressor of cytokine signaling 1 gene. Cancer Res 2010;70(8):3119–27.

[28] Dvinge H, Git A, Gräf S, Salmon-Divon M, Curtis C, Sottoriva A, et al. The shaping and functional consequences of the microRNA landscape in breast cancer. Nature 2013;497(7449):378–82.

[29] Chang S, Wang RH, Akagi K, Kim KA, Martin BK, Cavallone L, Kathleen Cuningham Foundation Consortium for Research into Familial Breast Cancer (kConFab), et al. Tumor suppressor BRCA1 epigenetically controls oncogenic microRNA-155. Nat Med 2011;17(10):1275–82.

[30] Stefansson OA, Esteller M. BRCA1 as a tumor suppressor linked to the regulation of epigenetic states: keeping oncomiRs under control. Breast Cancer Res 2012;14(2):304.

[31] Stefansson OA, Jonasson JG, Olafsdottir K, Hilmarsdottir H, Olafsdottir G, Esteller M, et al. CpG island hypermethylation of BRCA1 and loss of pRb as co-occurring events in basal/triple-negative breast cancer. Epigenetics 2011;6(5):638–49.

[32] Stefansson OA, Jonasson JG, Johannsson OT, Olafsdottir K, Steinarsdottir M, Valgeirsdottir S, et al. Genomic profiling of breast tumours in relation to BRCA abnormalities and phenotypes. Breast Cancer Res 2009;11(4):R47.

[33] Gasparini P, Lovat F, Fassan M, Casadei L, Cascione L, Jacob NK, et al. Protective role of miR-155 in breast cancer through RAD51 targeting impairs homologous recombination after irradiation. Proc Natl Acad Sci USA 2014;111(12):4536–41.

[34] Dinami R, Ercolani C, Petti E, Piazza S, Ciani Y, Sestito R, et al. miR-155 drives telomere fragility in human breast cancer by targeting TRF1. Cancer Res 2014;74(15):4145–56.

[35] Czyzyk-Krzeska MF, Zhang X. MiR-155 at the heart of oncogenic pathways. Oncogene 2014;33(6):677–8.

[36] Farazi TA, Horlings HM, Ten Hoeve JJ, Mihailovic A, Halfwerk H, Morozov P, et al. MicroRNA sequence and expression analysis in breast tumors by deep sequencing. Cancer Res 2011;71(13):4443–53.

[37] Si ML, Zhu S, Wu H, Lu Z, Wu F, Mo YY. miR-21-mediated tumor growth. Oncogene 2007;26(19): 2799–803.

[38] Lu Z, Liu M, Stribinskis V, Klinge CM, Ramos KS, Colburn NH, et al. MicroRNA-21 promotes cell transformation by targeting the programmed cell death 4 gene. Oncogene 2008;27(31):4373–9.

[39] Frankel LB, Christoffersen NR, Jacobsen A, Lindow M, Krogh A, Lund AH. Programmed cell death 4 (PDCD4) is an important functional target of the microRNA miR-21 in breast cancer cells. J Biol Chem 2008;283(2):1026–33.

[40] Lu Z, Liu M, Stribinskis V, Klinge CM, Ramos KS, Colburn NH, et al. MicroRNA-21 promotes cell transformation by targeting the programmed cell death 4 gene. Oncogene 2008;27(31):4373–9.

[41] Zhu S, Si ML, Wu H, Mo YY. MicroRNA-21 targets the tumor suppressor gene tropomyosin 1 (TPM1). J Biol Chem 2007;282(19):14328–36.

[42] Hamilton MP, Rajapakshe K, Hartig SM, Reva B, McLellan MD, Kandoth C, et al. Identification of a pan-cancer oncogenic microRNA superfamily anchored by a central core seed motif. Nat Commun 2013;4:2730.

[43] Calin GA, Dumitru CD, Shimizu M, Bichi R, Zupo S, Noch E, et al. Frequent deletions and down-regulation of micro-RNA genes miR15 and miR16 at 13q14 in chronic lymphocytic leukemia. Proc Natl Acad Sci USA 2002;99(24):15524–9.

[44] Zhang X, Wan G, Mlotshwa S, Vance V, Berger FG, Chen H, et al. Oncogenic Wip1 phosphatase is inhibited by miR-16 in the DNA damage signaling pathway. Cancer Res 2010;70(18):7176–86.

[45] Sachdeva M, Mo YY. MicroRNA-145 suppresses cell invasion and metastasis by directly targeting mucin 1. Cancer Res 2010;70(1):378–87.

[46] Götte M, Mohr C, Koo CY, Stock C, Vaske AK, Viola M, et al. miR-145-dependent targeting of junctional adhesion molecule A and modulation of fascin expression are associated with reduced breast cancer cell motility and invasiveness. Oncogene 2010;29(50):6569–80.

[47] Shimono Y, Zabala M, Cho RW, Lobo N, Dalerba P, Qian D, et al. Downregulation of miRNA-200c links breast cancer stem cells with normal stem cells. Cell 2009;138(3):592–603.

[48] Shen J, Ambrosone CB, DiCioccio RA, Odunsi K, Lele SB, Zhao H. A functional polymorphism in the miR-146a gene and age of familial breast/ovarian cancer diagnosis. Carcinogenesis 2008;29(10):1963–6.

[49] Jeon HS, Lee YH, Lee SY, Jang JA, Choi YY, Yoo SS, et al. A common polymorphism in pre-microRNA-146a is associated with lung cancer risk in a Korean population. Gene 2014;534(1):66–71.

[50] Xu T, Zhu Y, Wei QK, Yuan Y, Zhou F, Ge YY, et al. A functional polymorphism in the miR-146a gene is associated with the risk for hepatocellular carcinoma. Carcinogenesis 2008;29(11):2126–31.

[51] Yue C, Wang M, Ding B, Wang W, Fu S, Zhou D, et al. Polymorphism of the pre-miR-146a is associated with risk of cervical cancer in a Chinese population. Gynecol Oncol 2011;122(1):33–7.

[52] Hoffman AE, Zheng T, Yi C, Leaderer D, Weidhaas J, Slack F, et al. microRNA miR-196a-2 and breast cancer: a genetic and epigenetic association study and functional analysis. Cancer Res 2009;69(14): 5970–7.

[53] Catucci I, Yang R, Verderio P, Pizzamiglio S, Heesen L, Hemminki K, et al. Evaluation of SNPs in miR-146a, miR196a2 and miR-499 as low-penetrance alleles in German and Italian familial breast cancer cases. Hum Mutat 2010;31(1):E1052–7.

[54] Hu Z, Liang J, Wang Z, Tian T, Zhou X, Chen J, et al. Common genetic variants in pre-microRNAs were associated with increased risk of breast cancer in Chinese women. Hum Mutat 2009;30(1):79–84.

[55] Shen J, Ambrosone CB, Zhao H. Novel genetic variants in microRNA genes and familial breast cancer. Int J Cancer 2009;124(5):1178–82.

[56] Kontorovich T, Levy A, Korostishevsky M, Nir U, Friedman E. Single nucleotide polymorphisms in miRNA binding sites and miRNA genes as breast/ovarian cancer risk modifiers in Jewish high-risk women. Int J Cancer 2010;127(3):589–97.

[57] Yang R, Schlehe B, Hemminki K, Sutter C, Bugert P, Wappenschmidt B, et al. A genetic variant in the pre-miR-27a oncogene is associated with a reduced familial breast cancer risk. Breast Cancer Res Treat 2010;121(3):693–702.

[58] Dai ZJ, Shao YP, Wang XJ, Xu D, Kang HF, Ren HT, et al. Five common functional polymorphisms in microRNAs (rs2910164, rs2292832, rs11614913, rs3746444, rs895819) and the susceptibility to breast cancer: evidence from 8361 cancer cases and 8504 controls. Curr Pharm Des 2015;21(11):1455–63.

[59] Michailidou K, Hall P, Gonzalez-Neira A, Ghoussaini M, Dennis J, Milne RL, et al. Large-scale genotyping identifies 41 new loci associated with breast cancer risk. Nat Genet 2013;45(4):353–61. 361.e1–2.

[60] Rawlings-Goss RA, Campbell MC, Tishkoff SA. Global population-specific variation in miRNA associated with cancer risk and clinical biomarkers. BMC Med Genomics 2014;7:53.

[61] Tchatchou S, Jung A, Hemminki K, Sutter C, Wappenschmidt B, Bugert P, et al. A variant affecting a putative miRNA target site in estrogen receptor (ESR) 1 is associated with breast cancer risk in premenopausal women. Carcinogenesis 2009;30(1):59–64.

[62] Song F, Zheng H, Liu B, Wei S, Dai H, Zhang L, et al. An miR-502-binding site single-nucleotide polymorphism in the 3′-untranslated region of the SET8 gene is associated with early age of breast cancer onset. Clin Cancer Res 2009;15(19):6292–300.

[63] Yu Z, Li Z, Jolicoeur N, Zhang L, Fortin Y, Wang E, et al. Aberrant allele frequencies of the SNPs located in microRNA target sites are potentially associated with human cancers. Nucleic Acids Res 2007;35(13):4535–41.

[64] Jiang Y, Qin Z, Hu Z, Guan X, Wang Y, He Y, et al. Genetic variation in a hsa-let-7 binding site in RAD52 is associated with breast cancer susceptibility. Carcinogenesis 2013;34(3):689–93.

[65] Paranjape T, Heneghan H, Lindner R, Keane FK, Hoffman A, Hollestelle A, et al. 3′-untranslated region KRAS variant and triple-negative breast cancer: a case-control and genetic analysis. Lancet Oncol 2011;12(4):377–86.

[66] Khan S, Greco D, Michailidou K, Milne RL, Muranen TA, Heikkinen T, et al. MicroRNA related polymorphisms and breast cancer risk. PLoS One 2014;9(11):e109973.

[67] Calin GA, Sevignani C, Dumitru CD, Hyslop T, Noch E, Yendamuri S, et al. Human microRNA genes are frequently located at fragile sites and genomic regions involved in cancers. Proc Natl Acad Sci USA 2004;101(9):2999–3004.

[68] Persson H, Kvist A, Rego N, Staaf J, Vallon-Christersson J, Luts L, et al. Identification of new microRNAs in paired normal and tumor breast tissue suggests a dual role for the ERBB2/Her2 gene. Cancer Res 2011;71(1):78–86.

[69] Lehmann U. Aberrant DNA methylation of microRNA genes in human breast cancer – a critical appraisal. Cell Tissue Res 2014;356(3):657–64.

[70] Lehmann U, Hasemeier B, Christgen M, Müller M, Römermann D, Länger F, et al. Epigenetic inactivation of microRNA gene hsa-mir-9-1 in human breast cancer. J Pathol 2008;214(1):17–24.
[71] Li D, Zhao Y, Liu C, Chen X, Qi Y, Jiang Y, et al. Analysis of MiR-195 and MiR-497 expression, regulation and role in breast cancer. Clin Cancer Res 2011;17(7):1722–30.
[72] Xu Q, Jiang Y, Yin Y, Li Q, He J, Jing Y, et al. A regulatory circuit of miR-148a/152 and DNMT1 in modulating cell transformation and tumor angiogenesis through IGF-IR and IRS1. J Mol Cell Biol 2013;5(1):3–13.
[73] Zhang Y, Yang P, Sun T, Li D, Xu X, Rui Y, et al. miR-126 and miR-126* repress recruitment of mesenchymal stem cells and inflammatory monocytes to inhibit breast cancer metastasis. Nat Cell Biol 2013;15(3):284–94.
[74] Vrba L, Muñoz-Rodríguez JL, Stampfer MR, Futscher BW. miRNA gene promoters are frequent targets of aberrant DNA methylation in human breast cancer. PLoS One 2013;8(1):e54398.
[75] Lord CJ, Ashworth A. The DNA damage response and cancer therapy. Nature 2012;481(7381):287–94.
[76] Martin SA, Lord CJ, Ashworth A. DNA repair deficiency as a therapeutic target in cancer. Curr Opin Genet Dev 2008;18(1):80–6.
[77] Simone NL, Soule BP, Ly D, Saleh AD, Savage JE, Degraff W, et al. Ionizing radiation-induced oxidative stress alters miRNA expression. PLoS One 2009;4(7):e6377.
[78] van Jaarsveld MT, Wouters MD, Boersma AW, Smid M, van Ijcken WF, Mathijssen RH, et al. DNA damage responsive microRNAs misexpressed in human cancer modulate therapy sensitivity. Mol Oncol 2014;8(3):458–68.
[79] Rokavec M, Li H, Jiang L, Hermeking H. The p53/miR-34 axis in development and disease. J Mol Cell Biol 2014;6(3):214–30.
[80] Braun CJ, Zhang X, Savelyeva I, Wolff S, Moll UM, Schepeler T, et al. p53-Responsive microRNAs 192 and 215 are capable of inducing cell cycle arrest. Cancer Res 2008;68(24):10094–104.
[81] Georges SA, Biery MC, Kim SY, Schelter JM, Guo J, Chang AN, et al. Coordinated regulation of cell cycle transcripts by p53-inducible microRNAs, miR-192 and miR-215. Cancer Res 2008;68(24):10105–12.
[82] Aguda BD, Kim Y, Piper-Hunter MG, Friedman A, Marsh CB. MicroRNA regulation of a cancer network: consequences of the feedback loops involving miR-17-92, E2F, and Myc. Proc Natl Acad Sci USA. 2008;105(50):19678–83.
[83] Li Y, Choi PS, Casey SC, Dill DL, Felsher DW. MYC through miR-17-92 suppresses specific target genes to maintain survival, autonomous proliferation, and a neoplastic state. Cancer Cell 2014;26(2):262–72.
[84] Suzuki HI, Yamagata K, Sugimoto K, Iwamoto T, Kato S, Miyazono K. Modulation of microRNA processing by p53. Nature 2009;460(7254):529–33.
[85] Kawai S, Amano A. BRCA1 regulates microRNA biogenesis via the DROSHA microprocessor complex. J Cell Biol 2012;197(2):201–8.
[86] Zhang X, Wan G, Berger FG, He X, Lu X. The ATM kinase induces microRNA biogenesis in the DNA damage response. Mol Cell 2011;41(4):371–83.
[87] Wan G, Zhang X, Langley RR, Liu Y, Hu X, Han C, et al. DNA-damage-induced nuclear export of precursor microRNAs is regulated by the ATM-AKT pathway. Cell Rep 2013;3(6):2100–12.
[88] Aas T, Børresen AL, Geisler S, Smith-Søresnsen B, Jonsen H, Varhaug JE, et al. Specific P53 mutations are associated with de novo resistance to doxorubicin in breast cancer patients. Nat Med 1996;2:811–4.
[89] Andersson J, Larsson L, Klaar S, Holmberg L, Nilsson J, Inganäs M, et al. Worse survival for TP53 (p53)-mutated breast cancer patients receiving adjuvant CMF. Ann Oncol 2005;16:743–8.
[90] Santarius T, Shipley J, Brewer D, Stratton MR, Cooper CS. A census of amplified and overexpressed human cancer genes. Nat Rev Cancer 2010;10(1):59–64.
[91] Kumar M, Lu Z, Takwi AA, Chen W, Callander NS, Ramos KS, et al. Negative regulation of the tumor suppressor p53 gene by microRNAs. Oncogene 2011;30(7):843–53.
[92] Le MT, Teh C, Shyh-Chang N, Xie H, Zhou B, Korzh V, et al. MicroRNA-125b is a novel negative regulator of p53. Genes Dev 2009;23(7):862–76.
[93] Kato M, Paranjape T, Müller RU, Nallur S, Gillespie E, Keane K, et al. The mir-34 microRNA is required for the DNA damage response in vivo in *C. elegans* and in vitro in human breast cancer cells. Oncogene 2009;28(25):2419–24.
[94] Jiang H, Reinhardt HC, Bartkova J, Tommiska J, Blomqvist C, Nevanlinna H, et al. The combined status of ATM and p53 link tumor development with therapeutic response. Genes Dev 2009;23(16):1895–909.
[95] Millour J, de Olano N, Horimoto Y, Monteiro LJ, Langer JK, Aligue R, et al. ATM and p53 regulate FOXM1 expression via E2F in breast cancer epirubicin treatment and resistance. Mol Cancer Ther 2011;10(6):1046–58.
[96] Williamson CT, Kubota E, Hamill JD, Klimowicz A, Ye R, Muzik H, et al. Enhanced cytotoxicity of PARP inhibition in mantle cell lymphoma harbouring mutations in both ATM and p53. EMBO Mol Med 2012;4(6):515–27.
[97] Wang X, Simon R. Identification of potential synthetic lethal genes to p53 using a computational biology approach. BMC Med Genomics 2013;6:30.

[98] Sur S, Pagliarini R, Bunz F, Rago C, Diaz Jr LA, Kinzler KW, et al. A panel of isogenic human cancer cells suggests a therapeutic approach for cancers with inactivated p53. Proc Natl Acad Sci USA 2009;106(10):3964–9.
[99] Tessitore A, Cicciarelli G, Del Vecchio F, Gaggiano A, Verzella D, Fischietti M, et al. MicroRNAs in the DNA damage/repair network and cancer. Int J Genomics 2014;2014:820248.
[100] McCabe N, Turner NC, Lord CJ, Kluzek K, Bialkowska A, Swift S, et al. Deficiency in the repair of DNA damage by homologous recombination and sensitivity to poly(ADP-ribose) polymerase inhibition. Cancer Res 2006;66(16):8109–15.
[101] TCGA (Cancer Genome Atlas Network). Comprehensive molecular portraits of human breast tumours. Nature 2012;490(7418):61–70.
[102] Krishnan K, Steptoe AL, Martin HC, Wani S, Nones K, Waddell N, et al. MicroRNA-182-5p targets a network of genes involved in DNA repair. RNA 2013 Feb;19(2):230–42.
[103] Arbini AA, Guerra F, Greco M, Marra E, Gandee L, Xiao G, et al. Mitochondrial DNA depletion sensitizes cancer cells to PARP inhibitors by translational and post-translational repression of BRCA2. Oncogenesis 2013;2:e82.
[104] Bisso A, Faleschini M, Zampa F, Capaci V, De Santa J, Santarpia L, et al. Oncogenic miR-181a/b affect the DNA damage response in aggressive breast cancer. Cell Cycle 2013;12(11):1679–87.
[105] Shen J, Xu R, Mai J, Kim HC, Guo X, Qin G, et al. High capacity nanoporous silicon carrier for systemic delivery of gene silencing therapeutics. ACS Nano. 2013;7(11):9867–80.
[106] Song L, Lin C, Wu Z, Gong H, Zeng Y, Wu J, et al. miR-18a impairs DNA damage response through downregulation of ataxia telangiectasia mutated (ATM) kinase. PLoS One 2011;6(9):e25454.
[107] Neijenhuis S, Bajrami I, Miller R, Lord CJ, Ashworth A. Identification of miRNA modulators to PARP inhibitor response. DNA Repair (Amst) 2013;12(6):394–402.
[108] Choi YE, Pan Y, Park E, Konstantinopoulos P, De S, D'Andrea A, et al. MicroRNAs down-regulate homologous recombination in the G1 phase of cycling cells to maintain genomic stability. Elife 2014;3:e02445.
[109] Crosby ME, Kulshreshtha R, Ivan M, Glazer PM. MicroRNA regulation of DNA repair gene expression in hypoxic stress. Cancer Res 2009;69(3):1221–9.
[110] Heyn H, Engelmann M, Schreek S, Ahrens P, Lehmann U, Kreipe H, et al. MicroRNA miR-335 is crucial for the BRCA1 regulatory cascade in breast cancer development. Int J Cancer 2011;129(12):2797–806.
[111] Huan LC, Wu JC, Chiou BH, Chen CH, Ma N, Chang CY, et al. MicroRNA regulation of DNA repair gene expression in 4-aminobiphenyl-treated HepG2 cells. Toxicology 2014;322:69–77.
[112] Huang JW, Wang Y, Dhillon KK, Calses P, Villegas E, Mitchell PS, et al. Systematic screen identifies miRNAs that target RAD51 and RAD51D to enhance chemosensitivity. Mol Cancer Res 2013;11(12): 1564–73.
[113] Turner N, Tutt A, Ashworth A. Hallmarks of 'BRCAness' in sporadic cancers. Nat Rev Cancer 2004;4(10): 814–9.
[114] Murria Estal R, Palanca Suela S, de Juan Jiménez I, Egoavil Rojas C, García-Casado Z, Juan Fita MJ, et al. MicroRNA signatures in hereditary breast cancer. Breast Cancer Res Treat 2013;142(1):19–30.
[115] Tanic M, Yanowski K, Gómez-López G, Rodriguez-Pinilla MS, Marquez-Rodas I, Osorio A, et al. MicroRNA expression signatures for the prediction of BRCA1/2 mutation-associated hereditary breast cancer in paraffin-embedded formalin-fixed breast tumors. Int J Cancer 2015;136(3):593–602.
[116] Tanic M, Andrés E, Rodriguez-Pinilla SM, Marquez-Rodas I, Cebollero-Presmanes M, Fernandez V, et al. MicroRNA-based molecular classification of non-BRCA1/2 hereditary breast tumours. Br J Cancer 2013;109(10):2724–34.
[117] Allred DC, Wu Y, Mao S, Nagtegaal ID, Lee S, Perou CM, et al. Ductal carcinoma in situ and the emergence of diversity during breast cancer evolution. Clin Cancer Res 2008;14(2):370–8.
[118] Kristensen VN, Vaske CJ, Ursini-Siegel J, Van Loo P, Nordgard SH, Sachidanandam R, et al. Integrated molecular profiles of invasive breast tumors and ductal carcinoma in situ (DCIS) reveal differential vascular and interleukin signaling. Proc Natl Acad Sci USA 2012;109(8):2802–7.
[119] Sun EH, Zhou Q, Liu KS, Wei W, Wang CM, Liu XF, et al. Screening miRNAs related to different subtypes of breast cancer with miRNAs microarray. Eur Rev Med Pharmacol Sci 2014;18(19):2783–8.
[120] Volinia S, Galasso M, Sana ME, Wise TF, Palatini J, Huebner K, et al. Breast cancer signatures for invasiveness and prognosis defined by deep sequencing of microRNA. Proc Natl Acad Sci USA 2012;109(8):3024–9.
[121] Qi L, Bart J, Tan LP, Platteel I, Tv S, Huitema S, et al. Expression of miR-21 and its targets (PTEN, PDCD4, TM1) in flat epithelial atypia of the breast in relation to ductal carcinoma in situ and invasive carcinoma. BMC Cancer 2009;9:163.
[122] Perou CM, Børresen-Dale AL. Systems biology and genomics of breast cancer. Cold Spring Harb Perspect Biol 2011;3(2).
[123] Sørlie T, Perou CM, Tibshirani R, Aas T, Geisler S, Johnssen H, et al. Gene expression patterns of breast carcinomas distinguish tumor subclasses with clinical implications. Proc Natl Acad Sci USA 2001;98(19):10869–74.

[124] Dowsett M, Sestak I, Lopez-Knowles E, Sidhu K, Dunbier AK, Cowens JW, et al. Comparison of PAM50 risk of recurrence score with oncotype DX and IHC4 for predicting risk of distant recurrence after endocrine therapy. J Clin Oncol 2013;31(22):2783–90.
[125] Nielsen TO, Parker JS, Leung S, Voduc D, Ebbert M, Vickery T, et al. A comparison of PAM50 intrinsic subtyping with immunohistochemistry and clinical prognostic factors in tamoxifen-treated estrogen receptor-positive breast cancer. Clin Cancer Res 2010;16(21):5222–32.
[126] Sestak I, Cuzick J, Dowsett M, Lopez-Knowles E, Filipits M, Dubsky P, et al. Prediction of late distant recurrence after 5 years of endocrine treatment: a combined analysis of patients from the Austrian breast and colorectal cancer study group 8 and arimidex, tamoxifen alone or in combination randomized trials using the PAM50 risk of recurrence score. J Clin Oncol 2015;33(8):916–22.
[127] Paik S. Is gene array testing to be considered routine now? Breast 2011;20(Suppl. 3):S87–91.
[128] Glas AM, Floore A, Delahaye LJ, Witteveen AT, Pover RC, Bakx N, et al. Converting a breast cancer microarray signature into a high-throughput diagnostic test. BMC Genomics 2006;7:278.
[129] Enerly E, Steinfeld I, Kleivi K, Leivonen SK, Aure MR, Russnes HG, et al. miRNA-mRNA integrated analysis reveals roles for miRNAs in primary breast tumors. PLoS One 2011;6(2):e16915.
[130] Blenkiron C, Goldstein LD, Thorne NP, Spiteri I, Chin SF, Dunning MJ, et al. MicroRNA expression profiling of human breast cancer identifies new markers of tumor subtype. Genome Biol 2007;8(10):R214.
[131] Zheng SR, Guo GL, Zhang W, Huang GL, Hu XQ, Zhu J, et al. Clinical significance of miR-155 expression in breast cancer and effects of miR-155 ASO on cell viability and apoptosis. Oncol Rep 2012;27(4):1149–55.
[132] Zeng H, Fang C, Nam S, Cai Q, Long X. The clinicopathological significance of microRNA-155 in breast cancer: a meta-analysis. Biomed Res Int 2014;2014:724209.
[133] Kong W, Yang H, He L, Zhao JJ, Coppola D, Dalton WS, et al. MicroRNA-155 is regulated by the transforming growth factor beta/Smad pathway and contributes to epithelial cell plasticity by targeting RhoA. Mol Cell Biol 2008;28(22):6773–84.
[134] Mattie MD, Benz CC, Bowers J, Sensinger K, Wong L, Scott GK, et al. Optimized high-throughput microRNA expression profiling provides novel biomarker assessment of clinical prostate and breast cancer biopsies. Mol Cancer 2006;5:24.
[135] Chen J, Wang X. MicroRNA-21 in breast cancer: diagnostic and prognostic potential. Clin Transl Oncol 2014;16(3):225–33.
[136] Zhang Y, Yan LX, Wu QN, Du ZM, Chen J, Liao DZ, et al. miR-125b is methylated and functions as a tumor suppressor by regulating the ETS1 proto-oncogene in human invasive breast cancer. Cancer Res 2011;71(10):3552–62.
[137] Ma L, Teruya-Feldstein J, Weinberg RA. Tumour invasion and metastasis initiated by microRNA-10b in breast cancer. Nature 2007;449(7163):682–8.
[138] Gee HE, Camps C, Buffa FM, Colella S, Sheldon H, Gleadle JM, et al. MicroRNA-10b and breast cancer metastasis. Nature 2008;455(7216):E8–9.
[139] Ye F, Tang H, Liu Q, Xie X, Wu M, Liu X, et al. miR-200b as a prognostic factor in breast cancer targets multiple members of RAB family. J Transl Med 2014;12:17.
[140] Bojmar L, Karlsson E, Ellegård S, Olsson H, Björnsson B, Hallböök O, et al. The role of microRNA-200 in progression of human colorectal and breast cancer. PLoS One 2013;8(12):e84815.
[141] Baffa R, Fassan M, Volinia S, O'Hara B, Liu CG, Palazzo JP, et al. MicroRNA expression profiling of human metastatic cancers identifies cancer gene targets. J Pathol 2009;219(2):214–21.
[142] Feng X, Wang Z, Fillmore R, Xi Y. MiR-200, a new star miRNA in human cancer. Cancer Lett 2014;344(2):166–73.
[143] Song SJ, Poliseno L, Song MS, Ala U, Webster K, Ng C, et al. MicroRNA-antagonism regulates breast cancer stemness and metastasis via TET-family-dependent chromatin remodeling. Cell 2013;154(2):311–24.
[144] Valastyan S, Reinhardt F, Benaich N, Calogrias D, Szász AM, Wang ZC, et al. A pleiotropically acting microRNA, miR-31, inhibits breast cancer metastasis. Cell 2009;137(6):1032–46.
[145] Chou J, Lin JH, Brenot A, Kim JW, Provot S, Werb Z. GATA3 suppresses metastasis and modulates the tumour microenvironment by regulating microRNA-29b expression. Nat Cell Biol 2013;15(2):201–13.
[146] Hurst DR, Edmonds MD, Welch DR. Metastamir: the field of metastasis-regulatory microRNA is spreading. Cancer Res 2009;69(19):7495–8.
[147] Sempere LF, Christensen M, Silahtaroglu A, Bak M, Heath CV, Schwartz G, et al. Altered MicroRNA expression confined to specific epithelial cell subpopulations in breast cancer. Cancer Res 2007;67(24):11612–20.
[148] Agarwal S, Hanna J, Sherman ME, Figueroa J, Rimm DL. Quantitative assessment of miR34a as an independent prognostic marker in breast cancer. Br J Cancer 2015;112(1):61–8.
[149] Chen F, Hu SJ. Effect of microRNA-34a in cell cycle, differentiation, and apoptosis: a review. J Biochem Mol Toxicol 2012;26(2):79–86.
[150] Volinia S, Nuovo G, Drusco A, Costinean S, Abujarour R, Desponts C, et al. Pluripotent stem cell miRNAs and metastasis in invasive breast cancer. J Natl Cancer Inst 2014;106(12).

[151] Mitchell PS, Parkin RK, Kroh EM, Fritz BR, Wyman SK, Pogosova-Agadjanyan EL, et al. Circulating microRNAs as stable blood-based markers for cancer detection. Proc Natl Acad Sci USA 2008;105(30):10513–8.

[152] Kodahl AR, Lyng MB, Binder H, Cold S, Gravgaard K, Knoop AS, et al. Novel circulating microRNA signature as a potential non-invasive multi-marker test in ER-positive early-stage breast cancer: a case control study. Mol Oncol 2014;8(5):874–83.

[153] Mar-Aguilar F, Mendoza-Ramírez JA, Malagón-Santiago I, Espino-Silva PK, Santuario-Facio SK, Ruiz-Flores P, et al. Serum circulating microRNA profiling for identification of potential breast cancer biomarkers. Dis Markers 2013;34(3):163–9.

[154] Sochor M, Basova P, Pesta M, Dusilkova N, Bartos J, Burda P, et al. Oncogenic microRNAs: miR-155, miR-19a, miR-181b, and miR-24 enable monitoring of early breast cancer in serum. BMC Cancer 2014;14:448.

[155] Heneghan HM, Miller N, Lowery AJ, Sweeney KJ, Newell J, Kerin MJ. Circulating microRNAs as novel minimally invasive biomarkers for breast cancer. Ann Surg 2010;251(3):499–505.

[156] Chan M, Liaw CS, Ji SM, Tan HH, Wong CY, Thike AA, et al. Identification of circulating microRNA signatures for breast cancer detection. Clin Cancer Res 2013;19(16):4477–87.

[157] Pigati L, Yaddanapudi SC, Iyengar R, Kim DJ, Hearn SA, Danforth D, et al. Selective release of microRNA species from normal and malignant mammary epithelial cells. PLoS One 2010;5(10):e13515.

[158] Pradervand S, Weber J, Lemoine F, Consales F, Paillusson A, Dupasquier M, et al. Concordance among digital gene expression, microarrays, and qPCR when measuring differential expression of microRNAs. Biotechniques 2010;48(3):219–22.

[159] Devonshire AS, Whale AS, Gutteridge A, Jones G, Cowen S, Foy CA, et al. Towards standardisation of cell-free DNA measurement in plasma: controls for extraction efficiency, fragment size bias and quantification. Anal Bioanal Chem 2014;406(26):6499–512.

CHAPTER

30

MicroRNAs in Bone and Soft Tissue Sarcomas and Their Value as Biomarkers

Tomohiro Fujiwara[1,2,3], Yu Fujita[3], Yutaka Nezu[3], Akira Kawai[4], Toshifumi Ozaki[1], Takahiro Ochiya[3]

[1]Department of Orthopaedic Surgery, Okayama University Graduate School of Medicine, Dentistry, and Pharmaceutical Sciences, Okayama, Japan; [2]Center for Innovative Clinical Medicine, Okayama University Hospital, Okayama, Japan; [3]Division of Molecular and Cellular Medicine, National Cancer Center Research Institute, Tokyo, Japan; [4]Department of Musculoskeletal Oncology, National Cancer Center Hospital, Tokyo, Japan

OUTLINE

Epigenetic Biomarkers and Diagnostics
http://dx.doi.org/10.1016/B978-0-12-801899-6.00030-9

1. INTRODUCTION

Bone and soft tissue sarcomas are malignant neoplasms originating from transformed cells of mesenchymal origin and are different from carcinomas that are malignant neoplasms originating from epithelial cells. The word "sarcoma" is derived from the Greek word "sarkoma," meaning "fleshy outgrowth" [1]. Primary bone sarcomas constitute 0.2% of all malignancies in adults and approximately 5% of childhood malignancies [2]. Moreover, cancer registry data with histological stratification indicate that osteosarcoma is the most common primary malignant bone tumor, accounting for 35% of all cases, followed by chondrosarcoma (25%), Ewing sarcoma (16%), and chordoma (8%) [2]. Soft tissue sarcomas constitute fewer than 1% of all malignancies [2,3]. According to the results of the Surveillance, Epidemiology, and End Results (SEER) study, leiomyosarcoma was the most common sarcoma, accounting for 23.9% of all cases, followed by malignant fibrous histiocytoma (MFH; 17.1%), liposarcoma (11.5%), dermatofibrosarcoma (10.5%), rhabdomyosarcoma (RMS; 4.6%), and malignant peripheral nerve sheath tumor (MPNST; 4.0%) [4]. Although MFH was the second most common sarcoma in this series, the diagnostic term MFH is now reserved for undifferentiated sarcomas. Therefore, the incidence rates of MFH will be updated in future studies based on changes in diagnostic criteria.

Despite the development of combined modality treatments, including surgery, chemotherapy, and radiotherapy, a significant proportion of patients with sarcoma respond poorly to chemotherapy, leading to local relapse or distant metastasis. The main cause of death due to sarcoma is lung metastasis, for which prognosis is extremely poor [5,6]. Therefore, an early detection of recurrent or metastatic diseases and early decision making according to the tumor response to chemotherapy could improve patient prognosis. However, there are currently no effective biomarkers in such situations. Therefore, the discovery of novel biomarkers to detect tumors or predict drug sensitivity is one of the most important challenges in sarcoma management.

MicroRNAs (miRNAs) are small noncoding RNA molecules that modulate the expression of multiple target genes and play important roles in various physiological and pathological processes [7]. miRNAs are transcribed by RNA polymerase II as primary transcripts (pri-miRNAs) and are then processed into short 70-nucleotide precursor-miRNAs (pre-miRNAs) by an RNA-specific ribonuclease enzyme complex (Drosha). These processes take place within the cell nucleus. Then, pre-miRNAs are transported to the cytoplasm by exportin 5. These pre-miRNAs are further cleaved by the endonuclease (Dicer) generating mature miRNAs in the cytoplasm [7–9]. Mature miRNAs then regulate protein production in the cell by binding to complementary target mRNAs via the RNA-induced silencing complex (RISC) [8,9]. Over the past several years, multiple instances of miRNA dysregulation have been investigated in various human cancers [10]. Aberrant miRNA expression has been shown to contribute to cancer development through various mechanisms, including deletions, amplifications, and mutations, involving miRNA loci, epigenetic silencing, dysregulation of transcription factors that target specific miRNAs [11]. Growing evidence has revealed that miRNAs are frequently upregulated or downregulated in various tumors. These miRNAs include "oncomiRs" or "tumor suppressor miRs," which can function as oncogenes and tumor suppressors [12]. Interestingly, miRNAs, which have a pivotal role in epigenetic regulation, cellular senescence, or physiological development, such as miR-22, miR-34a, and let-7a, are also dysregulated in many cancers as well as bone and soft tissue sarcomas.

Recently, secreted miRNAs from tumor cells have been identified [13]. Data have demonstrated that serum circulating miRNAs remained stable despite the presence of RNase activity in blood.

Most circulating miRNAs cofractionate with complexes of AGO2, whereas a minority of specific circulating miRNAs are predominantly associated with vesicles [14,15]. Identification of such extracellular miRNA complexes in plasma raises the possibility that cells release a functional miRNA-induced silencing complex into the blood. One of the first studies measuring miRNA levels in serum was reported by Lawrie et al. [16] who demonstrated that the serum miR-21 levels were associated with relapse-free survival in patients with diffuse large B-cell lymphoma. Mitchell et al. [14] could distinguish patients with prostate cancer from healthy subjects by measuring the serum levels of miR-141. The analysis of circulating miRNA levels may present a novel approach for diagnosis of bone and soft tissue sarcomas, which has emerged in these several years.

2. miRNAs IN BONE SARCOMAS

Accumulated evidence of miRNA dysregulation in bone sarcomas has demonstrated their genetic complexity. Recent reports have mainly focused on miRNA dysregulation in osteosarcoma, and the various functions of each miRNA have suggested the genetic heterogeneity in osteosarcoma (Table 1). Ewing sarcoma has been also revealed its pattern of miRNA expression that is associated with *EWS-Fli1* gene, a unique chromosomal translocation of this sarcoma (Table 2). Although there have been few investigations for chondrosarcoma, several miRNAs have been documented as the regulators of important phenotypes including drug sensitivity (Table 2).

2.1 Osteosarcoma

Osteosarcoma (MIM #259500, ORPHA668) is a primary malignant bone tumor that is histologically characterized by the production of osteoid by malignant cells, mainly arising in adolescent and young adult populations. The most common primary sites are the distal femur, proximal tibia, and proximal humerus, with approximately 50% of cases originating around the knee. The survival rate has not improved for 20 years despite multiple clinical trials. Therefore, new diagnostic and therapeutic approaches must be sought for better prognosis.

There is accumulating evidence of miRNA dysregulation in osteosarcoma (Table 1) [6,17–113]. Many researchers have applied various approaches to profile miRNAs in osteosarcoma: (1) miRNAs identified from microarray assays using osteosarcoma cell lines and tissues, normal cell lines, and normal mesenchymal tissues; (2) miRNAs related to known genetic alterations such as p53; and (3) miRNAs related to cellular phenotypes such as chemoresistance, cell migration, or cell invasion [35,78]. In addition, the clinical relevance of miRNA dysregulation has emerged in recent years (Table 1).

Upregulated miRNAs in osteosarcoma include well-known oncomiRs in other cancers. miR-21, upregulated in all types of malignant tumors, was found to be overexpressed in osteosarcoma tissues and cell lines by Ziyan and Namløs [24,114]. Moreover, silencing of miR-21 decreased cell invasion and migration of osteosarcoma cells. Ziyan et al. [24] found that RECK, a tumor suppressor gene, was a direct target of miR-21. The clinical relevance of serum miR-21 has been recently reported, as described in the upregulation of the miR-17-92 cluster, encoding six miRNAs (miR-17, miR-18a, miR-19a, miR-20a, miR-19b, and miR-92) that were detected by several authors [33,91,115,116]. Baumhoer et al. [115] investigated six osteosarcoma cell lines, an osteoblast cell line, and mesenchymal stem cells (MSCs) for genome-wide miRNA expression and identified several miRNAs with oncogenic properties, including various members of the miR-17-92 cluster. Among these six miRNAs, Huang et al. [33] reported that miR-20a downregulated Fas expression, which inversely correlated with lung metastasis formation. Thus, this downregulation contributed

TABLE 1 Dysregulated miRNAs in Bone Sarcomas (Osteosarcoma)

Histology	Upregulated miRNAs	Downregulated miRNAs	Function, correlation	Clinical significance	miRNA Target (Predicted)	References
Osteosarcoma	miR-9		N/D	Prognosis, tumor size, metastasis		Xu et al. [102–106]
	miR-17-92		Cell proliferation, migration, invasion	Prognosis, clinical stage, recurrence	PTEN (RGMB, LRRC17, CCNE1, LIMA1, CAMK2N1)	Huang et al., Baumhoar et al., Gao et al., and Li et al.
	miR-20a		Metastasis		Fas	Huang et al. [33]
	miR-21		Cell invasion, migration		RECK	Ziyan et al. [24] and
	miR-25		Cell proliferation, tumor growth		p27	Wang et al. [100,101]
	miR-27a		Cell invasion, migration, tumorigenicity		N/D	Jones et al. [34]
	miR-29 family		N/D	Prognosis, tumor grade, recurrence, metastasis	N/D	Hong et al. [88]
	miR-31		Cell proliferation, apoptosis		E2F2	Creighton et al. [20]
	miR-33a		Chemoresistance		TWIST	Zhou et al. [113]
	miR-92a, -99b, -132, 193a-5p, -422a		Chemoresistance (IFO)		N/D	Gougelet et al. [182]
	miR-93		Cell proliferation, invasion		E2F1	Montanini et al. [37]
	miR-128		Cell proliferation		PTEN	Shen et al. [183]
	miR-133a (CSC)		Cell proliferation, invasion	Prognosis, metastasis	SGMS2, UBA2, SNX30, ANXA2	Fujiwara et al. [184]
	miR-140		Chemoresistance		HDAC4	Song et al. [19]
	miR-181a		Cell proliferation, apoptosis, invasion		(TIMP3, p21)	Jianwei et al. [50]
	miR-192, 215		Cell cycle arrest		CDKN1A/p21	Braun et al. [17]
	miR-195		Cell invasion, migration		FASN	Mao et al. [36]
	miR-199b-5p		N/D		(Notch signaling)	Won et al. [69]
	miR-210		N/D	Prognosis, tumor size, response to chemotherapy, metastasis	N/D	Cai et al. [45]

miR-214		Cell proliferation, invasion, tumor growth	Prognosis, tumor size, metastasis, response to chemotherapy	LZTS1	Wang et al. [100,101] and Xu et al. [102–106]
miR-215		Chemoresistance (MTX)		DHFR, TYMS	Song et al. [21]
miR-370, -654		N/D		IRS1	Stably et al. [185]
miR-7, -9/9*, -18a, -18b, -21*, -31/31*, -137, -301a, -301b, -503	miR-1, -126/126*, -133b, -142-3p, -144/144*, -150, -206, -223, -451, -486-5p	N/D		N/D	Namlos et al. [38]
	miR-1, -133b	Cell proliferation, apoptosis		MET	Novello et al. [186]
	miR-15a, -16-1	Cell proliferation, apoptosis		CCND1	Cai et al. [31]
	miR-16	Chemoresistance, tumorigenicity		IGF1R	Jones et al. [34] and Chen et al. [46]
	miR-22	Cell proliferation, migration, autophagy		HMGB1	Guo et al. [85]
	miR-23a	Cell proliferation, migration, invasion		RUNX2, CXCL 12	He et al. [87]
	miR-24	Cell proliferation		LPAATβ	Song et al. [63]
	miR-26a	Cell migration, invasion	Prognosis, clinical stage, metastasis	EZH2	Song et al. [97]
	miR-29 family (CSC)	Colony formation, apoptosis, CSC properties, cell proliferation, migration, invasion (miR-29b)		VEGF (CD133, N-Myc, CCDN2, E2F1, E2F2, Bcl-2, IAP-2)	Di Fiore et al. [187] and Zhang et al. [109–112,175]
	miR-32	Cell proliferation		Sox9	Xu et al. [102–106]
	miR-34a	Cell proliferation, apoptosis, migration, invasion		CDK6, E2F3, Cyclin E2, BCL2, cMet, CD44	He et al. [18], Yan et al. [40], Zhao et al. [76,77], and Tian et al. [99]
	miR-34c	Cell proliferation		RUNX2	van der Deen et al. [66]

Continued

TABLE 1 Dysregulated miRNAs in Bone Sarcomas (Osteosarcoma)—cont'd

Histology	Upregulated miRNAs	Downregulated miRNAs	Function, correlation	Clinical significance	miRNA Target (Predicted)	References
		miR-100	Cell proliferation		Cyr61	Huang et al. [89]
		miR-101	Cell migration, invasion		EZH2	Zhang et al. [109–112,175]
		miR-125b	Cell proliferation, migration		STAT3	Liu et al. [26]
		miR-126	Cell proliferation, migration, invasion		Sirt-1, Sox2	Xu et al. [71–73] and Yang et al. [107,108]
		miR-133a	Cell proliferation, apoptosis, tumorigenicity	Prognosis	Bcl-xL, Mcl-1	Ji et al. [49]
		miR-132	Cell proliferation	Prognosis, clinical stage, metastasis, response to chemotherapy	CCNE1	Yang et al. [74] and Wang et al. [100,101]
		miR-135b	Cell proliferation, migration, invasion		c-Myc	Liu et al. [92]
		miR-141	Cell proliferation, apoptosis		(ZEB1, ZEB2)	Xu et al. [71–73]
		miR-143	Cell invasion, apoptosis, tumorigenicity		MMP13	Osaki et al. [29] and Zhang et al. [23]
		miR-145	Cell proliferation, invasion	Prognosis, clinical stage, metastasis	ROCK1	Tang et al. [64] and Lei et al. [90]
		miR-183	Cell invasion, migration, proliferation	Prognosis, local recurrence, metastasis, response to chemotherapy	Ezrin	Zhu et al. [43], Zhao et al. [42], and Mu et al. [94]
		miR-194	Cell proliferation, migration, invasion, tumor growth, metastasis		CDH2, IGF1R	Han et al. [86]
		miR-199-3p	Cell invasion, migration, apoptosis		mTOR, STAT3	Duan et al. [25] and Tian et al. [99]
		miR-202	Cell proliferation, apoptosis, tumor growth		Gli2	Sun et al. [98]

miR-206	Cell viability, apoptosis, invasion, migration	Clinical stage, metastasis	NA	Bao et al. [44]
miR-217	Cell proliferation, migration, invasion	Metastasis	WASF3	Shen et al. [95,96]
miR-218	Cell proliferation, invasion, migration		(TIAM1, MMP2, MMP9)	Jin et al. [51,52]
miR-221	Cell proliferation, apoptosis, cisplatin resistance		PTEN	Zhao et al. [76–78]
miR-223	N/D	Prognosis, tumor grade, recurrence, metastasis, response to chemotherapy	ECT2	Zhang et al. [109–112,175]
miR-302b	Cell proliferation, apoptosis, induced by epirubcin		(Akt/pAkt, Bcl-2, Bim)	Zhang et al. [75]
miR-320	Cell proliferation, tumor growth		(FASN)	Cheng et al. [188]
miR-335	Cell invasion, migration		ROCK1	Wang et al. [67,68]
miR-340	Cell proliferation, migration, invasion	Prognosis, clinical stage, metastasis, response to chemotherapy	ROCK1	Zhou et al. [78,79], Cai et al. [81]
miR-376c	Cell proliferation, invasion		TGFA	Jin et al. [51,52]
miR-382, 369-3p, 544, 134	Apoptosis, cell proliferation, chemoresistance (miR-382)	Prognosis, response to chemotherapy (miR-382)	cMYC, KLF12, HIPK3 (miR-382)	Thayanithy et al. [39] and Xu et al. [102–106]
miRNAs in 14q32 (miR-134, -382, -544)	Survival period of mouse and dog, chemoresistance (miR-382)	Prognosis, metastasis (miR-3829)	KLF12 (miR-382), HIPK3 (miR-382)	Maire et al. [189], Sarver et al. [61], and Xu et al. [102–106]

N/D, not determined.

TABLE 2 Deregulated miRNAs in Bone Sarcomas (Ewing sarcoma and chondrosarcoma)

Histology	Upregulated miRNAs	Downregulated miRNAs	Function, correlation	Clinical significance	miRNA Target (Predicted)	References
Ewing sarcoma	miR-17-92, -106b-25, -106a-363		Colony formation		N/D	Dylla et al. [190]
	miR-125b		Chemoresistance		p53, Bak	Iida et al. [133]
	miR-106b, -150*, -371-5p, -557, -598	miR-22, -31, -31*, -145	N/D		IGF1 signaling, EWSR1, FLI1, EWS-FLI1	Mosakhani et al. [131]
		miR-22, -27a, -29a, -100, -125b, -221/222	Cell proliferation		IGF signaling	McKinsey et al. [124]
		miR-22	Clonogenic- and anchorage-independent cell growth		KDM3A	Parrish et al. [129]
		miR-30-5p	Cell proliferation, invasion		CD99	Franzetti et al. [128]
		miR-31	Cell proliferation, invasion, apoptosis		N/D	Karnuth et al. [130]
		miR-34a	Chemoresistance	Prognosis	N/D	Nakatani et al. [134]
		miR-143, -145	Tumor growth, CSC phenotypes		SOX2	De Vito et al. [126]
		miR-145	Function in a feedback loop with EWS/Fli1		EWS/Fli1, SOX2	Ban et al. [125]
		miR-708	Repressed by EWS/Fli1		EYA3	Robin et al. [132]
		let-7a	Tumor growth		HMGA2	De Vito et al. [127]
Chondrosarcoma	miR-96, 183	let-7a, miR-100, -136, -222, -335, -376a	N/D		N/D	Yoshitaka et al. [135]
		miR-100	Chemoresistance		mTOR	Zhu et al. [137]
		miR-145			SOX9	Mak et al. [191]

N/D, not determined; MAPK, mitogen-activated protein kinases; EGFR, epidermal growth factor receptor.

to the metastatic potential of osteosarcoma cells by altering the phenotype and allowing survival in the FasL+ lung microenvironment. Gao et al. [116] performed functional analysis of miR-17 and found that silencing of miR-17 substantially suppressed cell proliferation, migration, and invasion via its direct target, PTEN. The clinical relevance was reported by Li et al. [91] who demonstrated that higher expression of miR-17-92 cluster clearly predicted poorer recurrence-free and overall survival.

Downregulated miRNAs in osteosarcoma similarly include well-known tumor-suppressive miRs in other cancers. Osaki et al. [29] were the first to demonstrate that the expression of miR-143 decreased in metastatic osteosarcoma cells, similar to other cancers. They profiled the miRNA expression in a parental HOS cell line and its subclone 143B metastatic osteosarcoma cell line and found that miR-143 was the most downregulated in 143B cells. A significant inhibition of cell invasion was observed in miR-143-transfected 143B cells. Matrix metalloprotease-13 was one of the most probable targets of miR-143 and was positively identified in clinical specimens of lung metastasis-positive cases by immunohistochemistry, but not identified in those of at least three cases from the nonmetastatic group showing higher miR-143 expression levels [29]. Zhang et al. [23] additionally identified cell apoptosis by miR-143 and Bcl-2 as another direct target. Tumor-suppressive miRs in osteosarcoma include p53-related miRNAs. p53 is a well-known tumor suppressor gene involved in osteosarcoma [117], and is mutated in more than 20% of osteosarcomas. Further, p53 mutations have been demonstrated to be involved in osteosarcoma tumorigenesis [118]. The expression of the miR-34 family (miR-34a, 34b and 34c) was induced by p53 in response to DNA damage or oncogenic stress in multiple cancers [119]. miR-34 dysregulation in osteosarcoma has been reported by several authors. He et al. reported that the miR-34 family induced G1 arrest and apoptosis via their targets, CDK6, E2F3, Cyclin E2, and BCL2, in a p53-dependent manner in osteosarcoma cells. Yan et al. further performed an *in vivo* analysis and found inhibitory effects on tumor growth and pulmonary metastasis. Moreover, p53 also induced the expression of miR-192, -194, and -215 in U2OS cells carrying wild-type p53 [17]. Then, U2OS cells that were enhanced for miR-192 expression formed fewer colonies than those transfected with control miR or miR-34a [17].

No specific translocations or genetic abnormalities have been identified in osteosarcoma, unlike other sarcomas, such as synovial sarcoma and Ewing sarcoma [120]. Several detection modalities have provided a more accurate assessment of the complex cytogenetic aberrations in osteosarcoma; the most frequently detected amplifications include chromosomal regions 6p12–p21 (28%), 17p11.2 (32%), and 12q13–q14 (8%) [121]. However, analysis of genomic profiles of miRNA-gene expression in seven osteosarcoma tumors by Marie et al. has revealed that changes in miRNA expression were DNA copy number-correlated in most cases. However, copy number-independent miRNA expression in osteosarcoma was observed; a cluster of 23 miRNAs in cytoband 14q32.31 were downregulated in six samples, even though chromosomal gain was observed in this region, suggesting a position effect caused by rearrangements of chromosome 14 on miRNA expression from the 14q32.31 region. Subsequently, their analysis identified osteosarcoma-associated gene expression changes that are DNA copy number-correlated, DNA copy number-independent, mRNA-driven, and/or modulated by miRNA expression, suggesting that miRNAs provide a novel posttranscriptional mechanism for fine-tuning the expression of specific genes and pathways relevant to osteosarcoma [28].

Functional analyses using an animal model were conducted with the manipulation of miR-17-92, -25, -34a, -133a, -143, -202, and miRNAs located in 14q32 (Table 1). Among these reports, a systemic treatment model using oligonucleotides was developed by Osaki et al. They not only identified decreased miR-143 expression in highly metastatic 143B cell line as described above but also assessed the therapeutic potential of miR-143 against spontaneous lung metastasis in a model using systemic administration of a miR-143 mimic and miR-negative control (NC). Experimentally, 50 μg of miR-143 mimic or miR-NC was mixed with atelocollagen and intravenously administered into mice after inoculation of 143B cells [29]. Although miR-143 administration did not affect the growth of primary lesions, at 3 weeks after inoculation, 6 of the 8 mice exhibited lung metastasis by an *in vivo* imaging system and the remaining 2 mice died due to lung metastasis following miR-NC/atelocollagen treatment [29]. In contrast, only 2 of the 10 mice in the miR-143/atelocollagen-treated group showed lung metastasis [29]. This preclinical trial has highlighted the potential of miRNA therapeutics for osteosarcoma. Like this preclinical trial, further investigations of key miRNAs for other types of sarcomas and toxicological testing of miRNA mimics, along with the development of drug delivery system (DDS), would accelerate the therapeutic possibility of targeting miRNAs as novel sarcoma treatment options.

In recent years, the clinical relevance of dysregulated miRNAs in osteosarcoma has been reported. Upregulated miRNAs that are associated with poor prognosis include miR-9, -17–92, -29 family, -133a, -210, and -214 (Table 1). A clinical investigation based on the largest cohort was performed by Cai et al. Quantitative reverse transcription polymerase chain reaction (qRT-PCR) analysis was performed to detect the expression level of miR-210 and miR-214 in osteosarcoma and normal bone tissues from 92 children. As a result, the expression of both miR-210 and miR-214, which function as oncomiRs in other cancers, was significantly upregulated in osteosarcoma tissues compared with that in corresponding normal bone tissues [45]. Their clinical investigation revealed that miR-210 and miR-214 upregulation was associated with significantly decreased overall survival and progression-free survival [45]. Multivariate analysis confirmed that high expression was an independent prognostic factor of unfavorable survival in osteosarcoma [45]. Downregulated miRNAs that are associated with poor prognosis include miR-26a, miR-132, miR-133a, miR-145, miR-183, miR-206, miR-217, miR-223, miR-340, and miR-382 (Table 1). The largest cohort was investigated by Yang et al. [74]; they performed qRT-PCR using 166 pairs of osteosarcoma and normal bone tissues. miR-132 expression was significantly lower in osteosarcoma tissues than that in normal bones. The univariate and multivariate analyses showed that low miR-132 expression in osteosarcoma patients correlated with poorer overall and disease-free survival [74]. A treatment model targeting such miRNA dysregulation would provide novel insights for osteosarcoma biology.

2.2 Ewing Sarcoma

Ewing sarcoma (MIM #612219, ORPHA319) is the second most frequently occurring bone tumor, mostly affecting children and young adults. Ewing sarcoma is associated with a unique chromosomal translocation that generates a fusion protein comprising EWS (or, rarely, FUS/TLS) and a member of the Ets transcription factor family [122]. In 85–90% of cases, the t(11;22)(q24;q12) translocation fuses the 5′ end of the EWS gene to the 3′ end of the *Fli1* gene, giving rise to the EWS–Fli1 fusion protein [123]. EWS–Fli1 functions as an aberrant transcription factor, with both inducer and suppressor activities, which displays a distinct target gene specificity from those of its component parts [122]. miRNAs are of current interest as a mechanism of target gene regulation by EWS–Fli1. Indeed, several groups identified EWS–Fli1-regulated miRNAs (Table 2).

An miRNA microarray analysis using stable silencing of EWS–Fli1 in the Ewing sarcoma cell line A673, performed by McKinsey et al. [124] identified 30 upregulated miRNAs upon EWS–Fli1 depletion. As a result, seven miRNAs (miR-22, miR-100, miR-125b, miR-221, miR-222, miR-27a and miR-29) were highly changed and have predicted targets in the insulin-like growth factor (IGF) signaling pathway, a critical promoter of Ewing sarcoma oncogenesis. Ewing sarcoma cell growth was inhibited by the transfection of these miRNAs, revealing an oncogenic mechanism involving posttranscriptional derepression of IGF signaling by the EWS–Fli1 fusion oncoprotein via miRNAs. A similar approach was followed by Ban et al. [125] who identified miR-145 as a top EWS–Fli1-repressed miRNA. The upregulation of miR-145 in Ewing sarcoma cell lines reduced EWS–Fli1 protein levels, whereas the downregulation of miR-145 increased the EWS–Fli1 levels. Enhanced miR-145 expression inhibited cell line growth, indicating that feedback regulation between EWS–Fli1 and miR-145 is an important component in Ewing sarcoma [125]. An additional function of repressed miR-145, together with miR-143, in Ewing sarcoma has also been reported by De Vito et al. [126]. They compared miRNA profiles of $CD133^+$ cancer stem cell (CSC) populations and $CD133^-$ non-CSC populations from two Ewing sarcoma tumors, pediatric MSCs stably expressing ectopic EWS–Fli1, and an Ewing sarcoma cell line. As a result, the downregulation of miRNA processing factor TARBP2 was identified as a mechanism for miRNA depletion in $CD133^+$ cells [126]. A systemic administration of miR-143 and miR-145, two of the miRNAs downregulated in $CD133^+$ cells, resulted in the inhibition of tumor growth, supporting a role for miRNAs in the enhanced tumorigenicity of Ewing sarcoma CSC populations [126].

The other repressed miRNAs in Ewing sarcoma include let-7a, miR-22, miR-30-5p, and miR-31. De Vito et al. [127] reported that EWS–Fli1 directly represses let-7a expression and restoration of let-7a inhibits Ewing sarcoma tumor growth through the regulation of HMGA2 expression. Franzetti et al. [128] found 34 upregulated and 36 downregulated miRNAs by EWS–Fli1 depletion and identified miR-30a-5p as an EWS–Fli1-repressed miRNA, contributing to the regulation of CD99 expression by EWS–Fli1, which is highly expressed in the cell membrane of Ewing sarcoma cells. Parrish et al. [129] focused on EWS/Fli1-repressed miR-22 and identified the histone demethylase KDM3A as a miR-22-regulated tumor promoter in Ewing sarcoma. In addition, Karnuth et al. [130] profiled miRNA expression in 40 Ewing sarcoma biopsies, 6 Ewing sarcoma cell lines, and MSCs from 6 healthy donors. Among 16 lower expressed miRNAs, the expression of miR-31 was lowest relative to MSCs and the enhanced expression of miR-31 reduced cellular proliferation and invasion [130]. Interestingly, array comparative genomic hybridization and miRNA arrays by Mosakhani et al. [131] identified the upregulation of miR-106b, miR-150*, miR-371-5p, miR-557, and miR-598 and downregulation of miR-22, miR-31, and miR-31*, which were consistent with the results of other studies, indicating that these miRNAs are important miRNAs in Ewing sarcoma.

Several reports identified miRNAs that are associated with chemoresistance of Ewing sarcoma. Robin et al. [132] demonstrated that the DNA repair protein and transcriptional cofactor, EYA3, is highly expressed in Ewing sarcoma tumor samples and cell lines compared with MSCs and is modulated by EWS–Fli1. High expression levels of EYA3 significantly correlated with low levels of miR-708, while EWS/Fli1 increased EYA3 expression through the repression of miR-708 [132]. Furthermore, silencing of EYA3 in Ewing sarcoma cells resulted in sensitization to DNA-damaging chemotherapeutics, etoposide, and doxorubicin [132]. Moreover,

Iida et al. screened doxorubicin-resistant and parental Ewing sarcoma cells and identified miR-125b to be upregulated in two different doxorubicin-resistant Ewing sarcoma cell lines. Silencing of miR-125b showed an enhanced sensitivity to doxorubicin, whereas overexpression of miR-125b in parental EWS cells resulted in enhanced resistance to doxorubicin, etoposide, and vincristine [133].

miRNAs that are associated with disease prognosis have been reported. Nakatani et al. performed global miRNA microarray profiling on 34 primary Ewing sarcoma tumors, comparing the expression profiles of patients with early relapse to those without clinical relapse. Their analysis identified miR-34a, miR-23a, miR-92a, miR-490-3p, and miR-130b that significantly associated with both event-free and overall survival. Further analysis demonstrated that low levels of miR-34a emerged as a particularly robust predictor of early relapse, and the upregulation of miR-34a in EWS cell lines inhibited cell proliferation and, interestingly, sensitized cells to doxorubicin and vincristine [134].

2.3 Chondrosarcoma

Chondrosarcoma (MIM#215300, ORPHA55880) is the second most common primary malignant bone tumor [135]. Chondrosarcoma is unique among bone tumors as it shows resistance to chemotherapy and traditional radiotherapy. Surgical excision is the sole treatment option in most cases.

Yoshitaka et al. [135] were the first to report miRNA deregulation in chondrosarcoma (Table 2). They performed miRNA array and qRT-PCR using 20 clinical samples, two cell lines, and nontumorous adult articular chondrocytes. In their array analysis, 27 downregulated miRNAs and 2 upregulated miRNAs (miR-96 and miR-183) were identified. A validation study based on qRT-PCR demonstrated that let-7a, miR-100, miR-222, miR-136, miR-376a, and miR-335 were significantly downregulated in chondrosarcoma-derived samples [135]. Mak et al. [136] focused on miR-145, that inhibits SOX9, the master regulator of chondrogenesis. They identified high expression of SOX9 in patient samples and low expression of miR-145 in cell lines and patient samples [136]. The upregulation of miR-145 expression in chondrosarcoma cells contributed to the downregulation of SOX9 and then of ETS transcript variant 5 (ETV5) and matrix metalloproteinase-2 (MMP2). Functional miRNAs in chondrosarcoma cells have been identified by Zhu et al. The expression level of miR-100 was downregulated in cell lines and patient samples [137]. miR-100 expression was notably decreased in cisplatin-resistant chondrosarcoma cells compared with parental cells, and the upregulation of miR-100 increased the sensitivity to cisplatin in normal cell lines. mTOR was a direct target of miR-100 and recovery of the mTOR pathway desensitized the chondrosarcoma cells to cisplatin [137]. These data indicated that miR-100-mediated sensitization to cisplatin was dependent on the inhibition of mTOR in chondrosarcoma cells.

3. miRNAs IN SOFT TISSUE SARCOMAS

Soft tissue sarcomas encompass a wide range of tumor types. The first important study was reported by Subramanian et al. in 2008. They analyzed the miRNA expression profiles of 27 soft tissue sarcoma samples from 5 histological subtypes (synovial sarcoma, rhabdomyosarcoma, leiomyosarcoma, gastrointestinal stromal tumor (GIST), and liposarcoma) and 7 normal tissue samples [138]. In these expression profiles, different histological subtypes of sarcoma had distinct miRNA expression signatures, reflecting the apparent lineage and differentiation status of the tumors. Then, miRNA dysregulation in soft tissue sarcomas has gradually accumulated, revealing its functional importance and target genes (Table 3).

TABLE 3 Dysregulated miRNAs in Soft Tissue Sarcomas

Histology	Upregulated miRNAs	Downregulated miRNAs	Function	miRNA Target	References
Leiomyosarcoma	miR-1, -133a, -133b		N/D	N/D	Subramanian et al. [138]
	miR-1, 133a, -449a		N/D	N/D	Renner et al. [143]
	miR-17-92 cluster (uterine LMS)		Smooth muscle differentiation	N/D	Danielson et al. [142]
	miR-221 (uterine LMS)		N/D	N/D	Nuovo and Schmittgen [140]
	miR-320a		N/D	N/D	Guled et al. [198]
	let-7 (uterine LMS) (N/D)		Cell proliferation	*HMGA2*	Shi et al. [199]
		miR-483-5p, -656, -323-3p	N/D	N/D	Renner et al. [143]
Liposarcoma	miR-21, -26a (DDLS)		N/D	N/D	Ugras et al. [192]
	miR-26a-2 (DDLS, MLS)		Clonogenicity, adipocyte differentiation, cell apoptosis	*RCBTB1*	Lee et al. [150]
	miR-155 (DDLS)		Cell proliferation, colony formation, tumor growth	*CK1α*	Zhang et al. [146]
	miR-218-1* (DDLS)		N/D	N/D	Renner et al. [143]
	miR-9, -891a, -888 (MLS)		N/D	N/D	Renner et al. [143]
	miR-296-5p, -455-5p, -1249 (PLS)		N/D	N/D	Renner et al. [143]
		miR-143, -145 (DDLS)	Cell proliferation, apoptosis	*BCL2, Topoisomerase 2A, PRC1, PLK1*	Urgas et al. [198]
		miR-144, -1238 (DDLS)	N/D	N/D	Renner et al. [143]
		miR-193b (DDLS)	N/D (methylated)	N/D	Taylor et al. [147]
		miR-1257 (DDLS)	N/D	*CALR*	Hisaoka et al. [148]
		miR-486 (MLS)	Cell proliferation	*PAI-1*	Borjigin et al. [149]

Continued

TABLE 3 Dysregulated miRNAs in Soft Tissue Sarcomas—cont'd

Histology	Upregulated miRNAs	Downregulated miRNAs	Function	miRNA Target	References
		miR-486-3p, -1290 (MLS)	N/D	N/D	Renner et al. [143]
		miR-200b*, -200, -139-3p (PLS)	N/D	N/D	Renner et al. [143]
Rhabdoymyosarcoma	miR-9*		Cell migration	*E-cadherin*	Armeanu-Ebinger et al. [165]
	miR-17-92 cluster (ARMS)		Correlation with prognosis in 13q31 amplified ARMS	N/D	Reichek et al. [166]
	miR-183		Cell migration, cell invasion	*EGR1, PTEN*	Sarver et al. [164]
	miR-335 (ARMS)		N/D	*CHFR, HAND1, SP1*	Subramanian et al. [138]
	miR-485-3p (N/D)		Drug resistance	*NF-YB*	Chen et al. [193]
		miR-1, -133a/b	Myogenic differentiation, cell proliferation	*SRF, Cyclin D2*	Rao et al. [194]
		miR-26a	N/D	*Ezh2*	Ciarapica et al. [195]
		miR-29	Cell cycle arrest, muscle differentiation, tumor growth	*YY1*	Wang et al. [161] and Li et al. [157]
		miR-200c	Cell migration	N/D	Armeanu-Ebinger et al. [165]
		miR-203	Myogenic differentiation, cell proliferation, cell migration, tumor growth	*p63, LIFR*	Diao et al. [196]
		miR-206	Myogenic differentiation, cell growth, cell migration, tumor growth, correlation with prognosis	*c-Met, PAX3, PAX7, CCDN2, HDAC4*	Taulli et al. [155], Yan et al. [156], Li et al. [157], and Missiaglia et al. [197]
Synovial sarcoma	let-7e, miR-99b, miR-125a-3p		Cell proliferation	*HMGA2, SMARCA5*	Hisaoka et al. [168]

	miR-183		Cell migration, cell invasion	*EGR1*	Sarver et al. [164]
	miR-183, 200b*, -375		N/D	N/D	Renner et al. [143]
		miR-34b*, -142-5p, -34c-3p	N/D	N/D	Renner et al. [143]
		miR-143	N/D	*SSX1*	Subramanian et al. [138]
MPNST	miR-10b		Cell proliferation, migration and invasion	*NF1*	Chai et al. [174]
	miR-21		Apoptosis	*PDCD4*	Itani et al. [171]
	miR-204		Cell proliferation, migration and invasion	*HMGA2*	Gong et al. [173]
	miR-210, -339-5p		N/D	N/D	Presneau et al. [172]
		miR-29c	Cell invasion	*MMP2*	Presneau et al. [172]
		miR-30d	Apoptosis	*KPNB1*	Zhang et al. [109–112,175]
		miR-34a	Apoptosis	*MYCN*, *E2F2*, *CDK4*	Subramanian et al. [170]
UPS	miR-126, -223, -451, -1274b		N/D	N/D	Guled et al. [144]
		miR-100, -886-3p, -1260, -1274a, -1274b	N/D	*IMP3*	Guled et al. [144]
Epithelioid sarcoma	miR-206, -381, -671-5p		N/D	*SMARCB1 (INI1)*	Papp et al. [200]

DDLS, dedifferentiated liposarcoma; MLS, myxoid liposarcoma; PLS, pleomorphic liposarcoma; LMS, leiomyosarcoma; ARMS, alveolar rhabdomyosarcoma; AS, angiosarcoma; MPNST, malignant peripheral nerve sheath tumor; UPS, undifferentiated pleomorphic sarcoma; N/D, not determined.

3.1 Leiomyosarcoma

Leiomyosarcoma (LMS; ORPHA64720) is a malignant soft tissue tumor, demonstrating smooth muscle differentiation. Soft tissue LMS usually occurs in middle-aged or older individuals and originates in retroperitoneal lesions (40–45%), extremities (30–35%), skin (15–20%), and larger blood vessels (5%) [139]. For patients with LMS in the extremities, the local recurrence rate is 10–25%, whereas the 5-year survival rate is 64% [139].

miRNA dysregulation of LMS has been profiled in tumor tissues from the extremities and uterus [138,140–144]. All studies have demonstrated upregulated miRNAs in LMS relative to its benign counterparts, such as leiomyoma or other soft tissue sarcomas. Subramanian et al. [138] demonstrated that miR-1, miR-133a, and miR-133b, which play major roles in myogenesis and myoblast proliferation, were significantly overexpressed in LMS relative to normal smooth muscle. miR-206, a member of muscle-specific miRNAs with miR-1, and miR-133a, were underexpressed in both LMS and normal smooth muscle. miRNA dysregulation compared with other sarcomas has been demonstrated in recent reports. Guled et al. [144] profiled 10 high-grade LMS and UPS (undifferentiated pleomorphic sarcoma) samples each by microarray analysis and identified upregulated miR-320a in LMS relative to UPS. A similar approach was followed by Renner et al., who examined differentially expressed miRNAs in LMS compared with the other sarcoma subtypes. Their analysis identified upregulated miR-1, miR-133a, and miR-449a and downregulated miR-483-5p, miR-656, and miR-323-3p in LMS [143]. These results were partly consistent with those by Subramanian et al. [138,143].

3.2 Liposarcoma

Liposarcoma (MIM #613488, ORPHA69078) is one of the most common soft tissue sarcomas in adults and can be subdivided into four major types: atypical lipomatous tumor/well-differentiated liposarcoma (WDLS), myxoid liposarcoma (MLS), pleomorphic liposarcoma (PLS), and dedifferentiated liposarcoma (DDLS). DDLS is defined as a WDLS that shows an abrupt transition to a nonlipogenic sarcoma. MLS is relatively chemosensitive compared to the other subtypes.

Most investigations of miRNA profiling of liposarcoma have been focused on DDLS. Ugras et al. identified more than 40 miRNAs that were dysregulated in DDLS and not in normal adipose tissue or WDLS based on the deep sequencing of small RNA libraries and hybridization-based microarrays. Upregulated miR-21 and miR-26 and downregulated miR-143 and miR-145 have been identified [145]. The reexpression of miR-143 in DDLS cell lines inhibited cell proliferation and induced apoptosis through the downregulation of BCL2, topoisomerase 2A, protein regulator of cytokinesis 1, and polo-like kinase 1 [145]. A similar approach was adopted by Zhang et al., who performed miRNA profiling to compare WDLS/DDLS and normal adipose tissue. Their analysis identified upregulated miR-155 in DDLS [146], which had not been identified by Ugras et al. Silencing of miR-155 inhibited cell growth and colony formation, induced G1-S cell cycle arrest *in vitro*, and blocked tumor growth *in vivo* [146]. Renner et al. [143] identified upregulated miR-218-1* and HS_303_a and downregulated miR-144 and miR-1238 in DDLS relative to normal adipose tissues. Further, using unbiased, genome-wide methylation sequencing, Taylor et al. [147] demonstrated that miR-193b was downregulated in DDLS relative to normal adipose tissue and WDLS, whose putative miR-193b promoters were differentially methylated. A DDLS study by Hisaoka et al. [148] identified decreased expression of miR-1257, which targets calreticulin (CALR), an inhibitor of adipocyte differentiation.

MLS has a unique genomic abnormality characterized by a t(12;16)(q13;p11) translocation, which creates a TLS–CHOP chimeric oncoprotein.

Borijigin et al. [149] reported that miR-486 was downregulated in both TLS–CHOP-expressing fibroblasts and MLS, indicating that TLS–CHOP–miR-486–PAI-1 may be critical for MLS tumorigenesis and development. In miRNA profiling of MLS and normal adipose tissue, Renner et al. identified upregulated miR-9, miR-891a, and miR-888 and downregulated miR-486-3p and miR-1290, which was consistent with the report by Borijigin et al. [149] who also reported on dysregulated miRNAs in PLS relative to normal adipose tissue. Borijigin et al. also demonstrated that miR-1249, miR-296-5p, and miR-455-5p were upregulated and miR-200b*, miR-200, and miR-139-3p were downregulated.

Recent reports have demonstrated a clinical correlation with miRNA dysregulation in liposarcoma. In a single nucleotide polymorphism array of 75 liposarcoma samples, Lee et al. [150] identified frequent amplification of miR-26a-2c, which was upregulated in not only WDLS/DDLS but also in MLS. Importantly, high miR-26a-2 expression significantly correlated with poor patient survival, regardless of histological subtypes. An additional analysis revealed that the regulator of chromosome condensation and BTB domain-containing protein 1 (RCBTB1) was one of the targets of miR-26a-2, which regulates cellular apoptosis [150].

3.3 Rhabdomyosarcoma

Rhabdomyosarcoma (RMS; MIM #268210 #268220, ORPHA780) is the most common soft tissue sarcoma in children under 15 years of age (5–8% of all pediatric malignancies) and also arises in adolescents and young adults. Histopathologically, RMS is classified into the following four subtypes: embryonal RMS (ERMS), alveolar RMS (ARMS), pleomorphic RMS (PRMS), and spindle cell/sclerosing RMS.

Most studies have focused on miRNAs that are involved in skeletal muscle development ("muscle-specific miRNAs") as RMS has been predicted to originate from mesenchymal progenitor cells located in the muscle tissue [151–153]. Global miRNA expression analysis was conducted by Subramanian et al. [138] revealing that muscle-specific miRNAs (miR-1 and miR-133) were downregulated in PRMS muscle and miR-335 was upregulated in ARMS relative to normal skeletal muscle. Further, miR-335 resides in intron 2 of MEST, a downstream target of PAX3, the gene involved in the PAX3–FKHR fusion that is typical for ARMS. Rao et al. [154] also demonstrated that miR-1 and miR-133a were drastically reduced in ERMS and ARMS cell lines. Taulli et al. [155] and Yan et al. [156] examined the role of the muscle-specific miR-1 and miR-206 in RMS. The reexpression of these miRNAs in RMS cells targeted c-Met mRNA to promote myogenic differentiation, decreased cell growth and migration, and inhibited tumor growth in xenografted mice. Furthermore, Li et al. [157] reported PAX3 and CCND2 as additional important targets of miR-1, miR-206, and miR-29. Ciarapica et al. [158] also found that miR-26a was also downregulated in RMS cells and reported that it may have a role in RMS pathogenesis via the regulation of the expression of enhancer of zeste homologue 2 (Ezh2), which regulates embryonic development through the inhibition of homeobox gene expression. miR-203 was also found to be downregulated in RMS, which occurred due to promoter hypermethylation and could be reexpressed by DNA-demethylating agents [158]. Taulli et al. have recently further pursued the miR-206 target BAF53a, a subunit of the SWI/SNF chromatin remodeling complex, which is an important molecule during myogenic differentiation. Indeed, the BAF53a transcript was highly expressed in primary RMS tumors compared with normal muscle. Silencing of BAF53a in RMS cells inhibited cell proliferation and anchorage-independent growth, and inhibited ERMS and ARMS tumor growth, concluding that failure to downregulate the BAF53a subunit may contribute to RMS pathogenesis [159].

The clinical relevance of these muscle-specific miRNAs was demonstrated by Missiaglia et al. [160] using 163 primary RMS samples. The Kaplan–Meier curves showed a correlation between overall survival and miR-206 expression, whereas no correlation was observed with miR-1 or miR-133a/b. In particular, low miR-206 expression correlated with poor overall survival was an independent predictor of shorter survival times in metastatic ERMS and ARMS cases without PAX3/7–FOXO1 fusion genes [160].

Nonmuscle miRNAs also have been reported as key molecules that function in RMS. miR-29 was downregulated in RMS and acted as a tumor suppressor [138,157,161]. NF-κB and YY1 downregulation caused the derepression of miR-29 during myogenesis, whereas in RMS, miR-29 was epigenetically silenced by an activated NF-κB–YY1 pathway. Functional analysis revealed the inhibition of RMS tumor growth by the reexpression of miR-29 [161]. It has also been proposed that miR-29 can silence HDAC4 [162] or affect the Rybp epigenetic modifier [163], further promoting myogenic differentiation [151]. Sarver et al. [164] demonstrated that in addition to RMS, EGR1 is regulated by miR-183 in multiple tumor types, including synovial sarcoma and colon cancer. miR-183 silencing in RMS cells revealed the deregulation of an miRNA network comprising miR-183–EGR1–PTEN [164]. Armeanu-Ebinger et al. [165] analyzed miRNA expression in ARMS and malignant rhabdoid tumor (MRT) in tissue samples and cell lines. They demonstrated the overexpression of miR-9* in ARMS, whereas miR-200c was expressed at lower levels in ARMS than in MRT. Another important study on ARMS was reported by Reichek et al. [166], who investigated the 13q31 amplicon that contains the miR-17-92 cluster gene and observed its significant overexpression in RMS with the 13q31 amplicon. This was present in 23% of ARMS cases, particularly in PAX7–FKHR-positive cases compared to PAX3–FKHR-positive and fusion-negative cases. Importantly, high expression of the miR-17-92 cluster significantly correlated with poor prognosis in the 13q31-amplified group of patients, most of whom represented PAX7–FKHR-positive cases [166].

3.4 Synovial Sarcoma

Synovial sarcoma (MIM #300813, ORPHA3273) accounts for 10% of soft tissue sarcomas and includes two major histological subtypes, biphasic and monophasic [167]. Synovial sarcoma has a specific chromosomal translocation t(X;18)(p11;q11) that forms an SS18–SSX fusion gene.

In the first report on sarcoma miRNA profiling reported in 2008, Subramanian et al. [138] utilized microarray, cloning, and northern blot analysis to demonstrate that miR-143 was downregulated in synovial sarcoma relative to GIST and LMS. As SSX1 is predicted to be a target for miR-143 in databases, such as miR-Base and TargetScan, it is hypothesized that its decreased expression in synovial sarcoma enables the production of the SS18–SSX1 oncoprotein. Moreover, Sarver et al. [164] focused on the molecular feature of synovial sarcoma where the SS18–SSX fusion protein represses EGR1 expression through a direct association with the EGR1 promoter. The correlation between EGR1 and miR-183 was identified, and miR-183 was significantly overexpressed in synovial sarcoma. Through the functional analysis using many tumor cell lines, miR-183 was found to have an oncogenic role through the miR-183–EGR1–PTEN pathway [164]. Renner et al. also demonstrated that miR-183, miR-200b*, and miR-375 were upregulated in synovial sarcoma relative to other sarcomas. Hisaoka et al. [168] examined the global miRNA expression in synovial sarcoma and compared the results to Ewing sarcoma and normal skeletal muscle. They found 21 significantly upregulated miRNAs, including let-7e, miR-99b, and miR-125-3p. Functional analysis based on the silencing of let-7e and miR-99b resulted in the inhibition of cell proliferation and the expression of HMGA2 and SMARCA5, the putative targets of these miRNAs [168].

3.5 Malignant Peripheral Nerve Sheath Tumor

Malignant peripheral nerve sheath tumor (MPNST; ORPHA3148) originates from cells constituting the nerve sheath, such as Schwann and perineural cells. Approximately 50% of MPNSTs occur sporadically, with the remaining originating in neurofibromatosis type 1 (NF1) patients [169]. NF1 patients have a high risk of developing MPNSTs, and most are aggressive tumors with a poor prognosis.

Many reports have investigated the global miRNA profiling of MPNSTs compared with benign counterparts, such as neurofibromas. Subramanian et al. identified p53 inactivation in a majority of MPNSTs and revealed a relative downregulation of miR-34a expression in most tumors. They concluded that p53 inactivation and the subsequent loss of miR-34a expression may significantly contribute to MPNST development [170]. Itani et al. identified the overexpression of miR-21 in MPNSTs compared with neurofibromas. An *in silico* study predicted programmed cell death protein 4 (PDCD4) as a putative target of miR-21 [171]. Functional analysis using an MPNST cell line indicated that miR-21 silencing could induce apoptosis of MPNST cells [171]. Presneau et al. [172] compared miRNA profiling between MPNSTs and NFs and identified 14 downregulated and 2 upregulated miRNAs. The former included miR-29c, miR-30c, miR-139-5p, miR-195, miR-151-5p, miR-342-5p, miR-146a, miR-150, and miR-223, and the latter included miR-210 and miR-339-5p. Among them, enhanced miR-29c expression reduced cell invasion of MPNST cells, regulating the expression of its target, MMP2 [172]. In addition, Gong et al. [173] identified the downregulated expression of miR-204 in MPNSTs in the similar approach and reported Ras and HMGA2 as the target molecules of miR-204. Chai et al. [174] revealed that miR-10b was upregulated in primary Schwann cells isolated from NF1 neurofibromas and in cell lines and tumor tissues from MPNSTs. Importantly, NF1 mRNA was the target for miR-10b. Zhang et al. identified the unregulated expression in MPNSTs of polycomb group protein EZH2, an important regulator for various human malignancies, which inhibited miR-30d expression by binding to its promoter. Analysis of an *in silico* database identified KPNB1 as an miR-30d target, suggesting that Ezh2-miR-30d-KPNB1 signaling was critical for MPNST survival and tumorigenicity [175].

3.6 Undifferentiated/Unclassified Sarcoma

World Health Organization declassified malignant fibrous histiocytoma (MFH) as a formal diagnostic entity and renamed it as an undifferentiated pleomorphic sarcoma (UPS) not otherwise specified (NOS) in 2002 [176]. In 2013, UPS/MFH was categorized as an undifferentiated/unclassified sarcoma (ORPHA2023) [177]. Undifferentiated/unclassified sarcomas account for up to 20% of all sarcomas and have no clinical or morphological characteristics that would otherwise place them under specific types of sarcomas.

miRNA profiling on a series of LMS and UPS samples was performed by Guled et al. to identify specific signatures useful for differential diagnosis. Using 10 LMS, 10 UPS samples, and 2 cultured human mesenchymal stem cell samples as controls, they identified 38 upregulated miRNAs found in UPS compared to control samples [144]. In UPS samples, miR-126, miR-223, miR-451, and miR-1274b were significantly upregulated and miR-100, miR-886-3p, miR-1260, miR-1274a, and miR-1274b were significantly downregulated compared to control samples [144]. When comparing the profiles of LMS and UPS, miR-199-5p was highly expressed in UPS, while miR-320a was highly expressed in LMS [144]. For differential diagnosis, they also demonstrated that several genes, including IMP3, ROR2, MDM2, CDK4, and UPA, were targets of differentially expressed miRNAs and validated their expression in LMS and UPS by immunohistochemistry.

3.7 Epithelioid Sarcoma

Epithelioid sarcoma (ORPHA293202) represents approximately 1% of sarcomas and is most prevalent in adolescents and young adults between 10 and 35 years of age [178]. This sarcoma is the most common soft tissue sarcoma in the hand and wrist, followed by ARMS and synovial sarcoma. Two clinicopathological subtypes are recognized: (1) conventional or classic ("distal") form, characterized by its proclivity for acral sites and pseudogranulomatous growth pattern, and (2) proximal-type ("large-cell") variant that originates mainly in proximal/truncal regions and comprises nests and sheets of large epithelioid cells.

Proximal-type epithelioid sarcoma has similarities with MRT, including the lack of nuclear immunoreactivity of SMARCB1 (also known as INI1 or BAF47). Papp et al. hypothesized that miRNAs regulate SMARCB1 expression and analyzed eight candidate miRNAs selected from an *in silico* analysis. RT-PCR using tumor samples identified the overexpression of miR-206, miR-381, miR-671-5p, and miR-765 in epithelioid sarcomas [179]. Among them, three of the overexpressed miRNAs (miR-206, miR-381, and miR- 671-5p) could silence SMARCB1 mRNA expression in cell cultures. They concluded that the epigenetic mechanism of gene silencing by miRNAs caused the loss of SMARCB1 expression in epithelioid sarcoma [179].

4. CIRCULATING miRNAs IN BONE AND SOFT TISSUE SARCOMAS

Circulating miRNAs in sarcomas as potential diagnostic markers was first presented in 2010 by Miyachi et al. [180] who analyzed the expression levels of muscle-specific miRNAs in the sera of patients with rhabdomyosarcoma and healthy controls. To date, the evidence is restricted to only three types of sarcomas: osteosarcoma, RMS, and MPNST.

4.1 Osteosarcoma

To date, four miRNAs (miR-21, miR-34b, miR-143, and miR-199-3p) have been reported as potential osteosarcoma biomarkers. Yuan et al. [41] investigated serum miR-21 expression levels in 65 patients with osteosarcoma and 30 healthy controls by qRT-PCR and found that serum miR-21 expression levels were significantly higher in patients with osteosarcoma than in the controls. Moreover, upregulated serum miR-21 levels significantly correlated with Enneking stage and chemotherapeutic resistance. The upregulated expression of miR-21 in osteosarcoma had been previously reported by Kang et al. and Ziyan et al. (Table 1).

Ouyang et al. [59] evaluated the expression levels of 6 miRNAs (miR-34, miR-21, miR-199-3p, miR-143, miR-140, and miR-132) that had been reported as aberrantly expressed in osteosarcoma using plasma from 40 patients with osteosarcoma and 40 matched healthy controls by qRT-PCR. They found that plasma miR-21 levels were significantly higher in patients with osteosarcoma than in controls, consistent with the results from Yuan et al., whereas miR-199a-3p and miR-143 levels were decreased. Furthermore, plasma miR-21 and miR-143 levels were correlated with metastasis and histological subtype, whereas plasma miR-199a-3p levels correlated with histological subtype. The area under the curve value of the combined signature of three miRNAs (miR-21, miR-199-3p, and miR-143) was higher than that of bone-specific alkaline phosphatase (0.953 and 0.922, respectively), and the sensitivities and specificities of the combined miRNAs were 90.5% and 93.8%, respectively. Indeed, the aberrant expression of miR-143 and miR-199-3p in osteosarcoma has been previously reported by Osaki et al. and Duan et al. (Table 1).

Tian et al. investigated the association between osteosarcoma and the expression levels of plasma miR-34b/c, previously identified as a tumor suppressor in osteosarcoma by He et al., Yan et al., Zhao et al., and Honglin et al. (Table 1). Tian et al. [65] found that the plasma miR-34b level was significantly lower in patients with osteosarcoma than in the controls and correlated with its expression in osteosarcoma tissues. Moreover, plasma miR-34b expression levels significantly decreased in patients with metastatic disease compared with those with nonmetastatic diseases, while no significant difference in miR-34b levels was observed between patients with osteoblastic and non-osteoblastic diseases [65].

4.2 Rhabdomyosarcoma

Miyachi et al. were the first to suggest the use of circulating miRNAs for RMS diagnosis. They focused on muscle-specific miRNAs (miR-1, miR-133a, miR-133b, and miR-206) that were more abundantly expressed in myogenic tumors. First, the expression levels of these muscle-specific miRNAs were confirmed to be higher in RMS cell lines and culture supernatants than in other cell lines. Second, in their analysis of muscle-specific miRNA serum levels in RMS patients, normalized serum miR-206 showed the highest sensitivity and specificity among muscle-specific miRNAs [180]. Importantly, serum miR-206 expression decreased after the treatment of RMS [180]. Therefore, serum miR-206 expression may be used as a predictive biomarker of RMS aggressiveness and patient prognosis, which would be required to be confirmed by further studies with larger patient cohorts.

4.3 Malignant Peripheral Nerve Sheath Tumor

The use of serum miRNAs to distinguish MPNST patients with and without NF1 was investigated by Weng et al. [181]. Solexa sequencing was applied to screen for differentially expressed miRNA in pooled serum from 10 patients with NF1, 10 patients with sporadic MPNST, and 10 patients with NF1 MPNST. As a result, miR-801 and miR-214 showed higher expression levels in both sporadic MPNST and NF1 MPNST patients than in NF1 patients [181]. Furthermore, miR-24 was significantly upregulated in NF1 MPNST patients. Therefore, they concluded that the combination of the three miRNAs (miR-801, miR-214, and miR-24) could be used to distinguish NF1 MPNST patients from NF1 patients [181]. Subramanian et al. [170] have previously demonstrated that miR-214 was relatively upregulated in MPNST tissues compared with benign tumor tissues (Table 1), indicating that miRNA expression in the tumor tissues reflect the expression levels of circulating miRNA in the serum.

5. CONCLUSIONS

There is growing evidence of miRNA profiling not only in tumor cells and tissues but also in patient serum and plasma samples from patients with bone and soft tissue sarcomas. To date, there are few useful biomarkers to diagnose or monitor sarcoma development, which is one of the important problems in bone and soft tissue sarcoma treatment. Indeed, more than 100 sarcoma subtypes have been described. This variety can present a diagnostic challenge because the clinical and histopathological characteristics of each subtype are not always distinct. Therefore, the identification of miRNAs specific to histological subtypes would be a novel breakthrough for sarcoma research. In particular, circulating miRNAs in patient plasma would present a novel approach for the diagnosis and monitoring of sarcomas, which has been recently indicated in several reports [41,59,65,180,181]. Despite some exceptions, most of these findings have shown that aberrant expression of circulating miRNAs correlated with that of tumor cells and tissues, indicating that serum or plasma miRNA expression will serve as novel biomarkers for sarcomas. Further investigation of the correlation between tissue and serum miRNA dysregulation is required for the identification of important miRNAs for clinical applications. These studies would accelerate the potential of miRNA dysregulations to clinical applications as novel biomarkers for differential diagnosis, determination of tumor response to chemotherapy, tumor monitoring after treatment for primary lesions, and detection of micrometastasis, as well as novel therapeutics for combination therapy with neoadjuvant and/or adjuvant chemotherapy (Figure 1). Although some issues remain unresolved regarding the methodology of circulating miRNA levels and the effectiveness of miRNA therapeutics, we believe that a novel miRNA-based diagnosis and therapeutics will be a breakthrough for patients with bone and soft tissue sarcomas.

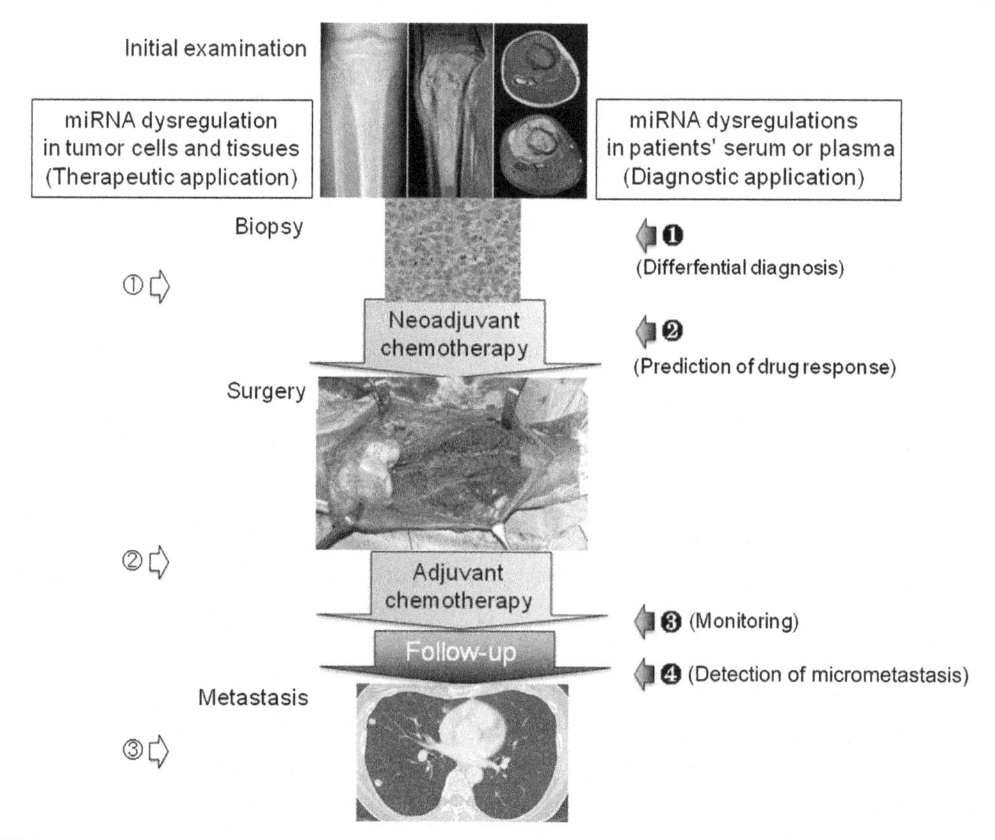

FIGURE 1 Examples of clinical applications of miRNA dysregulations as novel biomarkers and therapeutics for patients with bone and soft tissue sarcomas. As biomarkers: ❶ differential diagnosis, ❷ determine tumor response to chemotherapy, ❸ monitoring after treatment for primary lesions, ❹ detection of micrometastasis. As therapeutics: ① combination therapy with neoadjuvant chemotherapy, ② combination therapy with adjuvant chemotherapy, and ③ combination therapy with chemotherapy for metastasis.

LIST OF ABBREVIATIONS

ARMS Alveolar rhabdomyosarcoma
CSC Cancer stem cell
DDLS Dedifferentiated liposarcoma
ERMS Embryonal rhabdomyosarcoma
LMS Leiomyosarcoma
MFH Malignant fibrous histiocytoma
MLS Myxoid liposarcoma
MPNST Malignant peripheral nerve sheath tumor
MSC Mesenchymal stem cell
OS Osetosarcoma
RMS Rhabdomyosarcoma
UPS Undifferentiated pleomorphic sarcoma
WDLS Well-differentiated liposarcoma

Acknowledgment

The authors acknowledge a grant-in-aid for Scientific Research on Applying Health Technology from the Ministry of Health, Labour and Welfare of Japan.

References

[1] Misra A, Mistry N, Grimer R, Peart F. The management of soft tissue sarcoma. J Plast Reconstr Aesthet Surg 2009;62(2):161–74.
[2] Dorfman HD, Czerniak B. Bone cancers. Cancer 1995;75(S1):203–10.
[3] Gustafson P. Soft tissue sarcoma: epidemiology and prognosis in 508 patients. Acta Orthop 1994;65(S259):2–31.
[4] Toro JR, Travis LB, Wu HJ, Zhu K, Fletcher CD, Devesa SS. Incidence patterns of soft tissue sarcomas, regardless of primary site, in the surveillance, epidemiology and end results program, 1978–2001: an analysis of 26,758 cases. Int J Cancer 2006;119(12):2922–30.
[5] Tsuchiya H. Effect of timing of pulmonary metastases identification on prognosis of patients with osteosarcoma: the Japanese musculoskeletal oncology group study. J Clin Oncol 2002;20(16):3470–7.
[6] Fujiwara T, Kawai A, Yoshida A, Ozaki T, Ochiya T. Cancer stem cells of sarcoma. New Hampshire: CRC Press; 2013.
[7] Bartel DP. MicroRNAs: genomics, biogenesis, mechanism, and function. Cell 2004;116(2):281–97.
[8] Gregory RI, Shiekhattar R. MicroRNA biogenesis and cancer. Cancer Res May 1, 2005;65(9):3509–12.
[9] Nugent M. MicroRNA function and dysregulation in bone tumors: the evidence to date. Cancer Manag Res 2014;6:15–25.
[10] Croce CM. Causes and consequences of microRNA dysregulation in cancer. Nat Rev Genet October 2009;10(10):704–14.
[11] Kim VN, Han J, Siomi MC. Biogenesis of small RNAs in animals. Nat Rev Mol Cell Biol 2009;10(2):126–39.
[12] Esquela-Kerscher A, Slack FJ. Oncomirs—microRNAs with a role in cancer. Nat Rev Cancer April 2006;6(4): 259–69.
[13] Kosaka N, Iguchi H, Ochiya T. Circulating microRNA in body fluid: a new potential biomarker for cancer diagnosis and prognosis. Cancer Sci 2010; 101(10):2087–92.
[14] Mitchell PS, Parkin RK, Kroh EM, Fritz BR, Wyman SK, Pogosova-Agadjanyan EL, et al. Circulating microRNAs as stable blood-based markers for cancer detection. Proc Natl Acad Sci 2008;105(30):10513–8.
[15] Schwarzenbach H, Nishida N, Calin GA, Pantel K. Clinical relevance of circulating cell-free microRNAs in cancer. Nat Rev Clin Oncol March 2014;11(3):145–56.
[16] Lawrie CH, Gal S, Dunlop HM, Pushkaran B, Liggins AP, Pulford K, et al. Detection of elevated levels of tumour–associated microRNAs in serum of patients with diffuse large B–cell lymphoma. Br J Haematol 2008;141(5):672–5.
[17] Braun CJ, Zhang X, Savelyeva I, Wolff S, Moll UM, Schepeler T, et al. p53-Responsive microRNAs 192 and 215 are capable of inducing cell cycle arrest. Cancer Res 2008;68(24):10094–104.
[18] He C, Xiong J, Xu X, Lu W, Liu L, Xiao D, et al. Functional elucidation of MiR-34 in osteosarcoma cells and primary tumor samples. Biochem Biophys Res Commun 2009;388(1):35–40.
[19] Song B, Wang Y, Xi Y, Kudo K, Bruheim S, Botchkina GI, et al. Mechanism of chemoresistance mediated by miR-140 in human osteosarcoma and colon cancer cells. Oncogene 2009;28(46):4065–74.
[20] Creighton CJ, Fountain MD, Yu Z, Nagaraja AK, Zhu H, Khan M, et al. Molecular profiling uncovers a p53-associated role for microRNA-31 in inhibiting the proliferation of serous ovarian carcinomas and other cancers. Cancer Res 2010;70(5):1906–15.
[21] Song B, Wang Y, Titmus MA, Botchkina G, Formentini A, Kornmann M, et al. Research molecular mechanism of chemoresistance by miR-215 in osteosarcoma and colon cancer cells. Mol Cancer 2010;30(9):96.
[22] Stabley D, Kamara D, Holbrook J, Sol-Church K, Kolb E, McCahan S. Digital gene expression of miRNA in osteosarcoma xenografts: finding biological relevance in miRNA high throughput sequencing data. J Biomol Tech September 2010;21(3 Suppl):S25.
[23] Zhang H, Cai X, Wang Y, Tang H, Tong D, Ji F. MicroRNA-143, down-regulated in osteosarcoma, promotes apoptosis and suppresses tumorigenicity by targeting Bcl-2. Oncol Rep 2010;24(5):1363–9.
[24] Ziyan W, Shuhua Y, Xiufang W, Xiaoyun L. MicroRNA-21 is involved in osteosarcoma cell invasion and migration. Med Oncol 2011;28(4):1469–74.

[25] Duan Z, Choy E, Harmon D, Liu X, Susa M, Mankin H, et al. MicroRNA-199a-3p is downregulated in human osteosarcoma and regulates cell proliferation and migration. Mol Cancer Ther 2011;10(8):1337–45.
[26] Liu L, Li H, Li J, Zhong H, Zhang H, Chen J, et al. miR-125b suppresses the proliferation and migration of osteosarcoma cells through down-regulation of STAT3. Biochem Biophys Res Commun 2011;416(1–2).
[27] Lulla RR, Costa FF, Bischof JM, Chou PM, de FBM, Vanin EF, et al. Identification of differentially expressed microRNAs in osteosarcoma. Sarcoma 2011; 2011:732690.
[28] Maire G, Martin JW, Yoshimoto M, Chilton-MacNeill S, Zielenska M, Squire JA. Analysis of miRNA-gene expression-genomic profiles reveals complex mechanisms of microRNA deregulation in osteosarcoma. Cancer Genet 2011;204(3):138–46.
[29] Osaki M, Takeshita F, Sugimoto Y, Kosaka N, Yamamoto Y, Yoshioka Y, et al. MicroRNA-143 regulates human osteosarcoma metastasis by regulating matrix metalloprotease-13 expression. Mol Ther 2011;19(6):1123–30.
[30] Baumhoer D, Zillmer S, Unger K, Rosemann M, Atkinson MJ, Irmler M, et al. MicroRNA profiling with correlation to gene expression revealed the oncogenic miR-17-92 cluster to be up-regulated in osteosarcoma. Cancer Genet May 2012;205(5):212–9.
[31] Cai C-K, Zhao G-Y, Tian L-Y, Liu L, Yan K, Ma Y-L, et al. miR-15a and miR-16-1 downregulate CCND1 and induce apoptosis and cell cycle arrest in osteosarcoma. Oncol Rep 2012;28(5):1764–70.
[32] Dai N, Zhong Z, Cun Y, Qing Y, Chen C, Jiang P, et al. Alteration of the microRNA expression profile in human osteosarcoma cells transfected with APE1 siRNA. Neoplasma 2012;60(4):384–94.
[33] Huang G, Nishimoto K, Zhou Z, Hughes D, Kleinerman ES. miR-20a encoded by the miR-17–92 cluster increases the metastatic potential of osteosarcoma cells by regulating Fas expression. Cancer Res 2012;72(4): 908–16.
[34] Jones KB, Salah Z, Del Mare S, Galasso M, Gaudio E, Nuovo GJ, et al. miRNA signatures associate with pathogenesis and progression of osteosarcoma. Cancer Res 2012;72(7):1865–77.
[35] Kobayashi E, Hornicek FJ, Duan Z. MicroRNA involvement in osteosarcoma. Sarcoma 2012;2012: 359739.
[36] Mao JH, Zhou RP, Peng AF, Liu ZL, Huang SH, Long XH, et al. microRNA-195 suppresses osteosarcoma cell invasion and migration *in vitro* by targeting FASN. Oncol Lett November 2012;4(5):1125–9.
[37] Montanini L, Lasagna L, Barili V, Jonstrup SP, Murgia A, Pazzaglia L, et al. MicroRNA cloning and sequencing in osteosarcoma cell lines: differential role of miR-93. Cell Oncol 2012;35(1):29–41.
[38] Namlos HM, Meza-Zepeda LA, Baroy T, Ostensen IH, Kresse SH, Kuijjer ML, et al. Modulation of the osteosarcoma expression phenotype by microRNAs. PLoS One 2012;7(10):e48086.
[39] Thayanithy V, Sarver AL, Kartha RV, Li L, Angstadt AY, Breen M, et al. Perturbation of 14q32 miRNAs-cMYC gene network in osteosarcoma. Bone 2012;50(1):171–81.
[40] Yan K, Gao J, Yang T, Ma Q, Qiu X, Fan Q, et al. MicroRNA-34a inhibits the proliferation and metastasis of osteosarcoma cells both *in vitro* and *in vivo*. PLoS One 2012;7(3):e33778.
[41] Yuan J, Chen L, Chen X, Sun W, Zhou X. Identification of serum microRNA-21 as a biomarker for chemosensitivity and prognosis in human osteosarcoma. J Int Med Res 2012;40(6):2090–7.
[42] Zhao H, Guo M, Zhao G, Ma Q, Ma B, Qiu X, et al. miR-183 inhibits the metastasis of osteosarcoma via downregulation of the expression of Ezrin in F5M2 cells. Int J Mol Med 2012;30(5):1013.
[43] Zhu J, Feng Y, Ke Z, Yang Z, Zhou J, Huang X, et al. Down-regulation of miR-183 promotes migration and invasion of osteosarcoma by targeting Ezrin. Am J Pathol 2012;180(6):2440–51.
[44] Bao Y-P, Yi Y, Peng L-L, Fang J, Liu K-B, Li W-Z, et al. Roles of microRNA-206 in osteosarcoma pathogenesis and progression. Asian Pac J Cancer Prev 2013;14(6):3751–5.
[45] Cai H, Lin L, Cai H, Tang M, Wang Z. Prognostic evaluation of microRNA-210 expression in pediatric osteosarcoma. Med Oncol June 2013;30(2):499.
[46] Chen L, Wang Q, Wang GD, Wang HS, Huang Y, Liu XM, et al. miR-16 inhibits cell proliferation by targeting IGF1R and the Raf1-MEK1/2-ERK1/2 pathway in osteosarcoma. FEBS Lett May 2, 2013;587(9):1366–72.
[47] Di Fiore R, Fanale D, Drago-Ferrante R, Chiaradonna F, Giuliano M, De Blasio A, et al. Genetic and molecular characterization of the human osteosarcoma 3AB-OS cancer stem cell line: a possible model for studying osteosarcoma origin and stemness. J Cell Physiol June 2013;228(6):1189–201.
[48] Fujiwara T, Katsuda T, Hagiwara K, Kosaka N, Yoshioka Y, Takahashi R, et al. Clinical relevance and therapeutic significance of microRNA–133a expression profiles and functions in malignant osteosarcoma–initiating cells. Stem Cells 2013;32(4).
[49] Ji F, Zhang H, Wang Y, Li M, Xu W, Kang Y, et al. MicroRNA-133a, downregulated in osteosarcoma, suppresses proliferation and promotes apoptosis by targeting Bcl-xL and Mcl-1. Bone September 2013;56(1):220–6.
[50] Jianwei Z, Fan L, Xiancheng L, Enzhong B, Shuai L, Can L. MicroRNA 181a improves proliferation and invasion, suppresses apoptosis of osteosarcoma cell. Tumor Biol 2013:1–7.

[51] Jin J, Cai L, Liu Z-M, Zhou X-S. miRNA-218 inhibits osteosarcoma cell migration and invasion by down-regulating of TIAM1, MMP2 and MMP9. Asian Pac J Cancer Prev 2013;14(6):3681–4.

[52] Jin Y, Peng D, Shen Y, Xu M, Liang Y, Xiao B, et al. MicroRNA-376c inhibits cell proliferation and invasion in osteosarcoma by targeting to transforming growth factor-alpha. DNA Cell Biol 2013;32(6):302–9.

[53] Kelly AD, Haibe-Kains B, Janeway KA, Hill KE, Howe E, Goldsmith J, et al. MicroRNA paraffin-based studies in osteosarcoma reveal reproducible independent prognostic profiles at 14q32. Genome Med January 22, 2013;5(1):2.

[54] Kuijjer ML, Hogendoorn PC, Cleton-Jansen AM. Genome-wide analyses on high-grade osteosarcoma: making sense of a genomically most unstable tumor. Int J Cancer December 1, 2013;133(11):2512–21.

[55] Lauvrak SU, Munthe E, Kresse SH, Stratford EW, Namlos HM, Meza-Zepeda LA, et al. Functional characterisation of osteosarcoma cell lines and identification of mRNAs and miRNAs associated with aggressive cancer phenotypes. Br J Cancer October 15, 2013;109(8):2228–36.

[56] Long XH, Mao JH, Peng AF, Zhou Y, Huang SH, Liu ZL. Tumor suppressive microRNA-424 inhibits osteosarcoma cell migration and invasion via targeting fatty acid synthase. Exp Ther Med April 2013;5(4): 1048–52.

[57] Miao J, Wu S, Peng Z, Tania M, Zhang C. MicroRNAs in osteosarcoma: diagnostic and therapeutic aspects. Tumor Biol 2013;34(4):2093–8.

[58] Novello C, Pazzaglia L, Cingolani C, Conti A, Quattrini I, Manara MC, et al. miRNA expression profile in human osteosarcoma: role of miR-1 and miR-133b in proliferation and cell cycle control. Int J Oncol February 2013;42(2):667–75.

[59] Ouyang L, Liu P, Yang S, Ye S, Xu W, Liu X. A three-plasma miRNA signature serves as novel biomarkers for osteosarcoma. Med Oncol March 2013;30(1):340.

[60] Poos K, Smida J, Nathrath M, Maugg D, Baumhoer D, Korsching E. How microRNA and transcription factor co-regulatory networks affect osteosarcoma cell proliferation. PLoS Comput Biol 2013;9(8):e1003210.

[61] Sarver AL, Thayanithy V, Scott MC, Cleton-Jansen A-M, Hogendoorn P, Modiano JF, et al. MicroRNAs at the human 14q32 locus have prognostic significance in osteosarcoma. Orphanet J Rare Dis 2013;8(7).

[62] Shimada M. MicroRNA–Mediated regulation of apoptosis in osteosarcoma. J Carcinog Mutagen 2013. http://dx.doi.org/10.4172/2157-2518.S6-001.

[63] Song L, Yang J, Duan P, Xu J, Luo X, Luo F, et al. MicroRNA-24 inhibits osteosarcoma cell proliferation both *in vitro* and *in vivo* by targeting LPAATβ. Arch Biochem Biophys 2013;535(2).

[64] Tang M, Lin L, Cai H, Tang J, Zhou Z. MicroRNA-145 downregulation associates with advanced tumor progression and poor prognosis in patients suffering osteosarcoma. Onco Targets Ther 2013;6:833–8.

[65] Tian Q, Jia J, Ling S, Liu Y, Yang S, Shao Z. A causal role for circulating miR-34b in osteosarcoma. Eur J Surg Oncol September 11, 2013;40(1).

[66] van der Deen M, Taipaleenmaki H, Zhang Y, Teplyuk NM, Gupta A, Cinghu S, et al. MicroRNA-34c inversely couples the biological functions of the runt-related transcription factor RUNX2 and the tumor suppressor p53 in osteosarcoma. J Biol Chem July 19, 2013;288(29):21307–19.

[67] Wang Y, Zhao W, Fu Q. miR-335 suppresses migration and invasion by targeting ROCK1 in osteosarcoma cells. Mol Cell Biochem December 2013;384(1–2):105–11.

[68] Wang Z, Cai H, Lin L, Tang M, Cai H. Upregulated expression of microRNA-214 is linked to tumor progression and adverse prognosis in pediatric osteosarcoma. Pediatr Blood Cancer September 9, 2013;61(2).

[69] Won KY, Kim YW, Kim HS, Lee SK, Jung WW, Park YK. MicroRNA-199b-5p is involved in the Notch signaling pathway in osteosarcoma. Hum Pathol August 2013;44(8):1648–55.

[70] Wu X, Zhong D, Gao Q, Zhai W, Ding Z, Wu J. MicroRNA-34a inhibits human osteosarcoma proliferation by downregulating ether a go-go 1 expression. Int J Med Sci 2013;10(6):676–82.

[71] Xu H, Mei Q, Xiong C, Zhao J. Tumor-suppressing effects of miR-141 in human osteosarcoma. Cell Biochem Biophys 2013:1–7.

[72] Xu J, Yao Q, Hou Y, Xu M, Liu S, Yang L, et al. MiR-223/Ect2/p21 signaling regulates osteosarcoma cell cycle progression and proliferation. Biomed Pharmacother June 2013;67(5):381–6.

[73] Xu J-Q, Liu P, Si M-J, Ding X-Y. MicroRNA-126 inhibits osteosarcoma cells proliferation by targeting Sirt1. Tumor Biol 2013:3871–7.

[74] Yang J, Gao T, Tang J, Cai H, Lin L, Fu S. Loss of microRNA-132 predicts poor prognosis in patients with primary osteosarcoma. Mol Cell Biochem September 2013;381(1–2):9–15.

[75] Zhang Y, Hu H, Song L, Cai L, Wei R, Jin W. Epirubicin-mediated expression of miR-302b is involved in osteosarcoma apoptosis and cell cycle regulation. Toxicol Lett 2013;222(1):1–9.

[76] Zhao G, Cai C, Yang T, Qiu X, Liao B, Li W, et al. MicroRNA-221 induces cell survival and cisplatin resistance through PI3K/Akt pathway in human osteosarcoma. PLoS One 2013;8(1):e53906.

[77] Zhao H, Ma B, Wang Y, Han T, Zheng L, Sun C, et al. miR-34a inhibits the metastasis of osteosarcoma cells by repressing the expression of CD44. Oncol Rep 2013;29(3):1027–36.

[78] Zhou G, Shi X, Zhang J, Wu S, Zhao J. MicroRNAs in osteosarcoma: from biological players to clinical contributors, a review. J Int Med Res February 2013;41(1):1–12.
[79] Zhou X, Wei M, Wang W. MicroRNA-340 suppresses osteosarcoma tumor growth and metastasis by directly targeting ROCK1. Biochem Biophys Res Commun August 9, 2013;437(4):653–8.
[80] Almog N, Briggs C, Beheshti A, Ma L, Wilkie KP, Rietman E, et al. Transcriptional changes induced by the tumor dormancy-associated microRNA-190. Transcription 2014;4(4):177–91.
[81] Cai H, Lin L, Cai H, Tang M, Wang Z. Combined microRNA-340 and ROCK1 mRNA profiling predicts tumor progression and prognosis in pediatric osteosarcoma. Int J Mol Sci 2014;15(1):560–73.
[82] Cheng C, Chen Z-Q, Shi X-T. MicroRNA-320 inhibits osteosarcoma cells proliferation by directly targeting fatty acid synthase. Tumor Biol 2014;35(5):4177–83.
[83] Di Fiore R, Drago-Ferrante R, Pentimalli F, Di Marzo D, Forte IM, D'Anneo A, et al. MicroRNA-29b-1 impairs *in vitro* cell proliferation, self-renewal and chemoresistance of human osteosarcoma 3AB-OS cancer stem cells. Int J Oncol 2014;45(5):2013–23.
[84] Gao Y, Luo LH, Li S, Yang C. miR-17 inhibitor suppressed osteosarcoma tumor growth and metastasis via increasing PTEN expression. Biochem Biophys Res Commun February 7, 2014;444(2):230–4.
[85] Guo S, Bai R, Liu W, Zhao A, Zhao Z, Wang Y, et al. miR-22 inhibits osteosarcoma cell proliferation and migration by targeting HMGB1 and inhibiting HMGB1-mediated autophagy. Tumour Biol 2014;35(7):7025–34.
[86] Han K, Zhao T, Chen X, Bian N, Yang T, Ma Q, et al. microRNA-194 suppresses osteosarcoma cell proliferation and metastasis *in vitro* and *in vivo* by targeting CDH2 and IGF1R. Int J Oncol 2014;45(4):1437–49.
[87] He Y, Meng C, Shao Z, Wang H, Yang S. MiR-23a functions as a tumor suppressor in osteosarcoma. Cell Physiol Biochem October 8, 2014;34(5):1485–96.
[88] Hong Q, Fang J, Pang Y, Zheng J. Prognostic value of the microRNA-29 family in patients with primary osteosarcomas. Med Oncol August 2014;31(8):37.
[89] Huang J, Gao K, Lin J, Wang Q. MicroRNA-100 inhibits osteosarcoma cell proliferation by targeting Cyr61. Tumor Biol 2014;35(2):1095–100.
[90] Lei P, Xie J, Wang L, Yang X, Dai Z, Hu Y. microRNA-145 inhibits osteosarcoma cell proliferation and invasion by targeting ROCK1. Mol Med Rep 2014;10(1):155–60.
[91] Li X, Yang H, Tian Q, Liu Y, Weng Y. Upregulation of microRNA-17-92 cluster associates with tumor progression and prognosis in osteosarcoma. Neoplasma 2014;61(4).
[92] Liu Z, Zhang G, Li J, Liu J, Lv P. The tumor-suppressive MicroRNA-135b targets c-Myc in osteoscarcoma. PLoS One 2014;9(7):e102621.
[93] Lv H, Pei J, Liu H, Wang H, Liu J. A polymorphism site in the pre-miR-34a coding region reduces miR-34a expression and promotes osteosarcoma cell proliferation and migration. Mol Med Rep 2014;10(6):2912–6.
[94] Mu Y, Zhang H, Che L, Li K. Clinical significance of microRNA-183/Ezrin axis in judging the prognosis of patients with osteosarcoma. Med Oncol February 2014;31(2):821.
[95] Shen L, Chen X-D, Zhang Y-H. MicroRNA-128 promotes proliferation in osteosarcoma cells by downregulating PTEN. Tumor Biol 2014;35(3):2069–74.
[96] Shen L, Wang P, Yang J, Li X. MicroRNA-217 regulates WASF3 expression and suppresses tumor growth and metastasis in osteosarcoma. PLoS One 2014;9(10):e109138.
[97] Song Q-C, Shi Z-B, Zhang Y-T, Ji L, Wang KZ, Duan D-P, et al. Downregulation of microRNA-26a is associated with metastatic potential and the poor prognosis of osteosarcoma patients. Oncol Rep 2014;31(3):1263–70.
[98] Sun Z, Zhang T, Hong H, Liu Q, Zhang H. miR-202 suppresses proliferation and induces apoptosis of osteosarcoma cells by downregulating Gli2. Mol Cell Biochem December 2014;397(1–2):277–83.
[99] Tian Y, Zhang YZ, Chen W. MicroRNA-199a-3p and microRNA-34a regulate apoptosis in human osteosarcoma cells. Biosci Rep 2014;34(4).
[100] Wang J, Xu G, Shen F, Kang Y. miR-132 targeting cyclin E1 suppresses cell proliferation in osteosarcoma cells. Tumor Biol 2014;35(5):4859–65.
[101] Wang XH, Cai P, Wang MH, Wang Z. microRNA-25 promotes osteosarcoma cell proliferation by targeting the cell-cycle inhibitor p27. Mol Med Rep 2014;10(2).
[102] Xu J-Q, Zhang W-B, Wan R, Yang Y-Q. MicroRNA-32 inhibits osteosarcoma cell proliferation and invasion by targeting Sox9. Tumor Biol 2014;35(10):9847–53.
[103] Xu M, Jin H, Xu C-X, Sun B, Song Z-G, Bi W-Z, et al. miR-382 inhibits osteosarcoma metastasis and relapse by targeting Y Box-binding protein 1. Mol Ther 2014;23(1).
[104] Xu S-H, Yang Y-L, Han S-M, Wu Z-H. MicroRNA-9 expression is a prognostic biomarker in patients with osteosarcoma. World J Surg Oncol 2014;12(1):195.
[105] Xu W-G, Shang Y-L, Cong X-R, Bian X, Yuan Z. MicroRNA-135b promotes proliferation, invasion and migration of osteosarcoma cells by degrading myocardin. Int J Oncol 2014;45(5):2024–32.
[106] Xu Z, Wang T. miR-214 promotes the proliferation and invasion of osteosarcoma cells through direct suppression of LZTS1. Biochem Biophys Res Commun June 27, 2014;449(2):190–5.
[107] Yang C, Hou C, Zhang H, Wang D, Ma Y, Zhang Y, et al. miR-126 functions as a tumor suppressor in osteosarcoma by targeting Sox2. Int J Mol Sci 2014;15(1):423–37.
[108] Yang Y-Q, Qi J, Xu J-Q, Hao P. MicroRNA-142-3p, a novel target of tumor suppressor menin, inhibits osteosarcoma cell proliferation by down-regulation of FASN. Tumor Biol 2014;35(10):10287–93.

[109] Zhang C, Yao C, Li H, Wang G, He X. Serum levels of microRNA-133b and microRNA-206 expression predict prognosis in patients with osteosarcoma. Int J Clin Exp Pathol 2014;7(7):4194.
[110] Zhang H, Yin Z, Ning K, Wang L, Guo R, Ji Z. Prognostic value of microRNA-223/epithelial cell transforming sequence 2 signaling in patients with osteosarcoma. Hum Pathol July 2014;45(7):1430–6.
[111] Zhang K, Zhang C, Liu L, Zhou J. A key role of microRNA-29b in suppression of osteosarcoma cell proliferation and migration via modulation of VEGF. Int J Clin Exp Pathol 2014;7(9):5701.
[112] Zhang K, Zhang Y, Ren K, Zhao G, Yan K, Ma B. MicroRNA-101 inhibits the metastasis of osteosarcoma cells by downregulation of EZH2 expression. Oncol Rep 2014;32(5):2143–9.
[113] Zhou Y, Huang Z, Wu S, Zang X, Liu M, Shi J. miR-33a is up-regulated in chemoresistant osteosarcoma and promotes osteosarcoma cell resistance to cisplatin by down-regulating TWIST. J Exp Clin Cancer Res CR 2014;33(1):12.
[114] Namløs HM, Meza-Zepeda LA, Barøy T, Østensen IH, Kresse SH, Kuijjer ML, et al. Modulation of the osteosarcoma expression phenotype by microRNAs. PLoS One 2012;7(10):e48086.
[115] Baumhoer D, Zillmer S, Unger K, Rosemann M, Atkinson MJ, Irmler M, et al. MicroRNA profiling with correlation to gene expression revealed the oncogenic miR-17-92 cluster to be up-regulated in osteosarcoma. Cancer Genet 2012;205(5):212–9.
[116] Gao Y, Luo L-H, Li S, Yang C. miR-17 inhibitor suppressed osteosarcoma tumor growth and metastasis via increasing PTEN expression. Biochem Biophys Res Commun 2014;444(2):230–4.
[117] Marina N, Gebhardt M, Teot L, Gorlick R. Biology and therapeutic advances for pediatric osteosarcoma. Oncol 2004;9(4):422–41.
[118] Ta HT, Dass CR, Choong PF, Dunstan DE. Osteosarcoma treatment: state of the art. Cancer Metastasis Rev 2009;28(1–2):247–63.
[119] Hermeking H. The miR-34 family in cancer and apoptosis. Cell Death Differ 2009;17(2):193–9.
[120] Tang N, Song WX, Luo J, Haydon RC, He TC. Osteosarcoma development and stem cell differentiation. Clin Orthop Relat Res September 2008;466(9):2114–30.
[121] Sandberg AA, Bridge JA. Updates on the cytogenetics and molecular genetics of bone and soft tissue tumors: osteosarcoma and related tumors. Cancer Genet Cytogenet 2003;145(1):1–30.
[122] Riggi N, Stamenkovic I. The biology of Ewing sarcoma. Cancer Lett 2007;254(1):1–10.
[123] Delattre O, Zucman J, Plougastel B, Desmaze C, Melot T, Peter M, et al. Gene fusion with an ETS DNA-binding domain caused by chromosome translocation in human tumours. Nature 1992;359(6391):162–5.
[124] McKinsey EL, Parrish JK, Irwin AE, Niemeyer BF, Kern HB, Birks DK, et al. A novel oncogenic mechanism in Ewing sarcoma involving IGF pathway targeting by EWS/Fli1-regulated microRNAs. Oncogene December 8, 2011;30(49):4910–20.
[125] Ban J, Jug G, Mestdagh P, Schwentner R, Kauer M, Aryee DN, et al. Hsa-mir-145 is the top EWS-FLI1-repressed microRNA involved in a positive feedback loop in Ewing's sarcoma. Oncogene May 5, 2011;30(18):2173–80.
[126] De Vito C, Riggi N, Cornaz S, Suva ML, Baumer K, Provero P, et al. A TARBP2-dependent miRNA expression profile underlies cancer stem cell properties and provides candidate therapeutic reagents in Ewing sarcoma. Cancer Cell June 12, 2012;21(6):807–21.
[127] De Vito C, Riggi N, Suva ML, Janiszewska M, Horlbeck J, Baumer K, et al. Let-7a is a direct EWS-FLI-1 target implicated in Ewing's sarcoma development. PLoS One 2011;6(8):e23592.
[128] Franzetti GA, Laud-Duval K, Bellanger D, Stern MH, Sastre-Garau X, Delattre O. MiR-30a-5p connects EWS-FLI1 and CD99, two major therapeutic targets in Ewing tumor. Oncogene August 15, 2013;32(33): 3915–21.
[129] Parrish JK, Sechler M, Winn RA, Jedlicka P. The histone demethylase KDM3A is a microRNA-22-regulated tumor promoter in Ewing Sarcoma. Oncogene December 23, 2013;34(2).
[130] Karnuth B, Dedy N, Spieker T, Lawlor ER, Gattenlohner S, Ranft A, et al. Differentially expressed miRNAs in Ewing sarcoma compared to mesenchymal stem cells: low miR-31 expression with effects on proliferation and invasion. PLoS One 2014;9(3): e93067.
[131] Mosakhani N, Guled M, Leen G, Calabuig-Fariñas S, Niini T, Machado I, et al. An integrated analysis of miRNA and gene copy numbers in xenografts of Ewing's sarcoma. J Exp Clin Cancer Res 2012;31(1):24.
[132] Robin TP, Smith A, McKinsey E, Reaves L, Jedlicka P, Ford HL. EWS/FLI1 regulates EYA3 in Ewing sarcoma via modulation of miRNA-708, resulting in increased cell survival and chemoresistance. Mol Cancer Res 2012;10(8):1098–108.
[133] Iida K, J-i F, Matsumoto Y, Oda Y, Takahashi Y, Fujiwara T, et al. miR-125b develops chemoresistance in Ewing sarcoma/primitive neuroectodermal tumor. Cancer Cell Int 2013;13(1):21.
[134] Nakatani F, Ferracin M, Manara MC, Ventura S, Del Monaco V, Ferrari S, et al. miR-34a predicts survival of Ewing's sarcoma patients and directly influences cell chemo-sensitivity and malignancy. J Pathol April 2012;226(5):796–805.
[135] Yoshitaka T, Kawai A, Miyaki S, Numoto K, Kikuta K, Ozaki T, et al. Analysis of microRNAs expressions in chondrosarcoma. J Orthop Res December 2013;31(12):1992–8.

[136] Mak IW, Singh S, Turcotte R, Ghert M. The epigenetic regulation of SOX9 by miR-145 in human chondrosarcoma. J Cell Biochem January 2015;116(1):37–44.

[137] Zhu Z, Wang C-P, Zhang Y-F, Nie L. MicroRNA-100 resensitizes resistant chondrosarcoma cells to cisplatin through direct targeting of mTOR. Asian Pac J Cancer Prev 2014;15(2):917–23.

[138] Subramanian S, Lui WO, Lee CH, Espinosa I, Nielsen TO, Heinrich MC, et al. MicroRNA expression signature of human sarcomas. Oncogene March 27, 2008;27(14):2015–26.

[139] Farshid G, Pradhan M, Goldblum J, Weiss SW. Leiomyosarcoma of somatic soft tissues: a tumor of vascular origin with multivariate analysis of outcome in 42 cases. Am J Surg Pathol 2002;26(1):14–24.

[140] Nuovo GJ, Schmittgen TD. Benign metastasizing leiomyoma of the lung: clinicopathologic, immunohistochemical, and micro-RNA analyses. Diagn Mol Pathol 2008;17(3):145.

[141] Shi G, Perle MA, Mittal K, Chen H, Zou X, Narita M, et al. Let-7 repression leads to HMGA2 overexpression in uterine leiomyosarcoma. J Cell Mol Med 2009;13(9b):3898–905.

[142] Danielson LS, Menendez S, Attolini CS-O, Guijarro MV, Bisogna M, Wei J, et al. A differentiation-based microRNA signature identifies leiomyosarcoma as a mesenchymal stem cell-related malignancy. Am J Pathol 2010;177(2):908–17.

[143] Renner M, Czwan E, Hartmann W, Penzel R, Brors B, Eils R, et al. MicroRNA profiling of primary high-grade soft tissue sarcomas. Genes Chromosomes Cancer 2012;51(11):982–96.

[144] Guled M, Pazzaglia L, Borze I, Mosakhani N, Novello C, Benassi MS, et al. Differentiating soft tissue leiomyosarcoma and undifferentiated pleomorphic sarcoma: a miRNA analysis. Genes Chromosomes Cancer 2014;53(8).

[145] Ugras S, Brill E, Jacobsen A, Hafner M, Socci ND, Decarolis PL, et al. Small RNA sequencing and functional characterization reveals microRNA-143 tumor suppressor activity in liposarcoma. Cancer Res September 1, 2011;71(17):5659–69.

[146] Zhang P, Bill K, Liu J, Young E, Peng T, Bolshakov S, et al. MiR-155 is a liposarcoma oncogene that targets casein kinase-1α and enhances β-catenin signaling. Cancer Res 2012;72(7):1751–62.

[147] Taylor BS, DeCarolis PL, Angeles CV, Brenet F, Schultz N, Antonescu CR, et al. Frequent alterations and epigenetic silencing of differentiation pathway genes in structurally rearranged liposarcomas. Cancer Discov 2011;1(7):587–97.

[148] Hisaoka M, Matsuyama A, Nakamoto M. Aberrant calreticulin expression is involved in the dedifferentiation of dedifferentiated liposarcoma. Am J Pathol 2012;180(5).

[149] Borjigin N, Ohno S, Wu W, Tanaka M, Suzuki R, Fujita K, et al. TLS-CHOP represses miR-486 expression, inducing upregulation of a metastasis regulator PAI-1 in human myxoid liposarcoma. Biochem Biophys Res Commun 2012;427(2).

[150] Lee DH, Amanat S, Goff C, Weiss LM, Said JW, Doan NB, et al. Overexpression of miR-26a-2 in human liposarcoma is correlated with poor patient survival. Oncogenesis 2013;2:e47.

[151] Cieśla M, Dulak J, Józkowicz A. MicroRNAs and epigenetic mechanisms of *rhabdomyosarcoma* development. Int J Biochem Cell Biol 2014. http://dx.doi.org/10.1016/j.biocel.2014.05.003.

[152] Novák J, Vinklárek J, Bienertová–Vašků J, Slabý O. MicroRNAs involved in skeletal muscle development and their roles in rhabdomyosarcoma pathogenesis. Pediatr Blood Cancer 2013;60(11):1739–46.

[153] Rota R, Ciarapica R, Giordano A, Miele L, Locatelli F. MicroRNAs in rhabdomyosarcoma: pathogenetic implications and translational potentiality. Mol Cancer 2011;10:120.

[154] Rao PK, Missiaglia E, Shields L, Hyde G, Yuan B, Shepherd CJ, et al. Distinct roles for miR-1 and miR-133a in the proliferation and differentiation of rhabdomyosarcoma cells. FASEB J 2010;24(9):3427–37.

[155] Taulli R, Bersani F, Foglizzo V, Linari A, Vigna E, Ladanyi M, et al. The muscle-specific microRNA miR-206 blocks human rhabdomyosarcoma growth in xenotransplanted mice by promoting myogenic differentiation. J Clin Invest August 2009;119(8):2366–78.

[156] Yan D, Dong Xda E, Chen X, Wang L, Lu C, Wang J, et al. MicroRNA-1/206 targets c-Met and inhibits rhabdomyosarcoma development. J Biol Chem October 23, 2009;284(43):29596–604.

[157] Li L, Sarver AL, Alamgir S, Subramanian S. Downregulation of microRNAs miR-1, -206 and -29 stabilizes PAX3 and CCND2 expression in rhabdomyosarcoma. Lab Invest April 2012;92(4):571–83.

[158] Ciarapica R, De Salvo M, Carcarino E, Bracaglia G, Adesso L, Leoncini P, et al. The polycomb group (PcG) protein EZH2 supports the survival of PAX3-FOXO1 alveolar rhabdomyosarcoma by repressing FBXO32 (Atrogin1/MAFbx). Oncogene 2013;33(32).

[159] Taulli R, Foglizzo V, Morena D, Coda D, Ala U, Bersani F, et al. Failure to downregulate the BAF53a subunit of the SWI/SNF chromatin remodeling complex contributes to the differentiation block in rhabdomyosarcoma. Oncogene 2014;33(18):2354–62.

[160] Missiaglia E, Shepherd C, Patel S, Thway K, Pierron G, Pritchard-Jones K, et al. MicroRNA-206 expression levels correlate with clinical behaviour of rhabdomyosarcomas. Br J Cancer 2010;102(12):1769–77.

[161] Wang H, Garzon R, Sun H, Ladner KJ, Singh R, Dahlman J, et al. NF-κB–YY1–miR-29 regulatory circuitry in skeletal myogenesis and rhabdomyosarcoma. Cancer Cell 2008;14(5):369–81.

[162] Winbanks CE, Wang B, Beyer C, Koh P, White L, Kantharidis P, et al. TGF-β regulates miR-206 and miR-29 to control myogenic differentiation through regulation of HDAC4. J Biol Chem 2011;286(16):13805–14.

[163] Zhou L, Wang L, Lu L, Jiang P, Sun H, Wang H. A novel target of microRNA-29, Ring1 and YY1-binding protein (Rybp), negatively regulates skeletal myogenesis. J Biol Chem 2012;287(30):25255–65.

[164] Sarver AL, Li L, Subramanian S. MicroRNA miR-183 functions as an oncogene by targeting the transcription factor EGR1 and promoting tumor cell migration. Cancer Res 2010;70(23):9570–80.

[165] Armeanu-Ebinger S, Herrmann D, Bonin M, Leuschner I, Warmann SW, Fuchs J, et al. Differential expression of miRNAs in rhabdomyosarcoma and malignant rhabdoid tumor. Exp Cell Res 2012; 318(20):2567–77.

[166] Reichek JL, Duan F, Smith LM, Gustafson DM, O'Connor RS, Zhang C, et al. Genomic and clinical analysis of amplification of the 13q31 chromosomal region in alveolar rhabdomyosarcoma: a report from the Children's oncology group. Clin Cancer Res 2011;17(6):1463–73.

[167] Kawai A, Healey JH, Boland PJ, Lin PP, Huvos AG, Meyers PA. Prognostic factors for patients with sarcomas of the pelvic bones. Cancer 1998;82(5):851–9.

[168] Hisaoka M, Matsuyama A, Nagao Y, Luan L, Kuroda T, Akiyama H, et al. Identification of altered microRNA expression patterns in synovial sarcoma. Genes, Chromosomes Cancer 2011;50(3):137–45.

[169] Friedrich R, Kluwe L, Fünsterer C, Mautner V. Malignant peripheral nerve sheath tumors (MPNST) in neurofibromatosis type 1 (NF1): diagnostic findings on magnetic resonance images and mutation analysis of the NF1 gene. Anticancer Res 2005;25(3A):1699–702.

[170] Subramanian S, Thayanithy V, West RB, Lee CH, Beck AH, Zhu S, et al. Genome–wide transcriptome analyses reveal p53 inactivation mediated loss of miR–34a expression in malignant peripheral nerve sheath tumours. J Pathol 2010;220(1):58–70.

[171] Itani S, Kunisada T, Morimoto Y, Yoshida A, Sasaki T, Ito S, et al. MicroRNA-21 correlates with tumorigenesis in malignant peripheral nerve sheath tumor (MPNST) via programmed cell death protein 4 (PDCD4). J Cancer Res Clin Oncol 2012:1501–9.

[172] Presneau N, Eskandarpour M, Shemais T, Henderson S, Halai D, Tirabosco R, et al. MicroRNA profiling of peripheral nerve sheath tumours identifies miR-29c as a tumour suppressor gene involved in tumour progression. Br J Cancer 2013;108(4):964–72.

[173] Gong M, Ma J, Li M, Zhou M, Hock JM, Yu X. MicroRNA-204 critically regulates carcinogenesis in malignant peripheral nerve sheath tumors. Neuro Oncol 2012;14(8):1007–17.

[174] Chai G, Liu N, Ma J, Li H, Oblinger JL, Prahalad AK, et al. MicroRNA–10b regulates tumorigenesis in neurofibromatosis type 1. Cancer Sci 2010;101(9):1997–2004.

[175] Zhang P, Garnett J, Creighton CJ, Al Sannaa GA, Igram DR, Lazar A, et al. EZH2–miR–30d–KPNB1 pathway regulates malignant peripheral nerve sheath tumour cell survival and tumourigenesis. J Pathol 2014;232(3):308–18.

[176] Fletcher CD. The evolving classification of soft tissue tumours: an update based on the new WHO classification. Histopathology January 2006;48(1):3–12.

[177] Fletcher CD. The evolving classification of soft tissue tumours–an update based on the new 2013 WHO classification. Histopathology 2014;64(1):2–11.

[178] Chase DR, Enzinger FM. Epithelioid sarcoma: diagnosis, prognostic indicators, and treatment. Am J Surg Pathol 1985;9(4):241–63.

[179] Papp G, Krausz T, Stricker TP, Szendrői M, Sápi Z. SMARCB1 expression in epithelioid sarcoma is regulated by miR–206, miR–381, and miR–671–5p on both mRNA and protein levels. Genes Chromosomes Cancer 2014;53(2):168–76.

[180] Miyachi M, Tsuchiya K, Yoshida H, Yagyu S, Kikuchi K, Misawa A, et al. Circulating muscle-specific microRNA, miR-206, as a potential diagnostic marker for rhabdomyosarcoma. Biochem Biophys Res Commun 2010;400(1):89–93.

[181] Weng Y, Chen Y, Chen J, Liu Y, Bao T. Identification of serum microRNAs in genome-wide serum microRNA expression profiles as novel noninvasive biomarkers for malignant peripheral nerve sheath tumor diagnosis. Med Oncol June 2013;30(2):531.

[182] Gougelet A, Pissaloux D, Besse A, et al. Micro-RNA profiles in osteosarcoma as a predictive tool for ifosfamide response. Int J Cancer 2011;129(3):680–90.

[183] Shen L, Chen X-D, Zhang Y-H. MicroRNA-128 promotes proliferation in osteosarcoma cells by downregulating PTEN. Tumour Biol 2013:1–6.

[184] Fujiwara T, Katsuda T, Hagiwara K, et al. Clinical relevance and therapeutic significance of microRNA-133a expression profiles and functions in malignant osteosarcoma-initiating cells. Stem Cells 2014;32(4):959–73.

[185] Stabley D, Kamara D, Holbrook J, et al. Digital gene expression of miRNA in osteosarcoma xenografts: finding biological relevance in miRNA high throughput sequencing data. J Biomol Tech 2010;21(Suppl. 3):S25.

[186] Novello C, Pazzaglia L, Cingolani C, Conti A, Quattrini I, Manara MC, et al. miRNA expression profile in human osteosarcoma: role of miR-1 and miR-133b in proliferation and cell cycle control. Int J Oncol 2013;42(2):667–75.

[187] Di Fiore R, Drago-Ferrante R, Pentimalli F, et al. MicroRNA-29b-1 impairs in vitro cell proliferation, self-renewal and chemoresistance of human osteosarcoma 3AB-OS cancer stem cells. Int J Oncol 2014;45(5):2013–23.

[188] Cheng C, Chen ZQ, Shi XT. MicroRNA-320 inhibits osteosarcoma cells proliferation by directly targeting fatty acid synthase. Tumour Biol 2014;35(5):4177–83.

[189] Maire G, Martin JW, Yoshimoto M, et al. Analysis of miRNA-gene expression-genomic profiles reveals complex mechanisms of microRNA deregulation in osteosarcoma. Cancer Genet 2011;204(3):138–46.

[190] Dylla L, Jedlicka P. Growth-Promoting Role of the miR-106a~363 Cluster in Ewing Sarcoma. PLoS One 2013;8(4):e63032.

[191] Mak IW, Singh S, Turcotte R, et al. The epigenetic regulation of SOX9 by miR-145 in human chondrosarcoma. J Cell Biochem 2015;116(1):37–44.

[192] Ugras S, Brill E, Jacobsen A, et al. Small RNA sequencing and functional characterization reveals MicroRNA-143 tumor suppressor activity in liposarcoma. Cancer Res 2011;71(17):5659–69.

[193] Chen CF, He X, Arslan AD, et al. Novel regulation of nuclear factor-YB by miR-485-3p affects the expression of DNA topoisomerase IIα and drug responsiveness. Mol Pharmacol 2011;79(4):735–41.

[194] Rao PK, Missiaglia E, Shields L, et al. Distinct roles for miR-1 and miR-133a in the proliferation and differentiation of rhabdomyosarcoma cells. FASEB J 2010;24(9):3427–37.

[195] Ciarapica R, Russo G, Verginelli F, Raimondi L, Donfrancesco A, Rota R, et al. Deregulated expression of miR-26a and Ezh2 in rhabdomyosarcoma. Cell Cycle 2009;8(1):172–5.

[196] Diao Y, Guo X, Jiang L, et al. miR-203, a tumor suppressor frequently down-regulated by promoter hypermethylation in rhabdomyosarcoma. J Biol Chem 2014;289(1):529–39.

[197] Missiaglia E, Shepherd C, Patel S, et al. MicroRNA-206 expression levels correlate with clinical behaviour of rhabdomyosarcomas. Br J Cancer 2010;102(12):1769–77.

[198] Guled M, Pazzaglia L, Borze I, et al. Differentiating soft tissue leiomyosarcoma and undifferentiated pleomorphic sarcoma: A miRNA analysis. Genes Chromosomes Cancer August 2014;53(8):693–702.

[199] Shi G, Perle MA, Mittal K, et al. Let-7 repression leads to HMGA2 overexpression in uterine leiomyosarcoma. J Cell Mol Med 2009;13(9b):3898–905.

[200] Papp G, Krausz T, Stricker TP, et al. SMARCB1 expression in epithelioid sarcoma is regulated by miR-206, miR-381, and miR-671-5p on both mRNA and protein levels. Genes Chromosomes Cancer 2014;53(2):168–76.

CHAPTER

31

miRNAs in the Pathophysiology of Diabetes and Their Value as Biomarkers

Eoin Brennan[1], *Aaron McClelland*[1], *Shinji Hagiwara*[1], *Catherine Godson*[2], *Phillip Kantharidis*[1]

[1]Diabetic Complications Division, Baker IDI Heart and Diabetes Institute, Melbourne, VIC, Australia; [2]Conway Institute, Diabetes Complications Research Centre, University College Dublin, Dublin, Ireland

OUTLINE

Epigenetic Biomarkers and Diagnostics
http://dx.doi.org/10.1016/B978-0-12-801899-6.00031-0

1. INTRODUCTION

Diabetes mellitus is a disorder of dysregulated glucose homeostasis, characterized by either an inability to produce insulin (type 1 diabetes; MIM #222100) or an insufficient response to insulin coupled to an eventual decline in insulin production (type 2 diabetes; MIM #125853). Central to the pathophysiology of diabetes are the insulin-producing β-cells and glucagon-producing α-cells of the pancreatic islets of Langerhans. In response to elevations in blood glucose levels, insulin molecules released into the circulation bind to their cognate insulin receptors and promote the uptake of glucose into hepatic, muscle, and fat tissues. In contrast, low blood glucose-stimulated glucagon promotes hepatic glucose production, thereby raising blood glucose levels. Signaling among these organs/tissues is extremely important and modulation of this cross talk may lead to the development of diabetes. Diabetic microvascular and macrovascular complications may arise due to chronic elevation of blood glucose levels and hypertension, leading to damage of arteries (cardiovascular disease) and small blood vessels of the kidney (nephropathy), eye (retinopathy) and nervous system (neuropathy) and associated dyslipidemia.

Type 1 diabetes represents approximately 10% of all new cases of diabetes, affecting 20 million people worldwide. The majority of people are diagnosed with type 1 diabetes prior to 14 years of age with a recent rise seen in children diagnosed under 5 years [1,2]. The incidence of type 1 diabetes is increasing worldwide at an annual rate of ~3% and the lifetime risk of developing type 1 diabetes is much greater in siblings of type 1 diabetic subjects compared with the general population (6% vs ~0.4%) [3]. The concordance rate for type 1 diabetes in monozygotic twins (30–50%) is higher than that of dizygotic twins (~10%), indicating the contribution of heritable factors to the development of type 1 diabetes [4]. In type 1 diabetes, insulin secretion is insufficient due to a T cell-mediated autoimmune attack on the pancreatic β-cell population, with detectable disease onset occurring after 80–90% of the β-cell population is destroyed. A β-cell response is also initiated through the production of antibodies against β-cell autoantigens. Detection of serum autoantibodies against insulin, glutamic acid decarboxylase (GAD65), and insulinoma-associated antigen-2 (IA-2) are recognized as indicators of high risk of type 1 diabetes. In the case of type 1 diabetes, research efforts are focused on prevention or early intervention to halt the autoimmune process and preserve the β-cell function. Type 2 diabetes is characterized by impaired insulin secretion and/or cellular resistance to the metabolic effects of insulin. Approximately 90% of all cases of diabetes are diagnosed as type 2 diabetes, and these individuals differ from those with type 1 diabetes in that they are not entirely dependent on insulin replacement therapy. Monozygotic twin studies show high concordance for this disease in comparison with dizygotic twins, indicating a significant influence of heritable factors [5]. Familial aggregation of type 2 diabetes is observed, with a sibling risk ratio of 1.8 ($\lambda s = 1.8$) and an even greater risk if a parent and grandparent is affected ($\lambda s = 3.5$) [6].

Other forms of diabetes exist including diabetes arising from mutations in specific genes involved in regulating β-cell insulin secretion. Maturity-onset diabetes of the young (MODY) is a form of monogenic diabetes, where each different mutated gene causes a slightly different type of diabetes [7]. The most common forms are HNF1α-MODY (MODY3), and GCK-MODY (MODY2), due to mutations in the HNF1A and GCK genes. MODY is typically diagnosed in adolescence, or early adulthood. However, it has been known to develop in adults as late as their 50s. Many people with MODY are misdiagnosed as having type 1 or type 2 diabetes. Another subtype, gestational diabetes mellitus (GDM) occurs during pregnancy and affects approximately 4% of pregnancies each year [8]. Several risk factors for GDM confer a predisposition to type 2 diabetes in offspring, including maternal obesity prior to pregnancy, having a first degree relative with diabetes or previous history of GDM.

In 2000, it was estimated that approximately 170 million adults worldwide had diabetes. Further modeling predicts that the global prevalence of diabetes will increase to 366 million by 2030 [9], with greatest increases seen in countries undergoing rapid economic development where the number of people with diabetes is expected to increase from 84 to 228 million. Several factors indicative of modern society have been implicated in driving diabetes to epidemic levels, including longer life expectancy, and an increasing prevalence of obesity with an estimated 90% of type 2 diabetes cases attributable to excess weight [10]. Current antidiabetic drugs target the adipose and muscle tissue to reduce insulin resistance, act on liver to inhibit glucose production, or stimulate the pancreas to release insulin. However, while these drugs can help diminish the effects of diabetes, they cannot completely halt the progression of the disease. Therefore, the need for new diabetes drugs remains acute.

Diagnosis of diabetes typically involves several blood tests, including measurement of hemoglobin A1c (A1c test) which acts as an indicator of a person's blood glucose levels over the previous 3 months, a fasting plasma glucose test (FPG test) to measure blood glucose in a person who has fasted for at least 8 h, or an oral glucose tolerance test (OGTT) to determine blood glucose levels 2 h after a person ingests 75 g of glucose. By performing these tests, it is possible to determine whether a person has diabetes or is at the prediabetic stage where blood glucose levels are above normal but not high enough for a diagnosis of diabetes. In particular, there is a need to identify additional biomarkers of prediabetes, as several studies have shown that people at the prediabetic stage are able to reduce their risk of developing diabetes through weight management and dietary changes.

MicroRNAs (miRNAs) are short noncoding RNAs that directly bind to the 3′-untranslated region (3′-UTR) of complementary mRNA targets, leading to an inhibition of the target gene expression. More than 2000 miRNA genes have been reported in the human genome. Many miRNAs are evolutionarily conserved and ubiquitously expressed, but some of them display tissue-specific expression. A single miRNA can target multiple mRNAs, and a single mRNA may contain multiple miRNA-binding sites at its 3′-UTR, allowing a coordinated regulation by various miRNAs under different physiological conditions. From a therapeutics perspective, targeting disease-relevant miRNAs with either oligonucleotide inhibitors (anti-miRs) or miRNA mimics represents a novel strategy to modulate biological pathways. Recent reports indicate that miRNAs play important roles in controlling insulin biosynthesis and release, pancreatic β-cell development and survival, and glucose and lipid metabolism (Figure 1). Beyond diabetes, numerous miRNAs have been implicated in the development and progression of microvascular and macrovascular complications of diabetes, with several miRNAs identified as predictive biomarkers of these complications. Here, we outline important miRNAs that have been observed in β-cells, adipose, muscle, and liver of various diabetic models and diabetic individuals, and discuss their utility as biomarkers and targets for therapeutic intervention.

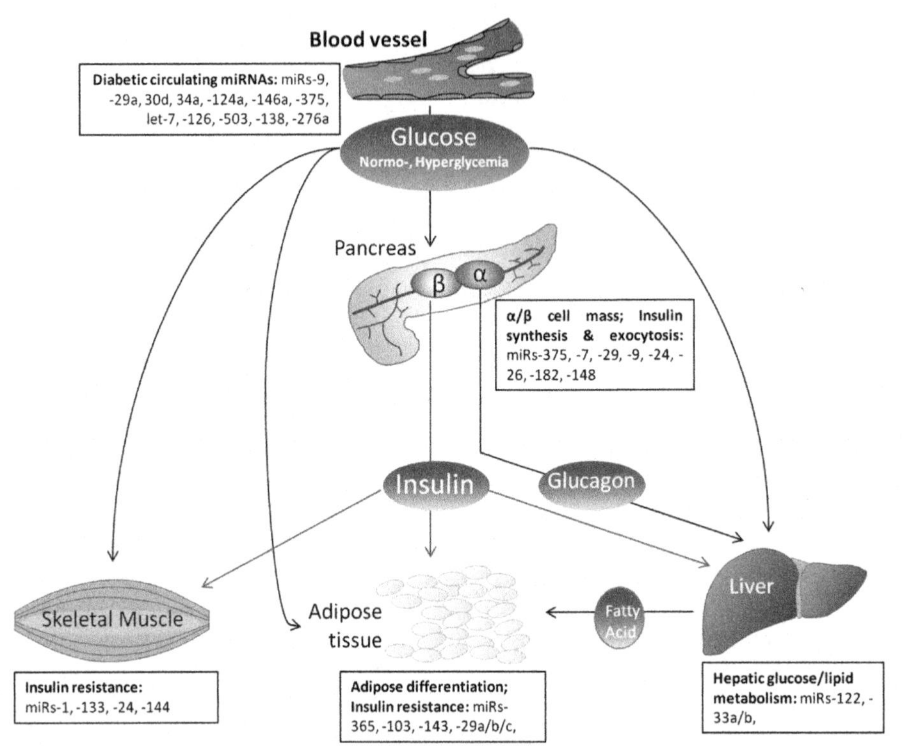

FIGURE 1 MicroRNAs (miRNAs) play important roles in controlling insulin biosynthesis and release, pancreatic β-cell development and survival, glucose and lipid metabolism, and secondary complications associated with diabetes.

2. miRNAs REGULATING PANCREATIC β-CELL FUNCTION AND INSULIN SECRETION

Central to the pathophysiology of diabetes are the insulin-producing β-cells of the pancreatic islets. Postprandial increases in blood glucose is associated with glucose uptake into the pancreatic β-cells via glucose transporter 2 (SLC2A2) and entry into the glycolytic pathway. As a result, adenosine triphosphate (ATP) production is elevated, leading to the closure of membrane-localized ATP-sensitive potassium channels, depolarization of the plasma membrane, and an influx of extracellular calcium. Elevated intracellular calcium levels mediate the release of insulin-containing granules into the circulation. Insulin molecules bind to their associated insulin receptors and initiate glucose uptake into various muscle and fat tissues via translocation of type 4 glucose transporters (SLC2A4) to the plasma membrane [11]. The maintenance of β-cell mass during development and adult life is a highly regulated process for normal glucose homeostasis, and β-cell loss and reduced production of insulin are hallmarks of both type 1 and type 2 diabetes. In recent years, miRNAs have been identified as key regulators of normal β-cell development, insulin biosynthesis, and release (Table 1). Indeed, deletion of Dicer, a key enzyme in miRNA biosynthesis pathway results in pancreatic developmental defects, including depleted β-cell mass [12]. Here, we review several specific miRNAs reported to play a key role in pancreatic β-cell function.

TABLE 1 MicroRNAs (miRNAs) Identified as Key Regulators of β-Cell Development, Insulin Biosynthesis, and Release

miRNAs	Gene targets	Proposed functions	Tissue expressions	References
miR-375	Insulin (*INS*); myotrophin (*MTPN*); phosphoinositide (PI)-dependent protein kinase-1 (*PDPK1*)	Pancreatic development α/β-cell mass Insulin synthesis/ exocytosis Regulates PI 3-kinase signaling	Abundant in pancreas ↑ in type 2 diabetes subjects ↑ in ob/ob obese mice ↑ in streptozotocin diabetic mice	[13–21]
miR-7	Paired box 6 (*PAX6*); transducer of regulated CREB1 (*TORC1*); MAPK-interacting kinases (*MKNK1*/ *MKNK2*); synuclein (*SNCA*)	Pancreatic development Regulates mTOR pathway Insulin secretion	Abundant in pancreas ↓ in ob/ob obese mice ↓ in type 2 diabetes subjects	[22–29]
miR-24/-26/-182/-148	Basic helix-loop-helix family, member e22 (*BHLHE22*); SRY (sex-determining region Y)-box 6 (*SOX6*)	Insulin exocytosis	Abundant in pancreas	[30]
miR-30d	Mitogen-activated protein 4 kinase 4 (*MAP4K4*)	Insulin synthesis	Abundant in pancreas ↓ in diabetic mice islets	[31,32]
miR-29/-9	One cut domain family member 2 (*ONECUT2*); sirtuin-1 (*SIRT1*)	Insulin exocytosis	Abundant in pancreas ↑ miR-29 in nonobese diabetic mice islets	[33–36]
miR-96	Rabphilin 3A-like, without C2 domains (*NOC2*)	Insulin exocytosis	Abundant in pancreas	[35]

2.1 miR-375

miR-375 was first identified as highly abundant in pancreatic islet cells acting as an important regulator of β-cell development and function [13]. Expression of miR-375 is under the control of Pdx-1 and NeuroD/Beta2, two transcription factors essential for pancreas development and insulin production [14]. Inhibition of miR-375 in zebrafish induces major defects in pancreatic islet development [15]. Furthermore, miR-375 knockout mice are hyperglycemic and exhibit a decrease in β-cell mass and concomitant increase in α-cell mass and plasma glucagon levels [16]. A recent study investigating islet transplantation for the treatment of type 1 diabetes through the generation of insulin-secreting cells from human-induced pluripotent stem cells reported that overexpression of miR-375 promotes pancreatic differentiation [17].

A role for miR-375 in insulin secretion has also been demonstrated. Inhibition of miR-375 leads to enhanced insulin secretion, whereas miR-375 overexpression attenuates insulin release by targeting myotrophin (Mtpn), a protein involved in insulin granule fusion [13,18]. miR-375 also represses insulin gene transcription through repression of phosphoinositide (PI)-dependent protein kinase-1, a key component in the PI 3-kinase signaling cascade [19]. High levels of miR-375 are found in the pancreatic islet of type 2 diabetic patients in comparison with nondiabetic patients and in an animal model of obesity and insulin resistance [16,20]. miR-375 expression is elevated in plasma samples of mice induced toward diabetes using the β-cell toxin streptozotocin, suggesting that circulating miR-375 levels could be utilized as biomarker [21]. Taken together, these studies suggest an important role for miR-375 in normal pancreatic islet development and function. In the setting of diabetes where miR-375 levels are elevated, inhibition of miR-375 to levels seen in healthy individuals may lead to improvements in β-cell insulin production and secretion.

2.2 miR-7

miR-7 is abundantly expressed in the developing and adult human pancreas, and is believed to play a crucial role in normal pancreatic development by targeting the developmental transcription factor PAX6 [22–24]. In adult tissues, miR-7 regulates β-cell proliferation through targeting the mammalian target of rapamycin (mTOR) pathway [25]. Previous studies demonstrate that activation of mTOR signaling promotes pancreatic β-cell replication, expansion of β-cell mass, and improved glucose tolerance [26–28]. A recent study utilizing miR-7-deficient/overexpressing mice revealed that miR-7 regulates β-cell function by directly regulating genes involved in insulin granule exocytosis in pancreatic β-cells [29]. Overall, these data indicate that therapeutic inhibition of miR-7 may act to enhance the insulin secretory capacity of β-cells, and also promote the expression of key transcription factors and signaling pathways important for β-cell differentiation and function.

Several additional miRNAs have been identified as important regulators of insulin synthesis and exocytosis in β-cells. For example, deficiency of miR-24, -26, -182, and -148 in islet β-cells results in upregulation of the transcriptional repressors, Bhlhe22 and Sox6, and subsequent downregulation of insulin gene expression [30]. miR-30d, the expression of which is reduced in islets of diabetic mice, also stimulates insulin production through repression of mitogen-activated protein 4 kinase 4 (MAP4K4), an inhibitor of insulin production [31,32]. miR-29 expression is increased in islet cells in response to high glucose and is increased in islets from nonobese diabetic (NOD) mice [33]. Overexpression of miR-29 leads to suppression of the transcription factor, one cut domain family member 2 (Onecut2), which in turn activates granuphilin, a known inhibitor of insulin exocytosis [33]. Likewise, miR-9 also negatively regulates Onecut2 expression, leading to increased granuphilin and impaired β-cell insulin release [34,35]. miR-9 also modulates insulin secretion by directly targeting and downregulating sirtuin-1 (SIRT1) expression, an NAD-dependent protein deacetylase involved in insulin secretion in response to glucose challenge [36]. miR-96 has shown similar effects on granuphilin through targeting of Noc2, which is required for insulin exocytosis [35].

3. miRNAs IN INSULIN TARGET TISSUES: LIVER, SKELETAL MUSCLE, AND ADIPOSE TISSUE

Insulin released into the circulation from pancreatic β-cells stimulates glucose uptake from the blood and utilization and storage in target tissues—liver, adipose, and skeletal muscles. Physiological and pathological factors such as diet, pregnancy, stress, and obesity can affect the sensitivity of these tissues toward insulin, such that they are resistant to the effects of insulin.

3.1 Liver

Once secreted from the pancreatic islet, insulin travels via the portal circulation to the liver, to control hepatic glucose and lipid metabolism. A key feature of type 2 diabetes is a fatty liver (hepatic steatosis), characterized by fat accumulation in the hepatocytes. This fat accumulation is thought to be caused by metabolic imbalances such as higher amounts of dietary lipids, increased trafficking of free fatty acids from adipose to liver, and increased lipogenesis. miRNAs are essential for normal liver development as evidenced by the fact that genetic deletion of Dicer in the liver affected protein localization, with protein expression more diffused throughout the liver rather than localized in specific compartments [37]. There are now several miRNAs identified as key regulators of hepatic pathophysiology in the development of diabetes, and these are summarized in Table 2.

3.2 miR-122

During development, several liver-enriched transcription factors (hepatocyte nuclear factors)

TABLE 2 MicroRNAs (miRNAs) Implicated in Maintaining Glucose Homeostasis and Insulin Sensitivity in Hepatic, Adipose, and Skeletal Muscle Tissues

miRNAs	Gene targets	Proposed functions	Tissue expressions	References
miR-122	Cut-like homeobox 1 (*CUTL1*); fatty acid synthase (*FASN*); HMG-CoA reductase (*HMGCR*)	Liver development Hepatic insulin sensitivity Hepatic fatty acid metabolism Cholesterol homeostasis	Abundant in liver ↑ during hepatic development ↓ in hepatic steatosis subjects ↓ in streptozotocin diabetic mice	[38–44]
miR-33a/-33b	Sterol regulatory element-binding protein 1 (*SREBP-1*); insulin receptor substrate 2 (*IRS2*); phosphoenolpyruvate carboxykinase (*PCK1*); glucose-6-phosphatase (*G6PC*)	Hepatic fatty acid metabolism Cholesterol homeostasis Hepatic insulin signaling Hepatic glucose production	Abundant in liver	[45–50]
miR-365	Runt-related transcription factor 1 Translocated to, 1 (*RUNX1T1*)	Adipose differentiation	Expressed in white/brown adipocyte tissue	[51–53]
miR-103/-143	Oxysterol-binding protein-related protein 8 (*ORP8*); mitogen-activated protein kinase 7 (*MAPK7*); mitogen-activated protein kinase 5 (*MAP2K5*)	Adipose differentiation Adipose insulin sensitivity	↑ during adipogenesis ↓ miR-103 in ob/ob obese mice adipocytes ↑ miR-143 in ob/ob obese mice adipocytes	[54–57]
miR-29a/b/c	Peroxisome proliferator-activated receptor gamma, coactivator 1 alpha (*PPARGC1A*)	Adipose insulin sensitivity	↑ in skeletal muscle/liver and adipose tissue in Zucker diabetic fatty rats ↑ in adipocytes in response to insulin/high glucose	[58,59]
miR-1/-133	Histone deacetylase 4 (*HDAC4*); serum response factor; Kruppel-like transcription factor 15 (*KLF15*)	Skeletal muscle differentiation Skeletal muscle insulin sensitivity	↓ by insulin in skeletal muscle biopsies ↓ in type 2 diabetic skeletal muscle	[62–66]
miR-24	P38 MAP kinase (*p38 MAPK*)	Skeletal muscle insulin sensitivity	↓ in diabetic rat muscle	[67]
miR-144	Insulin receptor substrate 1 (*IRS1*)	Skeletal muscle insulin sensitivity	↑ in type 2 diabetic skeletal muscle	[68]

activate miR-122, which in turn targets CUTL1, a transcriptional repressor of hepatocyte differentiation [38]. While the evidence for a role for miR-122 during embryogenesis is clear, there are conflicting data in the literature regarding the role of this miRNA during hepatic steatosis. Early studies demonstrated that inhibition of miR-122 in normal and diet-induced obese mice led to improved insulin sensitivity, decreased hepatic fatty acid, and cholesterol synthesis, along with a reduction in plasma cholesterol. Here, miR-122 was shown to regulate hepatic fatty acid metabolism by targeting several lipogenic enzymes, including fatty acid synthase and HMG-CoA reductase [39,40]. Similarly, in vivo inhibition of miR-122 in hepatocytes of nonhuman primates led to reduced plasma cholesterol levels [41]. However, in contrast to these transient inhibition studies, more recent efforts investigating the consequences of sustained loss of function of miR-122 in vivo suggest that genetic deletion of miR-122 leads to hepatosteatosis, hepatitis, and inflammation [42]. Measurement of miR-122 levels in human subjects revealed that miR-122 is downregulated in the hepatic tissues of patients of severe hepatic steatosis [43]. Interestingly, in the livers of streptozotocin-induced type 1 diabetic mice, miR-122 expression is reduced in comparison with control mice [44]. Taken together, these data suggest that while the short-term effects of anti-miR-122 therapy to the liver may result in reduced hepatic cholesterol synthesis and improved insulin sensitivity, the long-term consequences need to be considered.

3.3 miR-33a/33b

Another two miRNAs, miR-33a and miR-33b, have been shown to play an important role in cholesterol homeostasis, fatty acid metabolism, and insulin signaling. miR-33a and -33b are encoded within the introns of sterol regulatory element-binding proteins (SREBP) 2 and 1, respectively. SREBPs are the predominant transcription factors controlling the synthesis of cholesterol and fatty acids in the liver [45]. Genetic deletion of miR-33 in mice leads to obesity and liver steatosis, and this is believed to be mediated via enhanced expression of SREBP-1, demonstrating direct interaction between SREBP-1 and miR-33 [46]. Furthermore, transient inhibition of miR-33 in mice on a high-fat diet results in elevated plasma high-density lipoprotein (HDL), suggesting that inhibition of miR-33 may represent a therapeutic target for treating liver dysfunction in diabetes [47].

miR-33a and -33b are also believed to play an important role in insulin signaling and glucose metabolism in hepatocytes, with inhibition of endogenous miR-33a and -33b levels leading to increased expression of insulin receptor substrate 2 (IRS-2). In contrast, overexpression of miR-33a and -33b attenuates both fatty acid oxidation and insulin signaling in hepatocytes [48]. miR-33b is also believed to regulate hepatic glucose metabolism by targeting phosphoenolpyruvate carboxykinase (PCK1) and glucose-6-phosphatase (G6PC), key regulatory enzymes of hepatic gluconeogenesis [49]. However, a recent study has questioned the effects of long-term inhibition of miR-33. Here, persistent inhibition of miR-33 when mice are fed a high-fat diet led to hepatic steatosis and hypertriglyceridemia [50]. While it is clear that miR-33 plays an important role in hepatic cholesterol homeostasis and insulin signaling, further studies are required to assess the effect of chronic inhibition of miR-33 before translation to human therapeutics.

3.4 Adipose Tissue

Mammals have two principal types of fat—white adipose tissue, which stores extra energy as triglycerides, and brown adipose tissue, which functions to metabolize lipids for heat generation and energy expenditure. Alterations in fat mass and distribution can have a major impact on whole-body metabolism, leading to increased risk of type 2 diabetes and cardiovascular disease. Understanding the role of miRNAs in the development of adipocytes could identify new therapeutic targets for antidiabetic drugs. It is now known that miRNAs are important regulators of adipocyte differentiation and function, with a recent study demonstrating that mice with a fat-specific knockout of Dicer develop lipodystrophy and insulin resistance [51]. Table 2 highlights several miRNAs implicated in adipocyte development and dysfunction in diabetes.

3.5 miR-365

Knockout of Dicer in brown preadipocytes promotes a white adipocyte-like phenotype and leads to reduced expression of several miRNAs, including miR-365 [51]. Conflicting data exist for the precise role of miR-365 in promoting brown fat differentiation, with one study describing miR-365 as an essential miRNA required to promote brown fat differentiation [52], whereas a more recent study indicates that the development, differentiation, and function of brown adipose tissue do not require the presence of miR-365 [53]. Further investigations will be required to resolve these conflicting reports and determine the precise role of miR-365 in adipocyte differentiation and function.

3.6 miR-103 and miR-143

miR-103 and miR-143 expression is upregulated during adipogenesis of preadipocytes [54–57]. Overexpression of miR-103 in preadipocytes accelerates adipogenesis and induces an increase in triglyceride accumulation. Similarly, overexpression of miR-143 in preadipocytes increases the expression of adipocyte differentiation markers [54]. Interestingly, miR-103 levels were reduced in adipocytes from leptin-deficient ob/ob and diet-induced obese mice, and also in adipocytes in response to the proinflammatory cytokine tumor necrosis factor alpha [55,56]. Several reports have also revealed that miR-143 is upregulated in the liver of ob/ob mice or high-fat diet-treated mice, and increased miR-143 levels in adipocytes is associated with the development of obesity [55]. The mechanism through which overexpression of miR-143 in transgenic mice impairs insulin sensitivity is believed to be through repression of oxysterol-binding protein-related protein 8 (ORP8), thereby inhibiting insulin-stimulated protein kinase B activation. Importantly, the development of obesity is attenuated in miR-143 knockout mice, suggesting miR-143 as a potential therapeutic target for the treatment of obesity-associated diabetes.

3.7 miR-29a/b/c

As discussed previously, miR-29 expression is increased in islet cells in response to high glucose and is increased in islets from NOD mice [33]. The importance of this miRNA to the diabetic state extends beyond the pancreatic islets, with miR-29a/b/c expression also increased in skeletal muscle, liver, and fat tissues of diabetic rats [58]. miR-29 levels are increased in adipocytes in response to high glucose and insulin, and ectopic overexpression of miR-29 leads to insulin resistance in adipocytes [58]. A recent study exploring the functional relevance of elevated miR-29 levels in the liver of Zucker diabetic fatty rats suggests that miR-29 regulates FOXA2-mediated activation of lipid metabolism genes, including PPARGC1A [59]. Intriguingly, in this study, hepatic expression of miR-29 returned to normal upon treatment with the insulin-sensitizing agent pioglitazone, suggesting that existing therapeutic treatments may be utilized to modulate miR-29 levels.

3.8 Skeletal Muscle

Skeletal muscle is the primary site for glucose uptake, accounting for approximately 75% of insulin-stimulated glucose uptake. Insulin resistance in the skeletal muscle is a feature of type 2 diabetes, and many studies in both diabetic animal models and patients have identified several miRNAs dysregulated in diabetic skeletal muscle [60,61] (Table 2).

3.9 miR-1 and miR-133

miR-1 and miR-133 reside at the same chromosomal miRNA cluster and are coregulated in a tissue-specific manner during development. miR-1 and miR-133 regulate skeletal muscle proliferation and differentiation by targeting histone deacetylase 4 and serum response factor, respectively [62,63]. Muscle-specific expression of miR-1/133 is believed to be activated by myocyte enhancer factor 2C [64]. Interestingly, insulin represses miR-1/133 skeletal muscle expression by inhibiting myocyte enhancer factor 2C, and this effect is impaired in the skeletal muscle of type 2 diabetic patients [60]. A recent study has shown that the expression of miR-133a is downregulated in the skeletal muscle of type 2 diabetic patients, and altered levels of this miRNA are correlated with higher fasting glucose levels [65]. A role for miR-133 in glucose homeostasis in cardiac muscle cells has also been described. Here, miR-133 inhibited expression of its target Kruppel-like transcription factor KLF15, resulting in reduced expression of glucose transporter 4, and consequently a reduction in insulin-stimulated glucose uptake in these cells [66].

Other miRNAs identified as potential novel regulators of muscle cell insulin resistance in the diabetic state include miR-24 and miR-144. miR-24 expression is downregulated in diabetic rat muscle and it has been speculated that this loss is an adaptive response to high glucose conditions, whereby downregulation of miR-24 leads to increased expression of its target p38 MAPK, and increased expression of glucose transporter 4 to facilitate glucose uptake [67]. In contrast, miR-144 levels are upregulated in skeletal muscle of type 2 diabetic patients and this may result in suppression of insulin receptor substrate 1, a key component of the insulin signaling cascade, suggesting that targeted suppression of miR-144 in diabetic patients may be a novel therapeutic strategy to improve insulin sensitivity [68].

4. miRNA RESPONSES TO ANTIDIABETIC THERAPIES

Insulin replacement is the mainstay of therapy for type 1 diabetes. In type 2 diabetes, therapeutic interventions include those that impact target organ sensitivity and the pancreatic β-cells, including insulin sensitizers, sulfonylureas, incretin mimetics, and glitazones [69,70]. The insulin-sensitizing agent metformin is a first-line therapy which reduces hepatic glucose production and improves insulin resistance and glucose uptake into peripheral tissues [71]. Similarly, the thiazolidinediones/glitazones (TZD) improve insulin resistance and glucose uptake into peripheral tissues via activation of peroxisome proliferator-activated nuclear receptors [72]. There are also antidiabetic agents that stimulate pancreatic insulin secretion, thereby opposing insulin deficiency. Sulfonylureas stimulate insulin secretion by binding to the sulfonylurea receptor subunit of the ATP-sensitive potassium channel on pancreatic β-cells, resulting in channel closure, calcium influx, and exocytosis of insulin-containing secretory granules [73]. In recent years, several new agents have been approved for management of type 2 diabetes that stimulate pancreatic insulin secretion, including glucagon-like polypeptide 1 (GLP-1) agonists and dipeptidyl peptidase-4 (DPP4) inhibitors, which simulate the bioactions of endogenous incretins [74–76].

An important question arises regarding the ability of these antidiabetic treatments to modulate miRNA levels. These pharmacological agents may be acting through modulation of miRNAs known to play an important role in the pathogenesis of diabetes, or indeed through presently undescribed miRNA mechanisms highly specific to the agent used. While these studies have yet to address this in the clinical setting, there is evidence in the literature that miRNA levels are altered in diabetic individuals in response to antidiabetic agents. For example, as discussed previously, in vitro and in vivo data indicate that miR-29 expression is elevated in the diabetic state and hepatic expression of miR-29 can be reduced using the insulin-sensitizing agent pioglitazone [59]. The effect of pioglitazone on miR-29 levels in human type 2 diabetic patients has not yet been investigated. More recently it has been shown that levels of circulating let-7 miRs are reduced in patients with type 2 diabetes in comparison with healthy individuals, and following a 12-month period of antidiabetic treatment using metformin and DPP4 inhibitors, circulating levels of let-7 were increased in diabetic individuals and were comparable with healthy controls [77]. Interestingly, loss of let-7 miR levels has been reported in several studies to be associated with diabetic kidney disease [78,79]. These data suggest that measuring circulating let-7 levels in diabetic patients receiving this treatment may be a useful indicator for patient management, and metformin may prevent progression toward severe diabetic complications through upregulation of let-7 miRs. Similarly, circulating miR-192 levels were reduced in type 2 diabetes patients when compared with controls, and can be restored following a 3-month treatment with metformin [80].

5. PHARMACOLOGICAL TARGETING OF miRNAs IN DIABETES

Alteration of miRNA levels is complicated by the fact that the vast majority of miRNAs target a large number of genes, and a single gene 3′-UTR may be regulated by many miRNAs. Therefore, manipulating miRNA levels may elicit desirable effects on a disease-relevant pathway, but also may have consequences on additional "off-target" pathways unrelated to the disease pathogenesis. The goal now is to develop miRNA-based therapeutics to restore miRNA to normal levels in target cell populations. Strategies are being developed to generate miRNA mimics for miRNA replacement therapy, or miRNA inhibitors to deplete the expression of damaging miRNAs. Several classes of miRNA inhibitors exist including miRNA "sponge" vectors which rely on the expression of mRNAs containing multiple artificial miRNA-binding sites, thus acting as decoys, and anti-miRs that specifically bind to miRNAs, thereby preventing miRNA binding to mRNA targets. To prevent degradation by serum nucleases, a variety of chemical modifications have been incorporated into anti-miR oligonucleotides. Locked nucleic acid (LNA) anti-miRs offer high-affinity binding to target miRNAs and have proven to be efficient miRNA inhibitors in vivo [41]. Design of efficient delivery strategies to target miRNA mimics and anti-miRs to specific tissues and cell populations is perhaps the greatest hurdle that needs to be overcome. In the case of diabetes, delivery strategies will aim to deliver miRNA mimics/inhibitors to the pancreas and liver, as well as insulin target tissues. Current delivery strategies being developed include conjugation of cholesterol with anti-miR oligonucleotides to facilitate incorporation into lipoproteins and transport to the liver, gut, and kidney; incorporation into liposomes and nanoparticles; and linking anti-miRs to cell-targeting antibodies.

In recent years, several biopharmaceutical companies have been established to develop miRNA-based therapeutics. Currently there is a focus on miRNA replacement therapy in cancer patients, with a miR-34 mimic designated MRX34 developed by Mirna Therapeutics presently in Phase 1 clinical trial in patients with liver cancer. Mirna Therapeutics has also developed a let-7 mimic therapeutic which is currently at the preclinical stage. Given that circulating let-7 miR levels are reduced in patients with type 2 diabetes and restored following antidiabetic treatments [77], this miRNA mimic represents a promising candidate for miRNA replacement therapy in diabetes.

Beyond synthetic miRNA mimic/anti-miR delivery strategies, nontoxic natural agents used in disease treatment may exert their therapeutic effects through miRNA regulation. This is particularly evident in the cancer field where natural agents such as curcumin, resveratrol, genistein, isoflavone, and 3,3′-diindolylmethane (DIM) are believed to act via miRNA modulation [81]. Among these, isoflavone and DIM have been shown to upregulate miR-200 and let-7 levels in pancreatic cancer cells and reverse the mesenchymal phenotype observed in these cells [82]. These natural agents may be one such strategy to restore aberrant miRNA levels in diabetic patients.

6. CIRCULATING miRNAs AS PROMISING CLINICAL BIOMARKERS FOR DIABETES

Desirable characteristics of diabetes biomarkers include the capacity to detect these markers in body fluids such as blood and urine, ability to detect those patients with subclinical disease (prediabetes) and those at risk of progressing toward diabetic complications, and the capacity to monitor patient response to antidiabetic treatments. Currently, glycated hemoglobin (HbA1c) levels is the biomarker of choice for monitoring glycemic control, providing a measure of patient exposure to glucose over the 3-month lifespan of red blood cells. One notable limitation of HbA1c monitoring is the fact that this test is not suitable for assessing hyperglycemia in patients with hemoglobinopathies such as anemia or uremia, patients with end-stage renal disease treated by hemodialysis, or patients whose glucose levels fluctuate greatly. Other clinical and laboratory predictors of diabetes and the development of diabetes-related macrovascular complications include elevated blood lipid levels (cholesterol and triglycerides), hypertension, and body mass index, and there is abundant evidence in the literature demonstrating that management of lipid levels, blood pressure, and diet have beneficial effects on the progression of diabetes-associated cardiovascular disease. However, it is important to note that these biomarkers are generally deemed more reflective of established disease progress rather than predictive of disease onset. More recent efforts have focused on using novel technologies to identify additional biomarkers to improve diabetes detection and provide better indications of treatment effectiveness and the risk of developing complications. Among those identified, promising biomarker candidates include branched-chain amino acids, 2-aminoadipic acid, fructosamine, and glycated albumin [83–85].

Due to their relative stability in the circulation, miRNAs are currently being explored for their potential as biomarkers for the development and progression of diabetes. miRNAs are readily detectable in body fluids, such as blood, saliva, urine, and serum, with distinct expression profiles in different fluid types [86–89]. Unlike the majority of RNA molecules which are unstable in the extracellular environment due to circulating ribonucleases, miRNAs are believed to be protected from extracellular degradation by packaging into exosomes or by forming complexes with RNA-binding proteins or lipoprotein complexes [90,91]. Exosome-mediated transfer of miRNAs is one mechanism through which cells are believed to communicate. In recent years, altered circulating miRNAs have been reported in diverse diseases including cancer, heart failure, and liver injury. The biomarker potential of circulating miRNA is perhaps most advanced in the cancer field where several independent studies have reported the feasibility of using circulating miRNAs for disease detection, prognosis, and treatment response [92,93]. While these studies do show great promise, some limitations to overcome include the fact that circulating miRNAs offer little or no information on the cellular source of origin of the miRNA, nor it is clear what the degree of correlation is between diseased tissue and biofluid miRNA profiles.

Nevertheless, there are now several miRNAs identified as promising circulating biomarkers for diabetes (Table 3). Several studies have reported reduced expression of miR-126 levels in plasma from individuals with established type 2 diabetes or type 2 diabetes-susceptible individuals when compared with normal individuals, suggesting that circulating miR-126 levels may serve as a potential biomarker for early identification of susceptible individuals to type 2 diabetes [94,95]. Interestingly, reduced miR-15a, miR-29b, miR-126, miR-223, and elevated miR-28-3p levels preceded the onset of diabetes, highlighting the biomarker potential of these miRNAs [95]. Analysis of circulating levels of several diabetes-related miRNAs (miR-9, miR-29a, miR-30d, miR-34a, miR-124a, miR-146a, and miR-375) in patients with type 2 diabetes indicates that the expression levels of these miRNAs in patient serum is elevated in cases of newly diagnosed when compared with prediabetes individuals or type 2 diabetics with normal glucose tolerance [96]. An independent study in a Chinese Han population confirmed that circulating miR-146a levels are significantly elevated in newly diagnosed type 2 diabetes patients compared with healthy controls [97]. A more recent study has identified serum miRNAs as powerful predictive biomarkers for distinguishing type 2 diabetic patients from normal healthy controls (miR-138 or miR-376a), and type 2 diabetic

TABLE 3 MicroRNAs (miRNAs) Identified as Potential Circulating Biomarkers of Diabetes Diagnosis and Prognosis

miRNAs	Diabetes subtypes	Key findings	References
miR-126, -21, -24, -15a, -20b, -191, -197, -223, -320, -486, -28-3p	Type 2 diabetes	Reduced miR-15a, miR-29b, miR-126, miR-223, and elevated miR-28-3p levels precede disease onset	[94]
miR-126	Type 2 diabetes	Reduced expression in prediabetes and diabetes	[95]
miR-9, -29a, -30d, -34a, -124a, 146a, -375	Type 2 diabetes	Elevated in diabetes compared with prediabetes	[96]
miR-146a	Type 2 diabetes	Elevated in newly diagnosed type 2 diabetics	[97]
miR-138, -15b, -376a, -503	Type 2 diabetes	Distinguish obese diabetics from nonobese diabetics	[98]
miR-140-5p, -142-3p, -222, -423-5p, -125b, -192, -195, -130b, -532-5p, -126	Type 2 diabetes	Altered levels in type 2 diabetes patients miR-192 levels changed with 3-month metformin treatment	[80]
Let-7	Type 2 diabetes	Levels reduced in type 2 diabetes patients, and restored following metformin/DPP4 treatment	[77]
miR-152, -30a-5p, -181a, -24, -148a, -210, -27a, -29a, -26a, -27b, -25, -200a	Type 1 diabetes	Elevated in diabetes. miR-25 levels correlate with glycemic control HbA1c test	[99]
miR-103, -224	MODY	Elevated in MODY vs controls	[100]

DPP4, dipeptidyl peptidase-4; HbA1c, hemoglobin A1c; MODY, maturity-onset diabetes of the young.

from obese type 2 diabetic patients (miR-503 and miR-138) [98]. While the majority of studies have focused on serum miRNA biomarkers for type 2 diabetes, there is some evidence in the literature that supports the role of circulating miRNAs as predictive biomarkers for type 1 diabetes. A study comparing circulating miRNA levels in serum pools from new-onset type 1 diabetes children versus healthy controls identified a cohort of dysregulated miRNAs predicted to target β-cell networks and apoptosis, with miR-25 levels highly correlated with glycemic control 3 months after diagnosis in new-onset type 1 diabetes children [99]. Beyond type 1 and type 2 diabetes, serum levels of miR-103 have been shown to be elevated in individuals with MODY monogenic diabetes arising from a mutation in the HNFA1 gene [100].

7. CONCLUSION

Diabetes is a disorder of glucose homeostasis that involves a distortion of the homeostatic coordination between multiple tissues and organs. miRNAs are believed to play an important role in maintaining glucose homeostasis, with miRNA dysregulation observed in β-cells and insulin target tissues from diabetic subjects, while modulation of miRNA levels in cell model and in vivo systems indicate that restoration of miRNA to normal levels may improve insulin production and sensitivity. In the future, further studies will be required to determine the long-term effects of modifying miRNA levels, and how these interact with existing antidiabetic therapies. Efficient delivery systems targeting miRNA mimics/anti-miRs to specific cell populations will be required to minimize off-site effects.

Circulating miRNAs are also associated with diabetes pathogenesis and may prove to be effective clinical biomarkers for the development and progression of diabetes, and also for monitoring responses to antidiabetic agents. However, there are several key issues and considerations with respect to the biomarker potential of circulating miRNAs in diabetes [1]. First, while the studies described here have confirmed that miRNAs are detectable and display altered expression patterns in diabetic individuals, the majority of these studies have been performed in relatively small cohorts. Independent, large-scale biobank collections of serum samples from prediabetics, established diabetics, and matched control individuals will be required to accurately determine the predominant circulating miRNAs associated with diabetes [2]. Second, different antidiabetic drugs may elicit different miRNA responses in diabetic individuals and may confound results when determining circulating miRNA levels in a large cohort of diabetics undergoing a range of antidiabetic regimens [3]. Third, while a serum miRNA diabetes-associated biomarker has obvious diagnostic and prognostic potential, little is known about the cellular origin of the miRNA. This is an important question that needs to be addressed when these miRNAs are being targeted to specific tissues for therapeutic intervention using either miRNA mimics to replenish low levels of a specific miRNA or anti-miRs to reduce levels of a diabetes-associated miRNA [4]. Finally, there is ample evidence in the literature to suggest that miRNAs are highly stable in body fluids, and expression levels are typically quantified via the real-time quantitative reverse transcriptase-polymerase chain reaction. However, in the clinical setting, this may be complicated by the fact that the amount of RNA recovered from plasma or serum can often be low, and there is an absence of clear guidelines for correct use of endogenous controls and normalization approaches. Use of synthetic spiked-in control miRNAs and standardized RNA isolation methods are recommendations that would greatly enable the discovery of sensitive circulating miRNA biomarkers. By addressing these issues, the true potential of miRNA as both biomarkers and drug targets will be realized. Beyond diabetes, miRNA therapeutics and biomarkers will also be an important clinical strategy when diagnosing, monitoring, and treating micro- and macrovascular complications associated with diabetes.

LIST OF ABBREVIATIONS

3′-UTR 3′-Untranslated region
β-cell Pancreatic islet beta cell
Anti-miR MicroRNA inhibitor
ATP Adenosine triphosphate
DPP4 Dipeptidyl peptidase-4
FPG test Fasting plasma glucose test
GDM Gestational diabetes mellitus
GLP-1 Glucagon-like polypeptide 1
HbA1c test Hemoglobin A1c test
HDL High-density lipoprotein
LNA Locked nucleic acid
miRNA MicroRNA
MODY Maturity-onset diabetes of the young
mRNA Messenger RNA
mTOR pathway Mammalian target of rapamycin pathway
NOD mice Nonobese diabetic mice
OGTT Oral glucose tolerance test
SREBP Sterol regulatory element-binding proteins
Type 1 diabetes Insulin-dependent diabetes mellitus
Type 2 diabetes Noninsulin-dependent diabetes mellitus
TZDs Thiazolidinediones/glitazones

References

[1] Patterson CC, Dahlquist GG, Gyurus E, Green A, Soltesz G, Group ES. Incidence trends for childhood type 1 diabetes in Europe during 1989–2003 and predicted new cases 2005–20: a multicentre prospective registration study. Lancet June 13, 2009;373(9680):2027–33.

[2] Gardner SG, Bingley PJ, Sawtell PA, Weeks S, Gale EA. Rising incidence of insulin dependent diabetes in children aged under 5 years in the Oxford region: time trend analysis. The Bart's-Oxford Study Group. BMJ September 20, 1997;315(7110):713–7.

[3] Karvonen M, Viik-Kajander M, Moltchanova E, Libman I, LaPorte R, Tuomilehto J. Incidence of childhood type 1 diabetes worldwide. Diabetes Mondiale (DiaMond) Project Group. Diabetes Care October 2000;23(10):1516–26.

[4] Kaprio J, Tuomilehto J, Koskenvuo M, Romanov K, Reunanen A, Eriksson J, et al. Concordance for type 1 (insulin-dependent) and type 2 (non-insulin-dependent) diabetes mellitus in a population-based cohort of twins in Finland. Diabetologia November 1992;35(11):1060–7.

[5] Newman B, Selby JV, King MC, Slemenda C, Fabsitz R, Friedman GD. Concordance for type 2 (non-insulin-dependent) diabetes mellitus in male twins. Diabetologia October 1987;30(10):763–8.

[6] Weijnen CF, Rich SS, Meigs JB, Krolewski AS, Warram JH. Risk of diabetes in siblings of index cases with type 2 diabetes: implications for genetic studies. Diabet Med January 2002;19(1):41–50.

[7] Gardner DS, Tai ES. Clinical features and treatment of maturity onset diabetes of the young (MODY). Diabetes Metab Syndr Obes 2012;5:101–8.

[8] Hollander MH, Paarlberg KM, Huisjes AJ. Gestational diabetes: a review of the current literature and guidelines. Obstet Gynecol Surv February 2007;62(2):125–36.

[9] Wild S, Roglic G, Green A, Sicree R, King H. Global prevalence of diabetes: estimates for the year 2000 and projections for 2030. Diabetes Care May 2004;27(5):1047–53.

[10] Hossain P, Kawar B, El Nahas M. Obesity and diabetes in the developing world–a growing challenge. N Engl J Med January 18, 2007;356(3):213–5.

[11] Hedeskov CJ. Mechanism of glucose-induced insulin secretion. Physiol Rev April 1980;60(2):442–509.

[12] Lynn FC, Skewes-Cox P, Kosaka Y, McManus MT, Harfe BD, German MS. MicroRNA expression is required for pancreatic islet cell genesis in the mouse. Diabetes December 2007;56(12):2938–45.

[13] Poy MN, Eliasson L, Krutzfeldt J, Kuwajima S, Ma X, Macdonald PE, et al. A pancreatic islet-specific microRNA regulates insulin secretion. Nature November 11, 2004;432(7014):226–30.

[14] Keller DM, McWeeney S, Arsenlis A, Drouin J, Wright CV, Wang H, et al. Characterization of pancreatic transcription factor Pdx-1 binding sites using promoter microarray and serial analysis of chromatin occupancy. J Biol Chem November 2, 2007;282(44):32084–92.

[15] Kloosterman WP, Lagendijk AK, Ketting RF, Moulton JD, Plasterk RH. Targeted inhibition of miRNA maturation with morpholinos reveals a role for miR-375 in pancreatic islet development. PLoS Biol August 2007;5(8):e203.

[16] Poy MN, Hausser J, Trajkovski M, Braun M, Collins S, Rorsman P, et al. miR-375 maintains normal pancreatic alpha- and beta-cell mass. Proc Natl Acad Sci USA April 7, 2009;106(14):5813–8.

[17] Lahmy R, Soleimani M, Sanati MH, Behmanesh M, Kouhkan F, Mobarra N. miRNA-375 promotes beta pancreatic differentiation in human induced pluripotent stem (hiPS) cells. Mol Biol Rep April 2014;41(4):2055–66.

[18] Li Y, Xu X, Liang Y, Liu S, Xiao H, Li F, et al. miR-375 enhances palmitate-induced lipoapoptosis in insulin-secreting NIT-1 cells by repressing myotrophin (V1) protein expression. Int J Clin Exp Pathol 2010;3(3):254–64.

[19] El Ouaamari A, Baroukh N, Martens GA, Lebrun P, Pipeleers D, van Obberghen E. miR-375 targets 3′-phosphoinositide-dependent protein kinase-1 and regulates glucose-induced biological responses in pancreatic beta-cells. Diabetes October 2008;57(10):2708–17.

[20] Zhao H, Guan J, Lee HM, Sui Y, He L, Siu JJ, et al. Up-regulated pancreatic tissue microRNA-375 associates with human type 2 diabetes through beta-cell deficit and islet amyloid deposition. Pancreas August 2010;39(6):843–6.

[21] Erener S, Mojibian M, Fox JK, Denroche HC, Kieffer TJ. Circulating miR-375 as a biomarker of beta-cell death and diabetes in mice. Endocrinology February 2013;154(2):603–8.

[22] Bravo-Egana V, Rosero S, Molano RD, Pileggi A, Ricordi C, Dominguez-Bendala J, et al. Quantitative differential expression analysis reveals miR-7 as major islet microRNA. Biochem Biophys Res Commun February 22, 2008;366(4):922–6.

[23] Correa-Medina M, Bravo-Egana V, Rosero S, Ricordi C, Edlund H, Diez J, et al. MicroRNA miR-7 is preferentially expressed in endocrine cells of the developing and adult human pancreas. Gene Expr Patterns April 2009;9(4):193–9.

[24] Kredo-Russo S, Mandelbaum AD, Ness A, Alon I, Lennox KA, Behlke MA, et al. Pancreas-enriched miRNA refines endocrine cell differentiation. Development August 2012;139(16):3021–31.

[25] Wang Y, Liu J, Liu C, Naji A, Stoffers DA. MicroRNA-7 regulates the mTOR pathway and proliferation in adult pancreatic beta-cells. Diabetes March 2013;62(3):887–95.

[26] Bernal-Mizrachi E, Wen W, Stahlhut S, Welling CM, Permutt MA. Islet beta cell expression of constitutively active Akt1/PKB alpha induces striking hypertrophy, hyperplasia, and hyperinsulinemia. J Clin Invest December 2001;108(11):1631–8.

[27] Rachdi L, Balcazar N, Osorio-Duque F, Elghazi L, Weiss A, Gould A, et al. Disruption of Tsc2 in pancreatic beta cells induces beta cell mass expansion and improved glucose tolerance in a TORC1-dependent manner. Proc Natl Acad Sci USA July 8, 2008;105(27):9250–5.

[28] Hamada S, Hara K, Hamada T, Yasuda H, Moriyama H, Nakayama R, et al. Upregulation of the mammalian target of rapamycin complex 1 pathway by Ras homolog enriched in brain in pancreatic beta-cells leads to increased beta-cell mass and prevention of hyperglycemia. Diabetes June 2009;58(6):1321–32.

[29] Latreille M, Hausser J, Stutzer I, Zhang Q, Hastoy B, Gargani S, et al. MicroRNA-7a regulates pancreatic beta cell function. J Clin Invest June 2, 2014;124(6):2722–35.

[30] Melkman-Zehavi T, Oren R, Kredo-Russo S, Shapira T, Mandelbaum AD, Rivkin N, et al. miRNAs control insulin content in pancreatic beta-cells via downregulation of transcriptional repressors. EMBO J March 2, 2011;30(5):835–45.

[31] Zhao X, Mohan R, Ozcan S, Tang X. MicroRNA-30d induces insulin transcription factor MafA and insulin production by targeting mitogen-activated protein 4 kinase 4 (MAP4K4) in pancreatic beta-cells. J Biol Chem September 7, 2012;287(37):31155–64.

[32] Bouzakri K, Ribaux P, Halban PA. Silencing mitogen-activated protein 4 kinase 4 (MAP4K4) protects beta cells from tumor necrosis factor-alpha-induced decrease of IRS-2 and inhibition of glucose-stimulated insulin secretion. J Biol Chem October 9, 2009;284(41):27892–8.

[33] Roggli E, Gattesco S, Caille D, Briet C, Boitard C, Meda P, et al. Changes in microRNA expression contribute to pancreatic beta-cell dysfunction in prediabetic NOD mice. Diabetes July 2012;61(7):1742–51.

[34] Plaisance V, Abderrahmani A, Perret-Menoud V, Jacquemin P, Lemaigre F, Regazzi R. MicroRNA-9 controls the expression of Granuphilin/Slp4 and the secretory response of insulin-producing cells. J Biol Chem September 15, 2006;281(37):26932–42.

[35] Lovis P, Gattesco S, Regazzi R. Regulation of the expression of components of the exocytotic machinery of insulin-secreting cells by microRNAs. Biol Chem March 2008;389(3):305–12.

[36] Ramachandran D, Roy U, Garg S, Ghosh S, Pathak S, Kolthur-Seetharam U. Sirt1 and mir-9 expression is regulated during glucose-stimulated insulin secretion in pancreatic beta-islets. FEBS J April 2011;278(7):1167–74.

[37] Sekine S, Ogawa R, McManus MT, Kanai Y, Hebrok M. Dicer is required for proper liver zonation. J Pathol November 2009;219(3):365–72.

[38] Xu H, He JH, Xiao ZD, Zhang QQ, Chen YQ, Zhou H, et al. Liver-enriched transcription factors regulate microRNA-122 that targets CUTL1 during liver development. Hepatology October 2010;52(4):1431–42.

[39] Chang J, Nicolas E, Marks D, Sander C, Lerro A, Buendia MA, et al. miR-122, a mammalian liver-specific microRNA, is processed from hcr mRNA and may downregulate the high affinity cationic amino acid transporter CAT-1. RNA Biol July 2004;1(2):106–13.

[40] Esau C, Davis S, Murray SF, Yu XX, Pandey SK, Pear M, et al. miR-122 regulation of lipid metabolism revealed by in vivo antisense targeting. Cell Metab February 2006;3(2):87–98.

[41] Elmen J, Lindow M, Schutz S, Lawrence M, Petri A, Obad S, et al. LNA-mediated microRNA silencing in non-human primates. Nature April 17, 2008;452(7189):896–9.

[42] Hsu SH, Wang B, Kota J, Yu J, Costinean S, Kutay H, et al. Essential metabolic, anti-inflammatory, and anti-tumorigenic functions of miR-122 in liver. J Clin Invest August 1, 2012;122(8):2871–83.
[43] Cheung O, Puri P, Eicken C, Contos MJ, Mirshahi F, Maher JW, et al. Nonalcoholic steatohepatitis is associated with altered hepatic MicroRNA expression. Hepatology December 2008;48(6):1810–20.
[44] Li S, Chen X, Zhang H, Liang X, Xiang Y, Yu C, et al. Differential expression of microRNAs in mouse liver under aberrant energy metabolic status. J Lipid Res September 2009;50(9):1756–65.
[45] Brown MS, Goldstein JL. The SREBP pathway: regulation of cholesterol metabolism by proteolysis of a membrane-bound transcription factor. Cell May 2, 1997;89(3):331–40.
[46] Horie T, Nishino T, Baba O, Kuwabara Y, Nakao T, Nishiga M, et al. MicroRNA-33 regulates sterol regulatory element-binding protein 1 expression in mice. Nat Commun 2013;4:2883.
[47] Najafi-Shoushtari SH, Kristo F, Li Y, Shioda T, Cohen DE, Gerszten RE, et al. MicroRNA-33 and the SREBP host genes cooperate to control cholesterol homeostasis. Science June 18, 2010;328(5985):1566–9.
[48] Davalos A, Goedeke L, Smibert P, Ramirez CM, Warrier NP, Andreo U, et al. miR-33a/b contribute to the regulation of fatty acid metabolism and insulin signaling. Proc Natl Acad Sci USA May 31, 2011;108(22):9232–7.
[49] Ramirez CM, Goedeke L, Rotllan N, Yoon JH, Cirera-Salinas D, Mattison JA, et al. MicroRNA 33 regulates glucose metabolism. Mol Cell Biol August 2013;33(15):2891–902.
[50] Goedeke L, Salerno A, Ramirez CM, Guo L, Allen RM, Yin X, et al. Long-term therapeutic silencing of miR-33 increases circulating triglyceride levels and hepatic lipid accumulation in mice. EMBO Mol Med July 18, 2014;6.
[51] Mori MA, Thomou T, Boucher J, Lee KY, Lallukka S, Kim JK, et al. Altered miRNA processing disrupts brown/white adipocyte determination and associates with lipodystrophy. J Clin Invest August 1, 2014;124(8):3339–51.
[52] Sun L, Xie H, Mori MA, Alexander R, Yuan B, Hattangadi SM, et al. Mir193b-365 is essential for brown fat differentiation. Nat Cell Biol August 2011;13(8):958–65.
[53] Feuermann Y, Kang K, Gavrilova O, Haetscher N, Jang SJ, Yoo KH, et al. MiR-193b and miR-365-1 are not required for the development and function of brown fat in the mouse. RNA Biol December 1, 2013;10(12):1807–14.
[54] Esau C, Kang X, Peralta E, Hanson E, Marcusson EG, Ravichandran LV, et al. MicroRNA-143 regulates adipocyte differentiation. J Biol Chem December 10, 2004;279(50):52361–5.
[55] Takanabe R, Ono K, Abe Y, Takaya T, Horie T, Wada H, et al. Up-regulated expression of microRNA-143 in association with obesity in adipose tissue of mice fed high-fat diet. Biochem Biophys Res Commun November 28, 2008;376(4):728–32.
[56] Xie H, Lim B, Lodish HF. MicroRNAs induced during adipogenesis that accelerate fat cell development are downregulated in obesity. Diabetes May 2009;58(5):1050–7.
[57] Wang T, Li M, Guan J, Li P, Wang H, Guo Y, et al. MicroRNAs miR-27a and miR-143 regulate porcine adipocyte lipid metabolism. Int J Mol Sci 2011;12(11):7950–9.
[58] He A, Zhu L, Gupta N, Chang Y, Fang F. Overexpression of micro ribonucleic acid 29, highly up-regulated in diabetic rats, leads to insulin resistance in 3T3-L1 adipocytes. Mol Endocrinol November 2007;21(11):2785–94.
[59] Kurtz CL, Peck BC, Fannin EE, Beysen C, Miao J, Landstreet SR, et al. MicroRNA-29 fine-tunes the expression of key FOXA2-activated lipid metabolism genes and is dysregulated in animal models of insulin resistance and diabetes. Diabetes September 2014;63(9):3141–8.
[60] Granjon A, Gustin MP, Rieusset J, Lefai E, Meugnier E, Guller I, et al. The microRNA signature in response to insulin reveals its implication in the transcriptional action of insulin in human skeletal muscle and the role of a sterol regulatory element-binding protein-1c/myocyte enhancer factor 2C pathway. Diabetes November 2009;58(11):2555–64.
[61] Nielsen S, Scheele C, Yfanti C, Akerstrom T, Nielsen AR, Pedersen BK, et al. Muscle specific microRNAs are regulated by endurance exercise in human skeletal muscle. J Physiol October 15, 2010;588(Pt 20):4029–37.
[62] Chen JF, Mandel EM, Thomson JM, Wu Q, Callis TE, Hammond SM, et al. The role of microRNA-1 and microRNA-133 in skeletal muscle proliferation and differentiation. Nat Genet February 2006;38(2):228–33.
[63] Chen JF, Tao Y, Li J, Deng Z, Yan Z, Xiao X, et al. MicroRNA-1 and microRNA-206 regulate skeletal muscle satellite cell proliferation and differentiation by repressing Pax7. J Cell Biol September 6, 2010;190(5):867–79.
[64] Liu N, Williams AH, Kim Y, McAnally J, Bezprozvannaya S, Sutherland LB, et al. An intragenic MEF2-dependent enhancer directs muscle-specific expression of microRNAs 1 and 133. Proc Natl Acad Sci USA December 26, 2007;104(52):20844–9.
[65] Gallagher IJ, Scheele C, Keller P, Nielsen AR, Remenyi J, Fischer CP, et al. Integration of microRNA changes in vivo identifies novel molecular features of muscle insulin resistance in type 2 diabetes. Genome Med 2010;2(2):9.

[66] Horie T, Ono K, Nishi H, Iwanaga Y, Nagao K, Kinoshita M, et al. MicroRNA-133 regulates the expression of GLUT4 by targeting KLF15 and is involved in metabolic control in cardiac myocytes. Biochem Biophys Res Commun November 13, 2009;389(2):315–20.

[67] Huang B, Qin W, Zhao B, Shi Y, Yao C, Li J, et al. MicroRNA expression profiling in diabetic GK rat model. Acta Biochim Biophys Sin (Shanghai) June 2009;41(6):472–7.

[68] Karolina DS, Armugam A, Tavintharan S, Wong MT, Lim SC, Sum CF, et al. MicroRNA 144 impairs insulin signaling by inhibiting the expression of insulin receptor substrate 1 in type 2 diabetes mellitus. PloS One 2011;6(8):e22839.

[69] Inzucchi SE, Bergenstal RM, Buse JB, Diamant M, Ferrannini E, Nauck M, et al. Management of hyperglycaemia in type 2 diabetes: a patient-centered approach. Position statement of the American Diabetes Association (ADA) and the European Association for the Study of Diabetes (EASD). Diabetologia June 2012;55(6):1577–96.

[70] Bennett WL, Maruthur NM, Singh S, Segal JB, Wilson LM, Chatterjee R, et al. Comparative effectiveness and safety of medications for type 2 diabetes: an update including new drugs and 2-drug combinations. Ann Intern Med May 3, 2011;154(9):602–13.

[71] Janka HU, Plewe G, Riddle MC, Kliebe-Frisch C, Schweitzer MA, Yki-Jarvinen H. Comparison of basal insulin added to oral agents versus twice-daily premixed insulin as initial insulin therapy for type 2 diabetes. Diabetes Care February 2005;28(2):254–9.

[72] Lehmann JM, Moore LB, Smith-Oliver TA, Wilkison WO, Willson TM, Kliewer SA. An antidiabetic thiazolidinedione is a high affinity ligand for peroxisome proliferator-activated receptor gamma (PPAR gamma). J Biol Chem June 2, 1995;270(22):12953–6.

[73] Proks P, Reimann F, Green N, Gribble F, Ashcroft F. Sulfonylurea stimulation of insulin secretion. Diabetes December 2002;51(Suppl. 3):S368–76.

[74] Intensive blood-glucose control with sulphonylureas or insulin compared with conventional treatment and risk of complications in patients with type 2 diabetes (UKPDS 33). UK Prospective Diabetes Study (UKPDS) Group. Lancet September 12, 1998;352(9131):837–853.

[75] Pratley RE, Nauck M, Bailey T, Montanya E, Cuddihy R, Filetti S, et al. Liraglutide versus sitagliptin for patients with type 2 diabetes who did not have adequate glycaemic control with metformin: a 26-week, randomised, parallel-group, open-label trial. Lancet April 24, 2010;375(9724):1447–56.

[76] Gutniak M, Orskov C, Holst JJ, Ahren B, Efendic S. Antidiabetogenic effect of glucagon-like peptide-1 (7–36) amide in normal subjects and patients with diabetes mellitus. N Engl J Med May 14, 1992;326(20):1316–22.

[77] Santovito D, De Nardis V, Marcantonio P, Mandolini C, Paganelli C, Vitale E, et al. Plasma exosome microRNA profiling unravels a new potential modulator of adiponectin pathway in diabetes: effect of glycemic control. J Clin Endocrinol Metab June 17, 2014. jc20133843.

[78] Brennan EP, Nolan KA, Borgeson E, Gough OS, McEvoy CM, Docherty NG, et al. Lipoxins attenuate renal fibrosis by inducing let-7c and suppressing TGFbetaR1. J Am Soc Nephrol March 2013;24(4):627–37.

[79] Wang B, Jha JC, Hagiwara S, McClelland AD, Jandeleit-Dahm K, Thomas MC, et al. Transforming growth factor-beta1-mediated renal fibrosis is dependent on the regulation of transforming growth factor receptor 1 expression by let-7b. Kidney Int February 2014;85(2):352–61.

[80] Ortega FJ, Mercader JM, Moreno-Navarrete JM, Rovira O, Guerra E, Esteve E, et al. Profiling of circulating microRNAs reveals common microRNAs linked to type 2 diabetes that change with insulin sensitization. Diabetes Care May 2014;37(5):1375–83.

[81] Phuah NH, Nagoor NH. Regulation of microRNAs by natural agents: new strategies in cancer therapies. Biomed Res Int 2014;2014:804510.

[82] Li Y, VandenBoom 2nd TG, Kong D, Wang Z, Ali S, Philip PA, et al. Up-regulation of miR-200 and let-7 by natural agents leads to the reversal of epithelial-to-mesenchymal transition in gemcitabine-resistant pancreatic cancer cells. Cancer Res August 15, 2009;69(16):6704–12.

[83] Wang TJ, Larson MG, Vasan RS, Cheng S, Rhee EP, McCabe E, et al. Metabolite profiles and the risk of developing diabetes. Nat Med April 2011;17(4):448–53.

[84] Wang TJ, Ngo D, Psychogios N, Dejam A, Larson MG, Vasan RS, et al. 2-Aminoadipic acid is a biomarker for diabetes risk. J Clin Invest October 1, 2013;123(10):4309–17.

[85] Selvin E, Rawlings AM, Grams M, Klein R, Sharrett AR, Steffes M, et al. Fructosamine and glycated albumin for risk stratification and prediction of incident diabetes and microvascular complications: a prospective cohort analysis of the Atherosclerosis Risk in Communities (ARIC) study. Lancet Diabetes Endocrinol April 2014;2(4):279–88.

[86] Weber JA, Baxter DH, Zhang S, Huang DY, Huang KH, Lee MJ, et al. The microRNA spectrum in 12 body fluids. Clin Chem November 2010;56(11):1733–41.

[87] Wang K, Zhang S, Weber J, Baxter D, Galas DJ. Export of microRNAs and microRNA-protective protein by mammalian cells. Nucleic Acids Res November 2010;38(20):7248–59.

[88] Zubakov D, Boersma AW, Choi Y, van Kuijk PF, Wiemer EA, Kayser M. MicroRNA markers for forensic body fluid identification obtained from microarray screening and quantitative RT-PCR confirmation. Int J Legal Med May 2010;124(3):217–26.

[89] Hanson EK, Lubenow H, Ballantyne J. Identification of forensically relevant body fluids using a panel of differentially expressed microRNAs. Anal Biochem April 15, 2009;387(2):303–14.
[90] Gibbings DJ, Ciaudo C, Erhardt M, Voinnet O. Multivesicular bodies associate with components of miRNA effector complexes and modulate miRNA activity. Nat Cell Biol September 2009;11(9):1143–9.
[91] Valadi H, Ekstrom K, Bossios A, Sjostrand M, Lee JJ, Lotvall JO. Exosome-mediated transfer of mRNAs and microRNAs is a novel mechanism of genetic exchange between cells. Nat Cell Biol June 2007;9(6):654–9.
[92] Scholer N, Langer C, Kuchenbauer F. Circulating microRNAs as biomarkers – true blood? Genome Med 2011;3(11):72.
[93] Mo MH, Chen L, Fu Y, Wang W, Fu SW. Cell-free circulating miRNA biomarkers in cancer. J Cancer 2012;3:432–48.
[94] Zhang T, Lv C, Li L, Chen S, Liu S, Wang C, et al. Plasma miR-126 is a potential biomarker for early prediction of type 2 diabetes mellitus in susceptible individuals. Biomed Res Int 2013;2013:761617.
[95] Zampetaki A, Kiechl S, Drozdov I, Willeit P, Mayr U, Prokopi M, et al. Plasma microRNA profiling reveals loss of endothelial miR-126 and other microRNAs in type 2 diabetes. Circ Res September 17, 2010;107(6):810–7.
[96] Kong L, Zhu J, Han W, Jiang X, Xu M, Zhao Y, et al. Significance of serum microRNAs in pre-diabetes and newly diagnosed type 2 diabetes: a clinical study. Acta Diabetol March 2011;48(1):61–9.
[97] Rong Y, Bao W, Shan Z, Liu J, Yu X, Xia S, et al. Increased microRNA-146a levels in plasma of patients with newly diagnosed type 2 diabetes mellitus. PloS One 2013;8(9):e73272.
[98] Pescador N, Perez-Barba M, Ibarra JM, Corbaton A, Martinez-Larrad MT, Serrano-Rios M. Serum circulating microRNA profiling for identification of potential type 2 diabetes and obesity biomarkers. PloS One 2013;8(10):e77251.
[99] Nielsen LB, Wang C, Sorensen K, Bang-Berthelsen CH, Hansen L, Andersen ML, et al. Circulating levels of microRNA from children with newly diagnosed type 1 diabetes and healthy controls: evidence that miR-25 associates to residual beta-cell function and glycaemic control during disease progression. Exp Diabetes Res 2012;2012:896362.
[100] Bonner C, Nyhan KC, Bacon S, Kyithar MP, Schmid J, Concannon CG, et al. Identification of circulating microRNAs in HNF1A-MODY carriers. Diabetologia August 2013;56(8):1743–51.

Index

'*Note*: Page numbers followed by "f" indicate figures and "t" indicate tables.'

D

I

Q

R

T